A Complete Teaching and Learning Package

The following print, multi-media and web-based ancillaries are available to accompany the fifth edition of *Introduction to the Human Body*. Each of our supplementary products is specifically created with one goal in mind: to help teachers teach and students learn. *Please contact your Wiley sales representative for additional information about any of the following supplementary products.*

FOR INSTRUCTORS

- **Instructor's Resource Manual (0-471-38330-9)**—This on-line manual includes a chapter-by-chapter synopsis, suggested lecture outlines, learning objectives, and teaching tips. Each chapter also includes a description of what is new to this edition. The entire manual can be found on the text's dedicated website and is fully downloadable. The electronic format provides you with the opportunity to customize lecture outlines or activities for delivery either in print or electronically to your students.

- *NEW!* **Instructor's Presentation CD-ROM (0-471-38331-7)**—For lecture presentation purposes, this cross-platform CD-ROM includes all of the illustrations from the textbook in both labeled and unlabeled formats. An easy to use editing tool lets you modify each image to meet your individual instructional needs. Moreover, you can print images, use them in conjunction with *Microtest*, or export them to Microsoft PowerPoint software.

- **Transparency Acetates (0-471-38326-0)**—This set of full-color transparency acetates includes approximately 350 illustrations and photographs from the new edition. Each transparency displays the textbook figure number and caption and has enlarged labels for optimal visibility when projected in the classroom or lecture hall.

- **Printed Test Bank (0-471-38327-9)**—A test bank of approximately 2500 questions in a variety of formats—multiple choice, matching, true/false, short answer, and essay—in order to accommodate different testing preferences.

- *Microtest* **Computerized Testbank (Windows: 0-471-38328-7; Mac: 0-471-38332-5)**—Available for Windows and Macintosh this electronic version of the printed test bank features a user-friendly interface that lets you view and edit questions or add your own. The ability to link illustrations to the tests is included, as well as options for printing in a variety of styles.

- *NEW!* **Take Note! (0-471-38325-2)**—Students can organize their note taking and improve their understanding of anatomical structures and physiological processes by using this handy illustrated notebook. Following the illustration sequence in the textbook, each left-handed page displays an unlabeled, black-and-white copy of every text figure. Students can fill in the labels during lecture or lab at the instructor's directions and take additional notes on the blank right-hand pages.

- **Student Learning Guide (0-471-38329-5)**—This on-line study guide is designed to appeal to a broad range of student learning styles. It includes multiple activities for each chapter, including *Framework*—visual maps of the chapter content; *Wordbytes*—to help master vocabulary; *Checkpoints*—a series of study activities such as questions to answers, illustrations to label, tables to complete; *Critical Thinking Questions; Mastery Test*. Accessed through the text's dedicated website, this new format for the Learning Guide offers student's the opportunity to work on-line or print just those activities that they find most helpful and necessary, making this a more useful and affordable option.

- **A Photographic Atlas of the Human Body with Selected Cat, Sheep, and Cow Dissections (0-471-37487-3)**—Designed to support any course in anatomy and physiology, this new atlas also may be used in conjunction with, or in lieu or, a laboratory manual. Organized by body system, the clearly labeled photographs provide a stunning visual reference to gross anatomy. Histological micrographs are also included.

- **Anatomy and Physiology: A Companion Coloring Book (0-471-39515-3)**—This unique study tool is a breakthrough approach to learning and remembering the human body's anatomical structures and physiological processes. It features over 500 striking, original illustrations that give students a clear and enduring understanding of anatomy and physiology.

FOR STUDENTS

- *NEW!* **Student Companion Study CD-ROM**—This engaging study tool is attached to the inside front cover of each textbook, providing students with a wealth of activities to enhance their learning. Included are interactive activities such as feedback loop exercises, activities that trace complex physiological processes or challenge student understanding of anatomy. Animations of some key physiological processes are included along with entertaining interactive exercises based on those animations. Multiple choice quizzes for each chapter are included, as well as a Test feature that simulates mid-term or final exams. An audio pronunciation dictionary is provided as well.

DEDICATED WEB SITE

A new, dedicated web site to accompany this text is found at the Wiley home for anatomy and physiology—*Anatomy and Physiology Central*. The homepage gives you links to several other supportive sites, such as The Human Anatomy and Physiology Society (HAPS). Clicking on the book's cover will take you to the site dedicated to *Introduction of the Human Body 5e*. This site offers the students a variety of study options, including yet another new set of practice quizzes to test their understanding, as well as search engines for investigating homeostatic disorders linked to chapter content, and instructor reviewed websites that can enhance their understanding. The Web site can be accessed at www.wiley.com/college/apcentral.

Managing Editor   Bonnie Roesch
Marketing Manager   Clay Stone
Developmental Editor   Karen Trost
Production Director   Pamela Kennedy
Senior Production Editor   Kelly Tavares
Text and Cover Design   Karin Gerdes Kincheloe
Art Coordinator   Claudia Durrell
Photo Editor   Hilary Newman
Cover Photo   © Al Satterwhite\FPG International

This book was typeset by Progressive Information Technology and printed and bound by Replika Press Pvt. Ltd. The cover was also printed by Replika Press Pvt. Ltd.

The paper in this book was manufactured by a mill whose forest management programs include sustained yield harvesting of its timberlands. Sustained yield harvesting principles ensure that the number of trees cut each year does not exceed the amount of new growth.

This book is printed on acid-free paper. ∞

ISBN 0-471-36777-X

Printed in India.

10 9 8 7 6

INTRODUCTION TO THE HUMAN BODY
The Essentials of Anatomy and Physiology

Fifth Edition

Gerard J. Tortora

Bergen Community College

Sandra Reynolds Grabowski

Purdue University

JOHN WILEY & SONS, INC.
New York · Chichester · Weinheim · Brisbane · Singapore · Toronto

To my wife, Melanie, and to my children, Christopher, Anthony, and Andrew, who make it all worthwhile.—G.J.T.

To my students, whose questions continually nourish my love of teaching and writing.—S.R.G.

About the Authors

Gerard J. Tortora is Professor of Biology at Bergen Community College in Paramus, New Jersey, where he teaches human anatomy and physiology as well as microbiology. The author of several best-selling science textbooks and laboratory manuals, Jerry is devoted first and foremost to his students and their aspirations. In recognition of this commitment, he was named Distinguished Faculty Scholar at Bergen Community College. In 1996 Jerry received a National Institute for Staff and Organizational Development (NISOD) excellence award from the University of Texas, and he was selected to represent Bergen Community College in a campaign to increase awareness of the contributions of community colleges to higher education.

Jerry received his bachelor's degree in biology from Fairleigh Dickinson University and his master's degree in science education from Montclair State College. He is a member of many professional organizations, such as the Human Anatomy and Physiology Society (HAPS), the American Society of Microbiology (ASM), the American Association for the Advancement of Science (AAAS), the National Education Association (NEA), and the Metropolitan Association of College and University Biologists (MACUB).

Sandra Reynolds Grabowski is an instructor in the Department of Biological Sciences at Purdue University in West Lafayette, Indiana. For more than 20 years she has taught human anatomy and physiology to students in a wide range of academic programs. In 1992 students selected her as one of the top teachers in the School of Science at Purdue.

Sandy received her BS in biology and her PhD in neurophysiology from Purdue. She is a member of the Human Anatomy and Physiology Society (HAPS), served as editor of *HAPS News* from 1990 through 1992, and was elected to serve a 3-year term as President Elect, President, and Past President from 1992 to 1995. In addition, she is a member of the American Association for the Advancement of Science (AAAS), the Association for Women in Science (AWIS), the National Science Teachers Association (NSTA), and the Society for College Science Teachers (SCST).

Preface

Introduction to the Human Body: The Essentials of Anatomy and Physiology, Fifth Edition, is designed for courses in human anatomy and physiology or in human biology. It assumes no previous study of the human body. The successful approach of the previous editions—to provide students with a basic understanding of the structure and functions of the human body with an emphasis on homeostasis—has been retained. In the development of the fifth edition, we focused on improving the acknowledged strengths of the text as well as introducing several new and innovative features.

A New Co-Author

An important change in this edition is the addition of Sandy Grabowski as co-author, expanding our collaboration together as a writing team that began in 1993 with the publication of Principles of Anatomy and Physiology, seventh edition. The combination of our individual academic expertise (Jerry in anatomy and Sandy in physiology) uniquely enables us to fine-tune the balanced coverage between structure and function so that students can clearly see the relationship between these two integrated disciplines.

ORGANIZATION AND CONTENT IMPROVEMENTS

Like the previous edition of Introduction to the Human Body, the fifth edition of the book is divided into 24 chapters, approaches to the study of the human body system by system, beginning with the integumentary system in Chapter 5. Every chapter in the fifth edition incorporates a host of improvements to both the text and the art, many suggested by reviewers, educators, and students.

Some of the significant changes in selected chapters include:

- **Chapter 3 / Cells** Updates from front to back reflect the latest thinking in cell biology. Students will benefit in particular from improved coverage of the plasma membrane and membrane transport mechanisms. Virtually every figure has been redrawn in a new, vibrant, three-dimensional style.
- **Chapter 4 / Tissues** The much-imitated tissue tables in this chapter are redesigned to enhance student comprehension of histology. Now, every photomicrograph is accompanied by an interpretive illustration as well as an orientation diagram that indicates a location where a particular tissue

type occurs in the body. Newly added tables describe bone, blood, muscle tissue, and nervous tissue. There are also new sections dealing with tissue engineering, aging and tissues, common disorders, and medical terminology and conditions.

- **Chapter 7 / Joints** Improved coverage is further accentuated by clear and consistently stylized new range-of-motion photographs.
- **Chapter 8 / The Muscular System** The beginning of this chapter now provides clearer, simpler, and up-to-date explanations of the mechanics and physiology of muscle tissue. New, generously sized, three-dimensional illustrations of muscle tissue and its components complement our enhancements to this chapter's coverage. A new feature, Relating Muscles to Movements, encourages readers to group related muscles according to their common actions.
- **Chapter 9 / Nervous Tissue** New to this chapter is a discussion of neurotransmitters and added sections on common disorders and medical terminology and conditions.
- **Chapter 12 / Sensations** New three-dimensional drawings of major sensory organs by noted science illustrators Tomo Narashima and Steve Oh superbly enhance the content and utility of this chapter.
- **Chapter 15 / The Cardiovascular System: Heart** This chapter now features a section on exercise and the heart, as well as beautifully redrawn illustrations.
- **Chapter 16 / The Cardiovascular System: Blood Vessels** Several new pages of exhibits have been added to more completely show and describe blood vessels that serve all parts of the body. Meticulous new illustrations clearly reveal the relevant blood vessel anatomy.
- **Chapter 20 / Nutrition and Metabolism** This chapter has been reorganized so that the section on nutrition now comes before the metabolic fate of each type of nutrient is described. A new section describes the variety of lipoprotein particles that transport lipids in the bloodstream.

NEW OR ENHANCED FEATURES

Popular features of previous editions, such as Focus on Wellness Essays and Common Disorders, remain vital components of this new edition. In addition, several new features have been included to better meet the needs of students as they progress through the course.

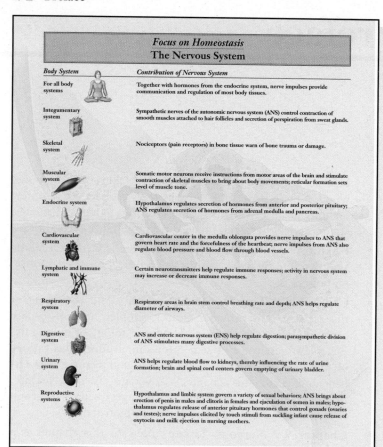

Focus on Homeostasis
The Nervous System

Body System	Contribution of Nervous System
For all body systems	Together with hormones from the endocrine system, nerve impulses provide communication and regulation of most body tissues.
Integumentary system	Sympathetic nerves of the autonomic nervous system (ANS) control contraction of smooth muscles attached to hair follicles and secretion of perspiration from sweat glands.
Skeletal system	Nociceptors (pain receptors) in bone tissue warn of bone trauma or damage.
Muscular system	Somatic motor neurons receive instructions from motor areas of the brain and stimulate contraction of skeletal muscles to bring about body movements; reticular formation sets level of muscle tone.
Endocrine system	Hypothalamus regulates secretion of hormones from anterior and posterior pituitary; ANS regulates secretion of hormones from adrenal medulla and pancreas.
Cardiovascular system	Cardiovascular center in the medulla oblongata provides nerve impulses to ANS that govern heart rate and the forcefulness of the heartbeat; nerve impulses from ANS also regulate blood pressure and blood flow through blood vessels.
Lymphatic and immune system	Certain neurotransmitters help regulate immune responses; activity in nervous system may increase or decrease immune responses.
Respiratory system	Respiratory areas in brain stem control breathing rate and depth; ANS helps regulate diameter of airways.
Digestive system	ANS and enteric nervous system (ENS) help regulate digestion; parasympathetic division of ANS stimulates many digestive processes.
Urinary system	ANS helps regulate blood flow to kidneys, thereby influencing the rate of urine formation; brain and spinal cord centers govern emptying of urinary bladder.
Reproductive systems	Hypothalamus and limbic system govern a variety of sexual behaviors; ANS brings about erection of penis in males and clitoris in females and ejaculation of semen in males; hypothalamus regulates release of anterior pituitary hormones that control gonads (ovaries and testes); nerve impulses elicited by touch stimuli from suckling infant cause release of oxytocin and milk ejection in nursing mothers.

• Focus on Homeostasis

An important new feature of this edition is the 10 **Focus on Homeostasis** pages, one each for the integumentary, skeletal, muscular, nervous, endocrine, cardiovascular, lymphatic and immune, respiratory, digestive, and urinary systems. Incorporating both graphic and narrative elements, these pages explain, clearly and succinctly, how the system under consideration contributes to the homeostasis of each of the other body systems. Use of this feature will enhance student understanding of the links between body systems and how interactions among systems contribute to the homeostasis of the body as a whole.

AGING AND THE RESPIRATORY SYSTEM

Objective: • **Describe the effects of aging on the respiratory system.**

With advancing age, the airways and tissues of the respiratory tract, including the alveoli, become less elastic; the chest wall becomes more rigid as well. The result is a decrease in lung capacity. Vital capacity can decrease as much as 35% by age 70. Moreover, a decrease in blood levels of O_2, decreased activity of alveolar macrophages, and diminished ciliary action of the epithelium lining the respiratory tract occur. Because of these changes, elderly people are more susceptible to pneumonia, bronchitis, emphysema, and other pulmonary disorders.

• Aging

Anatomy and physiology is not static. As the body ages, its structure and related functions change in subtle and not so subtle ways. Many students will go on to careers in health related fields, in which the average age of the client population is steadily advancing. For this reason, discussions of this professionally relevant topic have been added to chapters 1, 3, 4, 7, 8, 13, 16, 17, 18, 19, 22, and 23.

• Exercise

Physical exercise can produce favorable changes in some anatomical structures and enhance many physiological functions, most notably those associated with the muscular, skeletal and cardiovascular systems. This information is especially relevant to readers embarking on careers in physical education, sports training, and kinesiology. Hence, key chapters include brief discussions of exercise, signaled by a distinctive "running shoe" icon.

EXERCISE AND THE HEART

Objective: • **Explain the relationship between exercise and the heart.**

A person's level of fitness can be improved at any age with regular exercise. Some types of exercise are more effective than others for improving the health of the cardiovascular system. *Aerobic exercise,* any activity that works large body muscles for at least 20 minutes, elevates cardiac output and accelerates metabolic rate. Three to five such sessions a week are usually recommended for improving the health of the cardiovascular system. Brisk walking, running, bicycling, cross-country skiing, and swimming are examples of aerobic activities.

Sustained exercise increases the oxygen demand of the muscles. Whether the demand is met depends primarily on the adequacy of cardiac output and proper functioning of the respiratory system. After several weeks of training, a healthy person increases maximal cardiac output, thereby increasing the maximal rate of oxygen delivery to the tissues. Hemoglobin level increases and skeletal muscles develop more capillary networks, enhancing oxygen delivery. A physically fit person may even exhibit *bradycardia,* a resting heart rate under 60 beats per minute. A slowly beating heart is more energy efficient than one that beats more rapidly.

• Focus on Wellness

A popular feature of the last two editions has been the wellness essays, written by Barbara Brehm Curtis of Smith College. These essays increase students' appreciation of the relevancy of the concepts and details of anatomy and physiology presented in the text to the maintenance of good health. The wellness philosophy supports the notion that life-style choices that individuals make throughout the years have an important influence on their mental and physical well-being.

Several brand-new essays appear in this edition, along with revised and up-dated versions of the most popular essays from the previous edition. In addition, each essay includes a "Think It Over" concept application exercise. We believe that the information contained in these Focus on Wellness essays is timely and interesting; we hope students and instructors continue to feel this way, too.

Focus on Wellness

Neurotransmitters— Why Food Affects Mood

*E*veryone who has enjoyed the soothing relaxation of a good meal has experienced the effect of food on mood. Scientists have proposed that some of the relationship between food and mood can be explained by the effect of diet on the levels and regulation of neurotransmitters in the brain. As you learned in this chapter, neurotransmitters are the chemical messengers that allow neurons to "talk" to each other.

Neurons manufacture neurotransmitters from chemicals that come from food, so you could say that the story of the food–mood link begins with digestion. Many neurotransmitters are made from amino acids, which are the basic building blocks of proteins. Amino acids are made available when your body digests the protein in the food you eat. For example, the neurotransmitter serotonin is made from the amino acid tryptophan, and both dopamine and norepinephrine are synthesized from the amino acid tyrosine.

Mind-altering Food?
Regulation of neurotransmitter levels in the brain is quite complicated and depends not only on the availability of amino acid (and other) precursors, but also upon competition of these precursors for entry into the brain. Consider serotonin, one of the neurotransmitters that appears to have an important effect on mood. Serotonin leads to feelings of relaxation and sleepiness. Many antidepressant medications such as Prozac, relieve feelings of depression by increasing serotonin levels in the brain.

Although serotonin is manufactured from the amino acid tryptophan, protein foods do not lead to higher levels of tryptophan in the blood or brain. This is because after a high-protein meal, tryptophan must compete with more than 20 other amino acids for entry into the central nervous system, so its concentration in the brain remains relatively low. On the other hand, consumption of carbohydrate-rich foods, such as bread, pasta, potatoes, or sweets, is associated with an increase in the synthesis and release of serotonin in the brain. The result: Carbohydrates help us feel relaxed and sleepy.

▶ *Think It Over*

▶ Why might consuming a high-protein diet for several days lead to cravings for carbohydrate-rich foods? Why is depression often associated with weight gain?

• Exhibits and Tables

In previous editions, tabular materials were referred to as Exhibits. In this edition we have improved the utility and readability of tabular materials and refer to them as *Tables*. The term *Exhibit* is now reserved for a new, self-contained feature designed to give students the extra help that they need to learn the numerous structures that constitute certain body systems—most notably skeletal muscles, articulations, blood vessels, and nerves. Each Exhibit consists of an overview, a tabular summary of relevant anatomy, and an associated suite of illustrations. Students will find this clear and concise presentation to be the ideal study vehicle for learning anatomically complex body systems.

Exhibit 8.4 / Muscles That Act on the Anterior Abdominal Wall *(Figure 8.17)*

Objective: Describe the origin, insertion, and action of the muscles that act on the anterior abdominal wall.

• **Overview:** The anterior abdominal wall is composed of skin; fascia; and four pairs of flat, sheetlike muscles: rectus abdominis, external oblique, internal oblique, and transversus abdominis. The anterior surfaces of the rectus abdominis muscles are interrupted by three transverse fibrous bands of tissue called *tendinous intersections*. The *aponeuroses* of the external oblique, internal oblique, and transversus abdominis muscles meet at the midline to form the *linea alba* (white line), a tough fibrous band that extends from the xiphoid process of the sternum to the pubic symphysis. An *aponeurosis* (ap'-ō-noo-RO-sis) is a broad, flat tendon. The inferior free border of the external oblique aponeurosis, plus some collagen fibers, forms the *inguinal ligament*.

• **Relating muscles to movements:** Arrange the muscles in this exhibit according to the following actions on the vertebral column: (1) flexion, (2) lateral flexion, (3) extension, and (4) rotation. The same muscle may be mentioned more than once.

Muscle	Origin	Insertion	Action
Rectus abdominis (REK-tus ab-DOM-in-is; *rect-* = straight, fibers parallel to midline; *abdomin-* = abdomen)	Pubis and pubic symphysis.	Cartilage of fifth to seventh ribs and xiphoid process of sternum.	Flexes vertebral column, and compresses abdomen to aid in defecation, urination, forced expiration, and childbirth.
External oblique (ō-BLĒK; *external* = closer to surface; *oblique*-slanting; here, fibers that are diagonal to midline)	Lower eight ribs.	Crest of ilium and linea alba.	Contraction of both external obliques compresses abdomen and flexes vertebral column; contraction of one side alone bends vertebral column laterally and rotates it.
Internal oblique (*internal* = farther from surface)	Ilium, inguinal ligament, and thoracolumbar fascia.	Cartilage of last three or four ribs and linea alba.	Contraction of both internal obliques compresses abdomen and flexes vertebral column; contraction of one side alone bends vertebral column laterally and rotates it.
Transversus abdominis (trans-VER-sus; *transverse* = fibers that are perpendicular to midline)	Ilium, inguinal ligament, lumbar fascia, and cartilages of last six ribs.	Xiphoid process of sternum, linea alba, and pubis.	Compresses abdomen.

Figure 8.17 ■ Muscles of the male anterolateral abdominal wall.

The inguinal ligament separates the thigh from the body wall.

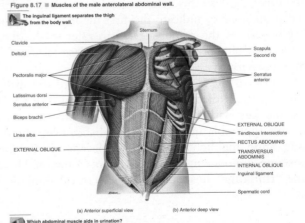

Clavicle
Deltoid
Pectoralis major
Latissimus dorsi
Serratus anterior
Biceps brachii
Linea alba
EXTERNAL OBLIQUE

Sternum
Scapula
Second rib
Serratus anterior
EXTERNAL OBLIQUE
Tendinous intersections
RECTUS ABDOMINIS
TRANSVERSUS ABDOMINIS
INTERNAL OBLIQUE
Inguinal ligament
Spermatic cord

(a) Anterior superficial view   (b) Anterior deep view

Which abdominal muscle aids in urination?

Table 19.1 / Major Hormones that Control Digestion

Hormone	Where Produced	Stimulant	Action
Gastrin	Stomach mucosa (pyloric region).	Stretching of stomach, partially digested proteins and caffeine in stomach, and high pH of stomach chyme.	Stimulates secretion of gastric juice, increases movement of GI tract, and relaxes pyloric sphincter.
Secretin	Intestinal mucosa.	Acidic chyme that enters the small intestine.	Inhibits secretion of gastric juice, stimulates secretion of pancreatic juice rich in bicarbonate ions, and stimulates secretion of bile.
Cholecystokinin (CCK)	Intestinal mucosa.	Amino acids and fatty acids in chyme in small intestine.	Inhibits gastric emptying, stimulates secretion of pancreatic juice rich in digestive enzymes, causes ejection of bile from the gallbladder, and induces a feeling of satiety (feeling full to satisfaction).

• Common Disorders

The Common Disorders sections, marked with a special stethoscope icon, are located at the end of appropriate chapters. The problems considered in these sections are selected to provide a review of normal body processes and demonstrate the importance of the study of anatomy and physiology to careers in health-related fields. These sections have been expanded, revised, and completely updated, providing resources to answer many questions students ask about medical disorders and diseases.

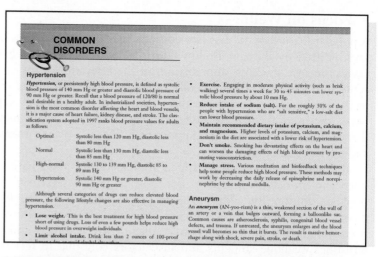

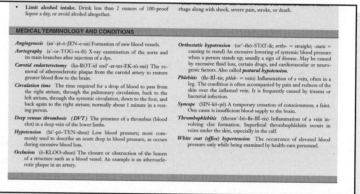

• Medical Terminology and Conditions

Vocabulary-building glossaries of selected medical terms and conditions appear at the end of appropriate chapters, as well. They have been expanded and updated for this edition.

• A New Design

The beautiful new design of the fifth edition has been carefully crafted to assist students in making the most of this text. Each page is carefully laid out to place related text, figures, and tables near one another, minimizing "page-turning" during the reading of a topic. New to this edition—and particularly helpful—is the red print used to indicate the first mention of a figure or table. Not only is the reader alerted to refer to the figure or table, but the color print also serves as a place locator for easy return to the narrative.

Also new to the design of this edition are the distinctive icons incorporated throughout to signal special features. These include the **key** with Key Concept Statements; the **question mark** with applicable questions that enhance every figure; the **running shoe** highlighting content relevant to exercise; the **stethoscope** announcing the section on Common Disorders; and the **"thinking" head** pointing out critical thinking questions.

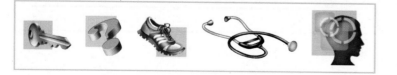

THE ILLUSTRATION PROGRAM

A fine illustration program is one of the signature features of this text. Beautiful artwork, carefully chosen photographs and photomicrographs, and unique pedagogical enhancements all combine to make the visual appeal and usefulness of the illustration program in *Introduction to the Human Body* distinctive. The fifth edition has been carefully reviewed, revised, and updated to uphold the standard of excellence that instructors and students alike have come to expect.

• Color Coding

Colors are used in a consistent and meaningful manner throughout the text to emphasize structural and functional relations. For example, sensory structures, sensory neurons, and sensory regions of the brain are shades of blue ■, whereas motor structures are red ■. Membrane phospholipids are gray ■ and aqua ■, the cytosol is sand ■, and extracellular fluid is blue ■. Illustrations negative and positive feedback loops also use color cues to aid the students in recognizing and understanding the concept. Stimulus and response are both orange ■ as they both alter the controlled condition. The controlled condition is green ■, the receptor is blue ■, the control center is purple ■, and the effector is red ■. Such color cues provide additional help for students who are trying to learn complex anatomical and physiological concepts.

• New Artwork

Many newly created figures have replaced older, less effective illustrations. Exciting new three-dimensional paintings grace the pages of Chapter 3, 5, 8, 9, 12, 15, 16, 19, and 21. Over 80 drawings have been newly rendered for this edition, and nearly every figure has been revised or improved in some way.

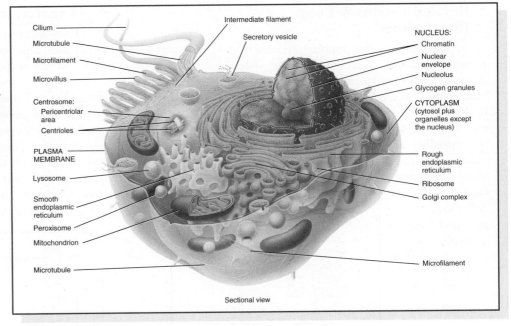

Sectional view

• New Histology-Based Art

As part of our plan of continuous improvement, many of the anatomical illustrations based on histological preparations have been replaced in this edition. For example, beautifully rendered artwork shows features of the skin and subcutaneous layer (see page 101), the interior of the eye (see page 283), layers of the gastrointestinal tract (see page 459), and structure of the small intestine (shown here).

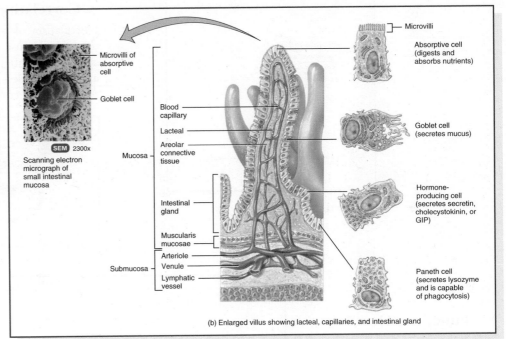

(b) Enlarged villus showing lacteal, capillaries, and intestinal gland

• New Histology Photographs

Many of the histology photographs are new for this edition. Dr. Michael Ross of the University of Florida photographed images specifically for this text. Dr. Ross' expertise in both histology and photographic technique has enhanced this important coverage in the text.

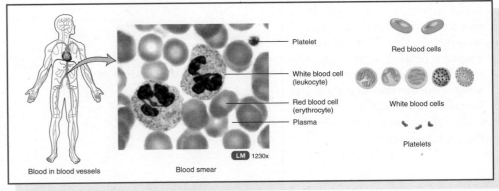

• Revised Feedback Loop Illustrations

As in past editions, this popular series of illustrations captures and clarifies the body's dynamic counterbalancing act in maintaining homeostasis. We have subtly revised the feedback loops for the fifth edition to visually accentuate the roles that receptors, control centers, and effectors play in modifying a controlled physiological condition.

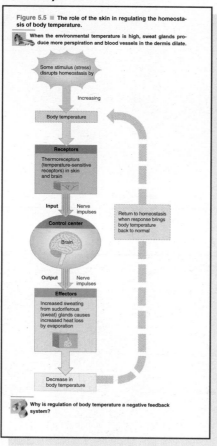

Figure 5.5 ▦ The role of the skin in regulating the homeostasis of body temperature.

When the environmental temperature is high, sweat glands produce more perspiration and blood vessels in the dermis dilate.

Some stimulus (stress) disrupts homeostasis by

Increasing

Body temperature

Receptors
Thermoreceptors (temperature-sensitive receptors) in skin and brain

Input — Nerve impulses

Control center
Brain

Return to homeostasis when response brings body temperature back to normal

Output — Nerve impulses

Effectors
Increased sweating from sudoriferous (sweat) glands causes increased heat loss by evaporation

Decrease in body temperature

Why is regulation of body temperature a negative feedback system?

• Orientation Diagrams

Students sometimes need help figuring out the perspective of structural illustrations—descriptions alone do not always suffice. An orientation diagram depicting and explaining the perspective of the view represented in the figure accompanies most anatomy and histology illustrations. The orientation diagrams are of three general types: (1) planes indicating where certain sections are made when a part of the body is cut; (2) diagrams containing a directional arrow and the word "View" to indicate the direction from which the body part is viewed, e.g. superior, inferior, posterior, anterior; (3) diagrams with arrows leading from or to them to direct attention to enlarged and detailed parts of illustrations.

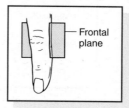 Frontal plane

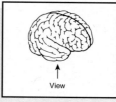 View

• Correlation of Sequential Processes

Correlation of sequential processes in text and art is achieved through the use of numbered lists in the narrative that correspond to numbered segments in the accompanying figure. This approach is used extensively throughout the book to lend clarity to the flow of complex processes.

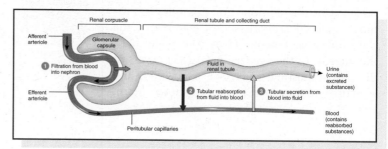

Renal corpuscle — Renal tubule and collecting duct

Afferent arteriole — Glomerular capsule

Fluid in renal tubule

① Filtration from blood into nephron

② Tubular reabsorption from fluid into blood

③ Tubular secretion from blood into fluid

Urine (contains excreted substances)

Efferent arteriole

Blood (contains reabsorbed substances)

Peritubular capillaries

• Functions Overview

The Functions Overview is a feature that juxtaposes the anatomical components and a brief functional overview for each body system. These function "boxes" accompany the first figure of chapters dealing with body systems. They help students to integrate visually the structure and function of a body system and make the connection between their interactions.

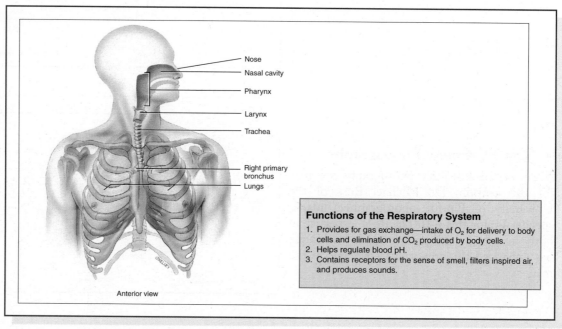

Nose
Nasal cavity
Pharynx
Larynx
Trachea
Right primary bronchus
Lungs

Functions of the Respiratory System

1. Provides for gas exchange—intake of O_2 for delivery to body cells and elimination of CO_2 produced by body cells.
2. Helps regulate blood pH.
3. Contains receptors for the sense of smell, filters inspired air, and produces sounds.

Anterior view

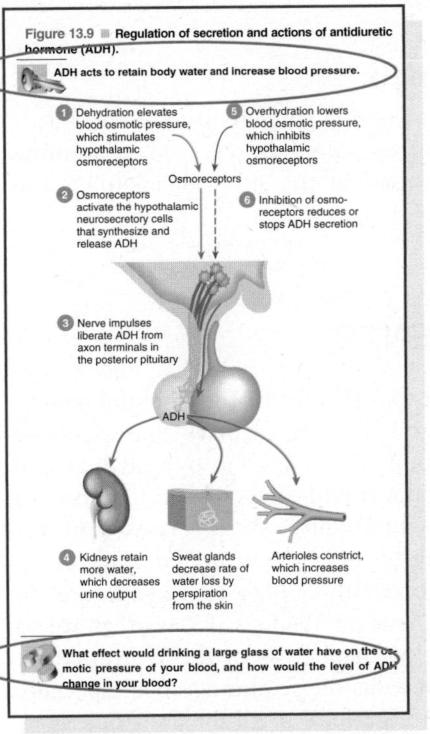

• Key Concept Statements

Included above every figure and denoted by the "key" icon, this feature summarizes an idea that is discussed in the text and demonstrated in a figure. A unique pedagogical feature to our text, these help students keep focused on the relevance of the figure to their understanding of specific content.

• Figure Questions

This highly applauded feature asks readers to synthesize verbal and visual information, think critically, or draw conclusions about what they see in a figure. Each Figure Question appears below its illustration and is highlighted by a distinctive Question Mark icon. Answers are located at the end of each chapter.

Figure 13.9 ■ Regulation of secretion and actions of antidiuretic hormone (ADH).

ADH acts to retain body water and increase blood pressure.

What effect would drinking a large glass of water have on the osmotic pressure of your blood, and how would the level of ADH change in your blood?

LEARNING AIDS

In response to users of the previous edition of *Introduction to the Human Body*, we have retained the learning aids that students and instructors find most useful and have tried to improve them wherever possible. Many—such as critical thinking questions and end of chapter quizzes—have been revised to reflect the enhancements to the text and art.

• Student Learning Objectives

Student Learning Objectives again appear at the beginning of each chapter and are also integrated into the body of the chapter, where they act as "checkpoints" for student reading. Chapter-opening objectives are also page-referenced.

• A Look Ahead

These outlines provide a quick overview of chapter topics and organization, with page references.

• Cross References

This new edition features cross-references that guide the reader to specific pages and figures. Most will help students relate new concepts to previously learned material. However, we acknowledge the really ambitious student by including some cross-references to material that has yet to be considered.

• Phonetic Pronunciations

We have carefully revised all phonetic pronunciations, which appear in parentheses after many anatomical and physiological terms. The pronunciations are given both in the text, where the term is introduced, and in the Glossary at the end of the book. Even more pronunciations have been added to this edition.

• Word Roots

These are derivations designed to provide an understanding of the meaning of new terms. They appear in parentheses when a term is introduced. Many new word roots have been added to this edition.

• Study Outline

The Study Outline at the end of each chapter summarizes major topics and includes specific page references so that students can easily turn to full text discussions.

• Self-Quizzes

Self-Quizzes at the end of every chapter include fill-in-the-blank, multiple choice, and matching questions. Many of the questions are new to this edition. These quizzes are meant not only for students to test their ability to memorize the facts presented in each chapter, but also to sharpen their critical thinking skills by applying the concepts and processes that are part of the way the human body is structured and how it functions. Answers to the Self-Quiz items are presented in an appendix at the end of the book.

• Critical Thinking Applications

In addition to the "Think It Over" questions with the Focus on Wellness in each chapter, Critical Thinking Applications are provided at the end of each chapter. These applications are essay-style problems that encourage students to think about and apply the concepts they have studied in each chapter. Suggested answers to these questions are available at the Web site (www.wiley/college/tortora).

• Glossary

A full glossary of terms with phonetic pronunciations appears at the end of the book. This edition includes over 1700 entries, with more than 250 new terms.

• **Inside Cover Materials**

The end-papers of this text provide students with useful information about prefixes, suffixes, word roots, and combining forms for the terminology used in the study of anatomy and physiology.

ACKNOWLEDGMENTS

We wish to especially thank four people for their helpful contributions to this edition. Caryl Tickner of Stark State College contributed to the completion of this edition by updating and revising the Self-Quizzes at the end of each chapter. Joan Barber of Delaware Technical & Community College revised or replaced the Critical Thinking Questions, also found at chapters' end. Barbara Brehm Curtis of Smith College once again took on the role of author for the Focus on Wellness Essays that are so well received. Happily, Dr. Michael Ross of the University of Florida agreed to provide new histology photographs especially to illustrate our work. We are grateful to all for their fine work in making this an even better edition for students to use.

The publication of the fifth edition of *Introduction to the Human Body* is the first with John Wiley & Sons Inc. We are delighted to be part of a truly outstanding publishing company that also shares our goals, vision, and aspirations. We are impressed! Our association with Wiley has also reunited us with our editor, Bonnie Roesch. Bonnie's multiplicity of talents was well known to us because we had the good fortune to work with her on previous editions of this book, as well as other textbooks. Under her leadership the books became even more successful best-sellers. Bonnie is one of the most professional, dedicated, talented, and knowledgble editors in publishing. As expected, she has been the source of expert guidance, great ideas, and innovative improvements. Her reunion with us is one of the most gratifying aspects of this edition. We look forward to working with Bonnie for many years. Karen Trost, our developmental editor, clearly understood the market for which the book is intended and directed her insightful comments, suggestions, and analysis in that direction. She shepherded the manuscript through two revision cycles and provided many perceptive suggestions, helping us to keep the book on target from beginning to end. The beautiful design for the book's interior is the brainchild of Karin Kincheloe, who also designed the book's cover. Moreover, Karin carefully laid out each page of the book to achieve the best possible arrangement of text, figures, and other elements. Both instructors and students will appreciate and benefit from the pedagogically effective and visually pleasing design elements that augment the changes made to this edition. Claudia Durell our Art Coordinator, like Bonnie Roesch, has been with us for many years. We are grateful that Wiley selected Claudia and has permitted us to maintain our collaboration with her. She remains a cornerstone of our projects. We owe a special thank you to Claudia for all her contributions. Her organizational skills, attention to detail, artistic ability, and understanding of our illustration preferences greatly enhance the visual appeal and style of the figures, surpassing our expectations. Kelly Tavares, Senior Production Editor, demonstrated her expertise and professionalism during each step of the production process. She coordinated all aspects of actually making and manufacturing the book; she also was "on press" as the book was being printed to ensure the highest possible quality. The many hours we spent on the telephone, including weekends, were worth the effort. Hillary Newman, Photo Editor, provided us with all of the photos we requested and did it with efficiency, accuracy, and professionalism. Mary O' Sullivan, Assistant Editor, coordinated the development of the many supplements that support this text. Her efforts have made their imprint on this book, and we are most appreciative of her input. Geraldine Osnato, Editorial Assistant, worked with Bonnie Roesch and helped her with all aspects of the project. She kept all the reviews and other information flowing smoothly to and from us. Her dedication and enthusiasm were obvious in all of her contacts with us. Thank you to everyone who helped during the revision and production of this book.

Finally, we would also like to thank all those who have corresponded with us to offer feedback on the usefulness of this text. Your input helps so much in revising. We particularly want to express our gratitude to the following reviewers who took the time to read and evaluate the draft manuscripts prior to the production of this edition.

Emma Jean Battles, Salt Lake Community College
Allen Billy, Douglas College
Daniel Gong, Seattle Central Community College
Robert Hyde, San Jose State University
Nancy Klepper-Kilgore, Mount Ida College
Dickson J. Phiri, Mesa College
Barbara Roller, Florida International University
Faiz Salehi, Pittsburgh State University
John Sheard, Eastern Michigan University
Emily Gay Williamsom, Mississippi State University

With this edition of *Introduction to the Human Body* we have had an opportunity to work with Wiley throughout the entire publishing process. Our journey has united us with some former professionals and introduced us to some new ones. Our association with the Wiley family has been most pleasant, satisfying, and rewarding. We have been given many reasons to "Celebrate the Difference." We invite all of our readers to celebrate the difference as well. In addition, we would like to invite all readers and users of the book to continue the tradition of sending comments and suggestions to us so that we can include them in the sixth edition.

Gerard J. Tortora
Department of Science and Health, S229
Bergen Community College
400 Paramus Road
Paramus, NJ 07652

Sandra Reynolds Grabowski
Department of Biological Sciences
1392 Lilly Hall of Life Sciences
Purdue University
West Lafayette, IN 47907-1392
email: SGrabows@bilbo.bio.purdue.edu

Note to Students

Your book has a variety of special features that will make your time studying anatomy and physiology a more rewarding experience. These have been developed based on feedback from students—like you—who have used previous editions of the text. Below are some hints for using some of these helpful aids. A review of the preface will give you insight, both visually and in narrative, to all of the text's distinctive features.

Begin your study by anticipating what is to be learned from each chapter. A brief, page referenced overview called **A Look Ahead** begins each chapter along with a list of **Student Learning Objectives**. You will encounter the learning objectives again as the topics are introduced in the body of the chapter. Use these features to help you focus on what is important as you read the material.

■ A Look Ahead

Endocrine Glands 305	Ovaries and Testes
Overview of Hormones 305	Pineal Gland 324
Hypothalamus and Pituitary Gland 307	Thymus 324
Thyroid Gland 313	Other Endocrine Tis
Parathyroid Glands 315	The Stress Response
Adrenal Glands 317	Aging and the Endoc
Pancreas 321	• *Focus on Homeostas*

■ Student Learning Objectives

1. Distinguish between exocrine and endocrine glands.   **305**
2. Define target cells and describe the role of hormone receptors.   **305**
3. Describe the two general mechanisms of action of hormones.   **306**
4. Explain how blood hormone levels are regulated.   **307**
7. Describe the locat and functions of t

LUNG VOLUMES AND CAPACITIES

Objective: • **Define the various lung volumes and capacities.**

The term for normal quiet breathing is *eupnea* (yoop-NĒ-a; *eu-* = normal; *-pnea* = breath). While at rest, a healthy adult averages 12 breaths a minute, with each inhalation and exhalation moving about 500 mL of air into and out of the lungs. The volume of one breath is called the *tidal volume*. The *minute venti-*

Studying the **Figures** (illustrations that include artwork and photographs) in this book is as important as reading the text. To get the most out of the visual parts of this book, use the tools we have added to the figures to help you understand the concepts being presented.

Start by reading the **legend,** which explains what the figure is about. Next, study the **key concept statement,** which reveals a basic idea portrayed in the figure. Added to many figures you will also find an **orientation diagram** to help you understand the perspective from which you are viewing a particular piece of anatomical art. Finally, at the bottom of each figure you will find a **Figure Question.** If you try to answer these questions as you go along, they will serve as self-checks to help you understand the material. Often it will be possible to answer a question by examining the figure itself. Other questions will encourage you to integrate the knowledge you've gained by carefully reading the text associated with the figure. Still other questions may prompt you to think critically about the topic at hand or predict a consequence in advance of its description in the text. You will find the answers to figure questions at the end of chapters.

legend

key concept statement

orientation diagram

figure question

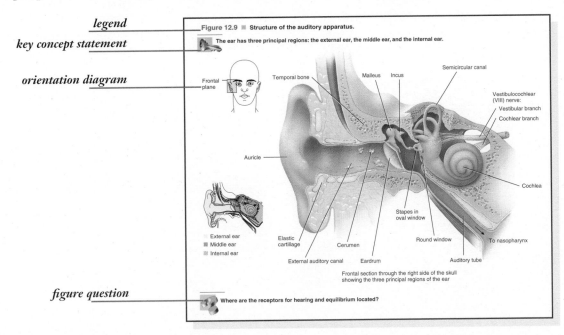

Figure 12.9 ■ Structure of the auditory apparatus.

The ear has three principal regions: the external ear, the middle ear, and the internal ear.

Frontal plane

Temporal bone — Malleus — Incus — Semicircular canal

Vestibulocochlear (VIII) nerve:
Vestibular branch
Cochlear branch

Auricle

Cochlea

Stapes in oval window

Round window

To nasopharynx

External ear
Middle ear
Internal ear

Elastic cartilage — Cerumen — Auditory tube

External auditory canal — Eardrum

Frontal section through the right side of the skull showing the three principal regions of the ear

Where are the receptors for hearing and equilibrium located?

At the end of each chapter are other resources that you will find useful. The **Study Outline** is a concise summary of important topics discussed in the chapter. Page numbers are listed next to key concepts so you can easily refer to specific passages in the text for clarification or amplification. The **Self-Quiz** is designed to help you evaluate your understanding of the chapter contents.

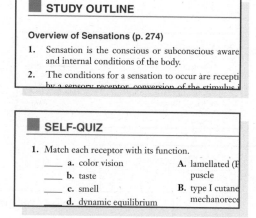

STUDY OUTLINE

Overview of Sensations (p. 274)

1. Sensation is the conscious or subconscious aware and internal conditions of the body.
2. The conditions for a sensation to occur are recepti by a sensory receptor, conversion of the stimulus

SELF-QUIZ

1. Match each receptor with its function.
   ____ **a.** color vision   **A.** lamellated (F
   ____ **b.** taste   puscle
   ____ **c.** smell   **B.** type I cutane
   ____ **d.** dynamic equilibrium   mechanorec

Critical Thinking Applications are word problems that allow you to apply the concepts you have studied in the chapter to specific situations.

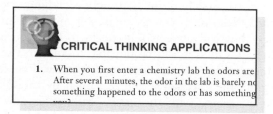

CRITICAL THINKING APPLICATIONS

1. When you first enter a chemistry lab the odors are After several minutes, the odor in the lab is barely n something happened to the odors or has something

Throughout the text we have included **Pronunciations** and, sometimes, **Word Roots,** for many terms that may be new to you. These appear in parentheses immediately following the new words, and the pronunciations are repeated in the glossary at the back of the book. Look at the words carefully and say them out loud several times. Learning to pronounce a new word will help you remember it and make it a useful part of your medical vocabulary. Take a few minutes now to read the following pronunciation key, so it will be familiar as you encounter new words. The key is repeated at the beginning of the Glossary, page G-1.

• Pronunciation Key

1. The most strongly accented syllable appears in capital letters, for example, bilateral (bī-LAT-er-al) and diagnosis (dī-ag-NŌ-sis).

2. If there is a secondary accent, it is noted by a prime ('), for example, physiology (fiz'-ē-OL-ō-jē). Any additional secondary accents are also noted by a prime, for example, decarboxylation (dē'-kar-bok'-si-LĀ-shun).

3. Vowels marked by a line above the letter are pronounced with the long sound as in the following common words:
   ā as in māke   ī as in īvy
   ē as in bē   ō as in pōle

4. Vowels not so marked are pronounced with the short sound as in the following words:
   e as in be   i as in sip
   o as in not   u as in bud

5. Other phonetic symbols are used to indicate the following sounds:
   a as in above   oo as in sue
   yoo as in cute   oy as in oil

Brief Contents

Contents

INTRODUCTION TO THE HUMAN BODY

Chapter 1

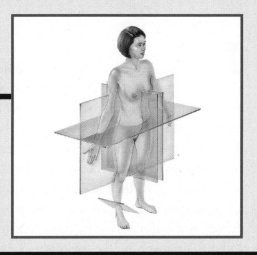

Organization of the Human Body

■ Student Learning Objectives

■ A Look Ahead

You are beginning a fascinating exploration of the human body in which you'll learn how it is organized and how it functions. First you will be introduced to the scientific disciplines of anatomy and physiology; we'll consider the levels of organization that characterize living things and the properties that all living things share. Then, we will examine how the body is constantly regulating its internal environment. This ceaseless process, called homeostasis, is a major theme in every chapter of this book. We will also discuss how the various individual systems that compose the human body cooperate with one another to maintain the health of the body as a whole. Finally, we will establish a basic vocabulary that allows us to speak about the body in a way that is understood by scientists and health-care professionals alike.

ANATOMY AND PHYSIOLOGY DEFINED

Objective: • **Define anatomy and physiology.**

The sciences of anatomy and physiology are the foundation for understanding the structures and functions of the human body. *Anatomy* (a-NAT-ō-mē; *ana-* = up; *tome* = a cutting) is the science of structure and the relationships among structures. *Physiology* (fiz′-ē-OL-ō-jē; *physis* = nature, *-ology* = study of) is the science of body functions, that is, how the body parts work. Because function can never be separated completely from structure, we can understand the human body best by studying anatomy and physiology together. We will look at how each structure of the body is designed to carry out a particular function and how the structure of a part often determines the functions it can perform. The bones of the skull, for example, are tightly joined to form a rigid case that protects the brain. The bones of the fingers, by contrast, are more loosely joined, which enables them to perform a variety of movements, such as turning the pages of this book.

LEVELS OF ORGANIZATION AND BODY SYSTEMS

Objectives: • **Describe the structural organization of the human body.**
• **Explain how body systems relate to one another.**

The structures of the human body are organized on several levels, similar to the way letters of the alphabet, words, sentences, and paragraphs make up language. Listed here, from smallest to largest, are the six levels of organization that are relevant to

anatomy and physiology: chemical, cellular, tissue, organ, system, and organismal (Figure 1.1).

1 The *chemical level* includes *atoms,* the smallest units of matter that participate in chemical reactions, and *molecules,* two or more atoms joined together. Certain atoms, such as carbon, hydrogen, oxygen, nitrogen, calcium, and others, are essential for maintaining life. Familiar examples of molecules found in the body are DNA (deoxyribonucleic acid), the genetic material passed on from one generation to another; hemoglobin, which carries oxygen in the blood; glucose, commonly known as blood sugar; and vitamins, which are needed for a variety of chemical processes. Chapters 2 and 20 focus on the chemical level of organization.

2 Molecules combine to form structures at the next level of organization—the *cellular level. Cells* are the basic structural and functional units of an organism and the smallest living units in the human body. Among the many types of cells in your body are muscle cells, nerve cells, and blood cells. Figure 1.1 shows a smooth muscle cell, one of three different types of muscle cells in your body. As you will see in Chapter 3, cells contain specialized structures called *organelles,* such as the nucleus, mitochondria, and lysosomes, that perform specific functions.

3 The *tissue level* is the next level of structural organization. *Tissues* are groups of cells and the materials surrounding them that work together to perform a particular function. The four basic types of tissue in your body are *epithelial tissue, connective tissue, muscle tissue, and nervous tissue.* The similarities and differences among the different types of tissues are the focus of Chapter 4. Note in Figure 1.1 that smooth muscle tissue consists of tightly packed smooth muscle cells.

4 At the *organ level,* different kinds of tissues join together to form body structures. *Organs* usually have a recognizable shape, are composed of two or more different types of tissues, and have specific functions. Examples of organs are the stomach, heart, liver, lungs, and brain. Figure 1.1 shows several tissues that make up the stomach. The *serous membrane* is a layer of epithelial tissue and connective tissue around the outside of the stomach that protects it and reduces friction when the stomach moves and rubs against other organs. Underneath the serous membrane are the *muscle tissue layers,* which contract to churn and mix food and push it on to the next digestive organ, the small intestine. The innermost lining of the stomach is an *epithelial tissue layer,* which contributes fluid and chemicals that aid digestion.

5 The next level of structural organization in the body is the *system level.* A *system* consists of related organs that have a common function. The example shown in Figure 1.1 is the digestive system, which breaks down and absorbs molecules in food. Its organs include the mouth, salivary glands, pharynx (throat), esophagus (food tube), stomach, liver, gallbladder, pancreas, small intestine, and large intestine. Sometimes an organ is part of more than one system. The pancreas, for example, is part of both the digestive system and the hormone-producing endocrine system.

Figure 1.1 ■ **Levels of structural organization in the human body.**

The levels of structural organization are chemical, cellular, tissue, organ, system, and organismal.

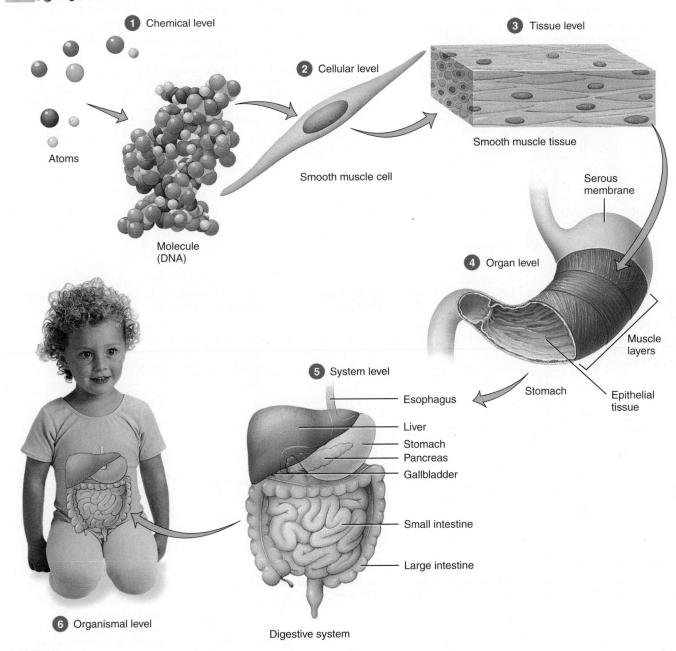

1 Chemical level

Atoms

Molecule (DNA)

2 Cellular level

Smooth muscle cell

3 Tissue level

Smooth muscle tissue

Serous membrane

4 Organ level

Muscle layers

Stomach

Epithelial tissue

5 System level

Esophagus

Liver

Stomach

Pancreas

Gallbladder

Small intestine

Large intestine

Digestive system

6 Organismal level

 Which level of structural organization usually has a recognizable shape and is composed of two or more different types of tissues?

In the chapters that follow, we will explore the anatomy and physiology of each of the body systems. Table 1.1 introduces the components and functions of these systems.

6 The *organismal level* is the largest level of organization. All the systems of the body combine to make up an *organism,* that is, one human being.

As you study the body systems, you will discover how they work together to maintain health, protect you from disease, and allow for reproduction of the species. As an example, consider how just two of the body systems—the integumentary and skeletal systems—cooperate to function at the organismal level. The integumentary system, which includes the skin, hair, and nails,

Table 1.1 / **Components and Functions of the Eleven Principal Systems of the Human Body**

Integumentary System

Components: Skin and structures derived from it, such as hair, nails, and sweat and oil glands.

Functions: Helps regulate body temperature; protects the body; eliminates some wastes; helps make vitamin D; detects sensations such as pressure, pain, heat, and cold.

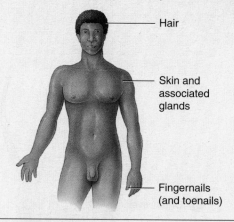

Hair

Skin and associated glands

Fingernails (and toenails)

Nervous System

Components: Brain, spinal cord, nerves, and sense organs such as the eye and ear.

Functions: Regulates body activities through nerve impulses by detecting changes in the environment, interpreting the changes, and responding to the changes by bringing about muscular contractions or glandular secretions.

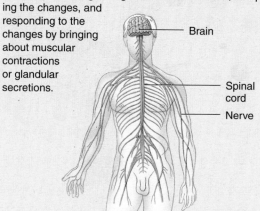

Brain

Spinal cord

Nerve

Skeletal System

Components: All the bones of the body, their associated cartilages, and joints.

Functions: Supports and protects the body, assists with body movements, stores cells that produce blood cells, stores minerals and lipids (fats).

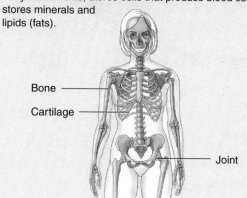

Bone

Cartilage

Joint

Endocrine System

Components: All glands and tissues that produce chemical regulators of body functions, called hormones.

Functions: Regulates body activities through hormones transported by the blood to various target organs.

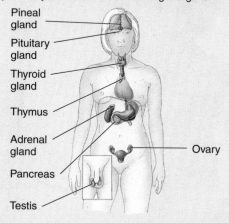

Pineal gland

Pituitary gland

Thyroid gland

Thymus

Adrenal gland

Pancreas

Testis

Ovary

Muscular System

Components: Specifically refers to skeletal muscle tissue, which is muscle usually attached to bones (other muscle tissues include smooth and cardiac).

Functions: Participates in bringing about movement, maintains posture, and produces heat.

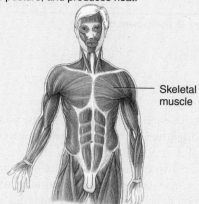

Skeletal muscle

Cardiovascular System

Components: Blood, heart, and blood vessels.

Functions: Heart pumps blood through blood vessels; blood carries oxygen and nutrients to cells and carbon dioxide and wastes away from cells, and helps regulate acidity, temperature, and water content of body fluids; blood components help defend against disease and mend damaged blood vessels.

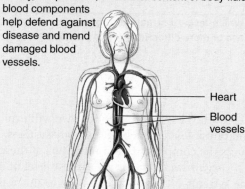

Heart

Blood vessels

Lymphatic and Immune System

Components: Lymphatic fluid and vessels; spleen, thymus, lymph nodes, and tonsils; cells that carry out immune responses (B cells, T cells, and others).

Functions: Returns proteins and fluid to blood; carries lipids from gastrointestinal tract to blood; contains sites of maturation and proliferation of B cells and T cells that protect against disease-causing organisms.

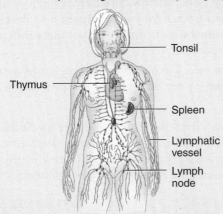

Tonsil

Thymus

Spleen

Lymphatic vessel

Lymph node

Digestive System

Components: Organs of gastrointestinal tract, including the mouth, esophagus, stomach, small and large intestines, rectum, and anus; also includes accessory digestive organs that assist in digestive processes, such as the salivary glands, liver, gallbladder, and pancreas.

Functions: Achieves physical and chemical breakdown of food; absorbs nutrients; eliminates solid wastes.

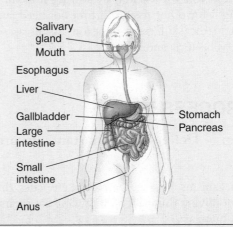

Salivary gland
Mouth
Esophagus
Liver
Gallbladder
Large intestine
Small intestine
Anus
Stomach
Pancreas

Respiratory System

Components: Lungs and air passageways such as the pharynx (throat), larynx (voice box), trachea (windpipe), and bronchial tubes leading into and out of them.

Functions: Transfers oxygen from inhaled air to blood and carbon dioxide from blood to exhaled air; helps regulate acidity of body fluids; air flowing out of lungs through vocal cords produces sounds.

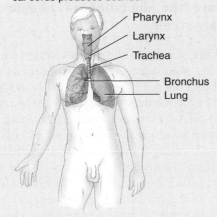

Pharynx
Larynx
Trachea
Bronchus
Lung

Urinary System

Components: Kidneys, ureters, urinary bladder, and urethra.

Functions: Produces, stores, and eliminates urine; eliminates wastes and regulates volume and chemical composition of blood; maintains body's mineral balance; helps regulate red blood cell production.

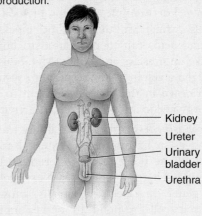

Kidney
Ureter
Urinary bladder
Urethra

Reproductive Systems

Components: Gonads (testes or ovaries) and associated organs: uterine tubes, uterus, and vagina in females, and epididymis, ductus (vas) deferens, and penis in males. Also, mammary glands in females.

Functions: Gonads produce gametes (sperm or oocytes) that unite to form a new organism and release hormones that regulate reproduction and other body processes; associated organs transport and store gametes. Mammary glands produce milk.

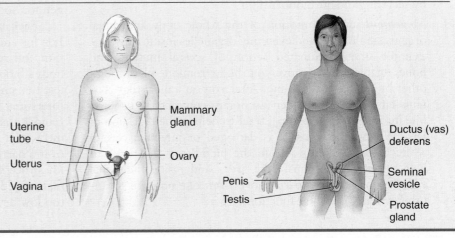

Uterine tube
Uterus
Vagina
Mammary gland
Ovary
Penis
Testis
Ductus (vas) deferens
Seminal vesicle
Prostate gland

protects all other body systems, including the skeletal system, which includes all the bones and joints of the body. The skin serves as a barrier between the outside environment and internal tissues and organs, such as those that make up the skeletal system. The skin also participates in the production of vitamin D, which the body needs to absorb the calcium in foods such as milk. Calcium is used to build bones and teeth. The skeletal system, in turn, provides support for the integumentary system, serves as a reservoir for calcium by storing it in times of plenty and releasing it for other tissues in times of need, and generates cells that help the skin resist invasion by disease-causing microbes such as bacteria and viruses.

LIFE PROCESSES

Objective: • **Define the important life processes of humans.**

All living organisms have certain characteristics that set them apart from nonliving things. The following are six important life processes of humans:

1. *Metabolism* (me-TAB-ō-lizm) is the sum of all the chemical processes that occur in the body. It includes the breakdown of large, complex molecules into smaller, simpler ones and the building of the body's structural and functional components. For example, proteins in food are split into amino acids, which are the building blocks of proteins. These amino acids can then be used to build new proteins that make up muscles and bones. Metabolism also involves using oxygen taken in by the respiratory system and nutrients broken down in the digestive system to provide the chemical energy needed to power cellular activities.

2. *Responsiveness* is the body's ability to detect and respond to changes in its internal or external environment. Different cells in the body detect different sorts of changes and respond in characteristic ways. Nerve cells respond to changes by generating electrical signals, known as nerve impulses. Muscle cells respond to nerve impulses by contracting, which generates force to move body parts. Endocrine cells in the pancreas respond to an elevated blood glucose level by secreting the hormone insulin.

3. *Movement* includes motion of the whole body, individual organs, single cells, and even tiny organelles inside cells. For example, the coordinated action of several muscles and bones enables you to move your body from one place to another by walking or running. After you eat a meal that contains fats, your gallbladder (an organ) contracts and squirts bile into the gastrointestinal tract to help in the digestion of fats. When a body tissue is damaged or infected, certain white blood cells move from the blood into the tissue to help clean up and repair the area. And inside individual cells, various cellular parts move from one position to another to carry out their functions.

4. *Growth* is an increase in body size. It may be due to an increase in (1) the size of existing cells, (2) the number of cells, or (3) the amount of material surrounding cells.

5. *Differentiation* is the process whereby unspecialized cells become specialized cells. Specialized cells differ in structure and function from the precursor cells that gave rise to them. For example, red blood cells and several types of white blood cells differentiate from the same unspecialized precursor cells in bone marrow. Similarly, a single fertilized egg cell undergoes tremendous differentiation to develop into a unique individual who is similar to, yet quite different from, either of the parents.

6. *Reproduction* refers to either the formation of new cells for growth, repair, or replacement or the production of a new individual.

Although not all of these processes are occurring in cells throughout the body all of the time, when they cease to occur properly cell death may occur. When cell death is extensive and leads to organ failure, the result is death of the organism.

HOMEOSTASIS: MAINTAINING LIMITS

Objectives: • **Define homeostasis and explain its importance.**
• **Describe the components of a feedback system.**
• **Compare the operation of negative and positive feedback systems.**

The trillions of cells of the various systems and organs of the human body need relatively stable conditions to function effectively and contribute to the survival of the body as a whole. The maintenance of relatively stable conditions, called *homeostasis* (hō′-mē-ō-STĀ-sis; *homeo-* = sameness; *-stasis* = standing still), is a condition in which the body's internal environment remains steady despite changes inside and outside the body. A large part of the internal environment consists of the fluid surrounding body cells, called *interstitial fluid*. Homeostasis ensures that the interstitial fluid remains at a proper temperature of 37° Celsius (98° to 99° Fahrenheit) and contains adequate nutrients and oxygen for body cells to flourish.

Each body system contributes to homeostasis in some way. For instance, in the cardiovascular system, alternating contraction and relaxation of the heart propels blood throughout the body's blood vessels. As blood flows through the blood capillaries, the smallest blood vessels, nutrients and oxygen move into interstitial fluid and wastes move into the blood. Cells, in turn, remove nutrients and oxygen from and release their wastes into interstitial fluid. Homeostasis is *dynamic;* that is, it can change over a narrow range that is compatible with maintaining cellular life processes. For example, the level of glucose in the blood is maintained within a narrow range. It normally does not fall too low between meals or rise too high even after eating a high-glucose meal. The brain needs a steady supply of glucose to keep functioning—a low blood glucose level may lead to unconsciousness or even death. A prolonged high blood glucose level, by contrast, can damage blood vessels and cause excessive loss of water in the urine.

Control of Homeostasis

Fortunately, every body structure, from cells to systems, has one or more homeostatic devices that work to keep the internal environment within normal limits. The homeostatic mechanisms of the body are mainly under the control of two systems, the nervous system and the endocrine system. The nervous system detects changes from the balanced state and sends messages in the form of *nerve impulses* to organs that can counteract the change. For example, when body temperature rises, nerve impulses cause sweat glands to release more sweat, which cools the body as it evaporates. The endocrine system corrects changes by secreting molecules called *hormones* into the blood. Hormones are targeted to specific body cells where they cause responses that restore homeostasis. For example, the hormone insulin reduces blood glucose level when it is too high. Nerve impulses typically cause rapid corrections, whereas hormones usually work more slowly.

Feedback Systems and Homeostasis

Homeostasis is maintained by means of many feedback systems. A *feedback system* is a cycle of events in which the status of a body condition is continually monitored, evaluated, changed, remonitored, reevaluated, and so on. Each monitored condition, such as body temperature, blood pressure, or blood glucose level, is termed a *controlled condition*. Any disruption that causes a change in a controlled condition is called a *stimulus*. Some disruptions come from the external environment (outside the body) in the form of physical insults such as intense heat or lack of oxygen. Other disruptions originate in the internal environment (within the body), for example, a blood glucose level that is too low. Homeostatic imbalances may also occur due to psychological stresses in our social environment—the demands of work and school, for example. In most cases, the disruption of homeostasis is mild and temporary, and the responses of body cells quickly restore balance in the internal environment. In other cases, the disruption of homeostasis may be intense and prolonged, as in poisoning, overexposure to temperature extremes, severe infection, or death of a loved one.

Three basic components compose a feedback system or *feedback loop:* a receptor, a control center, and an effector (Figure 1.2).

- A *receptor* is a body structure that monitors changes in a controlled condition and sends information called the *input* to a control center. Input is in the form of nerve impulses or chemical signals. Nerve endings in the skin that sense temperature are one of the hundreds of different kinds of receptors in the body.

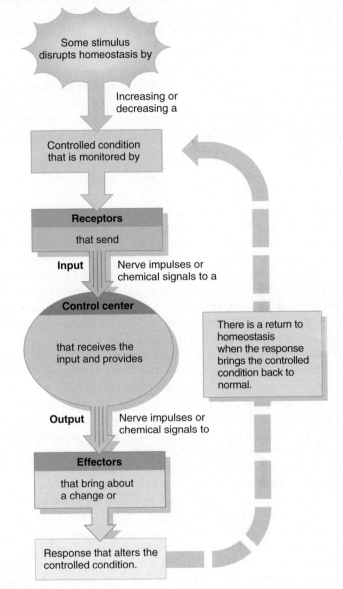

Figure 1.2 ■ **Components of a feedback system.** The dashed return arrow symbolizes negative feedback.

The three basic elements of a feedback system are the receptors, control center, and effectors.

Some stimulus disrupts homeostasis by

Increasing or decreasing a

Controlled condition that is monitored by

Receptors

that send

Input → Nerve impulses or chemical signals to a

Control center

that receives the input and provides

There is a return to homeostasis when the response brings the controlled condition back to normal.

Output → Nerve impulses or chemical signals to

Effectors

that bring about a change or

Response that alters the controlled condition.

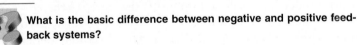

What is the basic difference between negative and positive feedback systems?

- A *control center* in the body sets the range of values within which a controlled condition should be maintained, evaluates the input it receives from receptors, and generates output commands when they are needed. *Output* is information, in the form of nerve impulses or chemical signals, that is relayed from the control center to an effector.

- An *effector* is a body structure that receives output from the control center and produces a *response* or effect that changes the controlled condition. Nearly every organ or tissue in the body can behave as an effector. For example, when your body temperature drops sharply, your brain (control center) sends nerve impulses to your skeletal muscles (effectors) that cause you to shiver, which generates heat and raises your temperature.

Feedback systems can produce either negative feedback or positive feedback. If the response reverses the original stimulus, as in the body temperature regulation example, the system is operating by *negative feedback*. If the response enhances or intensifies the original stimulus, the system is operating by *positive feedback*.

Negative Feedback Systems

A *negative feedback system* *reverses* a change in a controlled condition. First, a stimulus disrupts homeostasis by changing a controlled condition. The receptors that are part of the feedback system detect the change and send input to a control center. The control center evaluates the input and issues output commands to an effector. The effector produces a physiological response that reverses the change in the controlled condition. The activity of a negative feedback system slows and then stops as the controlled condition returns to its normal state.

Consider one negative feedback system that helps regulate blood pressure. Blood pressure (BP) is the force exerted by blood as it presses against the walls of blood vessels. When the heart beats faster or harder, BP increases. If a stimulus causes blood pressure (controlled condition) to rise, the following sequence of events occurs (Figure 1.3). The higher pressure is detected by *baroreceptors*, pressure-sensitive nerve cells located in the walls of certain blood vessels (the receptors). The baroreceptors send nerve impulses (input) to the brain (control center), which interprets the impulses and responds by sending nerve impulses (output) to the heart (the effector). Heart rate decreases, which causes blood pressure to decrease (response). This sequence of events returns the controlled condition—blood pressure—to normal, and homeostasis is restored. This loop is a negative feedback system because the activity of the effector produces a result, a drop in BP, that reverses the effect of the stimulus. Negative feedback systems tend to regulate conditions in the body that are held fairly stable over long periods of time.

Positive Feedback Systems

A *positive feedback system* operates similarly to a negative feedback system except for the effect of the response: the effector produces a physiological response that *reinforces* the initial change in the controlled condition. Normal positive feedback systems tend to reinforce conditions that don't happen very often, such as childbirth, ovulation, and blood clotting. Because a positive feedback system continually reinforces a change in a controlled condition, it must be shut off by some event outside

Figure 1.3 ■ **Homeostasis of blood pressure by a negative feedback system.** Note that the response is fed back into the system, and the system continues to lower blood pressure until there is a return to normal blood pressure (homeostasis).

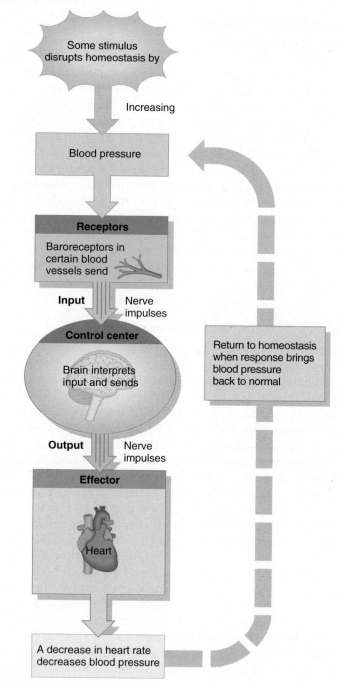

🔑 **If the response reverses the stimulus, a system is operating by negative feedback.**

What would happen to the heart rate if some stimulus caused blood pressure to decrease? Would this occur by positive or negative feedback?

Good Health— Homeostasis Is the Basis

You've seen *homeostasis* defined as a condition in which the body's internal environment remains relatively stable. What does this mean to you in your everyday life?

Homeostasis: The Power to Heal

The body's ability to maintain homeostasis gives it tremendous healing power and a remarkable resistance to abuse. The physiological processes responsible for maintaining homeostasis are in large part also responsible for your good health.

For most people, lifelong good health is not something that just happens. Two important factors in this balance called health are the environment and your own behavior. Your body's homeostasis is affected by the air you breathe, the food you eat, and even the thoughts you think. The way you live your life can either support or interfere with your body's ability to maintain homeostasis and recover from the inevitable stresses life throws your way.

Let's consider the common cold. You support your natural healing processes when you take care of yourself. Plenty of rest, fluids, and chicken soup allow the immune system to do its job. The cold runs its course, and you are soon back on your feet. If instead of taking care of yourself, you continue to smoke two packs of cigarettes a day, skip meals, and pull several all nighters studying for an anatomy and physiology exam, you interfere with the immune system's ability to fend off attacking microbes and bring the body back to homeostasis and good

health. Other infections take advantage of your weakened state, and pretty soon the cold has "turned into" bronchitis or pneumonia.

Homeostasis and Disease Prevention

Many diseases are the result of years of poor health behavior that interferes with the body's natural drive to maintain homeostasis. An obvious example is smoking-related illness. Smoking tobacco exposes sensitive lung tissue to a multitude of chemicals that cause cancer and damage the lung's ability to repair itself. Because diseases such as emphysema and lung cancer are difficult to treat and very rarely cured, it is much wiser to quit smoking—or never start—than to hope a doctor can fix you once you are diagnosed with a lung disease. Developing a lifestyle that works with, rather than against, your body's homeostatic processes helps you maximize your personal potential for optimal health and well-being.

▶ *Think It Over*

▶ What health habits have you developed over the past several years to prevent disease or enhance your body's ability to maintain health and homeostasis?

the system. If the action of a positive feedback system isn't stopped, it can "run away" and produce life-threatening changes in the body.

Homeostasis and Disease

As long as all of the body's controlled conditions remain within certain narrow limits, body cells function efficiently, homeostasis is maintained, and the body stays healthy. Should one or more components of the body lose their ability to contribute to homeostasis, however, the normal balance among all of the body's processes may be disturbed. If the homeostatic imbalance is

moderate, a disorder or disease may occur; if it is severe, death may result.

A *disorder* is any derangement or abnormality of function. *Disease* is a more specific term for an illness characterized by a recognizable set of signs and symptoms. *Symptoms* are subjective changes in body functions that are not apparent to an observer, for example, headache or nausea. *Signs* are objective changes that a clinician can observe and measure, such as bleeding, swelling, vomiting, diarrhea, fever, a rash, or paralysis. A *local disease* is one that affects one part or a limited region of the body. A *systemic disease* affects either several body parts or the entire body. Many diseases produce positive feedback loops that cause death by overwhelming the normal negative feedback loops that maintain homeostasis. Specific diseases alter body structure and function in characteristic ways, usually producing a recognizable cluster of signs and symptoms.

The science that deals with why, when, and where diseases occur and how they are transmitted in a human community is known as *epidemiology* (ep′-i-dē-mē-OL-ō-jē; *epi-* = upon; *demi-* = people). The science that deals with the effects and uses of drugs in the treatment of disease is called *pharmacology* (far′-ma-KOL-ō-jē; *pharmac-* = drug).

Diagnosis (dī′-ag-NŌ-sis; *dia-* = through; *-gnosis* = knowledge) is the identification of a disease or disorder based on a scientific evaluation of the patient's signs and symptoms, medical history, physical examination, and sometimes data from laboratory tests. Taking a *medical history* consists of collecting information about events that might be related to a patient's illness, including the chief complaint, history of present illness, past medical problems, family medical problems, and social history. A *physical examination* is an orderly evaluation of the body and its functions. This process includes inspection (looking at or into the body with various instruments), palpation (feeling body surfaces with the hands), auscultation (listening to body sounds, often using a stethoscope), percussion (tapping on body surfaces and listening to the resulting echo), and measuring vital signs (temperature, pulse, respiratory rate, and blood pressure). Some common laboratory tests include analyses of blood and urine.

ANATOMICAL TERMS

Objectives: • **Describe the anatomical position.**
• **Identify the major regions of the body and relate the common names to the corresponding anatomical terms for various parts of the body.**
• **Define the directional terms and the anatomical planes and sections used to locate parts of the human body.**

The language of anatomy and physiology has precisely defined meanings that allow unambiguous communication. When describing where the wrist is located, is it correct to say "the wrist is above the fingers"? This description is true if your arms are at your sides. But if you hold your hands up above your head, your fingers would be above your wrists. To prevent this kind of confusion, scientists and health-care professionals refer to one standard anatomical position and use a special vocabulary for relating body parts to one another.

Anatomical Position

In the study of anatomy, descriptions of any region or part of the human body assume that the body is in a specific stance called the *anatomical position.* In the anatomical position, the subject stands erect facing the observer, with the head level and the eyes facing forward. The feet are flat on the floor and directed forward, and the arms are at the sides with the palms turned forward (Figure 1.4).

Names of Body Regions

The human body is divided into several major regions that can be identified externally. The principal regions are the head, neck, trunk, upper limbs, and lower limbs (Figure 1.4). The *head* consists of the skull and face. The *skull* is the part of the head that encloses and protects the brain, and the *face* is the front portion of the head that includes the eyes, nose, mouth, forehead, cheeks, and chin. The *neck* supports the head and attaches it to the trunk. The *trunk* consists of the chest, abdomen, and pelvis. Each *upper limb* is attached to the trunk and consists of the shoulder, armpit, arm (portion of the limb from the shoulder to the elbow), forearm (portion of the limb from the elbow to the wrist), wrist, and hand. Each *lower limb* is also attached to the trunk and consists of the buttock, thigh (portion of the limb from the hip to the knee), leg (portion of the limb from the knee to the ankle), ankle, and foot. The *groin* is the area on the front surface of the body, marked by a crease on each side, where the trunk attaches to the thighs.

In Figure 1.4, the corresponding anatomical descriptive form (adjective) for each part of the body appears in parentheses next to the common name. For example, if you receive a tetanus shot in your *buttock*, it is a *gluteal* injection. The descriptive form of a body part is different from the common name because it is based on a Greek or Latin word or "root" for the same part or area. The Latin word for armpit is *axilla* (ak-SIL-a), for example, and thus one of the nerves passing within the armpit is called the axillary nerve. You will learn more about the word roots of anatomical and physiological terms as you read this book.

Directional Terms

To locate various body structures, anatomists use specific *directional terms,* words that describe the position of one body part relative to another. Several directional terms can be grouped in pairs that have opposite meanings, for example, anterior (front) and posterior (back). Study Exhibit 1.1 on page 12 and Figure 1.5 on page 13 to determine, among other things, whether your stomach is superior to your lungs.

Planes and Sections

You will also study parts of the body in various planes, that is, imaginary flat surfaces that pass through the body parts (Figure 1.6) on page 14. A *sagittal plane* (SAJ-i-tal; *sagitt-* = arrow) is a vertical plane that divides the body or an organ into right and left sides. More specifically, when such a plane passes through the midline of the body or organ and divides it into *equal* right and left sides, it is called a *midsagittal plane.* If the sagittal plane does not pass through the midline but instead divides the body or an organ into *unequal* right and left sides, it is called a

Figure 1.4 ■ **The anatomical position.** The common names and corresponding anatomical terms (in parentheses) indicate specific body regions. For example, the head is the cephalic region.

In the anatomical position, the subject stands erect facing the observer, with the head level and the eyes facing forward. The feet are flat on the floor and directed forward, and the arms are at the sides with the palms facing forward.

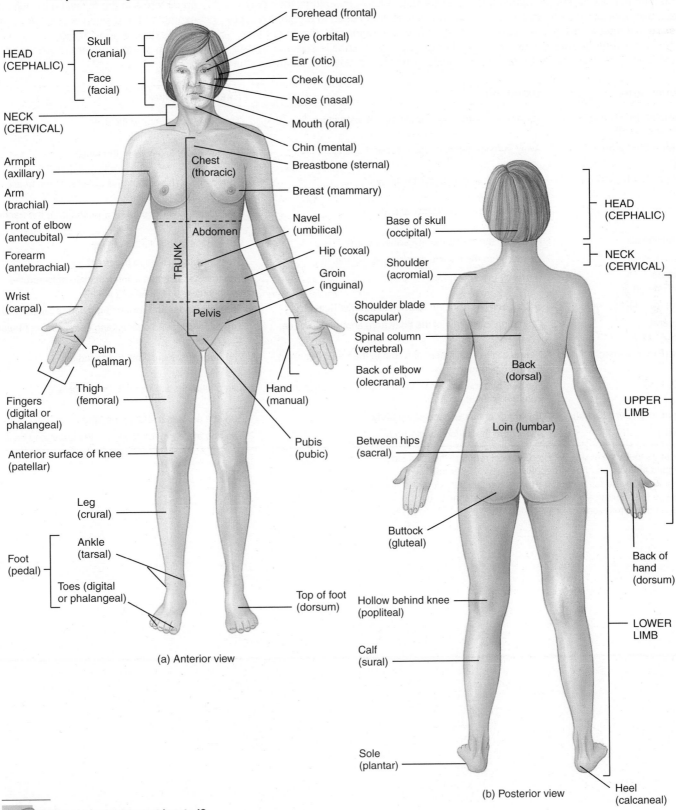

HEAD (CEPHALIC)
Skull (cranial)
Face (facial)

NECK (CERVICAL)

Armpit (axillary)

Arm (brachial)

Front of elbow (antecubital)

Forearm (antebrachial)

Wrist (carpal)

Palm (palmar)

Fingers (digital or phalangeal)

Thigh (femoral)

Anterior surface of knee (patellar)

Leg (crural)

Foot (pedal)
Ankle (tarsal)
Toes (digital or phalangeal)

Forehead (frontal)
Eye (orbital)
Ear (otic)
Cheek (buccal)
Nose (nasal)
Mouth (oral)
Chin (mental)
Breastbone (sternal)
Breast (mammary)

Chest (thoracic)

Abdomen

TRUNK

Pelvis

Navel (umbilical)
Hip (coxal)
Groin (inguinal)

Hand (manual)

Pubis (pubic)

Top of foot (dorsum)

(a) Anterior view

Base of skull (occipital)
Shoulder (acromial)
Shoulder blade (scapular)
Spinal column (vertebral)
Back of elbow (olecranal)

Back (dorsal)

Loin (lumbar)

Between hips (sacral)

Buttock (gluteal)

Hollow behind knee (popliteal)

Calf (sural)

Sole (plantar)

HEAD (CEPHALIC)
NECK (CERVICAL)

UPPER LIMB

Back of hand (dorsum)

LOWER LIMB

Heel (calcaneal)

(b) Posterior view

Where is a plantar wart located?

Exhibit 1.1 / Directional Terms Used to Describe the Human Body *(Figure 1.5)*

Objective: Define each directional term used to describe the human body.

Most of the directional terms used to describe the human body can be grouped into pairs that have opposite meanings. For example, **superior** means toward the upper part of the body, whereas **inferior** means toward the lower part of the body. Moreover, it is important to understand that directional terms have relative meanings; they only make sense when used to describe the position of one structure relative to another. For example, your knee is superior to your ankle, even though both are located in the inferior half of the body. Study the directional terms in the following exhibit and the example of how each is used. As you read each example, look at Figure 1.5 to see the location of the structures mentioned.

Directional Term	Definition	Example of Use
Superior (soo′-PEER-ē-or) (**cephalic** or **cranial**)	Toward the head, or the upper part of a structure.	The heart is superior to the liver.
Inferior (in′-FEER-ē-or) (**caudal**)	Away from the head, or the lower part of a structure.	The stomach is inferior to the lungs.
Anterior (an-TEER-ē-or) (**ventral**)*	Nearer to or at the front of the body.	The sternum (breastbone) is anterior to the heart.
Posterior (pos-TEER-ē-or) (**dorsal**)*	Nearer to or at the back of the body.	The esophagus (food tube) is posterior to the trachea (windpipe).
Medial (MĒ-dē-al)	Nearer to the midline† or midsagittal plane.	The ulna is medial to the radius.
Lateral (LAT-er-al)	Farther from the midline or midsagittal plane.	The lungs are lateral to the heart.
Intermediate (in′-ter-MĒ-dē-at)	Between two structures.	The transverse colon is intermediate between the ascending and descending colons.
Ipsilateral (ip′-si-LAT-er-al)	On the same side of the body as another structure.	The gallbladder and ascending colon are ipsilateral.
Contralateral (con′-tra-LAT-er-al)	On the opposite side of the body from another structure.	The ascending and descending colons are contralateral.
Proximal (PROK-si-mal)	Nearer to the attachment of a limb to the trunk; nearer to the point of origin.	The humerus is proximal to the radius.
Distal (DIS-tal)	Farther from the attachment of a limb to the trunk; farther from the point of origin.	The phalanges are distal to the carpals.
Superficial (soo′-per-FISH-al)	Toward or on the surface of the body.	The ribs are superficial to the lungs.
Deep (DĒP)	Away from the surface of the body.	The ribs are deep to the skin of the chest and back.

* Ventral refers to the belly side, whereas dorsal refers to the back side.
In four-legged animals anterior = cephalic (toward the head),
ventral = inferior, posterior = caudal (toward the tail), and dorsal = superior.
† The midline is an imaginary vertical line that divides the body
into equal right and left sides.

Figure 1.5 ■ Directional terms.

 Directional terms precisely locate various parts of the body in relation to one another.

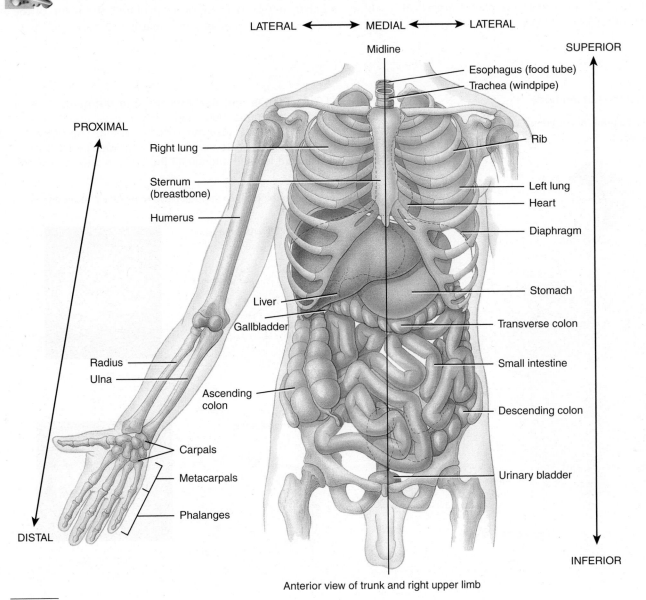

Anterior view of trunk and right upper limb

Is the radius proximal to the humerus? Is the esophagus anterior to the trachea? Are the ribs superficial to the lungs? Is the urinary bladder medial to the ascending colon? Is the sternum lateral to the descending colon?

parasagittal plane (*para* = near). A *frontal plane* or *coronal plane* divides the body or an organ into anterior (front) and posterior (back) portions. A *transverse plane* divides the body or an organ into superior (upper) and inferior (lower) portions. A transverse plane may also be termed a cross-sectional or horizontal plane. Sagittal, frontal, and transverse planes are all at right angles to one another. An *oblique plane,* by contrast, passes through the body or an organ at an angle between the transverse plane and a sagittal plane (Figure 1.6) or between the transverse plane and the frontal plane.

When you study a body region, you will often view it in *section,* meaning that you look at only one flat surface of the three-dimensional structure. It is important to know the plane of the section so you can understand the anatomical relationship of

Figure 1.6 ■ **Planes of the human body.**

 Frontal, transverse, sagittal, and oblique planes divide the body in specific ways.

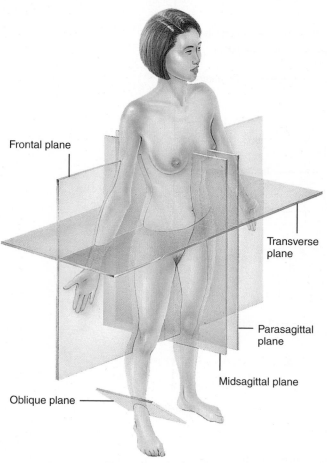

Frontal plane

Transverse plane

Parasagittal plane

Midsagittal plane

Oblique plane

Right anterolateral view

Which plane divides the heart into anterior and posterior portions?

Figure 1.7 ■ **Planes and sections through different parts of the brain.** The diagrams (left) show the planes and the photographs (right) show the resulting sections. (Note: The "view" arrows in the diagrams indicate the direction from which each section is viewed. This aid is used throughout the book to indicate viewing perspective.)

Planes divide the body in various ways to produce sections.

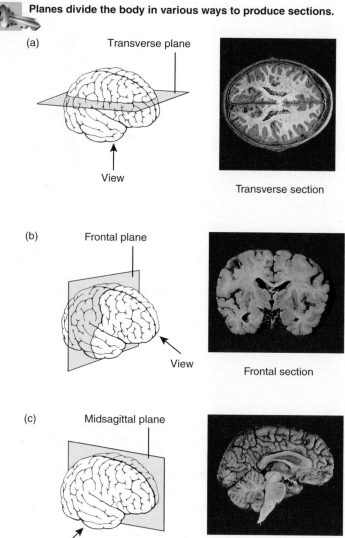

(a) Transverse plane

View

Transverse section

(b) Frontal plane

View

Frontal section

(c) Midsagittal plane

View

Midsagittal section

Which plane divides the brain into equal right and left sides?

one part to another. Figure 1.7 indicates how three different sections—a *transverse section*, a *frontal section*, and a *midsagittal section*—provide different views of the brain.

BODY CAVITIES

Objectives: • **Describe the principal body cavities and the organs they contain.**

• **Explain why the abdominopelvic cavity is divided into regions and quadrants.**

Spaces within the body that contain, protect, separate, and support internal organs are called **body cavities.** The two principal cavities are the dorsal and ventral body cavities (Figure 1.8). The **dorsal body cavity** is located near the dorsal (back) surface of the body. It consists of the **cranial cavity,** which is formed by the cranial (skull) bones and contains the brain, and the **vertebral (spinal) canal,** which is formed by the bones of the vertebral column (backbone) and contains the spinal cord.

The **ventral body cavity** is located on the ventral (front) aspect of the body and contains organs collectively called **viscera** (VIS-er-a). Like the dorsal body cavity, the ventral body cavity has two main subdivisions; the upper portion is the **thoracic cavity** (thor-AS-ik = chest) or chest cavity, and the lower portion is the **abdominopelvic cavity** (ab-dom′-i-nō-PEL-vik; *abdomino-* = belly; *-pelvic* = basin). The **diaphragm** (DĪ-a-fram = partition or wall), the large dome-shaped muscle that powers lung expansion during breathing, forms both the floor of the thoracic cavity and the roof of the abdominopelvic cavity.

Within the thoracic cavity are three smaller cavities: the **pericardial cavity** (per′-i-KAR-dē-al; *peri-* = around; *-cardial* = heart), a fluid-filled space that surrounds the heart, and two **pleural cavities** (PLOOR-al; *pleur-* = rib or side), each of which

Figure 1.8 ■ **Body cavities.** The ventral body cavity is divided into an upper thoracic cavity (blue) and a lower abdominopelvic cavity (green). The dashed lines indicate the border between the abdominal and pelvic cavities.

The two principal body cavities are the dorsal and ventral body cavities.

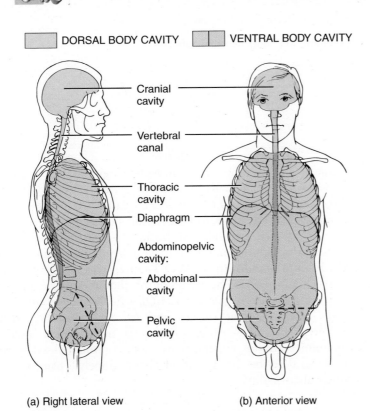

DORSAL BODY CAVITY   VENTRAL BODY CAVITY

Cranial cavity
Vertebral canal
Thoracic cavity
Diaphragm
Abdominopelvic cavity:
Abdominal cavity
Pelvic cavity

(a) Right lateral view   (b) Anterior view

CAVITY	COMMENTS
DORSAL	
Cranial	Formed by cranial bones and contains brain and its coverings.
Vertebral	Formed by vertebral column and contains spinal cord and the beginnings of spinal nerves.
VENTRAL	
Thoracic	Chest cavity; separated from abdominal cavity by diaphragm.
Pleural (right and left)	Contain lungs.
Pericardial	Contains heart.
Mediastinum	Region between the lungs from the breastbone to backbone that contains heart, thymus, esophagus, trachea, bronchi, and many large blood and lymphatic vessels.
Abdominopelvic	Subdivided into abdominal and pelvic cavities.
Abdominal	Contains stomach, spleen, liver, gallblader, small intestine, and most of large intestine.
Pelvic	Contains urinary bladder, portions of the large intestine, and internal female and male reproductive organs.

 In which cavities are the following organs located: urinary bladder, stomach, heart, small intestine, lungs, internal female reproductive organs, thymus, spleen, liver? Use the following symbols for your response: T = thoracic cavity, A = abdominal cavity, or P = pelvic cavity.

Figure 1.9 ■ **The thoracic cavity.** The dashed lines indicate the borders of the mediastinum. Notice that the pericardial cavity surrounds the heart, and that the pleural cavities surround the lungs.

The mediastinum is medial to the lungs; it extends from the sternum to the vertebral column and from the neck to the diaphragm.

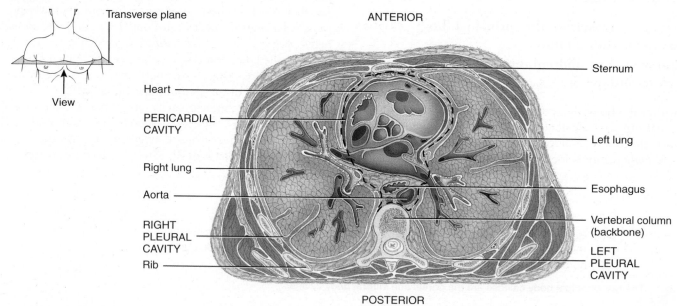

Inferior view of transverse section of thoracic cavity

Which of the following structures are contained in the mediastinum: right lung, heart, esophagus, spinal cord, aorta, left pleural cavity?

surrounds one lung and contains a small amount of fluid (Figure 1.9). The central portion of the thoracic cavity is called the *mediastinum* (mē′-dē-a-STĪ-num; *media-* = middle; *-stinum* = partition). It is a mass of tissues located between the pleural cavities and extending from the sternum (breastbone) to the vertebral column (backbone), and from the neck to the diaphragm (Figure 1.9). The mediastinum contains all thoracic viscera except the lungs themselves, including the heart, esophagus, trachea, and several large blood vessels.

As the name suggests, the abdominopelvic cavity is divided into two portions, although no wall separates them (see Figure 1.8). The upper portion, the *abdominal cavity,* contains the stomach, spleen, liver, gallbladder, small intestine, and most of the large intestine. The lower portion, the *pelvic cavity,* contains the urinary bladder, portions of the large intestine, and internal organs of the reproductive system. The pelvic cavity is located below the dashed line in Figure 1.8.

To describe the location of the many abdominal and pelvic organs more precisely, the abdominopelvic cavity may be divided into smaller compartments. In one method, two horizontal and two vertical lines, like a tick-tack-toe grid, partition the cavity into nine *abdominopelvic regions* (Figure 1.10). The names of the nine abdominopelvic regions are the *right hypochondriac* (hī′-pō-KON-drē-ak), *epigastric* (ep-i-GAS-trik), *left hypochondriac, right lumbar, umbilical* (um-BIL-i-kal), *left lumbar, right iliac* (IL-ē-ak), *hypogastric* (hī′-pō-GAS-trik), and *left iliac.* In another method, one horizontal and one vertical line passing through the umbilicus (navel) divide the abdominopelvic cavity into *quadrants* (KWOD-rantz; *quad-* = a one-fourth part) (Figure 1.11). The names of the abdominopelvic quadrants are the *right upper quadrant (RUQ), left upper quadrant (LUQ), right lower quadrant (RLQ),* and *left lower quadrant (LLQ).* Whereas the nine-region division is more widely used for anatomical studies, quadrants are more commonly used by clinicians to describe the site of an abdominopelvic pain, mass, or other abnormality.

• • •

We will next examine the chemical level of organization in Chapter 2. You will learn about the various groups of chemicals in your body, how they function, and how they contribute to homeostasis of your body.

Figure 1.10 ■ **The nine regions of the abdominopelvic cavity.** The internal reproductive organs in the pelvic cavity are shown in Figures 23.1 and 23.8.

 The nine-region designation is used for anatomical studies.

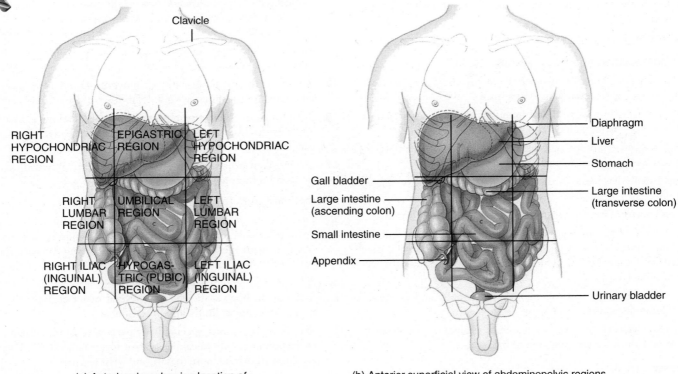

(a) Anterior view showing location of nine abdominopelvic regions

(b) Anterior superficial view of abdominopelvic regions

In which abdominopelvic region is each of the following found: most of the liver, ascending colon, urinary bladder, and appendix?

Figure 1.11 ■ **Quadrants of the abdominopelvic cavity.** The two lines cross at right angles at the umbilicus (navel).

 The quadrant designation is used to locate the site of pain, a mass, or some other abnormality.

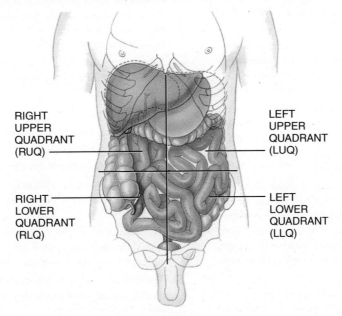

Anterior view

In which abdominopelvic quadrant would the pain from appendicitis (inflammation of the appendix) be felt?

■ STUDY OUTLINE

Anatomy and Physiology Defined (p. 2)

1. Anatomy is the science of structure and the relationships among structures.

2. Physiology is the science of how body structures function.

Levels of Organization and Body Systems (p. 2)

1. The human body consists of six levels of organization: chemical, cellular, tissue, organ, system, and organismal.

2. Cells are the basic structural and functional units of an organism and the smallest living units in the human body.

3. Tissues consist of groups of cells and the materials surrounding them that work together to perform a particular function.

4. Organs usually have recognizable shapes, are composed of two or more different types of tissues, and have specific functions.

5. Systems consist of related organs that have a common function.

6. Table 1.1 on page 4 introduces the eleven systems of the human body: integumentary, skeletal, muscular, nervous, endocrine, cardiovascular, lymphatic and immune, respiratory, digestive, urinary, and reproductive.

7. The human organism is a collection of structurally and functionally integrated systems.

8. Body systems work together to maintain health, protect against disease, and allow for reproduction of the species.

Life Processes (p. 6)

1. All living organisms have certain characteristics that set them apart from nonliving things.

2. Among the life processes in humans are metabolism, responsiveness, movement, growth, differentiation, and reproduction.

Homeostasis: Maintaining Limits (p. 6)

1. Homeostasis is a condition in which the internal environment of the body remains stable, within certain limits.

2. A large part of the body's internal environment is interstitial fluid, which surrounds all body cells.

3. Homeostasis is regulated by the nervous and endocrine systems acting together or separately. The nervous system detects body changes and sends nerve impulses to maintain homeostasis. The endocrine system regulates homeostasis by secreting hormones.

4. Disruptions of homeostasis come from external and internal stimuli and from psychological stresses. When disruption of homeostasis is mild and temporary, responses of body cells quickly restore balance in the internal environment. If disruption is extreme, the body's attempts to restore homeostasis may fail.

5. A feedback system consists of (l) receptors that monitor changes in a controlled condition and send input to (2) a control center that sets the value at which a controlled condition should be maintained, evaluates the input it receives, and generates output commands when they are needed, and (3) effectors that receive output from the control center and produce a response (effect) that alters the controlled condition.

6. If a response reverses the original stimulus, the system is operating by negative feedback. If a response enhances the original stimulus, the system is operating by positive feedback.

7. One example of negative feedback is the system that regulates blood pressure. If a stimulus causes blood pressure (controlled condition) to rise, baroreceptors (pressure-sensitive nerve cells, the receptors) in blood vessels send impulses (input) to the brain (control center). The brain sends impulses (output) to the heart (effector). As a result, heart rate decreases (response), and blood pressure drops back to normal (restoration of homeostasis).

8. Disruptions of homeostasis—homeostatic imbalances—can lead to disorders, disease, and even death.

9. A disorder is any abnormality of function. Disease is a more specific term for an illness with a definite set of signs and symptoms.

10. Symptoms are subjective changes in body functions that are not apparent to an observer, whereas signs are objective changes that can be observed and measured.

11. Diagnosis of disease involves a medical history, physical examination, and sometimes laboratory tests.

Anatomical Terms (p. 10)

1. Descriptions of any region of the body assume the body is in the anatomical position, in which the subject stands erect facing the observer, with the head level and the eyes facing forward, the feet flat on the floor and directed forward, and the arms at the sides, with the palms turned forward.

2. The human body is divided into several major regions: the head, neck, trunk, upper limbs, and lower limbs.

3. Within body regions, specific body parts have common names and corresponding anatomical descriptive forms (adjectives). Examples are chest (thoracic), nose (nasal), and wrist (carpal).

4. Directional terms indicate the relationship of one part of the body to another. Exhibit 1.1 on page 12 summarizes commonly used directional terms.

5. Planes are imaginary flat surfaces that divide the body or organs into two parts. A midsagittal plane divides the body or an organ into equal right and left sides. A parasagittal plane divides the body or an organ into unequal right and left sides. A frontal plane divides the body or an organ into anterior and posterior portions. A transverse plane divides the body or an organ into superior and inferior portions. An oblique plane passes through the body or an organ at an angle between a transverse plane and a sagittal plane, or between a transverse plane and a frontal plane.

6. Sections result from cuts through body structures. They are named according to the plane on which the cut is made: transverse, frontal, or sagittal.

Body Cavities (p. 15)

1. Spaces in the body that contain, protect, separate, and support internal organs are called body cavities.

2. The dorsal and ventral cavities are the two principal body cavities.

3. The dorsal cavity is subdivided into the cranial cavity, which contains the brain, and the vertebral canal, which contains the spinal cord.

4. The ventral body cavity is subdivided by the diaphragm into a superior thoracic cavity and an inferior abdominopelvic cavity. The viscera are organs within the ventral body cavity.

5. The thoracic cavity is subdivided into three smaller cavities: a pericardial cavity, which contains the heart, and two pleural cavities, which contain the lungs.

6. The central portion of the thoracic cavity is the mediastinum. It is located between the pleural cavities and extends from the sternum to the vertebral column and from the neck to the diaphragm. It contains all thoracic viscera except the lungs.

7. The abdominopelvic cavity is divided into a superior abdominal and an inferior pelvic cavity.

8. Viscera of the abdominal cavity include the stomach, spleen, liver, gallbladder, small intestine, and most of the large intestine.

9. Viscera of the pelvic cavity include the urinary bladder, portions of the large intestine, and internal organs of the reproductive system.

10. To describe the location of organs easily, the abdominopelvic cavity may be divided into nine abdominopelvic regions by two horizontal and two vertical lines.

11. The names of the nine abdominopelvic regions are right hypochondriac, epigastric, left hypochondriac, right lumbar, umbilical, left lumbar, right iliac, hypogastric, and left iliac.

12. The abdominopelvic cavity may also be divided into quadrants by passing one horizontal and one vertical line through the umbilicus (navel).

13. The names of the abdominopelvic quadrants are right upper quadrant (RUQ), left upper quadrant (LUQ), right lower quadrant (RLQ), and left lower quadrant (LLQ).

■ SELF-QUIZ

1. To properly reconnect the disconnected bones of a human body, you would need to have a good understanding of
   a. physiology   **b.** homeostasis   **c.** chemistry
   d. anatomy   **e.** feedback systems

2. Which of the following best illustrates the idea of increasing levels of organizational complexity?
   a. chemical → tissue → cellular → organ → organismal → system
   b. chemical → cellular → tissue → organ → system → organismal
   c. cellular → chemical → tissue → organismal → organ → system
   d. chemical → cellular → tissue → system → organ → organismal
   e. tissue → cellular → chemical → organ → system → organismal

3. Match the following:
   _____ **a.** transports oxygen, nutrients, and carbon dioxide
   _____ **b.** breaks down and absorbs food
   _____ **c.** functions in body movement, posture, and heat production
   _____ **d.** regulates body activities through hormones
   _____ **e.** supports and protects the body
   _____ **f.** eliminates wastes and regulates chemical composition and volume of blood
   _____ **g.** protects the body, detects sensations, and helps regulate body temperature

   A. urinary system
   B. digestive system
   C. endocrine system
   D. integumentary system
   E. muscular system
   F. skeletal system
   G. cardiovascular system

4. Fill in the missing blanks in the following table.

System	Major Organs	Functions
a	b	Regulates body activities by nerve impulses
c	Lymph vessels, spleen, thymus, tonsils, lymph nodes	d
e	f	Supplies oxygen to cells, eliminates carbon dioxide, regulates acid-base balance
Reproductive	g	h

5. Homeostasis is
   a. the sum of all of the chemical processes in the body
   b. the sign of a disorder or disease
   c. the combination of growth, repair, and energy release that is basic to life
   d. the tendency to maintain constant, favorable, internal body conditions
   e. caused by stress

6. Which of the following is NOT true concerning the life processes?
   a. The pupils of your eyes becoming smaller when exposed to strong light is an example of differentiation.
   b. The ability to walk to your car following class is a result of the life process called movement.
   c. The repair of injured skin would involve the life process of reproduction.
   d. Digesting and absorbing your breakfast food is an example of metabolism.
   e. Sweating on a hot summer day involves responsiveness.

7. In a negative feedback system,
   a. the controlled condition is never disrupted
   b. there tends to be a "runaway" body response
   c. the change in the controlled condition is reversed
   d. the body part that responds to the output is known as the receptor
   e. the response results in a reinforcement of the original stimulus

8. The part of a feedback system that receives the input and generates the output command is the
   a. effector   **b.** receptor   **c.** feedback loop   **d.** response
   e. control center

9. Match the following:
   _____ **a.** observable, measurable changes
   _____ **b.** abnormality of function
   _____ **c.** affects the entire body
   _____ **d.** subjective changes that aren't easily observed

   A. systemic disease
   B. symptom
   C. sign
   D. disorder

10. An itch in your axillary region would cause you to scratch
    a. your armpit   **b.** in front of your elbow   **c.** your neck
    d. the top of your head   **e.** your calf

11. If you were facing a person who is in the correct anatomical position, you could observe the
    a. antecubital region   **b.** lumbar region   **c.** gluteal region
    d. popliteal region   **e.** scapular region

12. Where would you look for the femoral artery?
    a. wrist   **b.** forearm   **c.** face   **d.** thigh   **e.** shoulder

13. The right hand is _____ in relation to the left hand.

    a. intermediate   **b.** inferior   **c.** contralateral   **d.** distal
    e. ipsilateral

14. Your chin is _____ in relation to your lips.

    a. lateral   **b.** superior   **c.** deep   **d.** posterior
    e. inferior

15. Your skull is _____ in relation to your brain.

    a. intermediate   **b.** superior   **c.** deep   **d.** superficial
    e. proximal

16. A magician is about to separate his assistant's body into superior and inferior portions. The plane through which he will pass his magic wand is

    a. midsagittal   **b.** frontal   **c.** transverse   **d.** parasagittal
    e. oblique

17. Which statement is NOT true of body cavities?

    a. The diaphragm separates the thoracic and pelvic cavities.
    b. The dorsal cavity consists of the cranial cavity and vertebral canal.
    c. The ventral cavity consists of the thoracic, abdominal, and pelvic cavities.
    d. The most superior body cavity is the cranial cavity.
    e. The ventral cavities are located anteriorly.

18. Match the following:

    ____ **a.** contains the urinary bladder and reproductive organs
    ____ **b.** contains the brain
    ____ **c.** cavity that contains the heart
    ____ **d.** region between the lungs, from the breastbone to the backbone
    ____ **e.** separates the thoracic and abdominal cavities
    ____ **f.** contains a lung
    ____ **g.** contains the spinal cord
    ____ **h.** contains the stomach and liver

    A. cranial cavity
    B. abdominal cavity
    C. vertebral canal
    D. pelvic cavity
    E. pleural cavity
    F. mediastinum
    G. diaphragm
    H. pericardial cavity

19. If Jamie is having her appendix removed, the surgeon would prepare which area for surgery?

    a. right upper quadrant   **b.** right lower quadrant   **c.** left upper quadrant   **d.** left lower quadrant   **e.** left hypochondriac region

20. To find the urinary bladder, you would look in the

    a. hypochondriac region   **b.** umbilical region   **c.** epigastric region   **d.** iliac region   **e.** hypogastric region

CRITICAL THINKING APPLICATIONS

1. Anthony was going for the playground record for the longest upside-down hang from the monkey bars. He didn't make it and may have broken his arm. The emergency room technician would like an x-ray film of Anthony's arm in the anatomical position. Use the proper anatomical terms to describe the position of Anthony's arm in the x-ray film.

2. Imagine that a manned space flight lands on Mars. The astronaut life specialist observes lumpy shapes that may be life forms. What are some characteristics of living organisms that may help the astronaut determine if these are life forms or mud balls?

3. Guy was trying to impress Jenna with a tale about his last rugby match. "The coach said I suffered a collateral injury to the dorsal antecubital in my groin." Jenna responded, "I think either you or your coach suffered a cephalic injury." Why wasn't Jenna impressed by Guy's athletic prowess?

4. There's a special fun-house mirror that hides half your body and doubles the image of your other side. In the mirror, you can do amazing feats such as lifting both legs off the ground. Along what plane is the mirror dividing your body?

ANSWERS TO FIGURE QUESTIONS

1.1 Organs have a recognizable shape and consist of two or more different types of tissues.

1.2 The basic difference between negative and positive feedback systems is that in negative feedback systems, the response reverses the change due to the original stimulus, whereas in positive feedback systems, the response enhances the change due to original stimulus.

1.3 If a stimulus caused blood pressure to decrease, then heart rate would increase due to operation of this negative feedback system.

1.4 A plantar wart is found on the sole.

1.5 No, No, Yes, Yes, No.

1.6 The frontal plane divides the heart into anterior and posterior portions.

1.7 The midsagittal plane divides the brain into equal right and left sides.

1.8 P, A, T, A, T, P, T, A, A.

1.9 Some structures in the mediastinum are the heart, esophagus, and aorta.

1.10 The liver is mostly in the epigastric region; the ascending colon is in the right lumbar region; the urinary bladder is in the hypogastric region; the appendix is in the right iliac region.

1.11 The pain associated with appendicitis would be felt in the right lower quadrant (RLQ).

Chapter 2

Introductory Chemistry

■ Student Learning Objectives

1. Define a chemical element, atom, ion, molecule, and compound. **22**

2. Explain how chemical bonds form. **24**

3. Define a chemical reaction and explain why it is important to the human body. **27**

4. Discuss the functions of water and inorganic acids, bases, and salts. **27**

5. Define pH and explain how the body attempts to keep pH within the limits of homeostasis. **28**

6. Discuss the functions of carbohydrates, lipids, and proteins. **29**

7. Explain the importance of deoxyribonucleic acid (DNA), ribonucleic acid (RNA), and adenosine triphosphate (ATP). **34**

■ A Look Ahead

Many common substances we eat and drink—water, sugar, table salt, proteins, starches, fats—play vital roles in keeping us alive. In this chapter, you will learn how these substances function in your body. Because your body is composed of chemicals and all body activities are chemical in nature, it is important to become familiar with the language and basic ideas of chemistry to understand human physiology.

INTRODUCTION TO CHEMISTRY

Objectives: • **Define a chemical element, atom, ion, molecule, and compound.**
• **Explain how chemical bonds form.**
• **Define a chemical reaction and explain why it is important to the human body.**

Chemistry (KEM-is-trē) is the science of the structure and interactions of *matter,* which is anything that occupies space and has mass. The amount of matter in any object is its *mass.* Both living and nonliving things consist of matter and may exist as solids, liquids, or gases.

Chemical Elements and Atoms

All forms of matter are made up of a limited number of building blocks called *chemical elements,* substances that cannot be broken down into a simpler form by ordinary chemical means. At present, scientists recognize 112 different elements. Each element is designated by a *chemical symbol,* one or two letters of the element's name in English, Latin, or another language. Examples are H for hydrogen, C for carbon, O for oxygen, N for nitrogen, K for potassium, Na for sodium, Fe for iron, and Ca for calcium.

Twenty-six different elements are found in your body. Oxygen, carbon, hydrogen, and nitrogen are the most abundant, making up about 96% of the body's mass. The elements calcium, phosphorus, potassium, sulfur, sodium, chlorine, magnesium, and iron contribute an additional 3.8% of the body's mass. Table 2.1 lists these 12 elements and provides an overview of their importance. An additional 14 elements, the *trace elements,* are present in tiny amounts—less than half a teaspoonful each—accounting for the remaining 0.2% of the body's mass. The trace elements are aluminum (Al), boron (B), chromium (Cr), cobalt (Co), copper (Cu), fluorine (F), iodine (I), manganese (Mn), molybdenum (Mo), selenium (Se), silicon (Si), tin (Sn), vanadium (V), and zinc (Zn).

Each element is made up of *atoms,* the smallest units of an element that retain its properties and characteristics. Each element is composed of one type of atom. For instance, a sample of

Table 2.1 / **Main Chemical Elements in the Body**		
Chemical Element (Symbol)	**% of Total Body Mass**	**Significance**
Oxygen (O)	65.0	Part of water and many organic (carbon-containing) molecules; used to generate ATP, a molecule used by cells to temporarily store chemical energy.
Carbon (C)	18.5	Forms backbone chains and rings of all organic molecules: carbohydrates, lipids (fats), proteins, and nucleic acids (DNA and RNA).
Hydrogen (H)	9.5	Constituent of water and most organic molecules; ionized form (H^+) makes body fluids more acidic.
Nitrogen (N)	3.2	Component of all proteins and nucleic acids.
Calcium (Ca)	1.5	Contributes to hardness of bones and teeth; ionized form (Ca^{2+}) needed for blood clotting, release of hormones, contraction of muscle, and many other processes.
Phosphorus (P)	1.0	Component of nucleic acids and ATP; required for normal bone and tooth structure.
Potassium (K)	0.35	Ionized form (K^+) is the most plentiful cation (positively charged particle) in the fluid inside cells; needed for nerve and muscle impulses.
Sulfur (S)	0.25	Component of some vitamins and many proteins.
Sodium (Na)	0.2	Ionized form (Na^+) is the most plentiful cation in the fluid outside cells; essential for maintaining water balance; needed for nerve and muscle impulses.
Chlorine (Cl)	0.2	Ionized form (Cl^-) is the most plentiful anion (negatively charged particle) in the fluid outside cells; essential for maintaining water balance.
Magnesium (Mg)	0.1	Ionized form (Mg^{2+}) needed for action of many enzymes, molecules that increase the rate of chemical reactions in organisms.
Iron (Fe)	0.005	Ionized forms (Fe^{2+} and Fe^{3+}) are part of hemoglobin (oxygen-carrying protein in blood) and some enzymes.

the element carbon, such as pure coal, contains only carbon atoms, and a tank of helium gas contains only helium atoms.

An atom consists of two basic parts: a nucleus and one or more electrons (Figure 2.1). The centrally located *nucleus* contains positively charged *protons* (p^+) and uncharged (neutral) *neutrons* (n^0). Because each proton has one positive charge, the nucleus is positively charged. The *electrons* (e^-) are tiny, negatively charged particles that move about in a large space surrounding the nucleus. The number of electrons in an atom equals the number of protons. Because each electron carries one

Figure 2.1 ■ **Structure of an atom.** In this simplified diagram of a carbon atom, note the centrally located nucleus. A carbon nucleus contains six neutrons and six protons, not all of which are visible in this view. The six electrons move about the nucleus in regions called electron shells, depicted here as circles.

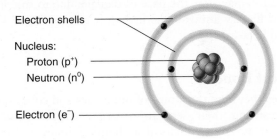

An atom is the smallest unit of matter that retains the properties and characteristics of its element.

Electron shells

Nucleus:
  Proton (p^+)
  Neutron (n^0)

Electron (e^-)

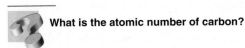

What is the atomic number of carbon?

negative charge, the negatively charged electrons and the positively charged protons balance each other. As a result, each atom is electrically neutral, meaning its total charge is zero.

The number of protons in the nucleus of an atom is called the atom's ***atomic number.*** The atoms of each different kind of element have a different number of protons in the nucleus: A hydrogen atom has 1 proton, a carbon atom has 6 protons, a sodium atom has 11 protons, a chlorine atom has 17 protons, and so on (Figure 2.2). Thus, each type of atom, or element, has a different atomic number. The total number of protons plus neutrons in an atom is its ***mass number.*** For instance, an atom of sodium, with 11 protons and 12 neutrons in its nucleus, has a mass number of 23.

The electrons move around the nucleus in regions called ***electron shells,*** which are depicted as circles in Figures 2.1 and 2.2. Each electron shell can hold only a specific number of electrons. For instance, the electron shell nearest the nucleus—the first electron shell—can hold only 2 electrons. The second electron shell can hold a maximum of 8 electrons, whereas the third can hold up to 18 electrons. Higher electron shells (there are as many as seven) can contain many more electrons. The electron shells are filled with electrons in a specific order, beginning with the first shell.

Figure 2.2 ■ **Atomic structures of several atoms that have important roles in the human body.**

The atoms of different elements have different atomic numbers because they have different numbers of protons.

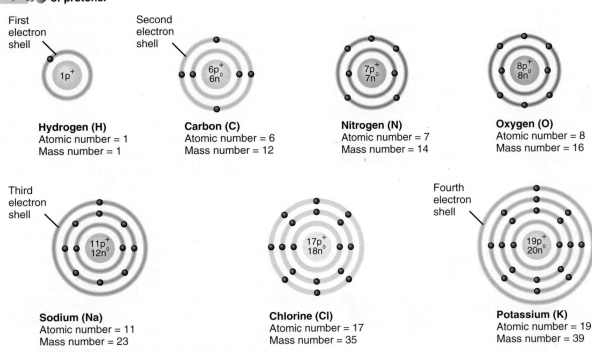

First electron shell

Second electron shell

Hydrogen (H)
Atomic number = 1
Mass number = 1

Carbon (C)
Atomic number = 6
Mass number = 12

Nitrogen (N)
Atomic number = 7
Mass number = 14

Oxygen (O)
Atomic number = 8
Mass number = 16

Third electron shell

Fourth electron shell

Sodium (Na)
Atomic number = 11
Mass number = 23

Chlorine (Cl)
Atomic number = 17
Mass number = 35

Potassium (K)
Atomic number = 19
Mass number = 39

Atomic number = number of protons in an atom
Mass number = number of protons and neutrons in an atom

Which four of these elements are most abundant in living organisms?

Ions, Molecules, and Compounds

The atoms of each element have a characteristic way of losing, gaining, or sharing their electrons when interacting with other atoms. The way that electrons behave enables atoms to exist as electrically charged forms called ions, or to join with each other to form molecules. If an atom either *gives up* or *gains* electrons, it becomes an **ion** (Ī-on), an atom that has a positive or negative charge due to unequal numbers of protons and electrons. An ion of an atom is symbolized by writing its chemical symbol followed by the number of its positive (+) or negative (−) charges. For example, Ca^{2+} stands for a calcium ion that has two positive charges because it has given up two electrons.

In contrast, when two or more atoms *share* electrons, the resulting combination of atoms is called a **molecule** (MOL-e-kyool). A molecule may consist of two or more atoms of the same element, such as an oxygen molecule or a hydrogen molecule, or of two or more atoms of different elements, such as a water molecule (Figure 2.3). A *molecular formula* indicates the number and type of atoms that make up a molecule. The molecular formula for a molecule of oxygen is O_2. The subscript 2 indicates there are two atoms of oxygen in the oxygen molecule. In the water molecule, H_2O, one atom of oxygen shares electrons with two atoms of hydrogen. Notice that two hydrogen molecules can combine with one oxygen molecule to form two water molecules (Figure 2.3).

A **compound** is a substance that can be broken down into two or more different elements by ordinary chemical means. Thus, a compound always contains atoms of two or more different elements. Most of the atoms in your body are joined into compounds, for example, water (H_2O). A molecule of oxygen (O_2) is *not* a compound because it consists of atoms of only one element.

A *free radical* is an electrically charged ion or molecule that has an unpaired electron in its outermost shell. (Most of an atom's electrons associate in pairs.) A common example of a free radical is *superoxide*, which is formed by the addition of an electron to an oxygen molecule. Having an unpaired electron makes a free radical unstable and destructive to nearby molecules. Free radicals break apart important body molecules by either giving

up their unpaired electron to or taking on an electron from another molecule. Among the many disorders and diseases linked to oxygen-derived free radicals are cancer, atherosclerosis, Alzheimer's disease, emphysema, diabetes mellitus, cataracts, macular degeneration, rheumatoid arthritis, and deterioration associated with aging. Consuming more **antioxidants**—substances that combine with and thus inactivate oxygen-derived free radicals—may slow the pace of damage due to free radicals. Important dietary antioxidants include vitamins E and C, selenium, and beta-carotene.

Chemical Bonds

The forces that bind the atoms of molecules and compounds together, resisting their separation, are **chemical bonds.** The chance that an atom will form a chemical bond with another atom depends on the number of electrons in its outermost shell, called the **valence shell.** The electrons in the valence shell are called **valence electrons.** An atom with a valence shell holding eight electrons is *chemically stable*, which means it is unlikely to form chemical bonds with other atoms. Neon, for example, has eight electrons in its valence shell, and for this reason it rarely forms bonds with other atoms.

The atoms of most biologically important elements do not have eight electrons in their valence shells. Given the right conditions, two or more such atoms can interact or bond in ways that produce a chemically stable arrangement of eight valence electrons for each atom. Three general types of chemical bonds occur:

1. In an *ionic bond*, an atom empties its partially filled valence shell by donating the valence electrons to another atom, or it fills its valence shell with valence electrons gained from another atom.

2. In a *covalent bond*, an atom shares its valence electrons with one or more other atoms.

3. In a *hydrogen bond*, a hydrogen atom that is covalently bonded to another atom is attracted to the partial negative charge of a nearby oxygen or nitrogen atom that is also covalently bonded to another atom.

Ionic Bonds

As you have already learned, an ion is formed when an atom loses or gains one or more valence electrons. Positively charged ions and negatively charged ions are attracted to one another. This force of attraction between ions of opposite charges is called an **ionic bond.** Ionic bonds always form between atoms of different elements. Consider sodium and chlorine atoms to see how an ionic bond forms (Figure 2.4). Sodium has one valence electron (Figure 2.4a). If a sodium atom *loses* this electron, it is left with the eight electrons in its second shell. However, the total number of protons (11) now exceeds the number of electrons (10). As a result, the sodium atom becomes a **cation,** a positively charged ion. A sodium ion has a charge of 1+ and is written Na^+. On the other hand, chlorine has seven valence electrons (Figure 2.4b), too many to lose. But if chlorine *accepts* an electron from a neighboring atom, it will have eight electrons in its

Figure 2.3 ■ **Molecules.**

A molecule may consist of two or more atoms of the same element or two or more atoms of different elements.

| 2 Hydrogen molecules ($2H_2$) | 1 Oxygen molecule (O_2) | Combine to form | 2 Water molecules ($2H_2O$) |

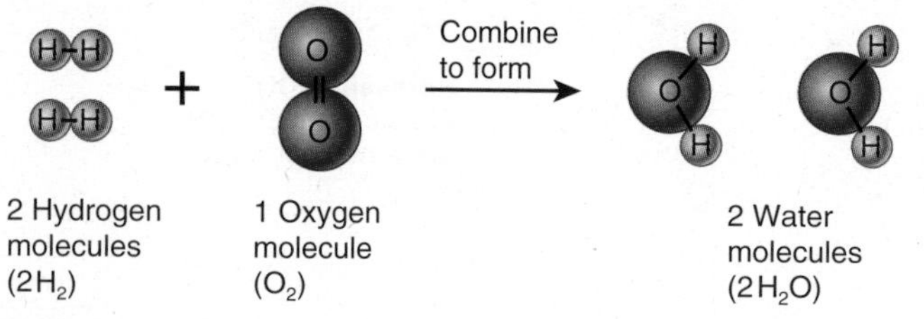

Which of the molecules shown here is a compound?

third electron shell. When this happens, the total number of electrons (18) exceeds the number of protons (17), and the chlorine atom becomes an ***anion,*** a negatively charged ion. The ionic form of chlorine is called a chloride ion. It has a charge of $1-$ and is written Cl^-. When an atom of sodium donates its sole valence electron to an atom of chlorine, the resulting positive and negative charges pull the ions tightly together into an ionic bond (Figure 2.4c). The resulting ionic compound is sodium chloride, written NaCl.

In the body, ionic bonds are found mainly in teeth and bones, where they give great strength to the tissue. Most other

Figure 2.4 ■ Ions and the formation of an ionic bond. (a) A sodium atom can attain the stability of eight electrons in its outermost shell by losing its one valence electron; it then becomes a sodium ion, Na^+. (b) A chlorine atom can attain the stability of eight electrons in its outermost shell by accepting one electron; it then becomes a chloride ion, Cl^-. (c) An ionic bond holds Na^+ and Cl^- together in the ionic compound sodium chloride, NaCl. The electron that is donated or accepted is colored red.

🔑 **An ionic bond is the force of attraction that holds together oppositely charged ions.**

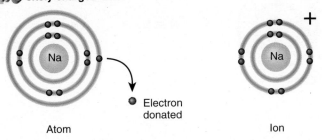

(a) Sodium: 1 valence electron

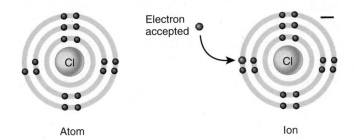

(b) Chlorine: 7 valence electrons

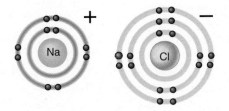

(c) Ionic bond in sodium chloride (NaCl)

 Will the element potassium (K) be more likely to form an anion or a cation? Why? (Hint: Look back to Figure 2.2 for the atomic structure of K.)

ions in the body are dissolved in body fluids. An ionic compound that breaks apart into cations and anions when dissolved is called an ***electrolyte*** (e-LEK-trō-līt) because the solution can conduct an electric current. As you will see in later chapters, electrolytes have many important functions. For example, they are critical for controlling water movement within the body, maintaining acid–base balance, and producing nerve impulses.

Covalent Bonds

When a ***covalent bond*** forms, neither of the combining atoms loses or gains electrons. Instead, the atoms form a molecule by *sharing* one, two, or three pairs of their valence electrons. Unlike ionic bonds, covalent bonds can form between atoms of the same element as well as between atoms of different elements. The greater the number of electron pairs shared between two atoms, the stronger the covalent bond. Covalent bonds are the most common chemical bonds in the body, and the compounds that result from them form most of the body's structures. Unlike ionic bonds, most covalent bonds do not break apart when the molecule is dissolved in water.

It is easiest to understand the nature of covalent bonds by considering those that form between atoms of the same element. As noted, covalently bonded atoms may share up to three pairs of electrons (Figure 2.5). A *single covalent bond* results when two atoms share one electron pair. For example, a molecule of hydrogen forms when two hydrogen atoms share their single valence electrons (Figure 2.5a), which allows both atoms to have a full valence shell. (Recall that the first electron shell holds only two electrons.) A *double covalent bond* (Figure 2.5b) or a *triple covalent bond* (Figure 2.5c) results when two atoms share two or three pairs of electrons. Notice the *structural formulas* for covalently bonded molecules in Figure 2.5. The number of lines between the chemical symbols for two atoms indicates whether the bond is a single (—), double (═), or triple (≡) covalent bond.

The same principles of covalent bonding that apply to atoms of the same element also apply to covalent bonds between atoms of different elements. Methane (CH_4), a gas, contains covalent bonds between the atoms of two different elements (Figure 2.5d). The valence shell of the carbon atom can hold eight electrons but has only four of its own. The single electron shell of a hydrogen atom can hold two electrons, but each hydrogen atom has only one of its own. A methane molecule is the product of four separate single covalent bonds; each hydrogen atom shares one pair of electrons with the carbon atom.

In some covalent bonds, atoms share the electrons equally—one atom does not attract the shared electrons more strongly than the other atom. This is called a *nonpolar covalent bond*. The bonds between two identical atoms always are nonpolar covalent bonds (Figure 2.5a–c). Another example of a nonpolar covalent bond is the single covalent bond that forms between carbon and each atom of hydrogen in a methane molecule (Figure 2.5d).

In a *polar covalent bond*, the sharing of electrons between atoms is unequal—one atom attracts the shared electrons more strongly than the other. When polar covalent bonds form, the resulting molecule has a partial negative charge, written δ^-, near the atom that attracts electrons more strongly. At least one other atom in the molecule then will have a partial positive charge,

Figure 2.5 ■ **Covalent bond formation.** The red electrons are shared. To the right are simpler ways to represent these molecules. In a structural formula, each covalent bond is denoted by a straight line between the chemical symbols for two atoms. In a molecular formula, the number of atoms in each molecule is noted by subscripts.

In a covalent bond, two atoms share one, two, or three pairs of valence electrons.

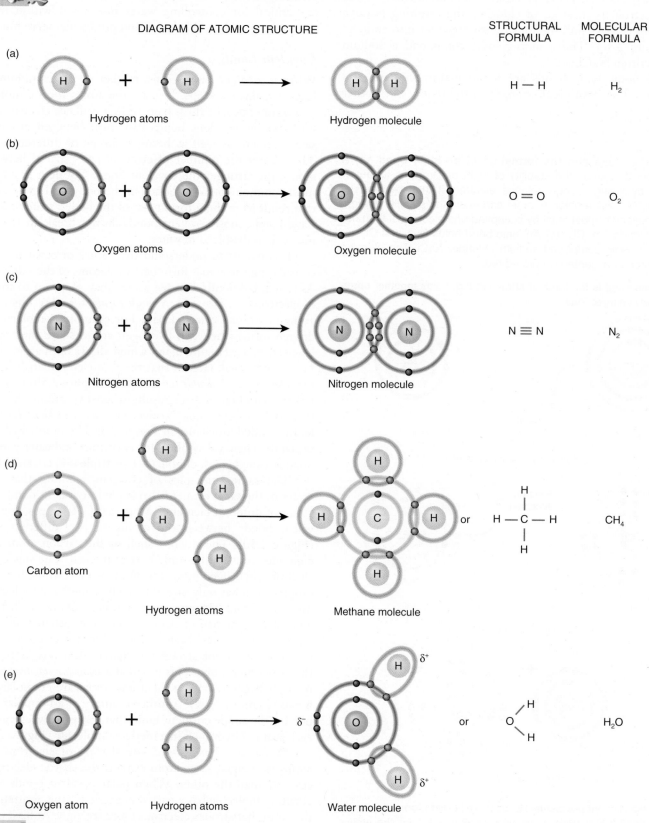

What is the main difference between an ionic bond and a covalent bond?

written δ^+. A very important example of a polar covalent bond in living systems is the bond between oxygen and hydrogen in a molecule of water (Figure 2.5e).

Hydrogen Bonds

The third type of chemical bond, a **hydrogen bond,** forms when the partial positive charge of a hydrogen atom attracts the partial negative charge of a neighboring oxygen or nitrogen atom, although both partners in the hydrogen bond are also covalently bonded to other atoms. Because hydrogen bonds are weak—only about 5% as strong as covalent bonds—they do not bind atoms into molecules. However, they do serve as links between different molecules or between various parts of a large, complex molecule. Very large molecules, such as proteins and nucleic acids, may contain hundreds of hydrogen bonds (see Figure 2.15). Collectively, they provide considerable strength and stability and help determine the three-dimensional shape of large molecules. As you will see later in this chapter, a large molecule's shape determines how it functions in the body.

Chemical Reactions

In a **chemical reaction** atoms combine with or break apart from other atoms. As a chemical reaction occurs, old bonds may be broken, and new bonds may be formed. The total number of atoms is the same after a chemical reaction. But new chemical substances with different properties arise because the atoms are rearranged. Through chemical reactions, body structures are built and body functions are carried out, processes that involve transfers of energy.

Forms of Energy and Chemical Reactions

Energy is the capacity to do work. The two main forms of energy are *potential energy,* energy stored by matter due to its *position,* and *kinetic energy,* the energy of matter *in motion.* For example, the energy stored in a battery or in a person poised to jump down some steps is potential energy. When the battery is used to run a clock or the person jumps, potential energy is converted into kinetic energy. **Chemical energy** is a form of potential energy that is stored in the bonds of molecules. In your body, chemical energy in the foods you eat is eventually converted into various forms of kinetic energy, such as mechanical energy, used to walk and talk, and heat energy, used to maintain body temperature. In chemical reactions, breaking old bonds requires an input of energy and forming new bonds releases energy. Because most chemical reactions involve both breaking old bonds and forming new bonds, the *overall reaction* may either release energy or require energy. The total amount of energy present at the beginning and end of a chemical reaction is the same—energy can neither be created nor be destroyed—although it may be converted from one form to another.

Synthesis Reactions

When two or more atoms, ions, or molecules combine to form new and larger molecules, the process is a **synthesis reaction.** The word *synthesis* means "to put together." Synthesis reactions can be expressed as follows:

$$\text{A} \quad + \quad \text{B} \quad \xrightarrow{\text{Combine to form}} \quad \text{AB}$$

Atom, ion, or molecule A   Atom, ion, or molecule B   New molecule AB

An example of a synthesis reaction is the synthesis of water from hydrogen and oxygen molecules (see Figure 2.3):

$$2\,\text{H}_2 \quad + \quad \text{O}_2 \quad \xrightarrow{\text{Combine to form}} \quad 2\,\text{H}_2\text{O}$$

Two hydrogen molecules   One oxygen molecule   Two water molecules

Decomposition Reactions

In a **decomposition reaction,** a molecule is split apart. The word *decompose* means to break down into smaller parts. Large molecules are split into smaller molecules, ions, or atoms. A decomposition reaction occurs in this way:

$$\text{AB} \quad \xrightarrow{\text{Breaks down into}} \quad \text{A} \quad + \quad \text{B}$$

Molecule AB   Atom, ion, or molelcule A   Atom, ion, or molecule B

For example, under the proper conditions, a methane molecule can decompose into one carbon atom and two hydrogen molecules:

$$\text{CH}_4 \quad \xrightarrow{\text{Breaks down into}} \quad \text{C} \quad + \quad 2\,\text{H}_2$$

One methane molecule   One carbon atom   Two hydrogen molecules

In general, energy-releasing reactions occur as nutrients, such as glucose, are broken down via decomposition reactions. Some of the energy released is temporarily stored in a special molecule called **adenosine triphosphate (ATP),** which will be discussed more fully later in this chapter. The energy transferred to the ATP molecules is then used to drive the energy-requiring synthesis reactions that lead to the building of body structures such as muscles and bones.

CHEMICAL COMPOUNDS AND LIFE PROCESSES

Objectives: • **Discuss the functions of water and inorganic acids, bases, and salts.**

• **Define pH and explain how the body attempts to keep pH within the limits of homeostasis.**

• **Discuss the functions of carbohydrates, lipids, and proteins.**

• **Explain the importance of deoxyribonucleic acid (DNA), ribonucleic acid (RNA), and adenosine triphosphate (ATP).**

Chemicals in the body can be divided into two main classes of compounds: inorganic and organic. In general, **inorganic compounds** lack carbon, are structurally simple, and are held together by ionic or covalent bonds. They include water; many salts, acids, and bases; and two carbon-containing compounds, carbon dioxide (CO_2) and bicarbonate ion (HCO_3^-). *Organic*

compounds, by contrast, always contain carbon, usually contain hydrogen, and always have covalent bonds. Examples include carbohydrates, lipids, proteins, nucleic acids, and adenosine triphosphate (ATP). Large organic molecules called *macromolecules* are formed by covalent bonding of many identical or similar building-block subunits termed *monomers*.

Inorganic Compounds

Water

Water is the most important and most abundant inorganic compound in all living systems, making up 55% to 60% of body mass in lean adults. With few exceptions, most of the volume of cells and body fluids is water. Several of its properties explain why water is such a vital compound for life.

1. **Water is an excellent solvent.** A *solvent* is a liquid or gas in which some other material, called a *solute,* has been dissolved. The combination of solvent plus solute is called a *solution.* Water is the solvent that carries nutrients, oxygen, and wastes throughout the body. The versatility of water as a solvent is due to its polar covalent bonds and its "bent" shape (see Figure 2.5), which allow each water molecule to interact with four or more neighboring ions or molecules. Solutes that contain ionic bonds or polar covalent bonds are *hydrophilic,* or "water loving," which means they dissolve easily in water. Common examples of hydrophilic solutes are sugar and salt. Molecules that contain mainly nonpolar covalent bonds, on the other hand, are *hydrophobic,* or "water fearing." They are not very water soluble. Examples of hydrophobic compounds include animal fats and vegetable oils.

2. **Water participates in chemical reactions.** Because water can dissolve so many different substances, it is an ideal medium for chemical reactions. Molecules dissolved in water continually collide, increasing their chance of undergoing a chemical reaction. Water also is an active participant in the decomposition of macromolecules such as proteins and carbohydrates, for example, during digestion of food, and it is produced in many synthesis reactions as macromolecules are built from monomers.

3. **Water absorbs and releases heat very slowly.** In comparison to most other substances, water can absorb or release a relatively large amount of heat with only a slight change in its own temperature. The large amount of water in the body thus moderates the effect of changes in the environmental temperature, thereby helping maintain the homeostasis of body temperature.

4. **Water requires a large amount of heat to change from a liquid to a gas.** When the water in sweat evaporates from the skin surface, it takes with it large quantities of heat and provides an excellent cooling mechanism.

5. **Water serves as a lubricant.** Water is a major part of saliva, mucus, and other lubricating fluids. Lubrication is especially necessary in the thoracic and abdominal cavities, where internal organs touch and slide over one another. It is

also needed at joints, where bones, ligaments, and tendons rub against one another.

Inorganic Acids, Bases, and Salts

Many inorganic compounds can be classified as acids, bases, or salts. An **acid** is a substance that breaks apart or *dissociates* (di-SŌ-sē-āts') into one or more *hydrogen ions* (H^+) and one or more anions when it dissolves in water (Figure 2.6a). A **base,** by contrast, dissociates into one or more *hydroxide ions* (OH^-) and one or more cations (Figure 2.6b). A **salt,** when dissolved in water, dissociates into cations and anions, neither of which is H^+ or OH^- (Figure 2.6c).

Acids and bases react with one another to form salts. For example, the reaction of hydrochloric acid (HCl) and potassium hydroxide (KOH), a base, produces the salt potassium chloride (KCl), along with water (H_2O). This reaction can be written as follows:

$$HCl + KOH \longrightarrow KCl + H_2O$$

Acid　　　Base　　　Salt　　　Water

Acid–Base Balance: The Concept of pH

To ensure homeostasis, body fluids must contain balanced quantities of acids and bases. The more hydrogen ions (H^+) dissolved in a solution, the more acidic the solution; conversely, the more hydroxide ions (OH^-), the more basic (alkaline) the solution. The chemical reactions that take place in the body are very sensitive to even small changes in the acidity or alkalinity of the body fluids in which they occur. Any departure from the narrow

Figure 2.6 ■ **Acids, bases, and salts.** (a) When placed in water, hydrochloric acid (HCl) ionizes into H^+ and Cl^-. (b) When the base potassium hydroxide (KOH) is placed in water, it ionizes into OH^- and K^+. (c) When the salt potassium chloride (KCl) is placed in water, it ionizes into positive and negative ions (K^+ and Cl^-), neither of which is H^+ or OH^-.

Ionization is the separation of inorganic acids, bases, and salts into ions in a solution.

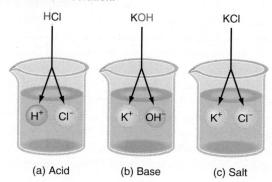

(a) Acid　　　(b) Base　　　(c) Salt

The compound CaCO₃ (calcium carbonate) dissociates into a calcium ion (Ca^{2+}) and a carbonate ion (CO_3^{2-}). Is it an acid, a base, or a salt? What about H_2SO_4, which dissociates into two H^+ and one SO_4^{2-}?

limits of normal H^+ and OH^- concentrations greatly disrupts body functions.

A solution's acidity or alkalinity is expressed on the **pH scale,** which extends from 0 to 14 (Figure 2.7). This scale is based on the number of hydrogen ions in a solution. The midpoint of the pH scale is 7, where the numbers of H^+ and OH^- are equal. A solution with a pH of 7, such as pure water, is neutral—neither acidic nor alkaline. A solution that has more H^+ than OH^- is **acidic** and has a pH below 7. A solution that has more OH^- than H^+ is **basic (alkaline)** and has a pH above 7. A change of one whole number on the pH scale represents a *10-fold* change in the number of H^+. At pH of 6, there are 10 times more H^+ than at a pH of 7. Put another way, a pH of 6 is 10 times more acidic than a pH of 7, and a pH of 9 is 100 times more alkaline than a pH of 7.

Maintaining pH: Buffer Systems

Although the pH of various body fluids may differ, the normal limits for each are quite narrow. Table 2.2 shows the pH

values for certain body fluids compared with those of common household substances. Homeostatic mechanisms maintain the pH of blood between 7.35 and 7.45, so that it is slightly more basic than pure water. Even though strong acids and bases may be taken into the body or be formed by body cells, the pH of fluids inside and outside cells remains almost constant. One important reason is the presence of **buffer systems,** in which chemical compounds called *buffers* convert strong acids or bases into weak acids or bases. (More will be said about buffers in Chapter 22.) Strong acids (or bases) dissociate easily and contribute many H^+ (or OH^-) to a solution. Therefore, they can change pH drastically. Weak acids (or bases) do not dissociate as much and contribute fewer H^+ (or OH^-). Hence, they have less effect on the pH.

Organic Compounds

Carbohydrates

Carbohydrates include sugars, glycogen, and starches. The elements present in carbohydrates are carbon, hydrogen, and oxygen. The ratio of hydrogen to oxygen atoms is usually $2:1$, as in water (H_2O), and the number of carbon and oxygen atoms is the same or nearly the same. For example, the molecular

Figure 2.7 ■ **The pH scale.** A pH below 7 indicates an acidic solution, or more H^+ than OH^-. The lower the numerical value of the pH, the more acidic the solution because the H^+ concentration becomes progressively greater. A pH above 7 indicates a basic (alkaline) solution; that is, there are more OH^- than H^+. The higher the pH, the more basic the solution.

At pH 7 (neutrality), the concentrations of H^+ and OH^- are equal.

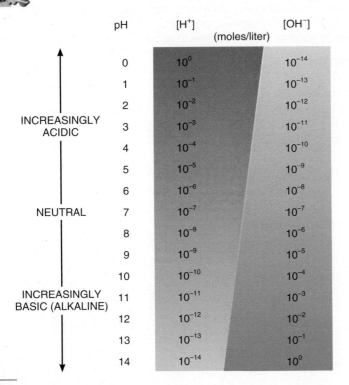

Which pH is more acidic, 6.82 or 6.91? Which pH is closer to neutral, 8.41 or 5.59?

Table 2.2 / pH Values of Selected Substances	
Substance*	**pH Value**
Gastric juice (digestive juice of the stomach)	1.2–3.0
Lemon juice	2.3
Grapefruit juice, vinegar, wine	3.0
Carbonated soft drink	3.0–3.5
Orange juice	3.5
Vaginal fluid	3.5–4.5
Tomato juice	4.2
Coffee	5.0
Urine	4.6–8.0
Saliva	6.35–6.85
Cow's milk	6.8
Distilled (pure) water	7.0
Blood	7.35–7.45
Semen (fluid containing sperm)	7.20–7.60
Cerebrospinal fluid (fluid associated with nervous system)	7.4
Pancreatic juice (digestive juice of the pancreas)	7.1–8.2
Bile (liver secretion that aids fat digestion)	7.6–8.6
Milk of magnesia	10.5
Lye	14.0

*Substances in the human body are highlighted by blue.

formula for the small carbohydrate glucose is $C_6H_{12}O_6$. Carbohydrates are divided into three major groups based on their size: monosaccharides, disaccharides, and polysaccharides. Monosaccharides and disaccharides are termed **simple sugars,** and polysaccharides are also known as **complex carbohydrates.**

1. **Monosaccharides** (mon'-ō-SAK-a-rīds; *mono-* = one; *sacchar-* = sugar) are the building blocks of carbohydrates. In your body, the principal function of the monosaccharide glucose is to serve as a source of chemical energy for generating the ATP that fuels metabolic reactions. Ribose and deoxyribose, two other monosaccharides, are used to make ribonucleic acid (RNA) and deoxyribonucleic acid (DNA), which are described on pages 34 to 37.

2. **Disaccharides** (dī-SAK-a-rīds; *di-* = two) are simple sugars that consist of two monosaccharides joined by a covalent bond. When two monosaccharides combine to form a disaccharide, a molecule of water is lost. This reaction is known as **dehydration synthesis** (*dehydration* = removal of water).

For example, the monosaccharides glucose and fructose combine to form the disaccharide sucrose (table sugar) as shown in Figure 2.8. Disaccharides can be split into monosaccharides by adding a molecule of water, a type of chemical reaction known as **hydrolysis** (*hydro-* = water; *-lysis* = to loosen). Sucrose, for example, may be hydrolyzed into its components of glucose and fructose by the addition of water (Figure 2.8).

3. **Polysaccharides** (pol'-ē-SAK-a-rīds; *poly-* = many) are large, complex carbohydrates that contain tens or hundreds of monosaccharides joined through dehydration synthesis reactions. The main polysaccharide in the human body is glycogen, which is made entirely of glucose units joined together in branching chains (Figure 2.9). Glycogen is stored in cells of the liver and in skeletal muscles. Like disaccharides, polysaccharides can be broken down into monosaccharides through hydrolysis reactions. Unlike simple sugars, however, polysaccharides usually are not soluble in water and do not taste sweet.

Figure 2.8 ■ **Dehydration synthesis and hydrolysis of a molecule of sucrose.** In the dehydration synthesis reaction (read from left to right), two smaller molecules, glucose and fructose, are joined to form a larger molecule of sucrose. Note the loss of a water molecule. In the hydrolysis reaction (read from right to left), the larger sucrose molecule is broken down into two smaller molecules, glucose and fructose. Here, a molecule of water is added to sucrose for the reaction to occur.

Monosaccharides are the building blocks of carbohydrates.

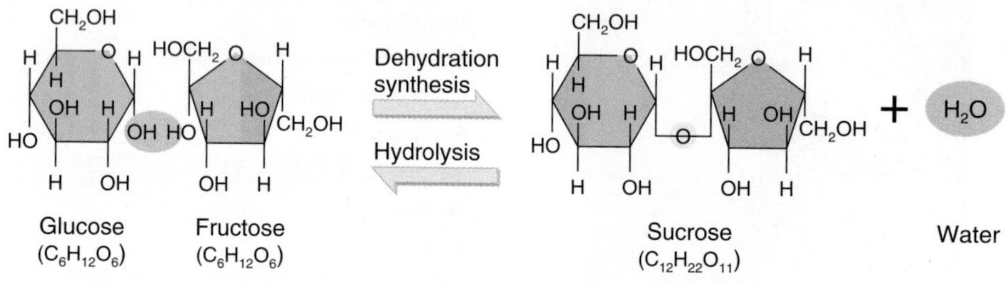

(a) Dehydration synthesis and hydrolysis of sucrose

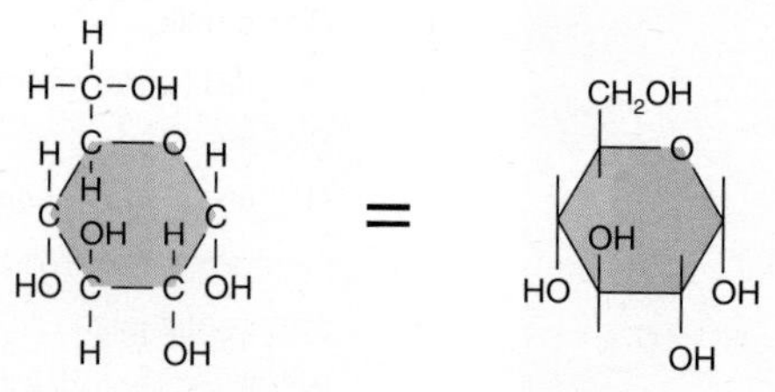

(b) Alternate chemical structures of organic molecules (shown here is glucose)

How many carbons are there in fructose? In sucrose?

Lipids

Like carbohydrates, *lipids* (*lipos* = fat) contain carbon, hydrogen, and oxygen. Unlike carbohydrates, they do not have a 2:1 ratio of hydrogen to oxygen. The proportion of oxygen atoms in lipids is usually smaller than in carbohydrates, so there are fewer polar covalent bonds. As a result, most lipids are hydrophobic; that is, they are insoluble in water (see page 28).

The diverse lipid family includes triglycerides (fats and oils), phospholipids, steroids, fatty acids, fat-soluble vitamins (vitamins A, D, E, and K), and eicosanoids (ī-KŌ-sa-noids). Fat-soluble vitamins are absorbed with dietary lipids. The eicosanoids include prostaglandins and leukotrienes, which contribute to regulation of a wide variety of body processes: inflammation, temperature regulation, platelet plug formation, allergic and immune responses, glandular secretions, and reproductive processes.

The most plentiful lipids in your body and in your diet are the *triglycerides* (trī-GLI-cer-īdes; *tri-* = three). At room temperature, triglycerides may be either solids (fats) or liquids (oils). They are the body's most highly concentrated form of chemical energy, storing more than twice as much chemical energy per gram as carbohydrates or proteins. Our capacity to store triglycerides in fat tissue, called *adipose tissue*, is unlimited, for all practical purposes. Excess dietary carbohydrates, proteins, fats, and oils all have the same fate: They are deposited in adipose tissue as triglycerides.

A triglyceride consists of two types of building blocks: a single glycerol molecule and three fatty acid molecules. A three-carbon *glycerol* molecule forms the backbone of a triglyceride (Figure 2.10). Three *fatty acids* are attached, by dehydration synthesis reactions, one to each carbon of the glycerol backbone. The fatty acid chains of a triglyceride may be saturated, monounsaturated, or polyunsaturated. *Saturated fats* contain only *single covalent bonds* between fatty acid carbon atoms. Because they do not contain any double bonds between fatty acid carbon atoms, each carbon atom is *saturated with hydrogen atoms* (see palmitic acid and stearic acid in Figure 2.10). Triglycerides with mainly saturated fatty acids are solid at room temperature and occur mostly in animal tissues. They also occur in a few tropical plants, such as cocoa, palm, and coconut. *Monounsaturated fats* (*mono-* = one) contain fatty acids with *one double covalent bond* between two fatty acid carbon atoms and thus are not completely saturated with hydrogen atoms. Olive oil and peanut oil are rich in triglycerides with monounsaturated fatty acids. *Polyunsaturated fats* (*poly-* = many) contain *more than one double covalent bond* between fatty acid carbon atoms. Canola oil, corn oil, safflower oil, sunflower oil, and soybean oil contain a high percentage of polyunsaturated fatty acids.

Like triglycerides, *phospholipids* have a glycerol backbone and two fatty acids attached to the first two carbons (Figure 2.11a). Attached to the third carbon is a phosphate group

Figure 2.9 ■ **Part of a glycogen molecule, the main polysaccharide in the human body.**

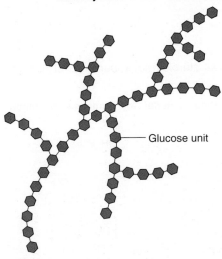

Glycogen is made up of glucose units and is the storage form of carbohydrate in the human body.

— Glucose unit

 Which body cells store glycogen?

Figure 2.10 ■ **Triglycerides consist of three fatty acids attached to a glycerol backbone.** The fatty acids vary in length and the number and location of double bonds between carbon atoms (C=C). Shown here is a triglyceride molecule that contains two saturated fatty acids and one monounsaturated fatty acid.

A triglyceride consists of two types of building blocks: a single glycerol molecule and three fatty acid molecules.

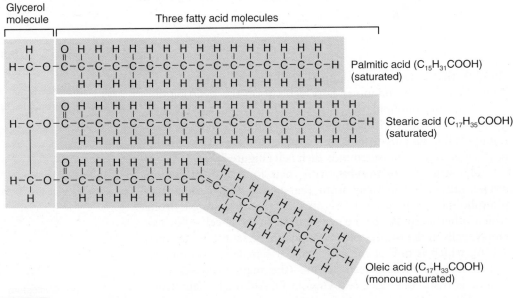

Palmitic acid ($C_{15}H_{31}COOH$) (saturated)

Stearic acid ($C_{17}H_{35}COOH$) (saturated)

Oleic acid ($C_{17}H_{33}COOH$) (monounsaturated)

How many double bonds are there in a monounsaturated fatty acid?

Figure 2.11 ■ **Phospholipids.** (a) In the synthesis of phospholipids, two fatty acids attach to the first two carbons of the glycerol backbone. A phosphate group links a small charged group to the third carbon in glycerol. In (b), the circle represents the polar head region, and the two wavy lines represent the two nonpolar tails.

Phospholipids are the main lipids in cell membranes.

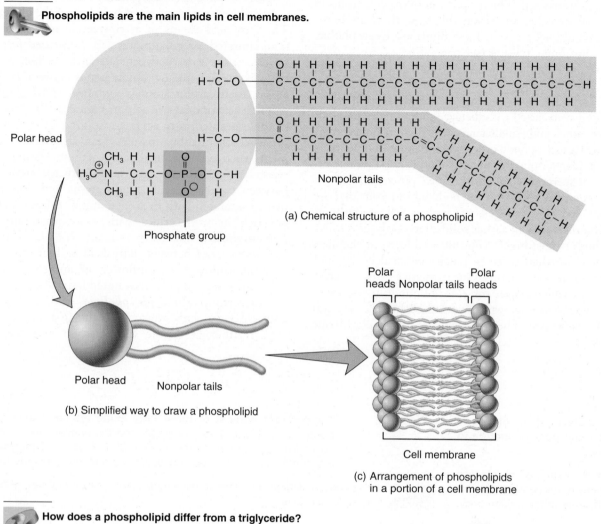

(a) Chemical structure of a phospholipid

Polar head

Phosphate group

Nonpolar tails

Polar heads Nonpolar tails Polar heads

Polar head   Nonpolar tails

(b) Simplified way to draw a phospholipid

Cell membrane

(c) Arrangement of phospholipids in a portion of a cell membrane

How does a phospholipid differ from a triglyceride?

(PO_4^{3-}) that links a small charged group to the glycerol backbone. Whereas the nonpolar fatty acids form the hydrophobic "tails" of a phospholipid, the polar phosphate group and charged group form the hydrophilic "head" (Figure 2.11b). Phospholipids line up tails-to-tails in a double row to make up much of the membrane that surrounds each cell (Figure 2.11c).

The structure of *steroids,* with their four rings of carbon atoms, differs considerably from that of the triglycerides and phospholipids. Cholesterol (Figure 2.12a) is the steroid from which other steroids may be synthesized by body cells. For example, cells in the ovaries of females synthesize estradiol (Figure 2.12b), which is one of the estrogens or female sex hormones. Other steroids include testosterone (the main male sex hormone), cortisol, bile salts, and vitamin D. Although their structures differ, excess dietary consumption of both saturated fats and cholesterol can contribute to the buildup of fatty plaques

Figure 2.12 ■ **Steroids.** All steroids have four rings of carbon atoms.

Cholesterol is the starting material for synthesis of other steroids in the body.

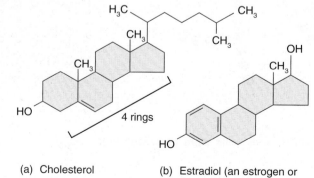

(a) Cholesterol

(b) Estradiol (an estrogen or female sex hormone)

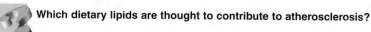
Which dietary lipids are thought to contribute to atherosclerosis?

that block blood vessels (atherosclerosis). In contrast, dietary monounsaturated fats such as olive oil and peanut oil are thought to offer some protection from atherosclerosis.

Proteins

Proteins are large molecules that contain carbon, hydrogen, oxygen, and nitrogen; some proteins also contain sulfur. Much more complex in structure than carbohydrates or lipids, proteins have many roles in the body and are largely responsible for the structure of body cells. For example, proteins termed enzymes speed up particular chemical reactions, other proteins are responsible for contraction of muscles, proteins called antibodies help defend the body against invading microbes, and some hormones are proteins.

Amino acids (a-MĒ-nō) are the building blocks of proteins. All amino acids have an *amino group* (—NH₂) at one end and an acidic *carboxyl group* (—COOH) at the other end. Each of the 20 different amino acids has a different *side chain* (R group) (Figure 2.13a). The covalent bonds that join amino acids together to form more complex molecules are called *peptide bonds* (Figure 2.13b).

The union of two or more amino acids produces a *peptide* (PEP-tīd). When two amino acids combine, the molecule is called a *dipeptide* (Figure 2.13b). Adding another amino acid to a dipeptide produces a *tripeptide*. A *polypeptide* contains a large number of amino acids. Proteins are polypeptides that contain as few as 50 or as many as 2000 amino acids. Because each varia-

tion in the number and sequence of amino acids produces a different protein, a great variety of proteins is possible. The situation is similar to using an alphabet of 20 letters to form words. Each letter would be equivalent to an amino acid, and each word would be a different protein.

A protein may consist of only one polypeptide or several intertwined polypeptides. A given type of protein has a unique three-dimensional shape because of the ways that each individual polypeptide twists and folds and associated polypeptides come together. If a protein encounters a hostile environment in which temperature, pH, or ion concentration is altered, it may unravel and lose its characteristic shape. This process is called *denaturation* (dē-nā′-chur-Ā-shun). Denatured proteins are no longer functional. A common example of denaturation is seen in frying an egg. In a raw egg the egg-white protein (albumin) is soluble and the egg white appears as a clear, viscous fluid. When heat is applied to the egg, however, the albumin denatures; it changes shape, becomes insoluble, and looks white.

Enzymes

As we have seen, chemical reactions occur when chemical bonds are made or broken as atoms, ions, or molecules collide with one another. At normal body temperature, such collisions occur too infrequently to maintain life. *Enzymes* (EN-zīms) are the living cell's solution to this problem. They speed up chemical reactions by increasing the frequency of collisions and by properly orienting the colliding molecules. Substances that can speed up chem-

Figure 2.13 ■ **Amino acids.** (a) In keeping with their name, amino acids have an amino group (shaded blue) and a carboxyl (acid) group (shaded red). The side chain (R group) is different in each type of amino acid. (b) When two amino acids are chemically united by dehydration synthesis (read from left to right), the resulting covalent bond between them is called a peptide bond. The peptide bond is formed at the point where water is lost. Here, the amino acids glycine and alanine are joined to form the dipeptide glycylalanine. Breaking a peptide bond occurs by hydrolysis (read from right to left).

Amino acids are the building blocks of proteins.

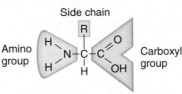

(a) Structure of an amino acid

(b) Protein formation

How many peptide bonds would there be in a tripeptide?

ical reactions without themselves being altered are called *cata-lysts* (KAT-a-lists). In living cells, most enzymes are proteins. The names of enzymes usually end in *-ase*. All enzymes can be grouped according to the types of chemical reactions they catalyze. For example, *oxidases* add oxygen, *kinases* add phosphate, *dehydrogenases* remove hydrogen, *anhydrases* remove water, *ATPases* split ATP, *proteases* break down proteins, and *lipases* break down lipids.

Enzymes catalyze selected reactions with great efficiency and with many built-in controls. Three important properties of enzymes are their specificity, efficiency, and control.

1. **Specificity.** Enzymes are highly specific. Each particular enzyme catalyzes a particular chemical reaction that involves specific **substrates,** the molecules on which the enzyme acts, and that gives rise to specific **products,** the molecules produced by the reaction. In some cases, the enzyme fits the substrate like a key fits in a lock. In other cases, the enzyme changes its shape to fit snugly around the substrate once the substrate and enzyme come together. Each of the more than 1000 known enzymes in your body has a characteristic three-dimensional shape with a specific surface configuration that allows it to fit specific substrates.

2. **Efficiency.** Under optimal conditions, enzymes can catalyze reactions at rates that are millions to billions of times more rapid than those of similar reactions occurring without enzymes. The rate at which a single enzyme molecule can convert substrate molecules to product molecules per second is generally between 1 and 10,000 per second and can be as high as 600,000 per second.

3. **Control.** Enzymes are subject to a variety of cellular controls. Their rate of synthesis and their concentration at any given time are under the control of a cell's genes. Substances within the cell may either enhance or inhibit activity of a given enzyme. Many enzymes exist in both active and inactive forms within the cell. The rate at which the inactive form becomes active or vice versa is determined by the chemical environment inside the cell. Many enzymes require a nonprotein substance, known as a **cofactor** or **coenzyme,** to operate properly. Ions of iron, zinc, magnesium, or calcium are cofactors whereas niacin or riboflavin, derivatives of B vitamins, act as coenzymes.

Figure 2.14 illustrates the actions of an enzyme.

❶ The substrates make contact with the active site on the surface of the enzyme molecule, forming a temporary intermediate compound called the **enzyme–substrate complex.** In this reaction, the substrates are the disaccharide sucrose and a molecule of water.

❷ The substrate molecules are transformed by the rearrangement of existing atoms, the breakdown of the substrate molecule, or the combination of several substrate molecules into products of the reaction. Here the products are two monosaccharides: glucose and fructose.

❸ After the reaction is completed and the reaction products move away from the enzyme, the unchanged enzyme is free to attach to another substrate molecule.

Sometimes a single enzyme may catalyze a reaction in either direction, depending on the abundance of the substrates and products. For example, the enzyme *carbonic anhydrase* catalyzes the reversible reaction (which is indicated by arrows pointing both directions):

$$\underset{\substack{\text{Carbon}\\\text{dioxide}}}{CO_2} + \underset{\text{Water}}{H_2O} \underset{}{\overset{\textit{Carbonic anhydrase}}{\rightleftharpoons}} \underset{\text{Carbonic acid}}{H_2CO_3}$$

During exercise, when more CO_2 is produced and released into the blood, the reaction flows to the right, increasing the amount of carbonic acid in the blood. Then as you exhale CO_2, its level in the blood falls and the reaction flows to the left, converting carbonic acid to CO_2 and H_2O.

Nucleic Acids: Deoxyribonucleic Acid (DNA) and Ribonucleic Acid (RNA)

Nucleic acids (noo-KLĒ-ic), so named because they were first discovered in the nuclei of cells, are huge organic molecules that contain carbon, hydrogen, oxygen, nitrogen, and phosphorus. The two kinds of nucleic acids are **deoxyribonucleic acid (DNA)** (dē-ok′-sē-rī′-bō-noo-KLĒ-ik) and **ribonucleic acid (RNA).**

A nucleic acid molecule is composed of repeating building blocks called **nucleotides.** Each nucleotide of DNA consists of three parts (Figure 2.15a on page 36):

Figure 2.14 ■ **How an enzyme works.**

An enzyme speeds up a chemical reaction without being altered or consumed.

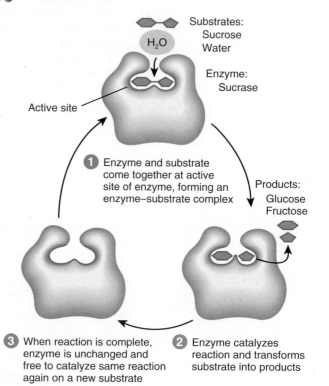

Substrates:
Sucrose
Water

Enzyme:
Sucrase

Active site

❶ Enzyme and substrate come together at active site of enzyme, forming an enzyme–substrate complex

Products:
Glucose
Fructose

❸ When reaction is complete, enzyme is unchanged and free to catalyze same reaction again on a new substrate

❷ Enzyme catalyzes reaction and transforms substrate into products

 What part of an enzyme combines with its substrate?

Herbal Supplements— They're Natural but Are They Safe?

Sales of herbal supplements are booming. Preparations of ginseng and echinacea stand next to bottles of vitamin C and aspirin in medicine cabinets across North America. But beware: Although some herbal supplements are helpful for specific problems, others are a waste of money, and many can be harmful to your health.

Does Natural Mean Safe?

Herbal supplements are preparations made from the leaves, flowers, bark, berries, or roots of plants. Herbal preparations have been used throughout the ages in cultures around the world to relieve pain, heal wounds, chase away evil spirits, and even to kill. Many of the active ingredients in drugs we use today were originally isolated from herbs. For example, the heart drug digitalis comes from the foxglove plant.

Anyone who understands even a little bit of chemistry can understand why "natural" does not necessarily mean "safe." Natural chemicals are still chemicals. They participate in chemical reactions in your body. They have chemical effects in the same way that manufactured drugs do.

Herbal products can't be effective and harmless at the same time because anything that has a physiological effect can be harmful at some dose. All drugs become toxic if you take too much of them.

Handle with Care

If you want to use herbal supplements, you must also use your head. Because regulation of these supplements is currently fairly loose in most countries, you can't believe everything the manufacturer says on the label or in advertising literature. If a product sounds too good to be true, beware!

Health-care professionals are especially concerned about the lack of data on long-term safety of many herbal products. Scientists are just beginning to investigate the use of herbs, and our understanding of these remedies is still in its infancy.

Talk to Your Doc

If you decide to try herbal supplements for an ailment, talk to your health-care provider to be sure you are not overlooking beneficial medical treatments. If you are taking any medications, ask your pharmacist whether you should be concerned about possible interactions between the supplement and your drugs. For example, it is dangerous to take ginkgo biloba and aspirin together, because both have potent blood-thinning effects that can lead to dangerous bleeding.

Women who are pregnant, intending to become pregnant, or nursing a baby should avoid supplements in the same way that they avoid drugs.

▶ *Think It Over*

▶ Your Aunt Mary tells you she is taking an herbal weight-loss supplement. "It's natural, so it's safe," she says. In fact, it's not working as well as it was two weeks ago, so she is now taking double the recommended dose. What do you say to her?

Figure 2.15 ■ **DNA molecule.** (a) A nucleotide consists of a nitrogenous base, a five-carbon sugar, and a phosphate group. (b) The paired nitrogenous bases project toward the center of the double helix. The structure is stabilized by hydrogen bonds (dotted lines) between each base pair. There are two hydrogen bonds between adenine and thymine and three between cytosine and guanine.

Nucleotides are the building blocks of nucleic acids.

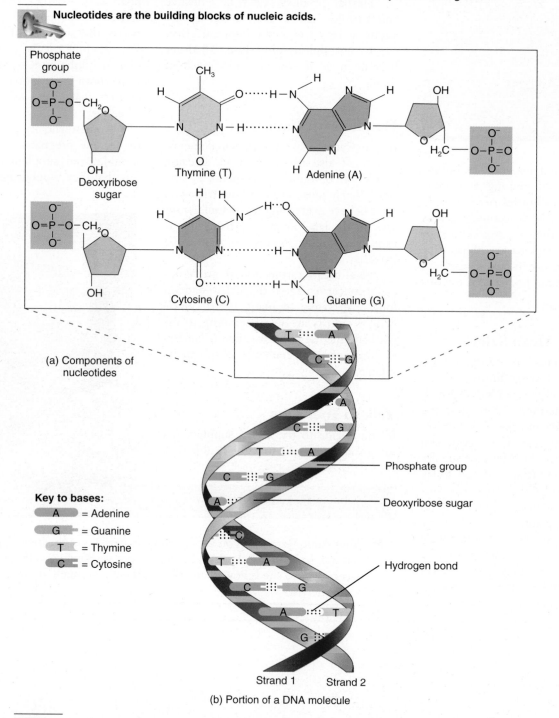

(a) Components of nucleotides

Key to bases:

A = Adenine
G = Guanine
T = Thymine
C = Cytosine

Phosphate group

Deoxyribose sugar

Hydrogen bond

Strand 1   Strand 2

(b) Portion of a DNA molecule

Which nitrogenous base is not present in RNA? Which nitrogenous base is not present in DNA?

- One of four different *nitrogenous bases*, ring-shaped molecules that contain atoms of C, H, O, and N.
- A five-carbon monosaccharide called *deoxyribose*.
- A *phosphate group* (PO_4^{3-}).

In DNA, the four bases are adenine (A), thymine (T), cytosine (C), and guanine (G). The nucleotides are named according to the base that is present. Thus, a nucleotide containing thymine is called a thymine nucleotide. One containing adenine is called an adenine nucleotide, and so on. Figure 2.15b shows the following structural characteristics of the DNA molecule:

1. The molecule consists of two strands with crossbars. The strands twist about each other in the form of a **double helix** so that the shape resembles a twisted rope ladder.

2. The uprights of the DNA ladder consist of alternating phosphate groups and the deoxyribose portions of the nucleotides.

3. The rungs of the ladder contain paired nitrogenous bases. Adenine always pairs with thymine, and cytosine always pairs with guanine.

About 1000 rungs of DNA comprise a **gene**, a portion of a DNA strand that performs a specific function, for example, providing instructions to synthesize the hormone insulin. Humans have about 100,000 functional genes. Genes determine which traits we inherit, and they control all the activities that take place in our cells throughout a lifetime. Any change that occurs in the nitrogenous base sequence of a gene is called a *mutation*. Some mutations can result in the death of a cell, cause cancer, or produce genetic defects in future generations.

RNA, the second kind of nucleic acid, is copied from DNA but differs from DNA in several respects. Whereas DNA is double stranded, RNA is single stranded. The sugar in the RNA nucleotide is ribose, and RNA contains the nitrogenous base uracil (U) rather than thymine. Cells contain three different kinds of RNA: messenger RNA, ribosomal RNA, and transfer RNA. Each has a specific role to perform in carrying out the instructions encoded in DNA, as will be described in Chapter 3.

Adenosine Triphosphate

Adenosine triphosphate (a-DEN-ō-sēn trī-FOS-fāt) or **ATP** is the "energy currency" of living organisms. It functions to transfer energy from energy-releasing reactions to energy-requiring reactions that maintain cellular activities. Among these cellular activities are contraction of muscles, movement of chromosomes during cell division, movement of structures within cells, transport of substances across cell membranes, and synthesis of larger molecules from smaller ones.

Structurally, ATP consists of three phosphate groups attached to adenosine, which is composed of adenine and ribose (Figure 2.16). The energy-transferring reaction occurs via hydrolysis: Removal of the last phosphate group (PO_4^{3-}), symbolized by Ⓟ in the following discussion, by addition of a water molecule liberates energy and leaves a molecule called **adenosine diphosphate (ADP)**. The enzyme that catalyzes the hydrolysis of

Figure 2.16 ■ **Structure of ATP and ADP.** The two phosphate bonds that can be used to transfer energy are indicated by "squiggles" (~). Most often energy transfer involves hydrolysis of the terminal phosphate bond of ATP.

🔑 **ATP transfers chemical energy to power cellular activities.**

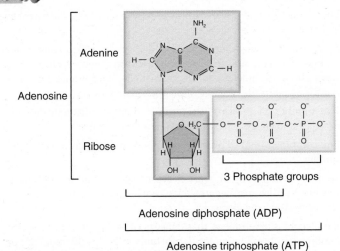

🦴 **What are some cellular activities that depend on energy supplied by ATP?**

ATP is called *ATPase*. This reaction may be represented as follows:

$$ \text{ATP} + \text{H}_2\text{O} \xrightarrow{\textit{ATPase}} \text{ADP} + \text{Ⓟ} + \text{E} $$

Adenosine triphosphate — Water — Adenosine diphosphate — Phosphate group — Energy

The energy supplied by the breakdown of ATP into ADP is constantly being used by the cell. As the supply of ATP at any given time is limited, a mechanism exists to replenish it: The enzyme *ATP synthase* catalyzes the addition of a phosphate group to ADP. The reaction may be represented as follows:

$$ \text{ADP} + \text{Ⓟ} + \text{E} \xrightarrow{\textit{ATP synthase}} \text{ATP} + \text{H}_2\text{O} $$

Adenosine diphosphate — Phosphate group — Energy — Adenosine triphosphate — Water

As you can see from this reaction, energy is required to produce ATP. The energy needed to attach a phosphate group to ADP is supplied mainly by the breakdown of glucose in a process called cellular respiration (Chapter 20).

• • •

In Chapter 1, you learned that the human body comprises various levels of organization and that the chemical level consists of atoms and molecules. Now that you have an understanding of the chemicals in the body, you will see in the next chapter how they are organized to form the structures of cells and perform the activities of cells that contribute to homeostasis.

STUDY OUTLINE

Introduction to Chemistry (p. 22)

1. Chemistry is the science of the structure and interactions of matter, which is anything that occupies space and has mass. Matter is made up of chemical elements.

2. The elements oxygen (O), carbon (C), hydrogen (H), and nitrogen (N) make up 96% of the body's mass.

3. Each element is made up of units called atoms, which consist of a nucleus that contains protons and neutrons, and electrons that move about the nucleus in electron shells. The number of electrons is equal to the number of protons in an atom.

4. The number of protons (the atomic number) distinguishes the atoms of one element from those of another element.

5. The combined total of protons and neutrons in an atom is its mass number.

6. An atom that *gives up* or *gains* electrons becomes an ion—an atom that has a positive or negative charge due to having unequal numbers of protons and electrons.

7. A molecule is a substance that consists of two or more chemically combined atoms. The molecular formula indicates the number and type of atoms that make up a molecule.

8. A compound is a substance that can be broken down into two or more different elements by ordinary chemical means.

9. A free radical is a destructive, electrically charged ion or molecule that has an unpaired electron in its outermost shell.

10. Chemical bonds hold the atoms of a molecule together.

11. Electrons in the valence (outermost) shell are the parts of an atom that participate in chemical reactions.

12. When valence electrons are transferred from one atom to another, the transfer forms ions, whose unlike charges attract each other and form ionic bonds. Positively charged ions are called cations; negatively charged ions are called anions.

13. In a covalent bond, one, two, or three pairs of valence electrons are shared between two atoms.

14. Hydrogen bonds are weak bonds between hydrogen and certain other atoms within large complex molecules such as proteins and nucleic acids. They add strength and stability and help determine the molecule's three-dimensional shape.

15. Energy is the capacity to do work. Potential energy is energy stored by matter due to its position. Kinetic energy is the energy of matter in motion. Chemical energy is a form of potential energy stored in the bonds of molecules.

16. In chemical reactions, breaking old bonds requires energy and forming new bonds releases energy.

17. In a synthesis reaction, two or more atoms, ions, or molecules combine to form a new and larger molecule. In a decomposition reaction, a molecule is split apart into smaller molecules, ions, or atoms.

18. When nutrients, such as glucose, are broken down via decomposition reactions, some of the energy released is temporarily stored in adenosine triphosphate (ATP) and then later used to drive energy-requiring synthesis reactions that build body structures, such as muscles and bones.

Chemical Compounds and Life Processes (p. 27)

1. Inorganic compounds usually are structurally simple and lack carbon. Organic substances always contain carbon, usually contain hydrogen, and always have covalent bonds.

2. Water is the most abundant substance in the body. It is an excellent solvent, participates in chemical reactions, absorbs and releases heat slowly, requires a large amount of heat to change from a liquid to a gas, and serves as a lubricant.

3. Inorganic acids, bases, and salts dissociate into ions in water. An acid ionizes into hydrogen ions (H^+); a base ionizes into hydroxide ions (OH^-). A salt ionizes into neither H^+ nor OH^- ions.

4. The pH of body fluids must remain fairly constant for the body to maintain homeostasis. On the pH scale, 7 represents neutrality. Values below 7 indicate acidic solutions, and values above 7 indicate alkaline solutions.

5. Buffer systems help maintain pH by converting strong acids or bases into weak acids or bases.

6. Carbohydrates include sugars, glycogen, and starches. They may be monosaccharides, disaccharides, or polysaccharides. Carbohydrates provide most of the chemical energy needed to generate ATP. Carbohydrates, and other large, organic molecules, are synthesized via dehydration synthesis reactions, in which a molecule of water is lost. In the reverse process, called hydrolysis, large molecules are broken down into smaller ones upon the addition of water.

7. Lipids are a diverse group of compounds that include triglycerides (fats and oils), phospholipids, steroids, fat-soluble vitamins (A, D, E, and K), and eicosanoids (prostaglandins and leukotrienes). Triglycerides protect, insulate, provide energy, and are stored in adipose tissue. Phospholipids are important membrane components. Steroids are synthesized from cholesterol.

8. Proteins are constructed from amino acids. They give structure to the body, regulate processes, provide protection, help muscles to contract, transport substances, and serve as enzymes.

9. Enzymes are highly specific catalysts that are subject to a variety of cellular controls.

10. Deoxyribonucleic acid (DNA) and ribonucleic acid (RNA) are nucleic acids consisting of nitrogenous bases, five-carbon sugars, and phosphate groups. DNA is a double helix and is the primary chemical in genes. RNA differs in structure and chemical composition from DNA; its main function is to carry out the instructions encoded in DNA.

11. Adenosine triphosphate (ATP) is the principal energy-transferring molecule in living systems. When it transfers energy, ATP is decomposed by hydrolysis to adenosine diphosphate (ADP) and Ⓟ. ATP is synthesized from ADP and Ⓟ using primarily the energy supplied by the breakdown of glucose.

SELF-QUIZ

1. A substance that dissociates in water to form H^+ and one or more anions is called

   a. a base   **b.** a salt   **c.** a buffer   **d.** an acid
   e. a nucleic acid

2. Ionic bonds are characterized by

   a. sharing electrons between atoms
   b. their ability to form strong, stable bonds
   c. atoms giving away and taking electrons
   d. the type of bonding formed in most organic compounds
   e. an attraction between water molecules

3. If an atom has two electrons in its second electron shell and its first electron shell is filled, it will tend to

   a. lose two electrons from its second electron shell
   b. lose the electrons from its first electron shell
   c. lose all of the electrons from its first and second electron shells
   d. gain six electrons in its second electron shell
   e. share two electrons in its second electron shell

4. Matter that cannot be broken down into simpler substances by chemical reactions is known as

   a. a molecule   **b.** an antioxidant   **c.** a compound   **d.** a buffer   **e.** a chemical element

5. Chlorine (Cl) has an atomic number of 17. An atom of chlorine may become a chloride ion (Cl^-) by

   a. losing one electron   **b.** losing one neutron   **c.** gaining one proton   **d.** gaining one electron   **e.** gaining two electrons

6. Which of the following is NOT true?

   a. A substance that separates in water to form some cation other than H^+ and some anion other than OH^- is known as a salt.
   b. A solution that has a pH of 9.4 is acidic.
   c. A solution with a pH of 5 is 100 times more acidic than distilled water, which has a pH of 7.
   d. Buffers help to make the body's pH more stable.
   e. Amino acids are linked by peptide bonds.

7. Which of the following organic compounds are NOT paired with their correct subunits (building blocks)?

   a. glycogen, glucose   **b.** proteins, monosaccharides
   c. DNA, nucleotides   **d.** lipids, glycerol and fatty acids
   e. ATP, ADP and ⓟ

8. The type of reaction by which a disaccharide is formed from two monosaccharides is known as a

   a. decomposition reaction   **b.** hydrolysis reaction   **c.** dehydration synthesis reaction   **d.** reversible reaction   **e.** dissociation reaction

9. Which of the following contains the genetic code in human cells?

   a. DNA   **b.** enzymes   **c.** RNA   **d.** glucose   **e.** ATP

10. What is the principal energy-transferring molecule in the body?

    a. ADP   **b.** RNA   **c.** DNA   **d.** ATP   **e.** NAD

11. Which of the following statements about water is NOT true?

    a. It is involved in many chemical reactions in the body.
    b. It is an important solvent in the human body.
    c. It helps lubricate a variety of structures in the body.
    d. It can absorb a large amount of heat without changing its temperature.
    e. It requires very little heat to change from a liquid to a gas.

12. The difference in H^+ concentration between solutions with a pH of 3 and a pH of 5 is that the solution with the pH of 3 has _____ H^+.

    a. 2 times more   **b.** 5 times more   **c.** 10 times more
    d. 100 times more   **e.** 200 times fewer

13. Which of the following is NOT a true statement about enzyme activity?

    a. Enzymes form a temporary complex with their substrates.
    b. Enzymes are not permanently altered by the chemical reaction they catalyze.
    c. Most enzymes work with a wide variety of substrates.
    d. Enzymes are considered organic catalysts.
    e. Enzymes are subject to cellular control.

14. For each item in the following list, place an R if it applies to RNA or a D if it refers to DNA; use R and D if it applies to both RNA and DNA.

    ____ **a.** composed of nucleotides   ____ **g.** contains the sugar deoxyribose
    ____ **b.** forms a double helix
    ____ **c.** contains thymine   ____ **h.** is single stranded
    ____ **d.** contains the sugar ribose   ____ **i.** contains adenosine
    ____ **e.** contains the nitrogenous base uracil   ____ **j.** contains phosphate groups
    ____ **f.** is the hereditary material of cells

15. An organic compound that consists of C, H, and O and that may be broken down into glycerol and fatty acids is a

    a. triglyceride   **b.** nucleic acid   **c.** monosaccharide
    d. carbohydrate   **e.** protein

16. Why is it important to consume foods that contain antioxidants?

    a. Antioxidants provide an energy source for the body.
    b. Antioxidants help inactivate damaging free radicals.
    c. Strands of antioxidants make up the body's genes.
    d. Antioxidants act as buffers to help maintain the blood's pH.
    e. Antioxidants are important solvents in the body.

17. If an enzyme is exposed to an extremely high temperature, it will

    a. divide   **b.** release energy   **c.** become an electrolyte
    d. form hydrogen bonds   **e.** denature

18. In what form are lipids stored in the adipose (fat) tissue of the body?

    a. triglycerides   **b.** glycogen   **c.** cholesterol
    d. polypeptides   **e.** disaccharides

19. Approximately 96% of your body's mass is composed of which of the following elements? Place an X beside each correct answer.

    ____ calcium   ____ iron   ____ nitrogen
    ____ phosphorus   ____ sodium   ____ chlorine
    ____ carbon   ____ oxygen   ____ sulfur
    ____ hydrogen   ____ potassium   ____ magnesium

20. Match the following:

    ____ **a.** inorganic compound   **A.** glycogen
    ____ **b.** monosaccharide   **B.** enzyme
    ____ **c.** polysaccharide   **C.** glucose
    ____ **d.** component of triglycerides   **D.** water
    ____ **e.** lipase   **E.** glycerol

CRITICAL THINKING APPLICATIONS

1. While having a tea party, your three-year-old cousin Sabrina added milk, lemon juice, and lots of sugar to her tea. The tea now has strange white lumps floating in it. What caused the milk to curdle?

2. When you eat a hamburger, you ingest the cow's DNA along with its protein and fat. So why don't you start to "moo"?

3. Joy is very proud of her healthy diet. "I drink only pure spring water and eat organic foods. I have a chemical-free body." Sonia replied, "Ever hear of H_2O?" Explain the error in Joy's reasoning.

4. Albert, Jr., was trying out the new Super Genius Home Chemistry Kit that he got for his birthday. He decided to check the pH of his secret formula: lemon juice and diet cola. The pH was 2.5. Next he added tomato juice for a really disgusting mixture with a pH of 3.5. "Wow! That's twice as strong!" Does Albert, Jr., have the makings of a "Super Genius"? Explain.

ANSWERS TO FIGURE QUESTIONS

2.1 The atomic number of carbon is 6.

2.2 The four most plentiful elements in living organisms are oxygen, carbon, hydrogen, and nitrogen.

2.3 Water is a compound because it contains atoms of both hydrogen and oxygen.

2.4 K is an electron donor; when it ionizes, it becomes a cation, K^+, because losing one electron from the fourth electron shell leaves eight electrons in the third shell.

2.5 An ionic bond involves the *loss* and *gain* of electrons; a covalent bond involves the *sharing* of pairs of electrons.

2.6 $CaCO_3$ is a salt, and H_2SO_4 is an acid.

2.7 A pH of 6.82 is more acidic than a pH of 6.91. Both pH = 8.41 and pH = 5.59 are 1.41 pH units from neutral (pH = 7).

2.8 There are 6 carbons in fructose, 12 in sucrose.

2.9 Glycogen is stored in liver and skeletal muscle cells.

2.10 A monounsaturated fatty acid has one double bond.

2.11 A triglyceride has three fatty acid molecules attached to a glycerol backbone, and a phospholipid has two fatty acid tails and a phosphate group attached to a glycerol backbone.

2.12 The dietary lipids thought to contribute to atherosclerosis are cholesterol and saturated fats.

2.13 A tripeptide would have two peptide bonds, each linking two amino acids.

2.14 The enzyme's active site combines with the substrate.

2.15 Thymine is present in DNA but not in RNA, and uracil is present in RNA but not in DNA.

2.16 A few cellular activities that depend on energy supplied by ATP are muscular contractions, movement of chromosomes, transport of substances across cell membranes, and synthesis reactions.

Chapter 3

Cells

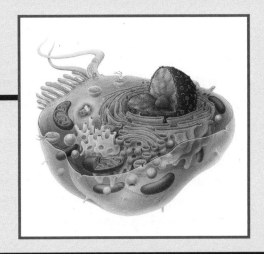

■ Student Learning Objectives

1. Name and describe the three main parts of a cell. **42**
2. Describe the structure and functions of the plasma membrane. **43**
3. Describe the processes that transport substances across the plasma membrane. **44**
4. Describe the structure and functions of cytoplasm, cytosol, and organelles. **51**
5. Describe the structure and functions of the nucleus. **56**
6. Outline the sequence of events involved in protein synthesis. **58**
7. Discuss the stages, events, and significance of somatic cell division. **61**
8. Explain the relationship of aging to cellular processes. **62**

■ A Look Ahead

A *cell* is the basic, living, structural and functional unit of the body. Cells are composed of characteristic parts that perform specific functions. *Cytology* (sī-TOL-ō-jē; *cyt-* = cell; *-ology* = study of) is the scientific study of cellular structure. *Cell physiology* is the study of cellular function. As you study the various parts of a cell and their relationships, you will appreciate the ways that cell structure and function depend on each other.

A GENERALIZED VIEW OF THE CELL

Objective: • **Name and describe the three main parts of a cell.**

Figure 3.1 is a generalized view of a cell that shows the main cellular subcomponents. Though some body cells lack some cellular structures shown in this diagram, such as secretory vesicles or a cilium, many body cells include most of these subcomponents. For ease of study, we can divide a cell into three main parts: the plasma membrane, cytoplasm, and nucleus.

- The *plasma membrane* forms a cell's sturdy yet flexible outer surface, separating the cell's internal environment (inside the cell) from its external environment (outside the cell). It is a selective barrier that regulates the flow of materials into and out of a cell to maintain the appropriate environment for normal cellular activities. The plasma membrane also plays a key role in communication among cells and between cells and their external environment.

- The *cytoplasm* (SĪ-tō-plazm; *-plasm* = formed or molded) consists of all the cellular contents between the plasma membrane and the nucleus. This compartment can be divided into two components: cytosol and organelles. *Cytosol* (SĪ-tō-sol) is the fluid portion of cytoplasm that consists mostly of water plus dissolved solutes and suspended particles. Surrounded by cytosol are several different types of *organelles* (or-ga-NELZ = little organs), each of which has a characteristic structure and specific functions.

Figure 3.1 ■ **Generalized view of a body cell.**

The cell is the basic, living, structural and functional unit of the body.

Cilium

Microtubule

Microfilament

Microvillus

Centrosome:
  Pericentriolar area
  Centrioles

PLASMA MEMBRANE

Lysosome

Smooth endoplasmic reticulum

Peroxisome

Mitochondrion

Microtubule

Intermediate filament

Secretory vesicle

NUCLEUS:
  Chromatin
  Nuclear envelope
  Nucleolus

Glycogen granules

CYTOPLASM (cytosol plus organelles except the nucleus)

Rough endoplasmic reticulum

Ribosome

Golgi complex

Microfilament

Sectional view

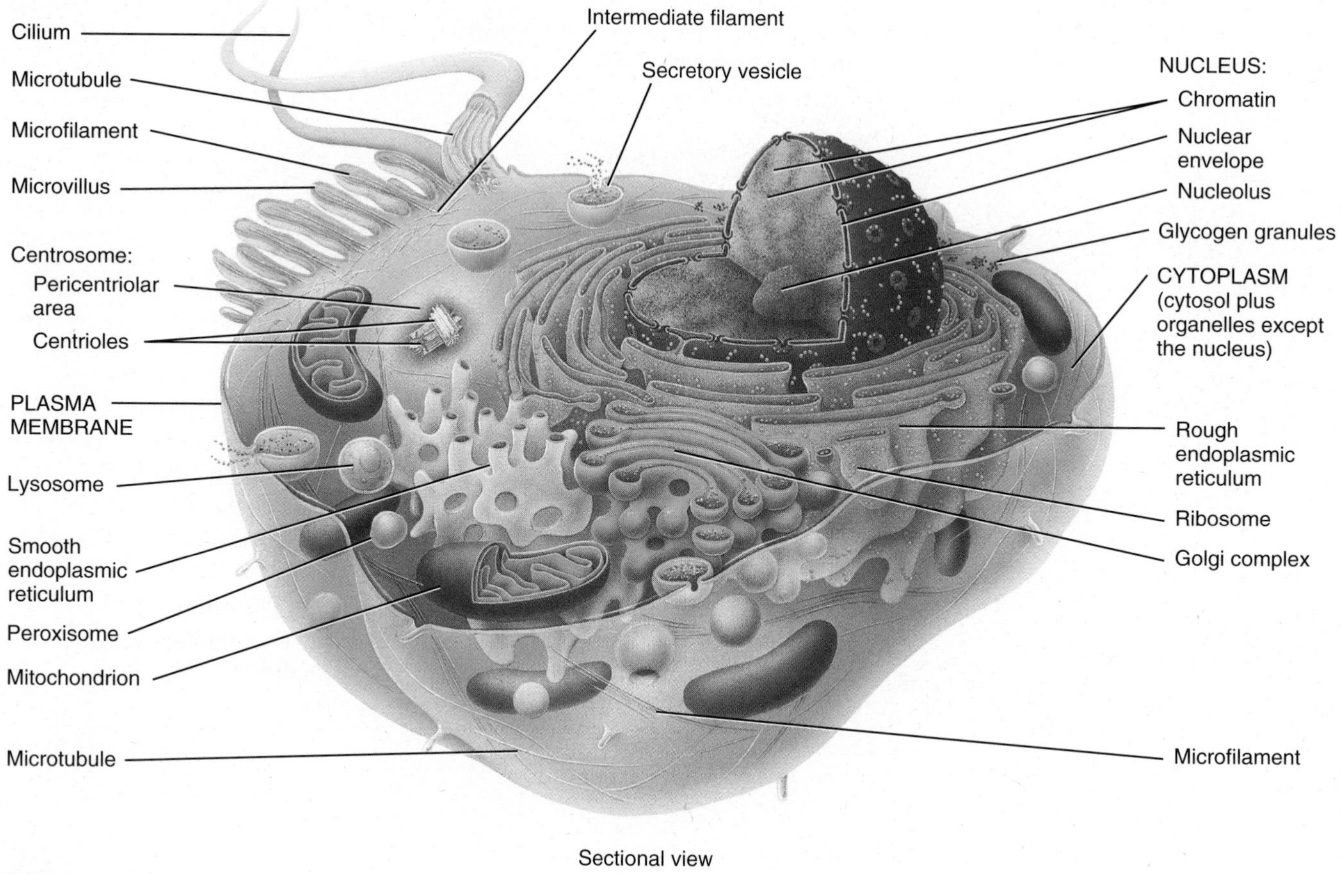

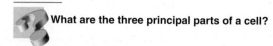

What are the three principal parts of a cell?

- The *nucleus* (NOO-klē-us = nut kernel) is the largest organelle of a cell. The nucleus acts as the control center for a cell because it contains the genes, which control cellular structure and most cellular activities.

THE PLASMA MEMBRANE

Objective: • **Describe the structure and functions of the plasma membrane.**

The *plasma membrane* consists mostly of phospholipids (lipids that contain phosphate groups) and proteins. Virtually all membrane proteins are *glycoproteins,* proteins with attached carbohydrate groups. Other molecules present in lesser amounts in the plasma membrane are cholesterol and glycolipids (lipids with attached carbohydrate groups). The basic framework of the plasma membrane is the *lipid bilayer,* two back-to-back layers made up of three types of lipid molecules: phospholipids, cholesterol, and glycolipids (Figure 3.2). The glycoproteins in a membrane are of two types—integral and peripheral (Figure 3.2). *Integral proteins* extend through the lipid bilayer

among the fatty acid tails. *Peripheral proteins* are loosely attached to the exterior or interior surface of the membrane. Although many of the proteins can float laterally in the lipid bilayer, each individual protein has a specific orientation with respect to the "inside" and "outside" faces of the membrane. Glycolipids appear only in the membrane layer that faces the extracellular fluid (the fluid outside body cells). The carbohydrate groups of glycoproteins also protrude into the extracellular fluid.

The plasma membrane allows some substances to move into and out of the cell but restricts the passage of other substances. This property of membranes is called *selective permeability* (per′-mē-a-BIL-i-tē). The lipid bilayer part of the membrane is permeable to water and to most lipid-soluble molecules, such as fatty acids, fat-soluble vitamins, steroids, oxygen, and carbon dioxide. The lipid bilayer is *not* permeable to ions and charged or polar molecules, such as glucose and amino acids. These small and medium-sized water-soluble materials may cross the membrane with the assistance of integral proteins. Some integral proteins form *channels* that have a *pore* (hole) through which specific substances can move into and out of cells (Figure 3.2). Other membrane proteins act as *transporters* (carriers), changing shape as they move a substance from one side of the membrane to the other. Large molecules such as proteins are

Figure 3.2 ■ **Chemistry and structure of the plasma membrane.**

The plasma membrane consists mostly of phospholipids, arranged in a bilayer, and proteins, virtually all of which are glycoproteins.

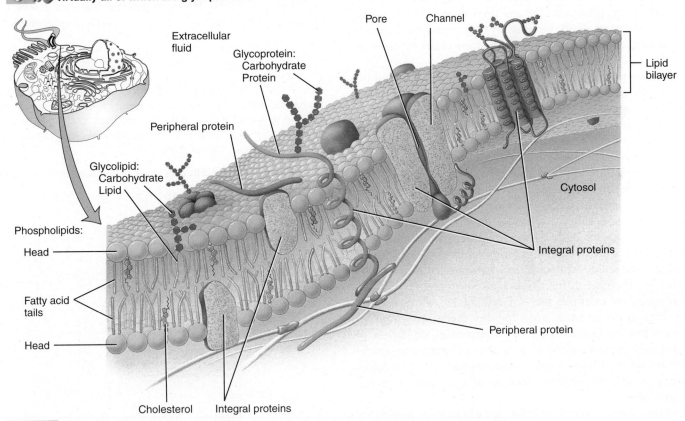

 Which components of the plasma membrane enable cells to recognize other similar cells or potentially dangerous foreign cells?

unable to pass through the plasma membrane except by vesicular transport (discussed later in this chapter).

Most functions of the plasma membrane depend on the types of proteins that are present. Integral proteins called **receptors** recognize and bind a specific molecule that governs some cellular function, for example, a hormone such as insulin. Different types of cells have different receptors for a variety of hormones and other substances. Some integral and peripheral proteins act as **enzymes**, catalyzing specific chemical reactions. Membrane glycoproteins and glycolipids often are **cell identity markers.** They enable a cell to recognize other cells of its own kind during tissue formation, or to recognize and respond to potentially dangerous foreign cells. Integral and peripheral proteins both may serve as **linkers,** which anchor proteins in the plasma membranes of neighboring cells to one another or to structures inside and outside the cell.

TRANSPORT ACROSS THE PLASMA MEMBRANE

Objective: • **Describe the processes that transport substances across the plasma membrane.**

Movement of materials across its plasma membrane is essential to the life of a cell. Certain substances must move into the cell to support metabolic reactions. Other materials must be moved out because they have been produced by the cell for export or are cellular waste products. Before discussing how materials move into and out of a cell, we need to understand what exactly is being moved as well as the form it needs to take to make its journey.

About two-thirds of the fluid in your body is contained inside body cells and is called **intracellular fluid** or **ICF** (*intra-* = within). Fluid outside body cells is called **extracellular fluid** or **ECF** (*extra-* = outside). The ECF in the microscopic spaces between the cells of tissues is **interstitial fluid** (in'-ter-STISH-al; *inter-* = between). The ECF in blood vessels is called **plasma,** and that in lymphatic vessels is called **lymph.**

Materials dissolved in body fluids include gases, nutrients, ions, and other substances needed to maintain life. Any material dissolved in a fluid is called a **solute,** and the fluid in which it is dissolved is the **solvent.** Body fluids are dilute solutions in which a variety of solutes are dissolved in a very familiar solvent, water. The amount of a solute in a solution is its **concentration.** A **concentration gradient** is a difference in concentration between two different areas, for example, the ICF and ECF. Solutes moving from a high-concentration area (where there are more of them) to a low-concentration area (where there are fewer of them) are said to move *down* or *with* the concentration gradient. Solutes moving from a low-concentration area to a high-concentration area are said to move *up* or *against* the concentration gradient.

Substances move across cellular membranes by passive or active processes. **Passive processes,** in which a substance moves down its concentration gradient through the membrane, using only its own energy of motion (kinetic energy), include diffusion

and osmosis. In **active transport,** cellular energy, usually in the form of ATP, is used to "push" the substance through the membrane "uphill" against its concentration gradient.

Another way that some substances may enter and leave cells is via tiny membrane sacs called **vesicles.** In **vesicular transport** (see Figure 3.11), vesicles either detach from the plasma membrane while bringing substances into the cell, or merge with the plasma membrane to release substances from the cell. Vesicular transport is the only way that large particles, such as whole bacteria and viruses, and large molecules, for example, polysaccharides and proteins, may enter and leave cells.

Passive Processes
Diffusion

Diffusion (di-FYOO-zhun; *diffus-* = to spread out) is the random mixing of substances that occurs in a solution due to the substances' kinetic energy. If a particular solute is present in high concentration in one area and in low concentration in another area, more solute particles diffuse from the region of high concentration to the region of low concentration than diffuse in the opposite direction. The difference in diffusion in the two directions is the overall or *net* diffusion. Solutes undergoing net diffusion move from a high to a low concentration, or *down their concentration gradient.* After some time, **equilibrium** (ē'-kwi-LIB-rē-um) is reached: The solute becomes evenly distributed throughout the solution and the concentration gradient disappears.

Placing a crystal of dye in a water-filled container provides an example of diffusion (Figure 3.3). At the beginning, the color

Figure 3.3 ■ **Principle of diffusion.** A crystal of dye placed in a cylinder of water dissolves (beginning), and there is net diffusion from the region of higher dye concentration to regions of lower dye concentration (intermediate). At equilibrium, dye concentration is uniform throughout the solution.

At equilibrium, net diffusion stops but random movement continues.

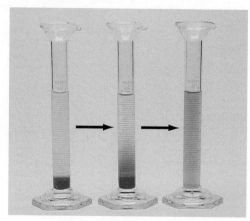

Beginning   Intermediate   Equilibrium

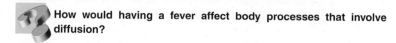

How would having a fever affect body processes that involve diffusion?

is most intense just next to the crystal because the dye concentration is greatest there. At increasing distances, the color is lighter and lighter because the dye concentration is lower and lower. The dye molecules undergo net diffusion, down their concentration gradient, until they are evenly mixed in the water. At equilibrium the solution has a uniform color.

In the example of dye diffusion, no membrane was involved. Substances may also diffuse across a membrane, if the membrane is permeable to them. Several factors influence the diffusion rate of substances across plasma membranes:

1. **Size of the concentration gradient.** The greater the concentration gradient, the faster the rate of diffusion.

2. **Temperature.** The higher the temperature, the faster the rate of diffusion. For example, in a person who has a fever, all of the body's diffusion processes occur more rapidly.

3. **Weight (mass) of the diffusing substance.** The larger the weight (mass) of the diffusing substance, the slower its diffusion rate. For example, if other factors are equal, oxygen (O_2) diffuses more rapidly than carbon dioxide (CO_2) because the molecular weight of O_2 is less than that of CO_2.

4. **Surface area.** The larger the membrane surface area available for diffusion, the faster the diffusion rate. For example, diffusion of oxygen from the air sacs of the lungs into the blood normally occurs rapidly because the air sacs have a large surface area. Lung diseases such as emphysema reduce this surface area, which slows the rate of diffusion and makes breathing more difficult.

5. **Diffusion distance.** The greater the distance over which diffusion must occur, the longer it takes. Diffusion across a plasma membrane takes only a fraction of a second because the membrane is so thin.

SIMPLE DIFFUSION In *simple diffusion,* substances diffuse across a membrane in one of two ways: lipid-soluble substances diffuse through the lipid bilayer and ions diffuse through pores of channels (Figure 3.4). Substances that move across membranes by simple diffusion through the lipid bilayer include oxygen, carbon dioxide, and nitrogen gases; fatty acids, steroids, and fat-soluble vitamins (A, D, E, and K); small alcohols; and ammonia. Simple diffusion through the lipid bilayer is important in the exchange of oxygen and carbon dioxide between blood and body cells and between blood and air within the lungs during breathing. It also is the transport method for absorption of lipid-soluble nutrients and release of some wastes from body cells.

An ion channel allows a specific type of ion to move across the membrane by simple diffusion through the channel's pore. In typical plasma membranes, the most common ion channels are selective for K^+ (potassium ions) or Cl^- (chloride ions); fewer channels are available for Na^+ (sodium ions) or Ca^{2+} (calcium ions). Many ion channels are gated; that is, a portion of the channel protein acts as a "gate," moving in one direction to open the pore and in another direction to close it (Figure 3.5). When the gates are open, ions diffuse into or out of cells, down their concentration gradient. Gated channels are important for the production of electrical signals by body cells.

Figure 3.4 ■ **Simple diffusion.** Lipid-soluble molecules may diffuse through the lipid bilayer, and ions may diffuse through channels in integral proteins. Plasma membranes have channels, formed by integral proteins, that are selective for potassium ions (K^+), sodium ions (Na^+), calcium ions (Ca^{2+}), and chloride ions (Cl^-).

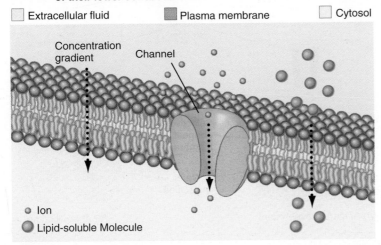

In simple diffusion there is a net (greater) movement of substances from a region of their higher concentration to a region of their lower concentration.

☐ Extracellular fluid   ■ Plasma membrane   ☐ Cytosol

Concentration gradient
Channel

○ Ion
◯ Lipid-soluble Molecule

What are some examples of substances that diffuse through the lipid bilayer?

Figure 3.5 ■ **Diffusion of K^+ through a gated channel.** A gated channel is one in which a portion of the channel protein acts as a gate to open or close the channel's pore to passage of ions.

Channels are integral membrane proteins that allow specific small, inorganic ions to pass across the membrane by simple diffusion.

☐ Extracellular fluid   ■ Plasma membrane   ☐ Cytosol

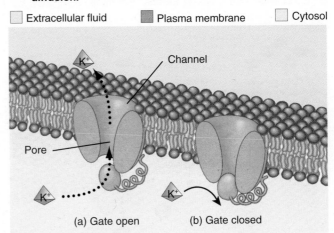

Channel
Pore
K^+
(a) Gate open   (b) Gate closed

Is the concentration of K^+ in body cells higher in the cytosol or in the extracellular fluid?

FACILITATED DIFFUSION Some substances that cannot diffuse through the lipid bilayer or through ion channels do cross the plasma membrane by *facilitated diffusion.* In this process, an integral membrane protein assists a specific substance across the membrane. The substance binds to a specific transporter on one side of the membrane and is released on the other side after the transporter undergoes a change in shape. As is true for simple diffusion, facilitated diffusion moves a substance down a concentration gradient—from a region of higher concentration to a region of lower concentration—and does not require cellular energy in the form of ATP.

Substances that move across plasma membranes by facilitated diffusion include glucose, fructose, galactose, urea, and some vitamins. Glucose enters many body cells by facilitated diffusion as follows (Figure 3.6)

❶ Glucose binds to a glucose transporter protein on the outside surface of the membrane.

❷ As the transporter undergoes a change in shape, glucose passes through the membrane.

❸ The transporter releases glucose on the other side of the membrane.

The selective permeability of the plasma membrane is often regulated to achieve homeostasis. For example, the hormone insulin promotes the insertion of many copies of a specific type of glucose transporter into the plasma membranes of certain cells. Thus, the effect of insulin is to increase entry of glucose into body cells by means of facilitated diffusion. With more trans-porters available, body cells can pick up glucose from the blood more rapidly.

Osmosis

Osmosis (oz-MŌ-sis) is the net movement of a solvent—water in living organisms—through a selectively permeable membrane. Water moves by osmosis from an area of *higher water concentration* to an area of *lower water concentration* (or from an area of *lower solute concentration* to an area of *higher solute concentration*). Water molecules pass through plasma membranes in two places: through the lipid bilayer and through integral membrane proteins that function as water channels. In osmosis, as in diffusion of solutes, the driving force is the kinetic energy of the water molecules themselves.

The device in Figure 3.7 demonstrates osmosis. A sac made of cellophane, a selectively permeable membrane that permits water but not sucrose (sugar) molecules to pass, is filled with a solution that is 20% sucrose and 80% water. The upper part of the cellophane sac is wrapped tightly about a stopper through which a glass tube is fitted. The sac is then placed into a beaker containing pure (100%) water (Figure 3.7a). Notice that the cel-

---

Figure 3.6 ■ Facilitated diffusion of glucose across a plasma membrane. The transporter binds to glucose in the extracellular fluid and releases it into the cytosol.

🔑 **Facilitated diffusion across a membrane requires a transporter but does not use ATP.**

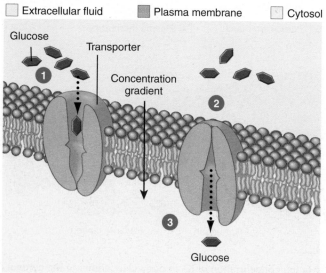

☐ Extracellular fluid   ☐ Plasma membrane   ☐ Cytosol

Glucose

Transporter

Concentration gradient

Glucose

❓ How does insulin alter glucose transport by facilitated diffusion?

Figure 3.7 ■ Principle of osmosis. (a) At the start of the experiment, a cellophane sac—a selectively permeable membrane that permits water but not sucrose molecules to pass—containing a 20% sucrose solution is immersed in a beaker of pure (100%) water. Osmosis begins (arrows) as water moves down its concentration gradient into the sac. **(b)** As the volume of the sucrose solution increases, the solution moves up the glass tubing. The added fluid in the tube exerts a pressure that drives some water molecules back into the beaker. At equilibrium, osmosis has stopped because the number of water molecules entering and the number leaving the cellophane sac are equal.

🔑 **Osmosis is the net movement of water molecules through a selectively permeable membrane.**

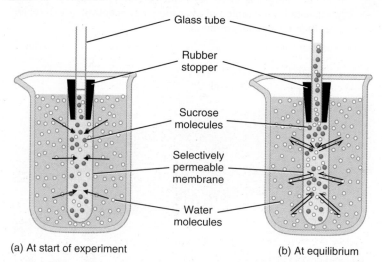

Glass tube

Rubber stopper

Sucrose molecules

Selectively permeable membrane

Water molecules

(a) At start of experiment   (b) At equilibrium

❓ **Will the fluid level in the tube continue to rise until the sucrose concentrations are the same in the beaker and in the sac?**

lophane now separates two fluids having different water concentrations. As a result, water begins to move by osmosis from the region where its concentration is higher (100% water in the beaker) through the cellophane to where its concentration is lower (80% water inside the sac). Because the cellophane is not permeable to sucrose, however, all the sucrose molecules remain inside the sac. As water moves into the sac, the volume of the sucrose solution increases and the fluid rises into the glass tube (Figure 3.7b). As the fluid rises in the tube, its water pressure forces some water molecules from the sac back into the beaker. At equilibrium, just as many water molecules are moving into the beaker due to the water pressure as are moving into the sac due to osmosis.

A solution containing solute particles that cannot pass through a membrane exerts a pressure on the membrane, called *osmotic pressure.* The osmotic pressure of a solution depends on the concentration of its solute particles—the higher the solute concentration, the higher the solution's osmotic pressure. Because the osmotic pressure of cytosol and interstitial fluid is the same, cell volume remains relatively constant. Cells neither shrink due to water loss by osmosis nor swell due to water gain by osmosis.

Any solution in which cells maintain their normal shape and volume is called an *isotonic solution* (iso- = same; tonic- = tension) (Figure 3.8a). This is a solution in which the concentrations of solutes that cannot pass through the plasma membrane are the *same* on both sides. For example, a 0.9% NaCl (sodium chloride, or table salt) solution, called a *normal saline solution,* is isotonic for red blood cells. When red blood cells are bathed in 0.9% NaCl, water molecules enter and exit the cells at the same rate, allowing the red blood cells to maintain their normal shape and volume.

If red blood cells are placed in a *hypotonic solution* (hypo- = less than), a solution that has a *lower* concentration of solutes than the cytosol inside the red blood cells (Figure 3.8b), water molecules enter the cells by osmosis faster than they leave. This situation causes the red blood cells to swell and eventually to burst. Rupture of red blood cells is called *hemolysis* (hē-MOL-i-sis). A *hypertonic solution* (hyper- = greater than) has a *higher* concentration of solutes than does the cytosol inside red blood cells (Figure 3.8c). When cells are placed in a hypertonic solution, water molecules move out of the cells by osmosis faster than they enter, causing the cells to shrink. Such shrinkage of red blood cells is called *crenation* (kri-NĀ-shun). Red blood cells and other body cells may be damaged or destroyed if exposed to hypertonic or hypotonic solutions. For this reason, most intravenous (IV) solutions, which are infused into the blood of patients via a vein, are isotonic.

Active Transport

In *active transport* cellular energy is used to transport substances across the membrane against a concentration gradient (from an area of low to an area of high concentration). The two types of active transport are (1) primary active transport, which uses energy obtained from splitting ATP, and (2) secondary active transport, which uses energy stored in a Na^+ concentration gradient.

Primary Active Transport

In *primary active transport,* energy derived from splitting ATP changes the shape of an integral protein, called a *pump,* which carries a substance across a cellular membrane against its concentration gradient. A typical body cell expends about 40% of its ATP on primary active transport. Drugs that turn off ATP production, such as the poison cyanide, are lethal because they shut down active transport in cells throughout the body. Substances transported across the plasma membrane by primary active transport are mainly ions, primarily Na^+, K^+, H^+, Ca^{2+}, I^-, and Cl^-.

The most important primary active transport pump expels sodium ions (Na^+) from cells and brings in potassium ions (K^+). The pump protein also acts as an enzyme to split ATP. Because of the ions it moves, this pump is called the *Na^+/K^+ pump* or, more simply, the *sodium pump.* All cells have thousands of sodium pumps in their plasma membranes. These pumps maintain a low concentration of sodium ions in the cytosol by pumping Na^+ into the extracellular fluid against the Na^+ concentration gradient. At the same time, the sodium pump moves potassium ions into cells against the K^+ concentration gradient. Because K^+ and Na^+ slowly leak back across the plasma membrane down their gradients, the sodium pumps must operate continually to maintain a low concentration of sodium ions and a high concentration of potassium ions in the cytosol. These differing concentrations of K^+ and Na^+ in cytosol and extracellular fluid are crucial for osmotic balance of the two fluids and also for the ability of some cells to generate electrical signals such as action potentials.

Figure 3.8 ▪ Principle of osmosis applied to red blood cells (RBCs). (a) In an isotonic solution, there is no water gain or loss by osmosis. (b) In a hypotonic solution, the RBC gains water rapidly by osmosis until it bursts, an event called hemolysis. (c) In a hypertonic solution, the RBC loses water by osmosis so rapidly that it shrinks, a process called crenation.

An isotonic solution is one in which cells maintain their normal shape and volume.

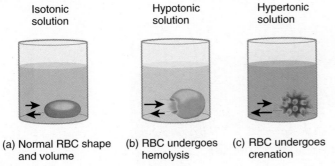

Isotonic solution	Hypotonic solution	Hypertonic solution
(a) Normal RBC shape and volume	(b) RBC undergoes hemolysis	(c) RBC undergoes crenation

Will a 2% solution of NaCl cause hemolysis or crenation of RBCs?

The sodium pump maintains a low intracellular concentration of Na⁺.

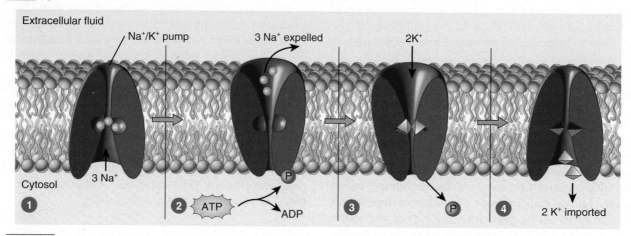

 What is the role of ATP in the operation of this pump?

Figure 3.9 shows how the sodium pump operates.

1 Three sodium ions (Na⁺) in the cytosol bind to the pump protein.

2 Na⁺ binding triggers the splitting of ATP into ADP plus a phosphate group (Ⓟ), which becomes attached to the pump protein. This chemical reaction changes the shape of the pump protein, releasing the three Na⁺ into the extracellular fluid. The changed shape of the pump protein then favors binding of two potassium ions (K⁺) in the extracellular fluid to the pump protein.

3 The binding of K⁺ causes the pump protein to release the phosphate group, which causes the pump protein to return to its original shape.

4 As the pump protein returns to its original shape, the two K⁺ are released into the cytosol. At this point, the pump is ready again to bind Na⁺, and the cycle repeats.

Secondary Active Transport

In *secondary active transport,* the energy stored in the Na⁺ concentration gradient is used to drive other substances across the

Figure 3.10 ■ **Secondary active transport mechanisms.** (a) Antiporters carry two substances across the membrane in opposite directions. (b) Symporters carry two substances across the membrane in the same direction.

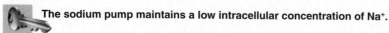 **Secondary active transport mechanisms use the energy stored in a Na⁺ concentration gradient.**

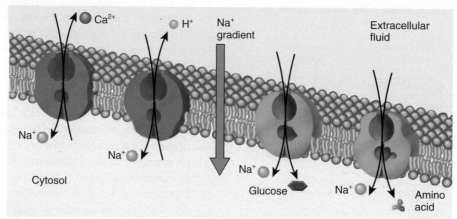

 What is the main difference between secondary and primary active transport mechanisms?

membrane against their own concentration gradients. It is the sodium pumps that maintain a steep concentration gradient of Na^+ across the plasma membrane. Because the Na^+ gradient is established by primary active transport, secondary active transport *indirectly* uses energy obtained from splitting ATP. The integral proteins that mediate secondary active transport work by simultaneously binding to Na^+ and another substance and then changing shape so that both substances cross the membrane at the same time. *Antiporters* move Na^+ and another substance in opposite directions across the membrane, and *symporters* move Na^+ and another substance in the same direction (Figure 3.10). Solutes transported across the plasma membrane by secondary active transport include ions, glucose, and amino acids.

Plasma membranes contain several types of antiporters and symporters that are powered by the Na^+ gradient. In most body cells, the concentration of calcium ions (Ca^{2+}) is low in the cytosol because Na^+/Ca^{2+} antiporters eject calcium ions (Figure 3.10a). Likewise, Na^+/H^+ antiporters help regulate the cytosol's pH (H^+ concentration) by expelling excess H^+. Dietary glucose and amino acids are absorbed into cells that line the small intestine by Na^+–glucose and Na^+–amino acid symporters (Figure 3.10b). In each case, sodium ions are moving down their concentration gradient while the other solutes move against their concentration gradients. Keep in mind that all these symporters and antiporters can do their job because the sodium pumps maintain a low concentration of Na^+ in the cytosol.

Vesicular Transport

A *vesicle* (VES-i-kul) is a small sac formed by budding off from an existing membrane. The membrane of the vesicle is selectively permeable, and the vesicle contains a small quantity of fluid plus dissolved or suspended particles. Vesicles transport substances from one structure to another within cells, take in substances from extracellular fluid, and release substances into extracellular fluid. Movement of vesicles requires energy supplied by ATP. The two main types of vesicular transport between a cell and the extracellular fluid that surrounds it are (1) *endocytosis* (endo- = within), in which materials move *into* a cell in a vesicle formed from the plasma membrane, and (2) *exocytosis* (exo- = out), in which materials move *out of* a cell by the fusion of a vesicle with the plasma membrane.

Endocytosis

As you have just learned, substances brought into the cell by means of endocytosis are surrounded by a piece of the plasma membrane, which buds off inside the cell to form a vesicle containing the ingested substances. The two main types of endocytosis are phagocytosis and pinocytosis.

PHAGOCYTOSIS In *phagocytosis* (fag'-ō-sī-TŌ-sis; *phago-* = to eat), large solid particles, such as whole bacteria or viruses or aged or dead cells, are taken in by the cell (Figure 3.11). Phagocytosis begins as the particle binds to a plasma membrane receptor, causing the cell to extend projections of its plasma membrane and cytoplasm, called *pseudopods* (SOO-dō-pods; *pseudo-* = false; *-pods* = feet). Two or more pseudopods surround the particle, and portions of their membranes fuse to

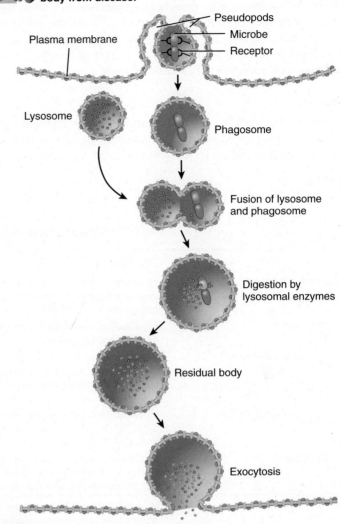

Figure 3.11 ■ **Endocytosis (phagocytosis) and exocytosis.**

Phagocytosis is a vital defense mechanism that helps protect the body from disease.

Plasma membrane · Pseudopods · Microbe · Receptor · Lysosome · Phagosome · Fusion of lysosome and phagosome · Digestion by lysosomal enzymes · Residual body · Exocytosis

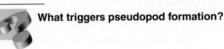

What triggers pseudopod formation?

form a vesicle called a *phagosome*, which enters the cytoplasm. The phagosome fuses with one or more lysosomes, organelles with powerful digestive enzymes; the ingested material is broken down; and useful substances are transported into the cytosol. The *residual body*, the vesicle that contains any undigested materials, may remain indefinitely in the cytoplasm or fuse with the plasma membrane and expel its contents from the cell by means of exocytosis.

Phagocytosis occurs only in *phagocytes*, cells that are specialized to engulf and destroy bacteria and other foreign substances. Phagocytes include certain types of white blood cells and macrophages, which are present in most body tissues. The process of phagocytosis is a vital defense mechanism that helps protect the body from disease.

PINOCYTOSIS In *pinocytosis* (pin'-ō-sī-TŌ-sis; *pino-* = to drink), cells take up vesicles containing tiny droplets of extracellular fluid. Pinocytosis occurs in most body cells and is not specific. It takes in any and all solutes dissolved in extracellular fluid. During pinocytosis, the plasma membrane folds inward and forms a ***pinocytic vesicle*** containing a droplet of extracellular fluid. The pinocytic vesicle detaches or "pinches off" from the plasma membrane and enters the cytosol. Within the cell, the pinocytic vesicle fuses with one or more lysosomes, where enzymes degrade the engulfed solutes. As in phagocytosis, undigested materials accumulate in a residual body.

Exocytosis

As shown in Figure 3.11, ***exocytosis*** involves the movement out of a cell of materials contained in vesicles that fuse with the plasma membrane. The ejected material may be either waste products, such as the undigested materials in Figure 3.11, or a useful secretory product, such as a protein hormone like insulin. All cells carry out exocytosis, but it is especially important in two types of cells. By means of exocytosis, nerve cells release substances called *neurotransmitters*, and secretory cells release digestive enzymes or hormones. During exocytosis in secretory cells, membrane-enclosed vesicles called *secretory vesicles* form inside the cell, fuse with the plasma membrane, and release their contents into the extracellular fluid (see Figure 3.1).

Table 3.1 summarizes the processes by which materials move through cellular membranes.

Table 3.1 / Movement of Materials Through Cellular Membranes

Process	Description	Substances Transported
Passive Processes	Movement of substances through the plasma membrane down a concentration gradient; do not require cellular energy in the form of ATP.	
Diffusion	Movement of a substance due to its kinetic energy down its concentration gradient.	
Simple diffusion	Diffusion of lipid-soluble molecules through the lipid bilayer of the plasma membrane, or of ions through channels in integral proteins.	*Lipid-soluble molecules:* oxygen, carbon dioxide, and nitrogen gases; fatty acids, steroids, and fat-soluble vitamins; small alcohols; and ammonia. *Ions:* Na^+, K^+, Ca^{2+}, and Cl^-.
Facilitated diffusion	Movement of substances across the plasma membrane aided by integral proteins that serve as transporters for particular substances.	Glucose, fructose, galactose, urea, and some vitamins.
Osmosis	Movement of water through a selectively permeable membrane from an area of higher to an area of lower water concentration until equilibrium is reached.	Water.
Active Transport	Movement of substances through the plasma membrane against their concentration gradient; requires expenditure of cellular energy in the form of ATP.	
Primary active transport	Transport of a substance through a cellular membrane by transporter proteins, termed pumps, that split ATP; the most important primary active transport pump expels Na^+ from cells and brings in K^+.	Na^+, K^+, Ca^{2+}, H^+, I^-, Cl^-, and other ions.
Secondary active transport	Simultaneous transport of Na^+ and a second substance across the membrane using energy stored in the Na^+ concentration gradient, which is maintained by primary active transport pumps.	
Antiporters	Move Na^+ and the second substance in opposite directions across the membrane.	Ca^{2+} and H^+ out of many cells.
Symporters	Move Na^+ and the second substance in the same direction.	Glucose into cells lining the small intestine and the kidney tubules; amino acids into most body cells.
Vesicular Transport	Movement of substances into or out of a cell within vesicles that bud from the plasma membrane; requires energy supplied by ATP.	
Endocytosis	Movement of substances into a cell in vesicles.	
Phagocytosis	"Cell eating"; formation of a phagosome around solid particles by pseudopods.	Bacteria, viruses, and aged or dead cells.
Pinocytosis	"Cell drinking"; formation of a pinocytic vesicle around an extracellular fluid droplet by infolding of the plasma membrane.	Most solutes in extracellular fluid.
Exocytosis	Movement of substances out of a cell in secretory vesicles that fuse with the plasma membrane and release their contents into the extracellular fluid.	Neurotransmitters, hormones, and digestive enzymes.

CYTOPLASM

Objective: • **Describe the structure and functions of cytoplasm, cytosol, and organelles.**

Cytoplasm consists of all of the cellular contents between the plasma membrane and the nucleus and includes both cytosol and organelles.

Cytosol

The *cytosol (intracellular fluid)* is the fluid portion of the cytoplasm that surrounds organelles and accounts for about 55% of the total cell volume. Although cytosol varies in composition and consistency from one part of a cell to another, typically it is 75% to 90% water plus various dissolved solutes and suspended particles. Among these are various ions, glucose, amino acids, fatty acids, proteins, lipids, ATP, and waste products. Some cells also contain *lipid droplets* that contain triglycerides and *glycogen granules*, clusters of glycogen molecules. The cytosol is the site of many of the chemical reactions that maintain cell structures and allow cellular growth.

Organelles

Organelles are specialized structures inside cells that have characteristic shapes and specific functions. Each type of organelle is a functional compartment where specific processes take place, and each has its own unique set of enzymes.

The Cytoskeleton

Extending throughout the cytosol, the cytoskeleton is a network of three different types of protein filaments: microfilaments, intermediate filaments, and microtubules (Figure 3.12).

Figure 3.12 ■ **Cytoskeleton.**

Extending throughout the cytoplasm, the cytoskeleton is a network of three kinds of protein filaments: microfilaments, intermediate filaments, and microtubules.

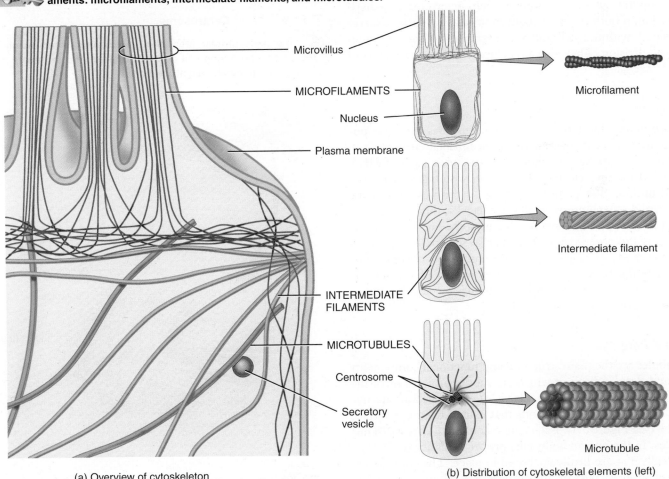

Microvillus

MICROFILAMENTS

Nucleus

Plasma membrane

Microfilament

INTERMEDIATE FILAMENTS

Intermediate filament

MICROTUBULES

Centrosome

Secretory vesicle

Microtubule

(a) Overview of cytoskeleton

(b) Distribution of cytoskeletal elements (left) and detail of structure (right)

Which cytoskeletal proteins help form the structure of centrioles, cilia, and flagella?

The thinnest elements of the cytoskeleton are the ***microfilaments,*** which are concentrated at the periphery of a cell and contribute to the cell's strength and shape. Microfilaments have two general functions: providing mechanical support and helping generate movements. They also anchor the cytoskeleton to integral proteins in the plasma membrane and provide support for microscopic, fingerlike projections of the plasma membrane called ***microvilli*** (*micro-* = small; *-villi* = tufts of hair; singular is *microvillus*). Because they greatly increase the surface area of the cell, microvilli are abundant on cells involved in absorption, such as the cells that line the small intestine. Some microfilaments extend beyond the plasma membrane and help cells attach to one another or to extracellular materials.

With respect to movement, microfilaments are involved in muscle contraction; cell division; and cell locomotion, such as the migration of embryonic cells during development, the invasion of tissues by white blood cells to fight infection, or the migration of skin cells during wound healing.

As their name suggests, ***intermediate filaments*** are thicker than microfilaments but thinner than microtubules (Figure 3.12). They are found in parts of cells subject to tension (such as stretching) and also help hold organelles such as the nucleus in place.

The largest of the cytoskeletal components, ***microtubules*** are long, hollow tubes. Microtubules help determine cell shape and function in both the movement of organelles, such as secretory vesicles, within a cell and the migration of chromosomes during cell division. They also are responsible for movements of cilia and flagella.

Centrosome

The ***centrosome,*** located near the nucleus, has two components—a pair of centrioles and a pericentriolar area (Figure 3.13). The two *centrioles* are cylindrical structures, each of which is composed of nine clusters of three microtubules (a triplet) arranged in a circular pattern. Although centrioles help in the formation and regeneration of cilia and flagella, their function in cells that lack these structures is still a mystery. Surrounding the centrioles is the *pericentriolar area* (per′-ē-sen′-trē-Ō-lar), containing hundreds of ring-shaped proteins called *tubulins*. The tubulins are the organizing centers for growth of the mitotic spindle, which plays a critical role in cell division, and for microtubule formation in nondividing cells.

Cilia and Flagella

Microtubules are the main structural and functional components of cilia and flagella, both of which are motile projections of the cell surface. ***Cilia*** (SIL-ē-a; singular is *cilium* = eyelash) are numerous, short, hairlike projections that extend from the surface of the cell (see Figure 3.1). Each cilium contains a core of 18 microtubules. In the human body, cilia propel fluids across the surfaces of cells that are firmly anchored in place. The coordinated movement of many cilia on the surface of a cell causes a steady movement of fluid along the cell's surface. Many cells of the res-

piratory tract, for example, have hundreds of cilia that help sweep foreign particles trapped in mucus away from the lungs. Their movement is paralyzed by nicotine in cigarette smoke. For this reason, smokers cough often to remove foreign particles from their airways. Cells that line the uterine (Fallopian) tubes also have cilia that sweep oocytes (egg cells) toward the uterus.

Flagella (fla-JEL-a; singular is *flagellum* = whip) are similar in structure to cilia but are much longer. Flagella usually move an entire cell. The only example of a flagellum in the human body is a sperm cell's tail, which propels the sperm toward its possible union with an oocyte.

Ribosomes

Ribosomes (RĪ-bō-sōms; *-somes* = bodies) are the sites of protein synthesis. Ribosomes are named for their high content of *ribo*nucleic acid (RNA). Besides ***ribosomal RNA (rRNA),*** these tiny organelles contain ribosomal proteins. Structurally, a ribosome consists of two subunits, large and small, one about half the size of the other (Figure 3.14). The two units are made separately in the nucleolus, a spherical body inside the nucleus. Once

Figure 3.13 ■ **Centrosome.**

 The pericentriolar area of a centrosome organizes the mitotic spindle during cell division, whereas the centrioles help form or regenerate flagella and cilia.

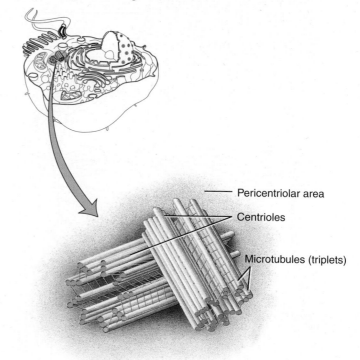

Pericentriolar area

Centrioles

Microtubules (triplets)

What are the components of the centrosome?

produced, the small and large subunits exit the nucleus and join together in the cytosol to participate in protein synthesis.

Some ribosomes, called *free ribosomes*, are not attached to other organelles. Free ribosomes synthesize proteins used inside the cell. Other ribosomes, called *membrane-bound ribosomes*, attach to the nuclear membrane and to an extensively folded membrane called the endoplasmic reticulum. These ribosomes synthesize proteins destined for insertion in the plasma membrane or for export from the cell.

Endoplasmic Reticulum

The **endoplasmic reticulum** (en′-dō-PLAS-mik re-TIK-yoo-lum; *-plasmic* = cytoplasm; *reticulum* = network) or **ER** is a network of folded membranes that form flattened sacs or tubules called **cisterns** (SIS-terns; cavities) (Figure 3.15). The ER extends throughout the cytoplasm and is so extensive that it constitutes more than half of the membranous surfaces within the cytoplasm of most cells.

Cells contain two distinct but interrelated forms of ER that differ in structure and function. **Rough ER** extends from the nu-

clear envelope (membrane around the nucleus) and usually is folded into a series of flattened sacs. The rough ER appears "rough" because its outer surface is studded with ribosomes. Proteins synthesized by ribosomes attached to rough ER enter cisterns within the ER for processing and sorting. These molecules may be incorporated into organelle membranes or the plasma membrane. Thus, rough ER is a factory for synthesizing secretory proteins and membrane molecules.

Smooth ER extends from the rough ER to form a network of membranous tubules (Figure 3.15). As you may already have guessed, smooth ER appears "smooth" because it lacks ribosomes. Although it is not the site of protein synthesis, smooth ER is where phospholipids, steroids such as estrogens and testosterone, and other lipids are synthesized. In liver cells, enzymes of the smooth ER also help release glucose into the bloodstream and inactivate or detoxify a variety of drugs and potentially harmful substances, including alcohol, pesticides, and carcinogens (cancer-causing agents). In muscle cells, calcium ions are released from the sarcoplasmic reticulum, a form of smooth ER, to trigger muscle contraction.

Figure 3.14 ■ **Ribosomes.**

Ribosomes, the sites of protein synthesis, consist of a large subunit and a small subunit

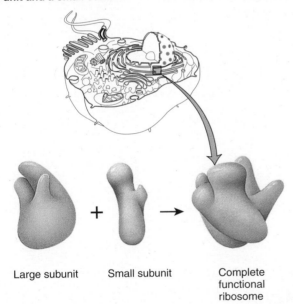

Large subunit   Small subunit   Complete functional ribosome

Where are ribosomal subunits synthesized and assembled?

Figure 3.15 ■ **Endoplasmic reticulum (ER).**

The ER is a network of cisterns—membrane-enclosed sacs and tubules—that extend throughout the cytoplasm and connect to the nuclear envelope.

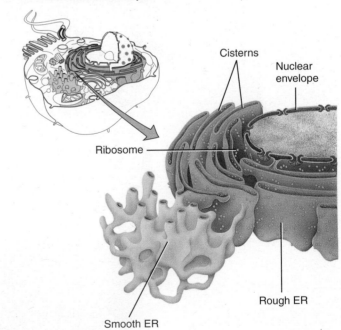

Cisterns
Nuclear envelope
Ribosome
Rough ER
Smooth ER

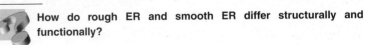

How do rough ER and smooth ER differ structurally and functionally?

Golgi Complex

After proteins are synthesized on a ribosome attached to rough ER, they usually are transported to another region of the cell. The first step in the transport pathway is through an organelle called the ***Golgi complex*** (GOL-jē). It consists of 3 to 20 ***cisterns,*** flattened membranous sacs with bulging edges, piled on each other like a stack of pita bread (Figure 3.16). The cisterns are often curved, giving the Golgi complex a cuplike shape. Most cells have only one Golgi complex, although some cells have several. The Golgi complex is more extensive in cells that secrete proteins into the extracellular fluid.

Proteins arriving at, passing through, and exiting the Golgi complex do so by means of vesicular transport in the following way (Figure 3.17):

❶ Proteins synthesized by ribosomes on the rough ER are surrounded by a piece of the ER membrane, which eventually buds from the ER to form a ***transport vesicle.***

❷ The transport vesicle moves toward the Golgi complex.

❸ The transport vesicle fuses with the side of the Golgi complex closest to the ER. As a result of this fusion, the proteins enter the Golgi complex.

❹ The proteins move from one cistern to another by way of ***transfer vesicles*** that bud from the edges of each cistern in the stack. As the proteins pass through the Golgi cisterns, enzymes in the cisterns modify them to form glycoproteins and lipoproteins.

❺ The modified proteins move by means of transfer vesicles into the last Golgi cistern.

❻ Within the last cistern, the products are sorted and packaged into vesicles.

❼ Some of the processed proteins leave the Golgi complex in ***secretory vesicles,*** which undergo exocytosis to release the proteins into the extracellular fluid.

❽ Other processed proteins leave the Golgi complex in ***membrane vesicles,*** vesicles that merge with the plasma membrane. Thus, the Golgi complex can modify the number and distribution of membrane molecules by adding new segments of plasma membrane.

❾ Finally, some processed proteins leave the Golgi complex in vesicles that are called ***storage vesicles.*** The major storage vesicle is a lysosome, whose structure and functions are discussed next.

Figure 3.16 ■ Golgi complex.

Most proteins synthesized by ribosomes attached to rough ER pass through the Golgi complex for processing.

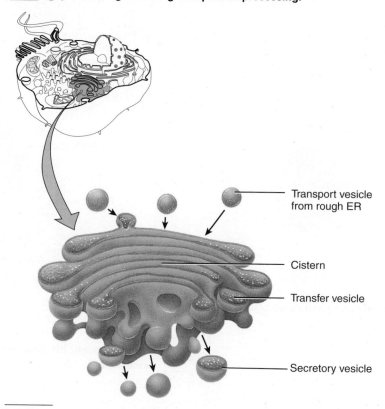

Transport vesicle from rough ER

Cistern

Transfer vesicle

Secretory vesicle

 What types of body cells are likely to have extensive Golgi complexes?

Lysosomes

Lysosomes (LĪ-sō-sōms; *lyso-* = dissolving) are membrane-enclosed vesicles that form in the Golgi complex (see Figure 3.17, step ❾). Inside are as many as 40 different digestive enzymes that can break down a wide variety of molecules once the lysosome fuses with a phagosome or pinocytic vesicle. The lysosomal membrane allows the final products of digestion, such as monosaccharides and amino acids, to be transported into the cytosol.

Lysosomal enzymes also help recycle the cell's own structures. A lysosome can engulf another organelle, digest it, and return the digested components to the cytosol for reuse. In this way, old organelles are continually replaced. The process by which worn-out organelles are digested is called **autophagy** (aw-TOF-a-jē; *auto-* = self; *-phagy* = eating). During autophagy, the organelle to be digested is enclosed by a membrane derived from the ER to create a vesicle called an *autophagosome*, which then fuses with a lysosome. In this way, a human liver cell, for example, recycles about half its contents every week. Lysosomal enzymes may also destroy their own cell, a process known as **autolysis** (aw-TOL-i-sis). Autolysis occurs in some pathological conditions and also is responsible for the tissue deterioration that occurs just after death.

Peroxisomes

Another group of organelles similar in structure to lysosomes, but smaller, are called **peroxisomes** (per-OK-si-sōms; *peroxi-* = peroxide; see Figure 3.1). Unlike lysosomes, which form in the Golgi complex, new peroxisomes arise through division of pre-existing peroxisomes. Peroxisomes contain one or more enzymes that can oxidize (remove hydrogen atoms from) various organic substances. For example, substances such as amino acids and fatty acids are oxidized in peroxisomes as part of normal metabolism. In addition, enzymes in peroxisomes oxidize toxic substances, such as alcohol. A byproduct of the oxidation reactions is hydrogen peroxide (H_2O_2) a potentially toxic compound. However, peroxisomes also contain an enzyme called *catalase* that decomposes the H_2O_2. Because the generation and degradation of H_2O_2 occurs within the same organelle, peroxisomes protect other parts of the cell from the toxic effects of H_2O_2.

Mitochondria

Because they are the site of most ATP production, the "powerhouses" of a cell are its **mitochondria** (mī-tō-KON-drē-a; singular is *mitochondrion* = threadlike granule). A cell may have as few

Figure 3.17 ■ Packaging of proteins by the Golgi complex.

Transport, transfer, and secretory vesicles all participate in the processing of proteins by the Golgi complex.

Ribosome
Synthesized protein
1 Transport vesicle
2
Cistern
3
9 Storage vesicle (lysosome)
4
Rough ER
Transfer vesicle
Protein packaging
5
6
7
8 Membrane vesicle
Transfer vesicle
Secretory vesicle
Proteins in vesicle membrane merge with plasma membrane
Proteins exported from cell by exocytosis
Plasma membrane

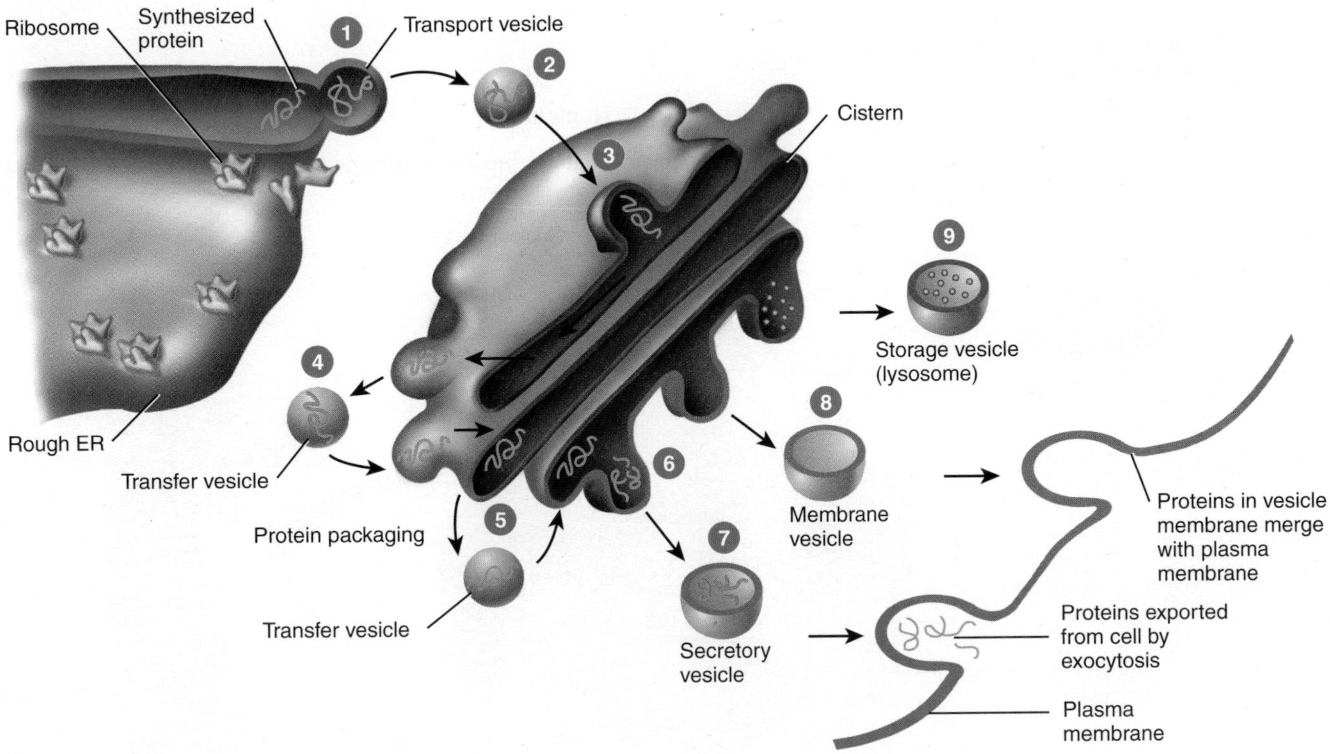

What are the three general destinations for proteins that leave the Golgi complex in vesicles?

Figure 3.18 ■ Mitochondrion.

 Within mitochondria, chemical reactions generate most of a cell's ATP.

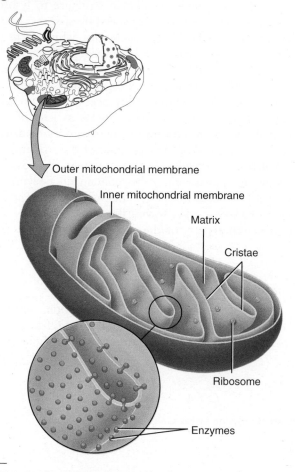

Outer mitochondrial membrane

Inner mitochondrial membrane

Matrix

Cristae

Ribosome

Enzymes

How do the cristae of a mitochondrion contribute to its ATP-producing function?

as one hundred or as many as several thousand mitochondria. Active cells, such as muscle, liver, and kidney tubule cells, have large numbers of mitochondria and use ATP at a high rate. A mitochondrion is bounded by two membranes, each of which is similar in structure to the plasma membrane (Figure 3.18). The *outer mitochondrial membrane* is smooth, but the *inner mitochondrial membrane* is arranged in a series of folds called **cristae** (KRIS-tē; singular is *crista* = ridge). The central fluid-filled cavity of a mitochondrion, enclosed by the inner membrane and cristae, is the **matrix.** The elaborate folds of the cristae provide an enormous surface area for a series of chemical reactions that provide most of a cell's ATP. Enzymes that catalyze these reactions are located in the matrix and on the cristae. Mitochondria also contain a small number of genes and a few ribosomes, enabling them to synthesize some proteins.

NUCLEUS

Objective: • **Describe the structure and functions of the nucleus.**

The **nucleus** is a spherical or oval structure that usually is the most prominent feature of a cell (Figure 3.19). Most body cells have a single nucleus, although some, such as mature red blood cells, have none. In contrast, skeletal muscle cells and a few other types of cells have several nuclei. A double membrane called the **nuclear envelope** separates the nucleus from the cytoplasm. Both layers of the nuclear envelope are lipid bilayers

Figure 3.19 ■ Nucleus.

The nucleus contains most of a cell's genes, which are located on chromosomes.

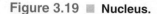

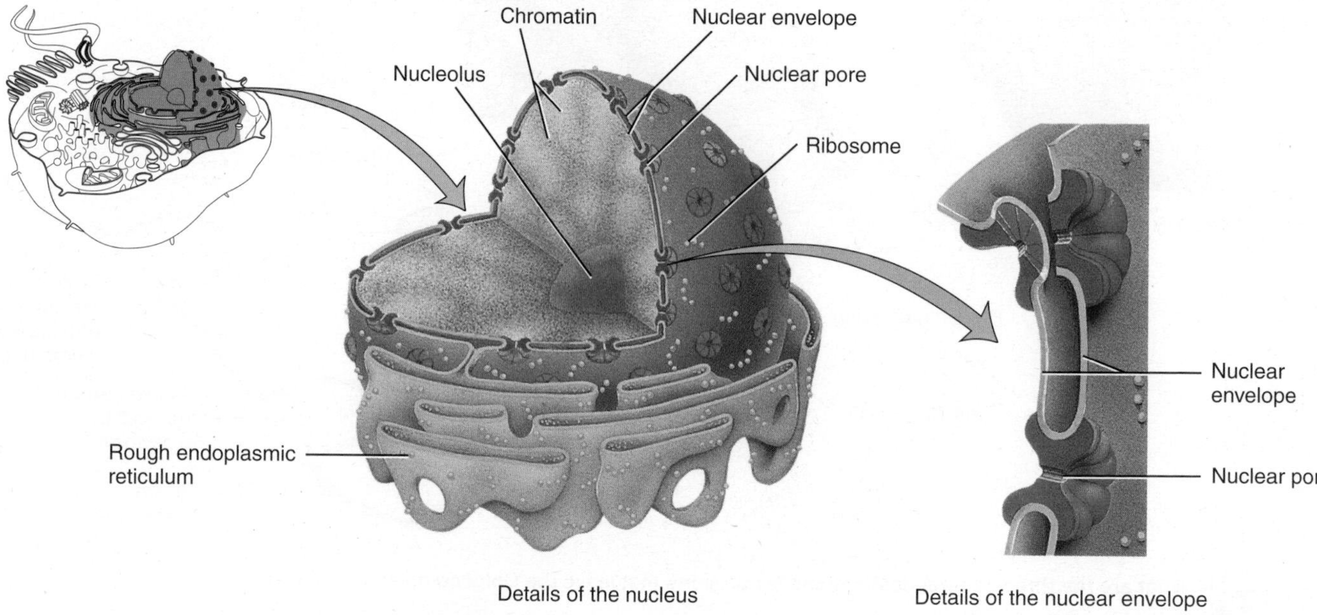

Chromatin

Nuclear envelope

Nucleolus

Nuclear pore

Ribosome

Rough endoplasmic reticulum

Nuclear envelope

Nuclear pore

Details of the nucleus

Details of the nuclear envelope

What are the functions of nuclear genes?

Table 3.2 / Cell Parts and Their Functions

Part	Structure	Functions
Plasma Membrane	Composed of a lipid bilayer consisting of phospholipids, cholesterol, and glycolipids with various proteins inserted; surrounds cytoplasm.	Protects cellular contents; makes contact with other cells; contains channels, transporters, receptors, enzymes, cell-identity markers, and linker proteins; mediates the entry and exit of substances.
Cytoplasm	Cellular contents between plasma membrane and nucleus, including cytosol and organelles.	
Cytosol	Composed of water, solutes, suspended particles, lipid droplets, and glycogen granules.	
Organelles	Specialized cellular structures with characteristic shapes and specific functions.	
Cytoskeleton	Network composed of three protein filaments: microfilaments, intermediate filaments, and microtubules.	Maintains shape and general organization of cellular contents; is responsible for cell movements.
Centrosome	Paired centrioles plus pericentriolar area.	Centrioles form and regenerate cilia and flagella; pericentriolar area is organizing center for microtubules and mitotic spindle.
Cilia and flagella	Motile cell surface projections with inner core of microtubules.	Cilia move fluids over a cell's surface; a flagellum moves an entire cell.
Ribosome	Composed of two subunits containing ribosomal RNA and proteins; may be free in cytosol or attached to rough ER.	Protein synthesis.
Endoplasmic reticulum (ER)	Membranous network of flattened sacs or tubules called cisterns. Rough ER is studded with ribosomes and is attached to nuclear membrane; smooth ER lacks ribosomes.	Rough ER is site of synthesis of secretory proteins and membrane molecules; smooth ER is site of phospholipid, fat, and steroid synthesis. Smooth ER also releases glucose into bloodstream, inactivates or detoxifies drugs and potentially harmful substances, and stores calcium ions for muscle contraction.
Golgi complex	A stack of 3–20 flattened membranous sacs called cisterns.	Accepts proteins from rough ER; forms glycoproteins and lipoproteins; stores, packages, and exports proteins.
Lysosome	Vesicle formed from Golgi complex; contains digestive enzymes.	Fuses with and digests contents of pinocytic vesicles and phagosomes; digests worn-out organelles (autophagy), entire cells (autolysis), and extracellular materials.
Peroxisome	Vesicle containing oxidative enzymes.	Detoxifies harmful substances.
Mitochondrion	Consists of outer and inner membranes, cristae, and matrix.	Site of reactions that produce most of a cell's ATP.
Nucleus	Consists of nuclear envelope with pores, nucleoli, and chromatin (or chromosomes).	Contains genes, which control cellular structure and most cellular activities.

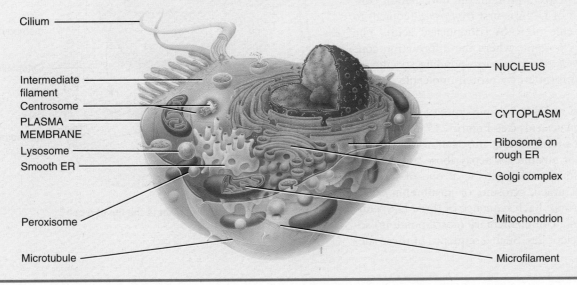

Cilium

Intermediate filament

Centrosome

PLASMA MEMBRANE

Lysosome

Smooth ER

Peroxisome

Microtubule

NUCLEUS

CYTOPLASM

Ribosome on rough ER

Golgi complex

Mitochondrion

Microfilament

similar to the plasma membrane. The outer membrane of the nuclear envelope is continuous with the rough endoplasmic reticulum and resembles it in structure. Many channels called **nuclear pores** pierce the nuclear envelope. Nuclear pores control the movement of substances between the nucleus and the cytoplasm.

Contained within the nucleus are one or more spherical bodies called **nucleoli** (noo-KLĒ-ō-lī; singular is *nucleolus*). These clusters of protein, DNA, and RNA are the sites of assembly of ribosomes, which exit the nucleus through nuclear pores and participate in protein synthesis in the cytoplasm. Cells that synthesize large amounts of protein, such as muscle and liver cells, have prominent nucleoli.

Also within the nucleus are most of the cell's hereditary units, called **genes,** which control cellular structure and direct most cellular activities. The nuclear genes are arranged in single file along **chromosomes** (*chromo-* = colored) (see Figure 3.24). Human somatic (body) cells have 46 chromosomes, 23 inherited from each parent. The two chromosomes of each pair—one contributed by the mother and one contributed by the father—are called *homologous chromosomes.* They have similar genes, which are usually arranged in the same order. In a cell that is not dividing, the 46 chromosomes appear as a diffuse, granular mass, which is called *chromatin* (Figure 3.19).

The main parts of a cell and their functions are summarized in Table 3.2 on page 57.

GENE ACTION: PROTEIN SYNTHESIS

Objective: • **Outline the sequence of events involved in protein synthesis.**

Although cells synthesize many chemicals to maintain homeostasis, much of the cellular machinery is devoted to protein production. Cells constantly synthesize large numbers of diverse proteins. The proteins, in turn, determine the physical and chemical characteristics of cells and, on a larger scale, of organisms.

The DNA contained in genes provides the instructions for making proteins. To synthesize a protein, the information contained in a specific region of DNA is first *transcribed* (copied) to produce a specific molecule of RNA (ribonucleic acid). The RNA then attaches to a ribosome, where the information contained in the RNA is *translated* into a corresponding specific sequence of amino acids to form a new protein molecule (Figure 3.20).

Information is stored in DNA in four types of nucleotides, the repeating units of nucleic acids (see Figure 2.15 on page 36). DNA nucleotides consist of one of four nitrogenous bases (adenine, thymine, cytosine, or guanine), a deoxyribose sugar, and a phosphate group. RNA nucleotides consist of one of four nitrogenous bases (adenine, uracil, cytosine, or guanine), a ribose sugar, and a phosphate group. Each sequence of three DNA nucleotides is transcribed as a complementary (corresponding) sequence of three RNA nucleotides. Such a sequence of three successive DNA nucleotides or three successive RNA nucleotides transcribed from DNA is called a **codon.** When translated, a given codon specifies a particular amino acid.

Transcription

During **transcription,** which occurs in the nucleus, the genetic information in DNA codons is copied into a complementary sequence of codons in a strand of RNA. Transcription of DNA is catalyzed by the enzyme *RNA polymerase,* which must be instructed where to start the transcription process and where to end it. The segment of DNA where RNA polymerase attaches to it is a special sequence of nucleotides called a **promoter,** located near the beginning of a gene (Figure 3.21a). Three kinds of RNA are made from DNA:

- *Messenger RNA (mRNA)* directs synthesis of a protein.
- *Ribosomal RNA (rRNA)* joins with ribosomal proteins to make ribosomes.
- *Transfer RNA (tRNA)* binds to an amino acid and holds it in place on a ribosome until it is incorporated into a protein during translation. Each of the more than 20 different types of tRNA binds to only one of the 20 different amino acids.

Figure 3.20 ■ **Overview of transcription and translation.**

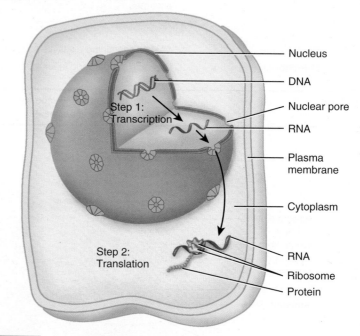

Whereas transcription occurs in the nucleus, translation takes place in the cytoplasm.

Nucleus

DNA

Nuclear pore

Step 1: Transcription

RNA

Plasma membrane

Cytoplasm

Step 2: Translation

RNA

Ribosome

Protein

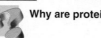

Why are proteins important in the life of a cell?

During transcription, nucleotides pair in a complementary manner: The nitrogenous base cytosine (C) in DNA dictates the complementary nitrogenous base guanine (G) in the new RNA strand, a G in DNA dictates a C in RNA, a thymine (T) in DNA dictates an adenine (A) in RNA, and an A in DNA dictates a uracil (U) in RNA. As an example, if a segment of DNA had the base sequence ATGCAT, the newly transcribed RNA strand would have the complementary base sequence UACGUA.

Transcription of DNA ends at another special nucleotide sequence on DNA called a ***terminator***, which specifies the end of the gene (Figure 3.21a). Upon reaching the terminator, RNA polymerase detaches from the transcribed RNA molecule and the DNA strand. Once synthesized, mRNA, rRNA (in ribosomes), and tRNA leave the nucleus of the cell by passing through a nuclear pore. In the cytoplasm, they participate in the next step in protein synthesis, translation.

Figure 3.21 ■ **Transcription.**

During transcription, the genetic information in DNA is copied to RNA.

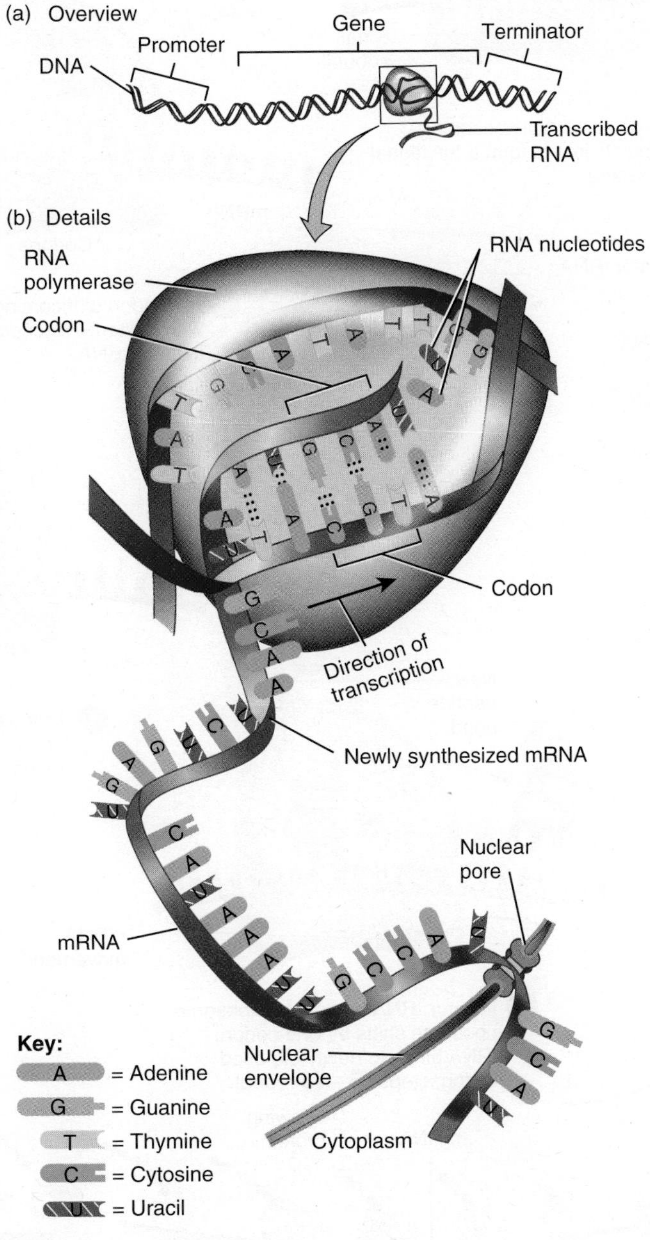

(a) Overview

(b) Details

Key:
A = Adenine
G = Guanine
T = Thymine
C = Cytosine
U = Uracil

What enzyme catalyzes transcription of DNA?

Translation

Translation is the process in which mRNA associates with ribosomes and directs synthesis of a protein by converting the sequence of nucleotides in mRNA into a specific sequence of amino acids. Translation occurs in the following way (Figure 3.22):

Figure 3.22 ■ **Protein elongation and termination of protein synthesis during translation.**

During protein synthesis the ribosomal subunits join, but they separate when the process is complete.

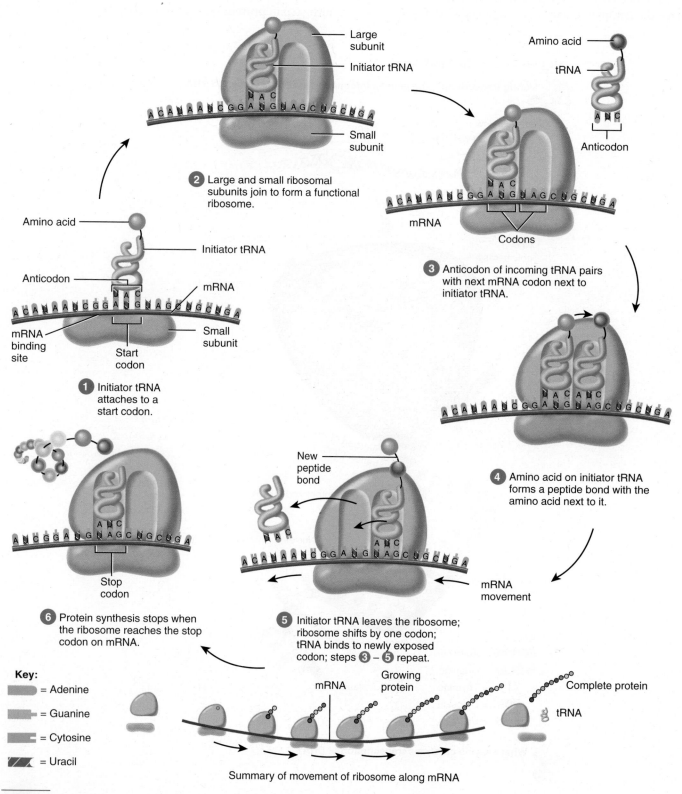

Large subunit
Initiator tRNA
Small subunit

2 Large and small ribosomal subunits join to form a functional ribosome.

Amino acid
tRNA
Anticodon

mRNA
Codons

3 Anticodon of incoming tRNA pairs with next mRNA codon next to initiator tRNA.

Amino acid
Initiator tRNA
Anticodon
mRNA
mRNA binding site
Start codon
Small subunit

1 Initiator tRNA attaches to a start codon.

4 Amino acid on initiator tRNA forms a peptide bond with the amino acid next to it.

mRNA movement

New peptide bond

Stop codon

6 Protein synthesis stops when the ribosome reaches the stop codon on mRNA.

5 Initiator tRNA leaves the ribosome; ribosome shifts by one codon; tRNA binds to newly exposed codon; steps **3** – **5** repeat.

Key:
= Adenine
= Guanine
= Cytosine
= Uracil

mRNA
Growing protein
Complete protein
tRNA

Summary of movement of ribosome along mRNA

What is the function of a stop codon?

1 An mRNA molecule binds to the small ribosomal subunit and a special tRNA, called *initiator tRNA*, binds to the start codon (AUG) on mRNA, where translation begins.

2 The large ribosomal subunit attaches to the small subunit, creating a functional ribosome. The initiator tRNA fits into position on the ribosome. One end of a tRNA carries a specific amino acid, and the opposite end consists of a sequence of three nucleotides called an *anticodon.* By pairing between complementary nitrogenous bases, the tRNA anticodon attaches to the mRNA codon. For example, if the mRNA codon is AUG, then a tRNA with the anticodon UAC would attach to it.

3 The anticodon of another tRNA with its amino acid attaches to the complementary mRNA codon next to the initiator tRNA.

4 A peptide bond is formed between the amino acids carried by the initiator tRNA and the tRNA next to it.

5 After the peptide bond forms, the tRNA detaches from the ribosome, and the ribosome shifts the mRNA strand by one codon. As the tRNA bearing the newly forming protein shifts, another tRNA with its amino acid binds to a newly exposed codon. Steps 3 through 5 repeat again and again as the protein lengthens.

6 Protein synthesis ends when the ribosome reaches a stop codon, at which time the completed protein detaches from the final tRNA. When the tRNA vacates the ribosome, the ribosome splits into its large and small subunits.

Protein synthesis progresses at a rate of about 15 amino acids per second. As the ribosome moves along the mRNA and before it completes synthesis of the whole protein, another ribosome may attach behind it and begin translation of the same mRNA strand. In this way, several ribosomes may be attached to the same mRNA, an assembly called a *polyribosome.* The simultaneous movement of several ribosomes along the same mRNA strand permits a large amount of protein to be produced from each mRNA.

SOMATIC CELL DIVISION

Objective: • **Discuss the stages, events, and significance of somatic cell division.**

As body cells become damaged, diseased, or worn out, they are replaced by *cell division,* the process whereby cells reproduce themselves. The two types of cell division are reproductive cell division and somatic cell division. *Reproductive cell division* or *meiosis* is the process that produces gametes—sperm and oocytes—the cells needed to form the next generation of sexually reproducing organisms. This is described in Chapter 23; here we will focus on somatic cell division.

All body cells, except those that produce gametes, are called somatic cells. In *somatic cell division,* a cell divides into two identical daughter cells. An important part of somatic cell division is replication (duplication) of the DNA sequences that make up genes and chromosomes so that the same genetic material can be passed on to the daughter cells. After somatic cell division, each daughter cell has the same number of chromosomes as the original cell. Somatic cell division replaces dead or injured cells and adds new ones for tissue growth. For example, skin cells are continually replaced by somatic cell divisions.

The *cell cycle* is the name for the sequence of changes that a cell undergoes from the time it forms until it duplicates its contents and divides into two cells. In somatic cells, the cell cycle consists of two major periods: interphase, when a cell is not dividing, and the mitotic (M) phase, when a cell is dividing.

Interphase

During *interphase* the cell replicates its DNA. It also manufactures additional organelles and cytosolic components in anticipation of cell division. Interphase is a state of high metabolic activity, and during this time the cell does most of its growing.

As DNA replication begins, DNA nucleotides are synthesized in the cytosol and imported into the nucleus. There, the helical structure of DNA partially uncoils, and the two strands separate at the points where hydrogen bonds connect base pairs (Figure 3.23). Each exposed base of the old DNA strand pairs with the complementary base of a newly synthesized nucleotide. A new DNA strand takes shape as chemical bonds form between neighboring nucleotides. The uncoiling and complementary base pairing continues until each of the two original DNA strands is joined with a newly formed complementary DNA strand. The original DNA molecule has become two identical DNA molecules.

A microscopic view of a cell during interphase shows a clearly defined nuclear envelope, a nucleolus, and a tangled mass of chromatin (Figure 3.24a). Once a cell completes its replication of DNA and other activities of interphase, the mitotic phase begins.

Mitotic Phase

The *mitotic phase* (mī-TOT-ik; *mito-* = thread) of the cell cycle consists of *mitosis*, division of the nucleus, followed by *cytokinesis*, division of the cytoplasm into two daughter cells. The events that take place during mitosis and cytokinesis are plainly visible under a microscope because chromatin condenses into chromosomes.

Nuclear Division: Mitosis

During *mitosis* (mī-TŌ-sis), the duplicated chromosomes become exactly segregated, one set into each of two separate nuclei. For convenience, biologists divide the process into four stages: prophase, metaphase, anaphase, and telophase. However, mitosis is a continuous process, with one stage merging imperceptibly into the next.

PROPHASE During early prophase, the chromatin fibers condense and shorten into chromosomes that are visible under the light microscope (Figure 3.24b). The condensation process may prevent entangling of the long DNA strands as they move during mitosis. Recall that DNA replication took place during interphase. Thus, each prophase chromosome consists of a pair of identical, double-stranded *chromatids.* A constricted region of the chromosome, called a **centromere,** holds the chromatid pair together.

Later in prophase, the nucleolus disappears, the nuclear envelope breaks down, and the two centrosomes start to form the **mitotic spindle,** a football-shaped assembly of microtubules (Figure 3.24b). Lengthening of the microtubules between centrosomes pushes the centrosomes to opposite poles (ends) of the cell. Finally, the spindle extends from pole to pole.

METAPHASE During metaphase, the centromeres of the chromatid pairs are aligned along the microtubules at the exact center of the mitotic spindle (Figure 3.24c). This midpoint region is called the **metaphase plate.**

ANAPHASE During anaphase the centromeres split, separating the two members of each chromatid pair, which move to opposite poles of the cell (Figure 3.24d). Once separated, the chromatids are called chromosomes. As the chromosomes are pulled by the microtubules during anaphase, they appear V-shaped because the centromeres lead the way and seem to drag the trailing arms of the chromosomes toward the pole.

TELOPHASE The final stage of mitosis, telophase, begins after chromosomal movement stops (Figure 3.24e). The identical sets of chromosomes, now at opposite poles of the cell, uncoil and revert to the threadlike chromatin form. A new nuclear envelope forms around each chromatin mass, nucleoli appear, and eventually the mitotic spindle breaks up.

Cytoplasmic Division: Cytokinesis

Division of a parent cell's cytoplasm and organelles is called **cytokinesis** (sī′-tō-ki-NĒ-sis; *-kinesis* = motion). This process begins late in anaphase or early in telophase with formation of a **cleavage furrow,** a slight indentation of the plasma membrane, that extends around the center of the cell (Figure 3.24d,e). Microfilaments in the cleavage furrow pull the plasma membrane progressively inward, constricting the center of the cell like a belt around a waist, and ultimately pinching it in two. After cytokinesis there are two separate daughter cells, each with equal portions of cytoplasm and organelles and identical sets of chromosomes. When cytokinesis is complete, interphase begins (Figure 3.24f).

Figure 3.23 ■ **Replication of DNA.** The two strands of the double helix separate by breaking the hydrogen bonds (shown as dotted lines) between nucleotides. New, complementary nucleotides attach at the proper sites, and a new strand of DNA is synthesized alongside each of the original strands. Arrows indicate hydrogen bonds forming again between pairs of bases.

🔑 **Replication doubles the amount of DNA.**

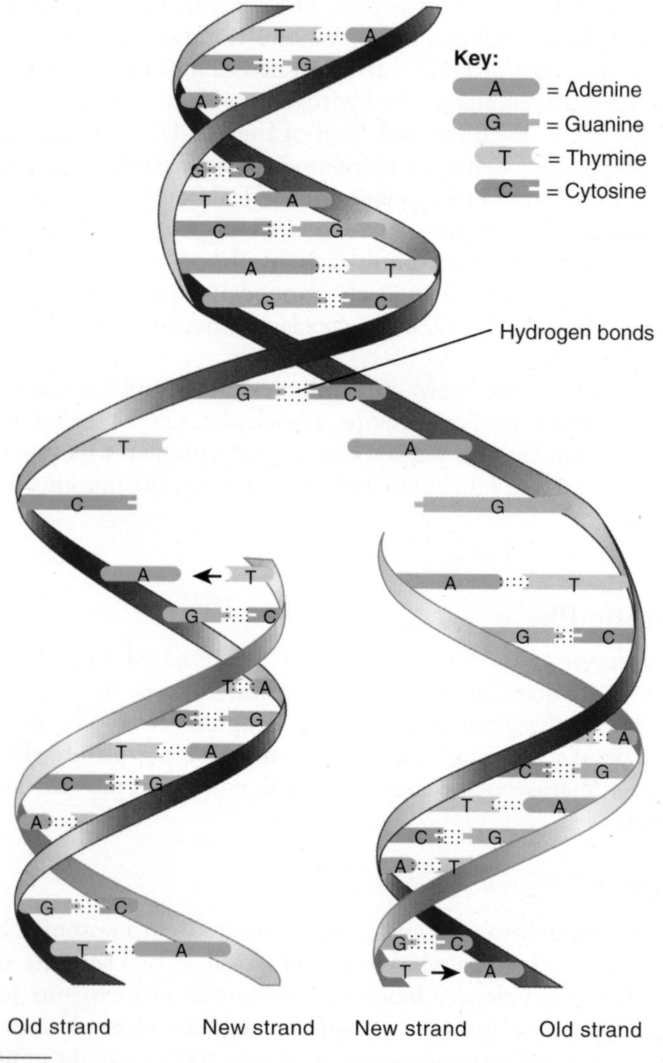

Key:
- A = Adenine
- G = Guanine
- T = Thymine
- C = Cytosine

Hydrogen bonds

Old strand   New strand   New strand   Old strand

🧩 **During which phase of the cell cycle does DNA replication occur?**

AGING AND CELLS

Objective: • **Explain the relationship of aging to cellular processes.**

Aging is a normal process accompanied by a progressive alteration of the body's homeostatic adaptive responses. It produces observable changes in structure and function and increases vulnerability to environmental stress and disease. The specialized branch of medicine that deals with the medical problems and care of elderly persons is called **geriatrics** (jer′-ē-AT-riks; *ger-* = old age; *-iatrics* = medicine).

Figure 3.24 ■ **Cell division: mitosis and cytokinesis.** Begin the sequence at (a) at the top of the figure and read clockwise until you complete the process.

In somatic cell division, a single diploid cell divides to produce two identical diploid daughter cells

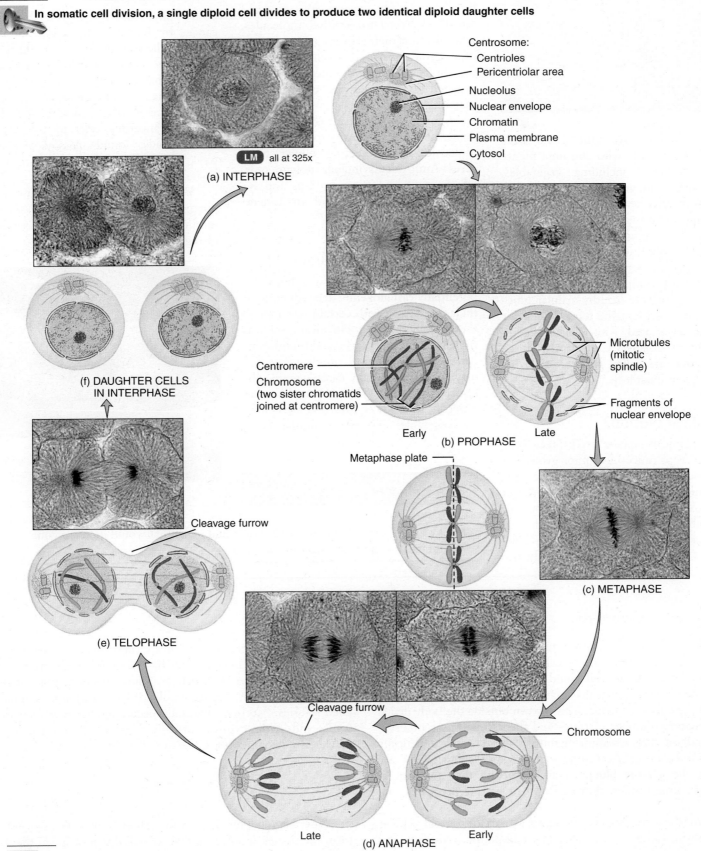

When does cytokinesis begin?

Phytochemicals— Protecting Cellular Function

Many studies over the years have shown that people who consume plenty of plant foods, including vegetables, beans, fruits, and grains, have a lower risk of cancer and heart disease than their meat-and-potato-eating peers. Scientists are just beginning to uncover the biochemical explanations for these associations. Their investigations have led to the discovery of compounds in plants that appear to promote healthy cellular function, and to prevent the types of cellular damage associated with cancer, aging, and heart disease. Collectively, these compounds are called *phytochemicals*, literally "plant chemicals."

A Radical Notion?

Phytochemicals appear to protect cells and interrupt cancerous tumor growth in a number of interesting ways. Some phytochemicals block chemicals that can cause oxidative damage to cells. You will learn about the process of oxidation in Chapter 20 when you read about metabolism. Oxidative damage commonly oc-curs in cells when byproducts of metabolism, known as oxygen free radicals, "steal" electrons from other molecules. This electron theft causes chain reactions of electron transfers that can damage cell membranes, the membranes of cellular organelles, and even the cells' genetic material.

Some phytochemicals act as antioxidants, donating electrons to free radical molecules, thus protecting cellular structures. Antioxidants include polyphenols, which are found in green tea, and lycopenes, which are found in tomato products.

Disabling the Opponent

Many substances entering the body are potentially carcinogenic, depending upon their interaction with certain enzymes in the liver. Some phytochemicals, such as the allyl sulfides in garlic and onions, enhance the production of enzymes that may render potentially carcinogenic substances harmless. The sulforaphane in broccoli, cauliflower, and other cruciferous vegetables performs a similar function.

Promoting Health

Some phytochemicals protect against cancer by blocking the action of substances called promoters. Promoters encourage the aggressive cellular division of cells that have undergone cancer-causing genetic changes. For example, estrogens are hormones that promote the division of cancerous cells in the breast. Isoflavonoids, found in soy products, weaken the action of estrogens in breast tissue.

Variety is the key to consuming more phytochemicals. Try to eat two to four servings of fruit and three to five servings of vegetables each day.

▶ *Think It Over*

▶ What are some dietary changes you could make that would increase your intake of helpful phytochemicals?

Although many millions of new cells normally are produced each minute, several kinds of cells in the body—heart muscle cells, skeletal muscle cells, and nerve cells—do not divide. Experiments have shown that many other cell types have only a limited capability to divide. Cells grown outside the body divide only a certain number of times and then stop. These observations suggest that cessation of mitosis is a normal, genetically programmed event. According to this view, "aging genes" are part of the genetic blueprint at birth, and they turn on at preprogrammed times, slowing down or halting processes vital to life.

Some free radicals are produced as part of normal cellular metabolism. Others are present in air pollution, radiation, and certain foods we eat. Free radicals cause oxidative damage in lipids, proteins, or nucleic acids by "stealing" an electron to accompany their unpaired electron. Some effects of this damage, which increase with age, are wrinkled skin, stiff joints, and hardened arteries. Although naturally occurring enzymes in peroxisomes and in the cytosol normally dispose of free radicals, the mechanism becomes less efficient with age.

• • •

Next, in Chapter 4, we will explore how cells associate to form the tissues and organs that we will discuss later in the text.

COMMON DISORDERS

Cancer

Cancer is a group of diseases characterized by uncontrolled division of abnormal cells. When cells in a part of the body divide without control, the excess tissue that develops is called a *tumor* or *neoplasm* (NĒ-ō-plazm; *neo-* = new). A cancerous neoplasm is called a *malignant tumor*. Malignant tumors exhibit destructive growth and have the ability to undergo *metastasis* (me-TAS-ta-sis), the spread of cancerous cells to other parts of the body. A *benign tumor* is a noncancerous growth. Although benign tumors do not metastasize, they may need to be surgically removed if they interfere with a normal body function or become disfiguring. The study of tumors is called *oncology* (on-KOL-ō-jē; *onco-* = swelling or mass), and a physician who specializes in this field is called an *oncologist.*

Types of Cancer

The name of the cancer is derived from the type of tissue in which it develops. Most human cancers are *carcinomas* (kar′-si-NŌ-maz; *carcin-* = cancer; *-omas* = tumors), malignant tumors that arise from epithelial cells. *Melanomas* (mel′-a-NŌ-maz; *melan-* = black), for example, are cancerous growths of melanocytes, skin epithelial cells that produce the pigment melanin. *Sarcoma* (sar-KŌ-ma) is a general term for any cancer arising from muscle or connective tissue cells. For example, *osteogenic sarcoma* (*osteo-* = bone; *-genic* = origin), the most frequent type of childhood cancer, destroys bone, which is one type of connective tissue. *Leukemia* (loo-KĒ-mē-a; *leuk-* = white; *-emia* = in the blood) is a cancer of blood-forming organs characterized by rapid growth of abnormal leukocytes (white blood cells).

Growth and Spread of Cancer

Cells of malignant tumors duplicate rapidly and continuously. As malignant cells invade surrounding tissues, they trigger *angiogenesis* (an′-jē-ō-GEN-e-sis), the growth of new networks of blood vessels. Proteins that trigger blood vessel growth in tumor tissue are called *tumor angiogenesis factors (TAFs)*. As the cancer grows, it begins to compete with normal tissues for space and nutrients. Eventually, the normal tissue decreases in size and dies. Malignant cells resist the antitumor defenses of the body. The pain that is often associated with cancer develops when a tumor presses on nerves or blocks a passageway in an organ so that secretions build up pressure.

Causes of Cancer

Several factors may trigger a normal cell to lose control and become abnormal. A chemical agent or radiation that produces cancer is called a *carcinogen* (car-SIN-ō-jen). A carcinogen can induce a *mutation,* a permanent structural change in the sequence of DNA nucleotides in a gene. The World Health Organization estimates that environmental carcinogens—substances in the air we breathe, the water we drink, and the food we eat—are associated with 60% to 90% of all human cancers.

Examples of carcinogens are hydrocarbons found in cigarette tar, radon gas from the earth, and ultraviolet (UV) radiation in sunlight.

A second cause of cancer lies within our genes. Abnormalities in genes that participate in regulation of the cell cycle are associated with many diseases. Some cancers are caused by loss of genes called *tumor-suppressor genes.* These genes produce proteins that normally inhibit cell division. Loss or alteration of a tumor-suppressor gene called *p53* is the most common genetic change in a wide variety of tumors, including breast and colon cancers. Other genes, called *oncogenes* (ON-kō-jēnz), have the ability to transform a normal cell into a cancerous cell when they are inappropriately activated. Oncogenes develop from normal genes, called *proto-oncogenes,* that regulate growth and development. Some proto-oncogenes are transformed into oncogenes by mutations. Others are activated by rearrangement of the DNA on the chromosomes. Rearrangement activates proto-oncogenes by placing them near genes that enhance their activity. Some viruses cause cancer by inserting their own oncogenes or proto-oncogenes into the host cell's DNA.

Carcinogenesis: A Multistep Process

Carcinogenesis (kar′-si-nō-JEN-e-sis), the development of a cancer, is a multistep process in which as many as 10 mutations may have to occur in a cell before it becomes cancerous. The progression of genetic changes leading to cancer is best understood for colon (colorectal) cancer. Such cancers, as well as lung and breast cancer, typically take years or decades to develop. In colon cancer, the tumor begins as an area of increased cell proliferation that results from a single mutation in the DNA of a cell in the lining of the large intestine (colon). When two or three additional mutations occur, including a mutation of *p53*, a carcinoma develops.

Treatment of Cancer

Many cancers are removed surgically. However, when cancer has metastasized or exists in organs such as the brain whose functioning would be greatly harmed by surgery, chemotherapy and radiation therapy may also be used. Chemotherapy involves administering drugs that harm rapidly dividing cells, thus affecting malignant cells more than normal cells. Many chemotherapeutic agents act at specific phases of the cell cycle and attack cells that are in the process of cell division. For example, taxol, a chemical first derived from the bark of the Pacific yew, prevents mitosis by disabling microtubules. Radiation therapy destroys chromosomes, which also prevents cell division. Because cancerous cells reproduce so rapidly, they are more vulnerable to the destructive effects of chemotherapy and radiation therapy than are normal cells. However, both chemotherapy and radiation therapy kill or disrupt the function of some normal cells as well.

Another approach to cancer treatment is *immunotherapy*, boosting the body's own defenses against disease. Agents that stimulate the body's natural defenses to attack cancer cells are called biological response modifiers. For example, some immunotherapy agents stimulate activity of natural killer cells, immune system cells that search out and destroy cancerous cells. Two immunotherapy agents are alpha-interferon, which is used for hairy cell leukemia and Kaposi's sarcoma associated with AIDS, and interleukin-2, which is used to treat kidney cancer.

MEDICAL TERMINOLOGY AND CONDITIONS

Note to the Student

Each chapter in this text that discusses cells, tissues, or a major system of the body is followed by a glossary of related *medical terminology and conditions.* Terms for both normal and pathological (disease) conditions are included in these glossaries.

Apoptosis (ap′-ō-TŌ-sis; a falling off, like dead leaves from a tree) An orderly, genetically programmed cell death in which "cell-suicide" genes become activated. Enzymes produced by these genes disrupt the cytoskeleton and nucleus; the cell shrinks and pulls away from neighboring cells; the DNA within the nucleus fragments; and the cytoplasm shrinks, although the plasma membrane remains intact. Phagocytes in the vicinity then ingest the dying cell. Apoptosis removes unneeded cells during development before birth and continues after birth both to regulate the number of cells in a tissue and to eliminate potentially dangerous cells such as cancer cells.

Atrophy (AT-rō-fē; *a-* = without; *-trophy* = nourishment) A decrease in the size of cells with subsequent decrease in the size of the affected tissue or organ; wasting away.

Biopsy (BĪ-op-sē; *bio-* = life; *-opsy* = viewing) The removal and microscopic examination of tissue from the living body for diagnosis.

Dysplasia (dis-PLĀ-zē-a; *dys-* = abnormal; *-plasia* = to shape) Alteration in the size, shape, and organization of cells due to chronic irritation or inflammation; may progress to a neoplasm (tumor formation, usually malignant) or revert to normal if the irritation is removed.

Hyperplasia (hī′-per-PLĀ-zē-a; *hyper-* = over) Increase in the number of cells of a tissue due to an increase in the frequency of cell division.

Hypertrophy (hī-PER-trō-fē) Increase in the size of cells in a tissue without cell division.

Metaplasia (met′-a-PLĀ-zē-a; *meta-* = change) The transformation of one type of cell into another.

Necrosis (ne-KRŌ-sis = death) A pathological type of cell death, resulting from tissue injury, in which many adjacent cells swell, burst, and spill their cytoplasm into the interstitial fluid; the cellular debris usually stimulates an inflammatory response, which does not occur in apoptosis.

Progeny (PROJ-e-nē; *pro-* = forward; *-geny* = production) Offspring or descendants.

STUDY OUTLINE

Introduction (p. 42)

1. A cell is the basic, living, structural and functional unit of the body.
2. Cytology is the scientific study of cellular structure. Cell physiology is the study of cellular function.

A Generalized View of the Cell (p. 42)

1. Figure 3.1 shows a generalized view of a cell that is a composite of many different cells in the body.
2. The principal parts of a cell are the plasma membrane; the cytoplasm, which consists of cytosol and organelles; and the nucleus.

The Plasma Membrane (p. 43)

1. The plasma membrane surrounds and contains the cytoplasm of a cell; it is composed of proteins and lipids.
2. The lipid bilayer consists of two back-to-back layers of phospholipids, cholesterol, and glycolipids.
3. Integral proteins extend through the lipid bilayer, whereas peripheral proteins associate with membrane lipids or integral proteins at the inner or outer surface of the membrane.
4. The membrane's selective permeability permits some substances to pass across it more easily than others. The lipid bilayer is permeable to water and to most lipid-soluble molecules. Small- and medium-sized water-soluble materials may cross the membrane with the assistance of integral proteins.
5. Membrane proteins have several functions. Channels and transporters are integral proteins that help specific solutes across the membrane; receptors serve as cellular recognition sites; some membrane proteins are enzymes; others are cell identity markers; and linkers anchor proteins in the plasma membranes to filaments inside and outside the cell.

Transport Across The Plasma Membrane (p. 44)

1. Fluid inside body cells is called intracellular fluid (ICF); fluid outside body cells is extracellular fluid (ECF). The ECF in the microscopic spaces between the cells of tissues is interstitial fluid. The ECF in blood vessels is plasma, and that in lymphatic vessels is lymph.
2. Any material dissolved in a fluid is called a solute, and the fluid that dissolves materials is the solvent. Body fluids are dilute solutions in which a variety of solutes are dissolved in the solvent water.
3. The selective permeability of the plasma membrane supports the existence of concentration gradients, which are differences in the concentration of chemicals between one side of the membrane and the other.
4. Materials move through cell membranes by passive processes or by active transport. In passive processes, a substance moves down its concentration gradient across the membrane. In active transport, cellular energy is used to drive the substance "uphill" against its concentration gradient.
5. In vesicular transport, tiny vesicles either detach from the plasma membrane while bringing materials into the cell or merge with the plasma membrane to release materials from the cell.
6. Diffusion is the movement of molecules or ions due to their kinetic energy. In net diffusion, substances move from an area of

higher concentration to an area of lower concentration until equilibrium is reached. At equilibrium the concentration is the same throughout the solution.

7. The rate of diffusion is faster when the concentration gradient is steeper, the temperature is higher, the weight (mass) of the diffusing substance is lower, the surface area available for diffusion is larger, and the distance over which diffusion is occurring is smaller.

8. In simple diffusion, substances move through the lipid bilayer or through channels in integral proteins. Ion channels selective for K^+, Cl^-, Na^+, and Ca^{2+} allow these ions to diffuse across the plasma membrane by simple diffusion. In facilitated diffusion, substances cross the membrane with the assistance of transporters, which bind to a specific substance on one side of the membrane and release it on the other side after the transporter undergoes a change in shape.

9. Osmosis is the movement of water molecules through a selectively permeable membrane from an area of higher to an area of lower water concentration.

10. In an isotonic solution, red blood cells maintain their normal shape; in a hypotonic solution, they gain water and undergo hemolysis; in a hypertonic solution, they lose water and undergo crenation.

11. With the expenditure of cellular energy, usually in the form of ATP, solutes can cross the membrane against their concentration gradient by means of active transport. Actively transported solutes include several ions such as Na^+, K^+, H^+, Ca^{2+}, I^-, and Cl^-; amino acids; and monosaccharides.

12. Two sources of energy are used to drive active transport. Energy from splitting of ATP is used in primary active transport, and energy stored in a Na^+ concentration gradient is used in secondary active transport.

13. The most important primary active transport pump is the Na^+/K^+ pump (the sodium pump) which expels Na^+ from cells and brings K^+ in.

14. Secondary active transport mechanisms include both symporters and antiporters that are powered by the Na^+ concentration gradient. Symporters move two substances in the same direction, whereas antiporters move two substances in opposite directions across the membrane.

15. Vesicular transport includes both endocytosis (phagocytosis and pinocytosis) and exocytosis.

16. Phagocytosis is the ingestion of solid particles. It is an important process used by some white blood cells to destroy bacteria that enter the body. Pinocytosis is the ingestion of extracellular fluid. In this process, the fluid becomes surrounded by a pinocytic vesicle.

17. Exocytosis involves movement of secretory or waste products out of a cell by fusion of vesicles with the plasma membrane.

Cytoplasm (p. 51)

1. Cytoplasm includes all the cellular contents between the plasma membrane and nucleus; it consists of cytosol and organelles.

2. The fluid portion of cytoplasm is cytosol, composed mostly of water, plus ions, glucose, amino acids, fatty acids, proteins, lipids, ATP, and waste products; cytosol is the site of many chemical reactions required for a cell's existence.

3. Organelles are specialized cellular structures with characteristic shapes and specific functions.

4. The cytoskeleton is a network of several kinds of protein filaments that extend throughout the cytoplasm; they provide a structural framework for the cell and generate movements. Components of the cytoskeleton are microfilaments, intermediate filaments, and microtubules.

5. The centrosome consists of two centrioles and a pericentriolar area. Centrioles function in the formation or regeneration of flagella and cilia. The pericentriolar area serves as a center for organizing microtubules in interphase cells and the mitotic spindle during cell division.

6. Cilia and flagella are motile projections of the cell surface. Cilia move fluid along the cell surface, whereas a flagellum moves an entire cell.

7. Ribosomes, composed of ribosomal RNA and ribosomal proteins, consist of two subunits and are the sites of protein synthesis.

8. Endoplasmic reticulum (ER) is a network of membranes that form flattened sacs or tubules called cisterns; it extends from the nuclear envelope throughout the cytoplasm.

9. Rough ER is studded with ribosomes. Proteins synthesized on the ribosomes enter the ER for processing and sorting. The ER is also where glycoproteins and phospholipids form and proteins attach to phospholipids.

10. Smooth ER lacks ribosomes. It is where phospholipids, steroids, and other lipids are synthesized. Smooth ER also participates in releasing glucose from the liver into the bloodstream, inactivating or detoxifying drugs and other potentially harmful substances, and releasing calcium ions that trigger contraction in muscle cells.

11. The Golgi complex consists of flattened sacs called cisterns that receive proteins synthesized in the rough ER. Within the Golgi cisterns the proteins are modified, sorted, and packaged into vesicles for transport to different destinations. Some processed proteins leave the cell in secretory vesicles, some are incorporated into the plasma membrane, and some enter lysosomes.

12. Lysosomes are membrane-enclosed vesicles that contain digestive enzymes. They function in digestion of worn-out organelles (autophagy) and even in digestion of their own cell (autolysis).

13. Peroxisomes are similar to lysosomes but smaller. They oxidize various organic substances such as amino acids, fatty acids, and toxic substances and, in the process, produce hydrogen peroxide. The hydrogen peroxide is degraded by an enzyme in peroxisomes called catalase.

14. Mitochondria consist of a smooth outer membrane, an inner membrane containing cristae, and a fluid-filled cavity called the matrix. They are called "powerhouses" of the cell because they produce most of a cell's ATP.

Nucleus (p. 56)

1. The nucleus consists of a double nuclear envelope; nuclear pores, which control the movement of substances between the nucleus and cytoplasm; nucleoli, which produce ribosomes; and genes arranged on chromosomes.

2. Most body cells have a single nucleus; some (red blood cells) have none, whereas others (skeletal muscle cells) have several.

3. Genes control cellular structure and most cellular functions.

Gene Action: Protein Synthesis (p. 58)

1. Most of the cellular machinery is devoted to protein synthesis.

2. Cells make proteins by transcribing and translating the genetic in-

formation encoded in the sequence of four types of nitrogenous bases in DNA.

3. In transcription, genetic information encoded in the DNA base sequence is copied into a complementary sequence of bases in a strand of messenger RNA (mRNA). Transcription begins on DNA in a region called a promoter.

4. Translation is the process in which mRNA associates with ribosomes and directs synthesis of a protein, converting the nucleotide sequence in mRNA into a specific sequence of amino acids.

5. In translation, mRNA binds to a ribosome, specific amino acids attach to tRNA, and anticodons of tRNA bind to codons of mRNA, bringing specific amino acids into position on a growing protein.

6. Translation begins at the start codon and terminates at the stop codon.

Somatic Cell Division (p. 61)

1. Cell division is the process by which cells reproduce themselves.

2. Cell division that results in an increase in the number of body cells is called somatic cell division; it involves a nuclear division called mitosis plus division of cytoplasm, called cytokinesis.

3. Cell division that results in the production of sperm and oocytes is called reproductive cell division.

4. The cell cycle is an orderly sequence of events in which a cell duplicates its contents and divides in two. It consists of interphase and a mitotic phase.

5. Before the mitotic phase, the DNA molecules, or chromosomes, replicate themselves so that identical chromosomes can be passed on to the next generation of cells.

6. A cell that is between divisions and is carrying on every life process except division is said to be in interphase.

7. Mitosis is the replication and distribution of two sets of chromosomes into separate and equal nuclei; it consists of prophase, metaphase, anaphase, and telophase.

8. Cytokinesis usually begins late in anaphase and ends in telophase.

9. A cleavage furrow forms and progresses inward, cutting through the cell to form two separate daughter cells, each with equal portions of cytoplasm, organelles, and chromosomes.

Aging and Cells (p. 62)

1. Aging is a normal process accompanied by progressive alteration of the body's homeostatic adaptive responses.

2. Many theories of aging have been proposed, including genetically programmed cessation of cell division, and the buildup of free radicals.

■ SELF-QUIZ

1. If the extracellular fluid contains a greater concentration of solutes than the cytosol of the cell, the extracellular fluid is said to be
   a. isotonic   **b.** hypertonic   **c.** hypotonic   **d.** allotonic
   e. epitonic

2. The proteins found in the plasma membrane
   a. are primarily glycoproteins
   b. allow the passage of many substances into the cell
   c. allow cells to recognize other cells
   d. help anchor cells to each other
   e. have all of the above functions

3. To enter many body cells, glucose must bind to a specific membrane transport protein, which assists glucose to cross the membrane without using ATP. This type of movement is known as
   a. facilitated diffusion   **b.** simple diffusion   **c.** vesicular transport   **d.** osmosis   **e.** active transport

4. A red blood cell placed in a hypotonic solution undergoes
   a. hemolysis   **b.** crenation   **c.** equilibrium   **d.** a decrease in osmotic pressure   **e.** shrinkage

5. Which of the following normally pass through the plasma membrane only by vesicular transport?
   a. water molecules   **b.** sodium ions   **c.** proteins   **d.** oxygen molecules   **e.** steroids

6. Which of the following statements concerning diffusion is NOT true?
   a. Diffusion speeds up as the body temperature rises.
   b. A small surface area slows down the rate of diffusion.
   c. A low-weight particle diffuses faster than a high-weight particle.
   d. It moves materials from an area of low concentration to an area of high concentration by kinetic energy.

   e. Diffusion over a greater distance takes longer than diffusion over a short distance.

7. Which of the following processes requires ATP?
   a. diffusion   **b.** active transport   **c.** osmosis   **d.** facilitated diffusion   **e.** net diffusion

8. Nicotine in cigarette smoke interferes with the ability of cells to rid the breathing passageways of debris. Which organelles are "paralyzed" by nicotine?
   a. flagella   **b.** ribosomes   **c.** microfilaments   **d.** cilia
   e. lysosomes

9. Many proteins found in the plasma membrane are formed by the _____ and packaged by the _____.
   a. ribosomes, Golgi complex
   b. smooth endoplasmic reticulum, Golgi complex
   c. Golgi complex, lysosomes
   d. mitochondria, Golgi complex
   e. nucleus, smooth endoplasmic reticulum

10. Match the following:
    _____ **a.** cellular movement
    _____ **b.** selective permeability
    _____ **c.** protein synthesis
    _____ **d.** lipid synthesis, detoxification
    _____ **e.** packages proteins and lipids
    _____ **f.** ATP production
    _____ **g.** digest bacteria and worn-out organelles
    _____ **h.** forms mitotic spindle

    A. centrosome
    B. cytoskeleton
    C. Golgi complex
    D. lysosomes
    E. mitochondria
    F. plasma membrane
    G. ribosomes
    H. smooth ER

11. If the smooth endoplasmic reticulum were destroyed, a cell would not be able to

a. form lysosomes **b.** synthesize certain proteins **c.** generate energy **d.** phagocytize bacteria **e.** synthesize phospholipids

12. Water moves into and out of red blood cells through the process of

a. endocytosis **b.** phagocytosis **c.** osmosis **d.** active transport **e.** facilitated diffusion

13. If a cell were missing its centrosomes, it would

a. be unable to undergo mitosis
b. have an abnormal shape due to the absence of the cytoskeleton
c. be missing its genetic material
d. not be capable of carrying out active transport
e. have to develop flagella to replace the missing centrosomes

14. Transcription involves

a. transferring information from the mRNA to tRNA
b. codon binding with anticodons
c. joining amino acids by peptide bonds
d. copying information contained in the DNA to mRNA
e. synthesizing the protein on the ribosome

15. If a DNA strand has a nitrogenous base sequence TACGA, then the sequence of bases on the corresponding mRNA would be

a. ATGCT **b.** AUGCU **c.** GUACU **d.** CTGAT
e. AUCUG

16. Place the following events of protein synthesis in the proper order.

1. DNA uncoils and mRNA is transcribed.
2. tRNA with an attached amino acid pairs with mRNA.
3. mRNA passes from the nucleus into the cytoplasm and attaches to a ribosome.
4. Protein is formed.
5. Two amino acids are linked by a peptide bond.

a. 1, 2, 3, 4, 5 **b.** 1, 3, 2, 5, 4 **c.** 1, 2, 3, 5, 4
d. 1, 5, 3, 2, 4 **e.** 2, 1, 3, 4, 5

17. Match the following descriptions with the phases of mitosis.

____ **a.** nuclear envelope (membrane) and nucleoli reappear

____ **b.** centromeres of the chromatid pairs line up in the center of the mitotic spindle

____ **c.** DNA duplicates

____ **d.** cleavage furrow splits cell into two daughter cells

____ **e.** chromosomes move toward opposite poles of cell

____ **f.** chromatids are attached at centromeres; mitotic spindle forms

A. prophase
B. cytokinesis
C. telophase
D. anaphase
E. metaphase
F. interphase

18. In which phase is a cell highly active and growing?

a. anaphase **b.** prophase **c.** metaphase **d.** telophase
e. interphase

19. If a virus were to enter a cell and destroy its ribosomes, how would the cell be affected?

a. It would be unable to undergo mitosis.
b. It could no longer produce ATP.
c. Movement of the cell would cease.
d. It would undergo autophagy.
e. It would be unable to synthesize proteins.

20. Which of the following statements concerning cancer is NOT true?

a. A benign tumor is noncancerous.
b. When a cancerous growth presses on nerves, it can cause pain.
c. Angiogenesis is the spread of cancerous cells to other parts of the body.
d. Ultraviolet radiation and radon gas are carcinogens.
e. Cancer is uncontrolled mitosis in abnormal cells.

CRITICAL THINKING APPLICATIONS

1. One of the functions of bones is to store minerals, especially calcium. The bone tissue must be dissolved to release the calcium for use by the body's systems. Which organelle would be involved in breaking down bone tissue?

2. You dream that you're floating on a raft in the middle of the ocean. The sun's hot, you're very thirsty, and you're surrounded by water. You want to take a long, cool drink of seawater, but something you learned in A & P (you knew that was coming) stops you from drinking and saves your life! Why shouldn't you drink seawater?

3. Mucin is a protein present in saliva. When mixed with water, mucin becomes the slippery substance known as mucus. Trace the route taken by mucin through the cells of the salivary glands, starting with the organelle where it is synthesized and ending with its release from the cells.

4. Your 65-year-old aunt has spent every sunny day at the beach for as long as you can remember. Her skin looks a lot like a comfortable lounge chair—brown and wrinkled. Recently her dermatologist removed a suspicious growth from the skin of her face. What would you suspect is the problem and its likely cause?

ANSWERS TO FIGURE QUESTIONS

3.1 The three main parts of a cell are the plasma membrane, cytoplasm, and nucleus.

3.2 Membrane glycolipids and glycoproteins are involved in cellular recognition.

3.3 Since a fever represents an increase in body temperature, all diffusion processes would be increased.

3.4 Oxygen, carbon dioxide, fatty acids, fat-soluble vitamins, steroids, ammonia, and small alcohols can cross the plasma membrane by simple diffusion through the lipid bilayer.

3.5 The concentration of K^+ is higher in the cytosol of body cells than in extracellular fluids.

3.6 Insulin promotes insertion of glucose transporters in the plasma membrane, which increases cellular glucose uptake by facilitated diffusion.

3.7 No, the water concentrations can never be the same because the beaker always contains pure (100%) water and the sac contains a solution that is less than 100% water.

3.8 A 2% solution of NaCl will cause crenation of RBCs because it is hypertonic.

3.9 ATP adds a phosphate group to the pump protein, which changes the pump's three-dimensional shape.

3.10 In secondary active transport, ATP is not split to drive the activity of symporter or antiporter proteins; ATP does directly power the pump protein in primary active transport.

3.11 The trigger that causes pseudopod extension is binding of a particle to a membrane receptor.

3.12 Clusters of microtubules form the structure of centrioles, cilia, and flagella.

3.13 The components of the centrosome are two centrioles and the pericentriolar area.

3.14 Large and small ribosomal subunits are synthesized in a nucleolus in the nucleus and then join together in the cytoplasm.

3.15 Rough ER has attached ribosomes where proteins that will be exported from the cell are synthesized whereas smooth ER lacks ribosomes and is associated with lipid synthesis and other metabolic reactions.

3.16 Cells that secrete proteins into extracellular fluid have extensive Golgi complexes.

3.17 After leaving the Golgi complex in vesicles, some proteins are discharged from the cell by exocytosis, some are incorporated into the plasma membrane, and some end up in lysosomes.

3.18 Mitochondrial cristae provide a large surface area for chemical reactions and contain enzymes needed for ATP production.

3.19 Nuclear genes control cellular structure and direct most cellular activities.

3.20 Proteins determine the physical and chemical characteristics of cells.

3.21 RNA polymerase catalyzes transcription of DNA.

3.22 When a ribosome encounters a stop codon in mRNA, the completed protein detaches from the final tRNA.

3.23 DNA replication occurs during interphase.

3.24 Cytokinesis begins late in anaphase or early in telophase.

Chapter 4

Tissues

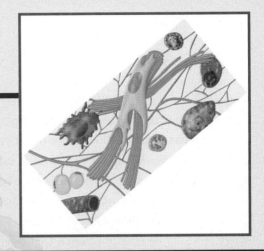

Student Learning Objectives

A Look Ahead

As you learned in the previous chapter, cells are highly organized living units, but they cannot function alone. Instead, cells work together in groups called tissues. A *tissue* is a group of similar cells, usually with a common embryonic origin, that function together to carry out specialized activities. *Histology* (hiss-TOL-ō-jē; *hist-* = tissue; *-ology* = study of) is the science that deals with the study of tissues. A *pathologist* (pa-THOL-ō-gist; *patho-* = disease) is a scientist who specializes in laboratory studies of cells and tissues to help physicians make accurate diagnoses. One of the principal functions of a pathologist is to examine tissues for any changes that might indicate disease.

TYPES OF TISSUES

Objective: • **Describe the characteristics of the four basic types of tissue that make up the human body.**

1. *Epithelial tissue* (ep′-i-THĒ-lē-al) covers body surfaces; lines body cavities, hollow organs, and ducts (tubes); and forms glands.

2. *Connective tissue* protects and supports the body and its organs, binds organs together, stores energy reserves as fat, and provides immunity.

3. *Muscle tissue* generates the force needed to make body structures move.

4. *Nervous tissue* initiates and transmits nerve impulses that coordinate body activities.

Epithelial tissue and connective tissue, except for bone tissue and blood, are discussed in detail in this chapter. The general features of bone tissue and blood will be introduced here, but their detailed discussion is presented in Chapters 6 and 14, respectively. The structure and function of muscle tissue and nervous tissue are examined in detail in Chapters 8 and 9, respectively.

EPITHELIAL TISSUE

Objective: • **Discuss the general features of epithelial tissue and the structure, location, and function of the various types of epithelia.**

Epithelial tissue, or more simply *epithelium* (plural is *epithelia*), may be divided into two types: (1) *covering and lining epithelium*

and (2) *glandular epithelium.* As its name suggests, covering and lining epithelium forms the outer covering of the skin and the outer covering of some internal organs. It also lines body cavities; blood vessels; ducts; and the interiors of the respiratory, digestive, urinary, and reproductive systems. It makes up, along with nervous tissue, the parts of the sense organs for hearing, vision, and touch. Glandular epithelium makes up the secreting portion of glands, such as sweat glands that secrete sweat.

General Features of Epithelial Tissue

As you will see shortly, there are many different types of epithelia, each with characteristic structure and functions. However, all of the different types of epithelial tissue also have features in common. General features of epithelial tissue include the following:

1. Epithelium consists largely or entirely of closely packed cells with little extracellular material between them.

2. Epithelial cells are arranged in continuous sheets, in either single or multiple layers.

3. Epithelial cells have a free (superficial) surface, which is exposed to a body cavity, lining of an internal organ, or the exterior of the body, and a basal surface, which is attached to a basement membrane. The **basement membrane** is an extracellular structure composed mostly of fibers. It is located between the epithelium and the underlying connective tissue layer and helps support the epithelium.

4. Epithelia are *avascular* (*a-* = without; *vascular* = blood vessels). The vessels that supply nutrients to and remove wastes from epithelia are located in adjacent connective tissues. The exchange of materials between epithelium and connective tissue occurs by means of diffusion.

5. Epithelia have a nerve supply.

6. Because epithelium is subject to a certain amount of wear and tear and injury, it has a high capacity for renewal by cell division.

7. Functions of epithelia include protection, secretion, absorption, excretion, sensory reception, and generation of gametes (sperm and oocytes).

Covering and Lining Epithelium
Classification by Cell Shape

Covering and lining epithelium, which covers or lines various parts of the body, contains cells with four basic shapes:

1. *Squamous* (SKWĀ-mus = flat) cells are flat and attach to each other like tiles. Their thinness allows for rapid passage of substances through them (see Chapter 3).

2. *Cuboidal* cells are thicker and shaped like cubes or hexagons. They produce several important body *secretions* (fluids that are produced and released by cells, such as mucus, sweat, and enzymes). They may also function in *absorption* (intake) of fluids and other substances, such as digested foods in the intestines.

3. *Columnar* cells are tall and cylindrical, thereby protecting underlying tissues. They may also be specialized for secretion and absorption. Some may also have cilia.

4. *Transitional* cells range in shape from flat to columnar and often change shape due to distention (stretching), expansion, or movement of body parts.

Classification by Arrangement of Layers

The cells of covering and lining epithelium typically are arranged in one or more layers depending on the function of the particular body part. The terms used to refer to their classification by arrangement of layers include the following:

1. *Simple epithelium* is a single layer of cells found in areas where diffusion, osmosis, filtration, secretion, and absorption occur.

2. *Stratified epithelium* (*stratum* = layer) contains two or more layers of cells used for protection of underlying tissues in areas where there is considerable wear and tear.

3. *Pseudostratified columnar epithelium* contains one layer of cells. The tissue appears to have several layers, but not all cells reach the surface; those that do are either ciliated or secrete mucus.

Combining the arrangements of layers and cell shapes provides the following classification scheme of covering and lining epithelium:

I. Simple epithelium
   A. Simple squamous epithelium
   B. Simple cuboidal epithelium
   C. Simple columnar epithelium

II. Stratified epithelium
   A. Stratified squamous epithelium*
   B. Stratified cuboidal epithelium*
   C. Stratified columnar epithelium*
   D. Transitional epithelium

III. Pseudostratified columnar epithelium

Table 4.1 illustrates all of these covering and lining epithelia. Most of the tables in this chapter contain figures consisting of a photomicrograph, a corresponding diagram, and an inset that identifies a principal location of the tissue in the body. Along with the illustrations are descriptions, locations, and functions of the tissues.

Simple Epithelium

SIMPLE SQUAMOUS EPITHELIUM This tissue consists of a single layer of flat cells that resembles a tiled floor when viewed from above (Table 4.1A). The nucleus of each cell is oval or spherical and centrally located. Simple squamous epithelium is found in parts of the body where filtration (kidneys) or diffusion (lungs) are priority processes. It is not found in body areas that are subjected to wear and tear.

   The simple squamous epithelium that lines the heart, blood vessels, and lymphatic vessels is known as *endothelium* (*endo-* =

---

* This classification is based on the shape of the superficial cells.

within; *-thelium* = covering); the type that forms the epithelial layer of serous membranes is called *mesothelium* (*meso-* = middle).

SIMPLE CUBOIDAL EPITHELIUM The cuboidal shape of the cells in this tissue (Table 4.1B) is obvious only when the tissue is sectioned and viewed from the side, as when a slice has been made at a right angle to the epithelial tissue layer. Cell nuclei are usually round and centrally located. Simple cuboidal epithelium performs the functions of secretion and absorption.

SIMPLE COLUMNAR EPITHELIUM When viewed from the side, the cells appear rectangular with oval nuclei near the base of the cells. Simple columnar epithelium exists in two forms: nonciliated simple columnar epithelium and ciliated simple columnar epithelium.

   Nonciliated simple columnar epithelium contains absorptive cells and goblet cells (Table 4.1C). *Absorptive cells* are columnar epithelial cells with *microvilli,* microscopic fingerlike projections that increase the surface area of the plasma membrane (see Figure 3.1 on page 42). Their presence increases the rate of absorption by the absorptive cell. *Goblet cells* are modified columnar cells that secrete mucus, a slightly sticky fluid. Before it is released, mucus accumulates in the upper portion of the cell, causing that area to bulge out. The whole cell then resembles a goblet or wine glass. Secreted mucus serves as a lubricant for the linings of the digestive, respiratory, reproductive, and most of the urinary tracts. Mucus also helps to trap dust entering the respiratory tract, and it prevents destruction of the stomach lining by digestive enzymes.

   Ciliated simple columnar epithelium (Table 4.1D) contains cells with cilia. In a few parts of the upper respiratory tract, ciliated columnar cells are interspersed with goblet cells. Mucus secreted by the goblet cells forms a film over the respiratory surface that traps inhaled foreign particles. The cilia wave in unison and move the mucus and any trapped foreign particles toward the throat, where it can be coughed up and swallowed or spit out. Cilia also help to move oocytes through the uterine tubes into the uterus.

Stratified Epithelium

In contrast to simple epithelium, stratified epithelium has at least two layers of cells. Thus, it is more durable and can better protect underlying tissues. Some cells of stratified epithelia also produce secretions. The name of the specific kind of stratified epithelium depends on the shape of the superficial cells.

STRATIFIED SQUAMOUS EPITHELIUM Cells in the superficial layers of this type of epithelium are flat, whereas in the deep layers, cells vary in shape from cuboidal to columnar (Table 4.1E). The basal (deepest) cells continually undergo cell division. As new cells grow, the cells of the basal layer are pushed upward toward the surface. As they move farther from the deep layer and from their blood supply in the underlying connective tissue, they become dehydrated, shrunken, and harder. At the surface, the cells lose their cell junctions and are sloughed off, but they are replaced as new cells continually emerge from the basal cells.

   Stratified squamous epithelium exists in both keratinized and nonkeratinized forms. In *keratinized stratified squamous*

Table 4.1 / Epithelial Tissues

Covering and Lining Epithelium

A. Simple squamous epithelium

Description: Single layer of flat cells; centrally located nucleus.

Location: Lines heart, blood vessels, lymphatic vessels, air sacs of lungs, glomerular (Bowman's) capsule of kidneys, and inner surface of the eardrum; forms epithelial layer of serous membranes.

Function: Filtration, diffusion, osmosis, and secretion in serous membranes.

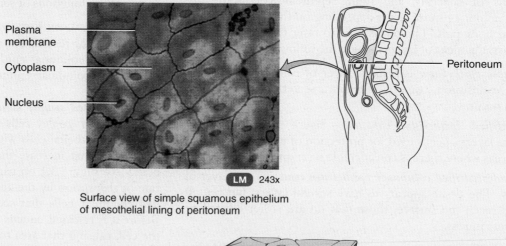

LM 243x

Surface view of simple squamous epithelium
of mesothelial lining of peritoneum

Simple squamous epithelium

B. Simple cuboidal epithelium

Description: Single layer of cube-shaped cells; centrally located nucleus.

Location: Covers surface of ovary, lines anterior surface of capsule of the lens of the eye, forms the pigmented epithelium at the back of the eye, lines kidney tubules and smaller ducts of many glands, and makes up the secreting portion of some glands such as the thyroid gland.

Function: Secretion and absorption.

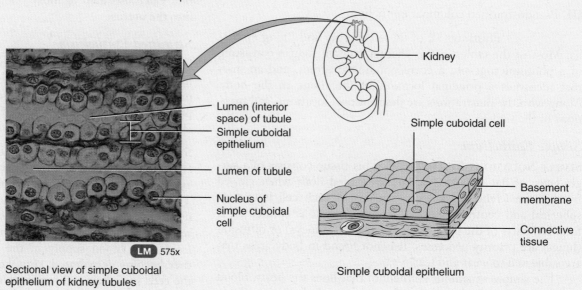

LM 575x

Sectional view of simple cuboidal
epithelium of kidney tubules

Simple cuboidal epithelium

Covering and Lining Epithelium

C. Nonciliated simple columnar

Description: Single layer of nonciliated rectangular cells; nucleus at base of cell; contains goblet cells and **epithelium**absorptive cells with microvilli in some locations.

Location: Lines the gastrointestinal, respiratory, reproductive, and urinary tracts, ducts of many glands, and gallbladder.

Function: Secretion and absorption.

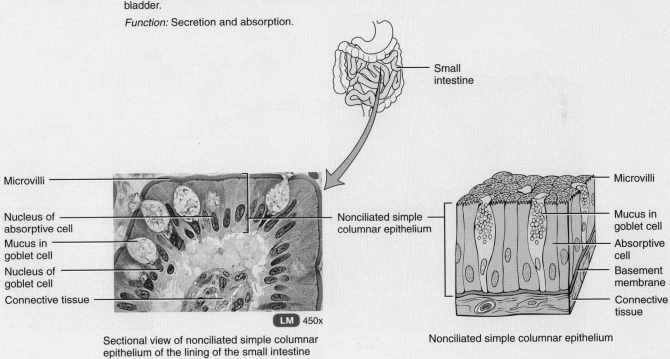

Small intestine

Microvilli

Nucleus of absorptive cell

Mucus in goblet cell

Nucleus of goblet cell

Connective tissue

Nonciliated simple columnar epithelium

Microvilli

Mucus in goblet cell

Absorptive cell

Basement membrane

Connective tissue

LM 450x

Sectional view of nonciliated simple columnar epithelium of the lining of the small intestine

Nonciliated simple columnar epithelium

D. Ciliated simple columnar epithelium

Description: Single layer of ciliated rectangular cells; nucleus at base of cell; contains goblet cells in some locations.

Location: Lines a few portions of upper respiratory tract, uterine (Fallopian) tubes, uterus, some paranasal sinuses, and central canal of spinal cord.

Function: Moves mucus and other substances by ciliary action.

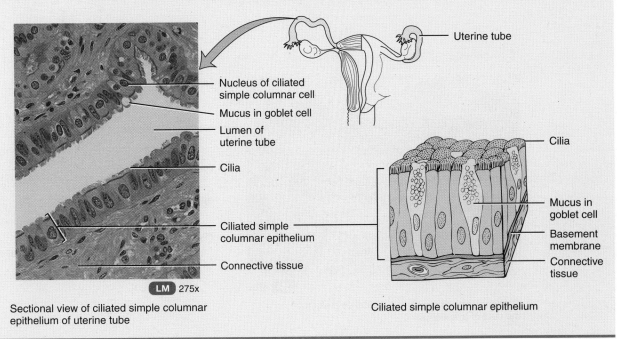

Uterine tube

Nucleus of ciliated simple columnar cell

Mucus in goblet cell

Lumen of uterine tube

Cilia

Ciliated simple columnar epithelium

Connective tissue

Cilia

Mucus in goblet cell

Basement membrane

Connective tissue

LM 275x

Sectional view of ciliated simple columnar epithelium of uterine tube

Ciliated simple columnar epithelium

(continues)

Table 4.1 / Epithelial Tissues (continued)

Covering and Lining Epithelium

E. Stratified squamous epithelium

Description: Several layers of cells; cuboidal to columnar shape in deep layers; squamous cells in superficial layers; basal (deepest) cells replace surface cells as they are lost.

Location: Keratinized variety forms superficial layer of skin; nonkeratinized variety lines wet surfaces, such as lining of the mouth, esophagus, part of epiglottis, and vagina, and covers the tongue.

Function: Protection.

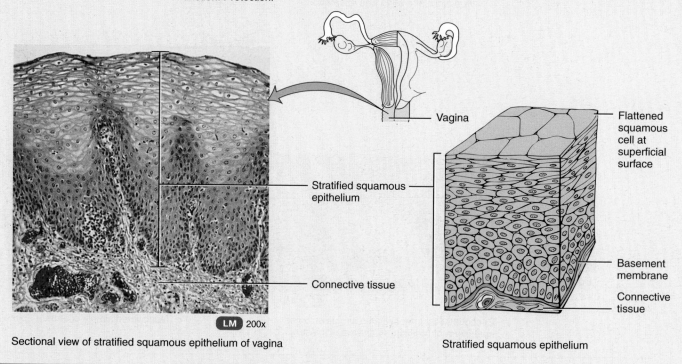

Vagina

Stratified squamous epithelium

Connective tissue

Flattened squamous cell at superficial surface

Basement membrane

Connective tissue

LM 200x

Sectional view of stratified squamous epithelium of vagina

Stratified squamous epithelium

F. Stratified cuboidal epithelium

Description: Two or more layers of cells in which the superficial cells are cube-shaped.

Location: Ducts of adult sweat glands and part of male urethra.

Function: Protection and limited secretion and absorbtion.

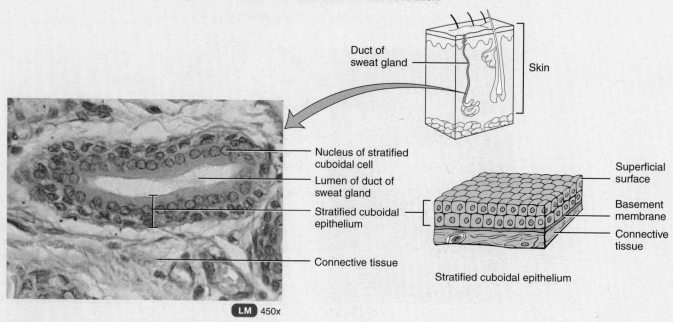

Duct of sweat gland

Skin

Nucleus of stratified cuboidal cell

Lumen of duct of sweat gland

Stratified cuboidal epithelium

Connective tissue

Superficial surface

Basement membrane

Connective tissue

LM 450x

Sectional view of stratified cuboidal epithelium of the duct of a sweat gland

Stratified cuboidal epithelium

Covering and Lining Epithelium

G. Stratified columnar epithelium

Description: Several layers of irregularly shaped cells; columnar cells are only in the superficial layer.

Location: Lines part of urethra, large excretory ducts of some glands, small areas in anal mucous membrane, and a part of the conjunctiva of the eye.

Function: Protection and secretion.

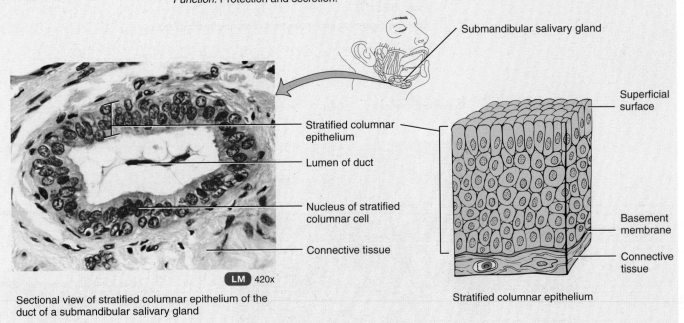

Sectional view of stratified columnar epithelium of the duct of a submandibular salivary gland

Stratified columnar epithelium

H. Transitional epithelium

Description: Appearance is variable (transitional); shape of superficial cells ranges from squamous (when stretched) to cuboidal (when relaxed).

Location: Lines urinary bladder and portions of ureters and urethra.

Function: Permits organs to stretch without rupturing.

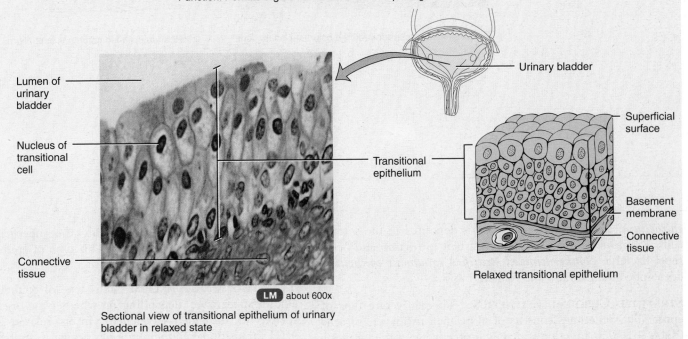

Sectional view of transitional epithelium of urinary bladder in relaxed state

Relaxed transitional epithelium

(continues)

Table 4.1 / Epithelial Tissues (continued)

Covering and Lining Epithelium

I. Pseudostratified columnar epithelium

Description: Not a true stratified tissue; nuclei of cells are at different levels; all cells are attached to basement membrane, but not all reach the surface.

Location: Pseudostratified ciliated columnar epithelium lines the airways of most of upper respiratory tract; pseudostratified nonciliated columnar epithelium lines larger ducts of many glands, epididymis, and part of male urethra.

Function: Secretion and movement of mucus by ciliary action.

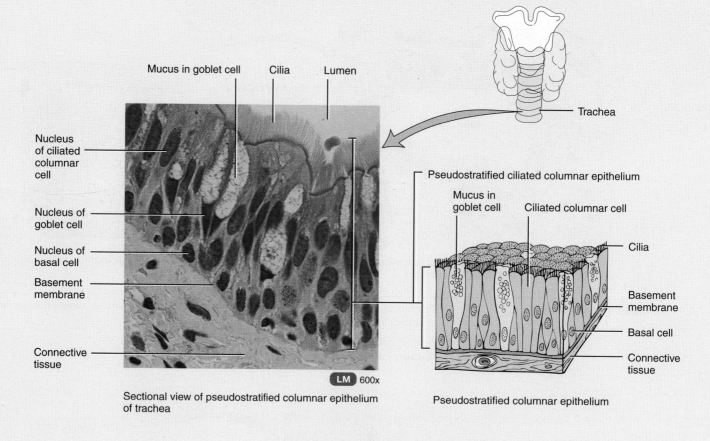

Sectional view of pseudostratified columnar epithelium of trachea

LM 600x

Pseudostratified columnar epithelium

epithelium, a tough layer of keratin is deposited in the surface cells. *Keratin* is a protein that is resistant to friction and helps repel bacteria. *Nonkeratinized stratified squamous epithelium* does not contain keratin and remains moist.

STRATIFIED CUBOIDAL EPITHELIUM   This fairly rare type of epithelium sometimes consists of more than two layers of cells (Table 4.1F). Its function is mainly protective; in some locations it also functions in secretion and absorption.

STRATIFIED COLUMNAR EPITHELIUM   This type of tissue also is uncommon. Usually the basal layer or layers consist of short-

ened, irregularly shaped cells; only the superficial cells are columnar in form (Table 4.1G). This type of epithelium functions in protection and secretion.

TRANSITIONAL EPITHELIUM   This type of epithelium is variable in appearance, depending on whether the organ it lines is relaxed or distended (stretched). In its relaxed state (Table 4.1H), transitional epithelium looks similar to stratified cuboidal epithelium, except that the superficial cells tend to be large and rounded. As the cells are stretched, they become flatter, giving the appearance of stratified squamous epithelium. Because of its elasticity, transitional epithelium lines hollow structures that are

Glandular Epithelium

J. Endocrine glands

Description: Secretions (hormones) released into blood.

Location: Examples include pituitary gland at base of brain, pineal gland in brain, thyroid and parathyroid glands near voice box, adrenal glands above kidneys, pancreas near stomach, ovaries in pelvic cavity, and testes in scrotum.

Function: Produce hormones that regulate various body activities.

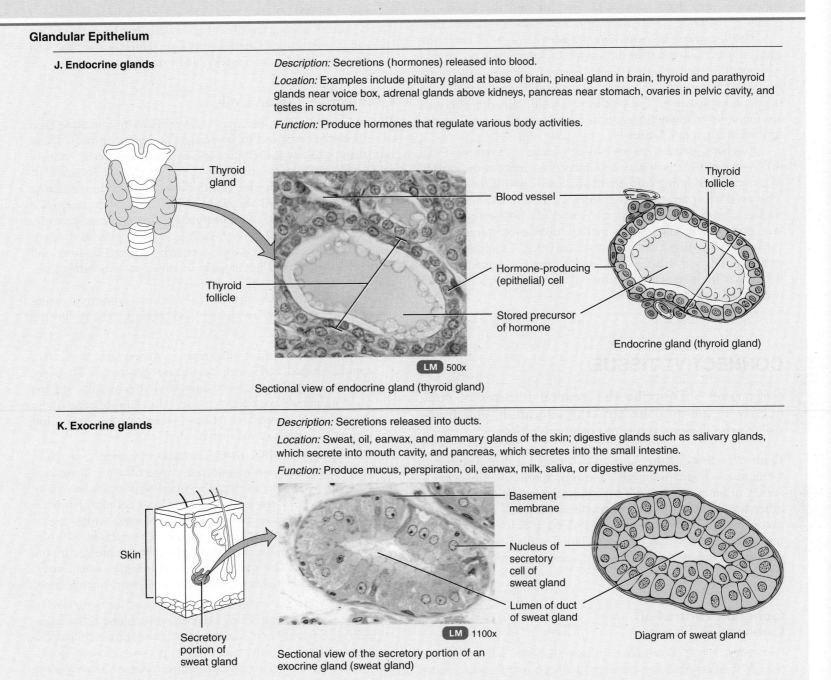

Thyroid gland

Blood vessel

Thyroid follicle

Thyroid follicle

Hormone-producing (epithelial) cell

Stored precursor of hormone

Endocrine gland (thyroid gland)

LM 500x

Sectional view of endocrine gland (thyroid gland)

K. Exocrine glands

Description: Secretions released into ducts.

Location: Sweat, oil, earwax, and mammary glands of the skin; digestive glands such as salivary glands, which secrete into mouth cavity, and pancreas, which secretes into the small intestine.

Function: Produce mucus, perspiration, oil, earwax, milk, saliva, or digestive enzymes.

Skin

Basement membrane

Nucleus of secretory cell of sweat gland

Lumen of duct of sweat gland

Secretory portion of sweat gland

LM 1100x

Sectional view of the secretory portion of an exocrine gland (sweat gland)

Diagram of sweat gland

subjected to expansion from within, such as the urinary bladder. It allows organs to stretch without rupturing.

Pseudostratified Columnar Epithelium

The third category of covering and lining epithelium is called pseudostratified columnar epithelium (Table 4.1I). The nuclei of the cells are at various depths. Even though all the cells are attached to the basement membrane in a single layer, some cells do not reach the surface. When viewed from the side, these features give the false impression of a multilayered tissue, thus the name *pseudo*stratified epithelium (*pseudo-* = false). In ***pseudostratified ciliated columnar epithelium,*** the cells that reach the surface either secrete mucus (goblet cells) or bear cilia that sweep away mucus and trapped foreign particles for eventual elimination from the body. ***Pseudostratified nonciliated columnar epithelium*** contains no cilia or goblet cells.

Glandular Epithelium

The function of glandular epithelium is secretion, which is accomplished by glandular cells that often lie in clusters deep to the covering and lining epithelium. A ***gland*** may consist of one cell or a group of highly specialized epithelial cells that secrete substances into ducts, onto a surface, or into the blood. All glands of the body are classified as either endocrine or exocrine.

The secretions of **endocrine glands** (Table 4.1J) enter the extracellular fluid and then diffuse directly into the bloodstream without flowing through a duct. These secretions, called *hormones*, regulate many metabolic and physiological activities to maintain homeostasis. The pituitary, thyroid, and adrenal glands are examples of endocrine glands. Endocrine glands will be described in detail in Chapter 13.

Exocrine glands (*exo-* = outside; *-crine* = secretion; Table 4.1K) secrete their products into ducts that empty at the surface of covering and lining epithelium or directly onto a free surface. The product of an exocrine gland may be released at the skin surface or into the lumen (interior space) of a hollow organ. The secretions of exocrine glands include mucus, perspiration, oil, earwax, milk, saliva, and digestive enzymes. Examples of exocrine glands are sweat glands, which produce perspiration to help lower body temperature, and salivary glands, which secrete mucus and digestive enzymes.

CONNECTIVE TISSUE

Objective: • **Describe the general features of connective tissue and the structure, location, and function of the various types of connective tissues.**

Connective tissue is the most abundant and widely distributed tissue in the body. In its various forms, connective tissue has a variety of functions. It binds together, supports, and strengthens other body tissues; protects and insulates internal organs; compartmentalizes structures such as skeletal muscles; is the major transport system within the body (blood, a fluid connective tissue); and is the major site of stored energy reserves (adipose, or fat, tissue).

General Features of Connective Tissue

Connective tissue consists of two basic elements: cells and a matrix. A **matrix** generally consists of fibers and a ground substance, the component of a connective tissue that occupies the space between the cells and the fibers. The matrix, which separates tissue cells from one another, may be fluid, semifluid, gelatinous, fibrous, or calcified. It is usually secreted by the connective tissue cells and determines the tissue's qualities. In blood, the matrix (which is not secreted by blood cells) is liquid. In cartilage, it is firm but pliable. In bone, it is hard and not pliable.

Other general features of connective tissue include these:

1. Connective tissue usually does not occur on free surfaces, such as the coverings or linings of internal organs or the external surface of the body.

2. Except for cartilage, connective tissue, like epithelium, has a nerve supply.

3. Connective tissue usually is highly vascular (has a rich blood supply). Exceptions include cartilage, which is avascular, and tendons, which have a scanty blood supply.

Connective Tissue Cells

Each major type of connective tissue contains an immature class of cells with a name ending in *-blast* (= to bud or sprout). These immature cells are called *fibroblasts* in loose and dense connective tissue, *chondroblasts* in cartilage, and *osteoblasts* in bone. Blast cells retain the capacity for cell division and secrete the matrix that is characteristic of the tissue. In cartilage and bone, once the matrix is produced, the fibroblasts differentiate into mature cells whose names end in *-cyte*, such as chondrocytes and osteocytes. Mature cells have reduced capacity for cell division and matrix formation and are mostly involved in maintaining the matrix.

The types of cells present in various connective tissues depend on the type of tissue and include the following (Figure 4.1):

1. **Fibroblasts** (FĪ-brō-blasts; *fibro-* = fibers) are large, flat, spindle-shaped cells with branching processes. They are present in all connective tissues, and they usually are the most numerous connective tissue cells. Fibroblasts migrate through the connective tissue, secreting the fibers and ground substance of the matrix.

2. **Macrophages** (MAK-rō-fā-jez; *macro-* = large; *-phages* = eaters) develop from monocytes, a type of white blood cell. Macrophages have an irregular shape with short branching projections and can engulf bacteria and cellular debris by phagocytosis. Some are **fixed macrophages,** which means they reside in a particular tissue—for example, alveolar macrophages in the lungs or spleen macrophages in the spleen. Others are **wandering macrophages,** which roam the tissues and gather at sites of infection or inflammation.

3. **Plasma cells** are small and either round or irregular in shape. They develop from a type of white blood cell. Plasma cells secrete *antibodies*, proteins that attack or neutralize foreign substances in the body. Thus, plasma cells are an important part of the body's immune system. Most plasma cells reside in connective tissues, especially in the gastrointestinal tract and the mammary glands.

4. **Mast cells** are abundant alongside the blood vessels that supply connective tissue. They produce *histamine*, a chemical that dilates small blood vessels as part of the body's reaction to injury or infection.

5. **Adipocytes,** also called fat cells, are connective tissue cells that store triglycerides (fats). They are found below the skin and around organs such as the heart and kidneys.

6. **White blood cells** are not found in significant numbers in normal connective tissue. However, in response to certain conditions, they migrate from blood into connective tissues,

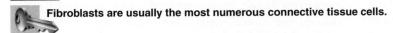

Figure 4.1 ■ **Representative cells and fibers present in connective tissues.**

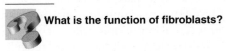 Fibroblasts are usually the most numerous connective tissue cells.

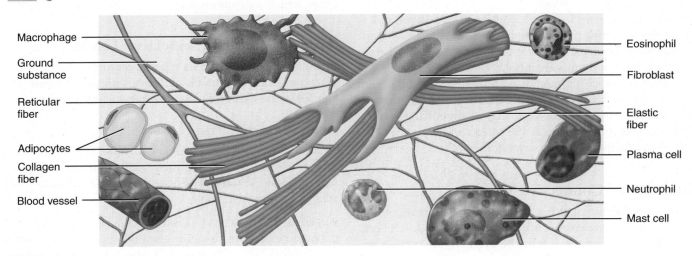

Macrophage

Ground substance

Reticular fiber

Adipocytes

Collagen fiber

Blood vessel

Eosinophil

Fibroblast

Elastic fiber

Plasma cell

Neutrophil

Mast cell

What is the function of fibroblasts?

where their numbers increase significantly. For example, *neutrophils* increase in number at sites of infection, and *eosinophils* increase in number in response to allergic conditions and parasitic invasions.

Connective Tissue Matrix

Each type of connective tissue has unique properties, due to accumulation of specific matrix materials between the cells. Matrix derives its properties from a fluid, gel, or solid ***ground substance*** plus its protein ***fibers.***

Ground Substance

Ground substance, the component of a connective tissue between the cells and fibers, supports cells, binds them together, and provides a medium through which substances are exchanged between the blood and cells. Until recently, ground substance was thought to function mainly as an inert scaffolding. Now it is clear, however, that the ground substance plays an active role in how tissues develop, migrate, proliferate, and change shape, and in how they carry out their metabolic functions.

Ground substance contains an assortment of large molecules, many of which are complex combinations of polysaccharides and proteins. For example, ***hyaluronic acid*** (hī′-a-loo-RON-ic) is a viscous, slippery substance that binds cells together, lubricates joints, and helps maintain the shape of the eyeballs. It also appears to play a role in helping phagocytes migrate through connective tissue during development and wound repair. White blood cells, sperm cells, and some bacteria produce *hyaluronidase*, an enzyme that breaks apart hyaluronic acid and causes the ground substance of connective tissue to become

watery. The ability to produce hyaluronidase enables white blood cells to move through connective tissues to reach sites of infection and sperm cells to penetrate the ovum during fertilization. It also accounts for how bacteria spread through connective tissues. Also present in the ground substance are ***adhesion proteins,*** which are responsible for linking components of the ground substance to each other and to the surfaces of cells. The principal adhesion protein of connective tissue is *fibronectin*, which binds to both collagen fibers (discussed shortly) and ground substance and cross-links them together. It also attaches cells to the ground substance.

Fibers

Fibers in the matrix strengthen and support connective tissues. Three types of fibers are embedded in the matrix between the cells: collagen fibers, elastic fibers, and reticular fibers.

Collagen fibers (*colla* = glue) are very strong and resist pulling forces, but they are not stiff, which promotes tissue flexibility. These fibers often occur in bundles lying parallel to one another (see Figure 4.1). The bundle arrangement affords great strength. Chemically, collagen fibers consist of the protein *collagen*. This is the most abundant protein in your body, representing about 25% of total protein. Collagen fibers are found in most types of connective tissues, especially bone, cartilage, tendons, and ligaments.

Elastic fibers, which are smaller in diameter than collagen fibers, branch and join together to form a network within a tissue. An elastic fiber consists of molecules of a protein called *elastin* surrounded by a glycoprotein named *fibrillin*, which is essential to the stability of an elastic fiber. Elastic fibers are strong but can be stretched up to one-and-a-half times their relaxed length without breaking. Equally important, elastic fibers have the ability to return to their original shape after being stretched, a property called *elasticity*. Elastic fibers are plentiful in skin, blood vessel walls, and lung tissue.

Reticular fibers (*reticul-* = net), consisting of *collagen* and a coating of glycoprotein, provide support in the walls of blood vessels and form branching networks around fat cells, nerve fibers, and skeletal and smooth muscle cells. Produced by fibroblasts, they are much thinner than collagen fibers. Like collagen fibers, reticular fibers provide support and strength and also form the *stroma* (= bed or covering) or supporting framework of many soft organs, such as the spleen and lymph nodes. These fibers also help form the basement membrane.

Classification of Connective Tissues

Classifying connective tissues can be challenging because of the diversity of cells and matrix that are present, and the differences in their relative proportions. Thus, the grouping of connective tissues into categories is not always clear-cut. We will classify them as follows:

I. Embryonic connective tissue
   A. Mesenchyme
   B. Mucous connective tissue
II. Mature connective tissue
   A. Loose connective tissue
     1. Areolar connective tissue
     2. Adipose tissue
     3. Reticular connective tissue
   B. Dense connective tissue
     1. Dense regular connective tissue
     2. Dense irregular connective tissue
     3. Elastic connective tissue
   C. Cartilage
     1. Hyaline cartilage
     2. Fibrocartilage
     3. Elastic cartilage
   D. Bone tissue
   E. Blood tissue
   F. Lymph

Note that our classification scheme has two major subclasses of connective tissue: embryonic and mature. *Embryonic connective tissue* is present primarily in the *embryo*, the developing human from fertilization through the first two months of pregnancy, and in the *fetus*, the developing human from the third month of pregnancy to birth.

One example of embryonic connective tissue found almost exclusively in the embryo is *mesenchyme* (MEZ-en-kīm), the tissue from which all other connective tissues eventually arise (Table 4.2A). Mesenchyme is composed of irregularly shaped cells, a semifluid ground substance, and delicate reticular fibers. Another kind of embryonic tissue is *mucous connective tissue (Wharton's jelly),* found primarily in the umbilical cord of the fetus. Mucous connective tissue is a form of mesenchyme that contains widely scattered fibroblasts, a more viscous jellylike ground substance, and collagen fibers (Table 4.2B).

The second major subclass of connective tissue, *mature connective tissue,* is present in the newborn and has cells produced from mesenchyme. Mature connective tissue is of several types, which we explore next.

Types of Mature Connective Tissue

As noted previously, the six types of mature connective tissue are (1) loose connective tissue, (2) dense connective tissue, (3) cartilage, (4) bone tissue, (5) blood tissue, and (6) lymph.

Loose Connective Tissue

As its name implies, the fibers in **loose connective tissue** are loosely arrayed among the many cells. The types of loose connective tissue are areolar connective tissue, adipose tissue, and reticular connective tissue.

AREOLAR CONNECTIVE TISSUE One of the most widely distributed connective tissues in the body is **areolar connective tissue** (a-RĒ-ō-lar; *areol-* = a small space). It contains several kinds of cells, including fibroblasts, macrophages, plasma cells, mast cells, adipocytes, and a few white blood cells (Table 4.3A). All three types of fibers—collagen, elastic, and reticular—are arranged randomly throughout the tissue. Combined with adipose tissue, areolar connective tissue forms the *subcutaneous layer,* the layer of tissue that attaches the skin to underlying tissues and organs.

ADIPOSE TISSUE Adipose tissue is a loose connective tissue in which the cells, called **adipocytes** (*adipo-* = fat), are specialized for storage of triglycerides (fats) (Table 4.3B). Adipocytes are derived from fibroblasts. Because the cell fills up with a single, large triglyceride droplet, the cytoplasm and nucleus are pushed to the periphery of the cell. Adipose tissue is found wherever areolar connective tissue is located. Adipose tissue is a good insulator and can therefore reduce heat loss through the skin. It is a major energy reserve and generally supports and protects various organs.

RECTICULAR CONNECTIVE TISSUE Reticular connective tissue consists of fine interlacing reticular fibers and reticular cells, cells that are connected to each other and form a network (Table 4.3C). Reticular connective tissue forms the stroma of certain organs and helps bind together smooth muscle cells.

Dense Connective Tissue

Dense connective tissue contains more numerous and thicker fibers arrayed densely among considerably fewer cells than loose connective tissue. There are three types: dense regular connective tissue, dense irregular connective tissue, and elastic connective tissue.

DENSE REGULAR CONNECTIVE TISSUE In this tissue, bundles of collagen fibers are arranged *regularly* in parallel patterns that confer great strength (Table 4.3D). The tissue structure withstands pulling along the axis of the fibers. Fibroblasts, which produce the fibers and ground substance, appear in rows between the fibers. The tissue is silvery white and tough, yet somewhat pliable. Examples are tendons and most ligaments.

Table 4.2 / Embryonic Connective Tissue

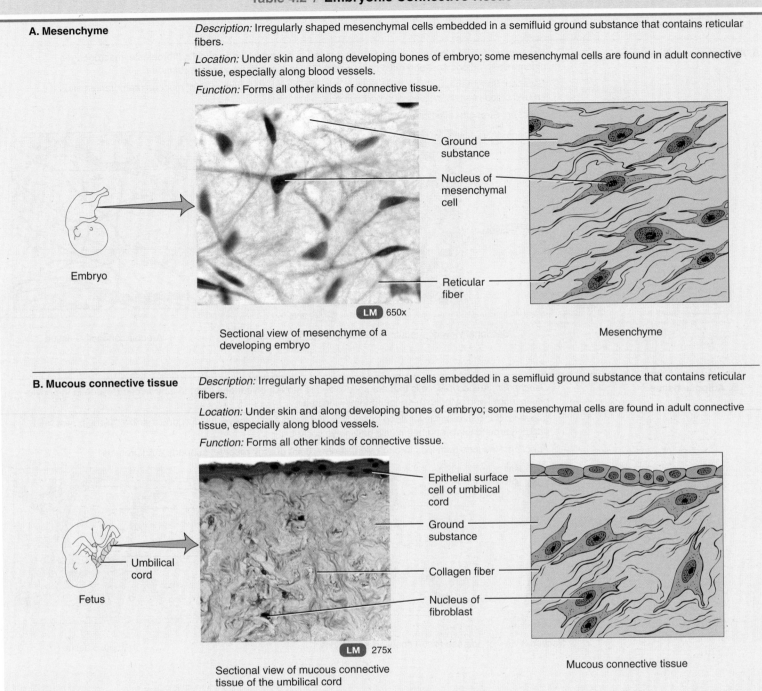

A. Mesenchyme

Description: Irregularly shaped mesenchymal cells embedded in a semifluid ground substance that contains reticular fibers.

Location: Under skin and along developing bones of embryo; some mesenchymal cells are found in adult connective tissue, especially along blood vessels.

Function: Forms all other kinds of connective tissue.

Embryo

Ground substance

Nucleus of mesenchymal cell

Reticular fiber

LM 650x

Sectional view of mesenchyme of a developing embryo

Mesenchyme

B. Mucous connective tissue

Description: Irregularly shaped mesenchymal cells embedded in a semifluid ground substance that contains reticular fibers.

Location: Under skin and along developing bones of embryo; some mesenchymal cells are found in adult connective tissue, especially along blood vessels.

Function: Forms all other kinds of connective tissue.

Fetus

Umbilical cord

Epithelial surface cell of umbilical cord

Ground substance

Collagen fiber

Nucleus of fibroblast

LM 275x

Sectional view of mucous connective tissue of the umbilical cord

Mucous connective tissue

DENSE IRREGULAR CONNECTIVE TISSUE   This tissue contains collagen fibers that are usually *irregularly* arranged (Table 4.3E), and it is found in parts of the body where pulling forces are exerted in various directions. The tissue usually occurs in sheets, such as in the dermis of the skin, which underlies the epidermis. Heart valves, the perichondrium (the membrane surrounding cartilage), and the periosteum (the membrane surrounding bone) are considered dense irregular connective tissues, despite a fairly orderly arrangement of their collagen fibers.

ELASTIC CONNECTIVE TISSUE   Freely branching elastic fibers predominate in elastic connective tissue (Table 4.3F), giving the unstained tissue a yellowish color. Fibroblasts are present in the spaces between the fibers. Elastic connective tissue is quite strong and can recoil to its original shape after being stretched. Elasticity is important to the normal functioning of lung tissue, which recoils as you exhale, and elastic arteries, whose recoil between heart beats helps maintain blood flow.

Table 4.3 / Mature Connective Tissue

Loose Connective Tissue

A. Areolar connective tissue

Description: Fibers (collagen, elastic, and reticular) and several kinds of cells (fibroblasts, macrophages, plasma cells, adipocytes, and mast cells) embedded in a semifluid ground substance.

Location: Subcutaneous layer deep to skin; superficial part of dermis of skin; mucous membranes; and around blood vessels, nerves, and body organs.

Function: Strength, elasticity, and support.

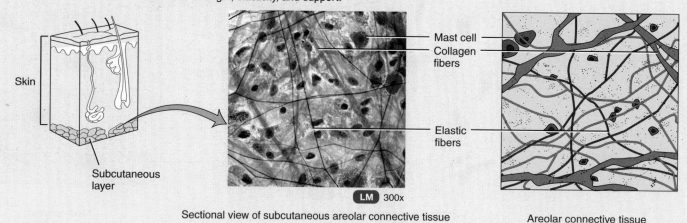

Skin

Subcutaneous layer

Mast cell
Collagen fibers

Elastic fibers

LM 300x

Sectional view of subcutaneous areolar connective tissue

Areolar connective tissue

B. Adipose tissue

Description: Adipocytes, cells that are specialized to store triglycerides (fats) in a large central area in their cytoplasm; nuclei and cytoplasm are peripherally located.

Location: Subcutaneous layer deep to skin, around heart and kidneys, yellow bone marrow of long bones, and padding around joints and behind eyeball in eye socket.

Function: Reduces heat loss through skin, serves as an energy reserve, supports, and protects.

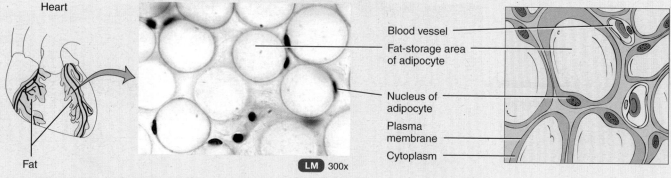

Heart

Fat

Blood vessel
Fat-storage area of adipocyte

Nucleus of adipocyte
Plasma membrane
Cytoplasm

LM 300x

Sectional view of adipose tissue showing adipocytes

Adipose tissue

Cartilage

Cartilage consists of a dense network of collagen fibers and elastic fibers firmly embedded in chondroitin sulfate, a rubbery component of the ground substance. Cartilage can endure considerably more stress than loose and dense connective tissues.

Whereas the strength of cartilage is due to its collagen fibers, its resilience (ability to assume its original shape after deformation) is due to chondroitin sulfate.

The cells of mature cartilage, called ***chondrocytes*** (KON-drō-sīts; *chondro-* = cartilage), occur singly or in groups within spaces called ***lacunae*** (la-KOO-nē = little lakes; singular is *la-*

Loose Connective Tissue

C. Reticular connective tissue

Description: A network of interlacing reticular fibers and reticular cells.

Location: Stroma (supporting framework) of liver, spleen, lymph nodes; portion of bone marrow that gives rise to blood cells; basement membrane; and around blood vessels and muscle.

Function: Forms stroma of organs; binds together smooth muscle tissue cells.

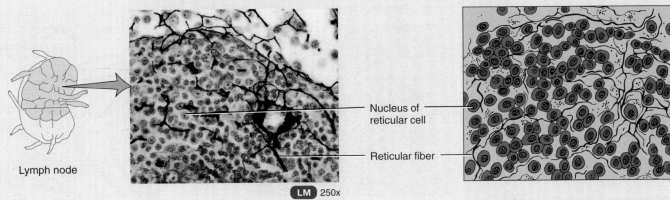

Lymph node

Nucleus of
reticular cell

Reticular fiber

LM 250x

Sectional view of reticular connective tissue
of a lymph node

Reticular connective tissue

Dense Connective Tissue

D. Dense regular connective tissue

Description: Matrix looks shiny white; consists predominantly of collagen fibers arranged in bundles; fibroblasts present in rows between bundles.

Location: Forms tendons (attach muscle to bone), most ligaments (attach bone to bone), and aponeuroses (sheet-like tendons that attach muscle to muscle or muscle to bone).

Function: Provides strong attachment between various structures.

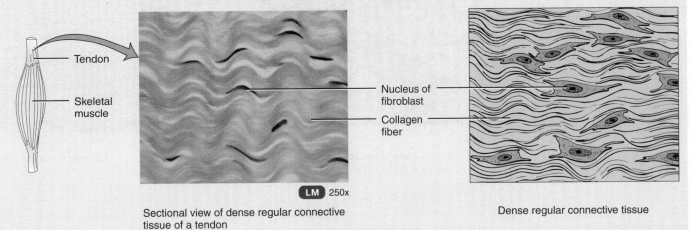

Tendon

Skeletal
muscle

Nucleus of
fibroblast

Collagen
fiber

LM 250x

Sectional view of dense regular connective
tissue of a tendon

Dense regular connective tissue

(continues)

cuna) in the matrix. The surface of most cartilage is surrounded by a membrane of dense irregular connective tissue called the ***perichondrium*** (per′-i-KON-drē-um; *peri-* = around). Unlike other connective tissues, cartilage has no blood vessels or nerves, except in the perichondrium. The three types of cartilage are hyaline cartilage, fibrocartilage, and elastic cartilage.

HYALINE CARTILAGE This type of cartilage contains a resilient gel as its ground substance and appears in the body as a bluish-white, shiny substance. The fine collagen fibers are not visible with ordinary staining techniques, and prominent chondrocytes are found in lacunae (Table 4.3G). Most hyaline cartilage is surrounded by a perichondrium. The exceptions are the articular

Table 4.3 / Mature Connective Tissue (continued)

Dense Connective Tissue

E. Dense irregular connective tissue

Description: Randomly arranged collagen fibers along with a few fibroblasts.

Location: Fasciae (tissue beneath skin and around muscles and other organs), deeper part of dermis of skin, periosteum (membrane) of bone, perichondrium of cartilage, joint capsules, membrane capsules around various organs (kidneys, liver, testes, lymph nodes), pericardium of the heart, and heart valves.

Function: Provides strength.

LM 275x

Sectional view of dense irregular connective tissue of inner region of dermis

Dense irregular connective tissue

F. Elastic connective tissue

Description: Freely branching elastic fibers with fibroblasts distributed in between.

Location: Lung tissue, walls of elastic arteries, trachea (windpipe), bronchial tubes, true vocal cords, suspensory ligament of penis, and ligaments between vertebrae.

Function: Allows stretching of various organs.

LM 335x

Sectional view of elastic connective tissue of aorta

Elastic connective tissue

cartilage in joints and at the epiphyseal plates, the regions where bones lengthen as a person grows. Hyaline cartilage is the most abundant cartilage in the body. It affords flexibility and support and, at joints, reduces friction and absorbs shock. Hyaline cartilage is the weakest of the three types of cartilage.

FIBROCARTILAGE Chondrocytes are scattered among clearly visible bundles of collagen fibers within the matrix of this type

of fibrocartilage (Table 4.3H). Fibrocartilage lacks a perichondrium. This tissue combines strength and rigidity and is the strongest of the three types of cartilage. One location of fibrocartilage is in the discs between vertebrae.

ELASTIC CARTILAGE In elastic cartilage, chondrocytes are located within a threadlike network of elastic fibers within the matrix (Table 4.3I). A perichondrium is present. Elastic cartilage

Cartilage

G. Hyaline cartilage

Description: Bluish-white, shiny ground substance with fine collagen fibers; contains numerous chondrocytes; most abundant type of cartilage.

Location: Ends of long bones, anterior ends of ribs, nose, parts of larynx (voice box), trachea, bronchi, bronchial tubes, and embryonic skeleton.

Function: Provides smooth surfaces for movement at joints, as well as flexibility and support.

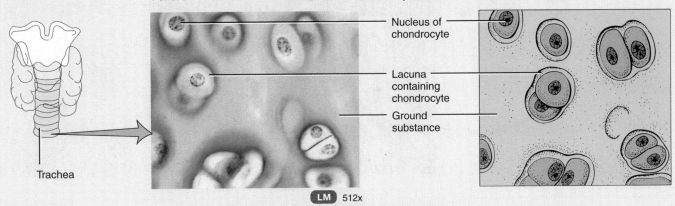

Trachea

LM 512x

Nucleus of chondrocyte

Lacuna containing chondrocyte

Ground substance

Sectional view of hyaline cartilage of trachea

Hyaline cartilage

H. Fibrocartilage

Description: Consists of chondrocytes scattered among bundles of collagen fibers within the matrix.

Location: Pubic symphysis (point where hip bones join anteriorly), intervertebral discs (discs between vertebrae), menisci (cartilage pads) of knee, and portions of tendons that insert into cartilage.

Function: Support and rigidity.

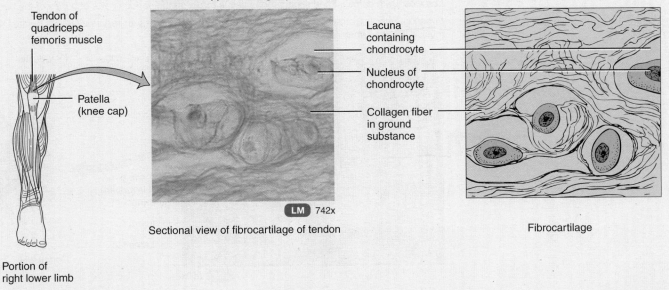

Tendon of quadriceps femoris muscle

Patella (knee cap)

Portion of right lower limb

LM 742x

Lacuna containing chondrocyte

Nucleus of chondrocyte

Collagen fiber in ground substance

Sectional view of fibrocartilage of tendon

Fibrocartilage

(continues)

provides strength and elasticity and maintains the shape of certain structures, such as the external ear.

Bone Tissue

Bones are organs composed of several different connective tissues, including **bone** or **osseous tissue** (OS-ē-us), the periosteum,

red and yellow bone marrow, and the endosteum (a membrane that lines a space within bone that stores yellow bone marrow). Bone tissue is classified as either compact or spongy, depending on how the matrix and cells are organized.

The basic unit of compact bone is an **osteon** or **Haversian system** (Table 4.3J). Each osteon is composed of four parts:

Table 4.3 / **Mature Connective Tissue (continued)**

Cartilage

I. Elastic cartilage

Description: Chondrocytes located in a threadlike network of elastic fibers within the matrix.

Location: Lid on top of larynx (epiglottis), external ear (auricle), and auditory (eustachian) tubes.

Function: Gives support and maintains shape.

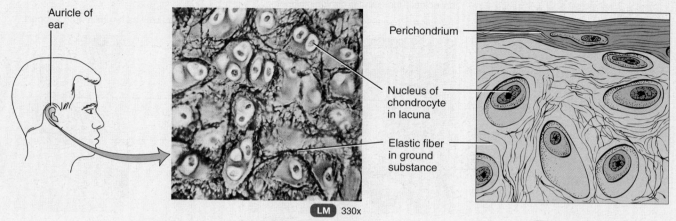

Sectional view of elastic cartilage of auricle of ear

Elastic cartilage

Bone Tissue

J. Compact bone

Description: Compact bone tissue consists of osteons (Haversian systems) that contain lamellae, lacunae, osteocytes, canaliculi, and central (Haversian) canals. By contrast, spongy bone tissue consists of thin columns called trabeculae; spaces between trabeculae are filled with red bone marrow.

Location: Both compact and spongy bone tissue make up the various parts of bones of the body.

Function: Support, protection, storage; houses blood-forming tissue; serves as levers that act together with muscle tissue to enable movement.

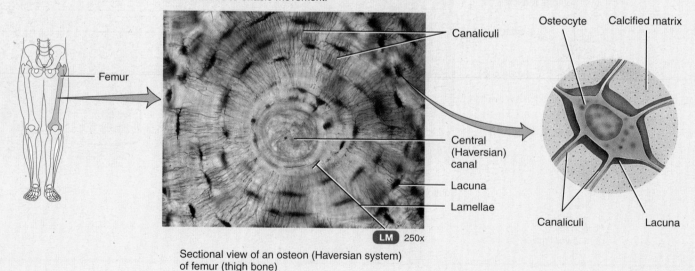

Sectional view of an osteon (Haversian system) of femur (thigh bone)

- The **lamellae** (la-MEL-lē = little plates) are concentric rings of matrix that consist of mineral salts (mostly calcium and phosphates), which give bone its hardness, and collagen fibers, which give bone its strength.

- **Lacunae** are small spaces between lamellae that contain mature bone cells called **osteocytes.**

- Projecting from the lacunae are **canaliculi** (CAN-a-lik′-ū-lī = little canals), networks of minute canals containing the processes of osteocytes. Canaliculi provide routes for nutrients to reach osteocytes and for wastes to leave them.

- A **central (Haversian) canal** contains blood vessels and nerves.

Blood Tissue

K. Blood

Description: Plasma and formed elements—red blood cells (erythrocytes), white blood cells (leukocytes), and platelets.

Location: Within blood vessels (arteries, arterioles, capillaries, venules, and veins) and within the chambers of the heart.

Function: Red blood cells transport oxygen and carbon dioxide; white blood cells carry on phagocytosis and are involved in allergic reactions and immune system responses; platelets are essential for the clotting of blood.

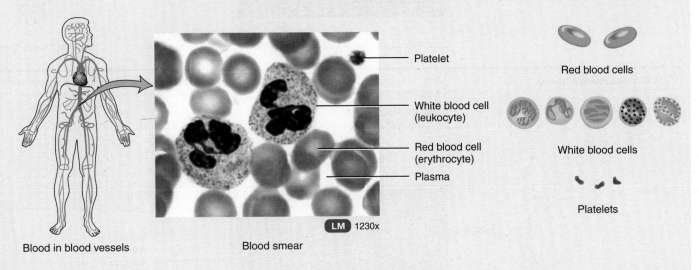

Blood in blood vessels

Blood smear

LM 1230x

Platelet

White blood cell (leukocyte)

Red blood cell (erythrocyte)

Plasma

Red blood cells

White blood cells

Platelets

Spongy bone has no osteons; instead, it consists of columns of bone called ***trabeculae*** (tra-BEK-ū-lē = little beams), which contain lamellae, osteocytes, lacunae, and canaliculi. Spaces between lamellae are filled with red bone marrow. Chapter 6 presents details of bone tissue histology.

The skeletal system has several functions. It supports soft tissues, protects delicate structures, and works with skeletal mus-cles to generate movement. Bone stores calcium and phosphorus; stores red bone marrow, which produces blood cells; and houses yellow bone marrow, a storage site for triglycerides.

Blood Tissue

Blood tissue (or simply blood) is a connective tissue with a liquid matrix called ***plasma,*** a pale yellow fluid that consists mostly of

Table 4.4 / **Muscle Tissue**

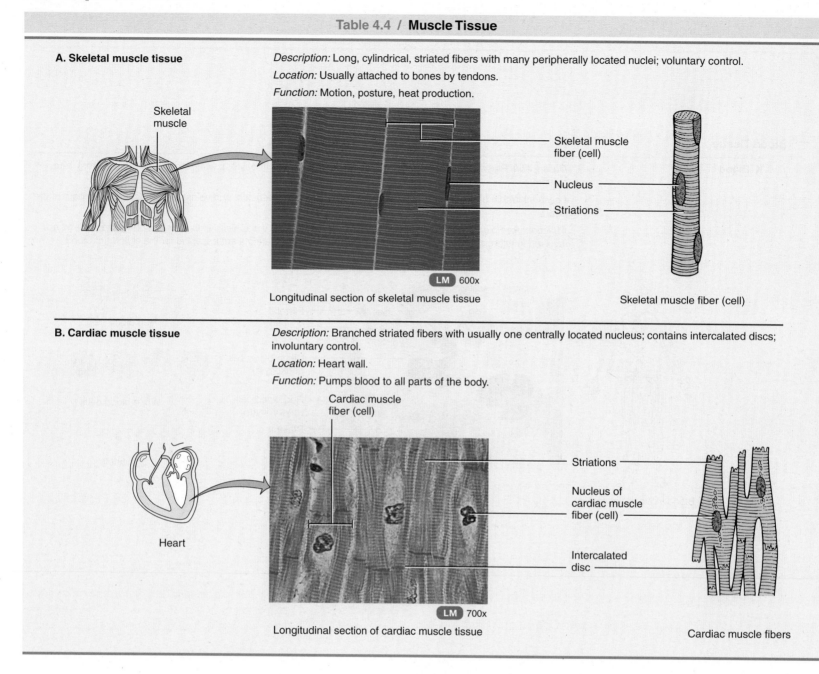

A. Skeletal muscle tissue

Description: Long, cylindrical, striated fibers with many peripherally located nuclei; voluntary control.
Location: Usually attached to bones by tendons.
Function: Motion, posture, heat production.

Skeletal muscle

Skeletal muscle fiber (cell)

Nucleus

Striations

LM 600x

Longitudinal section of skeletal muscle tissue

Skeletal muscle fiber (cell)

B. Cardiac muscle tissue

Description: Branched striated fibers with usually one centrally located nucleus; contains intercalated discs; involuntary control.
Location: Heart wall.
Function: Pumps blood to all parts of the body.

Cardiac muscle fiber (cell)

Heart

Striations

Nucleus of cardiac muscle fiber (cell)

Intercalated disc

LM 700x

Longitudinal section of cardiac muscle tissue

Cardiac muscle fibers

water with a wide variety of dissolved substances: nutrients, wastes, enzymes, hormones, respiratory gases, and ions (Table 4.3K). Suspended in the plasma are red blood cells, white blood cells, and platelets. ***Red blood cells*** transport oxygen to body cells and remove carbon dioxide from them. ***White blood cells*** are involved in phagocytosis, immunity, and allergic reactions. ***Platelets*** participate in blood clotting. The details of blood are considered in Chapter 14.

Lymph

Lymph is a fluid that flows in lymphatic vessels. It is a connective tissue that consists of a clear fluid similar to plasma but with much less protein. In the lymph are various cells and chemicals, whose composition varies from one part of the body to another.

For example, lymph leaving lymph nodes includes many lymphocytes, a type of white blood cell, whereas lymph from the small intestine has a high content of newly absorbed dietary lipids. The details of lymph are considered in Chapter 17.

MUSCLE TISSUE

Objective: • **Describe the general features of muscle tissue, and contrast the structure, location, and mode of control of the three types of muscle tissue.**

Muscle tissue consists of cells that are highly specialized to generate force. As a result of this characteristic, muscle tissue produces motion, maintains posture, and generates heat. Based on

C. Smooth muscle tissue

Description: Spindle-shaped, nonstriated fibers with one centrally located nucleus; involuntary control.

Location: Walls of hollow internal structures such as blood vessels, airways to the lungs, stomach, intestines, gallbladder, urinary bladder, and uterus.

Function: Motion: constriction of blood vessels and airways, propulsion of foods through gastrointestinal tract, contraction of urinary bladder and gallbladder.

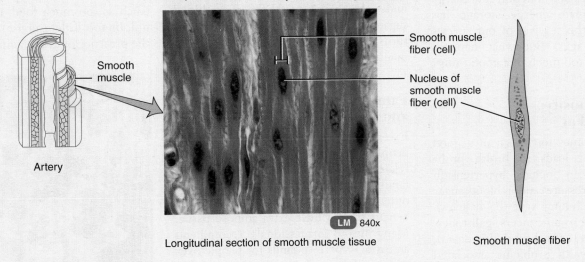

Artery

Smooth muscle

Smooth muscle fiber (cell)

Nucleus of smooth muscle fiber (cell)

LM 840x

Longitudinal section of smooth muscle tissue

Smooth muscle fiber

its location and certain structural and functional characteristics, muscle tissue is classified into three types: skeletal, cardiac, and smooth (Table 4.4).

Skeletal muscle tissue is named for its location—it is usually attached to the bones of the skeleton (Table 4.4A). It is also called *striated* muscle because the cells contain alternating light and dark bands called *striations* that are visible under a light microscope. Skeletal muscle is *voluntary* because it can be made to contract or relax by conscious control, for example, turning your head. A single skeletal muscle cell, called a **muscle fiber,** is very long, roughly cylindrical in shape, and has many nuclei located at its periphery.

Cardiac muscle tissue forms the bulk of the wall of the heart (Table 4.4B). Like skeletal muscle, it is striated. However, unlike skeletal muscle tissue, it is *involuntary:* its contraction is not con-

sciously controlled. Cardiac muscle fibers are branched (Y-shaped) and have one centrally located nucleus. Cardiac muscle fibers are attached end to end by transverse thickenings of the plasma membrane called **intercalated discs** (*intercalat-* = to insert between). The intercalated discs strengthen the tissue and hold the fibers together during vigorous contractions; they also provide a route for electrical signals to travel throughout the heart.

Smooth muscle tissue is located in the walls of hollow internal structures such as blood vessels, airways to the lungs, the stomach, intestines, gallbladder, and urinary bladder (Table 4.4C). Its contractions help constrict (narrow) the lumen of blood vessels, physically break down and move food along the gastrointestinal tract, move fluids through the body, and eliminate wastes. Smooth muscle fibers are usually *involuntary,* and

*A*dipose tissue contains adipocytes, cells specialized for the function of energy storage. Adequate energy storage has been vital to the survival of our species through the ages. But what happens when we have too much of a good thing?

Excess Adiposity: Too Much Fat

Adiposity becomes too much of a good thing when it leads to health problems, which can include hypertension (high blood pressure), poor blood sugar regulation (including type II diabetes), heart disease, certain cancers, gallstones, arthritis, and lower backaches. In general, the risk of health problems associated with excess adipose tissue increases in a dose-dependent fashion: the greater the excess weight, the greater the risk. People with a great deal of excess fat have a much higher risk for health problems than people who are only slightly too fat. But the health effects of excess fat de-

pend on several important factors besides quantity of adipose tissue.

Location of Adipose Tissue

People who carry extra fat on the torso are at greater risk for hypertension, type II diabetes, and artery disease than people whose extra fat resides in the hips and thighs. Especially risky is excess fat stored around the viscera (abdominal organs). Adipocytes in this area appear to be more "metabolically active" than those under the skin. Visceral fat affects blood sugar and blood fat regulation, which in turn can lead to the health problems mentioned above.

Family Medical History and Age

Excess body fat is especially risky for people who have already developed fat-related health problems, or who have a family history of these disorders. On the other hand, people over 70 years old may benefit from a little extra adipose tissue. Many health professionals recommend an extra 10 or 15 pounds for people over

70 to help them resist wasting if they should become ill.

Beware the Deadly Sins: Gluttony and Sloth

People with a moderate amount of excess adipose tissue who eat a healthful diet and exercise regularly have health risks similar to those of their leaner peers. This observation suggests that some of the health risks seen in overweight people may be caused by poor health habits (such as too much food, too much alcohol, or too little exercise) rather than by the presence of excess adipose tissue.

▶ **Think It Over**

▶ Why do you think body weight alone is not always a good measure of adiposity, or a good predictor of health risks associated with excess adipose tissue?

---

they are *nonstriated* (hence the term *smooth*). A smooth muscle fiber is small and spindle-shaped (thickest in the middle, with each end tapering), with one centrally located nucleus.

NERVOUS TISSUE

Objective: • **Describe the structural features and functions of nervous tissue.**

Despite the awesome complexity of the nervous system, it consists of only two principal kinds of cells: neurons and neuroglia. *Neurons,* or nerve cells, are sensitive to various stimuli. They

convert stimuli into nerve impulses (action potentials) and conduct these impulses to other neurons, to muscle fibers, or to glands. Most neurons consist of three basic portions: a cell body and two kinds of cell processes, dendrites and axons (Table 4.5). The *cell body* contains the nucleus and other organelles. Each neuron has multiple *dendrites* (*dendr-* = tree), cell processes that are tapering, highly branched, and usually short. Their function is to receive input from sensory receptors (see page 93) or from other neurons. The *axon* of a neuron is a single, thin, cylindrical process that may be very long. It is the output portion of a neuron, conducting nerve impulses toward another neuron or to some other tissue.

Neuroglia (noo-RŌG-lē-a; *-glia* = glue) do not generate or conduct nerve impulses, but they do have many other important

Table 4.5 / Nervous Tissue

Nervous tissue

Description: Neurons (nerve cells) and neuroglia. Neurons consist of a cell body and processes extending from the cell body, multiple dendrites, and a single axon. Neuroglia do not generate or conduct nerve impulses but have other important functions.

Location: Nervous system.

Function: Exhibits sensitivity to various types of stimuli; converts stimuli into nerve impulses; and conducts nerve impulses to other neurons, muscle fibers, or glands.

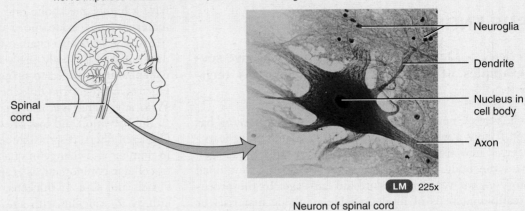

Neuron of spinal cord

functions (see Table 9.1 on page 222). The detailed structure and function of neurons and neuroglia are considered in Chapter 9.

MEMBRANES

Objective: • **Define a membrane, and describe the classification of membranes.**

Membranes are flat sheets of pliable tissue that cover or line a part of the body. The combination of an epithelial layer and an underlying connective tissue layer constitutes an **epithelial membrane.** The principal epithelial membranes of the body are mucous membranes; serous membranes; and the cutaneous membrane, or skin. (Skin is discussed in detail in Chapter 5 and will not be discussed here.) Another kind of membrane, a **synovial membrane,** lines joints and contains connective tissue but no epithelium.

Mucous Membranes

A **mucous membrane** or **mucosa** (myoo-KŌ-sa) lines a body cavity that opens directly to the exterior. Mucous membranes line the entire digestive, respiratory, and reproductive systems and much of the urinary system. Like all epithelial membranes, they consist of both a lining layer of epithelium and an underlying layer of connective tissue (see mucosa in Figure 19.2 on page 000). The epithelial layer of a mucous membrane secretes mucus, which prevents the cavities from drying out. It also traps

particles in the respiratory passageways, lubricates and absorbs food as it moves through the gastrointestinal tract, and secretes digestive enzymes. The connective tissue layer helps bind the epithelium to the underlying structures. It also provides the epithelium with oxygen and nutrients and removes wastes via its blood vessels.

Serous Membranes

A **serous membrane** (*serous* = watery) lines a body cavity that does not open directly to the exterior, and it also covers the organs that lie within the cavity. Serous membranes consist of two parts: a parietal layer and a visceral layer. The **parietal layer** (pa-RĪ-e-tal; *paries* = wall) is the part attached to the cavity wall, and the **visceral layer** (*viscer-* = body organ) is the part that covers and attaches to the organs inside these cavities. Each layer consists of areolar connective tissue sandwiched between two sheets of **mesothelium.** Mesothelium is a simple squamous epithelium. It secretes **serous fluid,** a watery lubricating fluid that allows organs to glide easily over one another or to slide against the walls of cavities.

The serous membrane lining the thoracic cavity and covering the lungs is the **pleura.** The serous membrane lining the heart cavity and covering the heart is the **pericardium.** The serous membrane lining the abdominal cavity and covering the abdominal organs is the **peritoneum.**

Synovial Membranes

Synovial membranes (si-NŌ-vē-al) line the cavities of some joints (see Figure 7.3 on page 156). They are composed of areolar connective tissue with elastic fibers and varying amounts of fat; they do not have an epithelial layer. Synovial membranes secrete *synovial fluid,* which lubricates the ends of bones as they move at joints and nourishes the articular cartilage covering the bones.

TISSUE ENGINEERING

Objective: • **Define tissue engineering and give several examples of tissues developed from this technology.**

In recent years a new technology called *tissue engineering* has emerged. Using a wide range of materials, scientists are learning to grow new tissues in the laboratory to replace damaged tissues in the body. Tissue engineers have already developed laboratory-grown versions of skin and cartilage. In the procedure, scaffolding beds of biodegradable synthetic materials or collagen are used as substrates that permit body cells such as skin cells or cartilage cells to be cultured. As the cells divide and assemble, the scaffolding degrades, and the new, permanent tissue is then implanted in the patient. Other structures being developed by tissue engineers include bones, tendons, heart valves, bone marrow, and intestines. Work is also underway to develop insulin-producing cells for the pancreases of diabetics, dopamine-producing cells for the brains of Parkinson's disease patients, and even entire livers and kidneys.

AGING AND TISSUES

Objective: • **Describe the effects of aging on tissues.**

Generally, tissues heal faster and leave less obvious scars in the young than in the aged. In fact, surgery performed on fetuses leaves no scars. The younger body is generally in a better nutritional state, its tissues have a better blood supply, and its cells have a higher metabolic rate. Thus, cells can synthesize needed materials and divide more quickly. The extracellular components of tissues also change with age. Glucose, the most abundant sugar in the body, plays a role in the aging process. Glucose is haphazardly added to proteins inside and outside cells, forming irreversible cross-links between adjacent protein molecules. With advancing age, more cross-links form, which contributes to the stiffening and loss of elasticity that occur in aging tissues. Collagen fibers, responsible for the strength of tendons, increase in number and change in quality with aging. Elastin, another extracellular component, is responsible for the elasticity of blood vessels and skin. It thickens, fragments, and acquires a greater affinity for calcium with age—changes that may also be associated with the development of atherosclerosis, the deposition of fatty materials in arterial walls.

• • •

Now that you have an understanding of tissues, we will look at the organization of tissues into organs and organs into systems. In the next chapter we will consider how the skin and other organs function as components of the integumentary system.

COMMON DISORDERS

Disorders of epithelial tissues are mainly specific to individual organs, for example, skin cancer, which involves cells in the epidermis, or peptic ulcer disease (PUD), which erodes the epithelial lining of the stomach or small intestine. For this reason, epithelial disorders are described together with the relevant body system throughout the text. The most prevalent disorders of connective tissues are *autoimmune diseases,* that is, diseases in which antibodies produced by the immune system fail to distinguish what is foreign from what is self and attack the body's own tissues. The most common autoimmune disorder is rheumatoid arthritis, which attacks the synovial membranes of joints. Because connective tissue is the most abundant and widely distributed of the four principal types of tissues, its disorders tend to affect multiple body systems. Common disorders of muscle tissue and nervous tissue are considered at the ends of Chapters 8 and 9, respectively.

Sjögren's Syndrome

Sjögren's syndrome (SHŌ-grenz) is a common autoimmune disorder that causes inflammation and destruction of glands, especially the lacrimal (tear) glands and salivary glands. It is characterized by dryness of the mucous membranes in the eyes and mouth and salivary gland enlargement. Systemic effects include arthritis, difficulty swallowing, pancreatitis, pleuritis, and migraine headaches. The disorder affects more females than males by a ratio of 9:1. About 20% of older adults experience some signs of Sjögren's syndrome, which also is associated with an increased risk of developing malignant lymphoma. Treatment is supportive, including using artificial tears to moisten the eyes and sipping fluids, chewing sugarless gum, and using a saliva substitute to moisten the mouth.

Systemic Lupus Erythematosus

Systemic lupus erythematosus (er-ith'-e-ma-TŌ-sus), *SLE,* or simply lupus, is a chronic inflammatory disease of connective tissue, occurring mostly in nonwhite women during their childbearing years. SLE is an autoimmune disease that can affect every body system. The ensuing inflammatory response causes tissue damage. The disease can range from a mild condition in most patients to a rapidly fatal disease. It is marked by periods of exacerbation and remission. The prevalence of SLE is about 1 in 2000 persons, with females more likely to be afflicted than males by a ratio of about 9:1. Although the cause of SLE is not known, genetic, environmental, and hormonal factors are implicated.

Signs and symptoms of SLE include painful joints, low-grade fever, fatigue, mouth ulcers, weight loss, enlarged lymph nodes and spleen, sensitivity to sunlight, rapid loss of large amounts of scalp hair, and loss of appetite. A distinguishing feature of lupus is an eruption across the bridge of the nose and cheeks called a "butterfly rash." Other skin lesions may occur with blistering and ulceration. The erosive nature of some SLE skin lesions was thought to resemble the damage in-flicted by the bite of a wolf, thus, the term *lupus* (= wolf). The most serious complications of the disease involve inflammation of the kidneys, liver, spleen, lungs, heart, brain, and gastrointestinal tract. Because there is no cure for SLE, treatment is supportive, including anti-inflammatory drugs, such as aspirin, and immunosuppressive drugs.

MEDICAL TERMINOLOGY AND CONDITIONS

Fibrosis   A process in which fibroblasts synthesize collagen and other matrix materials to form scar tissue to replace a damaged tissue. Because scar tissue is not specialized to perform the functions of the original tissue, the function of the tissue or organ is impaired.

Liposuction   (*lip-* = fat) or *suction lipectomy* (*-ectomy* = cut out) A surgical procedure that involves suctioning out small amounts of fat from various areas of the body; used as a body-contouring procedure in areas such as the thighs, buttocks, arms, breasts, and abdomen. Possible complications include fat emboli (clots), infection, fluid depletion, injury to internal structures, severe postoperative pain, or a result that is unsatisfactory to the client.

Papanicolaou smear   (pa′-pa-NI-kō-lō) or *Pap smear* A procedure that involves collecting and microscopically examining cells that have been sloughed off the surface of a tissue. A very common type of Pap smear involves examining the cells present in the secretions of the cervix and vagina. This type of Pap smear is performed mainly to detect early changes in the cells of the female reproductive system that may indicate cancer or a precancerous condition. Annual Pap smears are recommended for women over age 18 (or earlier if sexually active) as part of a routine pelvic exam.

Tissue transplantation   The replacement of a diseased or injured tissue or organ; the most successful transplants involve use of a person's own tissues or those from an identical twin.

Xenotransplantation   (zen′-ō-trans′-plan-TĀ-shun; *xeno-* = strange, foreign) The replacement of a diseased or injured tissue or organ with cells or tissues from an animal. Only a few cases of successful xenotransplantation exist to date.

■ STUDY OUTLINE

Types of Tissues (p. 72)

1. A tissue is a group of similar cells that usually has a similar embryological origin and is specialized for a particular function.

2. The various tissues of the body are classified into four basic types: epithelial, connective, muscle, and nervous.

Epithelial Tissue (p. 72)

1. The general types of epithelia include covering and lining epithelium and glandular epithelium.

2. Some general characteristics of epithelium: It consists mostly of cells with little extracellular material, is arranged in sheets, is attached to connective tissue by a basement membrane, is avascular (no blood vessels), has a nerve supply, and can replace itself.

3. Cell shapes include squamous (flat), cuboidal (cubelike), columnar (rectangular), and transitional (variable); layers are arranged as simple (one layer), stratified (several layers), and pseudostratified (one layer that appears to be several).

4. Simple squamous epithelium consists of a single layer of flat cells (Table 4.1A). It is found in parts of the body where filtration or diffusion are priority processes. One type, endothelium, lines the heart and blood vessels. Another type, mesothelium, forms the serous membranes that line the thoracic and abdominopelvic cavities and cover the organs within them.

5. Simple cuboidal epithelium consists of a single layer of cube-shaped cells that function in secretion and absorption (Table 4.1B). It is found covering the ovaries, in the kidneys and eyes, and lining some glandular ducts.

6. Nonciliated simple columnar epithelium consists of a single layer of nonciliated rectangular cells (Table 4.1C). It lines most of the gastrointestinal tract. Specialized cells containing microvilli perform absorption. Goblet cells secrete mucus.

7. Ciliated simple columnar epithelium consists of a single layer of ciliated rectangular cells (Table 4.1D). It is found in a few portions of the upper respiratory tract, where it moves foreign particles trapped in mucus out of the respiratory tract.

8. Stratified squamous epithelium consists of several layers of cells; superficial cells are flat (Table 4.1E). It is protective. A nonkeratinized variety lines the mouth; a keratinized variety forms the epidermis, the most superficial layer of the skin.

9. Stratified cuboidal epithelium consists of several layers of cells; superficial cells are cube-shaped (Table 4.1F). It is found in adult sweat glands and a portion of the male urethra.

10. Stratified columnar epithelium consists of several layers of cells; superficial cells are rectangular (Table 4.1G). It is found in a portion of the male urethra and large excretory ducts of some glands.

11. Transitional epithelium consists of several layers of cells whose appearance varies with the degree of stretching (Table 4.1H). It lines the urinary bladder.

12. Pseudostratified columnar epithelium has only one layer but gives the appearance of many (Table 4.1I). A ciliated variety contains goblet cells and lines most of the upper respiratory tract; a nonciliated variety has no goblet cells and lines ducts of many glands, the epididymis, and part of the male urethra.

13. A gland is a single cell or a group of epithelial cells adapted for secretion.

14. Endocrine glands secrete hormones into the blood (Table 4.1J).

15. Exocrine glands (mucous, sweat, oil, and digestive glands) secrete into ducts or directly onto a free surface (Table 4.1K).

Connective Tissue (p. 80)

1. Connective tissue is the most abundant body tissue.

2. Connective tissue consists of cells and a matrix of ground substance and fibers; it has abundant matrix with relatively few cells. It does not usually occur on free surfaces, has a nerve supply (except for cartilage), and is highly vascular (except for cartilage, tendons, and ligaments).

3. Cells in connective tissue include fibroblasts (secrete matrix), macrophages (perform phagocytosis), plasma cells (secrete antibodies), mast cells (produce histamine), adipocytes (store fat), and white blood cells (migrate from blood in response to infections).

4. The ground substance and fibers make up the matrix.

5. The ground substance supports and binds cells together, provides a medium for the exchange of materials, and is active in influencing cell functions.

6. Substances found in the ground substance include hyaluronic acid, chondroitin sulfate, and adhesion proteins.

7. The fibers in the matrix provide strength and support and are of three types: (a) collagen fibers (composed of collagen) are found in large amounts in bone, tendons, and ligaments; (b) elastic fibers (composed of elastin, fibrillin, and other glycoproteins) are found in skin, blood vessel walls, and lungs; and (c) reticular fibers (composed of collagen and glycoprotein) are found around fat cells, nerve fibers, and skeletal and smooth muscle cells.

8. The two major subclasses of connective tissue are embryonic connective tissue and mature connective tissue.

9. Mesenchyme is the embryonic connective tissue that gives rise to all mature connective tissues (Table 4.2A).

10. Mucous connective tissue is found in the umbilical cord of the fetus, where it gives support (Table 4.2B).

11. Mature connective tissue is subdivided into loose connective tissue, dense connective tissue, cartilage, bone tissue, blood tissue, and lymph.

12. Loose connective tissue includes areolar connective tissue, adipose tissue, and reticular connective tissue.

13. Areolar connective tissue consists of the three types of fibers, several cells, and a semifluid ground substance (Table 4.3A). It is found in the subcutaneous layer; in mucous membranes; and around blood vessels, nerves, and body organs.

14. Adipose tissue consists of adipocytes, which store triglycerides (Table 4.3B). It is found in the subcutaneous layer, around organs, and in the yellow bone marrow of long bones.

15. Reticular connective tissue consists of reticular fibers and reticular cells and is found in the liver, spleen, and lymph nodes (Table 4.3C).

16. Dense connective tissue includes dense regular connective tissue, dense irregular connective tissue, and elastic connective tissue.

17. Dense regular connective tissue consists of parallel bundles of collagen fibers and fibroblasts (Table 4.3D). It forms tendons, most ligaments, and aponeuroses.

18. Dense irregular connective tissue consists of usually randomly arranged collagen fibers and a few fibroblasts (Table 4.3E). It is found in fasciae, the dermis of skin, and membrane capsules around organs.

19. Elastic connective tissue consists of branching elastic fibers and fibroblasts (Table 4.3F). It is found in the walls of large arteries, lungs, trachea, and bronchial tubes.

20. Cartilage contains chondrocytes and has a rubbery matrix (chondroitin sulfate) containing collagen and elastic fibers.

21. Hyaline cartilage is found in the embryonic skeleton, at the ends of bones, in the nose, and in respiratory structures (Table 4.3G). It is flexible, allows movement, and provides support.

22. Fibrocartilage is found in the pubic symphysis, intervertebral discs, and menisci (cartilage pads) of the knee joint (Table 4.3H).

23. Elastic cartilage maintains the shape of organs such as the epiglottis of the larynx, auditory (Eustachian) tubes, and external ear (Table 4.3I).

24. Bone or osseous tissue consists of mineral salts and collagen fibers that contribute to the hardness of bone, and cells called osteocytes that are located in lacunae (Table 4.3J). It supports, protects, helps provide movement, stores minerals, and houses blood-forming tissue.

25. Blood tissue consists of plasma, red blood cells, white blood cells, and platelets (Table 4.3K). Its components function to transport oxygen and carbon dioxide, carry on phagocytosis, participate in allergic reactions, provide immunity, and bring about blood clotting.

26. Lymph flows in lymphatic vessels and is a clear fluid whose composition varies from one part of the body to another.

Muscle Tissue (p. 90)

1. Muscle tissue consists of cells (fibers) that are specialized for contraction. It provides motion, maintenance of posture, and heat production.

2. Skeletal muscle tissue is attached to bones and is striated; its action is voluntary (Table 4.4A).

3. Cardiac muscle tissue forms most of the heart wall and is striated; its action is involuntary (Table 4.4B).

4. Smooth muscle tissue is found in the walls of hollow internal structures (blood vessels and viscera) and is nonstriated; its action is involuntary (Table 4.4C).

Nervous Tissue (p. 92)

1. The nervous system is composed of neurons (nerve cells) and neuroglia (protective and supporting cells) (Table 4.5).

2. Neurons are sensitive to stimuli, convert stimuli into nerve impulses, and conduct nerve impulses.

3. Most neurons consist of a cell body and two types of processes called dendrites and axons.

Membranes (p. 93)

1. An epithelial membrane consists of an epithelial layer overlying a connective tissue layer. Examples are mucous, serous, and cutaneous membranes.

2. Mucous membranes line cavities that open to the exterior, such as the gastrointestinal tract.

3. Serous membranes line closed cavities (pleura, pericardium, peritoneum) and cover the organs in the cavities. These membranes consist of parietal and visceral layers.

4. Synovial membranes line joint cavities, bursae, and tendon sheaths and consist of areolar connective tissue instead of epithelium.

Tissue Engineeering (p. 94)

1. Tissue engineering refers to the growth of new tissues in the laboratory to replace damaged tissues in the body.

2. Laboratory versions of skin and cartilage have already been developed.

Aging and Tissues (p. 94)

1. Tissues heal faster and leave less obvious scars in the young than in the aged; surgery performed on fetuses leaves no scars.

2. The extracellular components of tissues, such as collagen and elastic fibers, also change with age.

■ SELF-QUIZ

1. Epithelial tissue functions in

a. conducting nerve impulses **b.** storing fat **c.** covering and lining the body and its parts **d.** movement **e.** storing minerals

2. Epithelial tissue is classified according to

a. its location **b.** its function **c.** the composition of the matrix **d.** the shape and arrangement of its cells **e.** whether it is under voluntary or involuntary control

3. Mucous membranes are

a. composed of two layers **b.** found in body cavities that open to the body's exterior **c.** located at the ends of bones **d.** found lining the thoracic cavity **e.** capable of producing synovial fluid

4. Which of the following is NOT a type of connective tissue?

a. blood **b.** adipose **c.** reticular **d.** cuboidal **e.** cartilage

5. Match the following tissue types with their descriptions.

____ **a.** fat storage
____ **b.** waterproofs the skin
____ **c.** forms the stroma (framework) of many organs
____ **d.** composes the intervertebral discs
____ **e.** stores red bone marrow, protects, supports
____ **f.** nonstriated, involuntary
____ **g.** involved in diffusion
____ **h.** found in kidney tubules, involved in absorption

A. simple cuboidal epithelium
B. simple squamous epithelium
C. adipose
D. fibrocartilage
E. reticular connective
F. smooth muscle
G. keratinized stratified squamous epithelium
H. bone

6. If you were going to design a hollow organ that needed to expand and have stretchability, which of the following epithelial and connective tissues might you use?

a. transitional epithelium and elastic connective tissue
b. stratified columnar epithelium and adipose tissue
c. simple columnar epithelium and dense regular connective tissue
d. simple squamous epithelium and hyaline cartilage
e. transitional epithelium and reticular connective tissue

7. Which of the following statements is NOT true concerning epithelial tissue?

a. The cells of epithelial tissue are closely packed.
b. The basal layer of cells rests on a basement membrane.
c. Epithelial tissue contains a nerve supply.
d. Epithelial tissue undergoes rapid rates of cell division.
e. Epithelial tissue is well supplied with blood vessels.

8. Which of the following is true concerning connective tissue?

a. Except for cartilage, connective tissue has a rich blood supply.
b. Connective tissue is classified according to cell shape and arrangement.
c. The cells of connective tissue are generally closely joined.
d. Loose connective tissue consists of many fibers arranged in a regular pattern.
e. The fibers in connective tissue are composed of lipids.

9. Where would you find smooth muscle tissue?

a. heart **b.** attached to the bones **c.** in joints **d.** the discs between the vertebrae **e.** in the walls of hollow organs

10. A connective tissue with a liquid matrix is

a. elastic cartilage **b.** blood **c.** areolar **d.** reticular **e.** osseous

11. The interior of your nose would be lined with

a. a mucous membrane **b.** smooth muscle tissue **c.** a synovial membrane **d.** keratinized stratified squamous epithelium **e.** a serous membrane

12. The four main types of tissue are

a. epithelial, embryonic, blood, nervous
b. blood, connective, muscle, nervous
c. connective, epithelial, muscle, nervous
d. stratified, muscle, striated, nervous
e. epithelial, connective, muscle, membranous

13. Which of the following materials would NOT be found in the matrix of connective tissue?

a. collagen fibers **b.** elastic fibers **c.** keratin **d.** reticular fibers **e.** fibronectin

14. Which connective tissue cells secrete antibodies?

a. mast cells **b.** adipocytes **c.** macrophages **d.** plasma cells **e.** chondrocytes

15. Modified columnar epithelial cells that secrete mucus are _____ cells.

a. microvilli **b.** keratinized **c.** mast **d.** fibroblast **e.** goblet

16. Which tissue is characterized by branching cells connected to each other by intercalated discs?

a. skeletal muscle **b.** nervous **c.** bone **d.** cardiac muscle **e.** smooth muscle

17. Stratified squamous epithelium functions in

a. protection and secretion **b.** contraction **c.** absorption **d.** stretching **e.** transport

18. What tissue type is found in tendons?
    a. dense irregular connective tissue
    b. elastic connective tissue
    c. dense regular connective tissue
    d. pseudostratified epithelium
    e. areolar tissue

19. In what tissue type would you find central (Haversian) canals, lacunae, and osteocytes?
    a. bone   b. hyaline cartilage   c. fibrocartilage   d. dense irregular connective tissue   e. elastic cartilage

20. Which of the following statements is true concerning glandular tissue?
    a. Endocrine glands are composed of connective tissue, whereas exocrine glands are composed of modified epithelium.
    b. Endocrine gland secretions diffuse directly into the bloodstream, whereas exocrine gland secretions enter ducts.
    c. A sweat gland is an example of an endocrine gland.
    d. Endocrine glands contain ducts, whereas exocrine glands do not.
    e. Exocrine glands produce substances known as hormones.

CRITICAL THINKING APPLICATIONS

1. The neighborhood kids are walking around with common pins and sewing needles stuck through their fingertips. There is no visible bleeding. What type of tissue have they pierced? (Be specific.) How do you know?

2. Tonia uses a line of cosmetics advertised as "all natural, no artificial anything." Her moisturizer includes "Wharton's jelly" in the ingredients list. What is Wharton's jelly? What kind of consistency would this add to the moisturizer?

3. Your lab partner Ned put a tissue slide labeled uterine tube under the microscope. He focused the slide and exclaimed, "Look! It's all hairy." Explain to Ned what the "hair" really is.

4. You've gone out to eat at your favorite fast-food joint, Goodbody's Fried Chicken Emporium. A health-food zealot grabs your chicken leg and declares, "This is all fat!" Using your knowledge of tissues, defend your dinner choice.

ANSWER TO FIGURE QUESTION

4.1 Fibroblasts secrete the fibers and ground substance of the connective tissue matrix.

Chapter 5

The Integumentary System

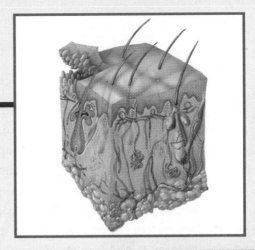

■ Student Learning Objectives

1. Describe the structure and functions of the skin. **100**
2. Explain the pigments involved in skin color. **103**
3. Describe the structure and functions of hair, skin glands, and nails. **103**
4. Describe the effects of aging on the integumentary system. **106**
5. Explain how the skin helps regulate body temperature. **107**

■ A Look Ahead

As you learned in Chapter 1, a group of tissues that performs a specific function is an **organ**. The next higher level of organization is a **system,** a group of organs operating together to perform specialized functions. The skin and its accessory structures, such as hair, nails, glands, and several specialized sensory receptors, constitute the **integumentary system** (in-teg′-yoo-MEN-tar-ē; *inte-* = whole; *-gument* = body covering).

Of all the body's organs, none is more easily inspected or more exposed to infection, disease, and injury than the skin. Because of its visibility, skin reflects our emotions and some aspects of normal physiology, as evidenced by frowning, blushing, and sweating. Changes in skin color or condition may indicate homeostatic imbalances in the body. For example, a skin rash such as occurs in chickenpox reveals a systemic infection, whereas a yellowing of the skin is an indication of jaundice, usually due to disease of the liver, an internal organ. Other disorders may be limited to the skin, such as warts, age spots, or pimples. The skin's location makes it vulnerable to damage from trauma, sunlight, microbes, or pollutants in the environment. Major damage to the skin, as occurs in third-degree burns, can be life threatening due to loss of protective skin functions.

Many interrelated factors may affect both the appearance and health of the skin, including nutrition, hygiene, circulation, age, immunity, genetic traits, psychological state, and drugs. So important is the skin to body image that people spend much time and money to restore it to a more youthful appearance.

SKIN

Objectives: • **Describe the structure and functions of the skin.**
• **Explain the pigments involved in skin color.**

Although it covers the entire body, the **skin** is an organ because it consists of different tissues that are joined to perform specific activities. It is one of the largest organs in surface area and weight. In adults, the skin covers an area of about 2 square meters (22 square feet). As you will see shortly, the skin is not merely a simple, thin covering. It also performs other essential

functions. **Dermatology** (der′-ma-TOL-ō-jē; *dermat-* = skin; *-ology* = study of) is the branch of medicine that specializes in diagnosing and treating skin disorders.

Structure of Skin

Structurally, the skin consists of two principal parts (Figure 5.1). The outer, thinner portion, which is composed of *epithelium*, is called the **epidermis** (ep′-i-DERM-is; *epi-* = above). The epidermis is attached to the deeper, thicker, *connective tissue* part, which is called the **dermis.** Deep to the dermis is a **subcutaneous (subQ) layer,** also called the **hypodermis,** which attaches the skin to underlying structures. Recall from Chapter 4 that the subcutaneous layer consists of adipose tissue and areolar connective tissue. The subcutaneous layer serves as a storage depot for fat and contains large blood vessels that supply the skin. The subcutaneous layer and the dermis also contain sensory nerve endings called *lamellated (Pacinian) corpuscles* (pa-SIN-ē-an) that are sensitive to pressure (Figure 5.1).

Functions of Skin

Among the many functions of the skin are the following:

1. **Regulation of body temperature.** In response to high environmental temperature or strenuous exercise, the evaporation of sweat from the skin surface helps lower an elevated body temperature to normal. Changes in the flow of blood in the skin also help regulate body temperature (described later in the chapter).

2. **Protection.** The skin covers the body and provides a physical barrier that protects underlying tissues from physical abrasion, bacterial invasion, dehydration, and ultraviolet (UV) radiation. Hair and nails also have protective functions, as described shortly.

3. **Sensation.** The skin contains abundant sensory nerve endings and sensory receptors that detect stimuli related to temperature, touch, pressure, and pain.

4. **Excretion.** Small amounts of water, salts, and several organic compounds (components of perspiration) are excreted by sweat glands.

5. **Immunity.** The Langerhans cells of the epidermis are an important component of the immune system, which fends off foreign invaders of the body.

6. **Synthesis of vitamin D.** Exposure of the skin to ultraviolet (UV) radiation initiates synthesis of the active form of vitamin D, a substance that aids in the absorption of calcium and phosphorus from the gastrointestinal tract into the blood.

Epidermis

The **epidermis** is composed of keratinized stratified squamous epithelium that contains four types of cells arranged in four or

Figure 5.1 ■ **Components of the integumentary system.** The skin consists of a thin, superficial epidermis and a deep, thicker dermis. Deep to the skin is the subcutaneous layer, which attaches the dermis to underlying organs and tissues.

The integumentary system includes the skin and its accessory structures—hair, nails, and glands—along with associated muscles and nerves.

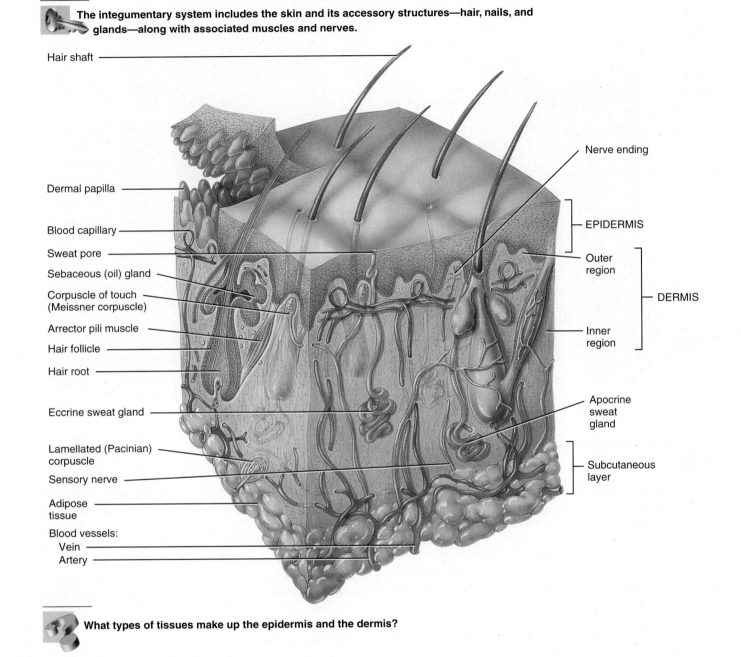

What types of tissues make up the epidermis and the dermis?

five layers (Figure 5.2). About 90% of epidermal cells are **keratinocytes** (ke-RAT-i-nō-sīts; *keratino-* = hornlike; *-cytes* = cells). These cells undergo a process termed **keratinization,** during which cells formed in the basal layers are pushed to the surface. As the cells move toward the surface they make **keratin,** a protein that helps protect the skin and underlying tissues from heat, microbes, and chemicals. At the same time, the cytoplasm, nucleus, and other organelles disappear, and the cells die. Eventually, the keratinized cells slough off and are replaced by underlying cells that, in turn, become keratinized. The whole process takes two to four weeks.

About 8% of the epidermal cells are **melanocytes** (MEL-a-nō-sīts; *melano-* = black), which produce the pigment melanin. Their long, slender projections extend between the keratinocytes and transfer melanin granules to them. **Melanin** is a brown-black pigment that contributes to skin color and absorbs damaging ultraviolet (UV) light. Once inside keratinocytes, the melanin granules cluster to form a protective veil over the

Figure 5.2 ■ Layers of the epidermis.

 The epidermis consists of keratinized stratified squamous epithelium.

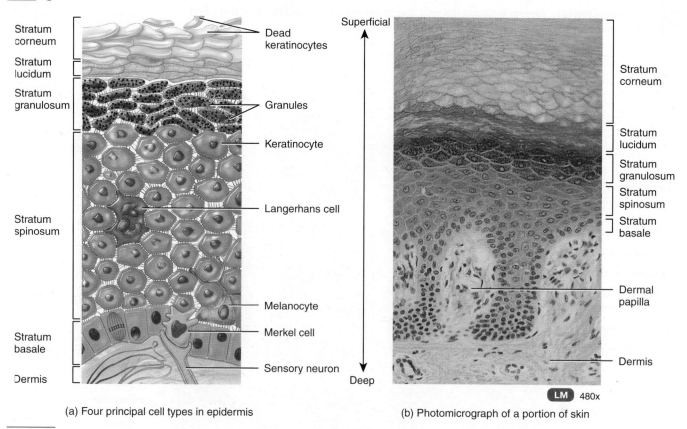

(a) Four principal cell types in epidermis

(b) Photomicrograph of a portion of skin

LM 480x

Which epidermal layer includes cells that continually undergo cell division?

nucleus, on the side toward the skin surface. In this way they shield the nuclear DNA from UV light.

Langerhans cells (LANG-er-hans) arise from red bone marrow and migrate to the epidermis, where they constitute a small proportion of the epidermal cells. They participate in immune responses mounted against microbes that invade the skin, and they are easily damaged by UV light.

Merkel cells are the least numerous of the epidermal cells. They are located in the deepest layer of the epidermis, where they contact the flattened process of a sensory neuron (nerve cell) and function in the sensation of touch.

In most regions of the body the epidermis has four layers and is called *thin skin. Thick skin* occurs where exposure to friction is greatest, such as in the palms and the soles. Five layers appear in thick skin (Figure 5.2). The five layers or strata (singular is stratum), from the deepest to the most superficial, are the following:

1. The *stratum basale* (ba-SA-lē = base) is a single layer of cuboidal or columnar keratinocytes that are capable of continued cell division. The cells multiply, push up toward the surface, and become part of the more superficial layers (keratinization). The stratum basale also contains melanocytes, Langerhans cells, and Merkel cells.

2. The *stratum spinosum* (spi-NŌ-sum = thornlike or prickly) consists of 8 to 10 layers of many-sided keratinocytes with spinelike projections. Langerhans cells and projections of melanocytes also appear in this stratum.

3. The *stratum granulosum* (gran-yoo-LŌ-sum = little grain) contains about five layers of flattened keratinocytes with granules that release a lipid secretion that functions as a water-repellent sealant. The secretion retards both loss of body fluid from the epidermis and entry of foreign materials into the epidermis.

4. The *stratum lucidum* (LOO-si-dum = clear) consists of about five layers of clear, flat, dead cells and is present only in the thick skin of the palms and soles.

5. The *stratum corneum* (COR-nē-um = horny) consists of about 30 layers of flat, dead keratinocytes. Constant exposure of thin or thick skin to friction or pressure stimulates formation of a callus, an abnormal thickening of the stratum corneum.

Dermis

The second principal part of the skin, the *dermis,* is composed of connective tissue containing collagen and elastic fibers (see

Figure 5.1). The combination of collagen and elastic fibers gives the skin its strength, *extensibility* (the ability to stretch), and *elasticity* (the ability to return to original shape after being stretched.) The ability of the skin to stretch can readily be seen during pregnancy, obesity, and tissue swelling (edema). Small tears in the skin due to extensive stretching that remain visible as silvery white streaks are called *striae* (STRĪ-ē = streaks). The types of cells in the dermis include fibroblasts, macrophages, and adipocytes. The dermis is very thick in the palms and soles and very thin in the eyelids and scrotum.

The outer region of the dermis consists of areolar connective tissue with fine elastic fibers. Its surface area is greatly increased by small, fingerlike projections called *dermal papillae* (pa-PIL- ē = nipples). Dermal papillae cause ridges in the epidermis, which produce fingerprints and help us to grip objects. Within dermal papillae are blood capillaries; *corpuscles of touch (Meissner corpuscles),* nerve endings that are sensitive to touch; and other nerve endings associated with thermal sensations, pain, tickling, and itching.

The inner region of the dermis consists of dense irregular connective tissue, adipose tissue, hair follicles, nerves, oil glands, and the ducts of sweat glands. It is attached to underlying bone and muscle by the subcutaneous layer.

Skin Color

Skin color is due to three pigments: melanin, carotene, and hemoglobin. The amount of *melanin* causes the skin's color to vary from pale yellow to tan to black. Because the number of melanocytes is about the same in all races, differences in skin color are due to the amount of pigment the melanocytes produce and transfer to keratinocytes. In some people, melanin tends to form in patches called *freckles.* When the skin is repeatedly exposed to ultraviolet rays, the amount and darkness of melanin increase, which darkens the skin and protects the body against UV radiation. Overexposure to UV rays, however, leads to skin cancer (see page 110).

An inherited inability to produce melanin causes *albinism* (AL-bin-izm; *albin-* = white), an absence of melanin pigment in the hair and eyes as well as the skin. In another condition, called *vitiligo* (vit′-i-LĪ-gō), the partial or complete loss of melanocytes from patches of skin produces irregular white spots.

Carotene (KAR-ō-tēn; *carot-* = carrot), a yellow-orange pigment, gives carrots and egg yolks their color. It is the precursor of vitamin A, which is used to synthesize pigments needed for vision. Carotene is found in the stratum corneum and fatty areas of the dermis and subcutaneous layer.

When little melanin or carotene is present, the epidermis appears translucent. The epidermis itself has no blood vessels, a characteristic of all epithelia. Thus, the skin of white people appears pink to red, depending on the amount and oxygen content of the blood moving through capillaries in the dermis. The red color is due to *hemoglobin,* the red, oxygen-carrying pigment in red blood cells.

The color of skin and mucous membranes can provide clues for diagnosing certain problems. When blood is not picking up an adequate amount of oxygen in the lungs, such as in someone who has stopped breathing, mucous membranes, nail beds, and light-colored skin appear bluish, or *cyanotic* (sī′-a-NOT-ic; *cyano-* = blue). Cyanosis occurs because hemoglobin that is depleted of oxygen looks deep, purplish blue. *Jaundice* (JON-dis; *jaund-* = yellow) is due to a buildup of the yellow pigment bilirubin in the blood. This condition, which usually indicates liver disease, gives a yellowed appearance to the whites of the eyes and to light-colored skin. *Erythema* (er′-e-THĒ-ma = red), redness of the skin, is caused by engorgement with blood of capillaries in the dermis due to exercise, exposure to heat, skin injury, infection, inflammation, or allergic reactions. All skin color changes are observed most readily in people who have light-colored skin and may be more difficult to discern in people with darker skin.

ACCESSORY STRUCTURES OF THE SKIN

Objective: • **Describe the structure and functions of hair, skin glands, and nails.**

Accessory structures of the skin that develop from the epidermis of an embryo—hair, glands, nails—perform vital functions. Hair and nails protect the body. Sweat glands help regulate body temperature.

Hair

Hairs, or *pili* (PI-lē), are present on most skin surfaces except the palms, palmar surfaces of the fingers, soles, and plantar surfaces of the toes. In adults, hair usually is most heavily distributed across the scalp, over the brows of the eyes, and around the external genitalia. Genetic and hormonal influences largely determine the thickness and pattern of distribution of hair. Hair on the head guards the scalp from injury and the sun's rays; eyebrows and eyelashes protect the eyes from foreign particles; hair in the nostrils protects against inhaling insects and foreign particles.

Each hair is a thread of fused, dead, keratinized cells that consists of a shaft and a root (Figure 5.3). The *shaft* is the superficial portion, most of which projects above the surface of the skin. The *root* is the portion below the surface that penetrates into the dermis and even into the subcutaneous layer. Surrounding the root is the *hair follicle,* which is composed of two layers of epidermal cells: *external* and *internal root sheaths* surrounded by a *connective tissue sheath.* Surrounding each hair follicle are nerve endings, called *hair root plexuses,* that are sensitive to touch. If their hair shaft is moved, the hair root plexuses respond.

The base of each follicle is enlarged into an onion-shaped structure, the *bulb.* This structure contains an indentation, the *papilla of the hair,* that contains many blood vessels and provides nourishment for the growing hair. The bulb also contains a region of cells called the *matrix,* which produces new hairs by cell division when older hairs are shed.

Figure 5.3 ■ **Hair.**

 Hairs are growths of epidermis composed of dead, keratinized cells.

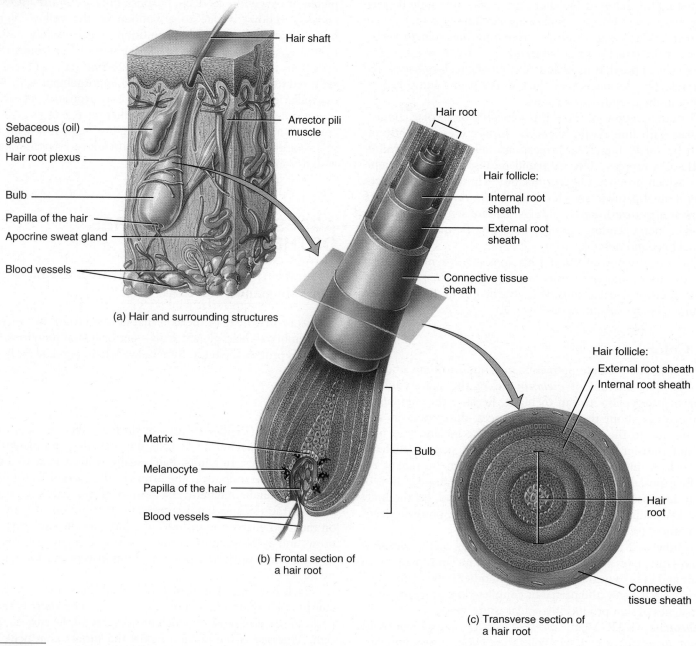

Hair shaft

Sebaceous (oil) gland

Hair root plexus

Bulb

Papilla of the hair

Apocrine sweat gland

Blood vessels

Arrector pili muscle

(a) Hair and surrounding structures

Hair root

Hair follicle:

Internal root sheath

External root sheath

Connective tissue sheath

Matrix

Melanocyte

Papilla of the hair

Blood vessels

Bulb

(b) Frontal section of a hair root

Hair follicle:

External root sheath

Internal root sheath

Hair root

Connective tissue sheath

(c) Transverse section of a hair root

Which part of a hair produces a new hair by cell division?

Sebaceous (oil) glands (discussed shortly) and a bundle of smooth muscle cells are also associated with hairs. The smooth muscle is called **arrector pili** (a-REK-tor PI-lē; *arrect* = to raise). It extends from the upper dermis to the side of the hair follicle. In its normal position, hair emerges at an angle to the surface of the skin. Under stress, such as cold or fright, nerve endings stimulate the arrector pili muscles to contract, which pulls the hair shafts perpendicular to the skin surface. This action causes "goose bumps" because the skin around the shaft forms slight elevations.

Each hair follicle goes through a **growth cycle,** which consists of a *growth stage* and a *resting stage.* During the growth stage, a hair is formed by cells of the matrix that differentiate, become keratinized, and die. As new cells are added at the base of the hair root, the hair grows longer. In time, the growth of the hair stops and the resting stage begins. After the resting stage, a new growth cycle begins in which a new hair replaces the old hair, and the old hair is pushed out of the hair follicle. In general, scalp hair grows for about three years and rests for one to two years. At any given time, most hair is in the growth stage.

Normal hair loss in the adult scalp is about 100 hairs per day. Both the rate of growth and the replacement cycle may be altered by any of the following factors: illness, radiation therapy and chemotherapy, diet, age, genetics, gender, and severe emotional stress. In women, the rate of shedding also increases for three to four months after childbirth.

At puberty, when their testes begin secreting significant quantities of androgens (masculinizing sex hormones), males develop the typical male pattern of hair growth, including a beard and a hairy chest. In females, both the ovaries and the adrenal glands produce small quantities of androgens. Surprisingly, androgens also must be present for the most common form of baldness, **male-pattern baldness,** to occur. In genetically predisposed males, androgens inhibit hair growth.

Minoxidil (Rogaine) is a potent vasodilator, that is, a drug that widens blood vessels and increases circulation. When applied to the scalp, it stimulates hair regrowth in some persons with thinning hair. For many, however, the hair growth is meager, and Minoxidil does not help people who already are bald.

The color of hair is due to melanin. It is synthesized by melanocytes in the matrix of the bulb and passes into cells of the root and shaft. Dark-colored hair contains mostly true melanin. Blond and red hair contain variants of melanin in which there is iron and more sulfur. Gray hair occurs with a decline in the synthesis of melanin. White hair results from accumulation of air bubbles in the hair shaft.

Glands

Three kinds of glands associated with the skin are sebaceous, sudoriferous, and ceruminous.

Sebaceous Glands

Sebaceous glands (se-BĀ-shus = greasy) or **oil glands,** with few exceptions, are connected to hair follicles (Figure 5.3). The secreting portions of the glands lie in the dermis and open into the necks of hair follicles or directly onto a skin surface. There are no sebaceous glands in the palms and soles.

Sebaceous glands secrete an oily substance called **sebum** (SĒ-bum). Sebum keeps hair from drying out, prevents excessive evaporation of water from the skin, keeps the skin soft, and inhibits the growth of certain bacteria.

When sebaceous glands of the face become enlarged because of accumulated sebum, **blackheads** develop. Because sebum is nutritive to certain bacteria, **pimples** or **boils** often result. The color of blackheads is due to melanin and oxidized oil, not dirt. Sebaceous gland activity increases during adolescence.

Sudoriferous Glands

Sudoriferous glands (soo'-dor-IF-er-us; *sudori-* = sweat; *-ferous* = bearing) or **sweat glands** are divided into two types. **Apocrine sweat glands** (*apo-* = from) are found in the skin of the axilla (armpit), pubic region, and pigmented areas (areolae) of the breasts. Apocrine sweat gland ducts open into hair follicles (see Figure 5.1). They begin to function at puberty and produce a sticky, viscous secretion. They are stimulated during emotional stresses and sexual excitement, and the secretions are commonly known as "cold sweat."

Eccrine sweat glands (*eccrine* = secreting outwardly) are distributed throughout the skin except for areas such as the margins of the lips, nail beds of the fingers and toes, and eardrums. Eccrine sweat glands are most numerous in the skin of the forehead, palms, and soles. Eccrine sweat gland ducts terminate at a **sweat pore** at the surface of the epidermis (see Figure 5.1) and function throughout life to produce a more watery secretion than that of apocrine sweat glands.

Perspiration, or **sweat,** is the substance produced by sudoriferous glands. Its principal function is to help regulate body temperature. It also helps eliminate wastes. Because the mammary glands are actually modified sudoriferous glands, they could be discussed here. However, because of their relationship to the reproductive system, they will be considered in Chapter 23.

Ceruminous Glands

Ceruminous glands (se-ROO-mi-nus; *cer-* = wax) are present in the external auditory canal, the outer ear canal. Their ducts open either directly onto the surface of the external auditory canal or into ducts of sebaceous glands. The combined secretion of the ceruminous and sebaceous glands is called **cerumen** or earwax. Cerumen and the hairs in the external auditory meatus provide a sticky barrier against foreign bodies.

Some people produce an abnormally large amount of cerumen in the external auditory canal. The cerumen may accumulate until it becomes impacted (firmly wedged), which prevents sound waves from reaching the eardrum. The treatment for impacted cerumen is usually periodic ear irrigation or removal of earwax with a blunt instrument by trained medical personnel. The use of cotton-tipped swabs or sharp objects is not recommended for this purpose because they may push the cerumen farther into the external auditory canal and damage the eardrum.

Figure 5.4 ■ **A fingernail.**

Nail cells arise by transformation of superficial cells of the nail matrix into nail cells.

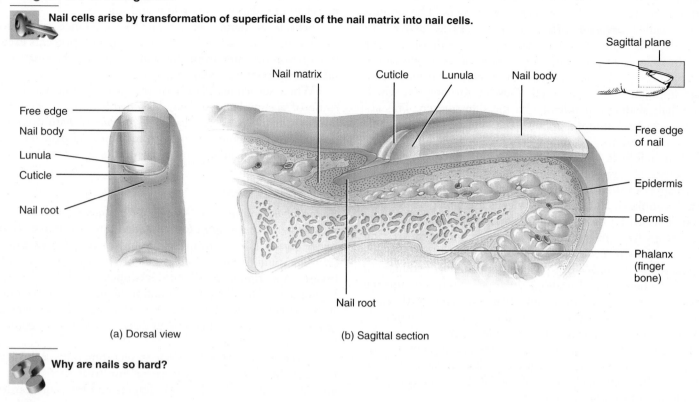

(a) Dorsal view

(b) Sagittal section

Why are nails so hard?

Nails

Nails are plates of tightly packed, hard, keratinized cells of the epidermis. Each nail (Figure 5.4) consists of a nail body, a free edge, and a nail root. The *nail body* is the portion of the nail that is visible; the *free edge* is the part that extends past the end of the finger or toe; the *nail root* is the portion that is not visible. Most of the nail body is pink because of the underlying blood capillaries. The whitish semilunar area near the nail root is called the *lunula* (LOO-nyoo-la = little moon). It appears whitish because the vascular tissue underneath does not show through due to the thickened stratum basale in the area. Nail growth occurs by the transformation of superficial cells of the *nail matrix* into nail cells. The average growth of fingernails is about 1 mm (0.04 inch) per week. The *cuticle* consists of stratum corneum.

Functionally, nails help us grasp and manipulate small objects, provide protection to the ends of the fingers and toes, and allow us to scratch various parts of the body.

AGING AND THE INTEGUMENTARY SYSTEM

Objective: • **Describe the effects of aging on the integumentary system.**

For most infants and children, relatively few problems are encountered with the skin as it ages. With the arrival of adoles-

cence, however, some teens develop acne. The pronounced effects of skin aging do not become noticeable until people reach their late forties. Most of the age-related changes occur in the dermis. Collagen fibers in the dermis begin to decrease in number, stiffen, break apart, and disorganize into a shapeless, matted tangle. Elastic fibers lose some of their elasticity, thicken into clumps, and fray, an effect that is greatly accelerated in the skin of smokers. Fibroblasts, which produce both collagen and elastic fibers, decrease in number. As a result, the skin forms the characteristic crevices and furrows known as wrinkles.

With further aging, Langerhans cells dwindle in number and macrophages become less-efficient phagocytes, thus decreasing the skin's immune responsiveness. Moreover, decreased size of sebaceous glands leads to dry and broken skin that is more susceptible to infection. Production of sweat diminishes, which probably contributes to the increased incidence of heat stroke in the elderly. There is a decrease in the number of functioning melanocytes, resulting in gray hair and atypical skin pigmentation. An increase in the size of some melanocytes produces pigmented blotching (liver spots). Walls of blood vessels in the dermis become thicker and less permeable, and subcutaneous fat is lost. Aged skin (especially the dermis) is thinner than young skin, and the migration of cells from the basal layer to the epidermal surface slows considerably. With the onset of old age, skin heals poorly and becomes more susceptible to pathological conditions such as skin cancer, itching, and pressure sores.

Growth of nails and hair begins to slow during the second and third decades of life. The nails also may become more brittle with age, often due to dehydration or repeated use of cuticle remover or nail polish.

Physical activity is good for your skin. During exercise, the body shunts blood to the skin to help release excess heat produced by the contracting muscles. This increased blood flow provides the skin with nutrients and gets rid of wastes.

Fun in the Sun

From your skin's point of view, the main problem with exercise is that it often occurs outdoors, where sun exposure over the years can lead to wrinkles, age spots, and cancers of the skin. To prevent these, do what you can to minimize sun exposure. The most effective skin protection is some form of sun block. Tightly woven clothing (hold it up to a light and see how much shines through) helps keep the sun's rays from reaching the skin, and wide-brimmed hats provide some protection. Zinc oxide blocks the sun and is good for noses and lips when long-term exposure is unavoidable.

When a sun block is not practical, a sunscreen should be used. These do not shield the skin completely, but they do reduce the damaging effects of the ultraviolet rays. Evidence suggests that the skin can repair some damage when sunscreens are consistently applied. But researchers warn that sunscreens can provide a false sense of security. Because they prevent burning, sunscreens may lull us into thinking the sun is not hurting us, while damage may still be occurring.

Barriers to skin protection

Chemists have yet to invent a sunscreen that is fun to wear. Many exercisers can't take the grease, especially, as an avid bicyclist put it, "as it mingles with sweat and dead bugs." Advice for heavy sweaters is to exercise in the early or late part of the day, take as shady a route as possible, wear a hat and protective clothing, and use as much sunscreen as you can tolerate.

Swimmers should note that "waterproof" sunscreen stays on for only about 30 minutes in the water and should be reapplied at that time.

Dry skin care

Although not life threatening, dry skin can be very uncomfortable. Frequent showers and water exposure can strip the skin of its natural protective oils. The only solution is frequent moisturizing. Use of a good moisturizing cream immediately after drying off will counteract the drying effect of a "wash-and-wear" lifestyle.

▶ **Think It Over**

▶ Imagine you have a friend who is training for a marathon and must exercise outdoors for an hour or more on most days. He has fair skin and a family history of skin cancer. What advice would you give him for minimizing sun exposure while continuing his training?

HOMEOSTASIS OF BODY TEMPERATURE

Objective: • **Explain how the skin helps regulate body temperature.**

One of the best examples of homeostasis in humans is the regulation of body temperature by the skin. (You may review the concept of homeostasis in Chapter 1 on page 7 and Figure 1.2 on page 7.) As warm-blooded animals, we maintain a remarkably constant body temperature close to 37°C (98.6°F) even though the environmental temperature varies greatly.

Suppose you are in an environment where the temperature is 38°C (101°F). Heat (the stimulus) continually flows from the environment to your body, raising body temperature. To counteract these changes in a controlled condition, a sequence of events is set into operation (Figure 5.5). Temperature-sensitive receptors (nerve endings) in the skin called ***thermoreceptors*** detect the stimulus and send nerve impulses (input) to your brain (control center). A temperature control region of the brain (called the hypothalamus, see Figure 10.10 on page 246) then sends nerve impulses (output) to the sweat glands (effectors), which produce perspiration more rapidly. As the sweat evaporates from the surface of your skin, heat is lost and your body

Figure 5.5 ■ **The role of the skin in regulating the homeostasis of body temperature.**

 When the environmental temperature is high, sweat glands produce more perspiration and blood vessels in the dermis dilate.

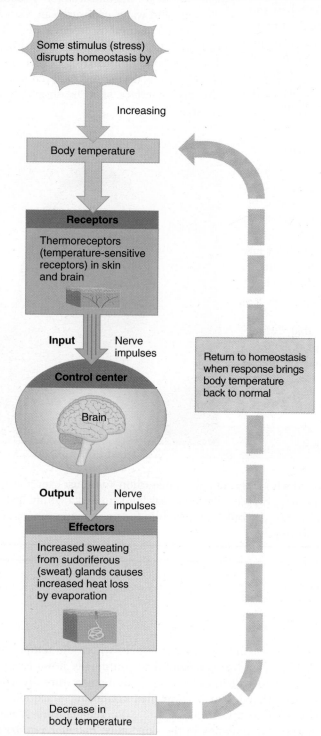

temperature drops to normal (return to homeostasis). When environmental temperature is low, sweat glands produce less perspiration.

Your brain also sends output to blood vessels (a second set of effectors), dilating (widening) those in the dermis so that skin blood flow increases. As more warm blood flows through capillaries close to the body surface, more heat can be lost to the environment, which lowers body temperature. Thus heat is lost from the body, and body temperature falls to the normal value to restore homeostasis.

Body temperature regulation is a *negative feedback system* because the response (cooling) is opposite to the stimulus (heating) that started the cycle. Also, the thermoreceptors continually monitor body temperature and feed this information back to the brain. The brain, in turn, continues to send impulses to the sweat glands and blood vessels until the temperature returns to 37°C (98.6°F).

Regulating the rate of sweating and changing dermal blood flow are only two mechanisms by which body temperature can be adjusted. Other mechanisms include regulating metabolic rate (a slower metabolic rate reduces heat production) and regulating skeletal muscle contractions (decreased muscle tone results in less heat production). These mechanisms are discussed in detail in Chapter 20.

If environmental temperature is low or some factor causes body temperature to decrease, a series of responses takes place to raise body temperature to normal. For example, blood vessels in the dermis constrict, thus reducing heat loss to the environment. Also, certain hormones are released that increase metabolic rate and therefore increase heat production. Finally, skeletal muscles contract involuntarily (shivering), and this too increases heat production. All of these responses act to raise body temperature back to normal.

• • •

To appreciate the many ways that skin contributes to homeostasis of other body systems, examine Focus on Homeostasis: The Integumentary System. This focus box is the first of ten, found at the end of selected chapters, that explain how the body system under consideration contributes to homeostasis of all the other body systems. The Focus on Homeostasis feature will help you understand how the individual body systems interact to contribute to the homeostasis of the entire body. Next, in Chapter 6, we will explore how bone tissue is formed and how bones are assembled into the skeletal system, which protects many of our internal organs.

Why is regulation of body temperature a negative feedback system?

Focus on Homeostasis
The Integumentary System

Body System		Contribution of Integumentary System
For all body systems		Skin and hair provide barriers that protect all internal organs from damaging agents in the external environment; sweat glands and skin blood vessels help regulate body temperature, needed for proper functioning of other body systems.
Skeletal system		Skin helps activate vitamin D, needed for proper absorption of dietary calcium and phosphorus to build and maintain bones.
Muscular system		Skin helps provide calcium ions, needed for muscle contraction, and rids the body of heat produced by muscular activity.
Nervous system		Nerve endings in skin and subcutaneous tissue provide input to the brain for touch, pressure, thermal, and pain sensations.
Endocrine system		Keratinocytes in skin help activate vitamin D, initiating its conversion to calcitriol, a hormone that aids absorption of dietary calcium and phosphorus.
Cardiovascular system		Local chemical changes in dermis cause widening and narrowing of skin blood vessels, which help adjust blood flow to the skin.
Lymphatic and immune system		Skin is "first line of defense" in immunity, providing mechanical barriers and chemical secretions that discourage penetration and growth of microbes; Langerhans cells in epidermis participate in immune responses by recognizing and processing foreign antigens.
Respiratory system		Hairs in nose filter dust particles from inhaled air; stimulation of pain nerve endings in skin may alter breathing rate.
Digestive system		Skin helps activate vitamin D to become the hormone calcitriol, which promotes absorption of dietary calcium and phosphorus in the small intestine.
Urinary system		Kidney cells receive partially activated vitamin D hormone from skin and convert it to calcitriol; some waste products are excreted from body in sweat, contributing to excretion by urinary system.
Reproductive systems		Nerve endings in skin and subcutaneous tissue respond to erotic stimuli, thereby contributing to sexual pleasure; suckling of a baby stimulates nerve endings in skin, leading to milk ejection; mammary glands (modified sweat glands) produce milk; skin stretches during pregnancy as fetus enlarges.

COMMON DISORDERS

Burns

The tissue damage from excessive heat, electricity, radioactivity, or corrosive chemicals is called a *burn.* Burns disrupt homeostasis because they destroy the protection afforded by the skin. They permit microbial invasion and infection, loss of fluid, and loss of body temperature regulation. A *first-degree burn* involves only the epidermis. It is characterized by mild pain and redness, but no blisters. A *second-degree burn* involves the epidermis and possibly part of the dermis. Some skin functions are lost. Redness, blister formation, edema, and pain are characteristic of such a burn. *A third-degree burn* destroys the epidermis, dermis, and accessory structures of the skin, and most skin functions are lost.

Acne

Acne is an inflammation of sebaceous glands that usually begins at puberty, when the glands grow in size and increase their production of sebum. Although testosterone, a male sex hormone, appears to play the greatest role in stimulating sebaceous glands, other steroid hormones from the ovaries and adrenal glands stimulate sebaceous secretions in females. Acne occurs predominantly in sebaceous follicles that have been colonized by bacteria, some of which thrive in the lipid-rich sebum.

Pressure Sores

Pressure sores are also called *decubitus ulcers* (dē-KYOO-bi-tus) or *bedsores.* They are caused by constant deficiency of blood to tissues over a bony projection that has been subjected to prolonged pressure against an object like a bed, cast, or splint. They occur most often in patients who are bedridden.

Skin Cancer

Excessive sun exposure causes three types of *skin cancer,* the most common cancer in white people. *Basal cell carcinomas (BCCs)* account for over 78% of all skin cancers. The tumors arise from the basal layer of the epidermis and rarely spread. *Squamous cell carcinomas (SCCs)* account for about 20% of all skin cancers. They arise from squamous cells of the epidermis and have a variable tendency to spread. Most SCCs arise from preexisting lesions on sun-exposed skin. Fortunately, most BCCs and SCCs can be removed surgically. *Malignant melanomas* arise from melanocytes and are the leading cause of death from all skin diseases because they spread rapidly. Malignant melanomas account for only about 2% of all skin cancers.

MEDICAL TERMINOLOGY AND CONDITIONS

Abrasion (a-BRĀ-shun; *ab-* = away; *-raison* = scraped) A portion of the epidermis that has been scraped away.

Alopecia (al′-o-PĒ-shē-a) Partial or complete lack of hair; may result from aging, endocrine disorders, chemotherapy for cancer, or skin disease.

Athlete's (ATH-lēts) *foot* A superficial fungus infection of the skin of the foot.

Cold sore A lesion, usually in oral mucous membrane, caused by type 1 herpes simplex virus (HSV) transmitted by oral or respiratory routes. The virus remains dormant until triggered by factors such as ultraviolet light, hormonal changes, and emotional stress. Also called a *fever blister.*

Contact dermatitis (der′-ma-TI-tis; *dermat-* = skin; *-itis* = inflammation) Inflammation of the skin characterized by redness, itching, and swelling and caused by exposure of the skin to chemicals that bring about an allergic reaction, such as poison ivy toxin.

Corn (KORN) A painful thickening of the stratum corneum of the epidermis found principally over toe joints and between the toes, often caused by friction or pressure. Corns may be hard or soft, depending on their location. Hard corns are usually found over toe joints, and soft corns are usually found between the fourth and fifth toes.

Dermabrasion (der′-ma-BRĀ-shun) Removal of acne, scars, tattoos, or moles by sandpaper or a high-speed brush.

Hemangioma (hē-man′-jē-Ō-ma; *hem-* = blood; *-angi-* = blood vessel; *-oma* = tumor) Localized tumor of the skin and subcutaneous layer that results from an abnormal increase in blood vessels. One type is a *port-wine stain,* a flat pink, red, or purple lesion present at birth, usually at the nape of the neck.

Hives (HĪVZ) Skin condition marked by reddened elevated patches that are often itchy. Most commonly caused by infections, physical trauma, medications, emotional stress, food additives, and certain food allergies.

Impetigo (im′-pe-TĪ-gō) Superficial skin infection caused by *Staphylococcus* bacteria; most common in children.

Intradermal (in-tra-DER-mal; *intra-* = within) Within the skin. Also called *intracutaneous.*

Keratosis (ker′-a-TŌ-sis; *kera-* = horn) Formation of a hardened growth of epidermal tissue, such as a *solar keratosis,* a premalignant lesion of the sun-exposed skin of the face and hands.

Laceration (las-er-Ā-shun; *lacer-* = torn) An irregular tear of the skin.

Nevus (NĒ-vus) A round, flat, or raised area of pigmented skin that may be present at birth or may develop later. Varies in color from yellow-brown to black. Also called a *mole* or *birthmark.*

Pruritus (proo-RĪ-tus; *pruri-* = to itch) Itching, one of the most common dermatological disorders. It may be caused by skin disorders (infections), systemic disorders (cancer, kidney failure), psychogenic factors (emotional stress), or allergic reactions.

Topical Refers to a medication applied to the skin surface rather than ingested or injected.

Wart Mass produced by uncontrolled growth of epithelial skin cells, caused by a papilloma virus. Most warts are noncancerous.

STUDY OUTLINE

Skin (p. 100)

1. The skin and its accessory structures (hair, glands, and nails) constitute the integumentary system.

2. The principal parts of the skin are the outer epidermis and inner dermis. The dermis overlies the subcutaneous layer.

3. The functions of the skin include regulation of body temperature, protection, sensation, excretion, immunity, and synthesis of vitamin D.

4. Epidermal cells include keratinocytes, melanocytes, Langerhans cells, and Merkel cells. The epidermal layers, from deepest to most superficial, are the stratum basale, stratum spinosum, stratum granulosum, stratum lucidum, and stratum corneum. The stratum basale undergoes continuous cell division and produces all other layers.

5. The dermis consists of two regions. The superficial region is areolar connective tissue containing blood vessels, nerves, hair follicles, dermal papillae, and corpuscles of touch (Meissner corpuscles). The deeper region is dense, irregularly arranged connective tissue containing adipose tissue, hair follicles, nerves, sebaceous (oil) glands, and ducts of sudoriferous (sweat) glands.

6. Skin color is due to the pigments melanin, carotene, and hemoglobin.

Accessory Structures of the Skin (p. 103)

1. Accessory structures of the skin develop from the epidermis of an embryo.

2. They include hair, skin glands (sebaceous, sudoriferous, and ceruminous), and nails.

3. Hairs are threads of fused, dead keratinized cells that function in protection.

4. Hairs consist of a shaft above the surface, a root that penetrates the dermis and subcutaneous layer, and a hair follicle.

5. Associated with hairs are bundles of smooth muscle called arrector pili and sebaceous (oil) glands.

6. New hairs develop from cell division of the matrix in the bulb; hair replacement and growth occur in a cyclic pattern that consists of a growth stage and a resting stage.

7. Sebaceous glands are usually connected to hair follicles; they are absent in the palms and soles. Sebaceous glands produce sebum, which moistens hairs and waterproofs the skin.

8. Sudoriferous glands are divided into apocrine and eccrine. They produce perspiration, which carries small amounts of wastes to the surface and assists in maintaining body temperature.

9. Ceruminous glands are modified sudoriferous glands that secrete cerumen. They are found in the external auditory canal.

10. Nails are hard, keratinized epidermal cells covering the terminal portions of the fingers and toes.

11. The principal parts of a nail are the body, free edge, root, lunula, cuticle, and matrix. Cell division of the matrix cells produces new nails.

Aging and the Integumentary System (p. 106)

1. Most effects of aging occur when an individual reaches the late forties.

2. Among the effects of aging are wrinkling, loss of subcutaneous fat, atrophy of sebaceous glands, and decrease in the number of melanocytes and Langerhans cells.

Homeostasis of Body Temperature (p. 107)

1. One of the functions of the skin is the regulation of normal body temperature of 37°C (98.6°F).

2. If environmental temperature is high, skin receptors sense the stimulus (heat) and generate impulses (input) that are transmitted to the brain (control center). The brain then sends impulses (output) to sweat glands and blood vessels (effectors) to produce perspiration and vasodilation. As the perspiration evaporates, the skin is cooled and the body temperature returns to normal.

3. The skin-cooling response is a negative feedback mechanism.

SELF-QUIZ

1. Hair follicles
   a. consist of dead cells   **b.** extend above the surface of the skin   **c.** can increase in number as you age   **d.** contain cells undergoing mitosis   **e.** are another name for arrector pili muscles

2. Skin coloration
   a. is due to melanin found in the subcutaneous layer   **b.** in whites is due mainly to carotene   **c.** is related to apocrine glands   **d.** is stimulated by exposure to the sun   **e.** is produced by Merkel cells

3. In which portion of the skin will you find dermal papillae?
   a. outer region of the dermis   **b.** epidermis   **c.** hypodermis   **d.** stratum spinosum   **e.** inner region of the dermis

4. If you pricked your fingertip with a needle, the first layer of epidermis that it would penetrate is the
   a. stratum basale   **b.** stratum spinosum   **c.** stratum granulosum   **d.** stratum lucidum   **e.** stratum corneum

5. A person with albinism has a defect in the production of
   a. carotene   **b.** keratin   **c.** collagen   **d.** cerumen   **e.** melanin

6. The red or pink tones seen in some skin are due to
   a. hemoglobin in the blood moving through capillaries in the dermis   **b.** the presence of carotene   **c.** the lack of oxygen   **d.** a buildup of bilirubin in the blood   **e.** an increased production of melanin

7. When you have your hair cut, scissors are cutting through the hair
   a. follicle   **b.** root   **c.** shaft   **d.** papilla   **e.** bulb

8. Which of the following is NOT true concerning eccrine sweat glands?
   a. They are most numerous on the palms and the soles.   **b.** They help regulate body temperature.   **c.** They produce a viscous secretion.   **d.** They function throughout life.   **e.** They terminate at pores on the skin's surface.

9. Which tissue is the main type found in the inner region of the dermis?

    a. dense irregular connective   b. stratified squamous epithelium   c. smooth muscle   d. nervous   e. cartilage

10. Which of the following is NOT a function of skin?

    a. calcium production   b. vitamin D synthesis   c. protection   d. immunity   e. temperature regulation

11. Which of the following is NOT true concerning hair?

    a. Hair growth cycles include a growth stage and a resting stage.   b. Normally, adults lose about 1000 hairs per day.   c. Hair color is due to melanin.   d. Sebaceous glands are associated with hair.   e. Contraction of the arrector pili muscles makes hair stand erect.

12. Sebaceous glands

    a. secrete an oily substance   b. are located on the palms and soles   c. are responsible for breaking out in a "cold sweat"   d. are involved in body temperature regulation   e. are found in the external auditory meatus

13. As keratinocytes in the stratum basale are pushed toward the skin's surface, they

    a. begin to divide more rapidly   b. become more elastic   c. begin to die   d. lose their melanin   e. begin to assume a columnar shape

14. To produce vitamin D, the skin cells need to be exposed to

    a. calcium and phosphorus   b. ultraviolet light   c. heat   d. pressure   e. keratin

15. To prevent an unwanted hair from growing back, you must destroy which structure?

    a. shaft   b. root sheath   c. lunula   d. matrix   e. arrector pili

16. Aging can result in

    a. an increase in collagen and elastic fibers in the skin   b. a steady increase in the activity of sudoriferous glands   c. a greater immune response from Langerhans cells   d. more efficient activity by macrophages   e. a decline in the activity of sebaceous glands

17. The portion of the nail that is responsible for nail growth is the

    a. cuticle   b. nail matrix   c. lunula   d. nail body   e. nail root

18. Match the following:

____ a. Langerhans cell	A.	earwax
____ b. Merkel cell	B.	silvery white streaks
____ c. keratin	C.	yellow-orange pigment
____ d. melanin	D.	function in immune responses
____ e. lamellated (Pacinian) corpuscles	E.	waterproofing protein of skin, hair
____ f. cerumen	F.	touch receptor found in epidermis
____ g. carotene	G.	yellow to black pigment
____ h. striae	H.	nerve endings sensitive to pressure
____ i. corpuscles of touch (Meissner corpuscles)	I.	touch receptors found in dermal papillae

19. Which of the following is NOT an accessory structure of the skin?

    a. dermal papillae   b. sudoriferous glands   c. sebaceous glands   d. ceruminous glands   e. nails

20. What is the response by effectors when the body temperature is elevated?

    a. Blood vessels to the skin's dermis constrict   b. Sweat glands increase production of sweat   c. Skeletal muscles begin to involuntarily contract   d. The body's metabolic rate increases   e. The ceruminous glands increase production

CRITICAL THINKING APPLICATIONS

1. Three-year-old Michael was having his first haircut. As the barber started to snip his hair, Michael cried, "Stop! You're killing it!" He then pulled his own hair, yelling, "Ouch! See! It's alive!" Is Michael right about his hair?

2. Michael's twin sister Michelle scraped her knee at the playground. She told her mother that she wanted "new skin that doesn't leak." Her mother promised that new skin would soon appear under the bandage. How does new skin grow?

3. An accident while slicing a bagel lands you in the emergency room. After examining the injury, the doctor states that you've cut through to the subQ layer. List the structures and functions of the skin layers that you have exposed.

4. Fifteen-year-old Jeremy has a bad case of "blackheads." According to his Aunt Emma, Jeremy's skin problems are from too much late-night TV, chicken nuggets, and cheddar popcorn. Explain the real cause of blackheads to Aunt Emma.

ANSWERS TO FIGURE QUESTIONS

5.1 The epidermis is made up of epithelial tissue, whereas the dermis is composed of connective tissue.

5.2 The stratum basale is the layer of the epidermis that contains stem cells that continually undergo cell division.

5.3 The matrix produces a new hair by cell division.

5.4 Nails are hard because they are composed of tightly packed, hard, keratinized epidermal cells.

5.5 Regulation of body temperature is a negative feedback system because the result of the effectors (lowering body temperature) is opposite to the initial stimulus (rising body temperature).

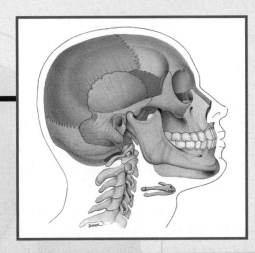

Chapter 6

The Skeletal System

■ Student Learning Objectives

1. Discuss the functions of bone and the skeletal system.   114
2. Classify bones on the basis of their shape and location.   114
3. Describe the parts of a long bone.   114
4. Describe the histological features of compact and spongy bone tissue.   116
5. Explain the steps involved in ossification.   118
6. Describe the factors involved in bone growth and maintenance and how hormones regulate calcium homeostasis.   120
7. Describe how exercise and mechanical stress affect bone tissue.   122

8. Describe the principal surface markings on bones and the function of each.   123
9. Classify the bones of the body into axial and appendicular divisions.   124
10. Name the cranial and facial bones and indicate their locations and major structural features.   124
11. Describe the relationship of the hyoid bone to the skull.   133
12. Identify the regions and normal curves of the vertebral column and describe its structural and functional features.   133
13. Identify the bones of the thorax and their principal markings.   138

14. Identify the bones of the pectoral (shoulder) girdle and their principal markings.   139
15. Identify the bones of the upper limb and their principal markings.   140
16. Identify the bones of the pelvic (hip) girdle and their principal markings.   142
17. List the skeletal components of the lower limb and their principal markings.   143
18. Identify the principal structural differences between female and male skeletons.   146
19. Describe the effects of aging on the skeletal system.   146

■ A Look Ahead

Despite its simple appearance, bone is a complex and dynamic living tissue that continuously is remodeled—new bone is built while old bone is broken down. Each individual bone is an organ because bone is composed of several different tissues working together: bone, cartilage, dense connective tissues, epithelium, various blood-forming tissues, adipose tissue, and nervous tissue. The entire framework of bones and their cartilages constitute the *skeletal system.* The study of bone structure and the treatment of bone disorders is termed *osteology* (os'-tē-OL-ō-jē; *oste-* = bone; *-ology* = study of).

FUNCTIONS OF BONE AND THE SKELETAL SYSTEM

Objective: • **Discuss the functions of bone and the skeletal system.**

Bone tissue and the skeletal system perform several basic functions:

1. **Support.** The skeleton provides a framework for the body by supporting soft tissues and providing points of attachment for most skeletal muscles.

2. **Protection.** The skeleton protects many internal organs from injury. For example, cranial bones protect the brain, vertebrae protect the spinal cord, and the rib cage protects the heart and lungs.

3. **Assisting in movement.** Because skeletal muscles attach to bones, when muscles contract, they pull on bones. Together, bones and muscles produce movement. This function is discussed in detail in Chapter 8.

4. **Storage of minerals.** Bone tissue stores several minerals, especially calcium and phosphorus. On demand, bone releases minerals into the blood to maintain critical mineral balances (homeostasis) and to distribute the minerals to other parts of the body.

5. **Production of blood cells.** Within certain bones a connective tissue called *red bone marrow* produces red blood cells, white blood cells, and platelets, a process called *hemopoiesis* (hēm'-ō-poy-Ē-sis; *hemo-* = blood; *poiesis* = making). Red bone marrow consists of developing blood cells, adipocytes, fibroblasts, and macrophages. It is present in developing bones of the fetus and in some adult bones, such as the pelvis, ribs, sternum (breastbone), vertebrae (backbones), skull, and ends of the arm bones and thigh bones.

6. **Storage of chemical energy.** Triglycerides stored in the adipose cells of *yellow bone marrow* are an important chemical energy reserve. In the newborn, all bone marrow is red and is involved in hemopoiesis. With increasing age, much of the bone marrow changes from red to yellow. Yellow bone marrow consists mainly of adipose cells and a few blood cells.

TYPES OF BONES

Objective: • **Classify bones on the basis of their shape and location.**

Almost all the bones of the body may be classified into five main types based on their shape: long, short, flat, irregular, or sesamoid. *Long bones* have greater length than width and consist of a shaft and a variable number of ends (extremities). They are usually somewhat curved for strength. Long bones include those in the thigh (femur), leg (tibia and fibula), arm (humerus), forearm (ulna and radius), and fingers and toes (phalanges).

Short bones are somewhat cube-shaped and nearly equal in length and width. Examples of short bones include most wrist and ankle bones.

Flat bones are generally thin, afford considerable protection, and provide extensive surfaces for muscle attachment. Bones classified as flat bones include the cranial bones, which protect the brain; the sternum (breastbone) and ribs, which protect organs in the thorax; and the scapulae (shoulder blades).

Irregular bones have complex shapes and cannot be grouped into any of the previous categories. Such bones include the vertebrae of the backbone and some facial bones.

Sesamoid bones (SES-a-moyd = like a sesame seed) develop in certain tendons where there is considerable friction, tension, and physical stress, such as the palms and soles. They vary in number from person to person. The exceptions are the two patellae (kneecaps), which are normally present in everyone. Functionally, sesamoid bones protect tendons from excessive wear and tear. They also may change the direction of pull of a tendon, thereby improving the efficiency of movement at the joint.

An additional type of bone is not included in this classification by shape, but instead is classified by location. *Sutural bones* (SOO-chur-al; *sutur-* = seam) are small bones located within joints called sutures that occur between certain cranial bones. Their number varies greatly from person to person.

PARTS OF A LONG BONE

Objective: • **Describe the parts of a long bone.**

A typical long bone, such as the humerus (arm bone) or femur (thigh bone), consists of the following parts (Figure 6.1):

Figure 6.1 ■ **Parts of a long bone: epiphysis, metaphysis, and diaphysis.** The spongy bone of the epiphysis and metaphysis contains red bone marrow, whereas the medullary cavity of the diaphysis contains yellow bone marrow in an adult.

🔑 **A long bone is covered by articular cartilage at its proximal and distal epiphyses and by periosteum around the diaphysis.**

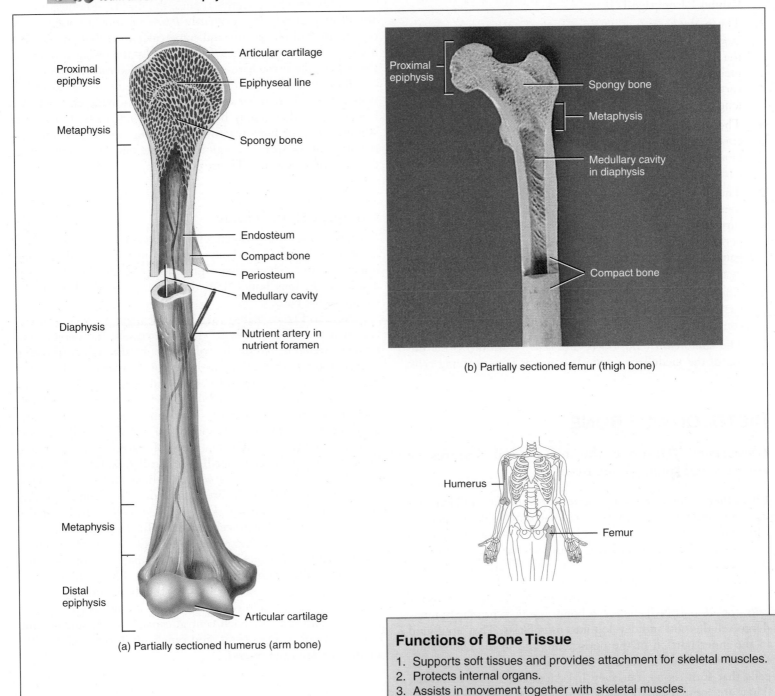

(a) Partially sectioned humerus (arm bone)

(b) Partially sectioned femur (thigh bone)

Functions of Bone Tissue

1. Supports soft tissues and provides attachment for skeletal muscles.
2. Protects internal organs.
3. Assists in movement together with skeletal muscles.
4. Stores and releases minerals.
5. Contains red bone marrow, which produces blood cells.
6. Contains yellow bone marrow, which stores triglycerides.

❓ **Which part of a bone reduces friction at joints? produces blood cells? lines the medullary cavity?**

1. The **diaphysis** (dī-AF-i-sis; *dia-* = through; *-physis* = growing) is the bone's body or shaft, the long, cylindrical main portion of the bone.

2. The **epiphyses** (ē-PIF-i-sēz; *epi-* = above or upon; singular is *epiphysis*) are the distal and proximal ends of the bone (see Exhibit 1.1 on page 12).

3. The **metaphyses** (me-TAF-i-sēz; *meta-* = between; singular is *metaphysis*) are the two regions in a mature bone where the diaphysis joins the epiphyses. In a growing bone, each metaphysis includes an **epiphyseal plate,** a layer of hyaline cartilage that allows the diaphysis of the bone to grow in length (described later in the chapter).

4. The **articular cartilage** is a thin layer of hyaline cartilage covering the epiphysis where the bone forms an articulation (joint) with another bone. The cartilage reduces friction and absorbs shock at freely movable joints.

5. The **periosteum** (per'-ē-OS-tē-um; *peri-* = around) is a tough, white fibrous membrane that surrounds the bone surface wherever it is not covered by articular cartilage. The periosteum is essential for bone growth (in diameter), repair, and nutrition. It also serves as a point of attachment for ligaments and tendons.

6. The **medullary cavity** (MED-yoo-lar'-ē; *medulla* = central part of a structure) is the space within the diaphysis that contains fatty yellow bone marrow in adults.

7. The **endosteum** (end-OS-tē-um; *endo-* = within) is the lining of the medullary cavity. It contains bone-forming cells.

HISTOLOGY OF BONE

Objective: • **Describe the histological features of compact and spongy bone tissue.**

Like other connective tissues, **bone tissue** contains an abundant matrix of intercellular materials that surround widely separated cells. Bone matrix consists of an inorganic component (mineral salts) that makes bone hard and an organic component (mostly collagen fibers) that gives bone its strength. There are four types of cells in bone tissue: osteogenic cells, osteoblasts, osteocytes, and osteoclasts (Figure 6.2). Throughout life, the **osteogenic cells** (os'-tē-ō-JEN-ik; *osteo-* = bone; *-genic* = producing) undergo cell division and develop into osteoblasts. They are found in the inner portion of the periosteum and in the endosteum. **Osteoblasts** (OS-tē-ō-blasts'; *-blasts* = buds or sprouts) are the cells that form bone, but they have lost the ability to divide. They are located on the surface of bone and secrete collagen and other organic components needed to build bone tissue. As osteoblasts surround themselves with matrix materials, they become trapped in their secretions and become osteocytes. **Osteocytes** (OS-tē-ō-sīts'; *-cytes* = cells), mature bone cells, are the main cells in bone tissue and maintain its daily metabolism, such as the exchange of nutrients and wastes with the blood. (NOTE: *Cytes* in bone or any other tissue means cells that maintain the tissue.) Like osteoblasts, osteocytes do not undergo

cell division. **Osteoclasts** (OS-tē-ō-clasts'; *-clast* = to break) are huge cells formed from the fusion of as many as 50 monocytes, one type of white blood cell. Osteoclasts have powerful lysosomal enzymes that function in destruction of bone matrix, a process known as **resorption.**

Unlike other connective tissues, the matrix of bone contains abundant mineral salts, primarily *hydroxyapatite,* a complex salt of calcium and phosphate, and some calcium carbonate. As salts are deposited in the framework formed by the collagen fibers of the matrix, the tissue hardens, a process called **calcification.**

Bone is not completely solid but has many small spaces between its hard components. Some spaces provide channels for blood vessels that supply bone cells with nutrients. Others are storage areas for bone marrow. Depending on the size and distribution of the spaces, a region of a bone may be categorized as compact or spongy (see Figure 6.1).

Compact Bone Tissue

Compact bone tissue contains few spaces. It forms the external layer of all bones of the body and the bulk of the body of long bones. Compact bone tissue provides protection and support and helps the long bones resist the stress placed on them by the weight of the body.

Note in Figure 6.2b-d that compact bone has a concentric-ring structure. Blood vessels, lymphatic vessels, and nerves from the periosteum penetrate the compact bone through **perforating canals** *(Volkmann's canals),* which extend transversely, across the width of bones. The blood vessels and nerves of these canals connect with those of the medullary cavity, periosteum, and **central canals** *(Haversian canals).* The central canals extend vertically, lengthwise through the bone. Around the canals are **concentric lamellae** (la-MEL-ē), rings of hard, calcified matrix. Between the lamellae are small spaces called **lacunae** (la-KOO-nē = little lakes; singular is *lacuna*), which contain osteocytes. Projecting outward in all directions from the lacunae are tiny channels called **canaliculi** (kan'-a-LIK-yoo-lī = small canals), which are filled with extracellular fluid. Inside the canaliculi are slender, fingerlike processes of osteocytes (see inset at top of Figure 6.2b). The canaliculi connect lacunae with one another and, eventually, with the central canals. Thus, an intricate, miniature canal system throughout the bone provides many routes for nutrients and oxygen to reach the osteocytes and for wastes to diffuse away. Each central canal, with its surrounding lamellae, lacunae, osteocytes, and canaliculi, forms a unit called an **osteon** *(Haversian system).*

Spongy Bone Tissue

In contrast to compact bone, **spongy bone** does not contain true osteons. As shown in Figure 6.2b-c, it consists of **trabeculae** (tra-BEK-yoo-lē = little beams), an irregular latticework of thin columns of bone. The spaces between the trabeculae of some bones are filled with red bone marrow. Within the trabeculae are osteocytes that lie in lacunae. Blood vessels from the periosteum penetrate through to the spongy bone, and osteocytes in

Figure 6.2 ■ **Histology of bone.** A photomicrograph of compact bone tissue is provided in Table 4.3J on page 88.

Osteocytes lie in lacunae arranged in concentric circles around a central (Haversian) canal in compact bone, and in lacunae arranged irregularly in the trabeculae of spongy bone.

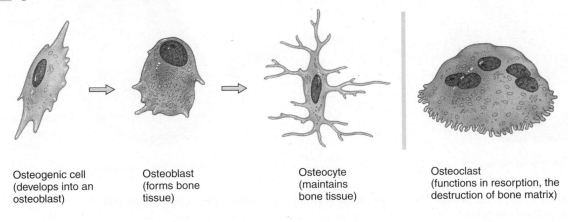

Osteogenic cell (develops into an osteoblast)

Osteoblast (forms bone tissue)

Osteocyte (maintains bone tissue)

Osteoclast (functions in resorption, the destruction of bone matrix)

(a) Types of cells in bone tissue

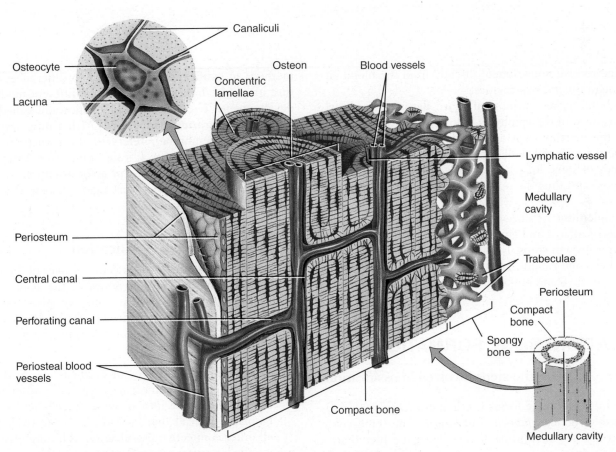

(b) Osteons (Haversian systems) in compact bone and trabeculae in spongy bone

(continues)

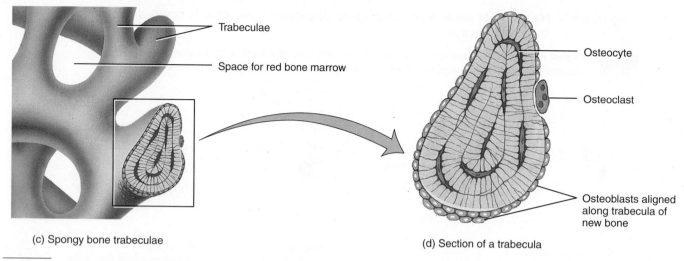

(c) Spongy bone trabeculae

(d) Section of a trabecula

Trabeculae

Space for red bone marrow

Osteocyte

Osteoclast

Osteoblasts aligned along trabecula of new bone

As people age, some central (Haversian) canals may become blocked. What effect would this have on the osteocytes?

the trabeculae receive nourishment directly from the blood circulating through the marrow cavities.

Spongy bone makes up most of short, flat, and irregularly shaped bones, most of the epiphyses of long bones, and a narrow rim around the marrow cavity of the diaphysis of long bones. Spongy bone tissue in the skull, hip bones, ribs, sternum, vertebrae, and ends of some long bones is the only site of red bone marrow storage and thus the only place where hemopoiesis occurs in adults.

Most people think of all bone as a very hard, rigid material. Yet the bones of infants and children are quite soft and become rigid only after growth stops. Even then, bone is constantly being broken down and rebuilt. Let us now see how this dynamic, living tissue is formed and how it grows.

OSSIFICATION: BONE FORMATION

Objective: • **Explain the steps involved in ossification.**

The process by which bone forms is called ***ossification*** (os′-i-fi-KĀ-shun; *ossi-* = bone; *-fication* = making). The embryonic skeleton is composed of fibrous connective tissue membranes and pieces of hyaline cartilage, which are shaped like bones and are the sites where ossification occurs. The first stage in the development of bone is the appearance of osteogenic cells that undergo cell division to produce osteoblasts, which secrete the matrix of bone. Ossification begins during the sixth or seventh week of embryonic life and continues throughout adulthood.

The two methods of bone formation involve the replacement of a preexisting connective tissue with bone. These two methods of ossification do not lead to differences in the structure of mature bones but are simply different methods of bone development. The first type of ossification, called ***intramembranous ossification*** (in′-tra-MEM-bra-nus; *intra-* = within; *membran-* = membrane), refers to the formation of bone directly on or within loose fibrous connective tissue membranes. The second kind, ***endochondral ossification*** (en′-dō-KON-dral; *endo-* = within; *-chondral* = cartilage), refers to the formation of bone within hyaline cartilage.

Intramembranous Ossification

Intramembranous ossification is the simpler of the two methods of bone formation. The flat bones of the skull and mandible (lower jawbone) are formed in this way. Also, the "soft spots" that help the fetal skull pass through the birth canal are replaced after birth by bone by means of intramembranous ossification, which occurs as follows (Figure 6.3):

1 **Development of center of ossification.** At the site where the bone develops, mesenchymal cells cluster together and differentiate, first into osteogenic cells and then into osteoblasts. (Recall that *mesenchyme* is the tissue from which all other connective tissues arise.) The site of such a cluster is called a ***center of ossification.*** Osteoblasts secrete the organic matrix of bone until they are completely surrounded by it.

2 **Calcification.** Then, secretion of matrix stops and the cells, now called osteocytes, lie in lacunae and extend their narrow cytoplasmic processes into canaliculi that radiate in all directions. Within a few days, calcium and other mineral salts are deposited and the matrix hardens or calcifies.

Figure 6.3 ■ **Intramembranous ossification.**

Intramembranous ossification involves the formation of bone directly on or within loose fibrous connective tissue membranes.

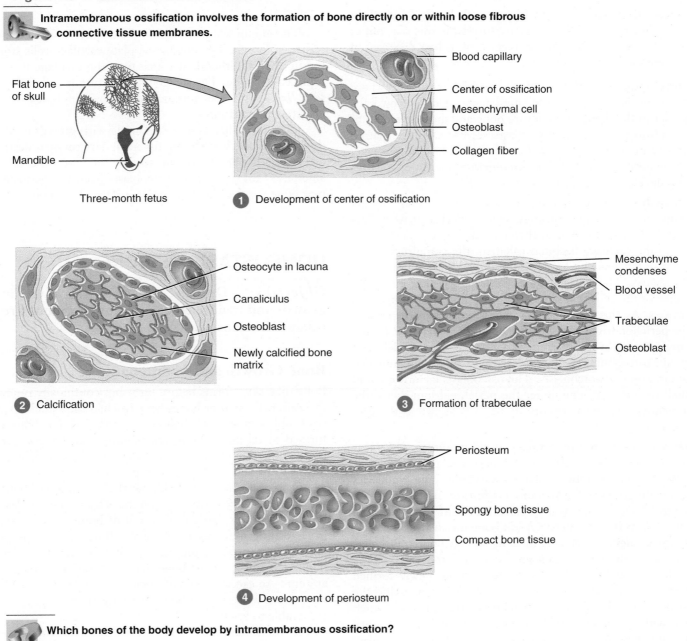

Flat bone of skull

Mandible

Three-month fetus

1 Development of center of ossification

Blood capillary

Center of ossification

Mesenchymal cell

Osteoblast

Collagen fiber

Osteocyte in lacuna

Canaliculus

Osteoblast

Newly calcified bone matrix

2 Calcification

Mesenchyme condenses

Blood vessel

Trabeculae

Osteoblast

3 Formation of trabeculae

Periosteum

Spongy bone tissue

Compact bone tissue

4 Development of periosteum

Which bones of the body develop by intramembranous ossification?

3 **Formation of trabeculae.** As the bone matrix forms, it develops into trabeculae that fuse with one another to form spongy bone. Blood vessels grow into the spaces between the trabeculae and the mesenchyme along the surface of newly formed bone. Connective tissue that is associated with the blood vessels in the trabeculae develops into red bone marrow.

4 **Development of periosteum.** At the periphery of the bone, the mesenchyme condenses and develops into the periosteum. Eventually, the surface layers of the spongy bone are replaced by compact bone, but spongy bone remains in the center. Much of the newly formed bone is remodeled (destroyed and reformed) as the bone is transformed into its adult size and shape.

Endochondral Ossification

The replacement of cartilage by bone is called *endochondral ossification.* Most bones of the body are formed in this way, but as shown in Figure 6.4, this type of ossification is best observed in a long bone such as the shin bone:

❶ Development of the cartilage model. At the site where the bone is going to form, mesenchymal cells crowd together in the shape of the future bone and then develop into chondroblasts. The chondroblasts secrete cartilage matrix, producing a *cartilage model* consisting of hyaline cartilage. A membrane called the *perichondrium* (per′-i-KON-drē-um) develops around the cartilage model.

❷ Growth of the cartilage model. As the cartilage model grows in length and thickness, chondrocytes in its midregion bring about chemical changes that trigger calcification. Once the cartilage becomes calcified, the chondrocytes die because nutrients no longer diffuse quickly enough through the matrix. The thin partitions between the now empty lacunae of the cells that have died break down, forming small cavities. As this is happening, a nutrient artery penetrates the perichondrium and then the bone through a hole (nutrient foramen) in the midregion of the cartilage model. As a result, osteogenic cells in the perichondrium are stimulated to develop into osteoblasts. The osteoblasts lay down a thin shell of compact bone under the perichondrium. Once the perichondrium starts to form bone, it is known as the *periosteum.*

❸ Development of the primary ossification center. Near the middle of the cartilage model, periosteal capillaries grow into the disintegrating calcified cartilage. The capillaries stimulate growth of a *primary ossification center,* a region where bone tissue will replace most of the cartilage. Osteoblasts then begin to deposit bone matrix over the remnants of calcified cartilage, forming spongy bone trabeculae. As the ossification center grows toward the ends of the bone, osteoclasts break down some of the newly formed spongy bone trabeculae. This activity leaves a cavity, the medullary cavity, in the core of the cartilage model, which fills with red bone marrow. This is how the diaphysis, once a solid mass of hyaline cartilage, is replaced by compact bone with a red bone marrow-filled core.

❹ Development of secondary ossification center. When branches of the epiphyseal artery enter an epiphysis, a *secondary ossification center* develops, usually around the time of birth, in much the same way that the primary ossification center develops.

❺ Formation of articular cartilage and epiphyseal plate. One difference between bone formation in the primary and secondary ossification centers is that spongy bone remains in the interior of the epiphyses (no medullary cavities are formed in the epiphyses). Also, hyaline cartilage is not replaced in the epiphyses. Instead, the hyaline cartilage remains, covering the epiphyses as *articular cartilage* and between the diaphysis and epiphysis as the *epiphyseal plate,* the area responsible for the lengthwise growth of long bones.

The epiphyseal plate allows the diaphysis of the bone to increase in length. The epiphyseal plate cartilage cells stop dividing in early adulthood, at which time the cartilage is eventually replaced by bone. The new bony structure is called the *epiphyseal line.* With the appearance of the epiphyseal line, bone growth in length stops.

Growth in diameter occurs along with growth in length. In this process, the bone lining the medullary cavity is destroyed by osteoclasts in the endosteum so that the cavity increases in diameter. At the same time, osteoblasts from the periosteum add new bone tissue around the outer surface of the bone.

HOMEOSTASIS OF BONE

Objective: • **Describe the factors involved in bone growth and maintenance and how hormones regulate calcium homeostasis.**

Bone Growth and Maintenance

Bone, like skin, forms before birth but continually renews itself thereafter. Even after bones have reached their adult shapes and sizes, old bone is continually destroyed, and new bone tissue is formed in its place. Bone *remodeling* is the ongoing replacement of old bone tissue by new bone tissue. Remodeling also removes worn and injured bone, replacing it with new bone tissue.

Osteoclasts are responsible for the resorption of bone tissue. A delicate balance exists between the actions of osteoclasts in removing minerals and collagen and of bone-making osteoblasts in depositing minerals and collagen. If too much mineral is deposited in the bone, the surplus bone tissue may form thick bumps, or spurs, on the bone that can interfere with movement at joints. An excessive loss of calcium or inadequate formation of new tissue weakens the bones, making them overly flexible or vulnerable to fracture.

Normal bone metabolism—growth in the young, bone remodeling in the adult, and repair of fractured bone—depends on several factors. These include (1) adequate minerals, most importantly calcium, phosphorus, and magnesium; (2) vitamins A, C, and D; (3) several hormones; and (4) weight-bearing exercise (exercise that places stress on bones). Before puberty, the main hormones that stimulate bone growth are human growth hormone (hGH), which is produced by the anterior lobe of the pituitary gland, and insulinlike growth factors (IGFs), which are produced locally by bone and also by the liver in response to hGH stimulation. Oversecretion of hGH produces giantism, in which a person becomes much taller and heavier than normal, whereas undersecretion of hGH produces dwarfism (short stature). Thyroid hormones, from the thyroid gland, and insulin, from the pancreas, also stimulate normal bone growth. At puberty, estrogens (sex hormones produced by the ovaries) and

Figure 6.4 ■ **Endochondral ossification of the tibia (shin bone).**

 During endochondral ossification, bone gradually replaces a cartilage model.

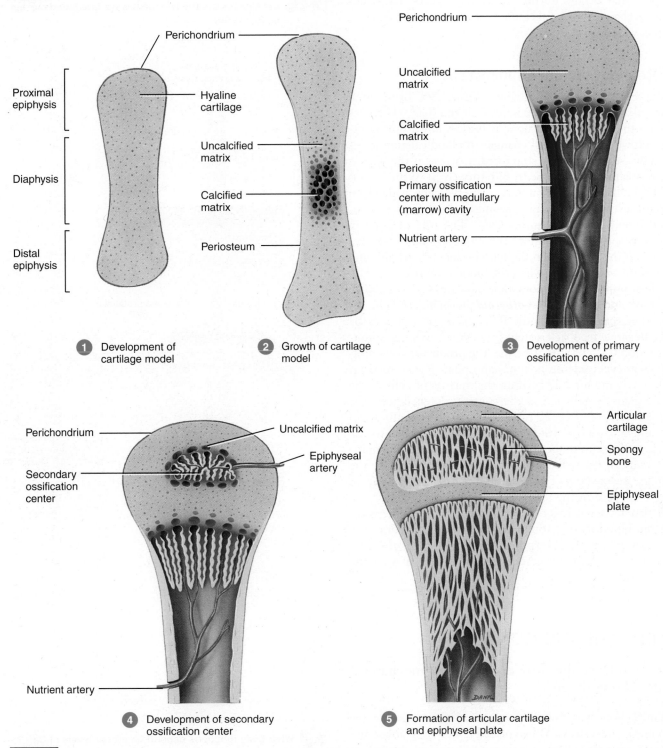

Proximal epiphysis

Diaphysis

Distal epiphysis

Perichondrium

Hyaline cartilage

Uncalcified matrix

Calcified matrix

Periosteum

1 Development of cartilage model

2 Growth of cartilage model

Perichondrium

Uncalcified matrix

Calcified matrix

Periosteum

Primary ossification center with medullary (marrow) cavity

Nutrient artery

3 Development of primary ossification center

Perichondrium

Secondary ossification center

Uncalcified matrix

Epiphyseal artery

Nutrient artery

4 Development of secondary ossification center

Articular cartilage

Spongy bone

Epiphyseal plate

5 Formation of articular cartilage and epiphyseal plate

Which structure signals that bone growth in length has stopped?

androgens (sex hormones produced by the testes in males and the adrenal glands in both sexes) start to be released in larger quantities. These hormones are responsible for the sudden growth spurt that occurs during the teenage years. Estrogens also promote changes in the skeleton that are typical of females, for example, widening of the pelvis.

Bone's Role in Calcium Homeostasis

Bone is the major reservoir of calcium, storing 99% of the total amount of calcium present in the body. Calcium (Ca^{2+}) becomes available to other tissues when bone is broken down during remodeling. However, even small changes in blood calcium levels can be deadly—the heart may stop (cardiac arrest) if the level is too high or breathing may cease (respiratory arrest) if the level is too low. In addition, most functions of nerve cells depend on just the right level of Ca^{2+}, many enzymes require Ca^{2+} as a cofactor, and blood clotting requires Ca^{2+}. The role of bone in calcium homeostasis is to "buffer" the blood calcium level, releasing Ca^{2+} to the blood when the blood calcium level falls and depositing Ca^{2+} back in bone when the blood level rises.

The most important hormone that regulates Ca^{2+} exchange between bone and blood is *parathyroid hormone (PTH)*, secreted by the parathyroid glands. PTH secretion operates by means of a negative feedback system (Figure 6.5). If some stimulus causes blood Ca^{2+} level to decrease, the production of a molecule known as cyclic adenosine monophosphate (cyclic AMP) is increased. The gene for PTH within the nucleus of cells in the parathyroid gland, which acts as the control center, detects the increased production of cyclic AMP, causing PTH synthesis to speed up and more PTH (the output) to be released into the blood. The presence of higher levels of PTH increases the number and activity of osteoclasts (the effectors), which step up the pace of bone resorption. The resulting release of Ca^{2+} from bone into blood returns the blood Ca^{2+} level to normal.

PTH also decreases loss of Ca^{2+} in the urine, so more is retained in the blood, and it stimulates formation of calcitriol, a hormone that promotes absorption of calcium from the gastrointestinal tract. Both of these effects also help elevate the blood Ca^{2+} level.

EXERCISE AND BONE

Objective: • **Describe how exercise and mechanical stress affect bone tissue.**

Within limits, bone has the ability to alter its strength in response to mechanical stress. When placed under stress, bone tissue becomes stronger with time, through increased deposition of mineral salts and production of collagen fibers. Without mechanical stress, bone does not remodel normally because resorption outpaces bone formation. The absence of mechanical stress weakens bone through decreased numbers of collagen fibers and *demineralization,* loss of bone minerals.

The main mechanical stresses on bone are those that result from the pull of skeletal muscles and the pull of gravity. If a per-

Figure 6.5 ■ **Negative feedback system for the regulation of blood calcium (Ca^{2+}) level.**

Release of calcium from bone matrix and retention of calcium by the kidneys are the two main ways that blood calcium level can be increased.

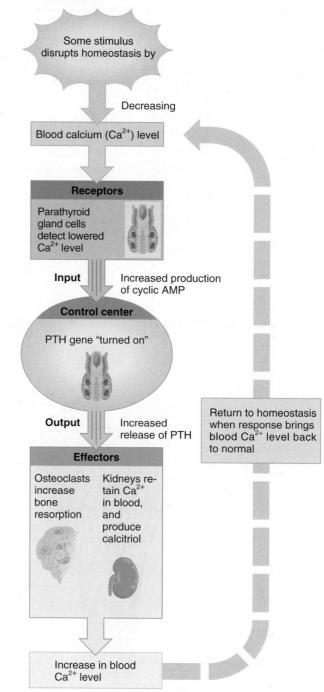

What body functions depend on proper levels of Ca^{2+}?

Table 6.1 / Summary of Factors that Influence Bone Metabolism

Factor	Comment
Minerals	
Calcium and phosphorus	Make bone matrix hard.
Magnesium	Needed for normal activity of osteoblasts.
Vitamins	
Vitamin A	Needed for activity of osteoclasts during remodeling of bone; deficiency stunts bone growth; toxic in high doses.
Vitamin C	Helps maintain bone matrix; deficiency leads to decreased collagen production, which retards bone growth and delays repair of broken bones.
Vitamin D	Active form (calcitriol) is formed in the skin and kidneys from vitamin D; helps build bone by increasing absorption of calcium from small intestine into blood; deficiency causes faulty calcification and retards bone growth; may reduce the risk of osteoporosis but is toxic if taken in high doses.
Hormones	
Human growth hormone (hGH)	Secreted by the anterior lobe of the pituitary gland; promotes general growth of all body tissues, including bone, mainly by stimulating production of insulinlike growth factors.
Insulinlike growth factors (IGFs)	Secreted by liver, bone, and other tissues upon stimulation by human growth hormone; stimulate uptake of amino acids and synthesis of proteins; promote tissue repair and bone growth.
Insulin	Secreted by the pancreas; promotes normal bone growth.
Thyroid hormones (thyroxine and triiodothyronine)	Secreted by thyroid gland; promote normal bone growth.
Parathyroid hormone (PTH)	Secreted by the parathyroid glands; promotes bone resorption by osteoclasts; enhances recovery of Ca^{2+} from urine; promotes formation of the active form of vitamin D (calcitriol).
Exercise	Weight-bearing activities help build thicker, stronger bones and retard the loss of bone mass that occurs as people age.

son is bedridden or has a fractured bone in a cast, the strength of the unstressed bones diminishes. Astronauts subjected to the weightlessness of space also lose bone mass. In both cases, the bone loss can be dramatic, as much as 1% per week. Bones of athletes, which are repetitively and highly stressed, become notably thicker than those of nonathletes. Weight-bearing activities, such as walking or moderate weightlifting, help build and retain bone mass. Adolescents and young adults should engage in regular weight-bearing exercise prior to the closure of the epiphyseal plates to help build total mass before its inevitable reduction with aging. However, the benefits of exercise do not end in young adulthood. Even elderly people can strengthen their bones by engaging in weight-bearing exercise.

Table 6.1 summarizes the factors that influence bone metabolism: growth, remodeling, and repair of fractured bones.

BONE SURFACE MARKINGS

Objective: • **Describe the principal surface markings on bones and the function of each.**

The surfaces of bones have various structural features adapted to specific functions. These features are called ***bone surface markings.*** For example, long weight-bearing bones have large, rounded ends that can form sturdy joints with other bones that have depressions to receive the rounded ends. Following are examples of bone surface markings:

1. **Depressions and openings.**

   A ***foramen*** (fō-RĀ-men = hole; plural is *foramina*) is an opening through which blood vessels, nerves, or ligaments pass, such as the foramen magnum of the occipital bone (see Figure 6.8).

   A ***meatus*** (mē-Ā-tus = passage) is a tubelike channel extending within a bone, such as the external auditory meatus of the temporal bone (see Figure 6.7b).

   A ***fossa*** (FOS-ah = ditch or trench) is a shallow depression in or on a bone, such as the mandibular fossa of the temporal bone (see Figure 6.8).

2. **Processes that form joints.**

   A ***condyle*** (KON-dīl = knucklelike process) is a large, rounded prominence that forms a joint, such as the medial condyle of the femur (see Figure 6.26).

   A ***head*** is a rounded projection that forms a joint and is supported on the constricted portion (neck) of a bone, such as the head of the femur (see Figure 6.26).

   A ***facet*** is a smooth, flat articular surface, such as the facet on a vertebra (see Figure 6.16).

3. **Processes to which tendons, ligaments, and other connective tissues attach.**

A *tuberosity* is a large, rounded projection, usually with a rough surface, such as the deltoid tuberosity of the humerus (see Figure 6.21).

A *spinous process* or *spine* is a sharp, slender projection, such as the spinous process of a vertebra (see Figure 6.17).

A *trochanter* (trō-KAN-ter) is a large, blunt projection found only on the femur, such as the greater trochanter (see Figure 6.26).

A *crest* is a prominent border or ridge, such as the iliac crest of the hip bone (see Figure 6.24).

DIVISIONS OF THE SKELETAL SYSTEM

Objective: • **Classify the bones of the body into axial and appendicular divisions.**

The adult human skeleton consists of 206 bones grouped in two principal divisions: 80 in the *axial skeleton* and 126 in the *appendicular skeleton* (Table 6.2 and Figure 6.6). The axial skeleton includes the bones of the skull, auditory ossicles (ear bones), hyoid bone, ribs, sternum, and vertebrae. The appendicular skeleton contains the bones of the upper and lower limbs plus the bone groups called girdles that connect the limbs to the axial skeleton.

SKULL

Objective: • **Name the cranial and facial bones and indicate their locations and major structural features.**

The *skull,* which contains 22 bones, rests on top of the vertebral column. It includes two sets of bones: cranial bones and facial bones. The eight *cranial bones* form the cranial cavity that encloses and protects the brain. They are the frontal bone, two parietal bones, two temporal bones, occipital bone, sphenoid bone, and ethmoid bone. Fourteen *facial bones* form the face: two nasal bones, two maxillae, two zygomatic bones, the mandible, two lacrimal bones, two palatine bones, two inferior nasal conchae, and the vomer. Figure 6.7 shows these bones from three different viewing directions to permit you to view all of the bones from their best perspective.

The cranial bones have functions besides protection of the brain. Their inner surfaces attach to membranes (meninges) that stabilize the positions of the brain, blood vessels, and nerves. Their outer surfaces provide large areas of attachment for muscles that move various parts of the head. Besides forming the framework of the face, the facial bones protect and provide support for the entrances to the digestive and respiratory systems. The facial bones also provide attachment for some muscles that are involved in producing various facial expressions. Together, the cranial and facial bones protect and support the delicate special sense organs for vision, taste, smell, hearing, and equilibrium (balance).

Table 6.2 / The Bones of the Adult Skeletal System

Division of the Skeleton	Structure	Number of Bones
Axial Skeleton		
	Skull	
	Cranium	8
	Face	14
	Hyoid	1
	Auditory ossicles	6
	Vertebral column	26
	Thorax	
	Sternum	1
	Ribs	24
		Subtotal = 80
Appendicular Skeleton		
	Pectoral (shoulder) girdles	
	Clavicle	2
	Scapula	2
	Upper limbs	
	Humerus	2
	Ulna	2
	Radius	2
	Carpals	16
	Metacarpals	10
	Phalanges	28
	Pelvic (hip) girdle	
	Hip, pelvic, or coxal bone	2
	Lower limbs	
	Femur	2
	Fibula	2
	Tibia	2
	Patella	2
	Tarsals	14
	Metatarsals	10
	Phalanges	28
		Subtotal = 126
		Total = 206

Figure 6.6 ■ **Divisions of the skeletal system.** The axial skeleton is indicated in blue. (Note the position of the hyoid bone in Figure 6.7b.)

 The adult human skeleton consists of 206 bones grouped into axial and appendicular divisions.

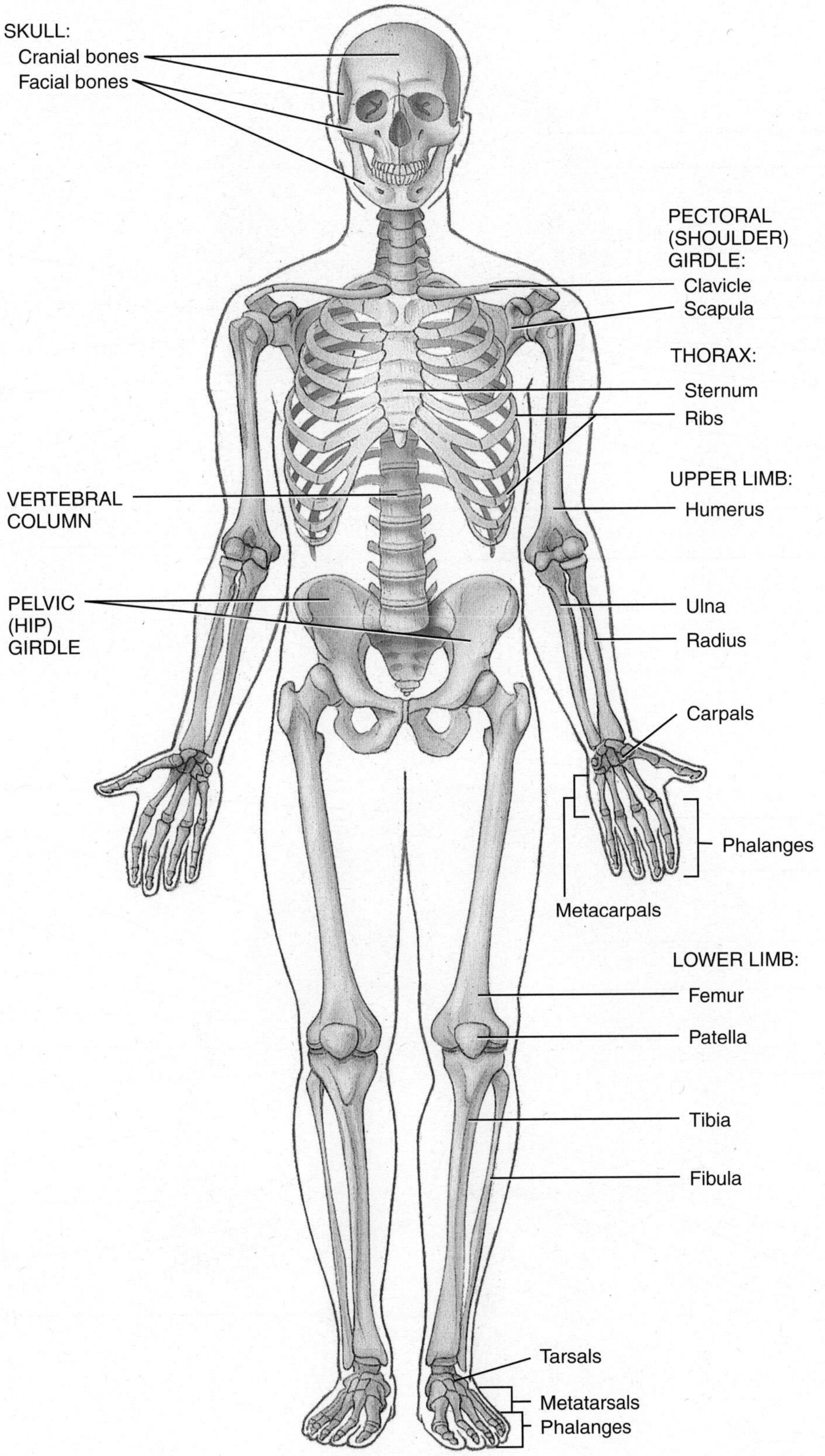

SKULL:
Cranial bones
Facial bones

PECTORAL
(SHOULDER)
GIRDLE:
Clavicle
Scapula

THORAX:
Sternum
Ribs

UPPER LIMB:
Humerus

VERTEBRAL
COLUMN

Ulna
Radius

PELVIC
(HIP)
GIRDLE

Carpals

Phalanges

Metacarpals

LOWER LIMB:
Femur
Patella

Tibia

Fibula

Tarsals
Metatarsals
Phalanges

Anterior view

 Identify each of the following bones as part of the axial skeleton or the appendicular skeleton: skull, clavicle, vertebral column, shoulder girdle, humerus, pelvic girdle, and femur.

Figure 6.7 ■ **Skull.** Although the hyoid bone is not part of the skull, it is included in (b) and (c) for reference.

🔑 The skull consists of two sets of bones: Eight cranial bones form the cranial cavity and fourteen facial bones form the face.

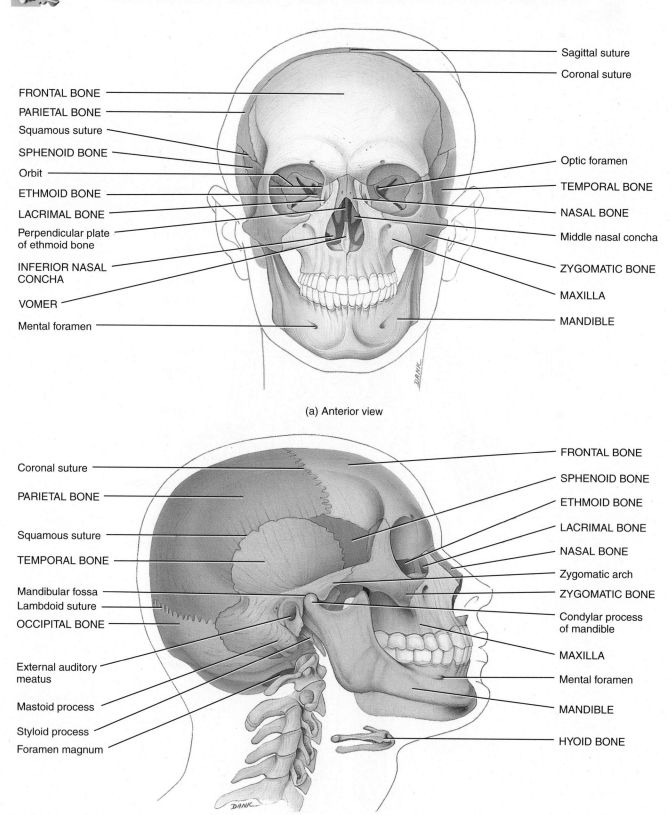

(a) Anterior view

(b) Right lateral view

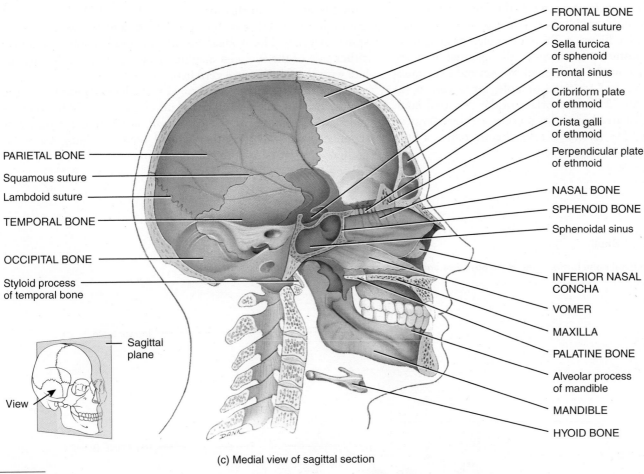

PARIETAL BONE

Squamous suture

Lambdoid suture

TEMPORAL BONE

OCCIPITAL BONE

Styloid process
of temporal bone

Sagittal
plane

View

FRONTAL BONE

Coronal suture

Sella turcica
of sphenoid

Frontal sinus

Cribriform plate
of ethmoid

Crista galli
of ethmoid

Perpendicular plate
of ethmoid

NASAL BONE

SPHENOID BONE

Sphenoidal sinus

INFERIOR NASAL
CONCHA

VOMER

MAXILLA

PALATINE BONE

Alveolar process
of mandible

MANDIBLE

HYOID BONE

(c) Medial view of sagittal section

What are the names of the cranial bones?

A *suture* (SOO-chur = seam) is an immovable joint that is found only between skull bones. Sutures hold skull bones together. Of the many sutures that are found in the skull, we will identify only four prominent ones (Figure 6.7):

1. The *coronal suture* (kō-RŌ-nal; *coron-* = crown) unites the frontal bone and two parietal bones.

2. The *sagittal suture* (SAJ-i-tal; *sagitt-* = arrow) unites the two parietal bones.

3. The *lambdoid suture* (LAM-doyd; so named because its shape resembles the Greek letter lambda, Λ) unites the parietal bones to the occipital bone.

4. The *squamous suture* (SKWĀ-mus; *squam-* = flat) unites the parietal bones to the temporal bones.

Cranial Bones

The *frontal bone* forms the forehead (the anterior part of the cranium), the roofs of the *orbits* (eye sockets; Figure 6.7a), and most of the anterior (front) part of the cranial floor. The *frontal sinuses* lie deep within the frontal bone (Figure 6.7c). These mucous membrane-lined cavities act as sound chambers that give the voice resonance. Other functions of the sinuses are given on page 132.

The two *parietal bones* (pa-RĪ-e-tal; *pariet-* = wall) form the greater portion of the sides and roof of the cranial cavity (Figure 6.7).

The two *temporal bones* (*tempor-* = temples) form the inferior (lower) sides of the cranium and part of the cranial floor. In the lateral view of the skull (Figure 6.7b), note that the temporal and zygomatic bones join to form the *zygomatic arch*. The *mandibular fossa* forms a joint with a projection on the mandible (lower jawbone) called the condylar process to form the *temporomandibular joint (TMJ)*. The mandibular fossa can be seen in Figure 6.8. The *external auditory meatus* is the canal in the temporal bone that leads to the middle ear. The *mastoid process (mastoid* = breast-shaped; see Figure 6.7b) is a rounded projection of the temporal bone posterior to (behind) the external auditory meatus. It serves as a point of attachment for several neck muscles. The *styloid process (styl-* = stake or pole; see Figure 6.7b) projects downward from the undersurface of the temporal bone and serves as a point of attachment for muscles and ligaments of the tongue and neck. The *carotid foramen* (Figure 6.8) is a hole through which the carotid artery passes.

The *occipital bone* (ok-SIP-i-tal; *occipit-* = back of head) forms the posterior part and most of the base of the cranium (Figures 6.7b,c and 6.8). The *foramen magnum* (*magnum* = large), the largest foramen in the skull, passes through the occipital bone. Within this foramen are the medulla oblongata of the brain, connecting to the spinal cord, and the vertebral and spinal arteries. The *occipital condyles* are oval processes, one on either side of the foramen magnum (see Figure 6.8), that articulate with the first cervical vertebra.

The *sphenoid bone* (SFE-noyd = wedge-shaped) lies at the middle part of the base of the skull (Figure 6.9). This bone is called the keystone of the cranial floor because it articulates (connects) with all the other cranial bones, holding them together. The shape of the sphenoid bone resembles a bat with

Figure 6.8 ■ **Skull.**

The occipital bone forms most of the posterior and inferior portion of the cranium.

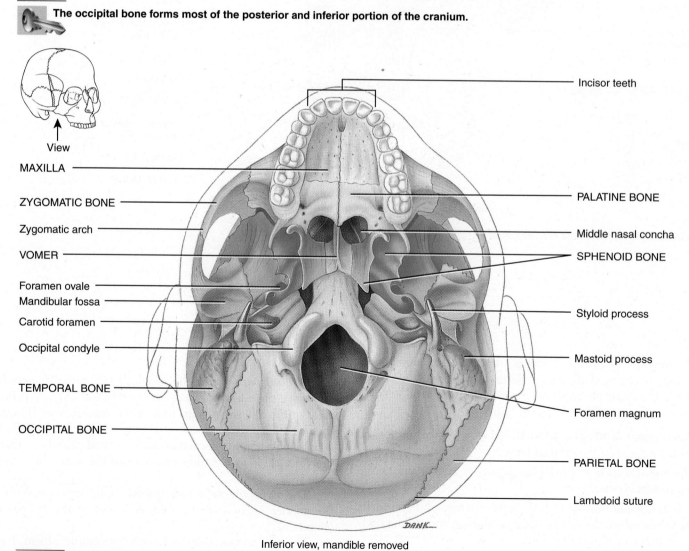

View

MAXILLA

ZYGOMATIC BONE

Zygomatic arch

VOMER

Foramen ovale

Mandibular fossa

Carotid foramen

Occipital condyle

TEMPORAL BONE

OCCIPITAL BONE

Incisor teeth

PALATINE BONE

Middle nasal concha

SPHENOID BONE

Styloid process

Mastoid process

Foramen magnum

PARIETAL BONE

Lambdoid suture

DANK

Inferior view, mandible removed

What is the largest foramen in the skull?

outstretched wings. The cubelike central portion of the sphenoid bone contains the *sphenoidal sinuses*, which drain into the nasal cavity (see Figure 6.12). On the superior surface of the sphenoid is a depression called the *sella turcica* (SEL-a TUR-si-ka = Turkish saddle), which cradles the pituitary gland. Two nerves pass through foramina in the sphenoid bone: the mandibular nerve through the *foramen ovale* and the optic nerve through the *optic foramen*.

The **ethmoid bone** (ETH-moid = sievelike) is a light, spongy bone located in the anterior part of the cranial floor between the orbits (Figure 6.10). It forms part of the anterior portion of the cranial floor, the medial wall of the orbits, the superior portions of the nasal septum, a partition that divides the nasal cavity into right and left sides, and most of the side walls of the nasal cavity. The ethmoid bone contains 3 to 18 air spaces, or "cells," that give this bone a sievelike appearance. The ethmoidal cells together form the *ethmoidal sinuses* (see Figure 6.12). The *perpendicular plate* forms the upper portion of the nasal septum (see Figure 6.11). The *cribriform plate* (KRIB-ri-form) forms the roof of the nasal cavity (Figure 6.10). It contains the *olfactory foramina* (*olfact-* = to smell) through which fibers of the olfactory nerve pass (see Figure 6.9). Projecting upward from the cribriform plate is a triangular process called the *crista galli* (= cock's comb), which serves as a point of attachment for the membranes (meninges) that cover the brain (Figure 6.10).

Also part of the ethmoid bone are two thin, scroll-shaped bones on either side of the nasal septum. These are called the *superior nasal concha* (KONG-ka; *conch-* = shell) and the *middle*

Figure 6.9 ■ **Sphenoid bone.**

The sphenoid bone is called the keystone of the cranial floor because it articulates with all other cranial bones, holding them together.

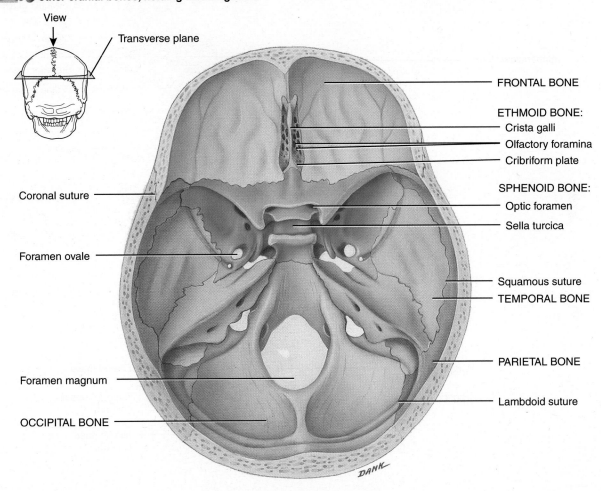

Viewed from above in floor of cranium

 Starting at the crista galli of the ethmoid bone and going in a clockwise direction, what are the names of the bones that articulate with the sphenoid bone?

nasal concha. The plural is *conchae* (KONG-kē). The conchae cause turbulence in inhaled air, which results in many inhaled particles striking and becoming trapped in the mucus that lines the nasal passageways. This turbulence thus cleanses the inhaled air before it passes into the rest of the respiratory tract.

Facial Bones

The shape of the face changes dramatically during the first two years after birth. The brain and cranial bones expand, the teeth form and erupt (emerge), and the paranasal sinuses increase in size. Growth of the face ceases at about 16 years of age.

The paired **nasal bones** form part of the bridge of the nose (see Figure 6.7a). The rest of the supporting tissue of the nose consists of cartilage.

The paired **maxillae** (mak-SIL-ē = jawbones; singular is *maxilla*) unite to form the upper jawbone and articulate with every bone of the face except the mandible (lower jawbone) (see Figure 6.7). Each maxilla contains a *maxillary sinus* that empties

Figure 6.10 ■ **Ethmoid bone.**

The ethmoid bone is the major supporting structure of the nasal cavity.

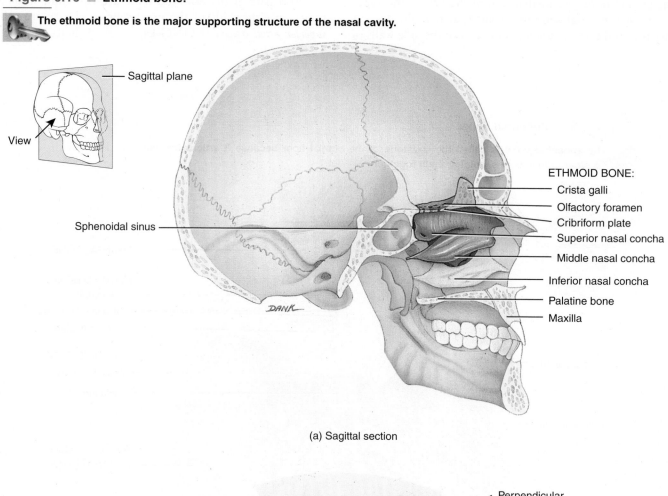

Sagittal plane

View

Sphenoidal sinus

ETHMOID BONE:
Crista galli
Olfactory foramen
Cribriform plate
Superior nasal concha
Middle nasal concha
Inferior nasal concha
Palatine bone
Maxilla

DANK

(a) Sagittal section

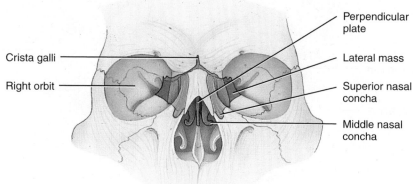

Perpendicular plate

Crista galli

Right orbit

Lateral mass

Superior nasal concha

Middle nasal concha

(b) Anterior view of position of ethmoid bone in skull

 What part of the ethmoid bone forms the top part of the nasal septum?

into the nasal cavity (see Figure 6.12). The *alveolar process* (al-VĒ-ō-lar; *alveol-* = small cavity) of the maxilla is an arch that contains the *alveoli* (sockets) for the maxillary (upper) teeth. The maxilla forms the anterior three-quarters of the hard palate. The fusion of the left and right maxillary bones is normally completed before birth. If fusion does not occur, a condition called *cleft palate* results. A *cleft lip*, in which the upper lip is split, often is associated with cleft palate. Depending on the extent and position of the cleft, speech and swallowing may be affected. Facial and oral surgeons recommend closure of a cleft lip during the first few weeks of life. Repair of a cleft palate is done between 12 and 18 months of age, ideally before the child begins to talk. The surgical results for both conditions usually are excellent.

The two L-shaped **palatine bones** (PAL-a-tīn; *palat-* = roof of mouth) are fused and form the posterior portion of the hard palate, part of the floor and lateral wall of the nasal cavity, and a small portion of the floors of the orbits (see Figure 6.8). In cleft palate, the palatine bones may also be incompletely fused.

The **mandible** (*mand-* = to chew), or lower jawbone, is the largest, strongest facial bone (see Figure 6.7b). It is the only movable skull bone. Recall from our discussion of the temporal bone that the mandible has a *condylar process* (KON-di-lar). This process articulates with the mandibular fossa of the temporal bone to form the temporomandibular joint. The mandible, like the maxilla, has an *alveolar process* containing the *alveoli* (sockets) for the mandibular (lower) teeth (see Figure 6.7c). The *mental foramen* (*ment-* = chin) is a hole in the mandible that can be used by dentists to reach the mental nerve when injecting anesthetics (see Figure 6.7a).

The two **zygomatic bones** (*zygo-* = like a yoke), commonly called cheekbones, form the prominences of the cheeks and part of the lateral wall and floor of each orbit (see Figure 6.7a,b). They articulate with the frontal, maxilla, sphenoid, and temporal bones.

The paired **lacrimal bones** (LAK-ri-mal; *lacrim-* = teardrop), the smallest bones of the face, are thin and roughly resemble a fingernail in size and shape. The lacrimal bones can be seen in the anterior and lateral views of the skull in Figure 6.7 (Figure 6.7a,b).

The two **inferior nasal conchae** are scroll-like bones that project into the nasal cavity below the superior and middle nasal conchae of the ethmoid bone (see Figures 6.7a and 6.10). They serve the same function as the other nasal conchae: the filtration of air before it passes into the lungs.

The **vomer** (VŌ-mer = plowshare) is a roughly triangular bone on the floor of the nasal cavity that articulates inferiorly with both the maxillae and palatine bones along the midline of the skull. The vomer, clearly seen in the anterior view of the skull in Figure 6.7a and the inferior view in Figure 6.8, is one of the components of the *nasal septum*, a partition that divides the nasal cavity into right and left sides.

The nasal septum is formed by the vomer, septal cartilage, and the perpendicular plate of the ethmoid bone (Figure 6.11a). The anterior border of the vomer articulates with the septal cartilage (hyaline cartilage) to form the anterior portion of the septum. The upper border of the vomer articulates with the perpendicular plate of the ethmoid bone to form the remainder of the nasal septum. A **deviated nasal septum** is one that bends sideways from the middle of the nose. If the deviation is severe, it may entirely block the nasal passageway. Even a partial blockage may lead to nasal congestion, blockage of the paranasal sinus openings, chronic sinusitis, headache, and nosebleeds. The condition can be corrected surgically.

Figure 6.11 ■ Maxillae.

🔑 **The maxillae articulate with every bone of the face, except the mandible.**

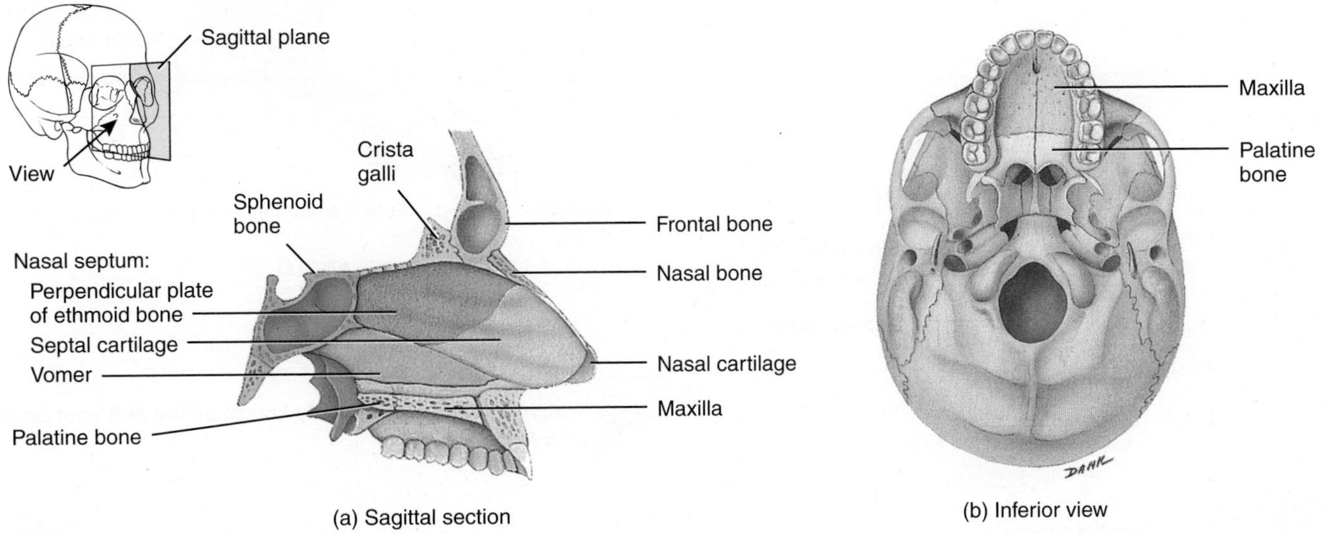

(a) Sagittal section

(b) Inferior view

Which bones form the hard palate?

Special Features of the Skull

Now that you are familiar with the names of the skull bones, we will take a closer look at two special features of the skull: paranasal sinuses and fontanels (present in the fetus and newborn babies).

Paranasal Sinuses

Paired cavities, the *paranasal sinuses* (*para-* = beside), are located in certain skull bones near the nasal cavity (Figure 6.12). The paranasal sinuses are lined with mucous membranes that are continuous with the lining of the nasal cavity. Skull bones containing paranasal sinuses include the frontal bone (frontal sinus), sphenoid bone (sphenoid sinus), ethmoid bone (ethmoidal sinuses), and maxillae (maxillary sinuses). Besides producing mucus, the paranasal sinuses serve as resonating chambers, producing the unique sounds of each of our speaking and singing voices.

Secretions produced by the mucous membranes of the paranasal sinuses drain into the nasal cavity. An inflammation of the membranes due to an allergic reaction or infection is called *sinusitis*. If the membranes swell enough to block drainage into the nasal cavity, fluid pressure builds up in the paranasal sinuses, resulting in a sinus headache.

Fontanels

Recall that the skeleton of a newly formed embryo consists of cartilage and fibrous connective tissue membrane structures shaped like bones. Gradually, ossification occurs—bone replaces the cartilage or fibrous connective tissue membranes. Included among the membrane-filled spaces called *fontanels* (fon-ta-NELZ = little fountains), found between cranial bones at birth (Figure 6.13), are the anterior fontanel (the "soft spot" on the top of an infant's head), the posterior fontanel, the anterolateral fontanel, and the posterolateral fontanel. These areas of fibrous connective tissue membranes will eventually be replaced with bone by intramembranous ossification and become sutures. Functionally, the fontanels enable the fetal skull to be compressed as it passes through the birth canal and permit rapid growth of the brain during infancy. The form and location of several fontanels are shown and described in Table 6.3.

Figure 6.12 ■ Paranasal sinuses.

 Paranasal sinuses are mucous membrane-lined spaces in the frontal, sphenoid, ethmoid, and maxillary bones that connect to the nasal cavity.

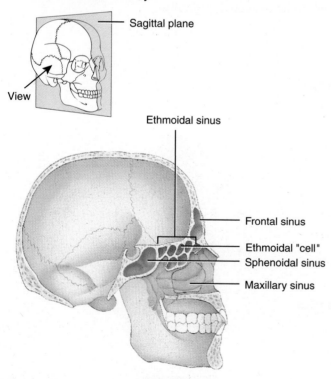

Sagittal plane

View

Ethmoidal sinus

Frontal sinus

Ethmoidal "cell"

Sphenoidal sinus

Maxillary sinus

Sagittal section

 What are two main functions of the paranasal sinuses?

Figure 6.13 ■ Fontanels of the skull at birth.

 Fontanels are membrane-filled spaces between cranial bones that are present at birth.

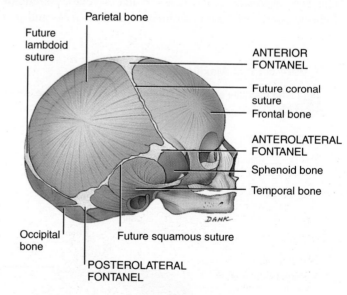

Parietal bone

Future lambdoid suture

ANTERIOR FONTANEL

Future coronal suture

Frontal bone

ANTEROLATERAL FONTANEL

Sphenoid bone

Temporal bone

DANK

Occipital bone

Future squamous suture

POSTEROLATERAL FONTANEL

Right lateral view

 Which fontanel is commonly known as the soft spot on the top of an infant's head?

Table 6.3 / **Fontanels (see also Figure 6.13)**			
Fontanel		**Location**	**Description**
Anterior		Between the two parietal bones and the frontal bone.	Roughly diamond-shaped, the largest of the fontanels; usually closes 18–24 months after birth.
Posterior		Between the two parietal bones and the occipital bone.	Diamond-shaped, considerably smaller than the anterior fontanel; generally closes about 2 months after birth.
Anterolateral		One on each side of the skull between the frontal, parietal, temporal, and sphenoid bones.	Small and irregular in shape; normally close about 3 months after birth.
Posterolateral		One on each side of the skull between the parietal, occipital, and temporal bones.	Irregularly shaped; begin to close 1 or 2 months after birth, but closure is generally not complete until 12 months.

HYOID BONE

Objective: • **Describe the relationship of the hyoid bone to the skull.**

The single *hyoid bone* (HĪ-oyd = U-shaped) is a unique component of the axial skeleton because it does not articulate with or attach to any other bone. Rather, it is suspended from the styloid processes of the temporal bones by ligaments and muscles. The hyoid bone is located in the neck between the mandible and larynx (see Figure 6.7b). It supports the tongue and provides attachment sites for some tongue muscles and for muscles of the neck and pharynx. The hyoid bone, as well as the cartilage of the larynx and trachea, are often fractured during strangulation. As a result, they are carefully examined in an autopsy when strangulation is suspected.

VERTEBRAL COLUMN

Objective: • **Identify the regions and normal curves of the vertebral column and describe its structural and functional features.**

The *vertebral column,* also called the *spine* or *backbone,* is composed of a series of bones called *vertebrae* (VER-te-brē; singular is *vertebra*). The vertebral column functions as a strong, flexible rod that can rotate and move forward, backward, and sideways. It encloses and protects the spinal cord, supports the head, and serves as a point of attachment for the ribs and the muscles of the back.

Regions of the Vertebral Column

The total number of vertebrae during early development is 33. Then, several vertebrae in the sacral and coccygeal regions fuse. As a result, the adult vertebral column, also called the spinal column, typically contains 26 vertebrae (Figure 6.14). These are distributed as follows:

- **7 *cervical vertebrae*** (*cervic-* = neck) in the neck region
- **12 *thoracic vertebrae*** (*thorax* = chest) posterior to the thoracic cavity
- **5 *lumbar vertebrae*** (*lumb-* = loin) supporting the lower back
- **1 *sacrum*** (SĀ-krum = sacred bone) consisting of five fused *sacral vertebrae*
- **1 *coccyx*** (KOK-siks = cuckoo, because the shape resembles the bill of a cuckoo bird) consisting of four fused *coccygeal vertebrae* (kok-SIJ-ē-al)

Whereas the cervical, thoracic, and lumbar vertebrae are movable, the sacrum and coccyx are immovable. Between adjacent vertebrae from the second cervical vertebra to the sacrum are *intervertebral discs* (*inter-* = between). Each disc has an outer ring of fibrocartilage and a soft, pulpy, highly elastic interior. The discs form strong joints, permit various movements of the vertebral column, and absorb vertical shock.

Normal Curves of the Vertebral Column

When viewed from the side, the vertebral column shows four slight bends called *normal curves* (Figure 6.14). Relative to the front of the body, the *cervical* and *lumbar curves* are convex

Figure 6.14 ■ **Vertebral column.**

The adult vertebral column typically contains 26 vertebrae.

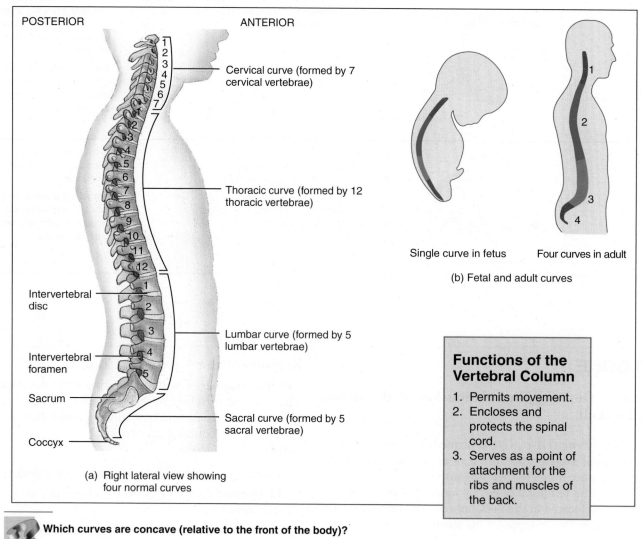

POSTERIOR   ANTERIOR

Cervical curve (formed by 7 cervical vertebrae)

Thoracic curve (formed by 12 thoracic vertebrae)

Intervertebral disc

Lumbar curve (formed by 5 lumbar vertebrae)

Intervertebral foramen

Sacrum

Sacral curve (formed by 5 sacral vertebrae)

Coccyx

(a) Right lateral view showing four normal curves

Single curve in fetus   Four curves in adult

(b) Fetal and adult curves

Functions of the Vertebral Column

1. Permits movement.
2. Encloses and protects the spinal cord.
3. Serves as a point of attachment for the ribs and muscles of the back.

Which curves are concave (relative to the front of the body)?

(bulging out), whereas the ***thoracic*** and ***sacral curves*** are concave (cupping in). The curves of the vertebral column increase its strength, help maintain balance in the upright position, absorb shocks during walking and running, and help protect the vertebrae from breaks.

In the fetus, there is a single concave curve (Figure 6.14b). At about the third month after birth, when an infant begins to hold its head erect, the cervical curve develops. Later, when the child sits up, stands, and walks, the lumbar curve develops.

Vertebrae

Vertebrae in different regions of the spinal column vary in size, shape, and detail, but they are similar enough that we can discuss the structure and functions of a typical vertebra (Figure 6.15).

1. The ***body,*** the thick, disc-shaped front portion, is the weight-bearing part of a vertebra.

2. The ***vertebral arch*** extends backwards from the body of the vertebra. It is formed by two short, thick processes, the *pedicles* (PED-i-kuls = little feet), which project backward from the body to unite with the laminae. The *laminae* (LAM-i-nē = thin layers) are the flat parts of the arch and end in a single sharp, slender projection *(spinous process).* The space between the vertebral arch and body contains the spinal cord and is known as the *vertebral foramen.* Together, the vertebral foramina of all vertebrae form the *vertebral canal.* When the vertebrae are stacked on top of one another, there is an opening between adjoining vertebrae on both sides of the column. Each opening, called an *intervertebral foramen,* permits the passage of a single spinal nerve.

3. Seven ***processes*** arise from the vertebral arch. At the point where a lamina and pedicle join, a *transverse process* extends laterally on each side. A single *spinous process (spine)* projects from the junction of the laminae. These three processes

Figure 6.15 ■ **Structure of a typical vertebra, as illustrated by a thoracic vertebra.** (Note the facets for the ribs, which vertebrae other than the thoracic vertebrae do not have.) In (b), only one spinal nerve has been included, and it has been extended beyond the intervertebral foramen for clarity.

A vertebra consists of a body, a vertebral arch, and several processes.

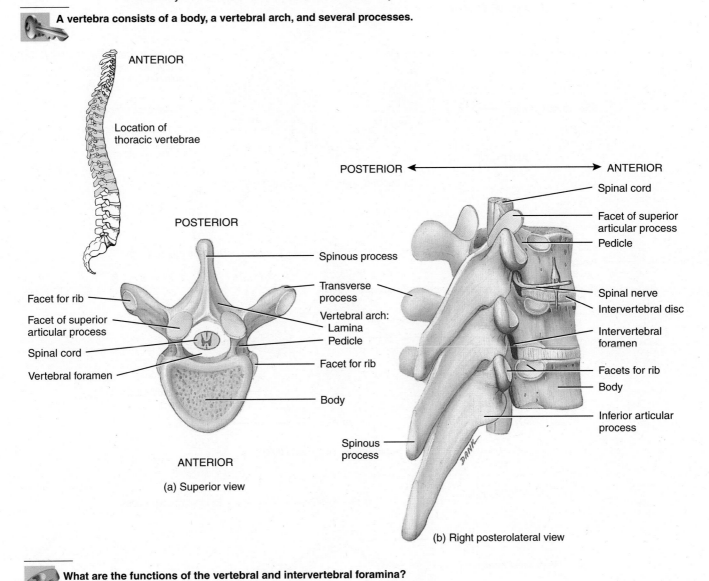

(a) Superior view

(b) Right posterolateral view

What are the functions of the vertebral and intervertebral foramina?

serve as points of attachment for muscles. The remaining four processes form joints with other vertebrae above or below. The two *superior articular processes* of a vertebra articulate with the vertebra immediately above them. The two *inferior articular processes* of a vertebra articulate with the vertebra immediately below them. The articulating surfaces of the articular processes are called *facets* (= little faces).

Vertebrae in each region are numbered in sequence from top to bottom. The seven **cervical vertebrae** are termed C1 through C7 (Figure 6.16). The spinous processes of the second through sixth cervical vertebrae are often *bifid*, or split into two parts (Figure 6.16b,c). All cervical vertebrae have three foramina: one vertebral foramen and two transverse foramina. Each

cervical transverse process contains a *transverse foramen* through which blood vessels and nerves pass.

The first two cervical vertebrae differ considerably from the others. The first cervical vertebra (C1), the **atlas,** supports the head and is named for the mythological Atlas who supported the world on his shoulders. The atlas lacks a body and a spinous process. The upper surface contains *superior articular facets* that articulate with the occipital bone of the skull. This articulation permits you to nod your head to indicate "yes." The inferior surface contains *inferior articular facets* that articulate with the second cervical vertebra.

The second cervical vertebra (C2), the **axis,** does have a body and a spinous process. A tooth-shaped process called the **dens** (= tooth) projects up through the vertebral foramen of the

Figure 6.16 ■ **Cervical vertebrae.**

 The cervical vertebrae are found in the neck region.

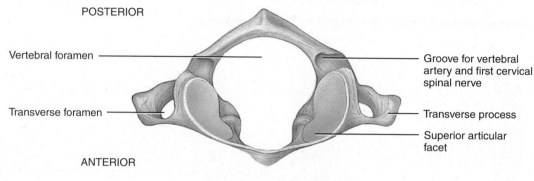

POSTERIOR

Vertebral foramen

Transverse foramen

ANTERIOR

Groove for vertebral
artery and first cervical
spinal nerve

Transverse process

Superior articular
facet

(a) Superior view of the atlas (C1)

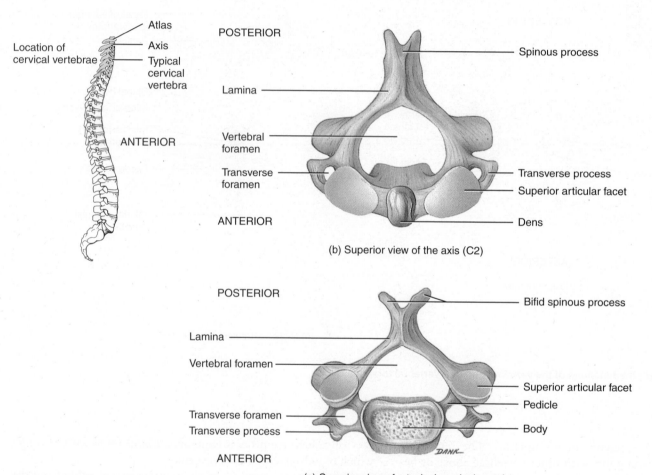

Location of
cervical vertebrae

Atlas

Axis

Typical
cervical
vertebra

ANTERIOR

POSTERIOR

Lamina

Vertebral
foramen

Transverse
foramen

ANTERIOR

Spinous process

Transverse process

Superior articular facet

Dens

(b) Superior view of the axis (C2)

POSTERIOR

Lamina

Vertebral foramen

Transverse foramen

Transverse process

ANTERIOR

Bifid spinous process

Superior articular facet

Pedicle

Body

(c) Superior view of a typical cervical vertebra

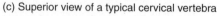 **Which bones permit the movement of the head to signify "no"?**

atlas. The dens is a pivot on which the atlas and head rotate, as in side-to-side rotation of the head to signify "no."

The third through sixth cervical vertebrae (C3 through C6), represented by the vertebra in Figure 6.16c, correspond to the structural pattern of the typical cervical vertebra described previously. The seventh cervical vertebra (C7), called the *vertebra prominens*, is somewhat different. It is marked by a single,

large spinous process that can be seen and felt at the base of the neck.

Thoracic vertebrae (T1 through T12) are considerably larger and stronger than cervical vertebrae. Distinguishing features of the thoracic vertebrae are their facets for articulating with the ribs (see Figure 6.15). Movements of the thoracic region are limited by the attachment of the ribs to the sternum.

Figure 6.17 ■ **Lumbar vertebrae.**

Lumbar vertebrae are found in the lower back.

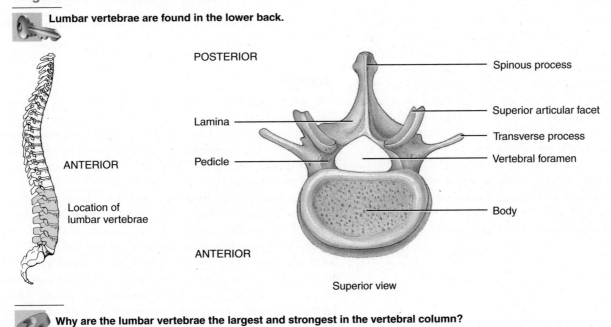

POSTERIOR

Spinous process

Superior articular facet

Lamina

Transverse process

Pedicle

Vertebral foramen

ANTERIOR

Location of lumbar vertebrae

Body

ANTERIOR

Superior view

Why are the lumbar vertebrae the largest and strongest in the vertebral column?

The *lumbar vertebrae* (L1 through L5) are the largest and strongest in the column (Figure 6.17). Their various projections are short and thick, and the spinous processes are well adapted for the attachment of the large back muscles.

The *sacrum* is a triangular bone formed by the fusion of five sacral vertebrae, indicated in Figure 6.18 as S1 through S5. The fusion of the sacral vertebrae begins between ages 16 and 18 years and is usually completed by age 30. The sacrum serves as a strong foundation for the pelvic girdle. It is positioned at the back of the pelvic cavity medial to the two hip bones.

Figure 6.18 ■ **Sacrum and coccyx.**

The sacrum is formed by the union of five sacral vertebrae, and the coccyx is formed by the union of four (usually) coccygeal vertebrae.

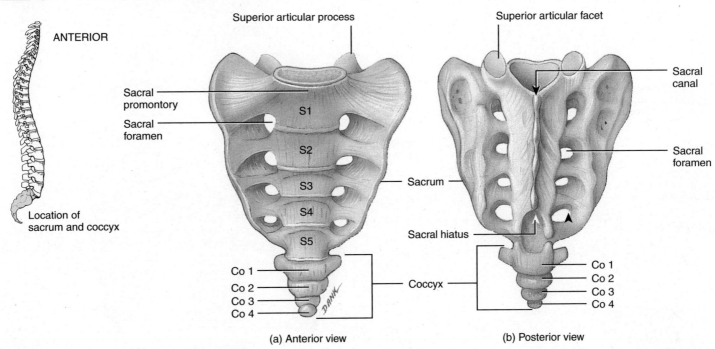

ANTERIOR

Superior articular process

Superior articular facet

Sacral canal

Sacral promontory

Sacral foramen

S1

S2

Sacral foramen

S3

Sacrum

Location of sacrum and coccyx

S4

S5

Sacral hiatus

Co 1

Co 1

Co 2

Coccyx

Co 2

Co 3

Co 3

Co 4

Co 4

(a) Anterior view

(b) Posterior view

What is the function of the sacral foramina?

The anterior and posterior sides of the sacrum contain four pairs of *sacral foramina*. Nerves and blood vessels pass through the foramina. The *sacral canal* is a continuation of the vertebral canal. The lower entrance is called the *sacral hiatus* (hī-Ā-tus = opening). Anesthetic agents are sometimes injected through the hiatus during childbirth in a procedure called caudal anesthesia (epidural block). The anterior top border of the sacrum has a projection, called the *sacral promontory* (PROM-on-tō′-rē), which is used as a landmark for measuring the pelvis prior to childbirth.

The *coccyx,* like the sacrum, is triangular in shape and is formed by the fusion of the four coccygeal vertebrae, which usually occurs between ages 20 and 30 years. These are indicated in Figure 6.18 as Co1 through Co4. The top of the coccyx articulates with the sacrum.

THORAX

Objective: • **Identify the bones of the thorax and their principal markings.**

The term *thorax* refers to the entire chest. The skeletal portion of the thorax, the *thoracic cage,* is a bony cage formed by the sternum, costal cartilages, ribs, and the bodies of the thoracic vertebrae (Figure 6.19). The thoracic cage encloses and protects the organs in the thoracic cavity and upper abdominal cavity. It also provides support for the bones of the shoulder girdle and upper limbs.

Sternum

The *sternum,* or breastbone, is a flat, narrow bone located in the center of the anterior thoracic wall and consists of three parts (Figure 6.19). The upper part is the *manubrium* (ma-NOO-brē-um = handlelike); the middle and largest part is the *body;* and the lowest, smallest part is the *xiphoid process* (ZĪ-foyd = sword-shaped).

The manubrium articulates with the clavicles and the first and second ribs. The body of the sternum articulates directly or indirectly with the second through tenth ribs. The xiphoid process consists of hyaline cartilage during infancy and childhood and does not ossify completely until about age 40. It has no ribs attached to it but provides attachment for some abdominal muscles. If the hands of a rescuer are incorrectly positioned during cardiopulmonary resuscitation (CPR), there is danger of fracturing the xiphoid process and driving it into internal organs.

Ribs

Twelve pairs of *ribs* make up the sides of the thoracic cavity (Figure 6.19). The ribs increase in length from the first through seventh ribs, then decrease in length to the twelfth rib. Each rib articulates posteriorly with its corresponding thoracic vertebra.

Figure 6.19 ■ **Skeleton of the thorax.**

The bones of the thorax enclose and protect organs in the thoracic cavity and upper abdominal cavity.

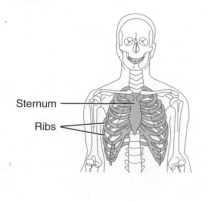
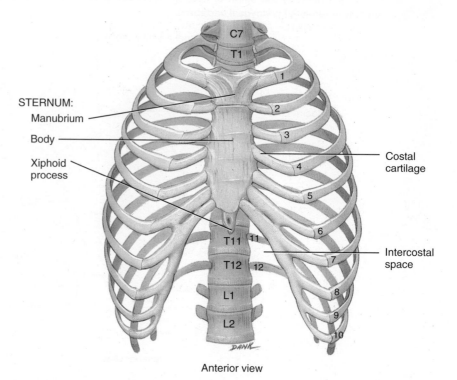

Anterior view

Which ribs are true ribs? false ribs? floating ribs?

The first through seventh pairs of ribs have a direct anterior attachment to the sternum by a strip of hyaline cartilage called *costal cartilage* (*cost-* = rib). These ribs are called *true ribs*. The remaining five pairs of ribs are termed *false ribs* because their costal cartilages either attach indirectly to the sternum or do not attach to the sternum at all. The cartilages of the eighth, ninth, and tenth pairs of ribs attach to each other and then to the cartilages of the seventh pair of ribs. The eleventh and twelfth false ribs are also known as *floating ribs* because the costal cartilage at their anterior ends does not attach to the sternum at all. Floating ribs attach only posteriorly to the thoracic vertebrae. Spaces between ribs, called *intercostal spaces*, are occupied by intercostal muscles, blood vessels, and nerves.

PECTORAL (SHOULDER) GIRDLE

Objective: • **Identify the bones of the pectoral (shoulder) girdle and their principal markings.**

The *pectoral girdles* (PEK-tō-ral) or *shoulder girdles* attach the bones of the upper limbs to the axial skeleton (Figure 6.20). The right and left pectoral girdles each consist of two bones: a clavicle and a scapula. The clavicle, the anterior component, articu-lates with the sternum, whereas the scapula, the posterior component, articulates with the clavicle and the humerus. The pectoral girdles do not articulate with the vertebral column. The joints of the shoulder girdles are freely movable and thus allow movements in many directions.

Clavicle

Each *clavicle* (KLAV-i-kul = key) or collarbone is a long, slender bone with two curves that is positioned horizontally above the first rib. The medial end of the clavicle articulates with the sternum, whereas the lateral end articulates with the acromion of the scapula (Figure 6.20). Because of its position, the clavicle transmits mechanical force from the upper limb to the trunk. If the force transmitted to the clavicle is excessive, as in falling on one's outstretched arm, a *fractured clavicle* may result.

Scapula

Each *scapula* (SCAP-yoo-la), or *shoulder blade,* is a large, flat, triangular bone situated in the posterior part of the thorax (Figure 6.20). A sharp ridge, the *spine,* runs diagonally across the posterior surface of the flattened, triangular *body* of the scapula. The lateral end of the spine, the *acromion* (a-KRŌ-mē-on; *acrom-* = topmost), is easily felt as the high point of the

Figure 6.20 ■ **Right pectoral (shoulder) girdle.**

The pectoral girdle attaches the bones of the upper limb to the axial skeleton.

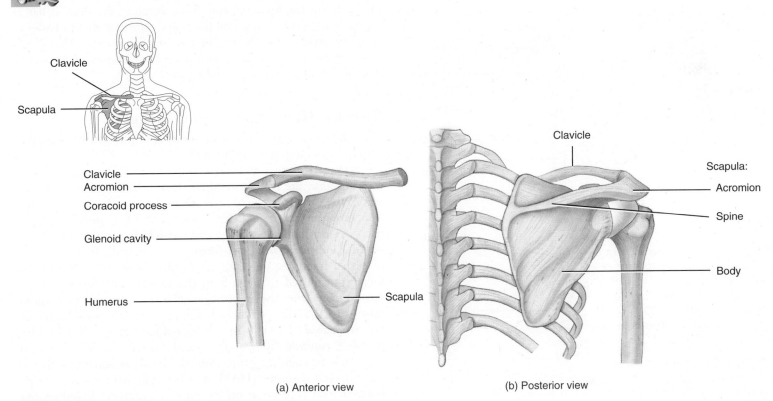

(a) Anterior view (b) Posterior view

Which bones make up a pectoral girdle?

shoulder and is the site of articulation with the clavicle. Inferior to the acromion is a depression called the *glenoid cavity*. This cavity articulates with the head of the humerus (arm bone) to form the shoulder joint. Also present on the scapula is a projection called the *coracoid process* (KOR-a-koyd = like a crow's beak) to which muscles attach.

UPPER LIMB

Objective: • **Identify the bones of the upper limb and their principal markings.**

The **upper limbs** consist of 60 bones. Each upper limb includes a humerus in the arm; ulna and radius in the forearm; and 8 carpals (wrist bones), 5 metacarpals (palm bones), and 14 phalanges (finger bones) in the hand (see Figure 6.6).

Humerus

The **humerus** (HYOO-mer-us), or arm bone, is the longest and largest bone of the upper limb (Figure 6.21). At the shoulder it

Figure 6.21 ■ **Right humerus in relation to the scapula, ulna, and radius.**

The humerus is the longest and largest bone of the upper limb.

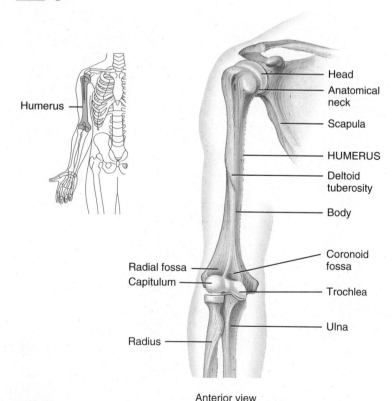

Anterior view

With which part of the scapula does the humerus articulate?

articulates with the scapula, and at the elbow it articulates with both the ulna and radius. The proximal end of the humerus consists of a *head* that articulates with the glenoid cavity of the scapula. It also has an *anatomical neck*, the former site of the epiphyseal plate, which is a groove just distal to the head. The *body* of the humerus contains a roughened, V-shaped area called the *deltoid tuberosity* where the deltoid muscle attaches. At the distal end of the humerus, the *capitulum* (ka-PIT-yoo-lum = small head), is a rounded knob that articulates with the head of the radius. The *radial fossa* is a depression that receives the head of the radius when the forearm is flexed (bent). The *trochlea* (TRŌK-lē-a) is a spool-shaped surface that articulates with the ulna. The *coronoid fossa* (KOR-o-noyd = crown-shaped) is a depression that receives part of the ulna when the forearm is flexed. The *olecranon fossa* (ō-LEK-ra-non) is a depression on the back of the bone that receives the olecranon of the ulna when the forearm is extended (straightened).

Ulna and Radius

The **ulna** is on the medial aspect (little-finger side) of the forearm and is longer than the radius (Figure 6.22). At the proximal end of the ulna is the *olecranon*, which forms the prominence of the elbow. The *coronoid process*, together with the olecranon, receives the trochlea of the humerus. The trochlea of the humerus also fits into the *trochlear notch*, a large curved area between the olecranon and the coronoid process. The *radial notch* is a depression for the head of the radius. A *styloid process* is at the distal end of the ulna.

The **radius** is located on the lateral aspect (thumb side) of the forearm. The proximal end of the radius articulates with the capitulum of the humerus and radial notch of the ulna. It has a raised, roughened area called the *radial tuberosity* that provides a point of attachment for the biceps brachii muscle. The distal end of the radius articulates with three carpal bones of the wrist. Also at the distal end is a *styloid process*.

Carpals, Metacarpals, and Phalanges

The **carpus (wrist)** of the hand contains eight small bones, the **carpals,** held together by ligaments (Figure 6.23). The carpals are arranged in two transverse rows, with four bones in each row, and they are named for their shapes. In the anatomical position, the carpals in the top row, from the lateral to medial position, are the **scaphoid** (SKAF-oid = boatlike), **lunate** (LOO-nāt = moon-shaped), **triquetrum** (trī-KWĒ-trum = three cornered), and **pisiform** (PĪ-si-form = pea-shaped). In about 70% of carpal fractures, only the scaphoid is broken because of the force transmitted through it to the radius. The carpals in the bottom row, from the lateral to medial position, are the **trapezium** (tra-PĒ-zē-um = four-sided figure with no two sides parallel), **trapezoid** (TRAP-e-zoid = four-sided figure with two sides parallel), **capitate** (KAP-i-tat = head-shaped; the largest carpal bone, whose rounded projection, the head, articulates with the lunate), and **hamate** (HAM-āt = hooked; named for a large hook-shaped projection on its anterior surface). Together, the concavity formed by the pisiform and hamate (on the ulnar side) and the scaphoid and trapezium (on the radial side) constitute a space called the **carpal tunnel.** Through it pass the long flexor tendons of the digits and thumb and the median nerve. Narrowing of the carpal tunnel gives rise to a condition called *carpal tunnel syndrome*, in which the median nerve is compressed. The nerve compression causes pain, numbness, tingling, and muscle weakness in the hand.

The **metacarpus (palm)** of the hand contains five bones called **metacarpals** (*meta-* = after or beyond). Each metacarpal bone consists of a proximal *base*, an intermediate *body*, and a distal *head*. The metacarpal bones are numbered I through V (or 1 to 5), starting with the lateral bone in the thumb or *pollex*. The heads of the metacarpals are commonly called the "knuckles" and are readily visible in a clenched fist.

The **phalanges** (fa-LAN-jēz = battle lines) are the bones of the fingers. They number 14 in each hand. Like the metacarpals, the phalanges are numbered I through V (or 1 to 5), beginning with the thumb. A single bone of a finger or toe is termed a *phalanx* (FĀ-lanks). Like the metacarpals, each phalanx consists of a proximal *base*, an intermediate *body*, and a distal *head*. There are two phalanges (proximal and distal) in the thumb and three phalanges (proximal, middle, and distal) in each of the other four digits. In order from the thumb, these other four digits are commonly referred to as the index finger, middle finger, ring finger, and little finger (Figure 6.23).

Figure 6.22 ■ **Right ulna and radius in relation to the humerus and carpals.**

In the forearm, the longer ulna is on the medial side, whereas the radius is on the lateral side.

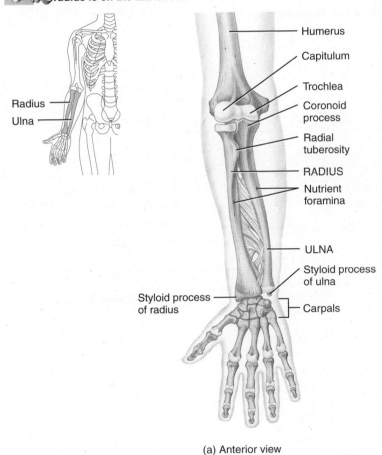

- Humerus
- Capitulum
- Trochlea
- Coronoid process
- Radial tuberosity
- RADIUS
- Nutrient foramina
- ULNA
- Styloid process of ulna
- Radius
- Ulna
- Styloid process of radius
- Carpals

(a) Anterior view

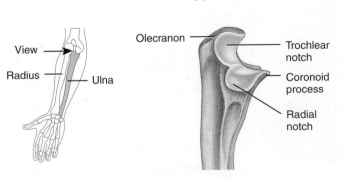

- View
- Radius
- Ulna
- Olecranon
- Trochlear notch
- Coronoid process
- Radial notch

(b) Lateral view of proximal end of ulna

What part of the ulna is called the "elbow"?

Figure 6.23 ■ **Right wrist and hand in relation to the ulna and radius.**

The skeleton of the hand consists of the carpals, metacarpals, and phalanges.

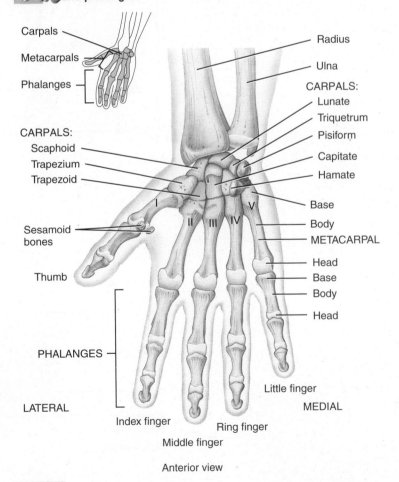

- Carpals
- Metacarpals
- Phalanges
- Radius
- Ulna
- CARPALS:
  - Lunate
  - Triquetrum
  - Pisiform
  - Capitate
  - Hamate
- CARPALS:
  - Scaphoid
  - Trapezium
  - Trapezoid
- Sesamoid bones
- Base
- Body
- METACARPAL
- Head
- Base
- Body
- Head
- Thumb
- PHALANGES
- LATERAL
- Index finger
- Middle finger
- Ring finger
- Little finger
- MEDIAL

Anterior view

What part of which bones are commonly called the "knuckles"?

PELVIC (HIP) GIRDLE

Objective: • **Identify the bones of the pelvic (hip) girdle and their principal markings.**

The ***pelvic (hip) girdle*** consists of the two ***hip bones,*** also called ***coxal bones*** (Figure 6.24). The pelvic girdle provides strong, stable support for the vertebral column, protects the pelvic viscera, and attaches the lower limbs to the axial skeleton. The hip bones are united to each other in front at a joint called the ***pubic sym-***

physis (PYOO-bik SIM-fi-sis); posteriorly they unite with the sacrum at the sacroiliac joint.

Together with the sacrum and coccyx, the two hip bones of the pelvic girdle form a basinlike structure called the ***pelvis*** (plural is *pelvises* or *pelves*). In turn, the bony pelvis is divided into upper and lower portions by a boundary called the ***pelvic brim*** (see Figure 6.24). The part of the pelvis above the pelvic brim is called the ***false (greater) pelvis.*** The false pelvis is actually part of the abdomen and does not contain any pelvic organs, except for the urinary bladder, when it is full, and the uterus

Figure 6.24 ■ **Female pelvic (hip) girdle.**

The hip bones are united in front at the pubic symphysis and in back at the sacrum.

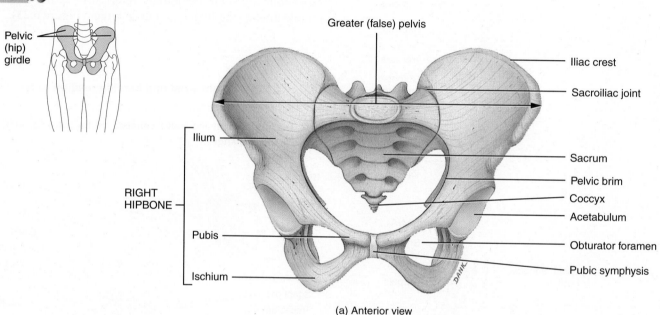

(a) Anterior view

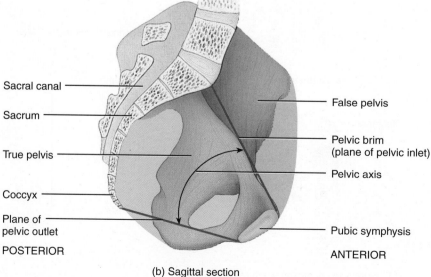

(b) Sagittal section

What part of the pelvis surrounds the pelvic organs in the pelvic cavity?

during pregnancy. The part of the pelvis below the pelvic brim is called the *true (lesser) pelvis*. The true pelvis surrounds the pelvic cavity (see Figure 1.8 on page 15). The upper opening of the true pelvis is called the *pelvic inlet*, whereas the lower opening of the true pelvis is called the *pelvic outlet*. The *pelvic axis* is an imaginary curved line passing through the true pelvis; it joins the central points of the planes of the pelvic inlet and outlet. During childbirth, the pelvic axis is the course taken by the baby's head as it descends through the pelvis.

Pelvimetry is the measurement of the size of the inlet and outlet of the birth canal, which may be done by ultrasonography or physical examination. Measurement of the pelvic outlet in pregnant females is important because it must become large enough for the fetus to pass through at birth.

Each of the two hip bones of a newborn is composed of three parts: the ilium, the pubis and the ischium (Figure 6.25). The *ilium* (= flank) is the largest of the three subdivisions of the hip bone. Its upper border is the *iliac crest*. On the lower surface is the *greater sciatic notch* (sī-AT-ik) through which the sciatic nerve, the longest nerve in the body, passes. The *ischium* (IS-kē-um = lip) is the lower, posterior portion of the hipbone. The *pubis* (PYOO-bis = pubic hair) is the anterior and inferior part of the hipbone. By age 23 years, the three separate bones have fused into one. The deep fossa (depression) where the three bones meet is the *acetabulum* (as-e-TAB-yoo-lum = vinegar cup). It is the socket for the head of the femur. The ischium joins with the pubis, and together they surround the *obturator foramen* (OB-too-rā-ter), the largest foramen in the skeleton.

LOWER LIMB

Objective: • List the skeletal components of the lower limb and their principal markings.

The two *lower limbs* are each composed of 30 bones: the femur in the thigh; the patella (kneecap); the tibia and fibula in the leg (the part of the lower limb between the knee and the ankle); and 7 tarsals (ankle bones), 5 metatarsals, and 14 phalanges (toes) in the foot (see Figure 6.6).

Femur

The *femur* (thigh bone) is the longest, heaviest, and strongest bone in the body (Figure 6.26). Its proximal end articulates with the hip bone, and its distal end articulates with the tibia and patella. The body of the femur bends medially, and as a result, the knee joints are brought nearer to the midline of the body.

Figure 6.26 ■ **Right femur in relation to the hip bone, patella, tibia, and fibula.**

The head of the femur articulates with the acetabulum of the hip bone to form the hip joint.

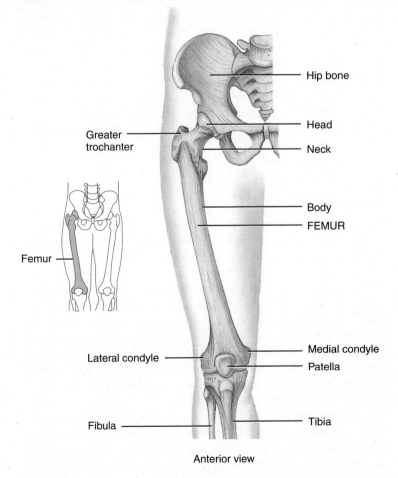

Anterior view

With which bones does the distal end of the femur articulate?

Figure 6.25 ■ **Right hip bone.** The lines of fusion of the ilium, ischium, and pubis are not always visible in an adult hip bone.

The two hip bones form the pelvic girdle, which attaches the lower limbs to the axial skeleton and supports the vertebral column and viscera.

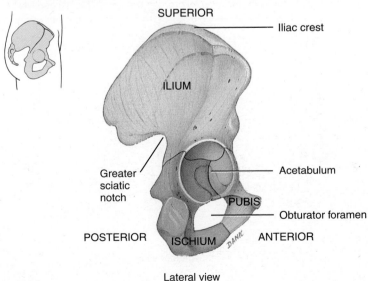

Lateral view

Which bone fits into the socket formed by the acetabulum?

The bend is greater in females because the female pelvis is broader.

The *head* of the femur articulates with the acetabulum of the hip bone to form the *hip joint*. The *neck* of the femur is a constricted region below the head. A fairly common fracture in the elderly occurs at the neck of the femur, which becomes so weak that it fails to support the weight of the body. Although it is actually the femur that is fractured, this condition is commonly known as a broken hip. The *greater trochanter* (trō-KAN-ter) is a projection felt and seen in front of the hollow on the side of the hip. It is where some of the thigh and buttock muscles attach and serves as a landmark for intramuscular injections in the thigh.

The distal end of the femur expands into the *medial condyle* and *lateral condyle*, which articulate with the tibia. The *patellar surface* is located on the anterior surface of the femur between the condyles.

Patella

The *patella* (= little dish), or kneecap, is a small, triangular bone in front of the joint between the femur and tibia, commonly known as the knee joint (Figure 6.26). The patella develops in the tendon of the quadriceps femoris muscle. Its functions are to increase the leverage of the tendon, maintain the position of the tendon when the knee is flexed, and protect the knee joint. During normal flexion and extension of the knee, the patella tracks (glides) up and down in the groove between the two femoral condyles. In "runner's knee," or *patellofemoral stress syndrome*, normal tracking does not occur. Instead, the patella tracks laterally, and the increased pressure of abnormal tracking causes the associated pain. A common cause of runner's knee is constantly walking, running, or jogging on the same side of the road. Because roads are high in the middle and slope down on the sides, the slope stresses the knee that is closer to the center of the road.

Tibia and Fibula

The *tibia,* or shin bone, is the larger, medial bone of the leg (Figure 6.27) and bears the weight of the body. The tibia articulates at its proximal end with the femur and fibula, and at its distal end with the fibula and talus of the ankle. The proximal end of the tibia expands into a *lateral condyle* and a *medial condyle*, which articulate with the condyles of the femur to form the *knee joint*. The *tibial tuberosity* is on the anterior surface below the condyles and is a point of attachment for the patellar ligament. The medial surface of the distal end of the tibia forms the *medial malleolus* (ma-LĒ-ō-lus = little hammer), which articulates with the talus of the ankle and forms the prominence that can be felt on the medial surface of your ankle.

Shin splints is the name given to soreness or pain along the tibia. Probably caused by inflammation of the periosteum brought about by repeated tugging of the muscles and tendons attached to the periosteum, it is often the result of walking or running up and down hills.

The *fibula* is parallel and lateral to the tibia (Figure 6.27) and is considerably smaller than the tibia. The *head* of the fibula articulates with the lateral condyle of the tibia below the knee joint. The distal end has a projection called the *lateral malleolus* that articulates with the talus of the ankle. This forms the prominence on the lateral surface of the ankle. As shown in Figure 6.27, the fibula also articulates with the tibia at the *fibular notch*.

Tarsals, Metatarsals, and Phalanges

The *tarsus* (*ankle*) of the foot contains seven bones, the *tarsals*, held together by ligaments (Figure 6.28). Of these, the *talus* (TĀ-lus = ankle bone) and *calcaneus* (kal-KĀ-nē-us = heel bone) are located on the posterior part of the foot. The anterior part of the ankle contains the **cuboid, navicular,** and three **cuneiform bones** called the *first, second,* and *third cuneiforms*. The talus is the only bone of the foot that articulates with the fibula and tibia. It articulates medially with the medial malleolus of the tibia and laterally with the lateral malleolus of the fibula. During walking, the talus initially bears the entire weight of the body. About half the weight is then transmitted to the calcaneus. The remainder is transmitted to the other tarsal bones. The calcaneus is the largest and strongest of the tarsals.

Figure 6.27 ■ **Right tibia and fibula in relation to the femur, patella, and talus.**

The tibia articulates with the femur and fibula proximally and with the fibula and talus distally, while the fibula articulates proximally with the tibia below the knee joint and distally with the talus.

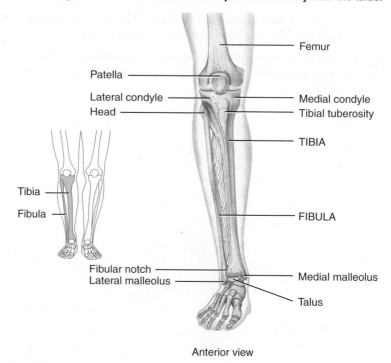

Anterior view

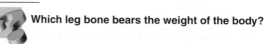 Which leg bone bears the weight of the body?

Figure 6.28 ■ Right foot.

 The skeleton of the foot consists of the tarsals, metatarsals, and phalanges.

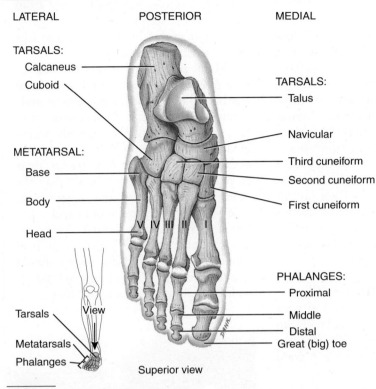

Superior view

 Which tarsal bone articulates with the tibia and fibula?

Five bones called *metatarsals* and numbered I to V (or 1 to 5) from the medial to lateral position form the skeleton of the *metatarsus*. Like the metacarpals of the palm, each metatarsal consists of a proximal *base*, an intermediate *body*, and a distal *head*. The first metatarsal, which is connected to the big toe, is thicker than the others because it bears more weight.

The *phalanges* of the foot resemble those of the hand both in number and arrangement. Each also consists of a proximal *base*, an intermediate *body*, and a distal *head*. The great or big toe *(hallux)* has two large, heavy phalanges—proximal and distal. The other four toes each have three phalanges—proximal, middle, and distal.

Arches of the Foot

The bones of the foot are arranged in two *arches* (Figure 6.29). These arches enable the foot to support the weight of the body, provide an ideal distribution of body weight over the hard and soft tissues of the foot, and provide leverage while walking. The arches are not rigid—they yield as weight is applied and spring back when the weight is lifted, thus helping to absorb shocks.

The *longitudinal arch* extends from the front to the back of the foot and has two parts, medial and lateral. The medial part of the longitudinal arch originates at the calcaneus, rises to the talus, and descends through the navicular, the three cuneiforms, and the heads of the three medial metatarsals. The lateral part of the longitudinal arch also begins at the calcaneus, then rises at the cuboid, and descends to the heads of the two lateral metatarsals. The *transverse arch* is formed by the navicular, three cuneiforms, and the bases of the five metatarsals.

The bones composing the arches are held in position by ligaments and tendons. If these ligaments and tendons are weakened by excess weight, postural abnormalities, or genetic predisposition, the height of the medial longitudinal arch may decrease or "fall." The result is a condition called *flatfoot.*

Figure 6.29 ■ Arches of the right foot.

Arches help the foot support and distribute the weight of the body and provide leverage during walking.

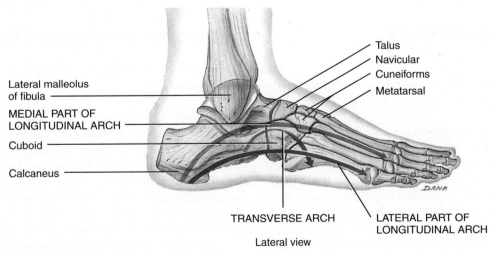

Lateral view

What structural aspect of the arches allows them to absorb shocks?

We take the structure and function of our feet for granted—until they start to hurt. And even then we often continue to mistreat them, cramming them into shoes that are too tight, and then walking on concrete sidewalks and taking long shopping expeditions. No wonder foot problems are such a common complaint! Fortunately, most foot problems are preventable by understanding the foot's structure and function and then using good footwear to support them in their work.

These Feet Were Made for Walking

Each time you take a step, your heel strikes the ground first. Then you roll through the arches, over the ball of your foot, and onto your toes. Your arches flatten slightly as they absorb the weight of your body. One foot continues to bear your weight until the heel of the other foot touches the ground. As you walk, your big toe maintains your balance while the other toes give your foot some resiliency. The two outer metatarsals move to accommodate uneven surfaces, while the inner three stay rigid for support.

The most common cause of foot problems is ill-fitting shoes, which stress the structure and interfere with the function of the foot. The high heel is a case in point, which explains why 80% of those suffering from foot problems are women. Although many people think high heels look good and are fun to wear, they should not be used for walking because they make the body's weight fall onto the forefoot. Thus, the arches of the foot are not allowed to absorb the force of the body's weight. This unnatural stress can injure soft-tissue structures, joints, and bones.

Good Shoes for Happy Feet

Choosing shoes that are "good" to your feet can prevent many foot problems, an especially important consideration if you are doing any amount of walking. A good shoe has a sole that is strong and flexible and provides a good gripping surface. Cushioned insoles help protect feet from hard surfaces. Arch supports help distribute weight over a broader area, just like the arches in your foot.

Many people spend a great deal of time researching which brand of shoes to buy but do not spend adequate time evaluating whether or not the shoes suit their feet. A high-quality shoe is only worth buying if it fits! Shop for shoes in the late afternoon when your feet are at their largest. One foot is often bigger than the other; always buy for the bigger foot. The shoes you try on should feel comfortable immediately—don't plan on shoes stretching with wear. The heel should fit snugly, and the instep should not gape open. The toe box should be wide enough to wiggle all your toes.

> ▶ **Think It Over**
>
> ▶ Why do you think excess body weight is associated with an increased risk of foot problems?

COMPARISON OF FEMALE AND MALE SKELETONS

Objective: • **Identify the principal structural differences between female and male skeletons.**

The bones of a male are generally larger and heavier than those of a female. The articular ends are thicker in relation to the shafts. In addition, because certain muscles of the male are larger than those of the female, the points of muscle attachment—tuberosities, lines, and ridges—are larger in the male skeleton.

Many significant structural differences between the skeletons of females and males are related to pregnancy and childbirth. Because the female's pelvis is wider and shallower than the male's, there is more space in the true pelvis of the female, especially in the pelvic inlet and pelvic outlet, which accommodate the passage of the infant's head at birth. Several of the significant differences between the female and male pelvis are shown in Table 6.4.

AGING AND THE SKELETAL SYSTEM

Objective: • **Describe the effects of aging on the skeletal system.**

From birth through adolescence, more bone is produced than is lost during bone remodeling. In young adults, the rates of bone production and loss are about the same. As the levels of sex steroids diminish during middle age, especially in women after menopause, a decrease in bone mass occurs because bone resorption outpaces bone formation. Because women's bones generally are smaller than men's bones to begin with, loss of bone mass in old age typically causes greater problems in women.

Aging has two main effects on bone: It becomes more brittle and it loses mass. Bone brittleness results from a decrease in the rate of protein synthesis and in the production of human growth hormone, which diminishes the production of the collagen fibers that give bone its strength and flexibility. As a result,

Table 6.4 / Comparison of the Pelvis in Females and Males

Point of Comparison	Female	Male
General structure	Light and thin.	Heavy and thick.
False (greater) pelvis	Shallow.	Deep.
Pelvic inlet	Larger and more oval.	Smaller and heart-shaped.
Pelvic outlet	Wider.	Narrower.
Acetabulum	Small and faces anteriorly.	Large and faces laterally.
Obturator foramen	Oval.	Round.
Pubic arch	Greater than 90° angle.	Less than 90° angle.

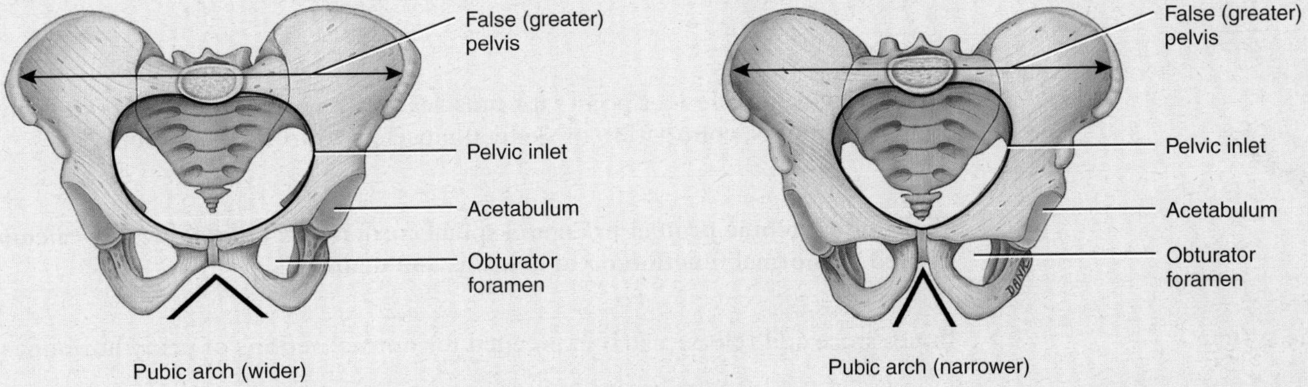

Anterior views

Iliac crest	Less curved.	More curved.
Ilium	Less vertical.	More vertical.
Greater sciatic notch	Wide.	Narrow.
Coccyx	More movable and more curved anteriorly.	Less movable and less curved anteriorly.
Sacrum	Short, wide (see anterior views), and more curved anteriorly.	Long, narrow (see anterior views), and less curved anteriorly.

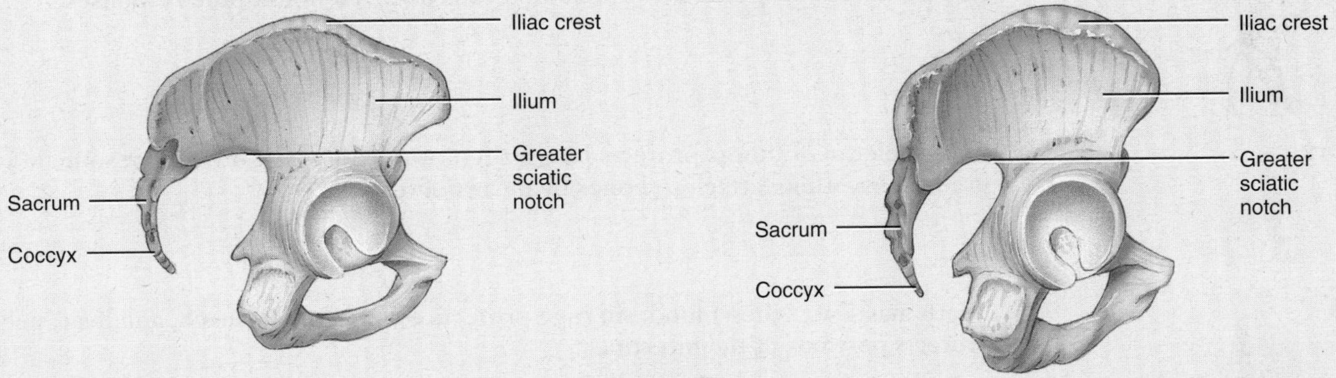

Right lateral views

inorganic minerals gradually constitute a greater proportion of the bone matrix. Loss of bone mass usually begins after age 30 in females, accelerates greatly around age 45 as levels of estrogens decrease, and continues until as much as 30% of the calcium in bones is lost by age 70. Once bone loss begins in females, about 8% of bone mass is lost every 10 years. In males, calcium loss from bone typically does not begin until after age 60, and about 3% of bone mass is lost every 10 years. The loss of calcium from bones is one of the problems in osteoporosis (described on page 149). Loss of bone mass also leads to bone deformity, pain, stiffness, some loss of height, and loss of teeth.

• • •

To appreciate the many ways that the skeletal system contributes to homeostasis of other body systems, examine Focus on Homeostasis: The Skeletal System on page 148. Next, in Chapter 7, we will see how joints both hold the skeleton together and permit it to participate in movements.

Focus on Homeostasis
The Skeletal System

Body System	*Contribution of the Skeletal System*
For all body systems	Bones provide support and protection for internal organs; bones store and release calcium, which is needed for proper functioning of most body tissues.
Integumentary system	Bones provide strong support for overlying muscles and skin, while joints provide flexibility that allows skin to bend.
Muscular system	Bones provide attachment points for muscles and leverage for muscles to bring about body movements; contraction of skeletal muscle requires calcium ions.
Nervous system	Skull and vertebrae protect brain and spinal cord; normal blood level of calcium is needed for normal functioning of neurons and neuroglia.
Endocrine system	Bones store and release calcium, needed for normal actions of many hormones.
Cardiovascular system	Red bone marrow carries out hemopoiesis (blood cell formation); rhythmical beating of the heart requires calcium ions.
Lymphatic and immune system	Red bone marrow produces white blood cells involved in immune responses.
Respiratory system	Axial skeleton of thorax protects lungs; rib movements assist breathing; some muscles used for breathing attach to bones by means of tendons.
Digestive system	Teeth masticate (chew) food; rib cage protects esophagus, stomach, and liver; pelvis protects portions of the intestines.
Urinary system	Ribs partially protect kidneys and pelvis protects urinary bladder and urethra.
Reproductive systems	Pelvis protects ovaries, uterine (Fallopian) tubes, and uterus in females and part of ductus (vas) deferens and accessory glands in males; bones are an important source of calcium needed for milk synthesis during lactation.

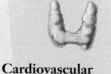

COMMON DISORDERS

Fractures

A *fracture* is any break in a bone. Although bone has a generous blood supply, healing sometimes takes months. Sufficient calcium and phosphorus to strengthen and harden new bone is deposited only gradually. Bone cells generally also grow and reproduce slowly. Types of bone fractures include the following:

1. *Partial:* an incomplete break across the bone, such as a crack.

2. *Complete:* a complete break across the bone; that is, the bone is broken into two or more pieces.

3. *Closed (simple):* the fractured bone does not break through the skin.

4. *Open (compound):* the broken ends of the bone protrude through the skin.

The healing process for bone varies with several factors such as the extent of the fracture, age, nutrition, the general health of the person, and the type of treatment received.

Osteoporosis

Osteoporosis (os′-tē-ō-ō-pō-RŌ-sis; *por-* = passageway; *-osis* = condition) is literally a condition of porous bones. The basic problem is that bone resorption outpaces bone deposition. As a result, bone mass becomes so depleted that the skeleton can no longer withstand the mechanical stresses of everyday living. For example, a hip fracture might result from simply sitting down too quickly. In the United States, osteoporosis causes more than a million fractures a year, mainly in the hip, wrist, and vertebrae. Osteoporosis afflicts the entire skeletal system. In addition to fractures, osteoporosis causes shrinkage of vertebrae and thus height loss, hunched backs, and bone pain.

Thirty million people in the United States suffer from osteoporosis. The disorder primarily affects middle-aged and elderly people, 80% of them women. Older women suffer from osteoporosis more often than men for two reasons: Women's bones are less massive than men's bones, and production of estrogens in women declines dramatically at menopause, whereas production of testosterone in older men wanes gradually and only slightly.

Besides gender, risk factors for developing osteoporosis include a family history of the disease, European or Asian ancestry, thin or small body build, an inactive lifestyle, cigarette smoking, a diet low in calcium and vitamin D, more than two alcoholic drinks a day, and the use of certain medications.

Calcium supplements and weight-bearing exercise help prevent or retard the development of osteoporosis. In postmenopausal women, treatment of osteoporosis may include estrogen replacement therapy (ERT; low doses of estrogens) or hormone replacement therapy (HRT; a combination of estrogens and progesterone, another sex steroid). A form of the hormone calcitonin (Miacalcin), which promotes bone deposition; raloxifene (Evista), which mimics the beneficial effects of estrogens on bone; and alendronate (Fosamax), which inhibits the activity of osteoclasts, are also used to treat osteoporosis. The most important aspect of treatment is prevention. Adequate calcium intake and exercise in her early years may be more beneficial to a woman than drugs and calcium supplements when she is older.

Rickets and Osteomalacia

Rickets and *osteomalacia* (os′-tē-ō-ma-LĀ-she-ah; *-malacia* = softness) are disorders in which bone calcification fails. The bones become soft or rubbery and are easily deformed. Whereas rickets affects the growing bones of children, osteomalacia affects the bones of adults.

Herniated Disc

If the ligaments of the intervertebral discs become injured or weakened, the resulting pressure may be great enough to rupture the surrounding fibrocartilage. When this occurs, the material inside may herniate (protrude). This condition is called a *herniated (slipped) disc.* It occurs most often in the lumbar region because that part of the vertebral column bears much of the weight of the body and is the region of the most bending.

Spina Bifida

Spina bifida (SPĪ-na BIF-i-da) is a congenital defect of the vertebral column in which laminae fail to unite at the midline. In serious cases, protrusion of the membranes (meninges) around the spinal cord or the spinal cord itself may produce partial or complete paralysis, partial or complete loss of urinary bladder control, and the absence of reflexes. Because an increased risk of spina bifida is associated with a low level of folic acid (one of the B vitamins) early in pregnancy, all women who might become pregnant are encouraged to take folic acid supplements.

MEDICAL TERMINOLOGY AND CONDITIONS

Bunion (BUN-yun) A deformity of the great toe that typically is caused by wearing tightly fitting shoes. The condition produces inflammation of bursae (fluid-filled sacs at the joint), bone spurs, and calluses.

Clawfoot A condition in which the medial part of the longitudinal arch is abnormally elevated. It is often caused by muscle deformities, such as may result from diabetes.

Kyphosis (kī-FŌ-sis; *kypho-* = bent) An exaggeration of the thoracic curve of the vertebral column. In the elderly, degeneration of the intervertebral discs leads to kyphosis; it may also be caused by rickets and poor posture.

Lordosis (lor-DŌ-sis; *lord-* = bent backward) An exaggeration of the lumbar curve of the vertebral column, also called *swayback.* It may result from increased weight of the abdomen as in pregnancy or extreme obesity, poor posture, rickets, or tuberculosis of the spine.

Osteogenic sarcoma (os′-tē-ō-Ō-JEN-ik sar-KŌ-ma; *sarcoma* = connective tissue tumor) Bone cancer that primarily affects osteoblasts and occurs most often in the femur, tibia, and humerus of teenagers during their growth spurt; metastases occur most often in the lungs. Treatment consists of multidrug chemotherapy and removal of the malignant growth or amputation of the affected limb.

Osteopenia (os′-tē-ō-PĒ-nē-a; *penia* = poverty) Reduced bone mass due to a decrease in the rate of bone production to a level insufficient to compensate for normal bone breakdown; any decrease in bone mass below normal. An example is osteoporosis.

Scoliosis (skō′-lē-Ō-sis; *scolio-* = crooked) A sideways bending of the vertebral column, usually in the thoracic region. It may result from congenitally (present at birth) malformed vertebrae, chronic sciatica, paralysis of muscles on one side of the vertebral column, poor posture, or one leg being shorter than the other.

STUDY OUTLINE

Functions of Bone and the Skeletal System (p. 114)

1. The skeletal system consists of all bones attached at joints and cartilage between joints.

2. The functions of the skeletal system include support, protection, movement, mineral storage, housing blood-forming tissue, and storage of energy.

Types of Bones (p. 114)

1. On the basis of shape, bones are classified as long, short, flat, irregular, or sesamoid.

2. Sutural bones are found between the sutures of certain cranial bones.

Parts of a Long Bone (p. 114)

1. Parts of a long bone include the diaphysis (shaft), epiphyses (ends), metaphysis, articular cartilage, periosteum, medullary (marrow) cavity, and endosteum.

2. The diaphysis is covered by periosteum.

Histology of Bone (p. 116)

1. Bone tissue consists of widely separated cells surrounded by large amounts of matrix (intercellular substance). The four principal types of cells are osteogenic cells, osteoblasts, osteocytes, and osteoclasts. The matrix contains collagen fibers (organic) and mineral salts that consist mainly of calcium phosphate (inorganic).

2. Compact (dense) bone tissue consists of osteons (Haversian systems) with little space between them. Compact bone lies over spongy bone and composes most of the bone tissue of the diaphysis. Functionally, compact bone protects, supports, and resists stress.

3. Spongy bone tissue consists of trabeculae surrounding many red bone marrow–filled spaces. It forms most of the structure of short, flat, and irregular bones and the epiphyses of long bones. Functionally, spongy bone stores some red and yellow bone marrow and provides some support.

Ossification: Bone Formation (p. 118)

1. Bone forms by a process called ossification.

2. The two types of ossification, intramembranous and endochondral, involve the replacement of preexisting connective tissue with bone.

3. Intramembranous ossification occurs within fibrous connective tissue membranes of the embryo and the adult.

4. Endochondral ossification occurs within a cartilage model. The primary ossification center of a long bone is in the diaphysis. Cartilage degenerates, leaving cavities that merge to form the medullary (marrow) cavity. Osteoblasts lay down bone. Next, ossification occurs in the epiphyses, where bone replaces cartilage, except for articular cartilage and the epiphyseal plate.

5. Because of the activity of the epiphyseal plate, the diaphysis of a bone increases in length.

6. Bone grows in diameter as a result of the addition of new bone tissue around the outer surface of the bone.

Homeostasis of Bone (p. 120)

1. The homeostasis of bone growth and development depends on a balance between bone formation and resorption.

2. Old bone is constantly destroyed by osteoclasts, while new bone is constructed by osteoblasts. This process is called remodeling.

3. Normal growth depends on minerals (calcium, phosphorus, magnesium), vitamins (A, C, D), and hormones (human growth hormone, insulinlike growth factors, insulin, thyroid hormones, sex hormones, and parathyroid hormone).

4. Under mechanical stress, bone increases in mass.

5. Bones store and release calcium and phosphate, controlled mainly by parathyroid hormone (PTH).

6. When placed under mechanical stress, bone tissue becomes stronger.

7. The important mechanical stresses result from the pull of skeletal muscles and the pull of gravity.

8. With aging, bone loses calcium and collagen fibers.

Exercise and Bone (p. 122)

1. Mechanical stress increases bone strength by increasing deposition of mineral salts and production of collagen fibers.

2. Removal of mechanical stress weakens bone through demineralization and collagen fiber reduction.

Bone Surface Markings (p. 123)

1. Bone surface markings are structural features visible on the surfaces of bones.

2. Each marking has a specific function: joint formation, muscle attachment, or passage of nerves and blood vessels.

Divisions of the Skeletal System (p. 124)

1. The axial skeleton consists of bones arranged along the longitudinal axis of the body. The parts of the axial skeleton are the skull, hyoid bone, auditory ossicles, vertebral column, sternum, and ribs.

2. The appendicular skeleton consists of the bones of the girdles and the upper and lower limbs. The parts of the appendicular skeleton are the pectoral (shoulder) girdles, bones of the upper limbs, pelvic (hip) girdle, and bones of the lower limbs.

Skull (p.124)

1. The skull consists of cranial bones and facial bones.

2. Sutures are immovable joints between bones of the skull. Examples are the coronal, sagittal, lambdoid, and squamous sutures.

3. The eight cranial bones include the frontal (1), parietal (2), temporal (2), occipital (1), sphenoid (1), and ethmoid (1).

4. The 14 facial bones are the nasal (2), maxillae (2), zygomatic (2), mandible (1), lacrimal (2), palatine (2), inferior nasal conchae (2), and vomer (1).

5. Paranasal sinuses are cavities in bones of the skull that communicate with the nasal cavity. They are lined by mucous membranes. Cranial bones containing paranasal sinuses are the frontal, sphenoid, ethmoid, and maxillae.

6. Fontanels are membrane-filled spaces between the cranial bones of fetuses and infants. The major fontanels are the anterior, posterior, anterolaterals, and posterolaterals.

Hyoid Bone (p. 133)

1. The hyoid bone is a U-shaped bone that does not articulate with any other bone.

2. It supports the tongue and provides attachment for some of its muscles as well as some neck muscles.

Vertebral Column (p. 133)

1. The bones of the adult vertebral column are the cervical vertebrae (7), thoracic vertebrae (12), lumbar vertebrae (5), the sacrum (5, fused), and the coccyx (4, fused).

2. The vertebral column contains normal curves that give strength, support, and balance.

3. The vertebrae are similar in structure, each consisting of a body, vertebral arch, and seven processes. Vertebrae in the different regions of the column vary in size, shape, and detail.

Thorax (p. 138)

1. The thoracic skeleton consists of the sternum, ribs, costal cartilages, and thoracic vertebrae.

2. The thoracic cage protects vital organs in the chest area.

Pectoral (Shoulder) Girdle (p. 139)

1. Each pectoral (shoulder) girdle consists of a clavicle and scapula.

2. Each attaches an upper limb to the trunk.

Upper Limb (p. 140)

1. There are 30 bones in each upper limb.

2. The upper limb bones include the humerus, ulna, radius, carpals, metacarpals, and phalanges.

Pelvic (Hip) Girdle (p. 142)

1. The pelvic (hip) girdle consists of two hip bones.

2. It attaches the lower limbs to the trunk at the sacrum.

3. Each hip bone consists of three fused components: ilium, pubis, and ischium.

Lower Limb (p. 143)

1. There are 30 bones in each lower limb.

2. The lower limb bones include the femur, patella, tibia, fibula, tarsals, metatarsals, and phalanges.

3. The bones of the foot are arranged in two arches, the longitudinal arch and the transverse arch, to provide support and leverage.

Comparison of Female and Male Skeletons (p. 146)

1. Male bones are generally larger and heavier than female bones and have more prominent markings for muscle attachment.

2. The female pelvis is adapted for pregnancy and childbirth. Differences in pelvic structure are listed in Table 6.4 on page 000.

Aging and the Skeletal System (p. 146)

1. The main effect of aging is a loss of calcium from bones, which may result in osteoporosis.

2. Another effect of aging is a decreased production of matrix proteins (mostly collagen fibers), which makes bones more brittle and thus more susceptible to fracture.

▓ SELF-QUIZ

1. Match the following cell types to their functions:

____ **a.** osteogenic cells	**A.**	mature bone cells
____ **b.** osteoclasts	**B.**	cells that form bone
____ **c.** mesenchymal cells	**C.**	can undergo mitosis and develop into osteoblasts
____ **d.** osteocytes	**D.**	differentiate into osteogenic cells
____ **e.** osteoblasts	**E.**	involved in bone resorption

2. When trying to locate a foramen in a bone, you would look for **a.** a large, rough projection **b.** a ridge **c.** a rounded projection **d.** a shallow depression **e.** an opening or hole

3. The ribs articulate with the **a.** thoracic vertebrae **b.** sacrum **c.** cervical vertebrae **d.** lumbar vertebrae **e.** atlas and axis

4. Match the following:

____ **a.** run lengthwise through bone	**A.**	lamellae
	B.	lacunae
____ **b.** connect central canals with lacunae	**C.**	perforating (Volkmann's) canal
____ **c.** concentric rings of matrix	**D.**	canaliculi
____ **d.** connect nutrient arteries and nerves from the periosteum to the central canals	**E.**	central (Haversian) canal
____ **e.** spaces that contain osteocytes		

5. The presence of an epiphyseal line in a long bone indicates that the bone **a.** is undergoing resorption **b.** has stopped growing in length **c.** is growing in diameter **d.** is still capable of growing in length **e.** is broken

6. The hyoid bone is unique because it **a.** is the smallest bone in the skull **b.** can malform causing a cleft palate **c.** forms the paranasal sinuses **d.** is often broken when an individual falls forward **e.** does not articulate with any other bone

7. The bones that form the pectoral girdle are the **a.** clavicle and scapula **b.** scapula and sternum **c.** humerus and scapula **d.** clavicle and humerus **e.** coxal bones

8. The main hormone that regulates the Ca^{2+} balance between bone and blood is **a.** parathyroid hormone **b.** insulin **c.** testosterone **d.** insulinlike growth factors **e.** human growth hormone

9. Spongy bone differs from compact bone because spongy bone **a.** is made up of numerous osteons **b.** is found primarily in the diaphyses of long bones **c.** has latticework walls known as trabeculae **d.** contains few, small spaces known as lacunae **e.** has lamellae arranged in concentric rings

10. In which of the following individuals might you expect to find the smallest bone mass? **a.** 20-year-old male weightlifter **b.** 45-year-old female weightlifter **c.** 45-year-old male astronaut **d.** 65-year-old bedridden female **e.** 65-year-old bedridden male

11. Place the following steps of endochondral ossification in the correct order:

1. Hyaline cartilage remains on the articular surfaces and epiphyseal plates

2. Chondroblasts produce a growing hyaline cartilage model surrounded by the perichondrium

3. Osteoblasts in perichondrium produce compact bone

4. Secondary ossification centers form

5. Primary ossification center and medullary cavity form

a. 2, 3, 4, 5, 1 **b.** 2, 3, 5, 4, 1 **c.** 5, 2, 1, 2, 4 **d.** 3, 2, 5, 4, 1 **e.** 5, 3, 2, 1, 4

12. Match each bone to its shape:

____ **a.** humerus	____ **d.** patella	**A.** flat	**D.** short
____ **b.** carpus	____ **e.** sternum	**B.** irregular	**E.** sesamoid
____ **c.** vertebra		**C.** long	

13. Where long bones form joints, the epiphyses are covered with **a.** yellow bone marrow **b.** osteoclasts **c.** periosteum **d.** endosteum **e.** hyaline cartilage

14. What substance in bone contributes to its strength?

    a. red bone marrow   **b.** collagen   **c.** yellow bone marrow
    d. hydroxyapatite   **e.** loose fibrous connective tissue

15. The skeletal system is responsible for

    a. protecting internal organs from injury   **b.** producing movement   **c.** providing a supporting framework for the body
    d. hemopoiesis   **e.** all of the above

16. For each of the following bones, place an AX in the blank if it belongs to the axial skeleton and an AP in the blank if it is part of the appendicular skeleton.

____ **a.** lacrimal   ____ **k.** ethmoid   ____ **u.** vertebrae

____ **b.** clavicle   ____ **l.** metatarsals   ____ **v.** coxal

____ **c.** radius   ____ **m.** temporal   ____ **w.** maxilla

____ **d.** mandible   ____ **n.** metacarpals   ____ **x.** frontal

____ **e.** patella   ____ **o.** vomer   ____ **y.** inferior nasal concha

____ **f.** carpals   ____ **p.** fibula

____ **g.** scapula   ____ **q.** palatine   ____ **z.** humerus

____ **h.** sternum   ____ **r.** hyoid   ____ **aa.** ulna

____ **i.** phalanges   ____ **s.** tibia   ____ **bb.** femur

____ **j.** tarsals   ____ **t.** spenoid   ____ **cc.** ribs

____ **dd.** occipital

CRITICAL THINKING APPLICATIONS

1. J.R. was riding his motorcycle across the Big Span Bridge when he had a collision with a near-sighted sea gull. In the resulting crash, J.R. crushed his left leg, fracturing both lower leg bones; snapped the pointy distal end of his lateral forearm bone; and broke the most lateral and proximal bone in his wrist. The sea gull flew off when the ambulance arrived. Name the bones that J.R. broke.

2. While investigating her new baby brother, a 4-year-old girl discovers a soft spot on the baby's skull and announces that the baby needs to go back because "it's not finished yet." Explain the presence of soft spots in the infant's skull and the lack of soft spots in yours.

3. Old Grandma Olga is a tiny, stooped woman with a big sense of humor. Her favorite movie line is from the *Wizard of Oz* when the wicked witch says "I'm melting." "That's me," laughs Olga, "melting away, getting shorter every year." What is happening to Grandma Olga?

4. Tyler fell from the top of a human pyramid during cheerleading practice and may have suffered a slipped disc. What is a disc? What is its composition and function?

ANSWERS TO FIGURE QUESTIONS

6.1 The articular cartilage reduces friction at joints; red bone marrow produces blood cells; and endosteum lines the medullary cavity.

6.2 Because the central canals are the main blood supply to the osteocytes, their blockage would lead to death of osteocytes.

6.3 Flat bones of the skull and mandible (lower jawbone) develop by intramembranous ossification.

6.4 The epiphyseal lines are indications of growth zones that have ceased to function.

6.5 Heartbeat, respiration, nerve cell functioning, enzyme functioning, and blood clotting all are processes that depend on proper levels of calcium.

6.6 Axial skeleton: skull, vertebral column. Appendicular skeleton: clavicle, shoulder girdle, humerus, pelvic girdle, femur.

6.7 The cranial bones are the frontal, parietal, occipital, sphenoid, ethmoid, and temporal bones.

6.8 The foramen magnum is the largest foramen in the skull.

6.9 Crista galli of ethmoid bone, frontal, parietal, temporal, occipital, temporal, parietal, frontal, crista galli of ethmoid bone all articulate with the sphenoid bone.

6.10 The perpendicular plate of the ethmoid bone forms the top part of the nasal septum.

6.11 The maxillae and palatine bones form the hard palate.

6.12 The paranasal sinuses produce mucus and serve as resonating chambers for vocalization.

6.13 The anterior fontanel is commonly known as the soft spot.

6.14 The thoracic and sacral curves are concave.

6.15 The vertebral foramina enclose the spinal cord, and the intervertebral foramina provide spaces for spinal nerves to exit the vertebral column.

6.16 The atlas and axis permit movement of the head to signify "no."

6.17 The body weight supported by vertebrae increases toward the inferior end of the vertebral column.

6.18 The sacral foramina are passageways for nerves and blood vessels.

6.19 The true ribs are pairs 1 through 7; the false ribs are pairs 8 through 12; and the floating ribs are pairs 11 and 12.

6.20 A pectoral girdle consists of a clavicle and a scapula.

6.21 The humerus articulates with the glenoid cavity of the scapula.

6.22 The "elbow" part of the ulna is the olecranon.

6.23 The knuckles are the heads of the metacarpals.

6.24 The true pelvis surrounds the pelvic organs in the pelvic cavity.

6.25 The femur fits into the acetabulum.

6.26 The distal end of the femur articulates with the tibia and the patella.

6.27 The tibia is the weight-bearing bone of the leg.

6.28 The talus articulates with the tibia and fibula.

6.29 The arches are not rigid, yielding when weight is applied and springing back when weight is lifted to allow them to absorb the shock of walking and running.

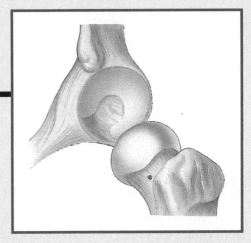

Chapter 7

Joints

■ Student Learning Objectives

1. Define a joint and describe how the structure of a joint determines its function. **154**
2. Describe the structural and functional classes of joints. **154**
3. Describe the structure and functions of the three types of fibrous joints. **154**
4. Describe the structure and functions of the two types of cartilaginous joints. **155**
5. Describe the structure and the six subtypes of synovial joints. **156**
6. Describe the types of movements that can occur at synovial joints. **157**
7. Describe the principal structures and functions of the knee joint. **162**
8. Explain the effects of aging on joints. **164**

■ A Look Ahead

*B*ones are too rigid to bend without being damaged. Fortunately, flexible connective tissues form joints that hold bones together while in most cases permitting some degree of movement. If you have ever damaged these areas you know how difficult it is to walk with a cast over your knee or to turn a doorknob with a splint on your finger. A few joints do not permit any movement at all but instead link bones to provide protection for the tissues they cover.

JOINTS

Objective: • **Define a joint and describe how the structure of a joint determines its function.**

A *joint* (also called an *articulation*) is a point of contact between bones, between cartilage and bones, or between teeth and bones. When we say one bone articulates with another bone, we mean that the two bones form a joint. *Arthrology* (ar-THROL-ō-jē; *arthr-* = joint; *-ology* = study of) is the scientific study of joints. Many joints of the body permit movement. The study of movement of the human body is called *kinesiology* (ki-nē′-sē-OL-ō-jē; *kinesi-* = movement). The field of medicine devoted to joint diseases and related conditions is *rheumatology* (roo′-ma-TOL-ō-jē; *rheumat-* = a watery flow).

A joint's structure determines its combination of strength and flexibility. At one end of the spectrum are joints that permit no movement and are thus very strong, but inflexible. In contrast, other joints afford fairly free movement and are thus flexible but not as strong. In general, the closer the fit at the point of contact, the stronger the joint. At tightly fitted joints, movement is obviously more restricted. The looser the fit, the greater the movement. However, loosely fitted joints are prone to *dislocation,* displacement of the articulating bones from their normal positions. Movement at joints is also determined by (1) the shape of the articulating bones, (2) the flexibility (tension or tautness) of the ligaments that bind the bones together, and (3) the tension of associated muscles and tendons. Joint flexibility may also be affected by hormones. For example, toward the end of pregnancy, a hormone called relaxin increases the flexibility of the fibrocartilage of the pubic symphysis and loosens the ligaments between the sacrum, hip bone, and coccyx. These changes enlarge the pelvic outlet, which assists in delivery of the baby.

CLASSIFICATION OF JOINTS

Objective: • **Describe the structural and functional classes of joints.**

Joints may be categorized into structural classes, based on anatomical characteristics, and into functional classes, based on the type of movement they permit.

Structural Classes of Joints

The structural classification of joints is based on (1) the presence or absence of a space, called a *synovial cavity* (si-NŌ-vē-al), between the articulating bones, and (2) the type of connective tissue that binds the bones together. A joint is classified structurally into one of the following categories:

- *Fibrous,* if there is no synovial cavity and the bones are held together by fibrous connective tissue that contains many collagen fibers. A suture is an example of a fibrous joint.
- *Cartilaginous* (kar-ti-LAJ-i-nus), if there is no synovial cavity and the bones are held together by cartilage. The epiphyseal plate is an example of a cartilaginous joint.
- *Synovial,* if there is a synovial cavity and the bones forming the joint are united by a surrounding articular capsule and often by accessory ligaments. The shoulder joint is an example of a synovial joint.

Functional Classes of Joints

The functional classification of joints takes into account the degree of movement they permit. Functionally, a joint is classified as follows:

- A *synarthrosis* (sin′-ar-THRŌ-sis; *syn-* = together; plural is *synarthroses*) is an immovable joint.
- An *amphiarthrosis* (am′-fē-ar-THRŌ-sis; *amphi-* = on both sides; plural is *amphiarthroses*) is a slightly movable joint.
- A *diarthrosis* (dī′-ar-THRŌ-sis = movable joint; plural is *diarthroses*) is a freely movable joint.

The following sections of this chapter present the joints of the body according to their structural classes. As we examine the structure of each type of joint, we will also discuss where it fits in these functional categories.

FIBROUS JOINTS

Objective: • **Describe the structure and functions of the three types of fibrous joints.**

Fibrous joints permit little or no movement. The three types of fibrous joints are (1) sutures, (2) syndesmoses, and (3) gomphoses.

Figure 7.1 ■ **Fibrous joints.**

 At a fibrous joint, the bones are held together by connective tis-sue containing many collagen fibers.

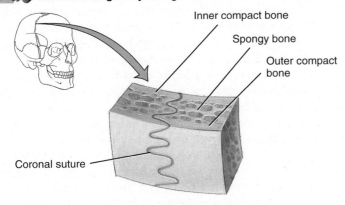

(a) Suture between skull bones

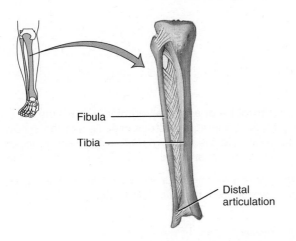

(b) Syndesmosis between tibia and fibula

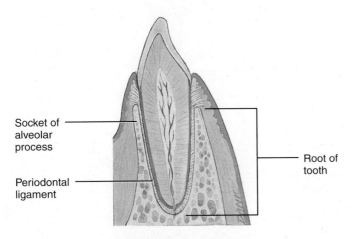

(c) Gomphosis between tooth and socket of alveolar process

Functionally, why are sutures classified as synarthroses and syndesmoses classified as amphiarthroses?

Sutures

A ***suture*** (SOO-cher; *sutur-* = seam) is a fibrous joint composed of a thin layer of dense fibrous connective tissue. Sutures unite the bones of the skull. An example is the coronal suture between the frontal and parietal bones (Figure 7.1a). The irregular, inter-locking edges of sutures give them added strength and decrease their chance of fracturing. Because a suture is immovable, it is classified functionally as a synarthrosis.

Syndesmoses

A ***syndesmosis*** (sin′-dez-MŌ-sis; *syndesmo-* = band or ligament) is a fibrous joint in which the distance between the articulating bones and the amount of fibrous connective tissue is greater than in a suture (Figure 7.1b). One example of a syndesmosis is the distal articulation between the tibia and fibula. Because it permits slight movement, a syndesmosis is classified functionally as an amphiarthrosis.

Gomphoses

A ***gomphosis*** (gom-FŌ-sis; *gompho-* = a bolt or nail; plural is *gomphoses*) is a type of fibrous joint in which a cone-shaped peg fits into a socket. The only gomphoses in the human body are the articulations of the roots of the teeth with the sockets of the alveolar processes of the maxillae and mandible (Figure 7.1c). The dense fibrous connective tissue between the root of a tooth and its socket is the periodontal ligament. A gomphosis is classi-fied functionally as a synarthrosis, an immovable joint.

CARTILAGINOUS JOINTS

Objective: • **Describe the structure and functions of the two types of cartilaginous joints.**

Like a fibrous joint, a ***cartilaginous joint*** allows little or no movement. Here the articulating bones are tightly connected by either fibrocartilage or hyaline cartilage. The two types of carti-laginous joints are synchondroses and symphyses.

Synchondroses

A ***synchondrosis*** (sin′-kon-DRŌ-sis; *chondro-* = cartilage) is a cartilaginous joint in which the connecting material is hyaline cartilage. An example of a synchondrosis is the epiphyseal plate that connects the epiphysis and diaphysis of an elongating bone (Figure 7.2a). Functionally, a synchondrosis is a synarthrosis, an immovable joint. When bone growth stops, bone replaces the hyaline cartilage.

Symphyses

A ***symphysis*** (SIM-fi-sis = growing together) is a cartilaginous joint in which the ends of the articulating bones are covered with hyaline cartilage, but the bones are connected by a broad, flat disc of fibrocartilage. The pubic symphysis between the anterior surfaces of the hip bones is one example of a symphysis (Figure 7.2b). This type of joint is also found at the intervertebral joints between bodies of vertebrae. Functionally, a symphysis is an amphiarthrosis, a slightly movable joint.

SYNOVIAL JOINTS

Objective: • **Describe the structure and the six subtypes of synovial joints.**

The distinguishing feature of a **synovial joint** is the space called a **synovial cavity** between the articulating bones (Figure 7.3). The structure of synovial joints allows them to be freely movable. Hence, all synovial joints are classified functionally as diarthroses.

Structure of Synovial Joints

The articulating surfaces of bones at a synovial joint are covered by articular cartilage (usually hyaline cartilage), which reduces friction between bones in the joint during movement and helps absorb shock. A sleevelike **articular capsule** surrounds a synovial joint, encloses the synovial cavity, and unites the articulating bones. The articular capsule is composed of two layers (Figure 7.3). The outer layer, the **fibrous capsule**, usually consists of dense, irregular connective tissue that attaches to the periosteum of the articulating bones. The flexibility of the fibrous capsule permits considerable movement at a joint, while its strength helps prevent the bones from dislocating. The fibers of some fibrous capsules are arranged in parallel bundles and are therefore highly adapted to resist strains. Such bundles of fibers are called **ligaments** (*liga-* = bound or tied). The inner layer of the articular capsule, the **synovial membrane,** is composed of areolar connective tissue with elastic fibers.

Figure 7.2 ■ **Cartilaginous joints.**

At a cartilaginous joint, the bones are held firmly together by cartilage.

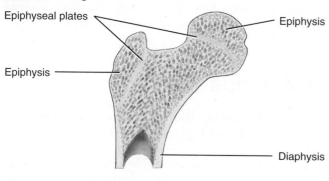

(a) Synchondrosis

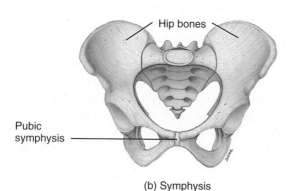

(b) Symphysis

What is the structural difference between a synchondrosis and a symphysis?

Figure 7.3 ■ **Structure of a typical synovial joint.** Note the two layers of the articular capsule: the fibrous capsule and the synovial membrane. Synovial fluid fills the synovial cavity, which is located between the synovial membrane and the hyaline articular cartilage.

The distinguishing feature of a synovial joint is the synovial (joint) cavity between the articulating bones.

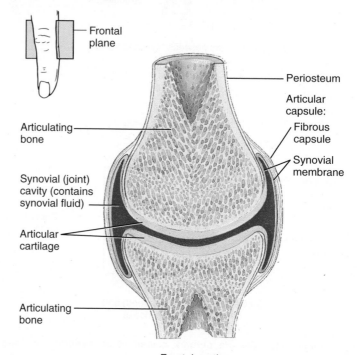

Frontal section

What is the functional classification of synovial joints?

The synovial membrane secretes *synovial fluid* (*ov-* = egg) into the synovial cavity, where the fluid forms a thin, viscous film over the surfaces within the articular capsule. Synovial fluid is similar in appearance and consistency to uncooked egg white. Its functions include lubricating and reducing friction in the joint as well as supplying nutrients and removing metabolic wastes. (Recall that cartilage lacks blood vessels.) Synovial fluid also contains phagocytic cells that remove microbes and debris resulting from wear and tear in the joint.

Many synovial joints also contain *accessory ligaments* that lie inside or outside the articular capsule. Inside some synovial joints, such as the knee, pads of fibrocartilage lie between the articular surfaces of the bones and are attached to the fibrous capsule. These pads are called *menisci* (me-NIS-ī; singular is *meniscus*). Figure 7.11d depicts the lateral and medial menisci in the knee joint, the largest and most complex joint in the body. The menisci allow two bones of different shapes to fit more tightly. A tearing of menisci in the knee, commonly called *torn cartilage,* occurs often among athletes. Such damaged cartilage may require surgical removal (meniscectomy). Repair of the torn cartilage may be assisted by *arthroscopy* (ar-THROS-kō-pē), a procedure that involves examination of the interior of a joint using an arthroscope, a lighted instrument about the size of a pencil in diameter.

Fluid-filled saclike structures called *bursae* (BER-sē = pouches or purses; singular is *bursa*) are strategically situated to reduce friction in some joints such as the shoulder and knee joints (see Figure 7.11c). Bursae cushion the movement of one part of the body over another. Lined by a synovial membrane and filled with a fluid similar to synovial fluid, bursae are located between the skin and bone in places where skin rubs over bone (such as in the knee or elbow) and also between tendons and bones, muscles and bones, and ligaments and bones and within articular capsules. An acute or chronic inflammation of a bursa is called *bursitis.* The condition may be caused by trauma, by an acute or chronic infection, or by rheumatoid arthritis. Repeated excessive friction often results in bursitis with local inflammation and the accumulation of fluid. *Bunions* are swellings that are often associated with a friction bursitis over the head of the first metatarsal bone. Symptoms include pain, swelling, tenderness, and limitation of motion.

Types of Synovial Joints

Although all synovial joints have a similar structure, the shapes of the articulating surfaces vary. Accordingly, synovial joints are divided into six subtypes: planar, hinge, pivot, condyloid, saddle, and ball-and-socket joints.

1. The articulating surfaces of bones in *planar joints* are flat or slightly curved (Figure 7.4a). Some examples of planar joints are the intercarpal (between carpal bones at the wrist), intertarsal (between tarsal bones at the ankle), sternoclavicular (between the sternum and the clavicle), and acromioclavicular (between the acromion of the scapula and the clavicle) joints. Planar joints permit side-to-side and back-and-forth gliding movements (described shortly).

2. In *hinge joints,* the convex surface of one bone fits into the concave surface of another bone (Figure 7.4b). Examples of hinge joints are the knee, elbow, ankle, and interphalangeal joints (between the phalanges of the fingers and toes). As the name implies, hinge joints produce an angular, opening-and-closing motion like that of a hinged door.

3. In *pivot joints,* the rounded or pointed surface of one bone articulates with a ring formed partly by another bone and partly by a ligament (Figure 7.4c). A pivot joint allows rotation around its own longitudinal axis. Examples of pivot joints are the atlantoaxial joint, in which the atlas rotates around the axis and permits you to turn your head from side to side as in signifying "no," and the radioulnar joints that allow you to move your palms forward and backward.

4. In *condyloid joints* (KON-di-loyd = knucklelike), the convex oval-shaped projection of one bone fits into the concave oval-shaped depression of another bone (Figure 7.4d). Examples are the wrist and metacarpophalangeal joints (between the metacarpals and phalanges) of the second through fifth digits. A condyloid joint permits up-and-down and side-to-side movements.

5. In *saddle joints,* the articular surface of one bone is saddle-shaped, and the articular surface of the other bone fits into the saddle like a rider sitting on a horse (Figure 7.4e). An example of a saddle joint is the carpometacarpal joint between the trapezium of the carpus and metacarpal of the thumb. Saddle joints permit side-to-side and up-and-down movements.

6. In *ball-and-socket joints,* the ball-like surface of one bone fits into a cuplike depression of another bone (Figure 7.4f). Ball-and-socket joints permit movement in several directions; the only examples in the human body are the shoulder and hip joints.

TYPES OF MOVEMENTS AT SYNOVIAL JOINTS

Objective: • **Describe the types of movements that can occur at synovial joints.**

Anatomists, physical therapists, and kinesiologists use specific terminology to designate specific types of movement that can occur at a synovial joint. These precise terms indicate the form of motion, the direction of movement, or the relationship of one body part to another during movement. Movements at synovial joints are grouped into four main categories: (1) gliding, (2) angular movements, (3) rotation, and (4) special movements. The last category includes movements that occur only at certain joints.

Figure 7.4 ■ **Types of synovial joints.** For each subtype, a drawing of the actual joint and a simplified diagram are shown.

 Synovial joints are classified into subtypes on the basis of the shapes of the articulating bone surfaces.

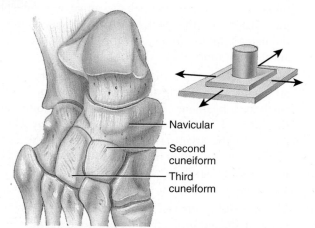

(a) Planar joint between the navicular and second and third cuneiforms of the tarsus in the foot

Navicular
Second cuneiform
Third cuneiform

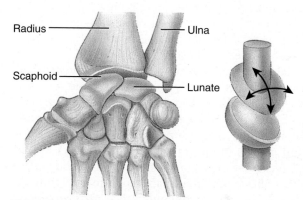

(d) Condyloid joint between radius and scaphoid and lunate bones of the carpus (wrist)

Radius
Ulna
Scaphoid
Lunate

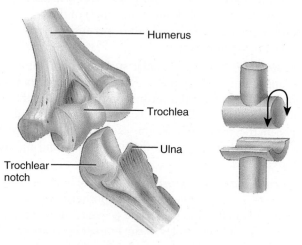

(b) Hinge joint between trochlea of humerus and trochlear notch of ulna at the elbow

Humerus
Trochlea
Ulna
Trochlear notch

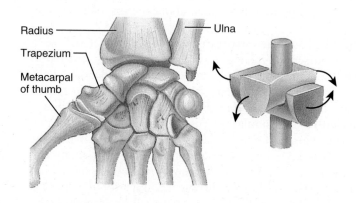

(e) Saddle joint between trapezium of carpus (wrist) and metacarpal of thumb

Radius
Ulna
Trapezium
Metacarpal of thumb

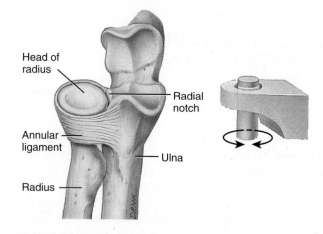

(c) Pivot joint between head of radius and radial notch of ulna

Head of radius
Radial notch
Annular ligament
Ulna
Radius

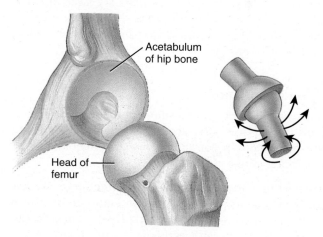

(f) Ball-and-socket joint between head of the femur and acetabulum of the hip bone

Acetabulum of hip bone
Head of femur

Which joints permit the greatest range of motion?

Figure 7.5 ■ **Gliding movements at synovial joints.**

 Gliding motion consists of side-to-side and back-and-forth movements.

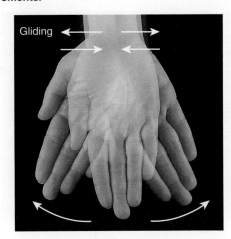

Joints between carpal bones

 At which subtype of synovial joints do gliding movements occur?

Gliding

Gliding is a simple movement in which relatively flat bone surfaces move back and forth and side to side relative to one another (Figure 7.5). Gliding movements are limited in range due to the loose-fitting structure of the articular capsule and associated ligaments and bones. Gliding occurs at planar joints.

Angular Movements

In *angular movements,* there is an increase or a decrease in the angle between articulating bones. The principal angular movements are flexion, extension, abduction, and adduction. In *flexion* (FLEK-shun = to bend), there is a decrease in the angle between articulating bones; in *extension* (eks-TEN-shun = to stretch out), there is an increase in the angle between articulating bones, often to restore a part of the body to the anatomical position after it has been flexed (Figure 7.6). Hinge, pivot, condyloid, saddle, and ball-and-socket joints all permit flexion

Figure 7.6 ■ **Angular movements at synovial joints: flexion, extension, and hyperextension.**

 In angular movements, there is an increase or decrease in the angle between articulating bones.

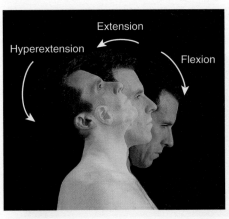

(a) Joints between atlas and occipital bone and between cervical vertebrae

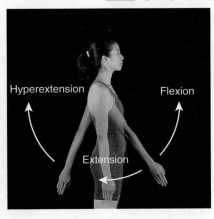

(b) Shoulder joint

(c) Elbow joint

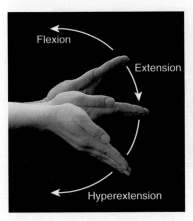

(d) Wrist joint

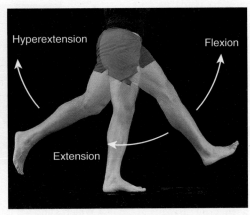

(e) Hip joint

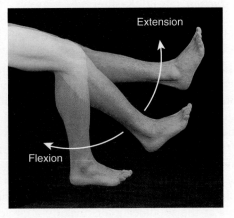

(f) Knee joint

 What prevents hyperextension at some synovial joints?

Figure 7.7 ■ **Angular movements at synovial joints: abduction and adduction.**

 Condyloid, saddle, and ball-and-socket joints permit abduction and adduction.

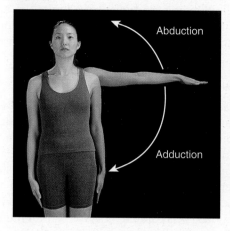

(a) Shoulder joint

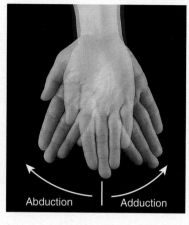

(b) Wrist joint

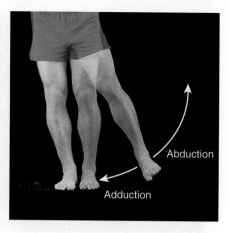

(c) Hip joint

One way to remember what adduction means is use of the phrase "adding your limb to your trunk." Why is this an effective learning device?

and extension. Examples of flexion include bending the head toward the chest (Figure 7.6a); moving the humerus forward at the shoulder joint as in swinging the arms forward while walking (Figure 7.6b); moving the forearm toward the shoulder (Figure 7.6c); moving the palm toward the forearm (Figure 7.6d); moving the femur forward, as in walking (Figure 7.6e); and bending the knee (Figure 7.6f).

Continuation of extension beyond the anatomical position is called **hyperextension** (*hyper-* = beyond or excessive). Examples of hyperextension include bending the head backward (Figure 7.6a); moving the humerus backward, as in swinging the arms backward while walking (Figure 7.6b); moving the palm backward at the wrist joint (Figure 7.6d); and moving the femur backward, as in walking (Figure 7.6e). Hyperextension of other joints, such as the elbow, interphalangeal (fingers and toes), and knee joints, is usually prevented by the arrangement of ligaments and bones.

Abduction (ab-DUK-shun; *ab-* = away; *-duct* = to lead) is the movement of a bone away from the midline, whereas **adduction** (ad-DUK-shun; *ad-* = toward) is the movement of a bone toward the midline. Condyloid, saddle, and ball-and-socket joints permit abduction and adduction. Examples of abduction include lateral movement of the humerus upward (Figure 7.7a), the palm away from the body (Figure 7.7b), and the femur away from the body (Figure 7.7c). Movement in the opposite direction (medially) in each case produces adduction (Figure 7.7).

Circumduction (ser-kum-DUK-shun; *circ-* = circle) is movement of the distal end of a part of the body in a circle (Figure 7.8). Ball-and-socket joints permit circumduction. Examples of joints that allow circumduction include the humerus at the shoulder joint (making a circle with your arm) and the femur at the hip joint (making a circle with your leg). Circumduction is more limited at the hip due to greater tension on the ligaments and muscles.

Figure 7.8 ■ **Angular movements at synovial joints: circumduction.**

Circumduction is the movement of the distal end of a body part in a circle.

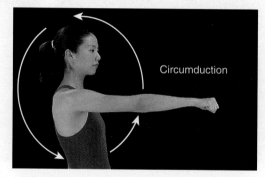

(a) Shoulder joint

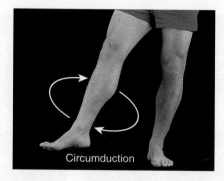

(b) Hip joint

 Which synovial joints permit circumduction?

Rotation

In *rotation* (rō-TĀ-shun; *rota-* = to revolve) a bone revolves around its own longitudinal axis. Pivot and ball-and-socket joints permit rotation. An example is turning the head from side to side, as in signifying "no" (Figure 7.9).

Figure 7.9 ■ **Rotation at synovial joints.**

 In rotation, a bone revolves around its own longitudinal axis.

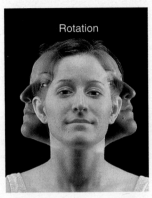

Rotation

Joint between atlas and axis

 Which synovial joints permit rotation?

Special Movements

The *special movements* that occur only at certain joints include elevation, depression, protraction, retraction, inversion, eversion, dorsiflexion, plantar flexion, supination, and pronation (Figure 7.10).

- *Elevation* (el'-e-VĀ-shun = to lift up) is the upward movement of a part of the body, such as closing the mouth to elevate the mandible (Figure 7.10a) or shrugging the shoulders to elevate the scapula.

- *Depression* (dē-PRESH-un = to press down) is the downward movement of a part of the body, such as opening the mouth to depress the mandible (Figure 7.10b) or returning shrugged shoulders to the anatomical position to depress the scapula.

- *Protraction* (prō-TRAK-shun = to draw forth) is the movement of a part of the body forward. You can protract your mandible by thrusting it outward (Figure 7.10c) or protract your clavicles by crossing your arms.

- *Retraction* (rē-TRAK-shun = to draw back) is the movement of a protracted part of the body back to the anatomical position (Figure 7.10d).

- *Inversion* (in-VER-zhun = to turn inward) is movement of the soles medially so that they face each other (Figure 7.10e).

Figure 7.10 ■ **Special movements at synovial joints.**

Special movements occur only at certain synovial joints.

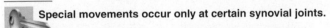

(a)   Temporomandibular joint   (b)   (c)   Temporomandibular joint   (d)

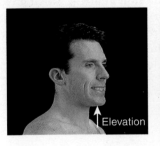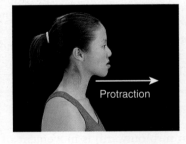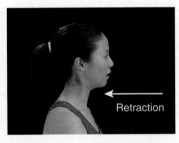

(e)   Joints between tarsal bones   (f)   (g) Ankle joint   (h) Joints between radius and ulna

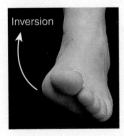 **What movement of the shoulder girdle is involved in bringing the arms forward until the elbows touch?**

Joint Care— Prevent Repetitive Motion Injury

Diarthroses (freely movable joints) allow extensive movement. But the human body is not a machine, and diarthroses were not designed to withstand the repetition of a given motion over and over and over again, all day long. When you repeat the same motion for extended periods of time, you may overstress the joint or joints responsible for that motion and the associated soft-tissue structures, such as the articular capsule, ligaments, bursae, muscles, tendons, and nerves. Repeated episodes of mechanical stress can lead to the development of *repetitive motion injuries.*

Repeat That?

Repetitive motion injuries are a type of *cumulative trauma disorder (CTD)*, which is a group of disorders characterized by ongoing damage to soft tissues. Repetitive motion injuries are the most common type of CTD, but CTDs may also involve trauma due to exposure to cold or hot temperatures, certain types of lighting, vibration, and so forth. Repetitive motion injuries are similar in many ways to the *overuse injuries* that athletes often experience. Just as tennis players may develop epicondylitis (tennis elbow), so too may construction workers who perform repeated elbow flexion and extension in their work and students who spend hours a day using their computer mouse ("mouse elbow").

Repetitive motions alone may cause repetitive motion injuries. Risk increases when repetitive motions are coupled with poor posture and biomechanics, which put excess strain on joints. Joint stress also increases when a person must apply force with the motion, such as when gripping or lifting heavy materials. The joints at highest risk are those that are the weakest. Wrists, backs, elbows, shoulders, and necks are the most common sites of repetitive motion injury.

Repetitive motion injuries usually develop slowly over a long period of time. They typically begin with mild to moderate discomfort in the affected joints, especially at night. Other symptoms include swelling in the joint, muscle fatigue, numbness, and tingling. Symptoms may come and go at first, but then become constant. Symptoms of more advanced damage include more intense pain, muscle weakness, and nerve problems. If left untreated, repetitive motion injuries can be extremely painful. They also may severely limit a joint's range of motion. Fortunately, because they develop slowly, most repetitive motion injuries are discovered early enough to be successfully treated.

▶ *Think It Over*

▶ Carpal tunnel syndrome is a repetitive motion injury in which pressure develops on the median nerve as it passes through the carpal tunnel, a narrow tunnel of bone and ligament at the wrist. Pressure on this nerve causes numbness, tingling, and pain in some or all of the fingers. What kind of workers do you think might be most at risk for the development of carpal tunnel syndrome?

- *Eversion* (ē-VER-zhun = to turn outward) is movement of the soles laterally so that they face away from each other (Figure 7.10f).

- *Dorsiflexion* (dor′-si-FLEK-shun) is bending of the foot in the direction of the dorsum (superior surface), as when you stand on your heels (Figure 7.10g).

- *Plantar flexion* involves bending of the foot in the direction of the plantar surface (Figure 7.10g), as when standing on your toes.

- *Supination* (soo′-pi-NĀ-shun) is movement of the forearm so that the palm is turned forward or upward (Figure 7.10h). Supination of the palms is one of the defining features of the anatomical position (see Figure 1.4 on page 000).

- *Pronation* (prō-NĀ-shun) is movement of the forearm so that the palm is turned backward or downward (Figure 7.10h).

KNEE JOINT

Objective: • **Describe the principal structures and functions of the knee joint.**

To give you an idea of the complexity of a synovial joint, we will examine some of the structural features of the knee joint, the largest and most complex joint in the body. Among the main structures of the knee joint are the following (Figure 7.11)

1. The *articular capsule* is strengthened by muscle tendons surrounding the joint.

2. The *patellar ligament* extends from the patella to the tibia and strengthens the anterior surface of the joint.

3. The *oblique popliteal ligament* (pop-LIT-ē-al) strengthens the posterior surface of the joint.

Figure 7.11 ■ **Structure of the right knee joint.**

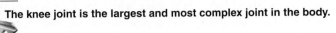

The knee joint is the largest and most complex joint in the body.

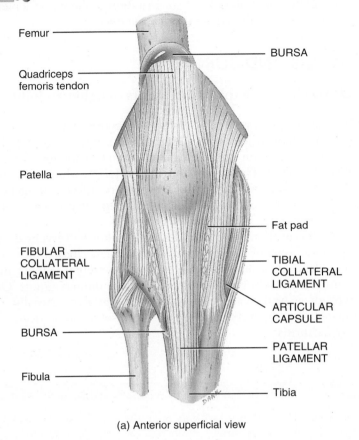

Femur

Quadriceps
femoris tendon

Patella

FIBULAR
COLLATERAL
LIGAMENT

BURSA

Fibula

BURSA

Fat pad

TIBIAL
COLLATERAL
LIGAMENT

ARTICULAR
CAPSULE

PATELLAR
LIGAMENT

Tibia

(a) Anterior superficial view

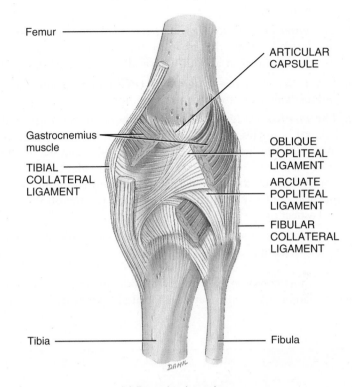

Femur

Gastrocnemius
muscle

TIBIAL
COLLATERAL
LIGAMENT

Tibia

ARTICULAR
CAPSULE

OBLIQUE
POPLITEAL
LIGAMENT

ARCUATE
POPLITEAL
LIGAMENT

FIBULAR
COLLATERAL
LIGAMENT

Fibula

(b) Posterior deep view

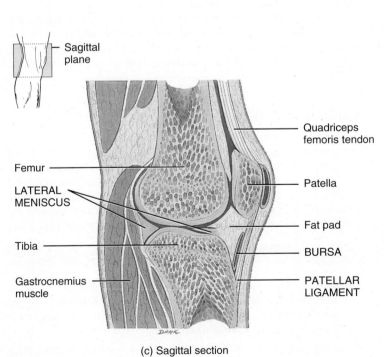

Sagittal
plane

Femur

LATERAL
MENISCUS

Tibia

Gastrocnemius
muscle

Quadriceps
femoris tendon

Patella

Fat pad

BURSA

PATELLAR
LIGAMENT

(c) Sagittal section

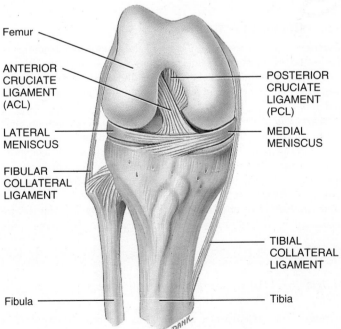

Femur

ANTERIOR
CRUCIATE
LIGAMENT
(ACL)

LATERAL
MENISCUS

FIBULAR
COLLATERAL
LIGAMENT

Fibula

POSTERIOR
CRUCIATE
LIGAMENT
(PCL)

MEDIAL
MENISCUS

TIBIAL
COLLATERAL
LIGAMENT

Tibia

(d) Anterior deep view (flexed)

What structures are damaged in the knee injury called torn cartilage?

4. The *arcuate popliteal ligament* strengthens the lower lateral part of the posterior surface of the joint.

5. The *tibial collateral ligament* strengthens the medial aspect of the joint.

6. The *fibular collateral ligament* strengthens the lateral aspect of the joint.

7. The *anterior cruciate ligament (ACL)* extends posteriorly and laterally from the tibia to the femur. The ACL is stretched or torn in about 70% of all serious knee injuries.

8. The *posterior cruciate ligament (PCL)* extends anteriorly and medially from the tibia to the femur.

9. The *menisci* are fibrocartilage discs between the tibial and femoral condyles. They help compensate for the irregular shapes of the articulating bones. The two menisci of the knee joint are the *medial meniscus*, a semicircular piece of fibrocartilage on the medial aspect of the knee, and the *lateral meniscus*, a nearly circular piece of fibrocartilage on the lateral aspect of the knee.

10. The *bursae* are saclike structures filled with fluid that help reduce friction.

The most common knee injury in football is rupture of the tibial collateral ligaments, often associated with tearing of the anterior cruciate ligament and medial meniscus. Usually, a hard blow to the lateral side of the knee while the foot is fixed on the ground causes the injury. Accumulation of synovial fluid in the knee joint, as a result of injury or irritation of the synovial membrane, is commonly called *water on the knee.*

AGING AND JOINTS

Objective: • **Explain the effects of aging on joints.**

The aging process usually results in decreased production of synovial fluid in joints. In addition, the articular cartilage becomes thinner with age, and ligaments shorten and lose some of their flexibility. The effects of aging on joints, which vary considerably from one person to another, are affected by genetic factors and by wear and tear. Although degenerative changes in joints may begin in individuals as young as 20 years of age, most do not occur until much later. By age 80, almost everyone develops some type of degeneration in the knees, elbows, hips, and shoulders. Degenerative changes in the vertebral column may cause a hunched-over posture and pressure on nerve roots. One type of arthritis, called osteoarthritis, is at least partially age related. Nearly everyone over age 70 displays some degree of osteoarthritis.

COMMON DISORDERS

Rheumatism and Arthritis

Rheumatism refers to any painful state of the supporting structures of the body—bones, ligaments, joints, tendons, or muscles. *Arthritis* is a form of rheumatism in which the joints are inflamed. Several different diseases can cause arthritis, and it afflicts about 40 million people in the United States.

Rheumatoid arthritis (RA) (ROO-ma-toyd) is an autoimmune disease in which the immune system attacks its own tissues, in this case its cartilage and joint linings. The primary symptom of rheumatoid arthritis is inflammation of the synovial membrane. RA is characterized by inflammation of the joint, which causes redness, warmth, swelling, pain, and loss of function.

Osteoarthritis (os′-tē-ō-ar-THRĪ-tis) is degenerative joint disease characterized by deterioration of articular cartilage. It is commonly known as "wear-and-tear" arthritis and is the leading cause of disability in older people. A major distinction between osteoarthritis and rheumatoid arthritis is that osteoarthritis strikes the larger joints (knees, hips) first, whereas rheumatoid arthritis first strikes smaller joints, such as those in the fingers.

In *gouty arthritis* (GOW-tē), sodium urate crystals are deposited in the soft tissues of the joints. The crystals irritate and erode the cartilage, causing inflammation, swelling, and acute pain. If the disorder is not treated, the ends of the articulating bones fuse, and the joint becomes immovable.

Sprain and Strain

A *sprain* is the forcible wrenching or twisting of a joint that stretches or tears its ligaments but does not dislocate the bones. It occurs when the ligaments are stressed beyond their normal capacity. The ankle joint is most often sprained; the lower back is another common location for sprains. A less serious injury is a *strain,* which is a stretched or partially torn muscle. It often occurs when a muscle contracts suddenly and powerfully, for example, in sprinters when they accelerate too quickly.

MEDICAL TERMINOLOGY AND CONDITIONS

Arthralgia (ar-THRAL-jē-a; *arthr-* = joint; *-algia* = pain) Pain in a joint.

Bursectomy (bur-SEK-tō-mē; *-ectomy* = to cut out) Removal of a bursa.

Chondritis (kon-DRĪ-tis; *chondro-* = cartilage) Inflammation of cartilage.

Dislocation (*dis-* = apart) or *luxation* (luks-A-shun; *lux-* = dislocation) The displacement of a bone from a joint with tearing of ligaments, tendons, and articular capsules. A partial or incomplete dislocation is called a *subluxation.*

Synovitis (sin′-ō-VĪ-tis) Inflammation of a synovial membrane in a joint.

STUDY OUTLINE

Joints (p. 154)

1. A joint (articulation) is a point of contact between two bones, cartilage and bone, or teeth and bone.

Classification of Joints (p. 154)

1. Structural classification is based on the presence or absence of a synovial cavity and the type of connecting tissue. Structurally, joints are classified as fibrous, cartilaginous, or synovial.

2. Functional classification of joints is based on the degree of movement permitted. Joints may be synarthroses (immovable), amphiarthroses (slightly movable), or diarthroses (freely movable).

Fibrous Joints (p. 154)

1. Bones held together by fibrous connective tissue are fibrous joints.

2. These joints include immovable sutures (found between skull bones), slightly movable syndesmoses (such as the distal joint between the tibia and fibula), and immovable gomphoses (roots of teeth in alveoli of the mandible and maxilla).

Cartilaginous Joints (p. 155)

1. Bones held together by cartilage are cartilaginous joints.

2. These joints include immovable synchondroses united by hyaline cartilage (epiphyseal plates) and slightly movable symphyses united by fibrocartilage (pubic symphysis).

Synovial Joints (p. 156)

1. A synovial joint contains a space between bones called the synovial cavity. All synovial joints are diarthroses.

2. Other characteristics of a synovial joint are the presence of articular cartilage and an articular capsule, made up of a fibrous capsule and a synovial membrane.

3. The synovial membrane secretes synovial fluid, which forms a thin, viscous film over the surfaces within the articular capsule.

4. Many synovial joints also contain accessory ligaments and menisci.

5. Bursae are saclike structures, similar in structure to joint capsules, that reduce friction in joints such as the shoulder and knee joints.

6. Subtypes of synovial joints are planar, hinge, pivot, condyloid, saddle, and ball-and-socket.

7. In a planar joint, the articulating surfaces are flat; examples are joints between carpals and tarsals.

8. In a hinge joint, the convex surface of one bone fits into the concave surface of another; examples are the elbow, knee, and ankle joints.

9. In a pivot joint, a round or pointed surface of one bone fits into a ring formed by another bone and a ligament; examples are the atlantoaxial and radioulnar joints.

10. In a condyloid joint, an oval-shaped projection of one bone fits into an oval cavity of another; examples are the wrist joint and metacarpophalangeal joints for the second through fifth digits.

11. In a saddle joint, the articular surface of one bone is shaped like a saddle, and the other bone fits into the "saddle" like a rider on a horse; an example is the carpometacarpal joint between trapezium and metacarpal of the thumb.

12. In a ball-and-socket joint, the ball-shaped surface of one bone fits into the cuplike depression of another; examples are the shoulder and hip joints.

Types of Movements at Synovial Joints (p. 157)

1. In a gliding movement, the nearly flat surfaces of bones move back and forth and side to side. Gliding movements occur at planar joints.

2. In angular movements, there is a change in the angle between bones. Examples are flexion–extension, hyperextension, abduction–adduction, and circumduction. Angular movements occur at hinge, pivot, condyloid, saddle, and ball-and-socket joints.

3. In rotation, a bone moves around its own longitudinal axis. Rotation can occur at pivot and ball-and-socket joints.

4. Special movements occur at specific synovial joints. Examples are elevation–depression, protraction–retraction, inversion–eversion, dorsiflexion–plantar flexion, and supination–pronation.

Knee Joint (p. 162)

1. The knee joint is a diarthrosis that illustrates the complexity of this type of joint.

2. It contains an articular capsule, several ligaments within and around the outside of the joint, menisci, and bursae.

Aging and Joints (p. 164)

1. With aging, a decrease in synovial fluid, thinning of articular cartilage, and decreased flexibility of ligaments occur.

2. Most individuals experience some degeneration in the knees, elbows, hips, and shoulders due to the aging process.

SELF-QUIZ

1. A joint that has a _____ fit offers a great amount of movement and is _____ likely to become dislocated. **a.** tight, less **b.** tight, more **c.** loose, less **d.** loose, more **e.** flexible, less

2. An example of a fibrous joint in which the bones are immovable is a **a.** suture **b.** syndesmosis **c.** gomphosis **d.** symphysis **e.** synchondrosis

3. Pulling out a tooth would disarticulate which type of joint?
   a. symphysis **b.** synovial **c.** gomphosis **d.** cartilaginous **e.** suture

4. Which of the following is NOT a function of synovial fluid?

a. It acts as a lubricant. **b.** It helps strengthen the joint. **c.** It removes microbes and debris from the joint. **d.** It provides nutrients to the tissues around the joints. **e.** It removes metabolic wastes.

5. Articular cartilage and bursae would most likely be found in which of the following? **a.** gomphosis **b.** a suture **c.** the pubic symphysis **d.** the knee **e.** a synchondrosis

6. Which of the following structures provides flexibility to a joint while also preventing dislocation? **a.** bursae **b.** articular cartilage **c.** synovial fluid **d.** muscles **e.** articular capsule

7. The joints between the vertebrae and the joint between the hip bones are examples of which joint type?   **a.** synovial **b.** symphysis   **c.** fibrous   **d.** synchondrosis   **e.** suture

8. Match the following:

_____ **a.** the joint between the atlas and axis

_____ **b.** allows gliding movements

_____ **c.** the joint between the carpal and metacarpal of the thumb

_____ **d.** hip joint

_____ **e.** knee joint

A. planar joint
B. hinge joint
C. ball-and-socket joint
D. pivot joint
E. saddle joint

9. Which of the following diarthrotic joints allows for the greatest degree of movement?   **a.** ball-and-socket   **b.** hinge **c.** condyloid   **d.** pivot   **e.** saddle

10. Moving the femur forward when walking is an example of **a.** abduction   **b.** circumduction   **c.** flexion   **d.** gliding **e.** inversion

11. When a gymnast performs the "splits," the primary movement at the hip joint is   **a.** rotation   **b.** adduction   **c.** extension **d.** gliding   **e.** abduction

12. In the anatomical position, the palms are   **a.** supinated **b.** flexed   **c.** inverted   **d.** pronated   **e.** protracted

13. A fluid-filled sac found between skin and bone that helps reduce friction between the skin and bone is a   **a.** meniscus   **b.** bursa **c.** ligament   **d.** articular capsule   **e.** synovial membrane

14. Nodding your head "yes" in response to a question involves **a.** abduction and adduction   **b.** circumduction   **c.** extension and hyperextension   **d.** rotation   **e.** flexion and extension

15. Match the following:

_____ **a.** movement of a bone around its own axis

_____ **b.** movement away from the midline of the body

_____ **c.** palm faces upward or forward

_____ **d.** downward movement of a body part

_____ **e.** movement toward the midline of the body

_____ **f.** movement of the mandible or shoulder backward

_____ **g.** turning the palm so it faces downward or backward

_____ **h.** upward movement of a body part

_____ **i.** movement of the distal end of a body part in a circle

_____ **j.** movement beyond the plane of extension

A. rotation
B. supination
C. depression
D. adduction
E. retraction
F. pronation
G. abduction
H. hyperextension
I. circumduction
J. elevation

CRITICAL THINKING APPLICATIONS

1. After your second A & P exam, you dropped to one knee, tipped your head back, raised one arm over your head, clenched your fist, pumped your arm up and down, and yelled "Yes!" Use the proper terms to describe the movements undertaken by the various joints.

2. Aunt Rosa's hip has been bothering her for years, and now she can hardly walk. Her doctor suggested a hip replacement. "It's one of those synonymous joints," Aunt Rosa explained. What type of joint is the hip joint? What types of movements can it perform?

3. John used to play football in high school, but his knees "gave out." Now he's going to have one knee "scoped" to clean out the torn cartilage. What structure did John injure, and what is the proper term for the surgery?

4. Hee Soo got slammed by a wave while bodysurfing. Now his arm feels useless, he's got an odd bump on his shoulder, and his shoulder really hurts! What happened to Hee Soo's shoulder?

ANSWERS TO FIGURE QUESTIONS

7.1   Sutures are synarthroses because they are immovable, whereas syndesmoses are classified as an amphiarthroses because they are slightly movable.

7.2   Hyaline cartilage holds a synchondrosis together, and fibrocartilage holds a symphysis together.

7.3   Synovial joints are diarthroses, freely movable joints.

7.4   Ball-and-socket joints permit the greatest degree of movement.

7.5   Gliding movements occur at planar joints.

7.6   The arrangement of ligaments and bones prevents hyperextension at some synovial joints.

7.7   When you adduct your arm or leg, you bring it closer to the midline of the body, thus "adding" it to the trunk.

7.8   Circumduction can occur at ball-and-socket joints.

7.9   Pivot and ball-and-socket joints permit rotation.

7.10   Bringing the arms forward until the elbows touch is an example of protraction.

7.11   In torn cartilage injuries of the knee, the menisci are damaged.

Chapter 8

The Muscular System

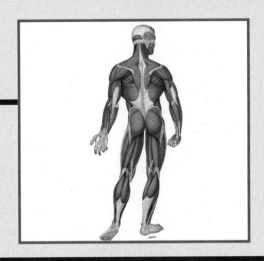

■ Student Learning Objectives

1. Describe the types, functions, and characteristics of muscle tissue. **168**
2. Explain the relation of connective tissue components, blood vessels, and nerves to skeletal muscles. **168**
3. Describe the histology of a skeletal muscle cell. **171**
4. Explain how skeletal muscle fibers contract and relax. **172**
5. Describe the sources of ATP and oxygen for muscle contraction. **177**
6. Define muscle fatigue and list its possible causes. **178**
7. List the reasons that oxygen consumption is higher after exercise than at rest. **178**

8. Explain the three phases of a twitch contraction. **178**
9. Describe how frequency of stimulation and motor unit recruitment affect muscle tension. **179**
10. Compare the three types of skeletal muscle fibers. **180**
11. Distinguish between isotonic and isometric contractions. **180**
12. Describe the effects of exercise on skeletal muscle tissue. **180**
13. Explain how aging affects skeletal muscle. **181**

14. Describe the structure and function of cardiac muscle tissue. **181**
15. Describe the structure and function of smooth muscle tissue. **181**
16. Describe how skeletal muscles cooperate to produce movement. **183**
17. List and describe several ways that skeletal muscles are named. **183**
18. Describe the location of skeletal muscles in various regions of the body and identify their functions. **187–209**

■ A Look Ahead

*A*lthough the bones and joints you learned about in the last two chapters form the framework of the body, they cannot move the body on their own. Motion results from the contraction and relaxation of muscles. When a muscle contracts, it generates tension that may result in shortening of the muscle. Muscle tissue constitutes about 40% to 50% of the total body weight and is composed of highly specialized cells. The scientific study of muscles is known as *myology* (mī-OL-ō-jē; *my-* = muscle; *-ology* = study of).

OVERVIEW OF MUSCLE TISSUE

Objective: • **Describe the types, functions, and characteristics of muscle tissue.**

Types of Muscle Tissue

Recall from Chapter 4 that the three types of muscle tissue are skeletal, cardiac, and smooth. As its name suggests, most *skeletal muscle tissue* is attached to bones and moves parts of the skeleton. It is *striated;* that is, *striations*, or alternating light and dark bands, are visible under a microscope. Because skeletal muscle can be made to contract and relax by conscious control, it is *voluntary.* Due to the presence of a small number of cells that can undergo cell division, skeletal muscle has a limited capacity for regeneration.

    Cardiac muscle tissue, found only in the heart, forms the bulk of the heart wall. The heart pumps blood through blood vessels to all parts of the body. Like skeletal muscle tissue, cardiac muscle tissue is *striated.* However, unlike skeletal muscle tissue, it is *involuntary:* its contractions are not under conscious control. Cardiac muscle cannot regenerate.

    Smooth muscle tissue is located in the walls of hollow internal structures, such as blood vessels, airways, the stomach, and the intestines. It participates in internal processes such as digestion and regulation of blood pressure. Smooth muscle is *nonstriated* (lacks striations) and *involuntary* (not under conscious control). Although smooth muscle tissue has considerable capacity to regenerate when compared with other muscle tissues, this capacity is limited when compared to other types of tissues, for example, epithelium.

Functions of Muscle Tissue

Through sustained contraction or alternating contraction and relaxation, muscle tissue has five key functions: producing body movements, stabilizing body positions, regulating organ volume, moving substances within the body, and generating heat.

1. **Producing body movements.** Body movements such as walking, running, writing, or nodding the head rely on the integrated functioning of bones, joints, and skeletal muscles.

2. **Stabilizing body positions.** Skeletal muscle contractions stabilize joints and help maintain body positions, such as standing or sitting. Postural muscles contract continuously when a person is awake; for example, sustained contractions of your neck muscles hold your head upright.

3. **Regulating organ volume.** Sustained contractions of ring-like bands of smooth muscles called *sphincters* prevent outflow of the contents of a hollow organ. Temporary storage of food in the stomach or urine in the urinary bladder is possible because smooth muscle sphincters close off the outlets of these organs.

4. **Moving substances within the body.** Cardiac muscle contractions pump blood through the body's blood vessels. Contraction and relaxation of smooth muscle in the walls of blood vessels helps adjust their diameter and thus regulate blood flow. Smooth muscle contractions also move food and other substances through the gastrointestinal tract, push gametes (sperm and oocytes) through the reproductive system, and propel urine through the urinary system. Skeletal muscle contractions aid the return of blood to the heart.

5. **Producing heat.** As muscle tissue contracts, it produces heat. Much of the heat released by muscle is used to maintain normal body temperature. Involuntary contractions of skeletal muscle, known as shivering, can help warm the body by greatly increasing the rate of heat production.

Characteristics of Muscle Tissue

Four characteristics of muscle tissue are important in understanding its functions and its contributions to homeostasis:

1. *Excitability* is the ability of muscle tissue (and nerve cells) to receive and respond to stimuli by producing electrical signals such as action potentials (described in Chapter 9).

2. *Contractility* is the ability of muscle tissue to contract (shorten and thicken) when stimulated by an action potential.

3. *Extensibility* is the ability of muscle tissue to stretch (extend) without being damaged.

4. *Elasticity* is the ability of muscle tissue to return to its original shape after contraction or extension.

SKELETAL MUSCLE TISSUE

Objectives: • **Explain the relation of connective tissue components, blood vessels, and nerves to skeletal muscles.**

• **Describe the histology of a skeletal muscle cell.**

Each skeletal muscle is a separate organ composed of hundreds to thousands of skeletal muscle cells called *muscle fibers* because of their elongated shapes. Connective tissues surround muscle fibers and whole muscles, and blood vessels and nerves penetrate muscle.

Connective Tissue Components

The term *fascia* (FASH-ē-a = bandage) is applied to a sheet or broad band of fibrous connective tissue beneath the skin or around muscles and other organs of the body. There are two types of fascia. (1) Immediately under the skin is the *superficial fascia (subcutaneous layer),* which is composed of areolar connective tissue and adipose tissue. (2) More important to the study of muscles is the *deep fascia,* which is composed of dense,

irregular connective tissue. Deep fascia holds muscles together and separates them into functional groups.

Several connective tissue coverings extend from the deep fascia (Figure 8.1). The entire muscle is wrapped in *epimysium* (ep'-i-MĪZ-ē-um; *epi-* = upon). Bundles of muscle fibers (cells)

Figure 8.1 ■ Organization of skeletal muscle and its connective tissue coverings.

A skeletal muscle consists of individual muscle fibers (cells) bundled into fascicles and surrounded by three connective tissue layers that are extensions of the deep fascia.

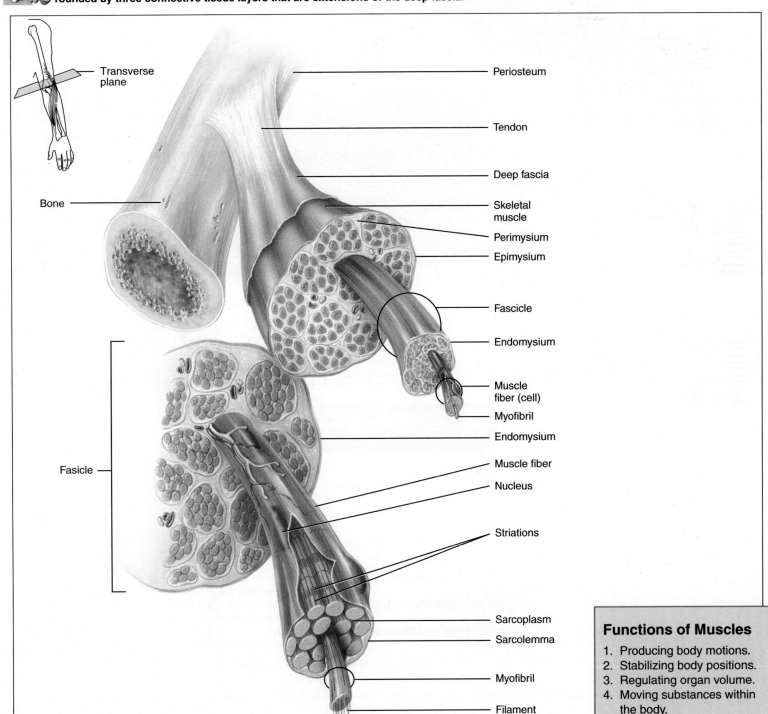

Labels: Transverse plane; Periosteum; Tendon; Deep fascia; Bone; Skeletal muscle; Perimysium; Epimysium; Fascicle; Endomysium; Muscle fiber (cell); Myofibril; Endomysium; Muscle fiber; Nucleus; Fasicle; Striations; Sarcoplasm; Sarcolemma; Myofibril; Filament

Functions of Muscles

1. Producing body motions.
2. Stabilizing body positions.
3. Regulating organ volume.
4. Moving substances within the body.
5. Producing heat.

Starting with the connective tissue that surrounds an individual muscle fiber (cell) and working toward the outside, list the connective tissue layers in order.

Figure 8.2 ■ **Organization of skeletal muscle from gross to molecular levels.** A photomicrograph of skeletal muscle tissue is shown in Table 4.4 on page 90.

The structural organization of a skeletal muscle from macroscopic to microscopic is as follows: skeletal muscle, fascicle (bundle of muscle fibers), muscle fiber, myofibril, and thin and thick filaments.

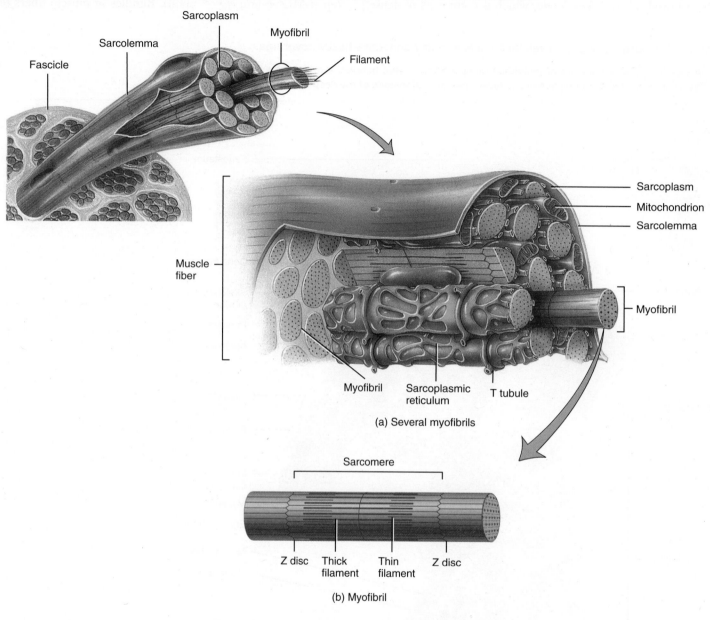

(a) Several myofibrils

(b) Myofibril

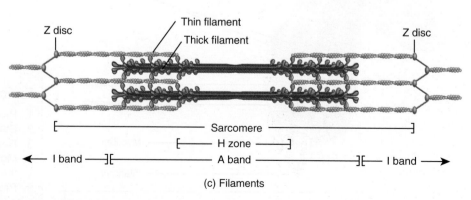

(c) Filaments

Which filaments are part of the A band and I band?

called *fascicles* (FAS-i-kuls = little bundles) are covered by *peri-mysium* (per'-i-MĪZ-ē-um; *peri-* = around). Finally, *endo-mysium* (en'-dō-MĪZ-ē-um; *endo-* = within) wraps each individual muscle fiber. Epimysium, perimysium, and endomysium extend beyond the muscle as a *tendon*—a cord of connective tissue that attaches a muscle to a bone. An example is the calcaneal (Achilles) tendon of the gastrocnemius muscle (see Figure 8.25a).

Nerve and Blood Supply

Skeletal muscles are well supplied with nerves and blood vessels (Figure 8.1), both of which are directly related to contraction, the chief characteristic of muscle. For a skeletal muscle fiber to contract, it must first be stimulated by an electrical signal called a *muscle action potential.* Muscle contraction also requires a good deal of ATP and therefore large amounts of nutrients and oxygen for ATP synthesis. Moreover, the waste products of these ATP-producing reactions must be eliminated. Thus prolonged muscle action depends on a rich blood supply to deliver nutrients and oxygen and remove wastes.

Generally, an artery and one or two veins accompany each nerve that penetrates a skeletal muscle. Within the endomysium, microscopic blood vessels called capillaries are distributed so that each muscle fiber is in close contact with one or more capillaries. Each skeletal muscle fiber also makes contact with the terminal portion of a neuron.

Histology

Microscopic examination of a skeletal muscle reveals that it consists of thousands of elongated, cylindrical cells called *muscle fibers* arranged parallel to one another (Figure 8.2a). Each muscle fiber is covered by a plasma membrane called the *sarcolemma* (*sarco-* = flesh; *-lemma* = sheath). *Transverse tubules* (*T tubules*), tunnel-like extensions of the sarcolemma, pass through the muscle fiber from side to side (transversely). Multiple nuclei lie at the periphery of the fiber, next to the sarcolemma. The muscle fiber's cytoplasm, called *sarcoplasm*, contains many mitochondria that produce large amounts of ATP during muscle contraction. Extending throughout the sarcoplasm is *sarcoplasmic reticulum* (sar'-kō-PLAZ-mik re-TIK-yoo-lum), a network of membrane-enclosed tubules (similar to smooth endoplasmic reticulum) that stores calcium ions required for muscle contraction. Also in the sarcoplasm are numerous molecules of *myoglobin* (mī'-ō-GLŌ-bin), a reddish pigment similar to hemoglobin in blood. In addition to the characteristic color it lends to skeletal muscle, myoglobin stores oxygen until needed by mitochondria to generate ATP.

Extending along the entire length of the muscle fiber are cylindrical structures called *myofibrils*. Each myofibril, in turn, consists of two types of protein filaments called *thin filaments* and *thick filaments* (Figure 8.2b), which do not extend the entire length of a muscle fiber. Filaments overlap in specific patterns and form compartments called *sarcomeres* (*-meres* = parts), the basic functional units of striated muscle fibers (Fig-

ure 8.2b,c). Sarcomeres are separated from one another by zig-zagging zones of dense material called *Z discs*. Within each sarcomere a darker area, called the *A band*, extends the entire length of the thick filaments. At the center of each A band is a narrow *H zone*, which contains only the thick filaments. At both ends of the A band, thick and thin filaments overlap. A lighter-colored area to either side of the A band, called the *I band,* is composed of thin filaments. Each I band extends into two sarcomeres, divided in half by a Z disc (see Figure 8.4a). The alternating darker A bands and lighter I bands give the muscle fiber its striated appearance.

Thick filaments are composed of the protein *myosin,* which is shaped like two golf clubs twisted together (Figure 8.3a). The *myosin tails* (handles of the golf clubs) are arranged parallel to each other, forming the shaft of the thick filament. The heads of the golf clubs project outward from the surface of the shaft. These projecting heads are referred to as *myosin heads* or *crossbridges.*

Thin filaments are anchored to the Z discs. Their main component is the protein *actin.* Individual actin molecules join to form an actin filament that is twisted into a helix (Figure 8.3b). Each actin molecule contains a *myosin-binding site* for a myosin head. The thin filaments contain two other protein molecules, *tropomyosin* and *troponin,* that cover the myosin-binding sites on actin in relaxed muscle fibers.

Figure 8.3 ■ **Detailed structure of filaments.** (a) About 200 myosin molecules compose a thick filament. The myosin tails all point toward the center of the sarcomere. (b) Thin filaments contain actin, troponin, and tropomyosin.

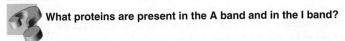
Myofibrils contain thick and thin filaments.

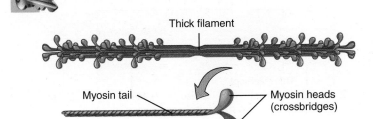

(a) One thick filament and a myosin molecule

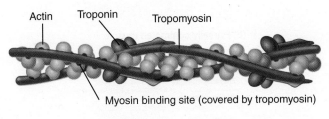

(b) Portion of a thin filament

What proteins are present in the A band and in the I band?

CONTRACTION AND RELAXATION OF SKELETAL MUSCLE

Objective: • **Explain how skeletal muscle fibers contract and relax.**

Sliding-Filament Mechanism

During muscle contraction, myosin heads of the thick filaments pull on the thin filaments, causing the thin filaments to slide toward the center of a sarcomere (Figure 8.4a, b). As the thin filaments slide, the I bands and H zones become narrower (Figure 8.4b) and eventually disappear altogether when the muscle is maximally contracted (Figure 8.4c).

The thin filaments slide past the thick filaments because the myosin heads move like the oars of a boat, pulling on the actin molecules of the thin filaments. Although the sarcomere shortens because of the increased overlap of thin and thick filaments, the lengths of the thin and thick filaments do not change. The sliding of filaments and shortening of sarcomeres in turn cause the shortening of the muscle fibers. This process, the *sliding-filament mechanism* of muscle contraction, occurs only when the level of calcium ions (Ca^{2+}) is high enough and ATP is available, for reasons you will see shortly.

Figure 8.4 ▦ **Sliding-filament mechanism of muscle contraction.**

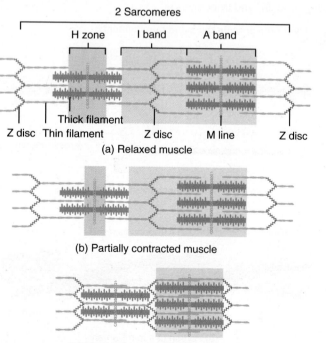

During muscle contraction, thin filaments move inward toward the H zone.

2 Sarcomeres

H zone   I band   A band

Thick filament

Z disc   Thin filament   Z disc   M line   Z disc
(a) Relaxed muscle

(b) Partially contracted muscle

(c) Maximally contracted muscle

What happens to the I bands as muscle contracts? Do the lengths of the thick and thin filaments change during contraction?

Neuromuscular Junction

Before a skeletal muscle fiber can contract, it must be stimulated by its neuron, called a ***motor neuron***. A single motor neuron along with all the muscle fibers it stimulates is called a ***motor unit***. Stimulation of one motor neuron causes all the muscle fibers in that motor unit to contract at the same time. Muscles that control small, precise movements, such as the muscles that move the eyes, have many motor units, each with fewer than 10 muscle fibers. Muscles of the body that are responsible for large, powerful movements, such as the biceps brachii in the arm and gastrocnemius in the leg, have fewer motor units, with as many as 2000 muscle fibers in each.

As the axon (long process) of a motor neuron enters a skeletal muscle, it divides into branches called ***axon terminals*** that approach—but do not touch—the sarcolemma of a muscle fiber (Figure 8.5a,b). The ends of the axon terminals enlarge into swellings known as ***synaptic end bulbs***, which contain ***synaptic vesicles*** filled with a chemical ***neurotransmitter***. The region of the sarcolemma near the axon terminal is called the ***motor end plate***. The space between the axon terminal and sarcolemma is the ***synaptic cleft***. The synapse formed between the axon terminals of a motor neuron and the motor end plate of a muscle fiber is known as the ***neuromuscular junction (NMJ)***. At the NMJ, a motor neuron excites a skeletal muscle fiber in the following way (Figure 8.5c):

① **Release of acetylcholine.** Arrival of the nerve impulse at the synaptic end bulbs triggers release of the neurotransmitter ***acetylcholine (ACh)*** (as′-e-til-KŌ-lēn). ACh then diffuses across the synaptic cleft between the motor neuron and the motor end plate.

② **Activation of ACh receptors.** Binding of ACh to its receptor in the motor end plate opens ion channels that allow small cations, especially sodium ions (Na^+), to flow across the membrane.

③ **Generation of muscle action potential.** The inflow of Na^+ (down its concentration gradient) generates a muscle action potential. The muscle action potential then travels along the sarcolemma and through the T tubules. Each nerve impulse normally elicits one muscle action potential. If another nerve impulse releases more acetylcholine, then steps **②** and **③** repeat. The details of nerve impulse generation are discussed in Chapter 9.

④ **Breakdown of ACh.** The effect of ACh lasts only briefly because the neurotransmitter is rapidly broken down in the synaptic cleft by an enzyme called ***acetylcholinesterase (AChE)*** (as′-e-til-kō′-lin-ES-ter-ās).

Several toxins and drugs can affect events at the NMJ. *Botulinum toxin*, produced by the bacterium *Clostridium botulinum*, blocks release of ACh. As a result, muscle contraction does not occur. The bacteria proliferate in improperly canned foods, and their toxin is one of the most lethal chemicals known. A tiny amount can cause death by paralyzing the diaphragm, the main muscle that powers breathing. The plant derivative *curare*, a poison used by South American Indians on arrows and blowgun darts, causes muscle paralysis by binding to and blocking

Figure 8.5 ■ Neuromuscular junction.

A neuromuscular junction includes the axon terminal of a motor neuron plus the motor end plate of a muscle fiber.

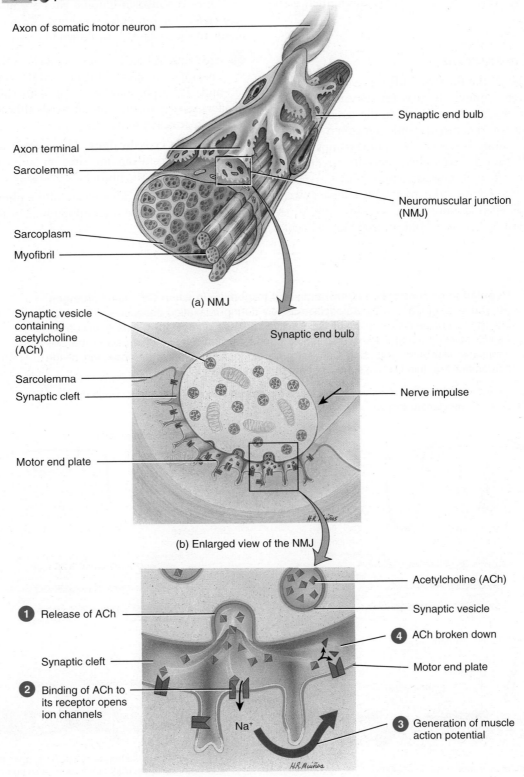

(a) NMJ

(b) Enlarged view of the NMJ

(c) Binding of acetylcholine to ACh receptors in the motor end plate

 What is the motor end plate?

ACh receptors. Curare-like drugs are often used during surgery to relax skeletal muscles. *Anticholinesterase agents* slow the enzymatic activity of AChE, thus slowing removal of ACh from the synaptic cleft. At low doses, these agents can strengthen weak muscle contractions.

Physiology of Contraction

Both Ca^{2+} and energy, in the form of ATP, are needed for muscle contraction. When a muscle fiber is relaxed (not contracting), there is a low concentration of Ca^{2+} in the sarcoplasm because the membrane of the sarcoplasmic reticulum contains Ca^{2+} active transport pumps (Figure 8.6a). These pumps continually transport Ca^{2+} from the sarcoplasm into the sarcoplasmic reticulum. However, when a muscle action potential travels along the sarcolemma and into the transverse tubule system, Ca^{2+} release channels open (Figure 8.6b), allowing Ca^{2+} to escape into the sarcoplasm. The Ca^{2+} binds to troponin molecules in the thin filaments, causing the troponin to change shape. This change in shape releases the troponin–tropomyosin complex from the myosin-binding sites on actin (Figure 8.6b).

Once the myosin-binding sites are uncovered, the ***contraction cycle***—the repeating sequence of events that causes the filaments to slide—begins, as shown in Figure 8.7:

1 **Splitting ATP.** The myosin heads include an ATPase, an enzyme that splits ATP into ADP (adenosine diphosphate) and Ⓟ (a phosphate group). This splitting reaction transfers energy to the myosin head, although ADP and Ⓟ remain attached to it.

2 **Forming crossbridges.** The energized myosin heads attach to the myosin-binding sites on actin, forming crossbridges and releasing the phosphate groups.

3 **Power stroke.** The release of the phosphate group triggers the ***power stroke*** of contraction. During the power stroke,

---

Figure 8.6 ▦ **Regulation of contraction by troponin and tropomyosin when Ca^{2+} level changes.** (a) The level of Ca^{2+} in the sarcoplasm is low during relaxation because it is pumped into the sarcoplasmic reticulum by Ca^{2+} active transport pumps. (b) A muscle action potential traveling along a transverse tubule opens calcium release channels in the sarcoplasmic reticulum, releasing Ca^{2+} into the sarcoplasm. Note the shortening of the sarcomere (the thin filaments are closer to the center than in part a).

An increase in the level of Ca^{2+} in the sarcoplasm starts the movement of thin filaments; when the level of Ca^{2+} declines, movement stops.

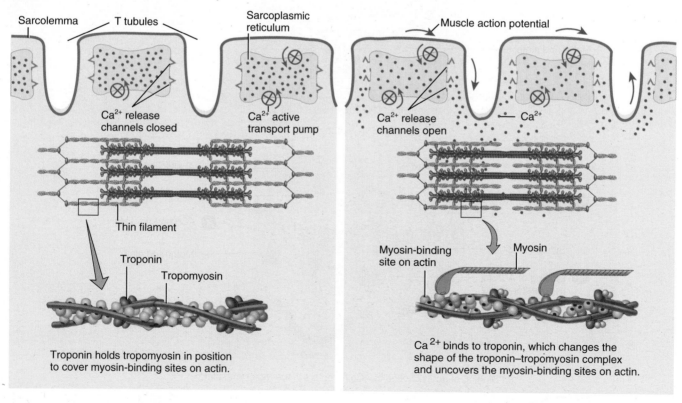

Sarcolemma   T tubules   Sarcoplasmic reticulum   Muscle action potential

Ca^{2+} release channels closed Ca^{2+} active transport pump Ca^{2+} release channels open Ca^{2+}

Thin filament

Troponin
Tropomyosin

Myosin-binding site on actin   Myosin

Troponin holds tropomyosin in position to cover myosin-binding sites on actin.

Ca^{2+} binds to troponin, which changes the shape of the troponin–tropomyosin complex and uncovers the myosin-binding sites on actin.

(a) Relaxation

(b) Contraction

 To which substance does Ca^{2+} bind when it is released from the sarcoplasmic reticulum?

the myosin head rotates or swivels and releases the ADP. The force produced as hundreds of myosin heads swivel slides the thin filament past the thick filament into the H zone.

④ **Binding ATP and detaching.** At the end of the power stroke, the myosin heads remain firmly attached to actin. When they bind another molecule of ATP, the myosin heads detach from actin.

As the myosin ATPase again splits ATP, the myosin head is reoriented and energized, ready to combine with another myosin-binding site farther along the thin filament. The contraction cycle repeats as long as ATP and Ca^{2+} are available in the sarcoplasm. At any one instant, some of the myosin heads are attached to actin and are producing force whereas others are detaching and preparing to bind again. During a maximal contraction, the sarcomere can shorten by as much as half its resting length.

Contraction is analogous to walking on a nonmotorized treadmill. One foot (myosin head) strikes the treadmill belt (thin filament) and pushes it backward (toward the H zone). Then the other foot comes down and imparts a second push. The belt soon moves smoothly while the walker (thick filament) remains stationary. Each myosin head progressively "walks" along a thin filament, coming closer to the Z disc with each "step," while the thin filament moves toward the H zone. And like the legs of the walker, the myosin heads need a constant supply of ATP to keep going!

After a person dies, Ca^{2+} begins to leak out of the sarcoplasmic reticulum and binds to troponin, causing the thin filaments to slide. ATP production has ceased, however, so the myosin heads cannot detach from actin. The resulting stiffness of the muscles is termed ***rigor mortis,*** rigidity of death. It lasts about 24 hours and then disappears as tissues disintegrate further.

Relaxation

Two changes permit a muscle fiber to relax after it has contracted. First, the neurotransmitter acetylcholine is rapidly broken down by the enzyme acetylcholinesterase (AChE). When nerve action potentials cease, release of ACh stops, and AChE rapidly breaks down the ACh already present in the synaptic cleft. This ends the generation of muscle action potentials, and the Ca^{2+} release channels in the sarcoplasmic reticulum membrane close.

Second, calcium ions are rapidly transported from the sarcoplasm into the sarcoplasmic reticulum. As the level of Ca^{2+} in the sarcoplasm falls, the tropomyosin–troponin complex slides back over the myosin-binding sites on actin. Once the myosin-binding sites are covered, the thin filaments slip back to their relaxed positions. Figure 8.8 summarizes the events of contraction and relaxation in a muscle fiber.

Figure 8.7 ■ **The contraction cycle.** Sarcomeres shorten through repeated cycles in which the myosin heads (crossbridges) attach to actin, rotate, and detach.

During the power stroke of contraction, myosin heads rotate and move the thin filaments past the thick filaments toward the center of the sarcomere.

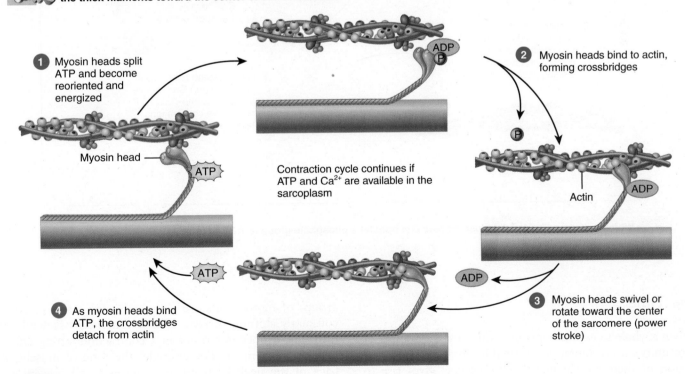

① Myosin heads split ATP and become reoriented and energized

Myosin head

ATP

Contraction cycle continues if ATP and Ca^{2+} are available in the sarcoplasm

② Myosin heads bind to actin, forming crossbridges

Actin

ADP

③ Myosin heads swivel or rotate toward the center of the sarcomere (power stroke)

④ As myosin heads bind ATP, the crossbridges detach from actin

 What causes myosin heads to detach from actin?

Figure 8.8 ■ **Summary of the events of contraction and relaxation in a skeletal muscle fiber.**

Acetylcholine released at the neuromuscular junction triggers a muscle action potential, which leads to muscle contraction.

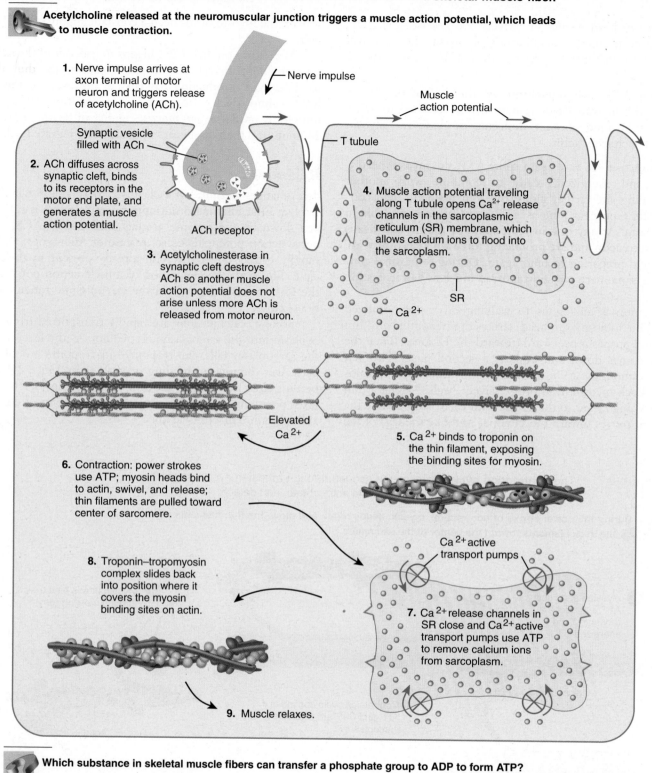

1. Nerve impulse arrives at axon terminal of motor neuron and triggers release of acetylcholine (ACh).

Nerve impulse

Muscle action potential

Synaptic vesicle filled with ACh

T tubule

2. ACh diffuses across synaptic cleft, binds to its receptors in the motor end plate, and generates a muscle action potential.

ACh receptor

4. Muscle action potential traveling along T tubule opens Ca^{2+} release channels in the sarcoplasmic reticulum (SR) membrane, which allows calcium ions to flood into the sarcoplasm.

3. Acetylcholinesterase in synaptic cleft destroys ACh so another muscle action potential does not arise unless more ACh is released from motor neuron.

SR

Ca^{2+}

Elevated Ca^{2+}

5. Ca^{2+} binds to troponin on the thin filament, exposing the binding sites for myosin.

6. Contraction: power strokes use ATP; myosin heads bind to actin, swivel, and release; thin filaments are pulled toward center of sarcomere.

Ca^{2+} active transport pumps

8. Troponin–tropomyosin complex slides back into position where it covers the myosin binding sites on actin.

7. Ca^{2+} release channels in SR close and Ca^{2+} active transport pumps use ATP to remove calcium ions from sarcoplasm.

9. Muscle relaxes.

Which substance in skeletal muscle fibers can transfer a phosphate group to ADP to form ATP?

Muscle Tone

Even when a whole muscle is not contracting, a small number of its motor units are involuntarily activated to produce a sustained contraction of their muscle fibers. This process gives rise to *muscle tone* (*tonos* = tension). To sustain muscle tone, small groups of motor units are alternately active and inactive in a constantly shifting pattern. Muscle tone keeps skeletal muscles firm, but it does not result in a contraction strong enough to produce movement. For example, the tone of muscles in the back of the neck keep the head upright and prevent it from slumping forward on the chest.

Effective Stretching Increases Muscle Flexibility

A certain degree of elasticity is an important attribute of skeletal muscles and their connective tissue attachments. Greater elasticity contributes to a greater degree of **flexibility,** increasing the range of motion of a joint. A joint's **range of motion (ROM)** is the maximum ability to move the bones about the joint through an arc of a circle. For example, a person may normally be able to extend the knee joint from 30° when it is maximally bent (flexed) to 170° when fully extended. The ROM or degree of flexibility is then $170° - 30° = 140°$. Physical therapists measure improvements in flexibility by increases in ROM.

Stretching It

When a relaxed muscle is physically stretched, its ability to lengthen is limited by connective tissue structures, such as fasciae. Regular stretching gradually lengthens these structures, but the process occurs very slowly. To see an im-

provement in flexibility, stretching exercises must be performed regularly—daily, if possible—for many weeks.

Tissues stretch best when slow, gentle force is applied at elevated tissue temperatures. An external source of heat, such as hot packs or ultrasound, can be used. But 10 or more minutes of muscular contraction is also a good way to raise muscle temperature. Exercise heats the muscle more deeply and thoroughly. That's where the term "warm-up" comes from. It's important to warm up *before* stretching, not vice versa. Stretching cold muscles does not increase flexibility and may even cause injury.

Just Relax . . .

The easiest and safest way to increase flexibility is with static stretching. A good static stretch is slow and gentle. After warming up, you get into a comfortable stretching position and relax. Continuing to relax and breathe deeply, you reach just a little farther, and a little farther, holding the stretch for at least 30 sec-

onds. If you have difficulty relaxing, you know you have stretched too far. Ease up until you feel a stretch but no strain.

When stretching, it is important to relax. Sounds simple, right? But if you ever visit an exercise class, you'll notice some people who are all tense, rigid, and hunched up, because the stretching positions are uncomfortable. As a result, their muscles tighten up in protest. These people figure they'd better push a little harder, and they tense up even more. They are unintentionally activating the motor neurons that initiate muscular contraction in the very muscles they are supposed to be relaxing, which of course interferes with the muscle's ability to elongate and stretch.

▶ *Think It Over*

▶ Using the information presented in this chapter, try to figure out which muscles are being stretched when you place one foot (keep that leg straight) up on a bar or chair.

METABOLISM OF SKELETAL MUSCLE TISSUE

Objectives: • **Describe the sources of ATP and oxygen for muscle contraction.**
• **Define muscle fatigue and list its possible causes.**
• **List the reasons that oxygen consumption is higher after exercise than at rest.**

Energy for Contraction

Unlike most cells of the body, skeletal muscle fibers often switch between virtual inactivity, when they are relaxed and using only

a modest amount of ATP, and great activity, when they are contracting and using ATP at a rapid pace. However, the ATP present inside muscle fibers is enough to power contraction for only a few seconds. If strenuous exercise is to continue, additional ATP must be synthesized. Muscle fibers have three sources for ATP production: (1) creatine phosphate, (2) glycolysis, and (3) aerobic cellular respiration.

While at rest, muscle fibers produce more ATP than they need. Some of the excess ATP is used to make **creatine phosphate,** an energy-rich molecule that is unique to muscle fibers. One of ATP's high-energy phosphate groups is transferred to creatine, forming creatine phosphate and ADP (adenosine diphosphate). While muscle is contracting, the high-energy phosphate group can be transferred from creatine phosphate back to ADP, quickly forming new ATP molecules. Together, creatine phosphate and ATP provide enough energy for muscles to contract maximally for about 15 seconds. This energy is sufficient for short bursts of intense activity, for example, running a 100-meter dash.

When muscle activity continues past the 15-second mark, the supply of creatine phosphate is depleted. The next source of ATP is *glycolysis,* a series of 10 reactions that produce 2 ATPs by breaking down a glucose molecule to pyruvic acid. Glucose passes easily from the blood into contracting muscle fibers and also is produced within muscle fibers by breakdown of glycogen. The reactions of glycolysis, which occur in the sarcoplasm, are *anaerobic* because they occur without using oxygen. Glycolysis can provide enough ATP for about 30 to 40 seconds of maximal muscle activity, for example, to run the last 300 meters of a 400-meter race.

Muscle activity that lasts longer than half a minute depends increasingly on *aerobic cellular respiration,* a series of ATP-producing reactions that occur in mitochondria and require oxygen. Muscle fibers have two sources of oxygen: (1) oxygen that diffuses into them from the blood and (2) oxygen released by myoglobin in the sarcoplasm. *Myoglobin* is an oxygen-binding protein found only in muscle fibers. It binds oxygen when oxygen is plentiful and releases oxygen when it is scarce. If enough oxygen is present, pyruvic acid enters the mitochondria, where it is completely oxidized in reactions that generate ATP, carbon dioxide, water, and heat. In comparison with glycolysis, aerobic cellular respiration yields much more ATP, about 36 molecules of ATP from each glucose molecule. In activities that last more than 10 minutes, aerobic cellular respiration provides most of the needed ATP.

During some activities, a lack of oxygen causes pyruvic acid to build up in the cytosol. Much of the pyruvic acid is then converted to lactic acid, which can diffuse out of muscle fibers into the blood. Liver cells pick up some of the lactic acid and convert it back to glucose. In this way, liver cells provide new glucose molecules and at the same time help reduce blood acidity.

Muscle Fatigue

The inability of a muscle to contract forcefully after prolonged activity is called *muscle fatigue.* One important factor in muscle fatigue is lowered release of calcium ions from the sarcoplasmic reticulum, resulting in a decline of Ca^{2+} level in the sarcoplasm. Other factors that contribute to muscle fatigue include depletion of creatine phosphate, insufficient oxygen, depletion of glycogen and other nutrients, buildup of lactic acid and ADP, and failure of nerve impulses in the motor neuron to release enough acetylcholine. Because increased lactic acid would cause a decrease in the pH of body fluids, muscle fatigue may be viewed as a homeostatic mechanism that prevents the pH from dropping below the normal acceptable range.

Oxygen Consumption After Exercise

During prolonged periods of muscle contraction, increases in breathing effort and blood flow enhance oxygen delivery to muscle tissue. After muscle contraction has stopped, heavy breathing continues for a period of time, and oxygen consumption remains above the resting level. In 1922, A. V. Hill coined the term *oxygen debt* for the added oxygen, over and above the oxygen consumed at rest, that is taken into the body after exercise. He proposed that this extra oxygen was used to "pay back" or restore metabolic conditions to the resting level in three ways: (1) to convert lactic acid back into glycogen stores in the liver, (2) to resynthesize creatine phosphate and ATP, and (3) to replace the oxygen removed from myoglobin.

The metabolic changes that occur *during exercise,* however, account for only some of the extra oxygen used *after exercise.* Only a small amount of resynthesis of glycogen occurs from lactic acid. Instead, glycogen stores are replenished much later from dietary carbohydrates. Much of the lactic acid that remains after exercise is converted back to pyruvic acid and used for ATP production via aerobic cellular respiration. Ongoing changes after exercise also boost oxygen use. First, the elevated body temperature after strenuous exercise increases the rate of chemical reactions throughout the body. Faster reactions use ATP more rapidly, and more oxygen is needed to produce ATP. Second, the heart and muscles used in breathing are still working harder than they were at rest, and thus they consume more ATP. Third, tissue repair processes are occurring at an increased pace. For these reasons, *recovery oxygen uptake* is a better term than oxygen debt for the elevated use of oxygen after exercise.

CONTROL OF MUSCLE TENSION

Objectives: • **Explain the three phases of a twitch contraction.**

• **Describe how frequency of stimulation and motor unit recruitment affect muscle tension.**

• **Compare the three types of skeletal muscle fibers.**

• **Distinguish between isotonic and isometric contractions.**

The contraction that results from a single muscle action potential, a muscle twitch, has significantly smaller force than the maximum force or tension the fiber is capable of producing. The total tension that a *single* muscle fiber can produce depends mainly on the rate at which nerve impulses arrive at its neuromuscular junction. The number of impulses per second is the *frequency of stimulation.* When considering the contraction of a *whole* muscle, the total tension it can produce depends on the number of muscle fibers that are contracting in unison.

Twitch Contraction

A *twitch contraction* is a brief contraction of all the muscle fibers in a motor unit in response to a single action potential in its motor neuron. A twitch also can be produced by direct electrical stimulation of a motor neuron or its muscle fibers. Figure 8.9 shows a recording of a muscle contraction, called a *myo-*

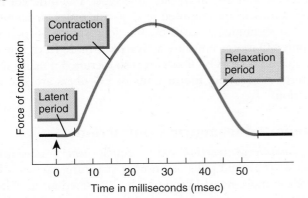

Figure 8.9 ■ **Myogram of a twitch contraction.** The arrow indicates the time at which the stimulus occurred.

A myogram is a record of a muscle contraction.

During which period do sarcomeres shorten?

binding to actin. During the second phase, the **contraction period** (upward tracing), repetitive powerstrokes are occurring, generating tension or force of contraction. In the third phase, the **relaxation period** (downward tracing), powerstrokes cease because the level of Ca^{2+} in the sarcoplasm is decreasing to the resting level. (Recall that calcium ions are actively transported back into the sarcoplasmic reticulum.)

Frequency of Stimulation

If a second stimulus occurs before a muscle fiber has completely relaxed, the second contraction will be stronger than the first (Figure 8.10a,b). This phenomenon, in which stimuli arriving one after the other cause larger contractions, is called **wave summation.** When a skeletal muscle is stimulated repeatedly, wave summation causes a sustained but wavering contraction called **unfused tetanus** (*tetan-* = rigid, tense; Figure 8.10c). Wave summation and tetanus result from the release of additional Ca^{2+} from the sarcoplasmic reticulum by the second, and subsequent, stimuli. The smooth, sustained voluntary muscle contractions that occur during most of your movements are achieved by out-of-synchrony unfused tetanus in different motor units. When the frequency of stimulation reaches about 90 stimuli per second, the result is **fused tetanus,** a sustained contraction that lacks even partial relaxation between stimuli (Figure 8.10d). For example, many motor units exhibit fused tetanus when you hold this book in an outstretched arm.

gram. Note that a brief delay, called the **latent period,** occurs between application of the stimulus (time zero on the graph) and the beginning of contraction. During the latent period, calcium ions are being released from the sarcoplasmic reticulum and binding to troponin, allowing the myosin heads to start

Figure 8.10 ■ **Myograms showing the effects of different frequencies of stimulation.** (a) Single twitch. (b) When a second stimulus occurs before the muscle has relaxed, wave summation occurs, and the second contraction is stronger than the first. (The dashed line indicates the force of contraction expected in a single twitch.) (c) In unfused tetanus, the curve looks jagged due to partial relaxation of the muscle between stimuli. (d) In fused tetanus, the contraction force is steady and sustained.

Due to wave summation, the tension produced during a sustained contraction is greater than during a single twitch.

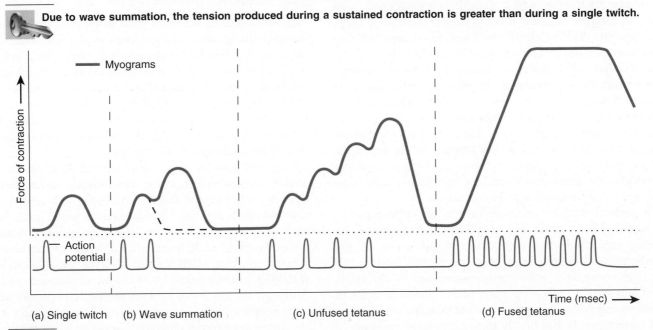

(a) Single twitch   (b) Wave summation   (c) Unfused tetanus   (d) Fused tetanus

What frequency of stimulation is needed to produce fused tetanus?

Motor Unit Recruitment

The process in which the number of contracting motor units is increased is called *motor unit recruitment*. Normally, the various motor neurons to a whole muscle fire *asynchronously:* While some motor units are contracting, others are relaxed. This pattern of motor unit activity delays muscle fatigue by allowing alternately contracting motor units to relieve one another, so that the contraction can be sustained for long periods.

Recruitment is one factor responsible for producing smooth movements rather than a series of jerky movements. Precise movements are brought about by small changes in muscle contraction. Typically, the muscles that produce precise movements are composed of small motor units. In this way, when a motor unit is recruited or turned off, only slight changes occur in muscle tension. On the other hand, large motor units are active where large tension is needed and precision is less important.

Types of Skeletal Muscle Fibers

Skeletal muscles contain three types of muscle fibers, which are present in varying proportions in different muscles of the body. The fiber types are (1) slow oxidative fibers, (2) fast oxidative–glycolytic fibers, and (3) fast glycolytic fibers.

Slow oxidative (SO) fibers are small in diameter and appear dark red because they contain a large amount of myoglobin. Because they have many large mitochondria, SO fibers generate ATP mainly by aerobic cellular respiration, which is why they are called oxidative fibers. These fibers are said to be "slow" because the contraction cycle proceeds at a slower pace than in "fast" fibers. SO fibers are very resistant to fatigue and are capable of prolonged, sustained contractions.

Fast oxidative–glycolytic (FOG) fibers are intermediate in diameter between the other two types. Like slow oxidative fibers, they contain a large amount of myoglobin, and thus appear dark red. FOG fibers can generate considerable ATP by aerobic cellular respiration, which gives them a moderately high resistance to fatigue. Because their glycogen content is high, they also generate ATP by glycolysis. These fibers are "fast" because they contract and relax more quickly than SO fibers.

Fast glycolytic (FG) fibers are largest in diameter, contain the most myofibrils, and generate the most powerful and most rapid contractions. They have a low myoglobin content and few mitochondria. FG fibers contain large amounts of glycogen and generate ATP mainly by glycolysis. They are used for intense movements of short duration, but they fatigue quickly. Strength-training programs that engage a person in activities requiring great strength for short times produce increases in the size, strength, and glycogen content of FG fibers.

Most skeletal muscles are a mixture of all three types of skeletal muscle fibers, about half of which are SO fibers. The proportions vary somewhat, depending on the action of the muscle, the person's training program, and genetic factors. For example, the continually active postural muscles of the neck, back, and legs have a high proportion of SO fibers. Muscles of the shoulders and arms, in contrast, are not constantly active but are used intermittently and briefly to produce large amounts of tension, such as in lifting and throwing. These muscles have a high proportion of FG fibers. Leg muscles, which not only support the body but are also used for walking and running, have large numbers of both SO and FOG fibers.

Even though most skeletal muscles are a mixture of all three types of skeletal muscle fibers, the skeletal muscle fibers of any given motor unit are all of the same type. The different motor units in a muscle are recruited in a specific order, depending on need. For example, if weak contractions suffice to perform a task, only SO motor units are activated. If more force is needed, the motor units of FOG fibers are also recruited. Finally, if maximal force is required, motor units of FG fibers are also called into action.

Isotonic and Isometric Contractions

In an *isotonic contraction* (*iso-* = equal; *-tonic* = tension), the muscle shortens and pulls on another structure, such as a bone, to produce movement. During such a contraction, muscle tension remains constant and energy is expended. Picking up a book from a desk is an example of an isotonic contraction.

In *isometric contractions,* considerable tension is generated without shortening of the muscle. An example would be holding a book steady in an outstretched arm. Isometric contractions are important because they stabilize some joints as others are moved. These contractions are important for maintaining posture and for supporting objects in a fixed position. Although isometric contractions do not result in body movement, energy is still expended. Most activities include both isotonic and isometric contractions.

EXERCISE AND SKELETAL MUSCLE TISSUE

Objective: • **Describe the effects of exercise on skeletal muscle tissue.**

Regular, repeated activities such as jogging or aerobic dancing increase the number of capillaries in skeletal muscles and thus improve the delivery of oxygen-rich blood for aerobic cellular respiration. By contrast, activities such as weightlifting stimulate synthesis of muscle proteins and result, over a period of time, in *muscle hypertrophy* (hī-PER-trō-fē), an increase in the diameters of muscle fibers. As a result, aerobic training builds endurance for prolonged activities, whereas weight training builds muscle strength for short-term feats.

Intense exercise also causes significant muscle damage, including torn sarcolemmas in some muscle fibers, damaged myofibrils, and disrupted Z discs. Microscopic muscle damage after exercise also is indicated by increases in blood levels of proteins, such as myoglobin, that are normally confined within muscle fibers. From 12 to 48 hours after a period of strenuous exercise, skeletal muscles often exhibit *delayed onset muscle soreness (DOMS),* accompanied by stiffness, tenderness, and swelling. Although the causes of DOMS are not completely understood, microscopic muscle damage seems to be a major factor.

In contrast to the changes that occur when muscles are contracting on a regular basis, muscles that contract rarely or not at all undergo *muscular atrophy* (A-trō-fē), a wasting away of muscles. Individual muscle fibers decrease in size due to a progressive loss of myofibrils. In *disuse atrophy,* muscles atrophy because they are not used. Bedridden individuals and people with casts that immobilize large muscle groups may experience disuse atrophy because the flow of nerve impulses to the inactive muscle is greatly reduced. If nerve impulses cease in its motor neurons, a muscle undergoes *denervation atrophy.* In six months to two years, the denervated muscle will be one-quarter its original size and the muscle fibers will be replaced by fibrous connective tissue. The transition to fibrous connective tissue, when complete, cannot be reversed. In people with spinal cord injuries, denervation atrophy occurs in skeletal muscles served by motor neurons below the level of injury.

AGING AND MUSCLE TISSUE

Objective: • **Explain how aging affects skeletal muscle.**

After birth, the *number* of skeletal muscle fibers does not increase significantly. During childhood, the increase in the *size* of muscle fibers is partially under the control of human growth hormone (hGH), produced by the anterior lobe of the pituitary gland. The hormone testosterone, from the testes in males and the adrenal gland in both sexes, causes a further increase in the size of skeletal muscle fibers. Because they have more testosterone, males generally have larger muscles than females.

Beginning at about 30 years of age, humans undergo a progressive loss of skeletal muscle mass that is replaced largely by fibrous connective tissue and adipose tissue. In part, this decline is due to decreasing activity. Accompanying the loss of muscle mass is a decrease in maximal strength and a slowing of muscle reflexes. Nevertheless, endurance and strength training programs can slow or even reverse the age-associated decline in muscular performance.

CARDIAC MUSCLE TISSUE

Objective: • **Describe the structure and function of cardiac muscle tissue.**

Most of the heart consists of *cardiac muscle tissue.* Like skeletal muscle, cardiac muscle is also *striated,* but its action is *involuntary:* its alternating contraction and relaxation are not consciously controlled. Cardiac muscle fibers often are branched, having a Y-shape; are shorter in length and larger in diameter than skeletal muscle fibers; and have a single, centrally located nucleus (see Figure 15.2b on page 353). Cardiac muscle fibers interconnect with one another by irregular transverse thickenings of the sarcolemma called *intercalated discs* (in-TER-ka-lāt-ed = to insert between). The intercalated discs hold the fibers together and contain *gap junctions,* which allow muscle action potentials to quickly spread from one cardiac muscle fiber to another.

A major difference between skeletal muscle and cardiac muscle is the source of stimulation. We have seen that skeletal muscle tissue contracts only when stimulated by acetylcholine

released by a nerve impulse in a motor neuron. In contrast, the heart beats because some of the cardiac muscle fibers act as a pacemaker to initiate each cardiac contraction. The built-in or intrinsic rhythm of heart contractions is called *autorhythmicity* (aw'-tō-rith-MIS-i-tē). Several hormones and neurotransmitters can increase or decrease adjust heart rate by speeding or slowing the heart's pacemaker.

Under normal resting conditions, cardiac muscle tissue contracts and relaxes an average of about 75 times a minute. Thus, cardiac muscle tissue requires a constant supply of oxygen and nutrients. The mitochondria in cardiac muscle fibers are larger and more numerous than in skeletal muscle fibers and produce most of the needed ATP via aerobic cellular respiration. Moreover, cardiac muscle fibers can use lactic acid, released by skeletal muscle fibers during exercise, to make ATP.

SMOOTH MUSCLE TISSUE

Objective: • **Describe the structure and function of smooth muscle tissue.**

Like cardiac muscle tissue, *smooth muscle tissue,* found in many internal organs and blood vessels, is *involuntary.* Smooth muscle fibers are considerably smaller in length and diameter than skeletal muscle fibers and are tapered at both ends. Within each fiber is a single, oval, centrally located nucleus (Figure 8.11). In addition to thick and thin filaments, smooth muscle fibers also

Figure 8.11 ■ **Histology of smooth muscle tissue.** A smooth muscle fiber is shown in the relaxed state (left) and the contracted state (right). A photomicrograph of smooth muscle tissue is shown in Table 4.4 on page 91.

Smooth muscle lacks striations—it looks "smooth"—because the thick and thin filaments and intermediate filaments are irregularly arranged.

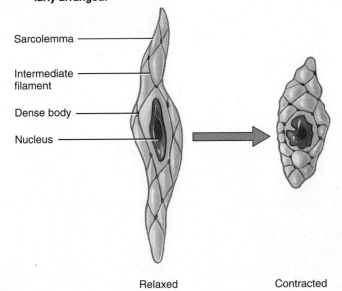

Sarcolemma

Intermediate filament

Dense body

Nucleus

Relaxed   Contracted

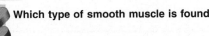
Which type of smooth muscle is found in the walls of hollow organs?

contain *intermediate filaments.* Because the various filaments have no regular pattern of overlap, smooth muscle lacks alternating dark and light bands and thus appears *nonstriated,* or smooth.

The intermediate filaments stretch between structures called *dense bodies,* which are similar to the Z discs found in striated muscle. Some dense bodies are attached to the sarcolemma; others are dispersed in the sarcoplasm. In contraction, those attached to the sarcolemma cause a lengthwise shortening of the fiber.

There are two kinds of smooth muscle tissue, visceral and multiunit. The more common type is *visceral (single-unit) muscle tissue.* It is found in sheets that wrap around to form part of the walls of small arteries and veins and hollow viscera such as the stomach, intestines, uterus, and urinary bladder. The fibers in visceral muscle tissue are tightly bound together in a continuous network. Like cardiac muscle, visceral smooth muscle is autorhythmic. Because the fibers connect to one another by gap junctions, muscle action potentials spread throughout the network. When a neurotransmitter, hormone, or autorhythmic signal stimulates one fiber, the muscle action potential spreads to neighboring fibers, which then contract in unison, as a single unit.

The second kind of smooth muscle tissue, *multiunit smooth muscle tissue,* consists of individual fibers, each with its own motor nerve endings. Whereas stimulation of a single visceral muscle fiber causes contraction of many adjacent fibers, stimulation of a single multiunit smooth muscle fiber causes contraction of that fiber only. Multiunit smooth muscle tissue is found in the walls of large arteries, in large airways to the lungs, in the arrector pili muscles attached to hair follicles, and in the internal eye muscles.

Smooth muscle tissue exhibits several important physiological differences from striated muscle tissue. Compared with contraction in a skeletal muscle fiber, contraction in a smooth muscle fiber starts more slowly and lasts much longer. Calcium ions enter smooth muscle fibers slowly and also move slowly out of the muscle fiber when excitation declines, which delays relaxation. The prolonged presence of Ca^{2+} in the cytosol provides for *smooth muscle tone,* a state of continued partial contraction. Smooth muscle tissue can thus sustain long-term tone, which is important in walls of blood vessels and in the walls of organs that maintain pressure on their contents. Finally, smooth muscle can both shorten and stretch to a greater extent than other muscle types. Stretchiness permits smooth muscle in the wall of hollow organs such as the uterus, stomach, intestines, and urinary bladder to expand as their contents enlarge, while still retaining the ability to contract.

Most smooth muscle fibers contract or relax in response to nerve impulses from the autonomic (involuntary) nervous system. In addition, many smooth muscle fibers contract or relax in response to stretching; hormones; or local factors such as changes in pH, oxygen and carbon dioxide levels, temperature, and ion concentrations. For example, the hormone epinephrine, released by the adrenal medulla, causes relaxation of smooth muscle in the airways and in some blood vessel walls.

Table 8.1 presents a summary of the major characteristics of the three types of muscle tissue. Now that you have a basic un-

Table 8.1 / Summary of the Principal Features of Muscle Tissue

Characteristics	Skeletal Muscle	Cardiac Muscle	Smooth Muscle
Cell appearance and features	Long cylindrical fiber with many peripherally located nuclei; striated; unbranched.	Branched cylinder usually with one centrally located nucleus; intercalated discs join neighboring fibers; striated.	Spindle-shaped fiber with one, centrally positioned nucleus; not striated.
Location	Primarily attached to bones by tendons.	Heart.	Walls of hollow viscera, airways, blood vessels, iris and ciliary body of eye, arrector pili of hair follicles.
Fiber diameter	Very large.	Large.	Small.
Fiber length	Very large.	Small.	Intermediate.
Sarcomeres	Yes.	Yes.	No.
Transverse tubules	Yes, aligned with each A–I band junction.	Yes, aligned with each Z disc.	No.
Speed of contraction	Fast.	Moderate.	Slow.
Nervous control	Voluntary.	Involuntary.	Involuntary.
Capacity for regeneration	Limited.	None.	Considerable compared with other muscle tissues, but limited compared with tissues such as epithelium.

derstanding of the structure and functions of muscle tissue, we will examine how skeletal muscles cooperate to produce various body movements.

HOW SKELETAL MUSCLES PRODUCE MOVEMENT

Objective: • **Describe how skeletal muscles cooperate to produce movement.**

Origin and Insertion

Based on the description of muscle tissue, we can define a ***skeletal muscle*** as an organ composed of several different types of tissues. These include skeletal muscle tissue, vascular tissue (blood vessels and blood), nervous tissue (motor neurons), and several types of connective tissues.

Skeletal muscles are not attached directly to bones; they produce movements by pulling on tendons, which, in turn, pull on bones. Most skeletal muscles cross at least one joint and are attached to the articulating bones that form the joint (Figure 8.12). When the muscle contracts, it draws one bone toward the other. The two bones do not move equally. One is held nearly in its original position; the attachment of a muscle (by means of a tendon) to the stationary bone is called the ***origin***. The other end of the muscle is attached by means of a tendon to the movable bone at a point called the ***insertion***. The fleshy portion of the muscle between the tendons of the origin and insertion is called the ***belly***. A good analogy is a spring on a door. The part of the spring attached to the door represents the insertion, the part attached to the frame is the origin, and the coils of the spring are the belly.

Group Actions

Most movements occur because several skeletal muscles are acting in groups rather than individually. Also, most skeletal muscles are arranged in opposing pairs at joints, that is, flexors–extensors, abductors–adductors, and so on. A muscle that causes a desired action is referred to as the ***prime mover*** or ***agonist*** (= leader). Often, another muscle, called the ***antagonist*** (*ant-* = against), relaxes while the prime mover contracts. The antagonist has an effect opposite to that of the prime mover; that is, the antagonist stretches and yields to the movement of the prime mover. When you bend (flex) your elbow, the biceps brachii is the prime mover. While the biceps brachii is contracting, the triceps brachii, the antagonist, is relaxing (see Figure 8.21). Do not assume, however, that the biceps brachii is always the prime mover and the triceps brachii is always the antagonist. For example, when straightening (extending) the elbow, the triceps brachii serves as the prime mover and the biceps brachii functions as the antagonist. If the prime mover and antagonist contracted together with equal force, there would be no movement, as in an isometric contraction.

Most movements also involve muscles called ***synergists*** (SIN-er-gists; *syn-* = together; *erg-* = work), which help the prime mover function more efficiently by reducing unnecessary

Figure 8.12 ■ **Relationship of skeletal muscles to bones.** Skeletal muscles produce movements by pulling on tendons attached to bones.

🔑 **In the limbs, the origin of a muscle is proximal and the insertion is distal.**

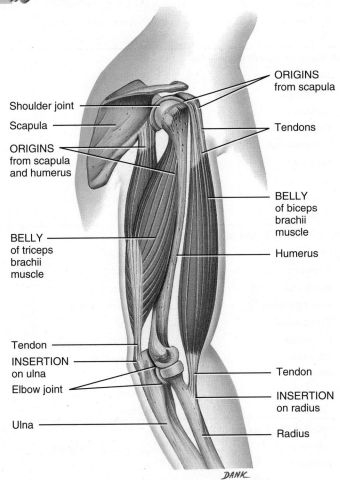

Origin and insertion of a skeletal muscle

Which muscle produces the desired action?

movement. Some muscles in a group also act as ***fixators***, stabilizing the origin of the prime mover so that the prime mover can act more efficiently. Under different conditions and depending on the movement, many muscles act at various times as prime movers, antagonists, synergists, or fixators.

NAMING SKELETAL MUSCLES

Objective: • **List and describe the ways that skeletal muscles are named.**

The names of most of the nearly 700 skeletal muscles are based on specific characteristics. Learning the terms used to indicate specific characteristics will help you remember the names of the muscles (Table 8.2).

Table 8.2 / Characteristics Used to Name Muscles

Name	Meaning	Example	Figure
DIRECTION: Orientation of muscle fibers relative to the body's midline			
Rectus	Parallel to midline	Rectus abdominis	8.17b
Transverse	Perpendicular to midline	Transversus abdominis	8.17b
Oblique	Diagonal to midline	External oblique	8.17a
SIZE: Relative size of the muscle			
Maximus	Largest	Gluteus maximus	8.24b
Minimus	Smallest	Gluteus minimus	8.24b
Longus	Longest	Adductor longus	8.24a
Latissimus	Widest	Latissimus dorsi	8.13b
Longissimus	Longest	Longissimus capitis	8.23
Magnus	Large	Adductor magnus	8.24b
Major	Larger	Pectoralis major	8.13a
Minor	Smaller	Pectoralis minor	8.19a
Vastus	Great	Vastus lateralis	8.24a
SHAPE: Relative shape of the muscle			
Deltoid	Triangular	Deltoid	8.17a
Trapezius	Trapezoid	Trapezius	8.13b
Serratus	Saw-toothed	Serratus anterior	8.17
Rhomboideus	Diamond-shaped	Rhomboideus major	8.19d
Orbicularis	Circular	Orbicularis oculi	8.14a
Pectinate	Comblike	Pectineus	8.24a
Piriformis	Pear-shaped	Piriformis	8.24b
Platys	Flat	Platysma	8.14a
Quadratus	Square	Quadratus lumborum	8.24a
Gracilis	Slender	Gracilis	8.24a
ACTION: Principal action of the muscle			
Flexor	Decreases joint angle	Flexor carpi radialis	8.22a
Extensor	Increases joint angle	Extensor carpi ulnaris	8.22b
Abductor	Moves bone away from midline	Abductor magnus	8.24b
Adductor	Moves bone closer to midline	Adductor longus	8.24a
Levator	Produces superior movement	Levator scapulae	8.19
Depressor	Produces inferior movement	Depressor labii inferioris	8.14b
Supinator	Turns palm superiorly or anteriorly	Supinator	
Pronator	Turns palm inferiorly or posteriorly	Pronator teres	8.22a
Sphincter	Decreases size of opening	External anal sphincter	
Tensor	Makes a body part rigid	Tensor fasciae latae	8.24a
NUMBER OF ORIGINS: Number of tendons of origin			
Biceps	Two origins	Biceps brachii	8.21a
Triceps	Three origins	Triceps brachii	8.21b
Quadriceps	Four origins	Quadriceps femoris	8.24a

LOCATION: Structure near which a muscle is found

Example: Frontalis, a muscle near the frontal bone (Figure 8.14a).

ORIGIN AND INSERTION: Sites where muscle originates and inserts

Example: Brachioradialis, orginating on the humerus and inserting on the radius (Figure 8.22a).

Figure 8.13 ■ **Principal superficial skeletal muscles.**

Most movements require contraction of several skeletal muscles acting in groups rather than individually.

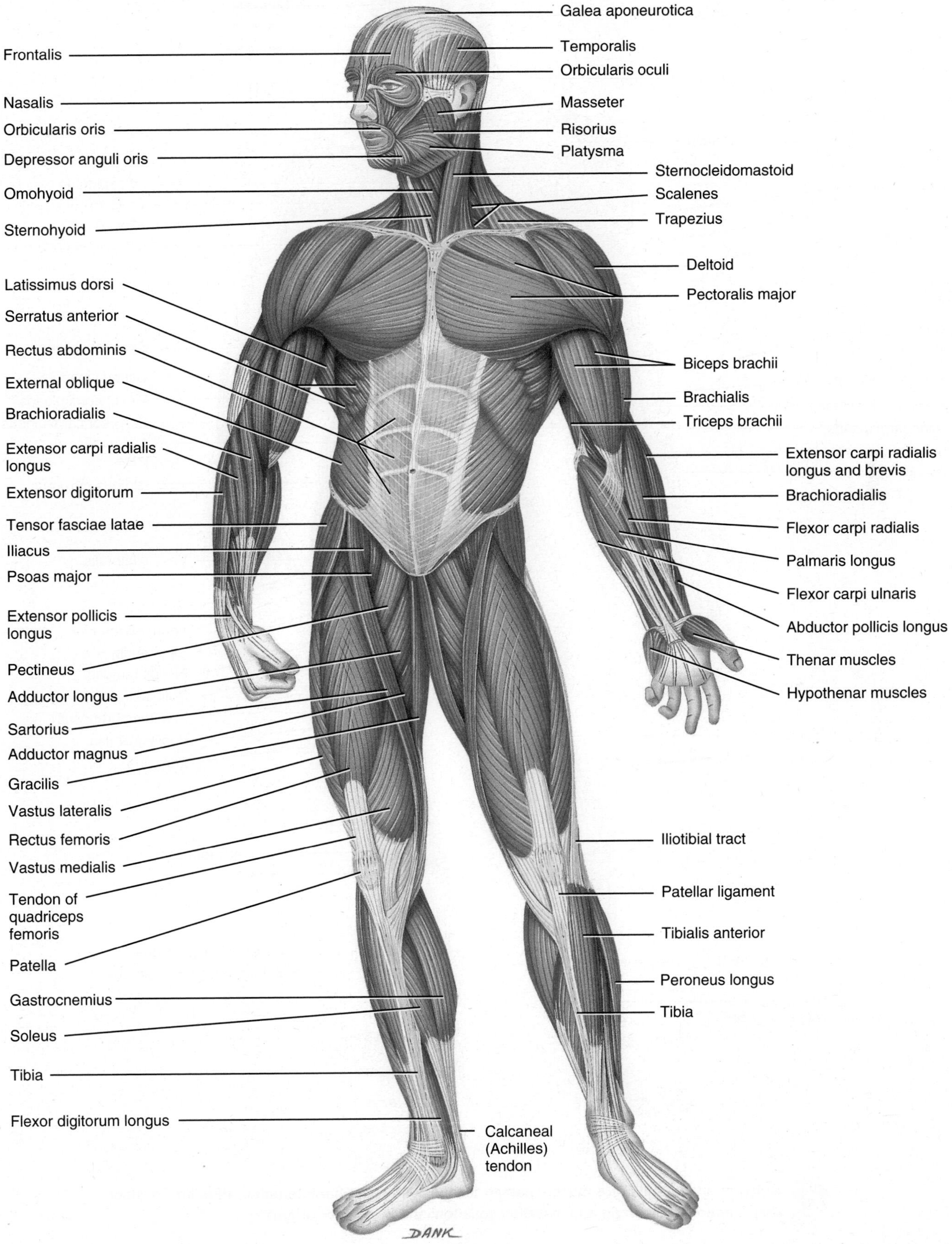

Galea aponeurotica

Frontalis

Temporalis

Orbicularis oculi

Nasalis

Masseter

Orbicularis oris

Risorius

Depressor anguli oris

Platysma

Omohyoid

Sternocleidomastoid

Sternohyoid

Scalenes

Trapezius

Latissimus dorsi

Deltoid

Serratus anterior

Pectoralis major

Rectus abdominis

Biceps brachii

External oblique

Brachialis

Brachioradialis

Triceps brachii

Extensor carpi radialis longus

Extensor carpi radialis longus and brevis

Extensor digitorum

Brachioradialis

Tensor fasciae latae

Flexor carpi radialis

Iliacus

Palmaris longus

Psoas major

Flexor carpi ulnaris

Extensor pollicis longus

Abductor pollicis longus

Pectineus

Thenar muscles

Adductor longus

Hypothenar muscles

Sartorius

Adductor magnus

Gracilis

Vastus lateralis

Rectus femoris

Iliotibial tract

Vastus medialis

Patellar ligament

Tendon of quadriceps femoris

Tibialis anterior

Patella

Peroneus longus

Gastrocnemius

Tibia

Soleus

Tibia

Flexor digitorum longus

Calcaneal (Achilles) tendon

DANK

(a) Anterior view

(Figure 8.13 continues)

Figure 8.13 (continued)

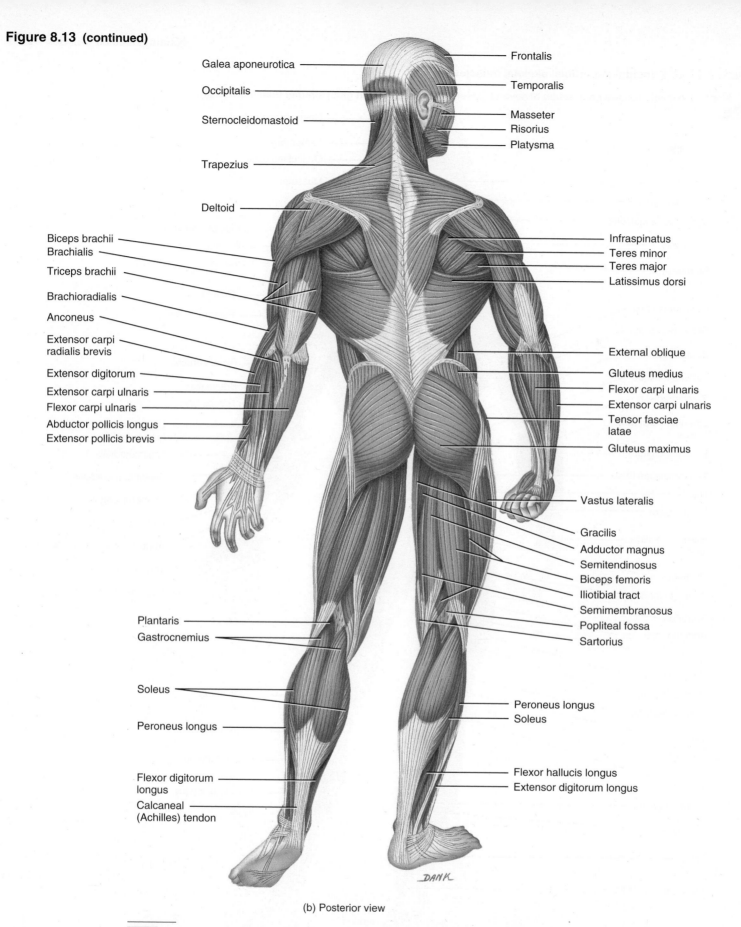

Galea aponeurotica

Occipitalis

Sternocleidomastoid

Trapezius

Deltoid

Biceps brachii
Brachialis

Triceps brachii

Brachioradialis

Anconeus

Extensor carpi
radialis brevis

Extensor digitorum

Extensor carpi ulnaris

Flexor carpi ulnaris

Abductor pollicis longus

Extensor pollicis brevis

Plantaris

Gastrocnemius

Soleus

Peroneus longus

Flexor digitorum
longus

Calcaneal
(Achilles) tendon

Frontalis

Temporalis

Masseter
Risorius
Platysma

Infraspinatus
Teres minor
Teres major
Latissimus dorsi

External oblique

Gluteus medius

Flexor carpi ulnaris

Extensor carpi ulnaris

Tensor fasciae
latae

Gluteus maximus

Vastus lateralis

Gracilis

Adductor magnus

Semitendinosus

Biceps femoris

Iliotibial tract

Semimembranosus

Popliteal fossa

Sartorius

Peroneus longus
Soleus

Flexor hallucis longus

Extensor digitorum longus

(b) Posterior view

 Which is an example of a muscle named for the following characteristics: direction of fibers, shape, action, size, origin and insertion, location, and number of origins?

PRINCIPAL SKELETAL MUSCLES

Objective: • **Describe the location of skeletal muscles in various regions of the body and identify their functions.**

Exhibits 8.1 through 8.13 list the principal muscles of the body with their origins, insertions, and actions. (By no means have all the muscles of the body been included.) For each exhibit, an overview section provides a general orientation to the muscles and their functions or unique characteristics. To make it easier for you to learn to say the names of skeletal muscles and understand how they are named, we have provided phonetic pronunciations and word roots that indicate how the muscles are named (refer also to Table 8.2). Once you have mastered the naming of the muscles, their actions will have more meaning and be easier to remember.

The muscles are divided into groups according to the part of the body on which they act. Figure 8.13 on pages 185 and 186 shows general anterior and posterior views of the muscular system. As you study groups of muscles in the following exhibits, refer to Figure 8.13 to see how each group is related to all others.

• • •

To appreciate the many ways that the muscular system contributes to homeostasis of other body systems, examine Focus on Homeostasis: The Muscular System on page 210. Next, in Chapter 9, we will see how the nervous system is organized, how neurons generate nerve impulses that activate muscle tissues as well as other neurons, and how synapses function.

Exhibit 8.1 / Muscles of Facial Expression (*Figure 8.14*)

Objective: Describe the origin, insertion, and action of the muscles of facial expression.

• **Overview:** The muscles of facial expression provide humans with the ability to express a wide variety of emotions, including displeasure, surprise, fear, and happiness. The muscles themselves lie within the layers of superficial fascia (connective tissue beneath the skin). As a rule, their origins are in the fascia or in the bones of the skull, with insertions into the skin. The "movable bone" in this case is the skin rather than a joint.

Bell's palsy, also known as *facial paralysis,* is a one-sided paralysis of the muscles of facial expression as a result of damage or disease of the facial cranial nerve (cranial nerve VII). Although the cause is unknown, a relationship between the herpes simplex virus and inflammation of the facial nerve has been suggested. In severe cases, the paralysis causes the entire side of the face to droop, and the person cannot wrinkle the forehead, close the eye, or pucker the lips on the affected side. Drooling and difficulty in swallowing also occur. Eighty percent of patients recover completely within a few weeks to a few months. For others, paralysis is permanent.

• **Relating muscles to movements:** Arrange the muscles in this exhibit into two groups: (1) those that act on the mouth and (2) those that act on the eyes.

Muscle	Origin	Insertion	Action
Frontalis (fron-TA-lis; *front-* = forehead)	Galea aponeurotica (flat tendon that attaches to the frontalis and occipitalis muscles).	Skin superior to orbit.	Draws scalp forward, raises eyebrows, and wrinkles skin of forehead horizontally.
Occipitalis (ok-si'-pi-TA-lis; *occipit-* = base of skull) (see Figure 8.13b)	Occipital and temporal bones.	Galea aponeurotica.	Draws scalp backward.
Orbicularis oris (or-bi'-kyoo-LAR-is OR-is; *orb* = circular; *or* = mouth)	Muscle fibers surrounding opening of mouth.	Skin at corner of mouth.	Closes and protrudes lips, compresses lips against teeth, and shapes lips during speech.
Zygomaticus major (zī-gō-MA-ti-kus; *zygomatic* = cheek bone; *major* = greater)	Zygomatic bone.	Skin at angle of mouth and orbicularis oris.	Draws angle of mouth upward and outward, as in smiling or laughing.
Buccinator (BUK-si-nā'-tor; *bucca* = cheek)	Maxilla and mandible.	Orbicularis oris.	Presses cheeks against teeth and lips, as in whistling, blowing, and sucking; draws corner of mouth laterally; assists in mastication (chewing) by keeping food between the teeth (and not between teeth and cheeks).
Platysma (pla-TIZ-ma; *platysma* = a flat plate)	Fascia over deltoid and pectoralis major muscles.	Mandible, muscles around angle of mouth, and skin of lower face.	Draws outer part of lower lip downward and backward as in pouting; depresses mandible.
Orbicularis oculi (OK-yoo-lī; *ocul-* = eye)	Medial wall of orbit.	Circular path around orbit.	Closes eye; wrinkles forehead vertically.
Levator palpebrae superioris (le-VĀ-tor PAL-pe-brē soo-per'-ē-OR-is; *palpebrae* = eyelids) (see also Figure 8.16)	Roof of orbit.	Skin of upper eyelid.	Elevates upper eyelid (opens eye).

(continues)

Exhibit 8.1 / continued *(Figure 8.14)*

Figure 8.14 ■ **Muscles of facial expression.** In this and subsequent figures in the chapter, the muscles indicated in all uppercase letters are the ones specifically referred to in the corresponding exhibit.

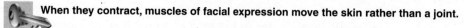 **When they contract, muscles of facial expression move the skin rather than a joint.**

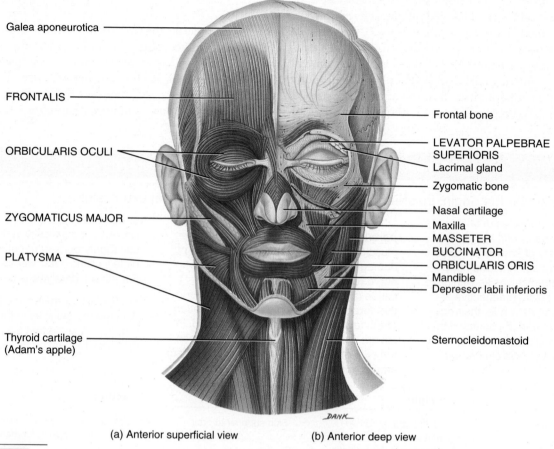

Galea aponeurotica

FRONTALIS

ORBICULARIS OCULI

ZYGOMATICUS MAJOR

PLATYSMA

Thyroid cartilage
(Adam's apple)

Frontal bone

LEVATOR PALPEBRAE
SUPERIORIS
Lacrimal gland

Zygomatic bone

Nasal cartilage
Maxilla
MASSETER
BUCCINATOR
ORBICULARIS ORIS
Mandible
Depressor labii inferioris

Sternocleidomastoid

DANK

(a) Anterior superficial view   (b) Anterior deep view

Which muscles of facial expression cause frowning, smiling, pouting, and squinting?

Exhibit 8.2 / Muscles that Move the Mandible (Lower Jaw) *(Figure 8.15)*

Objective: Describe the origin, insertion, and action of the muscles that move the mandible.

- **Overview:** Muscles that move the mandible (lower jaw) are also known as muscles of ***mastication*** (mas'-ti-KĀ-shun = to chew) because they are involved in biting and chewing. These muscles also assist in speech.

- **Relating muscles to movements:** Arrange the muscles in this exhibit according to their actions on the mandible: (1) elevation, (2) depression, (3) retraction, (4) protraction, and (5) side-to-side movement. The same muscle may be mentioned more than once.

Muscle	Origin	Insertion	Action
Masseter (MA-se-ter; *maseter* = chewer) See Figures 8.13 and 8.14.	Maxilla and zygomatic arch.	Mandible.	Elevates mandible as in closing mouth and retracts (draws back) mandible.
Temporalis (tem'-por-A-lis; *tempora* = temples)	Temporal bone.	Mandible.	Elevates and retracts mandible.
Medial pterygoid (TER-i-goid; *medial* = closer to midline; *pterygoid* = like a wing)	Sphenoid bone and maxilla.	Mandible.	Elevates and protracts (protrudes) mandible and moves mandible from side to side.
Lateral pterygoid (*lateral* = farther from midline)	Sphenoid bone.	Mandible and temporomandibular joint (TMJ).	Protracts mandible, depresses mandible as in opening mouth, and moves mandible from side to side.

Figure 8.15 ■ **Muscles that move the mandible (lower jaw).**

 The muscles that move the mandible are involved in mastication and speech.

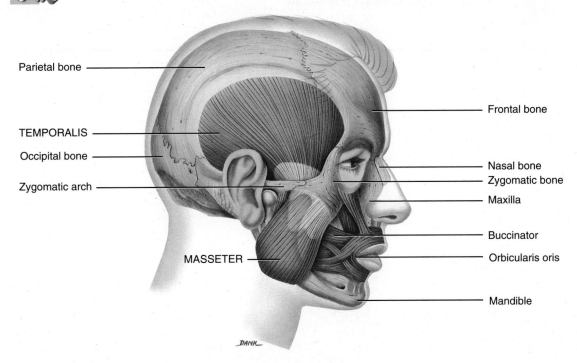

(a) Right lateral superficial view

Which is the strongest muscle of mastication?

Exhibit 8.2 / continued *(Figure 8.15)*

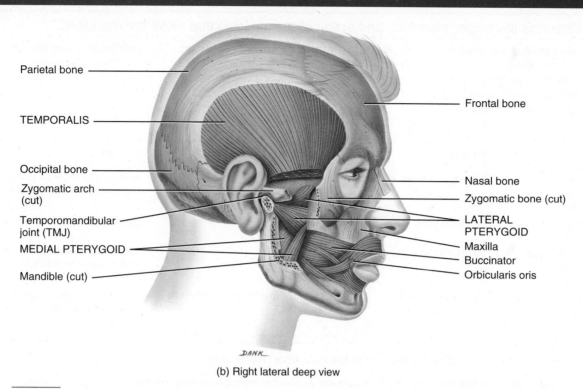

Parietal bone

TEMPORALIS

Occipital bone

Zygomatic arch (cut)

Temporomandibular joint (TMJ)

MEDIAL PTERYGOID

Mandible (cut)

Frontal bone

Nasal bone

Zygomatic bone (cut)

LATERAL PTERYGOID

Maxilla

Buccinator

Orbicularis oris

DANK

(b) Right lateral deep view

  Which is the strongest muscle of mastication?

Exhibit 8.3 / Muscles That Move the Eyeballs: Extrinsic Muscles (Figure 8.16)

Objective: Describe the origin, insertion, and action of the extrinsic muscles of the eyeballs.

- **Overview:** Two types of muscles are associated with the eyeball, extrinsic and intrinsic. *Extrinsic muscles* originate outside the eyeball and are inserted on its outer surface (sclera). They move the eyeballs in various directions. *Intrinsic muscles* originate and insert entirely within the eyeball. They move structures within the eyeballs, such as the iris and the lens.

Movements of the eyeballs are controlled by three pairs of extrinsic muscles. Two pairs of rectus muscles move the eyeball in the direction indicated by their respective names: superior, inferior, lateral, and medial. One pair of muscles, the oblique muscles—superior and inferior—rotate the eyeball on its axis. The extrinsic muscles of the eyeballs are among the fastest contracting and most precisely controlled skeletal muscles of the body.

- **Relating muscles to movements:** Arrange the muscles in this exhibit according to their actions on the eyeballs: (1) elevation, (2) depression, (3) abduction, (4) adduction, (5) medial rotation, and (6) lateral rotation. The same muscle may be mentioned more than once.

Muscle	Origin	Insertion	Action
Superior rectus (REK-tus; *superior* = above; *rect-* = straight; here, muscle fibers that are parallel to long axis of eyeball)	Tendinous ring attached to bony orbit around optic foramen.	Superior and central part of eyeball.	Moves eyeball upward (elevation) and medially (adduction), and rotates it medially.
Inferior rectus (*inferior* = below)	Same as above.	Inferior and central part of eyeball.	Moves eyeball downward (depression) and medially (adduction), and rotates it laterally.
Lateral rectus	Same as above.	Lateral side of eyeball.	Moves eyeball laterally (abduction).
Medial rectus	Same as above.	Medial side of eyeball.	Moves eyeball medially (adduction).
Superior oblique (ō-BLĒK; *oblique* = slanting; here, muscle fibers run diagonally to long axis of eyeball) The muscle moves through a ring of fibrocartilaginous tissue called the trochlea (*trochlea* = pulley).	Same as above.	Eyeball between superior and lateral recti.	Moves eyeball downward (depression) and laterally (abduction), and rotates it medially.
Inferior oblique	Maxilla.	Eyeball between inferior and lateral recti.	Moves eyeball upward (elevation) and laterally (abduction), and rotates it laterally.

Figure 8.16 ■ **Extrinsic muscles of the eyeballs.**

The extrinsic muscles of the eyeball are among the fastest contracting and most precisely controlled skeletal muscles in the body.

SUPERIOR OBLIQUE

Levator palpebrae superioris

SUPERIOR RECTUS
MEDIAL RECTUS

Optic (II) nerve

Sphenoid bone

Frontal bone

Trochlea

Eyeball

Maxilla

INFERIOR RECTUS   LATERAL RECTUS   INFERIOR OBLIQUE

Lateral view of right eyeball

 Which muscle passes through the trochlea?

Objective: Describe the origin, insertion, and action of the muscles that act on the anterior abdominal wall.

• **Overview:** The anterior abdominal wall is composed of skin; fascia; and four pairs of flat, sheetlike muscles: rectus abdominis, external oblique, internal oblique, and transversus abdominis. The anterior surfaces of the rectus abdominis muscles are interrupted by three transverse fibrous bands of tissue called ***tendinous intersections.*** The ***aponeuroses*** of the external oblique, internal oblique, and transversus abdominis muscles meet at the midline to form the ***linea alba*** (white line), a tough fibrous band that extends from the xiphoid process of the sternum to the pubic symphysis. An ***aponeurosis*** (ap'-ō-noo-RŌ-sis) is a broad, flat tendon. The inferior free border of the external oblique aponeurosis, plus some collagen fibers, forms the ***inguinal ligament.***

• **Relating muscles to movements:** Arrange the muscles in this exhibit according to the following actions on the vertebral column: (1) flexion, (2) lateral flexion, (3) extension, and (4) rotation. The same muscle may be mentioned more than once.

Muscle	Origin	Insertion	Action
Rectus abdominis (REK-tus ab-DOM-in-is; *rect-* = straight, fibers parallel to midline; *abdomin-* = abdomen)	Pubis and pubic symphysis.	Cartilage of fifth to seventh ribs and xiphoid process of sternum.	Flexes vertebral column, and compresses abdomen to aid in defecation, urination, forced expiration, and childbirth.
External oblique (ō-BLĒK; *external* = closer to surface; *oblique*-slanting; here, fibers that are diagonal to midline)	Lower eight ribs.	Crest of ilium and linea alba.	Contraction of both external obliques compresses abdomen and flexes vertebral column; contraction of one side alone bends vertebral column laterally and rotates it.
Internal oblique (*internal* = farther from surface)	Ilium, inguinal ligament, and thoracolumbar fascia.	Cartilage of last three or four ribs and linea alba.	Contraction of both internal obliques compresses abdomen and flexes vertebral column; contraction of one side alone bends vertebral column laterally and rotates it.
Transversus abdominis (trans-VER-sus; *transverse* = fibers that are perpendicular to midline)	Ilium, inguinal ligament, lumbar fascia, and cartilages of last six ribs.	Xiphoid process of sternum, linea alba, and pubis.	Compresses abdomen.

Figure 8.17 ■ **Muscles of the male anterolateral abdominal wall.**

The inguinal ligament separates the thigh from the body wall.

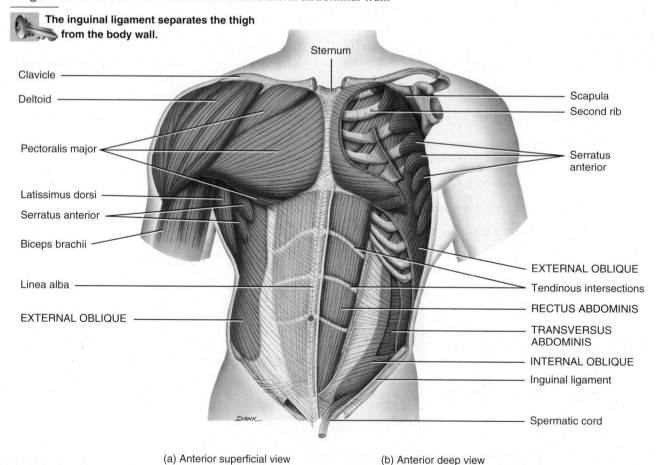

Sternum

Clavicle

Deltoid

Pectoralis major

Latissimus dorsi

Serratus anterior

Biceps brachii

Linea alba

EXTERNAL OBLIQUE

Scapula

Second rib

Serratus anterior

EXTERNAL OBLIQUE

Tendinous intersections

RECTUS ABDOMINIS

TRANSVERSUS ABDOMINIS

INTERNAL OBLIQUE

Inguinal ligament

Spermatic cord

DANK

(a) Anterior superficial view   (b) Anterior deep view

Which abdominal muscle aids in urination?

Exhibit 8.5 / Muscles Used in Breathing (Figure 8.18)

Objective: Describe the origin, insertion, and action of the muscles used in breathing.

- **Overview:** The muscles described here are attached to the ribs and by their contraction and relaxation alter the size of the thoracic cavity during breathing. ***Inhalation*** (breathing in) occurs when the thoracic cavity increases in size, and ***exhalation*** (breathing out) occurs when the thoracic cavity decreases in size. During normal quiet breathing, contraction of the diaphragm "lowers the floor" of the thoracic cavity and causes inhalation. Then, as the diaphragm relaxes, the highly elastic lungs recoil, causing exhalation. Deep, forceful breathing, as occurs during exercise or playing a wind instrument, involves accessory muscles—the sternocleidomastoid, external intercostals, pectoralis minor, and scalenes contribute to forceful inhalation, while the internal intercostals, external oblique, internal oblique, transversus abdominis, and rectus abdominis contribute to forceful exhalation.

- **Relating muscles to movements:** Arrange the muscles in this exhibit according to the following actions on the size of the thorax: (1) increase in vertical length, (2) increase in side-to-side (lateral) and front-to-back (anteroposterior) dimensions, and (3) decrease in side-to-side and front-to-back dimensions.

Muscle	Origin	Insertion	Action
Diaphragm (DĪ-a-fram; *dia* = across; *-phragm* = wall)	Xiphoid process of sternum, cartilages of last six ribs, and lumbar vertebrae.	Central tendon, a strong aponeurosis into which muscle fibers of the diaphragm insert.	Forms floor of thoracic cavity; pulls central tendon downward during inhalation, thus increasing vertical length of thorax.
External intercostals (in′-ter-KOS-tals; *external* = closer to surface; *inter-* = between; *costa-* = rib)	Inferior border of rib above.	Superior border of rib below.	Elevate ribs during forceful inhalation and thus increase side-to-side and front-to-back dimensions of thorax.
Internal intercostals (*internal* = farther from surface)	Superior border of rib below.	Inferior border of rib above.	Draw adjacent ribs together during forced exhalation and thus decrease side-to-side and front-to-back dimensions of thorax.

Figure 8.18 ▓ **Muscles used in breathing.**

The muscles used in breathing alter the size of the thoracic cavity.

Clavicle

INTERNAL INTERCOSTALS

EXTERNAL INTERCOSTALS

Pectoralis minor (cut)

Ribs

External oblique (cut)

Rectus abdominis (cut)

Transversus abdominis and aponeurosis

Rectus abdominis

Linea alba

Sternum

Ribs

EXTERNAL INTERCOSTALS

INTERNAL INTERCOSTALS

Central tendon of diaphragm

DIAPHRAGM

Quadratus lumborum

(a) Anterior superficial view   (b) Anterior deep view

Which muscle contracts during a normal quiet inhalation?

Exhibit 8.6 / Muscles That Move the Pectoral (Shoulder) Girdle (Figure 8.19)

Objective: Describe the origin, insertion, and action of the muscles that move the pectoral girdle.

- **Overview:** Muscles that move the pectoral girdle originate on the axial skeleton and insert on the clavicle or scapula. Based on their location, the muscles are divided into ***anterior*** and ***posterior thoracic muscles.*** The main action of the muscles is to hold the scapula in place so that it can function as a stable point of origin for most of the muscles that move the humerus (arm bone).

- **Relating muscles to movements:** Arrange the muscles in this exhibit according to the following actions on the scapula: (1) depression, (2) elevation, (3) lateral and forward movement, and (4) medial and backward movement. The same muscle may be mentioned more than once.

Muscle	Origin	Insertion	Action
Anterior Thoracic			
Pectoralis minor (pek'-tor-A-lis; *pect-* = breast, chest, thorax; *minor* = lesser)	Third through fifth ribs.	Scapula.	Depresses scapula, moves it laterally and forward, and rotates it downward; elevates third through fifth ribs during forced inspiration when scapula is fixed.
Serratus anterior (ser-Ā-tus; *serratus* = saw-toothed; *anterior* = before)	Upper eight or nine ribs.	Scapula.	Moves scapula laterally and forward, and rotates it upward; elevates ribs when scapula is fixed.
Posterior Thoracic			
Trapezius (tra-PĒ-zē-us; *trapezoides* = trapezoid-shaped)	Occipital bone and spines of seventh cervical and all thoracic vertebrae.	Clavicle and scapula.	Elevates clavicle; moves scapula medially and backward, rotates it upward, and elevates or depresses it; extends head.
Levator scapulae (le-VĀ-tor SKA-pyoo-lē; *levare* = to raise; *scapulae* = of the scapula)	Upper four or five cervical vertebrae.	Scapula.	Elevates scapula and rotates it downward.
Rhomboideus major (rom-BOID-ē-us; *rhomboides* = rhomboid or diamond-shaped)	Spines of second to fifth thoracic vertebrae.	Scapula.	Elevates scapula, moves it medially and backward, and rotates it downward.

Figure 8.19 ■ **Muscles that move the pectoral (shoulder) girdle.**

Muscles that move the pectoral girdle originate on the axial skeleton and insert on the clavicle or scapula.

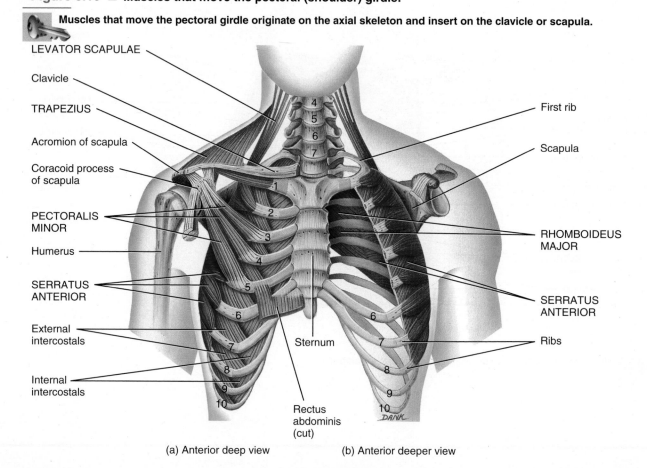

(a) Anterior deep view   (b) Anterior deeper view

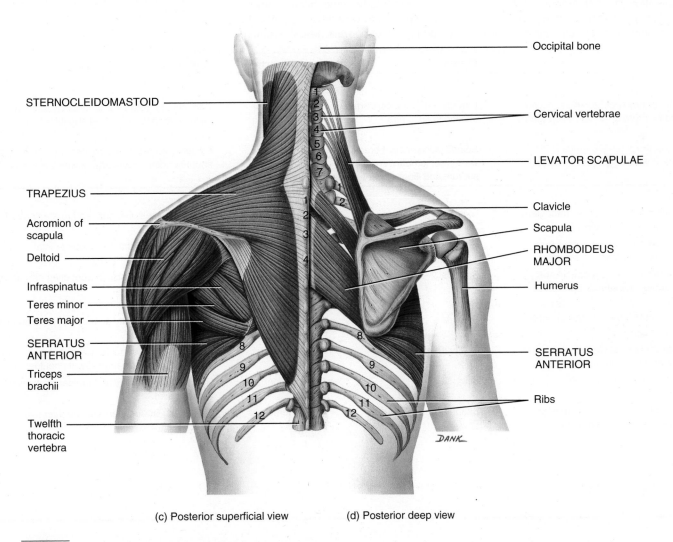

STERNOCLEIDOMASTOID

TRAPEZIUS

Acromion of
scapula

Deltoid

Infraspinatus

Teres minor

Teres major

SERRATUS
ANTERIOR

Triceps
brachii

Twelfth
thoracic
vertebra

Occipital bone

Cervical vertebrae

LEVATOR SCAPULAE

Clavicle

Scapula

RHOMBOIDEUS
MAJOR

Humerus

SERRATUS
ANTERIOR

Ribs

(c) Posterior superficial view   (d) Posterior deep view

Which muscles originate on the ribs? the vertebrae?

Exhibit 8.7 / Muscles That Move the Humerus (Arm Bone) *(Figure 8.20)*

Objective: Describe the origin, insertion, and action of the muscles that move the humerus.

• **Overview:** Of the nine muscles that cross the shoulder joint, only two of them (pectoralis major and latissimus dorsi) do not originate on the scapula. Because these muscles originate on the axial skeleton, they are termed *axial muscles.* The seven remaining muscles, the *scapular muscles,* have their origin on the scapula.

The strength and stability of the shoulder joint are provided by four deep muscles of the shoulder and their tendons: subscapularis, supraspinatus, infraspinatus, and teres minor. The tendons are arranged in a nearly complete circle around the joint, like the cuff on a shirt sleeve. This arrangement is called the *rotator cuff.*

One of the most common causes of shoulder pain and dysfunction in athletes is known as *impingement syndrome.* It results from repetitive overhead motions that are common in baseball, overhead racquet sports, volleyball (spiking), and swimming. Continual pinching of the supraspinatus tendon causes it to become inflamed and results in pain. If the condition persists, the tendon may degenerate near the attach-ment to the humerus and ultimately may tear away from the bone, a condition known as *rotator cuff injury.*

• **Relating muscles to movements:** Arrange the muscles in this exhibit according to the following actions on the humerus at the shoulder joint: (1) flexion, (2) extension, (3) abduction, (4) adduction, (5) medial rotation, and (6) lateral rotation. The same muscle may be mentioned more than once.

Muscle	Origin	Insertion	Action
Axial			
Pectoralis major (pek'-tor-A-lis; *pector-* = chest; *major* = greater) (see also Figure 8.17a)	Clavicle, sternum, cartilages of second to sixth ribs.	Humerus.	Adducts and rotates arm medially at shoulder joint; flexes and extends arm at shoulder joint.
Latissimus dorsi (la-TIS-i-mus DOR-sī; *latissimus* = widest; *dorsum* = back)	Spines of lower six thoracic vertebrae, lumbar vertebrae, sacrum and ilium, lower four ribs.	Humerus.	Extends, adducts, and rotates arm medially at shoulder joint; draws arm downward and backward.
Scapular			
Deltoid (DEL-toyd; *deltoid* = triangularly shaped) (see also Figure 8.17a)	Clavicle and scapula.	Humerus.	Abducts, flexes, extends, and rotates arm at shoulder joint.
Subscapularis (sub-scap'-yoo-LA-ris; *sub-* = below; *scapularis* = scapula)	Scapula.	Humerus.	Rotates arm medially at shoulder joint.
Supraspinatus (soo'-pra-spi-NA-tus; *supra* = above; *spinatus* = spine of scapula)	Scapula.	Humerus.	Assists deltoid muscle in abducting arm at shoulder joint.
Infraspinatus (in'-fra-spi-NA-tus; *infra* = below) (see Figure 8.19c)	Scapula.	Humerus.	Rotates arm laterally at shoulder joint; adducts arm at shoulder joint.
Teres major (TE-res) (*teres* = long and round) (see also Figure 8.19c)	Scapula.	Humerus.	Extends arm at shoulder joint; assists in adduction and rotation of arm medially at shoulder joint.
Teres minor (see Figure 8.19c)	Scapula.	Humerus.	Rotates arm laterally at shoulder joint; extends and adducts arm.
Coracobrachialis (kor'-a-kō-BRA-kē-a'-lis; *coraco* = coracoid process)	Scapula.	Humerus.	Flexes and adducts arm at shoulder joint.

Figure 8.20 ▦ **Muscles that move the humerus (arm).**

 The strength and stability of the shoulder joint are provided by the tendons of the muscles that form the rotator cuff.

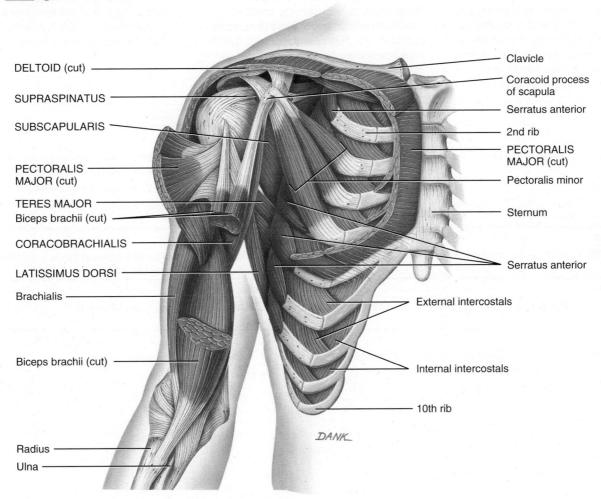

DELTOID (cut)

SUPRASPINATUS

SUBSCAPULARIS

PECTORALIS
MAJOR (cut)

TERES MAJOR

Biceps brachii (cut)

CORACOBRACHIALIS

LATISSIMUS DORSI

Brachialis

Biceps brachii (cut)

Radius

Ulna

Clavicle

Coracoid process
of scapula

Serratus anterior

2nd rib

PECTORALIS
MAJOR (cut)

Pectoralis minor

Sternum

Serratus anterior

External intercostals

Internal intercostals

10th rib

DANK

Anterior deep view (the intact pectoralis major muscle is shown in Figure 8.17a)

Which muscles that move the humerus are axial muscles, and why are they so named?

Exhibit 8.8 / Muscles That Move the Radius and Ulna (Forearm Bones) *(Figure 8.21)*

Objective: Describe the origin, insertion, and action of the muscles that move the radius and ulna.

• **Overview:** Most of the muscles that move the radius and ulna are divided into *forearm flexors* and *forearm extensors.* Recall that the elbow joint is a hinge joint, capable only of flexion and extension. The biceps brachii, brachialis, and brachioradialis are flexors of the elbow joint; the triceps brachii is an extensor. Other muscles that move the radius and ulna are concerned with supination and pronation.

In the limbs, functionally related skeletal muscles and their associated blood vessels and nerves are grouped together by deep fascia into regions called *compartments.* Thus, in the arm, the biceps brachii, brachialis, and coracobrachialis muscles constitute the *flexor compartment;* the triceps brachii muscle forms the *extensor compartment.*

• **Relating muscles to movements:** Arrange the muscles in this exhibit according to the following actions: (1) flexion and extension of the elbow joint; (2) supination and pronation of the forearm; and (3) flexion and extension of the humerus. The same muscle may be mentioned more than once.

Muscle	Origin	Insertion	Action
Forearm Flexors			
Biceps brachii (BĪ-ceps BRĀ- kē-ī; *biceps* = two heads of origin; *brach-* = arm)	Scapula.	Radius.	Flexes and supinates forearm at elbow joint; flexes arm at shoulder joint.
Brachialis (brā′-kē-A-lis)	Humerus.	Ulna.	Flexes forearm at elbow joint.
Brachioradialis (brā′-kē-ō-rā′-dē-A-lis; *radialis* = radius) (see Figure 8.22b)	Humerus.	Radius.	Flexes forearm at elbow joint.
Forearm Extensor			
Triceps brachii (TRĪ-ceps BRĀ-kē-ī; *triceps* = three heads of origin)	Scapula and humerus.	Ulna.	Extends forearm at elbow joint; extends arm at shoulder joint.
Forearm Supinator			
Supinator (SOO-pi-nā-tor; *supination* = turning palm upward or forward). Not illustrated.	Humerus and ulna.	Radius.	Supinates forearm.
Forearm Pronator			
Pronator teres (PRŌ-nā-tor TE-rēz; *pronation* = turning palm downward or backward) (see Figure 8.22a)	Humerus and ulna.	Radius.	Pronates forearm.

Figure 8.21 ■ **Muscles that move the radius and ulna (forearm).**

 Whereas the anterior arm muscles flex the forearm, the posterior arm muscles extend it.

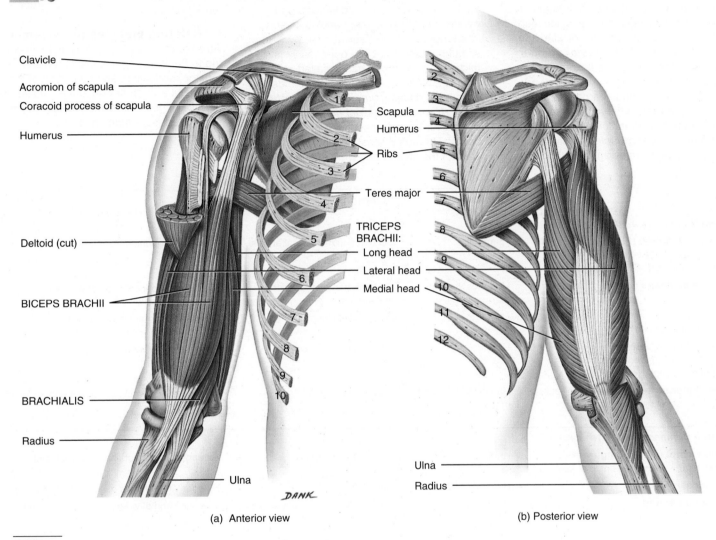

Clavicle

Acromion of scapula

Coracoid process of scapula

Humerus

Deltoid (cut)

BICEPS BRACHII

BRACHIALIS

Radius

Ulna

Scapula

Humerus

Ribs

Teres major

TRICEPS BRACHII:
Long head
Lateral head
Medial head

Ulna

Radius

DANK

(a) Anterior view

(b) Posterior view

What is a compartment?

Exhibit 8.9 / Muscles That Move the Wrist, Hand, and Fingers (Figure 8.22)

Objective: Describe the origin, insertion, and action of the muscles that move the wrist, hand, and fingers.

• **Overview:** Muscles that move the wrist, hand, and fingers are located on the forearm and are many and varied. However, as you will see, their names for the most part give some indication of their origin, insertion, or action. On the basis of location and function, the muscles are divided into two compartments. The **anterior compartment muscles** function as flexors. They originate on the humerus and typically insert on the carpals, metacarpals, and phalanges. The bellies of these muscles form the bulk of the proximal forearm. The **posterior compartment muscles** function as extensors. These muscles arise on the humerus and insert on the metacarpals and phalanges.

The tendons of the muscles of the forearm that attach to the wrist or continue into the hand, along with blood vessels and nerves, are held close to bones by fascia. The tendons are also surrounded by tendon sheaths. At the wrist, the deep fascia is thickened into fibrous bands called **retinacula** (re-ti-NAK-yoo-la = retaining bands; singular is *retinaculum*). The **flexor retinaculum (transverse carpal ligament)** is located over the palmar surface of the carpal bones. Through it pass the long flexor tendons of the fingers and wrist and the median nerve. The **extensor retinaculum (dorsal carpal ligament)** is located over the dorsal surface of the carpal bones. Through it pass the extensor tendons of the wrist and fingers.

• **Relating muscles to movements:** Arrange the muscles in this exhibit according to the following actions: (1) flexion, extension, abduction, and adduction of the wrist joint and (2) flexion and extension of the phalanges. The same muscle may be mentioned more than once.

Muscle	Origin	Insertion	Action
Anterior Compartment (Flexors)			
Flexor carpi radialis (FLEK-sor KAR-pē rā′-dē-A-lis; *flexor* = decreases angle at joint; *carpus* = wrist; *radialis* = radius)	Humerus.	Second and third metacarpals.	Flexes and abducts hand at wrist joint.
Flexor carpi ulnaris (ul-NAR-is; *ulnaris* = ulna)	Humerus and ulna.	Pisiform, hamate, and fifth metacarpal.	Flexes and adducts hand at wrist joint.
Palmaris longus (pal-MA-ris LON-gus; *palma* = palm; *longus* = long)	Humerus.	Flexor retinaculum.	Weakly flexes hand at wrist joint.
Flexor digitorum profundus (di′-ji-TOR-um pro-FUN-dus; *digit* = finger or toe; *profundus* = deep). Not illustrated.	Ulna.	Bases of distal phalanges.	Flexes hand at wrist joint; flexes phalanges of each finger.
Flexor digitorum superficialis (soo′-per-fish′-ē-A-lis; *superficialis* = closer to surface)	Humerus, ulna, and radius.	Middle phalanges.	Flexes hand at wrist joint; flexes phalanges of each finger.
Posterior Compartment (Extensors)			
Extensor carpi radialis longus (eks-TEN-sor; *extensor* = increases angle at joint	Humerus.	Second metacarpal.	Extends and abducts hand at wrist joint.
Extensor carpi ulnaris	Humerus and ulna.	Fifth metacarpal.	Extends and adducts hand at wrist joint.
Extensor digitorum	Humerus.	Second through fifth phalanges.	Extends hand at wrist joint; extends phalanges.

Figure 8.22 ■ **Muscles that move the wrist, hand, and fingers.**

 The anterior compartment muscles function as flexors, and the posterior compartment muscles function as extensors.

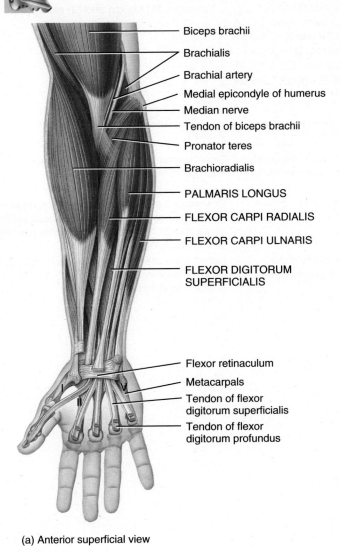

Biceps brachii

Brachialis

Brachial artery

Medial epicondyle of humerus

Median nerve

Tendon of biceps brachii

Pronator teres

Brachioradialis

PALMARIS LONGUS

FLEXOR CARPI RADIALIS

FLEXOR CARPI ULNARIS

FLEXOR DIGITORUM
SUPERFICIALIS

Flexor retinaculum

Metacarpals

Tendon of flexor
digitorum superficialis

Tendon of flexor
digitorum profundus

(a) Anterior superficial view

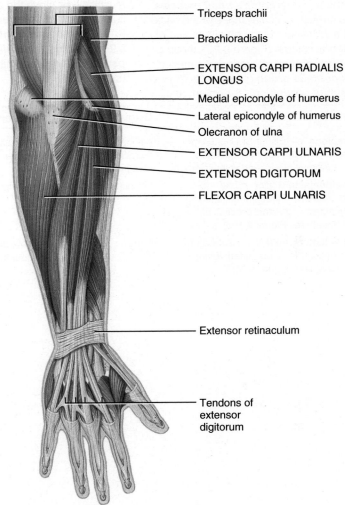

Triceps brachii

Brachioradialis

EXTENSOR CARPI RADIALIS
LONGUS

Medial epicondyle of humerus

Lateral epicondyle of humerus

Olecranon of ulna

EXTENSOR CARPI ULNARIS

EXTENSOR DIGITORUM

FLEXOR CARPI ULNARIS

Extensor retinaculum

Tendons of
extensor
digitorum

(b) Posterior superficial view

Which nerve is associated with the flexor retinaculum?

Exhibit 8.10 / Muscles That Move the Vertebral Column (Backbone) *(Figure 8.23)*

Objective: Describe the origin, insertion, and action of the muscles that move the vertebral column.

• **Overview:** The muscles that move the vertebral column are quite complex because they have multiple origins and insertions and there is considerable overlap among them. The ***erector spinae muscles*** form the largest muscular mass of the back, forming a prominent bulge on either side of the vertebral column (Figure 8.23). It consists of three groups of overlapping muscles: ***iliocostalis group*** (il′-ē-ō-kos-TĀ-lis), ***longissimus group*** (lon′-JI-si-mus), and ***spinalis group*** (spi-NĀ-lis). The iliocostalis group is laterally placed, the longissimus group is intermediate in placement, and the spinalis group is medially placed. Other muscles that move the vertebral column include the ***sternocleidomastoid*** (see Figure 8.19c), ***quadratus lumborum*** (see Figure 8.24a), ***rectus abdominis*** (see Exhibit 8.4), ***psoas major*** (see Exhibit 8.11), and ***iliacus*** (see Exhibit 8.11).

• **Relating muscles to movements:** Arrange the muscles in this exhibit according to the following actions on the vertebral column: (1) flexion and (2) extension.

Muscle	Origin	Insertion	Action
Erector spinae (e-REK-tor SPI-nē; *erector* = raise; *spinae* = of the spine) (iliocostalis group, longissimus group, and spinalis group)	All ribs plus cervical, thoracic, and lumbar vertebrae.	Occipital bone, temporal bone, ribs, and vertebrae.	Extends head; extends and laterally flexes vertebral column.
Sternocleidomastoid (ster′-nō-klī′-dō-MAS-toid; *sternum* = breastbone; *cleido-* = clavicle; *mastoid* = mastoid process of temporal bone) (see Figure 8.19c)	Sternum and clavicle.	Temporal bone.	Contractions of both muscles flex the cervical part of the vertebral column and extend the head; contraction of one muscle rotates head toward side opposite contracting muscle.
Quadratus lumborum (kwod-RĀ-tus lum-BOR-um; *quadratus* = four-sided; *lumbo* = lumbar region) (see Figure 8.24a)	Ilium.	Twelfth rib and upper four lumbar vertebrae.	Contractions of both muscles extend lumbar part of vertebral column; contraction of one muscle flexes lumbar part of vertebral column.

Figure 8.23 ■ **Major muscles that move the vertebral column.**

 The erector spinae muscles extend the vertebral column.

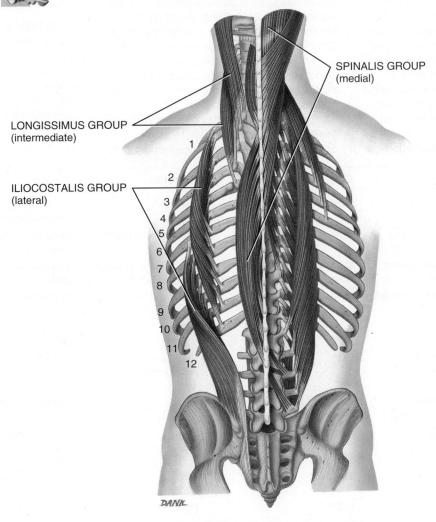

SPINALIS GROUP
(medial)

LONGISSIMUS GROUP
(intermediate)

ILIOCOSTALIS GROUP
(lateral)

1
2
3
4
5
6
7
8
9
10
11
12

DANK

Posterior view of erector spinae muscles

Which muscles constitute the erector spinae?

Exhibit 8.11 / Muscles That Move the Femur (Thigh Bone) *(Figure 8.24)*

Objective: Describe the origin, insertion, and action of the muscles that move the femur.

• **Overview:** Muscles of the lower limbs are larger and more powerful than those of the upper limbs to provide stability, locomotion, and maintenance of posture. In addition, muscles of the lower limbs often cross two joints and act equally on both. The majority of muscles that act on the femur originate on the pelvic (hip) girdle and insert on the femur. The anterior muscles are the psoas major and iliacus, together referred to as the *iliopsoas.* The remaining muscles (except for the pectineus, adductors, and tensor fasciae latae) are posterior muscles. Technically, the pectineus and adductors are components of the medial compartment of the thigh, but they are included in this exhibit because they act on the thigh. The tensor fasciae latae is laterally placed. The *fascia lata* is a deep fascia of the thigh that encircles the entire thigh. It is well developed laterally, where together with the tendons of the gluteus maximus and tensor fasciae latae it forms a structure called the *iliotibial tract.* The tract inserts into the lateral condyle of the tibia.

Certain activities may cause a *pulled groin,* a strain, stretching, or tearing of the distal attachments of the medial muscles of the thigh, psoas major, iliacus, and/or adductor muscles. This generally results from activities that involve quick sprints, such as soccer, tennis, football, and running.

• **Relating muscles to movements:** Arrange the muscles in this exhibit according to the following actions on the thigh at the hip joint: (1) flexion, (2) extension, (3) abduction, (4) adduction, (5) medial rotation, and (6) lateral rotation. The same muscle may be mentioned more than once.

Muscle	Origin	Insertion	Action
Psoas major (SŌ-as; *psoa* = muscle of loin)	Lumbar vertebrae.	Femur.	Flexes and rotates thigh laterally at the hip joint; flexes vertebral column.
Iliacus (il′-ē-AK-us; *iliac* = ilium)	Ilium.	With psoas major into femur.	Flexes and rotates thigh laterally at the hip joint; flexes vertebral column.
Gluteus maximus (GLOO-tē-us MAK-si-mus; *glut-* = buttock; *maximus* = largest)	Ilium, sacrum, coccyx, and aponeurosis of sacrospinalis.	Iliotibial tract of fascia lata and femur.	Extends and rotates thigh laterally at the hip joint.
Gluteus medius (MĒ-dē-us; *medius* = middle)	Ilium.	Femur.	Abducts and rotates thigh medially at the hip joint.
Gluteus minimus (MIN-i-mus; *minimus* = smallest)	Ilium.	Femur.	Abducts and rotates thigh medially at the hip joint.
Tensor fasciae latae (TEN-sor FA-shē-ē LĀ-tē; *tensor* = makes tense; *fascia* = band; *latus* = wide)	Ilium.	Tibia by means of the iliotibial tract.	Flexes and abducts thigh at the hip joint.
Adductor longus (LONG-us; *adductor* = moves part closer to midline; *longus* = long)	Pubis and pubic symphysis.	Femur.	Adducts, medially rotates, and flexes thigh at the hip joint.
Adductor magnus (MAG-nus; *magnus* = large)	Pubis and ischium.	Femur.	Adducts, flexes, medially rotates and extends thigh (anterior part flexes, posterior part extends) at the hip joint.
Piriformis (pir-i-FOR-mis; *pirum* = pear; *forma* = shape)	Sacrum.	Femur.	Rotates thigh laterally and abducts it at the hip joint.
Pectineus (pek-TIN-ē-us; *pecten* = comb-shaped)	Pubis.	Femur.	Flexes and adducts thigh at the hip joint.

Figure 8.24 ■ **Muscles that move the femur (thigh).**

 Most muscles that move the femur originate on the pelvic (hip) girdle and insert on the femur.

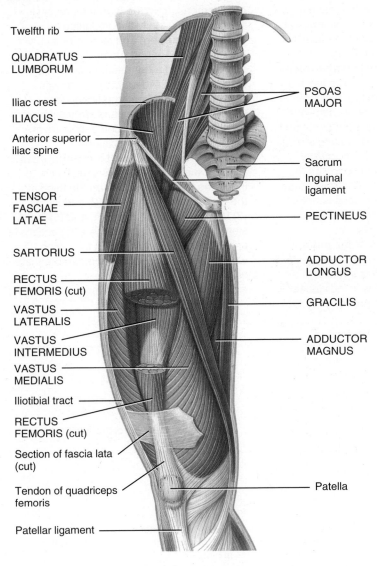

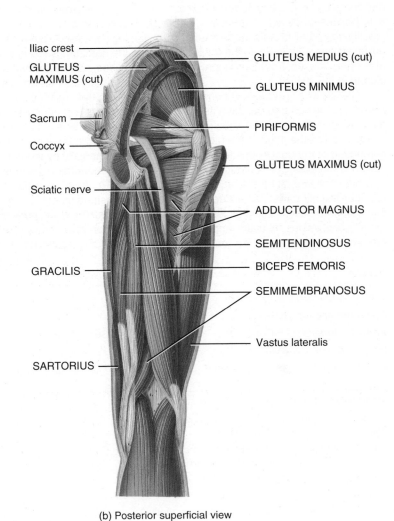

(a) Anterior superficial view

(b) Posterior superficial view

 Which muscles are part of the quadriceps femoris? the hamstrings?

Exhibit 8.12 / Muscles That Move the Femur (Thigh Bone), Tibia, and Fibula (Leg Bones) *(Figure 8.24)*

Objective: Describe the origin, insertion, and action of the muscles that move the femur, tibia, and fibula.

• **Overview:** The muscles that move the femur, tibia, and fibula originate in the hip and thigh and are separated into compartments by deep fascia. The **medial (adductor) compartment** is so named because its muscles adduct the thigh. The adductor magnus, adductor longus, and pectineus muscles, components of the medial compartment, are included in Exhibit 8.11 because they act on the femur. The gracilis, the other muscle in the medial compartment, not only adducts the thigh but also flexes the leg. For this reason, it is included in this exhibit.

The **anterior (extensor) compartment** is so designated because its muscles act to extend the leg at the knee joint, and some also flex the thigh at the hip joint. It is composed of the quadriceps femoris and sartorius muscles. The quadriceps femoris muscle is the largest muscle in the body but has four distinct parts, usually described as four separate muscles (rectus femoris, vastus lateralis, vastus medialis, and vastus intermedius). The common tendon for the four

muscles is the **quadriceps tendon,** which attaches to the patella. The tendon continues below the patella as the **patellar ligament** and attaches to the tibial tuberosity. The sartorius muscle is the longest muscle in the body, extending from the ilium of the hip bone to the medial side of the tibia. It moves both the thigh and the leg.

The **posterior (flexor) compartment** is so named because its muscles flex the leg (but also extend the thigh). Included are the hamstrings (biceps femoris, semitendinosus, and semimembranosus), so named because their tendons are long and string-like in the popliteal area. The **popliteal fossa** is a diamond-shaped space on the posterior aspect of the knee bordered laterally by the tendons of the biceps femoris and medially by the semitendinosus and semimembranosus muscles.

A strain or partial tear of the proximal hamstring muscles is referred to as **"pulled hamstrings"** or **hamstring strains.** They are common sports injuries in individuals who run very hard and/or are required to

perform quick starts and stops. Sometimes the violent muscular exertion required to perform a feat tears off part of the tendinous origins of the hamstrings, especially the biceps femoris, from the ischial tuberosity. This injury is usually accompanied by a contusion (bruising) and tearing of some of the muscle fibers and rupture of blood vessels, producing a hematoma (collection of blood) and pain. Adequate training with good balance between the quadriceps femoris and hamstrings and stretching exercises before running or competing are important in preventing this injury.

• **Relating muscles to movements:** Arrange the muscles in this exhibit according to the following actions on the thigh at the hip joint: (1) abduction, (2) adduction, (3) lateral rotation, (4) flexion, and (5) extension; and according to the following actions on the leg: (1) flexion and (2) extension. The same muscle may be mentioned more than once.

Muscle	Origin	Insertion	Action
Medial (Adductor) Compartment			
Adductor magnus (MAG-nus)			
Adductor longus (LONG-us)	See Exhibit 8.11.		
Pectineus (pek-TIN-ē-us)			
Gracilis (gra-SIL-is; *gracilis* = slender)	Pubic symphysis.	Tibia.	Adducts and medially rotates thigh at hip joint; flexes leg at knee joint.
Medial (Extensor) Compartment			
Quadriceps femoris (KWOD-ri-seps FEM-or-is; *quadriceps* = four heads of origin; *femoris* = femur)			
Rectus femoris (REK-tus FEM-or-is; *rectus* = straight; here, fibers run parallel to midline)	Ilium.	Patella by means of quadriceps tendon and then tibial tuberosity by means of patellar ligament.	All four heads extend leg at knee joint; rectus portion alone also flexes thigh at hip joint.
Vastus lateralis (VAS-tus lat′-er-A-lis; *vastus* = large; *lateralis* = lateral)	Femur.		
Vastus medialis (mē′-dē-A-lis; *medialis* = medial)	Femur.		
Vastus intermedius (in′-ter-MĒ-dē-us; *intermedius* = middle)	Femur.		
Sartorius (sar-TOR-ē-us; *sartor-* = tailor; refers to cross-legged position of tailors)	Ilium.	Tibia.	Flexes leg at knee joint; flexes, abducts, and laterally rotates thigh at hip joint, thus crossing leg.
Posterior (Flexor) Compartment			
Hamstrings			
Biceps femoris (BĪ-ceps FEM-or-is; *biceps* = two heads of origin)	Ischium and femur.	Fibula and tibia.	Flexes leg at knee joint; extends thigh at hip joint.
Semitendinosus (sem′-ē-TEN-di-nō′-sus; *semi-* = half; *tend-* = tendon)	Ischium.	Tibia.	Flexes leg at knee joint; extends thigh at hip joint.
Semimembranosus (sem′-ē-MEM-bra-nō′-sus; *membran-* = membrane)	Ischium.	Tibia.	Flexes leg at knee joint; extends thigh at hip joint.

Exhibit 8.13 / Muscles That Move the Foot and Toes *(Figure 8.25)*

Objective: Describe the origin, insertion, and action of the muscles that move the foot and toes.

• **Overview:** Muscles that move the foot and toes are located in the leg. The muscles of the leg, like those of the thigh, are divided into three compartments by deep fascia. The *anterior compartment* consists of muscles that dorsiflex the foot. In a situation like that at the wrist, the tendons of the muscles of the anterior compartment are held firmly to the ankle bones by thickenings of deep fascia called the *superior extensor retinaculum (transverse ligament of the ankle)* and *inferior extensor retinaculum (cruciate ligament of the ankle)*. The *lateral compartment* contains muscles that plantar flex and evert the foot. The *posterior compartment* consists of superficial

and deep muscles. The superficial muscles (gastrocnemius and soleus) share a common tendon of insertion, the calcaneal (Achilles) tendon, the strongest tendon of the body.

Shin splints refers to pain or soreness along the medial, distal two-thirds of the tibia. It may be caused by tendinitis of the tibialis posterior or toe flexors, inflammation of the periosteum around the tibia, or stress fractures of the tibia. The tendinitis usually occurs when poorly conditioned runners run on hard or banked surfaces with poorly supportive running shoes or walking or running up and down hills. The condition may also

occur as a result of vigorous activity of the legs following a period of relative inactivity. The muscles in the anterior compartment (mainly the tibialis anterior) can be strengthened to balance the stronger posterior compartment muscles.

• **Relating muscles to movements:** Arrange the muscles in this exhibit according to the following actions on the foot: (1) dorsiflexion, (2) plantar flexion, (3) inversion, and (4) eversion; and according to the following actions on the toes: (1) flexion and (2) extension. The same muscle may be mentioned more than once.

Muscle	Origin	Insertion	Action
Anterior Compartment			
Tibialis anterior (tib′-ē-A-lis; *tibialis* = tibia; *anterior* = front)	Tibia.	First metatarsal and first cuneiform.	Dorsiflexes and inverts foot.
Extensor digitorum longus (eks-TEN-sor di′-ji-TOR-um LON-gus; *extensor* = increases angle at joint; *digitorum* = finger or toe; *longus* = long)	Tibia and fibula.	Middle and distal phalanges of four outer toes.	Dorsiflexes and everts foot; extends toes.
Lateral Compartment			
Peroneus longus (per′-ō-NĒ-us LON-gus; *peroneus* = fibula)	Fibula and tibia.	First metatarsal and first cuneiform.	Plantar flexes and everts foot.
Posterior Compartment			
Gastrocnemius (gas′-trok-NĒ-mē-us; *gastro-* = belly; *-cnemi* = leg)	Femur.	Calcaneus by means of calcaneal (Achilles) tendon.	Plantar flexes foot; flexes leg at knee joint.
Soleus (SŌ-lē-us; *soleus* = sole of foot)	Fibula and tibia.	Calcaneus by means of calcaneal (Achilles) tendon.	Plantar flexes foot.
Tibialis posterior (*posterior* = back)	Tibia and fibula.	Second, third, and fourth metatarsals; navicular; all three cuneiforms, and cuboid.	Plantar flexes and inverts foot.
Flexor digitorum longus (FLEK-sor; *flexor* = decreases angle at joint)	Tibia.	Distal phalanges of four outer toes.	Plantar flexes foot; flexes toes.

(continues)

Exhibit 8.13 / continued (*Figure 8.25*)

Figure 8.25 ■ Muscles that move the foot and toes.

The superficial muscles of the posterior compartment share a common tendon of insertion, the calcaneal (Achilles) tendon, that inserts into the calcaneal bone of the ankle.

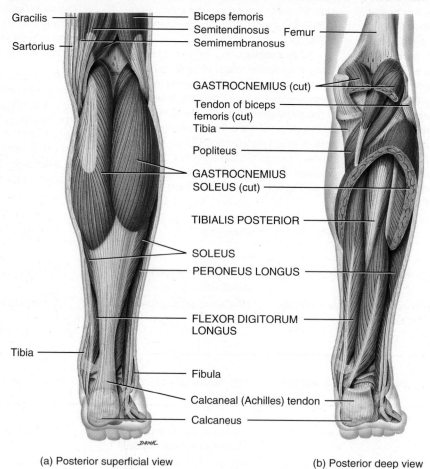

(a) Posterior superficial view

(b) Posterior deep view

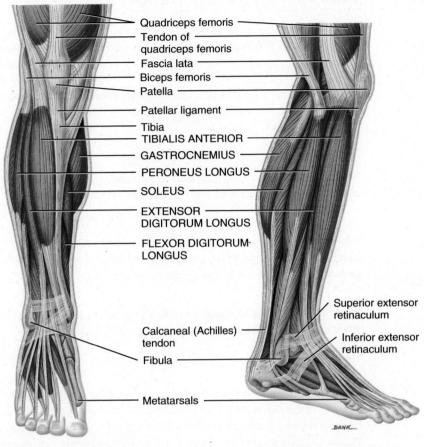

Quadriceps femoris

Tendon of
quadriceps femoris

Fascia lata

Biceps femoris

Patella

Patellar ligament

Tibia

TIBIALIS ANTERIOR

GASTROCNEMIUS

PERONEUS LONGUS

SOLEUS

EXTENSOR
DIGITORUM LONGUS

FLEXOR DIGITORUM
LONGUS

Calcaneal (Achilles)
tendon

Fibula

Metatarsals

Superior extensor
retinaculum

Inferior extensor
retinaculum

DANK

(c) Anterior superficial view

(d) Right lateral superficial view

 Which muscle is primarily affected in shin splints?

Focus on Homeostasis
The Muscular System

Body System	Contribution of the Muscular System
For all body systems	Muscles produce body movements, stabilize body positions, move substances within the body, and produce heat that helps maintain normal body temperature.
Integumentary system	Pull of skeletal muscles on attachments to skin of face cause facial expressions; muscular exercise increases skin blood flow.
Skeletal system	Skeletal muscle causes movement of body parts by pulling on attachments to bones; skeletal muscle provides stability for bones and joints.
Nervous system	Smooth, cardiac, and skeletal muscles carry out commands for the nervous system; shivering—involuntary contraction of skeletal muscles that is regulated by the brain—generates heat to raise body temperature.
Endocrine system	Regular activity of skeletal muscles (exercise) improves action of some hormones, such as insulin; muscles protect some endocrine glands.
Cardiovascular system	Cardiac muscle powers pumping action of heart; contraction and relaxation of smooth muscle in blood vessel walls help adjust the amount of blood flowing through various body tissues; contraction of skeletal muscles in the legs assist return of blood to the heart; regular exercise causes cardiac hypertrophy (enlargement) and increases heart's pumping efficiency; lactic acid produced by active muscles may be used for ATP production by the heart.
Lymphatic and immune system	Skeletal muscles protect some lymph nodes and lymphatic vessels; skeletal muscle contractions promote flow of lymph inside lymphatic vessels; exercise may increase or decrease some immune responses.
Respiratory system	Skeletal muscles involved with breathing cause air to flow into and out of the lungs; smooth muscles adjust size of airways; vibrations in skeletal muscles of larynx control air flowing past vocal cords, regulating voice production; coughing and sneezing, due to skeletal muscle contractions, help clear airways; regular exercise improves efficiency of breathing.
Digestive system	Skeletal muscles protect and support organs in the abdominal cavity; alternating contraction and relaxation of skeletal muscles power chewing and swallowing; smooth muscle sphincters control volume of organs of the gastrointestinal (GI) tract; smooth muscle in walls of GI tract mix and move its contents through the tract.
Urinary system	Skeletal and smooth muscle sphincters and smooth muscle in wall of urinary bladder control urination.
Reproductive systems	Skeletal and smooth muscle contractions eject semen from male; smooth muscle contractions propel oocyte along uterine (Fallopian) tube, help regulate flow of menstrual blood from uterus, and force baby from uterus during childbirth; during intercourse, skeletal muscle contractions are associated with orgasm and pleasurable sensations in both sexes.

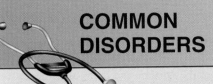

COMMON DISORDERS

Muscle function may be abnormal due to disease or damage of any of the components of a motor unit: the somatic motor neuron, neuromuscular junctions, or muscle fibers. The term *neuromuscular disease* encompasses problems at all three sites, whereas the term *myopathy* (mī-OP-a-thē; *-pathy* = disease) signifies a disease or disorder of skeletal muscle tissue itself.

Myasthenia Gravis

Myasthenia gravis (mī-as-THĒ-nē-a GRAV-is) is a chronic, progressive neuromuscular disease. People with myasthenia gravis inappropriately produce antibodies that bind to and block some of their acetylcholine receptors. As the disease progresses, more and more ACh receptors become dysfunctional. Muscles become increasingly weaker and may eventually cease to function.

Myasthenia gravis occurs in about 1 in 10,000 people and is more common in women. The muscles of the face and neck are most often affected. Initial symptoms include a weakness of the eye muscles, which may produce double vision, and difficulty in swallowing. Later, the individual has difficulty chewing and talking. Eventually, the muscles of the limbs may become involved. Death may result from paralysis of the respiratory muscles, but often the disorder does not progress to this stage.

Abnormal Contractions of Skeletal Muscle

One kind of abnormal muscular contraction is a *spasm*, a sudden involuntary contraction of a single muscle in a large group of muscles. A painful spasmodic contraction is known as a *cramp*. A *tic* is a spasmodic, involuntary twitching of muscles that are ordinarily under voluntary control. Twitching of the eyelid and facial muscles are examples of tics. A *tremor* is a rhythmic, involuntary, purposeless contraction that produces a quivering or shaking movement.

Running Injuries

Nearly 70% of those who jog or run sustain some type of running-related injury. Common sites of injury include the ankle, knee, calcaneal (Achilles) tendon, hip, groin, foot, and back. Of these, the knee often is the most severely injured area.

Running injuries often are caused by faulty training techniques such as improper (or lack of) warm-up routines, running too much, running too soon after an injury, or extended running on hard and/or uneven surfaces. Poorly constructed or worn-out running shoes can also contribute to injury. *Patellofemoral stress syndrome* ("runner's knee") is the most common problem in runners. A common cause of runner's knee is constantly walking, running, or jogging on the same side of the road. Because roads are high in the middle and slope down on the sides, the slope stresses the knee that is closer to the center of the road.

Most sports injuries should be treated initially with RICE therapy, which stands for **r**est, **i**ce, **c**ompression, and **e**levation. Immediately apply ice, and rest and elevate the injured part. Then apply an elastic bandage to compress the injured tissue. Continue using RICE for two to three days, and resist the temptation to apply heat, which may worsen the swelling. During the recovery period, it is important to keep active with an alternate fitness program that does not worsen the original injury. Finally, careful exercise is needed to rehabilitate the injured area itself.

MEDICAL TERMINOLOGY AND CONDITIONS

Charley horse   Tearing of a muscle due to forceful impact accompanied by bleeding and severe pain. It often occurs in contact sports and typically affects the quadriceps femoris muscle on the anterior surface of the thigh.

Electromyography   or *EMG* (e-lek′-trō-mī-OG-ra-fē; *electro-* = electricity; *myo-* = muscle; *-graphy* = to write) The recording and study of electrical changes that occur in muscle tissue.

Hypertonia   (*hyper-* = above) Increased muscle tone, characterized by increased muscle stiffness and sometimes associated with a change in normal reflexes.

Hypotonia   (*hypo-* = below) Decreased or lost muscle tone.

Muscular dystrophy   (*dystrophy* = degeneration) A group of inherited myopathies characterized by degeneration of individual muscle fibers, which leads to a progressive atrophy of the skeletal muscle.

Myalgia   (mī-AL-jē-a; *-algia* = painful condition) Pain in or associated with muscles.

Myositis   (mī′-ō-SĪ-tis; *-itis* = inflammation of) Inflammation of muscle fibers.

Myotonia   (mī′-ō-TŌ-nē-a; *-tonia* = tension) Increased muscular excitability and contractility with decreased power of relaxation; tonic spasm of the muscle.

Paralysis   (pa-RAL-a-sis; *para-* = beyond; *-lysis* = to loosen) Loss or impairment of voluntary movement in a muscle due to injury or disease of its nerve supply.

Volkmann's contracture   (FŌLK-manz kon-TRAK-tur; *contra* = against) Permanent shortening (contracture) of a muscle due to replacement of destroyed muscle fibers with fibrous connective tissue.

Wryneck or *torticollis*   (RĪ-neck; *torti-* = twisted; *-collis* = of the neck) Spasmodic contraction of several superficial and deep muscles of the neck that produces twisting of the neck and an unnatural position of the head.

STUDY OUTLINE

Overview of Muscle Tissue (p. 168)

1. The three types of muscle are skeletal muscle, cardiac muscle, and smooth muscle (summarized in Table 8.1 on page 000).

2. Skeletal muscle tissue is mostly attached to bones. It is striated and voluntary.

3. Cardiac muscle tissue forms most of the wall of the heart. It is striated and involuntary.

4. Smooth muscle tissue is located in viscera. It is nonstriated and involuntary.

5. Through contraction and relaxation, muscle tissue has five key functions: producing body movements, stabilizing body positions, regulating organ volume, moving substances within the body, and producing heat.

6. Muscle tissue has four characteristics: (1) excitability, the property of receiving and responding to stimuli by producing electrical signals; (2) contractility, the ability to contract (shorten and thicken); (3) extensibility, the ability to be stretched (extend); (4) elasticity, the ability to return to original shape after contraction or extension.

Skeletal Muscle Tissue (p. 168)

1. Fascia is a sheet or broad band of fibrous connective tissue underneath the skin or around muscles and organs of the body.

2. Other connective tissue components are epimysium, covering an entire muscle; perimysium, covering fascicles; and endomysium, covering individual muscle fibers.

3. Tendons and aponeuroses are extensions of connective tissue beyond muscle fibers that attach the muscle to bone or to another muscle.

4. Skeletal muscles are well supplied with nerves and blood vessels, which provide nutrients and oxygen for contraction.

5. Skeletal muscle consists of muscle fibers (cells) covered by a sarcolemma that features tunnel-like extensions, the transverse tubules. The fibers contain sarcoplasm, multiple nuclei, many mitochondria, myoglobin, and sarcoplasmic reticulum.

6. Each fiber also contains myofibrils that contain thin and thick filaments. The filaments are arranged in functional units called sarcomeres.

7. Thin filaments are composed of actin, tropomyosin, and troponin; thick filaments consist of myosin.

8. Projecting myosin heads are called crossbridges.

Contraction and Relaxation of Skeletal Muscle (p. 172)

1. Muscle contraction occurs when myosin heads attach to and "walk" along the thin filaments at both ends of a sarcomere, progressively pulling the thin filaments toward the center of a sarcomere. As the thin filaments slide inward, the Z discs come closer together, and the sarcomere shortens.

2. The neuromuscular junction (NMJ) is the synapse between a somatic motor neuron and a skeletal muscle fiber. The NMJ includes the axon terminals and synaptic end bulbs of a motor neuron plus the adjacent motor end plate of the muscle fiber sarcolemma.

3. A motor neuron and all of the muscle fibers it stimulates form a motor unit. A single motor unit may include as few as 10 or as many as 2000 muscle fibers.

4. When a nerve impulse reaches the synaptic end bulbs of a somatic motor neuron, it triggers exocytosis of the synaptic vesicles, releasing acetylcholine (ACh). ACh diffuses across the synaptic cleft and binds to ACh receptors, initiating a muscle action potential. Acetylcholineesterase then quickly destroys ACh.

5. An increase in the level of Ca^{2+} in the sarcoplasm, caused by the muscle action potential, starts the contraction cycle, whereas a decrease in the level of Ca^{2+} turns off the contraction cycle.

6. The contraction cycle is the repeating sequence of events that causes sliding of the filaments: (1) myosin ATPase splits ATP and becomes energized, (2) the myosin head attaches to actin, (3) the myosin head generates force as it swivels or rotates toward the center of the sarcomere (power stroke), and (4) binding of ATP to myosin detaches myosin from actin. The myosin head again splits ATP, returns to its original position, and binds to a new site on actin as the cycle continues.

7. Ca^{2+} active transport pumps continually remove Ca^{2+} from the sarcoplasm into the sarcoplasmic reticulum (SR). When the level of Ca^{2+} in the sarcoplasm decreases, the troponin–tropomyosin complexes slide back over and cover the myosin-binding sites, and the muscle fiber relaxes.

8. Continual involuntary activation of a small number of motor units produces muscle tone, which is essential for maintaining posture.

Metabolism of Skeletal Muscle Tissue (p. 177)

1. Muscle fibers have three sources for ATP production: creatine phosphate, glycolysis, and aerobic cellular respiration.

2. The transfer of a high-energy phosphate group from creatine phosphate to ADP forms new ATP molecules. Together, creatine phosphate and ATP provide enough energy for muscles to contract maximally for about 15 seconds.

3. Glucose is converted to pyruvic acid in the reactions of glycolysis, which yield two ATPs without using oxygen. These anaerobic reactions can provide enough ATP for about 30 to 40 seconds of maximal muscle activity.

4. Muscular activity that lasts longer than half a minute depends on aerobic cellular respiration, mitochondrial reactions that require oxygen to produce ATP. Aerobic cellular respiration yields about 36 molecules of ATP from each glucose molecule.

5. The inability of a muscle to contract forcefully after prolonged activity is muscle fatigue.

6. Elevated oxygen use after exercise is called recovery oxygen uptake.

Control of Muscle Tension (p. 178)

1. A twitch contraction is a brief contraction of all the muscle fibers in a motor unit in response to a single action potential.

2. A record of a contraction is called a myogram. It consists of a latent period, a contraction period, and a relaxation period.

3. Wave summation is the increased strength of a contraction that occurs when a second stimulus arrives before the muscle has completely relaxed after a previous stimulus.

4. Repeated stimuli can produce unfused tetanus, a sustained muscle contraction with partial relaxation between stimuli; more rapidly repeating stimuli will produce fused tetanus, a sustained contraction without partial relaxation between stimuli.

5. Motor unit recruitment is the process of increasing the number of active motor units.

6. On the basis of their structure and function, skeletal muscle fibers are classified as slow oxidative (SO), fast oxidative–glycolytic (FOG), and fast glycolytic (FG) fibers.

7. Most skeletal muscles contain a mixture of all three fiber types; their proportions vary with the typical action of the muscle.

8. The motor units of a muscle are recruited in the following order: first SO fibers, then FOG fibers, and finally FG fibers.

9. In an isotonic contraction, the muscle shortens to produce movement. Isometric contractions, in which tension is generated without muscle shortening, are important because they stabilize some joints as others are moved.

Exercise and Skeletal Muscle Tissue (p. 180)

1. Regular aerobic activities increase the number of capillaries in skeletal muscles and build endurance.

2. Weightlifting stimulates synthesis of muscle proteins and results in muscle hypertrophy, an increase in the diameters of muscle fibers.

3. Intense exercise causes microscopic muscle damage that is a major factor in delayed onset muscle soreness (DOMS).

4. Muscular atrophy refers to a state of wasting away of muscles that are not used.

Aging and Muscle Tissue (p. 181)

1. After birth, muscle fibers increase in size, stimulated by human growth hormone and testosterone, but the number of muscle fibers does not increase significantly.

2. Beginning at about 30 years of age, there is a progressive loss of skeletal muscle mass. As a result, muscle strength decreases and muscle reflexes slow.

3. Endurance and strength training programs can slow these age-related changes.

Cardiac Muscle Tissue (p. 181)

1. Cardiac muscle tissue, which is striated and involuntary, is found only in the heart.

2. Each cardiac muscle fiber usually contains a single centrally located nucleus and exhibits branching.

3. Cardiac muscle fibers are connected by means of intercalated discs, which hold the muscle fibers together and allow muscle action potentials to quickly spread from one cardiac muscle fiber to another.

4. Cardiac muscle tissue contracts when stimulated by its own autorhythmic fibers. Due to its continuous, rhythmic activity, cardiac muscle depends greatly on aerobic cellular respiration to generate ATP.

Smooth Muscle Tissue (p. 181)

1. Smooth muscle tissue is nonstriated and involuntary.

2. In addition to thin and thick filaments, smooth muscle fibers contain intermediate filaments and dense bodies.

3. Visceral (single-unit) smooth muscle is found in the walls of hollow viscera and of small blood vessels. Many visceral fibers form a network that contracts in unison.

4. Multiunit smooth muscle is found in large blood vessels, large airways to the lungs, arrector pili muscles, and the eye. The fibers contract independently rather than in unison.

5. The duration of contraction and relaxation is longer in smooth muscle than in skeletal muscle.

6. Smooth muscle fibers can be stretched considerably and still retain the ability to contract.

7. Smooth muscle fibers contract in response to nerve impulses, stretching, hormones, and local factors.

How Skeletal Muscles Produce Movement (p. 183)

1. Skeletal muscles produce movement by pulling on tendons attached to bones.

2. The attachment to the stationary bone is the origin. The attachment to the movable bone is the insertion.

3. The prime mover (agonist) produces the desired action. The antagonist produces an opposite action. The synergist assists the prime mover by reducing unnecessary movement. The fixator stabilizes the origin of the prime mover so that it can act more efficiently.

Naming Skeletal Muscles (p. 183)

1. The names of most skeletal muscles indicate specific characteristics.

2. The major descriptive categories are direction of fibers, location, size, number of origins (or heads), shape, origin and insertion, and action (see Table 8.2 on page 184).

Principal Skeletal Muscles (p. 187)

1. The principal skeletal muscles of the body are grouped according to region, as shown in Exhibits 8.1 through 8.13.

2. In studying muscle groups, refer to Figure 8.13 to see how each group is related to all others.

SELF-QUIZ

1. The characteristic of muscle tissue that allows it to return to its original shape after contraction is
   a. extensibility   b. excitability   c. fused tetanus   d. the latent period   e. elasticity

2. Match the connective tissue coverings with their locations:
   ____ a. wraps an entire muscle
   ____ b. lies immediately under the skin
   ____ c. separates muscle into functional groups
   ____ d. surrounds each individual muscle fiber
   ____ e. divides muscle fibers into fascicles

   A. endomysium
   B. deep fascia
   C. perimysium
   D. epimysium
   E. superficial fascia

3. Which of the following statements about skeletal muscle tissue is NOT true?
   a. Skeletal muscle requires a large blood supply.   b. Skeletal muscle fibers have many mitochondria.   c. The arrangement of thick and thin filaments produces the striations in skeletal muscle tissue.   d. Skeletal muscle fibers contain gap junctions that help conduct action potentials from one fiber to another.   e. A skeletal muscle fiber has many nuclei.

4. Match the following:
   ____ a. network of tubules that stores calcium
   ____ b. pigment used to store oxygen
   ____ c. composed of myosin
   ____ d. composed of actin, tropomyosin, and troponin
   ____ e. tunnel-like extensions of sarcolemma

   A. thick filaments
   B. transverse tubules
   C. sarcoplasmic reticulum
   D. myoglobin
   E. thin filaments

5. The sarcolemma is the equivalent of the
   a. cytoplasm   b. nucleus   c. plasma membrane   d. endoplasmic reticulum   e. mitochondria

6. You begin an intensive weightlifting plan because you want to enter a weightlifting contest. During the activity of weightlifting, your skeletal muscles will obtain energy (ATP) primarily through
   a. the conversion of lactic acid to glucose   b. the complete breakdown of pyruvic acid in the mitochondria   c. glycolysis   d. hypertrophy   e. aerobic cellular respiration

7. Which of the following events of skeletal muscle contraction does NOT occur during the latent period?
   a. Sarcomere shortens.   b. Action potentials conduct into the T tubules.   c. Concentration of calcium ions increases in the sarcoplasm.   d. Myosin-binding sites on the thin filaments are exposed.   e. Calcium release channels in the sarcoplasmic reticulum open.

8. For each of the following descriptions, indicate if it refers to skeletal muscle, cardiac muscle, or smooth muscle. Use the abbreviations SK for skeletal, CA for cardiac, and SM for smooth. The same response may be used more than once.
   ____ a. involuntary
   ____ b. multinucleated
   ____ c. striated
   ____ d. contain intercalated discs
   ____ e. elongated, cylindrical cells
   ____ f. voluntary
   ____ g. cells that taper at both ends
   ____ h. nonstriated
   ____ i. muscle fibers contract individually
   ____ j. autorhythmic

9. When ATP in the sarcoplasm is exhausted, the muscle must rely on _____ to quickly produce more ATP from ADP for contraction.
   a. acetylcholine   b. creatine phosphate   c. lactic acid   d. pyruvic acid   e. acetylcholinesterase

10. A motor unit consists of
    a. a transverse tubule and its associated sarcomeres   b. a motor neuron and all of the muscle fibers it stimulates   c. a muscle and all of its motor neurons   d. all of the filaments encased within a sarcomere   e. the motor end plate and the transverse tubules

11. Thick filaments
    a. include actin, troponin, and tropomyosin   b. compose the I band   c. stretch the entire length of a sarcomere   d. have binding sites for Ca^{2+} e. have myosin heads (crossbridges) used for the power stroke

12. The chemical that prevents the continuous stimulation of a muscle fiber is
    a. Ca^{2+} b. acetylcholinesterase   c. ATP   d. acetylcholine   e. troponin–tropomyosin

13. Which of the following is NOT associated with muscle fatigue?
    a. depletion of creatine phosphate   b. lack of oxygen   c. decrease in Ca^{2+} levels in the sarcoplasm   d. decrease in lactic acid levels   e. lack of glycogen

14. All of the following may result in an increase in muscle size EXCEPT
    a. denervation atrophy   b. weight training   c. human growth hormone   d. testosterone   e. isotonic contraction

15. Skeletal muscles are named using several characteristics. Which characteristic is NOT used to name skeletal muscles?
    a. direction of fibers   b. size   c. speed of contraction   d. location   e. shape

16. Arrange the following in the correct order for skeletal muscle fiber contraction.
    1. Sarcoplasmic reticulum releases Ca^{2+}.
    2. Ca^{2+} combines with troponin.
    3. Acetylcholine is released from the axon terminal.
    4. Action potential travels into transverse tubules.
    5. Energized myosin heads (crossbridges) attach to actin.
    6. Thin filaments slide towards the center of the sarcomere.
       a. 3, 4, 1, 2, 5, 6   b. 4, 3, 2, 1, 5, 6   c. 1, 2, 3, 4, 5, 6   d. 4, 1, 3, 5, 2, 6   e. 3, 1, 4, 5, 2, 6

17. Your instructor asks you to pick up a heavy box of books and carry them to the library in another building. Which of the following types of muscle contractions would you be utilizing?

a. hypertonic **b.** isotonic only **c.** spastic **d.** isometric only **e.** isometric and isotonic

18. Match the following:

_____ **a.** found on thick filaments

_____ **b.** contain myosin-binding site

_____ **c.** dense area that separates sarcomeres

_____ **d.** contain acetylcholine

_____ **e.** striated zone of the sarcomere composed of thick and thin filaments

_____ **f.** space between axon terminal and the sarcolemma

_____ **g.** striated zone of the sarcomere composed of thin filaments

_____ **h.** region of sarcolemma near the adjoining axon terminal

A. I band
B. synaptic vesicles
C. crossbridges
D. Z discs
E. motor end plate
F. actin molecules
G. A band
H. synaptic cleft

19. Matching the following:

_____ **a.** works with prime mover to reduce unnecessary movement

_____ **b.** muscle in a group that produces desired movement

_____ **c.** stationary end of a muscle

_____ **d.** muscle that has an action opposite to that of another muscle

_____ **e.** helps stabilize the origin of the prime mover

_____ **f.** the end of a muscle attached to the movable bone

A. insertion
B. origin
C. synergist
D. antagonist
E. prime mover
F. fixator

20. Match the following:

_____ **a.** extends and laterally rotates thigh at the hip joint

_____ **b.** adducts and medially rotates thigh at the hip joint

_____ **c.** compresses abdomen and flexes vertebral column

_____ **d.** flexes the neck

_____ **e.** flexes and abducts wrist joint

_____ **f.** extends phalanges

_____ **g.** adducts and rotates arm medially at shoulder joint

_____ **h.** extends leg at the knee and flexes thigh at hip joint

_____ **i.** plantar flexes foot at ankle joint and flexes leg at knee joint

_____ **j.** dorsiflexes and inverts foot

_____ **k.** abducts, flexes, extends, and rotates arm at shoulder joint

_____ **l.** elevates clavicle; depresses or elevates scapula

_____ **m.** elevates mandible; closes mouth

_____ **n.** draws scalp forward

_____ **o.** extends, adducts, and rotates arm medially at shoulder joint

A. trapezius
B. flexor carpi radialis
C. tibialis anterior
D. adductor longus
E. gluteus maximus
F. quadriceps group
G. rectus abdominis
H. sternocleidomastoid
I. frontalis
J. gastrocnemius
K. deltoid
L. masseter
M. extensor digitorum
N. latissimus dorsi
O. pectoralis major

CRITICAL THINKING APPLICATIONS

1. The newspaper reported several cases of botulism poisoning following a fund-raiser potluck dinner for the local clinic. The cause appeared to be three-bean salad "flavored" with the bacterium *Clostridium botulinum*. What would be the result of botulism poisoning on muscle function?

2. Latisha's nephew was squealing with laughter. She was entertaining him by sticking her thumb in her pursed lips, raising her eyebrows, pumping her arm up and down, and puffing her cheeks in and out. Name the muscles Latisha was using to maneuver her face.

3. Christopher tore some ligaments in his knee while skiing. He was in a toe-to-thigh cast for six weeks. When the cast was removed, the "healed" leg was noticeably thinner than the uncasted leg. What happened to his leg, and what should he do about it?

4. The coach of the track team has his athletes crosstraining. They ran 10 miles on Monday, then on Tuesday they lifted weights. How do these types of exercise affect the muscles?

ANSWERS TO FIGURE QUESTIONS

8.1   In order from the inside toward the outside, the connective tissue layers are endomysium, perimysium, epimysium, deep fascia, and superficial fascia.

8.2   The A band is thick filaments in its center and overlapping thick and thin filaments at each end; the I band is thin filaments.

8.3   A band: myosin, actin, troponin, and tropomyosin; I band: actin, troponin, and tropomyosin.

8.4   The I bands disappear. The lengths of the thick and thin filaments do not change.

8.5   The motor end plate is the region of the sarcolemma near the axon terminal.

8.6 Ca^{2+} binds to troponin.

8.7   Binding of ATP to the myosin heads detaches them from actin.

8.8   Creatine phosphate can transfer a phosphate group to ADP to form ATP.

8.9   Sarcomeres shorten during the contraction period.

8.10   Fused tetanus occurs when the frequency of stimulation reaches about 90 stimuli per second.

8.11   The walls of hollow organs contain visceral (single-unit) smooth muscle.

8.12   The prime mover or agonist produces the desired action.

8.13   The following are some possible responses (there are other correct answers): direction of fibers—external oblique; shape—del-toid; action—extensor digitorum; size—gluteus maximus; origin and insertion—sternocleidomastoid; location—tibialis anterior; number of origins—biceps brachii.

8.14   Frowning—frontalis; smiling—zygomaticus major; pouting—platysma; squinting—orbicularis oculi.

8.15   The strongest mastication muscle is the masseter.

8.16   The superior oblique passes through the trochlea.

8.17   The rectus abdominis aids in urination.

8.18   The diaphragm contracts during a normal quiet inhalation.

8.19   The subclavius, pectoralis major, and serratus anterior have origins on the ribs; the trapezius, levator scapulae, and rhomboideus major have origins on the vertebrae.

8.20   The pectoralis major and latissimus dorsi are termed axial muscles because they originate on the axial skeleton.

8.21   A compartment is a group of functionally related skeletal muscles in a limb, along with their blood vessels and nerves.

8.22   The median nerve is associated with the flexor retinaculum.

8.23   The iliocostalis, longissimus, and spinalis constitute the erector spinae.

8.24   Quadriceps femoris—rectus femoris, vastus lateralis, vastus medialis, and vastus intermedius; hamstrings—biceps femoris, semitendinosus, semimembranosus.

8.25   Shin splints affect the tibialis anterior.

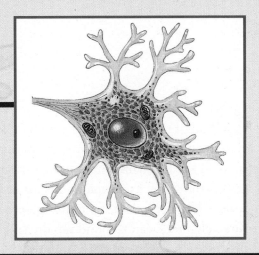

Chapter 9

Nervous Tissue

COMPONENTS OF THE NERVOUS SYSTEM

Objective: • **Describe the components of the nervous system.**

The nervous system is divided into two main subsystems: (1) the **central nervous system (CNS),** which consists of the brain and spinal cord, and (2) the **peripheral nervous system (PNS)** (pe-RIF-er-al), which includes all nervous tissue outside the central nervous system (Figure 9.1).

The nervous system has three basic functions:

- **Sensory functions.** The **sensory receptors** detect many different types of stimuli, both within your body, such as an increase in blood temperature, and outside your body, such as a touch on your arm. The neurons (nerve cells) that carry sensory information from the PNS to the CNS are called **sensory** or **afferent neurons** (AF-er-ent NOOR-onz; *af-* = toward; *-ferrent* = carried).

- **Integrative functions.** The nervous system *integrates* (processes) sensory information by *analyzing* and *storing*

*O*f all of the systems of the human body, the nervous system and the endocrine system share the greatest responsibility for maintaining homeostasis. Although their goal is the same, their operations differ in both form and speed. The nervous system, the subject of this and the next three chapters, can respond rapidly to help adjust body processes by means of nerve impulses. The endocrine system typically operates more slowly and influences the operation of specific cells throughout the body by releasing hormones into the bloodstream.

The branch of medical science that deals with the normal functioning and disorders of the nervous system is called **neurology** (noo-ROL-ō-jē; *neur-* = nerve or nervous system; *-ology* = study of).

Figure 9.1 ■ **Major structures of the nervous system.**

The central nervous system (CNS) includes the brain and spinal cord; the peripheral nervous system (PNS) includes all nervous tissue outside the CNS.

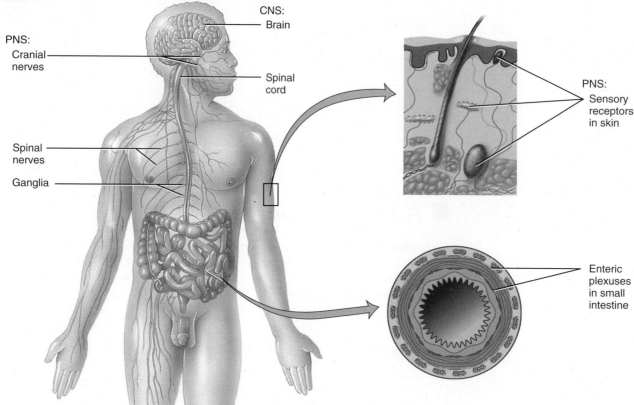

CNS:
Brain

PNS:
Cranial nerves

Spinal cord

Spinal nerves

Ganglia

PNS:
Sensory receptors in skin

Enteric plexuses in small intestine

 What is the total number of cranial and spinal nerves in your body?

218

some of it and by *making decisions* regarding appropriate responses. The neurons that carry out this function, called *interneurons,* are located entirely within the CNS and make up the vast majority of neurons in the body.

• *Motor functions.* The nervous system's motor functions involve *responding* to the integrative decisions of the interneurons. The neurons that serve this function, ***motor*** or ***efferent neurons*** (EF-er-ent; *ef-* = from), carry information from the CNS to the PNS. The cells and organs innervated by motor neurons are termed ***effectors.*** Examples of effectors are muscle fibers and cells in glands, for example, sweat glands.

The ***brain*** is housed within the skull and contains about 100 billion neurons. The ***spinal cord,*** which contains about 100 million neurons, is connected directly to the brain and is protected by the bones of the vertebral column. The CNS receives and correlates many different kinds of incoming sensory information and is the source of thoughts, emotions, and memories. It also is the source of most nerve impulses that stimulate muscles to contract and glands to secrete.

Communication to and from the CNS is by means of two components of the PNS: cranial nerves and spinal nerves. Twelve pairs of ***cranial nerves*** emerge from the base of the brain, and thirty-one pairs of ***spinal nerves*** emerge from the spinal cord (Figure 9.1). A ***nerve*** is a bundle containing hundreds to thousands of axons plus associated connective tissue and blood vessels. Each nerve follows a defined path and branches out to serve a specific region on the right or left side of the body. Most cranial and all spinal nerves are ***mixed nerves,*** which means that they contain axons of both sensory and motor

neurons. Signals traveling along cranial or spinal nerves may be relayed through ***ganglia,*** small clusters of neuronal cell bodies that are located outside the brain and spinal cord. In the walls of organs of the gastrointestinal tract are extensive networks of neurons, called ***enteric plexuses,*** that help regulate the digestive system.

The PNS may be subdivided further into a ***somatic nervous system (SNS)*** (*somat-* = body), an ***autonomic nervous system (ANS)*** (*auto-* = self; *-nomic* = law), and an ***enteric nervous system (ENS)*** (*enter-* = intestines) (Figure 9.2). The SNS consists of (1) sensory neurons that convey information from somatic receptors in the head, body wall, and limbs and from receptors for the special senses of vision, hearing, taste, and smell to the CNS and (2) motor neurons from the CNS, that conduct impulses to *skeletal muscles* only. Because these motor responses can be consciously controlled, the action of this part of the PNS is *voluntary.*

The ANS (the focus of Chapter 11) consists of (1) sensory neurons that convey information from autonomic sensory receptors, located primarily in visceral organs such as the stomach and lungs, to the CNS, and (2) motor neurons that conduct nerve impulses to *smooth muscle, cardiac muscle,* and *glands.* Because its motor responses are not normally under conscious control, the action of the ANS is *involuntary.* The motor portion of the ANS consists of two branches, the ***sympathetic division*** and the ***parasympathetic division.*** With a few exceptions, effectors are innervated by both divisions, and usually the two divisions have opposing actions. For example, sympathetic neurons speed the heartbeat, whereas parasympathetic neurons slow it down.

The ENS is the "brain of the gut" and its operation is involuntary. It consists of approximately 100 million neurons in

Figure 9.2 ■ **Organization of the nervous system.** Subdivisions of the PNS are the somatic nervous system (SNS), the autonomic nervous system (ANS), and the enteric nervous system (ENS).

The SNS is under voluntary control; the ANS and ENS are involuntary.

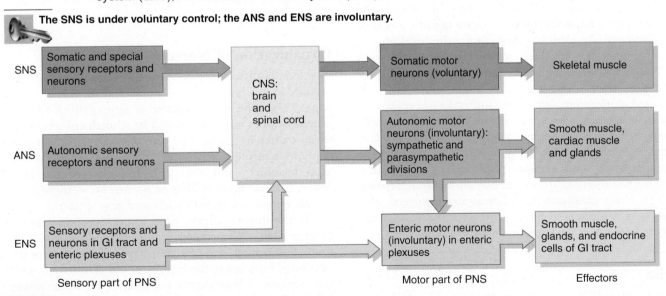

What terms are given to neurons that carry input to the CNS? that carry output from the CNS?

enteric plexuses that extend the entire length of the gastrointestinal (GI) tract. Sensory neurons of the ENS monitor chemical changes within the GI tract and stretching of its walls. Enteric motor neurons govern contraction of GI tract smooth muscle, secretions of the GI tract organs such as acid secretion by the stomach, and activity of GI tract endocrine cells.

CELLULAR ORGANIZATION OF NERVOUS TISSUE

Objective: • **Compare the structure and functions of neurons and neuroglia.**

The nervous system consists of two types of cells. *Neurons*, the basic information-processing units of the nervous system, are specialized for nerve impulse conduction. The function of *neuroglia* is the support and protection of the neurons.

Neurons

Neurons (nerve cells) usually have three parts: (1) a cell body, (2) dendrites, and (3) an axon (Figure 9.3). The **cell body** contains a nucleus surrounded by cytoplasm that includes typical organelles such as lysosomes, mitochondria, and a Golgi complex. The cell body also contains prominent clusters of rough endoplasmic reticulum, termed *Nissl bodies*. Newly synthesized proteins produced by the Nissl bodies are used to replace cellular components and for growth of neurons and regeneration of damaged axons in the PNS.

Two kinds of processes or extensions emerge from the cell body of a neuron: multiple dendrites and a single axon. The cell body and the **dendrites** (= little trees) are the receiving or input portion of a neuron. Usually, dendrites are short, tapering, and highly branched, forming a tree-shaped array of processes that emerge from the cell body. The second type of process, the **axon**, conducts nerve impulses toward another neuron, a muscle fiber, or a gland cell. An axon is a long, thin, cylindrical projection that often joins the cell body at a cone-shaped elevation called the **axon hillock** (= small hill). Impulses arise at the axon hillock and then travel along the axon. Some axons have side branches called **axon collaterals**. The axon and axon collaterals end by dividing into many fine processes called **axon terminals**.

The sites where two neurons or a neuron and an effector cell can communicate are termed **synapses**. The tips of most axon terminals swell into **synaptic end bulbs**. At chemical synapses, these bulb-shaped structures contain **synaptic vesicles**, tiny sacs that store chemicals called **neurotransmitters**. The neurotransmitter molecules released from synaptic vesicles are the means of communication from one neuron to another.

Neuroglia

The **neuroglia** (noo-RŌG-lē-a; *glia* = glue) or *glia*, which support and protect neurons, generally are smaller than neurons and are 5 to 50 times more numerous. The five main types of neuroglia are astrocytes, oligodendrocytes, microglia, and ependymal cells, found in the CNS, and Schwann cells, located in the PNS. Table 9.1 on page 222 shows the appearance of neuroglia and lists their functions.

Formation of Myelin

The axons of the majority of our neurons are surrounded by a **myelin sheath**, a many-layered covering composed of lipid and protein that is produced by neuroglia. Like insulation covering an electrical wire, the sheath insulates the axon of a neuron and increases the speed of nerve impulse conduction. Axons with such a covering are said to be **myelinated**, whereas those without it are said to be **unmyelinated**.

Two types of neuroglia produce myelin sheaths (see Table 9.1): Schwann cells in the PNS and oligodendrocytes in the CNS. Each **Schwann cell** wraps around a small segment of an axon's length many times. Eventually, multiple layers of Schwann cell plasma membrane surround the axon, with the Schwann cell's cytoplasm and nucleus located in the outermost layer (Figure 9.3b). The inner portion, consisting of up to 100 layers of Schwann cell membrane, is the myelin sheath. The outermost layer of the Schwann cell is called the **neurolemma**. Between adjacent Schwann cells are nodes of Ranvier (RON-vē-ā), small gaps in the myelin sheath.

Oligodendrocytes myelinate CNS axons in somewhat the same manner that Schwann cells myelinate PNS axons (see Table 9.1). Each oligodendrocyte produces myelin sheaths around many adjacent axons without forming a neurolemma.

Myelin sheaths begin to form around axons during the fourth month of fetal development. By the time a baby starts to talk, most myelin sheaths are partially formed, but myelination continues into the teenage years. Because some axons of immature motor neurons lack their myelin sheath, an infant's movements are not as rapid or coordinated as those of an older child or adult. Certain diseases such as multiple sclerosis (see page 229) destroy myelin sheaths and dramatically disrupt nervous system functions.

Organization of Nervous Tissue

In a freshly dissected section of the brain or spinal cord, some regions look white, whereas others appear gray (Figure 9.4 on page 223). The **white matter** is aggregations of myelinated processes from many neurons. The whitish color of myelin gives white matter its name. The **gray matter** of the nervous system contains neuronal cell bodies, dendrites, unmyelinated axons, axon terminals, and neuroglia. It looks grayish, rather than white, because there is little or no myelin in these areas. Blood vessels are present in both white and gray matter.

In the spinal cord, the white matter surrounds an inner core of gray matter shaped like a butterfly or the letter H. In the brain, a thin shell of gray matter covers the surface of the largest portions of the brain, the cerebrum and cerebellum. Several **nuclei**, regions of gray matter surrounded by white matter in the CNS, also lie deep within the brain. A **tract** is a bundle of axons located in the CNS that extends for some distance up or down the spinal cord or connects parts of the brain with each other and with the spinal cord. Spinal cord tracts that carry sensory

Figure 9.3 ■ **Structure of a typical neuron.** Arrows indicate the direction of information flow: from dendrites to the cell body to the axon to the axon terminals. The break in the figure indicates that the axon actually is longer than shown.

🔑 **The basic parts of a neuron are dendrites, a cell body, and an axon.**

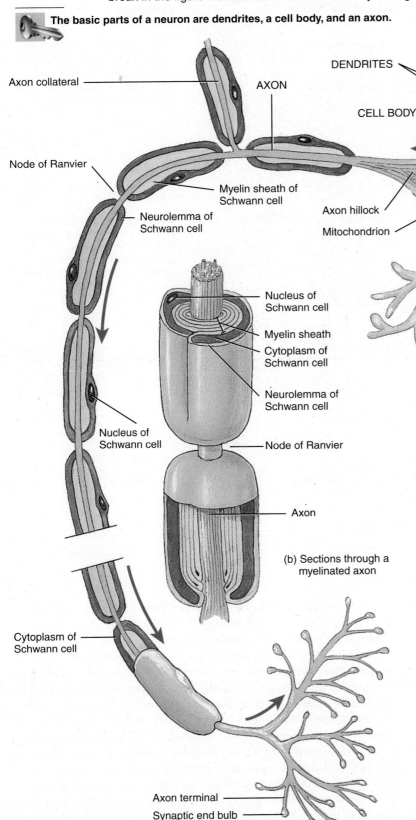

(a) Parts of a motor neuron

(b) Sections through a myelinated axon

🧩 **What roles do the axon and axon terminals play in neural communication?**

impulses toward the brain are called *ascending tracts*. Spinal cord tracts that carry motor impulses from the brain are called *descending tracts*.

Regeneration and Repair of Nervous Tissue

From birth to death, your nervous system exhibits ***plasticity,*** the capability to change based on experience. It is this capacity of the nervous system that allows you to learn. In the *hippocampus,* an area of the brain that is crucial for learning, significant numbers of new neurons arise, even in adults. At the level of individual neurons, the changes that can occur include the sprouting of new dendrites and synthesis of new proteins. Learning also depends on changes that occur at synaptic contacts between neurons. For example, your performance on exams will be determined by how much some of your synapses have been modified!

Despite plasticity, mammalian neurons have very limited powers of ***regeneration,*** the capability to replicate or repair themselves. In the PNS, axons and dendrites may undergo repair if the cell body is intact and if the Schwann cells are functional. The Schwann cells on either side of an injured site multiply by mitosis, grow toward each other, and may form a ***regeneration tube*** across the injured area. The tube guides growth of new processes from the proximal area across the injured area into the distal area previously occupied by the original axon. New axons cannot grow if the gap at the site of injury is too large or if the gap becomes filled with scar tissue.

Table 9.1 / Neuroglia in the CNS and PNS

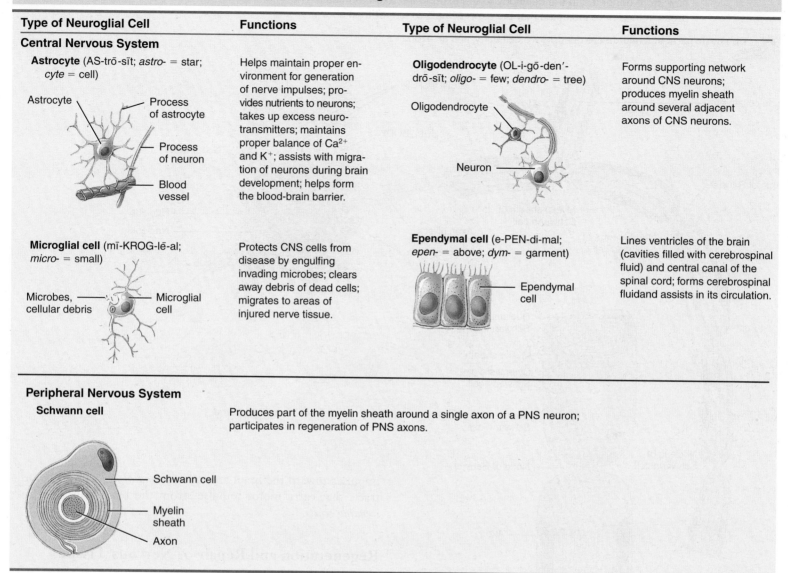

Type of Neuroglial Cell	Functions	Type of Neuroglial Cell	Functions
Central Nervous System			
Astrocyte (AS-trō-sīt; *astro-* = star; *cyte* = cell)	Helps maintain proper environment for generation of nerve impulses; provides nutrients to neurons; takes up excess neurotransmitters; maintains proper balance of Ca^{2+} and K^+; assists with migration of neurons during brain development; helps form the blood-brain barrier.	**Oligodendrocyte** (OL-i-gō-den′-drō-sīt; *oligo-* = few; *dendro-* = tree)	Forms supporting network around CNS neurons; produces myelin sheath around several adjacent axons of CNS neurons.
Microglial cell (mī-KROG-lē-al; *micro-* = small)	Protects CNS cells from disease by engulfing invading microbes; clears away debris of dead cells; migrates to areas of injured nerve tissue.	**Ependymal cell** (e-PEN-di-mal; *epen-* = above; *dym-* = garment)	Lines ventricles of the brain (cavities filled with cerebrospinal fluid) and central canal of the spinal cord; forms cerebrospinal fluidand assists in its circulation.
Peripheral Nervous System			
Schwann cell	Produces part of the myelin sheath around a single axon of a PNS neuron; participates in regeneration of PNS axons.		

In the CNS, even when the cell body remains intact, a cut axon is usually not repaired. Recall that, unlike the Schwann cells of the PNS, oligodendrocytes do not form neurolemmas. Interestingly, the presence of CNS myelin is one of the factors that actively inhibits regeneration of neurons, perhaps by means of the same mechanism that stops axonal growth once a target region has been reached during fetal development.

ACTION POTENTIALS

Objective: • **Describe how a nerve impulse is generated and conducted.**

Neurons communicate with one another by means of nerve action potentials, also called nerve impulses. Generation of action potentials depends on two basic features of the plasma membrane: the presence of special types of ion channels and the exis-tence of a resting membrane potential. Most body cells exhibit a ***membrane potential,*** a difference in electrical charge across the plasma membrane that is similar to voltage stored in a battery. A cell that has a membrane potential is said to be ***polarized.*** When muscle fibers and neurons are "at rest" (not conducting action potentials), the voltage across the plasma membrane is termed the ***resting membrane potential.*** If you connect the positive and negative terminals of a battery with a piece of metal (look in the battery compartment of your Walkman portable radio), an *electrical current* carried by electrons flows from the battery, allowing you to listen to your favorite music. In living tissues, the flow of *ions* (rather than electrons) constitutes the electrical current. Because the lipid bilayer of the plasma membrane is a good electrical insulator, the main paths for electrical current to flow across the membrane are through ion channels. Action potentials can occur in both muscle fibers and neurons because their plasma membranes contain two kinds of ion channels that open and close in response to changes in the membrane potential.

Figure 9.4 ■ Distribution of gray and white matter in the spinal cord and brain.

 White matter consists of myelinated processes of many neurons. Gray matter consists of neuron cell bodies, dendrites, axon terminals, bundles of unmyelinated axons, and neuroglia.

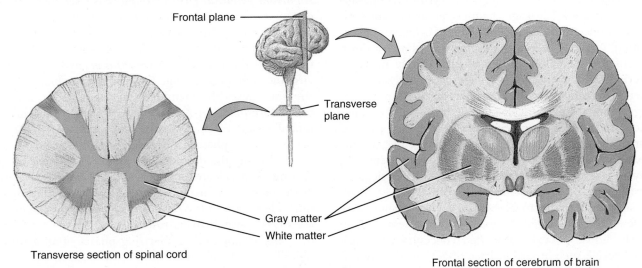

Frontal plane

Transverse plane

Gray matter

White matter

Transverse section of spinal cord

Frontal section of cerebrum of brain

What substance is responsible for the white appearance of white matter?

Ion Channels

When they are open, ion channels allow specific ions to diffuse across the plasma membrane from where the ions are more concentrated to where they are less concentrated. Similarly, positively charged ions will move toward a negatively charged area, and negatively charged ions will move toward a positively charged area. As ions diffuse across a plasma membrane to equalize differences in charge or concentration, the result is a flow of current that can change the membrane potential.

The two basic types of ion channels are leakage channels and gated channels. Leakage channels are always open, like a perforated garden hose that allows the constant flow of a small amount of water. Because plasma membranes typically have many more potassium ion (K^+) leakage channels than sodium ion (Na^+) leakage channels, the membrane's permeability to K^+ is much higher than its permeability to Na^+. Gated channels, in contrast, open and close on command (see Figure 3.5 on page 95). *Voltage-gated channels*—channels that open in response to a change in membrane potential—are used to generate and conduct action potentials.

Resting Membrane Potential

In a resting neuron, the outside surface of the plasma membrane has a positive charge and the inside surface has a negative charge. The separation of positive and negative electrical charges is a form of potential energy, which can be measured in volts. For example, two 1.5 volt batteries power an ordinary Walkman. Voltages produced by cells typically are much smaller and are measured in millivolts (1 millivolt = 1 mV = 1/1000 volt). In neurons, the resting membrane potential is about −70 mV. The minus sign indicates that the inside is negative relative to the outside.

The resting membrane potential arises from the unequal distributions of charges in cytosol and extracellular fluid (Figure 9.5). Extracellular fluid is rich in sodium ions (Na^+) and chloride ions (Cl^-). Inside cells, the main positively charged ions are

Figure 9.5 ■ The distribution of ions that produces the resting membrane potential.

The resting membrane potential is due to a small buildup of negatively charged ions, mainly organic phosphates (PO_4^{3-}) and proteins, in the cytosol just inside the membrane and an equal buildup of positively charged ions, mainly sodium ions (Na^+), in the extracellular fluid just outside the membrane.

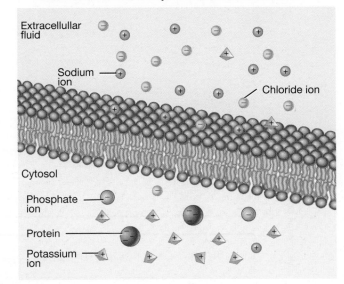

Extracellullar fluid

Sodium ion

Chloride ion

Cytosol

Phosphate ion

Protein

Potassium ion

 What is a typical value for the resting membrane potential of a neuron?

potassium ions (K⁺), and the two dominant negatively charged ions are organic phosphates, such as the three phosphates in ATP (adenosine triphosphate), and amino acids in proteins. Because the concentration of K⁺ is higher in cytosol and plasma membranes have many K⁺ leakage channels, potassium ions diffuse down their concentration gradient—out of cells into the extracellular fluid. As more and more positive potassium ions exit, the interior of the membrane becomes increasingly negative, and the exterior of the membrane becomes increasingly positive. Another factor contributes to the negativity inside: most negatively charged ions inside the cell are not free to leave. They cannot follow the K⁺ out of the cell because they are attached either to large proteins or to other large molecules. Because the negative charges inside attract K⁺ back into the cell, eventually, just as many K⁺ are entering the cell due to the interior negativity as are exiting the cell because of the concentration gradient.

Membrane permeability to Na⁺ is very low because there are only a few sodium leakage channels. Nevertheless, sodium ions do slowly diffuse inward, down their concentration gradient. Left unchecked, such inward leakage of Na⁺ would eventually destroy the resting membrane potential. The Na⁺/K⁺ pumps help maintain the resting membrane potential by pumping Na⁺ out of the cell as fast as it leaks in while at the same time pumping K⁺ into the cell.

Generation of Action Potentials

An *action potential (AP)* or *impulse* is a sequence of rapidly occurring events that decrease and reverse the membrane potential and then eventually restore it to the resting state. The ability of muscle fibers and neurons to convert stimuli into action potentials is called *excitability*. A *stimulus* is anything in the cell's environment that can change the resting membrane potential. If a stimulus causes the membrane to depolarize to a critical level, called *threshold* (typically, about −55 mV), then an action potential arises (Figure 9.6). An action potential (AP) or impulse has two main phases: a depolarizing phase and a repolarizing phase. During the *depolarizing phase,* a sequence of rapidly occurring events decrease and eventually reverse the polarization of the membrane, making the inside more *positive* than the outside. Then, during the *repolarizing phase,* the membrane polarization is restored to its resting state (Figure 9.6). In neurons, the depolarizing and repolarizing phases of an action potential typically last about 1 millisecond (1/1000 sec).

During an action potential, depolarization to threshold briefly opens two types of voltage-gated ion channels. In neurons, these channels are present mainly in the plasma membrane of the axon and axon terminals. First, a threshold depolarization

Figure 9.6 ■ **Action potential (AP).** When a stimulus depolarizes the membrane to threshold, an action potential is generated. The green outline in the inset indicates the presence of voltage-gated Na⁺ and K⁺ channels in the membrane of the axon and axon terminals.

An action potential consists of depolarizing and repolarizing phases.

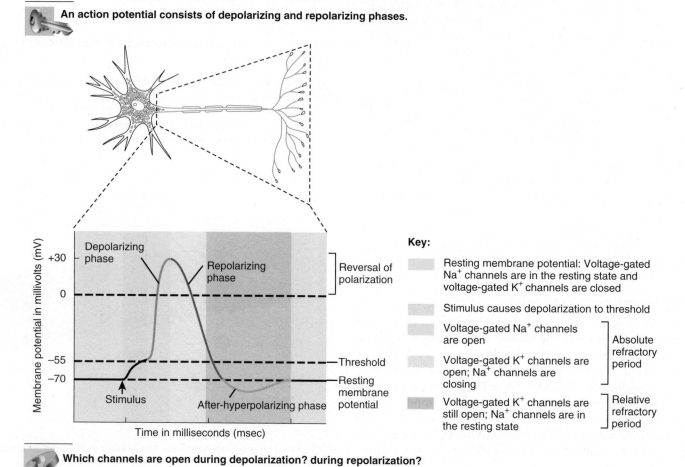

Which channels are open during depolarization? during repolarization?

opens voltage-gated Na^+ channels. As these channels open, about 20,000 sodium ions rush into the cell, causing the depolarizing phase. The inflow of Na^+ causes the membrane potential to pass 0 mV and finally reach +30 mV (Figure 9.6). Second, the threshold depolarization also opens voltage-gated K^+ channels. The voltage-gated K^+ channels open more slowly, so their opening occurs at about the same time the voltage-gated Na^+ channels are automatically closing. As the K^+ channels open, potassium ions flow out of the cell, producing the repolarizing phase.

While the voltage-gated K^+ channels are open, outflow of K^+ may be large enough to cause an *after-hyperpolarizing phase* of the action potential (Figure 9.6). During hyperpolarization, the membrane potential becomes even *more negative* than the resting level. Finally, as K^+ channels close, the membrane potential returns to the resting level of −70 mV.

Local anesthetics, such as procaine (Novocain) and lidocaine, act by blocking the opening of voltage-gated Na^+ channels. Nerve impulses cannot pass the obstructed region, so pain signals do not reach the CNS.

Action potentials arise according to the *all-or-none principle.* As long as a stimulus is strong enough to cause depolarization to threshold, the voltage-gated Na^+ and K^+ channels open, and an action potential occurs. A much stronger stimulus cannot cause a larger action potential because the size of an action potential is always the same. A weak stimulus that fails to cause a threshold-level depolarization does not elicit an action potential.

Refractory Period

The period of time during and after an action potential when a muscle fiber or neuron cannot generate another action potential is called the *refractory period* (see key in Figure 9.6). During the *absolute refractory period,* a second action potential *absolutely* cannot be initiated, even with a very strong stimulus. This period coincides with the time interval when Na^+ channels are either open or are closing and returning to their resting state.

The *relative refractory period* is the period of time during which a second action potential can be initiated, but only by a suprathreshold (larger than threshold) stimulus. It coincides with the period when Na^+ channels have returned to their resting state, but the voltage-gated K^+ channels are still open (Figure 9.6).

Conduction of Nerve Impulses

To communicate information from one part of the body to another, nerve impulses must travel from where they arise (trigger zone), usually the axon hillock, along the axon to the axon terminals. This special mode of impulse travel, which operates by positive feedback, is called *conduction* or *propagation.* Depolarization to threshold at the trigger zone opens voltage-gated Na^+ channels. The resulting inflow of sodium ions depolarizes the adjacent membrane to threshold, opening more voltage-gated Na^+ channels. Thus, depolarization in one segment of an axon causes further depolarization in the adjacent segment, a positive feedback effect. This situation is similar to pushing on the first domino in a long row: When the push on the first domino is strong enough, that domino falls against the second domino, and the entire row topples.

The type of action potential conduction that occurs in unmyelinated axons (and muscle fibers) is called *continuous conduction,* in which each adjacent segment of the plasma membrane depolarizes to threshold and generates an action potential that depolarizes the next patch of the membrane (Figure 9.7a). Note that the impulse has traveled only a relatively short distance after 10 milliseconds.

In myelinated axons, conduction is somewhat different. The voltage-gated Na^+ and K^+ channels are located primarily at the nodes of Ranvier, the gaps in the myelin sheath. When a nerve impulse conducts along a myelinated axon, current carried by Na^+ and K^+ flows through the extracellular fluid surrounding the myelin sheath and through the cytosol from one node to the next (Figure 9.7b). The nerve impulse at the first node generates ionic currents that open voltage-gated Na^+ channels at the second node and trigger a nerve impulse there. Then the nerve impulse from the second node generates an ionic current that opens voltage-gated Na^+ channels at the third node, and so on. Each node depolarizes and then repolarizes. Note the impulse has traveled much farther along the myelinated axon in Figure 9.7b in the same time period. Because current flows across the membrane only at the nodes, the impulse appears to leap from node to node as each nodal area depolarizes to threshold. This type of impulse conduction is called *saltatory conduction* (SAL-ta-tō-rē; *saltat-* = leaping).

The diameter of the axon and the presence or absence of a myelin sheath are the most important factors that determine the speed of nerve impulse conduction. Axons with large diameters conduct impulses faster than those with small diameters. Also, myelinated axons conduct impulses faster than do unmyelinated axons. Axons with the largest diameters are all myelinated and therefore capable of saltatory conduction. The smallest diameter axons are unmyelinated, so their conduction is continuous. Because axons conduct impulses at higher speeds when warmed and at lower speeds when cooled, localized cooling of a nerve can retard impulse conduction. Pain resulting from tissue injury such as that caused by a minor burn can be reduced by the application of ice because cooling slows conduction of nerve impulses along the axons of pain-sensitive neurons.

SYNAPTIC TRANSMISSION

Objective: • **Explain the events of synaptic transmission and the types of neurotransmitters used.**

Now that you know how action potentials arise and conduct along the axon of an individual neuron, we will explore how these signals are transmitted from one neuron to another. At synapses neurons communicate with other neurons or with effectors by a series of events known as *synaptic transmission.* In Chapter 8 we examined the events occurring at one type of synapse, the neuromuscular junction, the synapse between a somatic motor neuron and a skeletal muscle fiber. At a synapse between neurons, the neuron sending the signal is called the *presynaptic neuron* (*pre-* = before), and the neuron receiving the

Figure 9.7 ■ **Conduction of a nerve impulse after it arises at the trigger zone.** Dotted lines indicate ionic current flow. (a) In continuous conduction along an unmyelinated axon, ionic currents flow across each adjacent portion of the plasma membrane. (b) In saltatory conduction along a myelinated axon, the nerve impulse at the first node generates ionic currents in the cytosol and extracellular fluid that open voltage-gated Na⁺ channels at the second node, and so on at each subsequent node.

Unmyelinated axons exhibit continuous conduction, whereas myelinated axons exhibit saltatory conduction.

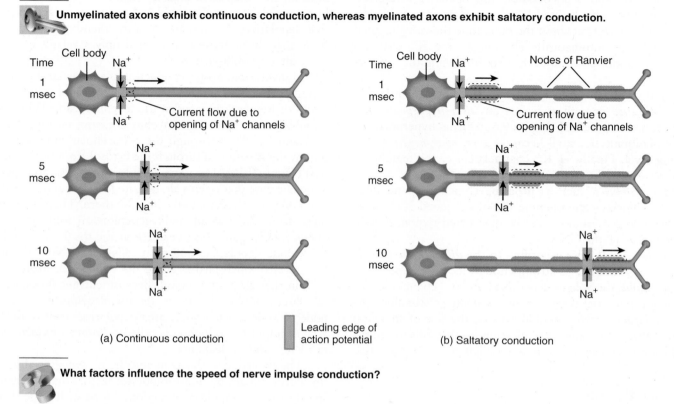

(a) Continuous conduction

Leading edge of action potential

(b) Saltatory conduction

What factors influence the speed of nerve impulse conduction?

message is called the ***postsynaptic neuron*** (*post-* = after). There are two types of synapses: electrical synapses and chemical synapses.

Electrical Synapses

At an ***electrical synapse,*** action potentials conduct directly from the presynaptic cell to the postsynaptic cell through ***gap junctions.*** Each gap junction contains a hundred or so tubules that connect the plasma membranes of the two cells. The tubules allow ions to flow from one cell to the next, providing a path for the flow of electrical current. Gap junctions are common in visceral smooth muscle, cardiac muscle, and the developing embryo. They also occur in the CNS.

Electrical synapses have two main advantages:

- They allow faster communication than chemical synapses because action potentials conduct directly through the gap junctions.

- They can synchronize the activity of a group of neurons or muscle fibers. In other words, a large number of neurons or muscle fibers can produce action potentials in unison if they are connected by gap junctions. The value of synchronized action potentials in the heart or in visceral smooth muscle is coordinated contraction of these muscle fibers.

Chemical Synapses

Although the presynaptic and postsynaptic neurons of a ***chemical synapse*** are in close proximity, their plasma membranes do not touch. They are separated by the ***synaptic cleft,*** a tiny space filled with extracellular fluid. Because nerve impulses cannot conduct across the synaptic cleft, an alternate, indirect form of communication occurs across this space. In response to a nerve impulse, the presynaptic neuron releases a neurotransmitter that diffuses across the synaptic cleft and binds to receptors in the plasma membrane of the postsynaptic neuron. The postsynaptic neuron receives the chemical signal and, in turn, generates an electrical signal. The time required for these events, a ***synaptic delay*** of about 0.5 milliseconds, is why chemical synapses relay signals more slowly than electrical synapses.

A typical chemical synapse operates as follows (Figure 9.8):

1 A nerve impulse arrives at a synaptic end bulb of a presynaptic axon.

2 The depolarizing phase of the nerve impulse opens ***voltage-gated Ca²⁺ channels,*** allowing calcium ions (Ca²⁺) to flow into the synaptic end bulb through the opened channels.

3 An increase in the level of Ca²⁺ inside the presynaptic neuron triggers exocytosis of some of the synaptic vesicles, releasing neurotransmitter molecules into the synaptic cleft.

Figure 9.8 ■ **Synaptic transmission at a chemical synapse.** Exocytosis of synaptic vesicles from a presynaptic neuron releases neurotransmitter molecules, which bind to receptors in the plasma membrane of the postsynaptic neuron.

At a chemical synapse, a presynaptic electrical signal (nerve impulse) is converted into a chemical signal (neurotransmitter release). The chemical signal is then converted back into an electrical signal (depolarization or hyperpolarization) in the postsynaptic cell.

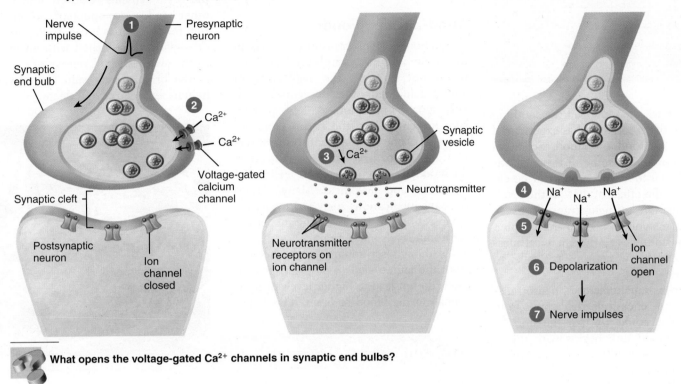

What opens the voltage-gated Ca^{2+} channels in synaptic end bulbs?

Each synaptic vesicle contains several thousand neurotransmitter molecules.

❹ The neurotransmitter molecules diffuse across the synaptic cleft and bind to ***neurotransmitter receptors*** on ion channels in the postsynaptic neuron's plasma membrane.

❺ Binding of neurotransmitter molecules to their receptors opens the ion channels and allows particular ions to flow across the membrane.

❻ Depending on which ions the channels admit, the ionic flow causes depolarization or hyperpolarization of the postsynaptic membrane. For example, opening of Na^+ channels allows inflow of Na^+, which causes depolarization. However, opening of Cl^- channels allows inflow of Cl^-, which causes hyperpolarization.

❼ If a depolarization in the postsynaptic neuron reaches threshold, one or more nerve impulses are generated.

At most chemical synapses, only *one-way information transfer* can occur, from a presynaptic neuron to a postsynaptic neuron or an effector, such as a muscle fiber or a gland cell. Only synaptic end bulbs of presynaptic neurons can release neurotransmitter, and only the postsynaptic neuron's membrane has the correct receptor proteins to recognize and bind that neurotransmitter.

The effect on the postsynaptic neuron of depolarization is excitatory; one or more nerve impulses are likely to occur. By contrast, the effect on the postsynaptic neuron of hyperpolarization is inhibitory; because hyperpolarization moves the membrane potential further from threshold, nerve impulses are less likely to arise. A typical neuron in the CNS receives input from 1000 to 10,000 synapses. Some of this input is excitatory and some is inhibitory. Integration of these inputs is known as ***summation***. The sum of all the excitatory and inhibitory effects at any given time determines the effect on the postsynaptic neuron, which may respond in the following ways:

- One or more nerve impulses are generated when the total excitatory effects are greater than the total inhibitory effects and the threshold level of stimulation is reached or surpassed.

- A nerve impulse does not arise when the total inhibitory effects are greater than the excitatory effects.

- If the total excitatory effects are greater than the total inhibitory effects but less than the threshold level of stimulation, the postsynaptic neuron is slightly depolarized. Subsequent excitatory stimuli, which might not be strong enough to reach threshold on their own, can more easily generate a nerve impulse.

*E*veryone who has enjoyed the soothing relaxation of a good meal has experienced the effect of food on mood. Scientists have proposed that some of the relationship between food and mood can be explained by the effect of diet on the levels and regulation of neurotransmitters in the brain. As you learned in this chapter, neurotransmitters are the chemical messengers that allow neurons to "talk" to each other.

Neurons manufacture neurotransmitters from chemicals that come from food, so you could say that the story of the food–mood link begins with digestion. Many neurotransmitters are made from amino acids, which are the basic building blocks of proteins. Amino acids are made available when your body digests the protein in the food you eat. For example, the neurotransmitter serotonin is made from the amino acid tryptophan, and both dopamine and norepinephrine are synthesized from the amino acid tyrosine.

Mind-altering Food?

Regulation of neurotransmitter levels in the brain is quite complicated and depends not only on the availability of amino acid (and other) precursors, but also upon competition of these precursors for entry into the brain. Consider serotonin, one of the neurotransmitters that appears to have an important effect on mood. Serotonin leads to feelings of relaxation and sleepiness. Many antidepressant medications such as Prozac, relieve feelings of depression by increasing serotonin levels in the brain.

Although serotonin is manufactured from the amino acid tryptophan, protein foods do not lead to higher levels of tryptophan in the blood or brain. This is because after a high-protein meal, tryptophan must compete with more than 20 other amino acids for entry into the central nervous system, so its concentration in the brain remains relatively low. On the other hand, consumption of carbohydrate-rich foods, such as bread, pasta, potatoes, or sweets, is associated with an increase in the synthesis and release of serotonin in the brain. The result: Carbohydrates help us feel relaxed and sleepy.

► *Think It Over*

► Why might consuming a high-protein diet for several days lead to cravings for carbohydrate-rich foods? Why is depression often associated with weight gain?

A neurotransmitter affects the postsynaptic neuron, muscle fiber, or gland cell as long as it remains in the synaptic cleft. Thus, removal of the neurotransmitter is essential for normal synaptic function. Neurotransmitter is removed from the synaptic cleft in three ways. (1) Some neurotransmitter molecules diffuse out of the synaptic cleft by moving down their concentration gradient. (2) Some neurotransmitters are destroyed by enzymes in the synaptic cleft. (3) Many neurotransmitters are actively transported back into the neuron that released them (reuptake) or are transported into neighboring neuroglia. Several therapeutically important drugs selectively block reuptake of specific neurotransmitters. For example, fluoxetine (Prozac), which is used to treat some forms of depression, is a *selective serotonin reuptake inhibitor (SSRI)*. Thus, it prolongs the synaptic activity of the neurotransmitter serotonin by interfering with its reuptake by the presynaptic neuron.

Neurotransmitters

The best-studied neurotransmitter is *acetylcholine (ACh)*, which is released by many PNS neurons and by some CNS neurons. ACh is an excitatory neurotransmitter at some synapses, such as the neuromuscular junction. It is also known to be an inhibitory neurotransmitter at other synapses; one example is the parasympathetic neurons of the vagus nerve (cranial nerve X) that innervate the heart. ACh slows heart rate at these inhibitory synapses.

Several amino acids serve as neurotransmitters in the CNS. The amino acid *glutamate* has powerful excitatory effects. Nearly all excitatory neurons in the CNS, perhaps half of the synapses in the brain, communicate using glutamate. A high level of glutamate in the interstitial fluid of the CNS causes *excitotoxicity*, that is, *destruction* of neurons through prolonged activation of excitatory synaptic transmission. The most common cause of excitotoxicity is oxygen deprivation of the brain, as happens during a stroke. Two other amino acids, *gamma aminobutyric acid (GABA)* and *glycine,* are important inhibitory neurotransmitters. Both cause hyperpolarization by opening Cl⁻ channels. As many as one-third of all brain synapses use GABA. In the spinal cord, about half of the inhibitory synapses use the amino acid glycine and half use GABA.

Some neurotransmitters are modified amino acids. These include norepinephrine, epinephrine, dopamine, and serotonin. *Norepinephrine (NE)* participates in awakening from deep sleep, dreaming, and regulating mood. A smaller number of neurons in the brain use *epinephrine* as a neurotransmitter. Both epinephrine and norepinephrine also serve as hormones. They are released by the adrenal medulla, the inner portion of the adrenal gland. Brain neurons containing the neurotransmitter *dopamine (DA)* are involved in regulating skeletal muscle tone and some aspects of movement due to contraction of skeletal muscles. Degeneration of dopamine-containing axons occurs in Parkinson's disease (see page 257). *Serotonin* is thought to be involved in sensory perception, temperature regulation, control of mood, and the onset of sleep.

Neuropeptides are neurotransmitters consisting of 3 to 40 amino acids linked by peptide bonds. Some neuropeptides, the *endorphins* (en-DOR-fins), have potent analgesic (pain-relieving) effects. It is thought that endorphins are the body's natural painkillers. Acupuncture may produce analgesia (loss of pain sensation) by increasing the release of endorphins. They have also been linked to improved memory and learning and to feelings of pleasure or euphoria.

An important newcomer to the ranks of recognized neurotransmitters is the simple gas *nitric oxide (NO),* which has widespread effects throughout the body. NO is different from all previously known neurotransmitters because it is not synthesized in advance and packaged into synaptic vesicles. Rather, it is formed on demand, diffuses out of cells that produce it and into neighboring cells, and acts immediately. Some research suggests that NO plays a role in learning and memory.

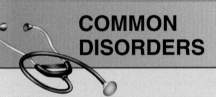

COMMON DISORDERS

Multiple Sclerosis

Multiple sclerosis (MS) is a progressive destruction of myelin sheaths of neurons in the CNS that afflicts about half a million people in the United States and 2 million people worldwide. It usually appears between the ages of 20 and 40 years, affecting females twice as often as males. MS is an autoimmune disease—the body's own immune system spearheads the attack. Myelin sheaths deteriorate to *scleroses,* which are hardened scars or plaques, in *multiple* regions. The destruction of myelin sheaths slows and then short-circuits conduction of nerve impulses.

The most common form of multiple sclerosis is called relapsing-remitting MS. Usually, the first symptoms, including a feeling of heaviness or weakness in the muscles, abnormal sensations, or double vision, occur in early adult life. An attack is followed by a period of remission during which the symptoms temporarily disappear. One attack follows another over the years, usually every year or two. The result is a progressive loss of function interspersed with remission periods, during which symptoms abate.

Although the cause of MS is unclear, both genetic susceptibility and exposure to some environmental factor (perhaps a herpesvirus) appear to contribute. Since 1993, many patients with relapsing-remitting MS have been treated with injections of beta-interferon, which lengthens the time between relapses, decreases the severity of relapses, and slows formation of new lesions in some cases. Unfortunately, not all MS patients can tolerate beta-interferon and therapy becomes less effective as the disease progresses.

Epilepsy

Epilepsy, which afflicts about 1% of the world's population, is characterized by short, recurrent attacks of motor, sensory, or psychological malfunction, although it almost never affects intelligence. The attacks, called *epileptic seizures,* are initiated by abnormal electrical discharges from millions of neurons in the brain. The discharges stimulate many of the neurons to send nerve impulses over their conduction pathways; as a result, lights, noise, or smells may be sensed when the eyes, ears, and nose have not been stimulated. Moreover, the skeletal muscles of a person undergoing a seizure may contract involuntarily. *Partial seizures* begin in a small focus on one side of the brain and produce milder symptoms, whereas *generalized seizures* involve larger areas on both sides of the brain and loss of consciousness. Epilepsy has many causes, including brain damage at birth (the most common cause), metabolic disturbances, infections, toxins (alcohol, tranquilizers, hallucinogens), head injuries, and brain tumors. However, most epileptic seizures have no demonstrable cause.

Epileptic seizures often can be eliminated or alleviated by antiepileptic drugs, such as phenytoin, carbamazepine, and valproate sodium. An implantable device that stimulates the vagus nerve (cranial nerve X) also has produced dramatic results in reducing seizures in some patients whose epilepsy was not well-controlled by drugs.

■ **STUDY OUTLINE**

Components of the Nervous System (p. 218)

1. Components of the nervous system include the brain, 12 pairs of cranial nerves and their branches, the spinal cord, 31 pairs of spinal nerves and their branches, sensory receptors, ganglia, and enteric plexuses.

2. The two main subsystems of the nervous system are the central nervous system (CNS)—the brain and spinal cord—and the peripheral nervous system (PNS)—all nervous tissues outside the brain and spinal cord.

3. The three basic functions of the nervous system are detecting stimuli (sensory functions); analyzing, integrating, and storing sensory information (integrative functions); and responding to integrative decisions (motor functions).

4. Sensory (afferent) neurons provide input to the CNS; motor (efferent) neurons carry output from the CNS to effectors.

5. The PNS also is subdivided into the somatic nervous system (SNS), autonomic nervous system (ANS), and enteric nervous system (ENS).

6. The SNS consists of (1) sensory neurons that conduct impulses from somatic and special sense receptors to the CNS, and (2) motor neurons from the CNS to skeletal muscles.

7. The ANS contains (1) sensory neurons from visceral organs and (2) motor neurons that convey impulses from the CNS to smooth muscle tissue, cardiac muscle tissue, and glands.

8. The ENS consists of neurons in two enteric plexuses that extend the length of the gastrointestinal (GI) tract; it monitors sensory changes and controls operation of the GI tract.

Cellular Organization of Nervous Tissue (p. 220)

1. Neurons (nerve cells) consist of a cell body, dendrites that receive stimuli, and a single axon that sends impulses to another neuron or to a muscle or gland.

2. Neuroglia are specialized tissue cells that support neurons, attach neurons to blood vessels, produce the myelin sheath around axons of the CNS, and carry out phagocytosis.

3. Neuroglia include astrocytes, oligodendrocytes, microglia, ependymal cells, and Schwann cells.

4. Myelin is formed by Schwann cells in the PNS and oligodendrocytes in the CNS.

5. White matter consists of aggregations of myelinated processes, whereas gray matter contains neuron cell bodies, dendrites, and axon terminals or bundles of unmyelinated axons and neuroglia.

6. In the spinal cord, gray matter forms an H-shaped inner core that is surrounded by white matter. In the brain, a thin, superficial shell of gray matter covers the cerebrum and cerebellum.

7. The nervous system exhibits plasticity—the capability to change based on experience—but it has very limited powers of regeneration—the capability to replicate or repair damaged neurons.

8. Axons and dendrites that are associated with a neurolemma in the PNS may undergo repair if the cell body is intact, the Schwann cells are functional, and scar tissue formation does not occur too rapidly.

Action Potentials (p. 222)

1. Neurons communicate with one another by means of nerve action potentials, also called nerve impulses.

2. Generation of action potentials depends on the presence of voltage-gated channels for Na^+ and K^+ and the existence of a resting membrane potential.

3. A typical value for the resting membrane potential (difference in electrical charge across the plasma membrane) is -70 mV. A cell that exhibits a membrane potential is said to be polarized.

4. The resting membrane potential arises due to unequal distribution of anions and cations across the plasma membrane and higher membrane permeability to K^+ than to Na^+. The level of K^+ is higher inside and the level of Na^+ is higher outside, a situation that is maintained by Na^+/K^+ pumps.

5. The ability of muscle fibers and neurons to respond to a stimulus and convert it into action potentials is called excitability.

6. During an action potential, voltage-gated Na^+ and K^+ channels open in sequence. Opening of voltage-gated Na^+ channels results in depolarization, the loss and then reversal of membrane polarization (from -70 mV to $+30$ mV). Then, opening of voltage-gated K^+ channels allows repolarization, recovery of the resting membrane potential.

7. According to the all-or-none principle, if a stimulus is strong enough to generate an action potential, the impulse generated is of a constant size. A stronger stimulus does not generate a larger impulse.

8. During the absolute refractory period, another impulse cannot be generated; during the relative refractory period, an impulse can be triggered only by a suprathreshold stimulus.

9. Nerve impulse conduction that occurs as a step-by-step process along an unmyelinated axon is called continuous conduction. In saltatory conduction, a nerve impulse "leaps" from one node of Ranvier to the next along a myelinated axon.

10. Axons with larger diameters conduct impulses faster than those with smaller diameters; myelinated axons conduct impulses faster than unmyelinated axons.

Synaptic Transmission (p. 225)

1. Neurons communicate with other neurons and with effectors at synapses in a series of events known as synaptic transmission.

2. The two types of synapses are electrical synapses, where gap junctions allow ions to flow from one cell to its neighbor, and chemical synapses, where neurotransmitter is released from a presynaptic neuron into the synaptic cleft and then binds to receptors on the postsynaptic plasma membrane.

3. An excitatory neurotransmitter depolarizes the postsynaptic neuron's membrane, bringing the membrane potential closer to threshold. An inhibitory neurotransmitter hyperpolarizes the membrane of the postsynaptic neuron.

4. The postsynaptic neuron integrates excitatory and inhibitory signals in a process called summation and then responds accordingly.

5. Neurotransmitter is removed from the synaptic cleft in three ways: diffusion, enzymatic degradation, and reuptake by neurons or neuroglia.

6. Important neurotransmitters include acetylcholine, glutamate, gamma amino butyric acid (GABA), glycine, norepinephrine, epinephrine, dopamine, serotonin, neuropeptides, and nitric oxide.

SELF-QUIZ

1. Which of the following are incorrectly matched?

   a. central nervous system, composed of the brain and spinal cord **b.** somatic nervous system, includes motor neurons to skeletal muscles **c.** sympathetic nervous system, includes motor neurons to skeletal, smooth, and cardiac muscles **d.** peripheral nervous system, includes cranial and spinal nerves **e.** autonomic nervous system, includes parasympathetic and sympathetic divisions

2. The portion of the nervous system that regulates the gastrointestinal (GI) tract is the

   a. somatic nervous system **b.** sympathetic division **c.** integrative division **d.** central nervous system **e.** enteric nervous system

3. Damage to dendrites would interfere with the ability to

   a. receive nerve impulses **b.** make proteins **c.** conduct nerve impulses to another neuron **d.** release neurotransmitters **e.** form myelin

4. The type of cell that produces myelin sheaths around axons in the CNS is the

   a. astrocyte **b.** myelinocyte **c.** Schwann cell **d.** oligodendrocyte **e.** microglia

5. A bundle of axons carrying sensory impulses up the spinal cord is

   a. an ascending tract **b.** a nucleus **c.** a mixed nerve **d.** a descending tract **e.** an enteric plexus

6. Which of the following is NOT true concerning the repair of nervous tissue?

   a. If the cell body is not damaged, neurons in the PNS may be able to repair themselves. **b.** In the CNS myelin inhibits neuronal regeneration. **c.** Injury to the CNS is usually permanent. **d.** Active Schwann cells contribute to the repair process in the PNS. **e.** The presence of a neurolemma in the CNS inhibits repair.

7. In a resting neuron

   a. there is a high concentration of K^+ outside the cell **b.** negatively charged ions move freely through the plasma membrane **c.** the Na^+/K^+ pumps help maintain the low concentration of Na^+ inside the cell **d.** the outside surface of the plasma membrane has a negative charge **e.** the plasma membrane is highly permeable to Na^+

8. The depolarizing phase of a nerve impulse is caused by a

   a. rush of Na^+ into the neuron **b.** rush of Na^+ out of the neuron **c.** rush of K^+ into the neuron **d.** rush of K^+ out of the neuron **e.** pumping of K^+ into the neuron

9. If a stimulus is strong enough to generate an action potential, the impulse generated is of a constant size. A stronger stimulus cannot generate a larger impulse. This is known as

   a. the principle of polarization–depolarization **b.** saltatory conduction **c.** the all-or-none principle **d.** the principle of reflex action **e.** the absolute refractory period

10. Place the following events in the correct order of occurrence:

    1. Voltage-gated Na^+ channels open and permit Na^+ to rush inside the neuron.
    2. The Na^+/K^+ pump restores the ions to their original sites.
    3. A stimulus of threshold strength is applied to the neuron.
    4. The membrane polarization changes from negative (-55 mV) to positive ($+30$ mV).
    5. Voltage-gated K^+ channels open, and K^+ flows out of the neuron.

    a. 4, 1, 2, 3, 5 **b.** 4, 3, 1, 2, 5 **c.** 3, 1, 4, 2, 5 **d.** 5, 3, 1, 4, 2 **e.** 3, 1, 4, 5, 2

11. Saltatory conduction occurs

    a. in unmyelinated axons **b.** at the nodes of Ranvier **c.** in the smallest diameter axons **d.** in skeletal muscle fibers **e.** in cardiac muscle fibers

12. The speed of nerve impulse conduction is increased by

    a. cold **b.** a very strong stimulus **c.** small diameter of the axon **d.** myelination **e.** astrocytes

13. For a signal to be transmitted by means of a chemical synapse from a presynaptic neuron to a postsynaptic neuron,

    a. the presynaptic neuron must be touching the postsynaptic neuron **b.** the postsynaptic neuron must contain neurotransmitter receptors **c.** there must be gap junctions present between the two neurons **d.** the postsynaptic neuron needs to release neurotransmitters from its synaptic vesicles **e.** the neurons must be myelinated

14. What would happen at the postsynaptic neuron if the total inhibitory effects of the neurotransmitters were greater than the total excitatory effects?

    a. There would be a nerve impulse generated. **b.** It would be easier to generate a nerve impulse when the next stimulus was received. **c.** The nerve impulse would be rerouted to another neuron. **d.** No nerve impulse would be generated. **e.** The neurotransmitter would be broken down more quickly.

15. Match the following neurotransmitters with their descriptions:

    _____ **a.** most common inhibitory neurotransmitter in the brain

    _____ **b.** a gaseous neurotransmitter that is not packaged into synaptic vesicles

    _____ **c.** excessive levels can result in excitotoxicity

    _____ **d.** body's natural painkillers

    _____ **e.** helps regulate mood

    _____ **f.** neurotransmitter that activates skeletal muscle fibers

    A. serotonin
    B. acetylcholine
    C. endorphins
    D. GABA
    E. nitric oxide
    F. glutamate

16. Match the following:

_____ **a.** the portion of a neuron containing the nucleus

_____ **b.** rounded structure at the distal end of an axon terminal

_____ **c.** highly branched, input part of a neuron

_____ **d.** sac in which neurotransmitter is stored

_____ **e.** neuron located entirely within the CNS

_____ **f.** long, cylindrical process that conducts impulses toward another neuron

_____ **g.** produces myelin sheath in PNS

_____ **h.** unmyelinated gaps in the myelin sheath

_____ **i.** substance that increases the speed of nerve impulse conduction

_____ **j.** neuron that conveys information from a receptor to the CNS

_____ **k.** neuron that conveys information from the CNS to an effector

_____ **l.** bundle of many axons in the PNS

_____ **m.** bundle of many axons in the CNS

_____ **n.** group of cell bodies in the PNS

_____ **o.** group of cell bodies in the CNS

_____ **p.** substance used for communication at chemical synapses

A. synaptic end bulb

B. motor neuron

C. sensory neuron

D. dendrite

E. interneuron

F. nucleus

G. myelin sheath

H. Schwann cell

I. cell body

J. node of Ranvier

K. ganglion

L. nerve

M. neurotransmitter

N. tract

O. synaptic vesicle

P. axon

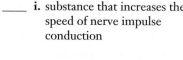

CRITICAL THINKING APPLICATIONS

1. The buzzing of the alarm clock awoke Rodrigo. He stretched, yawned, and started to salivate as he smelled the brewing coffee. He could feel his stomach rumble. List the divisions of the nervous system that are involved in each of these activities.

2. Angelina just figured out that her A & P class actually starts at 10:00 AM and not at 10:15, which has been her arrival time since the beginning of the term. One of the other students remarks that Angelina's "gray matter is pretty thin." Should Angelina thank him?

3. Sarah really looks forward to the great feeling she has after going for a nice long run on the weekends. By the end of her run, she doesn't even feel the pain in her sore feet. Sarah read in a magazine that some kind of natural brain chemical was responsible for the "runner's high" that she feels. Are there such chemicals in Sarah's brain?

4. The pediatrician was trying to educate the anxious new parents of a six-month-old baby. "No, don't worry about him not walking yet. The myelination of the baby's nervous system is not finished yet." Explain what the pediatrician means by this reassurance.

ANSWERS TO FIGURE QUESTIONS

9.1 The total number of cranial and spinal nerves in your body is $(12 \times 2) + (31 \times 2) = 86$.

9.2 Sensory or afferent neurons carry input to the CNS; motor or efferent neurons carry output from the CNS.

9.3 The axon conducts nerve impulses and transmits the message to another neuron or effector cell by releasing a neurotransmitter at its axon terminals.

9.4 Myelin gives white matter its characteristic color.

9.5 A typical value for the resting membrane potential in a neuron is -70 mV.

9.6 Voltage-gated Na^+ channels are open during the depolarizing phase, and voltage-gated K^+ channels are open during the repolarizing phase.

9.7 The two main factors that influence conduction speed of a nerve impulse are the axon diameter (larger axons conduct impulses more rapidly) and the presence or absence of a myelin sheath (myelinated axons conduct more rapidly than unmyelinated axons).

9.8 The depolarizing phase of the action potential opens the voltage-gated Ca^{2+} channels in synaptic end bulbs.

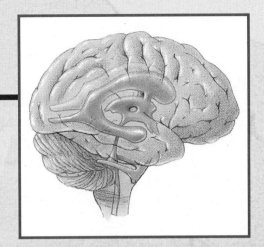

Chapter 10

Central and Somatic Nervous Systems

■ Student Learning Objectives

1. Describe how the spinal cord is protected. **234**

2. Describe the structure and functions of the spinal cord. **234**

3. Describe the composition, coverings, and branches of a spinal nerve. **239**

4. Discuss how the brain is protected and supplied with blood. **241**

5. Name the principal parts of the brain and explain the function of each part. **241**

6. Identify the 12 pairs of cranial nerves by name, number, type, location, and function. **252**

■ A Look Ahead

ow that you understand how the nervous system functions on the cellular level, in this chapter we will explore the structure and functions of the central nervous system, the brain and spinal cord. We will also examine spinal nerves and cranial nerves, which are part of the peripheral nervous system (see Figure 9.1 on page 218).

SPINAL CORD

Objectives: • **Describe how the spinal cord is protected.**

• **Describe the structure and functions of the spinal cord.**

Protection and Coverings:
Vertebral Canal and Meninges

The spinal cord is located in the vertebral canal of the vertebral column. Because the wall of the vertebral canal is essentially a ring of bone, the cord is well protected. Additional protection is provided by the meninges, cerebrospinal fluid, and vertebral ligaments.

The *meninges* (me-NIN-jēz; singular is *meninx*, MĒ-ninks) are three layers of connective tissue coverings that extend around the spinal cord and brain. The meninges that protect the spinal cord are called the *spinal meninges* (Figure 10.1), and those that protect the brain are called the *cranial meninges* (see Figure 10.8a). The outermost of the three layers of the meninges is called the *dura mater* (DOO-ra MĀ-ter = tough mother). Its tough, dense, irregular connective tissue helps protect the delicate structures of the CNS. The spinal dura mater is continuous with the cranial dura mater. The tube of spinal dura mater extends to the second sacral vertebra, well beyond the spinal cord, which ends at about the level of the second lumbar vertebra. The spinal cord is also protected by a cushion of fat and connective tissue located in the *epidural space,* a space between the dura mater and vertebral column (see Figure 10.6). Injection of an anesthetic into the epidural space results in a temporary sensory and motor paralysis called an *epidural block.* Such injections in the lower lumbar region are used to control pain during childbirth.

The middle layer of the meninges is called the *arachnoid* (a-RAK-noyd; *arachn-* = spider) because of its delicate spider's web arrangement of collagen and elastic fibers. Like the dura mater, the spinal arachnoid layer is continuous with the cranial arachnoid layer.

The inner layer, the *pia mater* (PĪ-a MĀ-ter = delicate mother), is a transparent layer of collagen and elastic fibers that adheres to the surface of the spinal cord and brain. It contains numerous blood vessels. Between the arachnoid and the pia mater is the *subarachnoid space,* where cerebrospinal fluid cir-

culates. Cerebrospinal fluid may be removed from the subarachnoid space between the third and fourth or fourth and fifth lumbar vertebrae by a *spinal tap (lumbar puncture).* A spinal tap may also be performed to introduce antibiotics and anesthetics and to administer chemotherapy.

General Features of the Spinal Cord

The length of the adult *spinal cord* ranges from 42 to 45 cm (16 to 18 inches). It extends from the foramen magnum of the occipital bone of the skull to the superior border of the second lumbar vertebra in the vertebral column (Figure 10.2). The spinal cord does not run the entire length of the vertebral column. Consequently, nerves arising from the lowest portion of the cord angle down the vertebral canal like wisps of flowing hair. They are appropriately named the *cauda equina* (KAW-da ē-KWĪ-na), meaning horse's tail. The spinal cord has two conspicuous enlargements: The *cervical enlargement* contains nerves that supply the upper limbs, and the *lumbar enlargement* contains nerves supplying the lower limbs. Each of 31 *spinal segments* of the spinal cord gives rise to a pair of spinal nerves (Figure 10.2).

Figure 10.1 ■ **Spinal meninges.**

Meninges are connective tissue coverings that surround the brain and spinal cord.

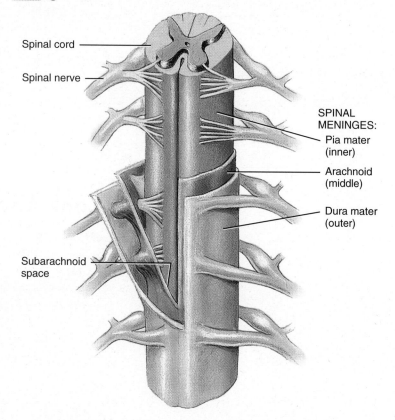

Spinal cord

Spinal nerve

SPINAL MENINGES:
Pia mater (inner)

Arachnoid (middle)

Dura mater (outer)

Subarachnoid space

Anterior view and transverse section through spinal cord

 In which space does cerebrospinal fluid circulate?

Figure 10.2 ■ **Spinal cord and spinal nerves.** Selected nerves are labeled on the left side of the figure. Together, the lumbar and sacral plexuses are called the lumbosacral plexus.

The spinal cord extends from the foramen magnum of the occipital bone to the superior border of the second lumbar vertebra.

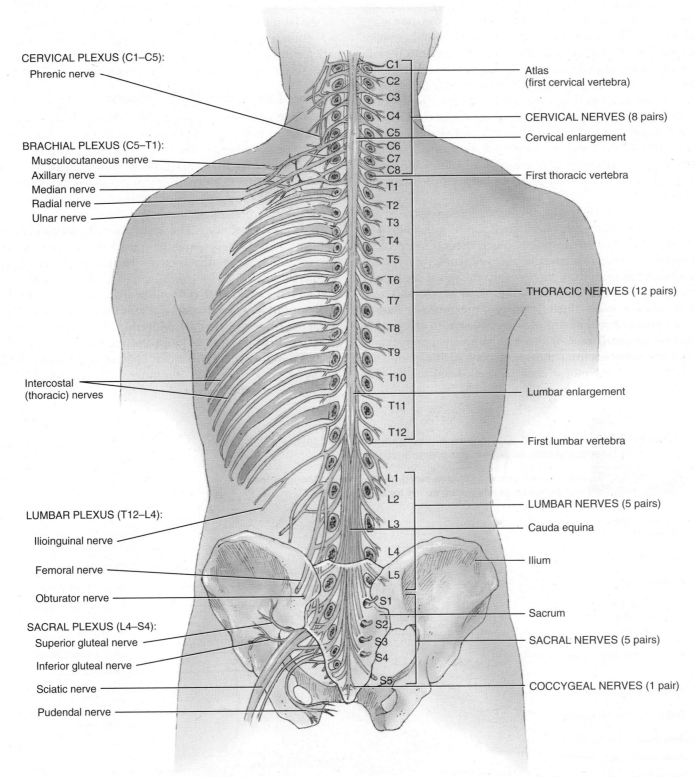

CERVICAL PLEXUS (C1–C5):
  Phrenic nerve

BRACHIAL PLEXUS (C5–T1):
  Musculocutaneous nerve
  Axillary nerve
  Median nerve
  Radial nerve
  Ulnar nerve

Intercostal (thoracic) nerves

LUMBAR PLEXUS (T12–L4):
  Ilioinguinal nerve
  Femoral nerve
  Obturator nerve

SACRAL PLEXUS (L4–S4):
  Superior gluteal nerve
  Inferior gluteal nerve
  Sciatic nerve
  Pudendal nerve

C1
C2
C3
C4
C5
C6
C7
C8
T1
T2
T3
T4
T5
T6
T7
T8
T9
T10
T11
T12
L1
L2
L3
L4
L5
S1
S2
S3
S4
S5

Atlas (first cervical vertebra)
CERVICAL NERVES (8 pairs)
Cervical enlargement
First thoracic vertebra
THORACIC NERVES (12 pairs)
Lumbar enlargement
First lumbar vertebra
LUMBAR NERVES (5 pairs)
Cauda equina
Ilium
Sacrum
SACRAL NERVES (5 pairs)
COCCYGEAL NERVES (1 pair)

Posterior view of entire spinal cord and portions of spinal nerves and their branches

 Spinal nerves are part of which division of the nervous system?

Structure in Cross Section

Two grooves, the deep *anterior median fissure* and the shallow *posterior median sulcus,* divide the spinal cord into right and left halves (Figure 10.3). As previously described, the spinal cord contains a centrally located H-shaped mass of gray matter surrounded by white matter. In the center of the gray matter is the *central canal,* which runs the length of the cord and contains cerebrospinal fluid. The sides of the H are divided into regions called *gray horns,* named relative to their location: anterior, lateral, and posterior. The gray matter consists mainly of interneurons and motor neurons that serve as relay stations for impulses (more on this later). The white matter of the spinal cord is organized into regions called anterior, lateral, and posterior *white columns.* The white columns consist of myelinated and unmyelinated axons organized into sensory (ascending) and motor (descending) *tracts* that convey impulses between the brain and spinal cord.

Spinal nerves are the paths of communication between the spinal cord tracts and the periphery. Each pair of spinal nerves is connected to the cord at two points called roots (Figure 10.3). The *posterior (dorsal) root* contains axons of sensory nerves and conducts impulses from the periphery into the spinal cord. Each posterior root also has a swelling, the *posterior (dorsal) root ganglion,* which contains the cell bodies of the sensory neurons. The other point of attachment of a spinal nerve to the cord is the *anterior (ventral) root.* It contains axons of motor nerves and conducts impulses from the spinal cord to the periphery. The cell bodies of the motor neurons are located in the gray matter of the cord.

Spinal Cord Functions

The spinal cord has two principal functions. (1) The white matter tracts in the spinal cord are highways for nerve impulse conduction. Along these highways, sensory impulses flow from the periphery to the brain, and motor impulses flow from the brain to the periphery. (2) The gray matter of the spinal cord receives and integrates incoming and outgoing information and is the site for integration of spinal cord reflexes. Both functions of the spinal cord are essential to maintaining homeostasis.

Impulse Conduction Along Pathways

The route that nerve impulses follow from their origin in the dendrites or cell body of a neuron in one part of the body to a destination elsewhere in the body is called a *pathway.* All pathways consist of several functionally related neurons. Somatic sensory information ascending from the spinal cord to the brain is conducted along two general pathways on each side of the cord: the anterolateral pathways and the posterior column–medial lemniscus pathway. The *anterolateral pathways* begin in two spinal cord tracts—anterior spinothalamic tract and lateral spinothalamic tract—that convey impulses for sensing pain, temperature (cold and warmth), crude (poorly localized) touch, pressure, tickle, and itch. Often, the name of a tract indicates its position in the white matter, where it begins and ends, and the direction of nerve impulse conduction. For example, the anterior spinothalamic tract is located in the *anterior* white column, it begins in the *spinal cord,* and it ends in the *thalamus* (a region of the brain). Figure 10.4a illustrates the anterolateral pathways as examples of sensory (ascending) tracts.

Figure 10.3 ■ **Spinal cord.** The organization of gray and white matter in the spinal cord as seen in transverse section.

The spinal cord conducts nerve impulses along tracts between the brain and periphery and serves as an integrating center for spinal reflexes.

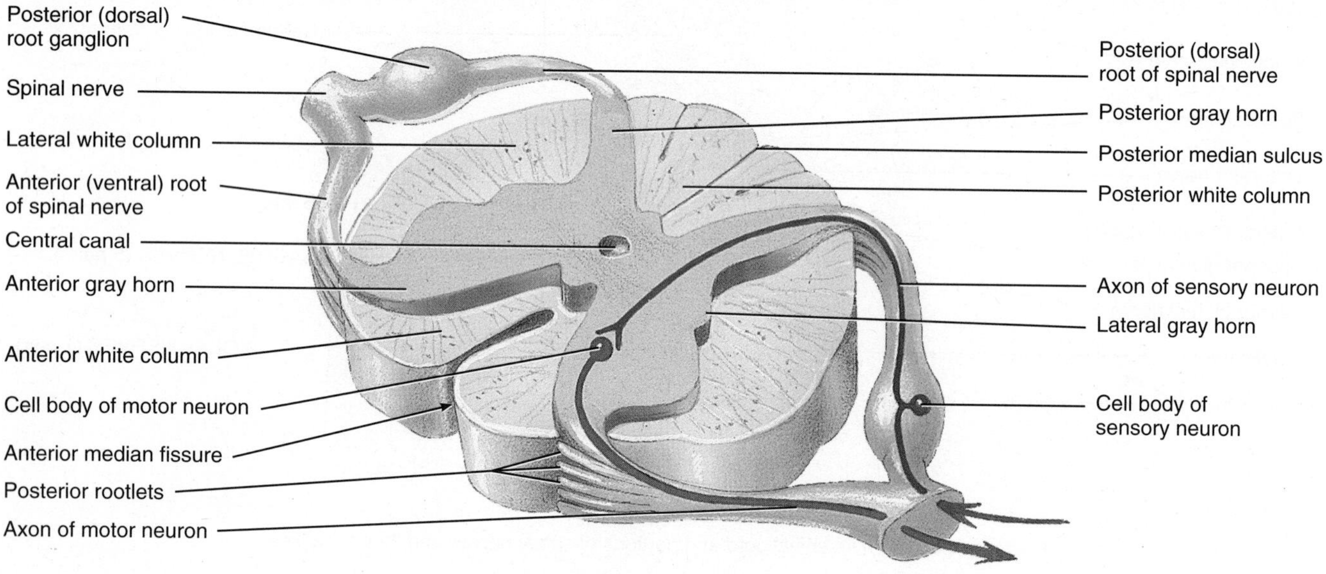

Posterior (dorsal) root ganglion

Spinal nerve

Lateral white column

Anterior (ventral) root of spinal nerve

Central canal

Anterior gray horn

Anterior white column

Cell body of motor neuron

Anterior median fissure

Posterior rootlets

Axon of motor neuron

Posterior (dorsal) root of spinal nerve

Posterior gray horn

Posterior median sulcus

Posterior white column

Axon of sensory neuron

Lateral gray horn

Cell body of sensory neuron

Transverse section of the thoracic spinal cord

Functionally, which type of neurons make up the posterior root? the anterior root?

The *posterior column–medial lemniscus pathway* carries nerve impulses for sensing (1) *proprioception,* awareness of the position of muscles, tendons, joints, and body movements; (2) *fine touch,* the ability to feel exactly what part of the body is touched; (3) *stereognosis,* the ability to recognize by feel the size, shape, and texture of an object; (4) *weight discrimination,* the ability to determine the weight of an object; and (5) *vibratory sensations,* the ability to detect vibration.

The sensory systems keep the central nervous system aware of the external and internal environments. Responses to this information are brought about by the motor systems, which enable us to move about and change our relationship to the world around us. As sensory information is conveyed to the central nervous system, it becomes part of a large pool of sensory input. We do not respond actively to every bit of input the central nervous system receives. Rather, each piece of incoming information is integrated with all the others. The integration process occurs at many places along the pathways of the central nervous system.

A part of the brain called the primary motor cortex assumes the major role for controlling precise, discrete, muscular movements. Other brain regions control semivoluntary movements like walking and laughing. Still other parts of the brain help make body movements smooth and coordinated. Nerve impulses that signal for the contraction of skeletal muscles descend in the spinal cord in two major motor pathways: the direct pathways and the indirect pathways.

Figure 10.4 ■ Sensory and motor pathways. (a) Sensory pathway composed of sets of three neurons. The first neurons (blue) synapse with the second neurons in the spinal cord gray matter. The second neurons (red) extend to the opposite thalamus, a region of the brain, and the third neurons (green) extend from the thalamus to the cerebral cortex, the outermost layer of gray matter in the brain. (b) Motor pathways composed of sets of two neurons (black and red).

Sensory (ascending) tracts convey nerve impulses from the periphery to the brain, and motor (descending) tracts convey nerve impulses from the brain to the periphery.

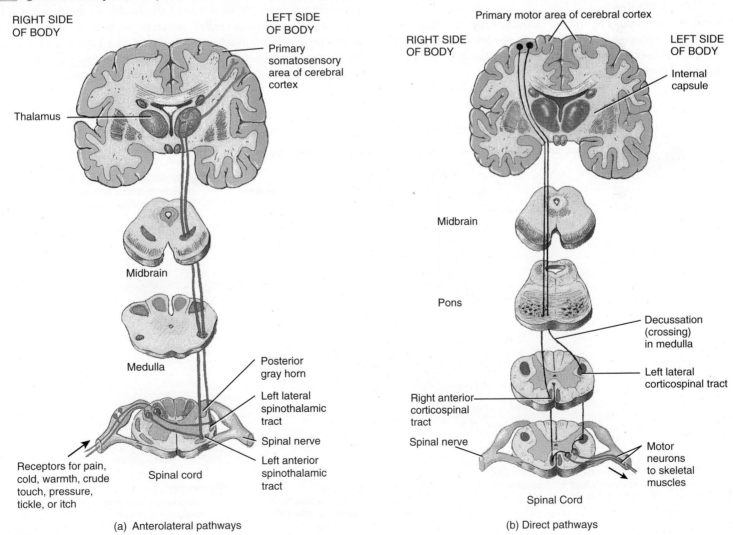

(a) Anterolateral pathways

(b) Direct pathways

Which spinal cord tracts shown here are sensory and which are motor?

Direct pathways—the lateral corticospinal, anterior corticospinal, and corticobulbar tracts—convey nerve impulses destined to cause precise, voluntary movements of skeletal muscles (Figure 10.4b). *Indirect pathways*—rubrospinal, tectospinal, and vestibulospinal tracts—convey nerve impulses that program automatic movements, help coordinate body movements with visual stimuli, maintain skeletal muscle tone and posture, and play a major role in equilibrium by regulating muscle tone in response to movements of the head.

Spinal Reflexes

The gray matter of the spinal cord serves as an integrating center for *spinal reflexes,* fast, automatic responses to sensory impulses that enter the spinal cord via spinal nerves. (Other reflexes occur by means of cranial nerves and are known as cranial reflexes.) The pathway followed by nerve impulses that produce a reflex is known as a *reflex arc.* The basic components of a spinal reflex arc are as follows (Figure 10.5):

1 **Sensory receptor.** The distal end of a sensory neuron serves as a *sensory receptor,* a part of the neuron that can respond to a specific type of stimulus by generating one or more nerve impulses.

2 **Sensory neuron.** The nerve impulses conduct from the sensory receptor along the axon of the *sensory neuron* to the axon terminals, which are located in the gray matter of the spinal cord.

3 **Integrating center.** One or more regions of gray matter within the spinal cord act as an *integrating center.* In the simplest type of reflex, the integrating center is a single synapse between a sensory neuron and a motor neuron. More often, the integrating center consists of one or more interneurons.

4 **Motor neuron.** Impulses triggered by the integrating center pass out of the spinal cord along a *motor neuron* to the part of the body that will respond.

5 **Effector.** The part of the body that responds to the motor nerve impulse, such as a muscle or gland, is the *effector.* Its action is a reflex.

When we discussed skeletal muscle physiology, we described skeletal muscle as voluntary. However, reflexes can cause the involuntary contraction of skeletal muscles. Reflexes involving skeletal muscles are termed *somatic reflexes.* One example is the *patellar reflex* (knee jerk reflex). This reflex is an extension of the knee by contraction of the quadriceps femoris muscle in response to tapping the patellar ligament. The reflex helps us to remain standing erect despite the effects of gravity, as well as providing an easy physical test of the health of the nervous system. The *withdrawal reflex* protects us from serious cuts or burns by causing immediate withdrawal of a limb from a source of injury, usually before we are even aware of any pain. Reflexes involving smooth muscle, cardiac muscle, and glands are called

Figure 10.5 ■ **General components of a reflex arc.** The arrows show the direction of nerve impulse conduction.

 Reflexes are fast, predictable, automatic responses to changes in the environment.

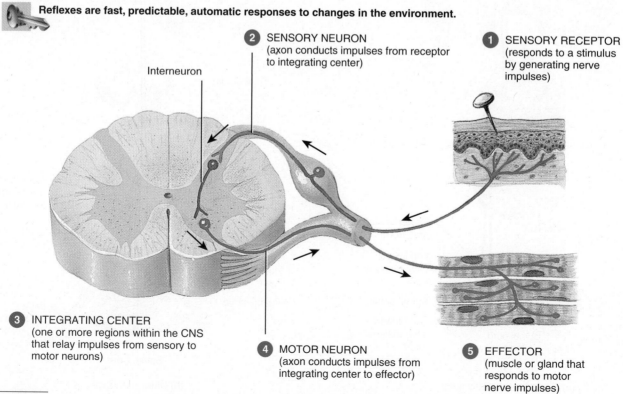

2 SENSORY NEURON
(axon conducts impulses from receptor to integrating center)

1 SENSORY RECEPTOR
(responds to a stimulus by generating nerve impulses)

Interneuron

3 INTEGRATING CENTER
(one or more regions within the CNS that relay impulses from sensory to motor neurons)

4 MOTOR NEURON
(axon conducts impulses from integrating center to effector)

5 EFFECTOR
(muscle or gland that responds to motor nerve impulses)

Which attachment of a spinal nerve to the spinal cord contains axons of motor neurons?

autonomic reflexes. For example, the acts of swallowing, urinating, and defecating all involve autonomic reflexes.

SPINAL NERVES

Objective: • Describe the composition, coverings, and branches of a spinal nerve.

Names

Spinal nerves and the nerves that branch from them to serve all parts of the body connect the CNS to sensory receptors, muscles, and glands and are part of the peripheral nervous system (PNS). The 31 pairs of spinal nerves are named and numbered according to the region and level of the vertebral column from which they emerge (see Figure 10.2). There are 8 pairs of cervical nerves, 12 pairs of thoracic nerves, 5 pairs of lumbar nerves, 5 pairs of sacral nerves, and 1 pair of coccygeal nerves. The first cervical pair emerges between the atlas and the occipital bone. All other spinal nerves leave the vertebral column by passing through the *intervertebral foramina,* the holes between vertebrae.

Composition and Coverings

A spinal nerve has two separate points of attachment to the cord, a posterior (dorsal) root that contains axons of sensory neurons and an anterior (ventral) root that contains axons of motor neurons (see Figure 10.3). These roots unite to form a spinal nerve at the intervertebral foramen. A spinal nerve thus contains both sensory and motor axons and therefore is a *mixed nerve.* Each spinal nerve consists of bundles of axons wrapped in connective tissue and supplied with blood vessels.

Distribution

Branches

After a spinal nerve exits the intervertebral foramen, it divides into several *rami* (RĀ-mī; singular is *ramus*), nerve branches that extend in different directions (Figure 10.6). The *posterior ramus* (RĀ-mus) supplies the deep muscles and skin of the back.

Figure 10.6 ■ **Branches of a spinal nerve.**

The various branches of a spinal nerve supply different regions of the body.

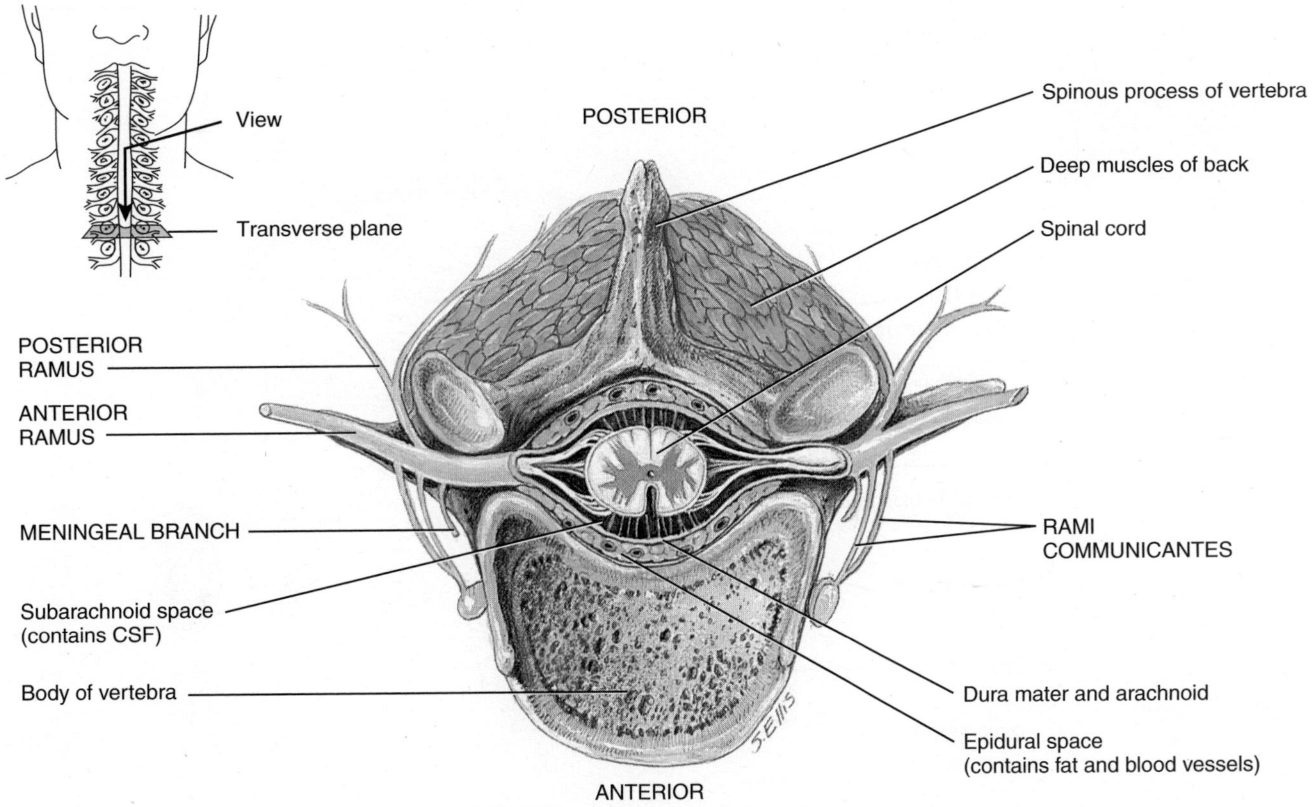

View
Transverse plane

POSTERIOR

Spinous process of vertebra

Deep muscles of back

Spinal cord

POSTERIOR RAMUS

ANTERIOR RAMUS

MENINGEAL BRANCH

Subarachnoid space (contains CSF)

Body of vertebra

RAMI COMMUNICANTES

Dura mater and arachnoid

Epidural space (contains fat and blood vessels)

ANTERIOR

Transverse section through vertebra and spinal cord

 Which branch of a spinal nerve supplies the deep muscles of the back?

The ***anterior ramus*** supplies the superficial back muscles, all the structures of the limbs, and the lateral and anterior trunk. The ***meningeal branch*** supplies the vertebrae, vertebral ligaments, blood vessels of the spinal cord, and the meninges. The ***rami communicantes*** (ko-myoo′-nī-KAN-tēz) are components of the autonomic nervous system and will be discussed in the next chapter.

Plexuses

The anterior rami of spinal nerves, except for thoracic nerves T2 to T11, do not extend directly to the body structures they supply. Instead, they form networks on either side of the body by joining with adjacent nerves. Such a network is called a ***plexus*** (= braid). The principal plexuses are the cervical plexus, brachial plexus, lumbar plexus, and sacral plexus (see Figure 10.2). Emerging from the plexuses are nerves bearing names that are often descriptive of the general regions they supply or the course they take. Each of the nerves, in turn, may have several branches named for the specific structures they supply.

The ***cervical plexus*** supplies the skin and muscles of the posterior head, neck, upper part of the shoulders, and the diaphragm. The phrenic nerves, which send motor impulses to the diaphragm, arise from the cervical plexus. Damage to the spinal cord above the origin of the phrenic nerves may cause respiratory failure. The ***brachial plexus*** constitutes the nerve supply for the upper limbs and a number of neck and shoulder muscles. Among the nerves that arise from the brachial plexus are the musculocutaneous, axillary, median, radial, and ulnar nerves. In a condition called ***carpal tunnel syndrome,*** the median nerve is compressed by trauma, edema, or repeated flexion of the wrist. The ***lumbar plexus*** supplies the abdominal wall, external genitals, and part of the lower limbs. Arising from this plexus are the ilioinguinal, femoral, and obturator nerves. The ***sacral plexus*** supplies the buttocks, perineum, and lower limbs. Among the nerves that arise from this plexus are the gluteal, sciatic, and pudendal nerves. The sciatic nerve is the longest nerve in the body.

Intercostal Nerves

Spinal nerves T2 to T11 do not form plexuses. They are known as ***intercostal nerves*** and extend directly to the structures they supply, including the muscles between ribs, abdominal muscles, and skin of the chest and back (see Figure 10.2).

Figure 10.7   ■   **Brain.** (Note: The infundibulum and pituitary gland are discussed together with the endocrine system in Chapter 13.)

The four principal parts of the brain are the brain stem, cerebellum, diencephalon, and cerebrum.

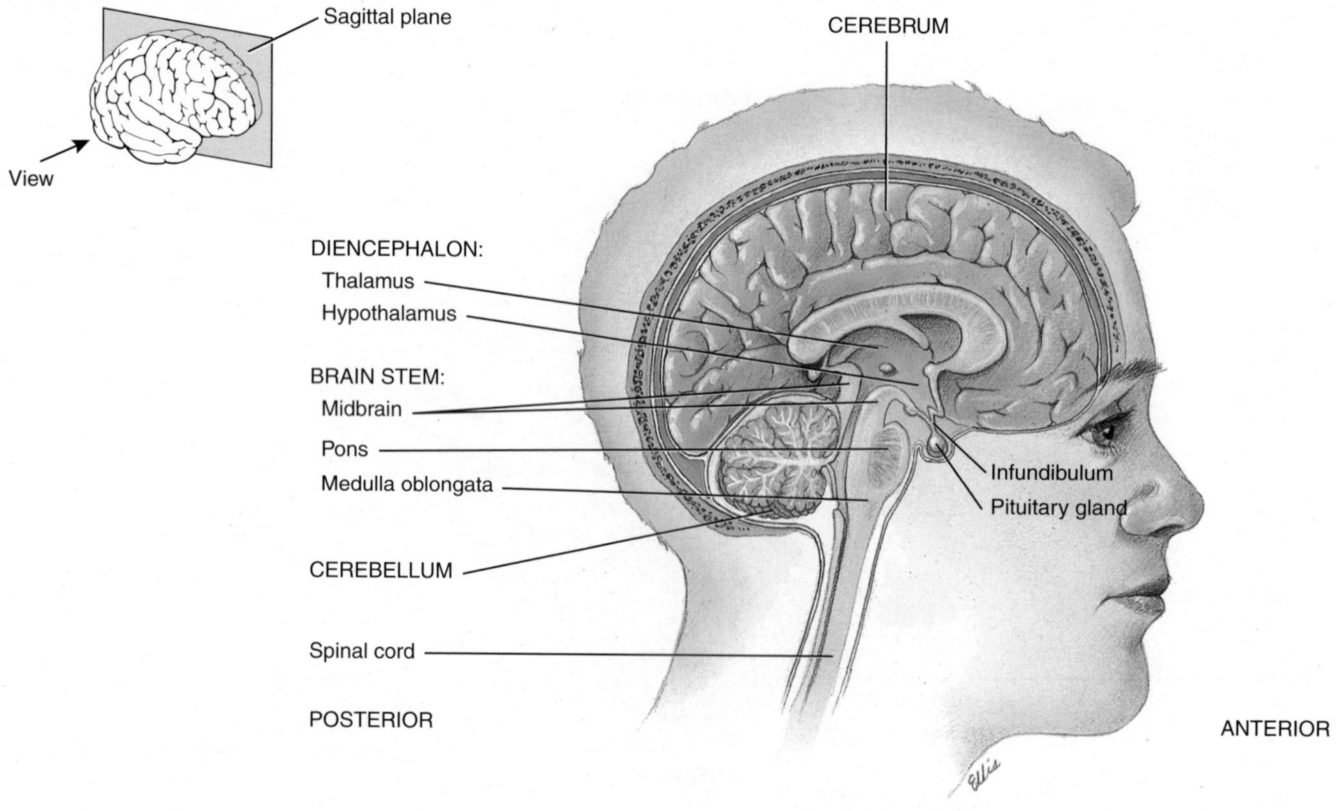

Sagittal plane

View

CEREBRUM

DIENCEPHALON:
Thalamus
Hypothalamus

BRAIN STEM:
Midbrain
Pons
Medulla oblongata

CEREBELLUM

Spinal cord

Infundibulum
Pituitary gland

POSTERIOR

ANTERIOR

(a) Sagittal section

Now we will consider the principal parts of the brain, how the brain is protected, and how it is related to the spinal cord and cranial nerves.

BRAIN

Objectives: • **Discuss how the brain is protected and supplied with blood.**
• **Name the principal parts of the brain and explain the function of each part.**

Principal Parts

The ***brain,*** made up of about 100 billion neurons and 1000 billion (1 trillion) neuroglia, is one of the largest organs of the body. Its mass is about 1300 g (almost 3 lb). The four principal parts are the brain stem, diencephalon, cerebrum, and cerebellum (Figure 10.7). The ***brain stem*** is continuous with the spinal cord and consists of the medulla oblongata, pons, and midbrain. Above the brain stem is the ***diencephalon*** (dī'-en-SEF-a-lon; *di-* = through; *-encephalon* = brain), consisting mostly of the thalamus and hypothalamus. Supported on the diencephalon and brain stem and forming the bulk of the brain is the ***cere-***

brum (se-RĒ-brum = brain). Its surface is composed of a thin layer of gray matter, the ***cerebral cortex*** (*cortex* = rind or bark), beneath which lies the cerebral white matter. Posterior to the brain stem is the ***cerebellum*** (ser'-e-BEL-um = little brain).

Brain Blood Supply

The brain is well supplied with oxygen and nutrients from a special circulatory route at the base of the brain called the ***cerebral arterial circle (circle of Willis)*** (see Figure 16.10c on page 387). Although the brain constitutes only about 2% of total body weight, it requires about 20% of the body's oxygen supply.

If blood flow to the brain is interrupted even briefly, unconsciousness may result. If brain neurons are totally deprived of oxygen for four or more minutes, many are permanently injured because lysosomes of brain cells are extremely sensitive to decreased oxygen. If the condition persists, the lysosomes break open and release enzymes that bring about self-destruction of brain cells.

Blood supplying the brain also contains glucose, the principal source of energy for brain cells. Because carbohydrate storage in the brain is limited, the supply of glucose must be continuous. If blood entering the brain has a low glucose level, mental confusion, dizziness, convulsions, and loss of consciousness may occur.

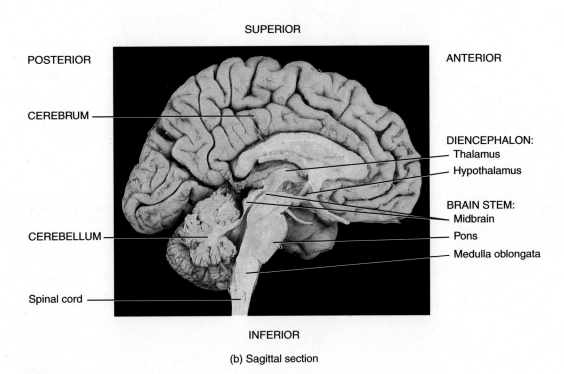

(b) Sagittal section

Which part of the brain attaches to the spinal cord?

Figure 10.8 ■ Meninges and ventricles of the brain.

Cerebrospinal fluid (CSF) protects the brain and spinal cord and delivers nutrients from the blood to the brain and spinal cord; CSF also removes wastes from the brain and spinal cord to the blood.

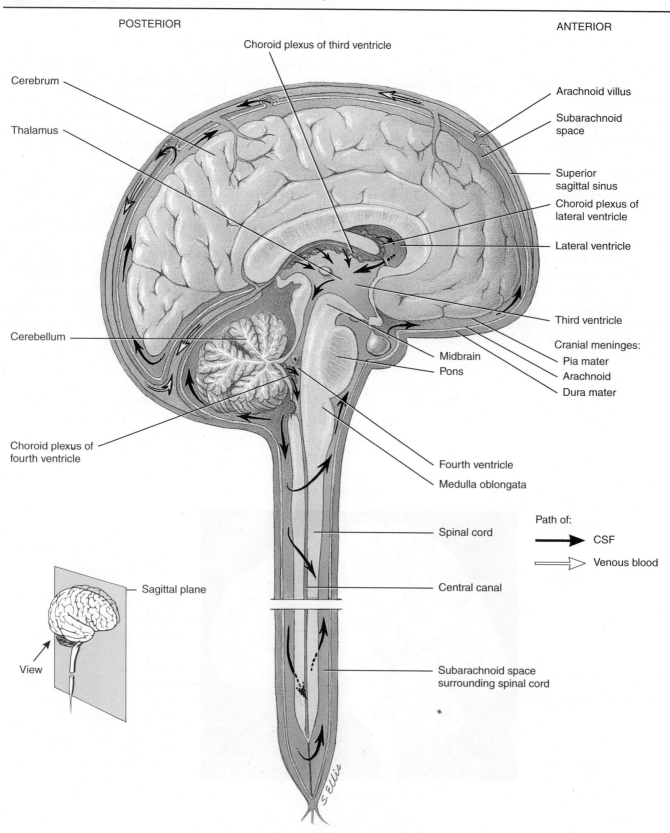

(a) Sagittal section of brain and spinal cord

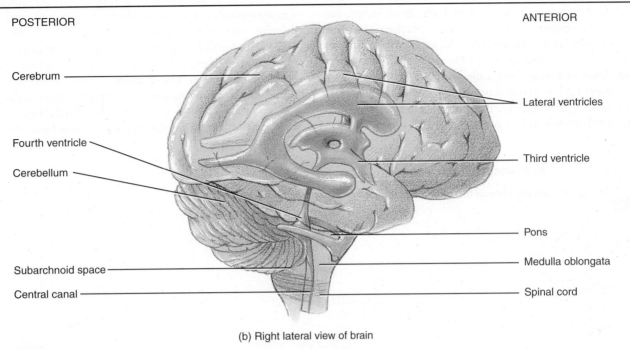

POSTERIOR

ANTERIOR

Cerebrum

Lateral ventricles

Fourth ventricle

Third ventricle

Cerebellum

Pons

Medulla oblongata

Subarachnoid space

Central canal

Spinal cord

(b) Right lateral view of brain

Where is CSF formed and absorbed?

Glucose, oxygen, and certain ions pass rapidly from the circulating blood into brain cells. Other substances enter slowly or not at all. The different rates of passage of certain materials from the blood into most parts of the brain are regulated by the ***blood-brain barrier (BBB)***. The cells of the brain capillary walls are tightly connected, have a thick basement membrane, and are surrounded by astrocytes (a type of neuroglial cell), thus forming a barrier to all but the smallest molecules or those selectively admitted through active transport. Although the blood-brain barrier protects brain cells from harmful substances, it also prevents entry of most antibiotics and therapeutic drugs for brain cancer. Trauma, inflammation, and certain toxins can cause a breakdown of the blood-brain barrier.

Protective Coverings of the Brain

The brain is protected by the cranium and cranial meninges. The ***cranial meninges*** have the same names as and are continuous with the spinal meninges: the outermost ***dura mater***, middle ***arachnoid***, and innermost ***pia mater*** (Figure 10.8)

Cerebrospinal Fluid

The spinal cord and brain are further protected against injury by ***cerebrospinal fluid (CSF)***. CSF circulates through the subarachnoid space, around the brain and spinal cord, and through cavities in the brain known as ***ventricles*** (VEN-tri-kuls = little cavities). There are four ventricles: two ***lateral ventricles***, one ***third***

ventricle, and one ***fourth ventricle*** (Figure 10.8). They are connected with one another, with the central canal of the spinal cord, and with the subarachnoid space.

The entire CNS contains 80 to 150 mL (3 to 5 oz) of cerebrospinal fluid. CSF is a clear, colorless liquid that contains proteins, glucose, urea, and salts. The two principal functions of CSF are related to homeostasis: protection and circulation. CSF protects the brain and spinal cord from jolts that would otherwise be traumatic, buoying the brain so that it "floats" in the cranial cavity. The CSF also delivers nutrients from the blood to the CNS and removes wastes and toxic substances produced by brain and spinal cord cells.

Cerebrospinal fluid, which circulates continuously, is formed by filtration and secretion from ***choroid plexuses*** (KŌ-royd = membranelike), specialized capillaries in the ventricles (Figure 10.8). From the ventricles, CSF flows into the subarachnoid space around the posterior part of the brain and downward around the posterior surface of the spinal cord, up the anterior surface of the spinal cord, and around the anterior part of the brain. From there it is gradually reabsorbed, primarily into a vein called the ***superior sagittal sinus***. The absorption occurs through ***arachnoid villi***, fingerlike projections of the arachnoid that extend into the superior sagittal sinus (Figure 10.8a). Normally, cerebrospinal fluid is absorbed as rapidly as it is formed so that the volume remains constant.

Abnormalities in the brain—tumors, inflammation, or developmental malformations—can interfere with the drainage of

CSF from the ventricles into the subarachnoid space. When excess CSF accumulates in the ventricles, the CSF pressure rises. Elevated CSF pressure causes a condition called **hydrocephalus** (hī'-drō-SEF-a-lus; *hydro-* = water; *cephal-* = head). In a baby in which the fontanels have not yet closed, the head bulges due to the increased pressure. If the condition persists, the fluid buildup compresses and damages the delicate nervous tissue. Hydrocephalus is relieved by draining the excess CSF. A neurosurgeon may implant a drain line, called a shunt, into the lateral ventricle to divert CSF into the superior vena cava or abdominal cavity. In adults, hydrocephalus may occur after head injury, meningitis, or subarachnoid hemorrhage.

Brain Stem

Medulla Oblongata

The **medulla oblongata** (me-DULL-la ob'-long-GA-ta), or simply **medulla,** is a continuation of the spinal cord. It forms the inferior part of the brain stem (Figure 10.9). See also Figure 10.7. The medulla contains all sensory (ascending) and motor (descending) tracts extending between the spinal cord and other parts of the brain. These tracts constitute the white matter of the medulla. Some of the white matter forms bulges on the anterior surface of the medulla, the **pyramids** (Figure 10.9), which contain the largest motor tracts that pass from the brain to the

Figure 10.9 ■ **Brain stem.**

The brain stem consists of the medulla oblongata, pons, and midbrain.

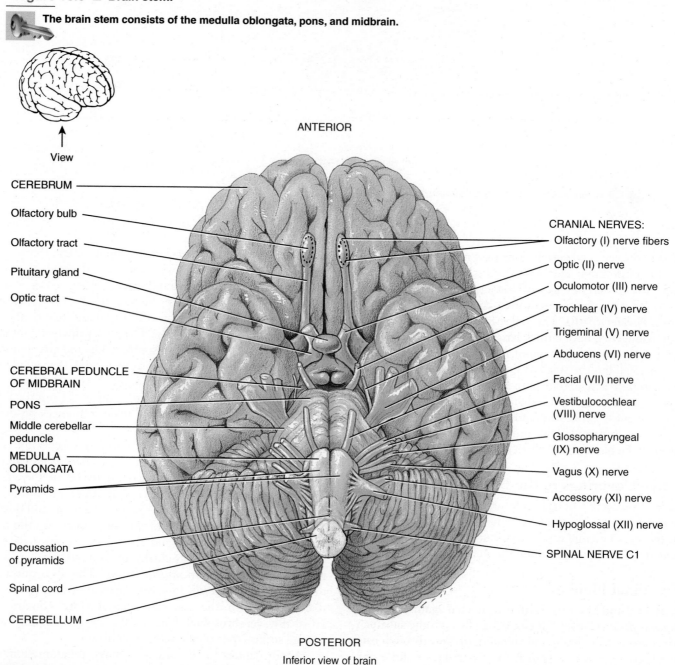

ANTERIOR

View

CEREBRUM

Olfactory bulb

Olfactory tract

Pituitary gland

Optic tract

CEREBRAL PEDUNCLE OF MIDBRAIN

PONS

Middle cerebellar peduncle

MEDULLA OBLONGATA

Pyramids

Decussation of pyramids

Spinal cord

CEREBELLUM

CRANIAL NERVES:

Olfactory (I) nerve fibers

Optic (II) nerve

Oculomotor (III) nerve

Trochlear (IV) nerve

Trigeminal (V) nerve

Abducens (VI) nerve

Facial (VII) nerve

Vestibulocochlear (VIII) nerve

Glossopharyngeal (IX) nerve

Vagus (X) nerve

Accessory (XI) nerve

Hypoglossal (XII) nerve

SPINAL NERVE C1

POSTERIOR

Inferior view of brain

Which part of the brain stem contains the cerebral peduncles?

Caffeine has been enjoyed by people around the world since the beginning of history. Found naturally in over 60 plants, caffeine is the most widely consumed drug in North America, primarily as a component of coffee, tea, cola, and other beverages. It is also found in many over-the-counter drugs and in small amounts in chocolate. The average coffee drinker consumes 3 cups a day, with some people (including many college students) consuming 10 cups or more a day. How is all of this caffeine affecting our health?

The Java Jitters

Caffeine's most obvious effect is on the nervous system. Caffeine is a *sympathomimetic* substance, which means its effects mimic those of the sympathetic division of the autonomic nervous system. In general, both caffeine and sympathetic arousal tend to wind you up. For example, both sympathetic arousal and caffeine make your heart beat faster and harder. (You will learn more about the

autonomic nervous system in Chapter 11.)

The immediate effects of caffeine vary greatly from person to person. Some people find that any amount of caffeine causes undesirable symptoms such as muscle twitches, anxiety, increased blood pressure, an irregular heartbeat, digestive complaints, headache, and difficulty sleeping. Other people get one or more of these symptoms only if their caffeine consumption exceeds a certain threshold.

How harmful is caffeine overload? If these symptoms are short-lived, and caffeine consumption is reduced or eliminated, no lasting harm seems to occur in otherwise healthy adults.

Studies on the long-term effect of high doses of caffeine are inconclusive at this point. Some animal studies suggest a link between caffeine and heart disease, cancer, and birth defects. Even though the evidence is only suggestive at this point, the U.S. Food and Drug Administration recommends that women avoid or greatly reduce caffeine intake during pregnancy.

How Much Is Too Much?

While caffeine tolerance varies, studies suggest that long-term consumption of moderate amounts of caffeine probably poses little or no risk to long-term health. A moderate amount of caffeine is equivalent to that contained in two cups of coffee per day. This guideline does not apply to people who experience negative symptoms with caffeine; people should avoid caffeine in any amount that leads to unhealthy symptoms.

▶ *Think It Over*

▶ Why might your caffeine tolerance go down during high-stress times, such as during final exam week? (Hint: Feelings of stress are associated with overactivation of the sympathetic division of the autonomic nervous system.)

spinal cord. Most of the fibers in the left pyramid cross to the right side, and most of the fibers in the right pyramid cross to the left. This crossing is called the ***decussation of pyramids*** (dē′-ka-SĀ-shun) and explains why one side of the brain controls the opposite side of the body. Motor fibers that originate in the right cerebral cortex activate muscles on the left side of the body and vice versa (see Figure 10.4b). Similarly, most sensory fibers also cross over in the medulla or spinal cord, so that nearly all sensory impulses generated on one side of the body are received in the opposite side of the cerebral cortex (see Figure 10.4a).

The medulla also contains nuclei that govern several autonomic functions. These nuclei include the ***cardiovascular center,*** which regulates the rate and force of the heartbeat and the diameter of blood vessels (see Figure 15.9 on page 364); the

medullary rhythmicity area of the respiratory center, which adjusts the basic rhythm of breathing (see Figure 18.14 on page 446); and other centers in the medulla that control reflexes for vomiting, coughing, and sneezing. Finally, the medulla contains the nuclei of origin for five pairs of cranial nerves: vestibulocochlear (VIII), glossopharyngeal (IX), vagus (X), accessory (XI), and hypoglossal (XII) (see Figure 10.9 and Table 10.2).

Given the many vital activities controlled by the medulla, it is not surprising that a hard blow to the back of the head or upper neck can be fatal. Damage to the medullary rhythmicity area is particularly serious and can rapidly lead to death. Symptoms of nonfatal injury to the medulla may include cranial nerve malfunctions, paralysis and loss of sensation, and irregularities in breathing or heart rhythm.

Pons

The **pons,** which means bridge, is superior to the medulla and anterior to the cerebellum (Figures 10.7, 10.8, and 10.9). Like the medulla, the pons consists of both nuclei and tracts. The pons connects the spinal cord with the brain and parts of the brain with each other. The nuclei for the trigeminal (V), abducens (VI), and facial (VII) nerves and the vestibular branch of the vestibulocochlear (VIII) nerve originate in the pons (see Figure 10.9 and Table 10.2). The pons also contains nuclei that help regulate breathing (see Figure 18.14 on page 446).

Midbrain

The **midbrain** extends from the pons to the lower portion of the diencephalon (Figures 10.7, 10.8, and 10.9). It contains the **cerebral peduncles** (pe-DUN-kulz), a pair of tracts containing motor fibers that connect the cerebral cortex to the pons and spinal cord, and sensory fibers that connect the spinal cord to the thalamus. The nuclei for the oculomotor (III) and trochlear (IV) nerves originate in the midbrain (see Table 10.2).

Reticular Formation

Extending throughout the brain stem and the diencephalon is the **reticular formation** (ret- = net), a netlike arrangement of small areas of gray matter interspersed among threads of white matter. The main sensory function of the reticular formation is alerting the cerebral cortex to incoming sensory signals. Its main motor function is to help regulate muscle tone, the slight degree of contraction that characterizes muscles at rest.

Diencephalon

The **diencephalon** consists principally of the thalamus and hypothalamus (see Figure 10.7).

Thalamus

The **thalamus** (THAL-a-mus = inner chamber) is an oval structure above the midbrain that consists mostly of paired masses of gray matter organized into nuclei (Figure 10.10). The thalamus is the principal relay station for sensory impulses from the spinal cord, brain stem, cerebellum, and other parts of the cerebrum to the cerebral cortex and allows crude perception of some sensations, such as pain, temperature, and pressure. The thalamus also plays an essential role in awareness and in the acquisition of knowledge, which is termed **cognition** (cogni- = to get to know).

Hypothalamus

The **hypothalamus** (hypo- = under) is the small portion of the diencephalon that lies below the thalamus and above the pituitary gland (see Figures 10.7 and 10.10). Don't be fooled by its small size; the hypothalamus controls many important body activities, most of them related to homeostasis. The chief functions of the hypothalamus are as follows:

1. **Control of the ANS.** The hypothalamus controls and integrates activities of the autonomic nervous system, which regulates contraction of smooth and cardiac muscle and the secretions of many glands. Through the ANS, the hypothalamus helps to regulate activities such as heart rate, move-

Figure 10.10 ■ **Diencephalon: thalamus and hypothalamus.** Also shown are the basal ganglia (discussed later).

🔑 The thalamus is the principal relay station for sensory impulses that reach the cerebral cortex from other parts of the brain and the spinal cord.

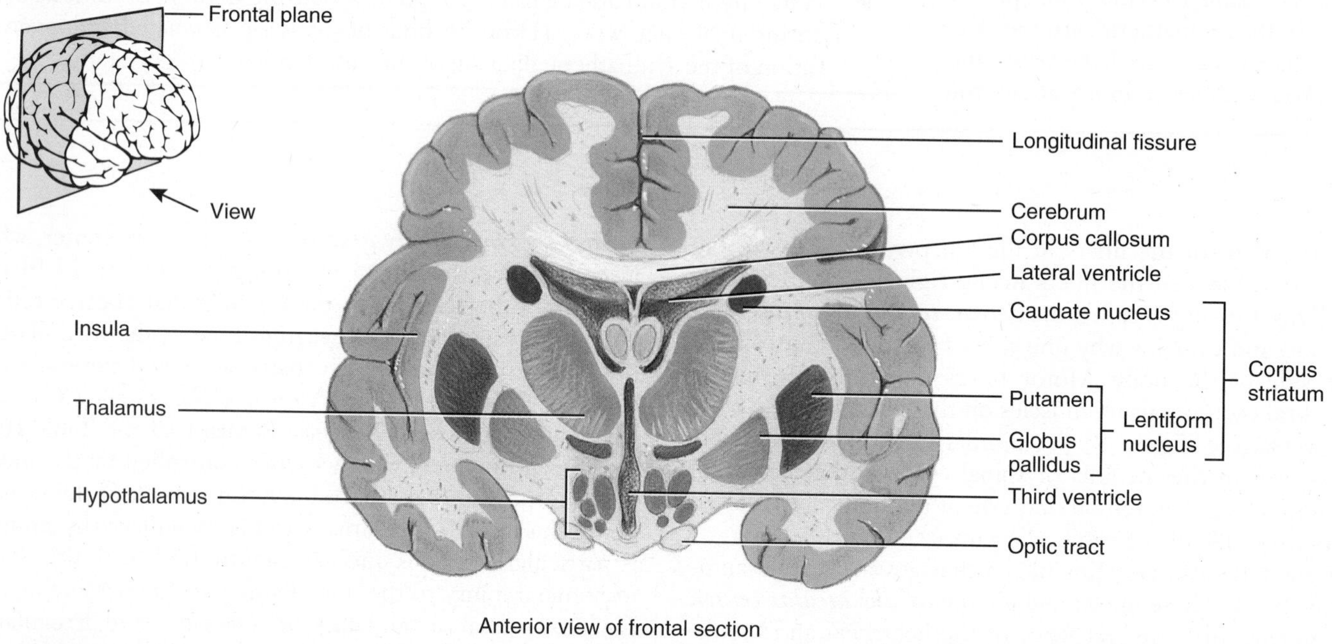

Frontal plane

View

- Longitudinal fissure
- Cerebrum
- Corpus callosum
- Lateral ventricle
- Caudate nucleus
- Putamen — Lentiform nucleus — Corpus striatum
- Globus pallidus
- Third ventricle
- Optic tract

Insula

Thalamus

Hypothalamus

Anterior view of frontal section

 In which principal part of the brain are the basal ganglia located, and what kind of tissue composes them?

ment of food through the gastrointestinal tract, and contraction of the urinary bladder.

2. **Control of the pituitary gland.** The hypothalamus controls the release of many hormones from the pituitary gland and thus serves as a primary connection between the nervous system and endocrine system. These two systems are the major control systems of the body.

3. **Regulation of emotional and behavioral patterns.** Together with the limbic system (described shortly), the hypothalamus regulates feelings of rage, aggression, pain, and pleasure and the behavioral patterns related to sexual arousal.

4. **Regulation of eating and drinking.** The hypothalamus regulates food intake through two centers. The *feeding center* is responsible for hunger sensations; when sufficient food has been ingested, the *satiety center* (sa-TĪ-e-tē; *sati-* = full, satisfied) is stimulated and sends out nerve impulses that inhibit the feeding center. The hypothalamus also contains a *thirst center.* When certain cells in the hypothalamus are stimulated by rising osmotic pressure of the interstitial fluid, they cause the sensation of thirst. The intake of water by drinking restores the osmotic pressure to normal, removing the stimulation and relieving the thirst.

5. **Control of body temperature.** If the temperature of blood flowing through the hypothalamus is above normal, the hypothalamus directs the autonomic nervous system to stimulate activities that promote heat loss. If, however, blood temperature is below normal, the hypothalamus generates impulses that promote heat production and retention.

6. **Regulation of circadian rhythms and states of consciousness.** Humans awaken and sleep in a fairly constant 24-hour rhythm called a *circadian rhythm* (ser-KĀ-dē-an), established by the hypothalamus.

Reticular Activating System, Consciousness, and Sleep

The reticular formation is responsible for the state of arousal of the brain. Because stimulating a portion of the reticular formation increases cortical activity, this area is also known as the *reticular activating system (RAS).* When the RAS is stimulated, many nerve impulses pass upward to widespread areas of the cerebral cortex. The result is a state of wakefulness called *consciousness.* Inactivation of the RAS produces *sleep,* a state of partial unconsciousness from which an individual can be aroused.

Consciousness may be altered by cocaine and amphetamines (which produce extreme alertness), meditation (which causes relaxed, focused consciousness), and alcohol and anesthetics (which cause various levels of unconsciousness called anesthesia). Damage or disease can produce a lack of consciousness called *coma.*

Cerebrum

During embryonic development, when there is a rapid increase in brain size, the gray matter of the cerebral cortex enlarges much faster than the underlying white matter. As a result, the cerebral cortex rolls and folds upon itself so that it can fit into the cranial cavity. The folds are called *gyri* (JĪ-rī = circles; singular is *gyrus*) (Figure 10.11)

The deep grooves between folds are *fissures;* the shallow grooves are *sulci* (SUL-sī = groove; singular is *sulcus,* SUL-kus). The *longitudinal fissure* separates the cerebrum into right and left halves called *cerebral hemispheres.* The hemispheres are connected internally by the *corpus callosum* (ka-LŌ-sum; *corpus* = body; *callosum* = hard), a broad band of white matter containing axons that extend between the hemispheres.

Lobes

Each cerebral hemisphere is subdivided into four lobes: *frontal lobe, parietal lobe, temporal lobe,* and *occipital lobe* (Figure 10.11). The *central sulcus* separates the frontal and parietal lobes. A major gyrus, the *precentral gyrus,* is located immediately anterior to the central sulcus. The precentral gyrus contains the primary motor area of the cerebral cortex. The *postcentral gyrus,* located immediately posterior to the central sulcus, contains the primary somatosensory area of the cerebral cortex, which is discussed shortly. A fifth part of the cerebrum, the *insula,* cannot be seen at the surface of the brain because it lies within the lateral cerebral fissure, deep to the parietal, frontal, and temporal lobes (see Figure 10.10).

White Matter

The *white matter* underlying the cerebral cortex consists of myelinated and unmyelinated axons that transmit impulses between gyri in the same hemisphere, from the gyri in one cerebral hemisphere to the corresponding gyri in the opposite cerebral hemisphere via the corpus callosum, and from the cerebrum to other parts of the brain and spinal cord.

Basal Ganglia

The *basal ganglia* are several paired masses of gray matter that lie within the white matter of each cerebral hemisphere. Although the term "ganglia" usually means collections of neuronal cell bodies outside the CNS, the basal ganglia are in the brain. The largest of the basal ganglia is the *corpus striatum* (strī-Ā-tum = striped), which consists of the *caudate nucleus* (*caud-* = tail) and the *lentiform nucleus* (*lentiform* = shaped like a lentil) (see Figure 10.10). Each lentiform nucleus, in turn, is subdivided into the *putamen* (pu-TĀ-men = shell) and the *globus pallidus* (GLŌ-bus PAL-li-dus; *globus* = ball; *pallidus* = pale) (see Figure 10.10).

The basal ganglia control large subconscious movements of skeletal muscles, such as swinging the arms while walking (such gross movements are also consciously controlled by the cerebral cortex), and they regulate muscle tone required for specific body movements. Damage to the basal ganglia results in uncontrollable shaking (tremor); muscular rigidity (stiffness); and involuntary muscle movements, such as occur in Parkinson's disease (see page 257).

Brain **247**

Figure 10.11 ■ **Cerebrum.** The inset in (a) indicates the differences among a gyrus, a sulcus, and a fissure.

The cerebrum is the seat of intelligence and provides us with the ability to read, write, and speak; make calculations and compose music; remember the past and plan for the future; and create works.

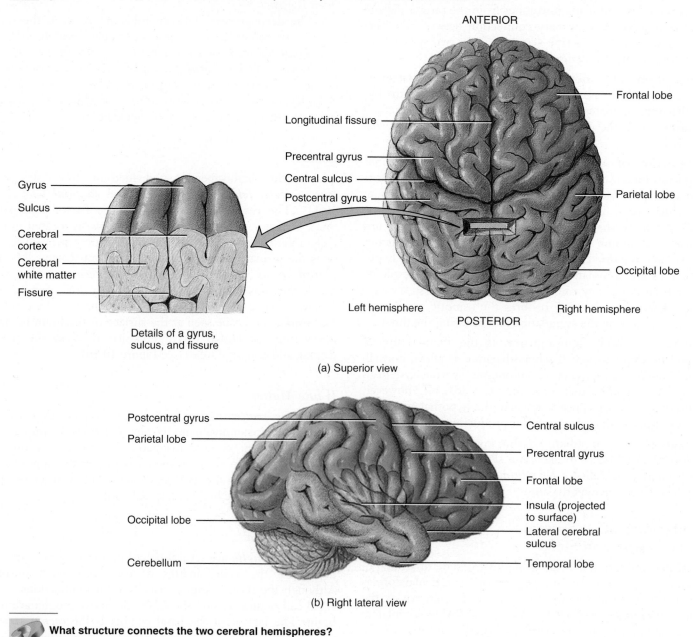

(a) Superior view

(b) Right lateral view

What structure connects the two cerebral hemispheres?

Limbic System

Certain parts of the cerebral hemispheres and diencephalon constitute the **limbic system** (*limbic* = border). It is a wishbone-shaped group of structures that encircles the brain stem and assumes a primary function in emotions such as pain, pleasure, anger, rage, fear, sorrow, sexual feelings, docility, and affection. It is therefore sometimes called the "emotional" brain. Although behavior is a function of the entire nervous system, the limbic system controls most of its involuntary aspects related to survival. Animal experiments suggest that it has a major role in controlling the overall pattern of behavior. Together with portions of the cerebrum, the limbic system also functions in memory; damage to the limbic system causes memory impairment.

Functional Areas of the Cerebral Cortex

The cerebral cortex is divided into three functional areas. The **sensory areas** receive and interpret sensory impulses, the **motor areas** control muscular movement, and the **association areas** are concerned with integrative functions such as memory, emotions, reasoning, will, judgment, personality traits, and intelligence. These areas are shown in Figure 10.12.

Figure 10.12 ■ **Functional areas of the cerebral cortex.** Most of the numbers in the circles are based on K. Brodmann's map of the cerebral cortex, first published in 1906. The numbers correlate specific regions of the brain with particular functions.

 Particular areas of the cerebral cortex process sensory, motor, and integrative signals.

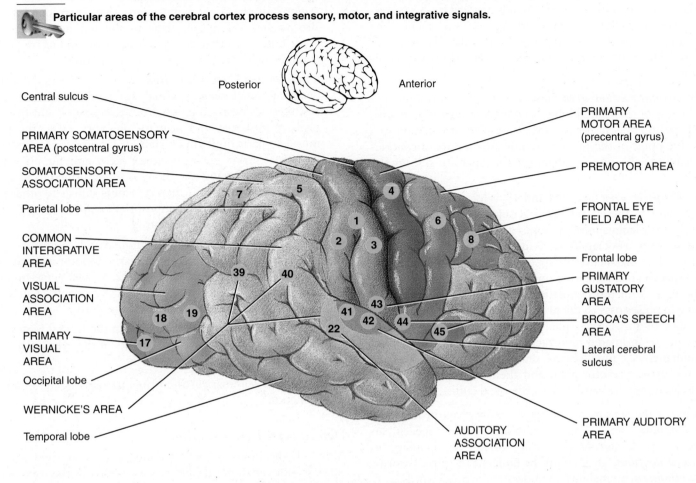

Lateral view of right cerebral hemisphere

Which part of the cerebrum localizes exactly where somatic sensations occur?

SENSORY AREAS Sensory input to the cerebral cortex flows mainly to the posterior half of the cerebral hemispheres, to regions behind the central sulci. In the cerebral cortex, primary sensory areas have the most direct connections with peripheral sensory receptors.

1. The ***primary somatosensory area*** (sō′-mat-ō-SEN-sō-rē; areas 1, 2, and 3 in Figure 10.12) is located directly posterior to the central sulcus of each cerebral hemisphere in the postcentral gyrus of the parietal lobe. It receives nerve impulses from somatic sensory receptors for touch, proprioception (joint and muscle position), pain, and temperature. Each point within the area receives nerve impulses from a specific part of the body, and the entire body is spatially represented in it. The size of the cortical area receiving impulses from a particular body part depends on the number of receptors present there rather than on the size of the part. For example, a larger region receives impulses from the lips and fingertips than from the thorax or hip. The ma-

jor function of the primary somatosensory area is to localize exactly the points of the body where sensations originate.

2. The ***primary visual area*** (area 17) is located in the occipital lobe and receives impulses that convey visual information concerning shape, color, and movement of visual stimuli.

3. The ***primary auditory area*** (areas 41 and 42) is located in the temporal lobe and receives impulses for hearing the characteristics of sound such as pitch and rhythm.

4. The ***primary gustatory area*** (area 43) is located at the base of the postcentral gyrus and receives impulses for taste.

5. The ***primary olfactory area*** (area 28) is located on the medial aspect of the temporal lobe (and thus is not visible in Figure 10.12). It receives impulses for smell.

MOTOR AREAS Just as the primary somatosensory area of the cortex has been mapped to reflect the amounts of sensory information coming from different body parts, the ***motor cortex*** has been mapped to indicate which specific cortical areas control particular muscle groups. There is a motor cortex in both the left and right cerebral hemispheres. The degree of representation is proportional to the precision of movement required of a particular body part. More cortical area is devoted to those muscles involved in skilled, complex, or delicate movement. The thumb, fingers, lips, tongue, and vocal cords have large representations. The trunk of the body has a relatively small representation.

1. The *primary motor area* (area 4) is located in the precentral gyrus of the frontal lobe. It consists of regions that control specific muscles or groups of muscles. Stimulation of a specific point of the primary motor area results in contraction of specific muscle fibers on the opposite side of the body.

2. *Broca's speech area* (areas 44 and 45) is located in the frontal lobe close to the lateral cerebral sulcus. Speaking and understanding language are complex activities that involve several sensory, association, and motor areas of the cerebral cortex. The production of speech occurs in Broca's (BRŌ-kaz) speech area, located in one frontal lobe—the *left* frontal lobe in 97% of the population. From Broca's speech area, nerve impulses pass to the premotor regions that control the muscles of the larynx, pharynx, and mouth, resulting in specific, coordinated contractions needed for speaking. Simultaneously, impulses are sent from Broca's speech area to the primary motor area, and then on to the breathing muscles to regulate airflow past the vocal cords.

Injury to the sensory or motor areas involved in language can cause *aphasia* (a-FĀ-zē-a; *a* = without; *-phasia* = speech), an inability to speak; *agraphia* (*graph* = write), an inability to write; *word deafness,* an inability to understand spoken words; or *word blindness,* an inability to understand written words.

ASSOCIATION AREAS   The *association areas* of the cerebral cortex consist of some motor and sensory areas, plus large areas on the lateral surfaces of the occipital, parietal, and temporal lobes and on the frontal lobes anterior to the motor areas. Association areas are connected with one another by tracts and include the following:

1. The *somatosensory association area* (areas 5 and 7) is just posterior to the primary somatosensory area. Its role is to integrate and interpret somatic sensations. This area permits you to determine the exact shape and texture of an object without looking at it, to determine the orientation of one object to another as they are felt, and to sense the relationship of one body part to another. Another role is the storage of memories of past sensory experiences.

2. The *visual association area* (areas 18 and 19) is located in the occipital lobe. It relates present and past visual experiences and is essential for recognizing and evaluating what is seen.

3. The *auditory association area* (area 22) is located below the primary auditory area in the temporal cortex. It allows one to realize if a sound is speech, music, or noise.

4. *Wernicke's area* (areas 22, 39, and 40) is a broad region in the temporal and parietal lobes. This area interprets the meaning of speech by translating words into thoughts. The regions in the *right* hemisphere that correspond to the Broca's and Wernicke's areas in the left hemisphere also contribute to verbal communication by adding emotional content such as anger or joy to spoken words.

5. The *common integrative area* (areas 5, 7, 39, and 40), bordered by somatosensory, visual, and auditory association areas, receives nerve impulses from these areas, and from the primary gustatory and primary olfactory areas, the thalamus, and parts of the brain stem. It integrates sensory interpretations from the association areas and impulses from other areas, allowing one thought to be formed from a variety of sensory inputs. It then transmits signals to other parts of the brain to cause the appropriate response to the interpretation of the sensory signals.

6. The *premotor area* (area 6), immediately anterior to the primary motor area, deals with learned motor activities of a complex and sequential nature. It generates nerve impulses that cause a specific group of muscles to contract in a specific sequence, for example, to write a word.

7. The *frontal eye field area* (area 8) in the frontal cortex controls voluntary scanning movements of the eyes, such as occur while reading this sentence.

Table 10.1 summarizes the principal parts of the brain and their functions.

Hemispheric Lateralization

Although the brain is fairly symmetrical on its right and left sides, subtle anatomical differences between the two hemispheres exist. For example, in about two-thirds of the population, a region of the temporal lobe that includes Wernicke's area is 50% larger on the left side than on the right side. They are also functionally different in some ways, with each hemisphere specializing in certain functions. This functional asymmetry is termed *hemispheric lateralization.*

The left hemisphere receives sensory signals from and controls the right side of the body, whereas the right hemisphere receives sensory signals from and controls the left side of the body. In most people, the left hemisphere is more important for spoken and written language, numerical and scientific skills, ability to use and understand sign language, and reasoning. Patients with damage in the left hemisphere, for example, often have difficulty speaking. The right hemisphere is more important for musical and artistic awareness; spatial and pattern perception; recognition of faces and emotional content of language; and for generating mental images of sight, sound, touch, taste, and smell. Patients with damage in the right hemispheric regions that correspond to Broca's and Wernicke's areas in the left hemisphere speak in a monotonous voice, having lost the ability to impart emotional inflection to what they say. In general, lateralization seems less pronounced in females than in males, both for language (left hemisphere) and for visual and spatial skills (right

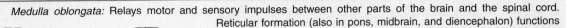

Table 10.1 / Summary of Functions of Principal Parts of the Brain

Part	Function
Brain stem 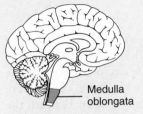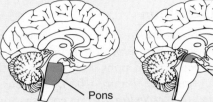 Medulla oblongata Pons Midbrain	*Medulla oblongata:* Relays motor and sensory impulses between other parts of the brain and the spinal cord. Reticular formation (also in pons, midbrain, and diencephalon) functions in consciousness and arousal. Vital centers regulate heartbeat, breathing (together with pons), and blood vessel diameter. Other centers coordinate swallowing, vomiting, coughing, sneezing, and hiccupping. Contains nuclei of origin for cranial nerves VIII, IX, X, XI, and XII. *Pons:* Relays impulses from one side of the cerebellum to the other and between the medulla and midbrain. Contains nuclei of origin for cranial nerves V, VI, VII, and VIII. Together with the medulla, helps control breathing. *Midbrain:* Relays motor impulses from the cerebral cortex to the pons and sensory impulses from the spinal cord to the thalamus. Contains nuclei of origin for cranial nerves III and IV.
Diencephalon 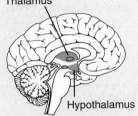 Thalamus Hypothalamus	*Thalamus:* Relays all sensory impulses to the cerebral cortex. Provides crude perception of touch, pressure, pain, and temperature. Also functions in cognition and awareness. *Hypothalamus:* Controls and integrates activities of the autonomic nervous system and pituitary gland. Regulates emotional and behavioral patterns and circadian rhythms. Controls body temperature and regulates eating and drinking behavior. Helps maintain waking state and establishes patterns of sleep.
Cerebrum 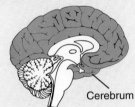 Cerebrum	Sensory areas interpret sensory impulses, motor areas control muscular movement, and association areas function in emotional and intellectual processes. Basal ganglia coordinate gross, automatic muscle movements and regulate muscle tone. Limbic system functions in emotional aspects of behavior related to survival.
Cerebellum 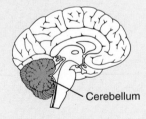 Cerebellum	Compares intended movements with what is actually happening to coordinate complex, skilled movements. Regulates posture and balance.

hemisphere). One possible reason for this is that females tend to have a larger anterior commissure and a larger posterior portion of the corpus callosum than males, both of which provide communication between the two hemispheres.

Memory

The portions of the brain known to be associated with memory include the association cortex of the frontal, parietal, occipital, and temporal lobes; parts of the limbic system; and the diencephalon. Although memory has been studied by scientists for decades, there is still no satisfactory explanation for how we remember. **Memory** is the ability to recall thoughts. For an experience to become part of memory, it must produce changes in the central nervous system.

Memory is divided into two categories, short-term and long-term. **Short-term memory** lasts only seconds or hours and is the ability to recall bits of information. One example is looking up a telephone number and remembering it long enough to dial. **Long-term memory,** on the other hand, lasts from days to years. If you frequently use a telephone number, it becomes part of long-term memory and can be retrieved for use for quite a long period.

Electroencephalogram (EEG)

At any instant, individual brain cells are generating millions of nerve impulses. Taken together, these electrical signals are called **brain waves.** Brain waves generated by neurons close to the brain surface, mainly neurons in the cerebral cortex, can be detected by placing sensors called electrodes on the forehead and scalp. A record of such waves is called an **electroencephalogram** (e-lek′-trō-en-SEF-a-lō-gram′) or **EEG.** Electroencephalograms are useful both in studying normal brain functions, such as changes that occur during sleep, and in diagnosing a variety of

brain disorders, such as epilepsy, tumors, metabolic abnormalities, sites of trauma, and degenerative diseases. An EEG may also be used to establish **brain death,** the complete absence of brain waves in two EEGs taken 24 hours apart.

Cerebellum

The **cerebellum** is the second-largest portion of the brain and consists of two **cerebellar hemispheres.** It is located posterior to the medulla and pons and inferior to the occipital lobes of the cerebrum (see Figure 10.7). The surface of the cerebellum, called the **cerebellar cortex,** consists of gray matter. Beneath the cortex are **white matter tracts** that resemble branches of a tree. Deep within the white matter are masses of gray matter, the **cerebellar nuclei.** The cerebellum is attached to the brain stem by three paired bundles of fibers called **cerebellar peduncles** (see Figure 10.9).

The cerebellum compares intended movements pro-

Table 10.2 / Summary of Cranial Nerves

Number and Name*	Type (Sensory or Mixed)	Location	Function
Cranial nerve I: olfactory (ol-FAK-tō-rē; *olfact-* = to smell) — Olfactory bulb — Olfactory (I) nerve — Olfactory tract	Sensory.	Consists of axons in lining of nose that terminate in olfactory bulb. The olfactory tract extends via two pathways to olfactory areas of cerebral cortex.	Smell.
Cranial nerve II: optic (OP-tik; *opti-* = the eye, vision) — Optic (II) nerve — Optic tract	Sensory.	Consists of axons from retina of the eye that terminate in thalamus. Neurons carrying impulses for vision extend from thalamus to the primary visual area (area 17) in occipital lobe of cerebral cortex.	Sight.

*A mnemonic device used to remember the names of the nerves is: "Oh, Oh, Oh, to touch and feel very green vegetables—AH!" The initial letter of each word (and both letters of "AH") corresponds to the initial letter of each pair of cranial nerves.

grammed by motor areas in the cerebral cortex with what is actually happening. It constantly receives sensory impulses from muscles, tendons, joints, equilibrium receptors, and visual receptors. The cerebellum helps to smooth and coordinate complex sequences of skeletal muscle contractions. It is the main brain region that regulates posture and balance and makes possible all skilled motor activities, from catching a baseball to dancing.

Damage to the cerebellum through trauma or disease disrupts muscle coordination, a condition called *ataxia* (*a-* = without; *-taxia* = order). Blindfolded people with ataxia cannot touch the tip of their nose with a finger because they cannot coordinate movement with their sense of where a body part is located. Another sign of ataxia is a changed speech pattern due to uncoordinated speech muscles. Cerebellar damage may also result in disturbances of gait (staggering or abnormal walking movements) and severe dizziness. Individuals who consume too much alcohol show signs of ataxia because alcohol inhibits activity of the cerebellum.

CRANIAL NERVES

Objective: • **Identify the 12 pairs of cranial nerves by name, number, type, location, and function.**

Cranial nerves, like spinal nerves, are part of the somatic nervous system (SNS) of the peripheral nervous system. Of the 12 pairs of ***cranial nerves,*** 10 originate from the brain stem (see Figure 10.9). The cranial nerves are designated with roman numerals and with names. The roman numerals indicate the order in which the nerves arise from the brain (anterior to posterior). The names indicate the distribution or function. Two cranial nerves contain only sensory fibers and thus are sensory nerves. The remainder contain both sensory and motor fibers and are mixed nerves. The cell bodies of sensory fibers are found outside the brain, whereas the cell bodies of motor fibers lie in nuclei within the brain. Table 10.2 lists each of the cranial nerves, along with its location and functions.

Number and Name	Type (Sensory or Mixed)	Location	Function
Cranial nerve III: oculomotor (ok′-ū-lō-MŌ-tor; *oculo-* = eye; *-motor* = mover) Oculomotor (III) nerve	Mixed, mainly motor.	*Sensory portion:* Consists of sensory axons from proprioceptors in eyeball muscles that terminate in midbrain. *Motor portion:* Originates in midbrain and passes to upper eyelid and the superior rectus, medial rectus, inferior rectus, and inferior oblique muscles (extrinsic eyeball muscles); parasympathetic axons of ANS pass to ciliary muscle of eyeball and sphincter muscle of iris.	Muscle sense (proprioception). Movement of eyelid and eyeball; alters lens for near vision and constricts pupil.
Cranial nerve IV: trochlear (TRŌK-lē-ar; *trochle-* = a pulley) Trochlear (IV) nerve	Mixed, mainly motor.	*Sensory portion:* Consists of sensory axons from proprioceptors in superior oblique muscles (extrinsic eyeball muscles) that terminate in midbrain. *Motor portion:* Originates in midbrain and passes to superior oblique muscles.	Muscle sense (proprioception). Movement of eyeball.

(continues)

Table 10.2 / Summary of Cranial Nerves (continued)

Number and Name	Type (Sensory or Mixed)	Location	Function
Cranial nerve V: trigeminal (trī-JEM-i-nal = triple, for its three branches) Trigeminal (V) nerve	Mixed.	*Sensory portion:* Consists of three branches—*ophthalmic, maxillary,* and *mandibular*—that terminate in pons. Sensory portion also consists of sensory axons from muscles used in chewing. *Motor portion:* Originates in pons and passes to muscles used in chewing.	Conveys sensations for touch, pain, and temperature; muscle sense (proprioception). Chewing.
Cranial nerve VI: abducens (ab-DŪ-senz; *ab-* = away; *-ducens* = to lead) Abducens (VI) nerve	Mixed, mainly motor.	*Sensory portion:* Consists of axons from proprioceptors in lateral rectus muscles (extrinsic eyeball muscles) that terminate in pons. *Motor portion:* Originates in pons and passes to lateral rectus muscles.	Muscle sense (proprioception). Movement of eyeball.
Cranial nerve VII: facial (FĀ-shal = face) Facial (VII) nerve	Mixed.	*Sensory portion:* Consists of axons from taste buds on tongue that terminate in pons. From there, axons extend to the thalamus and then to the gustatory areas in parietal lobes of cerebral cortex. Also contains sensory axons from proprioceptors in muscles of face and scalp. *Motor portion:* Originates in pons and passes to facial, scalp, and neck muscles; parasympathetic axons of ANS pass to lacrimal (tear) glands and salivary glands.	Taste and muscle sense (proprioception). Facial expressions; secretion of tears and saliva.

Number and Name	Type (Sensory or Mixed)	Location	Function
Cranial nerve VIII: vestibulocochlear (ves-tib′-ū-lō-KOK-lē-ar; *vestibulo-* = small cavity; *-cochlear* = a spiral, snail-like) Vestibulocochlear (VIII) nerve	Mixed, mainly sensory.	*Vestibular branch, sensory part:* Consists of axons from semicircular canals, saccule, and utricle (organs of equilibrium) that terminate in pons and cerebellum.	Equilibrium.
		Vestibular branch, motor part: Synapses with sensory receptors (hair cells) for equilibrium.	Adjusts sensitivity of hair cells.
		Cochlear branch, sensory part: Consists of axons from spiral organ (organ of hearing) that terminate in the thalamus; neurons then extend from thalamus to primary auditory area (areas 41 and 42) in temporal lobe of cerebral cortex.	Hearing.
		Cochlear branch, motor part: Synapses with sensory receptors (hair cells) for hearing.	Modifies responses of hair cells.
Cranial nerve IX: glossopharyngeal (glos′-ō-fah-RIN-jē-al; *glosso-* = tongue; *-pharyngeal* = throat) Glossopharyngeal (IX) nerve	Mixed.	*Sensory portion:* Consists of axons from taste buds and somatic sensory receptors on part of tongue, from proprioceptors in some swallowing muscles, and from stretch receptors in carotid sinus and chemoreceptors in carotid body that terminate in medulla.	Taste and somatic sensations (touch, pain, temperature) from tongue; muscle sense (proprioception); monitoring blood pressure; monitoring oxygen and carbon dioxide in blood for regulation of breathing.
		Motor portion: Originates in medulla and passes to swallowing muscles of throat; parasympathetic axons of ANS pass to a salivary gland.	Swallowing; secretion of saliva.
Cranial nerve X: vagus (VĀ-gus; *vagus* = vagrant or wandering) Vagus (X) nerve	Mixed.	*Sensory portion:* Consists of axons from proprioceptors in muscles of neck and throat, from stretch receptors and chemoreceptors in carotid sinus and carotid body, from chemoreceptors in aortic body, and from visceral sensory receptors in most organs of the thoracic and abdominal cavities that terminate in medulla and pons.	Somatic sensations (touch, pain, temperature) from throat and pharynx; monitoring of blood pressure; monitoring of oxygen and carbon dioxide in blood for regulation of breathing; sensations from visceral organs in thorax and abdomen.
		Motor portion: Originates in medulla. Somatic axons innervate skeletal muscles of throat and neck. Parasympathetic axons of the ANS supply smooth muscle in the airways, esophagus, stomach, small intestine, most of large intestine, and gallbladder; cardiac muscle in the heart; and glands of the gastrointestinal tract.	Swallowing, coughing, and voice production; smooth muscle contraction and relaxation in organs of the gastrointestinal tract; slowing of the heart rate; secretion of digestive fluids.

(continues)

Table 10.2 / Summary of Cranial Nerves (continued)

Number and Name	Type (Sensory or Mixed)	Location	Function
Cranial nerve XI: accessory (ak-SES-ō-rē = assisting)  Accessory (XI) nerve	Mixed, mainly motor.	*Sensory portion:* Consists of axons from proprioceptors in muscles of throat and voice box.	Muscle sense (proprioception).
		Motor portion: Cranial part (from medulla) supplies muscles of throat, and spinal part (from cervical spinal cord) supplies sternocleidomastoid and trapezius muscles.	Cranial part mediates swallowing movements, and spinal part governs movements of head and shoulders.
Cranial nerve XII: hypoglossal (hī′-pō-GLOS-al; hypo- = below; -glossal = tongue) Hypoglossal (XII nerve	Mixed, mainly motor.	*Sensory portion:* Consists of axons from proprioceptors in tongue muscles that terminate in the medulla.	Muscle sense (proprioception).
		Motor portion: Originates in medulla and supplies muscles of tongue.	Movement of tongue during speech and swallowing.

COMMON DISORDERS

Spinal Cord Injury

The spinal cord may be damaged by a tumor either within or adjacent to the spinal cord, herniated intervertebral discs, blood clots, penetrating wounds, or other trauma. Depending on the location and extent of spinal cord damage, paralysis may occur. *Monoplegia* (mono- = one; -plegia = blow or strike) is paralysis of a single limb. *Diplegia* (di- = two) is paralysis of both upper limbs or both lower limbs. *Paraplegia* (para- = beyond) is paralysis of both lower limbs and generally the lower trunk. *Hemiplegia* (hemi- = half) is paralysis of the upper limb, trunk, and lower limb on one side of the body, and *quadriplegia* (quad- = four) is paralysis of all four limbs.

Shingles

Shingles is an acute infection of the peripheral nervous system caused by *herpes zoster* (HER-pēz ZOS-ter), the virus that also causes chicken-

pox. After a person recovers from chickenpox, the virus retreats to a posterior root ganglion. If reactivated, the virus may leave the ganglion and travel down sensory neurons of the skin. The result is pain, discoloration of the skin, and a characteristic line of skin blisters. The line of blisters follows the path of the infected sensory nerves.

Cerebrovascular Accident

The most common brain disorder is a *cerebrovascular accident (CVA)*, also called a *stroke* or *brain attack*. CVAs affect 500,000 people a year in the United States and represent the third leading cause of death, behind heart attacks and cancer. A CVA is characterized by abrupt onset of persisting symptoms, such as paralysis or loss of sensation, that arise from destruction of brain tissue. Common causes of CVAs are hemorrhage from a blood vessel in the pia mater or brain, blood clots, and formation of cholesterol-containing atherosclerotic plaques that block brain blood flow.

The risk factors implicated in CVAs are high blood pressure, high blood cholesterol, heart disease, narrowed carotid arteries, transient ischemic attacks (discussed next), diabetes, smoking, obesity, and excessive alcohol intake.

Transient Ischemic Attack

A *transient ischemic attack (TIA)* is an episode of temporary cerebral dysfunction caused by impaired blood flow to part of the brain. Symptoms include dizziness, weakness, numbness, or paralysis in a limb or in one side of the body; drooping of one side of the face; headache; slurred speech or difficulty understanding speech; and a partial loss of vision or double vision. Sometimes nausea or vomiting also occurs. The onset of symptoms is sudden and reaches maximum intensity almost immediately. A TIA usually persists for 5 to 10 minutes and only rarely lasts as long as 24 hours; it leaves no persistent neurological deficits. The causes of TIAs include blood clots, atherosclerosis, and certain blood disorders.

Headache

One of the most common human afflictions is *headache*. Based on origin, two general types are distinguished: intracranial (within the brain) and extracranial (outside the brain). Serious headaches of intracranial origin are caused by brain tumors, blood vessel abnormalities, inflammation of the brain or meninges, decrease in oxygen supply to the brain, or damage to brain cells. Extracranial headaches are related to infections of the eyes, ears, nose, and sinuses. Tension headaches are extracranial headaches associated with stress, fatigue, and anxiety and usually occur in the occipital and temporal muscles. A *migraine* is usually a generalized headache but may affect only one side. It is accompanied by nausea, anorexia, and vomiting. Evidence suggests that a migraine is a genetic disorder that is related to regional alterations in cerebral blood flow.

Parkinson's Disease

Parkinson's disease (PD) is a progressive disorder of the CNS that typically affects its victims at around age 60. The cause is unknown, but toxic environmental factors are suspected, in part because only 5% of PD patients have a family history of the disease. Neurons that terminate in the basal ganglia and release the neurotransmitter dopamine (DA) degenerate in PD. Involuntary skeletal muscle contractions often interfere with voluntary movement in PD. For instance, the muscles of the upper limb may alternately contract and relax, causing the hand to shake. This shaking, called *tremor,* is the most common symptom of PD. Also, muscle tone may increase greatly, causing rigidity of the involved body part. Rigidity of the facial muscles gives the face a masklike appearance. The expression is characterized by a wide-eyed, unblinking stare and a slightly open mouth with uncontrolled drooling.

Motor performance is also impaired by *bradykinesia* (brady- = slow), in which activities such as shaving, cutting food, and buttoning a blouse take longer and become increasingly more difficult as the disease progresses. Muscular movements are performed not only slowly but with decreasing range of motion, or *hypokinesia* (hypo- = under). For example, handwritten letters get smaller, become poorly formed, and eventually become illegible. Often, walking is impaired; steps become shorter and shuffling, and arm swing diminishes. Even speech may be affected.

Alzheimer's Disease

Alzheimer's disease (ALTZ-hī-merz) *(AD)* is a disabling senile dementia that afflicts about 11% of the population over age 65. In the United States, AD afflicts 4 million people and claims over 100,000 lives a year, making it the fourth leading cause of death among the elderly, after heart disease, cancer, and stroke. A person with AD usually dies of some complication that afflicts bedridden patients, such as pneumonia.

Individuals with AD initially have trouble remembering recent events. They then become confused and forgetful, often repeating questions or getting lost while traveling to previously familiar places. Disorientation increases; memories of past events disappear; and episodes of paranoia, hallucination, or violent changes in mood may occur. As their minds continue to deteriorate, AD patients lose their ability to read, write, talk, eat, or walk. At autopsy, brains of AD victims show three distinct structural abnormalities: (1) loss of neurons that liberate acetylcholine from a brain region called the nucleus basalis, located below the globus pallidus; (2) beta-amyloid plaques, clusters of abnormal proteins deposited outside neurons; and (3) neurofibrillary tangles, abnormal bundles of protein filaments inside neurons in affected brain regions.

MEDICAL TERMINOLOGY AND CONDITIONS

Analgesia (an′-al-JĒ-zē-a; *an* = without; *algia* = painful condition) Pain relief.

Anesthesia (an′-es-THĒ-zē-a; *esthesia* = feeling) Loss of sensation.

Delirium (de-LIR-ē-um = off the track) Also called *acute confusional state (ACS).* A transient disorder of abnormal cognition and disordered attention accompanied by disturbances of the sleep-wake cycle and psychomotor behavior (hyperactivity or hypoactivity of movements and speech).

Dementia (de-MEN-sha; *de-* = away from; *-mentia* = mind) A mental disorder that results in permanent or progressive loss of intellectual abilities, including impairment of memory, judgment, and abstract thinking, and changes in personality.

Encephalitis (en′-sef-a-LĪ-tis) An acute inflammation of the brain caused by a direct viral attack or an allergic reaction to any of the many viruses that are normally harmless to the central nervous system. If the virus affects the spinal cord as well, the condition is called *encephalomyelitis.*

Lethargy (LETH-ar-jē) A condition of functional sluggishness.

Meningitis (men-in-JĪ-tis) Inflammation of the meninges.

Nerve block Loss of sensation due to injection of a local anesthetic; an example is local dental anesthesia.

Neuralgia (noo-RAL-ja; *neur-* = nerve; *-algia* = pain) Attacks of pain along the entire length or a branch of a peripheral sensory nerve.

Neuritis (*neur-* = nerve; *-itis* = inflammation) Inflammation of one or several nerves, resulting from irritation caused by bone fractures, contusions, or penetrating injuries. Additional causes include infections; vitamin deficiency (usually thiamin); and poisons such as carbon monoxide, carbon tetrachloride, heavy metals, and some drugs.

Reye's (RĪZ) *syndrome (RS)* Occurs after a viral infection, particularly chickenpox or influenza, most often in children or teens who have taken aspirin; characterized by vomiting and brain dysfunction (disorientation, lethargy, and personality changes) that may progress to coma and death.

Sciatica (sī-AT-i-ka) A type of neuritis characterized by severe pain along the path of the sciatic nerve or its branches; may be caused by a slipped disc, pelvic injury, osteoarthritis of the backbone, or pressure from an expanding uterus during pregnancy.

Stupor (STOO-por) Unresponsiveness from which a patient can be aroused only briefly and only by vigorous and repeated stimulation.

■ STUDY OUTLINE

Spinal Cord (p. 234)

1. The spinal cord is protected by the vertebral canal, meninges, cerebrospinal fluid, and vertebral ligaments.

2. The meninges are three connective tissue coverings of the spinal cord and brain: dura mater, arachnoid, and pia mater.

3. Removal of cerebrospinal fluid from the subarachnoid space is called a spinal tap. The procedure is used to remove CSF and to introduce antibiotics, anesthetics, and chemotherapy.

4. The spinal cord extends from the foramen magnum of the occipital bone to the top of the second lumbar vertebra.

5. The spinal cord contains cervical and lumbar enlargements that serve as points of origin for nerves to the limbs.

6. Nerves arising from the lowest portion of the cord are called the cauda equina.

7. The gray matter in the spinal cord is divided into horns and the white matter into columns. Parts of the spinal cord observed in cross section are the central canal; anterior, posterior, and lateral gray horns; anterior, posterior, and lateral white columns; and ascending and descending tracts.

8. The spinal cord conveys sensory and motor information by way of the sensory (ascending) and motor (descending) tracts.

9. One major function of the spinal cord is to convey sensory impulses from the periphery to the brain, and to conduct motor impulses from the brain to the periphery. The other major function is to serve as a reflex center.

10. The basic components of a reflex arc are a receptor, a sensory neuron, an integrating center, a motor neuron, and an effector.

11. A reflex is a quick, automatic response to a stimulus that passes along a reflex arc. Reflexes represent the body's principal mechanisms for responding to changes (stimuli) in the internal and external environments.

Spinal Nerves (p. 239)

1. The 31 pairs of spinal nerves are named and numbered according to the region and level of the spinal cord from which they emerge.

2. There are 8 pairs of cervical, 12 pairs of thoracic, 5 pairs of lumbar, 5 pairs of sacral, and 1 pair of coccygeal nerves.

3. Spinal nerves are attached to the spinal cord by means of a posterior root and an anterior root.

4. All spinal nerves are mixed nerves containing sensory and motor fibers.

5. Branches of a spinal nerve include the posterior ramus, anterior ramus, meningeal branch, and rami communicantes.

6. The anterior rami of spinal nerves, except for T2 to T11, form networks of nerves called plexuses.

7. The principal plexuses are called the cervical, brachial, lumbar, and sacral plexuses.

8. Nerves T2 to T11 do not form plexuses and are called intercostal nerves.

Brain (p. 241)

1. The principal parts of the brain are the brain stem, diencephalon, cerebrum, and cerebellum (See Table 10.1 on page 251).

2. The brain stem consists of the medulla oblongata, pons, and mid-brain. The diencephalon consists of the thalamus and hypothalamus.

3. The brain is supplied with oxygen and nutrients by the cerebral arterial circle, or circle of Willis.

4. Any interruption of the oxygen supply to the brain can weaken, permanently damage, or kill brain cells. Glucose deficiency may produce dizziness, convulsions, and unconsciousness.

5. The blood-brain barrier (BBB) limits the passage of certain material from the blood into the brain.

6. The brain is protected by cranial bones, meninges, and cerebrospinal fluid.

7. The cranial meninges are continuous with the spinal meninges and are named dura mater, arachnoid, and pia mater.

8. Cerebrospinal fluid is formed in the choroid plexuses and circulates continually through the subarachnoid space, ventricles, and central canal.

9. Cerebrospinal fluid protects by serving as a shock absorber. It also delivers nutritive substances from the blood and removes wastes.

10. The medulla oblongata, or medulla, is continuous with the upper part of the spinal cord. It contains regions for regulating heart rate, diameter of blood vessels, respiratory rate, swallowing, coughing, vomiting, sneezing, and hiccupping. The vestibulocochlear, accessory, vagus, and hypoglossal nerves originate at the medulla.

11. The pons connects the spinal cord with the brain and links parts of the brain with one another; it relays impulses related to voluntary skeletal movements from the cerebral cortex to the cerebellum, and it contains two regions that control respiration. The trigeminal, abducens, facial, and vestibular branches of the vestibulocochlear nerves originate at the pons.

12. The midbrain conveys motor impulses from the cerebrum to the cerebellum and cord, and sensory impulses from cord to thalamus.

13. The reticular formation is a netlike arrangement of gray and white matter extending throughout the brain stem that alerts the cerebral cortex to incoming sensory signals and helps regulate muscle tone.

14. The diencephalon consists of the thalamus and hypothalamus.

15. The thalamus contains nuclei that serve as relay stations for sensory impulses to the cerebral cortex. It also provides crude recognition of pain, temperature, touch, pressure, and vibration.

16. The hypothalamus, located below the thalamus, controls and integrates the autonomic nervous system and pituitary gland, functions in rage and aggression, controls body temperature, regulates food and fluid intake, and maintains consciousness and sleep patterns.

17. The reticular activating system functions in arousal (awakening from deep sleep) and consciousness (wakefulness).

18. Sleep is a state of partial unconsciousness from which an individual can be aroused.

19. The cerebrum is the largest part of the brain. Its cortex contains convolutions, fissures, and sulci.

20. The cerebral lobes are named the frontal, parietal, temporal, and occipital.

21. The white matter under the cerebral cortex consists of myelinated axons extending in three principal directions.

22. The basal ganglia are paired masses of gray matter in the cerebral hemispheres that help to control muscular movements.

23. The limbic system, found in the cerebral hemispheres and diencephalon, functions in emotional aspects of behavior and memory.

24. The sensory areas of the cerebral cortex receive and interpret sensory impulses. The motor areas govern muscular movement. The association areas are concerned with emotional and intellectual processes.

25. Although the brain is fairly symmetrical on its right and left sides, subtle anatomical differences exist between the two hemispheres, and each has unique functions.

26. The left hemisphere receives sensory signals from and controls the right side of the body; it also is more important for language, numerical and scientific skills, and reasoning.

27. The right hemisphere receives sensory signals from and controls the left side of the body. It also is more important for musical and artistic awareness; spatial and pattern perception; recognition of faces; emotional content of language; and generating mental images of sight, sound, touch, taste, and smell.

28. Memory, the ability to recall thoughts, is generally classified into two types: short-term memory and long-term memory.

29. Brain waves generated by the cerebral cortex are recorded as an electroencephalogram (EEG), which may be used to diagnose epilepsy, infections, and tumors.

30. The cerebellum occupies the inferior and posterior aspects of the cranial cavity. It consists of two cerebellar hemispheres with a cerebellar cortex of gray matter and an interior of white matter tracts.

31. It attaches to the brain stem by three pairs of cerebellar peduncles.

32. The cerebellum coordinates skeletal muscles and maintains normal muscle tone and body equilibrium.

Cranial Nerves (p. 253)

1. Twelve pairs of cranial nerves originate from the brain.

2. Like spinal nerves, cranial nerves are part of the PNS. See Table 10.2 on page 000 for the names, types, and functions of each of the cranial nerves.

▮ SELF-QUIZ

1. Trace a reflex arc from the stimulus to the response:

    1. effector   2. integrating center
    3. motor neuron   4. receptor
    5. sensory neuron

    a. 3, 1, 4, 5, 2   **b.** 1, 5, 2, 3, 4   **c.** 4, 3, 2, 5, 1
    d. 5, 2, 3, 4, 1   **e.** 4, 5, 2, 3, 1

2. Which of the following would carry sensory nerve impulses?

    a. anterior spinothalamic tract   **b.** anterior root   **c.** lateral corticospinal tract   **d.** direct pathways   **e.** pyramids

3. Inability to distinguish keys in your pocket by touch could indicate damage to the

    a. gray matter of the cerebellum   **b.** lateral spinothalamic tract   **c.** posterior column–medial lemniscus pathway   **d.** anterior ramus   **e.** primary motor cortex

4. Carpal tunnel syndrome is due to damage to a nerve in the

    a. lumbar plexus   **b.** cervical plexus   **c.** brachial plexus   **d.** cauda equina   **e.** sacral plexus

5. A needle used in a spinal tap would penetrate (in order):

    1. arachnoid
    2. dura mater
    3. epidural space
    4. subarachnoid space

    a. 1, 2, 3, 4   **b.** 2, 3, 1, 4   **c.** 3, 1, 4, 2   **d.** 3, 2, 1, 4
    e. 4, 1, 2, 3

6. The diencephalon is composed of the

    a. medulla, pons, and hypothalamus   **b.** midbrain, hypothalamus, and thalamus   **c.** cerebellum and midbrain
    d. medulla, pons, and midbrain   **e.** hypothalamus and thalamus

7. Which of the following statements about the blood supply to the brain is NOT true?

    a. The brain needs a constant supply of glucose delivered by the blood.   **b.** The structure of the brain capillaries allows selective passage of certain materials from the blood into the brain.   **c.** The oxygen brought to the brain can be stored for future use.   **d.** Lack of blood flow to the brain can cause lysosomes in the brain cells to break open and destroy the brain cells.   **e.** The cerebral arterial circle delivers blood to the brain.

8. After a car accident, Joe exhibits severe dizziness, difficulty walking, and slurred speech. He may have damaged his

    a. cerebellum   **b.** pons   **c.** reticular activating system   **d.** fifth cranial nerve   **e.** midbrain

9. Which of the following is NOT a function of cerebrospinal fluid?

    a. protection   **b.** circulation   **c.** conduction of nerve impulses   **d.** nutrition   **e.** shock absorption

10. Which part of the brain contains the centers that control the heart rate and breathing rhythm?

    a. medulla   **b.** midbrain   **c.** cerebellum   **d.** thalamus   **e.** pons

11. The part of the brain that serves as a link between the nervous and endocrine systems is the

    a. reticular formation   **b.** hypothalamus   **c.** pons   **d.** brain stem   **e.** cerebellum

12. Which of the following is NOT a function of the hypothalamus?

    a. regulates food intake   **b.** controls body temperature   **c.** regulates feelings of rage and aggression   **d.** helps establish sleep patterns   **e.** allows crude interpretation of pain and pressure

13. The part(s) of the brain concerned with memory, reasoning, judgment and intelligence is (are) the

    a. sensory areas   **b.** limbic system   **c.** motor areas   **d.** cerebellum   **e.** association areas

14. A broad band of white matter that connects the cerebral hemispheres is the

    a. corpus callosum   **b.** gyrus   **c.** insula   **d.** ascending tract   **e.** basal ganglia

15. The ringing of your clock alarm in the morning wakes you up by stimulating the
    a. thalamus   b. reticular activating system   c. Broca's area
    d. basal ganglia   e. spinal cord

16. Match the following functions to the primary lobe in which they are located:

    ____ a. contains primary visual area that allows interpretation of shape and color
    ____ b. receives impulses for smell
    ____ c. contains primary motor area that controls muscle movement
    ____ d. receives sensory impulses for touch, pain, and temperature

    A. frontal lobe
    B. parietal lobe
    C. occipital lobe
    D. temporal lobe

17. When entering a restaurant, you are bombarded with many different sensory stimuli. The part of the brain that combines all of those sensory inputs so that you can respond appropriately is the
    a. somatosensory association area   b. common integrative area
    c. premotor area   d. Wernicke's area   e. hypothalamus

18. Which cranial nerves contain only sensory fibers?
    a. olfactory, optic, and glossopharyngeal   b. optic and oculomotor   c. optic and trochlear   d. optic and olfactory
    e. vagus and facial

19. Which two of the following cranial nerves are NOT involved in controlling movement of the eyeball?
    a. oculomotor   b. trochlear   c. facial   d. abducens
    e. trigeminal

20. Match the following:

    ____ a. organization of white matter in the spinal cord
    ____ b. absorb cerebrospinal fluid
    ____ c. extension of nerves beyond the end of the spinal cord
    ____ d. folds of the cerebral cortex
    ____ e. contains the sensory fibers of a spinal nerve
    ____ f. contains the motor fibers of a spinal nerve
    ____ g. separates the cerebrum into right and left halves
    ____ h. divides spinal cord into right and left sides
    ____ i. brain cavities where CSF circulates
    ____ j. shallow grooves in the cerebrum
    ____ k. contains CSF in the spinal cord

    A. longitudinal fissure
    B. sulci
    C. ventricles
    D. anterior median fissure
    E. central canal
    F. posterior (dorsal) root
    G. columns
    H. arachnoid villi
    I. anterior (ventral) root
    J. gyri
    K. cauda equina

CRITICAL THINKING APPLICATIONS

1. While checking cranial nerve function in an emergency room patient, a nurse finds that the patient can see a light but cannot track the light with her eyes. What do these results tell the nurse about cranial nerve function in this patient?

2. Dennis was a little nervous. It was his first visit to the dentist in 10 years. "You won't feel a thing," said the dentist as she injected several doses of "numbing" medication. While having lunch right after the visit, soup drips out of Dennis' mouth because he still doesn't feel a thing in his lower lip and right upper lip. What happened to Dennis?

3. An elderly relative suffered a stroke and now has difficulty with the movement of her right arm. She is also working with a therapist due to some speech problems. What areas of the brain were damaged by the stroke?

4. Lynn flicked on the light when she heard her husband's yell. Kyle was bouncing on his left foot while holding his right foot with both hands. A pin was sticking out of the bottom of his right foot. Explain Kyle's response to stepping on the pin.

ANSWERS TO FIGURE QUESTIONS

10.1 CSF circulates in the subarachnoid space.

10.2 Spinal nerves are part of the peripheral nervous system.

10.3 Posterior root, sensory neurons; anterior root, motor neurons.

10.4 The lateral spinothalamic and anterior spinothalamic tracts are sensory, whereas the lateral corticospinal and anterior corticospinal tracts are motor.

10.5 Axons of motor neurons are part of the anterior root.

10.6 Posterior ramus supplies the deep muscles of the back.

10.7 The medulla oblongata of the brain attaches to the spinal cord.

10.8 CSF is formed in the choroid plexuses and reabsorbed into arachnoid villi.

10.9 The midbrain contains the cerebral peduncles.

10.10 The basal ganglia are located in the cerebrum and are composed of gray matter.

10.11 The corpus callosum connects the two cerebral hemispheres.

10.12 The primary somatosensory area localizes somatic sensations.

Chapter 11

Autonomic Nervous System

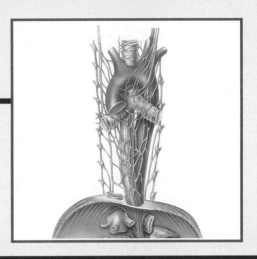

Student Learning Objectives

1. Compare the main structural and functional differences between the somatic and autonomic parts of the nervous system. **262**

2. Identify the structural features of the autonomic nervous system. **262**

3. Describe the functions of the sympathetic and parasympathetic divisions of the autonomic nervous system. **266**

A Look Ahead

The part of the nervous system that regulates smooth muscle, cardiac muscle, and certain glands is the **autonomic nervous system (ANS).** (Recall that together the ANS and somatic nervous system compose the peripheral nervous system; see Figure 9.1 on page 218.) The ANS was originally named *autonomic* (*auto-* = self; *-nomic* = law) because it was thought to function in a self-governing manner. Although the ANS usually does operate without conscious control from the cerebral cortex, it is regulated by other brain regions, mainly the hypothalamus and brain stem. In this chapter, we compare structural and functional features of the somatic and autonomic nervous systems; then we discuss the anatomy of the motor portion of the ANS and compare the organization and actions of its two major branches, the sympathetic and parasympathetic divisions.

COMPARISON OF SOMATIC AND AUTONOMIC NERVOUS SYSTEMS

Objective: • **Compare the main structural and functional differences between the somatic and autonomic parts of the nervous system.**

As you learned in Chapter 10, the somatic nervous system includes both sensory and motor neurons. The sensory neurons convey input from receptors for the special senses (vision, hearing, taste, smell, and equilibrium) and from receptors for somatic senses (pain, temperature, touch, and proprioceptive sensations). All these sensations normally are consciously perceived. In turn, somatic motor neurons synapse with skeletal muscle—the effector tissue of the somatic nervous system—and produce conscious, voluntary movements. When a somatic motor neuron stimulates a skeletal muscle, the muscle contracts. If somatic motor neurons cease to stimulate a muscle, the result is a paralyzed, limp muscle that has no muscle tone.

The main input to the ANS comes from **autonomic sensory neurons.** These neurons are associated with sensory receptors that monitor internal conditions, such as blood CO_2 level or the degree of stretching in the walls of internal organs (viscera) or blood vessels. When the viscera are functioning properly, these sensory signals are usually not consciously perceived. However, pain or nausea from damaged viscera or chest pain from inadequate blood flow to the heart can alert the conscious mind to potential problems.

Autonomic motor neurons regulate visceral activities by either exciting or inhibiting ongoing activities in their effector tissues: cardiac muscle, smooth muscle, and glands. Unlike skeletal muscle, these tissues generally function even if their nerve supply is interrupted. The heart continues to beat, for instance, when it is removed for transplantation into another person. Examples of autonomic responses are changes in the diameter of the pupil, dilation and constriction of blood vessels, and changes in the rate and force of the heartbeat. Because most autonomic responses cannot be consciously altered or suppressed to any great degree, they are the basis for polygraph ("lie detector") tests. However, practitioners of yoga or other techniques of meditation and those who employ biofeedback methods may learn how to modulate ANS activities. For example, they may be able to voluntarily decrease their heart rate or blood pressure.

Autonomic motor pathways consist of sets of two motor neurons (Figure 11.1a). The first neuron, called the **preganglionic neuron,** has its cell body in the CNS (brain or spinal cord). Its axon, which is *myelinated*, extends from the CNS through a cranial or a spinal nerve to an **autonomic ganglion,** where it synapses with the second neuron. (Recall that a ganglion is a collection of neuronal cell bodies outside the CNS.) The second neuron, the **postganglionic neuron,** lies entirely in the peripheral nervous system. Its cell body is located in an autonomic ganglion and its *unmyelinated* axon extends from the ganglion to the effector (smooth muscle, cardiac muscle, or a gland). The effect of the postganglionic neuron on the effector may be either excitation (causing contraction of smooth or cardiac muscle or increasing secretions of glands) or inhibition (causing relaxation of smooth or cardiac muscle or decreasing secretions of glands). In contrast, a single myelinated somatic motor neuron extends from the CNS and always excites its effector (causing contraction of skeletal muscle) (Figure 11.1b). Another difference between autonomic and somatic motor neurons is that all somatic motor neurons release acetylcholine (ACh) as their neurotransmitter. Some autonomic motor neurons release ACh, whereas others release norepinephrine (NE).

Table 11.1 summarizes the similarities and differences between the somatic and autonomic nervous systems. The output (motor) part of the ANS has two principal branches: the **sympathetic division** and the **parasympathetic division.** Most organs have **dual innervation;** that is, they receive impulses from both sympathetic and parasympathetic neurons. In general, nerve impulses from one division stimulate the organ to increase its activity (excitation), whereas impulses from the other division decrease the organ's activity (inhibition). For example, an increased rate of nerve impulses from the sympathetic division increases heart rate, whereas an increased rate of nerve impulses from the parasympathetic division decreases heart rate.

STRUCTURE OF THE AUTONOMIC NERVOUS SYSTEM

Objective: • **Identify the structural features of the autonomic nervous system.**

We will now examine the structure of preganglionic neurons, ganglia, and postganglionic neurons and how they relate to the activities of the autonomic nervous system.

Figure 11.1 ■ **Comparison of somatic and autonomic nervous systems.**

Stimulation by the autonomic nervous system either excites or inhibits its effectors: smooth muscle, cardiac muscle, and glands. Somatic nervous system stimulation always causes contraction of skeletal muscle.

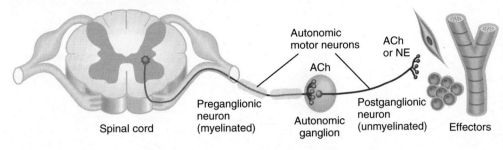

(a) Autonomic nervous system

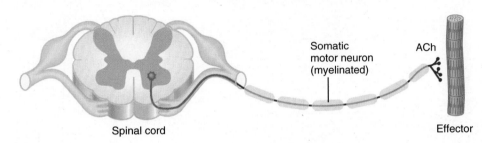

(b) Somatic nervous system

 What does "dual innervation" mean?

Preganglionic Neurons

Sympathetic Preganglionic Neurons

The sympathetic division of the ANS is also called the ***thoracolumbar division*** (thō′-ra-kō-LUM-bar) because the outflow of sympathetic nerve impulses is from the thoracic and lumbar segments of the spinal cord (Figure 11.2). The sympathetic preganglionic neurons have their cell bodies in the twelve thoracic and the first two lumbar segments of the spinal cord. The preganglionic axons emerge from the spinal cord through the anterior root of a spinal nerve along with axons of somatic motor neurons. After exiting the cord, the sympathetic preganglionic axons extend to the nearest sympathetic ganglion.

Parasympathetic Preganglionic Neurons

The parasympathetic division is also called the ***craniosacral division*** (krā′-nē-ō-SĀ-kral) because the outflow of parasympathetic nerve impulses is from cranial nerve nuclei and sacral segments of the spinal cord. The cell bodies of parasympathetic

Table 11.1 / Comparison of Somatic and Autonomic Nervous Systems		
Property	**Somatic**	**Autonomic**
Effectors	Skeletal muscles.	Cardiac muscle, smooth muscle, and glands.
Type of control	Voluntary.	Involuntary.
Neural pathway	One motor neuron extends from CNS and synapses directly with a skeletal muscle fiber.	One motor neuron extends from the CNS and synapses with another motor neuron in a ganglion; the second motor neuron synapses with a visceral effector.
Neurotransmitter	Acetylcholine.	Acetylcholine or norepinephrine.
Action of neurotransmitter on effector	Always excitatory (causing contraction of skeletal muscle).	May be excitatory (causing contraction of smooth muscle, increased heart rate, increased force of heart contraction, or increased secretions from glands) or inhibitory (causing relaxation of smooth muscle, decreased heart rate, or decreased secretions from glands).

Figure 11.2 ■ **Structure of the autonomic nervous system.**

Sympathetic and parasympathetic stimulation have opposing effects on organs that receive dual innervation.

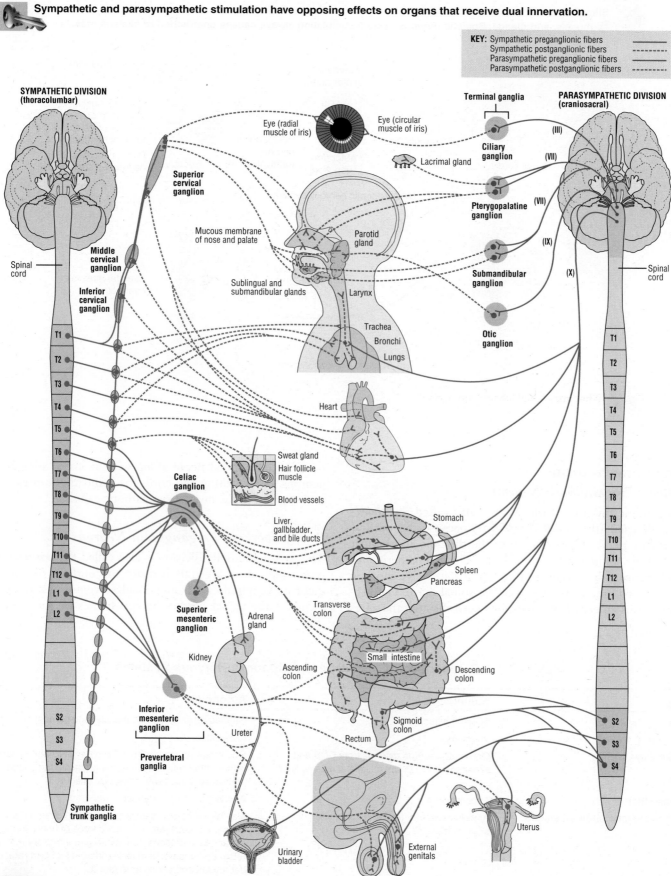

KEY: Sympathetic preganglionic fibers
Sympathetic postganglionic fibers
Parasympathetic preganglionic fibers
Parasympathetic postganglionic fibers

SYMPATHETIC DIVISION
(thoracolumbar)

Spinal cord

Middle cervical ganglion

Inferior cervical ganglion

Superior cervical ganglion

Eye (radial muscle of iris)

Eye (circular muscle of iris)

Terminal ganglia

PARASYMPATHETIC DIVISION
(craniosacral)

Ciliary ganglion   (III)

Lacrimal gland   (VII)

Pterygopalatine ganglion   (VII)

Mucous membrane of nose and palate

Parotid gland

Submandibular ganglion   (IX)

Sublingual and submandibular glands

Larynx

(X)

Spinal cord

Otic ganglion

Trachea
Bronchi
Lungs

Heart

Celiac ganglion

Sweat gland
Hair follicle muscle
Blood vessels

Liver, gallbladder, and bile ducts

Stomach

Spleen
Pancreas

Superior mesenteric ganglion

Adrenal gland

Transverse colon

Kidney

Ascending colon

Small intestine

Descending colon

Inferior mesenteric ganglion

Ureter

Prevertebral ganglia

Sigmoid colon

Rectum

Urinary bladder

External genitals

Uterus

Sympathetic trunk ganglia

T1 T2 T3 T4 T5 T6 T7 T8 T9 T10 T11 T12 L1 L2 S2 S3 S4

Which division has longer preganglionic axons? Why?

preganglionic neurons are located in the nuclei of four cranial nerves (III, VII, IX, and X) in the brain stem and in three sacral segments of the spinal cord (S2, S3, and S4) (Figure 11.2). Their axons emerge from the CNS as part of a cranial nerve or as part of the anterior root of a spinal nerve. Parasympathetic preganglionic axons of the vagus (X) nerve carry nearly 80% of the total parasympathetic outflow. In the thorax, axons of the vagus nerve extend to ganglia in the heart and the airways of the lungs. In the abdomen, axons of the vagus nerve extend to ganglia in the liver, gallbladder, bile ducts, stomach, pancreas, spleen, small intestine, transverse colon, and descending colon. Parasympathetic preganglionic axons exit the sacral spinal cord in the anterior roots of the second through fourth sacral nerves. The axons then extend to ganglia in the walls of the ascending colon, sigmoid colon, ureters, urinary bladder, and reproductive organs (Figure 11.2).

Autonomic Ganglia

The autonomic ganglia may be divided into three groups: *Sympathetic trunk ganglia* and *prevertebral ganglia* are part of the sympathetic division, whereas *terminal ganglia* are part of the parasympathetic division.

Sympathetic Ganglia

In the sympathetic ganglia, sympathetic preganglionic neurons synapse with postganglionic neurons. **Sympathetic trunk ganglia** lie in two vertical rows, one on either side of the vertebral column (Figure 11.3). Because the sympathetic trunk ganglia are near the spinal cord, most sympathetic preganglionic axons are short. The cervical portion of each sympathetic trunk is located in the neck and is subdivided into the *superior cervical ganglion,* *middle cervical ganglion,* and *inferior cervical ganglion* (see Figure

Figure 11.3 ■ **Sympathetic ganglia in the thorax, abdomen, and pelvis.**

The sympathetic trunk ganglia lie on either side of the vertebral column, whereas the prevertebral ganglia (celiac, superior mesenteric, and inferior mesenteric) lie anterior to the vertebral column, close to large abdominal arteries.

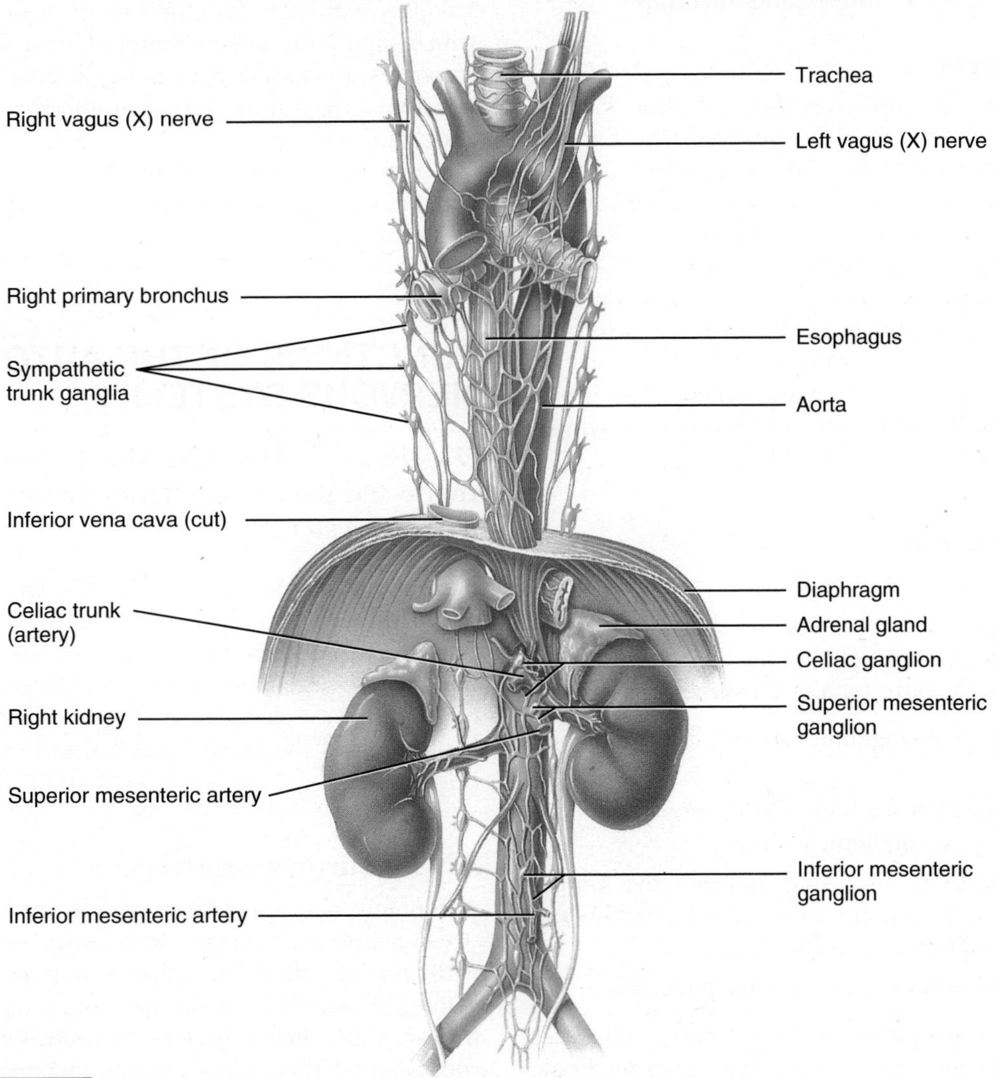

Right vagus (X) nerve —
Trachea

Left vagus (X) nerve

Right primary bronchus —

Sympathetic trunk ganglia —

Esophagus

Aorta

Inferior vena cava (cut) —

Diaphragm

Celiac trunk (artery) —

Adrenal gland

Celiac ganglion

Superior mesenteric ganglion

Right kidney —

Superior mesenteric artery —

Inferior vena cava

Inferior mesenteric ganglion

Inferior mesenteric artery —

 Which neurons synapse in a sympathetic trunk ganglion?

11.2). Most postganglionic axons emerging from sympathetic trunk ganglia supply organs above the diaphragm.

The second group of sympathetic ganglia, the **prevertebral ganglia,** lie anterior to the vertebral column and close to the large abdominal arteries. The prevertebral ganglia are the *celiac ganglion* (SĒ-lē-ak), on either side of the celiac artery just below the diaphragm; the *superior mesenteric ganglion,* near the beginning of the superior mesenteric artery; and the *inferior mesenteric ganglion,* near the beginning of the inferior mesenteric artery (see Figures 11.2 and 11.3). In general, postganglionic axons emerging from the prevertebral ganglia innervate organs below the diaphragm.

The sympathetic division of the ANS also includes part of the adrenal gland (Figure 11.3). The inner part of the adrenal gland, the **adrenal medulla** (me-DULL-a), develops from the same embryonic tissue as the sympathetic ganglia, and its cells are similar to sympathetic postganglionic neurons. Rather than extending to another organ, however, these cells release hormones into the blood. Upon stimulation by sympathetic preganglionic neurons, cells of the adrenal medulla release a mixture of hormones—about 80% **epinephrine** and 20% **norepinephrine.** These hormones circulate throughout the body and intensify responses elicited by sympathetic postganglionic neurons.

Parasympathetic Ganglia

Preganglionic axons of the parasympathetic division synapse with postganglionic neurons in **terminal ganglia,** which are located close to or actually within the wall of the innervated organ. Terminal ganglia in the head receive preganglionic axons from cranial nerves III (oculomotor), VII (facial), or IX (glossopharyngeal) and supply structures in the head (see Figure 11.2). Axons in cranial nerve X (vagus) extend to many terminal ganglia in the thorax and abdomen. Because the axons of parasympathetic preganglionic neurons extend from the brain stem or sacral spinal cord to a terminal ganglion in an innervated organ, they are longer than most of the axons of sympathetic preganglionic neurons (see Figure 11.2).

Postganglionic Neurons

Sympathetic Postganglionic Neurons

Once axons of preganglionic neurons of the sympathetic division enter a sympathetic trunk ganglion, they may follow one of several paths. Four possible paths for a given axon are as follows:

- It may synapse with postganglionic neurons in the ganglion it first reaches.

- It may ascend or descend to a higher or lower ganglion before synapsing with postganglionic neurons.

- It may continue, without synapsing, through the sympathetic trunk ganglion to end at a prevertebral ganglion and synapse with postganglionic neurons there.

- It may extend to and terminate in the adrenal medulla.

A single sympathetic preganglionic axon has many collaterals (branches) and may synapse with 20 or more postganglionic neurons. Thus, nerve impulses that arise in a single preganglionic neuron may activate many different postganglionic neurons that in turn synapse with several visceral effectors. This pattern helps explain why sympathetic responses can affect organs throughout the body almost simultaneously.

Most postganglionic axons leaving the superior cervical ganglion serve the head. They are distributed to sweat glands, smooth muscle of the eye, blood vessels of the face, nasal mucosa, and salivary glands. A few postganglionic axons from the superior cervical ganglion plus postganglionic axons from the middle and inferior cervical ganglia supply the heart. In the thoracic region, postganglionic axons from the sympathetic trunk serve the heart, lungs, and bronchi. Some axons from thoracic levels also supply sweat glands, blood vessels, and arrector pili muscles of hair follicles in the skin. In the abdomen, axons of postganglionic neurons leaving the prevertebral ganglia follow the course of various arteries to abdominal and pelvic visceral effectors.

Parasympathetic Postganglionic Neurons

Because the terminal ganglia are close to or in the walls of their visceral effectors, parasympathetic postganglionic axons are very short. In the ganglion, the preganglionic neuron usually synapses with only four or five postganglionic neurons, all of which supply the same visceral effector. Thus, parasympathetic responses are localized to a single effector, in contrast to the widespread activation of several effectors by a sympathetic preganglionic neuron.

Table 11.2 compares the organization of the sympathetic and parasympathetic divisions of the ANS.

FUNCTIONS OF THE AUTONOMIC NERVOUS SYSTEM

Objective: • **Describe the functions of the sympathetic and parasympathetic divisions of the autonomic nervous system.**

Neurotransmitters are chemical substances released by neurons at synapses. As you learned in Chapter 8, binding of the neurotransmitter acetylcholine to its receptors on skeletal muscle fibers causes excitation that leads to contraction. Some ANS neurons release acetylcholine, whereas others release norepinephrine, and the result is excitation in some cases and inhibition in others.

ANS Neurotransmitters

Autonomic neurons release neurotransmitters at synapses between neurons (preganglionic to postganglionic) and at synapses with visceral effectors (smooth muscle, cardiac muscle, and glands). Synapses between autonomic neurons and their effectors are called **neuroeffector junctions.** On the basis of the neurotransmitter they release, autonomic neurons may be classified as either cholinergic or adrenergic.

Table 11.2 / Organization of the Sympathetic and Parasympathetic Divisions of the ANS

Property	Sympathetic	Parasympathetic
Distribution	Bodywide: skin; sweat glands; arrector pili muscles of hair follicles; smooth muscle of blood vessels; and viscera of thorax, abdomen, and pelvis.	Limited mainly to head and to viscera of thorax, abdomen, and pelvis; some blood vessels.
Site of outflow	Thoracolumbar (T1–L2).	Craniosacral (cranial nerves III, VII, IX, and X; spinal nerves S2–S4).
Associated ganglia	Two types: sympathetic trunk ganglia and prevertebral ganglia.	One type: terminal ganglia.
Axon length and ganglia locations	Short preganglionic axons synapse with 20 or more postganglionic neurons in ganglia close to CNS; long postganglionic axons innervate many visceral effectors.	Long preganglionic axons usually synapse with 4–5 postganglionic neurons in ganglion near or within wall of visceral effectors; short postganglionic axons innervate a single visceral effector.

Cholinergic neurons (kō′-lin-ER-jik) release *acetylcholine (ACh)* and include the following: (1) all sympathetic and parasympathetic preganglionic neurons, (2) all parasympathetic postganglionic neurons, and (3) a few sympathetic postganglionic neurons. Because acetylcholine is quickly inactivated by the enzyme *acetylcholinesterase (AChE)*, the effects produced by cholinergic neurons are short-lived and localized. This feature enables the parasympathetic division to maintain precise control over its effectors.

Most sympathetic postganglionic neurons are **adrenergic** (ad′-ren-ER-jik); that is, they use the neurotransmitter *norepinephrine (NE)*. Because norepinephrine is inactivated much more slowly than acetylcholine and because norepinephrine is also released into the bloodstream by the adrenal medulla, the effects of activation of the sympathetic division are longer lasting and more widespread than those of the parasympathetic division. The consequence is your heart continues to pound for several minutes after a near miss at a busy intersection.

Activities of the ANS

As noted earlier, most body organs are innervated by both divisions of the ANS, which typically work in opposition to one another. This opposition is possible because the postganglionic neurons of the two divisions release different neurotransmitters. The balance between sympathetic and parasympathetic activity or "tone" is regulated by the hypothalamus. Typically, the hypothalamus turns up sympathetic tone at the same time it turns down parasympathetic tone, and vice versa. A few structures receive only sympathetic innervation: sweat glands, arrector pili muscles attached to hair follicles in the skin, the kidneys, most blood vessels, and the adrenal medulla (see Figure 11.2). In these cases there is no opposition from the parasympathetic division. Still, an *increase* in sympathetic tone has one effect, and a *decrease* in sympathetic tone produces the opposite effect.

Sympathetic Activities

During physical or emotional stress, high sympathetic tone favors body functions that can support vigorous physical activity and rapid production of ATP. At the same time, the sympathetic division reduces body functions that favor the storage of energy. Besides physical exertion, a variety of emotions—such as fear, embarrassment, or rage—stimulate the sympathetic division. Visualizing body changes that occur during "E situations" (exercise, emergency, excitement, embarrassment) will help you remember most of the sympathetic responses. Activation of the sympathetic division and release of hormones by the adrenal medulla result in a series of physiological responses collectively called the **fight-or-flight response**, in which the following occur:

1. The pupils of the eyes dilate.

2. Heart rate, force of heart contraction, and blood pressure increase.

3. The airways dilate, allowing faster movement of air into and out of the lungs.

4. The blood vessels that supply nonessential organs such as the kidneys and gastrointestinal tract constrict.

5. Blood vessels that supply organs involved in exercise or fighting off danger—skeletal muscles, cardiac muscle, liver, and adipose tissue—dilate, allowing greater blood flow.

6. Liver cells break down glycogen to glucose, and adipose cells break down triglycerides to fatty acids and glycerol, providing molecules that can be used by body cells for ATP production.

7. Release of glucose by the liver increases blood glucose level.

8. Processes that are not essential for meeting the stressful situation are inhibited. For example, muscular movements of the gastrointestinal tract and digestive secretions decrease or even stop.

Parasympathetic Activities

In contrast to the "fight-or-flight" activities of the sympathetic division, the parasympathetic division enhances "rest-and-digest" activities. Parasympathetic responses support body functions that conserve and restore body energy during times of rest and recovery. In the quiet intervals between periods of exercise,

Mind-Body Exercise— An Antidote to Stress

*W*hen we think of exercise, we usually think of toning up our muscles and maybe our hearts. But when some people think of exercise, their focus is on toning up neural input from the parasympathetic division of the autonomic nervous system. As you learned in this chapter, activation of the parasympathetic division helps restore homeostasis in many systems and is associated with feelings of relaxation.

Mind-Body Harmony

Mind-body exercise refers to exercise systems such as tai chi, hatha yoga, and many forms of the martial arts that couple muscular activity with an internally directed focus. These exercise systems exercise the mind as well as the body. Their internally directed focus usually includes an awareness of breathing, energy, and other physical sensations.

Practitioners often refer to this internal awareness as "mindful," meaning that the exerciser is open to physical and emotional sensations with an understanding, nonjudgmental attitude. A mindful attitude is typical of many kinds of meditation and relaxation practices. For example, when practicing a yoga pose, you would think something like "Deep, steady breathing; relax into the pose; shoulders pulling back, neck lengthening," rather than "That girl next to me sure is flexible; I'm really a failure at this stuff." Of course, in real life such external thoughts do sneak in, but we can redirect our attention back to a more neutral, nonjudgmental style.

Mind-Body Benefits

People practicing mind-body activities reap benefits from both the physical and mental activity. Hatha yoga, tai chi, and the martial arts increase muscular strength and flexibility, posture, balance, and coordination, and if performed vigorously, they can even improve cardiovascular health and endurance to some extent. In addition, the stress relief provided by the activity extends into both physical and psychological realms. Feelings of mental relaxation and emotional well-being translate into better resting blood pressure, a healthier immune system, and more relaxed muscles. Less stress can also mean an improvement in health habits. Those who practice mind-body exercise often improve their eating habits and reduce harmful behaviors such as cigarette smoking.

▶ *Think It Over*

▶ How could you make walking more of a mind-body activity?

parasympathetic impulses to the digestive glands and the smooth muscle of the gastrointestinal tract predominate over sympathetic impulses, allowing energy-supplying food to be digested and absorbed. At the same time, parasympathetic responses reduce body functions that support physical activity.

The acronym "SLUDD" can be helpful in remembering five responses that occur when parasympathetic tone rises: S for salivation, L for lacrimation (the secretion of tears), U for urination, D for digestion, and D for defecation. All of these activities are stimulated mainly by the parasympathetic division. Besides increasing "SLUDD" responses, the parasympathetic division causes three "decreases": decreased heart rate, decreased diameter of airways (bronchoconstriction), and decreased diameter (constriction) of the pupils.

Although moderate fear elicits a "fight-or-flight" sympathetic response, massive activation of the parasympathetic division occurs in *paradoxical fear*. This type of fear occurs when one is backed into a corner, with no way to win (so "fighting" is useless) and no escape route (so "flight" is not possible). It may happen to soldiers in a losing battle, to unprepared students tak-

Table 11.3 / Functions of the Autonomic Nervous System

Visceral Effector	Effect of Sympathetic Stimulation	Effect of Parasympathetic Stimulation
Glands		
Sweat	Increases secretion.	No known effect.
Lacrimal (tear)	Slightly stimulates secretion.	Stimulates secretion.
Adrenal medulla	Promotes secretion of epinephrine and norepinephrine.	No known effect.
Liver*	Promotes breakdown of glycogen into glucose, conversion of noncarbohydrates in the liver to glucose, and release of glucose into the blood; decreases bile secretion.	Promotes glycogen synthesis; increases bile secretion.
Kidney*	Stimulates secretion of renin (an enzyme), which helps raise blood pressure.	No known effect.
Pancreas	Inhibits secretion of digestive enzymes and insulin (hormone that lowers blood sugar level); promotes secretion of glucagon (hormone that raises blood sugar level).	Promotes secretion of digestive enzymes and insulin.
Adipose tissue*	Stimulates breakdown of triglycerides and release of fatty acids into blood.	No known effect.
Cardiac muscle		
Heart	Increases heart rate and increases strength of atrial and ventricular contraction.	Decreases heart rate and increases strength of atrial contraction.
Smooth muscle		
Radial muscle of iris of eye	Contraction → dilation of pupil.	No known effect.
Circular muscle of iris of eye	No known effect.	Contraction → constriction of pupil.
Ciliary muscle of eye	Relaxation for far vision.	Contraction for near vision.
Salivary glands (arterioles)	Decreases secretion.	Stimulates secretion.
Gastric glands (arterioles)	Inhibits secretion.	Promotes secretion.
Intestinal glands (arterioles)	Inhibits secretion.	Promotes secretion.
Gallbladder and ducts	Relaxation.	Contraction → increases release of bile into small intestine.
Stomach and intestines	Decreases motility (movement) and tone; contracts sphincters.	Increases motility and tone; relaxes sphincters.
Lungs (smooth muscle of bronchi)	Relaxation → airway widening (bronchodilation).	Contraction → airway narrowing (bronchoconstriction).
Urinary bladder	Relaxation of muscular wall; contraction of internal sphincter.	Contraction of muscular wall; relaxation of internal sphincter.
Spleen	Contraction and discharge of stored blood into general circulation.	No known effect.
Arrector pili of hair follicles	Contraction that results in erection of hairs.	No known effect.
Uterus	Inhibits contraction in nonpregnant women; stimulates contraction in pregnant women.	Minimal effect.
Sex organs	In men, produces ejaculation of semen.	Vasodilation; erection of clitoris (women) and penis (men).

*Listed with glands because they release substances into the blood.

ing an exam, or to athletes before competition. The high level of parasympathetic tone in such situations can cause loss of control over urination or defecation.

Table 11.3 summarizes the responses of glands, cardiac muscle, and smooth muscle to stimulation by the sympathetic and parasympathetic divisions of the ANS.

COMMON DISORDERS

Autonomic Dysreflexia

Autonomic dysreflexia is an exaggerated response of the sympathetic division of the ANS that occurs in about 85% of individuals with spinal cord injury at or above the sixth thoracic level. Symptoms include severe high blood pressure (hypertension); pounding headache; flushed, warm skin with profuse sweating above the injury level; and pale, cold, and dry skin below the injury level. Autonomic dysreflexia occurs when certain sensory impulses, particularly impulses from stretch receptors in the wall of a full urinary bladder, are unable to ascend the spinal cord because of the injury. Instead, the sensory impulses cause a massive stimulation of sympathetic nerves below the level of injury. One outcome is severe constriction of blood vessels below the level of the injury, which accounts for the pale and cold skin and also elevates blood pressure throughout the body to a dangerously high level. Autonomic dysreflexia is an emergency condition that requires immediate treatment, such as draining the urinary bladder and giving drugs to lower the blood pressure.

Horner's Syndrome

In *Horner's syndrome,* the sympathetic innervation to one side of the face is lost due to an inherited mutation, an injury, or a disease that affects sympathetic outflow through the superior cervical ganglion. Symptoms occur in the head on the affected side and include ptosis (drooping of the upper eyelid), miosis (constricted pupil), and anhidrosis (lack of sweating).

Raynaud's Disease

Raynaud's disease is excessive contraction of arterioles within the fingers and toes due to prolonged sympathetic stimulation. Blood flow is greatly diminished; the fingers and toes may be deprived of blood for minutes to hours and, in extreme cases, may become necrotic from lack of oxygen and nutrients. The disease is most common in young women and is worsened by cold climates, but its cause is unknown.

▮ STUDY OUTLINE

Comparison of Somatic and Autonomic Nervous Systems (p. 262)

1. The somatic nervous system produces conscious movement in skeletal muscles.

2. The autonomic nervous system regulates activities of smooth muscle, cardiac muscle, and glands, and it usually operates without conscious control.

3. The ANS is regulated by centers in the brain, in particular by the hypothalamus.

4. A single somatic motor neuron synapses on skeletal muscles; in the ANS, there are two motor neurons, one from the CNS to a ganglion (preganglionic) and one from a ganglion to a visceral effector (postganglionic).

5. Somatic motor neurons release acetylcholine (ACh), and autonomic motor neurons release either acetylcholine or norepinephrine (NE).

Structure of the Autonomic Nervous System (p. 262)

1. The autonomic nervous system consists of two principal divisions: sympathetic and parasympathetic.

2. The neurons of preganglionic autonomic neurons are myelinated; those of postganglionic autonomic neurons are unmyelinated.

3. Sympathetic ganglia are classified as sympathetic trunk ganglia (lateral to the vertebral column) or prevertebral ganglia (anterior to the vertebral column).

4. Parasympathetic ganglia are called terminal ganglia and are located near or within visceral effectors.

Functions of the Autonomic Nervous System (p. 266)

1. Cholinergic neurons release acetylcholine (ACh). Adrenergic neurons release norepinephrine (NE).

2. Activation of the sympathetic division causes widespread responses and is referred to as the "fight-or-flight" response.

3. Activation of the parasympathetic division produces more restricted responses that typically are concerned with "rest-and-digest" activities.

4. Table 11.3 on page 269 summarizes the main functions of the sympathetic and parasympathetic divisions of the ANS.

▮ SELF-QUIZ

1. In comparing the somatic nervous system with the autonomic nervous system, which of the following statements is true?
   a. The autonomic nervous system controls involuntary movement in skeletal muscle.   b. The somatic nervous system controls voluntary movement in glands and smooth muscle.   c. The autonomic nervous system controls involuntary movement in cardiac muscle, viscera, and glands.   d. The autonomic nervous system produces voluntary movement in smooth muscle and glands.

e. The somatic nervous system controls involuntary movement in smooth muscle, viscera, and glands.

2. Neurons in the autonomic nervous system include

a. two motor neurons and one ganglion   **b.** one motor neuron and two ganglia   **c.** two motor neurons and two ganglia   **d.** one motor and one sensory neuron and no ganglia   **e.** one motor and one sensory neuron and one ganglion

3. Which statement is NOT true?

a. Most sympathetic postganglionic neurons are adrenergic. **b.** Parasympathetic preganglionic neurons are cholinergic. **c.** Adrenergic effects are more localized and short-lived than cholinergic effects.   **d.** The effects from norepinephrine tend to be long lasting.   **e.** Collateral branches of a single postganglionic neuron in the sympathetic division extend to many organs.

4. Which of the following pairs is mismatched?

a. acetylcholine, parasympathetic nervous system   **b.** fight-or-flight, sympathetic nervous system   **c.** conserves body energy, parasympathetic nervous system   **d.** cholinergic, acetylcholine **e.** norepinephrine, parasympathetic nervous system

5. Which of the following statements is NOT true concerning the autonomic nervous system?

a. Most autonomic responses cannot be consciously controlled. **b.** In general, if the sympathetic division increases the activity in a specific organ, then the parasympathetic division decreases the activity of that organ.   **c.** Sensory receptors monitor internal body conditions.   **d.** Sensory neurons include pre- and postganglionic neurons.   **e.** Most visceral effectors receive dual innervation.

6. Which part of the central nervous system contains centers that regulate the autonomic nervous system?

a. hypothalamus   **b.** cerebellum   **c.** spinal cord **d.** basal ganglia   **e.** thalamus

7. Place the following structures in the correct order as they relate to an autonomic nervous system response from receipt of the stimulus to response:

1. visceral effector
2. centers in the CNS
3. autonomic ganglion
4. receptor and autonomic sensory neuron
5. preganglionic neuron
6. postganglionic neuron

a. 4, 5, 2, 3, 6, 1   **b.** 5, 6, 2, 3, 1, 4   **c.** 1, 6, 3, 5, 2, 4 **d.** 4, 2, 5, 3, 6, 1   **e.** 2, 4, 5, 6, 3, 1

8. Which of the following activities would NOT be monitored by autonomic sensory neurons?

a. carbon dioxide levels in the blood   **b.** hearing and equilibrium   **c.** blood pressure   **d.** stretching of the walls of visceral organs   **e.** nausea from damaged viscera

9. The autonomic ganglia associated with the parasympathetic division are the

a. trunk ganglia   **b.** prevertebral ganglia   **c.** posterior root ganglia   **d.** terminal ganglia   **e.** basal ganglia

10. Which of these statements about the parasympathetic division of the autonomic nervous system is NOT true? The parasympathetic division

a. arises from the cranial nerves in the brain stem and sacral spinal cord segments   **b.** is concerned with conserving and restoring energy   **c.** uses acetylcholine as its neurotransmitter   **d.** has ganglia near or within visceral effectors   **e.** initiates responses in preganglionic neurons that synapse with 20 or more postganglionic neurons

11. Which nerve carries most of the parasympathetic output from the brain?

a. spinal   **b.** vagus   **c.** oculomotor   **d.** facial **e.** glossopharyngeal

12. Which of the following would NOT be affected by the autonomic nervous system?

a. heart   **b.** intestines   **c.** urinary bladder   **d.** skeletal muscle   **e.** reproductive organs

13. Which of the following would be adrenergic?

a. somatic motor neurons   **b.** sympathetic postganglionic neurons   **c.** sympathetic preganglionic neurons   **d.** parasympathetic postganglionic neurons   **e.** parasympathetic preganglionic neurons

14. Match the following:

_____ **a.** cluster of cell bodies outside the CNS

_____ **b.** cell body is in ganglion; unmyelinated axon extends to effector

_____ **c.** cell body lies inside the CNS; myelinated axon extends to ganglion

_____ **d.** their postganglionic axons innervate organs below the diaphragm

_____ **e.** their postganglionic axons supply organs above the diaphragm

_____ **f.** contain the cell bodies and dendrites of parasympathetic postganglionic neurons

A. sympathetic trunk ganglia
B. prevertebral ganglia
C. ganglion
D. terminal ganglia
E. preganglionic neuron
F. postganglionic neuron

15. For each of the following, place a P if it refers to increased activity of the parasympathetic division or an S if it refers to increased activity of the sympathetic division.

_____ **a.** dilates pupils

_____ **b.** decreases heart rate

_____ **c.** causes bronchoconstriction

_____ **d.** stimulates breakdown of triglycerides

_____ **e.** occurs in "paradoxical fear"

_____ **f.** stimulates the gastrointestinal tract

_____ **g.** occurs during exercise

_____ **h.** causes release of glucose from the liver

_____ **i.** dilates blood vessels to cardiac muscle

CRITICAL THINKING APPLICATIONS

1. It's Thanksgiving and you've just eaten a huge turkey dinner with all the trimmings. Now you're headed for the football game on television. Which division of the nervous system will be handling your body's post-dinner activities? Give examples of some organs and the effects on their functions.

2. Anthony wanted a toy on top of the bookcase, so he climbed up the shelves. His mother ran in when she heard the crash and lifted the heavy bookcase with one arm while pulling her son out with the other. Later that day, she could not lift the bookcase back into position by herself. How do you explain the temporary "super-mom" effect?

3. The eight-year-old ball player was terrified when he fell down an abandoned well shaft in the outfield. He was soon rescued but he still needed a change of clothes—his bladder and bowels had emptied. Why did the child lose control of these functions?

4. Alicia was watching a scary late-night horror movie when she heard a door slam and a cat's yowl. The hair rose on her arms and she was covered with goose bumps. Trace the pathway taken by the impulses from her CNS to her arms.

ANSWERS TO FIGURE QUESTIONS

11.1 Dual innervation means that an organ receives impulses from both the sympathetic and parasympathetic divisions of the ANS.

11.2 Most parasympathetic preganglionic axons are longer than most sympathetic preganglionic axons because parasympathetic ganglia are located in the walls of visceral organs.

11.3 In the sympathetic trunk ganglia, sympathetic preganglionic axons form synapses with cell bodies and dendrites of sympathetic postganglionic neurons.

Chapter 12

SENSATIONS

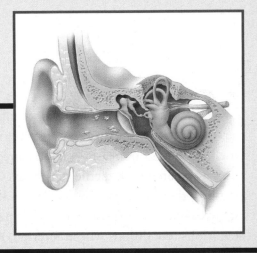

■ Student Learning Objectives

1. Define a sensation and describe the conditions necessary for a sensation to occur. **274**
2. Describe the location and function of the receptors for tactile, thermal, and pain sensations. **275**
3. Identify the receptors for proprioception and describe their functions. **277**
4. Describe the receptors for olfaction and the olfactory pathway to the brain. **279**
5. Describe the receptors for gustation and the gustatory pathway to the brain. **280**
6. Describe the accessory structures of the eye, the layers of the eyeball, the lens, the interior of the eyeball, image formation, and binocular vision. **282**
7. Describe the receptors for vision and the visual pathway to the brain. **285**
8. Describe the structures of the external, middle, and internal ear. **289**
9. Describe the receptors for hearing and equilibrium and their pathways to the brain. **292**

■ A Look Ahead

*M*ost of us are aware of sensory input to the central nervous system (CNS) from structures associated with smell, taste, vision, hearing, and balance. Input from structures associated with tactile sensations (touch, pressure, vibration), thermal sensations (warm and cold), pain, proprioceptive sensations (position of body parts), and visceral sensations (conditions within internal organs) are just as important to the maintenance of homeostasis.

OVERVIEW OF SENSATIONS

Objective: • **Define a sensation and describe the conditions necessary for a sensation to occur.**

Consider what would happen if you could not feel the pain of a hot pot handle or an inflamed appendix, or if you could not see, hear, smell, taste, or maintain your balance. In short, if you could not "sense" your environment and make the necessary homeostatic adjustments, you could not survive very well on your own.

Definition of Sensation

Sensation is the conscious or subconscious awareness of external or internal conditions of the body. For a sensation to occur, four conditions must be satisfied:

1. A *stimulus,* or change in the environment, capable of activating certain sensory neurons, must occur.
2. A *sensory receptor* must convert the stimulus to nerve impulses.
3. The nerve impulses must be *conducted* along a neural pathway from the sensory receptor to the brain.
4. A region of the brain must receive and *integrate* the nerve impulses into a sensation.

A stimulus that activates a sensory receptor may be in the form of light, heat, pressure, mechanical energy, or chemical energy. A sensory receptor responds to a stimulus by altering its membrane's permeability to small ions. In most types of sensory receptors, the resulting flow of ions across the membrane produces a depolarization called a *generator potential.* When a generator potential is large enough to reach the threshold level, it triggers one or more nerve impulses that are conducted along the sensory neuron toward the CNS.

Characteristics of Sensations

Conscious sensations or *perceptions* are integrated in the cerebral cortex. You seem to see with your eyes, hear with your ears, and feel pain in an injured part of your body. This is because sensory impulses from each part of the body arrive in a specific region of the cerebral cortex, which interprets the sensation as coming from the stimulated sensory receptors.

The distinct quality that makes one sensation different from others is its *modality.* A given sensory neuron carries information for one modality only. Neurons relaying impulses for touch, for example, do not also conduct impulses for pain. The specialization of sensory neurons enables nerve impulses from the eyes to be perceived as sight, and those from the ears to be perceived as sounds.

A characteristic of most sensory receptors is *adaptation,* a decrease in sensation during a prolonged stimulus. Adaptation is caused in part by a decrease in the responsiveness of sensory receptors. As a result of adaptation, the perception of a sensation decreases even though the stimulus persists. For example, when you first step into a hot shower, the water may feel very hot, but soon the sensation decreases to one of comfortable warmth even though the stimulus (the high temperature of the water) does not change. Receptors vary in how quickly they adapt. *Rapidly adapting receptors* adapt very quickly and are specialized for signaling *changes* in a particular stimulus. Receptors associated with pressure, touch, and smell are rapidly adapting. *Slowly adapting receptors,* in contrast, adapt slowly and continue to trigger nerve impulses as long as the stimulus persists. Slowly adapting receptors monitor stimuli associated with body positions and the chemical composition of the blood.

Classification of Sensations

The senses can be grouped into two classes: general senses and special senses.

1. The **general senses** include both **somatic senses** (*somat-* = of the body) and **visceral senses.** Somatic senses include tactile sensations (touch, pressure, and vibration); thermal sensations (warm and cold); pain sensations; and proprioceptive sensations, which allow perception of both the static positions of limbs and body parts (joint and muscle position sense) and movements of the limbs and head. Visceral senses provide information about conditions within body fluids and internal organs.
2. The **special senses** include smell, taste, vision, hearing, and equilibrium (balance).

Types of Sensory Receptors

Sensory receptors vary in their complexity. The simplest are *free nerve endings* that have no visible structural specializations (for example, pain receptors). Receptors for other general sensations, such as touch, pressure, and vibration, have *encapsulated nerve*

Table 12.1 / Classification of Sensory Receptors

I. Location

A. **Exteroceptors** (eks′-ter-ō-SEP-tors). Located at or near surface of body; provide information about *external* environment; transmit information for hearing, vision, smell, taste, touch, pressure, temperature, and pain.

B. **Interoceptors** (in′-ter-ō-SEP-tors). Located in blood vessels and viscera; provide information about *internal* environment; give rise to sensations such as pain, pressure, fatigue, hunger, thirst, and nausea.

C. **Proprioceptors** (prō′-prē-ō-SEP-tors). Located in muscles, tendons, joints, and the internal ear; provide information about body position and movement; transmit information related to muscle tension, position and tension of joints, and equilibrium (balance).

II. Type of stimulus

A. **Mechanoreceptors.** Detect pressure or stretching; stimuli are related to touch, pressure, proprioception, hearing, equilibrium, and blood pressure.

B. **Thermoreceptors.** Detect changes in temperature.

C. **Nociceptors** (nō′-sē-SEP-tors). Detect pain, usually as a result of physical or chemical damage to tissues.

D. **Photoreceptors.** Detect light in retina of eye.

E. **Chemoreceptors.** Detect taste in mouth; smell in nose; and chemicals such as oxygen, carbon dioxide, water, and glucose in body fluids.

III. Degree of complexity

A. **Simple receptors.** Simple structures and neural pathways that are associated with general senses (touch, pressure, heat, cold, and pain).

B. **Complex receptors.** Complex structures and neural pathways that are associated with special senses (smell, taste, vision, hearing, and equilibrium).

endings. Their dendrites are enclosed in a connective tissue capsule with a distinctive microscopic structure. Still other sensory receptors consist of specialized, separate cells that synapse with sensory neurons, for example, hair cells in the internal ear.

Sensory receptors are classified on the basis of their location (exteroceptors, interoceptors, and proprioceptors), the type of stimulus that activates them (mechanoreceptors, thermoreceptors, nociceptors, photoreceptors, and chemoreceptors), and their degree of complexity (simple receptors, complex receptors). Table 12.1 describes each of these categories.

SOMATIC SENSES

Objectives: • **Describe the location and function of the receptors for tactile, thermal, and pain sensations.** • **Identify the receptors for proprioception and describe their functions.**

Somatic sensations arise from stimulation of sensory receptors embedded in the skin or subcutaneous layer; in mucous membranes of the mouth, vagina, and anus; in muscles, tendons, and joints; and in the internal ear. The sensory receptors for somatic sensations are distributed unevenly. Some parts of the body surface are densely populated with receptors, whereas other parts contain only a few. The areas with the highest density of sensory receptors are the tip of the tongue, the lips, and the fingertips. Somatic sensations that result from stimulating the skin surface are called **cutaneous sensations** (kyoo-TĀ-nē-us; *cutane-* = skin).

Tactile Sensations

The **tactile sensations** (TAK-tīl; *tact-* = touch) are touch, pressure and vibration, and itch and tickle. Itch and tickle sensations are detected by free nerve endings. All other tactile sensations are detected by a variety of encapsulated mechanoreceptors. Tactile receptors in the skin or subcutaneous layer include corpuscles of touch, hair root plexuses, type I and II cutaneous mechanoreceptors, lamellated corpuscles, and free nerve endings (Figure 12.1).

Touch

Sensations of touch generally result from stimulation of tactile receptors in the skin or subcutaneous layer. *Crude touch* is the ability to perceive that something has contacted the skin, even though its exact location, shape, size, or texture cannot be determined. *Fine touch* provides specific information about a

Figure 12.1 ■ **Structure and location of sensory receptors in the skin and subcutaneous layer.**

The somatic sensations of touch, pressure, vibration, warmth, cold, and pain arise from sensory receptors in the skin, subcutaneous layer, and mucous membranes.

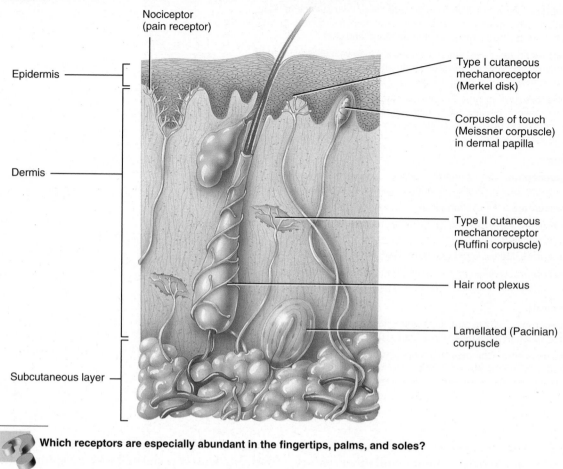

Which receptors are especially abundant in the fingertips, palms, and soles?

touch sensation, such as exactly what point on the body is touched plus the shape, size, and texture of the source of stimulation.

There are two types of rapidly adapting touch receptors. *Corpuscles of touch,* or *Meissner corpuscles* (MĪS-ner), are receptors for fine touch that are located in the dermal papillae of hairless skin. Each corpuscle is an egg-shaped mass of dendrites enclosed by a capsule of connective tissue. They are abundant in the fingertips, palms, eyelids, tip of the tongue, lips, nipples, soles, clitoris, and tip of the penis. *Hair root plexuses* are rapidly adapting touch receptors found in hairy skin; they consist of free nerve endings wrapped around hair follicles. Hair root plexuses detect movements on the skin surface that disturb hairs. For example, a flea landing on a hair causes movement of the hair shaft that stimulates the free nerve endings.

There are also two types of slowly adapting touch receptors. *Type 1 cutaneous mechanoreceptors,* also known as *Merkel disks,* function in fine touch. These saucer-shaped, flattened free nerve endings that contact Merkel cells of the stratum basale are plentiful in the fingertips, hands, lips, and external genitalia. *Type II cutaneous mechanoreceptors,* or *Ruffini corpuscles,* are elongated, encapsulated receptors located deep in the dermis, and in ligaments and tendons as well. Present in the hands and abundant on the soles, they are most sensitive to stretching that occurs as digits or limbs are moved.

Pressure and Vibration

Pressure is a sustained sensation that is felt over a larger area than touch. Receptors that contribute to sensations of pressure include corpuscles of touch, type I mechanoreceptors, and lamellated corpuscles. *Lamellated* or *Pacinian corpuscles* (pa-SIN-ē-an) are large oval structures composed of a multilayered connective tissue capsule that encloses a nerve ending (Figure 12.1). Like corpuscles of touch, lamellated corpuscles adapt rapidly. They are widely distributed in the body: in the dermis and subcutaneous layer; in tissues that underlie mucous and serous membranes; around joints, tendons, and muscles; in the periosteum; and in the mammary glands, external genitalia, and certain viscera, such as the pancreas and urinary bladder.

Sensations of *vibration* result from rapidly repetitive sensory signals from tactile receptors. The receptors for vibration sensations are corpuscles of touch and lamellated corpuscles.

Itch and Tickle

The *itch* sensation results from stimulation of free nerve endings by certain chemicals, such as bradykinin, often as a result of a local inflammatory response. Receptors for the *tickle* sensation are thought to be free nerve endings and lamellated corpuscles. This intriguing sensation typically arises only when someone else touches you, not when you touch yourself. The explanation of this puzzle seems to lie in the impulses that conduct to and from the cerebellum when you are moving your fingers and touching yourself that don't occur when someone else is tickling you.

Thermal Sensations

The sensory receptors for *thermal sensations* (sensations of heat and cold) consist of two types: cold receptors and warm receptors. *Cold receptors* are located in the stratum basale of the epidermis. Temperatures between 10° and 40°C (50° to 105°F) activate cold receptors. *Warm receptors,* located in the dermis, are activated by temperatures between 32° and 48°C (90° to 118°F). Both cold and warm receptors adapt rapidly at the onset of a stimulus but continue to generate some impulses throughout a prolonged stimulus. Temperatures below 10°C and above 48°C stimulate mainly pain receptors, rather than thermoreceptors, producing painful sensations.

Pain Sensations

The ability to perceive pain is indispensable for a normal life, providing us with information about tissue-damaging stimuli so we can protect ourselves from greater damage. Pain initiates our search for medical assistance, and our description and indication of the location of the pain may help pinpoint the underlying cause of disease.

The sensory receptors for pain, called *nociceptors* (nō′-sē-SEP-tors; *noci-* = harmful), are free nerve endings (Figure 12.1). Pain receptors are found in practically every tissue of the body except the brain, and they respond to several types of stimuli. Excessive stimulation of sensory receptors, excessive stretching of a structure, prolonged muscular contractions, inadequate blood flow to an organ, or the presence of certain chemical substances can all produce the sensation of pain.

During tissue irritation or injury, release of chemicals such as prostaglandins stimulates nociceptors. Nociceptors adapt only slightly or not at all to the presence of these chemicals, which are only slowly removed from the tissues following an injury. This situation explains why pain persists after the initial trauma. If there were adaptation to painful stimuli, irreparable tissue damage could result.

Recognition of the type and intensity of pain occurs primarily in the cerebral cortex. In most instances of somatic pain, the cortex projects the pain back to the stimulated area. If you burn your finger, you feel the pain in your finger, not in your cortex. In most instances of visceral pain, the sensation is not projected back to the point of stimulation. Rather, the pain is felt in the skin overlying the stimulated organ or in a surface area far from the stimulated organ. This phenomenon is called *referred pain*. It occurs because the area to which the pain is referred and the visceral organ involved are innervated by the same segment of the spinal cord. For example, sensory neurons from the heart as well as from the skin over the heart and left upper limb enter the left posterior (dorsal) roots of thoracic spinal cord segments T1 to T5. Thus the pain of a heart attack is typically felt in the skin over the heart and along the left arm.

A kind of pain often experienced by patients who have had a limb amputated is called *phantom pain*. They still experience sensations such as itching, pressure, tingling, or pain in the limb as if the limb were still there. One reason for these sensations is that the remaining proximal portions of the sensory nerves that previously received impulses from the limb are being stimulated by the trauma of the amputation. Stimuli from these nerves are interpreted by the brain as coming from the nonexistent (phantom) limb.

Some pain sensations are inappropriate; rather than warning of actual or impending damage, they occur out of proportion to minor damage or persist chronically for no obvious reason. In such cases, *analgesia* (*an-* = without; *-algesia* = pain) or pain relief is needed. Analgesic drugs such as aspirin and ibuprofen (for example, Advil) block formation of the chemicals that stimulate nociceptors. Local anesthetics, such as procaine (Novocain), provide short-term pain relief by blocking conduction of nerve impulses. Morphine and other opiate drugs alter the quality of pain perception in the brain; pain is still sensed, but it is no longer perceived as so unpleasant.

Proprioceptive Sensations

Proprioceptive sensations (*proprio-* = one's own) inform you, consciously and subconsciously, of the degree to which your muscles are contracted, the amount of tension present in your tendons, the positions of your joints, and the orientation of your head. The receptors for proprioception, called *proprioceptors,* adapt slowly and only slightly. Slow adaptation is advantageous because your brain must be aware of the status of different parts of your body at all times so that adjustments can be made. *Kinesthesia* (kin′-es-THĒ-zē-a; *kin-* = motion; *-esthesia* = perception), the perception of body movements, allows you to walk, type, or dress without using your eyes. Proprioceptive sensations also allow you to estimate the weight of objects and determine the amount of effort necessary to perform a task. For example, when you pick up a bag you quickly realize whether it contains feathers or books, and you then exert only the amount of effort needed to lift it.

Proprioceptors are located in skeletal muscles, in tendons, in and around synovial joints, and in the internal ear.

- *Muscle spindles* are delicate proprioceptors that are located among skeletal muscle fibers. When a muscle spindle is stretched, it sends impulses to the CNS, indicating how much and how fast the muscle is changing its length. This information is integrated within the CNS to coordinate muscle activity.

Pain is a useful sensation when it alerts us to an injury that needs attention. We pull our finger away from a hot stove, we take off shoes that are too tight, and we rest an ankle that has been sprained. We do what we can to help the injury heal and meanwhile take over-the-counter or prescription painkillers until the pain goes away.

Pain that persists for longer than two or three months despite appropriate treatment is known as *chronic pain.* The most common forms of chronic pain are low back pain and headache. Cancer, arthritis, fibromyalgia, and many other disorders are associated with chronic pain. People experiencing chronic pain often experience chronic frustration as they are sent from one specialist to another in search of a diagnosis. Their lives may turn into nightmares of fear and worry.

The goal of pain management programs, developed to help people with chronic pain, is to decrease pain as much as possible, and then help patients learn to cope with whatever pain remains. Because no single treatment works for everyone, pain management programs typically offer a wide variety of treatments from surgery and nerve blocks to acupuncture and exercise therapy. Following are some of the therapies that complement medical and surgical treatment for the management of chronic pain.

Counseling

Pain used to be regarded as a purely physical response to physical injury. Psychological factors are now understood to serve as important mediators in the perception of pain. Feelings such as fear and anxiety strengthen the pain perceptions. Pain may be used to avoid certain situations, or to gain attention. Depression and associated symptoms such as sleep disturbances can contribute to chronic pain. Psychological counseling techniques can help people with chronic pain confront issues such as these that may be worsening their pain.

Relaxation and Meditation

Relaxation and meditation techniques may reduce pain by decreasing anxiety and giving people a sense of personal control. Some of these techniques include deep breathing, visualization of positive images, and muscular relaxation. Others encourage people to become more aware of thoughts and situations that increase or decrease pain or provide a mental distraction from the sensations of pain.

Exercise

People with chronic pain tend to avoid movement because it hurts. Inactivity causes muscles and joint structures to atrophy, which may eventually cause the pain to worsen. Regular exercise and improved fitness helps to relieve pain. Why? Exercise stimulates the production of endorphins, chemicals produced by the body to relieve pain. It also improves self-confidence, can serve as a distraction from pain, and improves sleep quality, which is often a problem for people with chronic pain.

▶ *Think It Over*

▶ In what part of the nervous system do relaxation techniques have their effect?

- *Tendon organs (Golgi tendon organs)* are found at the junction of a tendon with a muscle. They protect tendons and their associated muscles from damage due to excessive tension by relaying information about the amount of tension to the CNS.

- *Joint kinesthetic receptors* are found in and around synovial joints, such as the shoulder, elbow, hip, and knee joints. They respond to pressure, acceleration and deceleration, and excessive strain on a joint.

- *Hair cells* of the internal ear are proprioceptors that provide information for maintaining balance and equilibrium. Their function is discussed in more detail later in the chapter.

Impulses for conscious proprioception pass along sensory tracts in the spinal cord and are relayed to the primary somatosensory area (postcentral gyrus) in the parietal lobe of the cerebral cortex (see areas 1, 2, and 3, in Figure 10.12 on page 249). Proprioceptive impulses also pass to the cerebellum along spinocerebellar tracts and contribute to subconscious proprioception.

278

SPECIAL SENSES

Receptors for the special senses—smell, taste, sight, hearing, and equilibrium—are structurally more complex than receptors for general sensations and are organized into familiar receptor organs (nose, tongue, eyes, and ears). The sense of smell is the least specialized, and the sense of sight, the most. Like the general senses, the special senses allow us to detect changes in our environment.

OLFACTION: SENSE OF SMELL

Objective: • **Describe the receptors for olfaction and the olfactory pathway to the brain.**

Smell and taste sensations arise from the interaction of molecules with chemoreceptors in the nose and mouth. The nose contains 10 million to 100 million receptors for the sense of smell, or **olfaction** (ol-FAK-shun; *olfact-* = smell). Because some nerve impulses for smell and taste propagate to the limbic system, certain odors and tastes can evoke strong emotional responses or a flood of memories.

Structure of the Olfactory Epithelium

The olfactory epithelium occupies the upper portion of the nasal cavity (Figure 12.2a) and consists of three types of cells: olfactory receptors, supporting cells, and basal cells (Figure 12.2b). Several cilia called **olfactory hairs** project from a knob-shaped tip on each **olfactory receptor**. The olfactory hairs are the parts of the olfactory receptor that respond to odors in the air. **Supporting cells** provide physical support, nourishment, and electrical insulation for the olfactory receptors, and they help detoxify chemicals that come in contact with the olfactory epithelium. **Basal cells** lie between the bases of the supporting cells

Figure 12.2 ■ Olfactory epithelium and olfactory receptors. (a) Location of olfactory epithelium in nasal cavity. (b) Anatomy of olfactory receptors, whose axons extend through the cribriform plate and terminate in the olfactory bulb.

The olfactory epithelium consists of olfactory receptors, supporting cells, and basal cells.

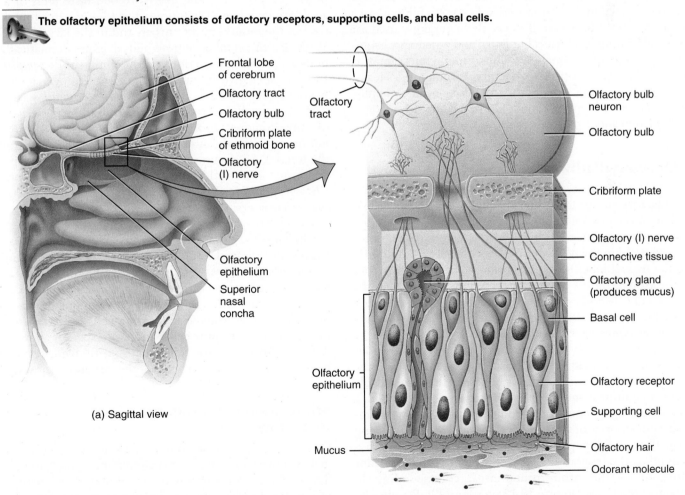

(a) Sagittal view

(b) Enlarged view of olfactory receptors

 Impulses from olfactory bulbs pass to which structures?

and are stem cells that continually undergo cell division to produce new olfactory receptors, which live for only a month or so before being replaced. This process is remarkable because olfactory receptors are neurons, and in general, mature neurons are not replaced. Within the connective tissue that supports the olfactory epithelium are mucus-producing **olfactory glands.** Mucus moistens the surface of the olfactory epithelium and serves as a solvent for inhaled odorants.

Stimulation of Olfactory Receptors

Many attempts have been made to distinguish among and classify "primary" sensations of smell. Genetic evidence now suggests that individual olfactory receptors respond to hundreds of different scents. Our ability to recognize about 10,000 different scents probably depends on patterns of activity in the brain that arise from activation of many different combinations of olfactory receptors.

Olfactory receptors react to odorant molecules by producing a generator potential that triggers one or more nerve impulses. Only a few molecules of certain substances need be present in air to be perceived as an odor. A good example is the chemical methylmercaptan, which smells like rotten cabbage and can be detected in concentrations as low as 1/25 billionth of a milligram per milliliter of air. Because the natural gas used for cooking and heating is odorless but lethal and potentially explosive if it accumulates, a small amount of methylmercaptan is added to natural gas to provide olfactory warning of gas leaks. Adaptation (decreasing sensitivity) to odors occurs rapidly. Olfactory receptors adapt by about 50% in the first second or so after stimulation and very slowly thereafter.

The Olfactory Pathway

On each side of the nose, bundles of slender, unmyelinated axons of olfactory receptors extend through holes in the cribriform plate of the ethmoid bone (Figure 12.2b). These bundles of axons form cranial nerve I, the **olfactory nerves.** They terminate in the brain in the **olfactory bulbs,** which are located inferior to the frontal lobes of the cerebrum. Within the olfactory bulbs, the axon terminals of olfactory receptors synapse with the dendrites and cell bodies of the next neurons in the olfactory pathway. The axons of the neurons extending from the olfactory bulb form the **olfactory tract.** The olfactory tract projects to the *primary olfactory area* in the temporal lobe, where conscious awareness of smells begins. Olfactory impulses also reach the limbic system and the hypothalamus. These pathways probably account for emotional and memory-evoked responses to odors, such as nausea upon smelling a food that once made you violently ill, or memories of your mother's kitchen at Thanksgiving upon smelling pumpkin pie. From the temporal lobe, pathways also extend to the frontal lobe. An important region for odor identification is the orbitofrontal area, corresponding to Brodmann's area 11 (see Figure 10.12 on page 249). People who suffer damage in this area have difficulty identifying different odors.

GUSTATION: SENSE OF TASTE

Objective: • **Describe the receptors for gustation and the gustatory pathway to the brain.**

Taste or **gustation** (GUS-tā-shun; *gust-* = taste), is much simpler than olfaction because only four major classes of stimuli can be distinguished: sour, sweet, bitter, and salty. All other "tastes," such as chocolate, pepper, and coffee, are combinations of these four, plus the accompanying olfactory sensations. Odors from food pass upward from the mouth into the nasal cavity, where they stimulate olfactory receptors. Because olfaction is much more sensitive than taste, foods may stimulate the olfactory system thousands of times more strongly than they stimulate the gustatory system. When persons with colds or allergies complain that they cannot taste their food, they are reporting blockage of olfaction, not of taste.

Structure of Taste Buds

The receptors for sensations of taste are located in the taste buds (Figure 12.3). The nearly 10,000 taste buds of a young adult are mainly on the tongue, but they are also found on the roof of the mouth, in the throat, and in the larynx (voice box). With age, the number of taste buds declines dramatically. Taste buds are found in elevations on the tongue called **papillae** (pa-PIL-ē), which give the upper surface of the tongue its rough appearance (Figure 12.3a,b). **Circumvallate papillae** (ser'-kum-VAL-āt) form an inverted V-shaped row at the posterior portion of the tongue. **Fungiform papillae** (FUN-ji-form) are mushroom-shaped elevations scattered over the entire surface of the tongue. All circumvallate and most fungiform papillae contain taste buds. In addition, the entire surface of the tongue has **filiform papillae** (FIL-i-form), pointed, threadlike structures that rarely contain taste buds.

Each **taste bud** is an oval body consisting of three types of epithelial cells: supporting cells, gustatory receptor cells, and basal cells (Figure 12.3c). The **supporting cells** surround about 50 **gustatory receptor cells.** A single, long microvillus, called a **gustatory hair,** projects from each gustatory receptor cell to the external surface through the **taste pore,** an opening in the taste bud. **Basal cells** produce supporting cells, which then develop into gustatory receptor cells with a life span of about 10 days. The gustatory receptor cells synapse with dendrites of sensory neurons that form the first part of the gustatory pathway.

Stimulation of Gustatory Receptors

Once a chemical is dissolved in saliva, it enters a taste pore and makes contact with the plasma membrane of the gustatory hairs. The result is a depolarizing potential that stimulates exocytosis of synaptic vesicles containing neurotransmitter from the gustatory receptor cell. Nerve impulses are triggered when these neurotransmitter molecules bind to their receptors on the dendrites

Figure 12.3 ■ **The relationship of gustatory receptors in taste buds to tongue papillae.**

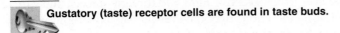 Gustatory (taste) receptor cells are found in taste buds.

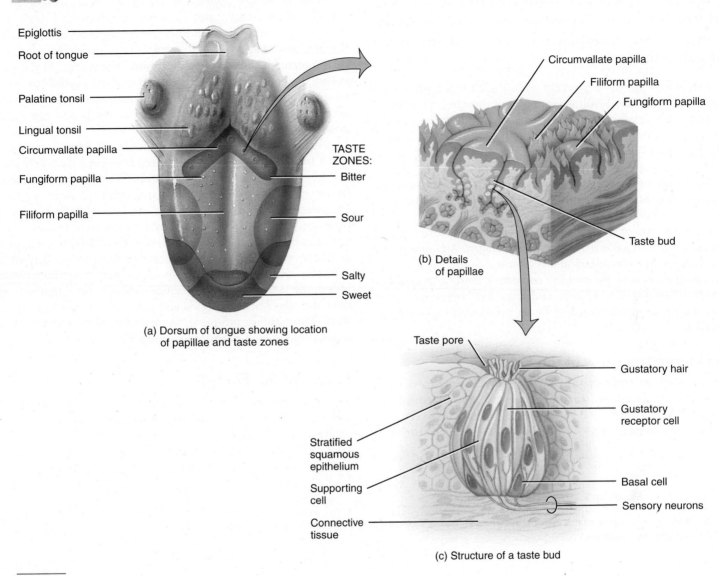

(a) Dorsum of tongue showing location of papillae and taste zones

Epiglottis

Root of tongue

Palatine tonsil

Lingual tonsil

Circumvallate papilla

Fungiform papilla

Filiform papilla

TASTE ZONES:

Bitter

Sour

Salty

Sweet

Circumvallate papilla

Filiform papilla

Fungiform papilla

Taste bud

(b) Details of papillae

Taste pore

Gustatory hair

Gustatory receptor cell

Stratified squamous epithelium

Supporting cell

Connective tissue

Basal cell

Sensory neurons

(c) Structure of a taste bud

In order, what structures form the gustatory pathway?

of the sensory neuron. The dendrites branch profusely and contact many gustatory receptors in several taste buds. Individual gustatory receptor cells may respond to more than one of the four primary tastes, but certain regions of the tongue are more sensitive to particular primary taste sensations (Figure 12.3a). Receptors in the tip of the tongue are highly sensitive to sweet and salty substances, receptors in the posterior portion of the tongue are highly sensitive to bitter substances, and those in the lateral areas of the tongue are most sensitive to sour substances. Complete adaptation (loss of sensitivity) to a specific taste can occur in 1 to 5 minutes of continuous stimulation.

The Gustatory Pathway

Three cranial nerves include axons of sensory neurons from taste buds: the facial (VII) nerve, the glossopharyngeal (IX) nerve, and the vagus (X) nerve. Impulses conduct along these cranial nerves to the medulla oblongata. From the medulla, some axons carrying impulses for taste extend to the limbic system and the hypothalamus, whereas others extend to the thalamus. From the thalamus, axons extend to the *primary gustatory area* in the parietal lobe of the cerebral cortex (see area 43 in Figure 10.12 on page 249), giving rise to the conscious perception of taste.

VISION

Objectives: • **Describe the accessory structures of the eye, the layers of the eyeball, the lens, the interior of the eyeball, image formation, and binocular vision.** • **Describe the receptors for vision and the visual pathway to the brain.**

More than half the sensory receptors in the human body are located in the eyes. As a result, a large part of the cerebral cortex is devoted to processing visual information. In this section of the chapter, we examine the accessory structures of the eye, the eyeball, the formation of visual images, the physiology of vision, and the visual pathway.

The study of the structure, function, and diseases of the eye is known as **ophthalmology** (of'-thal-MOL-ō-jē; *ophthalm-* = eye; *-ology* = study of). A physician who specializes in the diagnosis and treatment of eye disorders with drugs, surgery, and corrective lenses is known as an **ophthalmologist.** An **optometrist** has a doctorate in optometry and is licensed to test the eyes and treat visual defects by prescribing corrective lenses. An **optician** is a technician who fits, adjusts, and dispenses corrective lenses using the prescription supplied by an ophthalmologist or optometrist.

Accessory Structures of the Eye

The **accessory structures** of the eye include the eyebrows, eyelashes, eyelids, lacrimal (tearing) apparatus, and extrinsic eye muscles. The **eyebrows** and **eyelashes** help protect the eyeballs from foreign objects, perspiration, and direct rays of the sun.

The upper and lower **eyelids** shade the eyes during sleep, protect the eyes from excessive light and foreign objects, and spread lubricating secretions over the eyeballs (by blinking).

The **lacrimal apparatus** (*lacrima* = tear) refers to the glands, ducts, canals, and sacs that produce and drain tears (Figure 12.4). The right and left **lacrimal glands** are each about the size and shape of an almond. They secrete lacrimal fluid (tears) through the **lacrimal ducts** onto the surface of the upper eyelid. Tears then pass over the surface of the eyeball toward the nose into two **lacrimal canals** and a **nasolacrimal duct,** which allow the tears to drain into the nasal cavity (producing a runny nose when you cry).

Tears are a watery solution containing salts, some mucus, and a bacteria-killing enzyme called **lysozyme.** Tears clean, lubricate, and moisten the portion of the eyeball exposed to the air to prevent it from drying. Normally, tears are cleared away by evaporation or by passing into the nasal cavity as fast as they are produced. If, however, an irritating substance makes contact with the eye, the lacrimal glands are stimulated to oversecrete and tears accumulate. This protective mechanism dilutes and washes away the irritant. In response to parasympathetic stimulation, the lacrimal glands produce excessive tears that may spill over the edges of the eyelids and even fill the nasal cavity with fluid. Humans are unique in their ability to cry to express certain emotions such as happiness and sadness.

Layers of the Eyeball

The adult **eyeball** measures about 2.5 cm (1 inch) in diameter and is divided into three layers: fibrous tunic, vascular tunic, and retina (Figure 12.5a).

Figure 12.4 ▪ **Accessory structures of the eye.**

Accessory structures of the eye include the eyebrows, eyelashes, eyelids, the lacrimal apparatus, and extrinsic eye muscles.

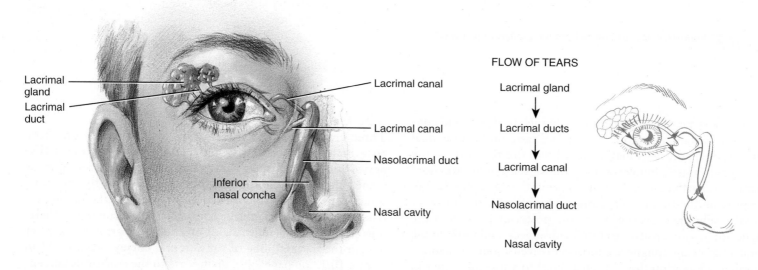

Lacrimal gland

Lacrimal duct

Lacrimal canal

Lacrimal canal

Nasolacrimal duct

Inferior nasal concha

Nasal cavity

Anterior view of the lacrimal apparatus

FLOW OF TEARS

Lacrimal gland
↓
Lacrimal ducts
↓
Lacrimal canal
↓
Nasolacrimal duct
↓
Nasal cavity

What substances are in tears and what are their functions?

Figure 12.5 ■ **Structure of the eyeball and responses of the pupil to light.**

The wall of the eyeball consists of three layers: the fibrous tunic, the vascular tunic, and the retina.

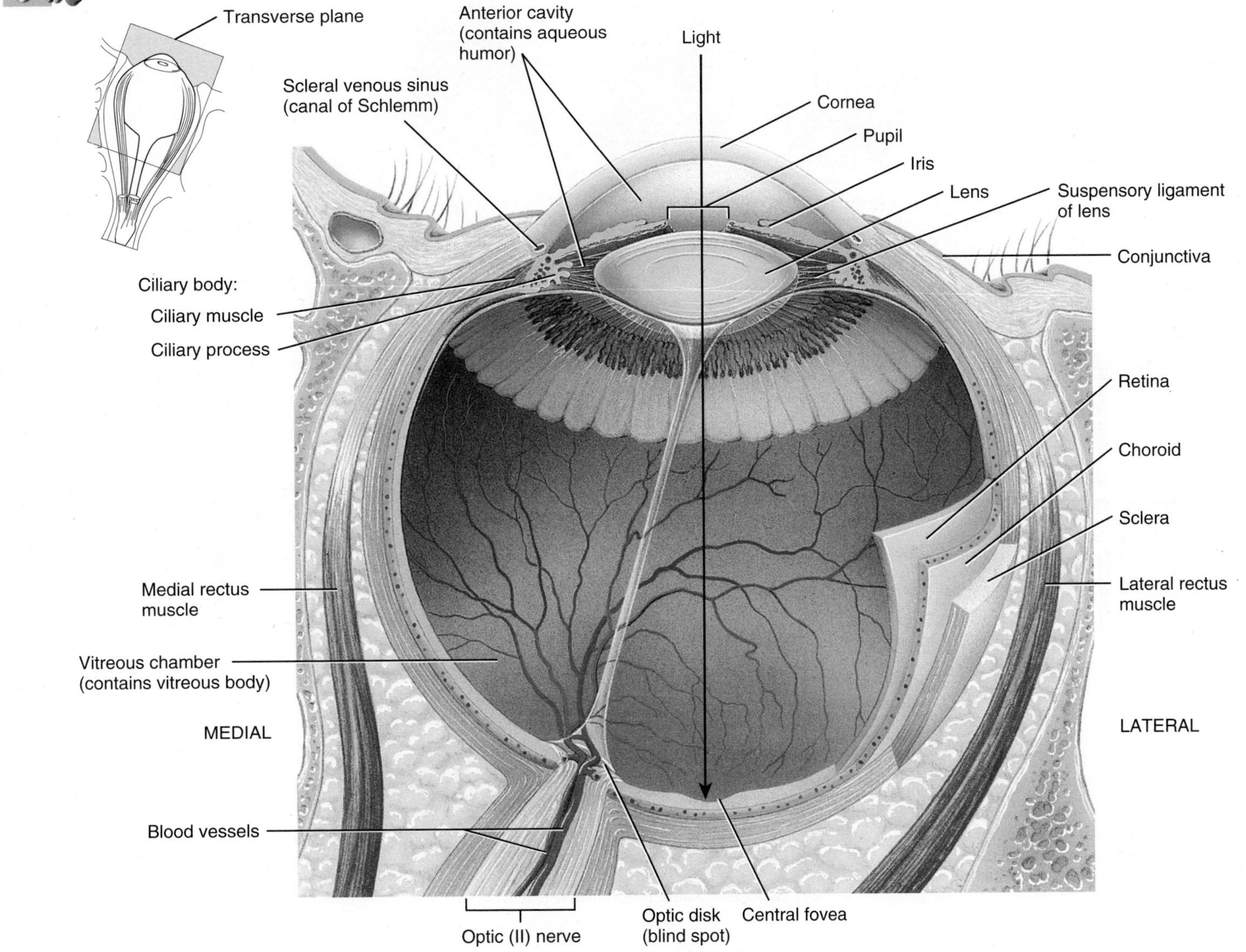

(a) Superior view of transverse section of right eyeball

Figure 12.5 (*continues*)

Fibrous Tunic

The ***fibrous tunic*** is the outer coat of the eyeball consisting of an anterior cornea and a posterior sclera. The ***cornea*** (KOR-nē-a) is a nonvascular, transparent fibrous coat that covers the colored part of the eyeball, the iris. The cornea's outer surface is covered by an epithelial layer called the ***conjunctiva,*** which also lines the eyelid. The ***sclera*** (SKLER-a = hard), the "white" of the eye, is a coat of dense connective tissue that covers all of the eyeball, except the cornea. The sclera gives shape to the eyeball, makes it more rigid, and protects its inner parts.

The cornea bends light rays entering the eyeball to produce a clear image. If the cornea is not curved properly, blurred vision results. A defective cornea can be removed and replaced with a donor cornea of similar diameter, a procedure called a ***corneal transplant.*** Corneal transplants are the most successful type of transplantation because corneas do not contain blood vessels and the body is unlikely to reject them.

Vascular Tunic

The ***vascular tunic*** is the middle layer of the eyeball and is composed of the choroid, ciliary body, and iris. The ***choroid*** (KOR-oyd) is a thin membrane that lines most of the internal surface of the sclera. It contains many blood vessels that help nourish the retina.

Figure 12.5 (*continued*)

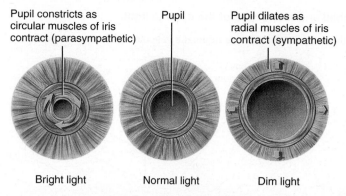

Pupil constricts as circular muscles of iris contract (parasympathetic)

Pupil

Pupil dilates as radial muscles of iris contract (sympathetic)

Bright light   Normal light   Dim light

(b) Anterior view of responses of the pupil to varying brightness of light

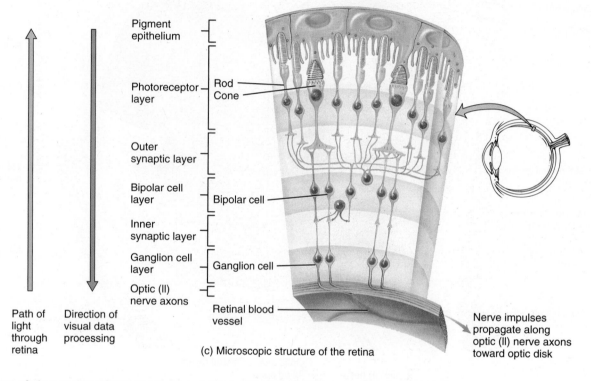

Pigment epithelium

Photoreceptor layer — Rod — Cone

Outer synaptic layer

Bipolar cell layer — Bipolar cell

Inner synaptic layer

Ganglion cell layer — Ganglion cell

Optic (II) nerve axons

Retinal blood vessel

Nerve impulses propagate along optic (II) nerve axons toward optic disk

Path of light through retina

Direction of visual data processing

(c) Microscopic structure of the retina

 Which type of photoreceptor is specialized for vision in dim light and allows us to see shapes and movement? Which type of photoreceptor is specialized for color vision and vision with high acuity?

At the front of the eye, the choroid becomes the *ciliary body* (SIL-ē-ar′-ē). The ciliary body consists of the *ciliary processes*, folds on the inner surface of the ciliary body whose capillaries secrete a watery fluid called aqueous humor, and the *ciliary muscle*, a smooth muscle that alters the shape of the lens for near or far vision. The *lens*, a transparent structure that focuses light rays onto the retina, is constructed of numerous layers of elastic protein fibers. *Suspensory ligaments* that attach the lens to the ciliary muscle hold the lens in position.

The *iris* (= colored circle) is the doughnut-shaped colored portion of the eyeball. It consists of circular and radial smooth muscle fibers. The hole in the center of the iris, through which light enters the eyeball, is the *pupil*. The iris regulates the amount of light passing though the lens. When the eye is stimulated by bright light, the circular muscles of the iris contract and decrease the size of the pupil (constriction). When the eye must adjust to dim light, the radial muscles contract, and the pupil increases in size (dilation) (Figure 12.5b). These muscles of the iris are controlled by the autonomic nervous system (see Table 11.3 on page 269).

Retina

The third and inner coat of the eyeball, the *retina*, lines the posterior three-quarters of the eyeball and is the beginning of the visual pathway. An ophthalmoscope allows an observer to peer through the pupil, providing a magnified image of the retina and the blood vessels that cross it. The surface of the retina is the only place in the body where blood vessels can be viewed directly and examined for pathological changes, such as those that occur with hypertension or diabetes mellitus.

The retina consists of a pigment epithelium (nonvisual portion) and a neural portion (visual portion). The *pigment epithelium* is a sheet of melanin-containing epithelial cells that lies between the choroid and the neural portion of the retina. Melanin in the choroid and in the pigment epithelium absorbs stray light rays, which prevents reflection and scattering of light within the eyeball. As a result, the image cast on the retina by the cornea and lens remains sharp and clear.

The neural portion of the retina is a multilayered structure that develops from brain tissue. It extensively processes visual data before transmitting nerve impulses to the thalamus. Three distinct layers of retinal neurons—the *photoreceptor layer*, the *bipolar cell layer*, and the *ganglion cell layer*—are separated by the outer and inner synaptic layers, where synaptic contacts are made (Figure 12.5c). The two types of cells located in the photoreceptor layer—rod and cone photoreceptors—are highly specialized for detecting light rays. *Rods* allow us to see in dim light, such as moonlight. Because they do not provide color vision, in dim light we see only shades of gray. Brighter lights stimulate the *cones*, giving rise to highly acute, color vision. The loss of cone vision causes a person to become legally blind. In contrast, a person who loses rod vision mainly has difficulty seeing in dim light and thus should not, for example, drive at night.

There are about 6 million cones and 120 million rods. Cones are most densely concentrated in the *central fovea*, a small depression in the center of the *macula lutea* (MAK-yoo-la LOO-tē-a), or yellow spot, in the exact center of the retina. The central fovea is the area of highest *visual acuity* or *resolution* (sharpness of vision) because of its high concentration of cones. The main reason that you move your head and eyes while looking at something, such as the words of this sentence, is to place images of interest on your fovea. Rods are absent from the central fovea and macula lutea and increase in numbers toward the periphery of the retina.

From photoreceptors, information flows through the outer synaptic layer to the *bipolar cells* of the bipolar cell layer, and then from bipolar cells through the inner synaptic layer to the *ganglion cells* of the ganglion cell layer. The axons of the ganglion cells extend posteriorly to a small area of the retina called the *optic disk (blind spot)*, where they all exit as the optic (II) nerve. Because the optic disk contains no rods or cones, we cannot see an image that strikes the blind spot.

A frequently encountered problem related to the retina is a *detached retina*, which may occur in trauma, such as a blow to the head. The actual detachment occurs between the neural part of the retina and the underlying choroid. Fluid accumulates between these layers, resulting in distorted vision and blindness. Often, it is possible to surgically reattach the retina.

Interior of the Eyeball

The lens divides the interior of the eyeball into two cavities, the anterior cavity and the vitreous chamber. The *anterior cavity* lies anterior to the lens and is filled with *aqueous humor* (Ā-kwē-us HYOO-mor; *aqua* = water), a watery fluid similar to cerebrospinal fluid. The aqueous humor is secreted into the anterior cavity from blood capillaries of the ciliary processes. It then drains into the *scleral venous sinus (canal of Schlemm)*, an opening where the sclera and cornea meet, and reenters the blood. The aqueous humor helps maintain the shape of the eye and nourishes the lens and cornea, neither of which has blood vessels.

The second, and larger, cavity of the eyeball is the *vitreous chamber*. It lies behind the lens and contains a clear, jellylike substance called the *vitreous body*. This substance helps prevent the eyeball from collapsing and holds the retina flush against the choroid. The vitreous body, unlike the aqueous humor, does not undergo constant replacement. It is formed during embryonic life and is not replaced thereafter.

The pressure in the eye, called *intraocular pressure*, is produced mainly by the aqueous humor with a smaller contribution from the vitreous body. Intraocular pressure maintains the shape of the eyeball and keeps the retina smoothly pressed against the choroid so the retina is well nourished and forms clear images. Normal intraocular pressure (about 16 mm Hg) is maintained by a balance between production and drainage of the aqueous humor.

Table 12.2 / Summary of Structures Associated with the Eyeball

Structure	Function	Structure	Function
Fibrous tunic	*Cornea:* Admits and refracts (bends) light. *Sclera:* Provides shape and protects inner parts.	**Lens**	Focuses light on the retina.
Vascular tunic	*Iris:* Regulates amount of light that enters eyeball. *Ciliary body:* Secretes aqueous humor and alters shape of lens for near or far vision. *Choroid:* Provides blood supply.	**Anterior cavity**	Contains aqueous humor that helps maintain shape of eyeball and supplies oxygen and nutrients to lens and cornea.
Retina	Converts light stimuli into nerve impulses. Output to brain is via the optic (II) nerve.	**Vitreous chamber**	Contains vitreous body that helps maintain shape of eyeball and keeps retina attached to choroid.

Table 12.2 summarizes the structures associated with the eyeball.

Image Formation and Binocular Vision

In some ways the eye is like a camera. Its optical elements focus an image of some object on a light-sensitive "film"—the retina—while ensuring the correct amount of light makes the proper "exposure." To understand how the eye forms clear images of objects on the retina, we must examine three processes: (1) the refraction or bending of light by the lens and cornea, (2) the change in shape of the lens, and (3) constriction of the pupil.

Refraction of Light Rays

When light rays traveling through a transparent substance (such as air) pass into a second transparent substance with a different density (such as water), they bend at the junction between the two substances. This bending is called *refraction* (Figure 12.6a). About 75% of the total refraction of light occurs at the anterior and posterior surfaces of the cornea. Both surfaces of the lens of the eye further refract the light rays so that they come into exact focus on the retina.

Images focused on the retina are inverted (upside down) (Figure 12.6b,c). They also undergo right-to-left reversal; that is, light from the right side of an object strikes the left side of the retina, and vice versa. The reason the world does not look inverted and reversed is that the brain "learns" early in life to coordinate visual images with the orientations of objects. The brain stores the inverted and reversed images we acquire when we first reach for and touch objects and interprets those visual images as being correctly oriented in space.

When an object is more than 6 meters (20 ft) away from the viewer, the light rays reflected from the object are nearly parallel to one another, and the curvatures of the cornea and lens exactly focus the image on the retina (Figure 12.6b). However, light rays from objects closer than 6 meters are divergent rather than parallel (Figure 12.6c). The rays must be refracted more if they are to be focused on the retina. This additional refraction is accomplished by changes in the shape of the lens, a process called accommodation.

Accommodation

A surface that curves outward, like the surface of a ball, is said to be *convex*. The convex surface of a lens refracts incoming light rays toward each other, so that they eventually intersect. The lens of the eye is convex on both its anterior and posterior surfaces, and its ability to refract light increases as its curvature becomes greater. When the eye is focusing on a close object, the lens becomes more convex and refracts the light rays more. This increase in the curvature of the lens for near vision is called *accommodation* (Figure 12.6c).

When you are viewing distant objects, the ciliary muscle is relaxed and the lens is fairly flat because it is stretched in all directions by taut suspensory ligaments. When you view a close

Figure 12.6 ■ **Refraction of light rays and accommodation.**

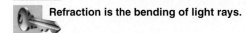
Refraction is the bending of light rays.

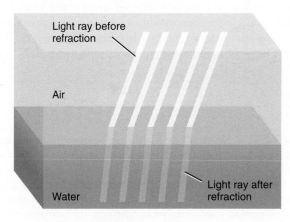

(a) Refraction of light rays

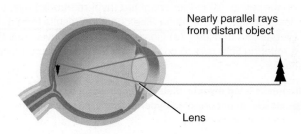

(b) Viewing distant object

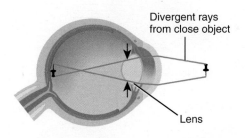

(c) Accommodation

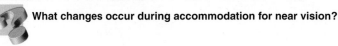
What changes occur during accommodation for near vision?

object, the ciliary muscle contracts, which pulls the ciliary process and choroid forward toward the lens. This action releases tension on the lens, allowing it to become rounder (more convex), which increases its focusing power and causes greater convergence of the light rays. With aging, the lens loses some of its elasticity so its ability to accommodate decreases. At about age 40, people who have not previously worn glasses begin to require them for near vision, such as reading. This condition is called *presbyopia* (prez-bē-Ō-pē-a; *presby-* = old; *-opia* = pertaining to the eye or vision).

The normal eye, known as an **emmetropic eye** (em'-e-TROP-ik; *emmetr-* = according to measure; *-opic* = eye), can sufficiently refract light rays so that a clear image is focused on the retina (Figure 12.7). Many people, however, lack this ability because of refraction abnormalities. For example, they may have an inability to clearly see distant objects, termed **myopia** (mī-Ō-pē-a; *my-* = to shut) or nearsightedness (Figure 12.7b,c). Or they may have an inability to clearly see near objects, termed **hy-**

permetropia (hī'-per-me-TRŌ-pē-a; *hyper-* = above, over) or farsightedness (Figure 12.7d,e). Another refraction abnormality is **astigmatism** (a-STIG-ma-tizm), an irregular curvature of either the cornea or the lens.

Constriction of the Pupil

Constriction of the pupil is a narrowing of the diameter of the hole through which light enters the eye. It occurs due to contraction of the circular muscles of the iris. This autonomic reflex occurs simultaneously with accommodation and prevents light rays from entering the eye through the periphery of the lens. Light rays entering at the periphery of the lens would not be brought to focus on the retina and would result in blurred vision. The pupil, as noted earlier, also constricts in bright light to limit the amount of light that strikes the retina.

Convergence

In humans, both eyes focus on a single object or set of objects, a characteristic called **binocular vision.** This feature of our visual system allows the perception of depth and an appreciation of the three-dimensional nature of objects. When you stare straight ahead at a distant object, the incoming light rays are aimed directly at the pupils of both eyes and are refracted to comparable spots on the two retinas. As you move closer to the object, your eyes must rotate toward the nose if the light rays from the object are to strike comparable points on both retinas. **Convergence** is the name for this automatic movement of the two eyeballs toward the midline, which is caused by the coordinated action of the extrinsic eye muscles. The nearer the object, the greater the convergence needed to maintain binocular vision.

Stimulation of Photoreceptors

After an image is formed on the retina by refraction, accommodation, constriction of the pupil, and convergence, light rays must be converted into neural signals. The initial step in this process is the absorption of light rays by the rods and cones of the retina. To understand how absorption occurs, it is necessary to understand the role of photopigments.

A **photopigment** is a substance that can absorb light and undergo a change in structure. The photopigment in rods is called **rhodopsin** (*rhodo-* = rose; *-opsin* = related to vision) and is composed of a protein called **opsin** and a derivative of vitamin A called **retinal.** Rhodopsin is a highly unstable compound in the presence of even very small amounts of light. Any amount of light in a darkened room causes some rhodopsin molecules to split into opsin and retinal and initiate a series of chemical changes in the rods. When the light level is dim, opsin and retinal recombine into rhodopsin so that production keeps pace with breakdown. Rods usually are nonfunctional in daylight, because rhodopsin is split apart faster than it can be re-formed. After going from bright sunlight into a dark room, it takes about 40 minutes before the rods function maximally.

Cones function in bright light and provide color vision. As in rods, absorption of light rays causes breakdown of photopigment molecules. The photopigments in cones also contain retinal, but there are three different opsin proteins. One type of cone photopigment responds best to yellow-orange light, the

Figure 12.7 ■ **Normal and abnormal refraction in the eyeball.** (a) In the normal (emmetropic) eye, light rays from an object are bent sufficiently by the cornea and lens and converge on the central fovea. A clear image is formed. (b) In the nearsighted (myopic) eye, the image is focused in front of the retina. The condition may result from an elongated eyeball or thickened lens. (c) Correction is by use of a concave lens that causes entering light rays to diverge so that they have to travel further through the eyeball and are focused directly on the retina. (d) In the farsighted (hypermetropic) eye, the image is focused behind the retina. The condition results from a shortened eyeball or a thin lens. (e) Correction is by a convex lens that causes entering light rays to converge so that they focus directly on the retina.

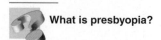

In uncorrected myopia, distant objects can't be seen clearly; in uncorrected hypermetropia, nearby objects can't be seen clearly.

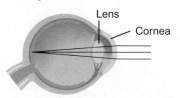

Lens
Cornea

(a) Normal (emmetropic) eye

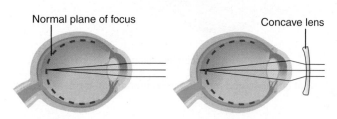

Normal plane of focus
Concave lens

(b) Nearsighted (myopic) eye, uncorrected

(c) Nearsighted (myopic) eye, corrected

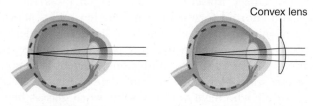

Convex lens

(d) Farsighted (hypermetropic) eye, uncorrected

(e) Farsighted (hypermetropic) eye, corrected

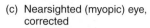

What is presbyopia?

Figure 12.8 ■ **Visual pathway.**

 The optic chiasm is the structure in which half of the retinal ganglion cell axons from each eye cross to the opposite side of the brain.

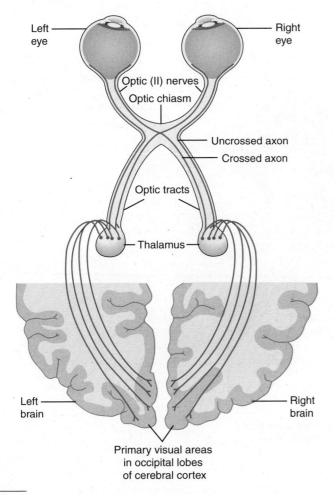

What is the correct order of structures that carry nerve impulses from the retina to the occipital lobe?

second to green, and the third to blue. An individual cone photoreceptor contains just one type of cone photopigment. The cone photopigments re-form much more quickly than the rod photopigment. If there are only three types of color photopigments, why don't we just see yellow orange, green, and blue? Just as an artist can obtain almost any color by mixing them on a palette, the cones can code for different colors by differential stimulation. If all three types of cones are stimulated, an object is perceived as white in color; if none is stimulated, the object looks black.

An individual with one type of cone missing from the retina cannot distinguish some colors from others and is said to be *colorblind*. In the most common type, *red–green colorblindness,* one cone photopigment is missing. Color blindness is an inherited condition that affects males far more often than females. The inheritance of the condition is discussed in Chapter 24 and illustrated in Figure 24.13 on page 591.

The Visual Pathway

After stimulation by light, the rods and cones trigger electrical signals in bipolar cells. Bipolar cells transmit both excitatory and inhibitory signals to ganglion cells. The ganglion cells become depolarized and generate nerve impulses. The axons of the ganglion cells exit the eyeball as cranial nerve II, the *optic nerve* (Figure 12.8) and extend posteriorly to the *optic chiasm* (KĪ-azm = a crossover, as in the letter X). In the optic chiasm, about half of the axons from each eye cross to the opposite side of the brain. After passing the optic chiasm, the axons, now part of the *optic tract,* terminate in the thalamus. Here the neurons of the optic tract synapse with thalamic neurons whose axons pass to the *primary visual areas* in the occipital lobes. Because of the crossing at the optic chiasm, the right side of the brain receives signals from both eyes for interpretation of visual sensations from the left side of an object, and the left side of the brain receives signals from both eyes for interpretation of visual sensations from the right side of an object.

HEARING AND EQUILIBRIUM

Objectives: • **Describe the structures of the external, middle, and internal ear.**
• **Describe the receptors for hearing and equilibrium and their pathways to the brain.**

The ear is a marvelously sensitive structure. Its sensory receptors can convert sound vibrations as small as the diameter of an atom of gold (0.3 nanometers) into electrical signals 1000 times faster than photoreceptors can respond to light. In addition to these incredibly sensitive receptors for sound waves, the ear also contains receptors for equilibrium (balance).

Structure of the Ear

The ear is divided into three main regions: the external ear, which collects sound waves and channels them inward; the middle ear, which conveys sound vibrations to the oval window; and the internal ear, which houses the receptors for hearing and equilibrium.

External Ear

The *external ear* collects sound waves and passes them inward (Figure 12.9). It consists of an auricle, external auditory canal, and eardrum. The *auricle,* the part of the ear that you can see, is a flap of elastic cartilage shaped like the flared end of a trumpet to direct sound waves into the external auditory canal. It is attached to the head by ligaments and muscles. The *external auditory canal* (audit- = hearing) is a curved tube that extends from the auricle to the eardrum. The canal contains a few hairs and *ceruminous glands* (se-ROO-mi-nus; cer- = wax), which secrete *cerumen* (se-ROO-min) (earwax). The hairs and cerumen help prevent foreign objects from entering the ear. The *eardrum,* also called the *tympanic membrane* (tim-PAN-ik = drum), is a thin, semitransparent partition between the external auditory canal and the middle ear.

Figure 12.9 ■ **Structure of the auditory apparatus.**

The ear has three principal regions: the external ear, the middle ear, and the internal ear.

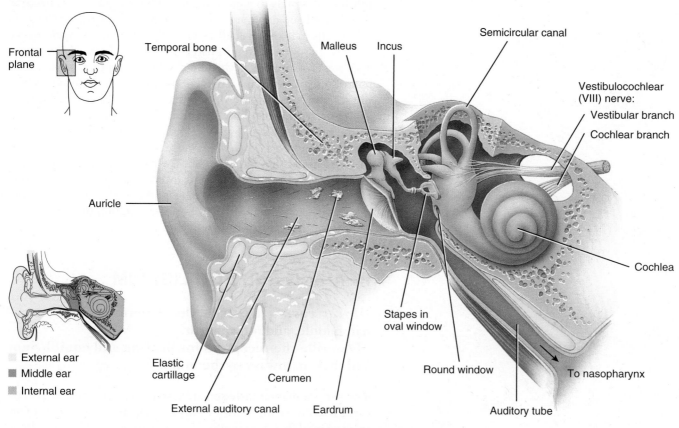

Frontal section through the right side of the skull
showing the three principal regions of the ear

 Where are the receptors for hearing and equilibrium located?

Middle Ear

The **middle ear** is a small, air-filled cavity between the eardrum of the external ear and internal ear (Figure 12.9). An opening in the anterior wall of the middle ear leads directly into the **auditory (Eustachian) tube,** which connects the middle ear with the nasopharynx, the upper part of the throat. Via the auditory tube, air pressure can equalize on both sides of the eardrum. Otherwise, abrupt changes in external or internal air pressure might cause the eardrum to rupture. During swallowing and yawning, the tube opens to relieve pressure, which explains why the sudden pressure change in an airplane may be equalized by swallowing or pinching the nose closed, closing the mouth, and gently forcing air up from the lungs.

Extending across the middle ear and attached to it by means of ligaments are three tiny bones called **auditory ossicles** (OS-si-kuls) that are named for their shapes: the malleus, incus, and stapes, commonly called the hammer, anvil, and stirrup (Figure 12.9). Equally tiny skeletal muscles control the amount of movement of these bones to prevent damage by excessively loud noises. The stapes fits into a small opening in the thin bony par-

tition between the middle and internal ear called the **oval window,** where the internal ear begins.

Internal Ear

The **internal ear** is divided into the outer bony labyrinth and inner membranous labyrinth (Figure 12.10). The **bony labyrinth** (LAB-i-rinth) is a series of cavities in the temporal bone, including the cochlea, vestibule, and semicircular canals. The cochlea is the sense organ for hearing, whereas the vestibule and semicircular canals are the sense organs for equilibrium and balance. The bony labyrinth contains a fluid called **perilymph**. This fluid surrounds the inner **membranous labyrinth,** a series of sacs and tubes with the same general shape as the bony labyrinth. The membranous labyrinth contains a fluid called **endolymph.**

The **vestibule** is the oval-shaped middle part of the bony labyrinth. The membranous labyrinth in the vestibule consists of two sacs called the **utricle** (YOO-tri-kul = little bag) and **saccule** (SAK-yool = little sac). Behind the vestibule are the three bony **semicircular canals.** The anterior and posterior semicircular canals are both vertical and the lateral canal is horizontal.

Figure 12.10 ■ **Details of the internal ear of the right ear.** (a) Relationship of the scala tympani, cochlear duct, and scala vestibuli. The arrows indicate the transmission of sound waves. (b) Details of the spiral organ (organ of Corti).

 The three channels in the cochlea are the scala vestibuli, scala tympani, and cochlear duct.

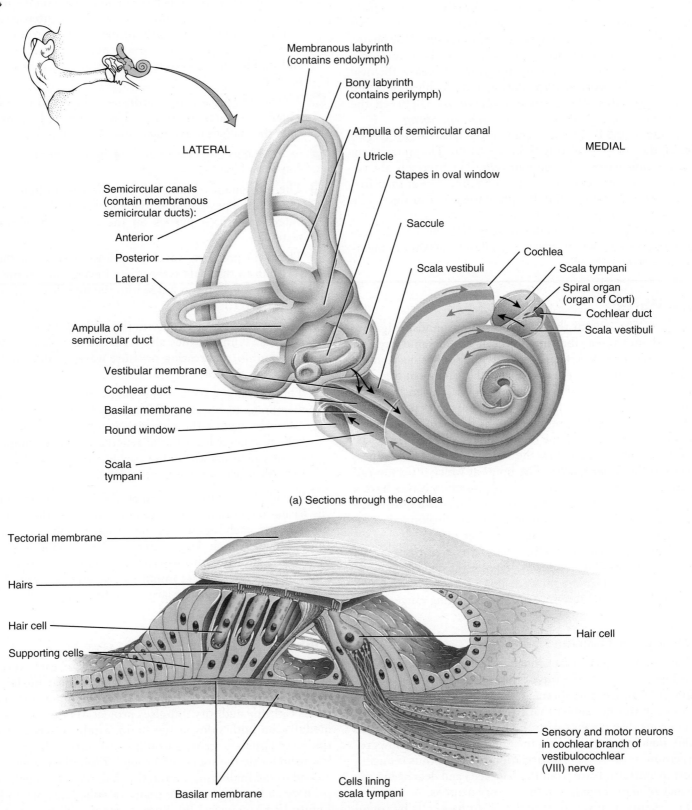

(a) Sections through the cochlea

(b) Enlargement of spiral organ (organ of Corti)

 What structures separate the external ear from the middle ear? the middle ear from the internal ear?

One end of each canal enlarges into a swelling called the *ampulla* (am-POOL-la = little jar). Inside the bony semicircular canals are the *semicircular ducts,* which connect with the utricle of the vestibule.

A transverse section through the *cochlea* (KOK-lē-a = snail's shell), a bony spiral canal that resembles a snail's shell, shows that it is divided into three channels: the *scala vestibuli* that begins near the oval window, the *scala tympani* that ends at the *round window* (a membrane-covered opening directly below the oval window), and the *cochlear duct.* Between the cochlear duct and the scala vestibuli is the *vestibular membrane.* Between the cochlear duct and scala tympani is the *basilar membrane.*

Resting on the basilar membrane is the *spiral organ (organ of Corti),* the organ of hearing (Figure 12.10b). The spiral organ consists of *supporting cells* and *hair cells.* The hair cells, the receptors for auditory sensations, have long processes at their free ends that extend into the endolymph of the cochlear duct. The hair cells form synapses with sensory and motor neurons in the cochlear branch of the vestibulocochlear (VIII) nerve. The *tectorial membrane* (tector- = covering), a flexible gelatinous membrane, covers the hair cells.

Sound Waves

Sound waves, the stimuli that we perceive as sounds, originate and spread out from a vibrating object much the same way that waves travel over the surface of water. The sounds heard most acutely by human ears are from sources that vibrate at frequencies between 500 and 5000 cycles per second. The entire range of human hearing extends from 20 to 20,000 cycles per second.

The *frequency* of a sound vibration is its *pitch.* The greater the frequency of vibration, the higher the pitch. The greater the *size* of the vibration, the louder the sound. Intensity or loudness is measured in *decibels (dB).* The point at which a person can just detect sound from silence is 0 dB. Rustling leaves have a decibel rating of 15, normal conversation 45, crowd noise 60, a vacuum cleaner 75, and a pneumatic drill 90. Prolonged exposure to sounds over 90 dB (loud music, jet planes, even some vacuum cleaners) damages hair cells of the spiral organ and causes hearing loss. The louder the sounds, the quicker the loss. If bystanders can hear the music you are listening to through headphones, the sound intensity is in the damaging range.

Physiology of Hearing

The hair cells convert a mechanical force into an electrical signal. When the hairs at the top of the cell are moved in one direction, the hair cell membrane depolarizes, causing release of neurotransmitter molecules that trigger nerve impulses in sensory neurons that synapse with the hair cell at its base (Figure 12.10b). Bending of the hairs in the opposite direction allows repolarization or even hyperpolarization to occur, thus reducing neurotransmitter release from the hair cells and decreasing the frequency of nerve impulses in the sensory neurons.

The events involved in stimulation of hair cells by sound waves are as follows (Figure 12.11):

1 The auricle directs sound waves into the external auditory canal.

2 Sound waves striking the eardrum cause it to vibrate. The distance and speed of its movement depend on the intensity and frequency of the sound waves. More intense (louder) sounds produce larger vibrations. The eardrum vibrates slowly in response to low-frequency (low-pitched) sounds and rapidly in response to high-frequency (high-pitched) sounds.

3 The central area of the eardrum connects to the malleus, which also starts to vibrate. The vibration is transmitted from the malleus to the incus and then to the stapes.

4 As the stapes moves back and forth, it pushes the oval window in and out.

5 The movement of the oval window sets up fluid pressure waves in the perilymph of the cochlea. As the oval window bulges inward, it pushes on the perilymph of the scala vestibuli.

6 The fluid pressure waves are transmitted from the scala vestibuli to the scala tympani and eventually to the round window, causing it to bulge into the middle ear. (See **9** in Figure 12.11.)

7 As the pressure waves deform the walls of the scala vestibuli and scala tympani, they also push the vestibular membrane back and forth, creating pressure waves in the endolymph inside the cochlear duct.

8 The pressure waves in the endolymph cause the basilar membrane to vibrate, which moves the hair cells of the spiral organ against the tectorial membrane. Bending of the hairs ultimately leads to the generation of nerve impulses in sensory neurons within the cochlear branch of the vestibulocochlear nerve (see Figure 12.10b).

Sound waves of various frequencies cause specific regions of the basilar membrane to vibrate more intensely than others. In other words, each segment of the basilar membrane is "tuned" for a particular pitch. High-intensity (loud) sound waves cause greater vibration of the basilar membrane, which leads to a higher frequency of nerve impulses reaching the brain. Louder sounds also may stimulate a larger number of hair cells.

Auditory Pathway

Sensory neurons in the cochlear branch of each vestibulocochlear (VIII) nerve terminate in the cochlear nuclei of the medulla oblongata on the same side of the brain. From there, axons carrying auditory impulses project to other nuclei in the medulla, on both sides of the brain. Slight differences in the timing of impulses arriving from the two ears at these nuclei allow us to locate the source of a sound. From the medulla, axons ascend to the midbrain, then to the thalamus, and finally to the primary auditory area in the temporal lobe (areas 41 and 42 in Figure 10.12 on page 249). Because many auditory axons cross to the opposite side, the right and left primary auditory areas receive nerve impulses from both ears.

Figure 12.11 ■ **Physiology of hearing shown in the right ear.** The numbers correspond to the events listed in the text. The cochlea has been uncoiled in order to more easily visualize the transmission of sound waves and their subsequent distortion of the vestibular and basilar membranes of the cochlear duct.

Sound waves originate from a vibrating object.

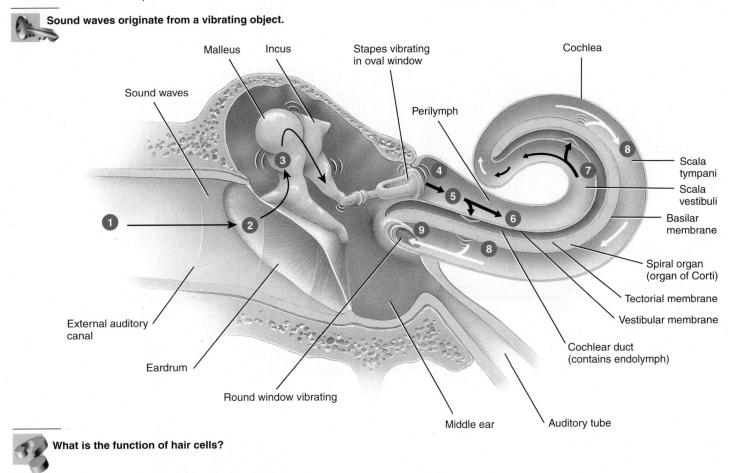

What is the function of hair cells?

Physiology of Equilibrium

You learned about the anatomy of the internal ear structures for equilibrium in the previous section. In this section we will cover the physiology of balance, or how you are able to stay on your feet after tripping over your roommate's shoes. There are two types of *equilibrium* (balance). One kind, called *static equilibrium,* refers to the maintenance of the position of the body relative to the force of gravity. The second kind, *dynamic equilibrium,* is the maintenance of body position in response to sudden movements such as rotation, acceleration, and deceleration. Collectively, the receptor organs for equilibrium, which include the saccule, utricle, and membranous semicircular ducts, are called the *vestibular apparatus* (ves-TIB-yoo-lar).

Saccule and Utricle

The walls of the saccule and the utricle contain small, thickened regions called *maculae* (MAK-yoo-lē; *macula* = spot). The two maculae, oriented perpendicular to one another, are the receptor organs for static equilibrium. They provide sensory information on the position of the head in space and help maintain appropriate posture and balance. They also contribute to some aspects of dynamic equilibrium by detecting linear acceleration and decel-

eration, such as the sensations you feel while in an elevator or a car that is speeding up or slowing down.

Like the spiral organ of the inner ear, the maculae consist of two kinds of cells: *hair cells* and *supporting cells* (Figure 12.12). Hair cells contain long, hairlike extensions of the cell membrane. Floating over the hair cells is a thick, jellylike substance called the **otolithic membrane.** Calcium carbonate crystals, called *otoliths* (*oto-* = ear; *-liths* = stones), are arranged in a layer over the entire surface of the otolithic membrane.

The otolithic membrane sits on top of the macula. If you tilt your head forward, the membrane (and the otoliths) is pulled by gravity and slides over the hair cells in the direction of the tilt. This stimulates the hair cells and triggers nerve impulses that conduct along the *vestibular branch* of the vestibulocochlear (VIII) nerve (see Figure 12.9).

Membranous Semicircular Ducts

The three membranous semicircular ducts lie at right angles to one another in three planes (see Figure 12.10a). This positioning permits detection of rotational acceleration or deceleration. The dilated portion of each duct, the ampulla, contains a small elevation called the *crista* (Figure 12.13). Each crista contains a

Figure 12.12 ■ **Location and structure of receptors in the maculae of the right ear.** Both sensory neurons (blue) and motor neurons (red) synapse with the hair cells.

Movements of the otolithic membrane stimulate the hair cells.

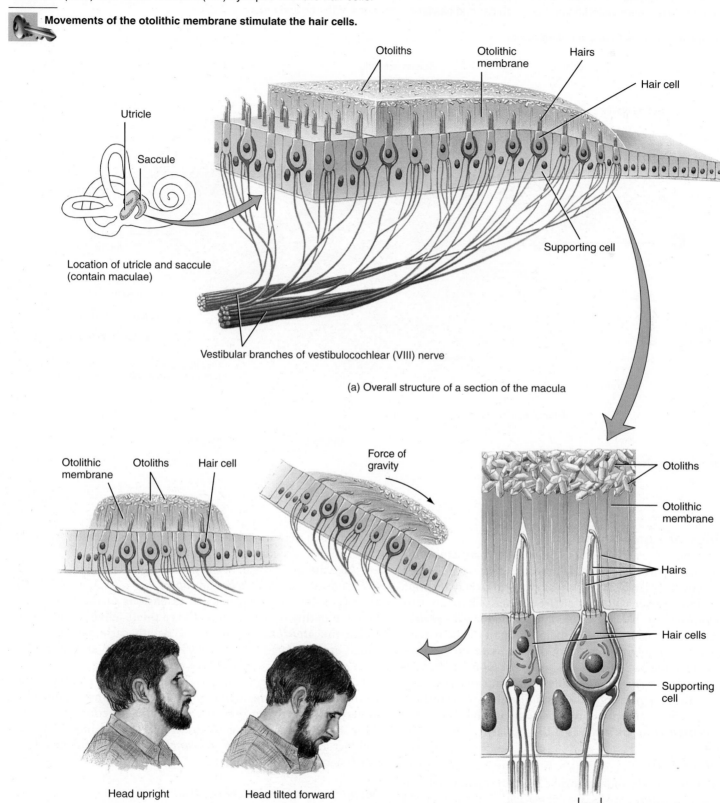

Otoliths

Otolithic membrane

Hairs

Hair cell

Utricle

Saccule

Location of utricle and saccule (contain maculae)

Supporting cell

Vestibular branches of vestibulocochlear (VIII) nerve

(a) Overall structure of a section of the macula

Otolithic membrane

Otoliths

Hair cell

Force of gravity

Otoliths

Otolithic membrane

Hairs

Hair cells

Supporting cell

Head upright

Head tilted forward

Vestibular branches of vestibulocochlear (VIII) nerve

(c) Position of macula with head upright (left) and tilted forward (right)

(b) Details of two hair cells

 What is the function of the maculae?

Figure 12.13 ■ **Location and structure of the membranous semicircular ducts of the right ear.**
Both sensory neurons (blue) and motor neurons (red) synapse with the hair cells. The ampullary nerves are branches of the vestibular division of the vestibulocochlear (VIII) nerve.

The positions of the membranous semicircular ducts permit detection of rotational movements.

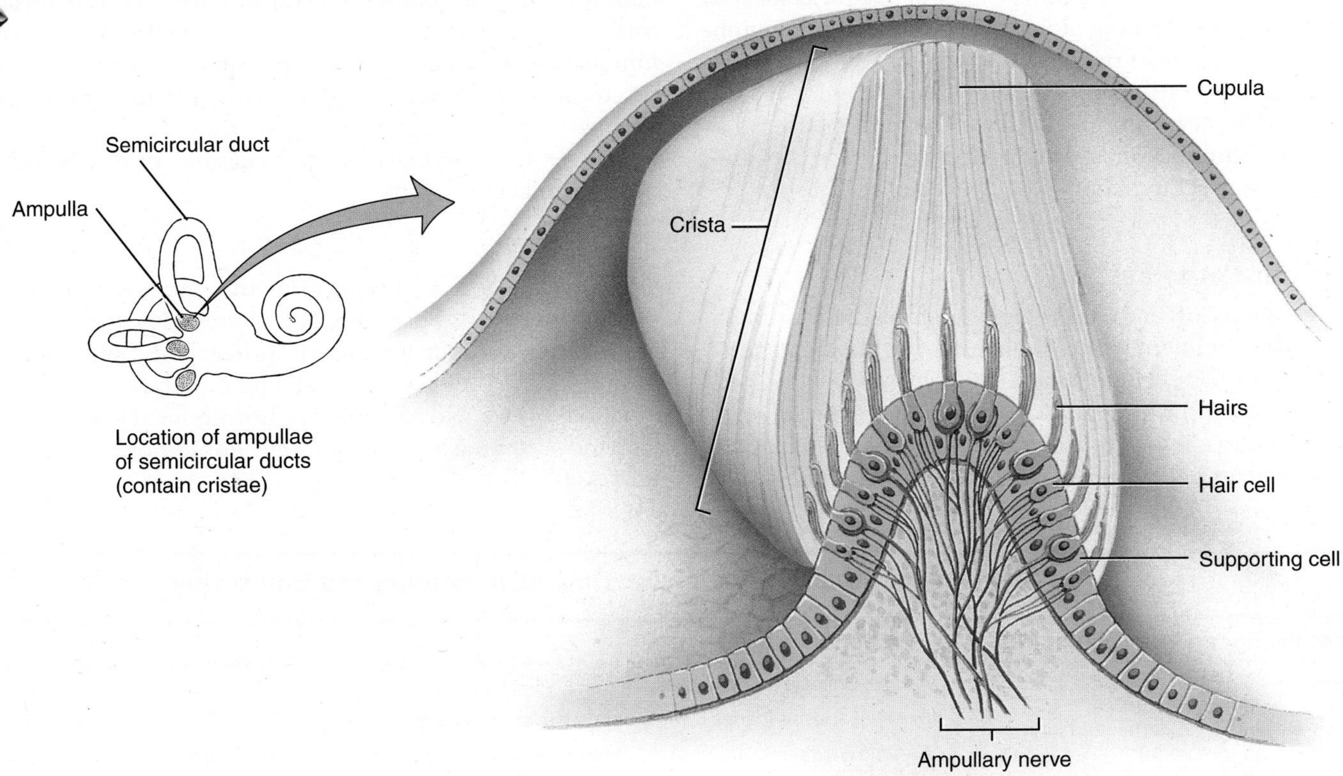

Semicircular duct

Ampulla

Location of ampullae of semicircular ducts (contain cristae)

Cupula

Crista

Hairs

Hair cell

Supporting cell

Ampullary nerve

(a) Details of a crista

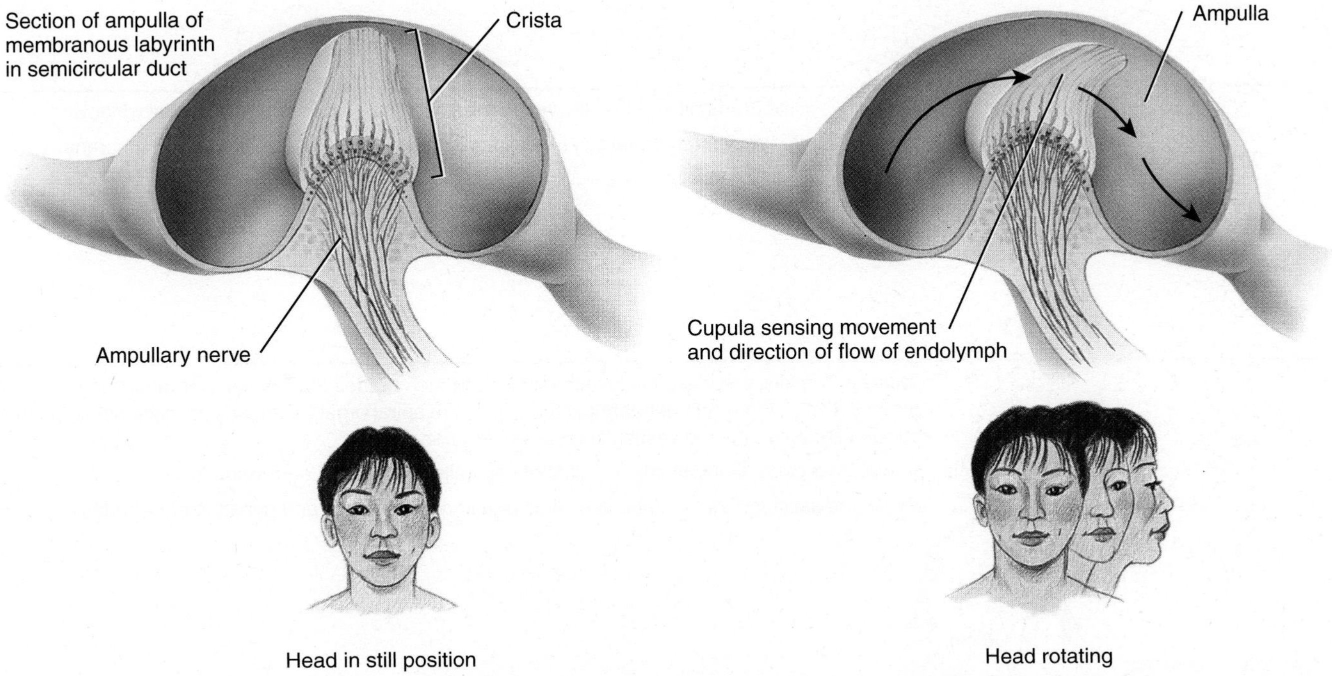

Section of ampulla of membranous labyrinth in semicircular duct

Crista

Ampulla

Ampullary nerve

Cupula sensing movement and direction of flow of endolymph

Head in still position

Head rotating

(b) Position of a crista with the head in the still position (left) and when the head rotates (right)

 With which type of equilibrium are the membranous semicircular ducts, the utricle, and the saccule associated?

group of *hair cells* and *supporting cells* covered by a mass of gelatinous material called the *cupula.* When the head moves, the attached membranous semicircular ducts and hair cells move with it. However, the endolymph within the membranous semicircular ducts is not attached and lags behind due to its inertia. As the moving hair cells drag along the stationary endolymph, the hairs bend. Bending of the hairs leads to nerve impulses that conduct along the vestibular branch of the vestibulocochlear (VIII) nerve. The nerve impulses follow the same pathways as those for static equilibrium and are eventually sent to the muscles that must contract to maintain body balance and posture.

Equilibrium Pathways

Most of the axons of the vestibular branch of the vestibulocochlear (VIII) nerve enter the brain stem and then extend to the medulla or the cerebellum, where they synapse with the next neurons in the equilibrium pathways. From the medulla, some axons conduct nerve impulses along the cranial nerves that control eye movements and head and neck movements. Other axons form a spinal cord tract that conveys impulses for regulation of muscle tone in response to head movements. Various pathways among the medulla, cerebellum, and cerebrum enable the cerebellum to play a key role in maintaining static and dynamic equilibrium. The cerebellum continuously receives sensory information from the utricle and saccule. In response, the cerebellum makes adjustments to the signals going from the motor cortex to specific skeletal muscles to maintain equilibrium and balance.

Table 12.3 summarizes the structures of the ear related to hearing and equilibrium.

• • •

Now that our exploration of the nervous system is completed, you can appreciate the many ways that this system contributes to homeostasis of other body systems by examining Focus on Homeostasis: The Nervous System. Next, in Chapter 13, we will see how the hormones released by the endocrine system also help maintain homeostasis of many body processes.

Table 12.3 / Summary of Structures of the Ear Related to Hearing and Equilibrium

Regions of the Ear and Key Structures	Functions
External ear	*Auricle*: Collects sound waves. *External auditory canal*: Directs sound waves to eardrum. *Eardrum (tympanic membrane)*: Sound waves cause it to vibrate, which, in turn, causes the malleus to vibrate.
Middle ear	*Auditory ossicles*: Transmit and amplify vibrations from tympanic membrane to oval window. *Auditory (Eustachian) tube*: Equalizes air pressure on both sides of the tympanic membrane.
Internal ear	*Cochlea*: Contains a series of fluids, channels, and membranes that transmit vibrations to the spiral organ (organ of Corti), the organ of hearing; hair cells in the spiral organ ultimately produce nerve impulses in the cochlear branch of the vestibulocochlear (VIII) nerve. *Semicircular ducts*: Contain cristae, sites of hair cells for dynamic equilibrium. *Utricle and saccule*: Contain maculae, sites of hair cells for static and dynamic equilibrium.

Focus on Homeostasis
The Nervous System

Body System	*Contribution of Nervous System*
For all body systems	Together with hormones from the endocrine system, nerve impulses provide communication and regulation of most body tissues.
Integumentary system	Sympathetic nerves of the autonomic nervous system (ANS) control contraction of smooth muscles attached to hair follicles and secretion of perspiration from sweat glands.
Skeletal system	Nociceptors (pain receptors) in bone tissue warn of bone trauma or damage.
Muscular system	Somatic motor neurons receive instructions from motor areas of the brain and stimulate contraction of skeletal muscles to bring about body movements; reticular formation sets level of muscle tone.
Endocrine system 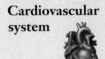	Hypothalamus regulates secretion of hormones from anterior and posterior pituitary; ANS regulates secretion of hormones from adrenal medulla and pancreas.
Cardiovascular system	Cardiovascular center in the medulla oblongata provides nerve impulses to ANS that govern heart rate and the forcefulness of the heartbeat; nerve impulses from ANS also regulate blood pressure and blood flow through blood vessels.
Lymphatic and immune system	Certain neurotransmitters help regulate immune responses; activity in nervous system may increase or decrease immune responses.
Respiratory system	Respiratory areas in brain stem control breathing rate and depth; ANS helps regulate diameter of airways.
Digestive system	ANS and enteric nervous system (ENS) help regulate digestion; parasympathetic division of ANS stimulates many digestive processes.
Urinary system	ANS helps regulate blood flow to kidneys, thereby influencing the rate of urine formation; brain and spinal cord centers govern emptying of urinary bladder.
Reproductive systems	Hypothalamus and limbic system govern a variety of sexual behaviors; ANS brings about erection of penis in males and clitoris in females and ejaculation of semen in males; hypothalamus regulates release of anterior pituitary hormones that control gonads (ovaries and testes); nerve impulses elicited by touch stimuli from suckling infant cause release of oxytocin and milk ejection in nursing mothers.

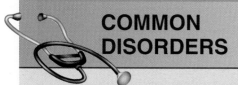

COMMON DISORDERS

Cataracts

A common cause of blindness is a loss of transparency of the lens known as a **cataract.** The lens becomes cloudy (less transparent) due to changes in the structure of the lens proteins. Cataracts often occur with aging but may also be caused by injury, excessive exposure to ultraviolet rays, certain medications (such as long-term use of steroids), or complications of other diseases (for example, diabetes). People who smoke also have increased risk of developing cataracts. Fortunately, sight can usually be restored by surgical removal of the old lens and implantation of an artificial one.

Glaucoma

In **glaucoma,** the most common cause of blindness in the United States, a buildup of aqueous humor within the anterior cavity causes an abnormally high intraocular pressure. Persistent pressure results in a progression from mild visual impairment to irreversible destruction of the retina, damage to the optic (II) nerve, and blindness. Because glaucoma is painless, and because the other eye initially compensates to a large extent for the loss of vision, a person may experience considerable retinal damage and loss of vision before the condition is diagnosed.

Macular Degeneration

Macular degeneration, the leading cause of blindness in individuals over age 75, is irreversible deterioration of the retina in the region of the macula, ordinarily the area of most acute vision. Initially, a person may experience blurring and distortion at the center of the visual field. Smokers have a threefold greater risk of developing macular degeneration than nonsmokers.

Deafness

Deafness is significant or total hearing loss. **Sensorineural deafness** is caused by either impairment of hair cells in the cochlea or damage of the cochlear branch of the vestibulocochlear (VIII) nerve. This type of deafness may be caused by atherosclerosis, which reduces blood supply to the ears; repeated exposure to loud noise, which destroys hair cells of the spiral organ; or certain drugs such as aspirin and streptomycin. **Conduction deafness** is caused by impairment of the external and middle ear mechanisms for transmitting sounds to the cochlea. It may be caused by otosclerosis, the deposition of new bone around the oval window; impacted cerumen; injury to the eardrum; or aging, which often results in thickening of the eardrum and stiffening of the joints of the auditory ossicles.

Ménière's Disease

Ménière's disease (mān-YĀRZ) results from an increased amount of endolymph that enlarges the membranous labyrinth. Among the symptoms are fluctuating hearing loss (caused by distortion of the basilar membrane of the cochlea) and roaring tinnitus (ringing). Vertigo (a sensation of spinning or whirling) is characteristic of Ménière's disease. Almost total destruction of hearing may occur over a period of years.

Otitis Media

Otitis media is an acute infection of the middle ear caused primarily by bacteria and associated with infections of the nose and throat. Symptoms include pain; malaise (discomfort or uneasiness); fever; and a reddening and outward bulging of the eardrum, which may rupture unless prompt treatment is received (this may involve draining pus from the middle ear). Bacteria from the nasopharynx passing into the auditory tube are the primary cause of all middle ear infections. Children are more susceptible than adults to middle ear infections because their auditory tubes are shorter, wider, and almost horizontal, which decreases drainage.

MEDICAL TERMINOLOGY AND CONDITIONS

Amblyopia (am′-blē-Ō-pē-a; ambly- = dull or dim) The loss of vision in a functionally normal eye that, because of muscle imbalance, cannot focus in synchrony with the other eye.

Blepharitis (blef′-a-RĪ-tis; blepharo = eyelid; itis = inflammation of) An inflammation of the eyelid.

Conjunctivitis (pinkeye) An inflammation of the conjunctiva; the type caused by bacteria such as pneumococci, staphylococci, or Hemophilus influenzae is very contagious and more common in children. Conjunctivitis may also be caused by irritants, such as dust, smoke, or pollutants in the air, in which case it is not contagious.

Keratitis (ker′-a-TĪ-tis; kerat- = cornea) An inflammation or infection of the cornea.

Labyrinthitis (lab′-i-rin-THĪ-tis) An inflammation of the internal ear.

Myringitis (mir′-in-JĪ-tis; myringa = eardrum) An inflammation of the eardrum; also called tympanitis.

Night blindness The lack of normal night vision; most often it is caused by vitamin A deficiency.

Nystagmus (nis-TAG-mus; nystagm- = nodding or drowsy) A rapid involuntary movement of the eyeballs, possibly caused by a disease of the central nervous system. It is associated with conditions that cause vertigo.

Otalgia (o-TAL-jē-a; oto = ear; algia = pain) Earache.

Ptosis (TŌ-sis; fall) Falling or drooping of the eyelid. (This term is also used for the slipping of any organ below its normal position.)

Retinoblastoma (ret′-i-nō-blas-TŌ-ma; blast = bud; oma = tumor) A tumor arising from immature retinal cells; it accounts for 2% of childhood cancers.

Scotoma (skō-TŌ-ma; scotoma = darkness) An area of reduced or lost vision in the visual field. Also called a **blind spot** (other than the normal blind spot or optic disk).

Strabismus (stra-BIZ-mus) An imbalance in the extrinsic eye muscles that cannot be controlled voluntarily. In **convergent strabismus (cross-eye),** the visual axes converge. In **divergent strabismus (walleye),** the visual axes diverge.

Trachoma (tra-KŌ-ma) A serious form of conjunctivitis and the greatest single cause of blindness in the world, caused by the bacterium Chlamydia trachomatis. The disease produces an excessive growth of subconjunctival tissue and invasion of blood vessels into the cornea, which progresses until the entire cornea is opaque.

■ STUDY OUTLINE

Overview of Sensations (p. 274)

1. Sensation is the conscious or subconscious awareness of external and internal conditions of the body.

2. The conditions for a sensation to occur are reception of a stimulus by a sensory receptor, conversion of the stimulus into a nerve impulse, conduction of the impulse to the brain, and integration of the impulse by a region of the brain.

3. When stimulated, most sensory receptors produce a depolarizing potential called a generator potential.

4. Sensory impulses from each part of the body arrive in specific regions of the cerebral cortex.

5. Adaptation is a decrease in sensation during a prolonged stimulus. Some receptors are rapidly adapting, whereas others are slowly adapting.

6. Modality is the distinct quality that makes one sensation different from others.

7. Two general classes of senses are general senses, which include somatic senses and visceral senses, and special senses, which include the modalities of smell, taste, vision, hearing, and equilibrium (balance).

8. Receptors can be classified by location as exteroceptors, interoceptors, and proprioceptors.

9. Receptors can be classified by the type of stimulus they detect as mechanoreceptors, thermoreceptors, nociceptors, photoreceptors, and chemoreceptors.

Somatic Senses (p. 275)

1. Somatic sensations that result from stimulating the skin surface are called cutaneous sensations. They include tactile sensations (touch, pressure, vibration, itch, and tickle), thermal sensations (heat and cold), and pain. Receptors for these sensations are located in the skin, subcutaneous layer, and mucous membranes of the mouth and anus.

2. Receptors for touch include corpuscles of touch (Meissner corpuscles), hair root plexuses, type I cutaneous mechanoreceptors (Merkel disks), and type II cutaneous mechanoreceptors (Ruffini corpuscles). Receptors for pressure and vibration are lamellated (Pacinian) corpuscles. Tickle and itch sensations result from stimulation of free nerve endings.

3. Thermoreceptors, free nerve endings in the epidermis and dermis, adapt to continuous stimulation.

4. Pain receptors (nociceptors) are free nerve endings that are located in nearly every body tissue.

5. Referred pain is felt in the skin near or away from the organ sending pain impulses.

6. Phantom pain is the sensation of pain in a limb that has been amputated.

7. Proprioceptors inform us of the degree to which muscles are contracted, the amount of tension present in tendons, the positions of joints, and the orientation of the head.

8. The proprioceptors include muscle spindles, tendon organs (Golgi tendon organs), joint kinesthetic receptors, and hair cells of the internal ear.

Olfaction: Sense of Smell (p. 279)

1. The olfactory epithelium in the upper portion of the nasal cavity contains olfactory receptors, supporting cells, and basal cells.

2. In olfactory reception, an odor molecule is dissolved in mucus and received by an olfactory receptor, which causes development of a generator potential and one or more nerve impulses.

3. Adaptation to odors occurs quickly.

4. Axons of olfactory receptors form the olfactory (I) nerves, which convey nerve impulses to the olfactory bulbs, olfactory tracts, limbic system, and cerebral cortex (temporal and frontal lobes).

Gustation: Sense of Taste (p. 280)

1. The receptors for gustation, the gustatory receptor cells, are located in taste buds.

2. To be tasted, substances must be dissolved in saliva.

3. The four primary tastes are salty, sweet, sour, and bitter.

4. Gustatory receptor cells trigger impulses in cranial nerves VII (facial), IX (glossopharyngeal), and X (vagus). Impulses for taste conduct to the medulla oblongata, limbic system, hypothalamus, thalamus, and the primary gustatory area in the parietal lobe of the cerebral cortex.

Vision (p. 282)

1. Accessory structures of the eyes include the eyebrows, eyelids, eyelashes, the lacrimal apparatus, and extrinsic eye muscles.

2. The lacrimal apparatus consists of structures that produce and drain tears.

3. The eyeball has three layers: (a) fibrous tunic (sclera and cornea), (b) vascular tunic (choroid, ciliary body, and iris), and (c) retina.

4. The retina consists of pigment epithelium and a neural portion (photoreceptor layer, bipolar cell layer, and ganglion cell layer).

5. The anterior cavity contains aqueous humor; the vitreous chamber contains the vitreous body.

6. Image formation on the retina involves refraction of light rays by the cornea and lens, which focus an inverted image on the central fovea of the retina.

7. For viewing close objects, the lens increases its curvature (accommodation), and the pupil constricts to prevent light rays from entering the eye through the periphery of the lens.

8. Improper refraction may result from myopia (nearsightedness), hypermetropia (farsightedness), or astigmatism (irregular curvature of the cornea or lens).

9. Movement of the eyeballs toward the nose to view an object is called convergence.

10. The first step in vision is the absorption of light rays by photopigments in rods and cones (photoreceptors). Stimulation of the rods and cones then activates bipolar cells, which in turn activate the ganglion cells.

11. Nerve impulses arise in ganglion cells and conduct along the optic (II) nerve, through the optic chiasm and optic tract to the thalamus. From the thalamus, the next axons in the visual pathway extend to the primary visual area in the occipital lobe of the cerebral cortex.

Hearing and Equilibrium (p. 289)

1. The external ear consists of the auricle and external auditory canal. The eardrum (tympanic membrane) separates the external ear from the middle ear.

2. The middle ear consists of the auditory (Eustachian) tube, ossicles, oval window, and round window.

3. The internal ear consists of the bony labyrinth and membranous labyrinth. The internal ear contains the spiral organ (organ of Corti), the organ of hearing.

4. Sound waves enter the external auditory canal, strike the eardrum, pass through the ossicles, strike the oval window, set up pressure waves in the perilymph, strike the vestibular membrane and scala tympani, increase pressure in the endolymph, vibrate the basilar membrane, and stimulate hair cells in the spiral organ.

5. Hair cells convert mechanical vibrations into depolarization of the hair cell membrane, which releases neurotransmitter that can initiate nerve impulses in sensory neurons.

6. Sensory neurons in the cochlear branch of the vestibulocochlear (VIII) nerve terminate in the medulla oblongata. Auditory signals then pass to the midbrain, thalamus, and temporal lobes.

7. Static equilibrium is the orientation of the body relative to the pull of gravity. The maculae of the utricle and saccule are the sense organs of static equilibrium.

8. Dynamic equilibrium is the maintenance of body position in response to rotation, acceleration, and deceleration. The maculae of the utricle and saccule and the cristae in the membranous semicircular ducts are the sense organs of dynamic equilibrium.

9. Most vestibular branch axons of the vestibulocochlear (VIII) nerve enter the brain stem and terminate in the medulla and pons; other axons extend to the cerebellum.

▪ SELF-QUIZ

1. You enter a sauna and it feels awfully hot, but soon the temperature feels comfortably warm. What have you have experienced?
   a. damage to your thermoreceptors   **b.** sensory adaptation
   c. a change in the temperature of the sauna   **d.** inactivation of your thermoreceptors   **e.** damage to the parietal lobe

2. The unique quality that makes one sensation different from others is its
   a. generator potential   **b.** modality   **c.** adaptability
   d. action potential   **e.** classification of receptors

3. Match each receptor with its function.
   _____ **a.** color vision
   _____ **b.** taste
   _____ **c.** smell
   _____ **d.** dynamic equilibrium
   _____ **e.** vision in dim light
   _____ **f.** stretch in a muscle
   _____ **g.** static equilibrium
   _____ **h.** pressure
   _____ **i.** fine touch
   _____ **j.** detects pain

   A. lamellated (Pacinian) corpuscle
   B. type I cutaneous mechanoreceptor
   C. rod photoreceptor
   D. nociceptor
   E. gustatory receptor cell
   F. olfactory receptor
   G. muscle spindle
   H. maculae
   I. cristae
   J. cones

4. The spiral organ (organ of Corti)
   a. contains hair cells
   b. is responsible for equilibrium
   c. is filled with perilymph
   d. is another name for the auditory (Eustachian) tube
   e. transmits auditory nerve impulses to the brain

5. Equilibrium and the activities of muscles and joints are monitored by
   a. olfactory receptors   **b.** nociceptors   **c.** tactile receptors
   d. proprioceptors   **e.** thermoreceptors

6. Which of the following pairs is NOT correctly matched?
   a. exteroceptors, monitor external environment
   b. proprioceptors, monitor body position
   c. nociceptors, detect pain
   d. mechanoreceptors, detect pressure
   e. interoceptors, located in ear

7. Which of the following is NOT required for a sensation to occur?
   a. the presence of a stimulus
   b. a receptor specialized to detect a stimulus
   c. the presence of slowly adapting receptors
   d. a sensory neuron to conduct an impulse
   e. a region of the brain for integration of the nerve impulse

8. For taste to occur
   a. the mouth must be dry
   b. the chemical must be in contact with the basal cells
   c. filiform papillae must be stimulated
   d. the limbic system needs to be activated
   e. the gustatory hair must be stimulated by the dissolved chemical

9. Which of the following characteristics of taste is NOT true?
   a. Olfaction can affect taste.
   b. Three cranial nerves conduct the impulses for taste to the brain.
   c. Taste adaptation occurs quickly.
   d. Humans can recognize about 10 primary tastes.
   e. Taste receptors are located in taste buds on the tongue and roof of the mouth.

10. You are seated at your desk and drop your pencil. As you lean over to retrieve it, what is occurring in your internal ear?
    a. The hair cells on the macula are responding to changes in static equilibrium.
    b. The hair cells in the cochlea are responding to changes in dynamic equilibrium.
    c. The cristae of each semicircular duct are responding to changes in dynamic equilibrium.

d. The cochlear branch of the vestibulocochlear (VIII) nerve begins to transmit nerve impulses to the brain.

e. The auditory (Eustachian) tube makes adjustments for varying air pressures.

11. A generator potential
a. results from the change in the receptor's membrane permeability to ions
b. is the same as an action potential
c. is a type of modality
d. results when there is a decreased sensitivity of receptors to a stimulus
e. is a region of the brain that integrates nerve impulses into sensations

12. Which of the following is NOT true concerning nociceptors?
a. They respond to stimuli that may cause tissue damage.
b. They consist of free nerve endings
c. They can be activated by excessive stimuli from other sensations.
d. They are found in virtually every body tissue except the brain.
e. They adapt very rapidly.

13. Match the following:

_____ **a.** focuses light rays onto the retina

_____ **b.** regulates the amount of light entering the eye

_____ **c.** contains aqueous humor

_____ **d.** contains blood vessels that help nourish the retina

_____ **e.** produce tears

_____ **f.** dense connective tissue that provides shape to the eye

_____ **g.** contains photoreceptors

A. sclera
B. choroid
C. lacrimal glands
D. lens
E. retina
F. iris
G. anterior cavity

14. Which of the following is NOT a function of tears?
a. moisten the eye
b. wash away eye irritants
c. destroy certain bacteria
d. lubricate the eye
e. provide nutrients to the cornea

15. Transmission of vibrations (sound waves) from the tympanic membrane to the oval window is accomplished by
a. nerve fibers
b. tectorial membrane
c. the auditory ossicles
d. the endolymph
e. the auditory (Eustachian) tube

16. Which of the following structures refracts light rays entering the eye?
a. cornea   **b.** sclera   **c.** pupil   **d.** retina   **e.** conjunctiva

17. Your 45-year-old neighbor has recently begun to have difficulty reading the morning newspaper. You explain that this condition is known as _____ and is due to _____ .
a. myopia, inability of his eyes to properly focus light on his retinas
b. night blindness, a vitamin A deficiency
c. binocular vision, the eyes focusing on two different objects
d. astigmatism, an irregularity in the curvature of the lens
e. presbyopia, the loss of elasticity in the lens

18. Damage to cells in the central fovea would interfere with
a. dynamic equilibrium
b. accommodation
c. visual acuity
d. ability to see in dim light
e. intraocular pressure

19. Place the following events concerning the visual pathway in the correct order:
1. Nerve impulses exit the eye via the optic (II) nerve.
2. Optic tract axons terminate in the thalamus.
3. Light reaches the retina.
4. Rods and cones are stimulated.
5. Synapses occur in the thalamus and continue to the primary visual area in the occipital lobe.
6. Ganglion cells generate nerve impulses.
    a. 4, 1, 2, 5, 6, 3   **b.** 5, 4, 1, 3, 2, 6   **c.** 3, 4, 6, 1, 5, 2
    d. 3, 4, 6, 1, 2, 5   **e.** 3, 4, 5, 6, 1, 2

20. Place the following events of the auditory pathway in the correct order:
1. Hair cells in the spiral organ bend as they rub against the tectorial membrane.
2. Movement in the oval window begins movement in the perilymph.
3. Nerve impulses exit the ear via the vestibulocochlear (VIII) nerve.
4. The eardrum and auditory ossicles transmit vibrations from sound waves.
5. Pressure waves from the perilymph cause bulging of the round window and formation of pressure waves in the endolymph.
    a. 4, 2, 5, 1, 3   **b.** 4, 5, 2, 3, 1   **c.** 5, 3, 2, 4, 1
    d. 3, 4, 5, 1, 2   **e.** 2, 4, 1, 5, 3

CRITICAL THINKING APPLICATIONS

1. When you first enter a chemistry lab the odors are quite strong. After several minutes, the odor in the lab is barely noticeable. Has something happened to the odors or has something happened to you?

2. Cliff works the night shift and sometimes falls asleep in A & P class. What is the effect on the structures in his internal ear when his head falls backward as he slumps in his seat?

3. A medical procedure used to improve vision involves shaving thin layers off the cornea. How could this procedure improve vision?

4. The optometrist put drops in Kate's eyes during her eye exam. When Kate looked in the mirror after the exam, her pupils looked very large, and her eyes were sensitive to the bright light. How did the eye drops produce this effect on Kate's eyes?

ANSWERS TO FIGURE QUESTIONS

12.1 Corpuscles of touch (Meissner corpuscles) are abundant in the fingertips, palms, and soles.

12.2 From the olfactory bulbs, impulses conduct into the olfactory tracts.

12.3 The gustatory pathway: gustatory receptors → cranial nerves VII, IX, and X → medulla oblongata → thalamus → primary gustatory area in the parietal lobe of the cerebral cortex.

12.4 Tears contain water, salts, some mucus, and lysozyme. Tears clean, lubricate, and moisten the eyeball.

12.5 Rods are specialized for vision in dim light and allow us to see shapes and movement; cones are specialized for color vision and acute vision.

12.6 During accommodation, the ciliary muscle contracts, suspensory ligaments slacken, and the lens becomes more rounded (convex) and refracts light more.

12.7 Presbyopia is the loss of elasticity in the lens that occurs with aging.

12.8 Structures carrying visual impulses from the retina: axons of ganglion cells → optic (II) nerve → optic chiasm → optic tract → thalamus → primary visual area in occipital lobe of the cerebral cortex.

12.9 The receptors for hearing and equilibrium are located in the internal ear: cochlea (hearing) and semicircular ducts (equilibrium).

12.10 The eardrum (tympanic membrane) separates the external ear from the middle ear. The oval and round windows separate the middle ear from the internal ear.

12.11 Hair cells convert a mechanical force (stimulus) into an electrical signal (depolarization and repolarization of the hair cell membrane).

12.12 The maculae are the receptors for static equilibrium and also contribute to dynamic equilibrium.

12.13 The membranous semicircular ducts, the utricle, and the saccule function in dynamic equilibrium.

Chapter 13

The Endocrine System

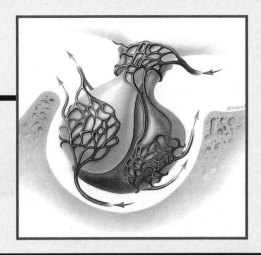

■ Student Learning Objectives

■ A Look Ahead

Together, the nervous and endocrine systems coordinate functions of all body systems. The nervous system controls body activities through nerve impulses, which trigger the release of neurotransmitters. In contrast, the glands of the *endocrine system* release molecules called *hormones* (*hormon* = to excite or get moving). Most hormones enter the bloodstream, and the circulating blood then delivers them to virtually all cells throughout the body.

The nervous system and endocrine system often work hand-in-hand. Certain parts of the nervous system stimulate or inhibit the release of hormones. The hormones may in turn promote or inhibit the generation of nerve impulses. The endocrine system not only helps regulate the activity of smooth muscle, cardiac muscle, and some glands, it affects virtually all other tissues as well. Hormones alter metabolism, regulate growth and development, and influence reproductive processes.

The nervous and endocrine systems typically respond to a stimulus at different rates. Nerve impulses most often produce an effect within a few milliseconds; some hormones can act within seconds, whereas others can take several hours or more to cause a response. Moreover, the effects of nervous system activation are generally briefer than those produced by the endocrine system.

In this chapter we examine the main endocrine glands and hormone-producing tissues and their roles in coordinating body activities. *Endocrinology* (en′-dō-kri-NOL-ō-jē; *endo-* = within; *-crin* = to secrete; *-ology* = study of) is the science of the structure and function of the endocrine glands and the diagnosis and treatment of disorders of the endocrine system.

Figure 13.1 ■ **Location of many endocrine glands.** Also shown are other organs that contain endocrine tissue, and associated structures.

Endocrine glands secrete hormones, which circulating blood delivers to target tissues.

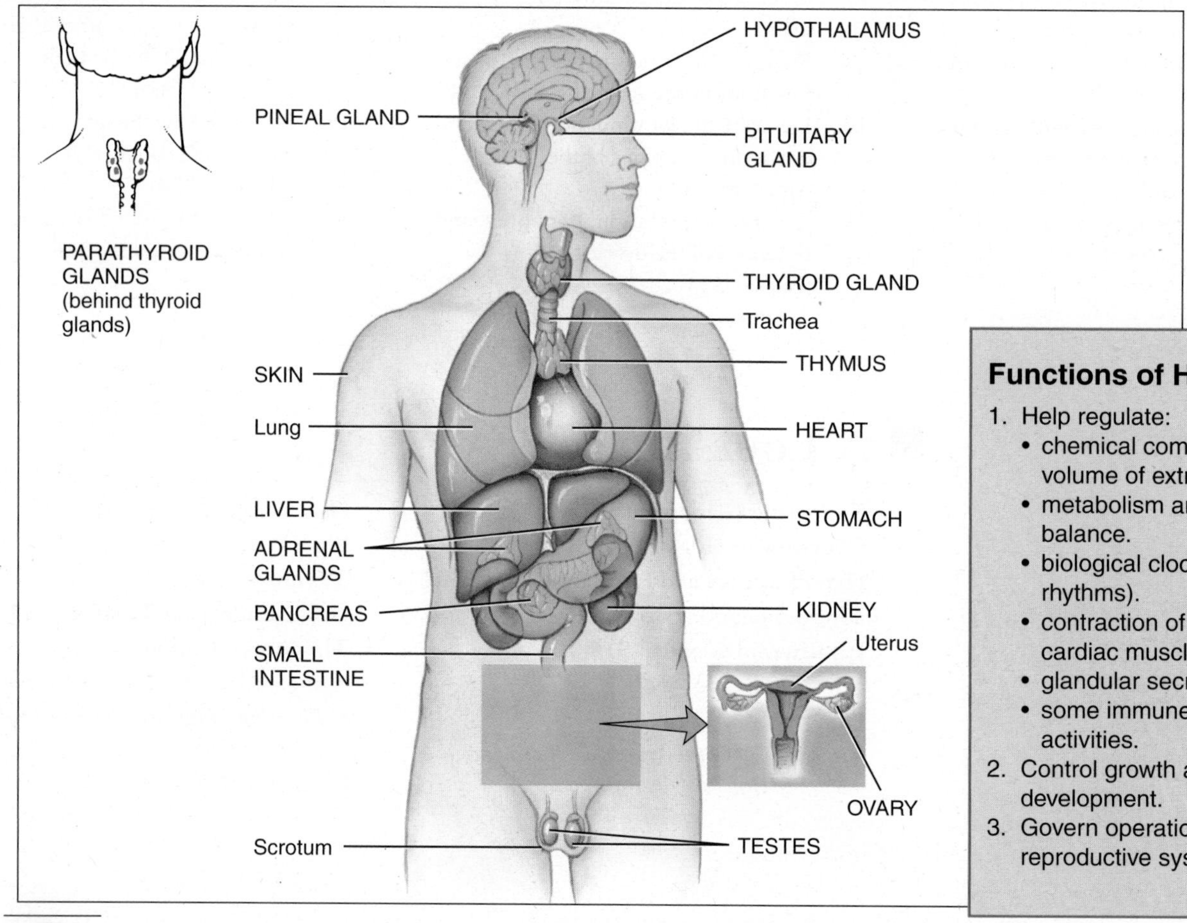

HYPOTHALAMUS

PINEAL GLAND

PITUITARY GLAND

PARATHYROID GLANDS (behind thyroid glands)

THYROID GLAND

Trachea

THYMUS

SKIN

Lung

HEART

LIVER

STOMACH

ADRENAL GLANDS

PANCREAS

KIDNEY

SMALL INTESTINE

Uterus

Scrotum

OVARY

TESTES

Functions of Hormones

1. Help regulate:
   - chemical composition and volume of extracellular fluid.
   - metabolism and energy balance.
   - biological clock (circadian rhythms).
   - contraction of smooth and cardiac muscle fibers.
   - glandular secretions.
   - some immune system activities.
2. Control growth and development.
3. Govern operation of reproductive systems.

 What is the basic difference between endocrine glands and exocrine glands?

ENDOCRINE GLANDS

Objective: • **Distinguish between exocrine and endocrine glands.**

The body contains two types of glands: exocrine glands and endocrine glands. *Exocrine glands* (*exo-* = outside) secrete their products into ducts that carry the secretions into body cavities, into the lumen of an organ, or to the outer surface of the body. Exocrine glands, which include sudoriferous (sweat), sebaceous (oil), mucous, and digestive glands, are not part of the endocrine system. *Endocrine glands* secrete their products (hormones) into the interstitial fluid surrounding the secretory cells, rather than into ducts. The endocrine glands of the body include the pituitary, thyroid, parathyroid, adrenal, and pineal glands (Figure 13.1). Several organs and tissues that are not considered endocrine glands also contain cells that secrete hormones. These include the hypothalamus, thymus, pancreas, ovaries, testes, kidneys, stomach, liver, small intestine, skin, heart, adipose tissue, and placenta.

Hormones have many functions (Figure 13.1). They help regulate most body functions, control growth and development, and govern operation of the male and female reproductive systems.

OVERVIEW OF HORMONES

Objectives: • **Define target cells and describe the role of hormone receptors.**
• **Describe the two general mechanisms of action of hormones.**
• **Explain how blood hormone levels are regulated.**

Circulating and Local Hormones

Most hormones are *circulating hormones*—they pass from the secretory cells that make them into interstitial fluid and then into the blood (Figure 13.2a). Although a circulating hormone travels throughout the body in the blood, it affects only specific *target cells* by chemically binding to its own specific *receptors.* Hormone receptors are protein molecules, and generally a target cell has 2000 to 100,000 receptors for a given hormone. Only the target cells for a given hormone have receptors that bind and recognize that hormone. For example, thyroid-stimulating hormone (TSH) binds to receptors on cells of the thyroid gland, but it does not bind to other body cells because only thyroid gland cells have TSH receptors.

Some hormones, termed *local hormones,* act locally on neighboring cells or on the same cell that secreted them without first entering the bloodstream (Figure 13.2b). Local hormones usually are inactivated quickly. Circulating hormones may linger in the blood and exert their effects for a few minutes or occasionally for a few hours. In time, circulating hormones are inactivated by the liver and excreted by the kidneys. In cases of kidney or liver failure, excessive levels of hormones may build up in the blood.

Chemistry of Hormones

Chemically, hormones can be divided into two broad classes—those that are soluble in lipids and those that are soluble in water. The lipid-soluble hormones include steroid hormones, thyroid hormones, and nitric oxide. *Steroid hormones* are derived from cholesterol; examples include cortisol, estrogens, and testosterone. The two *thyroid hormones* (T_3 and T_4) are synthesized by attaching iodine atoms to the amino acid tyrosine. The gas *nitric oxide (NO)* functions as both a hormone and a neurotransmitter.

The water-soluble hormones include amine hormones, peptide and protein hormones, and eicosanoid hormones. *Amine hormones* are synthesized by modifying certain amino acids. For instance, epinephrine and norepinephrine are formed by modifications to the amino acid tyrosine, and melatonin is made from the amino acid tryptophan. *Peptide hormones* and *protein hormones* consist of chains of amino acids; the smaller peptide hormones have 3 to 49 amino acids, whereas the larger protein hormones have 50 to 200. Examples of peptide hormones are antidiuretic hormone and oxytocin; protein hormones include human growth hormone and insulin. A more recently discovered group of chemical mediators are the *eicosanoid hormones* (ī-KŌ-sa-noid; *eicos-* = twenty forms; *-oid* = resembling), which are derived from a 20-carbon fatty acid. The two major types

Figure 13.2 ■ **Comparison of circulating and local hormones.**
Hormones exert their effects by binding to receptors on target cells.

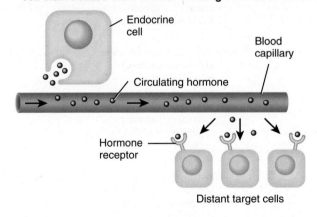

Circulating hormones enter the bloodstream and act on distant target cells; local hormones act on nearby cells or on the same cell that released them without entering the bloodstream.

Endocrine cell
Blood capillary
Circulating hormone
Hormone receptor
Distant target cells

(a) Circulating hormone

Hormone receptor
Local hormone
Endocrine cell
Nearby target cell

(b) Local hormone

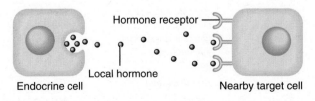

 Which organs inactivate and eliminate circulating hormones?

of eicosanoids are *prostaglandins* and *leukotrienes.* The eicosanoids are important local hormones, and they may act as circulating hormones as well.

Table 13.1 lists examples and sites of production of lipid-soluble hormones and water-soluble hormones.

Table 13.1 / Types of Hormones

Chemical Class	Hormone	Where Produced
Lipid Soluble		
Steroid hormones	Aldosterone, cortisol, and androgens.	Adrenal cortex.
	Calcitriol.	Kidneys.
	Testosterone.	Testes.
	Estrogens and progesterone.	Ovaries.
Thyroid hormones	T_3 (triiodothyronine) and T_4 (thyroxine).	Thyroid gland (follicular cells).
Gas	Nitric oxide.	Endothelial cells lining blood vessels.
Water Soluble		
Amines	Epinephrine and norepinephrine.	Adrenal medulla.
	Melatonin.	Pineal gland.
	Histamine.	Mast cells in connective tissues.
Peptides and proteins	All hypothalamic releasing and inhibiting hormones, oxytocin, antidiuretic hormone.	Hypothalamic neurosecretory cells.
	Human growth hormone, thyroid-stimulating hormone, adrenocorticotropic hormone, follicle-stimulating hormone, luteinizing hormone, prolactin, melanocyte-stimulating hormone.	Anterior pituitary.
	Insulin, glucagon, somatostatin, pancreatic polypeptide.	Pancreas.
	Parathyroid hormone.	Parathyroid glands.
	Calcitonin.	Thyroid gland (parafollicular cells).
	Gastrin, secretin, cholecystokinin, glucose-dependent insulinotropic peptide.	Stomach and small intestine.
	Erythropoietin.	Kidneys.
	Leptin.	Adipose tissue.
Eicosanoids	Prostaglandins, leukotrienes.	All cells except red blood cells.

Mechanisms of Hormone Action

The response to a hormone depends on both the hormone and the target cell. Various target cells respond differently to the same hormone. For example, insulin stimulates synthesis of glycogen in liver cells and synthesis of triglycerides in adipose cells. In addition to stimulating synthesis of new molecules, hormones can also change the permeability of the plasma membrane, stimulate transport of a substance into or out of its target cells, alter the rate of specific metabolic reactions, or cause contraction or relaxation of smooth muscle and cardiac muscle. The specific way in which a hormone exerts its effects depends on whether it is lipid soluble or water soluble.

Action of Lipid-Soluble Hormones

Lipid-soluble hormones diffuse through the lipid bilayer of the plasma membrane and bind to their receptors *within* target cells. They exert their effects in the following way (Figure 13.3):

1 A lipid-soluble hormone becomes a free hormone molecule when it detaches from its transport protein, which carries the hormone in the bloodstream. Then, the free hormone diffuses from blood into interstitial fluid, and through the plasma membrane into a cell.

2 If the cell is a target cell, the hormone binds to and activates receptors within the cell. The activated receptor–hormone

Figure 13.3 ■ **Mechanism of action of lipid-soluble hormones.**

Lipid-soluble hormones bind to their receptors inside target cells.

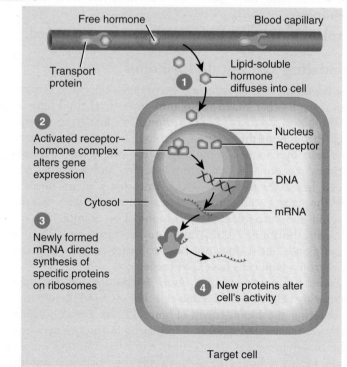

Free hormone

Blood capillary

Transport protein

1 Lipid-soluble hormone diffuses into cell

2 Activated receptor–hormone complex alters gene expression

Cytosol

Nucleus

Receptor

DNA

mRNA

3 Newly formed mRNA directs synthesis of specific proteins on ribosomes

4 New proteins alter cell's activity

Target cell

What type of molecules are synthesized after lipid-soluble hormones bind to their receptors?

complex then alters cell function by altering gene expression, turning specific genes on or off.

❸ As the DNA is transcribed, new messenger RNA (mRNA) forms, leaves the nucleus, and enters the cytosol; there it directs synthesis of new proteins, usually enzymes, on the ribosomes.

❹ The new proteins alter the cell's activity and cause the typical responses to that specific hormone.

Action of Water-Soluble Hormones

Certain hormones, such as peptide and protein hormones, cannot diffuse through the plasma membrane. The receptors for these *water-soluble hormones* are integral proteins in the plasma membrane. The hormone molecule that binds to its receptor in the plasma membrane is known as the ***first messenger***. Inside the target cell, another molecule, called the ***second messenger***, relays the signals for hormone-stimulated responses to take place. One common second messenger is ***cyclic AMP (cAMP)***, which is synthesized from ATP.

Water-soluble hormones exert their effects as follows (Figure 13.4):

❶ A water-soluble hormone (the first messenger) diffuses from the blood and binds to its receptor in a target cell's plasma

Figure 13.4 ■ Mechanism of action of water-soluble hormones.

Water-soluble hormones bind to receptors embedded in the plasma membrane of target cells.

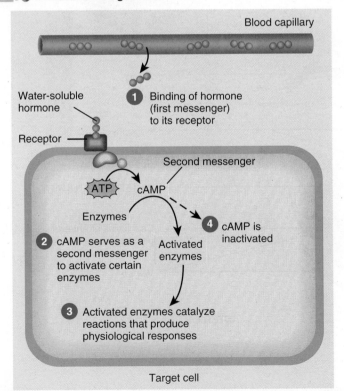

Blood capillary

Water-soluble hormone

Receptor

1 Binding of hormone (first messenger) to its receptor

ATP → cAMP

Second messenger

Enzymes

2 cAMP serves as a second messenger to activate certain enzymes

Activated enzymes

4 cAMP is inactivated

3 Activated enzymes catalyze reactions that produce physiological responses

Target cell

 Why is cAMP called a "second messenger"?

membrane. This binding starts a reaction that converts ATP into cyclic AMP in the cytosol of the cell.

❷ Cyclic AMP (the second messenger) causes the activation of several enzymes.

❸ Activated enzymes catalyze reactions that produce physiological responses.

❹ After a brief period of time, cyclic AMP is inactivated. Thus, the cell's response is turned off unless new hormone molecules continue to bind to their receptors in the plasma membrane.

Control of Hormone Secretions

The release of most hormones occurs in short bursts, with little or no secretion between bursts. When stimulated, an endocrine gland releases its hormone in more frequent bursts, increasing the concentration of the hormone in the blood. In the absence of stimulation, the blood level of the hormone decreases. Regulation of secretion normally prevents overproduction or underproduction of any given hormone.

Hormone secretion is regulated by (1) signals from the nervous system, (2) chemical changes in the blood, and (3) other hormones. For example, nerve impulses to the adrenal medullae regulate its release of epinephrine and norepinephrine; blood Ca^{2+} level regulates the secretion of parathyroid hormone; and a hormone from the anterior pituitary (ACTH) stimulates the release of cortisol by the adrenal cortex. Most systems that regulate secretion of hormones work by negative feedback, but a few operate by positive feedback. For example, during childbirth, the hormone oxytocin stimulates contractions of the uterus, and uterine contractions, in turn, stimulate more oxytocin release, a positive feedback effect.

Now that you have a general idea of the roles of hormones in the endocrine system, we turn to discussions of the various endocrine glands, the hormones they secrete, and how hormone secretion is controlled.

HYPOTHALAMUS AND PITUITARY GLAND

Objectives: • **Describe the locations of and relationship between the hypothalamus and pituitary gland.**
• **Describe the functions of each hormone secreted by the pituitary gland.**

For many years, the ***pituitary gland*** *(hypophysis)* was considered the "master" endocrine gland because it secretes several hormones that control other endocrine glands. We now know that the pituitary gland itself has a master—the ***hypothalamus***. This small region of the brain, below the thalamus, is the major link between the nervous and endocrine systems. Cells in the hypothalamus synthesize at least nine hormones, and the pituitary gland secretes seven. Together, these hormones play important roles in the regulation of virtually all aspects of growth, development, metabolism, and homeostasis.

The pituitary gland is a pea-sized mass composed of two lobes: a larger *anterior pituitary* or *anterior lobe* and a smaller *posterior pituitary* or *posterior lobe* (Figure 13.5). Both lobes of the pituitary gland rest in the *sella turcica*, a cup-shaped depression in the sphenoid bone. A stalklike structure, the *infundibulum,* attaches the pituitary gland to the hypothalamus. Within the infundibulum, blood vessels termed *hypophyseal portal veins* (hī′-pō-FIZ-ē-al) connect capillaries in the hypothalamus to capillaries in the anterior pituitary. Axons of hypothalamic neurons called *neurosecretory cells* end near the capillaries of the hypothalamus (inset of Figure 13.5), where they release several hormones into the blood. Axons of neurosecretory cells also extend from the hypothalamus to the posterior pituitary (see Figure 13.8), where they release hormones into capillaries of the posterior pituitary.

Anterior Pituitary Hormones

The anterior pituitary synthesizes and secretes hormones that regulate a wide range of bodily activities, from growth to repro-

Figure 13.5 ■ **The pituitary gland and its blood supply.** Note in the small figure to the right that releasing and inhibiting hormones synthesized by hypothalamic neurosecretory cells diffuse into capillaries and are carried by the hypophyseal portal veins to the anterior pituitary.

Hypothalamic releasing and inhibiting hormones are an important link between the nervous and endocrine systems.

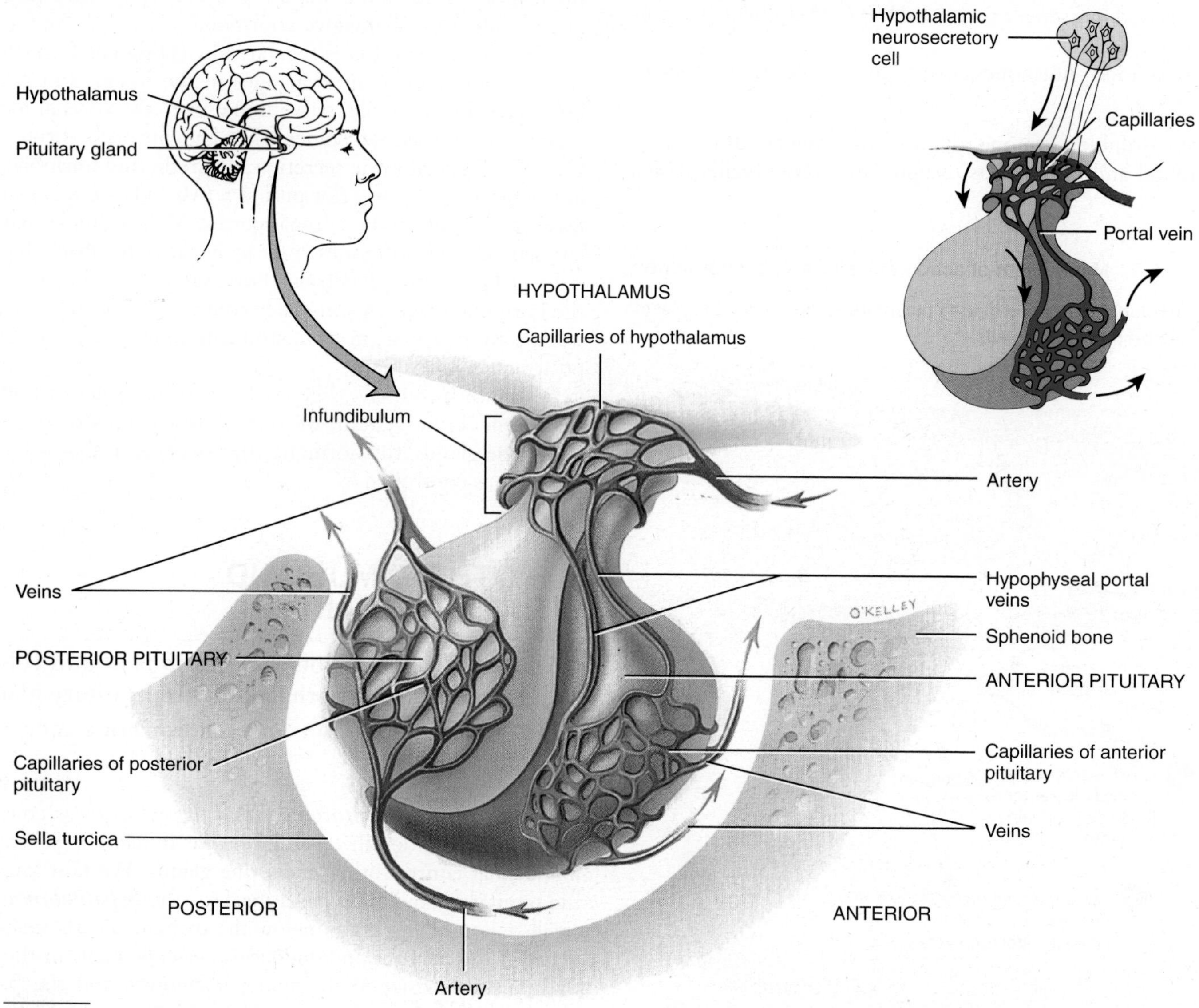

Which portion of the pituitary gland does not synthesize the hormones it releases? Where are its hormones produced?

duction. Some of these hormones influence other endocrine glands and are called *tropic hormones* (TRŌ-pik) or *tropins* (*tropin* = to change or nourish). Secretion of anterior pituitary hormones is stimulated by *releasing hormones* and suppressed by *inhibiting hormones,* both produced by neurosecretory cells of the hypothalamus. The hypothalamic releasing and inhibiting hormones are delivered to the anterior pituitary directly from the hypothalamus through the hypophyseal portal veins (Figure 13.5). This direct route allows the releasing and inhibiting hormones to act quickly on cells of the anterior pituitary before the hormones are diluted or destroyed in the general circulation. Figure 13.6 lists the seven hormones of the anterior pituitary and their target tissues and functions. With the exception of melanocyte-stimulating hormone (MSH), each hormone is described in the following sections. Although an excessive amount of MSH causes darkening of the skin, the function of normal levels of MSH is unknown.

Human Growth Hormone and Insulinlike Growth Factors

Human growth hormone (hGH) is the most plentiful anterior pituitary hormone. In response to human growth hormone, cells in the liver, skeletal muscle, cartilage, bone, and other tissues secrete ***insulinlike growth factors (IGFs),*** hormones that act to stimulate general body growth and regulate aspects of metabolism. IGFs are so named because several of their actions are similar to those of insulin. IGFs cause cells to grow and multiply, and they increase the rate of protein synthesis. As a result, human growth hormone increases the growth rate of the skeleton and skeletal muscles during childhood and the teenage years. In adults, human growth hormone and IGFs help maintain muscle and bone mass and promote healing of injuries and tissue repair.

IGFs also enhance lipolysis (breakdown of triglycerides) in adipose tissue, which results in increased use of the released fatty acids for ATP production by body cells. In addition, human

Figure 13.6 ■ Hormones produced by the anterior pituitary and their functions.

Hormones that influence other endocrine glands are called tropic hormones.

Hormone Produced	Target Tissues and Functions
Human growth hormone (hGH)	hGH stimulates bone, liver, and other tissues to release insulinlike growth factors, which stimulate general body growth and regulate metabolism.
Thyroid-stimulating hormone (TSH)	TSH stimulates thyroid gland to secrete its hormones.
Follicle-stimulating hormone (FSH)	FSH stimulates sperm production by testes. LH stimulates secretion of testosterone by testes.
Luteinizing hormone (LH)	FSH stimulates production of oocytes and secretion of estrogens by ovaries. LH stimulates secretion of estrogens and progesterone by ovaries and triggers ovulation.
Prolactin (PRL)	PRL promotes milk secretion in mammary glands.
Adrenocorticotropic hormone (ACTH)	ACTH stimulates adrenal cortex to secrete its hormones.
Melanocyte-stimulating hormone (MSH)	Excessive MSH darkens skin pigmentation.

Hypothalamus

Anterior pituitary

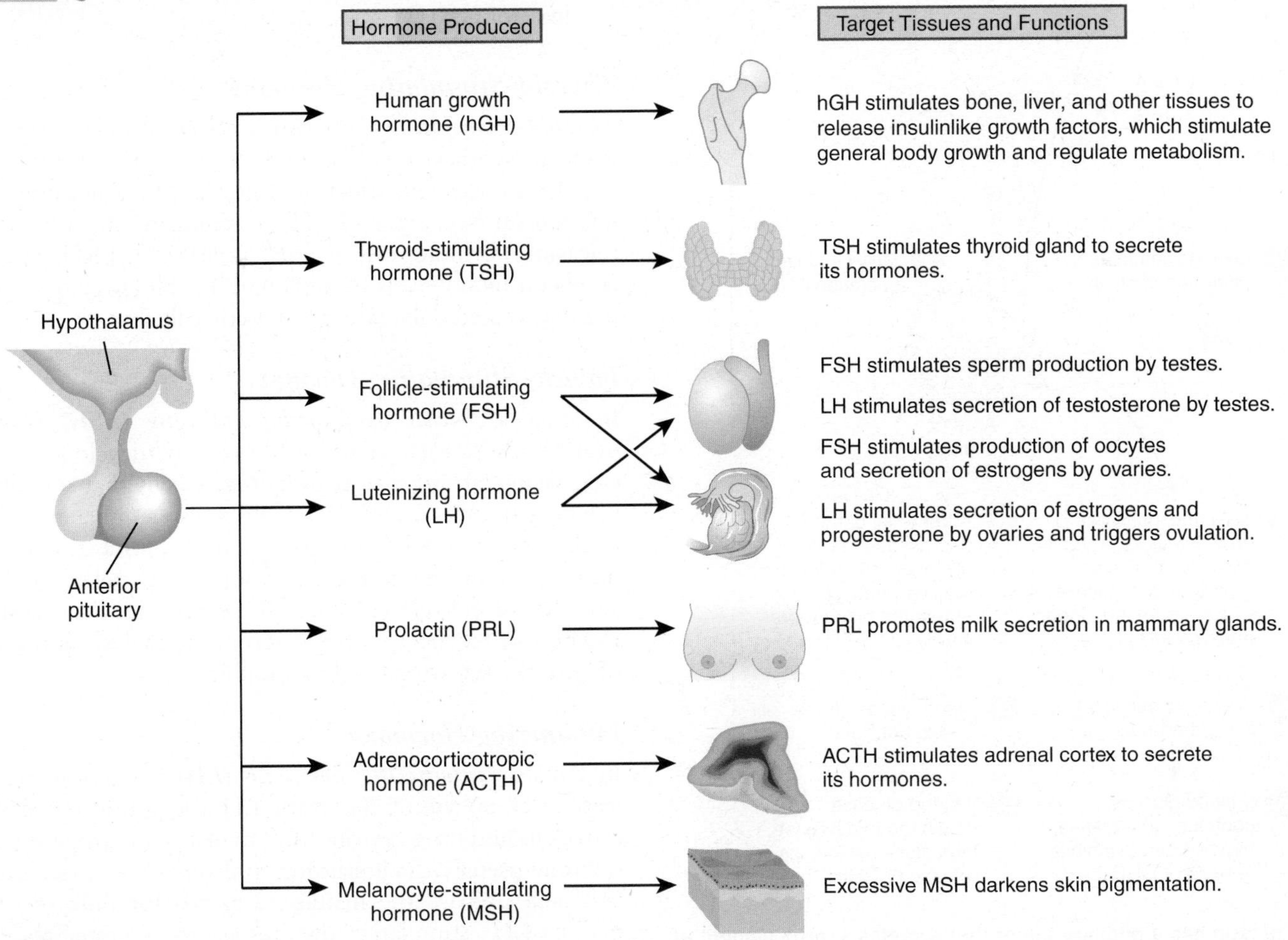

 Which endocrine glands are regulated by anterior pituitary hormones?

growth hormone and IGFs decrease glucose uptake by the body cells. This action spares glucose for use by neurons in ATP production.

The anterior pituitary releases bursts of human growth hormone every few hours, especially during sleep. Secretion of hGH is controlled by two hypothalamic hormones: *growth hormone-releasing hormone (GHRH)*, which promotes secretion of human growth hormone, and *growth hormone-inhibiting hormone (GHIH)*, which suppresses it. Blood glucose level is a major regulator of GHRH and GHIH secretion, as shown in Figure 13.7:

Figure 13.7 ■ Regulation of human growth hormone (hGH) secretion.

Secretion of hGH is stimulated by growth hormone-releasing hormone (GHRH) in response to hypoglycemia and is inhibited by growth hormone-inhibiting hormone (GHIH) in response to hyperglycemia.

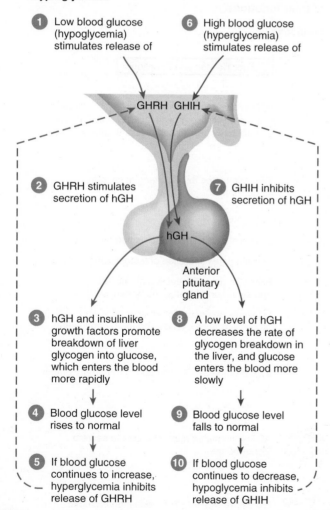

1 Low blood glucose (hypoglycemia) stimulates release of

6 High blood glucose (hyperglycemia) stimulates release of

GHRH   GHIH

2 GHRH stimulates secretion of hGH

7 GHIH inhibits secretion of hGH

hGH

Anterior pituitary gland

3 hGH and insulinlike growth factors promote breakdown of liver glycogen into glucose, which enters the blood more rapidly

8 A low level of hGH decreases the rate of glycogen breakdown in the liver, and glucose enters the blood more slowly

4 Blood glucose level rises to normal

9 Blood glucose level falls to normal

5 If blood glucose continues to increase, hyperglycemia inhibits release of GHRH

10 If blood glucose continues to decrease, hypoglycemia inhibits release of GHIH

If a person has a pituitary tumor that secretes a large amount of hGH and is not responsive to regulation by GHRH and GHIH, will hyperglycemia or hypoglycemia be more likely?

1 Low blood glucose level (hypoglycemia) stimulates the hypothalamus to secrete GHRH, which reaches the anterior pituitary in the hypophyseal portal veins.

2 Upon binding to its receptors in the anterior pituitary, GHRH stimulates release of human growth hormone (hGH).

3 Together, human growth hormone and insulinlike growth factors promote breakdown of liver glycogen into glucose, which enters the blood.

4 As a result, blood glucose concentration rises to the normal level.

5 By means of negative feedback, an increase in blood glucose concentration above the normal level (hyperglycemia) inhibits release of GHRH.

6 Hyperglycemia stimulates the hypothalamus to secrete GHIH (while reducing the secretion of GHRH).

7 Upon reaching the anterior pituitary, GHIH inhibits secretion of hGH.

8 A low level of hGH slows glycogen breakdown in the liver, and glucose is released into the blood more slowly.

9 Blood glucose falls to the normal level.

10 By means of negative feedback, hypoglycemia inhibits release of GHIH.

Thyroid-Stimulating Hormone

Thyroid-stimulating hormone (TSH) stimulates the thyroid gland to synthesize and secrete two thyroid hormones: T_3 and T_4. TSH secretion, which is controlled by a negative feedback mechanism (see Figure 13.11), is stimulated by a hypothalamic releasing hormone called *TRH*. Release of TRH, in turn, depends on blood levels of TSH and T_3, blood glucose level, and the body's metabolic rate, among other factors.

Follicle-Stimulating Hormone

In females, **follicle-stimulating hormone (FSH)** acts on the ovaries, where each month it initiates the development of follicles, saclike arrangements of secretory cells that surround a developing oocyte (future ovum). FSH also stimulates ovarian follicular cells to secrete estrogens. In males, FSH stimulates sperm production in the testes. A hypothalamic releasing hormone stimulates FSH release. Through negative feedback systems, estrogens (in females) and testosterone (in males) suppress release of both the releasing hormone and FSH.

Luteinizing Hormone

In females, **luteinizing hormone (LH)** triggers ovulation, release of an oocyte by the ovary. LH also stimulates secretion of estrogens and progesterone after ovulation. Estrogens and progesterone prepare the uterus for implantation of a fertilized ovum and help prepare the mammary glands for milk secretion. In males, LH stimulates the testes to secrete testosterone. Secretion of LH, like that of FSH, is controlled by a hypothalamic releasing hormone.

Prolactin

Prolactin (PRL), together with other hormones, initiates and maintains milk secretion by the mammary glands. After the mammary glands have been stimulated by other hormones, such as estrogens and progesterone, PRL brings about milk secretion. Ejection of milk from the mammary glands depends on the hormone oxytocin, which is released from the posterior pituitary. Together, milk secretion and ejection constitute *lactation*. The function of prolactin is unknown in males, but prolactin hypersecretion causes erectile dysfunction (impotence, the inability to have an erection of the penis).

The hypothalamus secretes both releasing and inhibiting hormones that regulate prolactin secretion. Most of the time, *prolactin-inhibiting hormone (PIH)* suppresses release of prolactin. Each month, just before menstruation begins, the secretion of PIH diminishes and the blood level of prolactin rises, but not enough to stimulate milk production. As the menstrual cycle begins anew, PIH is again secreted and the prolactin level drops. *Prolactin-releasing hormone (PRH)* stimulates a rise in prolactin level during pregnancy Nursing an infant causes a reduction in secretion of PIH.

Adrenocorticotropic Hormone

Adrenocorticotropic hormone (ACTH) controls the production and secretion of hormones called glucocorticoids by the cortex of the adrenal glands. A hypothalamic releasing hormone stimulates secretion of ACTH, as do stress-related stimuli, such as low blood glucose level or physical trauma. By means of negative feedback, glucocorticoids inhibit secretion of both the releasing hormone and ACTH.

Posterior Pituitary Hormones

Unlike the anterior pituitary and the hypothalamus, the posterior pituitary does not *synthesize* hormones. It is included among the endocrine glands because it *stores* and *releases* hormones. The posterior pituitary contains the axons and axon terminals of more than 10,000 neurosecretory cells with cell bodies in the hypothalamus (Figure 13.8). Within the cell bodies of different neurosecretory cells, the hormones **oxytocin (OT)** and **antidiuretic hormone (ADH)** are synthesized and packaged into secretory vesicles. The vesicles then move down the axons to the axon terminals in the posterior pituitary. Nerve impulses that

Figure 13.8 ■ **Axons of hypothalamic neurosecretory cells extend from the hypothalamus to the posterior pituitary.** Hormones synthesized in the cell body of a neurosecretory cell are packaged into secretory vesicles that move down to the axon terminals. Nerve impulses trigger release of the hormones from vesicles in the axon terminals in the posterior pituitary.

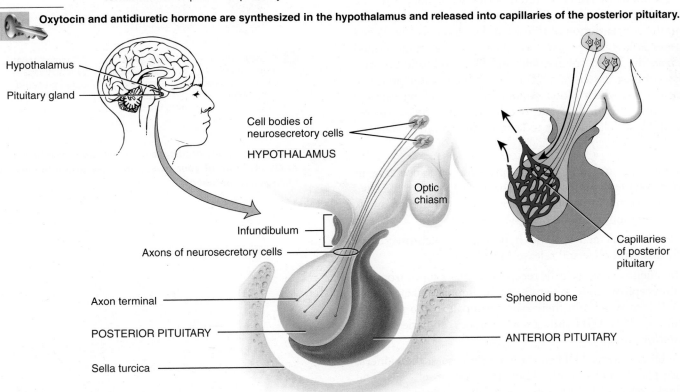

Oxytocin and antidiuretic hormone are synthesized in the hypothalamus and released into capillaries of the posterior pituitary.

 Where are the target cells of oxytocin located?

arrive at the axon terminals trigger release of these hormones into the capillaries of the posterior pituitary.

Oxytocin

During and after delivery of a baby, ***oxytocin*** (ok′-sē-TŌ-sin; *oxytoc-* = quick birth) has two target tissues: the mother's uterus and breasts. During delivery, oxytocin enhances contraction of smooth muscle cells in the wall of the uterus; after delivery, it stimulates milk ejection ("letdown") from the mammary glands in response to the mechanical stimulus provided by a suckling infant. The function of oxytocin in males and in nonpregnant females is not clear. Experiments with animals have suggested actions within the brain that foster parental caretaking behavior toward young offspring. Oxytocin also may be partly responsible for the feelings of sexual pleasure during and after intercourse.

Antidiuretic Hormone

An *antidiuretic* (*anti-* = against; *diuretic* = urine-producing agent) substance is something that decreases urine production. ***Antidiuretic hormone (ADH)*** causes the kidneys to return more water to the blood, thus decreasing urine volume. In the absence of ADH, urine output increases more than tenfold, from the normal 1 to 2 liters to about 20 liters a day. ADH also decreases the water lost through sweating and causes constriction of arterioles. This hormone's other name, ***vasopressin*** (*vaso-* = vessel; *pressin-* = pressing or constricting), reflects its effect on increasing blood pressure. Hyposecretion of ADH or nonfunctional ADH receptors causes diabetes insipidus (see page 329).

The amount of ADH secreted varies with blood osmotic pressure and blood volume. Blood osmotic pressure is proportional to the concentration of solutes in the blood plasma. When body water is lost faster than it is taken in, a condition termed *dehydration*, the blood volume falls and blood osmotic pressure rises. By contrast, faster gain than loss of body water produces *overhydration*, in which blood volume rises and blood osmotic pressure falls. Dehydration and overhydration influence secretion of ADH, which then acts to correct the hydration status as follows (Figure 13.9):

1 An increase in blood osmotic pressure or a decrease in blood volume because of hemorrhage, diarrhea, or excessive sweating stimulates ***osmoreceptors,*** neurons in the hypothalamus that monitor blood osmotic pressure.

2 Osmoreceptors activate the hypothalamic neurosecretory cells that synthesize and release ADH.

3 When neurosecretory cells receive excitatory input from the osmoreceptors, they generate nerve impulses that release ADH from vesicles in their axon terminals in the posterior pituitary. The ADH diffuses into blood capillaries of the posterior pituitary.

4 The blood carries ADH to three target tissues: the kidneys, sweat glands, and smooth muscle in blood vessel walls. The kidneys respond by retaining more water, which decreases urine output. Secretory activity of sweat glands decreases, which lowers the rate of water loss by perspiration from the

Figure 13.9 ■ **Regulation of secretion and actions of antidiuretic hormone (ADH).**

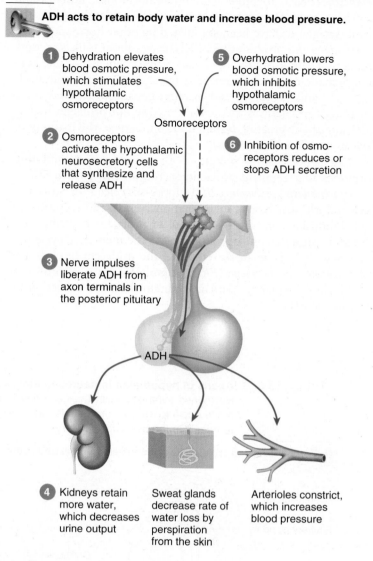

ADH acts to retain body water and increase blood pressure.

1 Dehydration elevates blood osmotic pressure, which stimulates hypothalamic osmoreceptors

5 Overhydration lowers blood osmotic pressure, which inhibits hypothalamic osmoreceptors

Osmoreceptors

2 Osmoreceptors activate the hypothalamic neurosecretory cells that synthesize and release ADH

6 Inhibition of osmo-receptors reduces or stops ADH secretion

3 Nerve impulses liberate ADH from axon terminals in the posterior pituitary

ADH

4 Kidneys retain more water, which decreases urine output

Sweat glands decrease rate of water loss by perspiration from the skin

Arterioles constrict, which increases blood pressure

What effect would drinking a large glass of water have on the osmotic pressure of your blood, and how would the level of ADH change in your blood?

skin. Smooth muscle in the walls of arterioles (small arteries) contracts in response to high levels of ADH, which constricts (narrows) these blood vessels and increases blood pressure.

5 Overhydration, which lowers blood osmotic pressure, or increased blood volume inhibits the osmoreceptors.

6 Inhibition of osmoreceptors reduces or stops ADH secretion. The kidneys then retain less water by forming a larger volume of urine, secretory activity of sweat glands increases, and arterioles dilate (widen). The blood volume and osmotic pressure of body fluids return to normal.

Secretion of ADH can also be altered in other ways. Pain, stress, trauma, and anxiety; acetylcholine; nicotine; and drugs such as morphine, tranquilizers, and some anesthetics stimulate ADH secretion. Alcohol inhibits ADH secretion, thereby increasing urine output. The resulting dehydration may cause both the thirst and the headache typical of a hangover.

Table 13.2 lists the pituitary gland hormones and summarizes their main actions.

THYROID GLAND

Objective: • **Describe the location, hormones, and functions of the thyroid gland.**

The butterfly-shaped *thyroid gland* is located just below the larynx (voice box). It is composed of right and left lobes, one on either side of the trachea, that are connected by an *isthmus* (IS-mus = a narrow passage) anterior to the trachea (Figure 13.10a).

Microscopic spherical sacs called *thyroid follicles* (Figure 13.10b) make up most of the thyroid gland. The wall of each follicle consists primarily of cells called *follicular cells,* which produce two hormones: *thyroxine* (thī-ROK-sēn), also called T_4 because it contains four atoms of iodine, and *triiodothyronine* (trī-ī′-ō-dō-THĪ-rō-nēn) (T_3), which contains three atoms of iodine. T_3 and T_4 are also known as *thyroid hormones.* The central cavity of each thyroid follicle contains stored thyroid hormones. As T_4 circulates in the blood and enters cells throughout the body, most of it is converted to T_3 by removal of one iodine. A smaller number of cells called *parafollicular cells* lie between the follicles (Figure 13.10b). They produce the hormone *calcitonin* (kal′-si-TŌ-nin), which normally has only a minor role in regulating calcium homeostasis.

Actions of Thyroid Hormones

Because most body cells have receptors for thyroid hormones, T_3 and T_4 exert their effects throughout the body. They regulate (1) basal metabolic rate and oxygen use, (2) cellular metabolism, and (3) growth and development.

Thyroid hormones increase *basal metabolic rate (BMR),* the rate of oxygen consumption at rest after an overnight fast, by

Table 13.2 / Summary of Pituitary Gland Hormones and Their Principal Actions

Hormone	Principal Actions
Anterior Pituitary	
Human growth hormone (hGH)	Stimulates liver, muscle, cartilage, bone, and other tissues to synthesize and secrete insulin-like growth factors (IGFs); IGFs promote growth of body cells, protein synthesis, tissue repair, lipolysis, and elevation of blood glucose concentration.
Thyroid-stimulating hormone (TSH)	Stimulates synthesis and secretion of thyroid hormones by thyroid gland.
Follicle-stimulating hormone (FSH)	In females, initiates development of oocytes and induces ovarian secretion of estrogens. In males, stimulates testes to produce sperm.
Luteinizing hormone (LH)	In females, triggers ovulation and stimulates secretion of estrogens and progesterone; in males, stimulates testes to produce testosterone.
Prolactin (PRL)	In females, promotes milk secretion by the mammary glands.
Adrenocorticotropic hormone (ACTH)	Stimulates secretion of glucocorticoids (mainly cortisol) by adrenal cortex.
Melanocyte-stimulating hormone (MSH)	Exact role in humans is unknown but can cause darkening of skin.
Posterior Pituitary	
Oxytocin (OT)	Stimulates contraction of smooth muscle cells of uterus during childbirth; stimulates milk ejection from mammary glands.
Antidiuretic hormone (ADH) or vasopressin	Conserves body water by decreasing urine volume; decreases water loss through sweating; raises blood pressure by constricting arterioles.

Figure 13.10 ■ Location, blood supply, and histology of the thyroid gland.

Thyroid hormones regulate (1) oxygen use and basal metabolic rate, (2) cellular metabolism, and (3) growth and development.

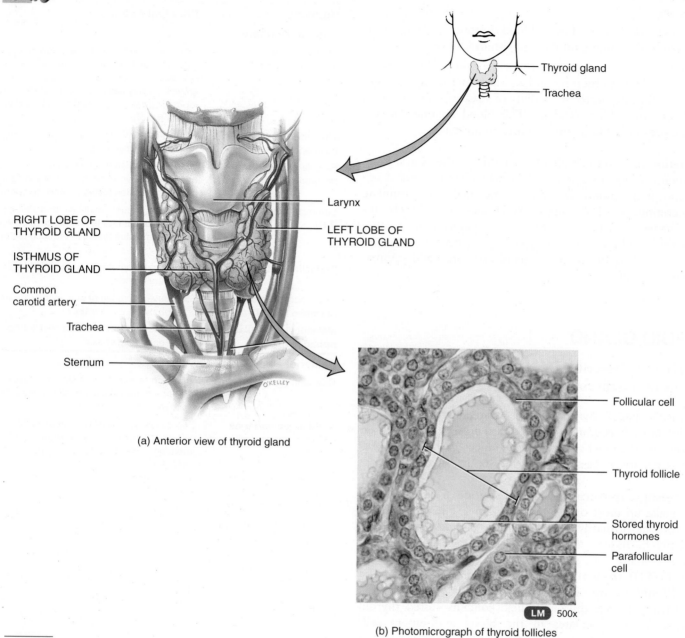

(a) Anterior view of thyroid gland

(b) Photomicrograph of thyroid follicles

 **Which cells secrete T$_3$ and T$_4$?**

stimulating synthesis of sodium pumps, which continually use ATP to eject sodium ions from the cytosol into the extracellular fluid. As cells use more oxygen to produce the ATP needed by these pumps, more heat is given off, and body temperature rises. In this way, thyroid hormones play an important role in the maintenance of normal body temperature. The thyroid hormones also stimulate protein synthesis, increase the use of glucose for ATP production, increase lipolysis, and enhance cholesterol excretion, thus reducing blood cholesterol level.

Excess secretion of thyroid hormones is known as *hyperthyroidism*. Symptoms of hyperthyroidism include increased heart rate and more forceful heartbeats, increased blood pressure, and increased nervousness. Together with human growth hormone and insulin, thyroid hormones stimulate body growth, particularly the growth and development of nervous tissue.

Control of Thyroid Hormone Secretion

The secretory activity of the thyroid gland is controlled in two main ways. First, an abnormally high blood iodine level suppresses secretion of thyroid hormones. Second, thyrotropin-releasing hormone (TRH) from the hypothalamus and TSH from the anterior pituitary stimulate synthesis and release of thyroid hormones as follows (Figure 13.11):

1 Low blood level of thyroid hormones or low metabolic rate stimulates the hypothalamus to secrete TRH.

2 TRH is carried to the anterior pituitary, where it stimulates secretion of thyroid-stimulating hormone (TSH).

3 TSH stimulates thyroid follicular cell activity, including thyroid hormone synthesis and secretion, and growth of the follicular cells.

4 The thyroid follicular cells release thyroid hormones into the blood until the metabolic rate returns to normal.

5 An elevated level of thyroid hormones inhibits release of TRH and TSH (negative feedback).

Conditions that increase ATP demand—a cold environment, low blood glucose, high altitude, and pregnancy—also increase secretion of the thyroid hormones.

Calcitonin

The hormone produced by the parafollicular cells of the thyroid gland, *calcitonin (CT)*, can decrease the level of calcium in the blood by inhibiting the action of osteoclasts, the cells that break down bone matrix. The secretion of CT is controlled by a negative feedback system (see Figure 13.13). Calcitonin's importance in normal physiology is unclear because it can be present in excess or completely absent without causing clinical symptoms. Miacalcin, a calcitonin extract derived from salmon, is an effective treatment for osteoporosis, a disorder in which the pace of bone matrix breakdown exceeds the pace of bone matrix rebuilding. It inhibits bone resorption (breakdown of bone matrix) and accelerates uptake of calcium and phosphates into bone matrix.

Figure 13.11 ■ **Regulation of thyroid hormone secretion.**

TSH promotes release of thyroid hormones.

1 Low blood level of thyroid hormones or low metabolic rate stimulates release of TRH

TRH

Hypothalamus

2 TRH, carried by hypophyseal portal veins to anterior pituitary, stimulates release of TSH

5 Elevated level of thyroid hormones inhibits release of TRH and TSH

TSH

Anterior pituitary gland

3 TSH released into blood stimulates thyroid follicular cells

4 Thyroid hormones released into blood by follicular cells

Thyroid follicle

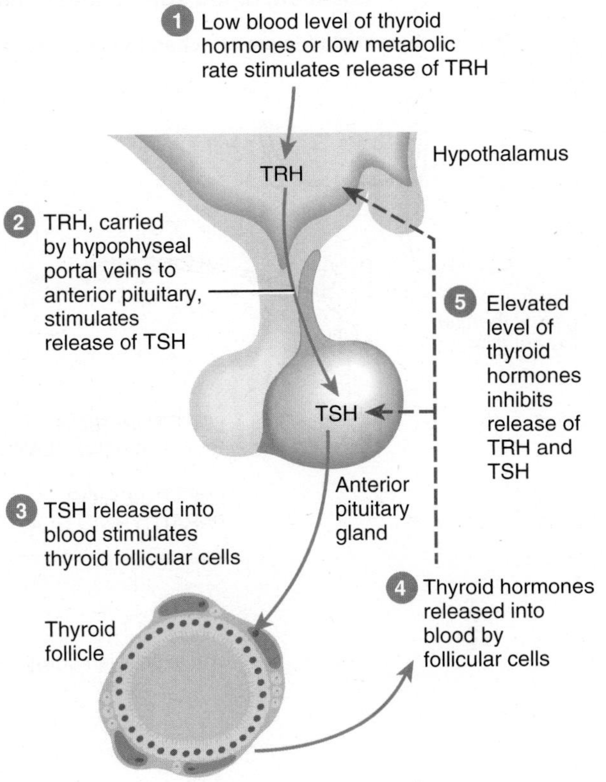

What is the effect of thyroid hormones on metabolic rate?

PARATHYROID GLANDS

Objective: • **Describe the location, hormones, and functions of the parathyroid glands.**

Attached to the posterior surface of the thyroid gland are small, round masses of tissue called the **parathyroid glands** (*para-* = beside). Usually there are four parathyroid glands, two attached to each thyroid lobe: the left superior, left inferior, right superior, and right inferior (Figure 13.12a). The most numerous cells in the parathyroid glands, the **principal cells,** produce **parathyroid hormone (PTH)**.

Parathyroid hormone is the major regulator of the levels of calcium (Ca^{2+}), magnesium (Mg^{2+}) and phosphate (HPO_4^{2-}) ions in the blood. PTH increases the number and activity of osteoclasts, which break down bone matrix and release Ca^{2+} and HPO_4^{2-} into the blood. PTH also produces a number of changes in the kidneys: it slows the rate at which Ca^{2+} and Mg^{2+} are lost from blood into the urine, and it increases loss of HPO_4^{2-} from blood in urine. Because more HPO_4^{2-} is lost in the urine than is gained from the bones, PTH decreases blood HPO_4^{2-} level and increases blood Ca^{2+} and Mg^{2+} levels. A third effect of PTH on the kidneys is to promote formation of the

Figure 13.12 ■ **Location, blood supply, and histology of the parathyroid glands.**

The parathyroid glands are attached to the posterior surface of the thyroid gland.

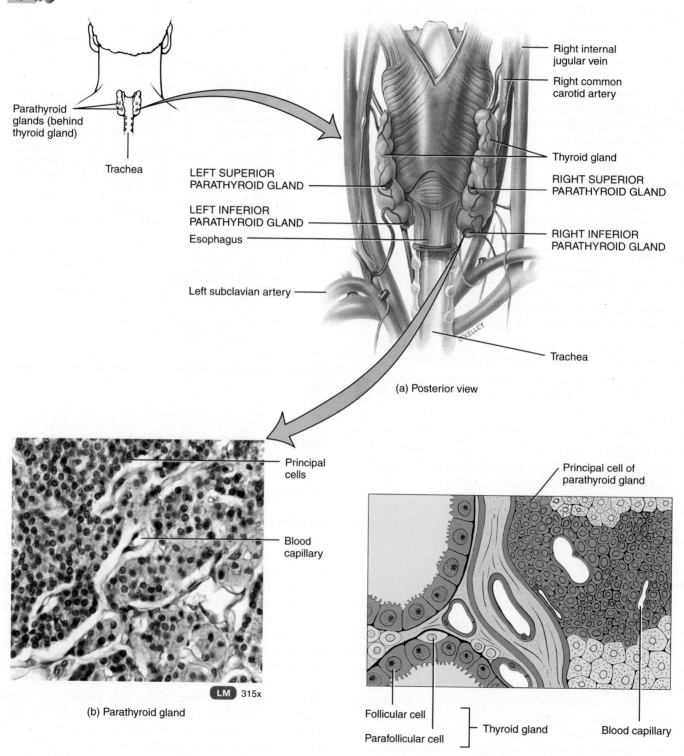

(a) Posterior view

(b) Parathyroid gland

(c) Diagram of a portion of the thyroid gland (left) and parathyroid gland (right)

What effect does PTH have on osteoclasts?

hormone *calcitriol,* the active form of vitamin D. Calcitriol increases the rate of Ca^{2+}, HPO_4^{2-}, and Mg^{2+} absorption from foods in the gastrointestinal tract into the blood.

The blood calcium level directly controls the secretion of calcitonin and parathyroid hormone via negative feedback, and the two hormones have opposite effects on blood Ca^{2+} level (Figure 13.13).

1 A higher than normal level of calcium ions (Ca^{2+}) in the blood stimulates parafollicular cells of the thyroid gland to release more calcitonin.

2 Calcitonin promotes movement of blood Ca^{2+} into bone matrix, thereby decreasing blood Ca^{2+} level.

3 A lower than normal level of Ca^{2+} in the blood stimulates principal cells of the parathyroid gland to release more PTH.

4 PTH promotes release of Ca^{2+} from bone matrix into the blood and slows loss of Ca^{2+} in the urine, raising the blood level of Ca^{2+}.

5 PTH also stimulates the kidneys to release calcitriol, the active form of vitamin D.

6 Calcitriol stimulates increased absorption of Ca^{2+} from foods in the gastrointestinal tract, which helps increase the blood level of Ca^{2+}.

ADRENAL GLANDS

Objective: • **Describe the location, hormones, and functions of the adrenal glands.**

The two *adrenal glands* are located superior to the kidneys (Figure 13.14). Each adrenal gland is composed of two regions: the outer *adrenal cortex,* which makes up 80% to 90% of the gland, and the inner *adrenal medulla.* The two regions produce different hormones.

Adrenal Cortex Hormones

The adrenal cortex is subdivided into three zones, each of which synthesizes and secretes different types of steroid hormones. The outer zone releases hormones called mineralocorticoids because they affect mineral homeostasis. The middle zone releases hormones called glucocorticoids because they affect glucose homeostasis. The inner zone releases androgens (male sex hormones).

Mineralocorticoids

Mineralocorticoids (min′-er-al-ō-KOR-ti-koyds) help control the homeostasis of water, sodium ions (Na^+), and potassium ions

Figure 13.13 ■ **The roles of calcitonin (green arrows), parathyroid hormone (blue arrows), and calcitriol (red arrows) in homeostasis of blood calcium level.**

PTH and calcitonin have opposite effects on the level of calcium ions in the blood.

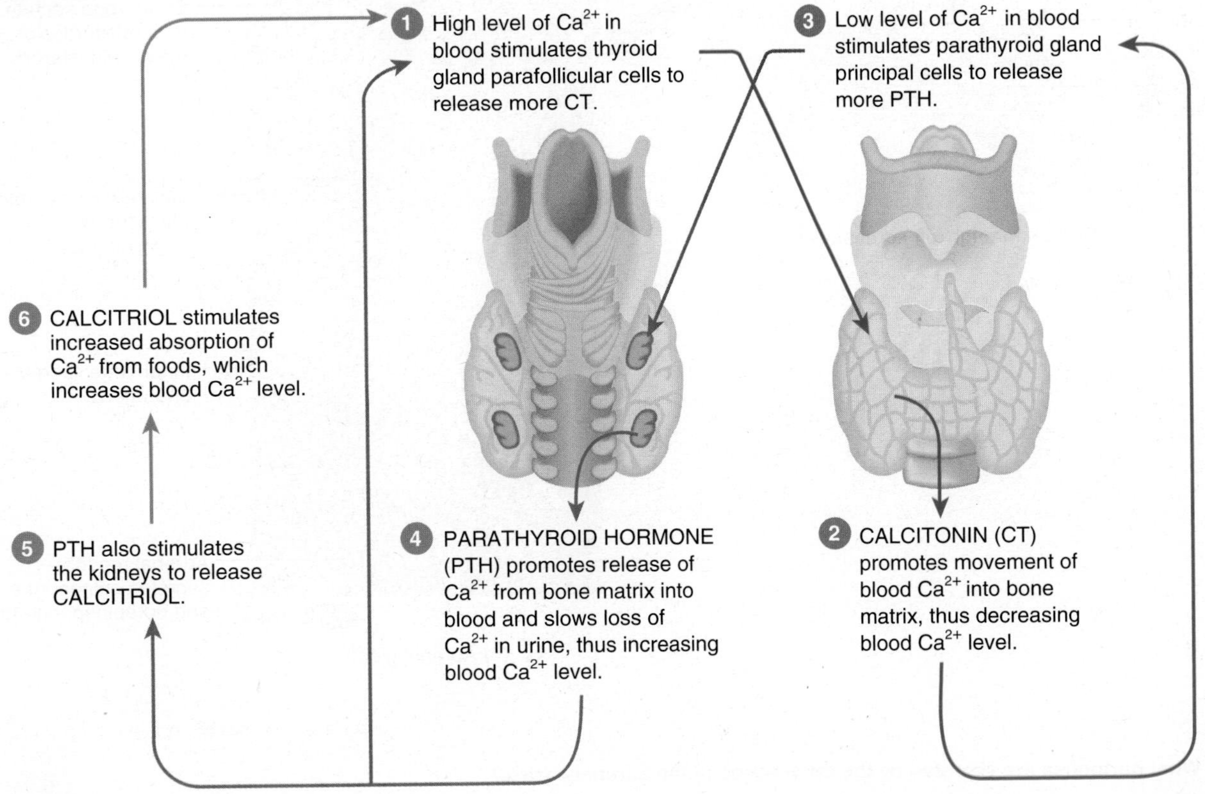

1 High level of Ca^{2+} in blood stimulates thyroid gland parafollicular cells to release more CT.

3 Low level of Ca^{2+} in blood stimulates parathyroid gland principal cells to release more PTH.

6 CALCITRIOL stimulates increased absorption of Ca^{2+} from foods, which increases blood Ca^{2+} level.

5 PTH also stimulates the kidneys to release CALCITRIOL.

4 PARATHYROID HORMONE (PTH) promotes release of Ca^{2+} from bone matrix into blood and slows loss of Ca^{2+} in urine, thus increasing blood Ca^{2+} level.

2 CALCITONIN (CT) promotes movement of blood Ca^{2+} into bone matrix, thus decreasing blood Ca^{2+} level.

Will a low level of calcium ions in blood cause increased secretions from the thyroid gland or parathyroid glands?

Figure 13.14 ■ **Location, blood supply, and histology of the adrenal glands.**

The adrenal cortex secretes steroid hormones, whereas the adrenal medulla secretes amine hormones.

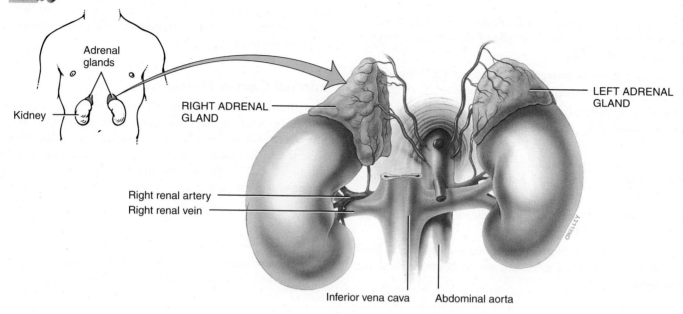

Adrenal glands

Kidney

RIGHT ADRENAL GLAND

LEFT ADRENAL GLAND

Right renal artery

Right renal vein

Inferior vena cava

Abdominal aorta

(a) Anterior view

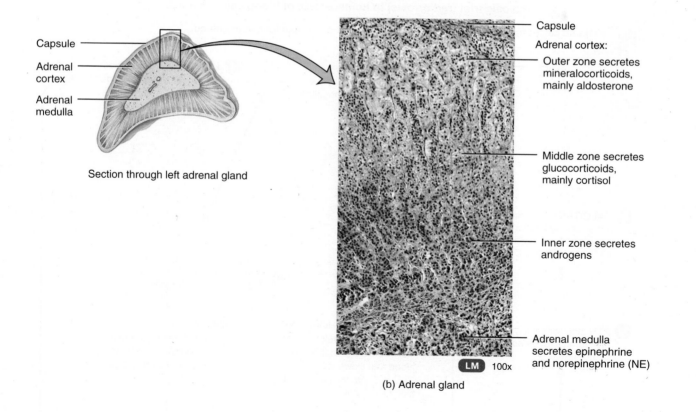

Capsule

Adrenal cortex

Adrenal medulla

Section through left adrenal gland

Capsule

Adrenal cortex:

Outer zone secretes mineralocorticoids, mainly aldosterone

Middle zone secretes glucocorticoids, mainly cortisol

Inner zone secretes androgens

Adrenal medulla secretes epinephrine and norepinephrine (NE)

LM 100x

(b) Adrenal gland

What hormones are secreted by the three zones of the adrenal cortex?

(K⁺). The hormone responsible for most mineralocorticoid activity is **aldosterone** (al-DOS-ter-ōn). Aldosterone increases reabsorption of Na⁺ from the urine into the blood, and it stimulates excretion of K⁺ into the urine.

Aldosterone secretion operates by the **renin–angiotensin–aldosterone pathway** (RĒ-nin an′-jē-ō-TEN-sin; Figure 13.15):

① Stimuli that initiate the renin–angiotensin–aldosterone pathway include dehydration, Na⁺ deficiency, or hemorrhage.

② These conditions cause a decrease in blood volume.

③ Decreased blood volume causes a decrease in blood pressure.

④ Lowered blood pressure stimulates certain kidney cells, called juxtaglomerular cells, to secrete the enzyme **renin.**

⑤ The level of renin in the blood increases.

⑥ Renin converts a plasma protein called **angiotensinogen** into **angiotensin I.**

⑦ Blood containing an increased level of angiotensin I circulates to the lungs.

⑧ As blood flows through the lungs, an enzyme called **angiotensin converting enzyme (ACE)** converts angiotensin I into the hormone **angiotensin II.**

⑨ Blood level of angiotensin II increases.

⑩ Angiotensin II acts on the adrenal cortex, stimulating it to secrete aldosterone.

⑪ Blood containing an increased level of aldosterone circulates to the kidneys.

⑫ In the kidneys, aldosterone increases Na⁺ and water reabsorption. Aldosterone also stimulates the kidneys to increase secretion of K⁺ into the urine.

⑬ As a result of increased water reabsorption by the kidneys, blood volume increases.

⑭ As blood volume increases, blood pressure increases to normal.

⑮ Angiotensin II also stimulates smooth muscle in the walls of arterioles to contract, producing vasoconstriction that increases blood pressure and thus helps raise blood pressure to normal.

Figure 13.15 ■ **Regulation of aldosterone secretion by the renin–angiotensin–aldosterone pathway.**

Aldosterone helps regulate blood volume, blood pressure, and levels of Na⁺ and K⁺ in the blood.

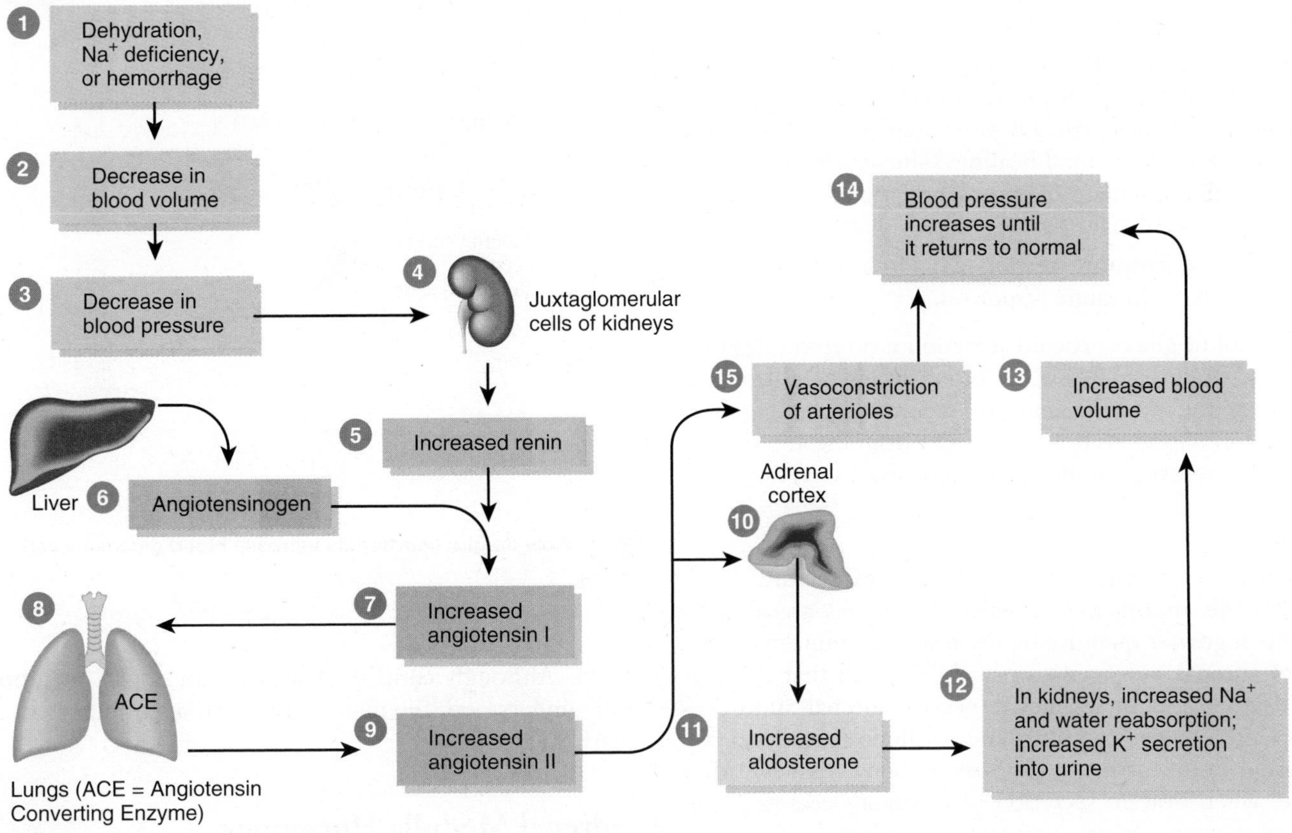

① Dehydration, Na⁺ deficiency, or hemorrhage

② Decrease in blood volume

③ Decrease in blood pressure

④ Juxtaglomerular cells of kidneys

Liver **⑥** Angiotensinogen

⑤ Increased renin

⑦ Increased angiotensin I

⑧ Lungs (ACE = Angiotensin Converting Enzyme)

ACE

⑨ Increased angiotensin II

⑪ Increased aldosterone

Adrenal cortex

⑩

⑮ Vasoconstriction of arterioles

⑬ Increased blood volume

⑭ Blood pressure increases until it returns to normal

⑫ In kidneys, increased Na⁺ and water reabsorption; increased K⁺ secretion into urine

In what two ways can angiotensin II increase blood pressure?

Glucocorticoids

The most abundant *glucocorticoid* (gloo′-kō-KOR-ti-koyd; *gluco-* = sugar; *cortic-* = the bark, shell) produced by the middle zone of the adrenal cortex is *cortisol*. Cortisol and the other glucocorticoids have the following effects on the body:

1. **Protein breakdown.** Glucocorticoids increase the rate of protein breakdown, mainly in muscle fibers, and thus increase the liberation of amino acids into the bloodstream. The amino acids may be used by body cells for synthesis of new proteins or for ATP production.

2. **Glucose formation.** Upon stimulation by glucocorticoids, liver cells may convert certain amino acids or lactic acid to glucose, which neurons and other cells can use for ATP production.

3. **Lipolysis.** Glucocorticoids stimulate *lipolysis,* the breakdown of triglycerides and release of fatty acids from adipose tissue.

4. **Resistance to stress.** Glucocorticoids work in many ways to provide resistance to stress. The additional glucose supplied by the liver cells provides tissues with a ready source of ATP to combat a range of stresses, including exercise, fasting, fright, temperature extremes, high altitude, bleeding, infection, surgery, trauma, and disease. Because glucocorticoids make blood vessels more sensitive to other hormones that cause vasoconstriction, they raise blood pressure. Not usually considered an advantage, increased blood pressure can lessen the effects of significant blood loss, which makes blood pressure fall.

5. **Anti-inflammatory effects.** Glucocorticoids inhibit white blood cells that participate in inflammatory responses. Unfortunately, glucocorticoids also retard tissue repair, and as a result they slow wound healing. Glucocorticoids are often used in the treatment of chronic inflammatory disorders such as rheumatoid arthritis.

6. **Depression of immune responses.** High doses of glucocorticoids depress immune responses.

The control of glucocorticoid secretion is a typical negative feedback loop (Figure 13.16). Low blood levels of glucocorticoids stimulate the hypothalamus to secrete a releasing hormone, which initiates the release of ACTH from the anterior pituitary. ACTH, in turn, stimulates glucocorticoid secretion.

Androgens

In both males and females, the adrenal cortex secretes small amounts of weak androgens. In males, androgens are also released in much greater quantity by the testes. In adult males the secretion of adrenal androgens is usually so small that their effects are insignificant. In females, however, adrenal androgens play important roles: They contribute to libido (sex drive) and are converted into estrogens by other body tissues. After menopause, when ovarian secretion of estrogens ceases, all female estrogens come from conversion of adrenal androgens. Adrenal androgens also stimulate growth of axillary and pubic hair in boys and girls and contribute to the prepubertal growth

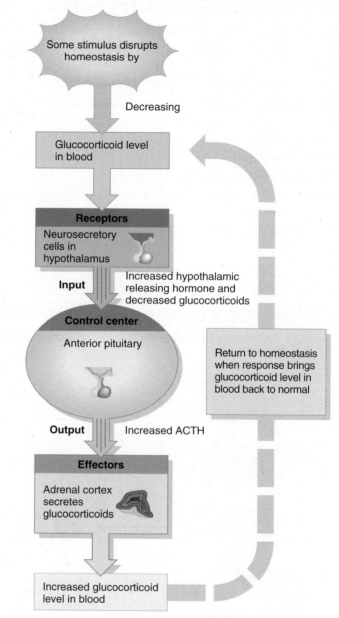

Figure 13.16 ■ Negative feedback regulation of glucocorticoid secretion.

A high level of a hypothalamic releasing hormone and a low level of glucocorticoids promote the release of ACTH, which stimulates glucocorticoid secretion by the adrenal cortex.

Some stimulus disrupts homeostasis by

Decreasing

Glucocorticoid level in blood

Receptors
Neurosecretory cells in hypothalamus

Input
Increased hypothalamic releasing hormone and decreased glucocorticoids

Control center
Anterior pituitary

Return to homeostasis when response brings glucocorticoid level in blood back to normal

Output   Increased ACTH

Effectors
Adrenal cortex secretes glucocorticoids

Increased glucocorticoid level in blood

 How do glucocorticoids increase blood glucose level?

spurt. Although control of adrenal androgen secretion is not fully understood, the main hormone that stimulates its secretion is ACTH.

Adrenal Medulla Hormones

The innermost region of the adrenal gland, the adrenal medulla, consists of sympathetic postganglionic cells of the autonomic

One of the most common endocrine disorders, type II diabetes, is characterized by high levels of insulin in the blood. Insulin levels appear to be high because of **insulin resistance,** which means the insulin receptors are not responding appropriately to insulin. In people with type II diabetes, these receptors gradually become less sensitive, and it takes more insulin than it should to accomplish the important task of moving glucose from the bloodstream into the cells. Despite a more than adequate amount of insulin, blood glucose level remains high, because the receptors are not letting insulin help the glucose across the membrane and into the cells.

A Little Riddle About Fat in the Middle

A large majority of people who develop type II diabetes also develop hypertension (high blood pressure) and high blood cholesterol levels. They also tend to be somewhat overweight and sedentary. This cluster of disorders—termed metabolic syndrome—may be related to excess adipose tissue around the abdominal viscera.

Why is visceral fat riskier than other adipose tissue? Adipocytes (fat cells) in the abdominal region, especially those located around the viscera, are metabolically "more active" than lower body fat cells; they are more responsive to hormones such as epinephrine. This means they release fatty acids into the bloodstream more readily, which, in the abdominal area, flows to the liver. The liver takes up the fatty acids and produces triglycerides that are packaged into very-low-density lipoprotein (VLDL) particles. Later, the VLDLs are converted into low-density lipoprotein (LDL) particles. Higher levels of LDLs are associated with the formation of artery-clogging atherosclerotic plaques.

The elevation in triglycerides may disrupt blood sugar regulation and trigger a rise in insulin. Elevated insulin levels in turn stimulate the sympathetic nervous system, which increases blood pressure. And there you have it, all in one package: high blood sugar, high blood lipids, hypertension, and visceral obesity, a package that significantly increases artery disease risk.

Type II diabetes appears to be related to both genetic and behavioral factors. Smoking, alcohol consumption, poor diet, and a sedentary lifestyle predispose a person to the development of this disorder. Both exercise and weight loss (in people who are overweight) increase the sensitivity of insulin receptors and improve transport of glucose into body cells.

► Think It Over

► Why do you think extra fat tends to be lost more easily in the abdominal region than in the hips and thighs? Are you at risk for metabolic syndrome? Use a tape measure to take the circumferences of your hips and waist. A waist–hips ratio greater than 1.0 for men or 0.8 for women is considered risky.

---

nervous system (ANS) that are specialized to secrete hormones. The two principal hormones of the adrenal medulla, **epinephrine** and **norepinephrine (NE),** also called adrenaline and noradrenaline, contribute greatly to the fight-or-flight response (see page 000). Like the glucocorticoids of the adrenal cortex, they also help the body resist stress.

When the body is under stress, impulses received by the hypothalamus are conveyed to sympathetic preganglionic neurons and then to the adrenal medulla, which rapidly increases its secretion of epinephrine and norepinephrine. Hypoglycemia (low blood sugar) also stimulates secretion of epinephrine and norepinephrine. By increasing heart rate and force of contraction, epinephrine and norepinephrine increase blood pressure. They also increase blood flow to the heart, liver, skeletal muscles, and adipose tissue; dilate airways to the lungs; and increase blood levels of glucose and fatty acids.

PANCREAS

Objective: • **Describe the location, hormones, and endocrine functions of the pancreas.**

The **pancreas** (*pan-* = all; *-creas* = flesh) is a flattened organ located in the curve of the duodenum, the first part of the small intestine (Figure 13.17a). It has both endocrine functions, discussed in this chapter, and exocrine functions, discussed in Chapter 19. The endocrine parts of the pancreas are clusters of cells called **pancreatic islets** or **islets of Langerhans** (LAHNG-er-hanz). The islets contain numerous blood capillaries and are surrounded by cells that form the exocrine part of the pancreas (Figure 13.17b,c). Four major types of cells are found in the islets: (1) **alpha cells** secrete the hormone glucagon; (2) **beta cells** secrete the hormone insulin; (3) **delta cells** secrete somatostatin,

Figure 13.17 ■ **Pancreas.**

The pancreas is both an endocrine gland and an exocrine gland.

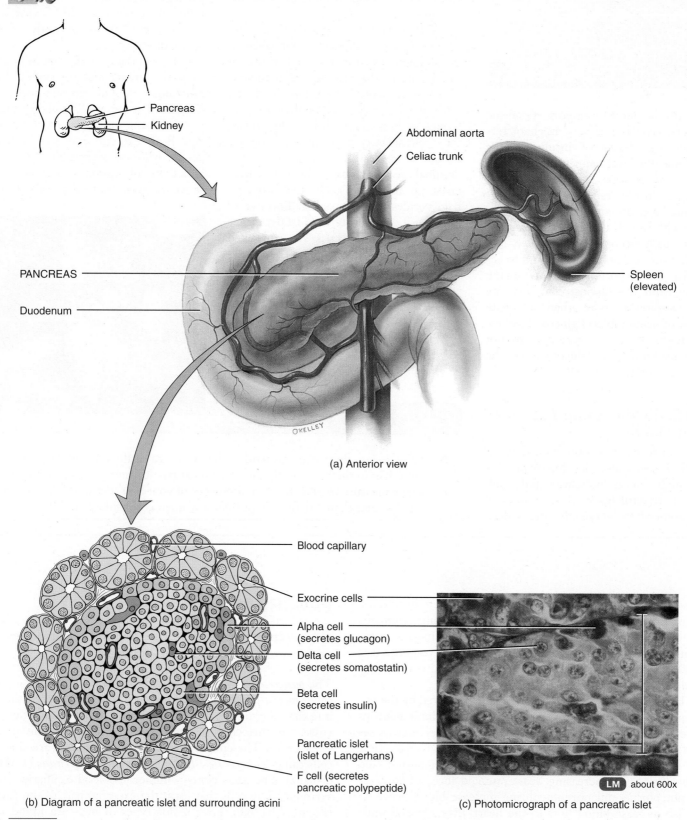

Pancreas

Kidney

Abdominal aorta

Celiac trunk

PANCREAS

Duodenum

Spleen (elevated)

O'KELLEY

(a) Anterior view

Blood capillary

Exocrine cells

Alpha cell (secretes glucagon)

Delta cell (secretes somatostatin)

Beta cell (secretes insulin)

Pancreatic islet (islet of Langerhans)

F cell (secretes pancreatic polypeptide)

(b) Diagram of a pancreatic islet and surrounding acini

LM about 600x

(c) Photomicrograph of a pancreatic islet

Which pancreatic hormones are the chief regulators of blood glucose level?

which inhibits secretion of insulin and glucagon; and (4) *F cells* secrete pancreatic polypeptide, which helps regulate the release of digestive enzymes from the pancreas. Glucagon and insulin are the chief regulators of blood glucose level.

Glucagon

Glucagon (GLOO-ka-gon) increases the blood glucose level when it falls below normal by (1) accelerating the conversion of glycogen in the liver into glucose, (2) promoting the conversion in the liver of amino acids and lactic acid into glucose, and (3) stimulating the release of glucose from the liver into the blood (Figure 13.18). Secretion of glucagon is controlled via a negative feedback system. When the blood glucose level falls, the alpha cells of the islets are stimulated to secrete glucagon. When

blood glucose rises, the cells are no longer stimulated and glucagon secretion decreases. Increased activity of the sympathetic division of the ANS also enhances glucagon release.

Insulin

When the blood glucose level is above normal, insulin decreases it by acting in several ways (Figure 13.18). It (1) accelerates the facilitated diffusion of glucose from blood into most body cells, especially skeletal muscle fibers; (2) accelerates the conversion of glucose into glycogen and fatty acids; (3) promotes uptake of amino acids into body cells and increases protein synthesis within cells; (4) slows the conversion of liver glycogen into glucose; and (5) slows formation of glucose by liver cells. The regulation of insulin secretion, like that of glucagon secretion, is directly regulated by the blood glucose level. High blood glucose level stimulates beta cells of the pancreatic islets to release insulin, whereas low blood glucose level inhibits insulin release.

Figure 13.18 ■ **Regulation of the secretion of glucagon and insulin.**

Low blood glucose level stimulates release of glucagon, whereas high blood glucose level stimulates secretion of insulin.

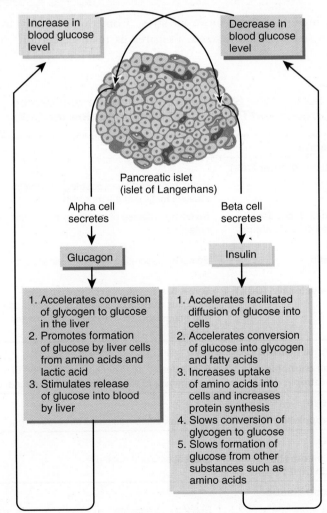

 Which cells of the pancreas produce a hormone that inhibits secretion of both insulin and glucagon?

OVARIES AND TESTES

Objective: • **Describe the location, hormones, and functions of the ovaries and testes.**

The female gonads, the *ovaries,* are paired oval bodies located in the pelvic cavity. They produce the female sex hormones *estrogens* and *progesterone.* These steroid hormones are responsible for the development and maintenance of feminine sexual characteristics. Along with FSH and LH from the anterior pituitary, the sex hormones regulate the menstrual cycle, maintain pregnancy, and prepare the mammary glands for lactation. The ovaries (and placenta) also produce two protein hormones: (1) *inhibin* inhibits secretion of FSH, and (2) *relaxin* helps enlarge the birth canal by increasing the flexibility of the pubic symphysis and dilating the uterine cervix toward the end of pregnancy.

The male gonads, the *testes,* are oval glands that lie in the scrotum. They produce *testosterone,* the primary androgen or male sex hormone. Testosterone regulates production of sperm and stimulates the development and maintenance of masculine characteristics such as beard growth. The testes also produce inhibin, which inhibits secretion of FSH. The detailed structure of the ovaries and testes and the specific roles of sex hormones will be discussed in Chapter 23.

The illegal use of *anabolic steroids* by athletes has received widespread attention in the print and broadcast media. These steroid hormones, similar to testosterone, are taken to increase muscle size and therefore to increase strength during athletic contests. The large doses needed to produce an effect, however, have damaging and devastating side effects. These include liver cancer, kidney damage, increased risk of heart disease, stunted growth, wide mood swings, increased irritability, and abnormally aggressive behavior. Females may experience atrophy of the breasts and uterus, menstrual irregularities, sterility, facial hair growth, and deepening of the voice. Males may experience diminished testosterone secretion, atrophy of the testes, and baldness.

PINEAL GLAND

Objective: • **Describe the location, hormone, and functions of the pineal gland.**

The *pineal gland* (PIN-ē-al = pinecone shape) is a small endocrine gland attached to the roof of the third ventricle of the brain at the midline (see Figure 13.1). Although many anatomical features of the pineal gland have been known for years, its physiological role is still unclear. One hormone secreted by the pineal gland is *melatonin*, an amine hormone. Melatonin contributes to setting the body's biological clock, which is controlled by the hypothalamus. More melatonin is released in darkness, and less melatonin is liberated in strong sunlight. In animals that breed during specific seasons, melatonin inhibits reproductive functions. Whether melatonin influences human reproductive function, however, is still unclear. Melatonin levels are higher in children and decline with age into adulthood, but there is no evidence that changes in melatonin secretion correlate with the onset of puberty and sexual maturation.

Seasonal affective disorder (SAD) is a type of depression that afflicts some people during the winter months, when day length is short. It is thought to be due, in part, to overproduction of melatonin. Bright light therapy—repeated exposure to artificial light—can provide relief.

THYMUS

Objective: • **Describe the location and functions of the thymus.**

The *thymus* is located behind the sternum between the lungs. Because of its role in immunity, the details of the structure and functions of the thymus are discussed in Chapter 17. The hormones produced by the thymus—*thymosin, thymic humoral factor (THF), thymic factor (TF),* and *thymopoietin*—promote the maturation of T cells (a type of white blood cell that destroys microbes and foreign substances) and may retard the aging process.

OTHER ENDOCRINE TISSUES

Objective: • **List the hormones secreted by cells in tissues and organs other than endocrine glands, and describe their functions.**

Hormones from Other Endocrine Cells

Some tissues and organs other than those described already contain endocrine cells that secrete hormones. Table 13.3 summarizes these hormones and their actions.

Eicosanoids

Two families of eicosanoid molecules, derived from 20-carbon fatty acids, the *prostaglandins* (pros'-ta-GLAN-dins), or *PGs*, and the *leukotrienes* (loo'-kō-TRĪ-ēns), or *LTs*, act as local hormones in most tissues of the body. Virtually all body cells except red blood cells release these local hormones in response to chemical and mechanical stimuli. The PGs and LTs act close to their sites of release so they appear in minute quantities in the blood.

Leukotrienes stimulate movement of white blood cells and mediate inflammation. The prostaglandins alter smooth muscle contraction, glandular secretions, blood flow, reproductive processes, platelet function, respiration, nerve impulse transmission, fat metabolism, and immune responses. PGs also have roles in inflammation, promoting fever, and intensifying pain.

Aspirin and related *nonsteroidal anti-inflammatory drugs (NSAIDs),* such as ibuprofen (Motrin), inhibit a key enzyme in prostaglandin synthesis without affecting synthesis of leukotrienes. They are used to treat a wide variety of inflammatory disorders, from rheumatoid arthritis to tennis elbow.

THE STRESS RESPONSE

Objective: • **Describe how the body responds to stress and how stress and disease are related.**

It is impossible to remove all stress from our everyday lives. Some stress, called *eustress*, prepares us to meet certain chal-

Table 13.3 / Summary of Hormones Produced by Organs and Tissues That Contain Endocrine Cells	
Hormone	**Principal Actions**
Gastrointestinal Tract	
Gastrin	Promotes secretion of gastric juice and increases motility of the stomach.
Glucose-dependent insulinotropic peptide (GIP)	Stimulates release of insulin by pancreatic beta cells.
Secretin	Stimulates secretion of pancreatic juice and bile.
Cholecystokinin (CCK)	Stimulates secretion of pancreatic juice, regulates release of bile from the gallbladder, and brings about a feeling of fullness after eating.
Placenta (during pregnancy)	
Human chorionic gonadotropin (hCG)	Stimulates ovary to continue production of estrogens and progesterone during pregnancy.
Kidneys	
Erythropoietin (EPO)	Increases rate of red blood cell production.
Heart	
Atrial natriuretic peptide (ANP)	Decreases blood pressure.
Adipose Tissue	
Leptin	Suppresses appetite.

lenges and thus is productive. Other stress, called *distress,* is harmful. Any stimulus that produces a stress response is called a *stressor.* A stressor may be almost any disturbance: heat or cold, environmental poisons, toxins given off by bacteria, heavy bleeding from a wound or surgery, or a strong emotional reaction. Stressors may be pleasant or unpleasant, and they vary among people and even within the same person at different times.

Homeostatic mechanisms attempt to counteract stress. When they are successful, the internal environment remains within normal physiological limits. If stress is extreme, unusual, or long lasting, the normal mechanisms may not be enough. A variety of stressful conditions or noxious agents elicit a similar sequence of bodily changes called the *stress response* or *general adaptation syndrome (GAS).* The stress response is controlled mainly by the hypothalamus and consists of neural and hormonal responses that help body systems cope with emergency situations. It occurs in three stages: (1) an initial fight-or-flight response, (2) a slower resistance reaction, and eventually (3) exhaustion.

The Fight-or-Flight Response

The *fight-or-flight response,* initiated by nerve impulses from the hypothalamus to the sympathetic division of the autonomic nervous system (ANS) and the adrenal medulla, quickly mobilizes the body's resources for immediate physical activity. It brings huge amounts of glucose and oxygen to the organs that are most active in warding off danger: the brain, which must become highly alert; the skeletal muscles, which may have to fight off an attacker or flee; and the heart, which must work vigorously to pump enough blood to the brain and muscles. During the fight-or-flight response, nonessential body functions such as digestive, urinary, and reproductive activities are inhibited. Reduction of blood flow to the kidneys, however, promotes release of renin, which sets into motion the renin–angiotensin–aldosterone pathway (see Figure 13.15). Aldosterone causes the kidneys to retain Na^+, which leads to water retention and elevated blood pressure. Water retention also helps preserve body fluid volume in the case of severe bleeding.

The Resistance Reaction

The second stage in the stress response is the *resistance reaction.* Unlike the short-lived fight-or-flight response, which is initiated by nerve impulses from the hypothalamus, the resistance reaction is initiated in large part by hypothalamic releasing hormones and is a longer-lasting response.

One hypothalamic releasing hormone stimulates the anterior pituitary to secrete ACTH, which in turn stimulates the adrenal cortex to release more cortisol. Cortisol then stimulates production of glucose by liver cells, breakdown of triglycerides into fatty acids (lipolysis), and breakdown of proteins into amino acids. Tissues throughout the body can use the resulting glucose, fatty acids, and amino acids to produce ATP or to repair damaged cells. Cortisol also reduces inflammation.

A second hypothalamic releasing hormone causes the ante-

rior pituitary to secrete human growth hormone (hGH). Acting by means of insulinlike growth factors, hGH stimulates lipolysis and the breakdown of glycogen to glucose. A third hypothalamic releasing hormone stimulates the anterior pituitary to secrete thyroid-stimulating hormone (TSH). TSH promotes secretion of thyroid hormones, which stimulate the increased use of glucose for ATP production. The combined actions of hGH and TSH thereby supply additional ATP for metabolically active cells.

The resistance stage helps the body continue fighting a stressor long after the fight-or-flight response dissipates. Generally, it is successful in seeing us through a stressful episode, and our bodies then return to normal. Occasionally, however, the resistance stage fails to combat the stressor, and the body moves into the state of exhaustion.

Exhaustion

The resources of the body may eventually become so depleted that they cannot sustain the resistance stage, and *exhaustion* ensues. Prolonged exposure to high levels of cortisol and other hormones involved in the resistance reaction causes wasting of muscle, suppression of the immune system, ulceration of the gastrointestinal tract, and failure of the pancreatic beta cells. In addition, pathological changes may occur because resistance reactions persist after the stressor has been removed.

Stress and Disease

Although the exact role of stress in human diseases is not known, it is clear that stress can lead to particular diseases by temporarily inhibiting certain components of the immune system. Stress-related disorders include gastritis, ulcerative colitis, irritable bowel syndrome, hypertension, asthma, rheumatoid arthritis, migraine headaches, anxiety, and depression. People under stress also are at a greater risk of developing chronic disease or dying prematurely.

AGING AND THE ENDOCRINE SYSTEM

Objective: • **Describe the effects of aging on the endocrine system.**

Although some endocrine glands shrink as we get older, their performance may or may not be compromised. With respect to the pituitary gland, production of human growth hormone decreases, which is one cause of muscle atrophy as aging proceeds. The thyroid gland often decreases its output of thyoid hormones with age, causing a decrease in metabolic rate, an increase in body fat, and hypothyroidism, which is seen more often in older people.

The thymus is largest in infancy. After puberty, its size begins to decrease, and thymic tissue is replaced by adipose and areolar connective tissue. In older adults, the thymus has atrophied significantly.

Focus on Homeostasis
The Endocrine System

Body System	Contribution of the Endocrine System

For all body systems

Together with the nervous system, circulating and local hormones of the endocrine system regulate activity and growth of target cells throughout the body; several hormones regulate metabolism, uptake of glucose, and molecules used for ATP production by body cells.

Integumentary system

Androgens stimulate growth of axillary and pubic hair and activation of sebaceous glands; melanocyte-stimulating hormone (MSH) can cause darkening of skin.

Skeletal system

Human growth hormone (hGH) and insulinlike growth factors (IGFs) stimulate bone growth; estrogens and testosterone cause closure of epiphyseal plates at the end of puberty and help maintain bone mass in adults; parathyroid hormone (PTH) and calcitonin regulate release versus uptake of calcium and other minerals between bone matrix and blood; thyroid hormones are needed for normal development and growth of the skeleton.

Muscular system

Epinephrine and norepinephrine help increase blood flow to exercising muscle; PTH maintains proper level of Ca^{2+}, needed for muscle contraction; glucagon, insulin, and other hormones regulate metabolism in muscle fibers; hGH, IGFs, thyroid hormones, and insulin help maintain muscle mass.

Nervous system

Several hormones, especially thyroid hormones, insulin, hGH, and IGFs, influence growth and development of the nervous system; PTH maintains proper level of Ca^{2+}, needed for generation and conduction of nerve impulses.

Cardiovascular system

Erythropoietin (EPO) promotes production of red blood cells; aldosterone and antidiuretic hormone (ADH) increase blood volume; epinephrine and norepinephrine increase heart rate and force of contraction; several hormones elevate blood pressure during exercise and other stresses.

Lymphatic and immune system

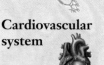

Glucocorticoids such as cortisol depress inflammation and immune responses; thymic hormones promote maturation of T cells (a type of white blood cell).

Respiratory system

Epinephrine and norepinephrine dilate (widen) airways during exercise and other stresses; erythropoietin regulates amount of oxygen carried in blood by adjusting number of red blood cells.

Digestive system

Epinephrine and norepinephrine depress activity of digestive system; gastrin, cholecystokinin, secretin, and GIP help regulate digestion; calcitriol promotes absorption of dietary calcium; leptin suppresses appetite.

Urinary system

ADH, aldosterone, and atrial natriuretic peptide (ANP) adjust the rate of fluid and ion loss in the urine, thereby regulating blood volume and ion content of the blood.

Reproductive systems

Hypothalamic releasing and inhibiting hormones, follicle-stimulating hormone (FSH), and luteinizing hormone (LH) regulate development, growth, and secretions of gonads (ovaries and testes); estrogens and testosterone contribute to development of oocytes and sperm and stimulate development of sexual characteristics; prolactin promotes milk secretion in mammary glands; oxytocin causes contraction of uterus and ejection of milk from mammary glands.

The adrenal glands contain increasingly more fibrous tissue and produce less cortisol and aldosterone with advancing age. However, production of epinephrine and norepinephrine remains normal. The pancreas releases insulin more slowly with age, and receptor sensitivity to glucose declines. As a result, blood glucose levels in older people increase faster and return to normal more slowly than in younger individuals.

The ovaries experience a drastic decrease in size with age, and they no longer respond to FSH and LH. The resultant decreased output of estrogen by the ovaries leads to conditions such as osteoporosis and atherosclerosis. Although testosterone production by the testes decreases with age, the effects are not usually apparent until very old age, and many elderly males can still produce active sperm in normal numbers.

• • •

To appreciate the many ways the endocrine system contributes to homeostasis of other body systems, examine Focus on Homeostasis: The Endocrine System. Next, in Chapter 14, we will begin to explore the cardiovascular system, starting with a description of the composition and functions of the blood.

COMMON DISORDERS

Disorders of the endocrine system often involve either *hyposecretion* (*hypo-* = too little or under), inadequate release of a hormone, or *hypersecretion* (*hyper-* = too much or above), excessive release of a hormone. In most cases the problem is faulty control of secretion, but in other cases the problem is faulty hormone receptors or an inadequate number of receptors.

Pituitary Gland Disorders

Several disorders of the anterior pituitary involve human growth hormone (hGH). Undersecretion of hGH during the growth years slows bone growth, and the epiphyseal plates close before normal height is reached. This condition is called *pituitary dwarfism*. Other organs of the body also fail to grow, and the body proportions are childlike.

Oversecretion of hGH during childhood results in *giantism (gigantism)*, an abnormal increase in the length of long bones. The person grows to be very tall, but body proportions are about normal. Oversecretion of hGH during adulthood is called *acromegaly* (ak′-rō-MEG-a-lē) (Figure 13.19a). Although hGH cannot produce further lengthening of the long bones because the epiphyseal plates are already closed, the bones of the hands, feet, cheeks, and jaws thicken and other tissues enlarge.

The most common abnormality associated with dysfunction of the posterior pituitary is *diabetes insipidus* (dī′-a-BĒ-tēs in-SIP-i-dus; *diabetes* = overflow; *insipidus* = tasteless). This disorder is due to defects in antidiuretic hormone (ADH) receptors or an inability to secrete ADH. Usually the disorder is caused by a brain tumor, head trauma, or brain surgery that damages the posterior pituitary or the hypothalamus. A common symptom is excretion of large volumes of urine, with resulting dehydration and thirst. Because so much water is lost in the urine, a person with diabetes insipidus may die of dehydration if deprived of water for only a day or so.

Thyroid Gland Disorders

Thyroid gland disorders affect all major body systems and are among the most common endocrine disorders. Hyposecretion of thyroid hormones during infancy results in *cretinism* (KRĒ-tin-izm) (Figure 13.19b). Cretins exhibit dwarfism because the skeleton fails to grow and mature, and they are severely mentally retarded because the brain fails to develop fully. Because maternal thyroid hormones cross the placenta and allow normal development, newborn babies usually are normal, even if they can't produce their own thyroid hormones. Most states require testing of all newborns to ensure adequate thyroid function. With early detection, cretinism can be prevented by giving thyroid hormone orally.

Hypothyroidism during the adult years produces *myxedema* (mix′-e-DĒ-ma), which occurs about five times more often in females than in males. A hallmark of this disorder is edema (accumulation of interstitial fluid) that causes the facial tissues to swell and look puffy. A person with myxedema has a slow heart rate, low body temperature, sensitivity to cold, dry hair and skin, muscular weakness, general lethargy, and a tendency to gain weight easily.

The most common form of hyperthyroidism is *Graves' disease,* which also occurs much more often in females than in males, usually before age 40. The disorder causes production of antibodies that mimic the action of thyroid-stimulating hormone (TSH). Thus, the thyroid gland is continually stimulated to grow and may enlarge to two to three times its normal size, a condition called *goiter* (GOY-ter; *guttur* = throat) (Figure 13.19c). Graves' patients often have a peculiar edema behind the eyes, called *exophthalmos* (ek′-sof-THAL-mos), which causes the eyes to protrude (Figure 13.19d). Goiter also occurs in other thyroid diseases and if dietary intake of iodine is inadequate.

Parathyroid Gland Disorders

Hypoparathyroidism—too little parathyroid hormone—leads to a deficiency of Ca^{2+}, which causes neurons and muscle fibers to depolarize and produce action potentials spontaneously. This leads to twitches, spasms, and *tetany* (maintained contraction) of skeletal muscle. The leading cause of hypoparathyroidism is accidental damage to the parathyroid glands or to their blood supply during surgery to remove the thyroid gland.

Adrenal Gland Disorders

Oversecretion of cortisol by the adrenal cortex produces *Cushing's syndrome.* The condition is characterized by breakdown of muscle proteins and redistribution of body fat, resulting in spindly arms and legs accompanied by a rounded "moon face" (Figure 13.19e), "buffalo hump" on the back, and pendulous (hanging) abdomen. The elevated level of cortisol causes hyperglycemia, osteoporosis, weakness, hypertension, increased susceptibility to infection, decreased resistance to stress, and mood swings.

328 Chapter 13 • The Endocrine System

COMMON DISORDERS (CONTINUED)

Undersecretion of glucocorticoids and aldosterone causes **Addison's disease.** Symptoms include mental lethargy, anorexia, nausea and vomiting, weight loss, hypoglycemia, and muscular weakness. Loss of aldosterone leads to elevated potassium and decreased sodium in the blood; low blood pressure; dehydration; and decreased cardiac output, cardiac arrhythmias, and even cardiac arrest.

Usually benign tumors of the adrenal medulla, called **pheochromocytomas** (fē′-ō-krō′-mō-si-TŌ-mas; *pheo-* = dusky; *chromo-* = color;

cyto- = cell), cause oversecretion of epinephrine and norepinephrine. The result is a prolonged version of the fight-or-flight response: rapid heart rate, headache, high blood pressure, high levels of glucose in blood and urine, an elevated basal metabolic rate (BMR), flushed face, nervousness, sweating, and decreased gastrointestinal motility.

Pancreatic Endocrine Disorders

Diabetes mellitus (MEL-i-tus) is a group of hereditary diseases that lead to elevated blood glucose (hyperglycemia) and excretion of glucose in the urine. Diabetes mellitus is also characterized by the three

Figure 13.19 ■ Photographs of various endocrine disorders.

Disorders of the endocrine system often involve hyposecretion or hypersecretion of hormones.

(a) Acromegaly (excess hGH during adulthood)

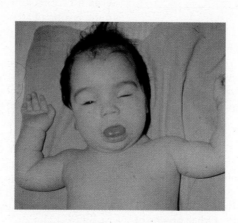

(b) Cretinism (lack of thyroid hormones during fetal life or infancy)

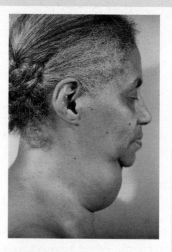

(c) Goiter (enlargement of thyroid gland as in Graves' disease or inadequate iodine intake)

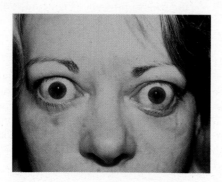

(d) Exophthalmos (excess thyroid hormones, as in Graves' disease)

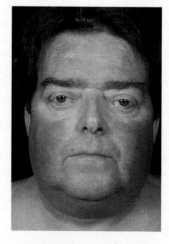

(e) Cushing's syndrome (excess glucocorticoids)

Which of the disorders is due to antibodies that mimic the action of TSH?

"polys": an inability to reabsorb water, resulting in increased urine production *(polyuria)*; excessive thirst *(polydipsia)*; and excessive eating *(polyphagia)*. There are two major types of diabetes mellitus: type I and type II. *Type I diabetes* is characterized by a deficiency of insulin because the person's immune system destroys the pancreatic beta cells. It is also called *insulin-dependent diabetes mellitus (IDDM)* because insulin injections are required to prevent death. Most commonly, IDDM develops in people younger than age 20, though it persists throughout life. Insulin deficiency speeds the breakdown of the body's fat reserves, which produces organic acids called ketones. Excess ketones cause a form of acidosis called *ketosis*, which lowers the pH of the blood and can result in death.

Type II diabetes, also called *non–insulin-dependent diabetes mellitus (NIDDM),* is much more common than type I. NIDDM most often occurs in people who are over 35 and overweight. The high glucose levels in the blood often can be controlled by diet, exercise, and weight loss. Sometimes, an antidiabetic drug such as *glyburide* (Diabeta) is used to stimulate secretion of insulin by beta cells of the pancreas. Although some type II diabetics need insulin, many have a sufficient amount (or even a surplus) of insulin in the blood. For these people, diabetes arises not from a shortage of insulin but because target cells become less sensitive to it.

Hyperinsulinism most often results when a diabetic injects too much insulin. The principal symptom is *hypoglycemia,* decreased blood glucose level, which occurs because the excess insulin stimulates excessive uptake of glucose by many cells of the body. When blood glucose falls, brain cells are deprived of the steady supply of glucose they need to function effectively. Severe hypoglycemia leads to mental disorientation, convulsions, unconsciousness, and shock and is termed *insulin shock.* Death can occur quickly unless blood glucose is raised.

◾ STUDY OUTLINE

Introduction (p. 306)

1. The nervous system controls homeostasis through nerve impulses; the endocrine system uses hormones.

2. The nervous system causes muscles to contract and glands to secrete; the endocrine system affects virtually all body tissues.

Endocrine Glands (p. 307)

1. Exocrine glands (sweat, oil, mucous, digestive) secrete their products through ducts into body cavities or onto body surfaces.

2. The endocrine system consists of endocrine glands and several organs that contain endocrine tissue.

Overview of Hormones (p. 307)

1. Endocrine glands secrete hormones into interstitial fluid. Circulating hormones enter the bloodstream; local hormones act locally on neighboring cells.

2. Hormones affect only specific target cells with receptors that bind a given hormone.

3. Chemically, hormones are either lipid soluble (steroids, thyroid hormones, and nitric oxide) or water soluble (amines, peptides and proteins, and eicosanoids).

4. Lipid-soluble hormones affect cell function by altering gene expression.

5. Water-soluble hormones alter cell function by activating plasma membrane receptors, which elicits production of a second messenger that activates various enzymes inside the cell.

6. Hormone secretion is controlled by signals from the nervous system, chemical changes in blood, and other hormones.

Hypothalamus and Pituitary Gland (p. 309)

1. The pituitary gland is attached to the hypothalamus and consists of two lobes: the anterior pituitary and the posterior pituitary.

2. Hormones of the pituitary gland are controlled by inhibiting and releasing hormones produced by the hypothalamus.

3. The hypophyseal portal veins carry hypothalamic releasing and inhibiting hormones from the hypothalamus to the anterior pituitary.

4. The anterior pituitary consists of cells that produce human growth hormone (hGH), prolactin (PRL), thyroid-stimulating hormone (TSH), follicle-stimulating hormone (FSH), luteinizing hormone (LH), adrenocorticotropic hormone (ACTH), and melanocyte-stimulating hormone (MSH).

5. Human growth hormone (hGH) stimulates body growth through insulinlike growth factors (IGFs) and is controlled by hypothalamic releasing and inhibiting hormones.

6. TSH regulates thyroid gland activities and is controlled by a hypothalamic releasing hormone.

7. FSH and LH regulate activities of the gonads—ovaries and testes—and are controlled by a hypothalamic releasing hormone.

8. PRL helps initiate milk secretion and is controlled by hypothalamic releasing and inhibiting hormones.

9. ACTH regulates activities of the adrenal cortex and is controlled by a hypothalamic releasing hormone.

10. The posterior pituitary contains axon terminals of neurosecretory cells whose cell bodies are in the hypothalamus.

11. Hormones made by the hypothalamus and stored in the posterior pituitary include oxytocin (OT), which stimulates contraction of the uterus and ejection of milk from the breasts, and antidiuretic hormone (ADH), which stimulates water reabsorption by the kidneys and constriction of arterioles.

12. Oxytocin secretion is stimulated by uterine stretching and by suckling during nursing; ADH secretion is controlled by osmotic pressure of the blood and blood volume.

Thyroid Gland (p. 315)

1. The thyroid gland is located below the larynx.

2. It consists of thyroid follicles composed of follicular cells, which secrete the thyroid hormones thyroxine (T_4) and triiodothyronine (T_3), and parafollicular cells, which secrete calcitonin.

3. Thyroid hormones regulate oxygen use and metabolic rate, cellular metabolism, and growth and development. Secretion is controlled by TRH from the hypothalamus and thyroid-stimulating hormone (TSH) from the anterior pituitary.

4. Calcitonin (CT) can lower the blood level of calcium; its secretion is controlled by the level of calcium in the blood.

Parathyroid Glands (p. 317)

1. The parathyroid glands are embedded on the posterior surfaces of the thyroid.

2. Parathyroid hormone (PTH) regulates the homeostasis of calcium, magnesium, and phosphate by increasing blood calcium and magnesium levels and decreasing blood phosphate level. PTH secretion is controlled by the level of calcium in the blood.

Adrenal Glands (p. 319)

1. The adrenal glands are located above the kidneys. They consist of an outer cortex and inner medulla.

2. The adrenal cortex is divided into three zones: The outer zone secretes mineralocorticoids; the middle zone secretes glucocorticoids; and the inner zone secretes androgens.

3. Mineralocorticoids (mainly aldosterone) increase sodium and water reabsorption and decrease potassium reabsorption. Secretion is controlled by the renin–angiotensin–aldosterone pathway.

4. Glucocorticoids (mainly cortisol) promote normal metabolism, help resist stress, and serve as anti-inflammatories. Secretion is controlled by ACTH.

5. Androgens secreted by the adrenal cortex stimulate growth of axillary and pubic hair, aid the prepubertal growth spurt, and contribute to libido.

6. The adrenal medulla secretes epinephrine and norepinephrine (NE), which are released under stress.

Pancreas (p. 323)

1. The pancreas lies in the curve of the duodenum. It has both endocrine and exocrine functions.

2. The endocrine portion consists of pancreatic islets or islets of Langerhans, made up of four types of cells: alpha, beta, delta, and F cells.

3. Alpha cells secrete glucagon, beta cells secrete insulin, delta cells secrete somatostatin, and F cells secrete pancreatic polypeptide.

4. Glucagon increases blood glucose level, whereas insulin decreases blood glucose level. Secretion of both hormones is controlled by the level of glucose in the blood.

Ovaries and Testes (p. 325)

1. The ovaries are located in the pelvic cavity and produce estrogens, progesterone, and inhibin. These sex hormones govern the development and maintenance of feminine sexual characteristics, reproductive cycles, pregnancy, lactation, and normal reproductive functions.

2. The testes lie inside the scrotum and produce testosterone and inhibin. These sex hormones govern the development and maintenance of masculine sexual characteristics and normal reproductive functions.

Pineal Gland (p. 326)

1. The pineal gland is attached to the roof of the third ventricle in the brain.

2. It secretes melatonin, which contributes to setting the body's biological clock.

Thymus (p. 326)

1. The thymus secretes several hormones related to immunity.

2. Thymosin, thymic humoral factor (THF), thymic factor (TF), and thymopoietin promote maturation of T cells.

Other Endocrine Tissues (p. 326)

1. Body tissues other than those normally classified as endocrine glands contain endocrine tissue and secrete hormones.

2. These include the gastrointestinal tract, placenta, kidneys, skin, and heart. (See Table 13.3 on page 000.)

3. Prostaglandins and leukotrienes are eicosanoids that act as local hormones in most body tissues.

The Stress Response (p. 326)

1. Stressors include surgical operations, poisons, infections, fever, and strong emotional responses.

2. Productive stress is termed eustress, and harmful, nonproductive stress is termed distress.

3. If stress is extreme, it triggers the stress response (general adaptation syndrome), which occurs in three stages: the fight-or-flight response, resistance reaction, and exhaustion.

4. The fight-or-flight response is initiated by nerve impulses from the hypothalamus to the sympathetic division of the autonomic nervous system and the adrenal medulla. This response rapidly increases circulation, promotes ATP production, and decreases nonessential activities.

5. The resistance reaction is initiated by releasing hormones secreted by the hypothalamus. Resistance reactions are longer lasting and accelerate breakdown reactions to provide ATP for counteracting stress.

6. Exhaustion results from depletion of body resources during the resistance stage.

7. Stress appears to trigger certain diseases by inhibiting the immune system.

■ SELF-QUIZ

1. Which of the following is NOT true concerning hormones?
   a. Responses by hormones are generally slower and longer lasting than the responses by the nervous system.   **b.** Hormones are generally controlled by negative feedback systems.   **c.** The hypothalamus inhibits the release of some hormones.   **d.** Most hormones are released steadily throughout the day.   **e.** Hormone secretion is determined by the body's needs to maintain homeostasis.

2. Which of the following statements concerning hormone action is NOT true?
   a. Hormones bring about changes in metabolic activities of cells.
   b. Target cells must have receptors for a hormone.   **c.** Lipid-

soluble hormones may directly enter target cells and activate the genes. **d.** A hormone that attaches to a membrane receptor is termed the first messenger. **e.** ATP is a common second messenger in target cells.

3. Which of the following statements is NOT true?

   a. The secretion of hormones by the anterior pituitary is controlled by hypothalamic releasing hormones. **b.** The pituitary is attached to the hypothalamus by the infundibulum. **c.** Hypophyseal portal veins connect the posterior pituitary to the hypothalamus. **d.** The anterior pituitary constitutes the majority of the pituitary gland. **e.** The posterior pituitary releases hormones produced by neurosecretory cells of the hypothalamus.

4. The hormone that promotes milk release from the mammary glands and that stimulates the uterus to contract is

   a. oxytocin **b.** prolactin **c.** relaxin **d.** calcitonin **e.** follicle-stimulating hormone

5. The gland that prepares the body to react to stress by releasing epinephrine is the

   a. posterior pituitary **b.** anterior pituitary **c.** adrenal cortex **d.** adrenal medulla **e.** pancreas

6. To help prevent rejection, organ transplant patients could be given

   a. glucocorticoids **b.** calcitonin **c.** mineralocorticoids **d.** thymopoietin **e.** melanocyte-stimulating hormone

7. A female who is sluggish, gaining weight, and has a low body temperature may be having problems with her

   a. pancreas **b.** parathyroid gland **c.** adrenal medulla **d.** ovaries **e.** thyroid gland

8. Destruction of the alpha cells of the pancreas might result in

   a. hypoglycemia **b.** seasonal affective disorder **c.** acromegaly **d.** hyperglycemia **e.** decreased urine output

9. Which of the following is NOT true concerning human growth hormone (hGH) and insulinlike growth factors?

   a. They stimulate protein synthesis. **b.** They have one primary target tissue in the body. **c.** They stimulate skeletal muscle growth. **d.** Hyposecretion in childhood results in dwarfism. **e.** Hypoglycemia can stimulate the release of hGH from the pituitary gland.

10. Follicle-stimulating hormone (FSH) acts on the _____ and luteinizing hormone (LH) acts on the _____ .

    a. the ovaries, the testes **b.** the testes, the ovaries **c.** the ovaries and testes, the ovaries and testes **d.** the ovaries, the mammary glands **e.** the ovaries and uterus, the testes

11. An injection of adrenocorticotropic hormone (ACTH) would

    a. stimulate the ovaries **b.** influence thyroid gland activity **c.** stimulate the release of cortisol **d.** cause uterine contractions **e.** decrease urine output

12. Which of the following is NOT true concerning glucocorticoids?

    a. They help to control electrolyte balance. **b.** They help provide resistance to stress. **c.** They help promote normal metabolism. **d.** They are anti-inflammatory hormones. **e.** They provide the body with energy.

13. Mineralocorticoids

    a. help prevent the loss of potassium from the body **b.** are secreted based upon the renin–angiotensin–aldosterone pathway **c.** increase the rate of sodium loss in the urine **d.** are involved in lowering the body's blood pressure **e.** increase water loss from the body by increasing urine production

14. A lack of iodine in the diet affects the production of which hormone?

    a. calcitonin **b.** parathyroid hormone **c.** aldosterone **d.** thyroxine **e.** glucagon

15. Which of the following hormones with opposite effects are correctly paired?

    a. parathyroid hormone, thyroid hormones **b.** parathyroid hormone, calcitonin **c.** oxytocin, glucocorticoids **d.** aldosterone, oxytocin **e.** thyroid hormones, thymosin

16. A patient exhibiting liver failure would tend to have

    a. higher than normal blood levels of circulating hormones **b.** higher than normal blood levels of local hormones **c.** lower than normal blood levels of circulating hormones **d.** lower than normal blood levels of local hormones **e.** excessive production of releasing hormones

17. Match the following:

    ____ **a.** produce thyroid hormones
    ____ **b.** secrete insulin
    ____ **c.** release hormones into capillaries of the posterior pituitary
    ____ **d.** produce parathyroid hormone
    ____ **e.** secrete glucagon
    ____ **f.** produce calcitonin
    ____ **g.** secrete somatostatin

    A. principal cells
    B. delta cells
    C. follicular cells
    D. alpha cells
    E. parafollicular cells
    F. beta cells
    G. neurosecretory cells

18. In a dehydrated person, you would expect to see an increased release of

    a. parathyroid hormone **b.** aldosterone **c.** insulin **d.** melatonin **e.** inhibin

19. Match the following:

    ____ **a.** diabetes insipidus
    ____ **b.** diabetes mellitus
    ____ **c.** myxedema
    ____ **d.** Cushing's syndrome
    ____ **e.** Addison's disease
    ____ **f.** tetany

    A. hypersecretion of glucocorticoids
    B. hyposecretion of antidiuretic hormone
    C. hyposecretion of insulin
    D. hyposecretion of parathyroid hormone
    E. hyposecretion of thyroid hormone
    F. hyposecretion of glucocorticoids

20. For each of the following, indicate at which stage they would occur as part of the general adaptation syndrome. Use **F** to indicate fight-or-flight response, **R** to indicate resistance reaction, and **E** to indicate exhaustion.

    ____ **a.** initiated by hypothalamic releasing hormones
    ____ **b.** initiated by the sympathetic division of the autonomic nervous system
    ____ **c.** immediately prepares the body for action
    ____ **d.** increases cortisol release
    ____ **e.** short-lived response
    ____ **f.** body resources become depleted
    ____ **g.** increased release of many hormones that ensure a continued ATP supply
    ____ **h.** failure of pancreatic beta cells
    ____ **i.** nonessential body functions inhibited

CRITICAL THINKING APPLICATIONS

1. Patrick was diagnosed with diabetes mellitus on his eighth birthday. His 65-year-old aunt was just diagnosed with diabetes also. Patrick is having a hard time understanding why he needs injections, while his aunt controls her blood sugar with diet and oral medication. Why is his aunt's treatment different from his?

2. Eddie, the tallest man in the world, suffered from an oversecretion of a pituitary hormone his entire life. Eddie died in early adulthood due to the effects of his condition. Name Eddie's condition and explain its cause.

3. Goiter was fairly common in regions of the American midwest before iodized salt was in common use. Explain the relationship of iodine to the thyroid gland.

4. Brian is in a 50-mile bike-a-thon on a hot summer day. He is sweating profusely, and he loses his water bottle. How will his hormones respond to this situation?

■ ANSWERS TO FIGURE QUESTIONS

13.1 Secretions of endocrine glands diffuse into extracellular spaces and then into the blood; exocrine secretions flow into ducts that lead into body cavities or to the body surface.

13.2 Hormones are inactivated by the liver and excreted by the kidneys.

13.3 RNA molecules are synthesized when genes are expressed (transcribed).

13.4 It brings the message of the first messenger, the water-soluble hormone, into the cell.

13.5 The posterior pituitary releases hormones synthesized in the hypothalamus.

13.6 Endocrine glands regulated by anterior pituitary hormones include the thyroid gland, adrenal cortex, ovaries, and testes.

13.7 Excess hGH causes hyperglycemia.

13.8 Oxytocin's target cells are in the uterus and mammary glands.

13.9 Absorption of a large glass of water in the intestines would decrease the osmotic pressure (concentration of solutes) of your blood plasma, turning off secretion of ADH and decreasing the ADH level in your blood.

13.10 Follicular cells secrete T_3 and T_4, also known as thyroid hormones.

13.11 Thyroid hormones increase metabolic rate.

13.12 PTH increases the number and activity of osteoclasts.

13.13 A low blood level of calcium ions causes increased secretion of PTH by the parathyroid gland.

13.14 The outer zone of the adrenal cortex secretes mineralocorticoids, the middle zone secretes glucocorticoids, and the inner zone secretes adrenal androgens.

13.15 Angiotensin II constricts blood vessels by causing contraction of smooth muscle, and it stimulates secretion of aldosterone (by cells of the adrenal cortex), which in turn causes the kidneys to conserve water and increase blood volume.

13.16 Glucocorticoids increase blood glucose level by stimulating liver cells to convert amino acids or lactic acid to glucose and release the glucose into the blood.

13.17 Insulin and glucagon are the chief regulators of blood glucose level.

13.18 Delta cells secrete somatostatin, which inhibits secretion of insulin and glucagon.

13.19 In Graves' disease antibodies are produced that mimic the action of TSH.

Chapter 14

The Cardiovascular System: Blood

Student Learning Objectives

1. List and describe the functions of blood. **334**
2. List the principal physical characteristics of whole blood. **334**
3. Discuss the formation, components, and functions of whole blood. **334**
4. Describe the various mechanisms that prevent blood loss. **342**
5. Describe the ABO and Rh blood groups. **345**

A Look Ahead

The *cardiovascular system* (*cardio-* = heart; *vascular* = blood or blood vessels) consists of three interrelated components: blood, the heart, and blood vessels. The focus of this chapter is blood; the next two chapters will cover the heart and blood vessels, respectively.

Functionally, the cardiovascular system transports substances to and from body cells. Oxygen breathed into the lungs and nutrients absorbed from food are transported by the blood to body tissues. The oxygen and nutrients move from blood into interstitial fluid, the fluid that bathes body cells, and then into body cells. Carbon dioxide and other wastes move in the reverse direction, from cells into interstitial fluid and then into blood. Blood then transports the wastes to various organs for elimination from the body. As we will see, blood not only transports various substances, it also helps regulate several life processes and affords protection against disease. To perform its functions, blood must circulate throughout the body. The heart serves as the pump for circulation, and blood vessels carry blood from the heart to body cells and from body cells back to the heart and lungs.

The branch of science concerned with the study of blood, blood-forming tissues, and the disorders associated with them is *hematology* (hēm-a-TOL-ō-jē; *hemo-* or *hemato-* = blood; *-ology* = study of).

FUNCTIONS OF BLOOD

Objective: • List and describe the functions of blood.

Blood is a liquid connective tissue that has three general functions: transportation, regulation, and protection.

1. **Transportation.** Blood transports oxygen from the lungs to cells throughout the body and carbon dioxide (a waste product of cellular respiration; see Chapter 20) from the cells to the lungs. It also carries nutrients from the gastrointestinal tract to body cells, heat and waste products away from cells, and hormones from endocrine glands to other body cells.

2. **Regulation.** Blood helps regulate the pH of body fluids. The heat-absorbing and coolant properties of the water in

blood plasma and its variable rate of flow through the skin help adjust body temperature. Blood osmotic pressure also influences the water content of cells.

3. **Protection.** Blood clots (becomes gel-like) in response to an injury, which protects against its excessive loss from the cardiovascular system. In addition, white blood cells protect against disease by carrying on phagocytosis and producing proteins called antibodies. Blood contains additional proteins, called interferons and complement, that also help protect against disease.

PHYSICAL CHARACTERISTICS OF WHOLE BLOOD

Objective: • List the principal physical characteristics of whole blood.

Blood is denser and more viscous (sticky) than water. The temperature of blood is about 38°C (100.4°F), which is slightly higher than normal body temperature. Its pH is slightly alkaline, ranging from 7.35 to 7.45. Blood constitutes about 8% of the total body weight. The blood volume is 5 to 6 liters (1.5 gal) in an average-sized adult male and 4 to 5 liters (1.2 gal) in an average-sized adult female.

COMPONENTS OF WHOLE BLOOD

Objective: • Discuss the formation, components, and functions of whole blood.

Whole blood is composed of two portions: (1) *plasma*, a liquid that contains dissolved substances, and (2) *formed elements*, which are cells and cell fragments. If a sample of blood is centrifuged (spun at high speed) in a small glass tube, the cells sink to the bottom of the tube and the lighter-weight plasma forms a layer on top (Figure 14.1a). Blood is about 45% formed elements and 55% plasma. Normally, more than 99% of the formed elements are red blood cells (RBCs). The percentage of total blood volume occupied by red blood cells is termed the *hematocrit* (he-MAT-ō-krit). Pale, colorless white blood cells (WBCs) and platelets occupy less than 1% of total blood volume. They form a very thin layer, called the *buffy coat*, between the packed RBCs and plasma in centrifuged blood. Figure 14.1b shows the composition of blood plasma and the numbers of the various types of formed elements in blood.

Plasma

When the formed elements are removed from blood, a straw-colored liquid called *plasma* remains. Plasma is about 91.5% water, 7% proteins, and 1.5% solutes other than proteins. Proteins in the blood, the *plasma proteins*, are synthesized mainly by the liver. The most plentiful plasma proteins are the *albumins*, which account for about 54% of all plasma proteins.

Blood is a connective tissue that consists of plasma (liquid) plus formed elements: red blood cells, white blood cells, and platelets.

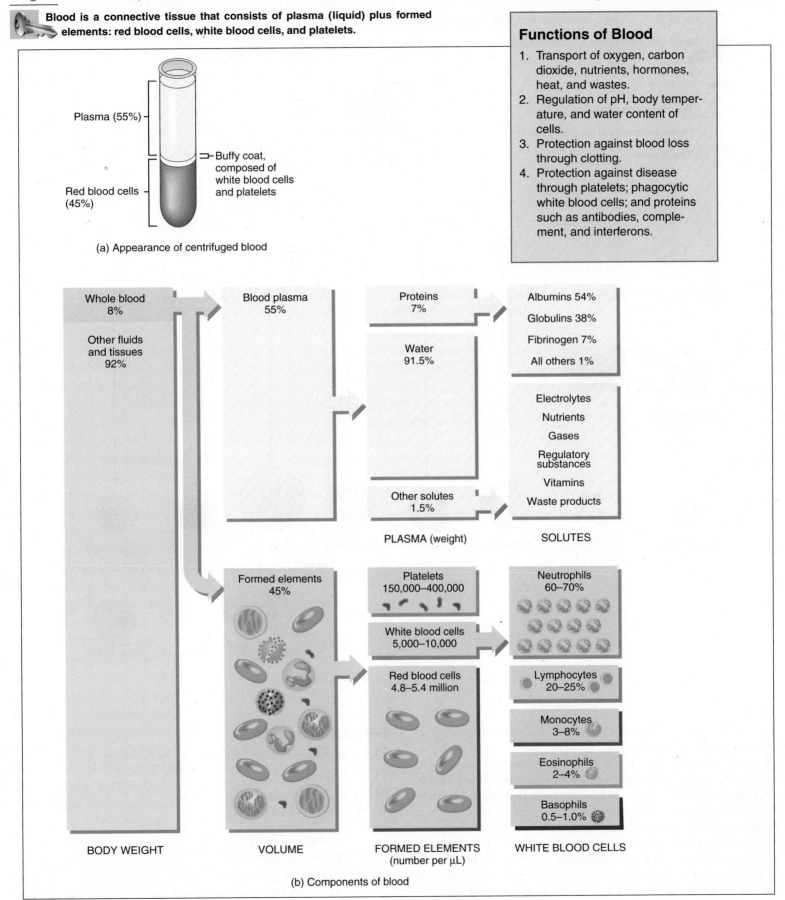

(a) Appearance of centrifuged blood

Plasma (55%)

Buffy coat, composed of white blood cells and platelets

Red blood cells (45%)

Functions of Blood

1. Transport of oxygen, carbon dioxide, nutrients, hormones, heat, and wastes.
2. Regulation of pH, body temperature, and water content of cells.
3. Protection against blood loss through clotting.
4. Protection against disease through platelets; phagocytic white blood cells; and proteins such as antibodies, complement, and interferons.

Whole blood 8%

Other fluids and tissues 92%

Blood plasma 55%

Proteins 7%

Albumins 54%
Globulins 38%
Fibrinogen 7%
All others 1%

Water 91.5%

Electrolytes
Nutrients
Gases
Regulatory substances
Vitamins
Waste products

Other solutes 1.5%

PLASMA (weight)   SOLUTES

Formed elements 45%

Platelets 150,000–400,000

White blood cells 5,000–10,000

Red blood cells 4.8–5.4 million

Neutrophils 60–70%

Lymphocytes 20–25%

Monocytes 3–8%

Eosinophils 2–4%

Basophils 0.5–1.0%

BODY WEIGHT   VOLUME   FORMED ELEMENTS (number per μL)   WHITE BLOOD CELLS

(b) Components of blood

Which formed elements of blood are most numerous?

Figure 14.2 ▪ **Origin, development, and structure of blood cells.**

Blood cell production is called hemopoiesis and occurs only in red bone marrow after birth.

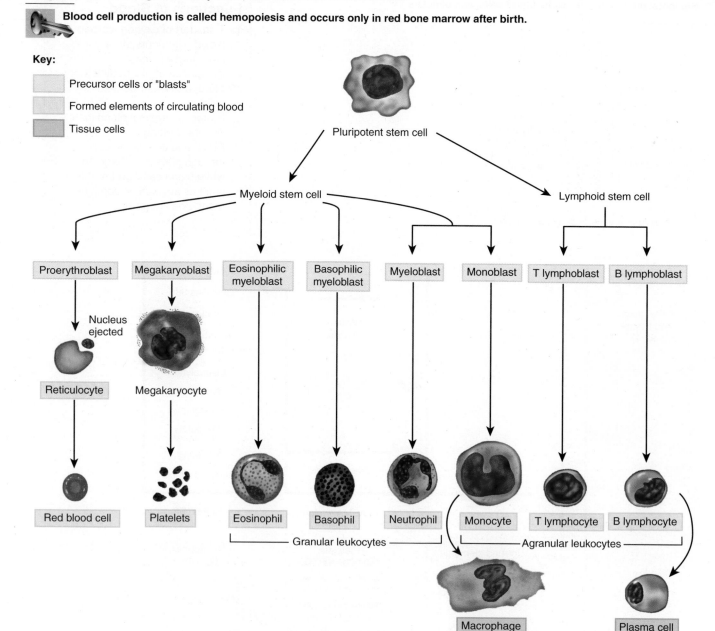

Key:

◻ Precursor cells or "blasts"

◻ Formed elements of circulating blood

◻ Tissue cells

Pluripotent stem cell

Myeloid stem cell   Lymphoid stem cell

Proerythroblast   Megakaryoblast   Eosinophilic myeloblast   Basophilic myeloblast   Myeloblast   Monoblast   T lymphoblast   B lymphoblast

Nucleus ejected

Reticulocyte   Megakaryocyte

Red blood cell   Platelets   Eosinophil   Basophil   Neutrophil   Monocyte   T lymphocyte   B lymphocyte

⎣————— Granular leukocytes —————⎦   ⎣————— Agranular leukocytes —————⎦

Macrophage   Plasma cell

(a) Origin of blood cells from pluripotent stem cells

Among other functions, albumins help maintain proper blood osmotic pressure, which is an important factor in the exchange of fluids across capillary walls. ***Globulins,*** which compose 38% of plasma proteins, include ***antibodies*** or ***immunoglobulins,*** proteins produced during certain immune responses (see page 417). For instance, different parts of foreign invaders such as bacteria and viruses stimulate production of many different antibodies. The antibodies then bind specifically to the substance, called an ***antigen,*** that provoked their production. ***Fibrinogen*** makes up about 7% of plasma proteins and is a key protein in formation of blood clots. Other solutes in plasma include electrolytes, nutrients, gases, regulatory substances such as enzymes and hormones, vitamins, and waste products.

Formed Elements

The ***formed elements*** of the blood are the following (Figure 14.2):

  I. Red blood cells

 II. White blood cells

    A. Granular leukocytes

        1. Neutrophils

        2. Eosinophils

        3. Basophils

    B. Agranular leukocytes

        1. T and B lymphocytes and natural killer cells

        2. Monocytes

III. Platelets

All **LM**

980x	910x	1260x
NEUTROPHIL	EOSINOPHIL	BASOPHIL

225x
BLOOD SMEAR

420x	1260x	1260x
RED BLOOD CELLS AND PLATELETS	MONOCYTE	LYMPHOCYTE

(b) Photomicrographs

 What percentage of body weight does blood represent?

Formation of Blood Cells

The process by which the formed elements of blood develop is called **hemopoiesis** (hē′-mō-poy-Ē-sis; *-poiesis* = making). During the embryonic and fetal stages of life, several tissues participate in producing the formed elements. After birth, hemopoiesis takes place only in red bone marrow, although small numbers of stem cells circulate in the bloodstream. Red bone marrow is found in long bones such as the humerus and femur; in flat bones such as the sternum, ribs, and cranial bones; in vertebrae; and in the pelvis.

About 0.1% of red bone marrow cells are **pluripotent stem cells** (Figure 14.2a, top), which give rise to all the formed elements of the blood. In response to stimulation by specific hormones, pluripotent stem cells generate two other types of stem cells: *myeloid stem cells* and *lymphoid stem cells* (Figure 14.2a). Myeloid stem cells begin their development in red bone marrow and differentiate into precursor "blast" cells from which red blood cells, platelets, eosinophils, basophils, neutrophils, and monocytes develop. Lymphoid stem cells begin their development in red bone marrow but complete it in lymphatic tissues. They differentiate into precursor cells from which the T and B lymphocytes develop.

Red Blood Cells

STRUCTURE   Red blood cells (RBCs) or **erythrocytes** (e-RITH-rō-sīts; *erythro-* = red; *-cyte* = cell) are biconcave (concave on both sides) disks averaging about 8 μm* in diameter (Figure 14.3a). Mature RBCs have a simple structure. Without a nucleus or other organelles, they cannot divide or carry on extensive

*1 μm = 1/25,000 of an inch or 1/10,000 of a centimeter (cm), which is 1/1000 of a millimeter (mm).

metabolic activities. Essentially, they consist of a selectively permeable plasma membrane, cytosol, and a red pigment called **hemoglobin.** Hemoglobin carries oxygen to body cells and is responsible for the red color of blood. As you will see later, certain glycolipids and glycoproteins on the surfaces of RBCs are responsible for the various blood groups such as ABO and Rh groups.

Figure 14.3 ■ **Shape of red blood cells and hemoglobin molecule.** In (b), the four polypeptides that make up the globin portion of the molecule are shown in blue; the heme (iron-containing) portions are shown in red. (Modified from R. E. Dickerson and I. Geis.)

🔑 Oxygen binds to the heme portion of hemoglobin for transportation.

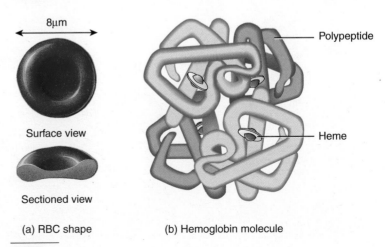

8μm

Surface view

Sectioned view

(a) RBC shape

Polypeptide

Heme

(b) Hemoglobin molecule

 How many molecules of O$_2$ can be transported by one hemoglobin molecule?

RBC FUNCTIONS　The hemoglobin molecules in RBCs consist of a protein called *globin,* composed of four polypeptide chains, plus four groups called *hemes* (Figure 14.3b). Each heme group is bound to one polypeptide chain and contains an iron ion that can combine with one oxygen molecule. The hemes bind the oxygen molecules picked up in the lungs and transport them in the blood. The hemes release oxygen, which diffuses into the interstitial fluid and then into cells. Because RBCs lack mitochondria, they generate ATP anaerobically (without using oxygen) and do not consume any of the oxygen that they transport.

RBC LIFE SPAN AND NUMBER　Red blood cells live only about 120 days because of wear and tear on their plasma membranes as they squeeze through blood capillaries. Worn-out red blood cells are phagocytized by macrophages in the spleen, liver, and red bone marrow (Figure 14.4, step 1). The red blood cell's hemoglobin is subsequently recycled. The globin is split from the heme groups and broken down into amino acids that may be reused by other cells for protein synthesis (step 2). The hemes are broken down into iron and biliverdin (step 3). The iron is carried by a plasma protein called *transferrin* to the liver for storage (steps 4 and 5) until it is needed, when it is released from storage and transported to red bone marrow to be reused for hemoglobin synthesis (steps 6 to 8). The noniron portion of heme is converted into *biliverdin,* a greenish pigment, and then into

bilirubin, which is yellow in color (step 9). Bilirubin enters the blood and is carried to the liver (step 10). Within the liver, bilirubin is secreted into bile, which passes into the small intestine (step 11). As bilirubin passes through the large intestine, bacteria convert it into *urobilinogen* (step 12). Most urobilinogen is converted into *stercobilin,* a brown pigment that gives feces its characteristic color (step 13). Some urobilinogen is absorbed back into the blood and converted to a yellow pigment called *urobilin* (yoo′-rō-BĪ-lin), which is excreted in urine and imparts a yellow color to urine (step 14).

　Jaundice is a yellowish coloration of the "white" of the eye (sclera), the skin, and mucous membranes due to a buildup of bilirubin in the body. Because the liver of a newborn functions poorly for the first week or so, many babies experience a mild form of jaundice called *neonatal jaundice* that disappears as the liver matures. It is usually treated by exposing the infant to blue light, which converts bilirubin into substances the kidneys can excrete.

PRODUCTION OF RBCs　The formation of new RBCs is termed *erythropoiesis* (e-rith′-rō-poy-Ē-sis). Near the end of erythropoiesis, an RBC precursor ejects its nucleus and becomes a *reticulocyte* (re-TIK-yoo-lō-sīt; see Figure 14.2a). Loss of the nucleus causes the center of the cell to indent, producing the RBC's distinctive biconcave shape. Reticulocytes, which are

Figure 14.4　■　**Formation and destruction of red blood cells and recycling of hemoglobin components.**

The rate of RBC formation by red bone marrow normally equals the rate of RBC destruction by macrophages.

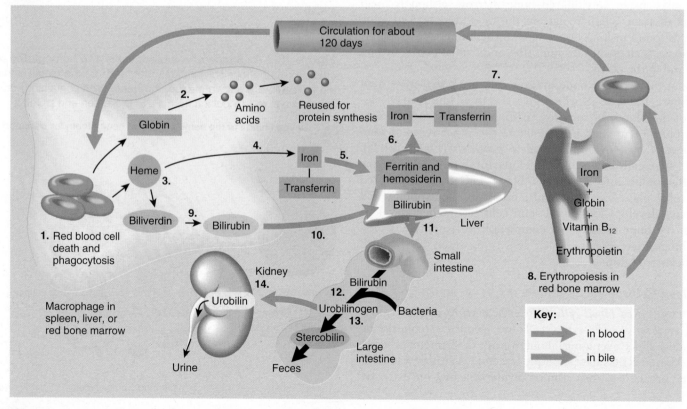

What substance is responsible for the brown color of feces?

about 34% hemoglobin and retain some mitochondria, ribosomes, and endoplasmic reticulum, pass from red bone marrow into the bloodstream. Reticulocytes usually develop into mature RBCs within 1 to 2 days after their release from bone marrow.

Normally, erythropoiesis and destruction of RBCs proceed at the same pace. If the oxygen-carrying capacity of the blood falls because erythropoiesis is not keeping up with RBC destruction, erythrocyte production increases (Figure 14.5). The controlled condition in this particular negative feedback loop is the amount of oxygen delivered to the kidneys (and thus to body tissues in general). *Hypoxia* (hī-POKS-ē-a), a deficiency of oxygen, stimulates increased release of *erythropoietin* (e-rith′-rō-POY-ē-tin), or *EPO,* a hormone made by the kidneys. EPO circulates through the blood to the red bone marrow, where it stimulates erythropoiesis. The larger the number of RBCs in the blood, the higher the oxygen delivery to the tissues (Figure 14.5). A person with prolonged hypoxia may develop a life-threatening condition called *cyanosis* (sī′-a-NŌ-sis), characterized by a bluish-purple skin coloration most easily seen in the nails and mucous membranes. Oxygen delivery may fall due to *anemia* (a lower-than-normal number of RBCs or reduced quantity of hemoglobin) or circulatory problems that reduce blood flow to tissues.

One medical use of EPO is to increase the amount of blood that can be produced and collected by individuals who choose to donate their own blood before surgery. Some athletes, seeking to improve their performance, use erythropoietin to increase their hematocrit, a practice termed *blood doping* that is banned by the International Olympic Committee. Blood doping is dangerous because it thickens the blood and thus increases both the heart's workload and the risk of clotting.

A test that measures the rate of erythropoiesis is called a reticulocyte count. This and several other tests related to red blood cells are explained in Table 14.1. A healthy male has about 5.4 million RBCs per microliter (μL), or cubic millimeter (mm³), of blood, and a healthy female about 4.8 million. (There are about 50 μL in a drop of blood.) The higher value in males is due to their higher level of testosterone, which stimulates the production of more red blood cells. To maintain normal quantities of RBCs, the body must be supplied with adequate amounts of iron, protein, folic acid, and vitamin B_{12} so that it can produce new mature cells at the astonishing rate of 2 million per second. Also, the body must produce enough *intrinsic factor* in the stomach to absorb vitamin B_{12} from the small intestine into the blood.

White Blood Cells

STRUCTURE AND TYPES Unlike red blood cells, *white blood cells (WBCs)* or *leukocytes* (LOO-kō-sīts; *leuko-* = white) have nuclei and do not contain hemoglobin. WBCs are classified as either granular or agranular, depending on whether they contain chemical-filled cytoplasmic vesicles (originally called *granules*). The *granular leukocytes* include *neutrophils* (NOO-trō-fils), *eosinophils* (ē′-ō-SIN-ō-fils), and *basophils* (BĀ-sō-fils). The *agranular leukocytes* include *lymphocytes* and *monocytes* (MON-ō-sīts). (See Table 14.2 for the sizes and microscopic characteristics of WBCs.)

Figure 14.5 ■ **Negative feedback regulation of erythropoiesis (red blood cell formation).**

The main stimulus for erythropoiesis is a decrease in the oxygen-carrying capacity of the blood.

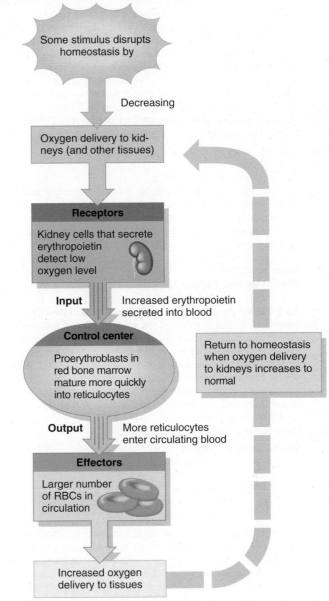

What is the term for oxygen deficiency?

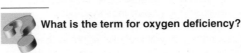

WBC FUNCTIONS The skin and mucous membranes of the body are continuously exposed to microbes (microscopic organisms), such as bacteria, some of which are capable of invading deeper tissues and causing disease. Once microbes enter the body, some WBCs combat them by *phagocytosis,* whereas others produce antibodies. Neutrophils respond first to bacterial invasion, carrying on phagocytosis and releasing enzymes such as lysozyme that destroy certain bacteria. Monocytes take longer to reach the site of infection than neutrophils, but they eventually arrive in larger numbers. Monocytes that migrate into

I. Obtaining Blood Samples

A. Venipuncture. This most frequently used procedure involves withdrawal of blood from a vein using a sterile needle and syringe. (Veins are used instead of arteries because they are closer to the skin, more readily accessible, and contain blood at a much lower pressure.) A commonly used vein is the median cubital vein in front of the elbow (see Figure 16.14 on page 385). A tourniquet is wrapped around the arm, which stops blood flow through the veins and makes the veins below the tourniquet stand out.

B. Fingerstick. Using a sterile needle or lancet, a drop or two of capillary blood is taken from a finger, earlobe, or heel.

C. Arterial stick. Sample is most often taken from radial artery in the wrist or femoral artery in the thigh (see Figure 16.9 on page 395).

II. Testing Blood Samples

A. Reticulocyte count
Normal value: 0.5% to 1.5%.
Indicates the rate of erythropoiesis. A high reticulocyte count might indicate the presence of bleeding or hemolysis (rupture of erythrocytes), or it may be the response of someone who is iron deficient to iron therapy. Low reticulocyte count in the presence of anemia might indicate a malfunction of the red bone marrow, owing to a nutritional deficiency, pernicious anemia, or leukemia.

B. Hematocrit or Hct.
Normal values: Females: 38 to 46 (average 42)
Males: 40 to 54 (average 47)
Hematocrit is the percentage of red blood cells in blood. A hematocrit of 40 means that 40% of the volume of blood is composed of red blood cells. The test is used to diagnose anemia, polycythemia (an increased percentage of red blood cells), and abnormal states of hydration. Anemia may vary from mild (hematocrit of 35) to severe (hematocrit of less than 15). Athletes often have a higher-than-average hematocrit, and the average hematocrit of persons living at high altitude is greater than that of persons living at sea level.

C. Differential white blood cell count

Normal values:	Type of WBC	Percentage
	neutrophils	60–70
	eosinophils	2–4
	basophils	0.5–1
	lymphocytes	20–25
	monocytes	3–8

A high neutrophil count might result from bacterial infections, burns, stress, or inflammation; a low neutrophil count might be caused by radiation, certain drugs, vitamin B_{12} deficiency, or systemic lupus erythematosus (SLE). A high eosinophil count could indicate allergic reactions, parasitic infections, autoimmune disease, or adrenal insufficiency; a low eosinophil count could be caused by certain drugs, stress, or Cushing's syndrome. Basophils could be elevated in some types of allergic responses, leukemias, cancers, and hyperthyroidism; decreases in basophils could occur during pregnancy, ovulation, stress, and hyperthyroidism. High lymphocyte counts could indicate viral infections, immune diseases, and some leukemias; low lymphocyte counts might occur as a result of prolonged severe illness, high steroid levels, and immunosuppression. A high monocyte count could result from certain viral or fungal infections, tuberculosis (TB), some leukemias, and chronic diseases; low monocyte levels rarely occur.

D. Complete blood count (CBC)
The most commonly ordered blood test. It is used to screen patients for anemia and various infections by assessing the red blood cell count, hematocrit, white blood cell count, and platelet count.

infected tissues develop into cells called *wandering macrophages* (*macro-* = large; *-phages* = eaters) that can phagocytize many more microbes than can neutrophils. They also clean up cellular debris following an infection.

Eosinophils leave the capillaries and enter interstitial fluid. They release enzymes that combat inflammation in allergic reactions. Eosinophils also phagocytize antigen–antibody complexes and are effective against certain parasitic worms. A high eosinophil count often indicates an allergic condition or a parasitic infection.

Basophils are also involved in inflammatory and allergic reactions. They leave capillaries, enter tissues, and can liberate heparin, histamine, and serotonin. These substances intensify the inflammatory reaction and are involved in allergic reactions.

Three types of lymphocytes—B cells, T cells, and natural killer cells—are the major combatants in immune responses, which are described in detail in Chapter 17. B cells develop into plasma cells, which produce antibodies that help destroy bacteria and inactivate their toxins. T cells attack viruses, fungi, transplanted cells, cancer cells, and some bacteria. Natural killer cells attack a wide variety of infectious microbes and certain spontaneously arising tumor cells.

White blood cells and other nucleated body cells have proteins, called *major histocompatibility (MHC) antigens,* protruding from their plasma membrane into the extracellular fluid. These "cell identity markers" are unique for each person (except identical twins). Although RBCs possess blood group antigens, they lack the MHC antigens. An incompatible tissue transplant is rejected by the recipient due, in part, to differences in donor and recipient MHC antigens. The MHC antigens are used to type tissues to identify compatible donors and recipients and thus reduce the chance of tissue rejection.

WBC NUMBER AND LIFE SPAN Red blood cells outnumber white blood cells about 700 to 1. There are normally about 5000 to 10,000 WBCs per μL of blood. Bacteria have continuous access to the body through the mouth, nose, and pores of the skin. Furthermore, many cells, especially those of epithelial tissue, age and die daily, and their remains must be removed. However, a WBC can phagocytize only a certain amount of material before it interferes with the WBC's own metabolic activities. Thus, the life span of most WBCs is only a few days. During a period of infection, many WBCs live only a few hours. However, some B and T cells remain in the body for years.

Leukocytosis (loo′-kō-sī-TŌ-sis), an increase in the number of WBCs, is a normal, protective response to stresses such as invading microbes, strenuous exercise, anesthesia, and surgery. Leukocytosis usually indicates an inflammation or infection. Because each type of white blood cell plays a different role, determining the percentage of each type in the blood assists in diagnosing the condition. This test, called a *differential white blood cell count,* measures the number of each kind of white cell in a sample of 100 white blood cells (see Table 14.1). An abnormally low level of white blood cells (below 5000 cells/μL), called *leukopenia* (loo′-kō-PĒ-nē-a), is never beneficial; it may be caused by exposure to radiation, shock, and certain chemotherapeutic agents.

Table 14.2 / Summary of the Formed Elements in Blood

Name and Appearance	Number	Characteristics*	Functions
Red blood cells (RBCs) or erythrocytes	4.8 million/μL (mm³) in females; 5.4 million/μL in males.	7–8 μm diameter, biconcave disks, without a nucleus; live for about 120 days.	Hemoglobin within RBCs transports most of the oxygen carried by the blood.
White blood cells (WBCs) or leukocytes	5000–10,000/μL	Most live for a few hours to a few days.†	Combat pathogens and other foreign substances that enter the body.
Granular leukocytes			
Neutrophils	60–70% of all WBCs.	10–12 μm diameter; nucleus has 2–5 lobes; cytoplasm has very fine, pale lilac granules.	Phagocytosis. Destruction of bacteria with lysozyme and strong oxidants, such as superoxide anion and hydrogen peroxide.
Eosinophils	2–4% of all WBCs.	10–12 μm diameter; nucleus has 2 or 3 lobes; large, red-orange granules fill the cytoplasm.	Combat the effects of histamine in allergic reactions, phagocytize antigen–antibody complexes, and destroy certain parasitic worms.
Basophils	0.5–1% of all WBCs.	8–10 μm diameter; nucleus has 2 lobes; large cytoplasmic granules appear deep blue-purple.	Liberate heparin, histamine, and serotonin in allergic reactions that intensify the overall inflammatory response.
Agranular leukocytes			
Lymphocytes (T cells, B cells, and natural killer cells)	20–25% of all WBCs.	Small lymphocytes are 6–9 μm in diameter; large lymphocytes are 10–14 μm in diameter; nucleus is round or slightly indented; cytoplasm forms a rim around the nucleus that looks sky blue; the larger the cell, the more cytoplasm is visible.	Mediate immune responses, including antigen–antibody reactions. B cells develop into plasma cells, which secrete antibodies. T cells attack invading viruses, cancer cells, and transplanted tissue cells. Natural killer cells attack a wide variety of infectious microbes and certain spontaneously arising tumor cells.
Monocytes	3–8% of all WBCs.	12–20 μm diameter; nucleus is kidney-shaped or horseshoe-shaped; cytoplasm is blue-gray and has foamy appearance.	Phagocytosis (after transforming into fixed or wandering macrophages).
Platelets	150,000–400,000/μL.	2–4 μm diameter cell fragments that live for 5–9 days; contain many granules but no nucleus.	Form platelet plug in hemostasis; release chemicals that promote vascular spasm and blood clotting.

*Colors are those seen when using Wright's stain.   †Some lymphocytes, called T and B memory cells, can live for many years once they are established.

PRODUCTION OF WBCS Leukocytes develop in red bone marrow. As shown in Figure 14.2a, monocytes and granular leukocytes develop from a myeloid stem cell. T and B cells develop from a lymphoid stem cell.

Platelets

Pluripotent stem cells also differentiate into cells that produce platelets (see Figure 14.2a). Some myeloid stem cells develop into cells called megakaryoblasts, which in turn transform into megakaryocytes, huge cells that splinter into 2000–3000 fragments in the red bone marrow and then enter the bloodstream. Each fragment, enclosed by a piece of the megakaryocyte cell membrane, is a *platelet.* Between 150,000 and 400,000 platelets are present in each μL of blood. Platelets are disk-shaped, have a diameter of 2–4 μm, and exhibit many granules but no nucleus. When blood vessels are damaged, platelets help stop blood loss by forming a platelet plug. Their granules also contain chemicals that promote blood clotting. After their short life span of 5–9 days, platelets are removed by macrophages in the spleen and liver.

Table 14.2 presents a summary of the formed elements in blood.

HEMOSTASIS

Objective: • **Describe the various mechanisms that prevent blood loss.**

Hemostasis (hē'-mō-STĀ-sis; *-stasis* = standing still) is a sequence of responses that stops bleeding when blood vessels are injured. (Be sure not to confuse the two words hemostasis and homeostasis.) The hemostatic response must be quick, localized to the region of damage, and carefully controlled. Three mechanisms can reduce loss of blood from blood vessels: (1) vascular spasm, (2) platelet plug formation, and (3) blood clotting (coagulation). When successful, hemostasis prevents *hemorrhage* (HEM-or-ij; *-rhage* = burst forth), the loss of a large amount of blood from the vessels. Hemostasis can prevent hemorrhage from smaller blood vessels, but extensive hemorrhage from larger vessels usually requires medical intervention.

Vascular Spasm

When a blood vessel is damaged, the smooth muscle in its wall contracts immediately, a response called a *vascular spasm.* Vascular spasm reduces blood loss for several minutes to several hours, during which time the other hemostatic mechanisms begin to operate. The spasm is probably caused by damage to the smooth muscle and by reflexes initiated by pain receptors. As platelets accumulate at the damaged site, they release chemicals that enhance vasoconstriction (narrowing of a blood vessel), thus maintaining the vascular spasm.

Platelet Plug Formation

When platelets come into contact with parts of a damaged blood vessel, their characteristics change drastically and they quickly come together to form a platelet plug that helps fill the gap in the injured blood vessel wall. Platelet plug formation occurs as follows (Figure 14.6):

1 Initially, platelets contact and stick to parts of a damaged blood vessel, such as collagen fibers of the connective tissue underlying the damaged endothelial cells. This process is called *platelet adhesion.*

2 As a result of adhesion, the platelets become activated, and their characteristics change dramatically. They extend many projections that enable them to contact and interact with one another, and they begin to liberate the chemicals contained in their granules. This phase is called the *platelet release reaction.* Liberated chemicals activate nearby platelets and sustain the vascular spasm, which decreases blood flow through the injured vessel.

3 The release of platelet chemicals makes other platelets in the area sticky, and the stickiness of the newly recruited and activated platelets causes them to stick to the originally activated platelets. This gathering of platelets is called *platelet aggregation.* Eventually, a large number of platelets forms a mass called a *platelet plug.* A platelet plug can stop blood loss completely if the hole in a blood vessel is small enough.

Because aspirin inhibits vascular spasms and platelet aggregation, low doses of aspirin are useful in reducing the risk of heart attacks and strokes (brain attacks).

Figure 14.6 ■ **Platelet plug formation.**

A platelet plug can stop blood loss completely if the hole in a blood vessel is small.

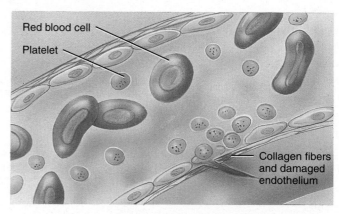

Red blood cell

Platelet

Collagen fibers and damaged endothelium

1 Platelet adhesion

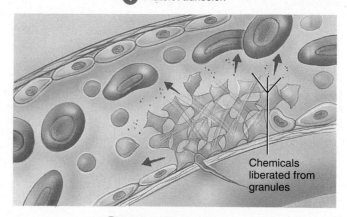

Chemicals liberated from granules

2 Platelet release reaction

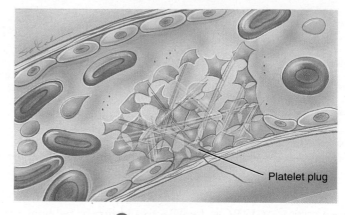

Platelet plug

3 Platelet aggregation

 What are the three mechanisms involved in hemostasis?

Clotting

Normally, blood remains in its liquid form as long as it stays within its vessels. If it is withdrawn from the body, however, it thickens and forms a gel. Eventually, the gel separates from the liquid. The straw-colored liquid, called *serum*, is simply plasma minus the clotting proteins. The gel is called a *clot* and consists of a network of insoluble protein fibers called *fibrin* in which the formed elements of blood are trapped (Figure 14.7)

The process of clot formation, called *clotting* or *coagulation*, is a series of chemical reactions that culminates in formation of fibrin threads. If blood clots too easily, the result can be *thrombosis*, clotting in an unbroken blood vessel. If the blood takes too long to clot, hemorrhage can result.

Clotting is a complex process in which various chemicals known as *clotting factors* (calcium ions, enzymes, and molecules associated with platelets or damaged tissues) activate each other. It occurs in three stages:

Stage 1. *Prothrombinase* is formed.

Stage 2. Prothrombinase converts *prothrombin* (a plasma protein formed by the liver) into the enzyme *thrombin.*

Stage 3. Thrombin converts soluble *fibrinogen* (another plasma protein formed by the liver) into insoluble fibrin. Fibrin forms the threads of the clot. (Cigarette smoke contains substances that interfere with fibrin formation.)

Prothrombinase can be formed in two ways, by either the extrinsic or the intrinsic pathway of blood clotting (Figure 14.8).

Figure 14.7 ■ **Scanning electron micrograph (SEM) of a portion of a blood clot, showing a platelet and red blood cells trapped by fibrin threads.**

A blood clot is a gel that contains formed elements of the blood entangled in fibrin threads.

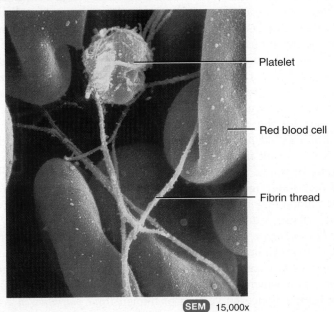

Platelet

Red blood cell

Fibrin thread

SEM 15,000x

What is serum?

Figure 14.8 ■ **Blood clotting.**

During blood clotting, the clotting factors activate each other, resulting in a cascade of reactions that includes positive feedback cycles.

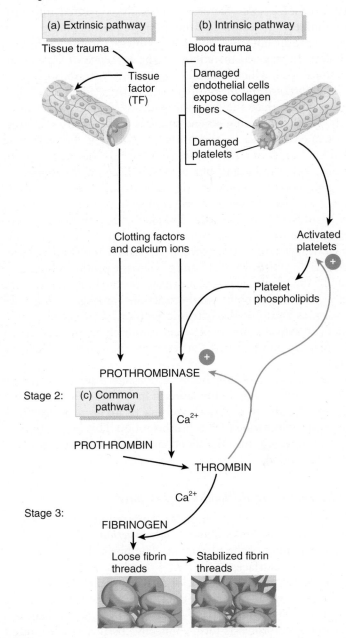

Stage 1:

(a) Extrinsic pathway

Tissue trauma

Tissue factor (TF)

(b) Intrinsic pathway

Blood trauma

Damaged endothelial cells expose collagen fibers

Damaged platelets

Clotting factors and calcium ions

Activated platelets

Platelet phospholipids

PROTHROMBINASE

Stage 2: (c) Common pathway

Ca^{2+}

PROTHROMBIN

THROMBIN

Ca^{2+}

Stage 3:

FIBRINOGEN

Loose fibrin threads → Stabilized fibrin threads

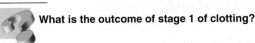

What is the outcome of stage 1 of clotting?

Extrinsic and Intrinsic Pathways

The *extrinsic pathway* of blood clotting occurs rapidly, within seconds. It is so named because damaged tissue cells release a tissue protein called *tissue factor (TF)* into the blood from *outside* (extrinsic to) blood vessels (Figure 14.8a). Following several additional reactions that require calcium ions (Ca^{2+}) and several clotting factors, tissue factor is eventually converted into prothrombinase. This completes the extrinsic pathway (stage 1). In stage 2, prothrombinase and Ca^{2+} convert prothrombin into thrombin. In stage 3, thrombin, in the presence of Ca^{2+}, converts fibrinogen, which is soluble, to fibrin, which is insoluble.

The *intrinsic pathway* of blood clotting (Figure 14.8b) is more complex than the extrinsic pathway, and it occurs more slowly, usually requiring several minutes. The intrinsic pathway is so named because its activators are either in direct contact with blood or contained *within* (intrinsic to) the blood. If endothelial cells lining the blood vessels become roughened or damaged, blood can come in contact with collagen fibers in the adjacent connective tissue. Such contact activates clotting factors. In addition, trauma to endothelial cells activates platelets, causing them to release phospholipids that can also activate certain clotting factors. After several additional reactions that require Ca^{2+} and several clotting factors, prothrombinase is formed. This completes the intrinsic portion of the pathway (stage 1). From this point on, the reactions for stages 2 and 3 are the same as those of the extrinsic pathway. Once thrombin is formed, it activates more platelets, resulting in the release of more platelet phospholipids, an example of a positive feedback cycle.

Clot formation occurs locally; it does not extend beyond the wound site into general circulation. One reason for this is that fibrin has the ability to absorb and inactivate up to nearly 90% of the thrombin formed from prothrombin. This helps stop the spread of thrombin into the blood and thus inhibits clotting except at the wound.

Clot Retraction and Blood Vessel Repair

Once a clot is formed, it plugs the ruptured area of the blood vessel and thus stops blood loss. *Clot retraction* is the consolidation or tightening of the fibrin clot. The fibrin threads attached to the damaged surfaces of the blood vessel gradually contract as platelets pull on them. As the clot retracts, it pulls the edges of the damaged vessel closer together, decreasing the risk of further damage. Permanent repair of the blood vessel can then take place. In time, fibroblasts form connective tissue in the ruptured area, and new endothelial cells repair the vessel lining.

Role of Vitamin K in Clotting

Normal clotting depends on adequate amounts of vitamin K in the body. Although vitamin K is not involved in actual clot formation, it is required for the synthesis of prothrombin and three other clotting factors. Vitamin K, which is normally produced by bacteria that inhabit the large intestine, is a fat-soluble vitamin, and it can normally be absorbed through the lining of the intestine and into the blood. People suffering from disorders that slow absorption of lipids (for example, inadequate release of bile into the small intestine) often experience uncontrolled bleeding as a consequence of vitamin K deficiency.

Hemostatic Control Mechanisms

Many times a day little clots start to form, often at a site of minor roughness inside a blood vessel. Usually, small, inappropriate clots dissolve in a process called *fibrinolysis* (fī'-bri-NOL-i-sis). When a clot is formed, an inactive plasma enzyme called *plasminogen* is incorporated into the clot. Both body tissues and blood contain substances that can activate plasminogen to *plasmin*, an active plasma enzyme. Once plasmin is formed, it can dissolve the clot by digesting fibrin threads. Plasmin also dissolves clots at sites of damage once the damage is repaired.

Several substances that inhibit clotting, called *anticoagulants*, are present in blood to keep it from clotting inappropriately. *Heparin*, an anticoagulant that is produced by mast cells and basophils, inhibits the conversion of prothrombin to thrombin, thereby preventing blood clot formation. Heparin extracted from animal tissues is often used to prevent clotting during and after open heart surgery. *Warfarin (Coumadin)* acts as an antagonist to vitamin K and thus blocks synthesis of four clotting factors. To prevent clotting in donated blood, blood banks and laboratories often add a substance that removes Ca^{2+}, for example, CPD (citrate phosphate dextrose).

Clotting in Blood Vessels

Despite fibrinolysis and the action of anticoagulants, blood clots sometimes form within blood vessels. The endothelial surfaces of a blood vessel may be roughened as a result of *atherosclerosis* (accumulation of fatty substances on arterial walls), trauma, or infection. These conditions also make the platelets that are attracted to the rough spots more sticky. Clots may also form in blood vessels when blood flows too slowly, allowing clotting factors to accumulate in high enough concentrations to initiate a clot.

Clotting in an unbroken blood vessel is called *thrombosis* (*thromb-* = clot; *-osis* = a condition of). The clot itself, called a *thrombus*, may dissolve spontaneously. If it remains intact, however, the thrombus may become dislodged and be swept away in the blood. A blood clot, bubble of air, fat from broken bones, or a piece of debris transported by the bloodstream is called an *embolus* (*em-* = in; *bolus* = a mass; plural is *emboli*). Because emboli often form in veins, where blood flow is slower, the most common site for the embolus to become lodged is in the lungs, a condition called *pulmonary embolism*. Massive emboli in the lungs may result in right ventricular failure and death in a few minutes or hours. An embolus that breaks away from an arterial wall may lodge in a smaller-diameter artery downstream. If it blocks blood flow to the brain, kidney, or heart, the embolus can cause a stroke, kidney failure, or heart attack, respectively.

Thrombolytic agents or "clot busters" are chemical substances that can be injected into the body to dissolve blood clots and restore circulation. Examples of thrombolytic agents include *streptokinase*, which is produced by streptococcal bacteria, and a genetically engineered version of human *tissue plasminogen activator (t-PA)*. These agents are useful in treating heart attacks and strokes that are due to blood clots.

BLOOD GROUPS AND BLOOD TYPES

Objective: • **Describe the ABO and Rh blood groups.**

The surfaces of red blood cells contain a genetically determined assortment of glycolipids and glycoproteins called *isoantigens* (ī′-sō-AN-ti-jenz; *iso-* = self) or *agglutinogens* (ag′-loo-TIN-ō-jenz). Based on the presence or absence of various isoantigens, blood is categorized into different *blood groups.* Within a given blood group there may be two or more different *blood types.* There are at least 24 blood groups and more than 100 isoantigens that can be detected on the surface of red blood cells. Here we discuss two major blood groups: ABO and Rh.

ABO Blood Group

The *ABO blood group* is based on two isoantigens called antigens A and B (Figure 14.9). People whose RBCs display only antigen A have type A blood. Those who have only antigen B are type B. Individuals who have both A and B antigens are type AB, whereas those who have neither antigen A nor B are type O. In about 80% of the population, soluble antigens of the ABO type appear in saliva and other body fluids, in which case blood type can be identified from a sample of saliva. The incidence of ABO blood types varies among different population groups, as indicated in Table 14.3.

Table 14.3 / Incidence of Blood Types in the United States

Population Group	Blood Type (percentage)				
	O	**A**	**B**	**AB**	**Rh⁺**
White	45	40	11	4	85
Black	49	27	20	4	95
Korean	32	28	30	10	100
Japanese	31	38	21	10	100
Chinese	42	27	25	6	100
Native American	79	16	4	1	100

In addition to isoantigens on RBCs, blood plasma usually contains *isoantibodies* or *agglutinins* (a-GLOO-ti-nins) that react with the A or B antigens if the two are mixed. These are *anti-A antibody,* which reacts with antigen A, and *anti-B antibody,* which reacts with antigen B. The antibodies present in each of the four ABO blood types are also shown in Figure 14.9. You do not have antibodies that react with your own isoantigens, but most likely you do have antibodies for any antigens that your RBCs lack. For example, if you have type A blood, it means that you have A antigens on the surfaces of your RBCs, but anti-B antibodies in your blood plasma. If you had anti-A antibodies in your blood plasma, they would attack your RBCs.

Figure 14.9 ■ **Antigens and antibodies involved in the ABO blood grouping system.**

Your plasma does not contain antibodies that could react with the antigens on your red blood cells.

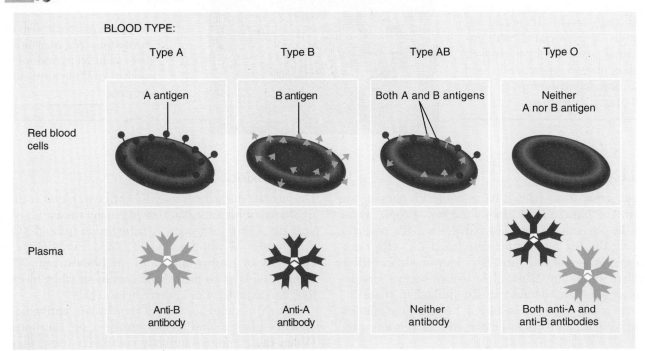

Which antibodies are found in type O blood?

Lifestyle and Blood Circulation— Let It Flow

Many people fear cholesterol, an evil substance that silently accumulates within artery walls, year after year, until it eventually kills its victim by shutting off blood flow to an important organ such as the heart or brain. But cholesterol is not the only villain in this atherosclerosis melodrama. Whereas cholesterol contributes to the formation of arterial plaques, the antagonist delivering the final blow is often a blood clot that forms in a blood vessel and subsequently blocks a narrowed artery, cutting off circulation to the tissues downstream. Fortunately, many of the things you can do to keep your arteries healthy also reduce your risk of blood clots.

Quit Smoking

If you need yet another reason to quit smoking, here it is: Smoking increases blood fibrinogen levels. Increased fibrinogen levels are associated with increased clotting risk. High fibrinogen levels increase platelet aggregation and fibrin deposition, contributing to both clotting and plaque deposition.

Exercise Regularly

Regular physical activity increases plasma volume. An increase in plasma volume means that the blood is more dilute, or "thinner," with a lower percentage of red blood cells and less fibrinogen, and consequently a reduced risk of blood clotting. Several studies have shown that vigorous exercise also reduces platelet stickiness and enhances fibrinolytic activity. These effects may help to explain why active people are at lower risk for heart disease and stroke. A sedentary lifestyle, by contrast, leads to increased clotting risk: Blood thickens as plasma volume decreases. Sedentary people have stickier platelets, which together with

higher levels of fibrinogen are more likely to form blood clots.

Cope Effectively with Stress

Prolonged mental stress impairs fibrinolysis by decreasing the activity of tissue plasminogen activator (t-PA), which helps break down fibrinogen.

Eat a Heart-healthy Diet

People with high blood cholesterol levels exhibit disturbances in coagulation, fibrinolysis, and platelet behavior. Lowering blood lipid levels by diet or drug therapy seems to reverse these disturbances and may be one way that a heart-healthy lifestyle reduces heart disease risk. An interesting study from Denmark found that volunteers who stuck to a low-fat, high-fiber diet showed increased fibrinolytic activity and thus a reduced risk of blood clot formation.

A moderate alcohol intake (one to two drinks per day) has been associated with a reduced heart disease risk. This risk reduction may be due in part to the increase in t-PA level observed in moderate drinkers.

> ▶ **Think It Over**
>
> ▶ Why are people at risk for clot formation told to avoid sitting for extended periods of time, such as on long airplane flights or car rides?

Rh Blood Group

The **Rh blood group** is so named because the Rh antigen was first found in the blood of the rhesus monkey. People whose RBCs have the Rh antigen are designated Rh$^+$ (Rh positive); those who lack the Rh antigen are designated Rh$^-$ (Rh negative). The percentages of Rh$^+$ and Rh$^-$ individuals in various populations are shown in Table 14.3. Under normal circumstances, plasma does not contain anti-Rh antibodies. If an Rh$^-$ person receives an Rh$^+$ blood transfusion, however, the immune system starts to make anti-Rh antibodies that do remain in the blood.

Transfusions

Despite the differences in RBC antigens, blood is the most easily shared of human tissues, saving many thousands of lives every year through transfusions. A **transfusion** (trans-FYOO-shun) is the transfer of whole blood or blood components (red blood cells only or plasma only) into the bloodstream. Most often a transfusion is given to alleviate anemia or when blood volume is low, for example, after a severe hemorrhage.

In an incompatible blood transfusion, antibodies in the recipient's plasma bind to the antigens on the donated RBCs. When these antigen–antibody complexes form, they cause hemolysis and release hemoglobin into the plasma. Consider what happens if a person with type A blood receives a transfusion of type B blood. In this situation, two things can happen. First, the anti-B antibodies in the recipient's plasma can bind to the B antigens on the donor's RBCs, causing hemolysis. Second, the anti-A antibodies in the donor's plasma can bind to the A antigens on the recipient's RBCs. The second reaction is usually not serious because the donor's anti-A antibodies become so diluted in the recipient's plasma that they do not cause any significant hemolysis of the recipient's RBCs.

People with type AB blood do not have any anti-A or anti-B antibodies in their plasma. They are sometimes called "universal recipients" because theoretically they can receive blood from donors of all four ABO blood types. People with type O blood have neither A nor B antigens on their RBCs and are sometimes called "universal donors." Theoretically, because there are no isoantigens on their RBCs for antibodies to attack, they can donate blood to all four ABO blood types. Type O persons requiring blood may receive only type O blood, as they have antibodies to both A and B antigens in their plasma. In practice, use of the terms *universal recipient* and *universal donor* is misleading and dangerous. Blood contains antigens and antibodies other than those associated with the ABO system, and they can cause transfusion problems. Thus, blood should always be carefully matched before transfusion.

• • •

We will next direct attention to the heart, the second major component of the cardiovascular system.

COMMON DISORDERS

Anemia

Anemia is a condition in which the oxygen-carrying capacity of blood is reduced. Many types of anemia exist; all are characterized by reduced numbers of RBCs or a decreased amount of hemoglobin in the blood. The person feels fatigued and is intolerant of cold, both of which are related to lack of oxygen needed for ATP and heat production. Also, the skin appears pale, due to the low content of red-colored hemoglobin circulating in skin blood vessels. Among the most important types of anemia are the following:

- *Iron-deficiency anemia*, the most prevalent kind of anemia, is caused by inadequate absorption of iron, excessive loss of iron, or insufficient intake of iron. Women are at greater risk for iron-deficiency anemia due to monthly menstrual blood loss.

- *Pernicious anemia* is caused by insufficient hemopoiesis resulting from an inability of the stomach to produce intrinsic factor (needed for absorption of dietary vitamin B_{12}).

- *Hemorrhagic anemia* is due to an excessive loss of RBCs through bleeding resulting from large wounds, stomach ulcers, or especially heavy menstruation.

- In *hemolytic anemia*, RBC plasma membranes rupture prematurely. The condition may result from inherited defects or from outside agents such as parasites, toxins, or antibodies from incompatible transfused blood.

- *Thalassemia* (thal'-a-SĒ-mē-a) is a group of hereditary hemolytic anemias in which there is an abnormality in one or more of the four polypeptide chains of the hemoglobin molecule. Thalassemia occurs primarily in populations from countries bordering the Mediterranean Sea.

- *Aplastic anemia* results from destruction of the red bone marrow caused by toxins, gamma radiation, and certain medications that inhibit enzymes needed for hemopoiesis.

Sickle Cell Disease

The RBCs of a person with *sickle cell disease (SCD)* contain an abnormal kind of hemoglobin termed hemoglobin S, or Hb S. When Hb S gives up oxygen to the interstitial fluid, it forms long, stiff, rodlike structures that bend the erythrocyte into a sickle shape. The sickled cells rupture easily. Even though the loss of RBCs stimulates erythropoiesis, it cannot keep pace with hemolysis; hemolytic anemia is the result. Prolonged oxygen reduction may eventually cause extensive tissue damage.

Hemophilia

Hemophilia (*-philia* = loving) is an inherited deficiency of clotting in which bleeding may occur spontaneously or after only minor trauma. Different types of hemophilia are due to deficiencies of different blood clotting factors and exhibit varying degrees of severity. Hemophilia is characterized by spontaneous or traumatic subcutaneous and intramuscular hemorrhaging, nosebleeds, blood in the urine, and hemorrhages in joints that produce pain and tissue damage. Treatment involves transfusions of fresh plasma or concentrates of the deficient clotting factor to relieve the tendency to bleed.

Hemolytic Disease of the Newborn

Hemolytic disease of the newborn (HDN) is a problem that results from Rh incompatibility between a mother and her fetus. Normally, no direct contact occurs between maternal and fetal blood while a woman is pregnant. However, if a small amount of Rh⁺ blood leaks from the fetus through the placenta into the bloodstream of an Rh⁻ mother, her body starts to make anti-Rh antibodies. Because the greatest possibility of fetal blood transfer occurs at delivery, the first-born baby typically is not affected. If the mother becomes pregnant again, however, her anti-Rh antibodies, made after delivery of the first baby, can cross the placenta and enter the bloodstream of the fetus. If the fetus is Rh⁻, there is no problem, because Rh⁻ blood does not have the Rh antigen. If, however, the fetus is Rh⁺, life-threatening *hemolysis* (rupture of RBCs) is likely to occur in the fetal blood. By contrast, ABO incompatibility between a mother and her fetus rarely causes problems because the anti-A and anti-B antibodies do not cross the placenta.

COMMON DISORDERS (CONTINUED)

HDN is prevented by giving all Rh⁻ women an injection of anti-Rh antibodies called anti-Rh gamma globulin (RhoGAM) soon after every delivery, miscarriage, or abortion. These antibodies destroy any Rh antigens that are present so the mother doesn't produce her own antibodies to them. In the case of an Rh⁺ mother, there are no complications, because she cannot make anti-Rh antibodies.

Leukemia

Acute leukemia is a malignant disease of blood-forming tissues characterized by uncontrolled production and accumulation of immature leukocytes. In *chronic leukemia*, mature leukocytes accumulate in the bloodstream because they do not die at the end of their normal life span, crowding out normally functioning WBCs, RBCs, and platelets. The *human T-cell leukemia-lymphoma virus-1 (HTLV-1)* is strongly associated with some types of leukemia. The abnormal accumulation of mature leukocytes may be reduced by radiation treatments and anti-leukemic drugs.

MEDICAL TERMINOLOGY AND CONDITIONS

Autologous preoperative transfusion (aw-TOL-o-gus trans-FYOO-zhun; *auto-* = self) Donating one's own blood in preparation for surgery; can be done up to six weeks before elective surgery. Also called *predonation.*

Blood bank A facility that collects and stores a supply of blood for future use by the donor or others. Because blood banks have now assumed additional and diverse functions (immunohematology reference work, continuing medical education, bone and tissue storage, and clinical consultation), they are more appropriately referred to as *centers of transfusion medicine.*

Hemochromatosis (hē′-mō-krō′-ma-TŌ-sis; *chroma* = color) Disorder of iron metabolism characterized by excess deposits of iron in tissues (especially the liver, heart, pituitary gland, gonads, and pancreas) that result in discoloration (bronzing) of the skin, cirrhosis, diabetes mellitus, and bone and joint abnormalities.

Phlebotomist (fle-BOT-ō-mist; *phlebo-* = vein; *-tom* = cut) A technician who specializes in withdrawing blood.

Polycythemia (pol′-ē-sī-THĒ-mē-a) An abnormal increase in the number of red blood cells in which hematocrit is above 55%, the upper limit of normal.

Septicemia (sep′-ti-SĒ-mē-a; *septic-* = decay; *-emia* = condition of blood) An accumulation of toxins or disease-causing bacteria in the blood. Also called *blood poisoning.*

Thrombocytopenia (throm′-bō-sī′-tō-PĒ-nē-a; *-penia* = poverty) Very low platelet count that results in a tendency to bleed from capillaries.

Venesection (vē′-ne-SEK-shun; *ven-* = vein) Opening of a vein for withdrawal of blood. Although *phlebotomy* (fle-BOT-ō-mē) is a synonym for venesection, in clinical practice, phlebotomy refers to therapeutic bloodletting, such as the removal of some blood from a patient with polycythemia to lower the blood's viscosity.

■ STUDY OUTLINE

Functions of Blood (p.334)

1. Blood transports oxygen, carbon dioxide, nutrients, wastes, and hormones.

2. It helps to regulate pH, body temperature, and water content of cells.

3. It prevents blood loss through clotting and combats microbes and toxins through the action of certain phagocytic white blood cells or specialized plasma proteins.

Physical Characteristics of Whole Blood (p. 334)

1. Physical characteristics of whole blood include a viscosity greater than that of water, a temperature of 38°C (100.4°F), and a pH range between 7.35 and 7.45.

2. Blood constitutes about 8% of body weight in an adult.

Components of Whole Blood (p. 334)

1. Blood consists of 55% plasma and 45% formed elements.

2. The formed elements in blood include red blood cells (erythrocytes), white blood cells (leukocytes), and platelets. Hematocrit (Hct) is the percentage of red blood cells in whole blood.

3. Plasma contains 91.5% water and 8.5% solutes.

4. Principal solutes include proteins (albumins, globulins, fibrinogen), nutrients, hormones, respiratory gases, electrolytes, and waste products.

5. Hemopoiesis, the formation of blood cells from pluripotent stem cells, occurs in red bone marrow.

6. Red blood cells (RBCs) are biconcave discs without nuclei that contain hemoglobin.

7. The function of the hemoglobin in red blood cells is to transport oxygen.

8. Red blood cells live about 120 days. A healthy male has about 5.4 million RBCs/μL of blood; a healthy female about 4.8 million RBCs/μL.

9. After phagocytosis of aged red blood cells by macrophages, hemoglobin is recycled.

10. RBC formation, called erythropoiesis, occurs in adult red bone marrow. It is stimulated by hypoxia, which stimulates release of erythropoietin by the kidneys.

11. A reticulocyte count is a diagnostic test that indicates the rate of erythropoiesis.

12. White blood cells (WBCs) are nucleated cells. The two principal

types are granular leukocytes (neutrophils, eosinophils, basophils) and agranular leukocytes (lymphocytes and monocytes).

13. The general function of WBCs is to combat inflammation and infection. Neutrophils and macrophages (which develop from monocytes) do so through phagocytosis.

14. Eosinophils combat the effects of histamine in allergic reactions, phagocytize antigen–antibody complexes, and combat parasitic worms; basophils develop into mast cells that liberate heparin, histamine, and serotonin in allergic reactions that intensify the inflammatory response.

15. B cells (lymphocytes) are effective against bacteria and other toxins. T cells (lymphocytes) are effective against viruses, fungi, and cancer cells. Natural killer cells attack microbes and tumor cells.

16. White blood cells usually live for only a few hours or a few days. Normal blood contains 5000 to 10,000 WBCs/μL.

17. Platelets are disk-shaped cell fragments without nuclei.

18. They are formed from megakaryocytes and take part in hemostasis by forming a platelet plug.

19. Normal blood contains 250,000 to 400,000 platelets/μL.

Hemostasis (p. 342)

1. Hemostasis refers to the stoppage of bleeding.

2. It involves vascular spasm, platelet plug formation, and blood clotting.

3. In vascular spasm, the smooth muscle of a blood vessel wall contracts.

4. Platelet plug formation is the aggregation of platelets to stop bleeding.

5. A clot is a network of insoluble protein fibers (fibrin) in which formed elements of blood are trapped.

6. The chemicals involved in clotting are known as clotting factors.

7. Blood clotting involves a series of reactions that may be divided into three stages: formation of prothrombinase (prothrombin activator) by either the extrinsic or intrinsic pathway, conversion of prothrombin into thrombin, and conversion of soluble fibrinogen into insoluble fibrin.

8. Normal coagulation requires vitamin K and also involves clot retraction (tightening of the clot) and fibrinolysis (dissolution of the clot).

9. Anticoagulants (for example, heparin) prevent clotting.

10. Clotting in an unbroken blood vessel is called thrombosis. A thrombus that moves from its site of origin is called an embolus.

Blood Groups and Blood Types (p. 345)

1. In the ABO system, the isoantigens on RBCs, termed antigens A and B, determine blood type. Plasma contains isoantibodies termed anti-A and anti-B antibodies.

2. In the Rh system, individuals whose erythrocytes have Rh antigens are classified as Rh$^+$. Those who lack the antigen are Rh$^-$.

◼ SELF-QUIZ

1. A hematocrit is
   a. used to measure the quantity of the five types of white blood cells   **b.** essential for determining a person's blood type   **c.** the percentage of red blood cells in whole blood   **d.** also known as a platelet count   **e.** involved in blood clotting

2. Match the following:
   ____ **a.** involved in certain immune responses
   ____ **b.** develop into mature red blood cells
   ____ **c.** required for vitamin B$_{12}$ absorption
   ____ **d.** most abundant plasma protein
   ____ **e.** blood after formed elements are removed
   ____ **f.** plasma without clotting proteins
   ____ **g.** needed for blood clotting

   A. albumin
   B. fibrinogen
   C. intrinsic factor
   D. immunoglobulins
   E. plasma
   F. serum
   G. reticulocytes

3. In adults, erythropoiesis primarily takes place in
   a. the liver   **b.** yellow bone marrow   **c.** red bone marrow   **d.** lymphatic tissue   **e.** the kidneys

4. Which of the following pigments contributes to the yellow color in urine?
   a. hemoglobin   **b.** stercobilin   **c.** biliverdin   **d.** urobilin   **e.** bilirubin

5. Which of the following statements is NOT true about red blood cells?
   a. The production of red blood cells is known as erythropoiesis.
   b. Red blood cells originate from pluripotent stem cells.
   c. Hypoxia increases the production of red blood cells.
   d. The liver takes part in the destruction and recycling of red blood cell components.   **e.** Red blood cells have a lobed nucleus and granular cytoplasm.

6. A primary function of erythrocytes is to
   a. maintain blood volume   **b.** help blood clot   **c.** provide immunity against some diseases   **d.** clean up debris following infection   **e.** deliver oxygen to the cells of the body

7. If a differential white blood cell count indicated higher than normal numbers of basophils, what may be occurring in the body?
   a. chronic infection   **b.** allergic reaction   **c.** leukopenia   **d.** initial response to invading bacteria   **e.** hemostasis

8. In a person with blood type A, the isoantibodies that would normally be present in the plasma is (are)
   a. anti-A antibody   **b.** anti-B antibody   **c.** both anti-A and anti-B antibodies   **d.** neither anti-A nor anti-B   **e.** anti-O antibodies

9. Hemolytic disease of the newborn (HDN) may result in the fetus of a second pregnancy if
   a. the mother is Rh$^+$ and the baby Rh$^-$ **b.** the mother is Rh$^+$ and the baby Rh$^+$ **c.** the mother is Rh$^-$ and the baby Rh$^-$ **d.** the mother is Rh$^-$ and the baby Rh$^+$ **e.** the father is Rh$^-$ and the mother is Rh$^+$

10. Place the following steps of hemostasis in the correct order.

   1. clot retraction
   2. prothrombinase formed
   3. fibrinolysis by plasmin
   4. vascular spasm
   5. conversion of prothrombin into thrombin
   6. platelet plug formation
   7. conversion of fibrinogen into fibrin

   a. 4, 6, 2, 5, 7, 1, 3   **b.** 1, 4, 7, 6, 2, 3, 1   **c.** 2, 5, 6, 7, 1, 4, 3
   d. 4, 6, 5, 2, 7, 1, 3   **e.** 4, 2, 6, 5, 3, 7, 1

11. In an individual with vitamin K deficiency,

   a. blood vessels undergo vascular spasms   **b.** thrombosis is stimulated   **c.** clotting is inhibited   **d.** hemoglobin cannot be produced   **e.** nutritional anemia develops

12. How does aspirin prevent thrombosis?

   a. It inhibits platelet aggregation.   **b.** It interferes with Ca^{2+} absorption.   **c.** It inhibits the conversion of prothrombin to thrombin.   **d.** It acts as an enzyme to dissolve the thrombus.   **e.** It prevents the accumulation of fatty substances on blood vessel walls.

13. Match the following:

   _____ **a.** become wandering macrophages   **A.** neutrophils

   _____ **b.** produce antibodies   **B.** eosinophils

   _____ **c.** are involved in allergic reactions **C.** basophils

   _____ **d.** first to respond to bacterial invasion **D.** lymphocytes

   _____ **e.** destroy antigen–antibody complexes; **E.** monocytes
             combat inflammation

14. Hemostasis is

   a. maintenance of a steady state in the body   **b.** an abnormal increase in leukocytes   **c.** a hereditary condition in which spontaneous hemorrhaging occurs   **d.** an anticoagulant produced by some leukocytes   **e.** a series of events that stop bleeding

15. Which of the following are mismatched?

   a. white blood cell count below 5000 cells/µL, leukopenia
   b. red blood cell count of 250,000 cells/µL, normal adult male
   c. white blood cell count above 10,000 cells/µL, leukocytosis
   d. platelet count of 300,000 cells/µL, normal adult   **e.** pH 7.4, normal blood

16. An individual with type A blood has _____ in the plasma membranes of red blood cells.

   a. antigen A   **b.** antigen B   **c.** major histocompatibility antigen A   **d.** antigen A and antigen Rh   **e.** antigen B and antigen Rh

17. Mrs. Smith arrives at a health clinic with her ill daughter Beth. It is suspected that Beth has recently developed a bacterial infection. It is likely that Beth's leukocyte count will be _____ cells per microliter of blood, a condition known as _____ . A differential white blood cell count shows an abnormally high percentage of _____ .

   a. 20,000, leukopenia, neutrophils   **b.** 5000, leukocytosis, monocytes   **c.** 7000, leukocytosis, basophils   **d.** 2000, leukopenia, platelets   **e.** 20,000, leukocytosis, neutrophils

18. Clot retraction

   a. draws torn edges of the damaged vessel closer together   **b.** dissolves clots   **c.** is also known as the intrinsic pathway   **d.** involves the formation of fibrin from fibrinogen   **e.** helps prevent the formation of an embolus

19. Persons with blood type AB are sometimes referred to as universal recipients because their blood

   a. lacks A and B antigens   **b.** lacks anti-A and anti-B antibodies   **c.** possesses type O antigens and anti-O antibodies   **d.** has natural immunity to disease   **e.** contains A and B antigens

20. A thrombus that is being transported by the bloodstream is called

   a. a plasma protein   **b.** a platelet   **c.** an embolus   **d.** a wandering macrophage   **e.** a reticulocyte

CRITICAL THINKING APPLICATIONS

1. Biliary atresia is a condition in which the ducts that transport bile out of the liver do not function properly. The whites of the eyes in a baby with this condition have a yellow color. What is the name of the yellow color and what is its cause?

2. A woman with blood type A Rh⁺ is married to a man with blood type B Rh⁻ and is pregnant with their second child. What is the chance the baby will have hemolytic disease of the newborn (HDN)?

3. The school nurse sighed, "I just can't get used to the blue nail polish the kids are wearing. I keep thinking there's a medical problem." What type of problem might result in blue fingernails?

4. Very small numbers of pluripotent stem cells occur normally in blood. If these cells could be isolated and grown in sufficient numbers, what medically useful products could they produce?

ANSWERS TO FIGURE QUESTIONS

14.1 Red blood cells are the most numerous formed element in blood.

14.2 Blood makes up 8% of body weight.

14.3 One hemoglobin molecule can transport four molecules of O_2, one attached to each heme group.

14.4 Stercobilin is responsible for the brown color of feces.

14.5 Hypoxia means cellular oxygen deficiency.

14.6 Hemostasis involves vascular spasm, platelet plug formation, and blood clotting.

14.7 Serum is blood plasma minus the clotting proteins.

14.8 Prothrombinase is formed during stage 1 of clotting.

14.9 Type O blood has anti-A and anti-B antibodies.

Chapter 15

The Cardiovascular System: Heart

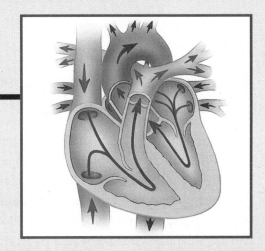

Student Learning Objectives

1. Describe the location of the heart and the structure and functions of the pericardium.   **352**
2. Describe the layers of the heart wall and the chambers of the heart.   **353**
3. Identify the major blood vessels that enter and exit the heart.   **356**
4. Describe the structure and functions of the valves of the heart.   **356**
5. Explain how blood flows through the heart.   **356**
6. Describe the clinical importance of the blood supply of the heart.   **356**
7. Explain how each heartbeat is initiated and maintained.   **358**
8. Describe the meaning and diagnostic value of an electrocardiogram.   **360**
9. Describe the phases of the cardiac cycle.   **361**
10. Define cardiac output, explain how it is calculated, and describe how it is regulated.   **362**
11. Explain the relationship between exercise and the heart.   **365**

A Look Ahead

In the last chapter we examined the composition and functions of blood. For blood to reach body cells and exchange materials with them, it must be constantly pumped by the heart through the body's blood vessels. The left side of the heart pumps blood through an estimated 100,000 km (60,000 mi) of blood vessels. The right side of the heart pumps blood through the lungs, enabling blood to pick up oxygen and unload carbon dioxide. Even while you are sleeping, your heart pumps 30 times its own weight each minute, which amounts to about 5 liters (5.3 qt) to the lungs and the same volume to the rest of the body. At this rate, the heart pumps more than 14,000 liters (3,600 gal) of blood in a day, or 10 million liters (2.6 million gal) in a year. You don't spend all your time sleeping, however, and your heart pumps more vigorously when you are active. Thus, the actual blood volume the heart pumps in a single day is much larger.

The scientific study of the normal heart and the diseases associated with it is **cardiology** (kar′-dē-OL-ō-jē; *cardio-* = heart; *-ology* = study of). This chapter explores the design of the heart and the unique properties that permit it to pump for a lifetime without a moment of rest.

LOCATION AND COVERINGS OF THE HEART

Objective: • **Describe the location of the heart and the structure and functions of the pericardium.**

The **heart** is situated between the two lungs in the thoracic cavity, with about two-thirds of its mass lying to the left of the body's midline (Figure 15.1). Shaped like a blunt, upside-down cone, your heart is about the size of your closed fist. The pointed end of the cone, the *apex*, is formed by the tip of the left ventricle, a lower chamber of the heart, and rests on the diaphragm. The major blood vessels enter and exit at the *base* of the heart, which is formed by the atria (upper chambers of the heart), mostly the left atrium.

Figure 15.1 ■ **Position of the heart and associated blood vessels in the thoracic cavity.** In this and subsequent illustrations, vessels that carry oxygenated blood are colored red; vessels that carry deoxygenated blood are colored blue.

The heart is located between the lungs, with about two-thirds of its mass to the left of the midline.

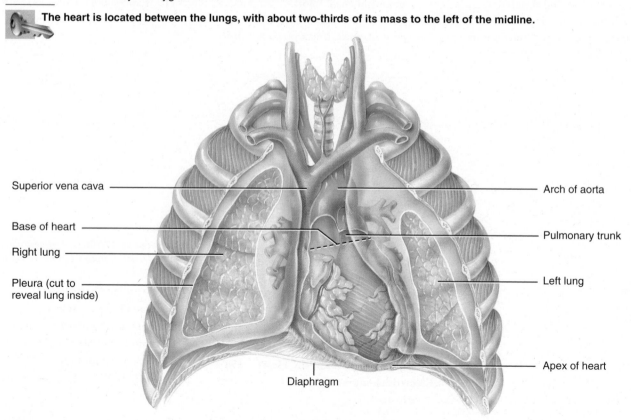

Superior vena cava — Arch of aorta

Base of heart — Pulmonary trunk

Right lung — Left lung

Pleura (cut to reveal lung inside) — Apex of heart

Diaphragm

Anterior view of the heart in the thoracic cavity

 What forms the base of the heart and where is it located?

The membrane that surrounds and protects the heart and holds it in place is the ***pericardium*** (*peri-* = around). It consists of two parts: the fibrous pericardium and the serous pericardium (Figure 15.2). The outer ***fibrous pericardium*** is a tough, inelastic, dense irregular connective tissue. It prevents overstretching of the heart, provides protection, and anchors the heart in place.

The inner ***serous pericardium*** is a thinner, more delicate membrane that forms a double layer around the heart. The outer ***parietal layer*** of the serous pericardium is fused to the fibrous pericardium, whereas the inner ***visceral layer*** of the serous pericardium, also called the ***epicardium*** (*epi-* = on top of), adheres tightly to the surface of the heart. Between the parietal and visceral layers of the serous pericardium is a thin film of fluid. This fluid, known as ***pericardial fluid,*** reduces friction between the membranes as the heart moves. The space that contains the pericardial fluid is the ***pericardial cavity.*** Inflammation of the pericardium is known as ***pericarditis.*** A buildup of pericardial fluid, which may occur in pericarditis, or extensive bleeding into the pericardium are both life-threatening conditions. Because the pericardium cannot stretch, the buildup of fluid or blood compresses the heart. This compression, known as ***cardiac tamponade*** (tam′-pon-ĀD = to plug up), can stop the beating of the heart.

HEART WALL AND CHAMBERS

Objective: • **Describe the layers of the heart wall and the chambers of the heart.**

The wall of the heart (Figure 15.2a) is composed of three layers: epicardium (external layer), myocardium (middle layer), and endocardium (inner layer). The ***epicardium,*** which is also known as the visceral layer of serous pericardium, is the thin, transparent outer layer of the wall. It is composed of mesothelium and connective tissue.

The ***myocardium*** (*myo-* = muscle) consists of cardiac muscle tissue, which constitutes the bulk of the heart. This tissue is found only in the heart and is specialized in structure and function. The myocardium is responsible for the pumping action of the heart. Cardiac muscle fibers (cells) are involuntary, striated, and branched, and the tissue is arranged in interlacing bundles of fibers (Figure 15.2b).

Cardiac muscle fibers form two separate networks—one atrial and one ventricular. Each cardiac muscle fiber connects with other fibers in the networks by thickenings of the sarcolemma (plasma membrane) called ***intercalated discs.*** Within the discs are ***gap junctions*** that allow action potentials to conduct

Figure 15.2 ■ Pericardium and heart wall.

The pericardium is a sac that surrounds and protects the heart.

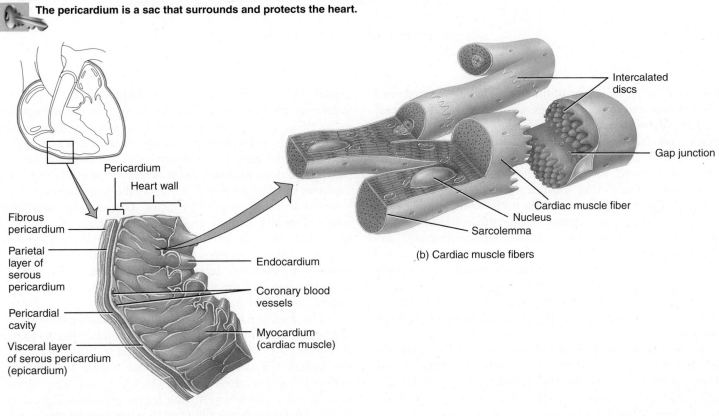

Intercalated discs

Gap junction

Cardiac muscle fiber

Nucleus

Sarcolemma

(b) Cardiac muscle fibers

Pericardium

Heart wall

Fibrous pericardium

Parietal layer of serous pericardium

Pericardial cavity

Visceral layer of serous pericardium (epicardium)

Endocardium

Coronary blood vessels

Myocardium (cardiac muscle)

(a) Portion of pericardium and right ventricular heart wall showing the divisions of the pericardium and layers of the heart wall

 Which layer is both a part of the pericardium and a part of the heart wall?

Figure 15.3 ▪ **Structure of the heart.**

The four chambers of the heart are the two upper atria and two lower ventricles.

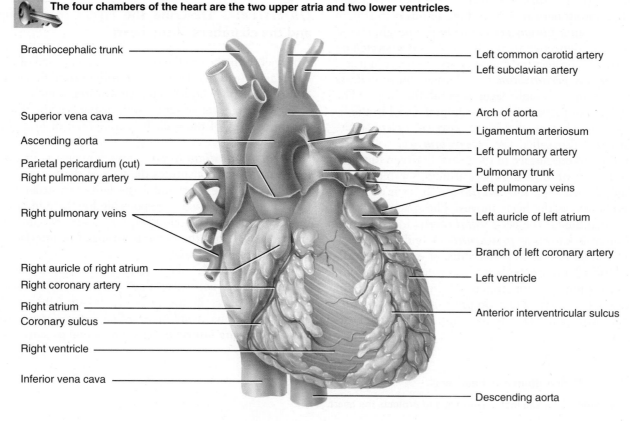

Brachiocephalic trunk
Left common carotid artery
Left subclavian artery
Superior vena cava
Arch of aorta
Ligamentum arteriosum
Ascending aorta
Left pulmonary artery
Parietal pericardium (cut)
Pulmonary trunk
Right pulmonary artery
Left pulmonary veins
Right pulmonary veins
Left auricle of left atrium
Branch of left coronary artery
Right auricle of right atrium
Left ventricle
Right coronary artery
Right atrium
Coronary sulcus
Anterior interventricular sulcus
Right ventricle
Inferior vena cava
Descending aorta

(a) Anterior external view showing surface features

from one cardiac muscle fiber to the next. The intercalated discs also link cardiac muscle fibers to one another so they do not pull apart. Each network contracts as a functional unit, so the atria contract separately from the ventricles. In response to a single action potential, cardiac muscle fibers develop a prolonged contraction, 10–15 times longer than the contraction observed in skeletal muscle fibers. Also, the refractory period of a cardiac fiber lasts longer than the contraction itself. Thus, another contraction of cardiac muscle cannot begin until relaxation is well underway. For this reason, tetanus (maintained contraction) cannot occur in cardiac muscle tissue.

The *endocardium* (*endo-* = within) is a thin layer of simple squamous epithelium that lines the inside of the myocardium and covers the valves of the heart and the tendons attached to the valves. It is continuous with the epithelial lining of the large blood vessels.

The heart contains four chambers (Figure 15.3). The two upper chambers are the *atria* (= entry halls or chambers), and the two lower chambers are the *ventricles* (= little bellies). Between the right atrium and left atrium is a thin partition called the *interatrial septum* (*inter-* = between; *septum* = a dividing wall or partition). Likewise, an *interventricular septum* separates the right ventricle from the left ventricle (Figure 15.3c). On the anterior surface of each atrium is a wrinkled pouchlike structure called an *auricle* (OR-i-kul; *auri-* = ear), so named because of its resemblance to a dog's ear. Each auricle

slightly increases the capacity of an atrium so that it can hold a greater volume of blood. Also on the surface of the heart are a series of grooves, called *sulci* (SUL-sē), that contain coronary blood vessels and a variable amount of fat. Each sulcus (SUL-kus) marks the external boundary between two chambers of the heart. The deep *coronary sulcus* (*coron-* = resembling a crown) encircles most of the heart and marks the boundary between the superior atria and inferior ventricles. The *anterior interventricular sulcus* is a shallow groove on the anterior surface of the heart that marks the boundary between the right and left ventricles. This sulcus continues around to the posterior surface of the heart as the *posterior interventricular sulcus,* which marks the boundary between the ventricles on the posterior aspect of the heart (Figure 15.3b).

The thickness of the myocardium of the chambers varies according to the amount of work each chamber has to perform. The walls of the atria are thin compared to those of the ventricles because the atria need only enough cardiac muscle tissue to deliver blood into the ventricles (Figure 15.3c). The right ventricle pumps blood only to the lungs (pulmonary circulation); the left ventricle pumps blood to all other parts of the body (systemic circulation). The left ventricle must work harder than the right ventricle to maintain the same rate of blood flow, so the muscular wall of the left ventricle is considerably thicker than the wall of the right ventricle.

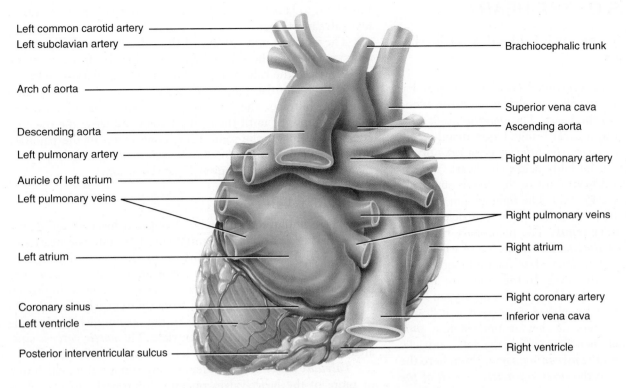

Left common carotid artery

Left subclavian artery

Arch of aorta

Descending aorta

Left pulmonary artery

Auricle of left atrium

Left pulmonary veins

Left atrium

Coronary sinus

Left ventricle

Posterior interventricular sulcus

Brachiocephalic trunk

Superior vena cava

Ascending aorta

Right pulmonary artery

Right pulmonary veins

Right atrium

Right coronary artery

Inferior vena cava

Right ventricle

(b) Posterior external view showing surface features

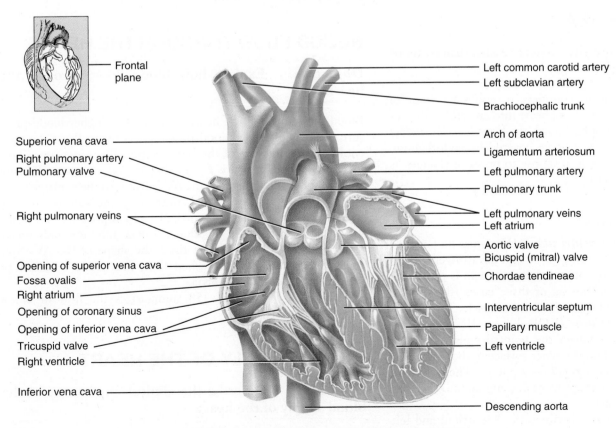

Frontal plane

Superior vena cava

Right pulmonary artery

Pulmonary valve

Right pulmonary veins

Opening of superior vena cava

Fossa ovalis

Right atrium

Opening of coronary sinus

Opening of inferior vena cava

Tricuspid valve

Right ventricle

Inferior vena cava

Left common carotid artery

Left subclavian artery

Brachiocephalic trunk

Arch of aorta

Ligamentum arteriosum

Left pulmonary artery

Pulmonary trunk

Left pulmonary veins

Left atrium

Aortic valve

Bicuspid (mitral) valve

Chordae tendineae

Interventricular septum

Papillary muscle

Left ventricle

Descending aorta

(c) Anterior view of frontal section showing internal anatomy

 Through which type of vessel does blood flow away from the heart?

GREAT VESSELS OF THE HEART

Objective: • **Identify the major blood vessels that enter and exit the heart.**

The right atrium receives *deoxygenated blood* (blood that has given up some of its oxygen to cells) through three *veins,* blood vessels that return blood to the heart. The *superior vena cava* (*vena* = vein; *cava* = hollow, a cave) brings blood mainly from parts of the body above the heart; the *inferior vena cava* brings blood mostly from parts of the body below the heart; and the *coronary sinus* drains blood from most of the vessels supplying the wall of the heart (Figure 15.3b,c). The right atrium then delivers the deoxygenated blood into the right ventricle, which pumps it into the *pulmonary trunk.* The pulmonary trunk divides into a *right* and *left pulmonary artery,* each of which carries blood to the corresponding lung. *Arteries* are blood vessels that carry blood away from the heart. In the lungs, the deoxygenated blood unloads carbon dioxide and picks up oxygen. This *oxygenated blood* (blood that has picked up oxygen as it flowed through the lungs) then enters the left atrium via four *pulmonary veins.* The blood then passes into the left ventricle, which pumps the blood into the *ascending aorta.* From here the oxygenated blood passes into the *coronary arteries, arch of the aorta, thoracic aorta,* and *abdominal aorta* (see Figure 16.9 on page 385). Branches from these blood vessels, the *systemic arteries,* carry blood to all parts of the body.

VALVES OF THE HEART

Objective: • **Describe the structure and functions of the valves of the heart.**

As each chamber of the heart contracts, it pushes a volume of blood into a ventricle or out of the heart into an artery. To prevent the blood from flowing backward, the heart has four *valves* composed of dense connective tissue covered by endothelium. These valves open and close in response to pressure changes as the heart contracts and relaxes.

Atrioventricular Valves

As their names imply, *atrioventricular (AV) valves* lie between the atria and ventricles (Figure 15.3c). The atrioventricular valve between the right atrium and right ventricle is called the *tricuspid valve* because it consists of three cusps (flaps). The pointed ends of the cusps project into the ventricle. Tendonlike cords, called *chordae tendineae* (KOR-dē ten-DIN-ē-ē; *chord-* = cord; *tend-* = tendon), connect the pointed ends to cardiac muscle projections located on the inner surface of the ventricles, called *papillary muscles* (*papill-* = nipple). The chordae tendineae prevent the valve cusps from pushing up into the atria when the ventricles contract.

The atrioventricular valve between the left atrium and left ventricle is called the *bicuspid (mitral) valve.* It has two cusps that work in the same way as the cusps of the tricuspid valve.

For blood to pass from an atrium to a ventricle, an atrioventricular valve must open.

The opening and closing of the valves are due to pressure differences across the valves. When blood moves from an atrium to a ventricle, the valve is pushed open, the papillary muscles relax, and the chordae tendineae slacken (Figure 15.4a). When a ventricle contracts, the pressure of the ventricular blood drives the cusps upward until their edges meet and close the opening (Figure 15.4b). At the same time, contraction of the papillary muscles and tightening of the chordae tendineae help prevent the cusps from swinging upward into the atrium.

Semilunar Valves

Near the origin of the pulmonary trunk and aorta are *semilunar valves* that prevent blood from flowing back into the heart (see Figure 15.3c). Each valve consists of three semilunar (half-moon-shaped) cusps that attach to the artery wall. Like the atrioventricular valves, the semilunar valves permit blood to flow in one direction only—in this case, from the ventricles into the arteries. The *pulmonary valve* lies in the opening where the pulmonary trunk leaves the right ventricle. The *aortic valve* is situated at the opening between the left ventricle and the aorta.

Valvular heart disease refers to any condition in which one or more of the heart valves operates improperly. Leaky valves can allow *regurgitation,* the backflow of blood through an incompletely closed valve. This could lead to congestive heart failure, in which the heart can no longer pump enough to supply the oxygen demands of the body.

BLOOD FLOW THROUGH THE HEART

Objective: • **Explain how blood flows through the heart.**

Blood flows through the heart from areas of higher blood pressure to areas of lower blood pressure. As the walls of the atria contract, the pressure of the blood within them increases. This increased blood pressure forces the AV valves open, allowing atrial blood to flow through the AV valves into the ventricles.

After the atria are finished contracting, the walls of the ventricles contract, increasing ventricular blood pressure and pushing blood through the semilunar valves into the pulmonary trunk and aorta. At the same time, the shape of the AV valve cusps causes them to be pushed shut, preventing backflow of ventricular blood into the atria.

Figure 15.5 on page 358 summarizes the flow of blood through the heart.

BLOOD SUPPLY OF THE HEART

Objective: • **Describe the clinical importance of the blood supply of the heart.**

The wall of the heart, like any other tissue, has its own blood vessels. The flow of blood through the numerous vessels in the myocardium is called **coronary (cardiac) circulation.** The principal coronary vessels are the **left** and **right coronary arteries,** which originate as branches of the ascending aorta (see Figure 15.3a). Each artery branches and then branches again to deliver oxygen and nutrients throughout the heart muscle. Most of the deoxygenated blood, which carries carbon dioxide and wastes, is collected by a large vein on the posterior surface of the heart, the **coronary sinus** (see Figure 15.3b), which empties into the right atrium.

Most heart problems result from faulty coronary circulation. Local anemia due to obstruction of blood vessels is called **ischemia** (is-KĒ-mē-a; *isch-* = to hold back; *-emia* = in the blood). Ischemia of the myocardium often causes chest pain, **angina pectoris** (an-JĪ-na, or AN-ji-na, PEK-to-ris; *angina* = pain; *pectoris* = chest), but some people with obstructed coro-

nary arteries have "silent ischemia," ischemia that occurs without the warning sensation of pain. Common causes include stress, strenuous exertion after a heavy meal, narrowing of coronary arteries, high blood pressure, and fever. Symptoms include chest pain, accompanied by tightness or pressure, and difficulty breathing. Sometimes weakness, dizziness, and perspiration occur.

A much more serious problem is **myocardial infarction** (in-FARK-shun), or **MI,** commonly called a heart attack. **Infarction** is the death of an area of tissue because of an interrupted blood supply. Myocardial infarction may result from a blood clot in one of the coronary arteries. Tissue beyond the blood clot dies and is replaced by noncontractile scar tissue, which causes the heart muscle to lose some of its strength. The aftereffects depend partly on the size and location of the infarct (dead area). In

Figure 15.4 ■ **Atrioventricular (AV) valves.** The bicuspid and tricuspid valves operate in a similar manner.

Heart valves open and close in response to pressure changes as the heart contracts and relaxes.

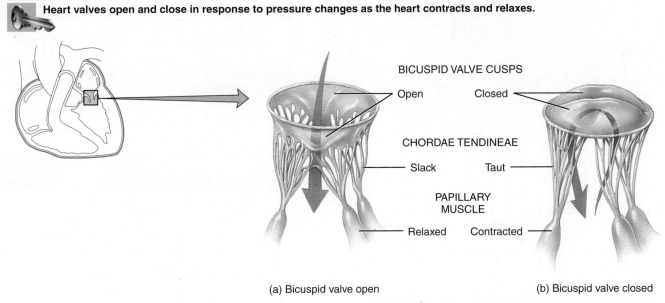

BICUSPID VALVE CUSPS

Open — Closed

CHORDAE TENDINEAE

Slack — Taut

PAPILLARY MUSCLE

Relaxed — Contracted

(a) Bicuspid valve open   (b) Bicuspid valve closed

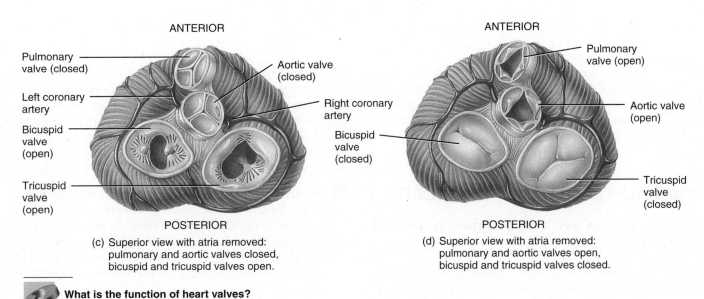

ANTERIOR

Pulmonary valve (closed)

Aortic valve (closed)

Left coronary artery

Right coronary artery

Bicuspid valve (open)

Tricuspid valve (open)

POSTERIOR

(c) Superior view with atria removed: pulmonary and aortic valves closed, bicuspid and tricuspid valves open.

ANTERIOR

Pulmonary valve (open)

Aortic valve (open)

Bicuspid valve (closed)

Tricuspid valve (closed)

POSTERIOR

(d) Superior view with atria removed: pulmonary and aortic valves open, bicuspid and tricuspid valves closed.

What is the function of heart valves?

Figure 15.5 ■ **Blood flow through the heart.**

The right and left coronary arteries deliver blood to the heart; the coronary veins drain blood from the heart into the coronary sinus.

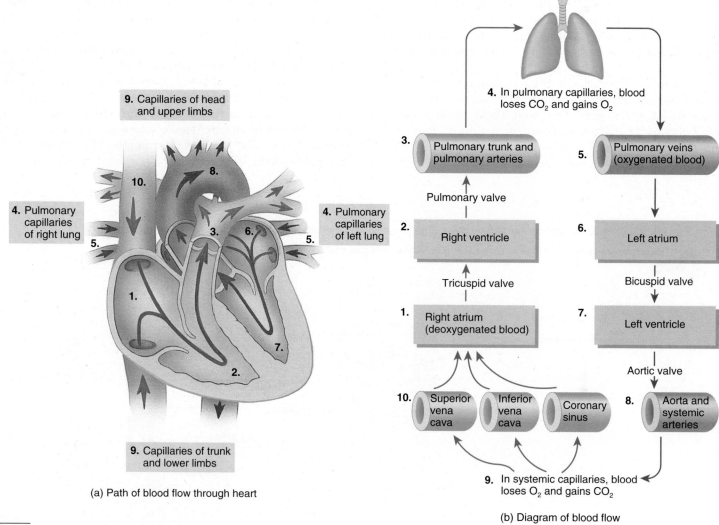

(a) Path of blood flow through heart

(b) Diagram of blood flow

Which veins deliver deoxygenated blood into the right atrium?

addition to destroying normal heart tissue, an infarction may disturb the conduction of action potentials through the heart that cause the heart to beat. An MI can cause sudden death but a person may be rescued by timely cardiopulmonary resuscitation (CPR).

CONDUCTION SYSTEM OF THE HEART

Objective: • **Explain how each heartbeat is initiated and maintained.**

About 1% of the cardiac muscle fibers repeatedly and rhythmically generate action potentials. These cells have two important

functions: They act as a *pacemaker*, setting the rhythm for the entire heart, and they form the *conduction system,* the route for propagating action potentials throughout the heart muscle. The conduction system ensures that cardiac chambers are stimulated to contract in a coordinated manner, which makes the heart an effective pump. Cardiac action potentials propagate through the following components of the conduction system (Figure 15.6):

❶ Normally, cardiac excitation begins in the *sinoatrial (SA) node,* located in the right atrial wall just inferior to the opening of the superior vena cava. An action potential spontaneously arises in the SA node and then conducts throughout both atria via gap junctions in the intercalated discs of atrial fibers. In the wake of the action potential, the atria contract.

2 By conducting along atrial muscle fibers, the action potential also reaches the ***atrioventricular (AV) node,*** located in the interatrial septum, just anterior to the opening of the coronary sinus. At the AV node, the action potential slows considerably, providing time for the atria to empty their blood into the ventricles.

3 From the AV node, the action potential enters the ***atrioventricular (AV) bundle*** (also known as the **bundle of His**), in the interventricular septum. The AV bundle is the only site where action potentials can conduct from the atria to the ventricles.

4 After conducting along the AV bundle, the action potential then enters both the ***right*** and ***left bundle branches*** that course through the interventricular septum toward the apex of the heart.

5 Finally, large-diameter ***conduction myofibers*** also known as ***Purkinje fibers*** (pur-KIN-jē) rapidly conduct the action potential, first to the apex of the ventricles and then upward to the remainder of the ventricular myocardium. Then, a fraction of a second after the atria contract, the ventricles contract.

The SA node initiates action potentials 90 to 100 times per minute, faster than any other region of the conducting system. Thus, the SA node sets the rhythm for contraction of the heart—it is the *pacemaker* of the heart. Various hormones and neurotransmitters can speed or slow pacing of the heart by SA node fibers. In a person at rest, for example, acetylcholine released by the parasympathetic division of the ANS typically slows SA node pacing to about 75 action potentials per minute, causing 75 heartbeats per minute. If the SA node becomes diseased or damaged, the slower AV node fibers can become the pacemaker. With pacing by the AV node, however, heart rate is slower, only 40 to 50 beats/min. If the activity of both nodes is suppressed, the heartbeat may still be maintained by the AV bundle, a bundle branch, or conduction myofibers. These fibers generate action potentials very slowly, about 20 to 40 times per minute. At such a low heart rate, blood flow to the brain is inadequate.

Figure 15.6 ■ **Conduction system of the heart.** The SA node, located in the right atrial wall, is the heart's pacemaker, initiating cardiac action potentials that cause contraction of the heart's chambers. The arrows indicate the flow of action potentials through the atria. The route of action potentials through the numbered components of the conduction system is described in the text.

The conduction system ensures that cardiac chambers contract in a coordinated manner.

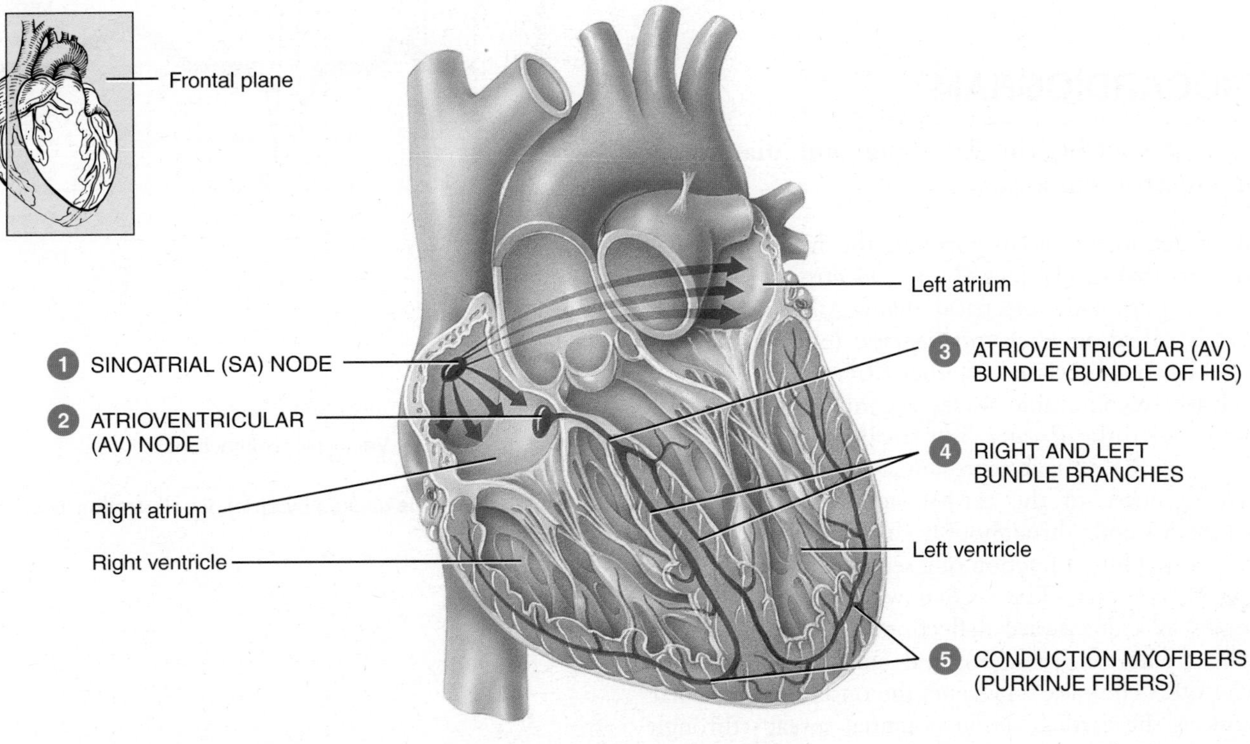

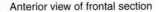

Anterior view of frontal section

 Which component of the conduction system provides the only route for action potentials to conduct between the atria and the ventricles?

Table 15.1 / Summary of Conduction System's Structures, Locations, and Functions

Structure	Location	Function
Sinoatrial (SA) node	Right atrial wall.	Initiates each heartbeat and sets basic pace for heart rate. Sends action potential to both atria, causing them to contract.
Atrioventricular (AV) node	Interatrial septum.	Picks up action potential from SA node and passes it to the atrioventricular (AV) bundle.
Atrioventricular (AV) bundle (bundle of His)	Interventricular septum.	Picks up action potential from AV node and passes it to right and left bundle branches.
Right and left bundle branches	Interventricular septum.	Picks up action potential and passes it to conduction myofibers.
Conduction myofibers (Purkinje fibers)	Ventricular myocardium.	Picks up action potential from bundle branches and passes it to ventricular myocardial cells, causing ventricles to contract.

Sometimes, a site other than the SA node develops abnormal self-excitability and becomes the pacemaker. Such a site is called an ***ectopic pacemaker*** (ek-TOP-ik; *ectop-* = displaced). An ectopic pacemaker may operate only occasionally, producing an irregular heartbeat, or it may pace the heart for some period of time. Triggers of ectopic activity include caffeine and nicotine, electrolyte imbalances, hypoxia, and toxic reactions to drugs such as digitalis. An irregular heart rhythm may be restored to normal with an ***artificial pacemaker,*** a device that sends out small electrical charges to the right atrium, right ventricle, or both, to stimulate the heart.

A summary of the conduction system is presented in Table 15.1.

ELECTROCARDIOGRAM

Objective: • **Describe the meaning and diagnostic value of an electrocardiogram.**

Conduction of action potentials through the heart generates electrical currents that can be picked up by electrodes placed on the skin. A recording of the electrical changes that accompany the heartbeat is called an ***electrocardiogram*** (e-lek′-trō-KAR-dē-ō-gram), which is abbreviated as either ***ECG*** or ***EKG.***

Three clearly recognizable waves accompany each heartbeat. The first, called the ***P wave,*** is a small upward deflection on the ECG (Figure 15.7); it represents atrial depolarization, the depolarizing phase of the cardiac action potential as it spreads from the SA node throughout both atria. Depolarization causes contraction. Thus, a fraction of a second after the P wave begins, the atria contract. The second wave, called the ***QRS complex,*** begins as a downward deflection (Q); continues as a large, upright, triangular wave (R); and ends as a downward wave (S). The QRS complex represents the onset of ventricular depolarization, as the cardiac action potential spreads through the ventricles. Shortly after the QRS complex begins, the ventricles start to contract. The third wave is a dome-shaped upward deflection called the ***T wave;*** it indicates ventricular repolariza-

Figure 15.7 ■ **Normal electrocardiogram (ECG) of a single heartbeat.** P wave = atrial depolarization; QRS complex = onset of ventricular depolarization; T wave = ventricular repolarization.

> An electrocardiogram is a recording of the electrical activity that initiates each heartbeat.

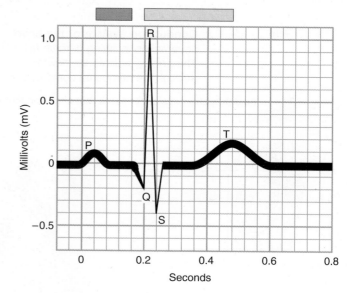

Key:

Atrial contraction

Ventricular contraction

What event is caused by atrial depolarization, seen as the P wave in the ECG?

Sudden Cardiac Death During Exercise— What's the Risk?

Regular physical activity helps keep hearts and arteries healthy. But very rarely, strenuous activity can precipitate a heart attack. In adults of middle age and older, a heart attack, or myocardial infarction, typically occurs because a blood clot lodges in an artery of the heart already narrowed by atherosclerotic plaque. During exercise, heart rate and blood pressure increase. Under this stress, an unstable plaque may rupture, stimulating the clotting process as the body tries to repair the damaged artery.

How Risky Is Exercise?

Many studies have been conducted in an attempt to quantify the risk imposed by strenuous exercise. Researchers have concluded that, in general, risk of heart attack is about two to six times higher during strenuous exercise than during light physical activity or rest. The statistical risk of heart attack varies considerably depending on a person's exercise history. Risk is lowest for those who exercise regularly and highest for people unaccustomed to exercise. Risk during exercise also rises with the number and severity of other cardiovascular risk factors. For example, people already diagnosed with heart disease are 10 times as likely to have a heart attack during exercise as apparently healthy individuals. Though this may sound discouraging, the risk can be seen from another point of view. Overall, the incidence of death during physical activity is very low, about 6 deaths per 100,000 middle-aged men per year.

Exercise is also safe for most people recovering from heart attacks. In 167 supervised cardiac rehabilitation exercise programs, the incidence of heart attack in 51,000 patients was 1 in 294,000 person-hours (number of people multiplied by number of hours each exercised). Incidence of death was only 1 in 784,000 person-hours.

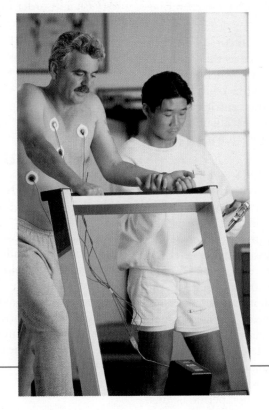

Who Is at Risk?

Although these figures illustrate that risk of heart attack and death are quite low, if the death occurs to someone you love, it happens 100% and is still a tragic loss. People with diagnosed or suspected artery disease are most at risk and should check with their physicians before starting an exercise program. Risk can be reduced by exercising regularly (several times per week) at a low to moderate intensity, and by heeding any warning signs of cardiovascular disease, such as chest pain or pressure, abnormal heart rhythms, or dizziness.

▶ *Think It Over*

▶ Why do you think exercise stress tests are used to help diagnose coronary artery disease?

---

tion and occurs just before the ventricles start to relax. Repolarization of the atria is not usually evident in an ECG because it is masked by the larger QRS complex.

Variations in the size and duration of the waves of an ECG are useful in diagnosing abnormal cardiac rhythms and conduction patterns and in following the course of recovery from a heart attack. An ECG can also reveal the presence of a living fetus.

THE CARDIAC CYCLE

Objective: • **Describe the phases of the cardiac cycle.**

A single *cardiac cycle* includes all the events associated with one heartbeat. In a normal cardiac cycle, the two atria contract while the two ventricles relax; then, while the two ventricles contract, the two atria relax. The term *systole* (SIS-tō-lē = contraction) refers to the phase of contraction; the phase of relaxation is *diastole* (dī-AS-tō-lē = dilation or expansion). A cardiac cycle consists of systole and diastole of both atria plus systole and diastole of both ventricles.

Phases

For the purposes of our discussion, we will divide the *cardiac cycle* into three phases (Figure 15.8):

❶ *Relaxation period.* The relaxation period begins at the end of a cardiac cycle when the ventricles start to relax and all four chambers are in diastole. Repolarization of the ventricular muscle fibers (T wave in the ECG) initiates relaxation. As the ventricles relax, pressure within them drops. When ventricular pressure drops below atrial pressure, the AV valves open and ventricular filling begins. About 75% of the ventricular filling occurs after the AV valves open and before the atria contract.

❷ *Atrial systole (contraction).* An action potential from the SA node causes atrial depolarization, noted as the P wave in the ECG. Atrial systole follows the P wave, which marks the end of the relaxation period. As the atria contract, they force the last 25% of the blood into the ventricles. At the end of atrial systole, each ventricle contains about 130 mL. The AV valves are still open and the semilunar valves are still closed.

❸ *Ventricular systole (contraction).* The QRS complex in the ECG indicates ventricular depolarization, which leads to contraction of the ventricles. Ventricular contraction pushes blood against the AV valves, forcing them shut. As ventricular contraction continues, pressure inside the chambers quickly rises. When left ventricular pressure surpasses aortic pressure at about 80 millimeters of mercury (mm Hg) and right ventricular pressure rises above the pressure in the pulmonary trunk (about 20 mm Hg), both semilunar valves open, and ejection of blood from the heart begins. Ejection continues until the ventricles start to relax. At rest, the volume of blood ejected from each ventricle during ventricular systole is about 70 mL (a little more than 2 oz). When the ventricles begin to relax, ventricular pressure drops, the semilunar valves close, and another relaxation period begins.

Timing

At rest, each cardiac cycle lasts about 0.8 sec. In one complete cycle, the first 0.4 sec of the cycle is the relaxation period, when all four chambers are in diastole. Then, the atria are in systole for 0.1 sec and in diastole for the next 0.7 sec. After atrial systole, the ventricles are in systole for 0.3 sec and in diastole for 0.5 sec. When the heart beats faster, for instance during exercise, the relaxation period is shorter.

Heart Sounds

The sound of the heartbeat comes primarily from turbulence in blood flow created by the closure of the valves, not from the contraction of the heart muscle. The first sound (S1, see Figure 15.8d), **lubb,** is a long, booming sound from the AV valves closing after ventricular systole begins. The second sound, a short, sharp sound (S2), **dupp,** is from the semilunar valves closing at the end of ventricular systole. There is a pause during the relaxation period. Thus, the cardiac cycle is heard as lubb, dupp, pause; lubb, dupp, pause; lubb, dupp, pause.

Heart sounds provide valuable information about the mechanical operation of the heart. A **heart murmur** is an abnormal sound consisting of a rushing or gurgling noise that is heard before, between, or after the normal heart sounds, or that may mask the normal heart sounds. Although some heart murmurs are not associated with a significant heart problem, most often a murmur indicates a valve disorder.

One example of a valve disorder is **mitral stenosis,** narrowing of the mitral valve by scar formation or congenital defect. Another cause of a heart murmur is **mitral valve prolapse (MVP),** a disorder in which a portion of a mitral valve is pushed back too far (prolapsed) during ventricular contraction and the cusps do not close properly. As a result, a small volume of blood may flow back into the left atrium during ventricular systole. Mitral valve prolapse often does not pose a serious threat and is found in up to 10% of otherwise healthy young persons.

CARDIAC OUTPUT

Objective: • **Define cardiac output, explain how it is calculated, and describe how it is regulated.**

The volume of blood ejected per minute from the left ventricle into the aorta is called the *cardiac output (CO).* (Note that the same amount of blood is also ejected from the right ventricle into the pulmonary trunk.) Cardiac output is determined by (1) the *stroke volume (SV),* the amount of blood ejected by the left ventricle during each beat (contraction), and (2) the number of heart beats per minute. In a resting adult, stroke volume averages 70 mL, and heart rate is about 75 beats per minute. Thus the average cardiac output in a resting adult is

$$
\begin{aligned}
\text{Cardiac output} &= \text{stroke volume} \times \text{beats per minute} \\
&= 70 \text{ mL/beat} \times 75 \text{ beats/min} \\
&= 5250 \text{ mL/min} \quad \text{or} \quad 5.25 \text{ liters/min}
\end{aligned}
$$

Factors that increase stroke volume or heart rate, such as exercise, increase cardiac output.

Regulation of Stroke Volume

Although some blood is always left in the ventricles at the end of their contraction, a healthy heart pumps out all the blood that has entered its chambers during the previous diastole. The more blood that returns to the heart during diastole, the more blood that is ejected during the next systole. Three factors regulate stroke volume and ensure that the left and right ventricles pump equal volumes of blood:

Figure 15.8 ■ **Cardiac cycle.** (a) ECG. (b) Left atrial, left ventricular, and aortic pressure changes along with the opening and closing of valves. (c) Left ventricular volume changes. (d) Heart sounds. (e) Phases of the cardiac cycle.

🔑 A cardiac cycle is composed of all the events associated with one heartbeat.

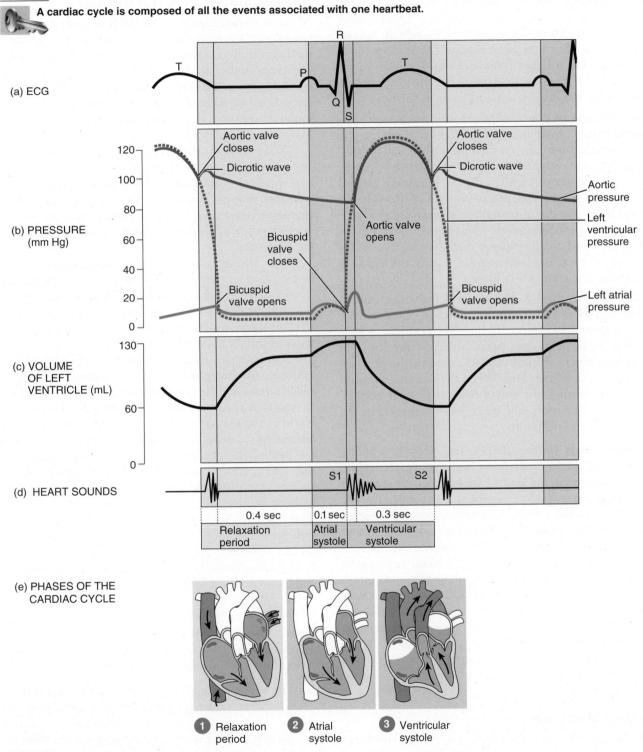

(a) ECG

(b) PRESSURE (mm Hg)

(c) VOLUME OF LEFT VENTRICLE (mL)

(d) HEART SOUNDS

(e) PHASES OF THE CARDIAC CYCLE

① Relaxation period ② Atrial systole ③ Ventricular systole

What is the term used for the contraction phase of the cardiac cycle? the relaxation phase?

1. **The degree of stretch in the heart before it contracts.** Within limits, the more the heart is stretched as it fills during diastole, the greater the force of contraction during systole, a relationship known as **Starling's law of the heart.** The situation is somewhat like stretching a rubber band: The more you stretch the heart, the more forcefully it contracts. If the left side of the heart pumps a little more blood than the right side, a larger volume of blood returns to the right ventricle. On the next beat the right ventricle contracts more forcefully, and the two sides are again in balance.

2. **The forcefulness of contraction of individual ventricular muscle fibers.** Even at a constant degree of stretch, the heart can contract more or less forcefully when certain substances are present. Stimulation of the sympathetic division of the autonomic nervous system (ANS), hormones such as epinephrine and norepinephrine, increased Ca^{2+} level in the interstitial fluid, and the drug digitalis all increase the force of contraction of cardiac muscle fibers. In contrast, inhibition of the sympathetic division of the ANS, anoxia, acidosis, some anesthetics, and increased K^+ level in the extracellular fluid decrease contraction force.

3. **The pressure required to eject blood from the ventricles.** The semilunar valves open and ejection of blood from the heart begins when pressure in the right ventricle exceeds the pressure in the pulmonary trunk (about 20 mm Hg), and when the pressure in the left ventricle exceeds the pressure in the aorta (about 80 mm Hg). When the required pressure is higher than normal, the valves open later than normal, stroke volume decreases, and more blood remains in the ventricles at the end of systole.

In **congestive heart failure (CHF)**, the heart is a failing pump. It pumps blood less and less effectively, leaving more blood in the ventricles at the end of each cycle. As the heart continues to be overstretched, it contracts less forcefully. The result is a positive feedback loop: Less-effective pumping leads to even lower pumping capability. Often, one side of the heart starts to fail before the other. If the left ventricle fails first, it can't pump out all the blood it receives, and blood backs up in the lungs. The result is *pulmonary edema*, fluid accumulation in the lungs that can lead to suffocation. If the right ventricle fails first, blood backs up in the systemic blood vessels. In this case, the resulting *peripheral edema* is usually most noticeable as swelling in the feet and ankles. Common causes of CHF are coronary artery disease (see page 366), long-term high blood pressure, myocardial infarctions, and valve disorders.

Regulation of Heart Rate

Adjustments to the heart rate are important in the short-term control of cardiac output and blood pressure. If left to itself, the sinoatrial node would set a constant heart rate of 90 to 100 beats/min. However, tissues require different volumes of blood flow under different conditions. During exercise, for example, cardiac output rises to supply working tissues with increased amounts of oxygen and nutrients. The most important factors in the regulation of heart rate are the autonomic nervous system and the hormones epinephrine and norepinephrine, released by the adrenal glands.

Autonomic Regulation of Heart Rate

Nervous system regulation of the heart originates in the **cardiovascular (CV) center** in the medulla oblongata. This region of the brain stem receives input from a variety of sensory receptors and from higher brain centers, such as the limbic system and cerebral cortex. The cardiovascular center then directs appropriate output by increasing or decreasing the frequency of nerve impulses sent out to both the sympathetic and parasympathetic branches of the ANS (Figure 15.9).

Neurons from the CV center extend down the spinal cord and provide input to sympathetic preganglionic neurons that pass out of the spinal cord. They then synapse with **cardiac accelerator nerves,** bundles of sympathetic postganglionic axons that innervate the conduction system, atria, and ventricles. The norepinephrine released by cardiac accelerator nerves increases the heart rate.

Also arising from the cardiovascular center are parasympathetic neurons that reach the heart via the **vagus (X) nerves.** These parasympathetic neurons extend to the conduction system and atria. The neurotransmitter they release—acetylcholine (ACh)—decreases the heart rate by slowing the pacemaking activity of the SA node.

Several types of sensory receptors help adjust heart rate. For example, **baroreceptors** (*baro-* = pressure), neurons sensitive to blood pressure changes, are strategically located in the arch of the aorta and carotid arteries (arteries in the neck that supply blood to the brain). If there is an increase in blood pressure, the baroreceptors send nerve impulses along sensory neurons that are part of the glossopharyngeal (IX) and vagus (X) nerves to the CV center (Figure 15.9). The cardiovascular center responds by putting out more nerve impulses along the parasympathetic (motor) neurons that are also part of the vagus nerves. The resulting decrease in heart rate lowers cardiac output and thus lowers blood pressure. If blood pressure falls, baroreceptors do not stimulate the cardiovascular center. As a result of this lack of stimulation, heart rate increases, cardiac output increases, and blood pressure increases to the normal level.

Chemical Regulation of Heart Rate

Certain chemicals influence both the basic physiology of cardiac muscle and its rate of contraction. Chemicals with major effects on the heart fall into one of two categories:

1. **Hormones.** Epinephrine and norepinephrine (from the adrenal medullae) enhance the heart's pumping effectiveness by increasing both heart rate and contraction force. Exercise, stress, and excitement cause the adrenal medullae to release more hormones. Thyroid hormones also increase heart rate. One sign of hyperthyroidism (excessive levels of thyroid hormone) is tachycardia (elevated heart rate at rest).

2. **Ions.** Elevated blood levels of K^+ or Na^+ decrease heart rate and contraction force. A moderate increase in extracellular and intracellular Ca^{2+} level increases heart rate and contraction force.

Figure 15.9 ■ **Autonomic nervous system regulation of heart rate.**

The cardiovascular center in the medulla oblongata controls both sympathetic and parasympathetic nerves that innervate the heart.

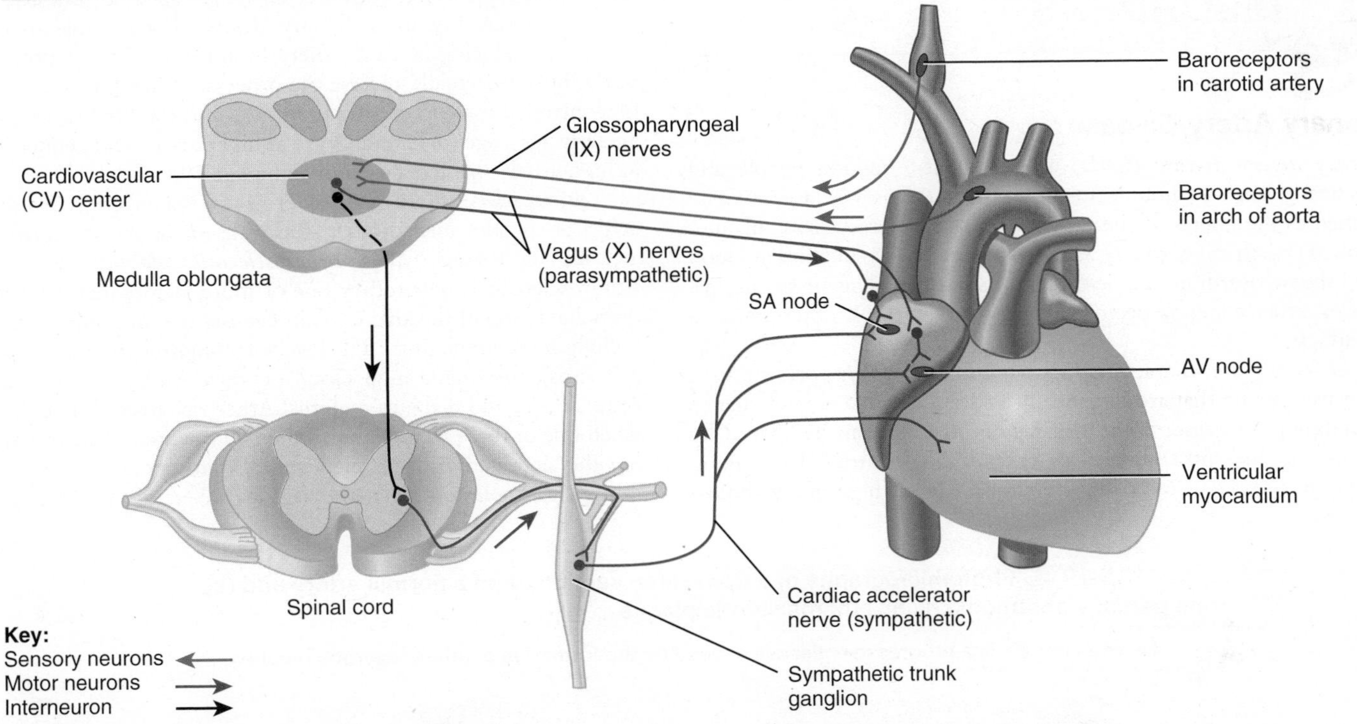

Cardiovascular (CV) center

Medulla oblongata

Glossopharyngeal (IX) nerves

Vagus (X) nerves (parasympathetic)

SA node

Baroreceptors in carotid artery

Baroreceptors in arch of aorta

AV node

Ventricular myocardium

Spinal cord

Cardiac accelerator nerve (sympathetic)

Sympathetic trunk ganglion

Key:
Sensory neurons
Motor neurons
Interneuron

What effect does acetylcholine, released by parasympathetic nerves, have on heart rate?

Other Factors in Heart Rate Regulation

Age, gender, physical fitness, and body temperature also influence resting heart rate. A newborn baby is likely to have a resting heart rate over 120 beats per minute; the rate then declines throughout childhood to the adult level of 75 beats per minute. Adult females generally have slightly higher resting heart rates than adult males, although regular exercise tends to bring resting heart rate down in both sexes. As adults age, their heart rate may increase.

Increased body temperature, such as occurs during fever or strenuous exercise, increases heart rate by causing the SA node to discharge more rapidly. Decreased body temperature decreases heart rate and force of contraction. During surgical repair of certain heart abnormalities, it is helpful to slow a patient's heart rate by deliberately cooling the body.

EXERCISE AND THE HEART

Objective: • **Explain the relationship between exercise and the heart.**

A person's level of fitness can be improved at any age with regular exercise. Some types of exercise are more effective than oth-

ers for improving the health of the cardiovascular system. *Aerobic exercise,* any activity that works large body muscles for at least 20 minutes, elevates cardiac output and accelerates metabolic rate. Three to five such sessions a week are usually recommended for improving the health of the cardiovascular system. Brisk walking, running, bicycling, cross-country skiing, and swimming are examples of aerobic activities.

Sustained exercise increases the oxygen demand of the muscles. Whether the demand is met depends primarily on the adequacy of cardiac output and proper functioning of the respiratory system. After several weeks of training, a healthy person increases maximal cardiac output, thereby increasing the maximal rate of oxygen delivery to the tissues. Hemoglobin level increases and skeletal muscles develop more capillary networks, enhancing oxygen delivery. A physically fit person may even exhibit *bradycardia,* a resting heart rate under 60 beats per minute. A slowly beating heart is more energy efficient than one that beats more rapidly.

During strenuous activity, a well-trained athlete can achieve cardiac output double that of a sedentary person, in part because training causes hypertrophy (enlargement) of the heart. Regular exercise also helps to reduce blood pressure, anxiety, and depression; control weight; and increase the body's ability to dissolve blood clots.

COMMON DISORDERS

Coronary Artery Disease

Coronary artery disease (CAD) affects about 7 million people and causes nearly half a million deaths in the United States each year. CAD is defined as the effects of the accumulation of atherosclerotic plaques (described shortly) in coronary arteries that lead to a reduction in blood flow to the myocardium. Some individuals have no signs or symptoms, others experience angina pectoris (chest pain), and still others suffer a heart attack.

Risk factors are characteristics, symptoms, or signs present in a disease-free person that are statistically associated with a greater chance of developing a disease. Some risk factors in CAD can be altered by changing diet and other habits, or they can be controlled by taking medications. Among these risk factors for CAD are high blood choles-terol level, high blood pressure, cigarette smoking, obesity, diabetes, "type A" personality, and sedentary lifestyle. For example, moderate exercise and reducing fat in the diet can not only stop the progression of CAD, but can actually reverse the process. Other risk factors that can't be controlled include genetic predisposition (family history of CAD at an early age), age (older adults are at higher risk), and gender (males are at higher risk than females until about age 70).

Atherosclerosis (ath'-er-ō-skler-Ō-sis) is a progressive disease characterized by the formation in the walls of large- and medium-sized arteries of lesions called *atherosclerotic plaques* (Figure 15.10). Atherosclerosis is initiated by one or more factors that damage the endothelial lining of the arterial wall. Factors that may initiate the process include high circulating LDL levels, cytomegalovirus (a common herpes virus), prolonged high blood pressure, carbon monoxide in cigarette smoke, and diabetes mellitus. Atherosclerosis is thought to begin when one of these factors injures the endothelium of an artery, promoting the aggregation of platelets and also attracting phagocytes. At the injury site, cholesterol and triglycerides collect in the inner layer of the

Figure 15.10 ■ **Photomicrographs of a transverse section of (a) a normal artery and (b) one partially obstructed by an atherosclerotic plaque.**

Atherosclerosis is a progressive disease caused by the formation of atherosclerotic plaques.

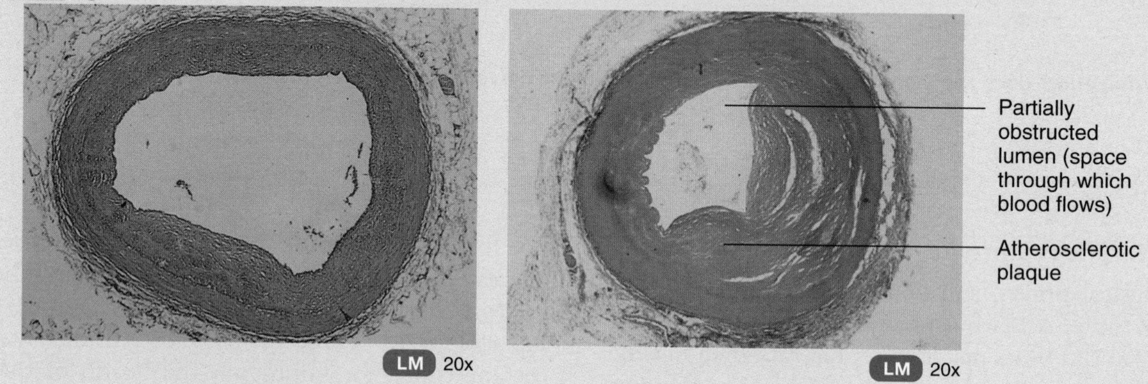

(a) Normal artery	(b) Obstructed artery
LM 20x	LM 20x

Partially obstructed lumen (space through which blood flows)

Atherosclerotic plaque

 What substances are part of an atherosclerotic plaque?

arterial wall. As part of the inflammatory process, macrophages also arrive at the site. Contact with platelets, lipids, and other components of blood stimulates smooth muscle cells and collagen fibers in the arterial wall to proliferate abnormally. In response, an atherosclerotic plaque develops, progressively obstructing blood flow as it enlarges. A plaque may provide a roughened surface that attracts platelets, initiating clot formation and further obstructing blood flow.

Treatment options for CAD include drugs (nitroglycerin, betablockers, and cholesterol-lowering and clot-dissolving agents) and various surgical and nonsurgical procedures designed to increase the blood supply to the heart.

Congenital Defects

A defect that exists at birth (and usually before) is a *congenital defect.* Among the several congenital defects that affect the heart are the following:

- In *patent ductus arteriosus,* the ductus arteriosus (temporary blood vessel) between the aorta and the pulmonary trunk, which normally closes shortly after birth, remains open.

- *Interventricular septal defect* is caused by an incomplete closure of the interventricular septum.

- *Valvular stenosis* is a narrowing of one of the valves regulating blood flow in the heart.

- *Tetralogy of Fallot* (te-TRAL-Ō-jē of fa-LŌ) is a combination of four defects: an interventricular septal defect, an aorta that emerges from both ventricles instead of from the left ventricle only, a narrowed pulmonary semilunar valve, and an enlarged right ventricle.

Arrhythmias

Arrhythmia (a-RITH-mē-a; *a-* = without) refers to an irregularity in heart rhythm resulting from a disturbance in the conduction system of the heart. Arrhythmias are caused by factors such as caffeine, nicotine, alcohol, anxiety, certain drugs, hyperthyroidism, potassium deficiency, and certain heart diseases. One serious arrhythmia is called a *heart block.* The most common heart block occurs in the atrioventricular node, which conducts impulses from the atria to the ventricles. This disturbance is called *atrioventricular (AV) block.*

In *atrial flutter,* the atrial rhythm averages between 240 and 360 beats per minute. The condition is essentially rapid atrial contractions accompanied by AV block. *Atrial fibrillation* is an uncoordinated contraction of the atrial muscles. When the muscle fibrillates, the muscle fibers of the atrium quiver individually instead of contracting together, canceling out the pumping of the atrium. *Ventricular fibrillation (VF)* is characterized by asynchronous, haphazard, ventricular muscle contractions. Ventricular ejection ceases, and circulatory failure and death occur.

MEDICAL TERMINOLOGY AND CONDITIONS

Cardiac angiography (an′-jē-OG-ra-fē) Procedure in which a cardiac catheter (plastic tube) is used to inject a radiopaque contrast medium into blood vessels or heart chambers. The procedure may be used to visualize blood vessels and the ventricles to assess structural abnormalities. Angiography can also be used to inject clot-dissolving drugs, such as streptokinase or tissue plasminogen activator (t-PA), into a coronary artery to dissolve an obstructing thrombus.

Cardiac arrest (KAR-dē-ak a-REST) A clinical term meaning cessation of an effective heartbeat. The heart may be completely stopped or in ventricular fibrillation.

Cardiac catheterization (kath′-e-ter-i-ZĀ-shun) Procedure that is used to visualize the heart's coronary arteries, chambers, valves, and great vessels. It may also be used to measure pressure in the heart and blood vessels; to assess cardiac output; and to measure the flow of blood through the heart and blood vessels, the oxygen content of blood, and the status of the heart valves and conduction system. The basic procedure involves inserting a catheter into a peripheral vein (for right heart catheterization) or artery (for left heart catheterization) and guiding it under fluoroscopy (x-ray observation).

Cardiomegaly (kar′-dē-ō-MEG-a-lē; *mega-* = large) Heart enlargement.

Cor pulmonale (CP) (kor pul-mōn-ALE; *cor-* = heart; *pulmon-* = lung) Right ventricular hypertrophy caused by hypertension (high blood pressure) in the pulmonary circulation.

Cardiopulmonary resuscitation (kar′-dē-ō-PUL-mō-ner-ē re-sus′-i-TĀ-shun) *(CPR)* The artificial establishment of normal or near-normal respiration and circulation. The *ABCs* of cardiopulmonary resuscitation are *Airway, Breathing,* and *Circulation,* meaning the rescuer must establish an airway, provide artificial ventilation if breathing has stopped, and reestablish circulation if there is inadequate cardiac action.

Incompetent valve (in-KOM-pe-tent VALV) Any valve that does not close properly, thus permitting a backflow of blood; also called *valvular insufficiency.*

Palpitation (pal′-pi-TĀ-shun) A fluttering of the heart or abnormal rate or rhythm of the heart.

Paroxysmal tachycardia (par′-ok-SIZ-mal tak′-e-KAR-dē-a) A period of rapid heartbeats that begins and ends suddenly.

Sudden cardiac death The unexpected cessation of circulation and breathing due to an underlying heart disease such as ischemia, myocardial infarction, or a disturbance in cardiac rhythm.

■ STUDY OUTLINE

Location and Coverings of the Heart (p. 352)

1. The heart is situated between the lungs, with about two-thirds of its mass to the left of the midline.

2. The pericardium consists of an outer fibrous layer and an inner serous pericardium.

3. The serous pericardium is composed of a parietal and visceral layer.

4. Between the parietal and visceral layers of the serous pericardium is the pericardial cavity, a space filled with pericardial fluid that reduces friction between the two membranes.

Heart Wall and Chambers (p. 353)

1. The wall of the heart has three layers: epicardium, myocardium, and endocardium.

2. The chambers include two upper atria and two lower ventricles.

Great Vessels of the Heart (p. 356)

1. The blood flows through the heart from the superior and inferior venae cavae and the coronary sinus to the right atrium, through the tricuspid valve to the right ventricle, through the pulmonary trunk to the lungs.

2. From the lungs blood flows through the pulmonary veins into the left atrium, through the bicuspid valve to the left ventricle, and out through the aorta.

Valves of the Heart (p. 356)

1. Four valves prevent backflow of blood in the heart.

2. Atrioventricular (AV) valves, between the atria and their ventricles, are the tricuspid valve on the right side of the heart and the bicuspid (mitral) valve on the left.

3. The atrioventricular valves, chordae tendineae, and their papillary muscles stop blood from flowing back into the atria.

4. The two arteries that leave the heart each have a semilunar valve.

Blood Flow Through the Heart (p. 356)

1. Blood flows through the heart from areas of higher pressure to areas of lower pressure.

2. The pressure is related to the size and volume of a chamber.

3. The movement of blood through the heart is controlled by the opening and closing of the valves and the contraction and relaxation of the myocardium.

Blood Supply of the Heart (p. 356)

1. Coronary circulation delivers oxygenated blood to the myocardium and removes carbon dioxide from it.

2. Deoxygenated blood returns to the right atrium via the coronary sinus.

3. Malfunctions of this system can result in angina pectoris or myocardial infarction (MI).

Conduction System of the Heart (p. 358)

1. The conduction system consists of specialized cardiac muscle tissue that generates and distributes action potentials.

2. Components of this system are the sinoatrial (SA) node (pacemaker), atrioventricular (AV) node, atrioventricular (AV) bundle (bundle of His), bundle branches, and conduction myofibers (Purkinje fibers).

Electrocardiogram (p. 360)

1. The record of electrical changes during each cardiac cycle is referred to as an electrocardiogram (ECG).

2. A normal ECG consists of a P wave (depolarization of atria), QRS complex (onset of ventricular depolarization), and T wave (ventricular repolarization).

3. The ECG is used to diagnose abnormal cardiac rhythms and conduction patterns.

Cardiac Cycle (p. 361)

1. A cardiac cycle consists of systole (contraction) and diastole (relaxation) of the chambers of the heart.

2. The phases of the cardiac cycle are (a) the relaxation period, (b) atrial systole, and (c) ventricular systole.

3. A complete cardiac cycle takes 0.8 sec at an average heartbeat of 75 beats per minute.

4. The first heart sound (lubb) represents the closing of the atrioventricular valves. The second sound (dupp) represents the closing of semilunar valves.

Cardiac Output (p. 362)

1. Cardiac output (CO) is the amount of blood ejected by the left ventricle into the aorta each minute: CO = stroke volume × beats per minute.

2. Stroke volume (SV) is the amount of blood ejected by a ventricle during ventricular systole. It is related to stretch on the heart before it contracts, forcefulness of contraction, and the amount of pressure required to eject blood from the ventricles.

3. Nervous control of the cardiovascular system originates in the cardiovascular center in the medulla oblongata.

4. Sympathetic impulses increase heart rate and force of contraction; parasympathetic impulses decrease heart rate.

5. Heart rate is affected by hormones (epinephrine, norepinephrine, thyroid hormones), ions (Na^+, K^+, Ca^{2+}), age, gender, physical fitness, and body temperature.

Exercise and the Heart (p. 365)

1. Sustained exercise increases oxygen demand on muscles.

2. Among the benefits of aerobic exercise are increased maximal cardiac output, decreased blood pressure, weight control, and increased ability to dissolve clots.

SELF-QUIZ

1. Match the following:
   ____ **a.** valve between the left atrium and left ventricle

   ____ **b.** valve between the right atrium and right ventricle

   ____ **c.** chamber that pumps blood to the lungs

   ____ **d.** chamber that pumps blood into aorta

   ____ **e.** chamber that receives oxygenated blood from lungs

   ____ **f.** chamber that receives deoxygenated blood from body

   ____ **g.** valve between the left ventricle and aorta

   ____ **h.** valve between right ventricle and pulmonary trunk

   A. aortic valve
   B. right atrium
   C. left atrium
   D. bicuspid (mitral) valve
   E. pulmonary valve
   F. right ventricle
   G. left ventricle
   H. tricuspid valve

2. Which of the following statements describes the pericardium?

   a. It is a layer of simple squamous epithelium. **b.** It lines the inside of the myocardium. **c.** It is continuous with the epithelial lining of the large blood vessels. **d.** It is responsible for the contraction of the heart. **e.** It anchors the heart to the diaphragm.

3. Which blood vessel delivers deoxygenated blood from the head and neck to the heart?

   a. pulmonary vein **b.** thoracic aorta **c.** pulmonary artery **d.** inferior vena cava **e.** superior vena cava

4. An embolus originating in the coronary sinus would first enter the

   a. right atrium **b.** pulmonary veins **c.** left atrium **d.** right ventricle **e.** aorta

5. The chordae tendineae and papillary muscles of the heart

   a. are responsible for connecting cardiac muscle fibers for the spread of action potentials **b.** can develop self-excitability and stimulate contraction **c.** help prevent the atrioventricular valves from protruding into the atria when the ventricles contract **d.** help anchor and protect the heart **e.** form the cusps (flaps) of the heart valves

6. Which chamber of the heart has the thickest layer of myocardium?

   a. right ventricle **b.** right atrium **c.** left ventricle **d.** left atrium **e.** coronary sinus

7. The normal "pacemaker" of the heart is the

   a. sinoatrial (SA) node **b.** atrioventricular (AV) node **c.** conduction myofibers (Purkinje fibers) **d.** atrioventricular (AV) bundle **e.** right bundle branch

8. In normal heart action,

   a. the right atrium and ventricle contract, followed by the contraction of the left atrium and ventricle **b.** the order of contraction is right atrium, then right ventricle, then left atrium, then left ventricle **c.** the two atria contract together, and then the two ventricles contract together **d.** the right atrium and left ventricle contract, followed by the contraction of the left atrium and right ventricle **e.** all four chambers of the heart contract and then relax simultaneously

9. Heart sounds are produced by

   a. contraction of the myocardium **b.** closure of the heart valves **c.** the flow of blood in the coronary arteries **d.** the flow of blood in the ventricles **e.** the transmission of action potentials through the conduction system

10. Heart rate and strength of contraction are controlled by the cardiovascular center, which is located in the

    a. cerebrum **b.** pons **c.** right atrium **d.** medulla **e.** atrioventricular node

11. The portion of the ECG that corresponds to atrial depolarization is the

    a. R peak **b.** space between the T wave and P wave **c.** T wave **d.** P wave **e.** QRS complex

12. The opening of the semilunar valves is due to the pressure in the

    a. ventricles exceeding the pressure in the aorta and pulmonary trunk **b.** ventricles exceeding the pressure in the atria **c.** atria exceeding the pressure in the ventricles **d.** atria exceeding the pressure in the aorta and pulmonary trunk **e.** aorta and pulmonary trunk exceeding the pressure in the ventricles

13. On the surface of the heart, the boundaries between the atria and ventricles can be distinguished by the

    a. anterior interventricular sulcus **b.** coronary sulcus **c.** auricles **d.** coronary arteries **e.** posterior interventricular sulcus

14. Starling's law of the heart

    a. is important in maintaining equal blood output from both ventricles **b.** is used in reference to the force of contraction of the atria **c.** results in a decreased heart rate **d.** causes blood to accumulate in the lungs **e.** is related to the stretching of the cardiac muscle cells in the atria

15. Which of the following sequences best represents the pathway of an action potential through the heart's conduction system?

    1. sinoatrial (SA) node
    2. conduction myofibers (Purkinje fibers)
    3. atrioventricular (AV) bundle
    4. atrioventricular (AV) node
    5. right and left bundle branches
    a. 1, 4, 3, 2, 5 **b.** 4, 1, 3, 5, 2 **c.** 3, 4, 1, 2, 5 **d.** 1, 4, 3, 5, 2 **e.** 2, 5, 3, 4, 1

16. Which of the following is NOT true concerning the ventricular filling parts of the cardiac cycle?

    a. The atrioventricular (AV) valves are open. **b.** The ventricles fill to 75% of their capacity before the atria contract. **c.** The remaining 25% of the ventricular blood is forced into the ventricles when the atria contract. **d.** The semilunar valves are open. **e.** Ventricular filling begins when the ventricular pressure drops below the atrial pressure, causing the AV valves to open.

17. Cardiac output

    a. equals stroke volume (SV) × blood pressure (BP) **b.** equals stroke volume (SV) × beats per minute **c.** is calculated using the formula for Starling's law of the heart **d.** is about 70 mL in the average adult male **e.** equals blood pressure (BP) × beats per minute

18. Most heart problems are due to
    a. old age   b. leakages at the valves   c. problems in the coronary circulation   d. the failure of the conduction system   e. infections in the heart coverings

19. Using the situations below, indicate if the heart rate would speed up **(A)** or slow down **(B)**.
    _____ a. sympathetic stimulation of the sinoatrial (SA) node
    _____ b. decrease in blood pressure
    _____ c. fever
    _____ d. parasympathetic stimulation of the heart's conduction system
    _____ e. release of epinephrine
    _____ f. elevated K^+ level
    _____ g. release of acetylcholine
    _____ h. strenuous exercise
    _____ i. stimulation by vagus nerve
    _____ j. fear, anger, stress
    _____ k. cooling the body
    _____ l. hypoxia
    _____ m. excessive thyroid hormones

20. Match the following:
    _____ a. may cause heart murmur
    _____ b. heart compression
    _____ c. inflammation of heart covering
    _____ d. heart chamber contraction
    _____ e. chest pain from ischemia
    _____ f. heart attack
    _____ g. heart chamber relaxation

    A. pericarditis
    B. mitral valve prolapse
    C. myocardial infarction
    D. angina pectoris
    E. diastole
    F. systole
    G. cardiac tamponade

CRITICAL THINKING APPLICATIONS

1. Your uncle had an artificial pacemaker inserted after his last bout with heart trouble. What is the function of a pacemaker? For which heart structure does the pacemaker substitute?

2. Deanna was strolling across a four-lane highway when a car suddenly appeared out of nowhere. As she finished sprinting across the road, she felt her heart racing. Trace the route of the signal from her brain to her heart.

3. Jean-Claude, a member of the college's cross-country ski team, volunteered to have his heart function evaluated by the exercise physiology class. His resting pulse rate was 40 beats per minute.

Assuming that he has an average cardiac output (CO), determine Jean-Claude's stroke volume (SV). Next, Jean-Claude rode an exercise bike until his heart rate had risen to 60 beats per minute. Assuming that his SV stayed constant, calculate Jean-Claude's CO during this moderate exercise.

4. After climbing to the top of the hill, Rosa popped a nitroglycerin tablet in her mouth to ease her chest pains. After a few minutes rest, she slowly continued her walk. What is causing Rosa's chest pain?

ANSWERS TO FIGURE QUESTIONS

15.1   The base of the heart consists mainly of the left atrium.

15.2   The visceral layer of the serous pericardium is also part of the heart wall (epicardium).

15.3   Blood flows away from the heart in arteries.

15.4   Heart valves prevent the backflow of blood.

15.5   Superior vena cava, inferior vena cava, and coronary sinus deliver deoxygenated blood into the right atrium.

15.6   The only electrical connection between the atria and the ventricles is the atrioventricular (AV) bundle.

15.7   Atrial depolarization causes contraction of the atria.

15.8   The contraction phase is called systole; the relaxation phase is called diastole.

15.9   Acetylcholine decreases heart rate.

15.10 Fatty substances, cholesterol, and smooth muscle fibers make up atherosclerotic plaques.

Chapter 16

The Cardiovascular System: Blood Vessels and Circulation

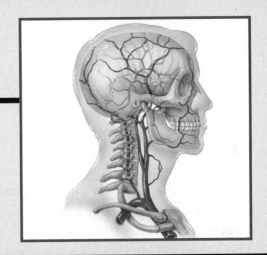

■ Student Learning Objectives

1. Compare the structure and function of the different types of blood vessels. **372**

2. Describe how substances enter and leave the blood in capillaries. **373**

3. Explain how venous blood returns to the heart. **375**

4. Define blood pressure and describe how it varies throughout the systemic circulation. **376**

5. Identify the factors that affect blood pressure and vascular resistance. **376**

6. Describe how blood pressure and blood flow are regulated. **377**

7. Explain how pulse and blood pressure are measured. **381**

8. Define shock and describe its common symptoms. **381**

9. Compare the major routes that blood takes through various regions of the body. **382**

10. Describe the effects of aging on the cardiovascular system. **382, 398**

■ A Look Ahead

371

veins of the abdominal organs (especially the liver and spleen) and the skin. Blood can be diverted quickly from its reservoirs to other parts of the body, for example, to skeletal muscles to support increased muscular activity.

Arteries and Arterioles

The walls of **arteries** have three layers of tissue surrounding a hollow space, the **lumen,** through which the blood flows (Figure 16.1a). The inner layer is composed of **endothelium,** a type of simple squamous epithelium; a basement membrane; and a type of elastic tissue called the elastic lamina. The middle layer consists of smooth muscle and elastic tissue. The outer layer is composed mainly of elastic and collagen fibers.

Sympathetic fibers of the autonomic nervous system innervate vascular smooth muscle. An increase in sympathetic stimulation typically stimulates the smooth muscle to contract, squeezing the vessel wall and narrowing the lumen. Such a decrease in the diameter of the lumen of a blood vessel is called **vasoconstriction.** In contrast, when sympathetic stimulation decreases, or in the presence of certain chemicals (such as nitric oxide or lactic acid), smooth muscle fibers relax. The resulting increase in lumen diameter is called **vasodilation.** Additionally, when an artery or arteriole is damaged, its smooth muscle contracts, producing vascular spasm of the vessel. Such a vasospasm limits blood flow through the damaged vessel and helps reduce blood loss if the vessel is small.

The arteries help propel blood onward while the ventricles are relaxing. As blood is ejected from the heart into arteries, their highly elastic walls stretch, accommodating the surge of blood. Then, while the ventricles are relaxing, the elastic fibers in the artery walls recoil, which forces blood onward through the smaller arteries.

An **arteriole** (= small artery) is a very small, almost microscopic, artery that delivers blood to capillaries. The smallest arterioles consist of little more than a layer of endothelium covered by a few smooth muscle fibers (see Figure 16.2a). Arterioles play a key role in regulating blood flow from arteries into capillaries. During vasoconstriction, blood flow from the arterioles to the capillaries is restricted; during vasodilation, the flow is significantly increased.

Capillaries

Capillaries (*capillar-* = hairlike) are microscopic vessels that connect arterioles to venules (Figure 16.1c). Because they are present near almost every body cell, capillaries permit the exchange of nutrients and wastes between the body's cells and the blood. The number of capillaries varies with the metabolic activity of the tissue they serve. Body tissues with high metabolic requirements, such as muscles, the liver, the kidneys, and the nervous system, have extensive capillary networks. Tissues with lower metabolic requirements, such as tendons and ligaments, contain fewer capillaries. A few tissues—all covering and lining epithelia, the cornea and lens of the eye, and cartilage—lack capillaries completely.

The cardiovascular system contributes to homeostasis of other body systems by transporting and distributing blood throughout the body to deliver materials such as oxygen, nutrients, and hormones and to carry away wastes. This transport is accomplished by blood vessels, which form closed circulatory routes for blood to travel from the heart to body organs and back again. In Chapters 14 and 15 we discussed the composition and functions of blood and the structure and function of the heart. In this chapter we examine the structure and functions of the different types of blood vessels that carry the blood to and from the heart, as well as factors that contribute to blood flow and regulation of blood pressure.

BLOOD VESSEL STRUCTURE AND FUNCTION

Objectives: • **Compare the structure and function of the different types of blood vessels.**

• **Describe how substances enter and leave the blood in capillaries.**

• **Explain how venous blood returns to the heart.**

There are five types of blood vessels: arteries, arterioles, capillaries, venules, and veins. **Arteries** (AR-ter-ēs) carry blood *away from the heart* to body tissues. Two large arteries—the aorta and the pulmonary trunk—emerge from the heart and branch out into medium-sized arteries that serve various regions of the body. These medium-sized arteries then divide into small arteries, which, in turn, divide into still smaller arteries called **arterioles** (ar-TER-ē-ōls). Arterioles within a tissue or organ branch into numerous microscopic vessels called **capillaries** (KAP-i-lar′-ēs). Groups capillaries within a tissue reunite to form small veins called **venules** (VEN-yoolz). These, in turn, merge to form progressively larger vessels called veins. **Veins** (VĀNZ) are the blood vessels that convey blood from the tissues *back to the heart.*

At any one time, systemic veins and venules contain about 60% of the total volume of blood in the system, systemic arteries and arterioles about 15%, systemic capillaries about 5%, pulmonary blood vessels about 12%, and the heart chambers about 8%. Because veins contain so much of the blood, certain veins function as **blood reservoirs.** The main blood reservoirs are the

Figure 16.1 ■ **Comparative structure of blood vessels.** The relative size of the capillary in (c) is enlarged for emphasis. Note the valve in the vein.

🔑 Arteries carry blood away from the heart to tissues. Veins carry blood from tissues to the heart.

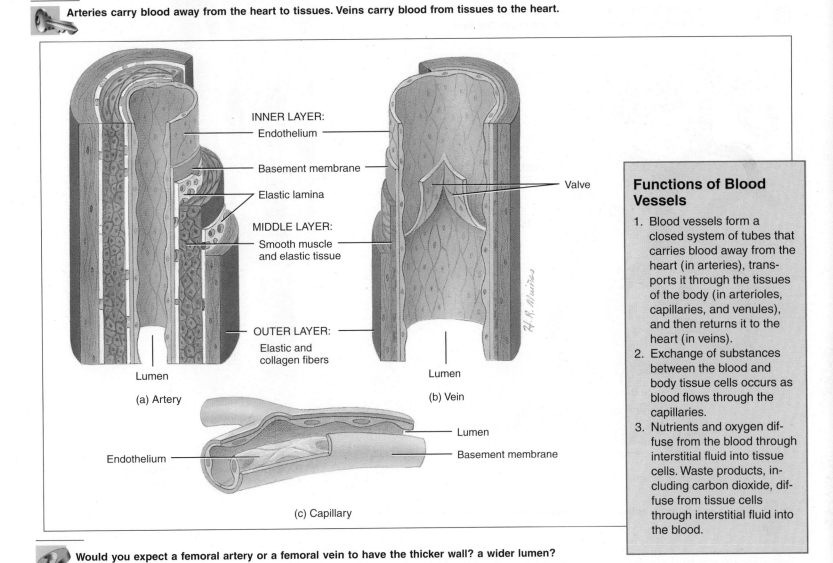

INNER LAYER:
Endothelium

Basement membrane

Elastic lamina

MIDDLE LAYER:
Smooth muscle
and elastic tissue

OUTER LAYER:
Elastic and
collagen fibers

Lumen

(a) Artery

Valve

Lumen

(b) Vein

Lumen

Endothelium

Basement membrane

(c) Capillary

Functions of Blood Vessels

1. Blood vessels form a closed system of tubes that carries blood away from the heart (in arteries), transports it through the tissues of the body (in arterioles, capillaries, and venules), and then returns it to the heart (in veins).
2. Exchange of substances between the blood and body tissue cells occurs as blood flows through the capillaries.
3. Nutrients and oxygen diffuse from the blood through interstitial fluid into tissue cells. Waste products, including carbon dioxide, diffuse from tissue cells through interstitial fluid into the blood.

Would you expect a femoral artery or a femoral vein to have the thicker wall? a wider lumen?

Structure of Capillaries

Because capillary walls are very thin, many substances easily pass through them to reach tissue cells from the blood or to enter the blood from interstitial fluid. The walls of all other blood vessels are too thick to permit the exchange of substances between blood and interstitial fluid. Depending on how tightly their endothelial cells are joined, different types of capillaries have varying degrees of permeability.

In some regions, capillaries link arterioles to venules directly. In other places, they form extensive branching networks (Figure 16.2). Blood flows through only a small part of a tissue's capillary network when metabolic needs are low. But when a tissue becomes active, the entire capillary network fills with blood. The flow of blood in capillaries is regulated by smooth muscle fibers in arteriole walls and by *precapillary sphincters,* rings of smooth muscle at the point where capillaries branch from arteri-

oles (Figure 16.2a). When precapillary sphincters relax, more blood flows into the connected capillaries; when precapillary sphincters contract, less blood flows through their capillaries.

Capillary Exchange

Because capillaries are so numerous, blood flows more slowly through them than through larger blood vessels. The slow flow aids the prime mission of the entire cardiovascular system: to keep blood flowing through capillaries so that *capillary exchange*—the movement of substances into and out of capillaries—can occur. The two main methods of capillary exchange are diffusion and bulk flow.

DIFFUSION Many substances, such as oxygen (O_2), carbon dioxide (CO_2), glucose, amino acids, and hormones, enter and leave capillaries by simple diffusion. Because O_2 and nutrients

Figure 16.2 ■ **Capillaries.**

Arterioles regulate blood flow into capillaries, where nutrients, gases, and wastes are exchanged between blood and interstitial fluid.

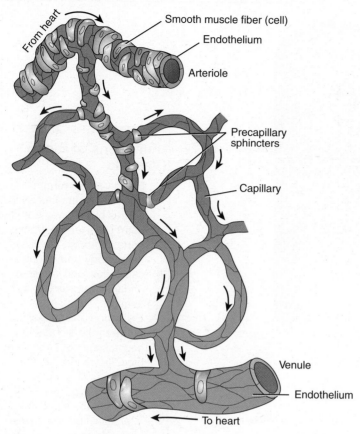

(a) Details of a capillary network

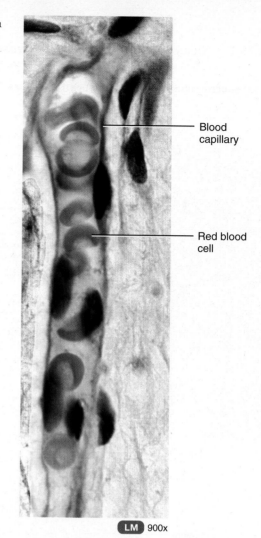

LM 900x

(b) Photomicrograph showing red blood cells squeezing through a blood capillary

Why do metabolically active tissues have extensive capillary networks?

normally are present in higher concentrations in blood, they diffuse down their concentration gradients into interstitial fluid and then into body cells. CO_2 and other wastes released by body cells are present in higher concentrations in interstitial fluid, so they diffuse into blood.

All plasma solutes except proteins pass easily across most capillary walls. The gaps between its endothelial cells are usually too small to allow proteins through the capillary wall. Thus, most plasma proteins remain in the blood. The gaps between the endothelial cells in the smallest blood vessels in the liver, called **sinusoids,** are so large, however, that they do allow proteins such as fibrinogen (the main clotting protein) and albumin to enter the bloodstream. Capillaries in most areas of the brain, by contrast, allow only a few substances to enter and leave because of the tightness of the endothelial layer, a property known as the *blood-brain barrier.*

BULK FLOW: FILTRATION AND REABSORPTION *Bulk flow* is a passive process by which *large* numbers of ions, molecules, or particles in a fluid move together in the same direction. Bulk flow is always from an area of higher pressure to an area of

lower pressure, and it continues as long as a pressure difference exists. Two opposing pressures—capillary blood pressure and blood colloid osmotic pressure—drive bulk flow in opposite directions across capillary walls.

Capillary blood pressure, the pressure of blood against the walls of capillaries, "pushes" fluid out of capillaries into interstitial fluid. An opposing pressure, termed **blood colloid osmotic pressure,** "pulls" fluid into capillaries. (Recall that an osmotic pressure is the pressure of a fluid due to its solute concentration. The higher the solute concentration, the greater the osmotic pressure.) Most solutes are present in nearly equal concentrations in blood and interstitial fluid. But the presence of proteins in plasma and their virtual absence in interstitial fluid gives blood the higher osmotic pressure. Blood colloid osmotic pressure is osmotic pressure due mainly to plasma proteins.

Capillary blood pressure is higher than blood osmotic pressure for about the first half of the length of a typical capillary. Thus, water and solutes flow out of the blood capillary into the surrounding interstitial fluid via bulk flow, a movement called **filtration** (Figure 16.3). Because capillary blood pressure decreases progressively as blood flows along a capillary, at about

Figure 16.3 ■ **Capillary exchange.**

 Capillary blood pressure pushes fluid out of capillaries (filtration), whereas blood colloid osmotic pressure pulls fluid into capillaries (reabsorption).

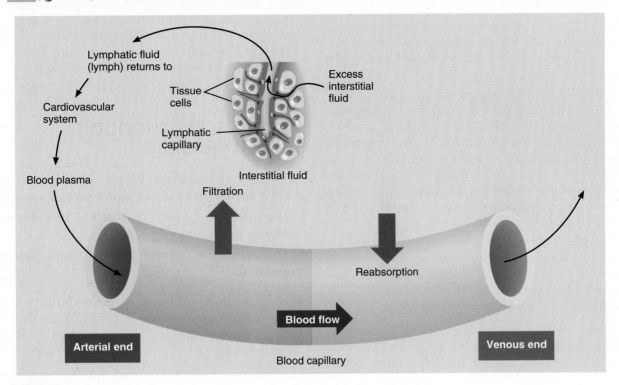

 What happens to excess filtered fluid and proteins that are not reabsorbed?

the capillary's midpoint, blood pressure drops below blood colloid osmotic pressure. Then, water and solutes move via bulk flow from interstitial fluid into the blood capillary, a process termed *reabsorption.* Normally, about 85% of the filtered fluid is reabsorbed. The excess filtered fluid and the few plasma proteins that do escape enter lymphatic capillaries and eventually are returned by the lymphatic system to the cardiovascular system.

Venules and Veins

When several capillaries unite, they form small veins called venules. Venules receive blood from capillaries and empty blood into veins, which return blood to the heart.

Structure of Venules and Veins

Venules (= little veins) are similar in structure to arterioles; their walls are thinner near the capillary end and thicker as they progress toward the heart. *Veins* are structurally similar to arteries, but their middle and inner layers are thinner (see Figure 16.1b). The outer layer of veins is the thickest layer. The lumen of a vein is wider than that of a corresponding artery.

In some veins, the inner layer folds inward to form *valves* that prevent backflow of blood. In people with weak venous

valves, gravity forces blood backwards through the valve. This increases venous blood pressure, which pushes the vein's wall outward. After repeated overloading, the walls lose their elasticity and become stretched and flabby, a condition called *varicose veins.*

By the time blood leaves the capillaries and moves into veins, it has lost a great deal of pressure. This can be observed in the blood leaving a cut vessel: Blood flows from a cut vein slowly and evenly, whereas it gushes out of a cut artery in rapid spurts. When a blood sample is needed, it is usually collected from a vein because pressure is low in veins and many of them are close to the skin surface.

Venous Return

Venous return, the volume of blood flowing back to the heart through systemic veins, occurs due to the pressure generated by contractions of the heart's left ventricle. Blood pressure is measured in millimeters of mercury, abbreviated mm Hg. The pressure difference from venules (averaging about 16 mm Hg) to the right ventricle (0 mm Hg), although small, normally is sufficient to cause venous return to the heart. When you stand, the pressure pushing blood up the veins in your lower limbs is barely enough to overcome the force of gravity pushing it back down. Besides the heart, two other mechanisms pump blood from the lower body back to the heart: (1) the skeletal muscle pump, and (2) the respiratory pump. Both pumps depend on the presence of valves in veins.

The *skeletal muscle pump* operates as follows (Figure 16.4):

1 While you are standing at rest, both the venous valve closer to the heart and the one farther from the heart in this part of the leg are open, and blood flows upward toward the heart.

2 Contraction of leg muscles, such as when you stand on tiptoes or take a step, compresses the vein. The compression pushes blood through the valve closer to the heart, an action called *milking*. At the same time, the valve farther from the heart in the uncompressed segment of the vein closes as some blood is pushed against it. People who are immobilized through injury or disease lack these contractions of leg muscles. As a result, their venous return is slower and they may develop circulation problems.

3 Just after muscle relaxation, pressure falls in the previously compressed section of vein, which causes the valve closer to the heart to close. The valve farther from the heart now opens because blood pressure in the foot is higher than in the leg, and the vein fills with blood from the foot.

The *respiratory pump* is also based on alternating compression and decompression of veins. During inhalation (breathing in) the diaphragm moves downward, which causes a decrease in pressure in the thoracic cavity and an increase in pressure in the abdominal cavity. As a result, abdominal veins are compressed, and a greater volume of blood moves from the compressed abdominal veins into the decompressed thoracic veins and then into the right atrium. When the pressures reverse during exhalation (breathing out), the valves in the veins prevent backflow of blood from the thoracic veins to the abdominal veins.

BLOOD FLOW THROUGH BLOOD VESSELS

Objectives: • **Define blood pressure and describe how it varies throughout the systemic circulation.**
• **Identify the factors that affect blood pressure and vascular resistance.**
• **Describe how blood pressure and blood flow are regulated.**

We saw in Chapter 15 that the entire cardiac output (CO) depends on stroke volume and heart rate. Two other factors influencing cardiac output and the proportion of blood that flows through specific circulatory routes are blood pressure and vascular resistance.

Blood Pressure

As you have just learned, blood flows from regions of higher pressure to regions of lower pressure; the greater the pressure difference, the greater the blood flow. Contraction of the ventricles generates *blood pressure (BP)*, the pressure exerted by blood on the walls of a blood vessel. BP is highest in the aorta and large systemic arteries, where in a resting, young adult, it rises to about 120 mm Hg during systole (contraction) and drops to about 80 mm Hg during diastole (relaxation). Blood pressure falls progressively as the distance from the left ventricle increases (Figure 16.5), from an average of 93 mm Hg in systemic arteries to about 35 mm Hg as blood passes into systemic capillaries. At the venous end of capillaries, blood pressure has dropped to about 16 mm Hg. Blood pressure continues to drop as blood enters systemic venules and then veins, and it reaches 0 mm Hg as blood returns to the right ventricle.

Blood pressure depends in part on the total volume of blood in the cardiovascular system. The normal volume of blood in an adult is about 5 liters (5.3 qt). Any decrease in this volume, as from hemorrhage, decreases the amount of blood that is circulated through the arteries. A modest decrease can be compensated by homeostatic mechanisms that help maintain blood pressure, but if the decrease in blood volume is greater than 10% of total blood volume, blood pressure drops, with potentially life-threatening results. Conversely, anything that increases blood volume, such as water retention in the body, tends to increase blood pressure.

Figure 16.4 ■ **Action of the skeletal muscle pump in returning blood to the heart.** Steps are described in the text.

 Milking refers to skeletal muscle contractions that drive venous blood toward the heart.

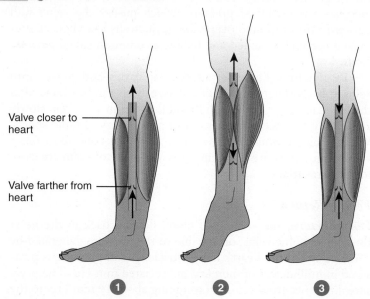

Valve closer to heart

Valve farther from heart

1   **2**   **3**

What mechanisms, besides cardiac contractions, act as pumps to boost venous return?

Figure 16.5 ■ **Blood pressure changes as blood flows through the systemic circulation.**

 Blood pressure falls progressively as blood flows from systemic arteries through capillaries and back to the right atrium.

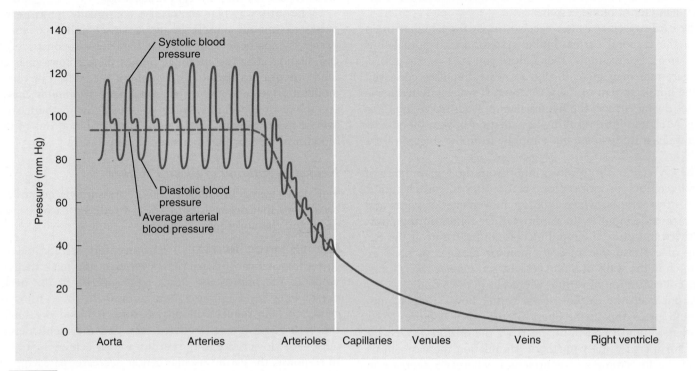

What is the relationship between blood pressure and blood flow?

Resistance

Vascular resistance is the opposition to blood flow due to friction between blood and the walls of blood vessels. An increase in vascular resistance increases blood pressure, whereas a decrease in vascular resistance has the opposite effect. Vascular resistance depends on (1) size of the blood vessel lumen, (2) blood viscosity, and (3) total blood vessel length.

1. **Size of the lumen.** The smaller the lumen of a blood vessel, the greater its resistance to blood flow. Vasoconstriction narrows the lumen, whereas vasodilation widens it. Normally, moment-to-moment fluctuations in blood flow through a given tissue are due to vasoconstriction and vasodilation of the tissue's arterioles.

2. **Blood viscosity.** The viscosity (thickness) of blood depends mostly on the ratio of red blood cells to plasma (fluid) volume, and to a smaller extent on the concentration of proteins in plasma. The higher the blood's viscosity, the higher the resistance. Any condition that increases the viscosity of blood, such as dehydration or polycythemia (an unusually high number of red blood cells), thus increases blood pressure. A depletion of plasma proteins or red blood cells, as a result of anemia or hemorrhage, decreases viscosity and thus decreases blood pressure.

3. **Total blood vessel length.** Resistance to blood flow increases when the total length of all blood vessels in the body increases. The longer the length, the greater the contact between the vessel wall and the blood. The greater the contact between the vessel wall and the blood, the greater the friction. An estimated 200 miles of additional blood vessels develop for each extra pound of fat, one reason why overweight individuals may have higher blood pressure.

Regulation of Blood Pressure and Blood Flow

Several interconnected negative feedback systems control blood pressure and blood flow by adjusting heart rate, stroke volume, vascular resistance, and blood volume. Some systems allow rapid adjustments to cope with sudden changes, such as the drop in blood pressure in the brain that occurs when you stand up; others provide long-term regulation. The body may also require adjustments to the distribution of blood flow. During exercise, for example, a greater percentage of blood flow is diverted to skeletal muscles.

Role of the Cardiovascular Center

In Chapter 15 we noted how the *cardiovascular (CV) center* in the medulla oblongata helps regulate heart rate and stroke volume. The CV center also controls the neural and hormonal negative feedback systems that regulate blood pressure and blood flow to specific tissues. Some of the CV center's neurons stimulate the heart, others inhibit the heart, and still others control the diameter of blood vessels.

The cardiovascular center receives input from higher brain regions: the cerebral cortex, limbic system, and hypothalamus (Figure 16.6). For example, even before you start to run a race, your heart rate may increase due to nerve impulses conveyed from the limbic system to the CV center. If your body temperature rises during a race, the hypothalamus sends nerve impulses to the CV center. The resulting vasodilation of skin blood vessels allows heat to dissipate more rapidly from the surface of the skin.

The CV center also receives input from three main types of sensory receptors: proprioceptors, baroreceptors, and chemoreceptors. *Proprioceptors*, which monitor movements of joints and muscles, provide input to the cardiovascular center during physical activity and cause the rapid increase in heart rate at the beginning of exercise. *Baroreceptors* monitor changes in pressure and stretch in the walls of blood vessels, and *chemoreceptors* monitor the concentration of various chemicals in the blood.

Output from the cardiovascular center flows along sympathetic and parasympathetic fibers of the ANS (Figure 16.6). Sympathetic impulses reach the heart via the *cardiac accelerator nerves*. An increase in sympathetic stimulation increases heart rate and the forcefulness of contraction, whereas a decrease in sympathetic stimulation decreases heart rate and contraction

force. Parasympathetic stimulation, conveyed along the *vagus (X) nerves*, decreases heart rate. Thus, autonomic control of the heart is the result of opposing sympathetic (stimulatory) and parasympathetic (inhibitory) influences.

The cardiovascular center also continually sends nerve impulses to smooth muscle in blood vessel walls via *vasomotor nerves*, sympathetic neurons that regulate vasoconstriction and vasodilation. The vasomotor region of the cardiovascular center sends impulses to arterioles throughout the body. The result is a moderate state of vasoconstriction, called *vasomotor tone*, that sets the resting level of vascular resistance. Sympathetic stimulation of most veins results in movement of blood out of venous blood reservoirs, which increases blood pressure.

Neural Regulation of Blood Pressure

The nervous system regulates blood pressure via negative feedback loops that occur as two types of reflexes: baroreceptor reflexes and chemoreceptor reflexes.

BARORECEPTOR REFLEXES Baroreceptors are located in the aorta, internal carotid arteries (arteries in the neck that supply blood to the brain), and other large arteries in the neck and chest. They send impulses continuously to the cardiovascular center to help regulate blood pressure. If blood pressure falls, the baroreceptors are stretched less, and they send nerve impulses at a slower rate to the cardiovascular center (Figure 16.7). In response, the cardiovascular center decreases parasympathetic stimulation of the heart via motor fibers of the vagus (X) nerves and increases sympathetic stimulation of the heart via cardiac

Figure 16.6 ■ **The cardiovascular (CV) center.** Located in the medulla oblongata, the CV center receives input from higher brain centers, proprioceptors, baroreceptors, and chemoreceptors. It provides output to both the sympathetic and parasympathetic divisions of the autonomic nervous system.

The cardiovascular center is the main region for nervous system regulation of heart rate, force of heart contractions, and vasodilation or vasoconstriction of blood vessels.

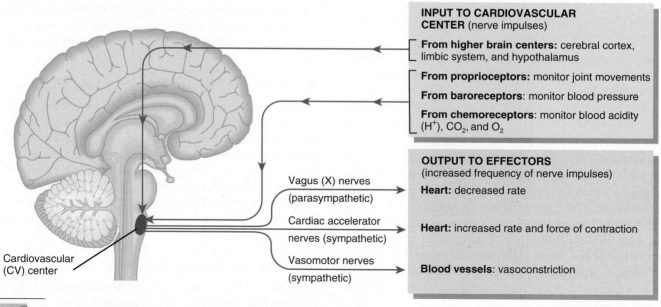

INPUT TO CARDIOVASCULAR CENTER (nerve impulses)

From higher brain centers: cerebral cortex, limbic system, and hypothalamus

From proprioceptors: monitor joint movements

From baroreceptors: monitor blood pressure

From chemoreceptors: monitor blood acidity (H^+), CO_2, and O_2

Vagus (X) nerves (parasympathetic)

Cardiac accelerator nerves (sympathetic)

Vasomotor nerves (sympathetic)

Cardiovascular (CV) center

OUTPUT TO EFFECTORS (increased frequency of nerve impulses)

Heart: decreased rate

Heart: increased rate and force of contraction

Blood vessels: vasoconstriction

How does vasoconstriction affect vascular resistance and blood flow?

Figure 16.7 ■ **Negative feedback regulation of blood pressure via baroreceptor reflexes.**

 The baroreceptor reflex is a neural mechanism for rapid regulation of blood pressure.

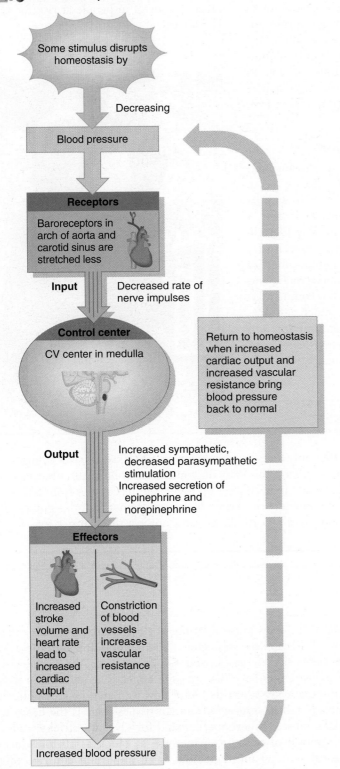

Does this negative feedback cycle happen when you lie down or when you stand up?

accelerator nerves. As the heart beats faster and more forcefully, and as vascular resistance increases, blood pressure increases to the normal level.

Conversely, when an increase in pressure is detected, the baroreceptors send impulses at a faster rate. The cardiovascular center responds by increasing parasympathetic stimulation and decreasing sympathetic stimulation. The resulting decreases in heart rate and force of contraction lower cardiac output. The cardiovascular center also slows the rate at which it sends sympathetic impulses along the vasomotor fibers that normally cause vasoconstriction. The resulting vasodilation lowers vascular resistance. Decreased cardiac output and decreased vascular resistance both lower blood pressure.

Moving from a prone (lying down) to an erect position decreases blood pressure and blood flow in the head and upper part of the body. The drop in pressure, however, is quickly counteracted by the baroreceptor reflexes. Sometimes these reflexes operate more slowly than normal, especially in older people. As a result, a person can faint due to reduced brain blood flow upon standing up too quickly.

CHEMORECEPTOR REFLEXES Chemoreceptors that monitor the chemical composition of blood are located in the two ***carotid bodies*** in the common carotid arteries and in the ***aortic body*** in the arch of the aorta. These chemoreceptors detect changes in blood levels of O_2, CO_2, and H^+. *Hypoxia* (lowered O_2 availability), *acidosis* (an increase in H^+ concentration), or *hypercapnia* (excess CO_2) stimulates the chemoreceptors to send impulses to the cardiovascular center. In response, the CV center increases sympathetic stimulation of arterioles and veins, producing vasoconstriction and an increase in blood pressure. These chemoreceptors also provide input to the respiratory center in the brain stem for adjustments in the rate of breathing.

Hormonal Regulation of Blood Pressure and Blood Flow

Several hormones help regulate blood pressure and blood flow by altering cardiac output, changing vascular resistance, or adjusting the total blood volume:

1. **Renin–angiotensin–aldosterone (RAA) system.** When blood volume falls or blood flow to the kidneys decreases, certain cells in the kidneys secrete *renin* into the bloodstream. Together, renin and angiotensin converting enzyme (ACE) produce the active hormone ***angiotensin II,*** which raises blood pressure by causing vasoconstriction. Angiotensin II also stimulates secretion of ***aldosterone,*** which increases reabsorption of sodium ions (Na^+) and water by the kidneys. The water reabsorption increases total blood volume, which increases blood pressure.

2. **Epinephrine and norepinephrine.** In response to sympathetic stimulation, the adrenal medulla releases epinephrine and norepinephrine. These hormones increase cardiac output by increasing the rate and force of heart contractions; they also cause vasoconstriction of arterioles and veins in the skin and abdominal organs.

3. **Antidiuretic hormone (ADH).** ADH is produced by the hypothalamus and released from the posterior pituitary in

Arterial Health— Undoing the Damage of Atherosclerosis

Not so long ago scientists believed that once plaque formed in an artery, it never went away. Medical researchers thought that lifestyle changes and drugs could slow the process of atherosclerosis, but they could not undo damage already done. In recent years, however, researchers have discovered that the body's own healing processes can reverse arterial plaque buildup. Lifestyle changes and drug treatments appear to stabilize the most dangerous atherosclerotic plaques and may even eliminate the need for surgical interventions, such as bypass surgery, in some people.

Stabilizing Dangerous Plaque

The health risk imposed by plaque that accumulates within the artery lining depends upon several factors. Some plaque is fairly stable: It has a low lipid content, is not growing much in size, and has a strong fibrous cap that keeps it from rupturing when blood pressure rises. Unstable plaque is characterized by a large accumulation of lipid in its core and only a thin fibrous cap. In addition, unstable plaques contain a large number of macrophages. In a misguided attempt to heal endothelial damage macrophages ingest plaque lipids; the net result is increased arterial injury and lipid accumulation. An unstable plaque is apt to rupture, triggering formation of a life-threatening blood clot at the plaque site.

Aggressive Prevention

The first step in preventing, slowing, and possibly reversing artery disease is to control the risk factors associated with its progression. Recommendations for a heart-healthy and artery-healthy lifestyle include no smoking, regular exercise (at least 30 minutes of moderate-intensity exercise per day), stress management, and a heart-healthy diet. Diet recommendations include limiting fat intake and dramatically increasing consumption of plant foods, such as grains, fruits, and vegetables. These recommendations help prevent arterial disease by reducing obesity, blood lipids, platelet stickiness, and blood pressure, and by improving blood glucose control in people at risk for type II diabetes.

▶ *Think It Over*

▶ Studies have found that only diets extremely low in fat (10% or fewer calories from fat) lead to plaque regression. Why do you think public health officials generally recommend a diet that supplies up to 30% of its calories from fat? Do you think this recommendation should be lower? Remember, this guideline is supposed to apply to all North Americans, not just those at risk for artery disease.

---

response to dehydration or decreased blood volume. Among other actions, ADH causes vasoconstriction, which increases blood pressure. For this reason ADH is also called *vasopressin.*

4. **Atrial natriuretic peptide (ANP).** Released by cells in the atria of the heart, ANP lowers blood pressure by causing vasodilation and by promoting the loss of salt and water in the urine, which reduces blood volume.

Local Regulation of Blood Flow

Localized changes in each capillary network can regulate vasodilation and vasoconstriction. When vasodilators are released by tissue cells, they cause dilation of nearby arterioles and relaxation of precapillary sphincters. Then, blood flow into the capillary networks increases, and O_2 delivery to the tissue rises. Vasoconstrictors have the opposite effect. The ability of a tissue to automatically adjust its blood flow to match its metabolic demands is called *autoregulation.* In tissues such as the heart and skeletal muscle, where the demand for O_2 and nutrients and for the removal of wastes can increase as much as 10-fold during physical activity, autoregulation is an important contributor to increased blood flow through the tissue. Autoregulation also controls regional blood flow in the brain; blood distribution to various parts of the brain changes dramatically for different mental and physical activities.

Two general types of stimuli contribute to autoregulation:

1. **Physical changes.** Warming promotes vasodilation, whereas cooling causes vasoconstriction. Also, stretching of vascular smooth muscle causes it to contract. When stretching lessens, the smooth muscle relaxes. For example, if blood flow through an arteriole decreases, then stretching of its walls decreases. As a result, the smooth muscle relaxes and produces vasodilation, which increases blood flow.

2. **Vasodilating and vasoconstricting chemicals.** Several types of cells, such as white blood cells, platelets, smooth muscle fibers, macrophages, and endothelial cells, release a wide variety of chemicals that alter blood vessel diameter. One important vasodilator released by endothelial cells is nitric oxide.

An important difference between the pulmonary and systemic circulations is their autoregulatory response to changes in O_2 level. The walls of blood vessels in the systemic circulation *dilate* in response to low O_2 concentration. With vasodilation, O_2 delivery increases, which restores the normal O_2 level. By contrast, the walls of blood vessels in the pulmonary circulation *constrict* in response to low levels of O_2. This response ensures that blood mostly bypasses those alveoli (air sacs) in the lungs that are not well ventilated by fresh air.

CHECKING CIRCULATION

Objective: • **Explain how pulse and blood pressure are measured.**

Pulse

The alternate expansion and elastic recoil of an artery after each contraction of the left ventricle is called a *pulse.* The pulse is strongest in the arteries closest to the heart. It becomes weaker as it passes through the arterioles, and it disappears altogether in the capillaries. The radial artery at the wrist is most commonly used to feel the pulse. Other sites where the pulse may be felt include the brachial artery along the medial side of the biceps brachii muscle; the common carotid artery, next to the voice box, which is usually monitored during cardiopulmonary resuscitation; the popliteal artery behind the knee; and the dorsalis pedis artery above the instep of the foot.

The pulse rate normally is the same as the heart rate, about 70 to 80 beats per minute at rest. *Tachycardia* (tak′-i-KAR-dē-a; *tachy-* = fast) is a rapid resting heart or pulse rate over 100 beats/min. *Bradycardia* (brād′-i-KAR-dē-a; *brady-* = slow) indicates a slow resting heart or pulse rate under 60 beats/min.

Measurement of Blood Pressure

In clinical use, the term *blood pressure* usually refers to the pressure in arteries generated by the left ventricle during systole and the pressure remaining in the arteries when the ventricle is in diastole. Blood pressure is usually measured in the brachial artery in the left arm (see Figure 16.10a). The device used to measure blood pressure is a *sphygmomanometer* (sfig′-mō-ma-NOM-e-ter; *sphygmo-* = pulse; *manometer* = instrument used to measure pressure). When the pressure cuff is inflated above the blood pressure attained during systole, the artery is compressed so that blood flow stops. The technician places a stethoscope below the cuff and then slowly deflates the cuff. When the cuff is deflated enough to allow the artery to open, a spurt of blood passes through, resulting in the first sound heard through the stethoscope. This sound corresponds to *systolic blood pressure (SBP)*—the force with which blood is pushing against arterial walls during ventricular contraction. As the cuff is deflated further, the sounds suddenly become faint. This level, called the *diastolic blood pressure (DBP),* represents the force exerted by the blood remaining in arteries during ventricular relaxation.

The normal blood pressure of a young adult male is 120 mm Hg systolic and 80 mm Hg diastolic, reported as "120 over 80" and written as 120/80. In young adult females, the pressures are 8 to 10 mm Hg less. People who exercise regularly and are in good physical condition may have even lower blood pressures.

SHOCK AND HOMEOSTASIS

Objective: • **Define shock and describe its common symptoms.**

Shock is a failure of the cardiovascular system to deliver enough O_2 and nutrients to meet cellular metabolic needs. The causes of shock are many and varied, but all are characterized by inadequate blood flow to body tissues. A common cause of shock is loss of body fluids, as occurs in hemorrhage, dehydration, burns, excessive vomiting, diarrhea, or sweating. If shock persists, cells and organs become damaged, and cells may die unless proper treatment begins quickly.

Although the symptoms of shock vary with the severity of the condition, the following are commonly observed:

- Systolic blood pressure lower than 90 mm Hg.
- Rapid resting heart rate due to sympathetic stimulation and increased blood levels of epinephrine.
- Weak, rapid pulse due to reduced cardiac output and fast heart rate.
- Cool, pale skin due to vasoconstriction of skin blood vessels.
- Sweating due to sympathetic stimulation.
- Reduced urine formation due to increased levels of aldosterone and antidiuretic hormone (ADH).
- Altered mental state due to reduced oxygen supply to the brain.
- Thirst due to loss of extracellular fluid.
- Nausea due to impaired circulation to digestive organs.

CIRCULATORY ROUTES

Objective: • **Compare the major routes that blood takes through various regions of the body.**

Blood vessels are organized into *circulatory routes* that carry blood throughout the body (Figure 16.8). As noted earlier, the two main circulatory routes are the systemic circulation and the pulmonary circulation.

Systemic Circulation

The *systemic circulation* includes the arteries and arterioles that carry blood containing oxygen and nutrients from the left ventricle to systemic capillaries throughout the body, plus the veins and venules that carry blood containing carbon dioxide and wastes to the right atrium. Blood leaving the aorta and traveling through the systemic arteries is a bright red color. As it moves through the capillaries, it loses some of its oxygen and takes on carbon dioxide, so that the blood in the systemic veins is a dark red color.

All systemic arteries branch from the *aorta,* which arises from the left ventricle of the heart (Figure 16.9). As the aorta emerges from the left ventricle, it passes upward and behind the pulmonary trunk. At this point, it is called the *ascending aorta.* The ascending aorta gives off two coronary arteries to the heart muscle. Then the ascending aorta turns to the left, forming the *arch of the aorta,* before descending as the *descending aorta.* The part of the descending aorta between the arch of the aorta and the diaphragm is the *thoracic aorta,* and the part between the diaphragm and the common iliac arteries is the *abdominal aorta.* Each part of the aorta gives off arteries that branch, leading to organs and finally into the arterioles and capillaries that service all the tissues of the body except the air sacs of the lungs. At the level of the fourth lumbar vertebra the abdominal aorta divides into two *common iliac arteries,* which carry blood to the lower limbs.

Deoxygenated blood returns to the heart through the systemic veins. All the veins of the systemic circulation empty into the *superior vena cava, inferior vena cava,* or the *coronary sinus,* which, in turn, empty into the right atrium. The principal blood vessels of the systemic circulation are described and illustrated in Exhibits 16.1 through 16.7 and Figures 16.9 through 16.15 starting on page 384.

Pulmonary Circulation

When deoxygenated blood returns to the heart from the systemic route, it is pumped out of the right ventricle into the lungs.

In the lungs, it loses carbon dioxide and picks up oxygen. Now bright red again, the blood returns to the left atrium of the heart and is pumped again into the systemic circulation. The flow of deoxygenated blood from the right ventricle to the air sacs of the lungs and the return of oxygenated blood from the air sacs to the left atrium is called the *pulmonary circulation* (Figure 16.16 on page 398). The *pulmonary trunk* emerges from the right ventricle and then divides into two branches. The *right pulmonary artery* runs to the right lung; the *left pulmonary artery* goes to the left lung. After birth, the pulmonary arteries are the only arteries that carry deoxygenated blood. On entering the lungs, the branches divide and subdivide until ultimately they form capillaries around the air sacs in the lungs. Carbon dioxide passes from the blood into the air sacs and is exhaled, while inhaled oxygen passes from the air sacs into the blood. The capillaries unite, venules and veins are formed, and, eventually, two *pulmonary veins* from each lung transport the oxygenated blood to the left atrium. (After birth, the pulmonary veins are the only veins that carry oxygenated blood.) Contractions of the left ventricle then send the blood into the systemic circulation.

Hepatic Portal Circulation

A vein that carries blood from one capillary network to another is called a *portal vein.* The hepatic portal vein, formed by the union of the splenic and superior mesenteric veins (Figure 16.17 on page 399), receives blood from capillaries of digestive organs and delivers it to capillary-like structures in the liver called sinusoids. In the *hepatic portal circulation* (*hepat-* = liver), venous blood from the gastrointestinal organs and spleen, rich with substances absorbed from the gastrointestinal tract, is delivered to the hepatic portal vein and enters the liver. The liver processes these substances before they pass into the general circulation. At the same time, the liver receives oxygenated blood from the systemic circulation via the hepatic artery. The oxygenated blood mixes with the deoxygenated blood in sinusoids. Ultimately, all blood leaves the sinusoids of the liver through the hepatic veins, which drain into the inferior vena cava.

Fetal Circulation

Fetal circulation, the circulatory route in a fetus, differs from an adult's because the lungs and digestive organs of a fetus are nonfunctional. The fetus derives its oxygen and nutrients from the maternal blood and eliminates its carbon dioxide and wastes into the maternal blood. The details of fetal circulation are discussed in Chapter 24.

Figure 16.8 ■ **Circulatory routes.** Red arrows indicate hepatic portal circulation. The details of the pulmonary circulation are shown in Figure 16.16, and the details of the hepatic portal circulation are shown in Figure 16.17.

 Blood vessels are organized into routes that deliver blood to various tissues of the body.

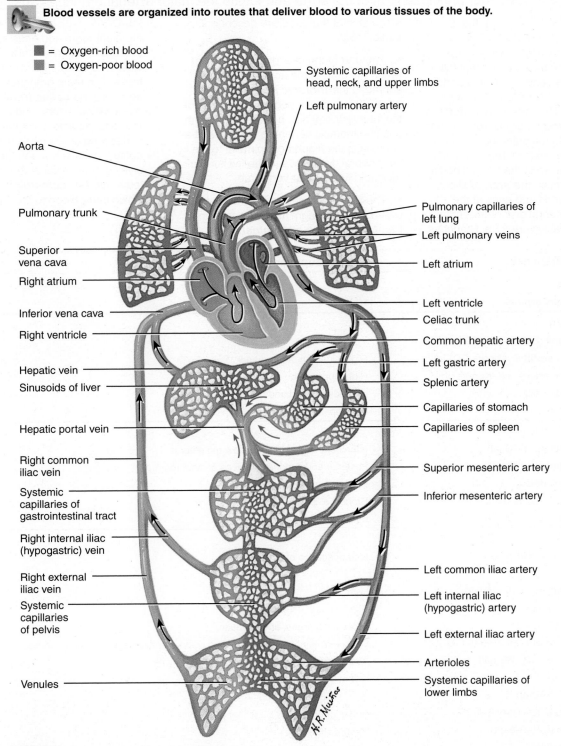

What are the two principal circulatory routes?

Exhibit 16.1 / The Aorta and Its Branches *(Figure 16.9)*

Objective: Identify the four principal parts of the aorta and locate the major arterial branches arising from each.

• The **aorta** (*aortae* = to lift up), the largest artery of the body, is 2 to 3 cm (about 1 in.) in diameter. Its four principal parts are the ascending aorta, arch of the aorta, thoracic aorta, and abdominal aorta. Recall that the thoracic aorta and abdominal aorta are part of the descending aorta. The **ascending aorta** emerges from the left ventricle posterior to the pulmonary trunk. It gives off two coronary artery branches that supply the myocardium of the heart. Then it turns to the left, forming the **arch of the aorta,** which descends and ends at the level of the intervertebral disk between the fourth and fifth thoracic vertebrae. Branches of the arch of the aorta are described in Exhibit 16.2.

• The part of the descending aorta between the arch of the aorta and the diaphragm, the **thoracic aorta,** is about 20 cm (8 in.) long. It begins at the level of the intervertebral disc between the fourth and fifth thoracic vertebrae and ends at an opening in the diaphragm parallel to the intervertebral disc between the twelfth thoracic and first lumbar vertebrae.

• The part of the descending aorta between the diaphragm and the common iliac arteries is the **abdominal aorta** (ab-DOM-i-nal). It begins at the opening in the diaphragm and ends at about the level of the fourth lumbar vertebra, where it divides into the right and left common iliac arteries. The main branches of the abdominal aorta are the **celiac trunk,** the **superior mesenteric artery,** and the **inferior mesenteric artery.** As the aorta continues to descend, it lies close to the vertebral bodies, passes through the diaphragm, and divides at the level of the fourth lumbar vertebra into two **common iliac arteries,** which carry blood to the lower limbs.

Aortic Parts and Branches	Region Supplied
ASCENDING AORTA	
Right and left coronary arteries	Heart
ARCH OF THE AORTA	
Bracheocephalic trunk (brā′-kē-ō-se-FAL-ik)	
Right common carotid artery (ka-ROT-id)	Right side of head and neck.
Right subclavian artery (sub-KLĀ-vē-an)	Right upper limb.
Left common carotid artery	Left side of head and neck.
Left subclavian artery	Left upper limb.
THORACIC AORTA (*thorac-* = chest)	
Intercostal arteries (in′-ter-KOS-tal)	Intercostal and chest muscles and pleurae.
Superior phrenic arteries (FREN-ik)	Posterior and superior surfaces of diaphragm.
Pericardial (per′-i-KAR-dē-al)	Pericardium.
Bronchial arteries (BRONG-kē-al)	Bronchi of lungs.
Esophageal arteries (e-sof′-a-JĒ-al)	Esophagus.
ABDOMINAL AORTA	
Inferior phrenic arteries (FREN-ik)	Inferior surface of diaphragm.
Celiac trunk (SĒ-lē-ak)	
Common hepatic artery (he-PAT-ik)	Liver.
Left gastric artery (GAS-trik)	Stomach and esophagus.
Splenic artery (SPLĒN-ik)	Spleen, pancreas, and stomach.
Superior mesenteric artery (MEZ-en-ter′-ik)	Small intestine, cecum, ascending colon, transverse colon, and pancreas.
Suprarenal arteries (soo′-pra-RĒ-nal)	Adrenal (suprarenal) glands.
Renal arteries (RĒ-nal)	Kidneys.
Gonadal arteries (gō-NAD-al)	
Testicular arteries (tes-TIK-yoo-lar)	Testes (male).
Ovarian arteries (ō-VAR-ē-an)	Ovaries (female).
Inferior mesenteric artery	Transverse colon, descending colon, sigmoid colon, and rectum.
Common iliac arteries (IL-ē-ak)	
External iliac arteries	Lower limbs.
Internal iliac (hypogastric) arteries	Uterus (female), prostate gland (male), muscles of buttocks, and urinary bladder.

Figure 16.9 ■ Aorta and its principal branches.

All systemic arteries branch from the aorta.

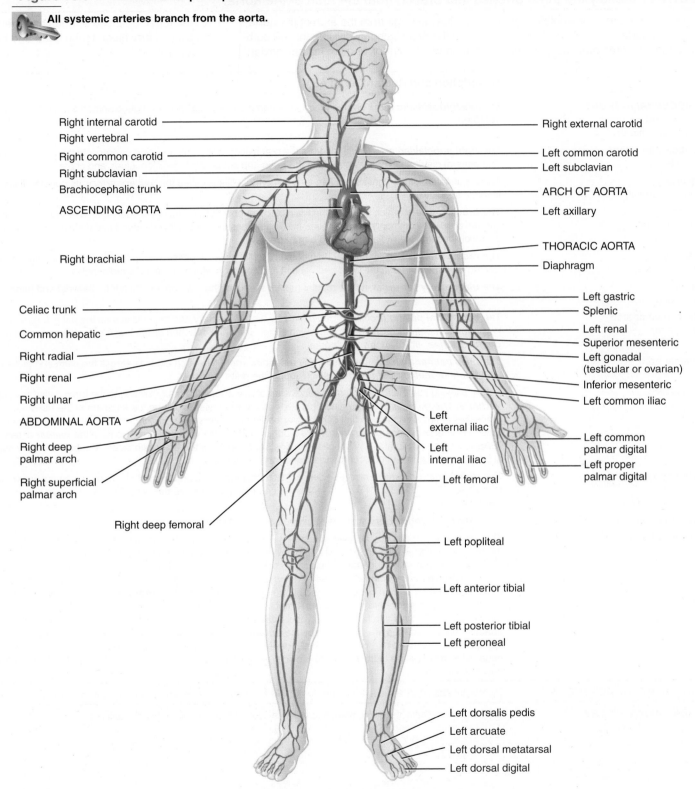

Right internal carotid

Right vertebral

Right common carotid

Right subclavian

Brachiocephalic trunk

ASCENDING AORTA

Right brachial

Celiac trunk

Common hepatic

Right radial

Right renal

Right ulnar

ABDOMINAL AORTA

Right deep
palmar arch

Right superficial
palmar arch

Right deep femoral

Right external carotid

Left common carotid

Left subclavian

ARCH OF AORTA

Left axillary

THORACIC AORTA

Diaphragm

Left gastric

Splenic

Left renal

Superior mesenteric

Left gonadal
(testicular or ovarian)

Inferior mesenteric

Left common iliac

Left
external iliac

Left
internal iliac

Left femoral

Left common
palmar digital

Left proper
palmar digital

Left popliteal

Left anterior tibial

Left posterior tibial

Left peroneal

Left dorsalis pedis

Left arcuate

Left dorsal metatarsal

Left dorsal digital

Anterior view of the principal branches of the aorta

 After blood is ejected from the heart, what are the names of the four parts of the aorta that it passes through?

Exhibit 16.2 / The Arch of the Aorta (Figure 16.10)

Objective: Identify the three arteries that branch from the arch of the aorta.

• The arch of the aorta, the continuation of the ascending aorta, is 4 to 5 cm (almost 2 in.) in length. Its three branches, in order as they emerge from the arch of the aorta, are the brachiocephalic trunk, the left common carotid artery, and the left subclavian artery. Together, these arteries and their branches carry blood to the head and both upper limbs.

Artery	Description and Region Supplied
BRACHIOCEPHALIC TRUNK (brā′-kē-ō-se-FAL-ik; *brachio-* = arm; *-cephalic* = head)	The *brachiocephalic trunk* divides to form the right subclavian artery and right common carotid artery (Figure 16.10a).
Right subclavian artery (sub-KLĀ-vē-an)	The *right subclavian artery* extends from the brachiocephalic trunk and then passes into the armpit (axilla). The general distribution of the artery is to the brain and spinal cord, neck, shoulder, and chest.
Axillary artery (AK-si-ler-ē = armpit)	The continuation of the right subclavian artery into the axilla is called the **axillary artery.** Its general distribution is to the shoulder.
Brachial artery (BRĀ-kē-al = arm)	The *brachial artery,* which provides the main blood supply to the arm, is the continuation of the axillary artery into the arm. It is commonly used to measure blood pressure. Just below the bend in the elbow, the brachial artery divides into the radial artery and ulnar artery.
Radial artery (RĀ-dē-al = radius)	The *radial artery* is a direct continuation of the brachial artery. It passes along the lateral (radial) aspect of the forearm and then through the wrist and hand; it is a common site for measuring radial pulse.
Ulnar artery (UL-nar = ulna)	The *ulnar artery* passes along the medial (ulnar) aspect of the forearm and then into the wrist and hand.
Superficial palmar arch (*palma* = palm)	The *superficial palmar arch* is formed mainly by the ulnar artery and extends across the palm. It gives rise to four **common palmar digital arteries,** which supply the palm and then divide into a pair of **proper palmar digital arteries,** which supply the fingers.
Deep palmar arch	The *deep palmar arch* is formed mainly by the radial artery. The arch extends across the palm and gives rise to the **palmar metacarpal arteries,** which supply the palm.
Vertebral artery (VER-te-bral)	Before passing into the axilla, the right subclavian artery gives off a major branch to the brain called the **right vertebral artery** (Figure 16.10b). The right vertebral artery passes through the foramina of the transverse processes of the sixth through first cervical vertebrae and enters the skull through the foramen magnum to reach the inferior surface of the brain. Here it unites with the left vertebral artery to form the **basilar artery** (BAS-i-lar). The vertebral artery supplies the posterior portion of the brain with blood. The basilar artery supplies the cerebellum and pons of the brain and the internal ear.
Right common carotid artery (ka-ROT-id)	The *right common carotid artery* begins at the branching of the brachiocephalic trunk and supplies structures in the head (Figure 16.10b). Near the larynx (voice box), it divides into the right external and right internal carotid arteries.
External carotid artery	The *external carotid artery* supplies structures *external* to the skull.
Internal carotid artery	The *internal carotid artery* supplies structures *internal* to the skull (the eyeball, ear, most of the cerebrum of the brain, pituitary gland, and external nose). Inside the cranium, the internal carotid arteries along with the basilar artery form an arrangement of blood vessels at the base of the brain near the sella turcica called the **cerebral arterial circle (circle of Willis).** From this circle (Figure 16.10c) arise arteries supplying most of the brain. The cerebral arterial circle is formed by the union of the **anterior cerebral arteries** (branches of internal carotids) and **posterior cerebral arteries** (branches of basilar artery). The posterior cerebral arteries are connected with the internal carotid arteries by the **posterior communicating arteries** (ko-MYOO-ni-kā′-ting). The anterior cerebral arteries are connected by the **anterior communicating arteries.** The *internal carotid arteries* are also considered part of the cerebral arterial circle. The functions of the cerebral arterial circle are to equalize blood pressure to the brain and provide alternate routes for blood flow to the brain, should the arteries become damaged.
LEFT COMMON CAROTID ARTERY	Divides into basically the same branches with the same names as the right common carotid artery.
LEFT SUBCLAVIAN ARTERY	Divides into basically the same branches with the same names as the right subclavian artery.

Figure 16.10 ■ **Arch of the aorta and its branches.**

 The arch of the aorta is the continuation of the ascending aorta.

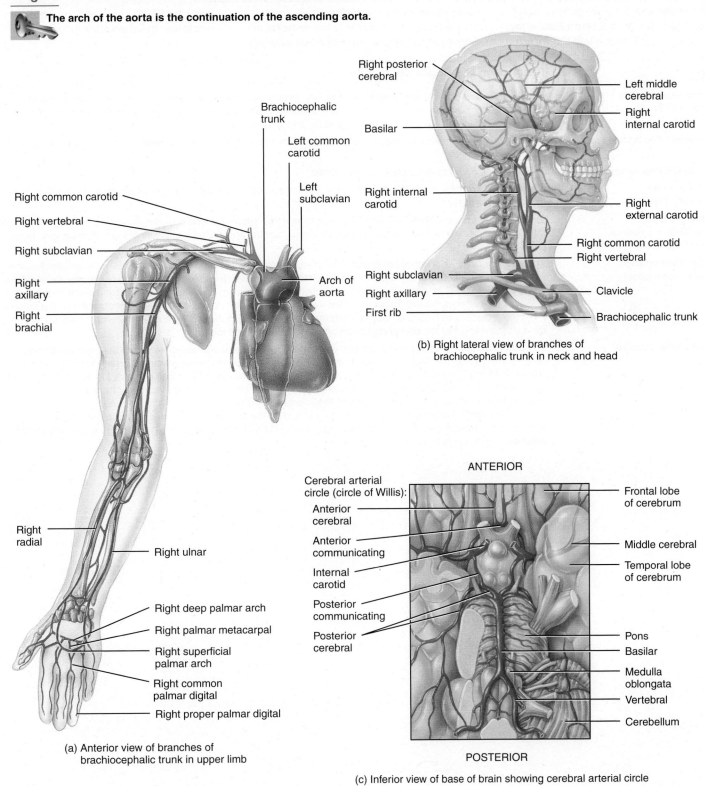

Right posterior cerebral

Left middle cerebral

Right internal carotid

Basilar

Right internal carotid

Right external carotid

Right common carotid

Right vertebral

Right subclavian

Right axillary

First rib

Clavicle

Brachiocephalic trunk

(b) Right lateral view of branches of brachiocephalic trunk in neck and head

Brachiocephalic trunk

Left common carotid

Left subclavian

Right common carotid

Right vertebral

Right subclavian

Right axillary

Right brachial

Arch of aorta

Right radial

Right ulnar

Right deep palmar arch

Right palmar metacarpal

Right superficial palmar arch

Right common palmar digital

Right proper palmar digital

(a) Anterior view of branches of brachiocephalic trunk in upper limb

ANTERIOR

Cerebral arterial circle (circle of Willis):

Anterior cerebral

Anterior communicating

Internal carotid

Posterior communicating

Posterior cerebral

Frontal lobe of cerebrum

Middle cerebral

Temporal lobe of cerebrum

Pons

Basilar

Medulla oblongata

Vertebral

Cerebellum

POSTERIOR

(c) Inferior view of base of brain showing cerebral arterial circle

What are the three major branches of the arch of the aorta, in order of their origination?

Exhibit 16.3 / Arteries of the Pelvis and Lower Limbs (Figure 16.11)

Objective: Identify the two major branches of the common iliac arteries.

• The abdominal aorta ends by dividing into the right and left ***common iliac arteries.*** These, in turn, divide into the ***internal*** and ***external iliac arteries.*** In sequence, the external iliacs become the ***femoral arteries*** in the thighs, the ***popliteal arteries*** posterior to the knee, and the ***anterior*** and ***posterior tibial arteries*** in the legs.

Artery	Description and Region Supplied
Common iliac arteries (IL-ē-ak = ilium)	At about the level of the fourth lumbar vertebra, the abdominal aorta divides into the right and left ***common iliac arteries.*** Each gives rise to two branches: internal iliac and external iliac arteries. The general distribution of the common iliac arteries is to the pelvis, external genitals, and lower limbs.
Internal iliac arteries	The ***internal iliac arteries*** are the primary arteries of the pelvis. They supply the pelvis, buttocks, external genitals, and thigh.
External iliac arteries	The ***external iliac arteries*** supply the lower limbs.
Femoral arteries (FEM-o-ral = thigh)	The ***femoral arteries,*** continuations of the external iliacs, supply the lower abdominal wall, groin, external genitals, and muscles of the thigh.
Popliteal arteries (pop'-li-TĒ-al = posterior surface of the knee)	The ***popliteal arteries,*** continuations of the femoral arteries, supply the adductor magnus and hamstring muscles and the skin on the posterior of the legs; the gastrocnemius, soleus, and plantaris muscles of the calf; knee joint; femur; patella; and fibula.
Anterior tibial arteries (TIB-ē-al = shin bone)	The ***anterior tibial arteries,*** which branch from the popliteal arteries, supply the knee joints, anterior muscles of the legs, skin on the anterior of the legs, and ankle joints. At the ankles, the anterior tibial arteries become the ***dorsalis pedis arteries*** ((PED-is; *ped* = foot), which supply the muscles, skin, and joints on the dorsal aspects of the feet. The dorsalis pedis arteries give off branches called the ***arcuate arteries*** (*arcuat-* = bowed) that extend over the metatarsals. From the arcuate arteries branch ***dorsal metatarsal arteries,*** which supply the feet. The dorsal metatarsal arteries terminate by dividing into the ***dorsal digital arteries,*** which supply the toes.
Posterior tibial arteries	The ***posterior tibial arteries,*** the direct continuations of the popliteal arteries, distribute to the muscles, bones, and joints of the leg and foot. Major branches of the posterior tibial arteries are the ***peroneal arteries*** (per'-o-NĒ-al; *perone* = fibula). They supply the peroneal, soleus, tibialis posterior, and flexor hallucis muscles; fibula; tarsus; and lateral aspect of the heel. Branching of the posterior tibial arteries gives rise to the medial and lateral plantar arteries. The ***medial plantar arteries*** (PLAN-tar = sole of foot) supply the muscles and skin of the feet and toes and medial aspect of the sole. The ***lateral plantar arteries*** unite with a branch of the dorsalis pedis arteries to form the ***plantar arch.*** The arch gives off ***plantar metatarsal arteries,*** which supply the feet. These terminate by dividing into ***plantar digital arteries,*** which supply the toes.

Figure 16.11 ■ Arteries of the pelvis and right lower limb.

The internal iliac arteries carry most of the blood supply to the pelvis, buttocks, external genitals, and thighs.

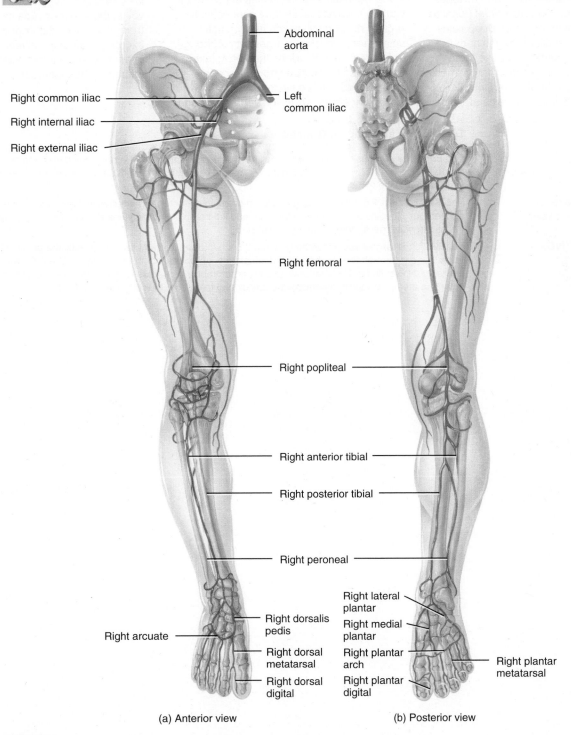

(a) Anterior view

(b) Posterior view

At what point does the abdominal aorta divide into the common iliac arteries?

Exhibit 16.4 / Veins of the Systemic Circulation (Figure 16.12)

Objective: Identify the three systemic veins that return deoxygenated blood to the heart.

• Whereas arteries distribute blood to various parts of the body, veins drain blood away from them. For the most part, arteries are deep. Veins may be **superficial** (located just beneath the skin) or **deep.** Deep veins generally travel alongside arteries and usually bear the same name. Because there are no large superficial arteries, the names of superficial veins do not correspond to those of arteries. Superficial veins are clinically important as sites for withdrawing blood or giving injections. Arteries usually follow definite pathways. Veins are more difficult to follow because they connect in irregular networks in which many smaller veins merge to form a larger vein. Although only one systemic artery, the aorta, takes oxygenated blood away from the heart (left ventricle), three systemic veins, the **coronary sinus, superior vena cava,** and ***inferior vena cava,*** deliver deoxygenated blood to the heart (right atrium). The coronary sinus receives blood from the cardiac veins; the superior vena cava receives blood from other veins superior to the diaphragm, except the air sacs (alveoli) of the lungs; the inferior vena cava receives blood from veins inferior to the diaphragm.

Vein	Description and Region Drained
Coronary sinus (KOR-ō-nar-ē; *corona* = crown)	The **coronary sinus** is the main vein of the heart; it receives almost all venous blood from the myocardium. It is located in the coronary sulcus (see Figure 15.3b) and opens into the right atrium between the opening of the inferior vena cava and the tricuspid valve.
Superior vena cava (SVC) (VĒ-na CĀ-va; *vena* = vein; *cava* = cavelike)	The **SVC** is about 7.5 cm (3 in.) long and 2 cm (1 in.) in diameter and empties its blood into the superior part of the right atrium. It begins by the union of the right and left brachiocephalic veins and enters the right atrium. The SVC drains the head, neck, chest, and upper limbs.
Inferior vena cava (IVC)	The **IVC** is the largest vein in the body, about 3.5 cm (1.5 in.) in diameter. It begins by the union of the common iliac veins, passes through the diaphragm, and enters the inferior part of the right atrium. The IVC drains the abdomen, pelvis, and lower limbs. The inferior vena cava is commonly compressed during the later stages of pregnancy by the enlarging uterus, producing edema of the ankles and feet and temporary varicose veins.

Figure 16.12 ■ **Principal veins.**

 Deoxygenated blood returns to the heart via the superior and inferior venae cavae and the coronary sinus.

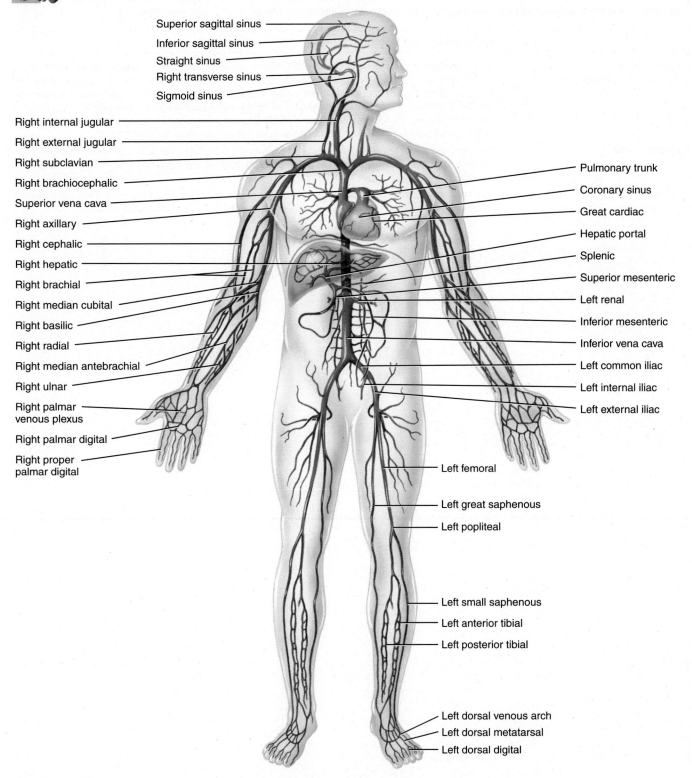

Superior sagittal sinus
Inferior sagittal sinus
Straight sinus
Right transverse sinus
Sigmoid sinus

Right internal jugular
Right external jugular
Right subclavian
Right brachiocephalic
Superior vena cava
Right axillary
Right cephalic
Right hepatic
Right brachial
Right median cubital
Right basilic
Right radial
Right median antebrachial
Right ulnar
Right palmar venous plexus
Right palmar digital
Right proper palmar digital

Pulmonary trunk
Coronary sinus
Great cardiac
Hepatic portal
Splenic
Superior mesenteric
Left renal
Inferior mesenteric
Inferior vena cava
Left common iliac
Left internal iliac
Left external iliac

Left femoral
Left great saphenous
Left popliteal
Left small saphenous
Left anterior tibial
Left posterior tibial

Left dorsal venous arch
Left dorsal metatarsal
Left dorsal digital

Overall anterior view

 **Which general regions of the body are drained by the superior vena cava and the inferior vena cava?**

Exhibit 16.5 / Veins of the Head and Neck *(Figure 16.13)*

Objective: Identify the three major veins that drain blood from the head.

• Most blood draining from the head passes into three pairs of veins: the ***internal jugular veins, external jugular veins,*** and ***vertebral veins.*** Within the brain, all veins drain into dural venous sinuses and then into the internal jugular veins. ***Dural venous sinuses*** are endothelium-lined venous channels between layers of the cranial dura mater.

Vein	Description and Region Drained
Internal jugular veins (JUG-yoo-lar; *jugular* = throat)	The dural venous sinuses (superior sagittal, inferior sagittal, straight, transverse, sigmoid, and cavernous) drain blood from the cranial bones, meninges, and brain. The right and left ***internal jugular veins*** pass inferiorly on either side of the neck lateral to the internal carotid and common carotid arteries. They then unite with the subclavian veins to form the right and left ***brachiocephalic veins*** (brā′-kē-ō-se-FAL-ik; *brachio-* = arm; *cephalic* = head). From here blood flows into the superior vena cava. The general structures drained by the internal jugular veins are the brain (through the dural venous sinuses), face, and neck.
External jugular veins	The right and left ***external jugular veins*** empty into the subclavian veins. The general structures drained by the external jugular veins are external to the cranium, such as the scalp and superficial and deep regions of the face.
Vertebral veins (VER-te-bral; *vertebra* = vertebrae)	The right and left ***vertebral veins*** empty into the brachiocephalic veins in the neck. They drain deep structures in the neck such as the cervical vertebrae, cervical spinal cord, and some neck muscles.

Figure 16.13 ■ **Principal veins of the head and neck.**

 Blood draining from the head passes into the internal jugular, external jugular, and vertebral veins.

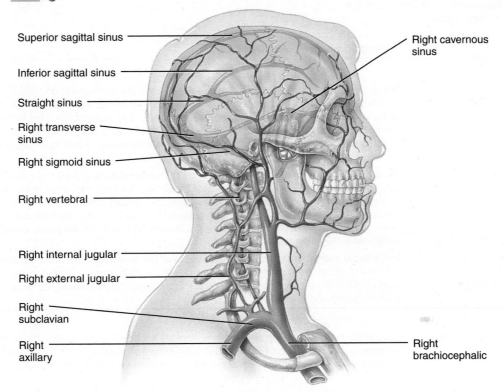

Superior sagittal sinus

Inferior sagittal sinus

Straight sinus

Right transverse sinus

Right sigmoid sinus

Right vertebral

Right internal jugular

Right external jugular

Right subclavian

Right axillary

Right cavernous sinus

Right brachiocephalic

Right lateral view

Into which veins in the neck does all venous blood from the brain drain?

Exhibit 16.6 / Veins of the Upper Limbs *(Figure 16.14)*

Objective: Identify the principal veins that drain the upper limbs.

• Blood from the upper limbs is returned to the heart by both superficial and deep veins. Both sets of veins have valves, which are more numerous in the deep veins.

Superficial veins are larger than deep veins and return most of the blood from the upper limbs.

Vein	Description and Region Drained
SUPERFICIAL VEINS	
Cephalic veins (se-FAL-ik = head)	The principal superficial veins that drain the upper limbs are the cephalic and basilic veins. They originate in the hand and convey blood from the smaller superficial veins into the axillary veins. The **cephalic veins** begin on the lateral aspect of the **dorsal venous arches** (VĒ-nus), networks of veins on the dorsum of the hands formed by the **dorsal metacarpal veins** (Figure 16.14a). These veins, in turn, drain the **dorsal digital veins,** which pass along the sides of the fingers. The cephalic veins drain blood from the lateral aspect of the upper limbs.
Basilic veins (ba-SIL-ik = royal)	The **basilic veins** begin on the medial aspects of the dorsal venous arches (Figure 16.14b) and drain blood from the medial aspects of the upper limbs. Anterior to the elbow, the basilic veins are connected to the cephalic veins by the **median cubital veins** (*cubitus* = elbow), which drain the forearm. If a vein must be punctured for an injection, transfusion, or removal of a blood sample, the median cubital vein is preferred. The basilic veins continue ascending until they join the brachial veins. As the basilic and brachial veins merge in the axillary area, they form the axillary veins.
Median antebrachial veins (an'-tē-BRĀ-kē-al; *ante-* = before, in front of; *brachium* = arm)	The **median antebrachial veins (median veins of the forearm)** begin in the **palmar venous plexuses,** networks of veins on the palms. The plexuses drain the **palmar digital veins** in the fingers. The median antebrachial veins ascend in the forearms to join the basilic or median cubital veins, sometimes both. They drain the palms and forearms.
DEEP VEINS	
Radial veins (RĀ-dē-al = radius)	The paired veins begin at the **deep palmar venous arches** (Figure 16.14c). These arches drain the **palmar metacarpal veins** in the palms. The radial veins drain the lateral aspects of the forearms and pass alongside the radial arteries. Just below the elbow joint, the radial veins unite with the ulnar veins to form the brachial veins.
Ulnar veins (UL-nar = ulna)	The paired **ulnar veins** begin at the **superficial palmar venous arches.** These arches drain the **common palmar digital veins** and the **proper palmar digital veins** in the fingers. The ulnar veins drain the medial aspect of the forearms, pass alongside the ulnar arteries, and join with the radial veins to form the brachial veins.
Brachial veins (BRĀ-kē-al)	The paired **brachial veins** accompany the brachial arteries. They drain the forearms, elbow joints, arms, and humerus. They join with the basilic veins to form the axillary veins.
Axillary veins (AK-si-ler'-ē; *axilla* = armpit)	The **axillary veins** ascend to become the subclavian veins. They drain the arms, axillae, and upper part of the chest wall.
Subclavian veins (sub-KLĀ-vē-an; *sub-* = under; *-clavian* = clavicle)	The **subclavian veins** are continuations of the axillary veins that unite with the internal jugular veins to form the brachiocephalic veins. The brachiocephalic veins unite to form the superior vena cava. The subclavian veins drain the arms, neck, and thoracic wall. The thoracic duct of the lymphatic system delivers lymph into the left subclavian vein at the junction with the internal jugular. The right lymphatic duct delivers lymph into the right subclavian vein at the corresponding junction (see Figure 17.1).

Figure 16.14 ■ **Principal veins of the right upper limb.**

Deep veins usually accompany arteries that have similar names.

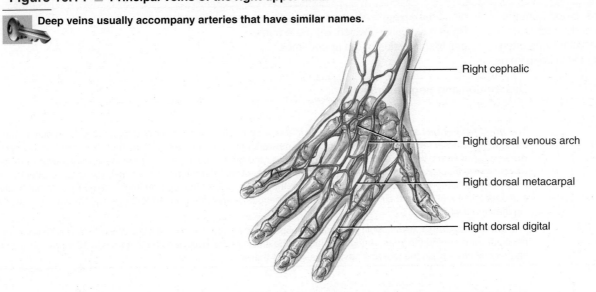

Right cephalic

Right dorsal venous arch

Right dorsal metacarpal

Right dorsal digital

(a) Posterior view of superficial veins of the hand

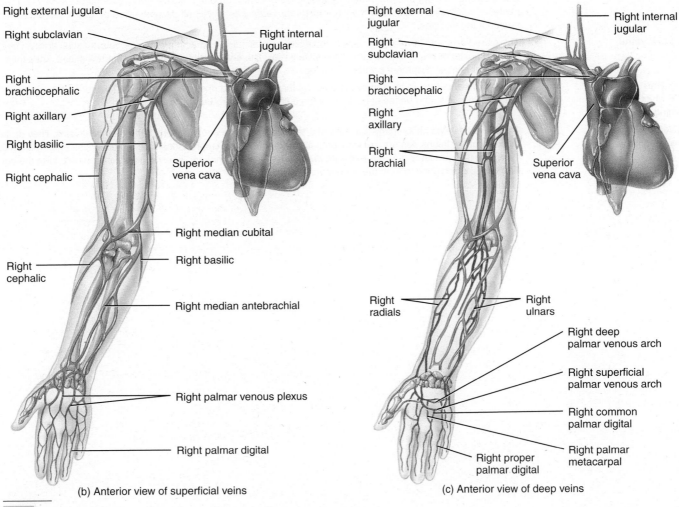

Right external jugular

Right subclavian

Right internal jugular

Right brachiocephalic

Right axillary

Right basilic

Right cephalic

Superior vena cava

Right median cubital

Right basilic

Right cephalic

Right median antebrachial

Right palmar venous plexus

Right palmar digital

(b) Anterior view of superficial veins

Right external jugular

Right subclavian

Right internal jugular

Right brachiocephalic

Right axillary

Right brachial

Superior vena cava

Right radials

Right ulnars

Right deep palmar venous arch

Right superficial palmar venous arch

Right common palmar digital

Right palmar metacarpal

Right proper palmar digital

(c) Anterior view of deep veins

From which vein in the upper limb is a blood sample often taken?

Exhibit 16.7 / Veins of the Lower Limbs *(Figure 16.15)*

Objective: Identify the principal veins that drain the lower limbs.

• As with the upper limbs, blood from the lower limbs is drained by both *superficial* and *deep veins.* The superficial veins often branch with each other and with deep veins along their length. All veins of the lower limbs have valves, which are more numerous than in veins of the upper limbs.

Vein	Description and Region Drained
SUPERFICIAL VEINS	
Great saphenous veins (sa-FĒ-nus = clearly visible)	The **great saphenous veins,** the longest veins in the body, begin at the medial end of the dorsal venous arches of the foot. The **dorsal venous arches** (VĒ-nus) are networks of veins on the top of the foot. They are formed by the **dorsal digital veins,** which collect blood from the toes, and then unite in pairs to form the **dorsal metatarsal veins,** which receive tributaries from superficial tissues and connect with the deep veins as well. The great saphenous veins empty into the femoral veins, and they mainly drain the medial side of the leg and thigh, the groin, external genitals, and abdominal wall. Along their length, the great saphenous veins have from 10 to 20 valves, with more located in the leg than the thigh.
Small saphenous veins	The **small saphenous veins** begin at the lateral aspect of the dorsal venous arches of the foot. They empty into the popliteal veins behind the knee. Along their length, the small saphenous veins have from 9 to 12 valves. The small saphenous veins drain the foot and posterior aspect of the leg.
DEEP VEINS	
Posterior tibial veins (TIB-ē-al)	The **plantar digital veins** on the plantar surfaces of the toes unite to form the **plantar metatarsal veins.** They unite to form the **deep plantar venous arches.** From each arch emerges the **medial** and **lateral plantar veins.** The paired **posterior tibial veins** are formed by the medial and lateral plantar veins. They accompany the posterior tibial artery through the leg and drain the foot and posterior compartment muscles. About two-thirds the way up the leg, the posterior tibial veins drain blood from the **peroneal veins** (per'-ō-NĒ-al; *perone* = fibula), which serve the lateral and posterior leg muscles.
Anterior tibial veins	The paired **anterior tibial veins** arise in the dorsal venous arch and accompany the anterior tibial artery. They unite with the posterior tibial veins to form the popliteal vein. The anterior tibial veins drain the ankle joint, knee joint, tibiofibular joint, and anterior portion of the leg.
Popliteal veins (pop'-li-TĒ-al; *popliteus* = hollow behind knee)	The **popliteal veins** are formed by the union of the anterior and posterior tibial veins. They drain the knee joint and the skin, muscles, and bones of portions of the calf and thigh around the knee joint.
Femoral veins (FEM-o-ral)	The **femoral veins** accompany the femoral arteries and are the continuations of the popliteal veins. They drain the muscles of the thighs, femurs, external genitals, and superficial lymph nodes. The femoral veins enter the pelvic cavity, where they are known as the **external iliac veins.** The external and internal iliac veins unite to form the common iliac veins, which unite to form the inferior vena cava.

Figure 16.15 ■ **Principal veins of the pelvis and lower limbs.**

 All veins of the lower limbs have valves.

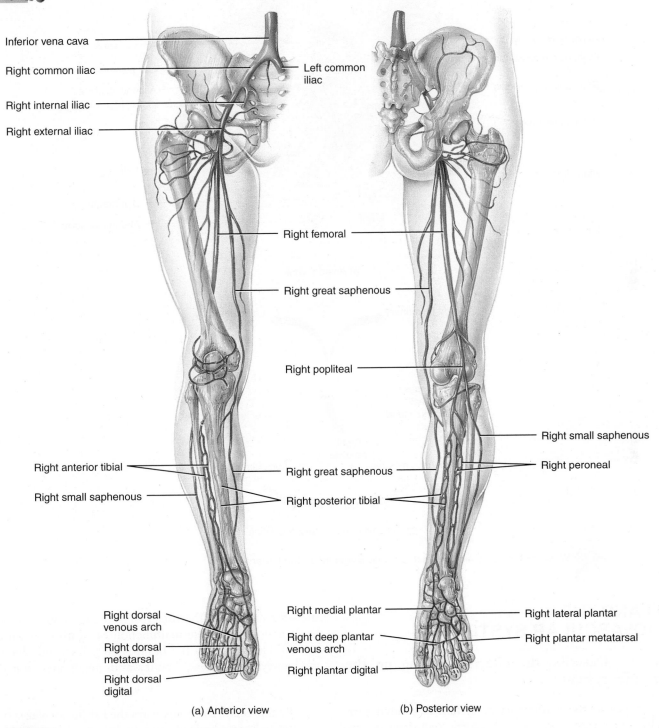

Inferior vena cava

Right common iliac

Right internal iliac

Right external iliac

Left common iliac

Right femoral

Right great saphenous

Right popliteal

Right small saphenous

Right anterior tibial

Right peroneal

Right great saphenous

Right small saphenous

Right posterior tibial

Right medial plantar

Right lateral plantar

Right dorsal venous arch

Right deep plantar venous arch

Right plantar metatarsal

Right dorsal metatarsal

Right plantar digital

Right dorsal digital

(a) Anterior view

(b) Posterior view

Which veins of the lower limb are superficial?

Figure 16.16 ■ **The pulmonary circulation.**

The pulmonary circulation brings deoxygenated blood from the right ventricle to the lungs and returns oxygenated blood from the lungs to the left atrium.

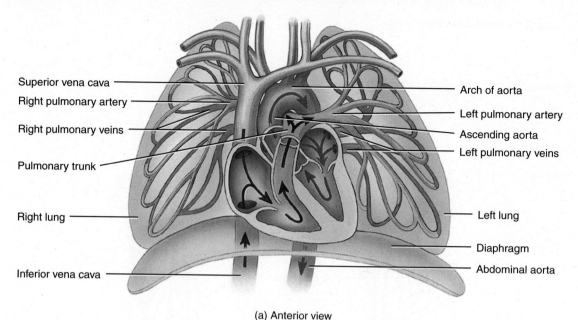

Superior vena cava

Right pulmonary artery

Right pulmonary veins

Pulmonary trunk

Right lung

Inferior vena cava

Arch of aorta

Left pulmonary artery

Ascending aorta

Left pulmonary veins

Left lung

Diaphragm

Abdominal aorta

(a) Anterior view

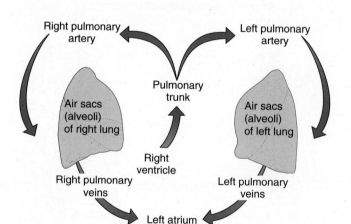

(b) Scheme of pulmonary circulation

Which are the only arteries that carry deoxygenated blood after birth?

AGING AND THE CARDIOVASCULAR SYSTEM

Objective: • **Describe the effects of aging on the cardiovascular system.**

General changes in the cardiovascular system associated with aging include increased stiffness of the aorta, reduction in cardiac muscle fiber size, progressive loss of cardiac muscular strength, reduced cardiac output, a decline in maximum heart rate, and an increase in systolic blood pressure. Coronary artery disease (CAD) is the major cause of heart disease and death in older Americans. Congestive heart failure, a set of symptoms associated with impaired pumping of the heart, is also prevalent in older individuals. Changes in blood vessels that serve brain tis-

sue—for example, atherosclerosis—reduce nourishment to the brain and result in the malfunction or death of brain cells. By age 80, blood flow to the brain is 20% less, and blood flow to the kidneys is 50% less, than it was in the same person at age 30.

• • •

To appreciate the many ways the cardiovascular system contributes to homeostasis of other body systems, examine Focus on Homeostasis: The Cardiovascular System on page 400. Next, in Chapter 17, we will examine the structure and function of the lymphatic system, seeing how it returns excess fluid filtered from capillaries to the cardiovascular system. We will also take a more detailed look at how some white blood cells function as defenders of the body by carrying out immune responses.

Figure 16.17 ■ **Hepatic portal circulation.**

The hepatic portal circulation delivers venous blood from the gastrointestinal organs and spleen to the liver.

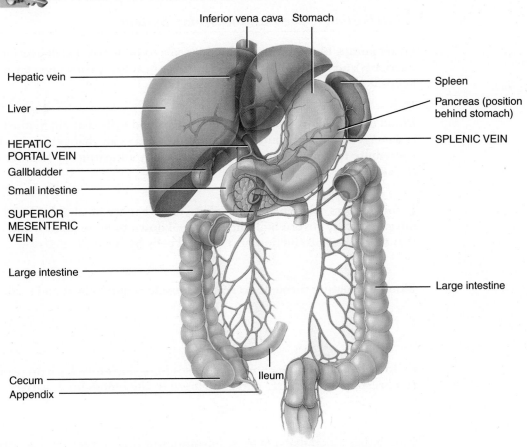

(a) Anterior view of veins draining into the hepatic portal vein

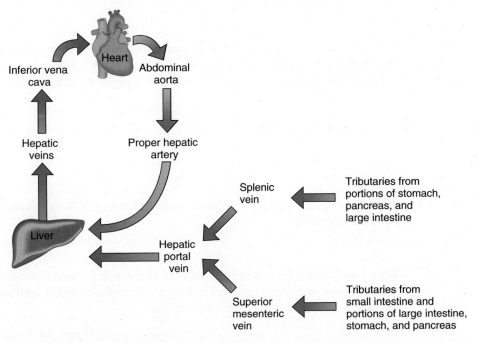

(b) Scheme of principal blood vessels of hepatic portal circulation and arterial supply and venous drainage of liver

 Which veins carry blood away from the liver?

Focus on Homeostasis
The Cardiovascular System

Body System	Contribution of the Cardiovascular System

For all body systems

Heart pumps blood through blood vessels to body tissues, delivering oxygen and nutrients and removing wastes by means of capillary exchange; circulating blood keeps body tissues at a proper temperature.

Integumentary system

Blood delivers clotting factors and white blood cells that aid hemostasis when skin is damaged and contribute to repair of injured skin; changes in skin blood flow contribute to body temperature regulation by adjusting the amount of heat loss via the skin.

Skeletal system

Blood delivers calcium and phosphate ions that are needed for building bone matrix, hormones that govern building and breakdown of bone matrix, and erythropoietin that stimulates production of red blood cells by red bone marrow.

Muscular system

Blood circulating through exercising muscle removes heat and lactic acid.

Nervous system

Endothelial cells lining choroid plexuses in brain ventricles help produce cerebrospinal fluid (CSF) and contribute to the blood–brain barrier.

Endocrine system

Circulating blood delivers most hormones to their target tissues; atrial cells secrete atrial natriuretic peptide.

Lymphatic and immune system

Circulating blood distributes cells and molecules (lymphocytes, macrophages, and antibodies) that carry out immune functions; lymph forms from excess interstitial fluid, which filters from blood plasma due to blood pressure generated by heart.

Respiratory system

Circulating blood transports oxygen from the lungs to body tissues and carbon dioxide to the lungs for exhalation.

Digestive system

Blood carries newly absorbed nutrients and water to liver; blood distributes hormones that aid digestion.

Urinary system

Heart and blood vessels deliver 20% of the resting cardiac output to the kidneys, where blood is filtered, needed substances are reabsorbed, and unneeded substances remain as part of urine, which is excreted.

Reproductive systems

Vasodilation of arterioles in penis and clitoris cause erection during sexual intercourse; blood distributes hormones that regulate reproductive functions.

COMMON DISORDERS

Hypertension

Hypertension, or persistently high blood pressure, is defined as systolic blood pressure of 140 mm Hg or greater and diastolic blood pressure of 90 mm Hg or greater. Recall that a blood pressure of 120/80 is normal and desirable in a healthy adult. In industrialized societies, hypertension is the most common disorder affecting the heart and blood vessels; it is a major cause of heart failure, kidney disease, and stroke. The classification system adopted in 1997 ranks blood pressure values for adults as follows:

Optimal	Systolic less than 120 mm Hg, diastolic less than 80 mm Hg
Normal	Systolic less than 130 mm Hg, diastolic less than 85 mm Hg
High-normal	Systolic 130 to 139 mm Hg, diastolic 85 to 89 mm Hg
Hypertension	Systolic 140 mm Hg or greater, diastolic 90 mm Hg or greater

Although several categories of drugs can reduce elevated blood pressure, the following lifestyle changes are also effective in managing hypertension.

- **Lose weight.** This is the best treatment for high blood pressure short of using drugs. Loss of even a few pounds helps reduce high blood pressure in overweight individuals.

- **Limit alcohol intake.** Drink less than 2 ounces of 100-proof liquor a day, or avoid alcohol altogether.

- **Exercise.** Engaging in moderate physical activity (such as brisk walking) several times a week for 30 to 45 minutes can lower systolic blood pressure by about 10 mm Hg.

- **Reduce intake of sodium (salt).** For the roughly 50% of the people with hypertension who are "salt sensitive," a low-salt diet can lower blood pressure.

- **Maintain recommended dietary intake of potassium, calcium, and magnesium.** Higher levels of potassium, calcium, and magnesium in the diet are associated with a lower risk of hypertension.

- **Don't smoke.** Smoking has devastating effects on the heart and can worsen the damaging effects of high blood pressure by promoting vasoconstriction.

- **Manage stress.** Various meditation and biofeedback techniques help some people reduce high blood pressure. These methods may work by decreasing the daily release of epinephrine and norepinephrine by the adrenal medulla.

Aneurysm

An *aneurysm* (AN-yoo-rizm) is a thin, weakened section of the wall of an artery or a vein that bulges outward, forming a balloonlike sac. Common causes are atherosclerosis, syphilis, congenital blood vessel defects, and trauma. If untreated, the aneurysm enlarges and the blood vessel wall becomes so thin that it bursts. The result is massive hemorrhage along with shock, severe pain, stroke, or death.

MEDICAL TERMINOLOGY AND CONDITIONS

Angiogenesis (an′-jē-ō-JEN-e-sis) Formation of new blood vessels.

Aortography (a′-or-TOG-ra-fē) X-ray examination of the aorta and its main branches after injection of a dye.

Carotid endarterectomy (ka-ROT-id end′-ar-ter-EK-tō-mē) The removal of atherosclerotic plaque from the carotid artery to restore greater blood flow to the brain.

Circulation time The time required for a drop of blood to pass from the right atrium, through the pulmonary circulation, back to the left atrium, through the systemic circulation, down to the foot, and back again to the right atrium; normally about 1 minute in a resting person.

Deep venous thrombosis (DVT) The presence of a thrombus (blood clot) in a deep vein of the lower limbs.

Hypotension (hī′-pō-TEN-shun) Low blood pressure; most commonly used to describe an acute drop in blood pressure, as occurs during excessive blood loss.

Occlusion (ō-KLOO-zhun) The closure or obstruction of the lumen of a structure such as a blood vessel. An example is an atherosclerotic plaque in an artery.

Orthostatic hypotension (or′-thō-STAT-ik; *ortho-* = straight; *-static* = causing to stand) An excessive lowering of systemic blood pressure when a person stands up; usually a sign of disease. May be caused by excessive fluid loss, certain drugs, and cardiovascular or neurogenic factors. Also called *postural hypotension.*

Phlebitis (fle-BĪ-tis; *phleb-* = vein) Inflammation of a vein, often in a leg. The condition is often accompanied by pain and redness of the skin over the inflamed vein. It is frequently caused by trauma or bacterial infection.

Syncope (SIN-kō-pē) A temporary cessation of consciousness; a faint. One cause is insufficient blood supply to the brain.

Thrombophlebitis (throm′-bō-fle-BĪ-tis) Inflammation of a vein involving clot formation. Superficial thrombophlebitis occurs in veins under the skin, especially in the calf.

White coat (office) hypertension The occurrence of elevated blood pressure only while being examined by health-care personnel.

■ STUDY OUTLINE

Blood Vessel Structure and Function (p. 372)

1. Arteries carry blood away from the heart. Their walls consist of three coats.

2. The structure of the middle coat gives arteries their two major properties, elasticity and contractility.

3. Arterioles are small arteries that deliver blood to capillaries.

4. Through constriction and dilation, arterioles play a key role in regulating blood flow from arteries into capillaries.

5. Capillaries are microscopic blood vessels through which materials are exchanged between blood and interstitial fluid.

6. Precapillary sphincters regulate blood flow through capillaries.

7. The two main methods of capillary exchange are diffusion and bulk flow (filtration and reabsorption).

8. Capillary blood pressure "pushes" fluid out of capillaries into interstitial fluid (filtration).

9. Blood colloid osmotic pressure "pulls" fluid into capillaries from interstitial fluid (reabsorption).

10. Venules are small vessels that emerge from capillaries and merge to form veins. They drain blood from capillaries into veins.

11. Veins consist of the same three coats as arteries but have less elastic tissue and smooth muscle. They contain valves that prevent backflow of blood.

12. Weak venous valves can lead to varicose veins.

13. Venous return, the volume of blood flowing back to the heart through systemic veins, occurs due to the pumping action of the heart, aided by skeletal muscle contractions (the skeletal muscle pump) and breathing (the respiratory pump).

Blood Flow Through Blood Vessels (p. 376)

1. Blood flow is determined by blood pressure and vascular resistance.

2. Blood flows from regions of higher pressure to regions of lower pressure.

3. Blood pressure is highest in the aorta and large systemic arteries; it drops progressively as distance from the left ventricle increases. Blood pressure in the right ventricle is close to 0 mm Hg.

4. An increase in blood volume increases blood pressure, and a decrease in blood volume decreases it.

5. Vascular resistance is the opposition to blood flow mainly as a result of friction between blood and the walls of blood vessels.

6. Vascular resistance depends on size of the blood vessel lumen, blood viscosity, and total blood vessel length.

7. Blood pressure and blood flow are regulated by neural and hormonal negative feedback systems and by autoregulation.

8. The cardiovascular center in the medulla oblongata helps regulate heart rate, stroke volume, and size of blood vessel lumen.

9. Vasomotor nerves (sympathetic) control vasoconstriction and vasodilation.

10. Baroreceptors—neurons sensitive to pressure—send impulses to the cardiovascular center to regulate blood pressure.

11. Chemoreceptors—neurons sensitive to concentrations of oxygen, carbon dioxide, and hydrogen ions—also send impulses to the cardiovascular center to regulate blood pressure.

12. Hormones such as angiotensin II, aldosterone, epinephrine, nor-

epinephrine, and antidiuretic hormone raise blood pressure, whereas atrial natriuretic peptide lowers it.

13. Autoregulation refers to local adjustments of blood flow in response to physical and chemical changes in a tissue.

Checking Circulation (p. 381)

1. Pulse is the alternate expansion and elastic recoil of an artery with each heartbeat. It may be felt in any artery that lies near the surface or over a hard tissue.

2. A normal pulse rate is about 75 beats per minute.

3. Blood pressure is the pressure exerted by blood on the wall of an artery when the left ventricle undergoes systole and then diastole. It is measured by a sphygmomanometer.

4. Systolic blood pressure (SBP) is the force of blood recorded during ventricular contraction. Diastolic blood pressure (DBP) is the force of blood recorded during ventricular relaxation. The normal blood pressure is 120/80 mm Hg.

Shock and Homeostasis (p. 381)

1. Shock is a failure of the cardiovascular system to deliver adequate amounts of oxygen and nutrients to meet the metabolic needs of cells.

2. Symptoms include low blood pressure; rapid resting heart rate; weak, rapid pulse; cool, pale skin; sweating; reduced urine formation; altered mental state; thirst; and nausea.

Circulatory Routes (p. 382)

1. The two major circulatory routes are the systemic circulation and the pulmonary circulation.

2. The systemic circulation takes oxygenated blood from the left ventricle through the aorta to all parts of the body and returns deoxygenated blood to the right atrium.

3. The parts of the aorta include the ascending aorta, the arch of the aorta, and the descending aorta (thoracic aorta and abdominal aorta). Each part gives off arteries that branch to supply the whole body.

4. Deoxygenated blood is returned to the heart through the systemic veins. All the veins of systemic circulation flow into either the superior or inferior vena cava or the coronary sinus, which empty into the right atrium.

5. The pulmonary circulation takes deoxygenated blood from the right ventricle to the air sacs of the lungs and returns oxygenated blood from the air sacs to the left atrium. It allows blood to be oxygenated for systemic circulation.

6. The hepatic portal circulation collects deoxygenated blood from the veins of the gastrointestinal tract and spleen and directs it into the hepatic portal vein of the liver. This routing allows the liver to extract and modify nutrients and detoxify harmful substances in the blood.

7. The liver also receives oxygenated blood from the hepatic artery.

Aging and the Cardiovascular System (p. 398)

1. General changes associated with aging include reduced elasticity of blood vessels, reduction in cardiac muscle size, reduced cardiac output, and increased systolic blood pressure.

2. The incidence of coronary artery disease (CAD), congestive heart failure (CHF), and atherosclerosis increase with age.

■ SELF-QUIZ

1. Sensory receptors that monitor changes in the blood pressure to the brain are

   a. chemoreceptors in the aorta   **b.** baroreceptors in the carotid arteries   **c.** the aortic bodies   **d.** precapillary sphincters in the arterioles   **e.** proprioceptors in the muscles

2. The blood vessels that allow the exchange of nutrients, wastes, oxygen, and carbon dioxide between the blood and tissues are the

   a. capillaries   **b.** arteries   **c.** venules   **d.** arterioles   **e.** veins

3. Substances undergo capillary exchange by means of

   a. simple diffusion and bulk flow   **b.** endocytosis, exocytosis, and active transport   **c.** simple diffusion and facilitated diffusion   **d.** simple diffusion and secondary active transport   **e.** filtration, reabsorption, and secretion

4. Blood flows through the blood vessels because of the

   a. establishment of a concentration gradient   **b.** elastic recoil of the veins   **c.** establishment of a pressure gradient   **d.** viscosity (stickiness) of the blood   **e.** thinness of the walls of capillaries

5. Which of the following represents pulmonary circulation as the blood flows from the right ventricle?

   a. pulmonary trunk → pulmonary veins → pulmonary capillaries → pulmonary arteries   **b.** pulmonary arteries → pulmonary capillaries → pulmonary trunk → pulmonary veins   **c.** pulmonary capillaries → pulmonary trunk → pulmonary arteries → pulmonary veins   **d.** pulmonary trunk → pulmonary arteries → pulmonary capillaries → pulmonary veins   **e.** pulmonary veins → pulmonary capillaries → pulmonary arteries → pulmonary trunk

6. The characteristic of arteries that allows them to stretch is

   a. contractility   **b.** vasoconstriction   **c.** excitability   **d.** vascular resistance   **e.** elasticity

7. Match the following descriptions to the appropriate blood vessel:

   ____ **a.** composed of a single layer of endothelial cells and a basement membrane

   ____ **b.** formed by reuniting capillaries

   ____ **c.** carry blood away from heart

   ____ **d.** regulate blood flow to capillaries

   ____ **e.** may contain valves

   A. arteries
   B. arterioles
   C. veins
   D. venules
   E. capillaries

8. Filtration of substances out of capillaries occurs when the capillary blood pressure

   a. is less than the blood colloid osmotic pressure   **b.** and the blood colloid osmotic pressure are equal   **c.** is high and the blood colloid osmotic pressure is high   **d.** is higher than the blood colloid osmotic pressure   **e.** is low and the blood colloid osmotic pressure is low

9. Weakened leg muscles would slow the

   a. blood flow out of the heart   **b.** respiratory pump   **c.** venous return   **d.** ability of arteries to vasodilate   **e.** pulse

10. Which of the following statements about blood vessels is true?

    a. Capillaries contain valves.   **b.** Walls of arteries are generally thicker and contain more elastic tissue than walls of veins.   **c.** Veins carry blood away from the heart.   **d.** Blood flows most rapidly through veins.   **e.** Blood pressure in arteries is always lower than in veins.

11. Why is it important that blood flows slowly through the capillaries?

    a. It allows time for the materials in the blood to pass through the thick capillary walls.   **b.** It prevents damage to the capillaries.   **c.** It permits the efficient exchange of nutrients and wastes between the blood and interstitial fluid.   **d.** It allows the heart time to rest.   **e.** It allows the blood pressure in capillaries to rise above the blood pressure in the veins.

12. Match the following:

    ____ **a.** source of all systemic arteries

    ____ **b.** supplies a lower limb

    ____ **c.** heart's blood system

    ____ **d.** returns blood to heart from lower limbs

    ____ **e.** carries blood to liver

    ____ **f.** leads to lungs

    ____ **g.** returns blood from lungs to heart

    ____ **h.** supplies blood to brain

    ____ **i.** returns blood to heart from head and upper body

    A. hepatic portal vein
    B. pulmonary trunk
    C. pulmonary vein
    D. common iliac artery
    E. coronary circulation
    F. inferior vena cava
    G. superior vena cava
    H. aorta
    I. cerebral arterial circle

13. For each of the following factors, indicate if it increases **(A)** or decreases **(B)** blood pressure:

    ____ **a.** an increase in cardiac output

    ____ **b.** hemorrhage

    ____ **c.** vasodilation

    ____ **d.** vasoconstriction

    ____ **e.** stimulation of the heart by the cardiac accelerator nerves

    ____ **f.** hypoxia

    ____ **g.** epinephrine

    ____ **h.** increase in blood volume

    ____ **i.** bradycardia

14. Aldosterone affects blood pressure by

    a. increasing heart rate   **b.** increasing vasoconstriction of arterioles   **c.** reducing blood volume   **d.** stimulating release of atrial natriuretic peptide by the heart   **e.** increasing reabsorption of sodium ions and water by the kidneys

15. In a blood pressure reading of 120/80,

    a. 120 represents the diastolic pressure   **b.** 80 represents the pressure of the blood against the arteries during ventricular relaxation   **c.** 120 represents the blood pressure and 80 represents the heart rate   **d.** 80 is the reading taken when the first sound is heard   **e.** the patient has a severe problem with hypertension

16. Which of the following statements is NOT true?

    a. Regulation of blood vessel diameter originates from the vasomotor region of the cerebral cortex.   **b.** The cerebral cortex can play a role in regulating blood pressure.   **c.** Baroreceptors may stimulate the cardiovascular center.   **d.** Activation of proprioceptors increases heart rate at the beginning of exercise.   **e.** Vasomotor nerves innervate smooth muscle in blood vessel walls.

17. Venous return to the heart is enhanced by all of the following EXCEPT

a. skeletal muscle "milking"   **b.** valves in veins   **c.** the pressure difference from venules to the right ventricle   **d.** vasodilation   **e.** inhalation during breathing

CRITICAL THINKING APPLICATIONS

1. The local anesthetic injected by a dentist often contains a small amount of epinephrine. What effect would epinephrine have on the blood vessels in the vicinity of the dental work? Why might this effect be desired?

2. In this chapter, you've read about varicose veins. Why didn't you read about varicose arteries?

3. As long as Gertie keeps taking her medication regularly, her blood pressure is pretty good—about 130 over 90. During her last annual checkup, she complained of feeling dizzy or lightheaded when getting out of bed in the morning and when standing up after her afternoon nap. Her doctor suspects the cause is low blood pressure

upon standing. After years of trying to keep her pressure down, Gertie was surprised to hear that this type of low BP is *not* good. What is Gertie's problem?

4. Julie was all flustered when she ran in late to her A & P lab. She had spilled a cup of coffee on herself while she was weaving in and out of traffic while trying to get around a traffic jam. Then she missed her exit while she was changing the station on the radio, couldn't find a place to park, and missed the lab quiz. The lab today is learning to take blood pressures, and Julie's is high! (It's normally 120 over 80.) What is the physiological explanation for Julie's elevated BP?

ANSWERS TO FIGURE QUESTIONS

16.1   The femoral artery has the thicker wall; the femoral vein has the wider lumen.

16.2   Metabolically active tissues have more capillaries because they use oxygen and produce wastes more rapidly than inactive tissues.

16.3   Excess filtered fluid and proteins that escape from plasma drain into lymphatic capillaries and are returned by the lymphatic system to the cardiovascular system.

16.4   The skeletal muscle pump and the respiratory pump help boost venous return.

16.5   As blood pressure increases, blood flow increases.

16.6   Vasoconstriction increases vascular resistance, which decreases blood flow through the vasoconstricted blood vessels.

16.7   It happens when you stand up because gravity causes pooling of blood in leg veins as you stand upright, decreasing the blood pressure in your upper body.

16.8   The principal circulations are the systemic and the pulmonary circulations.

16.9   The four parts of the aorta are the ascending aorta, arch of the aorta, thoracic aorta, and abdominal aorta.

16.10   Branches of the arch of the aorta are the brachiocephalic trunk, left common carotid artery, and left subclavian artery.

16.11   The abdominal aorta divides into the common iliac arteries at about the level of the fourth lumbar vertebra.

16.12   The superior vena cava drains regions above the diaphragm (except the cardiac veins and the alveoli of the lungs), and the inferior vena cava drains regions below the diaphragm.

16.13   All venous blood in the brain drains into the internal jugular veins.

16.14   The median cubital vein is often used for withdrawing blood.

16.15   Superficial veins of the lower limbs include the dorsal venous arch and the great saphenous and small saphenous veins.

16.16   The pulmonary arteries are the only arteries that carry deoxygenated blood after birth.

16.17   The hepatic veins carry blood away from the liver.

Chapter 17

The Lymphatic and Immune System

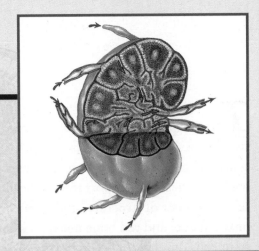

■ Student Learning Objectives

■ A Look Ahead

The *lymphatic and immune system* (lim-FAT-ik) assists in circulating body fluids and helps defend the body against disease-causing agents. It consists of a fluid called lymph; vessels to transport the fluid called lymphatic vessels; a number of structures and organs containing lymphatic tissue; and red bone marrow, where some of the stem cells develop into lymphocytes (Figure 17.1). *Lymphatic tissue* is a specialized form of reticular connective tissue (see Table 4.3C on page 85) that contains large numbers of lymphocytes. Recall from Chapter 14 that two types of lymphocytes participate in immune responses: *B cells* and *T cells*. B cells develop into plasma cells that protect us against disease by producing antibodies. T cells attack and destroy foreign cells and microbes.

FUNCTIONS OF THE LYMPHATIC AND IMMUNE SYSTEM

Objective: • **Describe the functions of the lymphatic and immune system.**

The lymphatic and immune system has several functions:

1. **Draining interstitial fluid.** Lymphatic vessels drain tissue spaces of excess interstitial fluid. This fluid forms as it filters out of the arterial ends of capillaries throughout the body (see Figure 16.3 on page 375). The lymphatic vessels return the fluid to venous blood in the subclavian veins (Figure 17.1).

2. **Transporting dietary lipids.** Lymphatic vessels transport triglycerides, cholesterol, and lipid-soluble vitamins (A, D, E, and K) from the gastrointestinal tract to the blood.

3. **Protecting against invasion.** Lymphocytes, with the aid of macrophages, protect the body from foreign cells, microbes, and cancer cells.

LYMPH AND INTERSTITIAL FLUID

Objective: • **Explain how lymph and interstitial fluid are related.**

The major difference between interstitial fluid and lymph is location. When the fluid bathes tissue cells, it is called *interstitial fluid* or *intercellular fluid*. When it flows through the lymphatic vessels, it is called *lymph* (= clear water). Both fluids are similar in composition to blood plasma. The main chemical difference is that interstitial fluid and lymph contain less protein than plasma because most plasma protein molecules are too large to filter through the capillary wall.

Each day, about 20 liters of fluid filter from blood into tissue spaces. This fluid must be returned to the cardiovascular system to maintain normal blood volume and functions. About 85% of the fluid filtered from the arterial end of a blood capillary is returned to the blood directly by reabsorption at the venous end of the capillary. The remaining 15% first passes into lymphatic vessels and then is returned to the blood.

LYMPHATIC VESSELS AND LYMPH CIRCULATION

Objective: • **Describe the organization of lymphatic vessels and the circulation of lymph.**

Lymphatic vessels originate as *lymphatic capillaries,* microscopic vessels in spaces between cells that are found throughout the body (Figure 17.2a). Tissues that lack lymphatic capillaries include avascular tissues, such as cartilage, the epidermis, and cornea of the eye; the central nervous system; portions of the spleen; and red bone marrow. Lymphatic capillaries are slightly larger than blood capillaries and have a unique structure that permits interstitial fluid to flow into them, but not out. The endothelial cells that make up the wall of a lymphatic capillary are not attached end to end, but rather, the ends overlap (Figure 17.2b). When pressure is greater in the interstitial fluid than in lymph, the cells separate slightly, like a one-way swinging door, and interstitial fluid enters the lymphatic capillary. When pressure is greater inside the lymphatic capillary, the cells adhere more closely and lymph cannot escape back into interstitial fluid.

Whereas blood capillaries link two larger blood vessels that form part of a circuit, lymphatic capillaries begin in the tissues and carry the lymph that forms there toward a larger lymphatic vessel. Just as blood capillaries converge to form venules and veins, lymphatic capillaries unite to form larger and larger *lymphatic vessels* (see Figure 17.1). Lymphatic vessels resemble veins in structure but have thinner walls and more valves. Located at intervals along lymphatic vessels are *lymph nodes,* encapsulated masses of B cells and T cells.

From the lymphatic vessels, lymph eventually passes into two main channels: the thoracic duct and the right lymphatic duct. The *thoracic duct,* the main lymph-collecting duct, receives lymph from the left side of the head, neck, and chest; the left upper limb; and the entire body below the ribs. The *right lymphatic duct* drains lymph from the upper right side of the body (see Figure 17.1).

Ultimately, the thoracic duct empties all its lymph into the junction of the left internal jugular vein and left subclavian vein, and the right lymphatic duct empties all its lymph into the junction of the right internal jugular vein and right subclavian vein. Thus, lymph drains back into the blood. In summary, the

Figure 17.1 ▥ **The lymphatic and immune system.**

 The lymphatic and immune system consists of lymph, lymphatic vessels, lymphatic tissue, and red bone marrow.

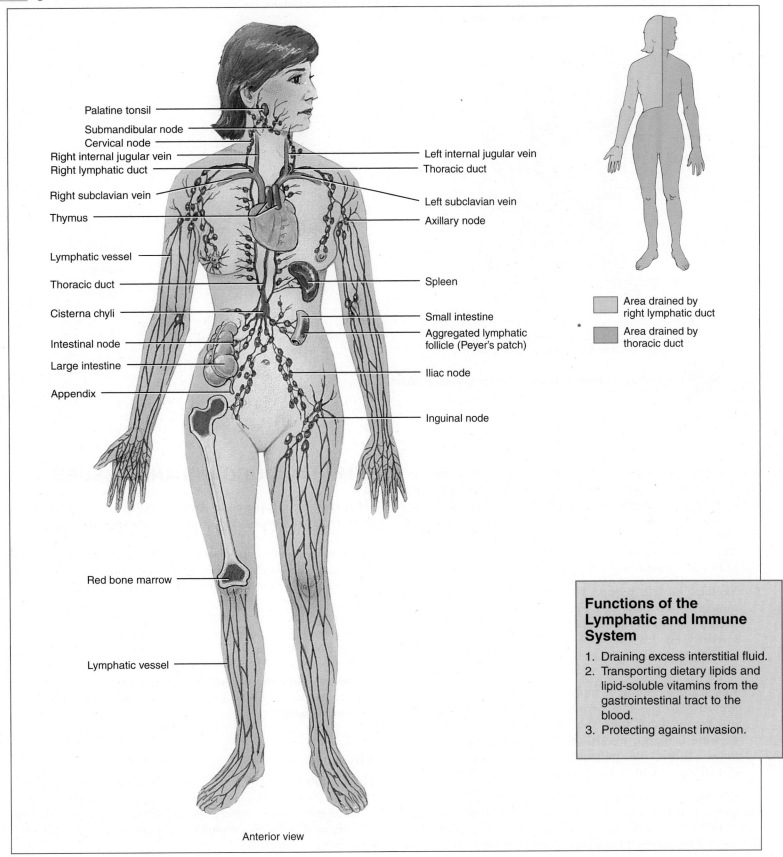

Palatine tonsil

Submandibular node

Cervical node

Right internal jugular vein

Right lymphatic duct

Right subclavian vein

Thymus

Lymphatic vessel

Thoracic duct

Cisterna chyli

Intestinal node

Large intestine

Appendix

Left internal jugular vein

Thoracic duct

Left subclavian vein

Axillary node

Spleen

Small intestine

Aggregated lymphatic follicle (Peyer's patch)

Iliac node

Inguinal node

Red bone marrow

Lymphatic vessel

Anterior view

Area drained by right lymphatic duct

Area drained by thoracic duct

Functions of the Lymphatic and Immune System

1. Draining excess interstitial fluid.
2. Transporting dietary lipids and lipid-soluble vitamins from the gastrointestinal tract to the blood.
3. Protecting against invasion.

What is lymphatic tissue?

Figure 17.2 ■ **Lymphatic capillaries.**

 Lymphatic capillaries are found throughout the body except in avascular tissues, the central nervous system, portions of the spleen, and red bone marrow.

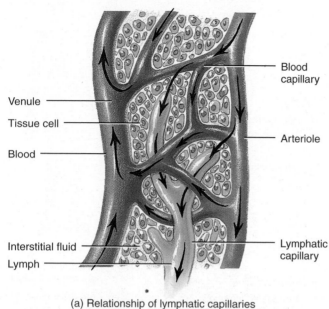

Blood capillary

Venule

Tissue cell

Blood

Interstitial fluid

Lymph

Arteriole

Lymphatic capillary

(a) Relationship of lymphatic capillaries to tissue cells and blood capillaries

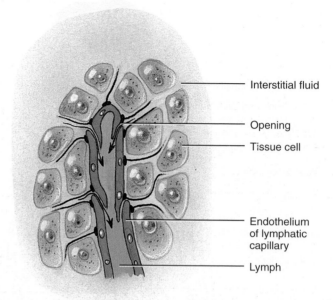

Interstitial fluid

Opening

Tissue cell

Endothelium of lymphatic capillary

Lymph

(b) Details of a lymphatic capillary

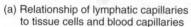 Is lymph more similar to blood plasma or interstitial fluid?

sequence of fluid flow is as follows: arteries (blood plasma) → blood capillaries (blood plasma) → interstitial spaces (interstitial fluid) → lymphatic capillaries (lymph) → lymphatic vessels and lymph nodes (lymph) → lymphatic ducts (lymph) → subclavian veins (blood plasma) (Figure 17.3).

The flow of lymph is maintained by the same two "pumps" that aid return of venous blood to the heart:

1. **Skeletal muscle pump.** The "milking action" of skeletal muscle contractions (see Figure 16.4 on page 376) compresses lymphatic vessels and forces lymph toward the subclavian veins. Like veins, lymphatic vessels contain valves, which ensure the one-way movement of lymph (Figure 17.3).

2. **Respiratory pump.** Lymph flow is also maintained by pressure changes that occur during inhalation (breathing in). Lymph flows from the abdominal region, where the pressure is higher, toward the thoracic region, where it is lower. When the pressures reverse during exhalation (breathing out), the valves prevent backflow of lymph.

Edema, an excessive accumulation of interstitial fluid in tissue spaces, may be caused by an obstruction, such as an infected lymph node, or by a blockage of lymphatic vessels. Edema may also result from increased capillary blood pressure, in which case interstitial fluid is formed faster than it can pass into lymphatic vessels.

LYMPHATIC ORGANS AND TISSUES

Objective: • **Compare the structure and functions of the various types of lymphatic organs and tissues.**

Lymphatic organs and tissues, which are widely distributed throughout the body, are classified into two groups based on their functions. *Primary lymphatic organs* are the sites where stem cells divide and mature into B cells and T cells. The primary lymphatic organs are the *red bone marrow* (in flat bones and the ends of the long bones of adults) and the *thymus.* Stem cells in red bone marrow give rise to mature B cells and to cells that migrate to the thymus, where they mature into T cells. The *secondary lymphatic organs* and *tissues,* the sites where most immune responses occur, include *lymph nodes,* the *spleen,* and *lymphatic nodules.*

Thymus

The *thymus* is a two-lobed organ located posterior to the sternum and medial to the lungs (see Figure 17.1). A *capsule,* a layer of connective tissue, covers each lobe. Internally, the thymus consists of T cells, macrophages, and epithelial cells that produce thymic hormones.

The thymus is large in the infant. After puberty, much of the thymic tissue is replaced by fat and connective tissue, but the

Figure 17.3 ■ **Relationship of lymphatic vessels to the cardiovascular system.**

 The flow of fluid is from arteries (blood plasma) to blood capillaries (blood plasma) to interstitial spaces (interstitial fluid) to lymphatic capillaries (lymph) to lymphatic vessels and lymph nodes (lymph) to lymphatic ducts (lymph) to subclavian veins (blood plasma).

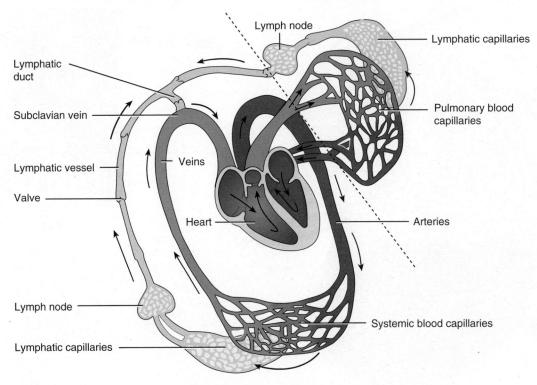

Arrows show direction of flow of lymph and blood

Which vessels of the cardiovascular system (arteries, veins, or capillaries) produce lymph?

gland continues to be functional throughout life. Recall from Chapter 13 that hormones produced by the thymus, such as thymosin and thymopoietin, promote the proliferation and maturation of T cells.

Lymph Nodes

The approximately 600 bean-shaped organs located along lymphatic vessels are called **lymph nodes.** They are scattered throughout the body, both superficially and deep, and usually occur in groups (see Figure 17.1). Lymph nodes are heavily concentrated near the mammary glands and in the axillae and groin. Each node is covered by a capsule of dense connective tissue (Figure 17.4). Internally, lymph nodes contain B cells that develop into plasma cells, which secrete antibodies, T cells, and macrophages.

Lymph nodes function as filters of lymphatic fluid. As lymph enters one end of a node through one of several **afferent lymphatic vessels** (af- = toward; -ferrent = to carry), foreign substances are trapped by **reticular fibers** within the **lymphatic**

sinuses. Then macrophages destroy some foreign substances by phagocytosis, and lymphocytes destroy others by a variety of immune responses. Filtered lymph then leaves the other end of the node through one or two **efferent lymphatic vessels** (ef- = away). Plasma cells and T cells that have proliferated within a lymph node can also leave the lymph nodes and circulate to other parts of the body. Both afferent and efferent lymphatic vessels contain valves that direct the flow of lymph inward through the afferent lymphatic vessels and outward through the efferent lymphatic vessels (Figure 17.4).

Lymphatic fluid is one vehicle for **metastasis** (me-TAS-ta-sis), the spread of cancer cells from a primary tumor to other sites in the body. Possible secondary tumor sites can be predicted by knowing the direction of lymph flow from a cancerous organ and the location of "downstream" lymph nodes.

Spleen

The **spleen** is the largest single mass of lymphatic tissue in the body (see Figure 17.1). It is covered by a capsule of dense

Figure 17.4 ■ **Structure of a lymph node.** Red arrows indicate direction of lymph flow.

Lymph nodes are present throughout the body, usually clustered in groups.

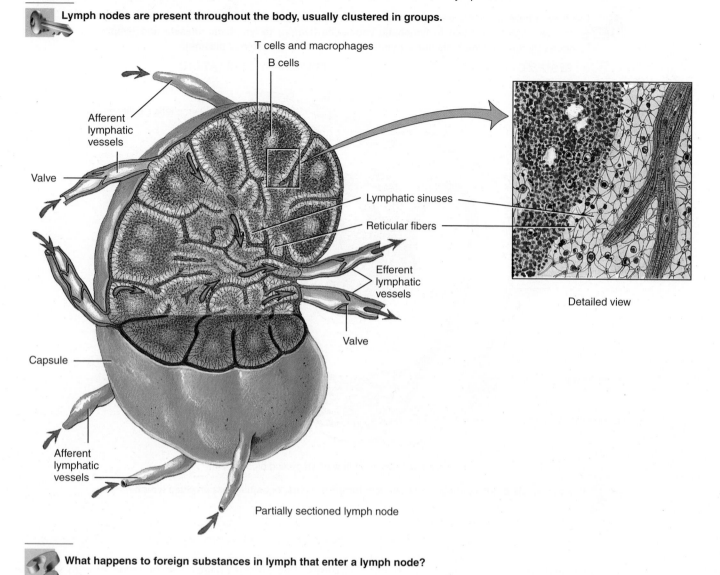

T cells and macrophages

B cells

Afferent lymphatic vessels

Valve

Lymphatic sinuses

Reticular fibers

Efferent lymphatic vessels

Valve

Capsule

Afferent lymphatic vessels

Partially sectioned lymph node

Detailed view

What happens to foreign substances in lymph that enter a lymph node?

connective tissue and lies between the stomach and diaphragm. The spleen contains two types of tissue called white pulp and red pulp. **White pulp** is lymphatic tissue, consisting mostly of lymphocytes and macrophages. **Red pulp** consists of blood-filled **venous sinuses** and cords of **splenic tissue** consisting of red blood cells, macrophages, lymphocytes, plasma cells, and granular leukocytes.

Blood flowing into the spleen through the splenic artery enters the white pulp. Within the white pulp, B cells and T cells carry out immune functions while macrophages destroy pathogens by phagocytosis. Within the red pulp, the spleen performs three functions related to blood cells: (1) removal by macrophages of worn out or defective blood cells and platelets; (2) storage of platelets, perhaps up to one-third of the body's

supply; and (3) production of blood cells (hemopoiesis) during fetal life.

The spleen is the organ most often damaged in cases of abdominal trauma. A ruptured spleen causes severe internal hemorrhage and shock. Prompt removal of the spleen *(splenectomy)* is needed to prevent death. After a splenectomy, other structures, particularly red bone marrow and the liver, can take over functions normally carried out by the spleen.

Lymphatic Nodules

Lymphatic nodules are oval-shaped concentrations of lymphatic tissue that are not surrounded by capsules. Because they are scattered throughout the connective tissue of mucous membranes lining the gastrointestinal, urinary, and reproductive tracts and the respiratory airways, lymphatic nodules are also referred to as *mucosa-associated lymphatic tissue (MALT)*.

Some lymphatic nodules occur as named masses in specific parts of the body. Among these are the tonsils in the pharyngeal region and the aggregated lymphatic follicles (Peyer's patches) in the ileum of the small intestine (see Figure 17.1). The five *tonsils*, which form a ring at the junction of the oral cavity, nasal cavity, and throat, are strategically positioned to participate in immune responses against inhaled or ingested foreign substances. The single *pharyngeal tonsil* (fa-RIN-jē-al) or *adenoid* is embedded in the posterior wall of the upper part of the throat (see Figure 18.2 on page 431). The two *palatine tonsils* (PAL-a-tīn) lie at the back of the mouth, one on either side; these are the tonsils commonly removed in a tonsillectomy. The paired *lingual tonsils* (LIN-gwal), located at the base of the tongue (see Figure 12.3a on page 281), may also require removal during a tonsillectomy.

NONSPECIFIC RESISTANCE TO DISEASE

Objective: • **Describe the mechanisms of nonspecific resistance to disease.**

A wide variety of defenses protect the human body and help maintain homeostasis by counteracting the activities of disease-producing organisms, called *pathogens* (PATH-ō-jenz), or their toxins (poisons). The ability to ward off disease is called *resistance*. Vulnerability or lack of resistance is called *susceptibility*. Defenses against disease may be grouped into two broad areas: nonspecific resistance and specific resistance. *Nonspecific resistance* comprises a wide variety of body reactions that provide immediate responses to fight invasion by a wide range of pathogens. Nonspecific resistance, as its name suggests, lacks specific responses to specific invaders; instead, its protective mechanisms function the same way, regardless of the type of invader. By contrast, the ability of the body to defend itself against specific invading agents such as bacteria, toxins, viruses, and foreign tissues is called *specific resistance* or *immunity*. In this section we consider mechanisms of nonspecific resistance to pathogens according to where they are encountered; specific resistance is the topic of the next section.

First Line of Defense: Skin and Mucous Membranes

The skin and mucous membranes of the body are the first line of defense against pathogens. Both physical and chemical barriers discourage pathogens and foreign substances from penetrating the body and causing disease.

With its many layers of closely packed, keratinized cells, the *epidermis* (the outer epithelial layer of the skin) provides a formidable physical barrier to the entrance of microbes (see Figure 5.1 on page 101). In addition, periodic shedding of epidermal cells helps remove microbes at the skin surface. Bacteria rarely penetrate the intact surface of healthy epidermis. But if this surface is broken by cuts, burns, or punctures, pathogens can penetrate the epidermis and then invade adjacent tissues or circulate in the blood to other parts of the body.

The epithelial layer of *mucous membranes*, which line body cavities, secretes a fluid called *mucus* that lubricates and moistens the cavity surface. Because mucus is slightly viscous, it traps many microbes and foreign substances. The mucous membrane of the nose has mucus-coated *hairs* that trap and filter microbes, dust, and pollutants from inhaled air. The mucous membrane of the upper respiratory tract contains *cilia*, microscopic hairlike projections on the surface of the epithelial cells. The waving action of cilia propels inhaled dust and microbes that have become trapped in mucus toward the throat. Coughing and sneezing accelerate movement of mucus and its entrapped pathogens out of the body.

The *lacrimal apparatus* (LAK-ri-mal) of the eyes (see Figure 12.4 on page 282) manufactures and drains away tears in response to irritants. Blinking spreads tears over the surface of the eyeball, and the continual washing action of tears helps to dilute microbes and keep them from settling on the surface of the eye. *Saliva*, produced by the salivary glands, washes microbes from the surfaces of the teeth and from the mucous membrane of the mouth, much as tears wash the eyes. The flow of saliva reduces colonization of the mouth by microbes. The cleansing of the urethra by the *flow of urine* retards microbial colonization of the urinary system. Vaginal secretions likewise move microbes out of the female body. *Defecation* and *vomiting* also expel microbes. For example, in response to some microbial toxins, the smooth muscle of the lower gastrointestinal tract contracts vigorously; the resulting *diarrhea* rapidly expels many of the microbes.

Certain chemicals also contribute to the resistance of the skin and mucous membranes to microbial invasion. Sebaceous (oil) glands of the skin secrete an oily substance called *sebum* that forms a protective film over the surface of the skin. The unsaturated fatty acids in sebum inhibit the growth of certain bacteria and fungi. The acidity of the skin (pH 3 to 5) is caused in part by the secretion of fatty acids and lactic acid. *Perspiration* helps flush microbes from the surface of the skin. It also contains *lysozyme*, an enzyme capable of breaking down the cell walls of certain bacteria. This same antimicrobial activity is exhibited by the lysozyme found in tears, saliva, nasal secretions, and tissue fluids. *Gastric juice*, produced by the glands of the stomach, is a mixture of hydrochloric acid, enzymes, and mucus. The strong acidity of gastric juice (pH 1.2 to 3.0) destroys many bacteria and most bacterial toxins. *Vaginal secretions* also are slightly acidic, which discourages bacterial growth.

Second Line of Defense: Internal Defenses

When pathogens penetrate the mechanical and chemical barriers of the skin and mucous membranes, they encounter a second line of defense: antimicrobial proteins, phagocytes, natural killer cells, inflammation, and fever.

Antimicrobial Proteins

Blood and interstitial fluids contain three main types of *antimicrobial proteins* that discourage microbial growth: interferons, complement, and transferrins.

INTERFERONS Virus-infected lymphocytes, macrophages, and fibroblasts produce proteins called *interferons* (in′-ter-FĒR-ons), or *IFNs*. Once released from infected cells, IFNs diffuse to uninfected neighboring cells, where they induce synthesis of antiviral proteins that interfere with viral replication. IFNs are an important defense against infection by many different viruses.

In clinical trials, IFNs have exhibited limited effects against some tumors. Intron A is an interferon that is used to treat several virus-associated disorders, including Kaposi's (KAP-ō-sēz) sarcoma, a cancer that often occurs in patients infected with HIV (the virus that causes AIDS, acquired immunodeficiency syndrome); genital warts caused by the human papilloma virus; hepatitis B and C, caused by the hepatitis B and C viruses; and hairy cell leukemia. Betaseron, another interferon drug, is used to slow the progression of multiple sclerosis (MS) and lessen the frequency and severity of MS attacks.

COMPLEMENT A family of proteins found in blood plasma and on plasma membranes is termed *complement* because, when activated, these proteins "complement" or enhance certain immune, allergic, and inflammatory reactions. Some complement proteins create holes in the plasma membrane of the microbe, causing the contents of the microbe to leak out, a process called *cytolysis.* Some cause *chemotaxis* (kē′-mō-TAK-sis), the chemical attraction of phagocytes to a site, and increase blood flow to the area by causing vasodilation of arterioles. Others promote *opsonization* (op′-son-i-ZĀ-shun), a process in which complement proteins bind to the surface of a microbe and promote phagocytosis.

TRANSFERRINS Iron-binding proteins called *transferrins* inhibit the growth of certain bacteria by reducing the amount of available iron.

Natural Killer Cells and Phagocytes

When microbes penetrate the skin and mucous membranes or bypass the antimicrobial proteins in blood, the next nonspecific line of defense consists of natural killer cells and phagocytes.

Figure 17.5 ■ **Phagocytosis.**

The stages of phagocytosis are chemotaxis, adherence, ingestion, digestion, and killing.

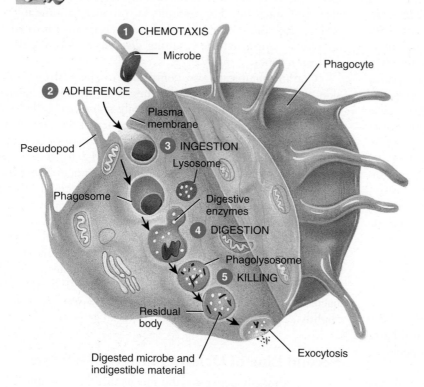

(a) Phases of phagocytosis

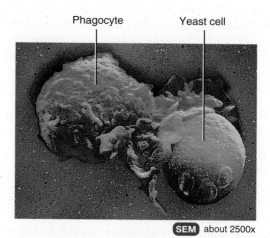

(b) Phagocyte engulfing a yeast cell

Which tissues and organs contain fixed macrophages?

Natural killer (NK) cells are lymphocytes that have the ability to induce cytolysis and thereby kill a wide variety of infectious microbes and certain tumor cells. The spleen, lymph nodes, red bone marrow, and blood all contain NK cells. Some cancer and AIDS patients have defective or decreased numbers of NK cells.

Phagocytes (*phago-* = eat; *-cytes* = cells) are specialized cells that perform *phagocytosis* (*-osis* = process), the ingestion of microbes or other particles such as cellular debris. There are two main types of phagocytes: neutrophils and macrophages. When an infection occurs, neutrophils and monocytes migrate to the infected area. During this migration, the monocytes enlarge and develop into actively phagocytic cells called *macrophages* (MAK-rō-fā-jez) (see Figure 14.2 on page 336). Some of these cells, called *wandering macrophages,* leave the blood and migrate to infected areas. Others, called *fixed macrophages,* remain in certain tissues and organs of the body, such as the skin and subcutaneous layer, liver, lungs, brain, spleen, lymph nodes, and red bone marrow.

Phagocytosis occurs in five stages: chemotaxis, adherence, ingestion, digestion, and killing (Figure 17.5).

1 **Chemotaxis.** Phagocytosis begins with chemotaxis. Chemicals that attract phagocytes might come from the microbes, white blood cells, damaged tissue cells, or activated complement.

2 **Adherence.** Attachment of the phagocyte to the microorganism or other foreign material, called *adherence,* sometimes occurs easily. Microorganisms can more readily be phagocytized if they first undergo opsonization, which as you have just learned is the binding of complement to the invading pathogen.

3 **Ingestion.** Following adherence, the plasma membrane of the phagocyte extends projections, called *pseudopods,* that engulf the microorganism in a process called *ingestion.* When the pseudopods meet, they fuse, surrounding the microorganism with a sac called a *phagosome.*

4 **Digestion.** The phagosome enters the cytoplasm and merges with lysosomes to form a single, larger structure called a *phagolysosome.* The lysosome contributes lysozyme, which breaks down microbial cell walls, and digestive enzymes, which degrade carbohydrates, proteins, lipids, and nucleic acids.

5 **Killing.** The chemical onslaught provided by lysozyme and digestive enzymes within a phagolysosome quickly kills many types of microbes. Any materials that cannot be degraded further remain in structures called *residual bodies,* which eventually undergo exocytosis.

Inflammation

Inflammation is a nonspecific, defensive response of the body to stress due to tissue damage. Among the conditions that may produce inflammation are pathogens, abrasions, chemical irritations, distortion or disturbances of cells, and extreme temperatures. Regardless of the cause, the injury may be viewed as a form of stress.

Inflammation is characterized by four symptoms: *redness, pain, heat,* and *swelling.* A fifth symptom can be *loss of function,* depending on the site and extent of the injury. Inflammation is an attempt to dispose of microbes, toxins, or foreign material at the site of injury; to prevent their spread to other organs; and to prepare the site for tissue repair.

There are three basic stages of the inflammatory response: (1) vasodilation and increased permeability of blood vessels, (2) phagocyte emigration, and (3) repair.

Vasodilation and Increased Permeability of Blood Vessels Two immediate changes occur in the blood vessels in a region of tissue injury: *vasodilation,* an increase in the diameter of the blood vessels, and *increased permeability* (Figure 17.6). Increased permeability means that substances normally retained in blood are permitted to pass out of the blood vessels. Vasodilation allows more blood to flow to the damaged area, and

Figure 17.6 ■ **Inflammatory response.** Several substances stimulate vasodilation, increased permeability of blood vessels, chemotaxis (attracting phagocytes to the area), emigration, and phagocytosis.

The three stages of inflammation are (1) vasodilation and increased permeability of blood vessels, (2) phagocyte emigration, and (3) tissue repair.

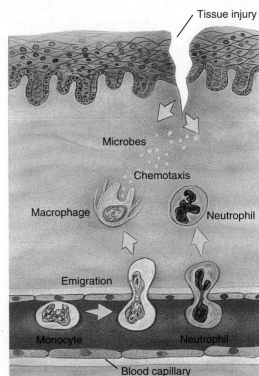

Phagocytes migrate from blood to site of tissue injury

 What causes redness at an inflammatory site?

increased permeability permits defensive substances such as antibodies and clot-forming chemicals to enter the injured area from the blood.

A substance called *histamine* contributes to vasodilation, increased permeability, and other aspects of the inflammatory response. It is released by mast cells in connective tissue and basophils and platelets in blood. Prostaglandins, membrane lipids that function as local hormones, intensify the effects of histamine.

Dilation of arterioles and increased permeability of capillaries produce three of the symptoms of inflammation: heat, redness, and swelling (edema). Heat and redness result from the large amount of blood that accumulates in the damaged area. The area swells due to an increased amount of interstitial fluid that has leaked out of the capillaries.

Pain results from injury to nerve fibers and from toxic chemicals released by microorganisms. Some of the substances that promote vasodilation and increased permeability, especially prostaglandins, also affect nerve endings and cause pain.

The increased permeability of capillaries causes leakage of clotting proteins into tissues. Fibrinogen is converted to an insoluble, thick network of fibrin threads, which traps the invading organisms and prevents their spread. The resulting clot isolates the invading microbes and their toxins.

EMIGRATION   Generally, within an hour after the inflammatory process starts, phagocytes appear on the scene. As large amounts of blood accumulate, neutrophils begin to stick to the inner surface of the endothelium (lining) of blood vessels (Figure 17.6). Then the neutrophils begin to squeeze through the wall of the blood vessel to reach the damaged area. This process, called *emigration*, depends on chemotaxis. Neutrophils attempt to destroy the invading microbes by phagocytosis. The neutrophils predominate in the early stages of infection but die off fairly rapidly.

As the inflammatory response continues, monocytes follow the neutrophils into the infected area. Once in the tissue, monocytes are transformed into wandering macrophages that add to the phagocytic activity of the fixed macrophages already present. True to their name, macrophages are much more potent phagocytes than are neutrophils. They are large enough to engulf damaged tissue, worn-out neutrophils, and invading microbes.

REPAIR   In all but very mild inflammations, pus is produced. *Pus* is a thick fluid that contains white blood cells and dead tissue debris. Pus formation continues until the infection subsides. The pus may push to the body surface or into an internal cavity for dispersal. If the pus cannot drain out of the body, an abscess develops. An *abscess* is an excessive accumulation of pus in a confined space. Common examples include pimples and boils.

Fever

Fever is an abnormally high body temperature that occurs because the hypothalamic thermostat is reset. Although its significance is still not fully understood, fever commonly occurs during infection and inflammation. Many bacterial toxins elevate body temperature, sometimes by triggering release of fever-causing substances such as interleukin-1. Elevated body temperature intensifies the effects of interferons, inhibits the growth of some microbes, and speeds up body reactions that aid repair. (Fever is discussed in more detail on page 506.)

Table 17.1 summarizes the factors that contribute to nonspecific resistance.

SPECIFIC RESISTANCE: IMMUNITY

Objectives: • **Define immunity and compare it with nonspecific resistance to disease.**

• **Explain the relationship between an antigen and an antibody.**

• **Compare the functions of cell-mediated immunity and antibody-mediated immunity.**

The various mechanisms of nonspecific resistance have one thing in common: They are not specifically directed against a particular type of invader. Specific resistance to disease, called *immunity*, involves the production of specific types of cells or specific antibodies to destroy a particular antigen. An *antigen* is any substance—such as microbes, foods, drugs, pollen, or tissue—that the immune system recognizes as foreign. The branch of science that deals with the responses of the body to antigens is called *immunology* (im′-yoo-NOL-ō-jē; *immun-* = free from service; *-ology* = study of). The *immune system* includes the cells and tissues that carry out immune responses.

Types of Immune Responses

Immunity consists of two types of closely allied responses, both triggered by antigens. In *cell-mediated immune responses,* certain T cells proliferate into cytotoxic T cells, which directly attack the invading antigen. In *antibody-mediated immune responses,* B cells change into plasma cells, which synthesize and secrete proteins called *antibodies.* A given antibody binds to and inactivates a specific antigen. Cell-mediated immunity augments nonspecific resistance and also plays a crucial role in the initiation of antibody-mediated immune responses.

Cell-mediated immunity is particularly effective against (1) intracellular pathogens that reside within host cells (primarily fungi, parasites, and viruses), (2) some cancer cells, and (3) foreign tissue transplants. Antibody-mediated immunity works mainly against (1) antigens present in body fluids and (2) extracellular pathogens that multiply in body fluids but rarely enter body cells (primarily bacteria). A given pathogen can provoke both types of immune responses.

Maturation of T Cells and B Cells

Both B cells and T cells develop from the pluripotent stem cells of red bone marrow (see Figure 14.2 on page 336). B cells complete their development in bone marrow, a process that continues throughout life. T cells develop from cells that migrate from

Table 17.1 / Summary of Nonspecific Resistance to Disease

Component	Functions
Skin and Mucous Membranes	
Mechanical factors	
Epidermis	Forms a physical barrier to the entrance of microbes.
Mucous membranes	Inhibit the entrance of many microbes.
Mucus	Traps microbes in respiratory and gastrointestinal tracts.
Hairs	Filter out microbes and dust in nose.
Cilia	Together with mucus, trap and remove microbes and dust from upper respiratory tract.
Lacrimal apparatus	Tears dilute and wash away irritating substances and microbes.
Saliva	Washes microbes from surfaces of teeth and mucous membranes of mouth.
Urine	Washes microbes from urethra.
Vaginal secretions	Move microbes out of the female body.
Defecation and vomiting	Expel microbes from body.
Chemical factors	
Unsaturated fatty acids	Antibacterial substances in sebum.
Acidic pH of skin	Discourages growth of many microbes.
Lysozyme	Antimicrobial substance in perspiration, tears, saliva, nasal secretions, and tissue fluids.
Gastric juice	Destroys bacteria and most toxins in stomach.
Vaginal secretions	Slight acidity discourages bacterial growth.
Internal Defenses	
Antimicrobial proteins	
Interferons (IFNs)	Protect uninfected host cells from viral infection.
Complement system	Causes rupture of microbes, promotes phagocytosis, and contributes to inflammation.
Transferrins	Inhibit the growth of certain bacteria by reducing iron levels.
Natural killer (NK) cells	Kill a wide variety of microbes and certain tumor cells.
Phagocytes	Ingest foreign particles and cellular debris.
Inflammation	Confines and destroys microbes and prepares for tissue repair.
Fever	Intensifies the effects of interferons, inhibits growth of some microbes, and speeds up some of the body's responses to infection.

bone marrow into the thymus, where they mature. Most T cells arise before puberty, but they continue to mature and leave the thymus throughout life.

Before T cells leave the thymus or B cells leave red bone marrow, they acquire several distinctive surface proteins. Some of these proteins function as *antigen receptors*—molecules capable of recognizing specific antigens. Other surface proteins allow the T cells to develop into either helper or cytotoxic T cells, which have very different functions.

Antigens and Antibodies

An antigen causes the body to produce specific antibodies and/or specific T cells that react with it. Antigens have two important characteristics: (1) the ability to stimulate the formation of specific antibodies and (2) the ability of the antigen to react specifically with the produced antibodies.

An entire microbe, such as a bacterium or virus, or just a part of it, such as the flagellum, capsule, or cell wall, may act as an antigen. Toxins secreted by bacteria are also highly antigenic. Nonmicrobial examples of antigens include foods, drugs, pollen, cancer cells, transplanted tissues or organs, and serum from other humans, animals, or insects.

Located at the plasma membrane surface of most body cells are "self-antigens," the *major histocompatibility complex (MHC)* proteins. These proteins are also called *human leukocyte antigens (HLA)* because they were first identified on white blood cells. Unless you have an identical twin, your MHC proteins are unique. Thousands to several hundred thousand MHC molecules mark the surface of all your body cells (except red blood cells). Although MHC proteins are the reason that tissues may be rejected when they are transplanted from one person to another, their normal function is to help T cells recognize that an antigen is foreign, not self. This recognition is an important first step in any immune response. The body's own molecules, recognized as self, do not normally act as antigens. Under certain conditions, however, the distinction between self and nonself antigens breaks down, leading to an *autoimmune disease* in which the immune system attacks body tissues (see page 425).

An *antibody (Ab)* is a protein produced by plasma cells in response to an antigen that combines specifically with the antigen that provoked its production. The combination of antibody and antigen depends on the size and shape of their combining sites (Figure 17.7), very much like the fit between a lock and key. The portion of an antibody (the key) that recognizes and com-

bines with an antigen (the lock) is called an *antigen-binding site.*

Chemically, antibodies belong to a group of plasma proteins called globulins, and for this reason they are also known as *immunoglobulins* (im'-yoo-nō-GLOB-yoo-lins), or *Igs*. Immunoglobulins are grouped in five different classes, designated IgG, IgA, IgM, IgD, and IgE. Each class has a distinct chemical structure and a specific role to play (Table 17.2). Because they appear first and are relatively short-lived, the presence of IgM antibodies indicates a recent invasion. In a sick patient, the responsible pathogen may be suggested by the presence of high levels of

Figure 17.7 ■ **Structure of an antibody (IgG) and relationship of an antigen to an antibody.**

An antigen induces the body to produce specific antibodies that combine with the antigen.

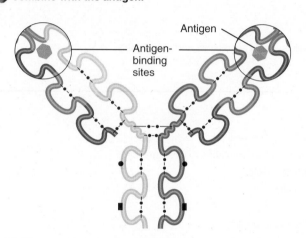

(a) Diagram of IgG

(b) Antibody symbol

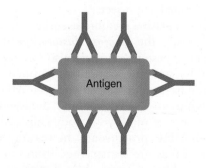

(c) Antibodies covering an antigen

 Chemically, antibodies are what type of molecules?

Table 17.2 / Classes of Antibodies, or Immunoglobulins (Igs)	
Name and Structure	**Characteristics and Functions**
IgG	Most abundant, about 80% of all antibodies in the blood; found in blood, lymph, and the intestines. Protects against bacteria and viruses by enhancing phagocytosis, neutralizing toxins, and triggering the complement system. It is the only class of antibody to cross the placenta from mother to fetus, thereby conferring considerable immune protection in newborns.
IgA	Makes up 10% to 15% of all antibodies in the blood. Found mainly in sweat, tears, saliva, mucus, milk, and gastrointestinal secretions. Smaller quantities are present in blood and lymph. Levels decrease during stress, lowering resistance to infection. On mucous membranes, provides localized protection against bacteria and viruses.
IgM	About 5% to 10% of all antibodies in the blood; first antibody class to be secreted by plasma cells after an initial exposure to any antigen; found in blood and lymph. Activates complement and causes agglutination and lysis of microbes. In blood plasma, the anti-A and anti-B antibodies of the ABO blood group, which bind to A and B antigens during incompatible blood transfusions, are also IgM antibodies (see Figure 14.9 on page 345).
IgD	About 0.2% of all antibodies in the blood; found in blood, in lymph, and on the surfaces of B cells as antigen receptors. Involved in activation of B cells.
IgE	Less than 0.1% of all antibodies in the blood; located on mast cells and basophils. Involved in allergic and hypersensitivity reactions; provides protection against parasitic worms.

Lifestyle, Immune Function, and Resistance to Disease

*I*f you want to observe the relationship between lifestyle and immune function, visit a college campus. As the semester progresses and the workload accumulates, an increasing number of students can be found in the waiting rooms of student health services.

Is Stress the Culprit?

Stress has been implicated as hazardous to immune function. Researchers in the field of *psychoneuroimmunology (PNI)* have found many communication pathways that link the nervous, endocrine, and immune systems. Chronic stress affects the immune system in several ways. For example, cortisol, a hormone secreted by the adrenal cortex in association with the stress response, inhibits immune system activity, perhaps one of its energy conservation effects. PNI research supports what many people have observed since the beginning of time: Your thoughts, feelings, moods, and beliefs influence your level of health and the course of disease. Especially toxic to the immune system are feelings of helplessness, hopelessness, fear, and social isolation.

People resistant to the negative health effects of stress are more likely to experience a sense of control over the future, a commitment to their work, expectations of generally positive outcomes for themselves, and feelings of social support. To increase your stress resistance, cultivate an optimistic outlook, get involved in your work, and build good relationships with others.

Or Is Lifestyle at Fault?

When work and stress pile up, health habits can change. Many people smoke or consume more alcohol when stressed, two habits detrimental to optimal immune function. Under stress, people are less likely to eat well or exercise regularly, two habits that enhance immunity.

Adequate sleep and relaxation are especially important for a healthy immune system. But when there aren't enough hours in the day, you may be tempted to steal some from the night. While skipping sleep may give you a few more hours of productive time in the short run, in the long run you end up even farther behind, especially if getting sick keeps you out of commission for several days, blurs your concentration, and blocks your creativity.

Even if you make time to get eight hours of sleep, stress can cause insomnia. If you find yourself tossing and turning at night, it's time to improve your stress management and relaxation skills! Be sure to "change the channel" and unwind from the day before going to bed.

▶ *Think It Over*

▶ Have you ever observed a connection between stress and illness in your own life? Do you feel it was caused by stress or other lifestyle factors?

IgM specific to a particular pathogen. Resistance of the fetus and newborn baby to infection stems mainly from maternal IgG antibodies that cross the placenta before birth and IgA antibodies in breast milk after birth.

Processing and Presenting Antigens

For an immune response to occur, B cells and T cells must recognize that a foreign antigen is present. Whereas B cells can recognize and bind to antigens in interstitial fluid, T cells only recognize fragments of antigenic proteins that are "presented" together with a major histocompatibility complex self-antigen.

A special class of cells, called *antigen-presenting cells (APCs)*, prepare antigens for display. APCs include macrophages, B cells, and dendritic cells (so named for their long, branchlike projections). APCs are strategically located in places where antigens are likely to penetrate nonspecific defenses and enter the body, such as the epidermis and dermis of the skin (Langerhans cells are a type of dendritic cell); the mucous membranes that line the respiratory, gastrointestinal, urinary, and reproductive tracts; and lymph nodes.

APCs are said to "process and present" antigens as follows (Figure 17.8). First, they ingest a foreign antigen, digest the antigen into fragments, combine a fragment of the antigen with an MHC protein, and insert the duo into their own plasma membrane. Then the APC enters a lymphatic vessel and migrates from the body tissue where it encountered the antigen to one of the lymphatic tissues. There, it meets a small number of T cells having the correct antigen receptors to fit a particular processed antigen fragment, and a cell-mediated immune response ensues.

T Cells and Cell-Mediated Immunity

The presentation of antigen by antigen-presenting cells informs T cells that intruders are present in the body and that combative action should begin. But a T cell becomes activated only if it binds to the foreign antigen and at the same time receives a second signal, a process known as *costimulation* (Figure 17.9). A common costimulator is *interleukin-1 (IL-1)*, a protein that is released by antigen-presenting cells. The need for two signals is a little like starting and driving a car. When you insert the correct key (antigen) in the ignition (T cell receptor) and turn it, the car starts (recognition of specific antigen), but it cannot move forward until you move the gear shift into drive (costimulation). The need for costimulation may prevent an immune response from occurring accidentally.

Once a T cell is activated, it enlarges and proliferates (divides many times), forming a population of identical cells, termed a *clone,* that can all recognize the same specific antigen. Before the first exposure to a given antigen, only a few T cells might be able to recognize it, but once an immune response has begun, there are thousands. Activation and proliferation of T cells occur in the secondary lymphatic organs and tissues. If you have ever noticed swollen tonsils or lymph nodes in your neck, the proliferation of lymphocytes participating in an immune response was likely the cause.

The three major types of T cells are helper T cells, cytotoxic T cells, and memory T cells. *Helper T cells,* also called *T4 cells,* help other cells of the immune system combat intruders. Helper T cells release *interleukin-2 (IL-2),* a protein that causes cytotoxic T cells (described next) to grow and divide. Other proteins released by helper T cells attract phagocytes and enhance the phagocytic ability of macrophages. Helper T cells also stimulate the development of B cells into antibody-producing plasma cells.

Cytotoxic T cells, also called *T8 cells,* are the soldiers that march forth to do battle with foreign invaders in cell-mediated immune responses. Cytotoxic T cells are especially effective against body cells infected by viruses, cancer cells associated with viral infection, and cells of a tissue transplant. After they proliferate, they leave secondary lymphatic organs and tissues and migrate to the site of invasion, infection, or tumor forma-

Figure 17.8 ■ **Processing and presenting of antigen by an antigen-presenting cell (APC).**
1. APC ingests a foreign antigen. **2.** It digests the antigen into fragments. **3.** The fragment of the foreign antigen is combined with an MHC protein. **4.** The MHC–processed antigen duo is inserted into the APC's plasma membrane.

An APC migrates to a lymphatic tissue where it presents a processed antigen to T cells having receptors that fit that particular processed antigen fragment.

4.

— Processed antigen
— MHC protein

3.

Migration via lymphatic vessels

T cell antigen receptor that fits processed antigen

1.

2.

Antigen-presenting cell (APC)

T cells in lymphatic tissue

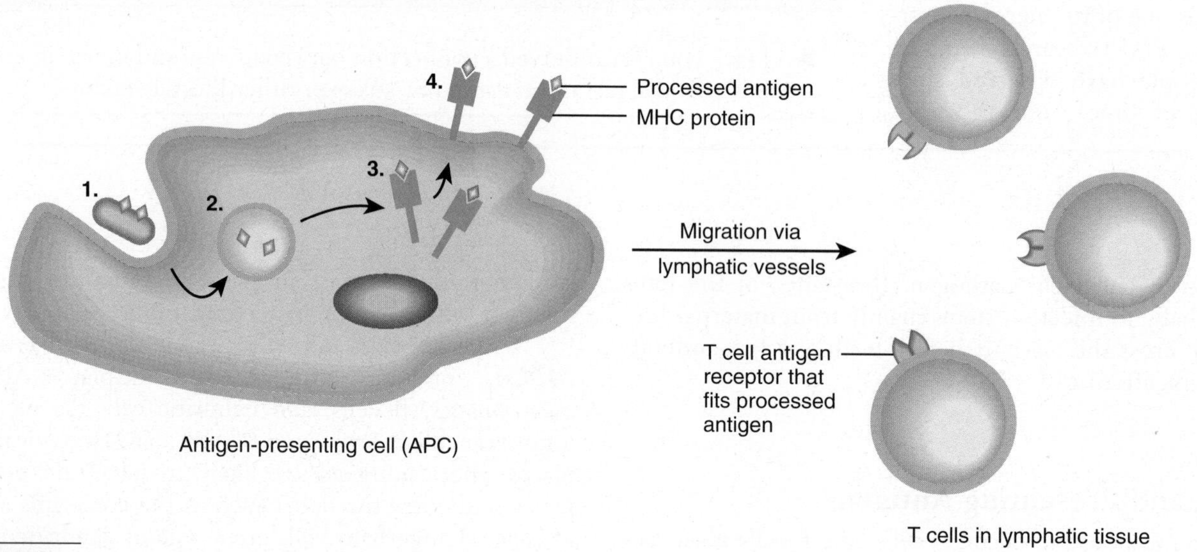

 Which types of cells are APCs?

Figure 17.9 ▪ **Activation and proliferation of T cells in cell-mediated immunity.**

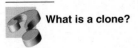 Two signals are needed for activation of a T cell: presentation of a processed antigen fragment by an antigen-presenting cell and costimulation.

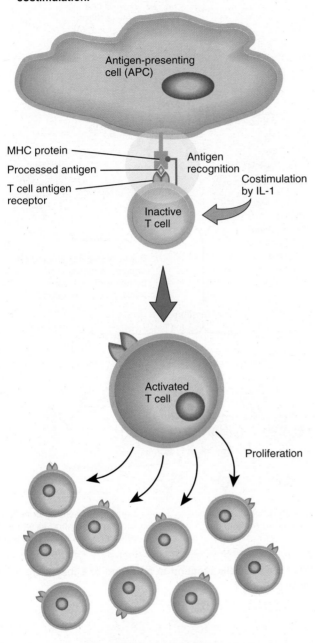

Antigen-presenting cell (APC)

MHC protein
Processed antigen
T cell antigen receptor

Antigen recognition

Costimulation by IL-1

Inactive T cell

Activated T cell

Proliferation

Clone of activated T cells in a secondary lymphatic organ

What is a clone?

tion. Cytotoxic T cells recognize and attach to cells bearing the antigen that stimulated their activation and proliferation, operating in the following way (Figure 17.10)

① Their antigen receptors recognize and attach to an antigen associated with an MHC protein at the surface of a virus-infected body cell.

② Then the cytotoxic T cells deliver "lethal hits" that kill the infected or intruder cells without damaging themselves. First, they release a protein called *perforin* that forms holes in the plasma membrane of the target cell, causing its cytolysis (rupture). Second, they release *lymphotoxin,* a toxic molecule that causes the target cell's DNA to fragment. As a result, the cell dies.

③ Finally, cytotoxic T cells secrete gamma-interferon, which activates macrophages at the scene of the battle. The macrophages help clean up the debris from dying cells. After detaching from its victim, a cytotoxic T cell can seek out and destroy another invader that displays the same antigen.

Figure 17.10 ▪ **Action of cytotoxic T cell.** After delivering a "lethal hit," a cytotoxic T cell can detach and attack another target cell displaying the same antigen.

Cytotoxic T cells kill their targets directly by secreting perforin and lymphotoxin.

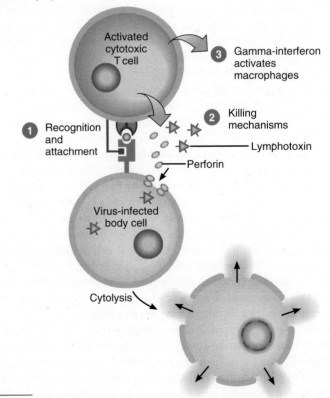

Activated cytotoxic T cell

③ Gamma-interferon activates macrophages

② Killing mechanisms

① Recognition and attachment

Lymphotoxin

Perforin

Virus-infected body cell

Cytolysis

 Besides cells infected by viruses, what other types of target cells are attacked by cytotoxic T cells?

Memory T cells remain in lymphatic tissue long after the original infection and remain able to recognize the original invading antigen. Should the same antigen invade the body at a later date, the memory T cells initiate a far swifter reaction than during the first invasion. In fact, the second response is so rapid that the pathogens are usually destroyed before any signs or symptoms of the disease occur, explaining why a person usually has the chickenpox only once. Memory T cells may provide immunity to a particular antigen for years.

When a normal cell transforms into a cancer cell, it may display cell surface components called *tumor antigens*. These are molecules that are rarely, if ever, displayed on the surface of normal cells. If the immune system can recognize tumor antigens as nonself, it can destroy the cancer cells carrying them. This type of immune response, called *immunologic surveillance*, is carried out by cytotoxic T cells, macrophages, and natural killer cells. It appears to be most effective in eliminating tumor cells that arise due to a cancer-causing virus. Despite immunologic surveillance, some cancer cells escape destruction, a phenomenon called *immunologic escape*. One possible explanation is that tumor cells shed their tumor antigens, thus evading recognition.

B Cells and Antibody-Mediated Immunity

The body contains not only millions of different T cells, but also millions of different B cells, each capable of responding to a specific antigen. Whereas cytotoxic T cells leave lymphatic tissues to seek out and destroy a foreign antigen, B cells stay put. In the presence of a foreign antigen, specific B cells in lymph nodes, the spleen, or mucosa-associated lymphatic tissue become activated. They then proliferate and develop into plasma cells that secrete specific antibodies, which in turn circulate in the lymph and blood to reach the site of invasion.

During activation of a B cell, antigen receptors on the cell surface bind to an antigen (Figure 17.11). B cell antigen receptors are chemically similar to the antibodies that eventually are secreted by the plasma cells. Although B cells can respond to an unprocessed antigen present in lymph or interstitial fluid, their response is much more intense when nearby antigen-presenting cells also process and present antigen to them. Some antigen is then taken into the B cell, broken down into peptide fragments and combined with MHC protein, and moved to the B cell surface. Helper T cells recognize the MHC–processed antigen duo and deliver the costimulation needed for B cell proliferation and differentiation. The helper T cell produces interleukin-2 and other proteins that function as costimulators to activate B cells. Interleukin-1, secreted by macrophages, also enhances B cell proliferation and development into plasma cells.

Some of the activated B cells enlarge, divide, and differentiate into a clone of antibody-secreting *plasma cells*. A few days after exposure to an antigen, a plasma cell is secreting hundreds of millions of antibodies daily, and secretion occurs for about four or five days, until the plasma cell dies. Most antibodies travel in lymph and blood to the invasion site. Some activated B

Figure 17.11 ■ **Activation and proliferation of B cells in antibody-mediated immunity.**

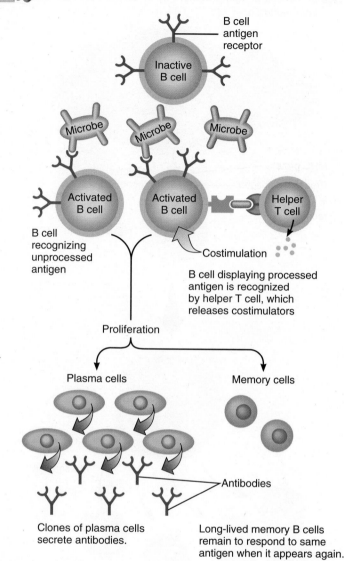

B cells develop into antibody-secreting plasma cells.

What type of cells respond to a second or subsequent invasion by an antigen?

cells do not differentiate into plasma cells but rather remain as *memory B cells* that are ready to respond more rapidly and forcefully should the same antigen reappear at a future time.

Although the functions of the five classes of antibodies differ somewhat, all attack antigens in several ways:

1. **Neutralizing antigen.** The reaction of antibody with antigen neutralizes some bacterial toxins and prevents attachment of some viruses to body cells.

2. **Immobilizing bacteria.** Some antibodies cause bacteria to lose their motility, which limits bacterial spread into nearby tissues.

3. **Agglutinating antigen.** The antigen–antibody reaction may connect pathogens to one another, causing agglutination (clumping together). Phagocytic cells ingest agglutinated microbes more readily.

4. **Activating complement.** Antigen–antibody complexes activate complement proteins, which then work to remove microbes as previously described.

5. **Enhancing phagocytosis.** Antibodies also enhance the activity of phagocytes by coating microbes so that they are more susceptible to phagocytosis (opsonization).

6. **Providing fetal and newborn immunity.** Resistance of the fetus and newborn baby to infection stems mainly from maternal IgG antibodies that pass across the placenta before birth and IgA antibodies in breast milk after birth.

An antibody-mediated response typically produces many different antibodies that recognize different features of an antigen or different antigens of a foreign cell. By contrast a *monoclonal antibody (MAb)* is pure antibody produced from a single clone of identical cells. Such cells can be obtained by fusing a specific B cell with a cultured cell that can divide endlessly. The resulting cell, called a *hybridoma* (hī-bri-DŌ-ma), is a long-term source of pure antibodies. One clinical use of monoclonal antibodies is for measuring levels of a drug in a patient's blood; other uses include the diagnosis of strep throat; pregnancy; allergies; and diseases such as hepatitis, rabies, and some sexually transmitted diseases. MAbs have also been used to detect cancer at an early stage and to determine the extent of metastasis. They may also be useful in preparing vaccines to counteract the rejection associated with transplants, to treat autoimmune diseases, and perhaps to treat AIDS.

Immunological Memory

A hallmark of immune responses is the lasting memory that remains for specific antigens that have triggered immune responses in the past. Immunological memory is due to the presence of long-lasting antibodies and very long lived lymphocytes that arise during proliferation of antigen-stimulated B cells and T cells.

Immune responses, whether cell-mediated or antibody-mediated, are much quicker and more intense after a second or subsequent exposure to an antigen than after the first exposure.

Initially, only a few cells have the correct specificity to respond, and the immune response may take several days to build to maximum intensity. Because thousands of memory cells exist after an initial encounter with an antigen, the next time the same antigen appears they can proliferate and differentiate into plasma cells or cytotoxic T cells within hours.

One measure of immunological memory is the amount of antibody in plasma. Following an initial contact with an antigen, no antibodies are present for a few days; then the antibody level slowly rises, first IgM and then IgG levels, followed by a gradual decline (Figure 17.12). This is called the *primary response*. Memory cells that are produced in the first response may remain for decades. Every new encounter with the same antigen results in a rapid proliferation of memory cells. Every time the antigen is contacted again, there is an immediate proliferation of memory B cells. The plasma level of antibodies (mainly IgG antibodies) rises to higher levels than were initially produced. This accelerated, more intense response is called the *secondary response*. When you recover from an infection without taking antimicrobial drugs, it is usually because of the primary response. If at a later time you are infected by the same microbe, the secondary response could be so swift that the microbes are destroyed before you exhibit any signs or symptoms of infection.

Immunological memory provides the basis for immunization by vaccination against certain diseases (for example, polio). When you receive the vaccine, which may contain weakened or killed whole microbes or portions of microbes, your B cells and T cells are activated. If you subsequently encounter the living pathogen as an infecting microbe, your secondary response provides successful protection. However, booster doses of some im-

Figure 17.12 ■ **Secretion of antibodies.** The primary response (after first exposure) is milder than the secondary response (after second or subsequent exposure) to a given antigen.

Immunological memory is the basis for successful immunization by vaccination.

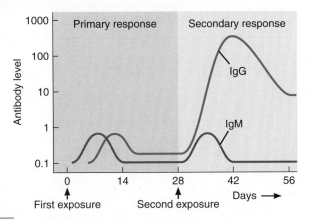

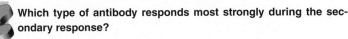

Which type of antibody responds most strongly during the secondary response?

munizing agents must be given periodically to maintain adequate protection against the pathogen.

Table 17.3 summarizes the functions of cells that participate in immune responses.

Graft Rejection

Organ transplantation involves the replacement of an injured or diseased organ, such as the heart, liver, kidney, lungs, or pancreas, with an organ donated by some other individual. Usually, the immune system recognizes the proteins in the transplanted organ as foreign and mounts both cell-mediated and antibody-mediated immune responses against them. This phenomenon is known as *graft rejection*. The more closely matched are the major histocompatibility complex proteins of the donor and recipient, the weaker the graft rejection response. To reduce the risk of graft rejection, organ transplant recipients receive immunosuppressive drugs. One such drug is *cyclosporine*, derived from a fungus, which inhibits secretion of interleukin-2 by helper T cells but has only a minimal effect on B cells. Thus, the risk of rejection is diminished while maintaining resistance to some diseases.

AGING AND IMMUNITY

Objective: **• Describe the effects of aging on immunity.**

With advancing age, elderly individuals become more susceptible to all types of infections and malignancies. Their response to vaccines is decreased, and they tend to produce more autoantibodies (antibodies against their body's own molecules). T cells become less responsive to antigens, and fewer T cells respond to infections. This may result from age-related atrophy of the thymus or decreased production of thymic hormones. Because the T cell population decreases with age, B cells are also less responsive. Consequently, antibody levels do not increase in response to a challenge by an antigen, resulting in increased susceptibility to various infections. It is for this key reason that elderly individuals are encouraged to get influenza (flu) vaccinations each year.

• • •

To appreciate the many ways that the lymphatic and immune systems contribute to homeostasis of other body systems, examine Focus on Homeostasis: The Lymphatic and Immune System. Next, in Chapter 18, we will explore the structure and function of the respiratory system and see how its operation is regulated by the nervous system. Most importantly, the respiratory system provides for gas exchange—taking in oxygen and blowing off carbon dioxide. The cardiovascular system aids gas exchange by transporting blood containing the gases between the lungs and tissue cells.

Table 17.3 / Summary of Functions of Cells Participating in Immune Responses	
Cell	**Functions**
Antigen-presenting cell	Processes and presents foreign antigens to T cells; secretes interleukin-1, which stimulates secretion of interleukin-2 by helper T cells and induces proliferation of B cells; secretes interferons that stimulate T cell growth.
Helper T cell (T4 cell)	Cooperates with B cells to amplify antibody production by plasma cells and stimulates proliferation of T cells and B cells.
Cytotoxic T cell (T8 cell)	Releases chemicals that cause cytolysis of target cells and damage to target cell DNA; attracts macrophages, enhances their phagocytic activity, and prevents them from leaving the site of action.
Memory T cell	Remains in lymphatic tissue and recognizes original invading antigen, even years after the first encounter.
B cell	Differentiates into antibody-producing plasma cell; processes and presents antigen to helper T cells.
Plasma cell	Descendant of B cell that produces and secretes antibodies.
Memory B cell	Ready to produce a more rapid and forceful secondary response should the same antigen enter the body in the future.

Focus on Homeostasis
The Lymphatic and Immune System

Body System	Contribution of Lymphatic and Immune System
For all body systems	B cells, T cells, and antibodies protect all body systems from attack by harmful foreign invaders (pathogens), foreign cells, and cancer cells.
Integumentary system	Lymphatic vessels drain excess interstitial fluid and leaked plasma proteins from dermis of skin; immune system cells (Langerhans cells) in skin help protect skin; lymphatic tissue provides IgA antibodies in sweat.
Skeletal system	Lymphatic vessels drain excess interstitial fluid and leaked plasma proteins from connective tissue around bones.
Muscular system	Lymphatic vessels drain excess interstitial fluid and leaked plasma proteins from connective tissue around bones.
Nervous system 	Lymphatic vessels drain excess interstitial fluid and leaked plasma proteins from the peripheral nervous system.
Endocrine system	Flow of lymph distributes some hormones; lymphatic vessels drain excess interstitial fluid and leaked plasma proteins from endocrine glands.
Cardiovascular system	Lymph returns excess fluid filtered from blood capillaries and leaked plasma proteins to venous blood; macrophages in spleen destroy aged red blood cells and remove debris in blood.
Respiratory system	Tonsils, alveolar macrophages, and MALT (mucosa-associated lymphatic tissue) help protect lungs from pathogens; lymphatic vessels drain excess interstitial fluid from lungs.
Digestive system	Tonsils and MALT help defend against toxins and pathogens that penetrate the body from the gastrointestinal tract; digestive system provides IgA antibodies in saliva and gastrointestinal secretions; lymphatic vessels pick up absorbed dietary lipids and fat-soluble vitamins from the small intestine and transport them to the blood; lymphatic vessels drain excess interstitial fluid and leaked plasma proteins from organs of the digestive system.
Urinary system	Lymphatic vessels drain excess interstitial fluid and leaked plasma proteins from organs of the urinary system; MALT helps defend against toxins and pathogens that penetrate the body via the urethra.
Reproductive systems	Lymphatic vessels drain excess interstitial fluid and leaked plasma proteins from organs of the reproductive system; MALT helps defend against toxins and pathogens that penetrate the body via the vagina and penis; in females, sperm deposited in the vagina are not attacked as foreign invaders due to inhibition of immune responses; IgG antibodies can cross the placenta to provide protection to a developing fetus; lymphatic tissue provides IgA antibodies in milk of nursing mother.

COMMON DISORDERS

AIDS: Acquired Immunodeficiency Syndrome

Acquired immunodeficiency syndrome (AIDS) is a condition in which a person experiences a telltale assortment of infections as a result of the progressive destruction of immune system cells by the *human immunodeficiency virus (HIV)*. AIDS represents the end stage of infection by HIV. A person who is infected with HIV may be symptom free for many years, even while the virus is actively attacking the immune system. HIV infection is serious because it takes control of and destroys the very cells that the body deploys to attack the virus.

Transmission

Because HIV is present in the blood and some body fluids, it is most effectively transmitted by actions or practices that involve the exchange of blood or body fluids. Within most populations, HIV is transmitted in semen or vaginal fluid during unprotected sexual intercourse or oral sex. HIV also is transmitted by direct blood-to-blood contact, such as occurs among intravenous drug users who share hypodermic needles. The sexual partners of HIV-infected individuals are at high risk of infection; those at lesser risk include health-care professionals who may be accidentally stuck by HIV-contaminated hypodermic needles. HIV may also be transmitted from a mother to her fetus or suckling infant. In the United States and Europe before 1985, HIV was unknowingly spread by the transfusion of blood and blood products containing the virus. Effective HIV screening of blood instituted after 1985 has largely eliminated this mode of HIV transmission in the United States and other developed nations.

In economically advantaged, industrialized nations, most people diagnosed with AIDS are either homosexual men or intravenous drug users. The rate of new HIV infections in the United States paints a different and disturbing picture: The greatest increases are among people of color, women, and teenagers. In developing nations, the primary mode of transmission is unprotected heterosexual intercourse; a secondary source is a contaminated blood supply. Of the 40 million people infected with HIV worldwide, about half are women and one-quarter are children.

HIV is a very fragile virus; it cannot survive for long outside the human body. It cannot be transmitted by insect bites, casual physical contact, or sharing of household items. The virus can be eliminated from personal care items and medical equipment by exposing them to heat (135°F for 10 minutes) or by cleaning them with common disinfectants: hydrogen peroxide, rubbing alcohol, Lysol, or bleach. Standard dish washing and clothes washing also kill HIV.

The chance of transmitting or of being infected by HIV during vaginal or anal intercourse can be greatly minimized—although not entirely eliminated—by the use of latex condoms. Public health programs aimed at encouraging intravenous drug users not to share needles have proved effective at checking the increase in new HIV infections among this population. Also, the prophylactic administration of certain drugs such as AZT to pregnant HIV-infected women has proved remarkably effective in minimizing the transmission of the virus to their newborn babies.

HIV: Structure and Infection

HIV consists of an inner core of ribonucleic acid (RNA) covered by a protein coat (capsid) surrounded by an outer layer, the envelope, composed of a lipid bilayer penetrated by proteins (Figure 17.13). Outside a living host cell, a virus has no life functions and is unable to replicate. However, once a virus makes contact with a host cell, the viral nucleic acid enters the host cell. Once inside, the viral nucleic acid uses the host cell's resources to make thousands of copies of the virus. The viruses eventually leave the infected cell to infect other cells.

HIV mainly damages helper T cells. Over 100 billion viral copies may be made each day. The viruses bud so rapidly from an infected cell's plasma membrane that cytolysis eventually occurs. In addition, the body's defenses attack the infected cells, killing them as well as the viruses they harbor.

As we saw in this chapter, helper T cells are instrumental in orchestrating the actions of the immune system. In most newly infected individuals, the body is able to replace the helper T cells at about the same rate that they are destroyed. After years of infection, the body's ability to replace helper T cells is slowly exhausted, and the number of helper T cells progressively declines.

Signs, Symptoms, and Diagnosis of HIV Infection

Immediately following infection with HIV, most people experience a brief flu-like illness. Common signs and symptoms are fever, fatigue, rash, headache, joint pain, sore throat, and swollen lymph nodes. In ad-

Figure 17.13 ■ Structure of the human immunodeficiency virus (HIV), the virus that causes AIDS.

HIV is most effectively transmitted by practices that involve the exchange of body fluids.

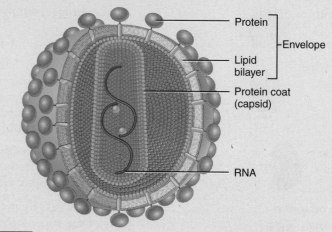

Protein
Envelope
Lipid bilayer
Protein coat (capsid)
RNA

Which cells of the immune system are attacked by HIV?

dition, about 50% of infected people have night sweats. After three to four weeks, plasma cells begin secreting antibodies to components of the HIV protein coat. These antibodies are detectable in blood plasma and form the basis for some of the screening tests for HIV. When people test "HIV positive," it usually means they have antibodies to HIV in their bloodstream. If an acute HIV infection is suspected but the antibody test is negative, laboratory tests based on detection of actual HIV components in blood plasma can confirm the presence of HIV.

Progression to AIDS

After a period of 2 to 10 years, the virus destroys enough helper T cells that most infected people begin to experience symptoms of immunodeficiency. HIV-infected people commonly have enlarged lymph nodes and experience persistent fatigue, involuntary weight loss, night sweats, skin rashes, diarrhea, and various lesions of the mouth and gums. In addition, the virus may begin to infect neurons in the brain, affecting memory and producing visual disturbances.

As the immune system slowly collapses, an HIV-infected person becomes susceptible to a host of *opportunistic infections*, diseases caused by microorganisms that are normally held in check but now proliferate because of the defective immune system. AIDS is diagnosed when the helper T cell count drops below 200 cells per microliter of blood, or when opportunistic infections arise, whichever occurs first. Typically, the opportunistic infections eventually cause death.

About 5% of individuals infected with HIV do not progress to AIDS; these people are called long-term nonprogressors. Their helper T cell count is stable, and they are symptom free. Their survival may be due to infection by a weak strain of HIV or the presence of potent natural killer cells and strong antibodies against HIV.

Treatment of HIV Infection

At present, infection with HIV cannot be cured, and despite intensive research, no effective vaccine is yet available to provide immunity against HIV. However, several drugs have proved successful in extending the life of many HIV-infected individuals.

Most HIV-infected individuals who receive *triple therapy*—a combination of three drugs—experience a drastic reduction in the number of viruses and an increase in the number of helper T cells. Not only does triple therapy delay the progression of HIV infection to AIDS, but many individuals with AIDS have seen the remission or disappearance of opportunistic infections and an apparent return to health. Unfortunately, triple therapy is very costly (exceeding $10,000 per year), the dosing schedule is grueling, and not all people can tolerate the harsh side effects of these drugs. It is currently thought that people who find the drugs helpful must keep taking them for a long time, perhaps for life.

Allergic Reactions

A person who is overly reactive to a substance that is tolerated by most other people is said to be **hypersensitive** or **allergic.** The antigens that induce an allergic reaction are called **allergens.** Common allergens include certain foods (milk, peanuts, shellfish, eggs), antibiotics (penicillin, tetracycline), vaccines (pertussis, typhoid), venoms (honeybee, wasp, snake), cosmetics, chemicals in plants such as poison ivy, pollens, dust, molds, iodine-containing dyes used in certain x-ray procedures, and even microbes.

Some allergic reactions, such as hives, eczema, swelling of the lips or tongue, abdominal cramps, and diarrhea, are *localized:* they affect one part or a limited area of the body. Others are *systemic,* affecting several parts or the entire body. An example is acute anaphylaxis (anaphylactic shock), in which the person develops wheezing and shortness of breath as bronchioles constrict, usually accompanied by cardiovascular failure due to vasodilation and fluid loss from blood.

Autoimmune Diseases

Under normal conditions, the body's immune mechanism can recognize its own tissues and chemicals, and it does not produce T cells or B cells against its own substances. Such recognition of self is called **immunological tolerance.**

At times, however, immunological tolerance breaks down, which leads to an **autoimmune disease.** For reasons still not understood, certain tissues undergo changes that cause the immune system to recognize them as foreign antigens and attack them. Among human autoimmune diseases are rheumatoid arthritis (RA), systemic lupus erythematosus (SLE), rheumatic fever, hemolytic and pernicious anemias, Addison's disease, Graves' disease, type I diabetes mellitus, myasthenia gravis, multiple sclerosis (MS), and ulcerative colitis.

Infectious Mononucleosis

Infectious mononucleosis or "mono" is a contagious disease caused by the *Epstein-Barr virus (EBV)*. It occurs mainly in children and young adults, and it affects more females than males by a 3 : 1 ratio. The virus most commonly enters the body through intimate oral contact such as kissing. It then multiplies in lymphatic tissues and spreads into the blood, where it infects and multiplies in B lymphocytes, the primary host cells. As a result of this infection, the B cells become enlarged and abnormal in appearance so that they resemble monocytes, hence the term *mononucleosis*. Signs and symptoms include an elevated white blood cell count with an abnormally high percentage of lymphocytes, fatigue, headache, dizziness, sore throat, enlarged and tender lymph nodes, and fever. There is no cure for infectious mononucleosis, but the disease usually runs its course in a few weeks.

Lymphomas

Lymphomas (lim-FŌ-mas; *lymph-* = clear water; *-oma* = tumor) are cancers of the lymphatic tissues, especially the lymph nodes. Most have no known cause. The two principal types of lymphomas are Hodgkin's disease and non-Hodgkin's lymphoma.

Hodgkin's disease (HD) is characterized by a painless, nontender enlargement of one or more lymph nodes, most commonly in the neck, chest, and axilla. If the disease has metastasized from these sites, fever, night sweats, weight loss, and bone pain also occur. HD primarily affects individuals between ages 15 and 35 and those over 60, and it is more common in males. If diagnosed early, HD has a 90% to 95% cure rate.

Non-Hodgkin's lymphoma (NHL), which is more common than HD, occurs in all age groups. NHL may start the same way as HD but may also include an enlarged spleen, anemia, and general malaise. Up to half of all individuals with NHL are cured or survive for a lengthy period. Treatment options for both HD and NHL include radiation therapy, chemotherapy, and bone marrow transplantation.

MEDICAL TERMINOLOGY AND CONDITIONS

Allograft (AL-ō-graft; *allo-* = other) A transplant between genetically different individuals of the same species. Skin transplants from other people and blood transfusions are allografts.

Autograft (AW-tō-graft; *auto-* = self) A transplant in which one's own tissue is grafted to another part of the body (such as skin grafts for burn treatment or plastic surgery).

Chronic fatigue syndrome (CFS) A disorder, usually occurring in young female adults, characterized by (1) extreme fatigue that impairs normal activities for at least six months and (2) the absence of other known diseases (cancer, infections, drug abuse, toxicity, or psychiatric disorders) that might produce similar symptoms.

Gamma globulin (GLOB-yoo-lin) Suspension of immunoglobulins from blood consisting of antibodies that react with a specific pathogen. It is prepared by injecting the pathogen into animals, removing blood from the animals after antibodies have been produced, isolating the antibodies, and injecting them into a human to provide short-term immunity.

Splenomegaly (splē′-nō-MEG-a-lē; *mega-* = large) Enlarged spleen.

Systemic lupus erythematosus (er-ith′em-a-TŌ-sus) *(SLE),* or *lupus* (*lupus* = wolf) An autoimmune, noncontagious, inflammatory disease of connective tissue, occurring mostly in young women. In SLE, damage to blood vessel walls results in the release of chemicals that mediate inflammation. Symptoms of SLE include joint pain, slight fever, fatigue, oral ulcers, weight loss, enlarged lymph nodes and spleen, photosensitivity, rapid loss of large amounts of scalp hair, and sometimes an eruption of the skin across the bridge of the nose and cheeks called a "butterfly rash."

Tonsillectomy (ton′-si-LEK-tō-mē; *-ectomy* = excision) Removal of a tonsil.

Xenograft (ZEN-ō-graft; *xeno-* = strange or foreign) A transplant between animals of different species. Xenografts from porcine (pig) or bovine (cow) tissue may be used in a human as a physiological dressing for severe burns.

■ STUDY OUTLINE

Functions of the Lymphatic and Immune System (p. 406)

1. The lymphatic and immune system consists of lymph, lymphatic vessels, structures and organs that contain lymphatic tissue (specialized reticular tissue containing large numbers of lymphocytes), and red bone marrow.

2. The lymphatic and immune system drains tissue spaces of excess fluid and returns proteins that have escaped from blood to the cardiovascular system; it transports lipids and lipid-soluble vitamins from the gastrointestinal tract to the blood, and it protects the body against invasion.

Lymph and Interstitial Fluid (p. 406)

1. Interstitial fluid and lymph differ mostly in location. When the fluid bathes body cells, it is called interstitial fluid; when it is found in lymph vessels, it is called lymph.

2. These fluids both contain less protein than plasma.

Lymphatic Vessels and Lymph Circulation (p. 406)

1. Lymphatic vessels begin as lymphatic capillaries in tissue spaces between cells.

2. Lymphatic capillaries merge to form larger vessels, called lymphatic vessels, which ultimately converge into the thoracic duct or right lymphatic duct.

3. Lymphatic vessels have thinner walls and more valves than veins.

4. The passage of lymph is from interstitial fluid, to lymphatic capillaries, to lymphatic vessels, to lymph trunks, to the thoracic duct or right lymphatic duct, to the subclavian veins.

5. Lymph flows as a result of skeletal muscle contractions, respiratory movements, and valves in the lymphatic vessels.

Lymphatic Organs and Tissues (p. 408)

1. The primary lymphatic organs are red bone marrow and the thymus.

2. Most immune responses occur in the secondary lymphatic organs and tissues—lymph nodes, spleen, and lymphatic nodules.

3. The thymus is the site of T cell maturation and produces hormones.

4. Lymph nodes are encapsulated, oval structures located along lymphatic vessels. Lymph enters nodes through afferent lymphatic vessels and exits through efferent lymphatic vessels.

5. Plasma cells and T cells also proliferate in lymph nodes.

6. The spleen is the single largest mass of lymphatic tissue in the body. It is a site where B cells proliferate into plasma cells and macrophages phagocytize worn-out red blood cells and platelets.

7. Lymphatic nodules are oval-shaped concentrations of lymphatic tissue that are not surrounded by a capsule. They are scattered throughout the mucosa of the gastrointestinal, respiratory, urinary, and reproductive tracts. This lymphatic tissue is termed mucosa-associated lymphatic tissue (MALT).

Nonspecific Resistance to Disease (p. 411)

1. The ability to ward off disease using a number of defenses is called resistance. Lack of resistance is called susceptibility.

2. Nonspecific resistance refers to a wide variety of body responses, including the skin and mucous membranes (first line of defense) as well as internal defenses (antimicrobial proteins, natural killer cells and phagocytes, inflammation, and fever), against a wide range of pathogens.

Specific Resistance: Immunity (p. 414)

1. Specific resistance to disease, called immunity, involves the production of specific types of cells or specific molecules (antibodies) to destroy a particular antigen.

2. B cells and T cells derive from stem cells in red bone marrow. T cells complete their maturation in the thymus.

3. Antigens are chemical substances that are recognized as foreign by the immune system.

4. The major histocompatibility complex (MHC) proteins are unique to each person's body cells. All cells except red blood cells display MHC molecules.

5. Antibodies (Abs) are proteins produced in response to antigens.

6. Antigen-presenting cells (APCs) process and present antigens to activate T cells, and they secrete substances that induce proliferation of T cells and B cells.

7. There are three main kinds of T cells: helper T cells, which stimulate growth and division of cytotoxic T cells, attract phagocytes, enhance phagocytosis by macrophages, and stimulate development of B cells; cytotoxic T cells, which destroy antigens on contact by causing cytolysis and DNA damage; and memory T cells, which recognize antigens at a later date.

8. Antibody-mediated immunity refers to destruction of antigens by antibodies, which are produced by descendants of B cells called plasma cells.

9. B cells develop into antibody-producing plasma cells under the influence of chemicals secreted by antigen-presenting cells and T cells.

10. Based on chemistry and structure, antibodies are grouped in five classes, each with specific biological roles: IgG, IgA, IgM, IgD, and IgE (see Table 17.2 on page 416). Functionally, antibodies neutralize antigens, immobilize bacteria, agglutinate antigens, activate complement, and enhance phagocytosis.

11. Immunization against certain microbes is possible because memory B cells and memory T cells remain after a primary response to an antigen. The secondary response provides protection should the same microbe enter the body again.

Aging and Immunity (p. 422)

1. With advancing age, individuals become more susceptible to infections and malignancies, respond less well to vaccines, and produce more autoantibodies.

2. T cell responses also diminish with age.

▪ SELF-QUIZ

1. Which of the following is NOT true concerning the lymphatic and immune system?

   a. Lymphatic vessels transport lipids from the gastrointestinal tract to the blood. **b.** Lymph is more similar to interstitial fluid than to blood. **c.** Lymphatic tissue is in only a few isolated organs in the body. **d.** The unique structure of lymphatic capillaries allows blood to flow into them but not out of them. **e.** The lymphatic vessels closely resemble veins in structure.

2. Which of the following are produced by virus-infected cells to protect uninfected cells from viral invasion?

   a. complement molecules **b.** prostaglandins **c.** fibrins **d.** interferons **e.** histamines

3. A blockage in the right lymphatic duct would interfere with lymph drainage from the

   a. left arm **b.** right leg **c.** lower abdomen **d.** left leg **e.** right arm

4. Which of the following best represents lymph flow from the interstitial spaces back to the blood?

   a. lymphatic capillaries → lymphatic ducts → lymphatic vessels → subclavian veins **b.** subclavian veins → lymphatic capillaries → lymphatic vessels → lymphatic ducts **c.** lymphatic capillaries → lymphatic vessels → lymphatic ducts → subclavian veins **d.** lymphatic ducts → lymphatic vessels → lymphatic capillaries → subclavian veins **e.** lymphatic capillaries → lymphatic vessels → subclavian veins → lymphatic ducts

5. Lymph nodes

   a. filter lymph **b.** are another name for tonsils **c.** produce lymph **d.** are a primary storage site for blood **e.** produce a protective mucus

6. Which of the following is NOT true concerning the role of skin in nonspecific resistance?

   a. Sebum inhibits the growth of certain bacteria.
   b. Epidermal cells produce interferons to destroy viruses.
   c. Shedding of epidermal cells helps remove microbes.
   d. Lysozyme in sweat destroys some bacteria. **e.** The skin forms a physical barrier to prevent entry of microbes.

7. Which of the following statements about B cells is true?

   a. They become functional while in the thymus gland.
   b. Some may develop into plasma cells that secrete antibodies.
   c. Nonactivated B cells become natural killer cells.
   d. Cytotoxic B cells travel in lymph and blood to react with foreign antigens. **e.** They kill virus-infected cells by secreting perforin.

8. The cells that attack and destroy foreign agents such as fungi, parasites, cancer cells, and foreign tissues are

   a. T cells **b.** plasma cells **c.** B cells **d.** natural killer cells **e.** memory cells

9. The secondary response in antibody-mediated immunity

   a. is characterized by a slow rise in antibody levels and then a gradual decline **b.** occurs when you first receive a vaccination against some disease **c.** produces fewer but more responsive antibodies than occur during the primary response **d.** is an intense response by memory cells to produce antibodies when an antigen is contacted again **e.** is rarely seen except in autoimmune disorders

10. The ability of the body's immune system to recognize its own tissue is known as

    a. immunologic escape **b.** autoimmunity **c.** nonspecific resistance **d.** hypersensitivity **e.** immunological tolerance

11. A disease that causes destruction of helper T cells would result in all of the following effects EXCEPT

    a. inability to produce cytotoxic T cells **b.** alteration of lymph flow **c.** lack of development of plasma cells **d.** decreased macrophage activity **e.** increased risk of developing infections

12. Which lymphatic organ functions in the production of T cells and hormones that promote the maturation of T cells?

    a. thyroid gland **b.** spleen **c.** thymus **d.** red bone marrow **e.** lymph node

13. Place the following steps involved in the process of inflammation in the correct order.

    1 arrival of large numbers of neutrophils

2. vasodilation and increased permeability of blood vessels
3. formation of pus
4. increased migration of monocytes
5. formation of fibrin network to form a clot
6. release of histamine

a. 6, 2, 4, 1, 5, 3   **b.** 3, 6, 1, 4, 2, 5   **c.** 5, 1, 4, 2, 6, 3
d. 6, 2, 5, 1, 4, 3   **e.** 4, 6, 1, 3, 2, 5

14. Match the following:

____ **a.** destroy antigens by cytolysis

____ **b.** stimulate other cells of
the immune system

____ **c.** are programmed to recognize
the original invading antigen;
allow immunity to last for years

____ **d.** function in nonspecific resistance

____ **e.** develop into plasma cells

A. natural killer
cells
B. helper T cells
C. B cells
D. memory T cells
E. cytotoxic T cells

15. Opsonization is the

a. engulfing of a microbe by a phagocyte   **b.** chemical attraction of a phagocyte   **c.** binding of complement to a microbe   **d.** attachment of a phagocyte to a microbe   **e.** breakdown of a microbe by enzymes

16. All of the following contribute to nonspecific resistance EXCEPT

a. complement   **b.** immunoglobulin   **c.** natural killer cells
d. lysozyme   **e.** interferons

17. Inflammation produces

a. redness due to bleeding   **b.** heat due to fever
c. swelling due to increased permeability of capillaries   **d.** pain due to histamine release   **e.** mucus due to phagocytosis

18. Place the phases of phagocytosis in the correct order.

1. adherence to foreign material
2. chemotaxis of phagocytes
3. exocytosis of indigestible materials
4. ingestion of foreign material

a. 1, 2, 3, 4   **b.** 2, 1, 4, 3   **c.** 1, 4, 3, 2   **d.** 4, 3, 2, 1
e. 2, 1, 3, 4

19. Antibodies attack antigens by all of the following methods EXCEPT

a. agglutination of antigens   **b.** activation of complement
c. opsonization to enhance phagocytosis   **d.** preventing attachment to body cells   **e.** producing acid secretions

20. What is the importance of tonsils in the body's defense mechanisms?

a. They help destroy microbes that are inhaled.   **b.** They contain ciliated cells that move trapped pathogens from the breathing passages.   **c.** They are needed for T cell maturation.
d. They are needed for B cell maturation.   **e.** They filter lymph.

CRITICAL THINKING APPLICATIONS

1. Marcia found a lump in her right breast during her monthly self-examination. The lump was found to be cancerous. The surgeon removed the breast lump, surrounding tissue, and some lymph nodes. Which nodes were probably removed and why?

2. Years ago, tonsillectomy was almost considered a "rite of passage" for children in elementary school. It seemed like all children were getting their tonsils removed! Why are tonsils frequently infected in young children?

3. Baby Carlos is due for his MMR "shot." The MMR vaccine is an immunization for measles, mumps, and rubella. How does immunization protect children against disease?

4. Every spring, the bees buzz, the flowers bloom, and Kathryn sneezes. Kathryn will suffer from itchy, watery eyes and nasal congestion for several weeks unless she takes her prescribed antihistamine. What is Kathryn's problem? How does an antihistamine help?

ANSWERS TO FIGURE QUESTIONS

17.1 Lymphatic tissue is reticular connective tissue that contains large numbers of lymphocytes.

17.2 Lymph is more similar to interstitial fluid, because its protein content is low.

17.3 Capillaries produce lymph.

17.4 Foreign substances in lymph may be phagocytized by macrophages or destroyed by T cells or antibodies produced by plasma cells.

17.5 The skin and subcutaneous layer, liver, lungs, brain, spleen, lymph nodes, and red bone marrow contain fixed macrophages.

17.6 Redness is caused by increased blood flow due to vasodilation.

17.7 Antibodies are proteins.

17.8 APCs include macrophages, B cells, and dendritic cells.

17.9 A clone is a population of identical cells.

17.10 Cytotoxic T cells attack some tumor cells and transplanted tissue cells.

17.11 Memory cells respond to a second invasion by the same antigen.

17.12 IgG is the antibody secreted in greatest amount during a secondary response.

17.13 HIV attacks helper T cells.

Chapter **18**

The Respiratory System

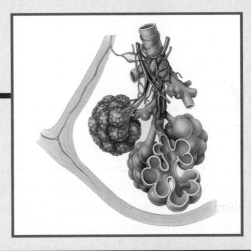

Student Learning Objectives

1. Describe the structure and functions of the nose, pharynx, larynx, trachea, bronchi, and lungs. **431**
2. Define the various lung volumes and capacities. **438**
3. Explain how inhalation and exhalation take place. **438**
4. Explain how oxygen and carbon dioxide are exchanged between alveolar air and blood (external respiration) and between blood and body cells (internal respiration). **442**
5. Describe how the blood transports oxygen and carbon dioxide. **444**
6. Explain how the nervous system controls breathing and list the factors that can alter the rate and depth of breathing. **445**
7. Describe the effects of exercise on the respiratory system. **449**
8. Describe the effects of aging on the respiratory system. **449**

A Look Ahead

Cells continually use oxygen (O_2) for the metabolic reactions that release energy from nutrient molecules and produce ATP. These same reactions produce carbon dioxide (CO_2), which must be eliminated quickly and efficiently. The two systems that cooperate to supply O_2 and eliminate CO_2 are the cardiovascular and respiratory systems. The *respiratory system* provides for intake of O_2 and elimination of CO_2, whereas the cardiovascular system transports blood containing these gases between the lungs and tissue cells. Failure of either system causes rapid cell death from oxygen starvation and the buildup of waste products. The respiratory system also helps regulate blood pH, contains receptors for the sense of smell, filters inspired air, produces sounds, and rids the body of some water and heat in exhaled air.

Respiration, the exchange of O_2 and CO_2 between atmospheric air, blood, and tissue cells, involves four processes: (1) ventilation (breathing), the inhalation (inflow) and exhalation (outflow) of air between the atmosphere and the lungs; (2) external respiration, the exchange of O_2 and CO_2 between air in the lungs and blood; (3) transport of O_2 from the lungs to tissue cells and CO_2 back to the lungs via the bloodstream; and (4) internal respiration, the exchange of O_2 and CO_2 between the blood and tissue cells. The chemical reactions whereby cells use O_2 to produce ATP and liberate CO_2, a process called *cellular respiration,* is described in detail in Chapter 20.

The branch of medicine that deals with the diagnosis and treatment of diseases of the ears, nose, and throat is called **otorhinolaryngology** (ō′-tō-rī′-nō-lar′-in-GOL-ō-jē; *oto-* = ear; *rhino-* = nose; *laryng-* = voice box; *-ology* = study of). A **pulmonologist** (*pulmon-* = lung) is a specialist in the diagnosis and treatment of lung diseases.

Figure 18.1 ■ **Organs of the respiratory system in relation to surrounding structures.**

The upper respiratory system includes the nose, pharynx, and associated structures. The lower respiratory system includes the larynx, trachea, bronchi, and lungs.

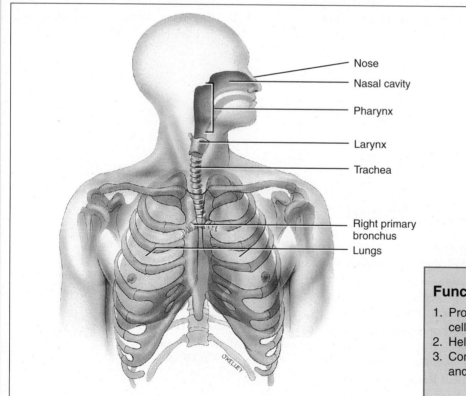

Nose
Nasal cavity
Pharynx
Larynx
Trachea
Right primary bronchus
Lungs

Anterior view

Functions of the Respiratory System

1. Provides for gas exchange—intake of O_2 for delivery to body cells and elimination of CO_2 produced by body cells.
2. Helps regulate blood pH.
3. Contains receptors for the sense of smell, filters inspired air, and produces sounds.

 Which structures are part of the conducting portion of the respiratory system?

ORGANS OF THE RESPIRATORY SYSTEM

Objective: • **Describe the structure and functions of the nose, pharynx, larynx, trachea, bronchi, and lungs.**

The organs of the respiratory system include the nose, pharynx (throat), larynx (voice box), trachea (windpipe), bronchi, and lungs (Figure 18.1). Structurally, the respiratory system consists of two portions: the **upper respiratory system** includes the nose, pharynx, and associated structures; the **lower respiratory system** consists of the larynx, trachea, bronchi, and lungs. The respiratory system can also be divided into two parts based on function. The **conducting portion** consists of a series of interconnecting cavities and tubes—nose, pharynx, larynx, trachea, bronchi, bronchioles, and terminal bronchioles—that conduct air into the lungs. The **respiratory portion** consists of those portions of the respiratory system where gas exchange occurs—the respiratory bronchioles, alveolar ducts, alveolar sacs, and alveoli.

Nose

Structure

The **nose** has a visible external portion and an internal portion inside the skull (Figure 18.2). The external portion consists of bone and cartilage covered with skin and lined with mucous membrane. It has two openings called the **external nares** (NA-rēz; singular is **naris**) or **nostrils.**

The internal portion of the nose, a large cavity in the skull that lies below the cranium and above the mouth, is connected to the throat (pharynx) through two openings called the **internal nares.** Four paranasal sinuses (frontal, sphenoidal, maxillary, and ethmoidal) and the nasolacrimal ducts also connect to the internal nose. The cavity inside the external and internal portions of the nose is called the **nasal cavity** and is divided into right and left sides by a partition called the **nasal septum.** The septum consists of the perpendicular plate of the ethmoid bone, vomer, and cartilage (see Figure 6.11 on page 131).

Functions

The interior structures of the nose are specialized for three basic functions: (1) warming and moistening incoming air to facilitate

Figure 18.2 ■ Respiratory organs in the head and neck.

As air passes through the nose, it is warmed, filtered, and moistened.

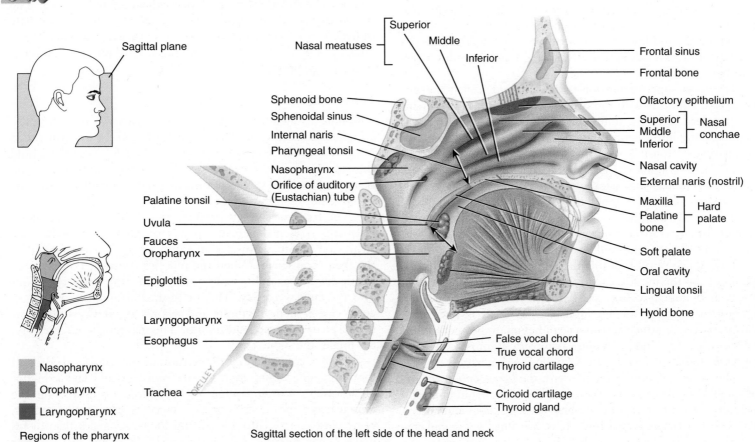

Regions of the pharynx

Sagittal section of the left side of the head and neck

What is the path taken by air molecules into and through the nose?

gas exchange in the lungs, and filtering incoming air to remove some inhaled particles; (2) receiving olfactory stimuli; and (3) providing a resonating chamber for speech sounds.

When air enters the nostrils, it passes through coarse hairs that filter out large dust particles. The air then passes by three shelves formed by the superior, middle, and inferior **nasal conchae** that extend out of the wall of the cavity, increasing the surface area for warming, moistening, and filtering incoming air. Mucous membrane lines the nasal cavity and the three conchae. Above the superior nasal concha is the **olfactory epithelium,** which receives olfactory stimuli.

The mucous membrane of the nasal cavity contains pseudostratified ciliated columnar epithelium with many goblet cells and an extensive blood supply in the underlying connective tissue. As the air whirls around the conchae, it is warmed by the blood in capillaries of the connective tissue. Mucus secreted by the goblet cells moistens the air and traps dust particles, forming mucus–dust packages. The cilia move the mucus–dust packages toward the pharynx, where they can be swallowed or eliminated from the body. Substances in cigarette smoke inhibit movement of cilia. When this happens, only coughing can remove mucus–dust packages from the airways. This is the origin of the distinctive smoker's cough.

Pharynx

The **pharynx** (FAIR-inks), or throat, is a tube that starts at the internal nares and extends part way down the neck (see Figure 18.2). It lies just posterior to the nasal and oral cavities and just anterior to the cervical (neck) vertebrae. Its wall is composed of skeletal muscle and lined with mucous membrane. The pharynx functions as a passageway for air and food and provides a resonating chamber for speech sounds.

The most superior portion of the pharynx, called the **nasopharynx,** connects with the two internal nares, has two openings that lead into the auditory (Eustachian) tubes, and has a single opening into the oropharynx. The posterior wall contains the pharyngeal tonsil (called the adenoid when it becomes inflamed). The nasopharynx exchanges air with the nasal cavities and receives mucus–dust packages. The cilia of its pseudostratified ciliated columnar epithelium move the mucus–dust packages toward the mouth. The nasopharynx also exchanges small amounts of air with the auditory tubes to equalize air pressure between the pharynx and middle ear.

The middle portion of the pharynx, the **oropharynx,** has an opening from the mouth, called the **fauces** (FAW-sēz = throat), as well as the opening from the nasopharynx, and it serves as a common passageway for air, food, and drink. It is lined with nonkeratinized stratified squamous epithelium to protect it from abrasion by coarse food particles. Two pairs of tonsils, the *palatine tonsils* and *lingual tonsils*, are found in the oropharynx.

The lowest portion of the pharynx, the **laryngopharynx** (la-rin′-gō-FAIR-inks), extends downward from the hyoid bone and connects posteriorly with the esophagus (food tube) and anteriorly with the larynx (voice box). It is also lined with nonkeratinized stratified squamous epithelium. Like the oropharynx, the laryngopharynx is both a respiratory and a digestive pathway.

Larynx

The **larynx** (LAR-inks), or voice box, is a short passageway that connects the pharynx with the trachea. It lies in the midline of the neck anterior to the fourth, fifth, and sixth cervical vertebrae (C4 to C6).

Structure

The wall of the larynx is composed of nine pieces of cartilage (Figure 18.3), three single and three paired. The three single pieces are the thyroid cartilage, epiglottis, and cricoid cartilage.

The **thyroid cartilage (Adam's apple),** which consists of hyaline cartilage, forms the anterior wall of the larynx and gives it its triangular shape. It gets its common name because it is often larger in males than in females due to the influence of male sex hormones during puberty.

The **epiglottis** (epi- = over; -glottis = tongue) is a large, leaf-shaped piece of elastic cartilage that is covered by epithelium and is the most superior part of the larynx (see also Figure 18.2). The "stem" of the epiglottis is attached to the thyroid cartilage. The "leaf" of the epiglottis is unattached and free to move up and down like a trap door. During swallowing, the larynx elevates, causing the leaf portion of the epiglottis to seal off the air passageways below it. Thus, liquids and foods are routed into the esophagus, part of the digestive system, and kept out of the larynx. When anything but air passes into the larynx, a cough reflex attempts to expel the material.

The **cricoid cartilage** (KRĪ-koyd) is a ring of hyaline cartilage attached to the first ring of cartilage of the trachea. The paired **arytenoid cartilages** (ar′-i-TĒ-noyd), consisting mostly of hyaline cartilage, are located above the cricoid cartilage. They attach to the true vocal cords and pharyngeal muscles and function in voice production.

Voice Production

The mucous membrane of the larynx forms two pairs of folds: an upper pair called the **false vocal cords** and a lower pair called the **true vocal cords** (see Figure 18.2). The false vocal cords hold the breath against pressure in the thoracic cavity when you strain to lift a heavy object, such as a backpack filled with textbooks. They do not produce sound.

The true vocal cords produce sound. They contain elastic ligaments stretched between pieces of rigid cartilage like the strings on a guitar. Muscles attach to both the cartilage and to the true vocal cords. When the muscles contract, they pull the elastic ligaments tight, which moves the true vocal cords out into the air passageway. The air pushed against the true vocal cords causes them to vibrate and sets up sound waves in the air in the pharynx, nose, and mouth. The greater the air pressure, the louder the sound.

Pitch is controlled by the tension of the true vocal cords. If they are pulled taut, they vibrate more rapidly and a higher pitch results. Lower sounds are produced by decreasing the muscular tension. Due to the influence of male sex hormones, vocal cords are usually thicker and longer in males than in females. They therefore vibrate more slowly, giving men a lower range of pitch than women.

Figure 18.3 ■ **Larynx.**

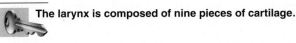

The larynx is composed of nine pieces of cartilage.

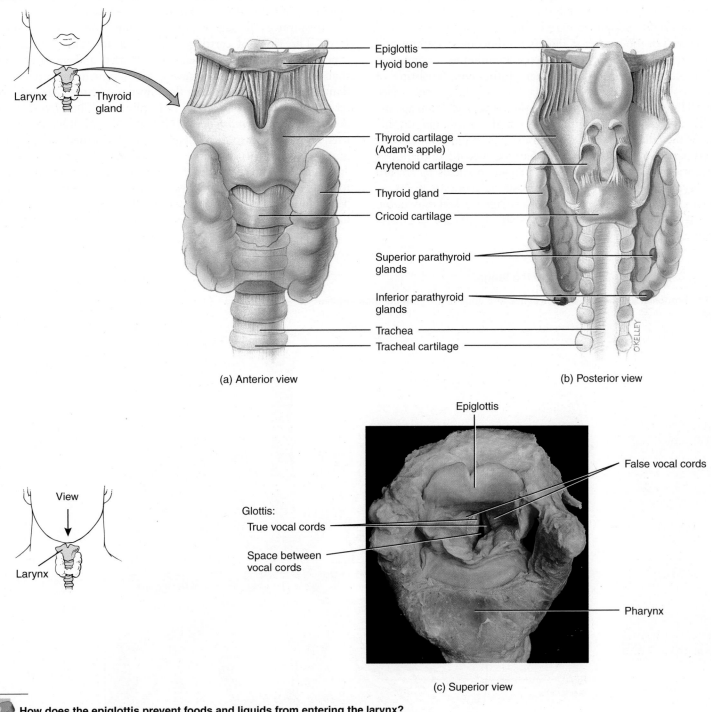

Larynx — Thyroid gland

Epiglottis
Hyoid bone

Thyroid cartilage (Adam's apple)
Arytenoid cartilage
Thyroid gland
Cricoid cartilage

Superior parathyroid glands

Inferior parathyroid glands

Trachea
Tracheal cartilage

(a) Anterior view

(b) Posterior view

View
Larynx

Epiglottis

False vocal cords

Glottis:
True vocal cords

Space between vocal cords

Pharynx

(c) Superior view

How does the epiglottis prevent foods and liquids from entering the larynx?

Laryngitis is an inflammation of the larynx, most often caused by a respiratory infection, respiratory irritants, or excessive shouting or coughing, and it results in hoarseness or loss of voice. Inflammation interferes with the contraction of the cords or causes them to swell to the point where they cannot vibrate freely. Long-term smokers often acquire a permanent hoarseness from the damage done by chronic inflammation of the larynx.

Trachea

The *trachea* (TRĀ-kē-a), or windpipe, is a passageway for air that is located anterior to the esophagus. It extends from the larynx to the upper part of the fifth thoracic vertebra (T5), where it divides into right and left primary bronchi (Figure 18.4).

The wall of the trachea is lined with mucous membrane and is supported by cartilage. The mucous membrane is composed of pseudostratified ciliated columnar epithelium, consisting of ciliated columnar cells, goblet cells, and basal cells (see Table 4.1I on page 78), and provides the same protection against dust as the membrane lining the nasal cavity and larynx. Whereas the cilia in the upper respiratory tract move mucus and trapped particles *down* toward the pharynx, the cilia in the lower respiratory tract move mucus and trapped particles *up* toward the pharynx. The cartilage layer consists of about 16 to 20 incomplete, C-shaped rings of hyaline cartilage stacked one on top of another. The open part of each C-shaped cartilage ring faces

the esophagus and permits it to expand slightly into the trachea during swallowing. The solid parts of the C-shaped cartilage rings provide a rigid support so the tracheal wall does not collapse inward and obstruct the air passageway. The rings of cartilage may be felt under the skin below the larynx.

Respiratory passageways may become obstructed in several ways. The rings of cartilage may be accidentally crushed, the mucous membrane may become inflamed and swell so much that it closes off the passageway, excess mucus secreted by inflamed membranes may clog the lower respiratory passages, or a large object may be breathed in (aspirated). The **Heimlich (abdominal thrust) maneuver** (HĪM-lik ma-NOO-ver) may be used to expel an aspirated object. It is performed by applying a quick upward thrust that causes sudden elevation of the diaphragm and forceful, rapid expulsion of air from the lungs, forcing air out of the trachea to eject the obstructing object. The Heimlich maneuver is also used to expel water from the lungs of near-drowning victims before resuscitation is begun. If

Figure 18.4 ■ Air passageways to the lungs.

The bronchial tree begins at the trachea and ends at the terminal bronchioles.

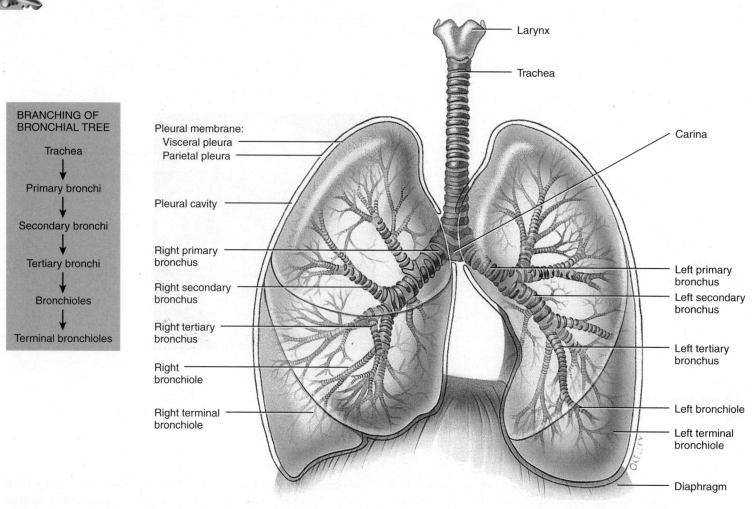

BRANCHING OF
BRONCHIAL TREE

Trachea
↓
Primary bronchi
↓
Secondary bronchi
↓
Tertiary bronchi
↓
Bronchioles
↓
Terminal bronchioles

Larynx

Trachea

Carina

Pleural membrane:
  Visceral pleura
  Parietal pleura

Pleural cavity

Right primary bronchus

Right secondary bronchus

Right tertiary bronchus

Right bronchiole

Right terminal bronchiole

Left primary bronchus

Left secondary bronchus

Left tertiary bronchus

Left bronchiole

Left terminal bronchiole

Diaphragm

Anterior view

How many lobes and secondary bronchi are present in each lung?

the Heimlich maneuver is not successful and the obstruction is above the level of the chest, a *tracheostomy* (trā′-kē-OS-tō-mē) may be performed. In this procedure, an incision is made in the trachea below the cricoid cartilage and a tracheal tube is inserted to create an emergency air passageway. Another method is *intubation.* A tube is inserted into the mouth or nose and passed down through the larynx and trachea. The tube pushes back any flexible obstruction and provides a passageway for air. If mucus is clogging the trachea, it can be suctioned out through the tube.

Bronchi

The trachea divides into a *right primary bronchus* (BRON-kus = windpipe), which goes to the right lung, and a *left primary bronchus,* which goes to the left lung (Figure 18.4). The right primary bronchus is more vertical, shorter, and wider than the left. As a result, foreign objects are more likely to enter and lodge in the right primary bronchus than in the left. Like the trachea, the primary bronchi (BRONG-kī) contain incomplete rings of cartilage and are lined by pseudostratified ciliated columnar epithelium.

On entering the lungs, the primary bronchi divide to form the *secondary bronchi,* one for each lobe of the lung. The right lung has three lobes; the left lung has two. The secondary bronchi continue to branch, forming still smaller bronchi, called *tertiary bronchi,* that divide several times, ultimately giving rise to smaller *bronchioles.* Bronchioles, in turn, branch into even smaller tubes called *terminal bronchioles.* Because of its resemblance to a tree trunk with many branches, this arrangement is commonly referred to as the *bronchial tree. Bronchoscopy* is the visual examination of the bronchi through a *bronchoscope,* an illuminated, tubular instrument that is passed through the trachea into the bronchi.

As the branching becomes more extensive in the bronchial tree, several structural changes occur. First, the rings of cartilage in the primary bronchi are replaced by strips of cartilage in the

Figure 18.5 ■ **Lungs.**

The subdivisions of the lungs are lobes, bronchopulmonary segments, and lobules.

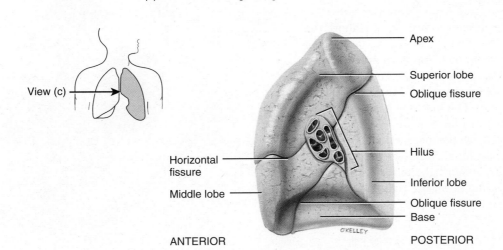

(a) Lateral view of right lung

(b) Lateral view of left lung

(c) Medial view of right lung

Why are the right and left lungs slightly different in size and shape?

secondary and tertiary bronchi. Cartilage disappears in the bronchioles. As the amount of cartilage decreases, the amount of smooth muscle encircling the lumen increases. The mucous membrane in the bronchial tree changes from pseudostratified ciliated columnar epithelium in the primary bronchi, secondary bronchi, and tertiary bronchi to ciliated simple columnar epithelium with some goblet cells in larger bronchioles, to mostly ciliated simple cuboidal epithelium with no goblet cells in smaller bronchioles, to mostly nonciliated simple cuboidal epithelium in terminal bronchioles.

The parasympathetic division of the autonomic nervous system and mediators of allergic reactions, such as histamine, cause narrowing of the lumen of bronchioles (bronchoconstriction), whereas the sympathetic division and epinephrine cause bronchodilation. During an *asthma attack* (see page 451), the smooth muscle goes into spasm. Because there is no supporting cartilage, the spasms can close off the air passageways. Movement of air through constricted bronchial tubes causes breathing to be more labored.

Lungs

The *lungs* (= lightweights, because they float) are paired, cone-shaped organs lying in the thoracic cavity. They are separated from each other by the heart and other structures in the mediastinum (see Figure 15.1 on page 350). The *pleural membrane* is a double-layered serous membrane that encloses and protects each lung (Figure 18.4). The outer layer is attached to the wall of the thoracic cavity and diaphragm and is called the *parietal pleura.* The inner layer, the *visceral pleura,* covers the lungs. Between the visceral and parietal pleurae is a narrow space, the *pleural cavity,* which contains a lubricating fluid secreted by both membranes. This fluid reduces friction between the membranes and allows them to move easily during breathing.

Pleurisy, or inflammation of the pleural membranes, causes friction during breathing that can be quite painful when the swollen membranes rub against each other.

Structure

The lungs extend from the diaphragm to just slightly above the clavicles and lie against the ribs. The broad bottom portion of each lung is its *base,* whereas the narrow top portion is the *apex* (Figure 18.5 on page 435). The *hilus* is an area on the medial side through which bronchi, pulmonary blood vessels, lymphatic vessels, and nerves enter and exit. The left lung has a concavity, the *cardiac notch,* in which the heart lies.

Due to the space occupied by the heart, the left lung is about 10% smaller than the right lung. Although the right lung is thicker and broader, it is also somewhat shorter than the left lung because the diaphragm is higher on the right side, accommodating the liver that lies below it.

Lobes, Fissures, and Lobules

Each lung is divided into lobes by one or more deep grooves called fissures. The left lung has one fissure (oblique) and two

lobes (superior and inferior); the right lung has two fissures (oblique and horizontal) and three lobes (superior, middle, and inferior) (Figure 18.5). Each lobe receives its own secondary bronchus. Thus the right primary bronchus branches into three secondary bronchi called the *superior, middle,* and *inferior secondary bronchi.* The left primary bronchus branches into a *superior* and an *inferior secondary bronchus.*

Each lobe of the lungs is divided into regions called *bronchopulmonary segments,* each supplied by its own tertiary bronchus. The segments, in turn, are divided into many small compartments called *lobules* (Figure 18.6). Each lobule contains a lymphatic vessel, an arteriole, a venule, and a branch from a terminal bronchiole wrapped in elastic connective tissue. Terminal bronchioles subdivide into microscopic branches called *respiratory bronchioles.* Respiratory bronchioles, in turn, subdivide into 2 to 11 *alveolar ducts.* Respiratory bronchioles

Figure 18.6 ■ **Lobule of the lung.**

Alveolar sacs are two or more alveoli that share a common opening.

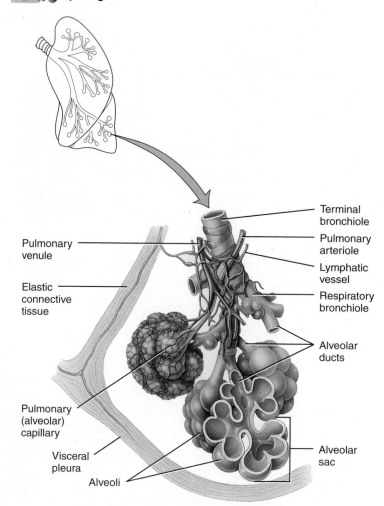

Portion of a lobule of the lung

What are the major parts of a lobule of a lung?

contain mostly nonciliated simple cuboidal epithelium. From the trachea to the alveolar ducts, there are about 25 levels of branching of the respiratory passageways. That is, the trachea divides into primary bronchi (first level), the primary bronchi divide into secondary bronchi (second level), and so on.

Alveoli

Numerous alveoli and alveolar sacs surround each alveolar duct. An *alveolus* (al-VĒ-ō-lus; plural is *alveoli*) is a cup-shaped outpouching of an alveolar sac. The two or more alveoli that share a common opening to the alveolar duct are called *alveolar sacs* (Figure 18.6). The walls of alveoli consist of two types of alveolar epithelial cells (Figure 18.7). *Type I alveolar cells,* the most numerous (about 95%), are simple squamous epithelial cells. They form a mostly continuous lining of the alveolar wall that is interrupted by occasional *type II alveolar cells.* The thin type I alveolar cells are the main sites of gas exchange. Type II alveolar cells, which are rounded or cuboidal epithelial cells with microvilli, secrete *alveolar fluid,* which keeps the surface between the cells and the air moist. Included in the alveolar fluid is *surfactant* (sur-FAK-tant), a mixture of phospholipids and lipoproteins. Surfactant reduces the tendency of alveoli to collapse. Associated with the alveolar wall are *alveolar macrophages,* wandering phagocytes that remove fine dust particles and other

Figure 18.7 ■ **Structure of an alveolus.**

The exchange of respiratory gases occurs by diffusion across the respiratory membrane.

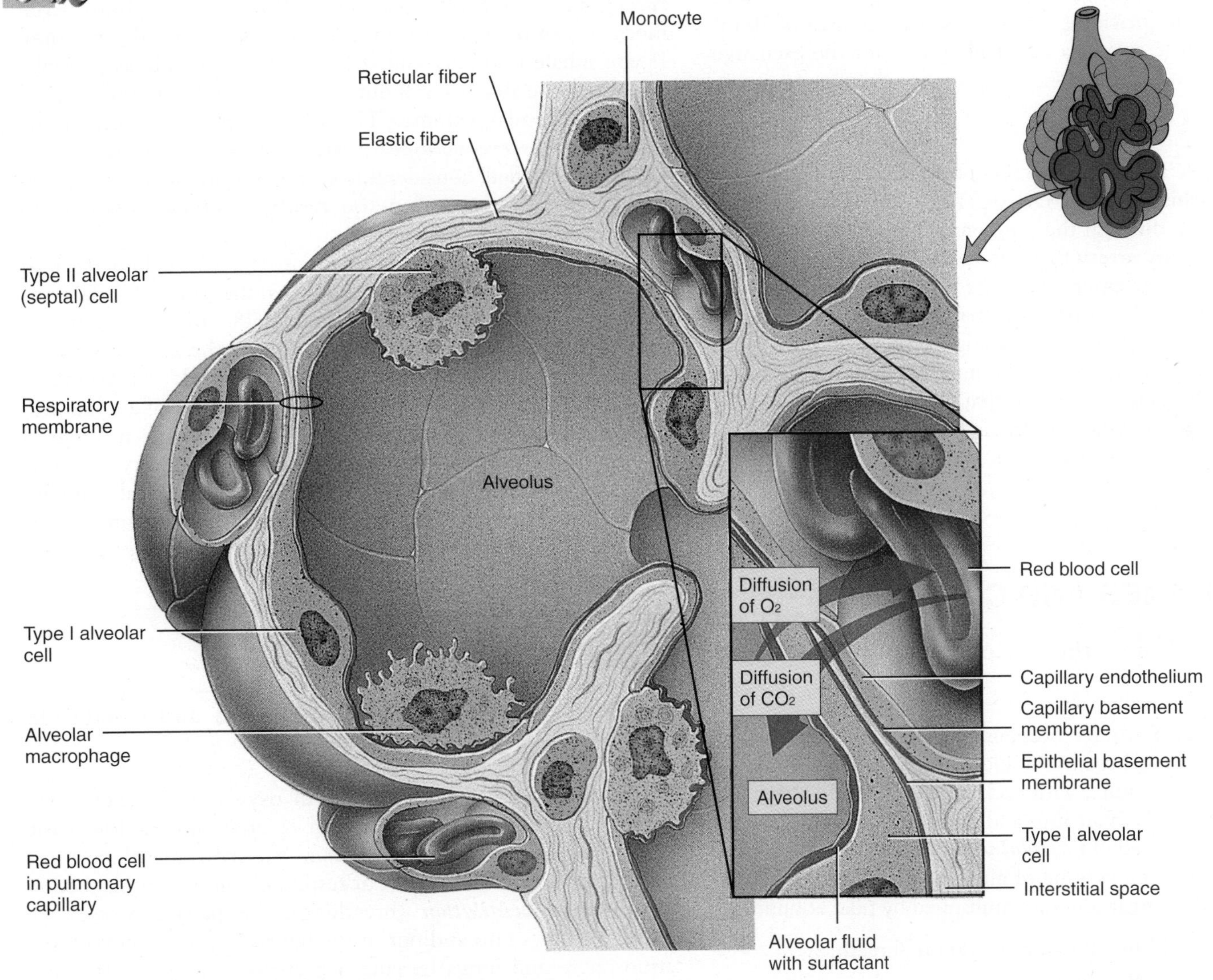

Monocyte

Reticular fiber

Elastic fiber

Type II alveolar (septal) cell

Respiratory membrane

Type I alveolar cell

Alveolar macrophage

Red blood cell in pulmonary capillary

Alveolus

Diffusion of O_2

Diffusion of CO_2

Alveolus

Red blood cell

Capillary endothelium

Capillary basement membrane

Epithelial basement membrane

Type I alveolar cell

Interstitial space

Alveolar fluid with surfactant

(a) Section through an alveolus showing its cellular components

(b) Details of respiratory membrane

What is the function of type II alveolar cells?

debris in the alveolar spaces. Also present are fibroblasts that produce reticular and elastic fibers. Underlying the layer of type I alveolar cells is an elastic basement membrane. Around the alveoli, the lobule's arteriole and venule form a network of blood capillaries. The exchange of O_2 and CO_2 between the air spaces in the lungs and the blood takes place by diffusion across alveolar and capillary walls. The gases diffuse through the **respiratory membrane,** which consists of four layers (Figure 18.7b):

1. The **wall** of an alveolus, comprised of a layer of type I and type II alveolar cells and associated alveolar macrophages.

2. An **epithelial basement membrane** underlying the alveolar wall.

3. A **capillary basement membrane** that is often fused to the epithelial basement membrane.

4. The **capillary endothelium.**

The respiratory membrane is very thin, only 0.5 μm* thick, which allows rapid diffusion of gases. The lungs contain about 300 million alveoli, providing an immense surface area of 70 m² (750 ft²)—about the size of a handball court—for the exchange of gases.

Blood Supply

The lungs receive blood via two sets of arteries: pulmonary arteries and bronchial arteries. Deoxygenated blood passes from the right ventricle through the pulmonary trunk, which divides into a left pulmonary artery that enters the left lung and a right pulmonary artery that enters the right lung. The return of the oxygenated blood to the heart is by way of the pulmonary veins, which drain into the left atrium (see Figure 16.16 on page 398).

Oxygenated blood supplying the lungs' own tissues is delivered through bronchial arteries that branch off the aorta. Deoxygenated venous blood is drained from the tissues of the lungs by the bronchial veins, which ultimately empty into the superior vena cava.

LUNG VOLUMES AND CAPACITIES

Objective: • **Define the various lung volumes and capacities.**

The term for normal quiet breathing is **eupnea** (yoop-NĒ-a; *eu-* = normal; *-pnea* = breath). While at rest, a healthy adult averages 12 breaths a minute, with each inhalation and exhalation moving about 500 mL of air into and out of the lungs. The volume of one breath is called the **tidal volume.** The **minute ventilation (MV)**—the total volume of air inhaled and exhaled each minute—is equal to breathing rate multiplied by tidal volume:

$$MV = 12 \text{ breaths/min} \times 500 \text{ mL/breath}$$
$$= 6 \text{ liters/min}$$

*1 μm (micrometer) = 1/25,000 of an inch or 1/1,000,000 of a meter.

Tidal volume varies considerably from one person to another and in the same person at different times. In an average adult, about 70% of the tidal volume (350 mL) actually reaches the respiratory portion of the respiratory system and participates in respiration; the other 30% (150 mL) remains in the conducting airways of the nose, pharynx, larynx, trachea, bronchi, bronchioles, and terminal bronchioles. Collectively, these conducting airways are known as the **anatomic dead space.** The part of the tidal volume that remains in the anatomic dead space cannot be used in gas exchange.

The apparatus commonly used to measure respiratory rate and the volume of air inhaled and exhaled during breathing is a **spirometer** (*spiro-* = breathe; *meter* = measuring device). The record is called a **spirogram.** Inhalation is recorded as an upward deflection, and exhalation is recorded as a downward deflection (Figure 18.8).

By taking a very deep breath, you can inhale a good deal more than 500 mL. This additional inhaled air, called the **inspiratory reserve volume,** is about 3100 mL (Figure 18.8). Even more air can be inhaled if inspiration follows forced expiration. If you inhale normally and then exhale as forcibly as possible, you should be able to push out 1200 mL of air in addition to the 500 mL of tidal volume. The extra 1200 mL is called the **expiratory reserve volume.** Even after the expiratory reserve volume is expelled, considerable air remains in the lungs and airways. This volume, called the **residual volume,** amounts to about 1200 mL.

Lung *capacities* are combinations of specific lung *volumes* (Figure 18.8). **Inspiratory capacity** is the sum of tidal volume and inspiratory reserve volume (500 mL + 3100 mL = 3600 mL). **Functional residual capacity** is the sum of residual volume and expiratory reserve volume (1200 mL + 1200 mL = 2400 mL). **Vital capacity,** the maximum amount of air that can be expired after maximal inspiration, is the sum of inspiratory reserve volume, tidal volume, and expiratory reserve volume (4800 mL). Finally, **total lung capacity** is the sum of all volumes (6000 mL). The values given here are averages for young adults.

Table 18.1 summarizes lung volumes and capacities.

VENTILATION

Objective: • **Explain how inhalation and exhalation take place.**

Respiration supplies body cells with oxygen and removes the carbon dioxide produced by cellular respiration. The four basic events of respiration are ventilation, external respiration, internal respiration, and transport of respiratory gases. In this section we describe **ventilation** (breathing), the passive process by which air flows into and out of the lungs. Air flows between the atmosphere and lungs because a pressure difference exists between them. We inhale or breathe in when the pressure inside the lungs is less than the air pressure in the atmosphere. We exhale or breathe out when the pressure inside the lungs is greater than the pressure in the atmosphere.

Figure 18.8 ■ **Spirogram of lung volumes and capacities.**

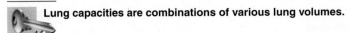

Lung capacities are combinations of various lung volumes.

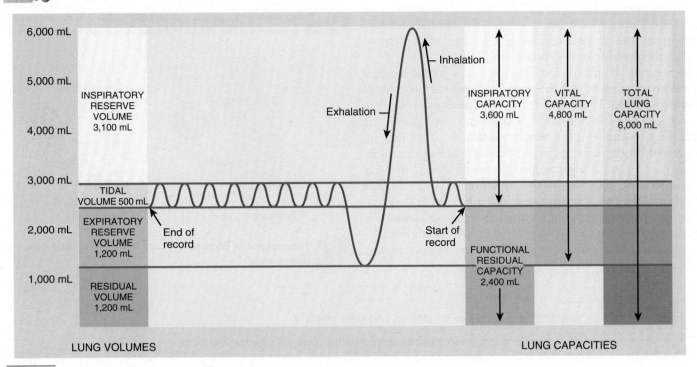

Breathe in as deeply as possible and then exhale as much air as you can. Which lung capacity have you demonstrated?

Table 18.1 / Summary of Lung Volumes and Capacities

Volume or Capacity	Value	Definition
Volumes		
Tidal volume (TV)*	500 mL	Amount of air inhaled or exhaled with each breath during normal quiet breathing.
Inspiratory reserve volume (IRV)	3100 mL	Amount of air that can be forcefully inhaled over and above tidal volume.
Expiratory reserve volume (ERV)	1200 mL	Amount of air that can be forcefully exhaled over and above tidal volume.
Residual volume (RV)	1200 mL	Amount of air remaining in the lungs after a forced exhalation.
Capacities		
Inspiratory capacity (IC)	3600 mL	Maximum inspiratory capacity of the lungs following a normal tidal volume exhalation (IC = TV + IRV).
Functional residual capacity (FRC)	2400 mL	Volume of air remaining in lungs after a normal tidal volume exhalation (FRC = ERV + RV).
Vital capacity (VC)	4800 mL	Maximum amount of air that can be expired following maximum inhalation (VC = TV + IRV + ERV).
Total lung capacity (TLC)	6000 mL	Maximum amount of air in lungs following maximum inhalation (TLC = TV + IRV + ERV + RV).

*Only 350 mL of the tidal volume reach the alveoli; 150 mL remain in the ***anatomic dead space.***

Inhalation

Breathing in is called *inhalation* or *inspiration*. Just before each inhalation, the air pressure inside the lungs equals the pressure of the atmosphere, which is about 760 mm Hg (millimeters of mercury), or 1 atmosphere (atm), at sea level. For air to flow into the lungs, the air pressure inside the lungs must become lower than the pressure in the atmosphere.

The pressure of any volume of gas depends on the number of gas molecules present and the space (volume) they occupy. If a container's volume is decreased, then the pressure of the gas inside the container increases. If a container's volume is increased, the pressure of a gas inside it decreases. This may be demonstrated by placing a gas in a container that has a movable piston (Figure 18.9). The initial pressure is created by the gas molecules striking the wall of the container. If the piston is pushed down, the gas is compressed into a smaller volume, so that the same number of gas molecules strike less wall area. The pressure doubles as the gas is compressed to half its original volume. Conversely, if the piston is raised to increase the volume, the pressure decreases.

Figure 18.9 ■ **The pressure of a gas varies inversely with its volume.**

🔑 **The pressure of a gas is due to the number of molecules of the gas present and the volume of space that they occupy.**

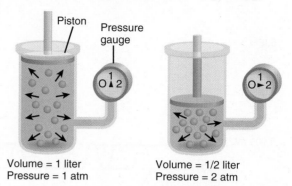

Volume = 1 liter
Pressure = 1 atm

Volume = 1/2 liter
Pressure = 2 atm

❓ **If the volume is decreased from 1/2 to 1/4 liter, how does the pressure change?**

Figure 18.10 ■ **Ventilation: muscles of inhalation and exhalation.**

🔑 **During deep, labored inhalation, accessory muscles of inhalation (external intercostals, sterno-cleidomastoids, and scalenes) participate.**

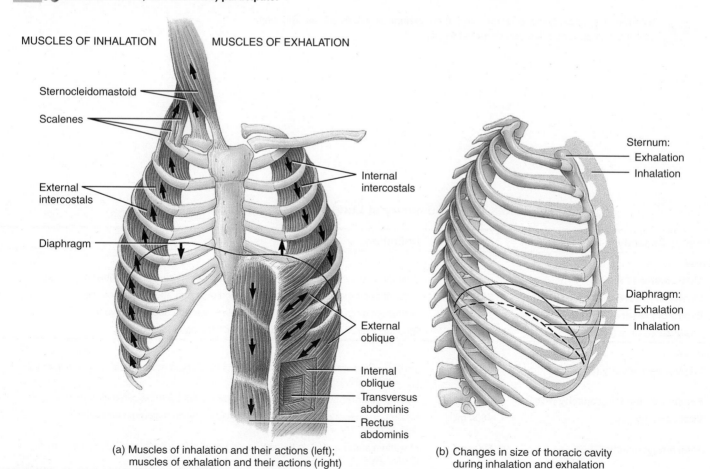

MUSCLES OF INHALATION   MUSCLES OF EXHALATION

Sternocleidomastoid

Scalenes

External
intercostals

Diaphragm

Internal
intercostals

External
oblique

Internal
oblique

Transversus
abdominis

Rectus
abdominis

Sternum:
Exhalation
Inhalation

Diaphragm:
Exhalation
Inhalation

(a) Muscles of inhalation and their actions (left);
muscles of exhalation and their actions (right)

(b) Changes in size of thoracic cavity
during inhalation and exhalation

❓ **What muscle contracts and relaxes during eupnea?**

For inhalation to occur, the lungs must expand, which increases lung volume and decreases the pressure in the lungs to below atmospheric pressure. The first step in expanding the lungs is contraction of the diaphragm, the main muscle of inhalation (Figure 18.10a). The diaphragm is a dome-shaped skeletal muscle that forms the floor of the thoracic cavity. Contraction of the diaphragm flattens it (it normally curves up), which increases the size of the thoracic cavity from top to bottom (Figure 18.10b). Advanced pregnancy, obesity, confining clothing, or increased size of the stomach after eating a large meal can impede complete descent of the diaphragm and may cause shortness of breath.

During deep, forceful inhalations, accessory muscles of inhalation also participate in increasing the size of the thoracic cavity (Figure 18.10a) in the following ways: The external inter-costals elevate the ribs, the sternocleidomastoid elevates the sternum, and the scalenes elevate the two uppermost ribs. As the ribs and sternum are elevated, the size of the thoracic cavity increases in the anterior–posterior direction (Figure 18.10b).

Expansion of the lungs is aided by movement of the pleural membrane. Normally, the parietal and visceral pleurae are tightly attached to each other due to surface tension created by their moist adjoining surfaces. As the diaphragm contracts and the thoracic cavity expands, the parietal pleura is pulled outward, pulling the visceral pleura along with it. As the volume of the lungs increases, the pressure inside the lungs, called the **alveolar pressure,** drops from 760 to 758 mm Hg (Figure 18.11), creating a pressure difference between the atmosphere and the alveoli and causing air to flow into the lungs. Inhalation continues as long as the pressure difference exists.

Figure 18.11 ▪ Pressure changes during ventilation. At the beginning of inhalation, the diaphragm contracts, the chest expands, the volume of the lungs increases, and alveolar pressure decreases. As the diaphragm relaxes, the lungs recoil inward. Alveolar pressure rises, forcing air out until alveolar pressure equals atmospheric pressure.

Air moves into the lungs when alveolar pressure is less than atmospheric pressure and out of the lungs when alveolar pressure is greater than atmospheric pressure.

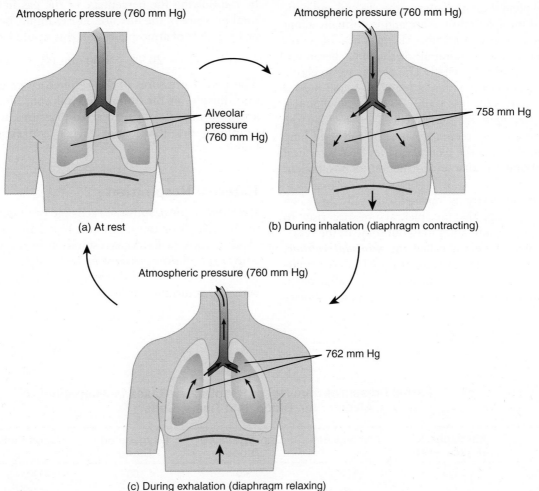

Atmospheric pressure (760 mm Hg)

Atmospheric pressure (760 mm Hg)

Alveolar pressure (760 mm Hg)

758 mm Hg

(a) At rest

(b) During inhalation (diaphragm contracting)

Atmospheric pressure (760 mm Hg)

762 mm Hg

(c) During exhalation (diaphragm relaxing)

What is alveolar pressure during inhalation? during exhalation?

Exhalation

Breathing out, called **exhalation** or **expiration,** occurs when the pressure in the lungs is greater than the pressure of the atmosphere. Because no muscular contractions are involved, normal exhalation, unlike inhalation, is a *passive process.* Exhalation results from **elastic recoil** of the chest wall and lungs, both of which have a natural tendency to spring back after they have been stretched.

Exhalation begins when the diaphragm relaxes and again curves upward, which decreases the volume of the thoracic cavity. Lung volume decreases and alveolar pressure increases, to about 762 mm Hg. Air then flows from the area of higher pressure in the alveoli to the area of lower pressure in the atmosphere (Figure 18.11).

Exhalation becomes active only during more forceful breathing, such as in playing a wind instrument or during exercise. During these times, muscles of exhalation—the abdominals (external oblique, internal oblique, transversus abdominis, and rectus abdominis) and internal intercostals—contract to move the lower ribs downward and compress the abdominal viscera, thus forcing the diaphragm upward (see Figure 18.10a).

Collapsed Lung

The pressure in the pleural cavity between the visceral and parietal pleurae is called intrapleural pressure and is normally about 4 mmHg *below* atmospheric pressure. Maintenance of a low pressure here is vital to the functioning of the lungs because it helps keep the alveoli slightly inflated. The alveoli are so elastic that at the end of an exhalation they attempt to recoil inward and collapse on themselves like the walls of a deflated balloon. A collapsed lung is prevented by the "suction" created by the slightly lower pressure in the pleural cavities. A collapsed lung may be caused by air entering the pleural cavities from a surgical incision or chest wound, an airway obstruction, or insufficient surfactant.

When the thoracic cavity is opened surgically or when trauma creates an opening, the intrapleural pressure rises to equal the atmospheric pressure and forces out some of the residual volume. The remaining air is called the **minimal volume.** Minimal volume provides a medical and legal tool for determining whether a baby was born dead (*stillborn*) or died after birth. The presence of minimal volume can be demonstrated by placing a piece of lung in water and observing if it floats. Fetal lungs contain no air, and so the lung of a stillborn baby does not float.

EXCHANGE OF OXYGEN AND CARBON DIOXIDE

Objective: • **Explain how oxygen and carbon dioxide are exchanged between alveolar air and blood (external respiration) and between blood and body cells (internal respiration).**

The exchange of oxygen and carbon dioxide between alveolar air and pulmonary blood occurs by passive diffusion. Air is a mixture of gases—nitrogen, oxygen, carbon dioxide, water vapor, and a number of others—each of which has its own pressure and behaves as if no other gases are present. This **partial pressure** is denoted as P. The total pressure of air, the atmospheric pressure, is calculated by adding all the partial pressures:

Atmospheric pressure (760 mm Hg)
$$= P_{N_2} + P_{O_2} + P_{H_2O} + P_{CO_2} + P_{Other\ gases}$$

We can determine the partial pressure of each gas in the mixture by multiplying the percentage of the gas in the mixture by the total pressure of the mixture. For example, because oxygen composes 20.9% of atmospheric air, this would be our calculation:

$$P_{O_2} = 20.9\% \times 760\ mm\ Hg = 159\ mm\ Hg$$

The partial pressures of the respiratory gases (oxygen and carbon dioxide) in the atmosphere, alveoli, blood, and tissue cells are shown in Table 18.2. Partial pressures are important in determining the diffusion of oxygen and carbon dioxide in the body.

External Respiration

External respiration is the exchange of O_2 and CO_2 between air in the alveoli of the lungs and blood in pulmonary capillaries (Figure 18.12a). External respiration in the lungs results in the conversion of **deoxygenated blood** (depleted of some O_2) coming from the right side of the heart to **oxygenated blood** (saturated with O_2) returning to the left side of the heart. As blood flows

Table 18.2 / **Partial Pressures (mm Hg) of Respiratory Gases in Atmospheric Air, Alveolar Air, Blood, and Tissue Cells**				
	Atmospheric Air (sea level)	**Alveolar Air**	**Deoxygenated Blood**	**Oxygenated Blood**	**Tissue Cells**
P_{O_2}	159	105	40	100	40
P_{CO_2}	0.3	40	45	40	45

through the pulmonary capillaries, it picks up O_2 from alveolar air and unloads CO_2 into alveolar air. Although this process is commonly called an "exchange" of gases, each gas diffuses independently from the area where its partial pressure is higher to the area where its partial pressure is lower.

Compared with inspired air, alveolar air has less O_2 (20.9% versus 13.6%) and more CO_2 (0.04% versus 5.2%) for two reasons. First, gas exchange in the alveoli increases the CO_2 content and decreases the O_2 content of alveolar air. Second, when air is inhaled it becomes humidified as it passes along the moist mucosal linings. As water vapor content of the air increases, the relative percentage that is O_2 decreases.

As Figure 18.12a shows, O_2 diffuses from alveolar air, where its partial pressure is 105 mm Hg, into the capillary blood, where P_{O_2} is only 40 mm Hg (in a person at rest). Diffusion continues until the P_{O_2} values in both areas are equal. Because blood leaving capillaries near alveoli mixes with a small volume of blood that has flowed through capillaries where gas exchange does not occur, the P_{O_2} of blood in the pulmonary veins is slightly less than the P_{O_2} in pulmonary capillaries—about 100 mm Hg. At rest, the P_{CO_2} of capillary blood is 45 mm Hg when it arrives at the alveoli. CO_2 diffuses into the alveolar air, where the P_{CO_2} is only 40 mm Hg, until the P_{CO_2} values in both areas are equal. Exhalation constantly keeps alveolar P_{CO_2} at 40 mm Hg. Blood returning to the left side of the heart thus has a P_{CO_2} of 40 mm Hg.

External respiration is enhanced by several structural features of the lungs that we have already discussed. The respiratory membrane is very thin, so diffusion occurs quickly, and the surface area of the alveoli is huge. In addition, many capillaries surround each alveolus, so many that as much as 900 mL of blood is able to participate in gas exchange at any given time. Finally, the capillaries are so narrow that the red blood cells must flow through them in single file, giving each red blood cell maximum exposure to the available oxygen.

The efficiency of external respiration depends on several factors. One is altitude. Alveolar P_{O_2} must be higher than blood P_{O_2} for oxygen to diffuse from alveolar air into the blood. As a person ascends in altitude, atmospheric P_{O_2} decreases, alveolar P_{O_2} decreases correspondingly, and less oxygen diffuses into the blood. The common symptoms of **high altitude sickness**—shortness of breath, nausea, dizziness—are due to a lower level of oxygen in the blood. Another factor that affects external respiration is the total surface area available for O_2–CO_2 exchange. Any pulmonary disorder that decreases the functional surface area of the respiratory membrane, for example, emphysema (see page 451), decreases the rate of external respiration. A third factor is the minute ventilation. Certain drugs, such as morphine, slow the rate of respiration, thereby decreasing the amount of O_2 and CO_2 that can be exchanged between the alveoli and the blood. A final factor is the amount of O_2 that reaches the alveoli, which depends on intact and clear air passageways and sufficient oxygen.

Internal Respiration

The movement of O_2 and CO_2 between tissue capillaries and tissue cells is called *internal respiration* (Figure 18.12b). As O_2 leaves the bloodstream, oxygenated blood is converted into deoxygenated blood. Unlike external respiration, which occurs only in the lungs, internal respiration occurs in tissues throughout the body.

The P_{O_2} of blood delivered to systemic tissue cells is higher (100 mm Hg) than the P_{O_2} in the cells (40 mm Hg) because the cells constantly use up O_2 to produce ATP. Due to this pressure difference, oxygen diffuses out of the capillaries into tissue cells, and blood P_{O_2} drops to 40 mm Hg by the time the blood reaches the venules.

Because cells are constantly producing CO_2, the P_{CO_2} of tissue cells (45 mm Hg) is higher than that of capillary blood (40 mm Hg). Carbon dioxide continuously diffuses into the capillaries so that the P_{CO_2} of blood returning to the right side of the heart rises to 45 mm Hg. The deoxygenated blood is pumped through the heart to the lungs for another cycle of external respiration.

Figure 18.12 ■ **Changes in partial pressures (in mm Hg) during external (a) and internal (b) respiration.**

Gases diffuse from areas of higher partial pressure to areas of lower partial pressure.

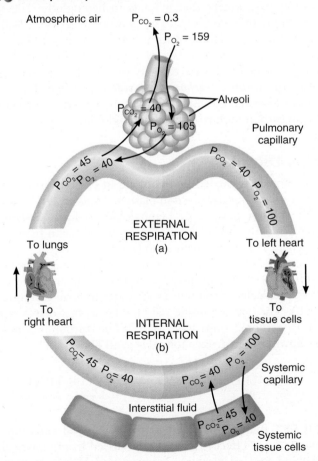

Atmospheric air
P_{CO_2} = 0.3
P_{O_2} = 159

Alveoli
P_{CO_2} = 40
P_{O_2} = 105

Pulmonary capillary

P_{CO_2} = 45 P_{O_2} = 40
P_{CO_2} = 40 P_{O_2} = 100

To lungs

To left heart

EXTERNAL RESPIRATION
(a)

To right heart

To tissue cells

INTERNAL RESPIRATION
(b)

P_{CO_2} = 45 P_{O_2} = 40
P_{CO_2} = 40 P_{O_2} = 100

Systemic capillary

Interstitial fluid

P_{CO_2} = 45 P_{O_2} = 40

Systemic tissue cells

What causes oxygen (O_2) to enter pulmonary capillaries from alveolar air and to enter tissue cells from systemic capillaries?

TRANSPORT OF RESPIRATORY GASES

Objective: • **Describe how the blood transports oxygen and carbon dioxide.**

When O_2 and CO_2 enter the blood, certain physical and chemical changes occur that aid in gas transport and exchange.

Oxygen

Oxygen does not dissolve well in water, and therefore only about 1.5% of O_2 absorbed from the lungs into the blood is dissolved in plasma, which is mostly water. About 98.5% of blood O_2 is carried in chemical combination with hemoglobin in red blood cells (Figure 18.13).

The heme portion of hemoglobin contains four atoms of iron, each capable of combining with a molecule of O_2 (see Figure 14.3b on page 337). Oxygen and hemoglobin bind in an easily reversible reaction to form *oxyhemoglobin* (Hb-O_2):

$$Hb + O_2 \underset{\text{Release of } O_2}{\overset{\text{Binding of } O_2}{\rightleftharpoons}} Hb\text{-}O_2$$

When blood P_{O_2} is high, hemoglobin binds with large amounts of O_2 and is *fully saturated*; that is, every available iron atom has combined with a molecule of O_2. When blood P_{O_2} is low, hemoglobin releases O_2. Therefore, in systemic capillaries, where the P_{O_2} is lower, hemoglobin does not hold as much O_2, and the O_2 is released for diffusion into the tissue cells (Figure 18.13b).

Besides P_{O_2}, several other factors influence the amount of O_2 released by hemoglobin. One factor is the pH: In an acidic environment, hemoglobin releases O_2 more readily. Another factor is temperature: Within limits, as temperature increases, so does the amount of O_2 released from hemoglobin.

Carbon monoxide (CO) is a colorless and odorless gas found in exhaust fumes from automobiles, gas furnaces, and space heaters and in tobacco smoke. CO binds to hemoglobin just as O_2 does, except that it binds to hemoglobin 200 times more strongly than does O_2. In concentrations as small as 0.1%, carbon monoxide combines with half the hemoglobin molecules of the blood, blocking access of oxygen to those molecules. When the oxygen-carrying capacity of the blood is reduced in this way, the result is *carbon monoxide poisoning*. The condition may be treated by administering pure O_2, which hastens the release of carbon monoxide from hemoglobin.

Figure 18.13 ■ **Transport of oxygen and carbon dioxide in the blood.**

Most O_2 is transported by hemoglobin as oxyhemoglobin within red blood cells; most CO_2 is transported in blood plasma as bicarbonate ions.

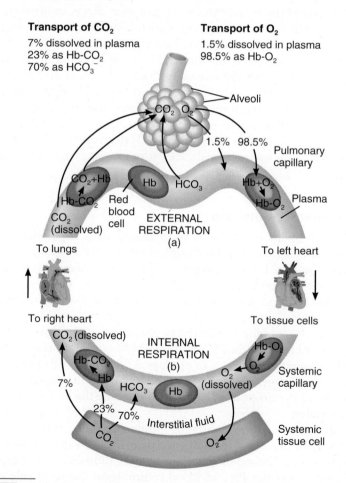

Transport of CO_2
7% dissolved in plasma
23% as Hb-CO_2
70% as HCO_3^-

Transport of O_2
1.5% dissolved in plasma
98.5% as Hb-O_2

Alveoli

EXTERNAL RESPIRATION (a)

To lungs

To right heart

INTERNAL RESPIRATION (b)

To left heart

To tissue cells

Pulmonary capillary

Plasma

Systemic capillary

Interstitial fluid

Systemic tissue cell

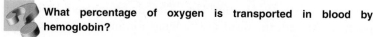

What percentage of oxygen is transported in blood by hemoglobin?

Carbon Dioxide

Carbon dioxide is carried by the blood in several forms (Figure 18.13). About 7% is dissolved in plasma. Another 23% combines with the globin portion of hemoglobin to form *carbaminohemoglobin* (Hb-CO_2). CO_2 and globin combine in tissue capillaries, where P_{CO_2} is relatively high (Figure 18.13b).

About 70% of CO_2 is transported in plasma as bicarbonate ions (HCO_3^-). As CO_2 diffuses into tissue capillaries and enters the red blood cells, it combines with water to form carbonic acid (H_2CO_3). The enzyme inside red blood cells that drives this reaction is carbonic anhydrase (CA). The carbonic acid then breaks down into hydrogen ions (H^+) and HCO_3^-:

$$\underset{\substack{\text{Carbon}\\\text{dioxide}}}{CO_2} + \underset{\text{Water}}{H_2O} \overset{CA}{\rightleftharpoons} \underset{\substack{\text{Carbonic}\\\text{acid}}}{H_2CO_3} \rightleftharpoons \underset{\substack{\text{Hydrogen}\\\text{ion}}}{H^+} + \underset{\substack{\text{Bicarbonate}\\\text{ion}}}{HCO_3^-}$$

HCO_3^- then diffuses out of the red blood cells into the plasma. The net effect is that CO_2 is carried from tissue cells as bicarbonate ions in plasma. Deoxygenated blood returning to the lungs, then, contains CO_2 dissolved in plasma, CO_2 com-

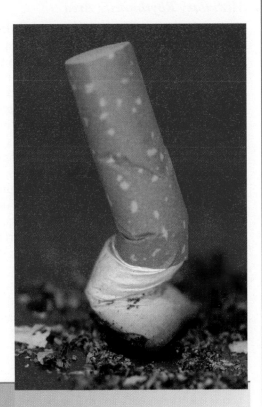
bined with globin, and CO_2 incorporated in bicarbonate ions in plasma (Figure 18.13b).

In the pulmonary capillaries, the events reverse. The CO_2 dissolved in plasma diffuses into alveolar air. The CO_2 combined with globin splits and diffuses into the alveoli. The bicarbonate ions (HCO_3^-) reenter the red blood cells and recombine with H^+ to form H_2CO_3, which splits into CO_2 and H_2O. The CO_2 leaves the red blood cells and diffuses into the alveoli (Figure 18.13a).

CONTROL OF RESPIRATION

Objective: • **Explain how the nervous system controls breathing and list the factors that can alter the rate and depth of breathing.**

At rest, about 200 mL of O_2 are used each minute by body cells. During strenuous exercise, however, O_2 use typically increases 15- to 20-fold in normal healthy adults, and as much as 30-fold in elite endurance-trained athletes. Several mechanisms help match respiratory effort to metabolic demand.

Respiratory Center

The basic rhythm of respiration is controlled by groups of neurons in the brain stem. The area from which nerve impulses are sent to respiratory muscles is called the **respiratory center** and consists of groups of neurons functionally divided into three areas: the medullary rhythmicity area, the pneumotaxic area, and the apneustic area (Figure 18.14).

Medullary Rhythmicity Area

The **medullary rhythmicity area** (rith-MIS-i-tē) in the medulla oblongata controls the basic rhythm of respiration. During quiet breathing, inhalation usually lasts for about 2 seconds and exhalation for about 3 seconds. Within the medullary rhythmicity area are both inspiratory and expiratory areas.

Nerve impulses generated in the **inspiratory area** generate the basic rhythm of breathing. The nerve impulses last for about 2 seconds and reach the diaphragm via the phrenic nerves. When the nerve impulses reach the diaphragm, it contracts and inhalation occurs. At the end of 2 seconds, nerve impulses cease in the inspiratory area, and the diaphragm relaxes for about 3 seconds, allowing passive elastic recoil of the lungs and thoracic wall; then the cycle repeats (Figure 18.15a).

The neurons of the **expiratory area** remain inactive during most normal, quiet respirations. However, during forceful ventilation, nerve impulses from the inspiratory area activate the expiratory area. Impulses from the expiratory area then cause contraction of the internal intercostals and abdominal muscles, which decreases the size of the thoracic cavity and causes forceful exhalation (Figure 18.15b).

Pneumotaxic Area

The **pneumotaxic area** (noo-mō-TAK-sik; *pneumo-* = breath; *-taxic* = arrangement) in the upper pons (see Figure 18.14) transmits inhibitory impulses to the inspiratory area. The major effect of these nerve impulses is to help turn off the inspiratory area before the lungs become too full of air. In other words, the impulses limit the duration of inhalation. When the pneumotaxic area is more active, breathing rate is more rapid.

Apneustic Area

The **apneustic area** (ap-NOO-stik) in the lower pons (see Figure 18.14) sends stimulatory impulses to the inspiratory area that activate it and prolong inhalation, producing a long, deep inhalation. When the pneumotaxic area is active, it overrides the apneustic area.

Regulation of the Respiratory Center

Although the basic rhythm of respiration is set and coordinated by the inspiratory area, the rhythm can be modified in response to inputs from other brain regions, receptors in the peripheral nervous system, and other factors.

Cortical Influences on Respiration

Because the cerebral cortex has connections with the respiratory center, we can voluntarily alter our pattern of breathing. We can even refuse to breathe at all for a short time. Voluntary control is protective because it enables us to prevent water or irritating gases from entering the lungs. The ability to not breathe, however, is limited by the buildup of CO_2 and H^+ in the body. When P_{CO_2} and H^+ concentrations increase to a certain level, the inspiratory area is strongly stimulated and breathing resumes, whether the person wants it or not. It is impossible for people to kill themselves by voluntarily holding their breath. Even if breath is held long enough to cause fainting, breathing resumes when consciousness is lost. Nerve impulses from the hypothalamus and limbic system also stimulate the respiratory center, allowing emotional stimuli to alter respirations (as, for example, in crying).

Chemoreceptor Regulation of Respiration

Certain chemical stimuli modulate how quickly and how deeply we breathe. The respiratory system functions to maintain proper levels of CO_2 and O_2 and is very responsive to changes in the levels of either in body fluids. Sensory neurons that are responsive to chemicals are termed chemoreceptors. **Central chemoreceptors** located within the medulla oblongata respond to changes in H^+ level or P_{CO_2}, or both, in cerebrospinal fluid. The **peripheral chemoreceptors,** located within the arch of the aorta and common carotid arteries, are especially sensitive to changes in P_{O_2}, H^+, and P_{CO_2} in the blood.

Because CO_2 is lipid soluble, it easily diffuses into cells, where it combines with water (H_2O) to form carbonic acid (H_2CO_3), which quickly breaks down into H^+ and HCO_3^-. Any increase in CO_2 in the blood thus causes an increase in H^+ inside cells, and any decrease in CO_2 causes a decrease in H^+.

Under normal circumstances, the P_{CO_2} in arterial blood is 40 mm Hg. If even a slight increase in P_{CO_2} occurs—a condition called **hypercapnia**—the central chemoreceptors are stimulated and respond vigorously to the resulting increase in H^+ level. The peripheral chemoreceptors also are stimulated by both the high P_{CO_2} and the rise in H^+. In addition, the peripheral chemoreceptors respond to severe **hypoxia,** a deficiency of O_2. If P_{O_2} in arterial blood falls from a normal level of 100 mm Hg to about 50 mm Hg, the peripheral chemoreceptors are strongly stimulated.

The chemoreceptors participate in a negative feedback system that regulates the levels of CO_2, O_2, and H^+ in the blood (Figure 18.16 on page 448). As a result of increased P_{CO_2}, increased H^+, and decreased P_{O_2}, input from the central and peripheral chemoreceptors causes the inspiratory area to become highly active, and the rate and depth of breathing increase. Rapid and deep breathing, called **hyperventilation,** allows the exhalation of more CO_2 until P_{CO_2} and H^+ are lowered to normal.

If arterial P_{CO_2} is lower than 40 mm Hg—a condition called **hypocapnia**—the central and peripheral chemoreceptors are not stimulated, and stimulatory impulses are not sent to the inspiratory area. Consequently, the area sets its own moderate pace un-

Figure 18.14 ■ **Areas of the respiratory center.**

The respiratory center is located in the medulla oblongata and pons of the brain stem

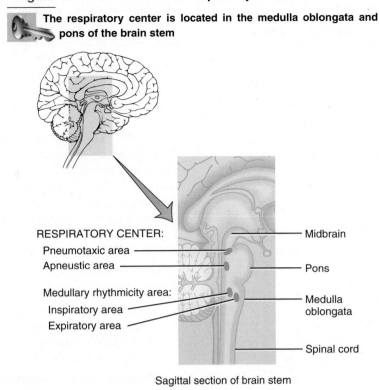

RESPIRATORY CENTER:
Pneumotaxic area
Apneustic area

Medullary rhythmicity area:
Inspiratory area
Expiratory area

Midbrain
Pons
Medulla oblongata
Spinal cord

Sagittal section of brain stem

In normal, quiet breathing, which of the areas shown contains cells that set the basic rhythm of respiration?

Figure 18.15 ■ **Roles of the medullary rhythmicity area in controlling (a) the basic rhythm of respiration and (b) labored breathing.**

During normal, quiet respirations, the expiratory area is inactive; during forceful respirations, the expiratory area is activated by the inspiratory area.

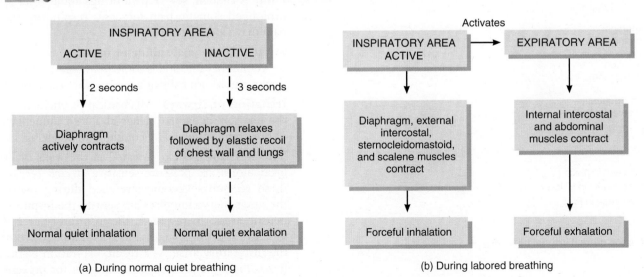

INSPIRATORY AREA

ACTIVE   INACTIVE

2 seconds   3 seconds

Diaphragm actively contracts

Diaphragm relaxes followed by elastic recoil of chest wall and lungs

Normal quiet inhalation

Normal quiet exhalation

(a) During normal quiet breathing

Activates

INSPIRATORY AREA ACTIVE → EXPIRATORY AREA

Diaphragm, external intercostal, sternocleidomastoid, and scalene muscles contract

Internal intercostal and abdominal muscles contract

Forceful inhalation

Forceful exhalation

(b) During labored breathing

Which nerves convey impulses from the respiratory center to the diaphragm?

Figure 18.16 ■ Negative feedback control of breathing in response to changes in blood P_{CO_2}, H^+ level, and P_{O_2}.

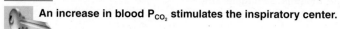

An increase in blood P_{CO_2} stimulates the inspiratory center.

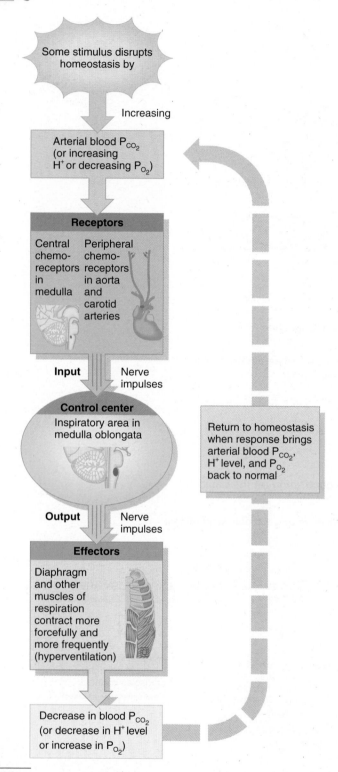

Some stimulus disrupts homeostasis by

Increasing

Arterial blood P_{CO_2} (or increasing H^+ or decreasing P_{O_2})

Receptors

Central chemoreceptors in medulla

Peripheral chemoreceptors in aorta and carotid arteries

Input   Nerve impulses

Control center

Inspiratory area in medulla oblongata

Return to homeostasis when response brings arterial blood P_{CO_2}, H^+ level, and P_{O_2} back to normal

Output   Nerve impulses

Effectors

Diaphragm and other muscles of respiration contract more forcefully and more frequently (hyperventilation)

Decrease in blood P_{CO_2} (or decrease in H^+ level or increase in P_{O_2})

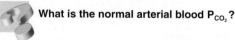

What is the normal arterial blood P_{CO_2}?

til CO_2 accumulates and the P_{CO_2} rises to 40 mm Hg. People who hyperventilate voluntarily and cause hypocapnia can hold their breath for an unusually long period of time. Swimmers were once encouraged to hyperventilate just before diving in to compete, but this practice is risky because the O_2 level may fall dangerously low and cause fainting before the P_{CO_2} rises high enough to stimulate inhalation. A person who faints on land may suffer bumps and bruises, but one who faints in the water may drown.

Severe deficiency of O_2 depresses activity of the central chemoreceptors and inspiratory area, which then do not respond well to any inputs and send fewer impulses to the muscles of respiration. As the respiration rate decreases or breathing ceases altogether, P_{O_2} falls lower and lower, thereby establishing a positive feedback cycle with a possibly fatal result.

Other Influences on Respiration

Other factors that contribute to regulation of respiration include the following:

- **Limbic system stimulation.** Anticipation of activity or emotional anxiety may stimulate the limbic system, which then sends excitatory input to the inspiratory area, increasing the rate and depth of ventilation.

- **Proprioceptor stimulation of respiration.** As soon as you start exercising, your rate and depth of breathing increase, even before changes in P_{O_2}, P_{CO_2}, or H^+ level occur. The main stimulus for these quick changes in respiratory effort is input from proprioceptors, which monitor movement of joints and muscles. Nerve impulses from the proprioceptors stimulate the inspiratory area of the medulla oblongata.

- **Temperature.** An increase in body temperature, as during a fever or vigorous muscular exercise, increases the rate of respiration; a decrease in body temperature decreases respiratory rate. A sudden cold stimulus (such as plunging into cold water) causes **apnea** (AP-nē-a; *a-* = without; *-pnea* = breath), a temporary cessation of breathing.

- **Pain.** A sudden, severe pain brings about brief apnea, but a prolonged somatic pain increases respiratory rate. Visceral pain may slow respiratory rate.

- **Stretching the anal sphincter muscle.** This action increases the respiratory rate and is sometimes used to stimulate respiration in a newborn baby or a person who has stopped breathing.

- **Irritation of airways.** Mechanical or chemical irritation of the pharynx or larynx brings about an immediate cessation of breathing followed by coughing or sneezing.

- **The inflation reflex.** Located in the walls of bronchi and bronchioles are pressure-sensitive **stretch receptors.** When these receptors become stretched during overinflation of the lungs, nerve impulses are sent to the inspiratory and apneustic areas. In response, the inspiratory area is inhibited directly, and the apneustic area is inhibited from activating the inspiratory area. As a result, expiration begins. This reflex is mainly a protective mechanism for preventing excessive inflation of the lungs rather than a key component in the normal regulation of respiration.

Table 18.3 / Summary of Stimuli That Affect Ventilation Rate and Depth

Stimuli That Increase Ventilation Rate and Depth	Stimuli That Decrease Ventilation Rate and Depth
Voluntary hyperventilation controlled by cerebral cortex and anticipation of activity by stimulation of the limbic system.	Voluntary hypoventilation controlled by cerebral cortex.
Increase in arterial blood P_{CO_2} above 40 mm Hg (causes an increase in H^+) detected by peripheral and central chemoreceptors.	Decrease in arterial blood P_{CO_2} below 40 mm Hg (causes a decrease in H^+) detected by peripheral and central chemoreceptors.
Decrease in arterial blood P_{O_2} from 100 mm Hg to 50 mm Hg.	Decrease in arterial blood P_{O_2} below 50 mm Hg.
Increased activity of proprioceptors.	Decreased activity of proprioceptors.
Increase in body temperature.	Decrease in body temperature decreases rate of respiration, and sudden cold stimulus causes apnea.
Prolonged pain.	Severe pain causes apnea.
Decrease in blood pressure.	Increase in blood pressure.
Stretching anal sphincter.	Irritation of pharynx or larynx by touch or chemicals causes brief apnea followed by coughing or sneezing.

- **Blood pressure.** The baroreceptors in arteries detect changes in blood pressure. Although these baroreceptors are active mainly in the control of blood pressure, they have a small effect on respiration. A sudden rise in blood pressure decreases the rate of respiration, and a drop in blood pressure increases the respiratory rate.

Table 18.3 summarizes the stimuli that affect the rate and depth of ventilation.

EXERCISE AND THE RESPIRATORY SYSTEM

Objective: • **Describe the effects of exercise on the respiratory system.**

During exercise, the respiratory and cardiovascular systems make adjustments in response to both the intensity and duration of the exercise. The effects of exercise on the heart are discussed in Chapter 15; here we focus on how exercise affects the respiratory system.

Recall that the heart pumps the same amount of blood to the lungs as to all the rest of the body. Thus, as cardiac output rises, the rate of blood flow through the lungs also increases. If blood flows through the lungs twice as fast as at rest, it picks up twice as much oxygen. In addition, the rate at which O_2 diffuses from alveolar air into the blood increases during maximal exercise because blood flows through a larger percentage of the pulmonary capillaries, providing a greater surface area for diffusion of O_2 into the blood.

The skeletal muscles need this additional oxygen. When they contract during exercise, muscles consume large amounts of O_2 and produce large amounts of CO_2, forcing the respiratory system to work harder to maintain normal blood gas levels. During vigorous exercise, ventilation increases dramatically as well. At the onset of exercise, an abrupt increase in ventilation, due to activation of proprioceptors, is followed by a more gradual increase. With moderate exercise, the depth of ventilation rather than breathing rate is increased. When exercise is more strenuous, breathing rate also increases.

At the end of an exercise session, an abrupt decrease in pulmonary ventilation is followed by a more gradual decline to the resting level. The initial decrease is due mainly to decreased stimulation of proprioceptors when movement stops or slows. The more gradual decrease reflects the slower return of blood chemistry to resting levels.

AGING AND THE RESPIRATORY SYSTEM

Objective: • **Describe the effects of aging on the respiratory system.**

With advancing age, the airways and tissues of the respiratory tract, including the alveoli, become less elastic; the chest wall becomes more rigid as well. The result is a decrease in lung capacity. Vital capacity can decrease as much as 35% by age 70. Moreover, a decrease in blood levels of O_2, decreased activity of alveolar macrophages, and diminished ciliary action of the epithelium lining the respiratory tract occur. Because of these changes, elderly people are more susceptible to pneumonia, bronchitis, emphysema, and other pulmonary disorders.

• • •

To appreciate the many ways that the respiratory system contributes to homeostasis of other body systems, examine Focus on Homeostasis: The Respiratory System on page 450. Next, in Chapter 19, we will see how the digestive system makes nutrients available to body cells so that oxygen provided by the respiratory system can be used for ATP production.

Focus on Homeostasis
The Respiratory System

Body System

Contribution of Respiratory System

For all body systems

Provides oxygen and removes carbon dioxide; helps adjust pH of body fluids through exhalation of carbon dioxide.

Muscular system

Increased rate and depth of breathing support increased activity of skeletal muscles during exercise.

Nervous system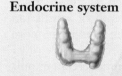

Nose contains receptors for sense of smell (olfaction); vibrations of air flowing across vocal cords produces sounds for speech.

Endocrine system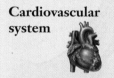

Angiotensin converting enzyme (ACE) in lungs promotes production of the hormone angiotensin II.

Cardiovascular system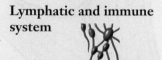

Respiratory pump (during breathing) aids return of venous blood to the heart.

Lymphatic and immune system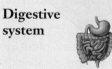

Hairs in nose; cilia and mucus in trachea, bronchi, and smaller airways; and alveolar macrophages contribute to nonspecific resistance to disease; pharynx (throat) contains lymphatic tissue (tonsils); respiratory pump (during breathing) promotes flow of lymph.

Digestive system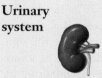

Forceful contraction of respiratory muscles can assist in defecation.

Urinary system

Together, respiratory and urinary systems regulate pH of body fluids.

Reproductive systems

Increased rate and depth of breathing support activity during sexual intercourse; internal respiration provides oxygen to developing fetus.

COMMON DISORDERS

Asthma

Asthma (AZ-ma = panting) is a disorder characterized by chronic airway inflammation, airway hypersensitivity to a variety of stimuli, and airway obstruction. The airway obstruction may be due to smooth muscle spasms in the walls of smaller bronchi and bronchioles, swelling of the mucosa of the airways, increased mucus secretion, or damage to the epithelium of the airway.

Individuals with asthma typically react to low concentrations of stimuli that do not normally affect people without asthma. Sometimes the trigger is an allergen such as pollen, dust mites, molds, or a particular food. Other common triggers include emotional upset, aspirin, sulfiting agents (used in wine and beer and to keep greens fresh in salad bars), exercise, and breathing cold air or cigarette smoke. Symptoms include difficult breathing, coughing, wheezing, chest tightness, tachycardia, fatigue, moist skin, and anxiety. Asthma is at least partially reversible, either spontaneously or with treatment. It is more common in children than in adults.

Chronic Obstructive Pulmonary Disease

Chronic obstructive pulmonary disease (COPD) is a type of respiratory disorder characterized by chronic obstruction of airflow. The principal types of COPD are emphysema and chronic bronchitis. In most cases, COPD is preventable; its most common cause is cigarette smoking or breathing secondhand smoke. Other causes include air pollution, pulmonary infection, occupational exposure to dusts and gases, and genetic factors.

Emphysema

Emphysema (em′-fi-SĒ-ma = blown up or full of air) is a disorder characterized by destruction of the walls of the alveoli, which produces abnormally large air spaces that remain filled with air during exhalation. With less surface area for gas exchange, O_2 diffusion across the respiratory membrane is reduced. Blood O_2 level is somewhat lowered, and any mild exercise that raises the O_2 requirements of the cells leaves the patient breathless. As increasing numbers of alveolar walls are damaged, lung elastic recoil decreases due to loss of elastic fibers, and an increasing amount of air becomes trapped in the lungs at the end of exhalation. Over several years, added respiratory exertion increases the size of the chest cage, resulting in a "barrel chest."

Chronic Bronchitis

Chronic bronchitis is a disorder characterized by excessive secretion of bronchial mucus accompanied by a productive cough (sputum is raised). Inhaled irritants lead to chronic inflammation with an increase in the size and number of mucous glands and goblet cells in the airway epithelium. The thickened and excessive mucus produced narrows the airway and impairs the action of cilia. Thus, inhaled pathogens become embedded in airway secretions and multiply rapidly. Besides a productive cough, symptoms of chronic bronchitis are shortness of breath, wheezing, cyanosis, and pulmonary hypertension.

Lung Cancer

In the developed world, *lung cancer* is the leading cause of cancer death in both males and females. At the time of diagnosis, lung cancer is usually well advanced; most people with lung cancer die within a year of the initial diagnosis, and the overall survival rate is only 10–15%. About 85% of lung cancer cases are due to smoking, and the disease is 10 to 30 times more common in smokers than nonsmokers. Exposure to secondhand smoke also causes lung cancer.

Symptoms of lung cancer are related to the location of the tumor and may include a chronic cough, spitting blood from the respiratory tract, wheezing, shortness of breath, chest pain, hoarseness, difficulty swallowing, weight loss, anorexia, fatigue, bone pain, confusion, problems with balance, headache, anemia, low blood platelet count, and jaundice.

Pneumonia

Pneumonia or *pneumonitis* is an acute infection or inflammation of the alveoli. It is the most common infectious cause of death in the United States, where an estimated 4 million cases occur annually. When certain microbes enter the lungs of susceptible individuals, they release damaging toxins, stimulating inflammation and immune responses that have damaging side effects. The toxins and immune response damage alveoli and bronchial mucous membranes; inflammation and edema cause the alveoli to fill with debris and fluid, interfering with ventilation and gas exchange. The most common cause is the pneumococcal bacterium *Streptococcus pneumoniae*, but other bacteria and even viruses may also cause pneumonia.

Tuberculosis

The bacterium *Mycobacterium tuberculosis* produces an infectious, communicable disease called *tuberculosis (TB)* that most often affects the lungs and the pleurae but may involve other parts of the body. Once the bacteria are inside the lungs, they multiply and cause inflammation, which stimulates neutrophils and macrophages to migrate to the area and engulf the bacteria to prevent their spread. If the immune system is not impaired, the bacteria may remain dormant for life. Impaired immunity may enable the bacteria to escape into blood and lymph to infect other organs. In many people, symptoms—fatigue, weight loss, lethargy, anorexia, a low-grade fever, night sweats, cough, dyspnea, chest pain, and spitting blood (hemoptysis)—do not develop until the disease is advanced.

Coryza and Influenza

Hundreds of viruses, especially the rhinoviruses, are responsible for *coryza* (ko-RĪ-za) or the ***common cold.*** Typical symptoms include sneezing, excessive nasal secretion, dry cough, and congestion. The uncomplicated common cold is not usually accompanied by a fever. Complications include sinusitis, asthma, bronchitis, ear infections, and

COMMON DISORDERS (CONTINUED)

laryngitis. Recent investigations suggest an association between emotional stress and the common cold. The higher the stress level, the greater the frequency and duration of colds.

Influenza (flu) is also caused by a virus. Its symptoms include chills, fever (usually higher than 101°F, or 38°C), headache, and muscular aches. Coldlike symptoms appear as the fever subsides.

Pulmonary Edema

Pulmonary edema is an abnormal accumulation of interstitial fluid in the interstitial spaces and alveoli of the lungs. The most common symptom is painful or labored breathing. Other symptoms include wheezing, rapid breathing rate, restlessness, a feeling of suffocation, cyanosis, paleness, and excessive perspiration.

Cystic Fibrosis

Cystic fibrosis (CF) is an inherited disease of secretory epithelia that affects the airways, liver, pancreas, small intestine, and sweat glands. The cause of cystic fibrosis is a genetic mutation affecting a transporter protein that carries chloride ions across the plasma membranes of many epithelial cells. The mutation also disrupts the normal functioning of several organs by causing their ducts to become obstructed by thick mucus secretions. Buildup of these secretions leads to inflammation and replacement of injured cells with connective tissue that further blocks the ducts. Obstruction of bile ducts in the liver interferes with digestion and disrupts liver function; clogging of pancreatic ducts prevents digestive enzymes from reaching the small intestine. Clogging and infection of the airways leads to difficulty in breathing and eventual destruction of lung tissue. Lung disease accounts for most deaths from CF.

MEDICAL TERMINOLOGY AND CONDITIONS

Acute respiratory distress syndrome (ARDS)   A form of respiratory failure characterized by excessive leakiness of the respiratory membranes and severe hypoxia. Situations that can cause ARDS include near-drowning, aspiration of acidic gastric juice, drug reactions, inhalation of an irritating gas such as ammonia, allergic reactions, various lung infections such as pneumonia or TB, and pulmonary hypertension.

Asphyxia   (as-FIK-sē-a; *sphyxia* = pulse) Oxygen starvation due to low atmospheric oxygen or interference with ventilation, external respiration, or internal respiration.

Atelectasis   (at'-ē-LEK-ta-sis; *atel-* = incomplete; *ectasis* = expansion) Incomplete expansion of a lung or a portion of a lung caused by airway obstruction, lung compression, or inadequate pulmonary surfactant.

Bronchography   (bron-KOG-ra-fē) An imaging technique used to visualize the bronchial tree using x rays. After an opaque contrast medium is inhaled through an intratracheal catheter, radiographs of the chest in various positions are taken, and the developed film, a *bronchogram* (BRON-kō-gram), provides a picture of the bronchial tree.

Cheyne-Stokes respiration   (CHĀN STŌKS res'-pi-RĀ-shun) A repeated cycle of irregular breathing that begins with shallow breaths that increase in depth and rapidity and then decrease and cease altogether for 15 to 20 seconds. Cheyne-Stokes is normal in infants; it is also often seen just before death from pulmonary, cerebral, cardiac, and kidney disease.

Dyspnea   (DISP-nē-a; *dys-* = painful, difficult) Painful or labored breathing.

Epistaxis   (ep'-i-STAK-sis) Loss of blood from the nose due to trauma, infection, allergy, malignant growths, or bleeding disorders. It can be arrested by cautery with silver nitrate, electrocautery, or firm packing. Also called *nosebleed*.

Hypoxia   (hī-POK-sē-a; *hypo-* = below or under) A deficiency of O_2 at the tissue level that may be caused by a low P_{O_2} in arterial blood, as from high altitudes; too little functioning hemoglobin in the blood, as in anemia; inability of the blood to carry O_2 to tissues fast enough to sustain their needs, as in heart failure; or inability of tissues to use O_2 properly, as in cyanide poisoning.

Rales   (RĀLS) Sounds sometimes heard in the lungs that resemble bubbling or rattling. Different types are due to the presence of an abnormal type or amount of fluid or mucus within the bronchi or alveoli, or to bronchoconstriction that causes turbulent airflow.

Respiratory failure   A condition in which the respiratory system either cannot supply sufficient O_2 to maintain metabolism or cannot eliminate enough CO_2 to prevent respiratory acidosis (a higher-than-normal H^+ level in extracellular fluid).

Rhinitis   (rī-NĪ-tis; *rhin-* = nose) Chronic or acute inflammation of the mucous membrane of the nose.

Sudden infant death syndrome (SIDS)   Death of infants between the ages of 1 week and 12 months thought to be due to hypoxia while sleeping in a prone position (on the stomach) and the rebreathing of exhaled air trapped in a depression of the mattress. It is now recommended that normal newborns be placed on their backs for sleeping (remember: "back to sleep").

Tachypnea   (tak'-ip-NĒ-a; *tachy-* = rapid) Rapid breathing rate.

STUDY OUTLINE

Organs of the Respiratory System (p. 431)

1. Respiratory organs include the nose, pharynx, larynx, trachea, bronchi, and lungs, and they act with the cardiovascular system to supply oxygen and remove carbon dioxide from the blood.

2. The external portion of the nose is made of cartilage and skin and is lined with mucous membrane. Openings to the exterior are the external nares.

3. The internal portion of the nose, divided from the external portion by the septum, communicates with the paranasal sinuses and nasopharynx through the internal nares.

4. The nose is adapted for warming, moistening, filtering air; olfaction; and serving as a resonating chamber for special sounds.

5. The pharynx (throat), a muscular tube lined by a mucous membrane, is divided into the nasopharynx, oropharynx, and laryngopharynx.

6. The nasopharynx functions in respiration. The oropharynx and laryngopharynx function both in digestion and in respiration.

7. The larynx connects the pharynx and the trachea. It contains the thyroid cartilage (Adam's apple), the epiglottis, the cricoid cartilage, three paired cartilages, false vocal cords, and true vocal cords. Taut true vocal cords produce high pitches; relaxed ones produce low pitches.

8. The trachea (windpipe) extends from the larynx to the primary bronchi. It is composed of smooth muscle and C-shaped rings of cartilage and is lined with pseudostratified ciliated columnar epithelium.

9. The bronchial tree consists of the trachea, primary bronchi, secondary bronchi, tertiary bronchi, bronchioles, and terminal bronchioles.

10. Lungs are paired organs in the thoracic cavity enclosed by the pleural membrane. The parietal pleura is the outer layer; the visceral pleura is the inner layer.

11. The right lung has three lobes separated by two fissures; the left lung has two lobes separated by one fissure plus a depression, the cardiac notch.

12. Each lobe consists of lobules, which contain lymphatic vessels, arterioles, venules, terminal bronchioles, respiratory bronchioles, alveolar ducts, alveolar sacs, and alveoli.

13. Exchange of gases (oxygen and carbon dioxide) occurs across the respiratory membrane of the alveoli.

Lung Volumes and Capacities (p. 438)

1. The minute ventilation is the total air taken in during 1 minute (breathing rate per minute multiplied by tidal volume).

2. The lung volumes are tidal volume, inspiratory reserve volume, expiratory reserve volume, and residual volume.

3. Lung capacities, the sum of two or more lung volumes, include inspiratory, functional residual, vital, and total.

Ventilation (p. 438)

1. Ventilation (breathing) consists of inhalation and exhalation, the movement of air into and out of the lungs. Air flows from higher to lower pressure.

2. Inhalation occurs when alveolar pressure falls below atmospheric pressure. Contraction of the diaphragm increases the size of the thorax so that the lungs expand. Expansion of the lungs decreases alveolar pressure, so that air moves along the pressure gradient from the atmosphere into the lungs.

3. Exhalation occurs when alveolar pressure is higher than atmospheric pressure. Relaxation of the diaphragm decreases the size of the thorax and lung volume, and alveolar pressure increases so that air moves from the lungs to the atmosphere.

4. During forced inhalation, accessory muscles of inhalation are also used.

5. Forced exhalation involves contraction of the internal intercostals and abdominal muscles.

6. A collapsed lung may be caused by air in the pleural cavities, airway obstruction, or decreased surfactant.

7. When the thoracic cavity is opened, the air that then remains in the lungs is the minimal volume.

Exchange of Oxygen and Carbon Dioxide (p. 442)

1. The partial pressure of a gas (P) is the pressure exerted by that gas in a mixture of gases.

2. Each gas in a mixture of gases exerts its own pressure and behaves as if no other gases are present.

3. In external and internal respiration, O_2 and CO_2 move from areas of higher partial pressure to areas of lower partial pressure.

4. External respiration is the exchange of gases between alveolar air and pulmonary blood capillaries. It is aided by a thin respiratory membrane, a large alveolar surface area, and a rich blood supply.

5. Internal respiration is the exchange of gases between systemic tissue capillaries and systemic tissue cells.

Transport of Respiratory Gases (p. 444)

1. Most oxygen, 98.5%, is carried by the iron atoms of the heme in hemoglobin; the rest is dissolved in plasma.

2. The association of O_2 and hemoglobin is affected by P_{O_2}, pH, temperature, and P_{CO_2}.

3. Hypoxia refers to O_2 deficiency at the tissue level.

4. Carbon dioxide is transported in three ways. About 7% is dissolved in plasma, 23% combines with the globin of hemoglobin, and 70% is converted to bicarbonate ions (HCO_3^-) that travel in plasma.

Control of Respiration (p. 445)

1. The respiratory center consists of a medullary rhythmicity area (inspiratory and expiratory areas), pneumotaxic area, and apneustic area.

2. The inspiratory area sets the basic rhythm of respiration.

3. The pneumotaxic and apneustic areas coordinate the transition between inspiration and expiration.

4. Respirations may be modified by several factors, including cortical influences; chemical stimuli, such as levels of O_2, CO_2, and H^+; limbic system stimulation; proprioceptor input; temperature; pain; the inflation reflex; blood pressure changes; and irritation to the airways.

Exercise and the Respiratory System (p. 449)

1. The rate and depth of ventilation change in response to both the intensity and duration of exercise.

2. An increase in pulmonary blood flow and O_2 diffusing capacity occurs during exercise.

3. The abrupt increase in ventilation at the start of exercise is due to neural changes that send excitatory impulses to the inspiratory area in the medulla oblongata. The more gradual increase in ventilation during moderate exercise is due to chemical and physical changes in the bloodstream.

Aging and the Respiratory System (p. 449)

1. Aging results in decreased vital capacity, decreased blood level of O_2, and diminished alveolar macrophage activity.

2. Elderly people are more susceptible to pneumonia, emphysema, bronchitis, and other pulmonary disorders.

◼ SELF-QUIZ

1. Which of the following is NOT true concerning the pharynx?
   a. Food, drink, and air pass through the oropharynx and laryngopharynx. **b.** The auditory (Eustachian) tubes have openings in the nasopharynx. **c.** The pseudostratified ciliated epithelium of the nasopharynx helps move dust-laden mucus toward the mouth. **d.** The palatine and lingual tonsils are located in the laryngopharynx. **e.** The wall of the pharynx is composed of skeletal muscle lined with mucous membranes.

2. During speaking, you raise your voice's pitch. This is possible because
   a. the epiglottis vibrates rapidly **b.** you have increased the air pressure pushing against the vocal cords **c.** you have increased the tension on the true vocal cords **d.** your true vocal cords have become thicker and longer **e.** the true vocal cords begin to vibrate more slowly

3. Johnny is having an asthma attack and feels as if he cannot breathe. Why?
   a. His diaphragm is not contracting. **b.** Spasms in the bronchiole smooth muscle have blocked airflow to the alveoli. **c.** Excess mucus production is interfering with airflow into the lungs. **d.** The epiglottis has closed and air is not entering the lungs. **e.** Insufficient surfactant is being produced by the type II alveolar cells.

4. Which sequence of events best describes inhalation?
   a. contraction of diaphragm → increase in size of thoracic cavity → decrease in alveolar pressure **b.** relaxation of diaphragm → decrease in size of thoracic cavity → increase in alveolar pressure **c.** contraction of diaphragm → decrease in size of thoracic cavity → decrease in alveolar pressure **d.** relaxation of diaphragm → increase in size of thoracic cavity → increase in alveolar pressure **e.** contraction of diaphragm → decrease in size of thoracic cavity → increase in alveolar pressure

5. Which of the following does NOT help keep air passages clean?
   a. nostril hairs **b.** alveolar macrophages **c.** capillaries in the nasal cavities **d.** cilia in the upper and lower respiratory tracts **e.** mucus

6. If the total pressure of a mixture of gases is 760 mm Hg and gas Z makes up 20% of the total mixture, then the partial pressure of gas Z would be:
   a. 152 mm Hg **b.** 175 mm Hg **c.** 225 mm Hg **d.** 608 mm Hg **e.** 760 mm Hg

7. How does hypercapnia affect respiration?
   a. It increases the rate of respiration. **b.** It decreases the rate of respiration. **c.** It causes hypoventilation. **d.** It does not change the rate of respiration. **e.** It activates stretch receptors in the lungs.

8. Air would flow into the lungs along the following route:
   1. bronchioles
   2. primary bronchi
   3. secondary bronchi
   4. terminal bronchioles
   5. tertiary bronchi
   6. trachea
   a. 6, 1, 2, 3, 5, 4 **b.** 6, 5, 3, 4, 2, 1 **c.** 6, 2, 3, 5, 4, 1
   d. 6, 2, 3, 5, 1, 4 **e.** 6, 1, 4, 5, 3, 2

9. Match the following:
   _____ **a.** normally inactive; when activated, causes contraction of internal intercostals and abdominal muscles and forced exhalation
   _____ **b.** located in pons; stimulates inspiratory area to prolong inhalation
   _____ **c.** sets basic rhythm of respiration; located in medulla
   _____ **d.** transmits inhibitory impulses to inspiratory area; located in pons
   _____ **e.** allows voluntary alteration of breathing patterns

   A. inspiratory area
   B. expiratory area
   C. pneumotaxic area
   D. apneustic area
   E. cerebral cortex

10. Under normal body conditions, hemoglobin releases oxygen more readily when
    a. body temperature increases **b.** blood acidity decreases **c.** blood pH increases **d.** blood oxygen partial pressure is high **e.** blood CO_2 is low

11. Match the following:
    _____ **a.** decreased carbon dioxide levels
    _____ **b.** normal, quiet breathing
    _____ **c.** rapid breathing
    _____ **d.** exchange of gases between the blood and lungs
    _____ **e.** inhalation and exhalation
    _____ **f.** increased carbon dioxide levels
    _____ **g.** exchange of gases between blood and tissue cells
    _____ **h.** temporary cessation of breathing

    A. external respiration
    B. apnea
    C. hypercapnia
    D. eupnea
    E. internal respiration
    F. hypocapnia
    G. pulmonary ventilation
    H. hyperventilation

12. Which of the following statements is NOT true concerning the lungs?

    a. Pulmonary blood vessels enter the lungs at the hilus. **b.** The left lung is thicker and broader because the liver lies below it. **c.** The right lung is composed of three lobes. **d.** The top portion of the lung is the apex. **e.** The lungs are surrounded by a serous membrane.

13. Exhalation

    a. occurs when alveolar pressure reaches 758 mm Hg **b.** is normally considered an active process requiring muscle contraction **c.** occurs when alveolar pressure is greater than atmospheric pressure **d.** involves the expansion of the pleural membranes **e.** occurs when the atmospheric pressure is equal to the pressure in the lungs

14. In which structures would you find simple squamous epithelium?

    a. secondary bronchi **b.** larynx and pharynx **c.** tertiary bronchi **d.** primary bronchi **e.** terminal bronchioles

15. Overinflation of the lungs is prevented by

    a. the inflation reflex **b.** pain in the pleural membranes **c.** nerve impulses from proprioceptors **d.** control from the cerebral cortex **e.** controlling blood pressure

16. The function of goblet cells in the nasal cavities is to

    a. warm the air entering the nose **b.** produce mucus to trap inhaled dust **c.** increase the surface area inside the nose **d.** help produce speech **e.** exchange O_2 and CO_2 within the nasal cavities

17. Decreasing the surface area of the respiratory membrane would affect

    a. internal respiration **b.** inhalation **c.** speech **d.** external respiration **e.** mucus production

18. In which form is carbon dioxide NOT carried in the blood?

    a. bicarbonate ion **b.** bound to globin **c.** oxyhemoglobin **d.** carbaminohemoglobin **e.** dissolved in plasma

19. Fill in the blanks in the following chemical reactions.

    $$CO_2 = \underline{\hspace{1.5cm}} \rightleftharpoons H_2CO_3 \rightleftharpoons H^+ + \underline{\hspace{1.5cm}}$$

    a. HCO_3^-, O_2 **b.** HCO_3^-, H_2O **c.** H^+, H_2O **d.** O_2, HCO_3^- **e.** H_2O, HCO_3^-

20. Of the following, which would have the highest partial pressure of oxygen?

    a. Alveolar air at the end of exhalation **b.** Rapidly contracting skeletal muscle fibers **c.** Alveolar air immediately after inhalation **d.** Blood flowing into the lungs from the right side of the heart **e.** Blood returning to the heart from the tissue cells.

21. Match the following:

    ____ **a.** forceful exhalation of air

    ____ **b.** inspiratory reserve volume + tidal volume + expiratory reserve volume

    ____ **c.** volume of air moved during normal quiet breathing

    ____ **d.** air remaining after forced exhalation

    ____ **e.** forceful inhalation of air

    A. vital capacity

    B. inspiratory reserve volume

    C. residual volume

    D. expiratory reserve volume

    E. tidal volume

CRITICAL THINKING APPLICATIONS

1. Your three-year-old nephew Levi likes to get his own way all the time! Right now, Levi wants to eat 20 chocolate kisses (1 for each finger and toe), but you'll only give him one for each year of his age. He is at this moment "holding my breath until I turn blue and won't you be sorry!" Is he in danger of death?

2. Tracey's mother caught her smoking a cigarette. "All the rock stars do it!" Tracey cried. "If all the rock stars jumped off a cliff, would you do that too?" answered Mom. (You knew that was coming.) What should her mom tell Tracey about the effects of smoking?

3. Brianna has a flare for being dramatic. "I can't come to work today," she whispered, "I've got laryngitis and a horrible case of coryza." What is wrong with Brianna?

4. A child collapses at a party after inhaling an uninflated balloon. What procedures may be used to try to dislodge or remove the balloon? If the balloon has reached the end of the trachea, where is it likely to lodge?

ANSWERS TO FIGURE QUESTIONS

18.1   The conducting part of the respiratory system includes the nose, pharynx, larynx, trachea, bronchi, and bronchioles (except the respiratory bronchioles).

18.2   Air molecules flow through the external nares, the nasal cavity, and then the internal nares.

18.3   During swallowing, the epiglottis closes over the larynx to stop food and liquids from entering.

18.4   There are two lobes and two secondary bronchi in the left lung and three lobes and three secondary bronchi in the right lung.

18.5   Because two-thirds of the heart lies to the left of the midline, the left lung contains a cardiac notch to accommodate the position of the heart. The right lung is shorter to accommodate the liver lying below it.

18.6   A lung lobule includes a lymphatic vessel, arteriole, venule, and branch of a terminal bronchiole wrapped in elastic connective tissue.

18.7   Type II alveolar cells secrete alveolar fluid, which includes surfactant.

18.8   Vital capacity.

18.9   The pressure would increase to 4 atm.

18.10   Eupnea, normal breathing at rest, occurs due to contraction and relaxation of the diaphragm.

18.11   Alveolar pressure during inhalation is 758 mm Hg; alveolar pressure during exhalation is 762 mm Hg.

18.12   Oxygen enters pulmonary capillaries from alveolar air and enters tissue cells from systemic capillaries due to a difference in P_{O_2}.

18.13   Hemoglobin transports about 98.5% of the oxygen carried in blood.

18.14   Neurons in the inspiratory area of the medullary rhythmicity area establish the basic rhythm of quiet respiration.

18.15   The phrenic nerves stimulate the diaphragm to contract.

18.16   Normal arterial blood P_{CO_2} is 40 mm Hg.

Chapter **19**

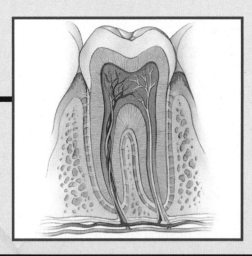

The Digestive System

■ Student Learning Objectives

1. Identify the organs of the digestive system and their basic functions. **458**
2. Describe the layers that form the wall of the gastrointestinal tract. **459**
3. Describe the structure and functions of the tongue. **461**
4. Identify the locations of the salivary glands, and describe the functions of their secretions. **462**
5. Identify the parts of a typical tooth, and compare deciduous and permanent dentitions. **462**

6. Describe the location, structure, and function of the pharynx. **463**
7. Describe the location, structure, and function of the esophagus. **463**
8. Describe the location, structure, and functions of the stomach. **465**
9. Describe the location, structure, and functions of the pancreas. **468**

10. Describe the location, structure, and functions of the liver and gallbladder. **468**
11. Describe the location, structure, and functions of the small intestine. **472**
12. Describe the location, structure, and functions of the large intestine. **476**
13. List the factors that regulate food intake. **480**
14. Describe the effects of aging on the digestive system. **481**

■ A Look Ahead

The food we eat contains a variety of nutrients, the molecules needed for building new body tissues, repairing damaged tissues, and sustaining chemical reactions. To make use of these nutrients, our bodies must break down food into molecules small enough for cells to use, a process known as **digestion**. The passage of these smaller molecules through cells into the blood and lymph is termed **absorption**. Collectively, the organs that perform these functions are known as the **digestive system**.

The medical specialty that deals with the structure, function, diagnosis, and treatment of diseases of the stomach and intestines is **gastroenterology** (gas′-trō-en′-ter-OL-o-jē; *gastro-* = stomach; *enter-* = intestines; *-ology* = study of). The medical specialty that deals with the diagnosis and treatment of disorders of the rectum and anus is **proctology** (prok-TOL-ō-jē; *proct-* = rectum).

OVERVIEW OF THE DIGESTIVE SYSTEM

Objective: • **Identify the organs of the digestive system and their basic functions.**

The digestive system (Figure 19.1) is composed of two groups of organs: the gastrointestinal tract and the accessory digestive organs. The **gastrointestinal (GI) tract** is a continuous tube that extends from the mouth to the anus. The GI tract contains food from the time it is eaten until it is digested and absorbed or eliminated from the body. Organs of the gastrointestinal tract include the mouth, most of the pharynx, esophagus, stomach, small intestine, and large intestine. The teeth, tongue, salivary glands, liver, gallbladder, and pancreas serve as **accessory digestive organs.** Teeth aid in the physical breakdown of food, and the tongue assists in chewing and swallowing. The other accessory digestive organs never come into direct contact with food. The secretions that they produce or store flow into the GI tract through ducts and aid in the chemical breakdown of food.

Overall, the digestive system has six basic functions:

Figure 19.1 ■ **Organs of the digestive system and related structures.**

Organs of the gastrointestinal (GI) tract include the mouth, pharynx, esophagus, stomach, small intestine, and large intestine. Accessory digestive organs include the teeth, tongue, salivary glands (parotid, sublingual, and submandibular), liver, gallbladder, and pancreas.

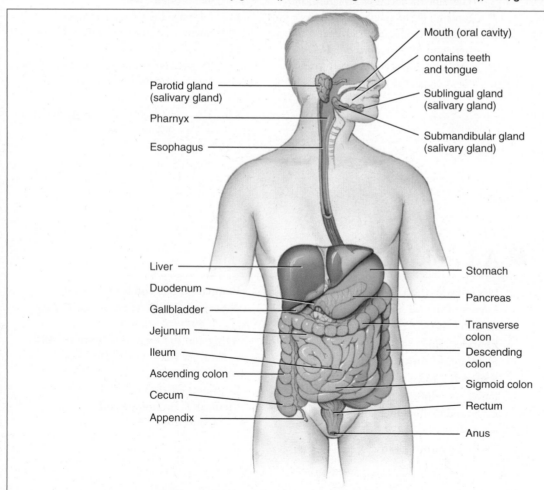

Mouth (oral cavity)

contains teeth and tongue

Parotid gland (salivary gland)

Pharnyx

Esophagus

Sublingual gland (salivary gland)

Submandibular gland (salivary gland)

Liver

Duodenum

Gallbladder

Jejunum

Ileum

Ascending colon

Cecum

Appendix

Stomach

Pancreas

Transverse colon

Descending colon

Sigmoid colon

Rectum

Anus

Functions of the Digestive System

1. Ingestion: taking food into the mouth.
2. Secretion: release of water, acid, buffers, and enzymes into the lumen of the GI tract.
3. Mixing and propulsion: churning and propulsion of food through the GI tract.
4. Digestion: mechanical and chemical breakdown of food.
5. Absorption: passage of digested products from the GI tract into the blood and lymph.
6. Defecation: the elimination of feces from the GI tract.

 What type of movement propels food along the GI tract?

1. **Ingestion.** This process involves taking foods and liquids into the mouth (eating).

2. **Secretion.** Each day, cells within the walls of the GI tract and accessory organs secrete a total of about 7 liters of water, acid, buffers, and enzymes into the lumen of the tract.

3. **Mixing and propulsion.** Alternating contraction and relaxation of smooth muscle in the walls of the GI tract mix food and secretions and propel them toward the anus. The ability of the GI tract to mix and move material along its length is termed *motility*.

4. **Digestion.** Mechanical and chemical processes break down ingested food into small molecules. In *mechanical digestion* the teeth cut and grind food before it is swallowed, and then smooth muscles of the stomach and small intestine churn the food. As a result, food molecules become dissolved and thoroughly mixed with digestive enzymes. In *chemical digestion* the large carbohydrate, lipid, protein, and nucleic acid molecules in food are broken down into smaller molecules.

5. **Absorption.** The entrance of ingested and secreted fluids, ions, and the small molecules that are products of digestion into the epithelial cells lining the lumen of the GI tract is called *absorption*. The absorbed substances pass into blood or lymph and circulate to cells throughout the body.

6. **Defecation.** Wastes, indigestible substances, bacteria, cells sloughed from the lining of the GI tract, and digested materials that were not absorbed leave the body through the anus in a process called *defecation*. The eliminated material is termed *feces*.

LAYERS OF THE GI TRACT

Objective: • **Describe the layers that form the wall of the gastrointestinal tract.**

The wall of the GI tract, from the lower esophagus to the anal canal (organs below the diaphragm), has the same basic, four-layered arrangement of tissues. The four layers of the tract, from the inside out, are the mucosa, submucosa, muscularis, and serosa (Figure 19.2).

Figure 19.2 ■ **Layers of the gastrointestinal tract.** Variations in this basic plan may be seen in the stomach (Figure 19.8a), small intestine (Figure 19.12a), and large intestine (Figure 19.15a).

The four layers of the GI tract from inside to outside are the mucosa, submucosa, muscularis, and serosa.

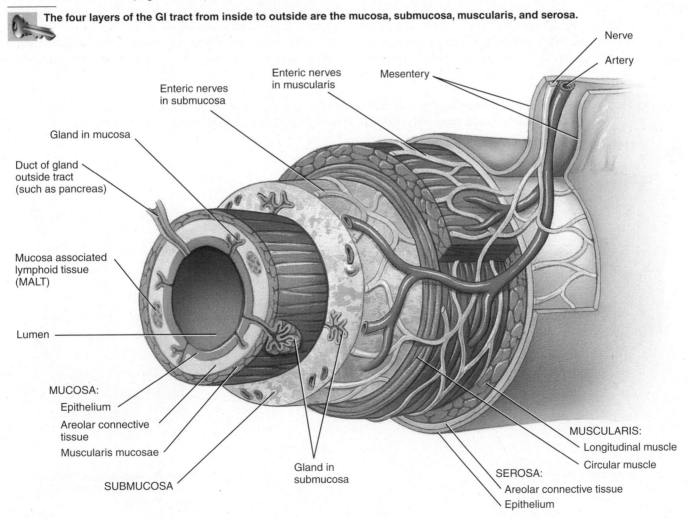

What is the function of the nerves in the wall of the gastrointestinal tract?

Mucosa

The *mucosa,* or inner lining of the tract, is a mucous membrane. It is composed of a layer of epithelium in direct contact with the contents of the GI tract, areolar connective tissue, and a thin layer of smooth muscle (muscularis mucosae). Contractions of the muscularis mucosae create folds in the mucosa that increase the surface area for digestion and absorption. The mucosa also contains *mucosa-associated lymphatic tissue (MALT),* prominent lymphatic nodules that protect against entry of pathogens through the GI tract.

Submucosa

The *submucosa* consists of areolar connective tissue that binds the mucosa to the muscularis. It contains numerous blood and lymphatic vessels for the absorption of the breakdown products of digestion and networks of neurons that control GI tract secretions and motility. The neurons are part of the *enteric nervous system (ENS),* the "brain of the gut." Enteric neurons in the submucosa regulate movements of the mucosa and vasoconstriction of blood vessels of the GI tract. Because its neurons also innervate secretory cells of mucosal and submucosal glands, the ENS is important in controlling secretions by the GI tract.

Muscularis

As its name implies, the *muscularis* of the GI tract is a thick layer of muscle. In the mouth, pharynx, and upper esophagus, it consists in part of skeletal muscle that produces voluntary swallowing. Skeletal muscle also forms the external anal sphincter, which permits voluntary control of defecation. Recall that a sphincter is a thick circle of muscle around an opening. In the rest of the tract, the muscularis consists of smooth muscle, usually arranged as an inner sheet of circular fibers and an outer sheet of longitudinal fibers. Involuntary contractions of these smooth muscles help break down food physically, mix it with digestive secretions, and propel it through the tract. Enteric neurons within the muscularis control the frequency and strength of contraction of the muscularis.

Serosa and Peritoneum

The *serosa,* the outermost layer around organs of the GI tract below the diaphragm, is a membrane composed of simple squamous epithelium and connective tissue. The serosa secretes slippery, watery fluid that allows the tract to glide easily against other organs.

The serosa is also called the *visceral peritoneum.* Recall from Chapter 4 that the *peritoneum* (per′-i-tō-NĒ-um = to stretch over) is the largest serous membrane of the body. Whereas the parietal peritoneum lines the wall of the abdominal cavity, the visceral peritoneum covers organs in the cavity.

In addition to binding the organs to each other and to the walls of the abdominal cavity, the peritoneal folds contain blood vessels, lymphatic vessels, and nerves that supply the abdominal organs. The *greater omentum* (ō-MENT-um = fat skin) drapes over the transverse colon and coils of the small intestine like a "fatty apron" (Figure 19.3a,b). The many lymph nodes of the

Figure 19.3 ■ **Views of the abdomen and pelvis.** The relationship of the parts of the peritoneum (greater omentum, falciform ligament, and mesentery) to each other and to organs of the digestive system is shown.

The peritoneum is the largest serous membrane in the body.

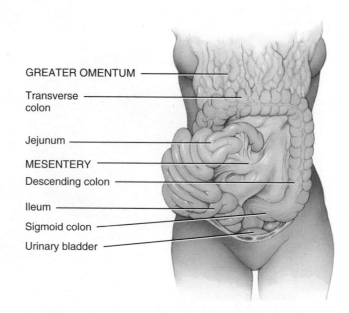

FALCIFORM LIGAMENT
Liver
Stomach
Transverse colon

GREATER OMENTUM

Urinary bladder

(a) Anterior view

GREATER OMENTUM

Transverse colon

Jejunum

MESENTERY

Descending colon

Ileum

Sigmoid colon

Urinary bladder

(b) Anterior view (greater omentum lifted and small intestine reflected to right side)

Which part of the peritoneum binds the small intestine to the posterior abdominal wall?

greater omentum contribute macrophages and antibody-producing plasma cells that help combat and contain infections of the GI tract. The *falciform ligament* (FAL-si-form; *falc-* = sickle-shaped) attaches the liver to the anterior abdominal wall and diaphragm (Figure 19.3a). A part of the peritoneum, the *mesentery* (MEZ-en-ter′-ē; *mes-* = middle), binds the small intestine to the posterior abdominal wall (Figure 19.3b).

MOUTH

Objectives: • **Identify the locations of the salivary glands, and describe the functions of their secretions.**
• **Describe the structure and functions of the tongue.**
• **Identify the parts of a typical tooth, and compare deciduous and permanent dentitions.**

The *mouth* or *oral cavity* is formed by the cheeks, hard and soft palates, and tongue (Figure 19.4). The *lips* are fleshy folds around the opening of the mouth. They are covered on the outside by skin and on the inside by a mucous membrane. During chewing, the lips and cheeks help keep food between the upper and lower teeth. They also assist in speech.

The *hard palate*, consisting of the maxillae and palatine bones, forms most of the roof of the mouth. The rest is formed by the muscular *soft palate*. Hanging from the soft palate is a projection called the **uvula** (YOO-vyoo-la). During swallowing, the uvula moves upward with the soft palate, which prevents entry of swallowed foods and liquids into the nasal cavity. At the back of the soft palate, the mouth opens into the oropharynx through an opening called the **fauces** (FAW-sēz). The *palatine tonsils* are just posterior to the fauces.

Tongue

The *tongue* forms the floor of the oral cavity. It is an accessory digestive organ composed of skeletal muscle covered with mucous membrane (see Figure 12.3 on page 281).

Figure 19.4 ■ Structures of the mouth (oral cavity).

The mouth is formed by the cheeks, hard and soft palates, and tongue.

Superior lip (pulled upward)

Gingivae (gums)

Fauces

Hard palate

Soft palate

Uvula

Cheek

Palatine tonsil

Tongue

Molars

Lingual frenulum

Premolars

Opening of duct of submandibular gland

Cuspid (canine)

Gingivae (gums)

Incisors

Inferior lip (pulled down)

Anterior view

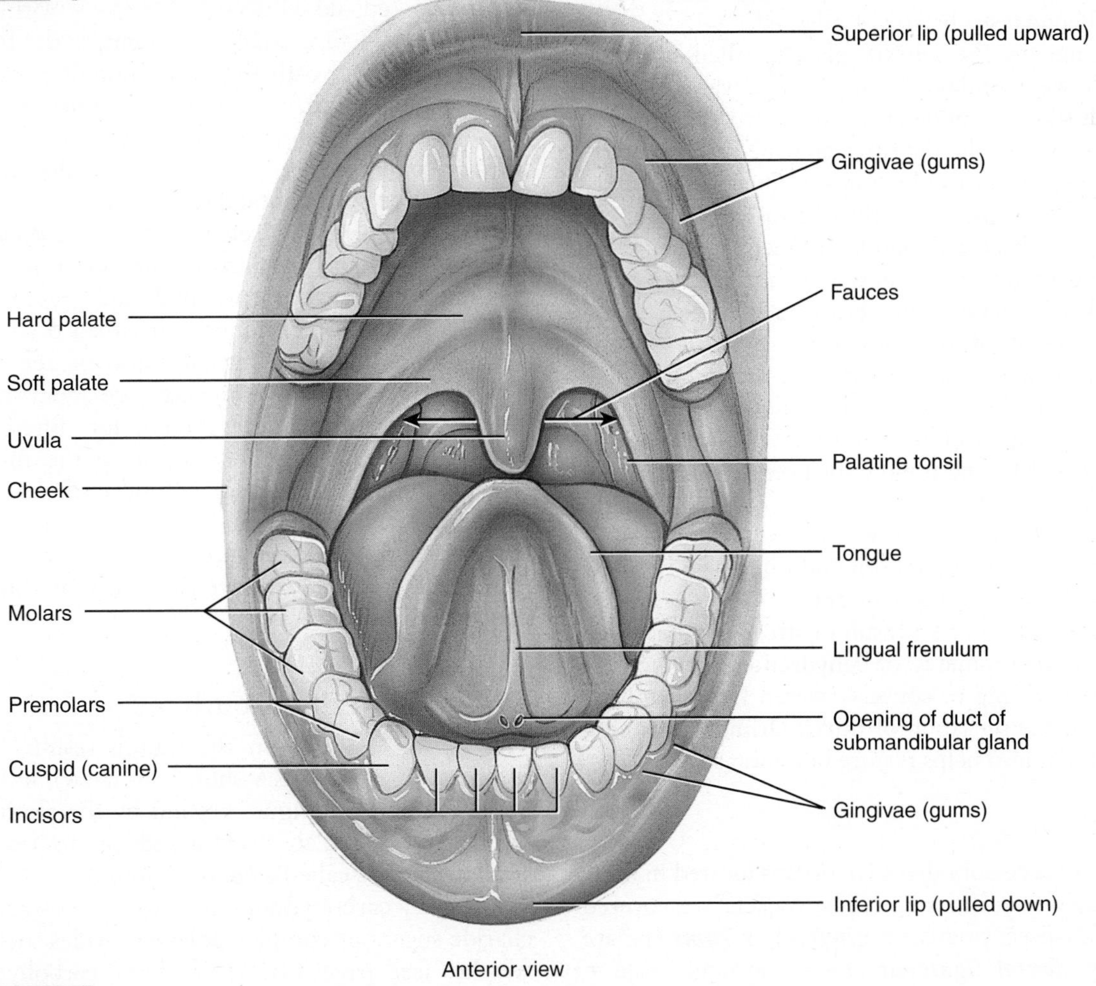

 What are the functions of the muscles of the tongue?

The muscles of the tongue maneuver food for chewing, shape the food into a rounded mass, force the food to the back of the mouth for swallowing, and alter the shape and size of the tongue for swallowing and speech. The *lingual frenulum* (LING-gwal FREN-yoo-lum; *lingua* = tongue; *frenum* = bridle), a fold of mucous membrane in the midline of the undersurface of the tongue, limits the movement of the tongue posteriorly (Figure 19.4). The lingual tonsils lie at the base of the tongue (see Figure 12.3a). The upper surface and sides of the tongue are covered with projections called *papillae* (pa-PIL-ē), some of which contain taste buds.

Salivary Glands

Fluid is secreted by three pairs of *salivary glands*, accessory organs of digestion that lie outside the mouth and pour their contents into ducts emptying into the oral cavity: the parotid, submandibular, and sublingual (see Figure 19.1). The *parotid glands* are located inferior and anterior to the ears between the skin and the masseter muscle. The *submandibular glands* lie inferior to the base of the tongue in the floor of the mouth. The *sublingual glands* lie anterior to the submandibular glands.

Mumps is an inflammation and enlargement of the parotid glands accompanied by fever, malaise, a sore throat, and swelling on one or both sides of the face. In about 30% of males past puberty, a testis may also become inflamed; sterility rarely results because usually only one testis becomes inflamed.

The fluid secreted by the salivary glands, called *saliva,* is composed of 99.5% water and 0.5% solutes. The water in saliva helps dissolve foods so they can be tasted and starts digestive reactions. One of the solutes, the digestive enzyme *salivary amylase,* begins the digestion of carbohydrates in the mouth. Mucus in saliva lubricates food so it can easily be swallowed. The enzyme lysozyme destroys bacteria, thereby protecting the mouth's mucous membrane from infection and the teeth from decay.

Salivation (sal'-i-VĀ-shun), the secretion of saliva, is entirely under the control of the autonomic nervous system. Parasympathetic stimulation arriving via the facial (VII) and glossopharyngeal (IX) nerves promotes continuous secretion of saliva to keep the tongue and lips moist. Whenever food is tasted, smelled, touched by the tongue, or even thought about, parasympathetic impulses increase secretion of saliva. Saliva continues to be secreted heavily for some time after food is swallowed. This flow washes out the mouth and dilutes and buffers chemical remnants of any irritating substances.

Dryness of the mouth can be a result of stress, during which sympathetic stimulation dominates, or dehydration, in which the salivary glands stop secreting to conserve water. Dryness of the mouth contributes to the sensation of thirst. Drinking not only moistens the mouth but also helps restore homeostasis.

Teeth

The *teeth (dentes)* are accessory digestive organs located in bony sockets of the mandible and maxillae. The sockets are covered by the *gingivae* (JIN-ji-vē; singular is *gingiva*) or *gums* and are lined with the *periodontal ligament (peri-* = around; *odous* = tooth)*, dense fibrous connective tissue that anchors the teeth to

bone, keeps them in position, and acts as a shock absorber during chewing (Figure 19.5).

A typical tooth has three principal parts. The *crown* is the exposed portion above the level of the gums. The *root* consists of one to three projections embedded in the socket. The *neck* is the junction line of the crown and root, near the gum line.

Teeth are composed primarily of *dentin,* a calcified connective tissue that gives the tooth its basic shape and rigidity. The dentin encloses the *pulp cavity,* a space in the crown filled with *pulp,* a connective tissue containing blood vessels, nerves, and lymphatic vessels. Narrow extensions of the pulp cavity run through the root of the tooth and are called *root canals.* Each root canal has an opening at its base through which blood vessels bring nourishment, lymphatic vessels offer protection, and nerves provide sensation. The dentin of the crown is covered by *enamel* that consists primarily of calcium phosphate and calcium carbonate. Enamel, the hardest substance in the body and the richest in calcium salts (about 95% of its dry weight), protects the tooth from the wear of chewing. It is also a barrier against acids that easily dissolve the dentin. The dentin of the root is covered by *cementum,* a bonelike substance that attaches the root to the periodontal ligament.

The branch of dentistry that is concerned with the prevention, diagnosis, and treatment of diseases that affect the pulp, root, periodontal ligament, and alveolar bone is known as *endodontics* (en'-dō-DON-tiks; *endo-* = within). *Orthodontics* (or'-thō-DON-tiks; *ortho-* = straight) is the branch of dentistry that is concerned with the prevention and correction of abnormally aligned teeth, whereas *periodontics* (per'-ē-ō-DON-tiks) is the branch of dentistry concerned with the treatment of abnormal conditions of the tissues immediately surrounding the teeth, such as periodontal diseases.

Humans have two sets of teeth. The *deciduous teeth* begin to erupt at about 6 months of age, and one pair appears about each month thereafter until all 20 are present. They are generally lost in the same sequence between 6 and 12 years of age. The *permanent teeth* appear between age 6 and adulthood. There are 32 teeth in a complete permanent set.

Humans have different teeth for different functions (see Figure 19.4). *Incisors* are closest to the midline, are chisel-shaped, and are adapted for cutting into food; *cuspids* (canines) are next to the *incisors* and have one pointed surface (cusp) to tear and shred food; *premolars* have two cusps to crush and grind food; and *molars* have three or more blunt cusps to crush and grind food.

Digestion in the Mouth

Mechanical digestion in the mouth results from chewing, or *mastication* (mas'-ti-KĀ-shun = to chew), in which food is manipulated by the tongue, ground by the teeth, and mixed with saliva. As a result, the food is reduced to a soft, flexible, easily swallowed mass called a *bolus* (= lump).

Dietary carbohydrates are either monosaccharide and disaccharide sugars or complex polysaccharides such as glycogen and starches (see page 30). Most of the carbohydrates we eat are starches from plant sources, but only monosaccharides can be

Figure 19.5 ■ Parts of a typical tooth.

 There are 20 teeth in a complete deciduous set and 32 teeth in a complete permanent set.

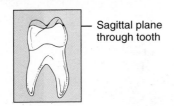

— Sagittal plane through tooth

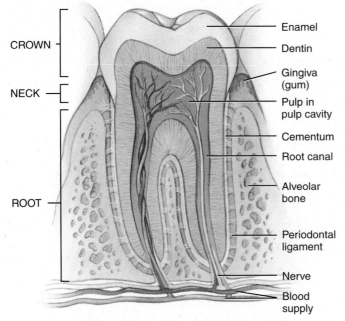

CROWN

NECK

ROOT

— Enamel

— Dentin

— Gingiva (gum)

— Pulp in pulp cavity

— Cementum

— Root canal

— Alveolar bone

— Periodontal ligament

— Nerve

— Blood supply

Sagittal section of a mandibular (lower) molar

What type of tissue is the main component of teeth?

absorbed into the bloodstream. Thus, ingested starches must be broken down into monosaccharides. *Salivary amylase* begins the breakdown of starch and glycogen by breaking particular chemical bonds between the glucose subunits. The resulting products include the disaccharide maltose (2 glucose subunits), the trisaccharide maltotriose (3 glucose subunits), and larger fragments called dextrins (5 to 10 glucose subunits). Salivary amylase in the swallowed food continues to act for about an hour until it is inactivated by stomach acids.

PHARYNX

Objective: • **Describe the location, structure, and function of the pharynx.**

The *pharynx* (FAIR-inks) is a tube that is composed of skeletal muscle and lined by mucous membrane. It extends from the internal nares to the esophagus posteriorly and the larynx anteriorly (Figure 19.6a), and it is divided into two sections important to digestion: the oropharynx and the laryngopharynx. As you learned in Chapter 18, the nasopharynx is involved in respiration (see Figure 18.2 on page 429). Food that is swallowed passes from the mouth into the oropharynx and laryngopharynx before passing into the esophagus. Muscular contractions of the oropharynx and laryngopharynx help propel food into the esophagus.

ESOPHAGUS

Objective: • **Describe the location, structure, and function of the esophagus.**

The *esophagus* (e-SOF-a-gus = eating gullet) is a muscular tube that lies posterior to the trachea. It begins at the end of the laryngopharynx, passes through the mediastinum and diaphragm, and connects to the superior aspect of the stomach. It transports food to the stomach and secretes mucus.

    Swallowing, the movement of food from the mouth to the stomach, involves the mouth, pharynx, and esophagus and is helped by saliva and mucus. Swallowing is divided into three stages: the voluntary, pharyngeal, and esophageal stages.

    In the *voluntary stage* of swallowing, the bolus is forced to the back of the mouth cavity and into the oropharynx by the movement of the tongue upward and backward against the soft palate. With the passage of the bolus into the oropharynx, the involuntary *pharyngeal stage* of swallowing begins (Figure 19.6b). Breathing is temporarily interrupted when the soft palate and uvula move upward to close off the nasopharynx, the epiglottis seals off the larynx, and the vocal cords come together. After the bolus passes through the oropharynx, the respiratory passageways reopen and breathing resumes.

    In the *esophageal stage,* food is pushed through the esophagus by a process called *peristalsis* (Figure 19.6c):

Figure 19.6 ■ **Swallowing.** During the pharyngeal stage of swallowing (b), the tongue rises against the palate, the nasopharynx is closed off, the larynx rises, the epiglottis seals off the larynx, and the bolus is passed into the esophagus.

 Swallowing is a mechanism that moves food from the mouth into the stomach.

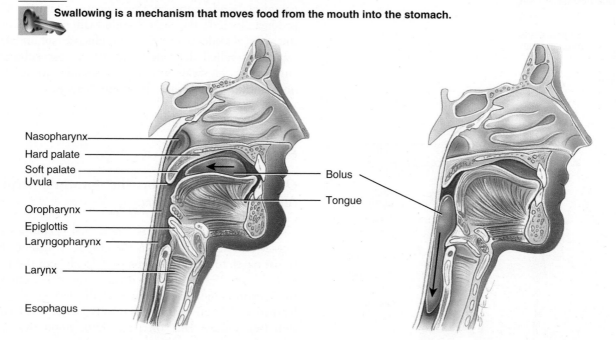

Nasopharynx

Hard palate

Soft palate

Uvula

Oropharynx

Epiglottis

Laryngopharynx

Larynx

Esophagus

Bolus

Tongue

(a) Position of structures before swallowing

(b) During the pharyngeal stage of swallowing

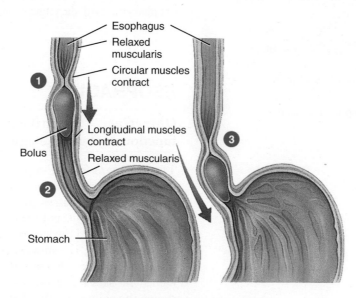

Esophagus

Relaxed muscularis

Circular muscles contract

Longitudinal muscles contract

Relaxed muscularis

Bolus

Stomach

(c) Esophageal stage of swallowing

Is swallowing a voluntary or involuntary action?

1 The circular muscle fibers in the section of esophagus above the bolus contract, constricting the wall of the esophagus and squeezing the bolus downward.

2 Longitudinal muscle fibers around the bottom of the bolus contract, shortening the section of the esophagus below the bolus and pushing its walls outward.

3 After the bolus moves into the new section of the esophagus, the circular muscles above it contract, and the cycle repeats. The contractions move the bolus down the esophagus and into the stomach.

Sometimes, after food has entered the stomach, the stomach contents can back up or reflux into the lower esophagus, a condition known as *gastroesophageal reflux disease (GERD)*. Reflux of hydrochloric acid (HCl) from the stomach contents can irritate the esophageal wall, resulting in a burning sensation commonly called *heartburn*. Although it is experienced in a region very near the heart, it is unrelated to any cardiac problem.

STOMACH

Objective: • **Describe the location, structure, and functions of the stomach.**

The *stomach* is a J-shaped region of the GI tract directly below the diaphragm. The superior part of the stomach is connected to the esophagus. The inferior part empties into the duodenum, the first section of the small intestine.

Structure of the Stomach

The stomach is divided into four main areas: cardia, fundus, body, and pylorus (Figure 19.7). The *cardia* (CAR-dē-a) surrounds the superior opening of the stomach. The stomach then curves upward. The portion superior and to the left of the cardia is the *fundus* (FUN-dus). Inferior to the fundus is the large central portion of the stomach, called the *body.* The narrow, most inferior region is the *pylorus* (pī-LOR-us; *pyl-* = gate; *-orus* = guard). Between the pylorus and duodenum is the *pyloric sphincter,* also called the *pyloric valve.*

The stomach wall is composed of the same four basic layers as the rest of the GI tract (mucosa, submucosa, muscularis, serosa), with certain modifications. When the stomach is empty, the mucosa lies in large folds, called *rugae* (ROO-jē = wrinkles). The surface of the mucosa is a layer of nonciliated simple columnar epithelial cells called *mucous surface cells* (Figure 19.8). Epithelial cells also extend downward and form columns of secretory cells called *gastric glands* that line narrow channels called *gastric pits.* Secretions from the gastric glands flow into the gastric pits and then into the lumen of the stomach. The glands contain three types of *exocrine gland* cells that secrete their products into the stomach: mucous neck cells, chief cells, and parietal cells. In addition, gastric glands include one type of hormone-producing cell (G cell).

Figure 19.7 ■ **External and internal anatomy of the stomach.** The dashed lines indicate the approximate borders of the regions of the stomach.

The four regions of the stomach are the cardia, fundus, body, and pylorus.

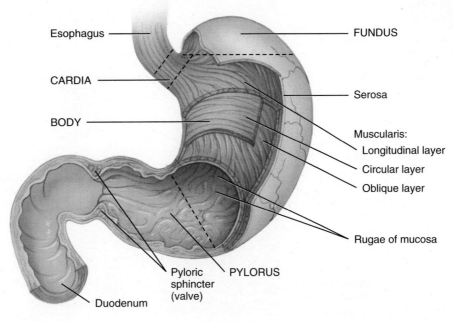

Anterior view

 Does your stomach have rugae after a big Thanksgiving dinner?

Figure 19.8 ■ **Histology of the stomach.**

The secretions of the mucous, chief, and parietal cells are referred to as gastric juice.

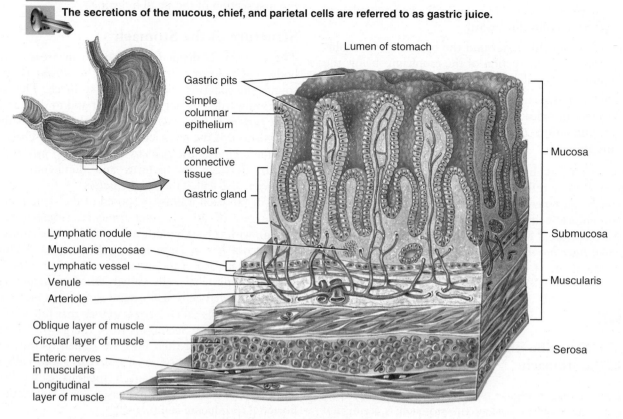

Lumen of stomach

Gastric pits

Simple columnar epithelium

Areolar connective tissue

Gastric gland

Lymphatic nodule

Muscularis mucosae

Lymphatic vessel

Venule

Arteriole

Oblique layer of muscle

Circular layer of muscle

Enteric nerves in muscularis

Longitudinal layer of muscle

Mucosa

Submucosa

Muscularis

Serosa

(a) Three dimensional view of layers of the stomach

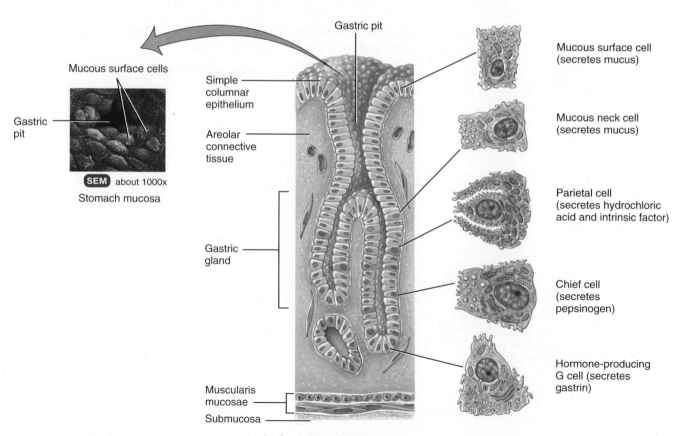

Gastric pit

Mucous surface cells

Gastric pit

SEM about 1000x

Stomach mucosa

Simple columnar epithelium

Areolar connective tissue

Gastric gland

Muscularis mucosae

Submucosa

Mucous surface cell (secretes mucus)

Mucous neck cell (secretes mucus)

Parietal cell (secretes hydrochloric acid and intrinsic factor)

Chief cell (secretes pepsinogen)

Hormone-producing G cell (secretes gastrin)

(b) Sectional view of the stomach mucosa showing gastric glands

What are the functions of gastrin?

Both mucous surface cells and ***mucous neck cells*** secrete mucus. The ***chief cells*** secrete an inactive gastric enzyme called ***pepsinogen. Parietal cells*** produce hydrochloric acid, which helps convert pepsinogen to the active digestive enzyme ***pepsin,*** and *intrinsic factor,* which is involved in the absorption of vitamin B_{12}. Inadequate production of intrinsic factor can result in pernicious anemia because vitamin B_{12} is needed for red blood cell production. The secretions of the mucous, chief, and parietal cells are collectively called ***gastric juice.*** The ***G cells*** secrete the hormone gastrin into the bloodstream. Gastrin stimulates secretion of gastric juice by the three types of exocrine gland cells, increases motility (contractions) of the GI tract, and relaxes the pyloric sphincter.

The submucosa of the stomach is composed of areolar connective tissue that connects the mucosa to the muscularis. The muscularis has three rather than two layers of smooth muscle: an outer longitudinal layer, a middle circular layer, and an inner oblique layer (see Figure 19.7). Contractions of the muscularis break food into small particles, mix it with gastric juice, and pass it to the duodenum. The serosa covering the stomach, composed of simple squamous epithelium and areolar connective tissue, is part of the visceral peritoneum.

Digestion and Absorption in the Stomach

Mechanical digestion in the stomach begins several minutes after food enters, when gentle, rippling, peristaltic movements called ***mixing waves*** start passing over the stomach about every 20 seconds. These waves macerate food, mix it with the secretions of the gastric glands, and change it to a thin liquid called ***chyme*** ($K\bar{I}M$ = juice). Each mixing wave forces a small amount of chyme through the pyloric sphincter into the duodenum, a process called ***gastric emptying.*** Most of the chyme is forced back into the body of the stomach, where it is subjected to further mixing. The next wave pushes it forward again and forces a little more into the duodenum. The forward and backward movements of the chyme are responsible for almost all mixing in the stomach.

The main event of chemical digestion in the stomach is the beginning of protein digestion by pepsin. As its name implies, pepsin breaks peptide bonds between the amino acids of proteins, fragmenting the proteins into ***peptides,*** smaller strings of amino acids. Pepsin is most effective in the very acidic environment (pH 2) of the stomach.

What keeps pepsin from digesting the protein in stomach cells along with the food? First, recall that chief cells secrete pepsin in an inactive form (pepsinogen). It is not converted into active pepsin until it comes in contact with the hydrochloric acid secreted by the parietal cells. Second, mucus secreted by mucous cells coats the mucosa, forming a thick barrier between the cells of the stomach lining and the gastric juices.

The epithelial cells of the stomach are impermeable to most materials, so little absorption occurs. However, mucous cells of the stomach absorb some water, ions, and short-chain fatty acids, as well as certain drugs (especially aspirin). Because it is lipid soluble and thus can diffuse across the lipid bilayer of plasma membranes, alcohol absorption begins in the stomach.

Much of the alcohol absorbed through the stomach wall never reaches the bloodstream, however, because an enzyme known as alcohol dehydrogenase breaks it down inside stomach cells. Because their stomach cells have much less alcohol dehydrogenase, females and people of Asian heritage often experience higher blood alcohol levels and therefore become more drunk than white or black males after drinking the same amount of alcohol.

Regulation of Gastric Secretion and Motility

Parasympathetic nerve impulses from the medulla oblongata, conducted along the vagus (X) nerves, activate enteric nervous system (ENS) neurons that promote peristalsis and stimulate secretions from the gastric glands. Just as salivation may begin before food enters the mouth, gastric secretions may occur in response to the sight, smell, taste, or thought of food before food even enters the stomach. The cerebral cortex and feeding center in the hypothalamus send nerve impulses to the medulla oblongata, which then sends impulses via parasympathetic neurons to the ENS in the stomach.

Once food reaches the stomach, neural and hormonal mechanisms ensure that gastric secretion continues. Food of any kind causes stretching, which stimulates receptors in the wall of the stomach to send nerve impulses to the medulla and from there back to the gastric glands. Partially digested proteins, caffeine, and a high pH of stomach chyme also stimulate the secretion of gastric juice by triggering the release of gastrin.

Within 2 to 6 hours after food is ingested, the stomach has emptied its contents into the duodenum, a small quantity at a time. Foods rich in carbohydrate spend the least time in the stomach. Proteins move more slowly, and emptying is slowest after a meal containing large amounts of triglycerides (fats). As the amount of chyme in the small intestine increases, it starts to inhibit gastric secretion and motility by means of reflexes that inhibit parasympathetic activity and stimulate sympathetic activity. These reflexes slow the exit of chyme from the stomach and prevent overloading of the small intestine with more chyme than it can handle. Stimulation of the sympathetic nervous system by emotions such as anger, fear, and anxiety may also slow digestion in the stomach.

When chyme leaves the stomach and enters the small intestine, it triggers cells in the small intestinal mucosa to release into the blood two hormones that affect the stomach—***secretin*** (se-KRĒ-tin) and ***cholecystokinin*** (kō′-lē-sis′-tō-KĪN-in) or ***CCK.*** Secretin mainly decreases gastric secretions, whereas CCK mainly inhibits stomach emptying. Both hormones have other important effects on the pancreas, liver, and gallbladder (explained shortly) that contribute to the regulation of digestive processes. Table 19.1 summarizes the effects of gastrin, secretin, and cholecystokinin.

Vomiting is the forcible expulsion of the contents of the upper GI tract (stomach and sometimes duodenum) through the mouth. The strongest stimuli for vomiting are irritation and stretching of the stomach such as that caused by overeating. Other stimuli include general anesthesia, unpleasant sights, dizziness, and certain drugs such as morphine. Nerve impulses

Hormone	Where Produced	Stimulant	Action
Gastrin	Stomach mucosa (pyloric region).	Stretching of stomach, partially digested proteins and caffeine in stomach, and high pH of stomach chyme.	Stimulates secretion of gastric juice, increases movement of GI tract, and relaxes pyloric sphincter.
Secretin	Intestinal mucosa.	Acidic chyme that enters the small intestine.	Inhibits secretion of gastric juice, stimulates secretion of pancreatic juice rich in bicarbonate ions, and stimulates secretion of bile.
Cholecystokinin (CCK)	Intestinal mucosa.	Amino acids and fatty acids in chyme in small intestine.	Inhibits gastric emptying, stimulates secretion of pancreatic juice rich in digestive enzymes, causes ejection of bile from the gallbladder, and induces a feeling of satiety (feeling full to satisfaction).

Table 19.1 / Major Hormones that Control Digestion

from the vomiting center in the medulla oblongata travel to the upper GI tract organs, diaphragm, and abdominal muscles to bring about the vomiting act. Basically, vomiting involves squeezing the stomach between the diaphragm and abdominal muscles and expelling the contents into the esophagus and through the mouth. Prolonged vomiting, especially in infants and elderly people, can be serious because the loss of gastric juice and fluids leads to disturbances in fluid and acid–base balance.

The next organ of the GI tract involved in the breakdown of food is the small intestine. Chemical digestion in the small intestine depends not only on its own secretions, but also on activities of three accessory structures: the pancreas, liver, and gallbladder.

PANCREAS

Objective: • **Describe the location, structure, and functions of the pancreas.**

Structure of the Pancreas

The *pancreas* (*pan-* = all; *-creas* = flesh) lies behind the stomach (see Figure 19.1). Secretions pass from the pancreas to the duodenum via the *pancreatic duct,* which unites with the common bile duct from the liver and gallbladder, forming the common duct to the duodenum (Figure 19.9).

The pancreas is made up of small clusters of glandular epithelial cells, about 99% of which are arranged in clusters called *acini* (AS-i-nī). The acini constitute the *exocrine* portion of the organ (see Figure 13.17b on page 000). The cells within acini secrete a mixture of fluid and digestive enzymes called *pancreatic juice.* The remaining 1% of the cells are organized into clusters called *pancreatic islets (islets of Langerhans),* the *endocrine* portion of the pancreas. These cells secrete the hormones glucagon, insulin, somatostatin, and pancreatic polypeptide, which are discussed in Chapter 13.

Pancreatic Juice

Pancreatic juice is a clear, colorless liquid that consists mostly of water, some salts, sodium bicarbonate, and enzymes. The sodium bicarbonate gives pancreatic juice a slightly alkaline pH (7.1 to 8.2), which inactivates pepsin from the stomach and creates the optimal environment for activity of enzymes in the small intestine. The enzymes in pancreatic juice include a carbohydrate-digesting enzyme called *pancreatic amylase;* several protein-digesting enzymes called *trypsin* (TRIP-sin), *chymotrypsin* (kī'-mō-TRIP-sin), *carboxypeptidase* (kar-bok'-sē-PEP-ti-dās), and *elastase* (ē-LAS-tās); the main triglyceride-digesting enzyme in adults, called *pancreatic lipase;* and nucleic acid-digesting enzymes called *ribonuclease* and *deoxyribonuclease.* The protein-digesting enzymes are produced in inactive form, which prevents them from digesting the pancreas itself. Upon reaching the small intestine, the inactive form of trypsin is activated by an enzyme called *enterokinase.* In turn, trypsin activates the other protein-digesting pancreatic enzymes.

Regulation of Pancreatic Secretions

Pancreatic secretion, like gastric secretion, is regulated by both neural and hormonal mechanisms. The secretion of pancreatic enzymes is stimulated by parasympathetic fibers from the vagus (X) nerves and by the hormones secretin and cholecystokinin (see Table 19.1).

LIVER AND GALLBLADDER

Objective: • **Describe the location, structure, and functions of the liver and gallbladder.**

In an average adult, the *liver* weighs 1.4 kg (about 3 lb) and, after the skin, is the second largest organ of the body. It is located below the diaphragm, mostly on the right side of the body. A connective tissue capsule covers the liver, which in turn is covered by peritoneum, the serous membrane that covers all the

Figure 19.9 ■ **Relation of the pancreas to the liver, gallbladder, and duodenum.** The inset shows details of the common bile duct and pancreatic duct forming the common duct that empties into the duodenum.

Pancreatic juice in the pancreatic duct and bile in the common bile duct both flow into the common duct to the duodenum.

Anterior view

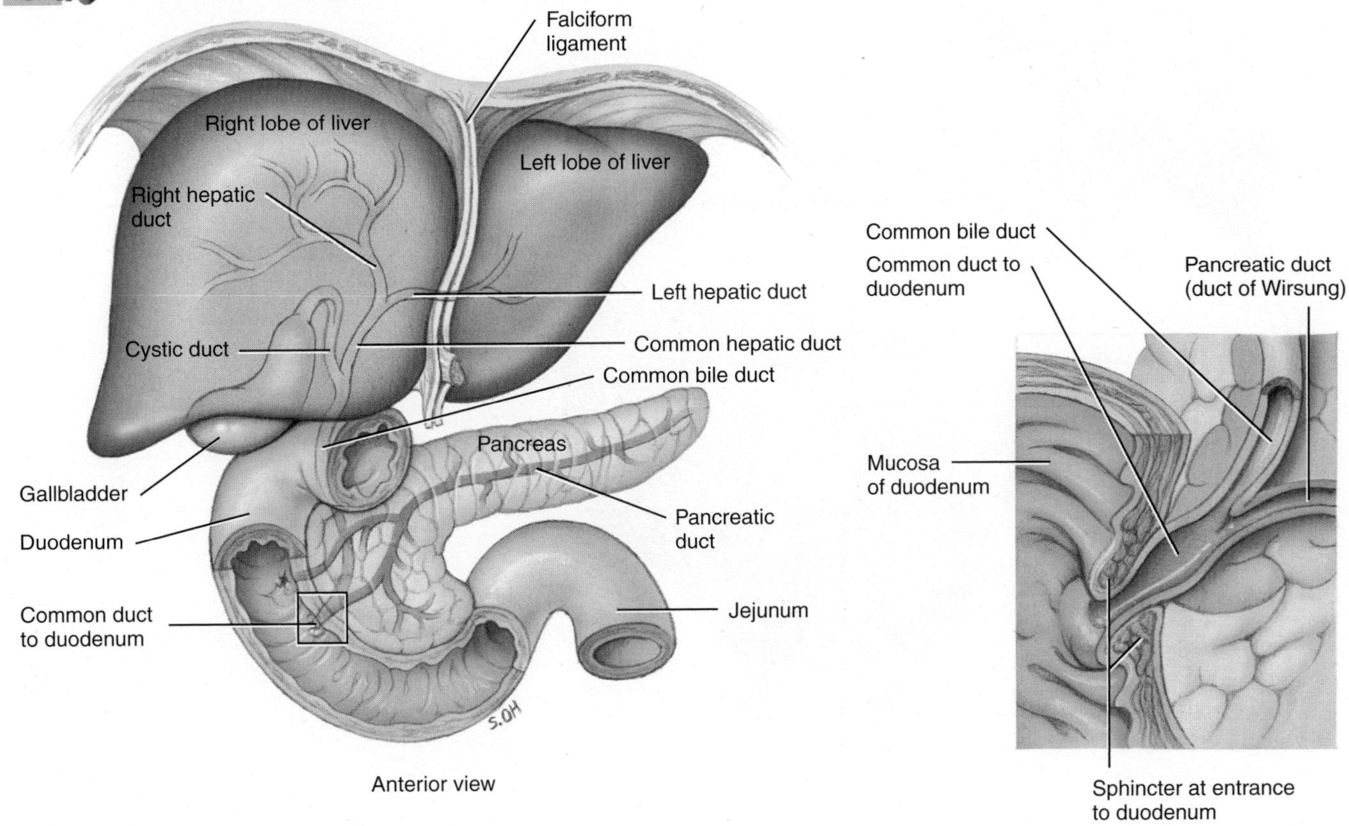

What type of fluid would you find in the pancreatic duct?

viscera. The **gallbladder** (*gall-* = bile) is a pear-shaped sac that hangs from the lower front margin of the liver (Figure 19.9). The gallbladder stores and concentrates bile made by the liver until the bile is needed for absorption of dietary fats in the small intestine.

Structure of the Liver and Gallbladder

The liver is divided by the *falciform ligament* into two principal lobes: the **right lobe** and the **left lobe** (Figure 19.9). Each lobe of the liver is made up of many functional units called **lobules** (Figure 19.10). A lobule consists of **hepatocytes** (*hepat-* = liver; *-cytes* = cells) arranged in irregular, branching, interconnected plates around a **central vein.** Hepatocytes secrete **bile,** a yellow, brownish, or olive-green liquid with a pH of 7.6 to 8.6. Bile consists of water, bile salts, cholesterol, a phospholipid called lecithin, bile pigments, and several ions. It is produced in the liver but stored in the gallbladder. In the liver, blood passes

through larger, endothelium-lined spaces called **sinusoids** instead of capillaries. Phagocytes in the sinusoids, called **stellate reticuloendothelial (Kupffer's) cells** (STEL-āt rē-TIK-yoo-en'-dō-thē'-lē-al; *stellate* = star-shaped; *reticulo-* = netlike), destroy worn-out white blood cells and red blood cells, bacteria, and other foreign matter in the venous blood draining from the gastrointestinal tract.

Bile produced by the hepatocytes enters the **right** and **left hepatic ducts,** which unite to leave the liver as the **common hepatic duct** (see Figure 19.9). The contraction of the smooth muscle fibers of the middle, muscular coat of the wall of the gallbladder ejects bile from the gallbladder into the **cystic duct.** The cystic duct joins the common hepatic duct to become the **common bile duct.** The common bile duct and pancreatic duct merge, forming the common duct to the duodenum, which enters the duodenum. When the small intestine is empty, the sphincter around the common duct at the entrance to the duodenum closes, and bile backs up into the cystic duct to the gallbladder for storage.

Figure 19.10 ■ **Histology of the liver.**

The lobes of the liver are composed of many functional units called lobules; a lobule consists of hepatocytes arranged around a central vein.

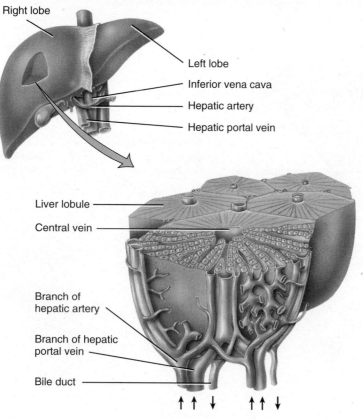

Right lobe

Left lobe

Inferior vena cava

Hepatic artery

Hepatic portal vein

Liver lobule

Central vein

Branch of hepatic artery

Branch of hepatic portal vein

Bile duct

(a) Relationship of several liver lobules

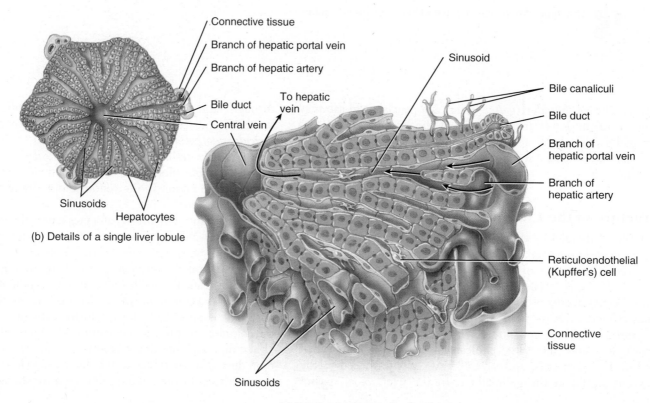

Connective tissue

Branch of hepatic portal vein

Branch of hepatic artery

Bile duct

Central vein

To hepatic vein

Sinusoid

Bile canaliculi

Bile duct

Branch of hepatic portal vein

Branch of hepatic artery

Reticuloendothelial (Kupffer's) cell

Connective tissue

Sinusoids

Hepatocytes

(b) Details of a single liver lobule

Sinusoids

(c) Details of a portion of a liver lobule

Blood Supply of the Liver

The liver receives oxygenated blood from the hepatic artery and deoxygenated blood containing newly absorbed nutrients from the hepatic portal vein (see Figure 16.17 on page 399). Branches of both the hepatic artery and the hepatic portal vein carry blood to the sinusoids of the lobules, where oxygen, most of the nutrients, and toxins are extracted by the hepatocytes. Stellate reticuloendothelial cells remove microbes and foreign or dead matter from the blood. Nutrients are stored or used to make new materials. The toxins are stored or inactivated. Hepatic cell products and nutrients needed by other cells are secreted back into the blood. The blood then drains into the central vein, the hepatic vein, and the inferior vena cava, and finally returns to the heart.

Function of Bile

Bile is partially an excretory product and partially a digestive secretion. Bile salts in bile aid in *emulsification,* the conversion of big triglyceride (fat) globules into a suspension of tiny triglyceride droplets, and in absorption of triglycerides following their digestion. The tiny triglyceride droplets present a very large surface area so that pancreatic and intestinal lipase can digest them rapidly. The principal bile pigment is *bilirubin,* which is derived from heme. When worn-out red blood cells are broken down, iron, globin, and bilirubin are released. The iron and globin are recycled, but some of the bilirubin is excreted in bile.

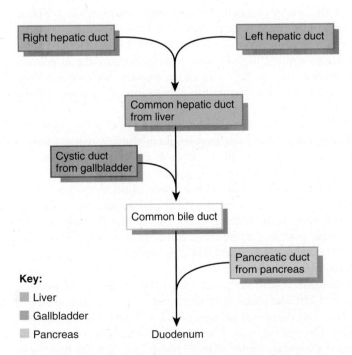

Key:
■ Liver
■ Gallbladder
■ Pancreas

(d) Ducts carrying bile from liver and gallbladder and pancreatic juice from pancreas to the duodenum

From the hepatic portal vein, what is the path of blood flow through the liver?

Bilirubin eventually is broken down in the intestine, and one of its breakdown products (stercobilin) gives feces their normal color (see Figure 14.4 on page 338).

The rate at which bile is secreted is determined by both nervous and hormonal factors. Parasympathetic stimulation by way of the vagus (X) nerves increases the production of bile. Secretin also stimulates the secretion of bile (see Table 19.1). When triglycerides enter the small intestine, cholecystokinin (CCK) is released. CCK stimulates contraction of the gallbladder's muscular layer, sending bile into the small intestine (see Table 19.1).

The components of bile sometimes crystallize and form *gallstones.* As they grow in size and number, gallstones may cause intermittent or complete obstruction to the flow of bile from the gallbladder into the duodenum. Treatment consists of using gallstone-dissolving drugs or lithotripsy, a shock-wave therapy that smashes the gallstones into particles small enough to pass through the ducts. For people with recurrent gallstones or for whom drugs or lithotripsy is not indicated, *cholecystectomy*—the removal of the gallbladder and its contents—is necessary.

Functions of the Liver

The liver performs many other vital functions in addition to the secretion of bile and bile salts and the phagocytosis of bacteria and dead or foreign material by the Kupffer's cells described above. Many of these are related to metabolism and are discussed in Chapter 20. Briefly, however, other functions of the liver include the following:

1. **Carbohydrate metabolism.** The liver is especially important in maintaining a normal blood glucose level. When blood glucose is low, the liver can break down glycogen to glucose and release glucose into the bloodstream. The liver can also convert certain amino acids and lactic acid to glucose, and it can convert other sugars, such as fructose and galactose, into glucose. When blood glucose is high, as occurs just after eating a meal, the liver converts glucose to glycogen and triglycerides for storage.

2. **Lipid metabolism.** Hepatocytes store some triglycerides; break down fatty acids to generate ATP; synthesize lipoproteins, which transport fatty acids, triglycerides, and cholesterol to and from body cells; synthesize cholesterol; and use cholesterol to make bile salts.

3. **Protein metabolism.** Hepatocytes remove the amino group ($—NH_2$) from amino acids so that the amino acids can be used for ATP production or converted to carbohydrates or fats. They also convert the resulting toxic ammonia (NH_3) into the much less toxic urea, which is excreted in urine. Hepatocytes also synthesize most plasma proteins, such as globulins, albumin, prothrombin, and fibrinogen.

4. **Processing of drugs and hormones.** The liver can detoxify substances such as alcohol or excrete drugs such as penicillin, erythromycin, and sulfonamides into bile. It can also chemically alter or excrete thyroid hormones and steroid hormones such as estrogens and aldosterone.

5. **Excretion of bilirubin.** As previously noted, bilirubin, derived from the heme of aged red blood cells, is absorbed by the liver from the blood and secreted into bile. Most of the bilirubin in bile is metabolized in the small intestine by bacteria and eliminated in feces.

6. **Storage of vitamins and minerals.** In addition to storing glycogen, the liver stores certain vitamins (A, B$_{12}$, D, E, and K) and minerals (iron and copper), which are released from the liver when needed elsewhere in the body.

7. **Activation of vitamin D.** The skin, liver, and kidneys participate in synthesizing the active form of vitamin D.

SMALL INTESTINE

Objective: • **Describe the location, structure, and functions of the small intestine.**

The major events of digestion and absorption occur in a long tube called the *small intestine.* The small intestine averages 2.5 cm (1 in.) in diameter; its length is about 3 m (10 ft) in a living person and about 6.5 m (21 ft) in a cadaver due to the loss of smooth muscle tone after death.

Structure of the Small Intestine

The small intestine is divided into three segments (Figure 19.11): the duodenum, the jejunum, and the ileum. The *duode-num* (doo′-ō-DĒ-num), the shortest part (about 25 cm or 10 in.), originates at the pyloric sphincter of the stomach. *Duodenum* means "twelve"; the structure is so named because it is about as long as the width of 12 fingers. The next section, the *jejunum* (jē-JOO-num = empty) is about 1 m (3 ft) long. It is so named because it is empty at death. The final portion of the small intestine, the *ileum* (IL-ē-um = twisted), measures about 2 m (6 ft) and joins the large intestine at the *ileocecal sphincter* (il′-ē-ō-SĒ-kal).

The structure of the small intestine is specially adapted for the absorption of nutrients. Its length provides a large surface area for absorption, and that area is further increased by modifications in the structure of its wall. The wall of the small intestine is composed of the same four layers that make up most of the GI tract. However, special features of both the mucosa and the submucosa allow the small intestine to complete the processes of digestion and absorption (Figure 19.12). The mucosa forms a series of fingerlike *villi* (= tuft of hair; singular is *villus*). These projections are 0.5 to 1 mm long and give the intestinal mucosa a velvety appearance (Figure 19.12a,b). The large number of villi (20 to 40 per square millimeter) vastly increases the surface area of the epithelium available for absorption and digestion. Each villus has a core of connective tissue that contains an arteriole, a venule, a capillary network, and a *lacteal* (LAK-tē-al = milky), which is a lymphatic capillary. Nutrients absorbed by the epithelial cells covering the villus pass through the wall of a capillary or a lacteal to enter blood or lymph, respectively.

The epithelium of the mucosa consists of simple columnar epithelium that contains absorptive cells, mucus-secreting goblet cells, hormone-producing cells, and Paneth cells. The membrane of the absorptive cells features *microvilli* (mī′-krō-VIL-ī; *micro-* = small). Each microvillus is a tiny, cylindrical, membrane-covered projection. Each square millimeter of small intestine contains about 200 million microvilli. Because the microvilli greatly increase the surface area of the plasma membrane, larger amounts of digested nutrients can diffuse into absorptive cells in a given period of time.

The mucosa contains many cavities lined with glandular epithelium. Cells lining the cavities form the *intestinal glands,* which secrete intestinal juice. *Paneth cells* are found in the deepest parts of the intestinal glands; they secrete lysozyme, a bactericidal enzyme, and are also capable of phagocytosis. They may have a role in regulating the microbial population in the intestines. Hormone-producing cells, also in the deepest part of the intestinal glands, secrete three hormones: secretin, cholecystokinin (see Table 19.1), and GIP, which stimulates release of insulin from the pancreas.

In addition to the microvilli and villi, a third set of projections called *circular folds* further increases the surface area for absorption and digestion (Figure 19.12c). The folds are permanent ridges in the mucosa. They enhance absorption by causing the chyme to spiral, rather than to move in a straight line, as it passes through the small intestine. The folds and villi decrease in size in the ileum; most absorption occurs in the duodenum and jejunum. The absorptive surface area of the small intestine is increased about 600 times by the villi, microvilli, and circular folds.

Figure 19.11 ■ **Divisions of the small intestine.**

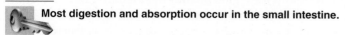

Most digestion and absorption occur in the small intestine.

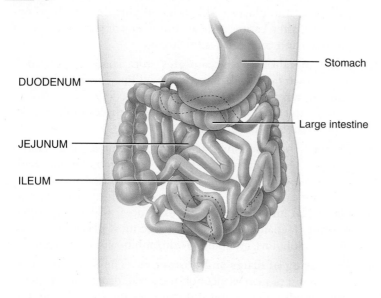

Anterior view

Which portion of the small intestine is the longest?

The areolar connective tissue of the mucosa of the small intestine has an abundance of mucosa-associated lymphatic tissue (MALT). **Solitary lymphatic nodules** are most numerous in the ileum. Groups of lymphatic nodules, referred to as **aggregated lymphatic follicles (Peyer's patches),** are also numerous in the ileum. The submucosa of the duodenum contains **duodenal glands** that secrete an alkaline mucus. It helps neutralize gastric acid in the chyme.

The muscularis of the small intestine consists of two layers of smooth muscle. The outer, thinner layer contains longitudinally arranged fibers. The inner, thicker layer contains circularly arranged fibers. The serosa completely surrounds the small intestine, except for a major portion of the duodenum, which is surrounded by areolar connective tissue (adventitia).

Intestinal Juice

Intestinal juice, secreted by the intestinal glands, is a watery clear yellow fluid with a slightly alkaline pH of 7.6 that contains some mucus. Rapidly reabsorbed by the villi, intestinal juice allows the small intestine to absorb substances from chyme as they come in contact with the villi. Intestinal enzymes are synthesized in the epithelial cells that line the villi. Most digestion by enzymes of the small intestine occurs in or on the surface of these epithelial cells.

Digestion in the Small Intestine
Mechanical Digestion

Two types of movements contribute to intestinal motility in the small intestine: segmentations and peristalsis. **Segmentations** are localized contractions that slosh chyme back and forth, mixing it with digestive juices and bringing food particles into contact with the mucosa for absorption. The movements are similar to alternately squeezing the middle and the ends of a tube of toothpaste. They do not push the intestinal contents along the tract.

After most of a meal has been absorbed, segmentation stops and peristalsis begins. The type of peristalsis that occurs in the small intestine, termed a **migrating motility complex (MMC),** begins in the lower portion of the stomach and pushes chyme forward along a short stretch of small intestine. The MMC slowly migrates down the small intestine, reaching the end of the ileum in 90 to 120 minutes. Then another MMC begins in the stomach. Altogether, chyme remains in the small intestine for 3 to 5 hours.

Chemical Digestion

The chyme entering the small intestine contains partially digested carbohydrates and proteins. The completion of digestion in the small intestine is a collective effort of pancreatic juice, bile, and intestinal juice. Once digestion is completed, the final products of digestion are ready for absorption.

Figure 19.12 ■ **Small intestine.** Shown are various structures that adapt the small intestine for digestion and absorption.

Villi, microvilli, and circular folds increase the surface area of the small intestine for digestion and absorption.

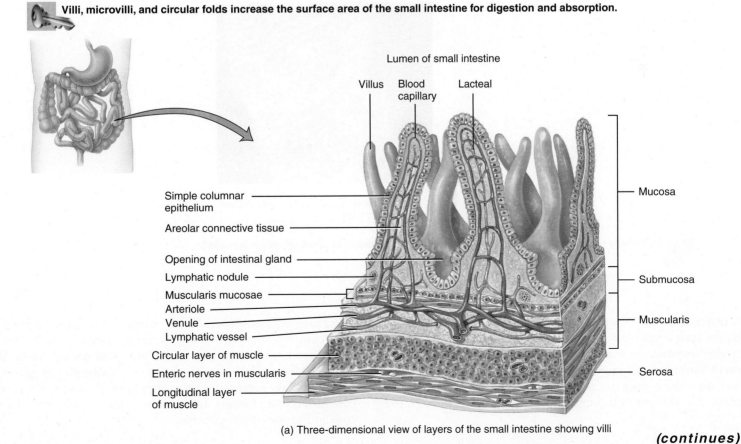

(a) Three-dimensional view of layers of the small intestine showing villi

(continues)

Figure 19.12 (continued)

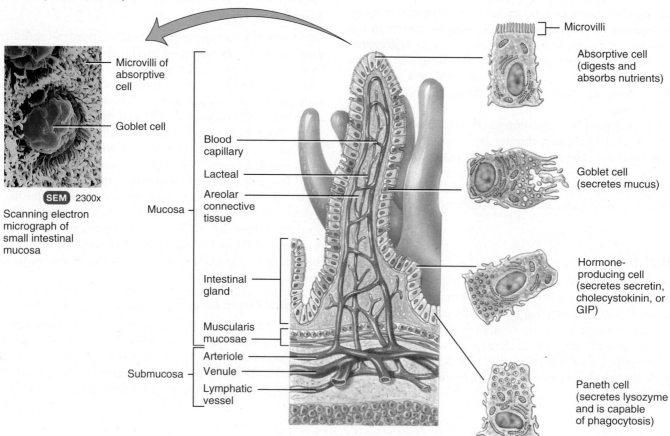

Scanning electron micrograph of small intestinal mucosa

Microvilli of absorptive cell

Goblet cell

SEM 2300x

Mucosa

Blood capillary

Lacteal

Areolar connective tissue

Intestinal gland

Muscularis mucosae

Submucosa

Arteriole

Venule

Lymphatic vessel

Microvilli

Absorptive cell (digests and absorbs nutrients)

Goblet cell (secretes mucus)

Hormone-producing cell (secretes secretin, cholecystokinin, or GIP)

Paneth cell (secretes lysozyme and is capable of phagocytosis)

(b) Enlarged villus showing lacteal, capillaries, and intestinal gland

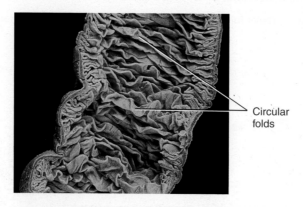

Circular folds

(c) Photograph of jejunum cut open to expose the circular folds

  How many villi and microvilli are present in a square millimeter of small intestinal mucosa?

CARBOHYDRATES Starches and dextrins not reduced to maltose by the time chyme leaves the stomach are broken down by *pancreatic amylase,* an enzyme in pancreatic juice that acts in the small intestine. Although amylase acts on glycogen, starches, and dextrins, it does not act on the polysaccharide cellulose, a plant fiber that passes through undigested.

Three enzymes located at the surface of intestinal epithelial cells complete the digestion of carbohydrates, breaking them down into monosaccharides, which are small enough to be absorbed. *Maltase* splits maltose into two molecules of glucose. *Sucrase* breaks sucrose into a molecule of glucose and a molecule of fructose. *Lactase* digests lactose into a molecule of glucose and a molecule of galactose.

PROTEINS Enzymes in pancreatic juice (trypsin, chymotrypsin, elastase, and carboxypeptidase) continue the digestion of proteins begun in the stomach, though their actions differ somewhat because each splits the peptide bond between different amino acids. Protein digestion is completed by *peptidases,* enzymes produced by epithelial cells that line the villi. The final products of protein digestion are amino acids, dipeptides, and tripeptides.

LIPIDS In an adult, most lipid digestion occurs in the small intestine. In the first step of lipid digestion, bile salts emulsify large globules of triglycerides and lipids into small droplets, giving the enzymes easy access. Recall that triglycerides consist of a molecule of glycerol with three attached fatty acids (see Figure 2.10 on page 000). In the second step, *pancreatic lipase,* found in pancreatic juice, breaks down each triglyceride molecule by removing two of the three fatty acids from glycerol; the third remains attached to the glycerol. Thus, fatty acids and monoglycerides are the end products of triglyceride digestion.

NUCLEIC ACIDS Both intestinal juice and pancreatic juice contain *nucleases* that digest nucleotides into their constituent pentoses and nitrogenous bases. *Ribonuclease* acts on ribonucleic acid nucleotides, and *deoxyribonuclease* acts on deoxyribonucleic acid nucleotides.

Table 19.2 summarizes the enzymes that contribute to digestion.

Regulation of Intestinal Secretion and Motility

The most important means for regulating small intestinal secretion and motility are local reflexes in the ENS in response to the presence of chyme. Segmentation movements are stimulated by intestinal stretching. Parasympathetic nerve impulses intensify intestinal motility, whereas sympathetic nerve impulses decrease it. The first remnants of a meal reach the beginning of the large intestine in about 4 hours.

Absorption in the Small Intestine

All the chemical and mechanical phases of digestion from the mouth down through the small intestine are directed toward changing food into molecules that can pass through the epithelial cells of the mucosa into the underlying blood and lymphatic vessels. Molecules that are absorbed include monosaccharides from digested carbohydrates; single amino acids, dipeptides, and tripeptides from digested proteins; and fatty acids and monoglycerides from digested triglycerides.

About 90% of all absorption takes place in the small intestine. The other 10% occurs in the stomach and large intestine. Absorption in the small intestine occurs by diffusion, facilitated diffusion, osmosis, and active transport. Any undigested or unabsorbed material left in the small intestine is passed on to the large intestine.

Table 19.2 / Summary of Digestive Enzymes

Enzyme	Source	Substrate	Product
Carbohydrate Digesting			
Salivary amylase	Salivary glands.	Starches, glycogen, and dextrins.	Maltose (disaccharide), maltotriose (trisaccharide), and dextrins.
Pancreatic amylase	Pancreas.	Starches, glycogen, and dextrins.	Maltose, maltotriose, and dextrins.
Maltase	Small intestine.	Maltose.	Glucose.
Sucrase	Small intestine.	Sucrose.	Glucose and fructose.
Lactase	Small intestine.	Lactose.	Glucose and galactose.
Protein Digesting			
Pepsin	Stomach (chief cells).	Proteins.	Peptides.
Trypsin	Pancreas.	Proteins.	Peptides.
Chymotrypsin	Pancreas.	Proteins.	Peptides.
Elastase	Pancreas.	Proteins.	Peptides.
Carboxypeptidase	Pancreas.	Last amino acid at carboxyl (acid) end of peptides.	Peptides and amino acids.
Peptidases	Small intestine.	Last amino acid at amino end of peptides and dipeptides.	Peptides and amino acids.
Lipid Digesting			
Pancreatic lipase	Pancreas.	Triglycerides (fats) that have been emulsified by bile salts.	Fatty acids and monoglycerides.
Nucleases			
Ribonuclease	Pancreas and small intestine.	Ribonucleic acid nucleotides.	Pentoses and nitrogenous bases.
Deoxyribonuclease	Pancreas and small intestine.	Deoxyribonucleic acid nucleotides.	Pentoses and nitrogenous bases.

Absorption of Monosaccharides

Essentially all carbohydrates are absorbed as monosaccharides. Glucose and galactose are transported into epithelial cells of the villi by active transport. Fructose is transported by facilitated diffusion (Figure 19.13a). After absorption, monosaccharides are transported out of the epithelial cells by facilitated diffusion into the capillaries, which drain into venules of the villi. From here, monosaccharides are carried to the liver via the hepatic portal vein, then through the heart and to the general circulation (Figure 19.13b).

Absorption of Amino Acids, Dipeptides, and Tripeptides

Most dietary proteins are absorbed as amino acids, dipeptides, and tripeptides by means of active transport processes that occur mainly in the duodenum and jejunum. About half of the absorbed amino acids are present in food, whereas the other half come from proteins in digestive juices and dead cells that slough off the mucosal surface! Amino acids move out of the epithelial cells by means of diffusion and enter capillaries of the villus (Figure 19.13a). Like the monosaccharides, amino acids are carried in the blood to the liver by way of the hepatic portal vein (Figure 19.13b). If not removed by hepatocytes, amino acids enter the general circulation.

Absorption of Lipids

All dietary lipids are absorbed by means of diffusion. As a result of their digestion, triglycerides are broken down into monoglycerides and fatty acids. The small amount of short-chain fatty acids (those having fewer than 10 to 12 carbon atoms) in the diet passes into the epithelial cells by diffusion and follows the same route taken by monosaccharides and amino acids into a blood capillary of a villus (see Figure 19.13a,b).

Most dietary triglycerides, however, contain long-chain fatty acids. They and monoglycerides reach the bloodstream by a different route and require bile for adequate absorption. Bile salts emulsify lipids, forming tiny droplets called *micelles* (mī-SELZ = small morsels) that include 20 to 50 bile salt molecules plus dietary lipids. Close to the mucosal epithelial cells, long-chain fatty acids and monoglycerides, liberated from triglycerides by lipases, diffuse out of micelles into the cells, leaving the bile salts behind. When chyme reaches the ileum, 90–95% of the bile salts are reabsorbed and returned by the blood to the liver for recycling. Insufficient bile salts, due either to obstruction of the bile ducts or removal of the gallbladder, can result in the loss of up to 40% of dietary lipids in feces due to diminished lipid absorption.

Within the epithelial cells, many monoglycerides are further digested by lipase to glycerol and fatty acids. The fatty acids and glycerol are then recombined to form triglycerides, which are packaged along with phospholipids and cholesterol into large spherical particles, called *chylomicrons,* that become coated with proteins (Figure 19.13a). Chylomicrons leave the epithelial cells by means of exocytosis. Because they are so large and bulky, chylomicrons cannot enter blood capillaries in the small intestine; instead, they enter the much leakier lacteals. From there, they are transported by way of lymphatic vessels to the thoracic duct and enter the blood at the left subclavian vein (see Figure 19.13b). Within 10 minutes after absorption, about half of the chylomicrons are removed from the blood as they pass through blood capillaries in the liver and adipose tissue. Two or three hours after a meal, few chylomicrons remain in the blood.

Absorption of Water

The total volume of the fluids that are ingested and secreted into the GI tract each day is about 9.3 liters (9.9 qt). Most of the water in these fluids is absorbed by means of osmosis—about 8.3 liters in the small intestine and 0.9 liters in the large intestine. The remaining 0.1 liter (100 mL) is excreted in the feces.

Absorption of Electrolytes

Many of the electrolytes absorbed by the small intestine come from gastrointestinal secretions. Others are part of ingested foods and liquids. After sodium ions (Na^+) have moved into epithelial cells by diffusion and active transport from the lumen, they are actively transported into blood capillaries on the other side. Most of the Na^+ in gastrointestinal secretions is reclaimed and not lost in the feces. Chloride, iodide, and nitrate ions can passively follow sodium ions into the epithelial cells or can be actively transported. Calcium ions are absorbed actively in a process stimulated by calcitriol, the active form of vitamin D. Other electrolytes such as iron, potassium, magnesium, and phosphate ions also are absorbed by active transport mechanisms.

Absorption of Vitamins

Fat-soluble vitamins (A, D, E, and K) are absorbed along with ingested dietary triglycerides in micelles. In fact, these vitamins are not adequately absorbed unless some triglycerides are present. Most water-soluble vitamins, such as the B vitamins and C, are absorbed by diffusion. Vitamin B_{12} must be combined with intrinsic factor produced by the stomach for its absorption in the small intestine by active transport.

LARGE INTESTINE

Objective: • Describe the location, structure, and functions of the large intestine.

The large intestine is the last part of the GI tract. Its overall functions are the completion of absorption, the production of certain vitamins, the formation of feces, and the expulsion of feces from the body.

Structure of the Large Intestine

The *large intestine* averages about 6.5 cm (2.5 in.) in diameter and about 1.5 m (5 ft) in length. It extends from the ileum to the anus and is attached to the posterior abdominal wall by its mesentery (see Figure 19.3b). The large intestine is divided into four principal regions: cecum, colon, rectum, and anal canal (Figure 19.14 on page 478).

Figure 19.13 ◼ **Absorption of digested nutrients in the small intestine.** For simplicity, all digested foods are shown in the lumen of the small intestine, even though some nutrients are digested at the surface of or in epithelial cells of the villi.

Long-chain fatty acids and monoglycerides are absorbed into lacteals; other products of digestion enter blood capillaries.

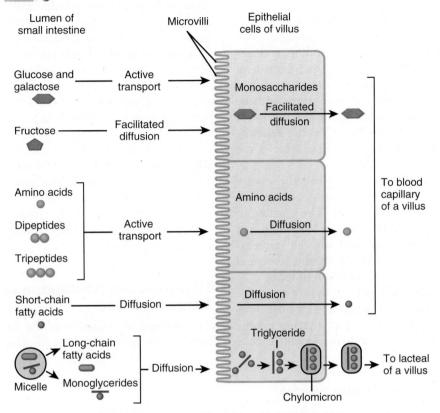

(a) Mechanisms for movement of nutrients through epithelial cells of the villi

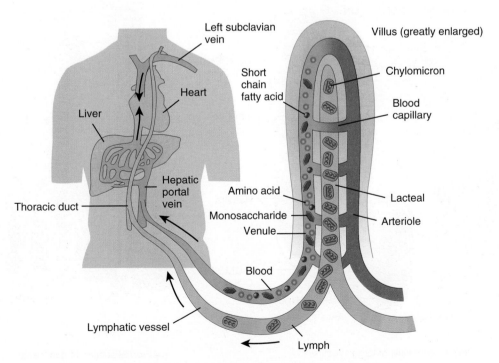

(b) Movement of absorbed nutrients into the blood and lymph

 How are fat-soluble vitamins (A, D, E, and K) absorbed?

At the opening of the ileum into the large intestine is a valve called the *ileocecal sphincter.* It allows materials from the small intestine to pass into the large intestine. Inferior to the ileocecal sphincter is the first segment of large intestine, called the *cecum.* Attached to the cecum is a twisted coiled tube called the *appendix.*

The open end of the cecum merges with the longest portion of the large intestine, called the *colon* (= food passage). The colon is divided into ascending, transverse, descending, and sigmoid portions. The *ascending colon* ascends on the right side of the abdomen, reaches the undersurface of the liver, and turns to the left. The colon continues across the abdomen to the left side as the *transverse colon.* It curves beneath the lower border of the spleen on the left side and passes downward as the *descending colon.* The *sigmoid colon* begins near the iliac crest of the left hip bone and ends as the *rectum.*

The last 2 to 3 cm (1 in.) of the rectum is called the *anal canal.* The mucous membrane of the anal canal is arranged in longitudinal folds containing arteries and veins. The opening of the anal canal to the exterior is called the *anus.* It has an internal sphincter of smooth (involuntary) muscle and an external sphincter of skeletal (voluntary) muscle. Normally, the anal sphincters are closed except during the elimination of feces.

The wall of the large intestine differs from that of the small intestine in several respects. No villi or permanent circular folds are found in the mucosa. The epithelium of the mucosa is simple columnar epithelium that contains mostly absorptive cells and goblet cells (Figure 19.15). The cells form long tubes called intestinal glands. The absorptive cells function primarily in water absorption. The goblet cells secrete mucus that lubricates the contents of the colon. Individual lymphatic nodules also are found in the mucosa. The muscularis consists of an external layer of longitudinal muscles and an internal layer of circular muscles. Unlike other parts of the gastrointestinal tract, the outer longitudinal layer of the muscularis is bundled into three longitudinal bands. These bands, the *teniae coli* (TĒ-nē-ē KŌ-lī; *tenia* = flat band), run the length of most of the large intestine (see Figure 19.14a) and are clearly visible on the surface of the colon. Contractions of the teniae coli gather the colon into a series of pouches called *haustra* (HAWS-tra; singular is *haustrum*), which give the colon its puckered appearance.

Varicosities in any veins involve inflammation and enlargement. Varicosities of the rectal veins are known as *hemorrhoids.* Initially contained within the anus (first degree), they gradually enlarge until they prolapse or extend outward on defecation (second degree) and finally remain prolapsed through the anal orifice (third degree). Hemorrhoids may be caused by constipation and pregnancy.

Digestion and Absorption in the Large Intestine

The passage of chyme from the ileum into the cecum is regulated by the ileocecal sphincter. The sphincter normally remains

---

Figure 19.14 ■ **Anatomy of the large intestine.**

The subdivisions of the large intestine are the cecum, colon, rectum, and anal canal.

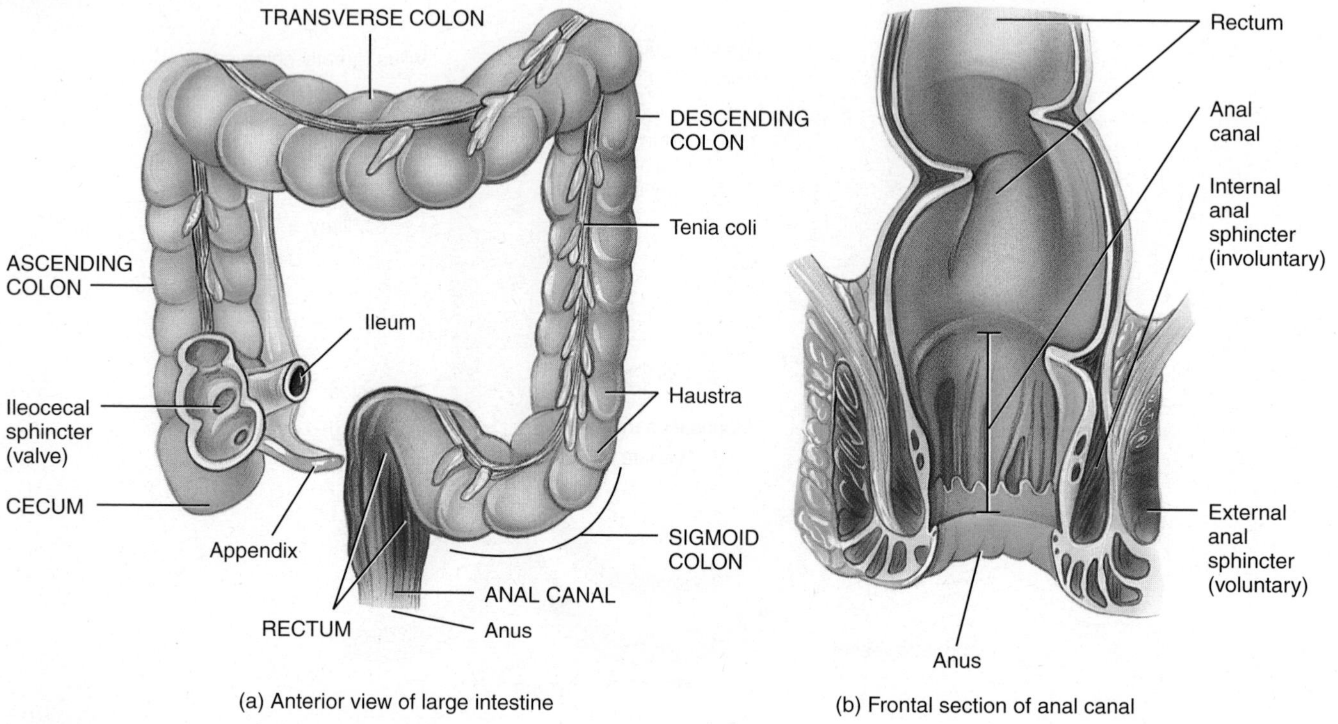

(a) Anterior view of large intestine

(b) Frontal section of anal canal

What are the functions of the large intestine?

Figure 19.15 ■ **Histology of the large intestine.**

Intestinal glands formed by absorptive cells and goblet cells extend the full thickness of the mucosa.

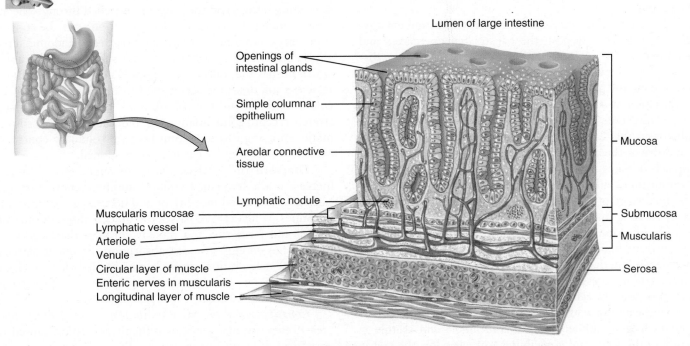

(a) Three-dimensional view of layers of the large intestine

Lumen of large intestine

Openings of intestinal glands

Simple columnar epithelium

Areolar connective tissue

Lymphatic nodule

Muscularis mucosae

Lymphatic vessel

Arteriole

Venule

Circular layer of muscle

Enteric nerves in muscularis

Longitudinal layer of muscle

Mucosa

Submucosa

Muscularis

Serosa

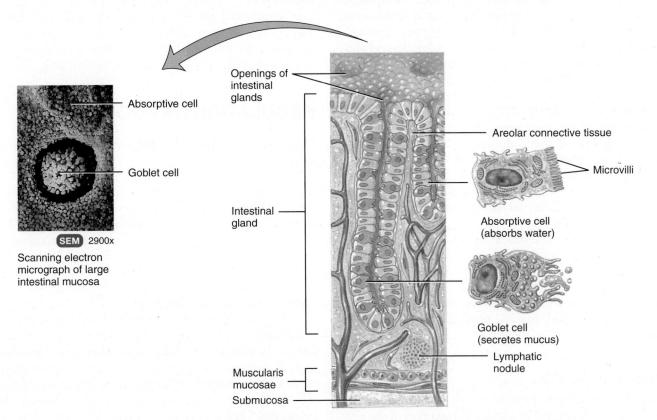

Absorptive cell

Goblet cell

SEM 2900x

Scanning electron micrograph of large intestinal mucosa

Openings of intestinal glands

Areolar connective tissue

Microvilli

Absorptive cell (absorbs water)

Intestinal gland

Goblet cell (secretes mucus)

Lymphatic nodule

Muscularis mucosae

Submucosa

(b) Sectional view of the large intestinal mucosa showing intestinal glands

How does the muscularis of the large intestine differ from that of other parts of the GI tract?

mildly contracted so the passage of chyme is usually a slow process, but immediately following a meal, there is a reflex action that intensifies peristalsis, forcing any chyme in the ileum into the cecum.

One movement characteristic of the large intestine is *haustral churning.* In this process, the haustra remain relaxed and distended (stretched) while they fill with chyme. When the distention reaches a certain point, the walls of the haustra contract so that the contents pass into the next haustrum. *Peristalsis* also occurs, although at a slower rate than in other portions of the tract. The third type of movement is *mass peristalsis,* a strong peristaltic wave that begins in about the middle of the transverse colon and drives the colonic contents into the rectum. Food in the stomach initiates mass peristalsis, which usually takes place three or four times a day, during or immediately after a meal.

The final stage of digestion occurs in the colon through the activity of bacteria that normally inhabit the lumen. The glands of the large intestine secrete mucus but no enzymes. Bacteria ferment any remaining carbohydrates and release hydrogen, carbon dioxide, and methane gases. These gases contribute to flatus (gas) in the colon; excessive gas in the colon is termed *flatulence.* Bacteria also convert remaining proteins to amino acids and break down the amino acids into simpler substances: indole, skatole, and hydrogen sulfide. Some of the indole and skatole is eliminated in the feces and contributes to their odor; the rest is absorbed and transported to the liver, where these compounds are converted to less toxic compounds and excreted in the urine. Bacteria also decompose bilirubin to simpler pigments, including stercobilin, which give feces their brown color. Several vitamins needed for normal metabolism, including some B vitamins and vitamin K, are bacterial products that are absorbed in the colon. When bacteria leave the large intestine because of a perforation or ruptured appendix, they can cause *peritonitis,* an acute inflammation of the peritoneum.

Although 90% of all water absorption occurs in the small intestine, the large intestine also absorbs a significant amount. Of the 0.9 liter of water that enters the large intestine, all but about 100 mL is absorbed by means of osmosis. The large intestine also absorbs electrolytes, including sodium and chloride, and some dietary vitamins.

By the time chyme has remained in the large intestine 3 to 10 hours, it has become solid or semisolid as a result of water absorption and is now called *feces.* Chemically, feces consist of water, inorganic salts, sloughed-off epithelial cells from the mucosa of the gastrointestinal tract, bacteria, products of bacterial decomposition, unabsorbed digested materials, and indigestible parts of food.

The Defecation Reflex

Mass peristaltic movements push fecal material from the sigmoid colon into the rectum. The resulting distention of the rectal wall stimulates stretch receptors, which initiates a *defecation reflex* that empties the rectum. Impulses from the spinal cord travel along parasympathetic nerves to the descending colon, sigmoid colon, rectum, and anus. The resulting contraction of the longitudinal rectal muscles shortens the rectum, thereby in-

creasing the pressure within it. This pressure plus parasympathetic stimulation opens the internal sphincter. The external sphincter is voluntarily controlled. If it is voluntarily relaxed, defecation occurs and the feces are expelled through the anus; if it is voluntarily constricted, defecation can be postponed. Voluntary contractions of the diaphragm and abdominal muscles aid defecation by increasing the pressure within the abdomen, which pushes the walls of the sigmoid colon and rectum inward. If defecation does not occur, the feces back up into the sigmoid colon until the next wave of mass peristalsis stimulates the stretch receptors. In infants, the defecation reflex causes automatic emptying of the rectum because voluntary control of the external anal sphincter has not yet developed.

Diarrhea (dī'-a-RĒ-a; *dia-* = through; *rhea* = flow) is an increase in the frequency, volume, and fluid content of the feces caused by increased motility of and decreased absorption by the intestines. When chyme passes too quickly through the small intestine and feces pass too quickly through the large intestine, there is not enough time for absorption. Frequent diarrhea can result in dehydration and electrolyte imbalances. Excessive motility may be caused by stress or by microbes that irritate the gastrointestinal mucosa.

Constipation (kon'-sti-PĀ-shun; *con-* = together; *stip-* = to press) refers to infrequent or difficult defecation caused by decreased motility of the intestines. Because the feces remain in the colon for prolonged periods of time, excessive water absorption occurs, and the feces become dry and hard. Constipation may be caused by poor habits (delaying defecation), spasms of the colon, insufficient fiber in the diet, inadequate fluid intake, lack of exercise, emotional stress, or certain drugs.

REGULATION OF FOOD INTAKE

Objective: • **List the factors that regulate food intake.**

Two groups of neurons in the hypothalamus are related to food intake. When the *feeding center* is stimulated experimentally, animals begin to eat heartily, even if they are already full. When the *satiety center* (sa-TĪ-i-tē) is stimulated, the animals stop eating, even if they have been starved for days. *Satiety* is a feeling of fullness accompanied by lack of desire to eat. Other parts of the brain that function in hunger and satiety are the cerebral cortex, brain stem, and limbic system.

How do neurons in the hypothalamus sense whether a person is well-fed or in need of food? One factor is changes in the chemical composition of the blood after eating and during fasting, for example, increases and decreases in blood glucose level. Also, signaling molecules in the blood act on the hypothalamus. Those that decrease appetite include several hormones (glucagon, cholecystokinin, epinephrine, and leptin, a protein hormone that is released by adipocytes as they synthesize triglycerides). Others increase appetite, such as growth hormone-releasing hormone (GHRH), glucocorticoids, epinephrine, insulin, progesterone, and somatostatin (growth hormone-inhibiting hormone).

Emotional Eating— Consumed by Food

In addition to keeping us alive, eating serves countless psychological, social, and cultural purposes. We eat to celebrate, punish, comfort, defy, and deny. Eating in response to emotional drives, such as feeling stressed, bored, or tired, rather than in response to true physical hunger, is called emotional eating.

Food as Emotional Rescue

Emotional eating is so common that, within limits, it is considered well within the range of normal behavior. Who hasn't at one time or another headed for the refrigerator after a bad day? Problems arise when emotional eating becomes so excessive that it interferes with health. Physical health problems include obesity and associated disorders such as hypertension and heart disease. Psychological health problems include poor self-esteem; an inability to cope effectively

with feelings of stress; and in extreme cases, eating disorders.

For emotional eaters, the drive to eat often masks unpleasant feelings such as boredom, loneliness, depression, anxiety, anger, or fatigue. Eating provides comfort and solace, numbing pain and "feeding the hungry heart." Some emotional overeaters say that stuffing themselves with food becomes a metaphor for suppressing undesirable feelings.

Eating may provide a biochemical "fix" as well. Emotional eaters typically overeat carbohydrate foods (sweets and starches), which may raise brain serotonin levels and lead to feelings of relaxation. Food becomes a way to self-medicate when negative emotions arise.

Consumed by Food

In extreme cases, eating becomes an addiction, and the drive to consume excessive amounts of food begins to take over a person's life. People with bulimia or binge-eating disorder have an over-

whelmingly urgent and totally uncontrollable drive to eat, causing them to consume huge volumes of food several times a week, sometimes several times a day. People with bulimia try to purge the calories they have consumed by vomiting, exercising excessively, or using laxatives and diuretics, whereas people with binge-eating disorder usually do not.

Eating disorders can be very dangerous and even lethal, requiring prompt, comprehensive, and in-depth professional treatment that helps people cope with the underlying psychological issues. Therapy for emotional eaters requires addressing the emotions that trigger overeating and devising effective coping strategies that eliminate the need to deal with stress by overeating.

▶ *Think It Over*

▶ Why might repeated attempts to lose weight with very restrictive diets lead to emotional overeating?

Food intake is also regulated by stretching of the GI tract, particularly the stomach and duodenum. The stretching of these organs initiates a reflex that activates the satiety center and depresses the feeding center. Psychological factors, such as those that accompany anorexia nervosa and obesity, may override the usual regulators of appetite and satiety.

AGING AND THE DIGESTIVE SYSTEM

Objective: • **Describe the effects of aging on the digestive system.**

Changes in the digestive system associated with aging include decreased secretory mechanisms, decreased motility of the digestive organs, loss of strength and tone of the muscular tissue and its supporting structures, changes in neurosensory feedback regarding enzyme and hormone release, and diminished response to pain and internal sensations. In the upper portion of

the GI tract, common changes include reduced sensitivity to mouth irritations and sores, loss of taste, periodontal disease, difficulty in swallowing, hiatal hernia, gastritis, and peptic ulcer disease. Changes that may appear in the small intestine include duodenal ulcers, appendicitis, and malabsorption. Other pathologies that increase in incidence with age are gallbladder problems, jaundice, cirrhosis of the liver, and acute pancreatitis. Changes in the large intestine such as constipation, hemorrhoids, and diverticular disease may also occur. The incidence of cancer of the colon or rectum increases with age.

• • •

Now that our exploration of the digestive system is completed, you can appreciate the many ways that this system contributes to homeostasis of other body systems by examining Focus on Homeostasis: The Digestive System on page 482. Next, in Chapter 20, you will discover how the nutrients absorbed by the GI tract are utilized in metabolic reactions by the body tissues.

Focus on Homeostasis
The Digestive System

Body System	Contribution of Digestive System
For all body systems	Digestive system breaks down dietary nutrients into forms that can be absorbed and used by body cells for ATP production and building of body tissues; absorbs water, minerals, and vitamins needed for growth and function of body tissues; and eliminates wastes from body tissues in feces.
Integumentary system	Small intestine absorbs vitamin D, which skin and kidneys modify to produce the hormone calcitriol; excess dietary calories are stored as triglycerides in adipose cells in dermis and subcutaneous layer.
Skeletal system	Small intestine absorbs dietary calcium and phosphorus salts needed to build bone matrix.
Muscular system	Liver can convert lactic acid produced by muscles during exercise to glucose.
Nervous system	Gluconeogenesis (synthesis of new glucose molecules) in liver plus digestion and absorption of dietary carbohydrates provide glucose, needed for ATP production by neurons.
Endocrine system	Liver inactivates some hormones, ending their activity; pancreatic islets release insulin and glucagon; hormones that regulate digestive activities are released by cells in mucosa of stomach and small intestine; liver produces angiotensinogen.
Cardiovascular system	GI tract absorbs water that helps maintain blood volume and iron that is needed for synthesis of hemoglobin in red blood cells; bilirubin from hemoglobin breakdown is partially excreted in feces; liver synthesizes most plasma proteins.
Lymphatic and immune system 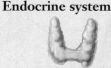	Acidity of gastric juice destroys bacteria and most toxins in stomach.
Respiratory system	Pressure of abdominal organs against the diaphragm helps expel air quickly during a forced exhalation.
Urinary system 	Absorption of water by GI tract provides water needed to excrete waste products in urine.
Reproductive systems	Digestion and absorption provides adequate nutrients, including fats, for normal development of reproductive structures, for production of gametes (oocytes and sperm), and for fetal growth and development during pregnancy.

COMMON DISORDERS

Dietary Fiber and the Digestive System

Dietary fiber consists of indigestible plant substances, such as cellulose, lignin, and pectin, found in fruits, vegetables, grains, and beans. *Insoluble fiber,* which does not dissolve in water, includes the woody or structural parts of plants such as fruit and vegetable skins and the bran coating around wheat and corn kernels. Insoluble fiber passes through the GI tract largely unchanged and speeds up the passage of material through the tract. *Soluble fiber,* which does dissolve in water, is found in abundance in beans, oats, barley, broccoli, prunes, apples, and citrus fruits. It tends to slow the passage of material through the tract.

People who choose a fiber-rich diet may reduce their risk of developing obesity, diabetes, atherosclerosis, gallstones, hemorrhoids, diverticulitis, appendicitis, and colon cancer. There is also evidence that insoluble fiber may help protect against colon cancer and that soluble fiber may help lower blood cholesterol level.

Dental Caries

Dental caries, or tooth decay, involves a gradual demineralization (softening) of the enamel and dentin by bacterial acids. If untreated, various microorganisms may invade the pulp, causing inflammation and infection with subsequent death of the pulp. Such teeth are treated by root canal therapy.

Periodontal Disease

Periodontal disease refers to a variety of conditions characterized by inflammation and degeneration of the gums, bone, periodontal ligament, and cementum. Periodontal diseases are often caused by poor oral hygiene; by local irritants, such as bacteria, impacted food, and cigarette smoke; or by a poor "bite."

Peptic Ulcer Disease

Five to ten percent of the U.S. population develops *peptic ulcer disease* *(PUD)* each year. An *ulcer* is a craterlike lesion in a membrane; ulcers that develop in areas of the GI tract exposed to acidic gastric juice are called *peptic ulcers.* The most common complication of peptic ulcers is bleeding, which can lead to anemia. In acute cases, peptic ulcers can lead to shock and death. Three distinct causes of PUD are recognized: (1) the bacterium *Helicobacter pylori,* (2) nonsteroidal anti-inflammatory drugs (NSAIDs) such as aspirin, and (3) hypersecretion of HCl.

Appendicitis

Appendicitis is an inflammation of the appendix. Appendectomy (surgical removal of the appendix) is recommended in all suspected cases because it is safer to operate than to risk gangrene, rupture, and peritonitis.

Colorectal Cancer

Colorectal cancer is among the deadliest of malignancies. An inherited predisposition contributes to more than half of all cases of colorectal cancer. Intake of alcohol and diets high in animal fat and protein are associated with increased risk of colorectal cancer, whereas dietary fiber, retinoids, calcium, and selenium may be protective. Signs and symptoms of colorectal cancer include diarrhea, constipation, cramping, abdominal pain, and rectal bleeding. Screening for colorectal cancer includes testing for blood in the feces, digital rectal examination, sigmoidoscopy, colonoscopy, and barium enema.

Diverticulitis

Diverticulosis is the development of diverticula, saclike outpouchings of the wall of the colon in places where the muscularis has become weak. Many people who develop diverticulosis have no symptoms and experience no complications. About 15% of people with diverticulosis eventually develop an inflammation known as *diverticulitis,* characterized by pain, either constipation or increased frequency of defecation, nausea, vomiting, and low-grade fever. Patients who change to high-fiber diets often show marked relief of symptoms.

Hepatitis

Hepatitis is an inflammation of the liver caused by viruses; drugs; or chemicals, including alcohol.

Hepatitis A (infectious hepatitis), caused by the hepatitis A virus, is spread by fecal contamination of food, clothing, toys, eating utensils, and so forth (fecal–oral route). It does not cause lasting liver damage.

Hepatitis B, caused by the hepatitis B virus, is spread primarily by contaminated syringes and transfusion equipment. It can also be spread by any secretion of fluid by the body (tears, saliva, semen). Hepatitis B can produce chronic liver inflammation. Vaccines are available for hepatitis B and are required for certain individuals, such as health-care providers.

Hepatitis C is a form of hepatitis that cannot be traced to either the hepatitis A or hepatitis B viruses. It is clinically similar to hepatitis B and is often spread by blood transfusions. The hepatitis C virus can cause cirrhosis and liver cancer.

Hepatitis D is caused by the hepatitis D virus. It is transmitted like hepatitis B. A person must be infected with hepatitis B to contract hepatitis D. Hepatitis D results in severe liver damage and has a fatality rate higher than that of people infected with hepatitis B virus alone.

Hepatitis E is caused by the hepatitis E virus and is spread like hepatitis A. Although it does not cause chronic liver disease, the hepatitis E virus is responsible for a very high mortality rate in pregnant women.

MEDICAL TERMINOLOGY AND CONDITIONS

Anorexia nervosa A chronic disorder characterized by self-induced weight loss, negative perception of body image, and physiological changes that result from nutritional depletion. Patients have a fixation on weight control and often abuse laxatives, which worsens their fluid and electrolyte imbalances and nutrient deficiencies. The disorder is found predominantly in young, single females, and it may be inherited. Individuals may become emaciated and may ultimately die of starvation or one of its complications.

Borborygmus (bor′-bō-RIG-mus) A rumbling noise caused by the propulsion of gas through the intestines.

Bulimia (*bu-* = ox; *limia* = hunger) or *binge-purge syndrome* A disorder characterized by overeating at least twice a week followed by purging by self-induced vomiting, strict dieting or fasting, vigorous exercise, or use of laxatives or diuretics; it occurs in response to fears of being overweight, stress, depression, and physiological disorders such as hypothalamic tumors.

Canker sore (KANG-ker) Painful ulcer on the mucous membrane of the mouth that affects females more often than males, usually between ages 10 and 40; it may be an autoimmune reaction or result from a food allergy.

Cholecystitis (kō′-lē-sis-TĪ-tis; *chole-* = bile; *cyst-* = bladder; *-itis* = inflammation of) In some cases, an autoimmune inflammation of the gallbladder; other cases are caused by obstruction of the cystic duct by bile stones.

Cirrhosis Distorted or scarred liver as a result of chronic inflammation due to hepatitis, chemicals that destroy hepatocytes, parasites that infect the liver, or alcoholism; the hepatocytes are replaced by fibrous or adipose connective tissue. Symptoms include jaundice, edema in the legs, uncontrolled bleeding, and increased sensitivity to drugs.

Colitis (kō-LĪ-tis) Inflammation of the mucosa of the colon and rectum in which absorption of water and salts is reduced, producing watery, bloody feces and, in severe cases, dehydration and salt depletion. Spasms of the irritated muscularis produce cramps. It is thought to be an autoimmune condition.

Colostomy (kō-LOS-tō-mē; *-stomy* = provide an opening) The diversion of the fecal stream through an opening in the colon, creating a surgical "stoma" (artificial opening) that is affixed to the exterior of the abdominal wall. This opening serves as a substitute anus through which feces are eliminated into a bag worn on the abdomen.

Gastroscopy (gas-TROS-kō-pē; *-scopy* = to view with a lighted instrument) Endoscopic examination of the stomach in which the examiner can view the interior of the stomach directly to evaluate an ulcer, tumor, inflammation, or source of bleeding.

Hernia (HER-nē-a) Protrusion of all or part of an organ through a membrane or cavity wall, usually the abdominal cavity. *Diaphragmatic (hiatal) hernia* is the protrusion of the lower esophagus, stomach, or intestine into the thoracic cavity through the esophageal hiatus. In *inguinal hernia*, the intestine may protrude through a weakened area of the abdominal wall and may extend into the scrotal compartment in males, causing strangulation of the herniated part.

Inflammatory bowel disease (in-FLAM-a-tō′-rē BOW-el) Disorder that exists in two forms: (1) Crohn's disease, an inflammation of the gastrointestinal tract, especially the distal ileum and proximal colon, in which the inflammation may extend from the mucosa through the serosa, and (2) ulcerative colitis, an inflammation of the mucosa of the gastrointestinal tract, usually limited to the large intestine and usually accompanied by rectal bleeding.

Irritable bowel syndrome (IBS) Disease of the entire gastrointestinal tract in which a person reacts to stress by developing symptoms (such as cramping and abdominal pain) associated with alternating patterns of diarrhea and constipation. Excessive amounts of mucus may appear in feces; other symptoms include flatulence, nausea, and loss of appetite.

Malocclusion (mal′-ō-KLOO-zhun; *mal-* = bad; *occlusion* = to fit together) Condition in which the surfaces of the maxillary (upper) and mandibular (lower) teeth fit together poorly.

Nausea (NAW-sē-a = seasickness) Discomfort characterized by a loss of appetite and the sensation of impending vomiting. Its causes include local irritation of the gastrointestinal tract, a systemic disease, brain disease or injury, overexertion, or the effects of medication or drug overdose.

Traveler's diarrhea Infectious disease of the gastrointestinal tract that results in loose, urgent bowel movements; cramping; abdominal pain; malaise; nausea; and occasionally fever and dehydration. It is acquired through ingestion of food or water contaminated with fecal material typically containing bacteria (especially *Escherichia coli*); viruses or protozoan parasites are a less common cause.

STUDY OUTLINE

Introduction (p. 458)

1. The breakdown of larger food molecules into smaller molecules is called digestion; the passage of these smaller molecules into blood and lymph is termed absorption.

2. The organs that collectively perform digestion and absorption constitute the digestive system.

Overview of the Digestive System (p. 458)

1. The GI tract is a continuous tube extending from the mouth to the anus.

2. The accessory digestive organs include the teeth, tongue, salivary glands, liver, gallbladder, and pancreas.

3. Digestion includes six basic processes: ingestion, secretion, mixing and propulsion, mechanical and chemical digestion, absorption, and defecation.

4. Mechanical digestion consists of mastication and movements of the gastrointestinal tract that aid chemical digestion.

5. Chemical digestion is a series of hydrolysis reactions that break down large carbohydrates, lipids, proteins, and nucleic acids in foods into smaller molecules that are usable by body cells.

Layers of the GI Tract (p. 459)

1. The basic arrangement of layers in most of the gastrointestinal tract, from deep to superficial, is the mucosa, submucosa, muscularis, and serosa.

2. The mucosa contains extensive patches of lymphatic tissue called mucosa-associated lymphatic tissue (MALT).

3. Parts of the peritoneum include the mesentery, falciform ligament, and greater omentum.

Mouth (p. 461)

1. The mouth is formed by the cheeks, palates, lips, and tongue, which aid mechanical digestion.

2. The opening from the mouth to the throat is the fauces.

3. The tongue forms the floor of the oral cavity. It is composed of skeletal muscle covered with mucous membrane. The superior surface and lateral areas of the tongue are covered with papillae. Some papillae contain taste buds.

4. Most saliva is secreted by the salivary glands, which lie outside the mouth and pour their contents into ducts that empty into the oral cavity. There are three pairs of salivary glands: the parotid, submandibular, and sublingual. Saliva lubricates food and starts the chemical digestion of carbohydrates. Salivation is entirely under nervous control.

5. The teeth, or dentes, project into the mouth and are adapted for mechanical digestion. A typical tooth consists of three principal portions: crown, root, and neck. Teeth are composed primarily of dentin and are covered by enamel, the hardest substance in the body. Humans have two dentitions: deciduous and permanent.

6. Through mastication, food is mixed with saliva and shaped into a bolus.

7. Salivary amylase begins the digestion of starches in the mouth.

Pharynx (p. 463)

1. Food that is swallowed passes from the mouth into the oropharynx.

2. From the oropharynx, food passes into the laryngopharynx.

Esophagus (p. 363)

1. The esophagus is a muscular tube that connects the pharynx to the stomach.

2. It passes a bolus into the stomach by peristalsis.

3. Swallowing moves a bolus from the mouth to the stomach. It consists of a voluntary stage, pharyngeal stage (involuntary), and esophageal stage (involuntary).

Stomach (p. 465)

1. The stomach attaches to the esophagus and ends at the pyloric sphincter.

2. The anatomic subdivisions of the stomach are the cardia, fundus, body, and pylorus.

3. Adaptations of the stomach for digestion include rugae; glands that produce mucus, hydrochloric acid, a protein-digesting enzyme (pepsin), intrinsic factor, and gastrin; and a three-layered muscularis for efficient mechanical movement.

4. Mechanical digestion consists of mixing waves.

5. Chemical digestion consists of the conversion of proteins into peptides by pepsin.

6. Mixing waves and gastric secretions reduce food to chyme.

7. Gastric secretion and motility are regulated by nervous and hormonal mechanisms. Parasympathetic impulses and gastrin cause secretion of gastric juices.

8. The presence of food in the small intestine, secretin, and cholecystokinin inhibit gastric secretion.

9. Gastric emptying is stimulated in response to stretching, and gastrin is released in response to the presence of certain types of food. Gastric emptying is inhibited by reflex action and hormones (secretin and cholecystokinin).

10. The stomach wall is impermeable to most substances. Among the substances the stomach can absorb are water, certain ions, drugs, and alcohol.

Pancreas (p. 468)

1. The pancreas is connected to the duodenum by the pancreatic duct.

2. Pancreatic islets (islets of Langerhans) secrete hormones and constitute the endocrine portion of the pancreas.

3. Acinar cells, which secrete pancreatic juice, constitute the exocrine portion of the pancreas.

4. Pancreatic juice contains enzymes that digest starch, glycogen, and dextrins (pancreatic amylase); proteins (trypsin, chymotrypsin, and carboxypeptidase); triglycerides (pancreatic lipase); and nucleic acids (nucleases).

5. Pancreatic secretion is regulated by nervous control (parasympathetic fibers from the vagus nerves) and hormonal mechanisms (secretin and cholecystokinin).

Liver and Gallbladder (p. 468)

1. The liver has left and right lobes. The gallbladder is a sac located in a depression under the liver that stores and concentrates bile produced by the liver.

2. The lobes of the liver are made up of lobules that contain hepatocytes (liver cells), sinusoids, stellate reticuloendothelial (Kupffer's) cells, and a central vein.

3. Hepatocytes produce bile that is carried by a duct system to the gallbladder for concentration and temporary storage. Cholecystokinin (CCK) stimulates ejection of bile into the common bile duct.

4. Bile's contribution to digestion is the emulsification of dietary lipids.

5. The liver also functions in carbohydrate, lipid, and protein metabolism; processing of drugs and hormones; excretion of bilirubin; synthesis of bile salts; storage of vitamins and minerals; phagocytosis; and activation of vitamin D.

6. Bile secretion is regulated by neural and hormonal mechanisms.

Small Intestine (p. 472)

1. The small intestine extends from the pyloric sphincter to the ileocecal sphincter. It is divided into the duodenum, the jejunum, and the ileum.

2. The small intestine is highly adapted for digestion and absorption. Its glands produce enzymes and mucus, and the microvilli, villi, and circular folds of its wall provide a large surface area for digestion and absorption.

3. Mechanical digestion in the small intestine involves segmentations and migrating motility complexes.

4. Intestinal enzymes in intestinal juice, pancreatic juice, and bile break down disaccharides to monosaccharides; protein digestion is completed by peptidase enzymes; triglycerides are broken down into fatty acids and monoglycerides by pancreatic lipase; and nucleases break down nucleic acids to pentoses and nitrogenous bases.

5. The most important regulators of intestinal secretion and motility are local reflexes and digestive hormones. Parasympathetic impulses increase motility; sympathetic impulses decrease motility.

6. Absorption is the passage of nutrients from digested food in the gastrointestinal tract into the blood or lymph. Absorption, which occurs mostly in the small intestine, occurs by means of diffusion, facilitated diffusion, osmosis, and active transport.

7. Monosaccharides, amino acids, and short-chain fatty acids pass into the blood capillaries.

8. Long-chain fatty acids and monoglycerides are absorbed as part of micelles, resynthesized to triglycerides, and transported in chylomicrons to the lacteal of a villus.

9. The small intestine also absorbs water, electrolytes, and vitamins.

Large Intestine (p. 476)

1. The large intestine extends from the ileocecal sphincter to the anus. Its subdivisions include the cecum, colon, rectum, and anal canal.

2. The mucosa contains numerous absorptive and goblet cells, and the muscularis contains taeniae coli.

3. Mechanical movements of the large intestine include haustral churning, peristalsis, and mass peristalsis.

4. In the large intestine, substances are further broken down, and some vitamins are synthesized through bacterial action.

5. The large intestine absorbs water, electrolytes, and vitamins.

6. Feces consist of water, inorganic salts, epithelial cells, bacteria, and undigested foods.

7. The elimination of feces from the rectum is called defecation. Defecation is a reflex action aided by voluntary contractions of the diaphragm and abdominal muscles and relaxation of the external anal sphincter.

Regulation of Food Intake (p. 480)

1. Two centers in the hypothalamus regulate food intake: the feeding center and satiety center.

2. Signaling molecules in the blood that decrease appetite include several hormones and leptin. Signaling molecules that increase appetite include growth hormone–releasing hormone (GHRH), glucocorticoids, epinephrine, insulin, progesterone, and somatostatin (GHIH).

3. Food intake is also regulated by stretching of the GI tract, which initiates a reflex that activates the satiety center and depresses the feeding center.

Aging and the Digestive System (p. 481)

1. General changes with age include decreased secretory mechanisms, decreased motility, and loss of tone.

2. Specific changes may include loss of taste, hernias, peptic ulcer disease, constipation, hemorrhoids, and diverticular diseases.

SELF-QUIZ

1. Which of the following is NOT an accessory digestive organ?

   a. teeth   **b.** salivary glands   **c.** liver   **d.** pancreas
   e. esophagus

2. Chewing food is an example of

   a. absorption   **b.** mechanical digestion   **c.** secretion
   d. chemical digestion   **e.** ingestion

3. Which of the following is mismatched?

   a. submucosa, enteric nervous system (ENS)   **b.** muscularis,
   mucosa-associated lymphatic tissue (MALT)   **c.** serosa, greater
   omentum   **d.** mucosa, villi   **e.** serosa, visceral peritoneum

4. Most chemical digestion occurs in the

   a. liver   **b.** stomach   **c.** duodenum   **d.** colon
   e. pancreas

5. Absorption is defined as

   a. the elimination of solid wastes from the digestive system
   b. a reflex action controlled by the autonomic nervous system
   c. the breakdown of foods by enzymes   **d.** the passage of nutrients from the gastrointestinal tract into the bloodstream
   e. the mechanical breakdown of triglycerides

6. The exposed portions of the teeth that you clean with a toothbrush
   are the

   a. crowns   **b.** periodontal ligaments   **c.** roots   **d.** pulp
   cavities   **e.** gingivae

7. The smell of your favorite food cooking makes "your mouth water"; this is due to

   a. sympathetic stimulation of the salivary glands   **b.** mastication   **c.** parasympathetic stimulation of the salivary glands
   d. increased mucus secretion by the pharynx   **e.** the enteric
   nervous system

8. Match the following:

   _____ **a.** pouches in the colon
   _____ **b.** proteins combined with
            triglycerides and cholesterol
   _____ **c.** surrounds the opening
            between the stomach and
            duodenum
   _____ **d.** secrete pancreatic juice
   _____ **e.** increase surface area in
            small intestine
   _____ **f.** bile salts combined with
            partially digested lipids
   _____ **g.** located between the opening
            of the small and large intestine
   _____ **h.** large mucosal folds in stomach

   A. pyloric sphincter
   B. circular folds
   C. micelles
   D. haustra
   E. ileocecal sphincter
   F. rugae
   G. chylomicrons
   H. acini

9. Which of the following correctly describes the esophagus?

   a. Food enters the esophagus from the pyloric region of the stomach.   **b.** The movement of food through the entire esophagus
   is under voluntary control.   **c.** It allows the passage of chyme.
   d. It produces several enzymes that aid in the digestion of food.
   e. It is a muscular tube extending from the pharynx to stomach.

10. If an incision were made into the stomach, the tissue layers would
    be cut in what order?

    a. mucosa, muscularis, serosa, submucosa   **b.** mucosa, muscu-
    laris, submucosa, serosa   **c.** serosa, muscularis, mucosa, submu-
    cosa   **d.** muscularis, submucosa, mucosa, serosa   **e.** serosa,
    muscularis, submucosa, mucosa

11. Most water absorption in the digestive tract occurs in the

    a. small intestine   **b.** stomach   **c.** mouth   **d.** liver
    e. large intestine

12. Which of the following would NOT result in secretion of gastric
    juices in the stomach?

    a. activation of the feeding center in the hypothalamus
    b. stimulation by the vagus nerves   **c.** the presence of partially
    digested proteins   **d.** stretching of the stomach   **e.** stimula-
    tion by the sympathetic nervous system

13. Bile

    a. is produced in the gallbladder   **b.** is an enzyme that breaks
    down carbohydrates   **c.** emulsifies triglycerides   **d.** is re-
    quired for the absorption of amino acids   **e.** enters the small
    intestine through the right hepatic duct

14. Which of the following is NOT a function of the liver?

    a. processing newly absorbed nutrients   **b.** producing enzymes
    that digest proteins   **c.** breaking down old red blood cells
    d. detoxifying certain poisons   **e.** producing bile

15. The purpose of villi in the small intestine is to

    a. aid in the movement of food through the small intestines
    b. phagocytize microbes   **c.** produce digestive enzymes
    d. increase the surface area for absorption of digested nutrients
    e. produce acidic secretions

16. Which of the following is NOT produced in the stomach?

    a. sodium bicarbonate ($NaHCO_3$)   **b.** gastrin   **c.** pepsino-
    gen   **d.** mucus   **e.** hydrochloric acid (HCl)

17. Which of the following is NOT correctly paired?

    a. esophagus, peristalsis   **b.** mouth, mastication   **c.** large
    intestine, mass peristalsis   **d.** small intestine, segmentations
    e. stomach, migrating mobility complex

18. The enzyme pancreatic lipase digests triglycerides into

    a. glucose   **b.** amino acids   **c.** fatty acids and monoglyc-
    erides   **d.** nucleic acids   **e.** amylase

19. Place the following in the correct order as food passes from the
    small intestine:

    1. sigmoid colon
    2. transverse colon
    3. ascending colon
    4. rectum
    5. cecum
    6. descending colon

    a. 1, 3, 2, 6, 5, 4   **b.** 5, 1, 6, 2, 3, 4   **c.** 4, 1, 6, 2, 3, 5
    d. 2, 3, 5, 6, 4, 1   **e.** 5, 3, 2, 6, 1, 4

20. Lacteals function

    a. in the absorption of lipids in chylomicrons   **b.** to produce
    bile in the liver   **c.** in the absorption of electrolytes   **d.** in
    the fermentation of carbohydrates in the large intestine   **e.** to
    produce salivary amylase

CRITICAL THINKING APPLICATIONS

1. Four out of five dentists think that you should chew sugarless gum, but all five think that you should brush your teeth. Why?

2. In a classic experiment, Pavlov conditioned dogs to salivate at the sound of a bell in anticipation of being fed. You probably won't salivate at the sound of a bell, but you may salivate in anticipation of a delicious meal. Explain why.

3. Tomo put a plastic spider in his sister's drink as a joke. Unfortunately, his mother doesn't think the joke was very funny because his sister swallowed it and now they're all at the ER (emergency room). The doctor suspects that the spider may have lodged at the junction of the stomach and the duodenum. Name the sphincter at this junction. Trace the path taken by the plastic spider on its journey to its new home. What procedure could the doctor use to view the interior of the stomach? What structures may be viewed in the stomach (besides the spider)?

4. Jerry hadn't eaten all day when he bought a dried out, lukewarm hot dog from a street vendor for lunch. A few hours later, he was a victim of food poisoning and was desperately seeking a bathroom. After vomiting several times, Jerry noticed that he was expelling a greenish-yellow liquid. The hot dog may have been a bit shriveled, but it wasn't green! What is the source of this colored fluid?

ANSWERS TO FIGURE QUESTIONS

19.1 Peristalsis propels food along the GI tract.

19.2 Nerves in its wall help regulate secretions and contractions of the gastrointestinal tract.

19.3 Mesentery binds the small intestine to the posterior abdominal wall.

19.4 Muscles of the tongue maneuver food for chewing, shape food into a bolus, force food to the back of the mouth for swallowing, and alter the shape of the tongue for swallowing and speech production.

19.5 The main component of teeth is a connective tissue called dentin.

19.6 Swallowing is both voluntary and involuntary. Initiation of swallowing, carried out by skeletal muscles, is voluntary. Completion of swallowing—moving a bolus along the esophagus and into the stomach—involves peristalsis of smooth muscle and is involuntary.

19.7 After a very large meal the stomach probably does not have rugae because as the stomach fills, the rugae stretch out.

19.8 Gastrin stimulates secretion of gastric juice, increases motility of the GI tract, and relaxes the pyloric sphincter.

19.9 The pancreatic duct contains pancreatic juice, composed of fluid and digestive enzymes.

19.10 The path of blood flow is hepatic portal vein → branch of hepatic portal vein → sinusoid → central vein → hepatic vein → inferior vena cava.

19.11 The ileum is the longest portion of the small intestine.

19.12 A square millimeter of small intestinal mucosa contains 20 to 40 villi and 200 million microvilli.

19.13 Fat-soluble vitamins are absorbed by diffusion from micelles.

19.14 Functions of the large intestine include completion of absorption, synthesis of certain vitamins, formation of feces, and elimination of feces.

19.15 The muscularis of the large intestine contains teniae coli that form haustra.

Chapter **20**

Nutrition and Metabolism

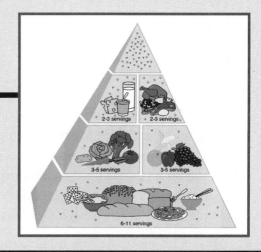

■ Student Learning Objectives

■ A Look Ahead

The food we eat is our only source of energy for performing biological work. Many molecules needed to maintain cells and tissues can be made from building blocks within the body; others must be obtained in food because we cannot make them. Food molecules absorbed by the gastrointestinal (GI) tract have three main fates:

1. Most food molecules are used to *supply energy* for sustaining life processes, such as active transport, DNA replication, protein synthesis, muscle contraction, and mitosis.

2. Some food molecules *serve as building blocks* for the synthesis of more complex structural or functional molecules, such as muscle proteins, hormones, and enzymes.

3. Other food molecules are *stored for future use.* For example, glycogen is stored in liver cells, and triglycerides are stored in adipose cells.

In this chapter we will discuss the major groups of nutrients; guidelines for healthy eating; how each group of nutrients is used for ATP production, growth, and repair of the body; and how various factors affect the body's metabolic rate.

NUTRIENTS

Objectives: • **Define a nutrient and identify the six main classes of nutrients.**

• **List the guidelines for healthy eating.**

Nutrients are chemical substances in food that provide energy, form new body components, or assist in various body processes. The six main classes of nutrients are carbohydrates, lipids, proteins, water, minerals, and vitamins. The structures and functions of carbohydrates, proteins, lipids, and water were discussed in Chapter 2. In this chapter we discuss minerals and vitamins. Some minerals and many vitamins are part of enzyme systems that catalyze the breakdown and synthesis of carbohydrates, lipids, and proteins.

Minerals

Minerals are inorganic elements that constitute about 4% of the total body weight and are concentrated most heavily in the skeleton. Minerals with known functions in the body include calcium, phosphorus, potassium, sulfur, sodium, chloride, fluoride, magnesium, iron, iodide, manganese, cobalt, copper, zinc, selenium, and chromium. Others—aluminum, boron, silicon, and molybdenum—are present but may have no functions. Typical diets supply adequate amounts of potassium, sodium, chlorine, and magnesium. Some attention must be paid to eating foods that provide enough calcium, phosphorus, iron, and iodine. Excess amounts of most minerals are excreted in the urine and feces.

A major role of minerals is to help regulate enzymatic reactions. Calcium, iron, magnesium, and manganese are part of some coenzyme compounds. Magnesium also serves as a catalyst for the conversion of ADP to ATP. Minerals such as sodium and phosphorus work in buffer systems, which help control the pH of body fluids. Sodium also helps regulate the osmosis of water and, with other ions, is involved in the generation of nerve impulses. Table 20.1 describes the roles of several minerals in various body functions.

Vitamins

Organic nutrients required in small amounts to maintain growth and normal metabolism are called *vitamins*. Unlike carbohydrates, lipids, or proteins, vitamins do not provide energy or serve as the body's building materials. Most vitamins with known functions serve as coenzymes.

Most vitamins cannot be synthesized by the body and must be ingested. Other vitamins, such as vitamin K, are produced by bacteria in the GI tract and then absorbed. The body can assemble some vitamins if the raw materials, called *provitamins,* are provided. For example, vitamin A is produced by the body from the provitamin beta-carotene, a chemical present in orange and yellow vegetables such as carrots and in dark green vegetables such as spinach. No single food contains all the vitamins required by the body—one of the best reasons to eat a varied diet.

Vitamins are divided into two main groups: fat soluble and water soluble. The *fat-soluble vitamins* are vitamins A, D, E, and K. They are absorbed along with dietary lipids in the small intestine and packaged into chylomicrons. They cannot be absorbed in adequate quantity unless they are ingested with lipids. Fat-soluble vitamins may be stored in cells, particularly in the liver. The *water-soluble vitamins* include several B vitamins and vitamin C. They are dissolved in body fluids. Excess quantities of these vitamins are not stored but instead are excreted in the urine.

Besides their other functions, three vitamins—C, E, and beta-carotene (a provitamin)—are termed *antioxidant vitamins* because they inactivate oxygen free radicals. Recall that free radicals are highly reactive ions or molecules that carry an unpaired electron in their outermost electron shell. Free radicals damage cell membranes, DNA, and other cellular structures and contribute to the formation of atherosclerotic plaques. Some free radicals arise naturally in the body, and others derive from environmental hazards such as tobacco smoke and radiation. Antioxidant vitamins are thought to play a role in protecting

Table 20.1 / Minerals and Their Functions in the Body

Mineral	Distribution in Body and Sources	Functions
Calcium	Most abundant mineral in body. Appears in combination with phosphates. About 99% is stored in bones and teeth. Calcitriol is needed for absorption of dietary calcium. Sources are milk, egg yolk, shellfish, and green leafy vegetables.	Formation of bones and teeth, blood clotting, normal muscle and nerve activity, endocytosis and exocytosis, cellular motility, chromosome movement before cell division, and synthesis and release of neurotransmitters.
Phosphorus	About 80% is found in bones and teeth. Sources are dairy products, meat, fish, poultry, nuts.	Formation of bones and teeth, component of many enzymes, involved in energy transfer (ATP), component of DNA and RNA.
Potassium	Principal cation (K^+) in intracellular fluid. Present in most foods (meats, milk, fruits, vegetables, grains).	Needed for generation and conduction of action potentials in neurons and muscle fibers.
Sulfur	Component of many proteins, electron carriers in electron transport chain, and some vitamins (thiamine and biotin). Sources include beef, liver, lamb, fish, poultry, eggs, cheese, beans.	As component of hormones and vitamins, regulates various body activities. Needed for ATP production by electron transport chain.
Sodium	Most abundant cation (Na^+) in extracellular fluids; some found in bones. Sources include salt, soy sauce, processed foods.	Affects distribution of water by osmosis. Needed for generation and conduction of action potentials in neurons and muscle fibers.
Chloride	Principal anion (Cl^-) in extracellular fluid. Sources include salt, soy sauce, processed foods.	Plays role in water balance and formation of HCl in stomach.
Magnesium	Second most abundant cation (Mg^{2+}) in intracellular fluid. Sources include green leafy vegetables, seafood, and whole-grain cereals.	Required for normal functioning of muscle and nervous tissue, participates in bone formation, constituent of many coenzymes.
Iron	About 66% found in hemoglobin of blood. Sources are meat, liver, shellfish, egg yolk, beans, legumes, dried fruits, nuts, cereals.	As component of hemoglobin, reversibly binds O_2. Component of electron carriers in electron transport chain.
Iodide	Essential component of thyroid hormones. Sources are seafood, iodized salt, and vegetables grown in iodine-rich soils.	Needed for synthesis of thyroid hormones, which regulate metabolic rate.
Manganese	Some stored in liver and spleen. Sources are nuts, legumes, whole-grain cereals, fruits, and leafy green vegetables.	Activates several enzymes. Needed for hemoglobin synthesis, urea formation, growth, reproduction, lactation, and bone formation.
Copper	Some stored in liver and spleen. Sources include eggs, whole-wheat flour, beans, beets, liver, fish, spinach, asparagus.	Needed for synthesis of hemoglobin, component of coenzymes in electron transport chain.
Cobalt	Constituent of vitamin B_{12}. Sources include meat, milk, and milk products.	Needed for synthesis of hemoglobin.
Zinc	Important component of certain enzymes. Widespread in many foods, especially meats.	As a component of carbonic anhydrase, important in carbon dioxide metabolism. As a component of peptidases, is involved in protein digestion. Necessary for normal tissue repair, normal taste and smell sensations, and normal sperm counts in males.
Fluoride	Components of bones, teeth, other tissues. Found in tea, seafoods, and fluoridated water.	Improves tooth structure and inhibits tooth decay.
Selenium	Found in seafood, meat, chicken, grain cereals, egg yolk, milk, mushrooms, and garlic.	An antioxidant. Prevents chromosome breakage and may play a role in preventing some cancers and birth defects.
Chromium	Present in most body cells. Sources are meats, brewers' yeast, wine, and some beers.	Needed for normal activity of insulin in carbohydrate and lipid metabolism.

against some kinds of cancer, reducing the buildup of atherosclerotic plaque, delaying some effects of aging, and decreasing the chance of cataract formation in the lenses of the eyes. Table 20.2 lists the principal vitamins, their sources, their functions, and related deficiency disorders.

Most nutritionists recommend eating a balanced diet that includes a variety of foods rather than taking vitamin or mineral supplements, except in special circumstances. Common examples of necessary supplementations include iron for women who have excessive menstrual bleeding; iron and calcium for women who are pregnant or breast-feeding; folic acid (folate) for all women who may become pregnant, to reduce the risk of fetal neural tube defects; calcium for most adults, because they do not receive the recommended amount in their diets; and vitamin B_{12} for strict vegetarians, who eat no meat. Because most North Americans do not ingest in their food the high levels of antioxidant vitamins thought to have beneficial effects, some experts recommend supplementing vitamins C and E. More is not always better; larger doses of vitamins or minerals can be very harmful.

Guidelines for Healthy Eating

Each gram of dietary protein or carbohydrate in food provides about 4 Calories, whereas a gram of fat (lipids) provides about 9 Calories. (The number of Calories in a food is a measure of the

Table 20.2 / The Principal Vitamins

Vitamin	Comment and Source	Functions	Deficiency Symptoms and Disorders
Fat Soluble	All require bile salts and some dietary lipids for adequate absorption.		
A	Formed from provitamin beta-carotene (and other provitamins) in GI tract. Stored in liver. Sources of carotene and other provitamins include orange, yellow, and green vegetables; sources of vitamin A include liver and milk.	Maintains general health and vigor of epithelial cells. Beta-carotene acts as an antioxidant to inactivate free radicals.	Deficiency results in atrophy and keratinization of epithelium, leading to dry skin and hair; increased incidence of ear, sinus, respiratory, urinary, and digestive system infections; inability to gain weight; drying of cornea; and skin sores.
		Essential for formation of light-sensitive pigments in photoreceptors of retina.	Night blindness or decreased ability for dark adaptation.
		Aids in growth of bones and teeth by helping to regulate activity of osteoblasts and osteoclasts.	Slow and faulty development of bones and teeth.
D	In the presence of sunlight, the skin, liver, and kidneys form active form of vitamin D (calcitriol). Stored in tissues to slight extent. Most excreted in bile. Dietary sources include fish-liver oils, egg yolk, fortified milk.	Essential for absorption and utilization of calcium and phosphorus. Works with parathyroid hormone (PTH) to maintain Ca^{2+} homeostasis.	Defective utilization of calcium by bones leads to rickets in children and osteomalacia in adults. Possible loss of muscle tone.
E (tocopherols)	Stored in liver, adipose tissue, and muscles. Sources include fresh nuts and wheat germ, seed oils, green leafy vegetables.	Inhibits catabolism of certain fatty acids that help form cell structures, especially membranes. Involved in formation of DNA, RNA, and red blood cells. May promote wound healing, contribute to the normal structure and functioning of the nervous system, and prevent scarring. May help protect liver from toxic chemicals such as carbon tetrachloride. Acts as an antioxidant to inactivate free radicals.	May cause the oxidation of monounsaturated fats, resulting in abnormal structure and function of mitochondria, lysosomes, and plasma membranes. A possible consequence is hemolytic anemia.
K	Produced by intestinal bacteria. Stored in liver and spleen. Dietary sources include spinach, cauliflower, cabbage, liver.	Coenzyme essential for synthesis of several clotting factors by liver, including prothrombin.	Delayed clotting time results in excessive bleeding.
Water Soluble	Dissolved in body fluids. Not stored in body. Excess intake eliminated in urine.		
B₁ (thiamin)	Rapidly destroyed by heat. Sources include whole-grain products, eggs, pork, nuts, liver, yeast.	Acts as coenzyme for many different enzymes that break carbon-to-carbon bonds and are involved in carbohydrate metabolism of pyruvic acid to CO_2 and H_2O. Essential for synthesis of the neurotransmitter acetylcholine.	Improper carbohydrate metabolism leads to buildup of pyruvic and lactic acids and insufficient production of ATP for muscle and nerve cells. Deficiency leads to: (1) beriberi, partial paralysis of smooth muscle of GI tract, causing digestive disturbances; skeletal muscle paralysis; and atrophy of limbs; (2) polyneuritis, due to degeneration of myelin sheaths; impaired reflexes, impaired sense of touch, stunted growth in children, and poor appetite.

Vitamin	Comment and Source	Functions	Deficiency Symptoms and Disorders
B₂ (riboflavin)	Small amounts supplied by bacteria of GI tract. Dietary sources include yeast, liver, beef, veal, lamb, eggs, whole-grain products, asparagus, peas, beets, peanuts.	Component of certain coenzymes (for example, FAD) in carbohydrate and protein metabolism, especially in cells of eye, integument, mucosa of intestine, blood.	Deficiency may lead to improper utilization of oxygen resulting in blurred vision, cataracts, and corneal ulcerations. Also, dermatitis and cracking of skin, lesions of intestinal mucosa, and one type of anemia.
Niacin (nicotinamide)	Derived from amino acid tryptophan. Sources include yeast, meats, liver, fish, whole-grain products, peas, beans, nuts.	Essential component of coenzymes NAD and NADP. In lipid metabolism, inhibits production of cholesterol; assists in triglyceride breakdown.	Principal deficiency is pellagra, characterized by dermatitis, diarrhea, and psychological disturbances.
B₆ (pyridoxine)	Synthesized by bacteria of GI tract. Other sources include salmon, yeast, tomatoes, yellow corn, spinach, whole-grain products, liver, yogurt.	Essential coenzyme for normal amino acid metabolism. Assists production of circulating antibodies. May function as coenzyme in triglyceride metabolism.	Most common deficiency symptom is dermatitis of eyes, nose, and mouth. Other symptoms are retarded growth and nausea.
B₁₂ (cyano-cobalamin)	Only B vitamin not found in vegetables; only vitamin containing cobalt. Absorption from GI tract depends on intrinsic factor secreted by stomach. Sources include liver, kidney, milk, eggs, cheese, meat.	Coenzyme necessary for red blood cell formation, formation of the amino acid methionine, entrance of some amino acids into Krebs cycle, and manufacture of choline (used to synthesize acetylcholine).	Pernicious anemia, ataxia, memory loss, weakness, personality and mood changes, abnormal sensations, and impaired osteoblast activity.
Pantothenic acid	Some produced by bacteria of GI tract. Other sources include kidney, liver, yeast, green vegetables, cereal.	Constituent of coenzyme A essential for transfer of acetyl group from pyruvic acid into Krebs cycle, conversion of lipids and amino acids into glucose, and synthesis of cholesterol and steroid hormones.	Fatigue, muscle spasms, insufficient production of adrenal steroid hormones, vomiting, insomnia.
Folic acid (folate, folacin)	Synthesized by bacteria of GI tract. Dietary sources include legumes, green leafy vegetables, broccoli, asparagus, fortified breads, and citrus fruits.	Component of enzyme systems synthesizing nitrogenous bases of DNA and RNA. Essential for normal production of red and white blood cells.	Production of abnormally large red blood cells. Higher risk of neural tube defects in babies born to folate-deficient mothers.
Biotin	Synthesized by bacteria of GI tract. Dietary sources include yeast, liver, egg yolk, kidneys.	Essential coenzyme for conversion of pyruvic acid to oxaloacetic acid (in Krebs cycle) and synthesis of fatty acids.	Mental depression, muscular pain, dermatitis, fatigue, nausea.
C (ascorbic acid)	Sources include citrus fruits, tomatoes, green vegetables. Rapidly destroyed by heat.	Promotes protein synthesis, including laying down of collagen in formation of connective tissue. As coenzyme, may combine with poisons, rendering them harmless until excreted. Works with antibodies, promotes wound healing, and functions as an antioxidant.	Scurvy; anemia; many symptoms related to poor connective tissue growth and repair, including tender swollen gums, loosening of teeth, poor wound healing, bleeding (vessel walls are fragile because of connective tissue degeneration), and retardation of growth.

heat it releases upon oxidation, which is described later in the chapter.) On a daily basis, many women and older people require about 1600 Calories; children, teenage girls, active women, and most men need about 2200 Calories; and teenage boys and active men need about 2800 Calories.

We do not know with certainty what levels and types of carbohydrate, fat, and protein are optimal in the diet, for different populations the world over eat radically different diets adapted to their particular lifestyles. Experts recommend the following distribution of calories: 50-60% from carbohydrates, with less than 15% from simple sugars; less than 30% from fats (triglycerides are the main type of dietary fat), with no more than 10% as saturated fats; and about 12-15% from proteins.

The guidelines for healthy eating are as follows:

• Eat a variety of foods.
• Maintain healthy weight.
• Choose foods low in fat, saturated fat, and cholesterol.
• Eat plenty of vegetables, fruits, and grain products.

• Use sugars in moderation only.
• Use salt and sodium in moderation (less than 6 grams daily).
• If you drink alcoholic beverages, do so in moderation (less than 1 ounce of the equivalent of pure alcohol per day).

To help people achieve a good balance of vitamins, minerals, carbohydrates, fats, and proteins in their food, the U.S. Department of Agriculture developed the food guide pyramid (Figure 20.1). The sections of the pyramid indicate how many daily servings of each of the five major food groups to eat. The smallest number of servings corresponds to a 1600 Calorie per day diet, whereas the largest number of servings corresponds to a 2800 Calorie per day diet. Because they should be consumed in largest quantity, foods rich in complex carbohydrates—the bread, cereal, rice, and pasta group—form the base of the pyramid. Vegetables and fruits form the next level. The health benefits of eating generous amounts of these foods are well documented. Foods on the next level—the milk, yogurt, and cheese group, and the meat, poultry, fish, dry beans, eggs, and nuts

Figure 20.1 ■ **The food guide pyramid.** The smallest number of servings corresponds to 1600 Calories per day, whereas the largest number of servings corresponds to 2800 Calories per day. Each example given equals one serving.

The sections of the pyramid show how many servings of five major food groups to eat each day.

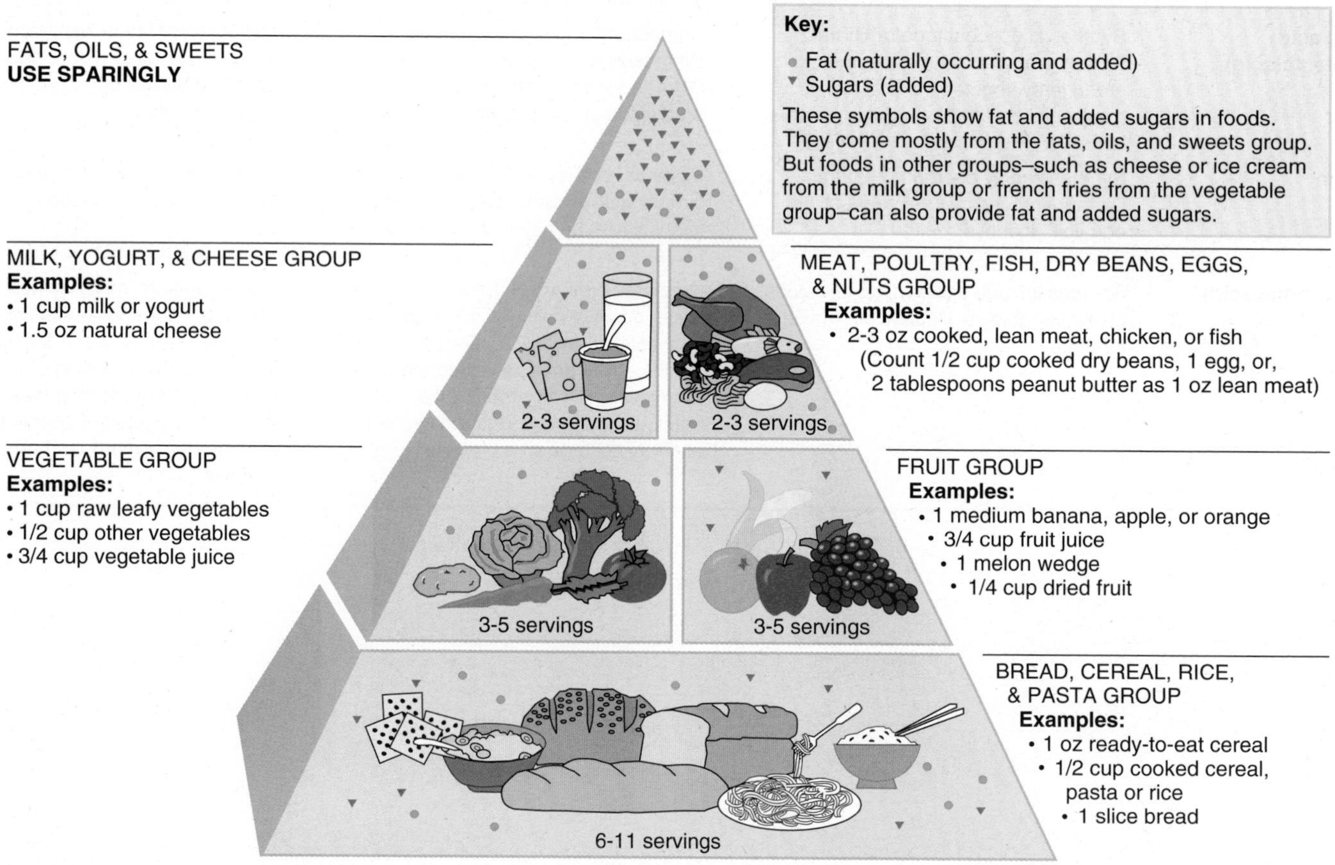

FATS, OILS, & SWEETS
USE SPARINGLY

Key:
• Fat (naturally occurring and added)
▼ Sugars (added)

These symbols show fat and added sugars in foods. They come mostly from the fats, oils, and sweets group. But foods in other groups–such as cheese or ice cream from the milk group or french fries from the vegetable group–can also provide fat and added sugars.

MILK, YOGURT, & CHEESE GROUP
Examples:
• 1 cup milk or yogurt
• 1.5 oz natural cheese

2-3 servings

MEAT, POULTRY, FISH, DRY BEANS, EGGS, & NUTS GROUP
Examples:
• 2-3 oz cooked, lean meat, chicken, or fish
(Count 1/2 cup cooked dry beans, 1 egg, or, 2 tablespoons peanut butter as 1 oz lean meat)

2-3 servings

VEGETABLE GROUP
Examples:
• 1 cup raw leafy vegetables
• 1/2 cup other vegetables
• 3/4 cup vegetable juice

3-5 servings

FRUIT GROUP
Examples:
• 1 medium banana, apple, or orange
• 3/4 cup fruit juice
• 1 melon wedge
• 1/4 cup dried fruit

3-5 servings

BREAD, CEREAL, RICE, & PASTA GROUP
Examples:
• 1 oz ready-to-eat cereal
• 1/2 cup cooked cereal, pasta or rice
• 1 slice bread

6-11 servings

Which food group(s) shown contain cholesterol and most of the saturated fatty acids in the diet?

group—should be eaten in smaller quantities. These two food groups have higher fat and protein content than the food groups below them. To lower your daily intake of fats, choose low-fat foods from these groups—nonfat milk and yogurt, low-fat cheese, fish, and skinless poultry.

The apex of the pyramid is not a food group but rather a caution to use fats, oils, and sweets sparingly. The food guide pyramid does not distinguish among the different types of fatty acids—saturated, polyunsaturated, and monounsaturated—in dietary fats. Atherosclerosis and coronary artery disease are prevalent in populations that consume large amounts of saturated fats and cholesterol. By contrast, populations living around the Mediterranean Sea, who eat a diet with considerable olive oil, which is rich in monounsaturated fatty acids and has no cholesterol, have low rates of coronary artery disease despite eating up to 40% of their Calories as fat. Canola oil, avocados, nuts, and peanut oil are also rich in monounsaturated fatty acids.

OVERVIEW OF METABOLISM

Objective: • **Define metabolism and describe its importance in homeostasis.**

Metabolism (me-TAB-ō-lizm; *metabol-* = change) refers to all the chemical reactions of the body. The body's metabolism may be thought of as an energy-balancing act between anabolic (synthesis) and catabolic (decomposition) reactions.

Anabolism

Chemical reactions that combine simple substances into more complex molecules are collectively known as ***anabolism*** (a-NAB-ō-lizm; *ana-* = upward). Overall, anabolic reactions use more energy than they produce. The energy they use is supplied by catabolic reactions (Figure 20.2). One example of an anabolic process is the formation of peptide bonds between amino acids, combining the amino acids into proteins.

Catabolism

The chemical reactions that break down complex organic compounds into simple ones are collectively known as ***catabolism*** (ka-TAB-ō-lizm; *cata-* = downward). Catabolic reactions produce more energy than they use. These reactions release the potential energy stored in the bonds of organic molecules. This energy is transferred to molecules of ATP and then used to power anabolic reactions. Chemical digestion is an example of a catabolic process. Another example is oxidation, to be described shortly.

Metabolism and Enzymes

Recall that chemical reactions occur when chemical bonds between substances are formed or broken, and that enzymes serve

Figure 20.2 ■ **Catabolism, anabolism, and ATP.** The breakdown of ATP provides energy for the formation of complex compounds from simple ones (anabolism). When large compounds are split apart (catabolism), some of the energy is transferred to and trapped in ATP and then utilized to drive anabolic reactions. Both the formation and breakdown of ATP give off heat.

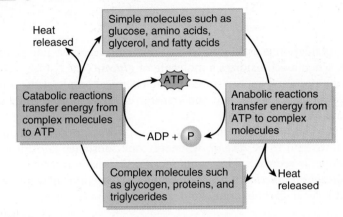

The body's metabolism is an energy-balancing act between anabolism and catabolism.

Is the process of making digestive enzymes anabolic or catabolic?

as catalysts to speed up chemical reactions. You may wish to review the characteristics and functions of enzymes in Chapter 2 before reading on.

Some enzymes require the presence of an ion such as calcium, iron, or zinc that must be attached to the enzyme before a substrate can bind to the enzyme's active site. Other enzymes work together with ***coenzymes,*** which function as temporary carriers of atoms being removed from or added to a substrate during a reaction. Many coenzymes are derived from vitamins. Examples include the coenzyme *NAD⁺*, derived from the B vitamin niacin, and the coenzyme *FAD,* derived from vitamin B_2 (riboflavin).

Oxidation–Reduction Reactions

Oxidation refers to the *removal* of electrons (e^-) and hydrogen ions (H^+) from a molecule. This is equivalent to the removal of an atom of hydrogen: $e^- + H^+ = H$. An example of an oxidation reaction is the conversion of lactic acid into pyruvic acid:

$$
\begin{array}{ccc}
\text{COOH} & & \text{COOH} \\
| & & | \\
\text{H}-\text{C}-\text{OH} & \xrightarrow[\text{(remove 2 H)}]{\text{Oxidation}} & \text{C}=\text{O} \\
| & & | \\
\text{CH}_3 & & \text{CH}_3 \\
\text{Lactic acid} & & \text{Pyruvic acid}
\end{array}
$$

A coenzyme picks up the removed hydrogen atoms and immediately transfers them to another compound.

Reduction refers to the *addition* of electrons and hydrogen ions (hydrogen atoms) to a molecule. Reduction is the opposite of oxidation. An example of a reduction reaction is the conversion of pyruvic acid to lactic acid:

$$
\begin{array}{ccc}
\text{COOH} & & \text{COOH} \\
| & \xrightarrow[\text{(add 2 H)}]{\text{Reduction}} & | \\
\text{O}=\text{O} & & \text{H}-\text{C}-\text{OH} \\
| & & | \\
\text{CH}_3 & & \text{CH}_3 \\
\text{Pyruvic acid} & & \text{Lactic acid}
\end{array}
$$

Oxidation is usually an energy-releasing reaction. For example, when a cell oxidizes a molecule of **glucose** ($C_6H_{12}O_6$), some of the potential energy stored in the chemical bonds of the glucose molecule is released and transferred to adenosine triphosphate (ATP).

Oxidation and reduction reactions always occur together and are said to be *coupled;* whenever one substance is oxidized, another is reduced. This coupling of reactions is referred to as *oxidation–reduction.* Oxidation–reduction reactions play important roles in carbohydrate, lipid, and protein metabolism.

CARBOHYDRATE METABOLISM

Objective: • **Explain how the body uses carbohydrates.**

During digestion, polysaccharide and disaccharide carbohydrates are digested to monosaccharides—glucose, fructose, and galactose—which are absorbed in the small intestine. Shortly after their absorption, however, fructose and galactose are converted to glucose. Thus, the story of carbohydrate metabolism is really the story of glucose metabolism.

Fate of Carbohydrates

Because glucose is the body's preferred source for synthesizing ATP, the fate of glucose absorbed from the diet depends on the needs of body cells. If the cells require ATP immediately, the glucose is oxidized by the cells. Glucose not needed for immediate ATP production may be converted to glycogen (glycogenesis) for storage by liver cells and skeletal muscle fibers. If these glycogen stores are full, the liver cells can transform the glucose to triglycerides (lipogenesis) for storage in adipose tissue. At a later time, when the cells need more ATP, the glycogen and triglycerides can be converted back to glucose.

Before glucose can be used by body cells, it must pass through the plasma membrane by facilitated diffusion and enter the cytosol. Insulin increases the rate of facilitated diffusion of glucose.

Glucose Catabolism

The **oxidation** of glucose, also known as **cellular respiration** (Figure 20.3), involves four sets of reactions:

① **Glycolysis.** One glucose molecule is oxidized and two molecules of pyruvic acid are produced in a set of reactions known as **glycolysis** (glī-KOL-i-sis; *glyco-* = sugar; *-lysis* = breakdown). The reactions also produce two molecules of ATP and two energy-containing molecules of NADH (an H^+ is produced with each NADH). Because glycolysis does not require oxygen, it is a way to produce ATP anaerobically (without oxygen) and is known as **anaerobic cellular respiration.**

② **Formation of acetyl coenzyme A.** In a transition step that prepares pyruvic acid for entrance into the Krebs cycle, a molecule of acetyl coenzyme A is formed. This step also produces energy-containing NADH plus carbon dioxide (CO_2).

③ **Krebs cycle.** Acetyl coenzyme A enters the Krebs cycle, a series of reactions that oxidize acetyl coenzyme A and produce CO_2, ATP, NADH, and $FADH_2$.

④ **Electron transport chain.** The reactions of the electron transport chain oxidize NADH and $FADH_2$ and transfer their electrons to a series of electron carriers. The Krebs cycle and the electron transport chain require oxygen to produce ATP and together are known as **aerobic cellular respiration.**

Glycolysis

Glycolysis takes place in the cytosol of most body cells. The fate of the pyruvic acid produced during glycolysis depends on the availability of oxygen. If oxygen is scarce (anaerobic conditions)—for example, in skeletal muscle fibers during strenuous exercise—then pyruvic acid is converted to lactic acid. As lactic acid is produced, it rapidly diffuses out of muscle fibers and enters the blood. Liver cells remove lactic acid from the blood and convert it back to pyruvic acid.

When oxygen is plentiful (aerobic conditions), most cells convert pyruvic acid to acetyl coenzyme A. This molecule links glycolysis, which occurs in the cytosol, with the Krebs cycle, which occurs in mitochondria. Because they lack mitochondria, red blood cells can produce ATP only through glycolysis.

Formation of Acetyl Coenzyme A

Each step in the oxidation of glucose requires a different enzyme, and often a coenzyme as well. The coenzyme used at this point in cellular respiration is **coenzyme A (CoA),** which is derived from a B vitamin. During the transitional step between glycolysis and the Krebs cycle, pyruvic acid is first oxidized and converted to a two-carbon fragment by removing a molecule of carbon dioxide (Figure 20.3). Then, the coenzyme NAD^+ picks up a hydrogen atom from pyruvic acid. Finally, the remaining two-carbon fragment, called an **acetyl group,** is attached to coenzyme A, to form **acetyl coenzyme A (acetyl CoA).**

Krebs Cycle

During the **Krebs cycle** a series of oxidation–reduction reactions breaks down acetyl coenzyme A; releases CO_2; produces a sub-

Figure 20.3 ■ **Oxidation of glucose: cellular respiration.**

The oxidation of glucose involves glycolysis, the Krebs cycle, and the electron transport chain.

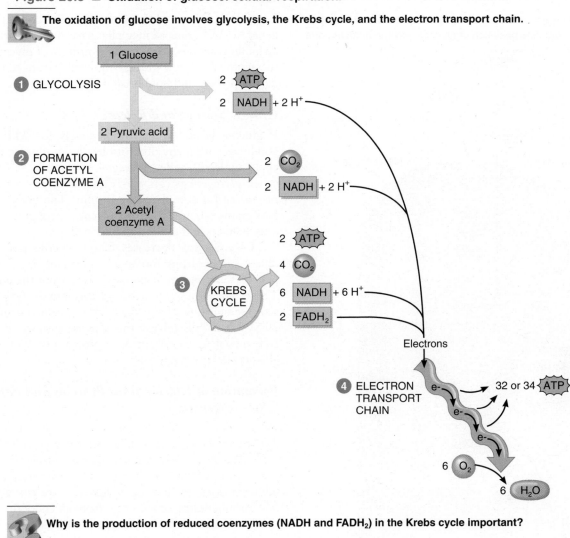

Why is the production of reduced coenzymes (NADH and FADH$_2$) in the Krebs cycle important?

stance called guanosine triphosphate (GTP); and reduces two energy-containing coenzymes, NADH and FADH$_2$. (FAD is the oxidized form; FADH$_2$ is the reduced form.) The CO$_2$ is transported by the blood to the lungs, where it is exhaled. The GTP is a high-energy compound that is used to change ADP into ATP (GTP is the energy equivalent of ATP). The reduced coenzymes now contain the potential energy that was in glucose, pyruvic acid, and acetyl coenzyme A. To make use of this energy for cellular activities, the reduced coenzymes must first go through the electron transport chain.

Electron Transport Chain

The **electron transport chain** is a series of oxidation–reduction reactions that take place in mitochondria (see Figure 20.3). This series of reactions transfers the energy that was stored in the reduced coenzymes (NADH and FADH$_2$) to 32 or 34 molecules of ATP (depending on which electron carriers are used in the electron transport chain). As NADH and FADH$_2$ are oxidized, they pass their electrons to a series of electron carriers. This electron transfer generates the ATP. The energy used to produce ATP represents only 40% of the energy originally stored in glucose; the rest of the energy is released as heat. In the final step of the electron transport chain, the electrons are passed to oxygen in a reaction that produces water.

The complete catabolism of a molecule of glucose can be summarized as follows:

1 Glucose + 6 Oxygen $\longrightarrow$
36 or 38 ATP + 6 Carbon dioxide + 6 Water

A summary of the sites of the principal events of glucose oxidation is shown in Figure 20.4. Glycolysis, the Krebs cycle, and the electron transport chain provide the ATP for all cellular activities. Because the Krebs cycle and electron transport chain are aerobic processes, the cells cannot carry on their activities for long without sufficient oxygen.

Figure 20.4 ■ Sites of the principal events of glucose oxidation.

 Except for glycolysis, which occurs in cytosol, all other reactions involving the complete oxidation of glucose occur in mitochondria.

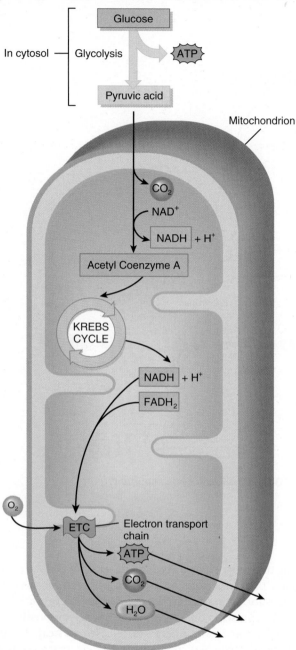

How many molecules of ATP are produced during the complete oxidation of one molecule of glucose?

Glucose Anabolism

Even though most of the glucose in the body is catabolized to generate ATP, glucose may take part in or be formed via several anabolic reactions. One is the synthesis of glycogen; another is the synthesis of new glucose molecules from some of the products of protein and lipid breakdown.

Glucose Storage and Release

If glucose is not needed immediately for ATP production, it combines with many other molecules of glucose to form a long-chain molecule called **glycogen** (Figure 20.5). This process is called **glycogenesis** (glī′-kō-JEN-e-sis; glyco- = sugar; -genesis = origin) and is stimulated by insulin. The body can store about 500 grams (about 1.1 lb) of glycogen, roughly 75% in skeletal muscle fibers and the rest in liver cells.

Glycogenesis decreases blood glucose level. When blood glucose level drops too much, glucagon is released from the pancreas and epinephrine is released from the adrenal medulla. These hormones stimulate **glycogenolysis** (glī′-kō-je-NOL-i-sis), the splitting of glycogen into its glucose subunits (Figure 20.5). Liver cells release this glucose into the blood, and body cells pick it up to use for ATP production. Glycogenolysis usually occurs between meals.

Formation of Glucose from Proteins and Fats: Gluconeogenesis

When your liver runs low on glycogen, it is time to eat. If you don't, your body starts catabolizing triglycerides (fats) and proteins. Actually, the body normally catabolizes some of its triglycerides and proteins, but large-scale triglyceride and protein catabolism does not happen unless you are starving, eating very few carbohydrates, or suffering from an endocrine disorder.

The glycerol part of triglycerides, lactic acid, and certain amino acids can be converted in the liver to glucose (Figure 20.5). The process by which glucose is formed from these noncarbohydrate sources is called **gluconeogenesis** (gloo′-kō-nē′-ō-JEN-e-sis; neo = new). An easy way to distinguish this term from glycogenesis or glycogenolysis is to remember that in this case glucose is not converted back from glycogen, but is instead *newly formed*. Gluconeogenesis occurs when the liver is stimulated by cortisol from the adrenal cortex and glucagon from the pancreas.

LIPID METABOLISM

Objectives: • **Describe the lipoproteins that transport lipids in the blood.**

• **Explain how the body uses lipids.**

Transport of Lipids by Lipoproteins

Most lipids, such as triglycerides, do not dissolve in water. To be transported in blood, such molecules first must be made water

Figure 20.5 ■ **Reactions of glucose anabolism: synthesis of glycogen (glycogenesis), breakdown of glycogen (glycogenolysis), and synthesis of glucose from amino acids, lactic acid or glycerol (gluconeogenesis).**

About 500 grams (1.1 lb) of glycogen are stored in skeletal muscles and the liver.

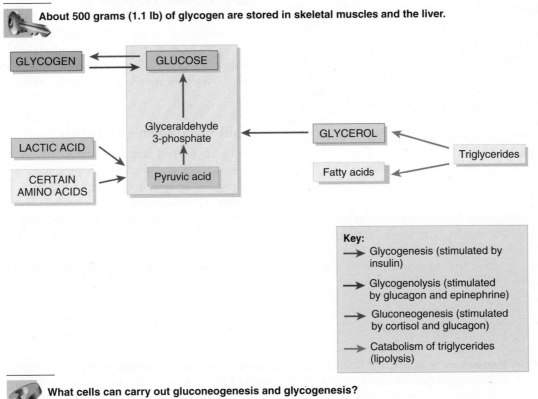

What cells can carry out gluconeogenesis and glycogenesis?

soluble by combining them with proteins produced by the liver and small intestine. These water-soluble combinations of proteins and lipids are *lipoproteins,* spherical particles with an outer shell of proteins, phospholipids, and cholesterol molecules surrounding an inner core of triglycerides and other lipids. The proteins in the outer shell, called *apoproteins,* help the lipoprotein particles dissolve in body fluids and also have specific functions.

Lipoproteins are essentially transport vehicles: They provide delivery and pick-up services so that lipids can be available when cells need them or be removed from circulation when they are not needed. Lipoproteins are categorized and named mainly according to their size and density. From largest and lightest to smallest and heaviest, the four major classes of lipoproteins are chylomicrons, very-low-density lipoproteins (VLDLs), low-density lipoproteins (LDLs), and high-density lipoproteins (HDLs).

Chylomicrons form in mucosal epithelial cells of the small intestine and transport dietary lipids to adipose tissue for storage. They contain about 1–2% proteins, 85% triglycerides, 7% phospholipids, and 6–7% cholesterol, plus a small amount of fat-soluble vitamins. As chylomicrons circulate through the capillaries of adipose tissue, one of their apoproteins activates an enzyme that removes fatty acids from chylomicron triglycerides. The free fatty acids are then taken up by adipose cells for syn-

thesis and storage of triglycerides and by muscle cells for ATP production. Liver cells remove chylomicron remnants from the blood.

Very-low-density lipoproteins (VLDLs) contain about 10% proteins, 50% triglycerides, 20% phospholipids, and 20% cholesterol. VLDLs transport triglycerides made in liver cells to adipose cells for storage. As they deposit some of their triglycerides in adipose cells, VLDLs are converted to LDLs.

Low-density lipoproteins (LDLs) contain 25% proteins, 5% triglycerides, 20% phospholipids, and 50% cholesterol. They carry about 75% of the total cholesterol in blood and deliver it to cells throughout the body for use in repair of cell membranes and synthesis of steroid hormones and bile salts. After LDL particles enter body cells by means of receptor-mediated endocytosis, they are broken down, and their cholesterol is released to serve the cell's needs.

When present in excessive numbers, LDLs also deposit cholesterol in and around smooth muscle fibers in arteries, forming fatty plaques that increase the risk of coronary artery disease (see page 366). For this reason, the cholesterol in LDLs, called LDL cholesterol, is known as "bad" cholesterol. Eating a high-fat diet increases the production of VLDLs, which elevates the LDL level and increases the formation of fatty plaques.

High-density lipoproteins (HDLs), which contain 40–45% proteins, 5–10% triglycerides, 30% phospholipids, and 20%

cholesterol, remove excess cholesterol from body cells and transport it to the liver for elimination. Because HDLs prevent accumulation of cholesterol in the blood, a high HDL level is associated with decreased risk of coronary artery disease. For this reason, HDL cholesterol is known as "good" cholesterol.

Desirable levels of blood cholesterol in adults are total cholesterol under 200 mg/dL, LDL under 130 mg/dL, and HDL over 40 mg/dL. The ratio of total cholesterol to HDL cholesterol predicts the risk of developing coronary artery disease. A person with a total cholesterol of 180 mg/dL and HDL of 60 mg/dL has a risk ratio of 3. Ratios above 4 are considered undesirable; the higher the ratio, the greater the risk of developing coronary artery disease.

Fate of Lipids

Lipids, like carbohydrates, may be oxidized to produce ATP. If the body has no immediate need to use lipids in this way, they are stored in adipose tissue throughout the body and in the liver. A few lipids are used as structural molecules or to synthesize other essential substances, such as steroid hormones.

Triglyceride Storage

Triglycerides stored in adipose tissue constitute 98% of all body energy reserves. The major function of adipose tissue is to remove triglycerides from chylomicrons and VLDLs and store them until they are needed for ATP production in other parts of the body. Adipose tissue also insulates and protects the body. About 50% of stored triglycerides are deposited in the subcutaneous layer beneath the dermis of the skin.

Lipid Catabolism: Lipolysis

Muscle, liver, and adipose tissue routinely catabolize fatty acids derived from triglycerides to produce ATP. First the triglycerides must be split into glycerol and fatty acids, a process called *lipolysis* (li-POL-i-sis) (Figure 20.6). The hormones epinephrine and norepinephrine enhance lipolysis.

The glycerol and fatty acids that result from lipolysis are catabolized via different pathways. Glycerol is converted by many cells of the body to glyceraldehyde 3-phosphate. If ATP supply in a cell is high, glyceraldehyde 3-phosphate is converted into

Figure 20.6 ■ **Metabolism of lipids.** The breakdown of triglycerides into glycerol and fatty acids is called lipolysis. Glycerol may be converted to glyceraldehyde 3-phosphate, which can then be converted to glucose or enter the Krebs cycle for oxidation. Fatty acids undergo beta oxidation and enter the Krebs cycle as acetyl coenzyme A. Fatty acids also can be converted into ketone bodies (ketogenesis). Lipogenesis is the synthesis of lipids from glucose or amino acids.

Glycerol and fatty acids are catabolized in separate pathways.

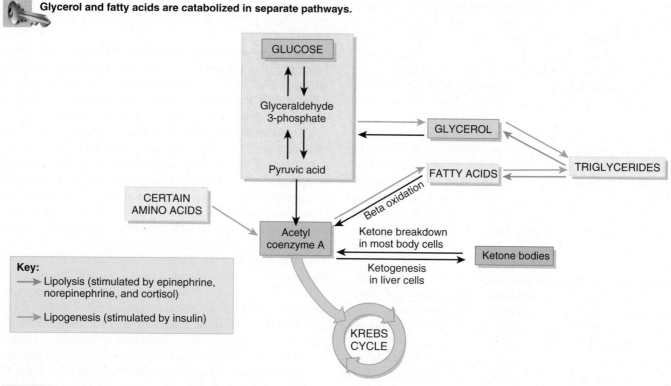

Which cells carry out beta oxidation and ketogenesis?

glucose, an example of gluconeogenesis. If ATP supply in a cell is low, glyceraldehyde 3-phosphate enters the catabolic pathway to pyruvic acid.

Fatty acid catabolism begins with several reactions, collectively called **beta oxidation,** that occur in mitochondria. Enzymes remove two carbon atoms at a time from the fatty acid and attach them to molecules of coenzyme A, forming acetyl coenzyme A (acetyl CoA). Then the acetyl CoA enters the Krebs cycle (Figure 20.6). A 16-carbon fatty acid such as palmitic acid can yield as many as 129 ATPs via oxidation, the Krebs cycle, and the electron transport chain.

As part of normal fatty acid catabolism, the liver converts some acetyl CoA molecules into substances known as **ketone bodies,** a process known as **ketogenesis** (kē′-tō-JEN-e-sis) (Figure 20.6). Ketone bodies then leave the liver to enter body cells, where they are broken down into acetyl CoA, which enters the Krebs cycle.

The level of ketone bodies in the blood normally is very low because other tissues use them for ATP production as fast as they are generated. When the concentration of ketone bodies in the blood rises above normal—a condition called **ketosis**—the ketone bodies, most of which are acids, must be buffered. If too many accumulate, blood pH falls. Prolonged ketosis can lead to **acidosis,** an abnormally low blood pH that can result in death.

Lipid Anabolism: Lipogenesis

Liver cells and adipose cells can synthesize lipids from glucose or amino acids through **lipogenesis** (Figure 20.6), which is stimulated by insulin. Lipogenesis occurs when more calories are consumed than are needed to satisfy ATP needs. Excess dietary carbohydrates, proteins, and fats all have the same fate: They are converted into triglycerides. Certain amino acids can undergo the following reactions: amino acids → acetyl CoA → fatty acids → triglycerides. The use of glucose to form lipids takes place via two pathways: (1) glucose → glyceraldehyde 3-phosphate → glycerol and (2) glucose → glyceraldehyde 3-phosphate → acetyl CoA → fatty acids. The resulting glycerol and fatty acids can undergo anabolic reactions to become stored triglycerides, or they can go through a series of anabolic reactions to produce other lipids. Examples are phospholipids, which are part of plasma membranes; lipoproteins, which are used to transport cholesterol throughout the body; thromboplastin, which is needed for blood clotting; and myelin sheaths, which speed up nerve impulse conduction.

PROTEIN METABOLISM

Objective: • **Explain how the body uses proteins.**

During digestion, proteins are broken down into amino acids. Unlike carbohydrates and triglycerides, proteins are not warehoused for future use. Instead, their amino acids are either oxidized to produce ATP or used to synthesize new proteins for growth and repair of body tissues. Excess dietary amino acids are converted into glucose (gluconeogenesis) or triglycerides (lipogenesis).

Fate of Proteins

The active transport of amino acids into body cells is stimulated by insulinlike growth factors (IGFs) and insulin. Almost immediately after digestion, amino acids are reassembled into proteins. Many proteins function as enzymes; other proteins are involved in transportation (hemoglobin) or serve as antibodies, clotting factors (fibrinogen), hormones (insulin), or contractile elements in muscle fibers (actin and myosin). Several proteins serve as structural components of the body (collagen, elastin, and keratin).

Protein Catabolism

A certain amount of protein catabolism occurs in the body each day. Proteins from worn-out cells (such as red blood cells) are broken down into amino acids. The main hormone that promotes protein catabolism is cortisol from the adrenal cortex. Some amino acids are converted into other amino acids, peptide bonds are re-formed, and new proteins are made (protein anabolism) as part of the recycling process. Other amino acids are converted by liver cells to fatty acids, ketone bodies, or glucose. See Figure 20.5 for the details of the conversion of amino acids into glucose (gluconeogenesis). Figure 20.6 shows the conversion of amino acids into fatty acids (lipogenesis) or ketone bodies (ketogenesis).

Amino acids also are oxidized to generate ATP. Before amino acids can enter the Krebs cycle, however, their amino group ($—NH_2$) must first be removed, a process called **deamination** (dē-am′-i-NĀ-shun). Deamination occurs in liver cells and produces ammonia (NH_3). Liver cells then convert the highly toxic ammonia to urea, a relatively harmless substance that is excreted in the urine.

Protein Anabolism

Protein anabolism, the formation of peptide bonds between amino acids to produce new proteins, is carried out on the ribosomes of almost every cell in the body, directed by the cells' DNA and RNA. Insulinlike growth factors, thyroid hormones (T_3 and T_4) insulin, estrogen, and testosterone stimulate protein synthesis. Because proteins are a main component of most cell structures, adequate dietary protein is especially essential during the growth years, during pregnancy, and when tissue has been damaged by disease or injury. Once dietary intake of protein is adequate, eating more protein does not increase bone or muscle mass; only a regular program of forceful, weight-bearing muscular activity accomplishes that goal.

Of the 20 amino acids in the human body, 10 are **essential amino acids:** they must be present in the diet because they cannot be synthesized in the body in adequate amounts. Humans can obtain the essential amino acids by eating plants or the

Figure 20.7 ■ **Summary of catabolism.**

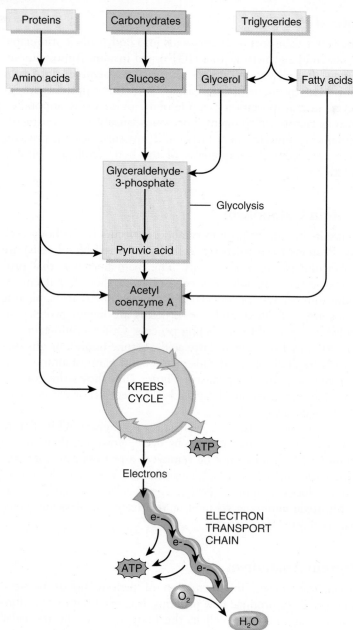

Breakdown products of carbohydrates, triglycerides, and proteins undergo glycolysis or enter the Krebs cycle at various points.

Why is acetyl coenzyme A important in metabolism?

animals that eat plants. *Nonessential amino acids* are those synthesized by the body. They are formed by the transfer of an amino group from an amino acid to pyruvic acid or to an acid in the Krebs cycle. Once the appropriate essential and nonessential amino acids are present in cells, protein synthesis occurs rapidly.

Figure 20.7 summarizes carbohydrate, lipid, and protein catabolism, and Table 20.3 summarizes the processes occurring in both catabolism and anabolism of carbohydrates, lipids, and proteins.

METABOLISM AND BODY HEAT

Objectives: • **Explain how body heat is produced and lost.**
• **Describe how body temperature is regulated.**

We now consider the relationship of foods to body heat, heat production and loss, and the regulation of body temperature.

Measuring Heat

Heat is a form of energy that can be measured as *temperature* and expressed in units called calories. A *calorie (cal)* is defined as the amount of heat required to raise the temperature of 1 gram of water 1°C. One-thousand calories is a *Calorie* (spelled with an upper case C), or *kilocalorie (kcal),* the unit used to express the caloric value of foods and to measure the body's metabolic rate. Thus, when we say that a particular food item contains 500 Calories, we are actually referring to kilocalories. Knowing the caloric value of foods is important. If we know the amount of energy the body uses for various activities, we can adjust our food intake by taking in only enough kilocalories to sustain our activities.

Production and Loss of Body Heat

The body produces more or less heat depending on the rates of metabolic reactions. Homeostasis of body temperature can be maintained only if the rate of heat production by metabolism equals the rate of heat loss from the body. Thus, it is important to understand the ways in which heat can be produced and lost.

Body Heat Production

Most of the heat produced by the body comes from oxidation of the food we eat. The rate at which this heat is produced, the *metabolic rate,* is measured in kilocalories. Because many factors affect metabolic rate, it is measured under certain standard conditions. These conditions of the body are called the *basal state,* and the measurement obtained is the *basal metabolic rate (BMR).* BMR is 1200 to 1800 Calories per day in adults, which breaks down to about 24 Calories per kilogram of body mass in adult males and 22 Calories per kilogram in adult females.

The added Calories needed to support daily activities, such as digestion and walking, range from 500 Calories for a small,

Table 20.3 / Summary of Metabolism

Process	Comment
Carbohydrate Metabolism	
Glucose catabolism	Complete oxidation of glucose (cellular respiration) is the chief pathway for ATP production in most cells. It consists of glycolysis, Krebs cycle, and electron transport chain. Complete oxidation of one molecule of glucose yields 36 or 38 molecules of ATP.
Glycolysis	Conversion of glucose into pyruvic acid, with net production of two ATP per glucose molecule; reactions do not require oxygen (anaerobic cellular respiration).
Krebs cycle	Series of oxidation–reduction reactions in which coenzymes (NAD^+ and FAD) pick up hydrogen atoms from oxidized organic acids; some ATP is produced. CO_2, H_2O, and heat are byproducts. Reactions are aerobic.
Electron transport chain	Third set of reactions in glucose catabolism is another series of oxidation–reduction reactions, in which electrons are passed from one carrier to the next, and most of the ATP is produced. Reactions are aerobic (aerobic cellular respiration).
Glucose anabolism	Some glucose is converted into glycogen (glycogenesis) for storage if not needed immediately for ATP production. Glycogen can be converted back to glucose (glycogenolysis) for use in ATP production. Gluconeogenesis is the synthesis of glucose from amino acids, glycerol, or lactic acid.
Lipid Metabolism	
Triglyceride catabolism	Triglycerides are broken down into glycerol and fatty acids. Glycerol may be converted into glucose (gluconeogenesis) or catabolized via glycolysis. Fatty acids are catabolized via beta oxidation into acetyl CoA that can enter the Krebs cycle for ATP production or be converted into ketone bodies (ketogenesis).
Triglyceride anabolism	Synthesis of triglycerides from glucose and amino acids is lipogenesis. Triglycerides are stored in adipose tissue.
Protein Metabolism	
Catabolism	Amino acids are deaminated to enter the Krebs cycle and then are oxidized. Amino acids may be converted into glucose (gluconeogenesis), fatty acids, or ketone bodies.
Anabolism	Protein synthesis is directed by DNA and uses the cell's RNA and ribosomes.

relatively sedentary person to over 3000 Calories for a person in training for Olympic-level competitions or mountain climbing. The following factors affect metabolic rate:

1. **Exercise.** During strenuous exercise the metabolic rate increases by as much as 15 to 20 times the BMR.

2. **Hormones.** Thyroid hormones (T_3 and T_4) are the main regulators of BMR, which increases as the blood levels of thyroid hormones rise. Testosterone, insulin, and human growth hormone can increase the metabolic rate by 5–15%.

3. **Nervous system.** During exercise or in a stressful situation, the sympathetic division of the autonomic nervous system releases norepinephrine, and it stimulates release of the hormones epinephrine and norepinephrine by the adrenal medulla. Both epinephrine and norepinephrine increase the metabolic rate of body cells.

4. **Body temperature.** The higher the body temperature, the higher the metabolic rate. As a result, metabolic rate is substantially increased during a fever.

5. **Ingestion of food.** The ingestion of food, especially proteins, can raise metabolic rate by 10–20%.

6. **Age.** The metabolic rate of a child, in relation to its size, is about double that of an elderly person due to the high rates of growth-related reactions in children.

7. **Other factors.** Other factors that affect metabolic rate are gender (lower in females, except during pregnancy and lactation), climate (lower in tropical regions), sleep (lower), and malnutrition (lower).

Body Heat Loss

Because body heat is continuously produced by metabolic reactions, heat must also be removed continuously or body temperature would rise steadily. The principal routes of heat loss are radiation, conduction, convection, and evaporation.

RADIATION   The transfer of heat in the form of infrared rays between a warmer object and a cooler one without physical contact is known as *radiation*. Your body loses heat by radiating more infrared waves than it absorbs from cooler objects. In a room at 21°C (70°F), about 60% of heat loss occurs by means of radiation in a resting person. If surrounding objects are warmer than you are, you absorb more heat by radiation than you lose.

CONDUCTION   The heat exchange that occurs between two materials that are in direct contact is termed *conduction*. At rest, about 3% of body heat is lost by conduction to solid materials in contact with the body, such as a your chair, clothing, and jewelry. Heat can also be gained by conduction, for example, while soaking in a hot tub.

*A*thletes spend hours a day training for their sports. Many physiological changes occur as a result of all this training, including an increased ability to produce ATP for muscle contraction. These improvements are specific to the metabolic pathways that are used during the training. Athletes design their training programs to challenge the ATP production system or systems most vital to their sports.

Metabolic Power

Some sports require a short burst of power, high energy output that lasts only a few seconds. Such events include the 100-meter sprint, the shot put, the discus throw, and the 25-meter swim. Muscle contraction for these events is supplied primarily by existing ATP and creatine phosphate. (Recall from Chapter 8 that creatine phosphate can donate a phosphate group to ADP to restore ATP.)

Athletes improve their ability to generate power by practicing their events, over and over. Power lifting—forcefully lifting very heavy weights—also challenges the muscles to produce more ATP faster. In response to such training, the concentration of enzymes required for these ATP-production pathways as well as the levels of ATP and creatine phosphate increase in trained muscles.

Going for the Burn

Many athletic events require well-trained glycolytic pathways. Anaerobic glycolysis, combined with ATP and creatine phosphate, provides most of the energy for high-intensity exercise lasting up to 90 seconds, such as a 400-meter run or a 100-meter swim. Many sports, such as basketball, soccer, and tennis, require bursts of high-ATP production by anaerobic glycolysis interspersed with somewhat lower energy output.

Athletes train the anaerobic glycolytic pathway by exercising at high intensities for periods of a minute or longer. Interval training includes both high-intensity work and periods of lower-intensity work or rest. In response to high-intensity exercise, the concentration of enzymes required for anaerobic glycolysis increases.

Going the Distance

Most athletic events, including all events lasting longer than a few minutes, require aerobic cellular respiration. Athletes improve these ATP-production pathways by exercising at moderate to vigorous intensities for extended periods of time, with or without high-intensity intervals. Aerobic training increases the size and number of mitochondria as well as the concentration of enzymes required for ATP production by aerobic pathways.

▶ *Think It Over*

▶ Do you think that the burning sensation that can develop with high-intensity exercise, producing ATP via anaerobic glycolysis, represents the breakdown of adipose tissue?

CONVECTION The transfer of heat by the movement of a gas or a liquid between areas of different temperature is known as *convection.* The contact of air or water with your body results in heat transfer by both conduction and convection. When cool air makes contact with the body, it becomes warmed and is carried away by convection currents. The faster the air moves—for example, by a breeze or a fan—the faster the rate of convection. At rest, about 15% of body heat is lost to the air by conduction and convection.

EVAPORATION The conversion of a liquid to a vapor is called *evaporation.* Under typical resting conditions, about 22% of heat loss occurs through evaporation of water—a daily loss of about 300 mL in exhaled air and 400 mL from the skin surface.

The higher the relative humidity, the lower the rate of evaporation. At 100% humidity, heat is gained by condensation of water on the skin surface as fast as heat is lost by evaporation. Evaporation provides the main defense against overheating during exercise. Under extreme conditions, a maximum of about 3 liters of sweat can be produced each hour, removing more than 1700 kcal of heat if all of it evaporates. Sweat that drips off the body rather than evaporating removes very little heat.

Homeostasis of Body Temperature Regulation

If the amount of heat production equals the amount of heat loss, you maintain a nearly constant body temperature near 37°C

Figure 20.8 ■ **Negative feedback mechanisms that increase heat production.**

When stimulated, the heat-promoting center in the hypothalamus raises body temperature.

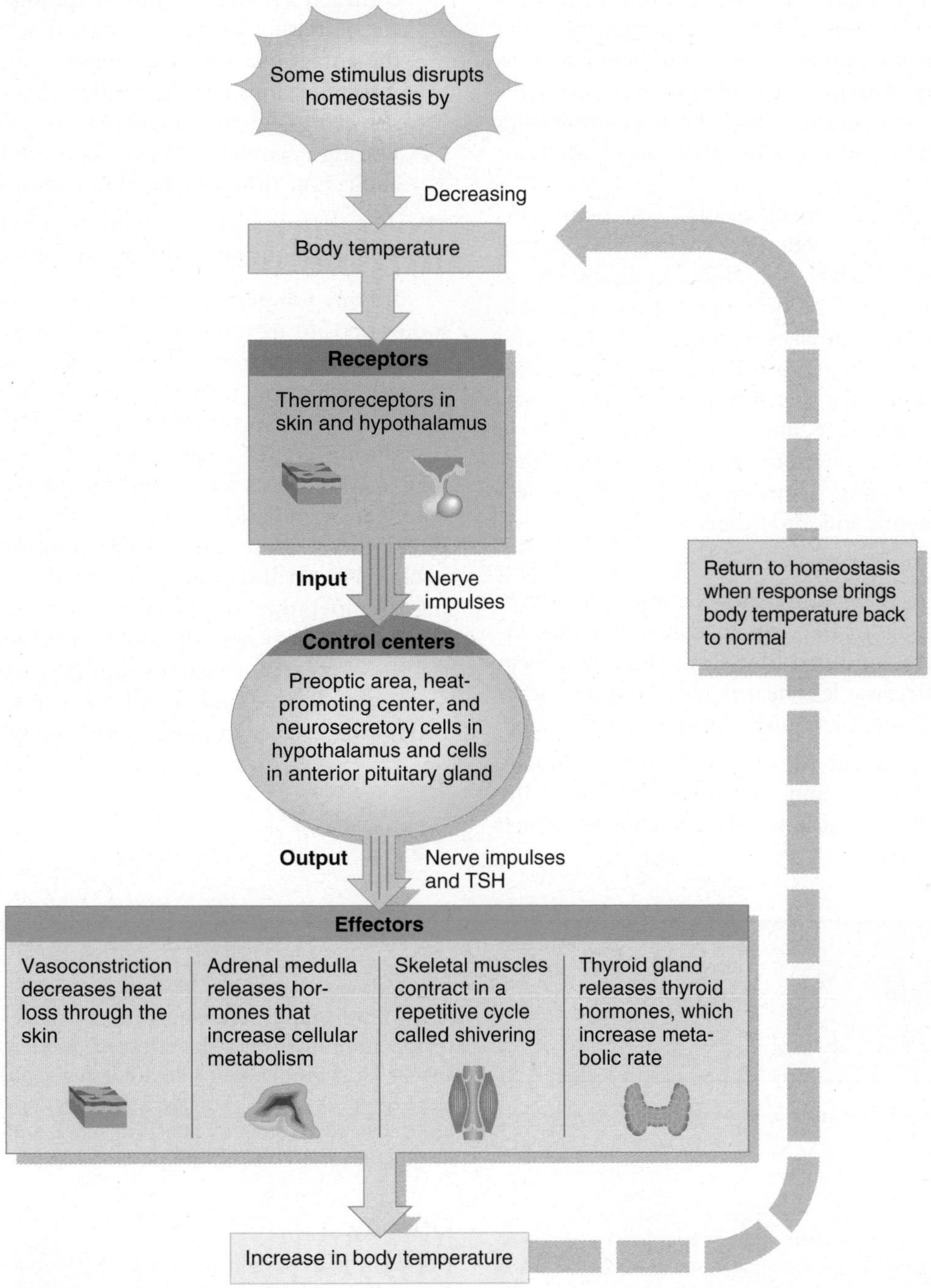

Some stimulus disrupts homeostasis by

Decreasing

Body temperature

Receptors

Thermoreceptors in skin and hypothalamus

Input | Nerve impulses

Control centers

Preoptic area, heat-promoting center, and neurosecretory cells in hypothalamus and cells in anterior pituitary gland

Output | Nerve impulses and TSH

Effectors

| Vasoconstriction decreases heat loss through the skin | Adrenal medulla releases hormones that increase cellular metabolism | Skeletal muscles contract in a repetitive cycle called shivering | Thyroid gland releases thyroid hormones, which increase metabolic rate |

Increase in body temperature

Return to homeostasis when response brings body temperature back to normal

What factors can increase your metabolic rate and thus increase your rate of heat production?

(98.6°F). If your heat-producing mechanisms generate more heat than is lost by your heat-losing mechanisms, your body temperature rises. For example, strenuous exercise and some infections elevate body temperature. If you lose heat faster than you produce it, your body temperature falls. Immersion in icy water, certain diseases such as hypothyroidism, and some drugs such as alcohol and antidepressants can cause body temperature to fall. An elevated temperature may destroy body proteins, whereas a depressed temperature may cause cardiac arrhythmias; both can lead to death.

The balance between heat production and heat loss is a function of a group of neurons in the hypothalamus called the **preoptic area.** Neurons of the preoptic area generate more nerve impulses when blood temperature increases and fewer impulses when blood temperature decreases. Nerve impulses from the preoptic area can activate two other areas of the hypothalamus known as the **heat-losing center**, which initiates a series of responses that lower body temperature, and the **heat-promoting center**, which initiates a series of responses that raise body temperature.

If body temperature falls, mechanisms that help conserve heat and increase heat production act by means of several negative feedback loops to raise the body temperature to normal (Figure 20.8 on page 505). Thermoreceptors in the skin and hypothalamus send nerve impulses to the preoptic area and the heat-promoting center in the hypothalamus, and to neurosecretory cells in the hypothalamus that produce a releasing hormone called thyrotropin-releasing hormone (TRH). TRH in turn stimulates the anterior pituitary gland to release thyrotropin, or thyroid-stimulating hormone (TSH). Nerve impulses from the hypothalamus and TSH then activate several effectors:

- The heat-promoting center stimulates sympathetic nerves that cause blood vessels of the skin to constrict (vasoconstriction). The decrease of blood flow slows the rate of heat loss from the skin. Because less heat is lost, body temperature increases even if the metabolic rate remains the same.

- Sympathetic nerves leading to the adrenal medulla stimulate the release of epinephrine and norepinephrine into the blood. These hormones increase cellular metabolism, which increases heat production.

- The heat-promoting center stimulates parts of the brain that increase muscle tone and hence heat production. As muscle tone increases in one muscle (the agonist), the small contractions stretch muscle spindles in its antagonist muscle, initiating a stretch reflex. The resulting contraction in the antagonist stretches muscle spindles in the agonist, and it too develops a stretch reflex. This repetitive cycle—called **shivering**—greatly increases the rate of heat production. During maximal shivering, body heat production can rise to about four times the basal rate in just a few minutes.

- The thyroid gland responds to TSH by releasing more thyroid hormones into the blood, increasing the metabolic rate.

If body temperature rises above normal, a negative feedback loop opposite to the one depicted in Figure 20.8 goes into action. The higher temperature of the blood stimulates the preoptic area, which in turn activates the heat-losing center and inhibits the heat-promoting center. Nerve impulses from the heat-losing center cause dilation of blood vessels in the skin. The skin becomes warm, and the excess heat is lost to the environment by radiation and conduction as an increased volume of blood flows from the warmer interior of the body into the cooler skin. At the same time, metabolic rate decreases, and the high temperature of the blood stimulates sweat glands of the skin by means of hypothalamic activation of sympathetic nerves. As the water in sweat evaporates from the surface of the skin, the skin is cooled. All these responses counteract heat-promoting effects and help return body temperature to normal.

COMMON DISORDERS

Fever

A *fever* is an elevation of body temperature that results from a resetting of the hypothalamic thermostat. The most common causes of fever are viral or bacterial infections and bacterial toxins; other causes are ovulation, excessive secretion of thyroid hormones, tumors, and reactions to vaccines. When phagocytes ingest certain bacteria, they are stimulated to secrete a *pyrogen* (PĪ-rō-gen; *pyro-* = fire; *-gen* = produce), a fever-producing substance. The pyrogen circulates to the hypothalamus and induces neurons of the preoptic area to secrete prostaglandins. Some prostaglandins can reset the hypothalamic thermostat at a higher temperature, and temperature-regulating reflex mechanisms then act to bring body temperature up to this new setting. *Antipyretics* are agents that relieve or reduce fever. Examples include aspirin, acetaminophen (Tylenol), and ibuprofen (Advil), all of which reduce fever by inhibiting synthesis of certain prostaglandins.

Although death results if core temperature rises above 44–46°C (112–114°F), up to a point, fever is beneficial. For example, a higher temperature intensifies the effect of interferon and the phagocytic activities of macrophages while hindering replication of some pathogens. Because fever increases heart rate, infection-fighting white blood cells are delivered to sites of infection more rapidly. In addition, antibody production and T cell proliferation increase.

Obesity

Obesity is body weight more than 20% above a desirable standard due to an excessive accumulation of adipose tissue; it affects one-third of the adult population in the United States. (An athlete may be *overweight* due to a higher-than-normal amount of muscle tissue without being obese.) Even moderate obesity is hazardous to health; it is implicated as a risk factor in cardiovascular disease, hypertension, pulmonary disease, non–insulin-dependent diabetes mellitus, arthritis, certain cancers (breast, uterus, and colon), varicose veins, and gallbladder disease.

In a few cases, obesity may result from trauma to or tumors in the food-regulating centers in the hypothalamus. In most cases of obesity, no specific cause can be identified. Contributing factors include genetic factors, eating habits taught early in life, overeating to relieve tension, and social customs.

MEDICAL TERMINOLOGY AND CONDITIONS

Heat cramps Cramps that result from profuse sweating. The salt lost in sweat causes painful contractions of muscles; such cramps tend to occur in muscles used while working but do not appear until the person relaxes once the work is done. Drinking salted liquids usually leads to rapid improvement.

Heatstroke (sunstroke) A severe and often fatal disorder caused by exposure to high temperatures, especially when the relative humidity is high, which makes it difficult for the body to lose heat. Blood flow to the skin is decreased, perspiration is greatly reduced, and body temperature rises sharply because of failure of the hypothalamic thermostat. Body temperature may reach 43°C (110°F). Treatment, which must be undertaken immediately, consists of cooling the body by immersing the victim in cool water and by administering fluids and electrolytes.

Kwashiorkor (kwash′-ē-OR-kor) A disorder in which protein intake is deficient despite normal or nearly normal caloric intake, characterized by edema of the abdomen, enlarged liver, decreased blood pressure, low pulse rate, lower than normal body temperature, and sometimes mental retardation. Because the main protein in corn lacks two essential amino acids, which are needed for growth and tissue repair, many African children whose diet consists largely of cornmeal develop kwashiorkor.

Malnutrition (*mal-* = bad) An imbalance of total caloric intake or intake of specific nutrients, which can be either inadequate or excessive.

Phenylketonuria (fen′-il-kē′-tō-NOO-rē-a) or ***PKU*** A genetic error of protein metabolism characterized by elevated blood levels of the amino acid phenylalanine. Most children with phenylketonuria have a mutation in the gene that codes for the enzyme phenylalanine hydroxylase; this enzyme is needed to convert phenylalanine into the amino acid tyrosine, which can enter the Krebs cycle. PKU can be treated if caught early by restricting dietary intake of phenylalanine.

◼ STUDY OUTLINE

Introduction (p. 490)

1. Our only source of energy for performing biological work is the food we eat. Food also provides essential substances that we cannot synthesize.

2. Most food molecules absorbed by the gastrointestinal tract are used to supply energy for life processes, serve as building blocks during synthesis of complex molecules, or are stored for future use.

Nutrients (p. 490)

1. Nutrients include carbohydrates, lipids, proteins, water, minerals, and vitamins.

2. Some minerals known to perform essential functions include calcium, phosphorus, potassium, sodium, chloride, magnesium, iron, manganese, copper, and zinc. Their functions are summarized in Table 20.1 on page 491.

3. Vitamins are organic nutrients that maintain growth and normal metabolism. Many function as coenzymes.

4. Fat-soluble vitamins are absorbed with fats and include vitamins A, D, E, and K; water-soluble vitamins are absorbed with water and include the B vitamins and vitamin C.

5. The functions of the principal vitamins and their deficiency disorders are summarized in Table 20.2 on page 492.

6. Most teens and adults need between 1600 and 2800 Calories per day.

7. Nutrition experts suggest dietary calories be 50–60% from carbohydrates, 30% or less from fats, and 12–15% from proteins.

8. The food guide pyramid indicates how many servings of five food groups are recommended each day to attain the number of calories and variety of nutrients needed for wellness.

Overview of Metabolism (p. 495)

1. Metabolism refers to all chemical reactions of the body and has two phases: catabolism and anabolism.

2. Anabolism consists of a series of synthesis reactions whereby small molecules are built up into larger ones that form the body's structural and functional components.

3. Catabolism refers to decomposition reactions that provide energy.

4. Anabolic reactions require energy, which is supplied by catabolic reactions.

5. Metabolic reactions are catalyzed by enzymes, proteins that speed up chemical reactions without being changed.

6. Oxidation is the removal of electrons and hydrogen ions from a molecule; oxidation releases energy.

7. Reduction is the addition of electrons and hydrogen ions to a molecule; oxidation and reduction reactions are always coupled.

Carbohydrate Metabolism (p. 496)

1. During digestion, polysaccharides and disaccharides are converted to monosaccharides (glucose, fructose, and galactose), which are transported to the liver.

2. Carbohydrate metabolism is primarily concerned with glucose metabolism.

3. Some glucose is oxidized by cells to provide energy. Glucose moves into cells by facilitated diffusion, which is stimulated by insulin.

4. Excess glucose can be stored by the liver and skeletal muscles as glycogen or converted to fat.

5. Glucose oxidation is also called cellular respiration. The complete oxidation of glucose to CO_2 and H_2O involves glycolysis, the Krebs cycle, and the electron transport chain.

6. Glycolysis is also called anaerobic respiration because it occurs without oxygen. During glycolysis, which occurs in the cytosol, glucose is broken down into two molecules of pyruvic acid. Glycolysis yields a net of two molecules of ATP and two molecules of NADH.

7. When oxygen is in short supply, pyruvic acid is converted to lactic acid; under aerobic conditions, pyruvic acid enters the Krebs cycle.

8. Pyruvic acid is prepared for entrance into the Krebs cycle by conversion to a two-carbon acetyl group followed by the addition of coenzyme A to form acetyl coenzyme A.

9. The Krebs cycle occurs in mitochondria. A series of oxidations and reductions of various organic acids take place; coenzymes (NAD^+ and FAD) are reduced. The potential energy originally contained in glucose and pyruvic acid is now stored in the reduced coenzymes (NADH and $FADH_2$).

10. The electron transport chain is a series of oxidation–reduction reactions that occur in mitochondria in which the energy in the coenzymes is liberated and transferred to ATP.

11. The complete oxidation of glucose can be represented as follows: 1 Glucose + 6 Oxygen $\rightarrow$ 36 or 38 ATP + 6 Carbon dioxide + 6 Water.

12. The conversion of glucose to glycogen for storage in the liver and skeletal muscle is called glycogenesis. It occurs extensively in liver and skeletal muscle fibers (cells) and is stimulated by insulin. The body can store about 500 g of glycogen.

13. The conversion of glycogen back to glucose is called glycogenolysis; it occurs between meals.

14. Gluconeogenesis is the conversion of fat and protein molecules into glucose.

Lipid Metabolism (p. 498)

1. Lipoproteins transport lipids in the bloodstream. Types of lipoproteins include chylomicrons, which carry dietary lipids to adipose tissue; very-low-density lipoproteins (VLDLs), which carry triglycerides from the liver to adipose tissue; low-density lipoproteins (LDLs), which deliver cholesterol to body cells; and high-density lipoproteins (HDLs), which remove excess cholesterol from body cells and transport it to the liver for elimination.

2. During digestion, triglycerides are ultimately broken down into fatty acids and glycerol.

3. Some triglycerides may be oxidized to produce ATP, whereas others are stored in adipose tissue.

4. Triglycerides must be split into fatty acids and glycerol before they can be catabolized. Glycerol can be converted into glucose by conversion into glyceraldehyde 3-phosphate. Fatty acids are catabolized through beta oxidation, yielding acetyl coenzyme A, which can enter the Krebs cycle.

5. Other lipids are used as structural molecules or to synthesize essential molecules.

6. The formation of ketone bodies by the liver is a normal phase of fatty acid catabolism, but an excess of ketone bodies, called ketosis, may cause acidosis.

7. The conversion of glucose or amino acids into lipids, called lipogenesis, is stimulated by insulin.

Protein Metabolism (p. 501)

1. Amino acids, under the influence of insulinlike growth factors and insulin, enter body cells by means of active transport.

2. Inside cells, amino acids are reassembled into proteins that function as enzymes, hormones, structural elements, and so forth; stored as fat or glycogen; or used for energy.

3. Before amino acids can be catabolized, they must deaminated. Liver cells convert the resulting ammonia to urea, which is excreted in urine.

4. Amino acids may also be converted into glucose, fatty acids, and ketone bodies.

5. Protein synthesis is stimulated by insulinlike growth factors, thyroid hormones, insulin, estrogen, and testosterone.

6. Protein synthesis is directed by DNA and RNA and carried out on ribosomes.

7. Table 20.3 on page 503 summarizes carbohydrate, lipid, and protein metabolism.

Metabolism and Body Heat (p. 502)

1. A calorie, is the amount of energy required to raise the temperature of 1 gram of water 1°C.

2. The Calorie is the unit of heat used to express the caloric value of foods and to measure the body's metabolic rate. One Calorie equals 1000 calories, or 1 kilocalorie.

3. Most body heat is a result of oxidation of the food we eat. The rate at which this heat is produced is known as the metabolic rate and is affected by exercise, hormones, the nervous system, body temperature, ingestion of food, age, gender, climate, sleep, and malnutrition.

4. Measurement of the metabolic rate under basal conditions is called the basal metabolic rate (BMR).

5. Mechanisms of heat loss are radiation, conduction, convection, and evaporation.

6. Radiation is the transfer of heat from a warmer object to a cooler object without physical contact.

7. Conduction is the transfer of heat between two objects in contact with each other.

8. Convection is the transfer of heat by a liquid or gas between areas of different temperatures.

9. Evaporation is the conversion of a liquid to a vapor; in the process, heat is lost.

10. The hypothalamic thermostat is in the preoptic area. Neurons there stimulate the heat-losing and heat-promoting centers, also in the hypothalamus. A normal body temperature is maintained by a delicate balance between heat-producing and heat-losing mechanisms.

11. Responses that produce or retain heat when body temperature falls are vasoconstriction; release of epinephrine, norepinephrine, and thyroid hormones; and shivering.

12. Responses that increase heat loss when body temperature increases include vasodilation, decreased metabolic rate, and evaporation of sweat.

SELF-QUIZ

1. Creating a protein from amino acids is an example of
   a. oxidation   **b.** anabolism   **c.** gluconeogenesis
   d. catabolism   **e.** cellular respiration

2. Free radicals
   a. are a type of provitamin   **b.** are essential amino acids
   c. can cause damage to cellular structures   **d.** help regulate enzymatic reactions   **e.** are a form of energy

3. Which of the following statements about vitamins is NOT true?
   a. Most vitamins are synthesized by the body cells.
   b. Vitamins can act as coenzymes.   **c.** Vitamin K is produced by bacteria in the GI tract.   **d.** Fat-soluble vitamins may be stored in the liver.   **e.** Excess water-soluble vitamins are excreted in urine.

4. Match the following:
   _____ **a.** precursor for vitamin A
   _____ **b.** form in which lipids are transported in the blood plasma
   _____ **c.** needed to convert ADP to ATP
   _____ **d.** derived from vitamin B_2 (riboflavin)

   A. lipoproteins
   B. FAD
   C. beta-carotene
   D. magnesium

5. Body temperature is controlled by the
   a. pons   **b.** thyroid gland   **c.** hypothalamus
   d. adrenal medulla   **e.** autonomic nervous system

6. The removal of an amino group ($-NH_2$) from amino acids entering the Krebs cycle is known as
   a. deamination   **b.** convection   **c.** ketogenesis
   d. beta oxidation   **e.** aerobic respiration

7. If your diet is low in carbohydrates, which compound(s) does your body begin to catabolize next for ATP production?
   a. vitamins   **b.** lipids   **c.** minerals   **d.** cholesterol
   e. amino acids

8. Cellular respiration includes the following steps in order:
   a. Krebs cycle, glycolysis, electron transport chain   **b.** Krebs cycle, electron transport chain, glycolysis   **c.** glycolysis, electron transport chain, Krebs cycle   **d.** electron transport chain, Krebs cycle, glycolysis   **e.** glycolysis, Krebs cycle, electron transport chain

9. Which of the following is most often used to synthesize ATP?
   a. galactose   **b.** triglycerides   **c.** amino acids
   d. glucose   **e.** glycerol

10. How does glucose enter the cytosol of cells?
    a. facilitated diffusion   **b.** simple diffusion   **c.** active transport   **d.** osmosis   **e.** electron transport

11. In a person at rest in a room at 21°C (70°F), loss of body heat is greatest by means of
    a. radiation   **b.** conduction   **c.** convection
    d. evaporation   **e.** vasodilation

12. Glycolysis
    a. requires the presence of oxygen   **b.** produces two ATP molecules per glucose molecule   **c.** takes place in mitochondria
    d. is also known as the Krebs cycle   **e.** is the conversion of glucose to glycogen

13. Which of the following statements is NOT true?
    a. Triglycerides are stored in adipose tissue.   **b.** Chylomicrons enter the blood by way of lacteals in the intestines.   **c.** Most of the body's cholesterol is carried in low-density lipoproteins.
    d. Lipids can be stored in the liver.   **e.** High-density lipoproteins contribute to the formation of fatty plaques.

14. Which of the following equations summarizes the complete catabolism of a molecule of glucose?
    a. Glucose + 6 Water $\rightarrow$ 36 or 38 ATP + 6 CO_2 + 6 O_2
    b. Glucose + 6 O_2 $\rightarrow$ 36 or 38 ATP + 6 CO_2 + 6 Water
    c. Glucose + ATP $\rightarrow$ 31 or 38 CO_2 + 6 Water
    d. Glucose + Pyruvic acid $\rightarrow$ 36 or 38 ATP + 6 O_2
    e. Glucose + Citric acid $\rightarrow$ 31 or 38 ATP + 6 CO_2

15. Those amino acids that cannot by synthesized by the body and must be obtained from the diet are known as
    a. coenzymes   **b.** ketones   **c.** essential amino acids
    d. nonessential amino acids   **e.** polypeptides

16. Which of the following would NOT increase the metabolic rate?
    a. increased levels of thyroid hormones   **b.** epinephrine
    c. old age   **d.** fever   **e.** exercise

17. A protein that speeds up the rate of a chemical reaction without being altered is
    a. a nutrient   **b.** an antioxidant   **c.** an oxidizer
    d. an enzyme   **e.** an apoprotein

18. The process by which glucose is formed from amino acids is
    a. gluconeogenesis   **b.** glycogenesis   **c.** glycogenolysis
    d. lipogenesis   **e.** glycolysis

19. All of the following can contribute to an increase in body temperature EXCEPT
    a. shivering   **b.** release of thyroid hormones   **c.** sympathetic stimulation of the adrenal medulla   **d.** vasodilation of blood vessels in the skin   **e.** activation of the heat-promoting center in the hypothalamus

20. Match the following:
    _____ **a.** removal of electrons and hydrogen ions from a molecule
    _____ **b.** the oxidation of glucose
    _____ **c.** building simple molecules into more complex ones
    _____ **d.** addition of electrons and hydrogen ions to a molecule
    _____ **e.** the breakdown of organic compounds

    A. catabolism
    B. anabolism
    C. oxidation
    D. cellular respiration
    E. reduction

CRITICAL THINKING APPLICATIONS

1. Some people will try anything to loose weight, including a high protein–zero carbohydrate diet. Intake of carbohydrates in this diet are severely restricted while eating large amounts of protein and fat is encouraged. What effect would this diet have on the body's metabolism?

2. It's noon on a hot summer day, the sun is directly overhead, and a group of sunbathers roasts on the beach. What mechanism causes their body temperature to increase? Several of the sunbathers jump into the cool water. What mechanisms decrease their body temperature?

3. Shannon is a morning person but her roommate Darla is not. In fact, Shannon teases Darla for being a classic example of "BMR" (barely mentally responsive) during her 8 AM class. What does BMR really mean? How is metabolism measured?

4. Rob takes a multivitamin tablet and extra vitamin C each day. What are the functions of vitamin C in the body? Does Rob need to worry about his body storing toxic levels of vitamin C?

ANSWERS TO FIGURE QUESTIONS

20.1 The food groups that contain cholesterol and most of the saturated fatty acids in the diet are the milk, yogurt, and cheese group and the meat, etc. group.

20.2 The formation of digestive enzymes is anabolic.

20.3 The reduced coenzymes from the Krebs cycle later yield ATP in the electron transport chain.

20.4 Complete oxidation of glucose yields 36 or 38 ATP, depending on which carriers are used in the electron transport chain.

20.5 Liver cells can carry out gluconeogenesis and glycogenesis.

20.6 Liver cells carry out beta oxidation and ketogenesis.

20.7 Formation of acetyl CoA is a transition step that prepares pyruvic acid to enter the Krebs cycle.

20.8 Exercise, the sympathetic nervous system, hormones (epinephrine, norepinephrine, thyroid hormones, testosterone, human growth hormone), elevated body temperature, and ingestion of food increase metabolic rate.

Chapter **21**

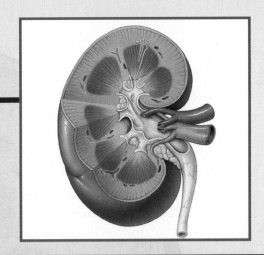

The Urinary System

■ Student Learning Objectives

■ A Look Ahead

The **urinary system** consists of two kidneys, two ureters, one urinary bladder, and one urethra (Figure 21.1). After the kidneys filter blood and return most of the water and many of the solutes to the bloodstream, the remaining water and solutes are excreted from the kidneys as **urine.** Urine passes through the ureters and is stored in the urinary bladder until it is expelled from the body through the urethra. **Nephrology** (nef-ROL-ō-jē; *nephr-* = kidney; *-ology* = study of) is the scientific study of the anatomy, physiology, and pathology of the kidney. The branch of medicine that deals with the male and female urinary systems and the male reproductive system is **urology** (yoo-ROL-ō-jē; *uro-* = urine).

OVERVIEW OF KIDNEY FUNCTIONS

Objective: • **List the functions of the kidneys.**

The kidneys do the major work of the urinary system. The other parts of the system are primarily passageways and storage areas. Functions of the kidneys include the following:

- **Regulation of blood ionic composition.** The kidneys help regulate the blood levels of several ions, most importantly sodium ions (Na^+), potassium ions (K^+), calcium ions (Ca^{2+}), chloride ions (Cl^-), and phosphate ions (HPO_4^{2-}).

- **Regulation of blood volume.** The kidneys adjust the volume of blood in the body by conserving or eliminating water.

- **Regulation of blood pressure.** Besides adjusting blood volume, the kidneys help regulate blood pressure by secreting the enzyme renin, which activates the renin–angiotensin–aldosterone pathway (see Figure 13.15 on page 319), and by adjusting blood flow into and out of the kidneys.

- **Regulation of blood pH.** The kidneys excrete H^+ into the urine and conserve bicarbonate ions (HCO_3^-), an important buffer of H^+.

- **Red blood cell production.** The kidneys release the hormone erythropoietin, which stimulates red blood cell production (see Figure 14.5 on page 339).

- **Vitamin D synthesis.** Together with the skin and the liver, the kidneys synthesize calcitriol, the active form of vitamin D.

- **Excretion of wastes and foreign substances.** By forming urine, the kidneys help excrete wastes—substances that no

Figure 21.1 ■ **Organs of the female urinary system in relation to surrounding structures.**

Urine formed by the kidneys passes first into the ureters, then to the urinary bladder for storage, and finally through the urethra for elimination from the body.

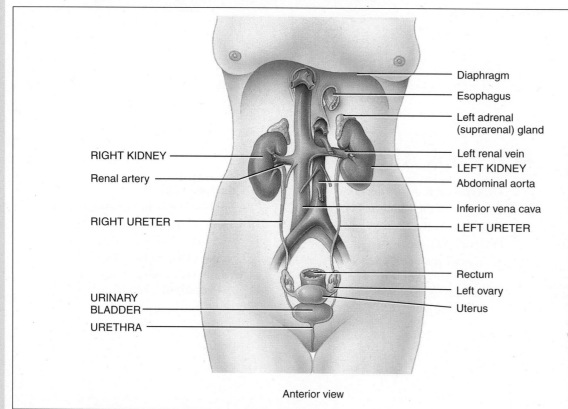

RIGHT KIDNEY
Renal artery

RIGHT URETER

URINARY
BLADDER
URETHRA

Diaphragm
Esophagus
Left adrenal
(suprarenal) gland
Left renal vein
LEFT KIDNEY
Abdominal aorta
Inferior vena cava
LEFT URETER

Rectum
Left ovary
Uterus

Anterior view

Functions of the Urinary System

1. The kidneys regulate blood volume and composition, help regulate blood pressure and pH, participate in red blood cell production and vitamin D synthesis, and excrete wastes and foreign substances.
2. The ureters transport urine from the kidneys to the urinary bladder.
3. The urinary bladder stores urine and expels urine into the urethra.
4. The urethra discharges urine from the body.

 Which organ of the urinary system does most of the work?

longer have useful functions in the body. Some wastes result from metabolic reactions in the body, such as ammonia and urea from the breakdown of amino acids; bilirubin from the breakdown of hemoglobin; and creatinine from the breakdown of creatine phosphate in muscle fibers. Foreign substances, such as drugs and environmental toxins, may also be excreted in urine.

STRUCTURE OF THE KIDNEYS

Objective: • **Describe the structure and blood supply of the kidneys.**

The paired *kidneys* are reddish organs shaped like kidney beans. They lie just superior to the waist against the back wall of the abdominal cavity. Because they are behind the peritoneal lining of the abdominal cavity, their position is described as *retroperitoneal* (re′-trō-per′-i-tō-NĒ-al; *retro-* = behind). The ureters

and adrenal glands are also retroperitoneal. The kidneys are partially protected by the 11th and 12th pairs of ribs. The right kidney is slightly lower than the left because the liver occupies a large area above the kidney on the right side.

External Anatomy

An adult kidney is 10 to 12 cm (4 to 5 in.) long, 5 to 7 cm (2 to 3 in.) wide, and 3 cm (1 in.) thick—about the size of a bar of bath soap. Near the center of the concave border is an indentation called the *renal hilus* (RĒ-nal HĪ-lus; *ren-* = kidney), through which the ureter leaves the kidney and blood vessels, lymphatic vessels, and nerves enter and exit.

Each kidney is enclosed in a *renal capsule,* a smooth, transparent, fibrous membrane that helps maintain the shape of the kidney and serves as a barrier against trauma (Figure 21.2). A mass of fatty tissue surrounds the renal capsule and cushions the kidney. Along with a thin layer of dense irregular connective tissue, the fatty tissue anchors the kidney to the posterior abdominal wall.

Figure 21.2 ■ **Structure of the kidney.**

A renal capsule covers the kidney; internally, the two main regions are the renal cortex and the renal medula which contains the renal pyramids.

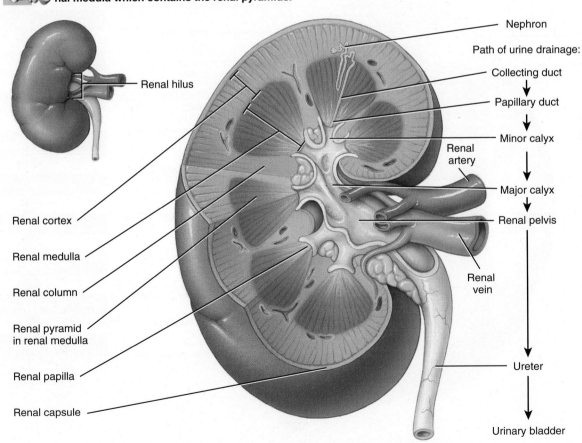

Frontal section of right kidney

 Through which structure do blood vessels, lymphatic vessels, and nerves enter the kidneys?

Internal Anatomy

The outer part of the kidney is a reddish area called the ***renal cortex*** (*cortex* = rind or back) and the inner, striated (striped) part is a red-brown region called the ***renal medulla*** (*medulla* = inner portion) (Figure 21.2). Within the renal medulla are 8 to 18 striated, triangular structures, the ***renal pyramids***. The striations are straight tubules and blood vessels. The bases of the re-

nal pyramids face the renal cortex, and their tips, called ***renal papillae***, face the center of the kidney. The portions of the renal cortex that extend between renal pyramids are called ***renal columns***.

The functional parts of each kidney are approximately 1 million microscopic units called nephrons (described shortly). Urine formed by the nephrons drains into collecting ducts and then into larger ***papillary ducts***. A large, funnel-shaped basin or

Figure 21.3 ■ **Blood supply of the right kidney.**

The renal arteries deliver about 25% of the resting cardiac output to the kidneys.

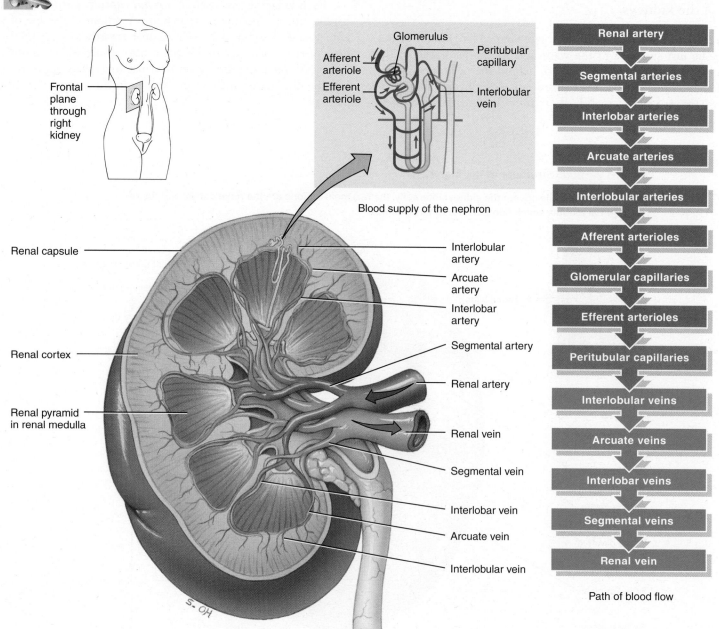

Blood supply of the nephron

Frontal section of right kidney

Path of blood flow

cavity called the **renal pelvis** (*pelv-* = basin) receives the urine. The edge of the renal pelvis has cuplike extensions called **major** and **minor calyces** (KĀ-li-sēz = cups; singular is **calyx**). Urine drains from papillary ducts into a minor calyx and from there through a major calyx into the renal pelvis, which connects to a ureter.

Blood Supply

The nephrons are largely responsible for removing wastes from the blood and regulating its fluid and electrolyte content. Thus, they are abundantly supplied with blood vessels. About 25% of the resting cardiac output—1200 mL of blood per minute—flows into the kidneys via the right and left **renal arteries** (Figure 21.3).

Close to the renal hilus, the renal artery divides into several branches that enter the kidney. The arteries divide into smaller and smaller vessels (*segmental, interlobar, arcuate, interlobular arteries*) that eventually terminate as **afferent arterioles** (*af-* = toward; *-ferre* = to carry).

One afferent arteriole is distributed to each nephron, where the arteriole divides into a tangled capillary network called the **glomerulus** (glō-MER-yoo-lus = little ball; plural is **glomeruli**). The capillaries of the glomerulus permit the passage of some substances but restrict the passage of others. The resulting fluid, **glomerular filtrate,** consists of water and smaller solutes such as glucose, vitamins, amino acids, very small proteins, nitrogenous wastes, and ions. Once blood is filtered by the glomeruli, they unite to form an **efferent arteriole** (*ef-* = out) that drains blood out of the glomerulus. Because the efferent arteriole is smaller in diameter than the afferent arteriole, it helps raise the blood pressure in the glomerulus. Glomerular capillaries are unique because they are positioned between two arterioles, rather than between an arteriole and a venule.

Upon leaving the glomerulus, each efferent arteriole divides to form a network of capillaries, called the **peritubular capillaries** (*peri-* = around), which reclaim useful substances from glomerular filtrate and eliminate other substances into it. These capillaries eventually reunite to form *peritubular veins*, which merge into *interlobular, arcuate, interlobar,* and *segmental veins.* Ultimately, all the veins deliver their blood to the **renal vein.**

Nephron

The functional unit of the kidney is the **nephron** (NEF-ron; Figure 21.4). Each nephron consists of two portions: a **renal corpuscle** (KOR-pus-ul = tiny body) where blood plasma is filtered and a **renal tubule** into which the filtered fluid passes. The glomerular filtrate is modified as it moves through the renal tubules: Wastes and excess substances are added, and useful materials are returned to the blood.

The renal corpuscles of the nephron lie in the renal cortex. Each renal corpuscle has two parts: the **glomerulus** (capillary network) and the **glomerular (Bowman's) capsule,** a double-

Figure 21.4 ■ **Renal corpuscle.**

A renal corpuscle consists of a glomerular (Bowman's) capsule and a glomerulus.

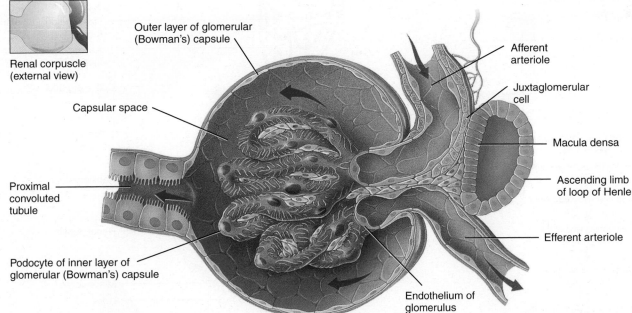

Renal corpuscle (external view)

Outer layer of glomerular (Bowman's) capsule

Capsular space

Proximal convoluted tubule

Podocyte of inner layer of glomerular (Bowman's) capsule

Afferent arteriole

Juxtaglomerular cell

Macula densa

Ascending limb of loop of Henle

Efferent arteriole

Endothelium of glomerulus

Renal corpuscle (internal view)

 Which structures make up the filtration membrane?

walled epithelial cup that surrounds the glomerulus. Think of the glomerulus as a fist punched into a limp balloon (the glomerular capsule) until the fist is covered by two layers of balloon with a space in between, the ***capsular space.*** The cells that make up the inner wall of the glomerular capsule, called ***podocytes,*** adhere closely to the endothelial cells of the glomerulus. Together, they form a ***filtration membrane*** that permits the passage of water and solutes from the blood into the capsular space.

From the renal corpuscle, glomerular filtrate passes into the renal tubule. The first portion of the renal tubule is called the

proximal convoluted tubule (PCT) (Figure 21.5). The wall of the proximal convoluted tubule consists of cuboidal epithelial cells with numerous microvilli. From the proximal convoluted tubules, water and solutes pass into the ***loop of Henle.*** Water and solutes pass from the first portion of this loop, called the ***descending limb of the loop of Henle,*** into the second portion, called the ***ascending limb of the loop of Henle.*** Back in the cortex of the kidney, the water and solutes pass into a ***distal convoluted tubule.*** The distal convoluted tubules of several nephrons empty into a single ***collecting duct.*** Collecting ducts then unite and converge until eventually, deep in the renal pyramids, there

Figure 21.5 ■ **Parts of a nephron and its blood supply.**

Nephrons are the functional units of the kidneys.

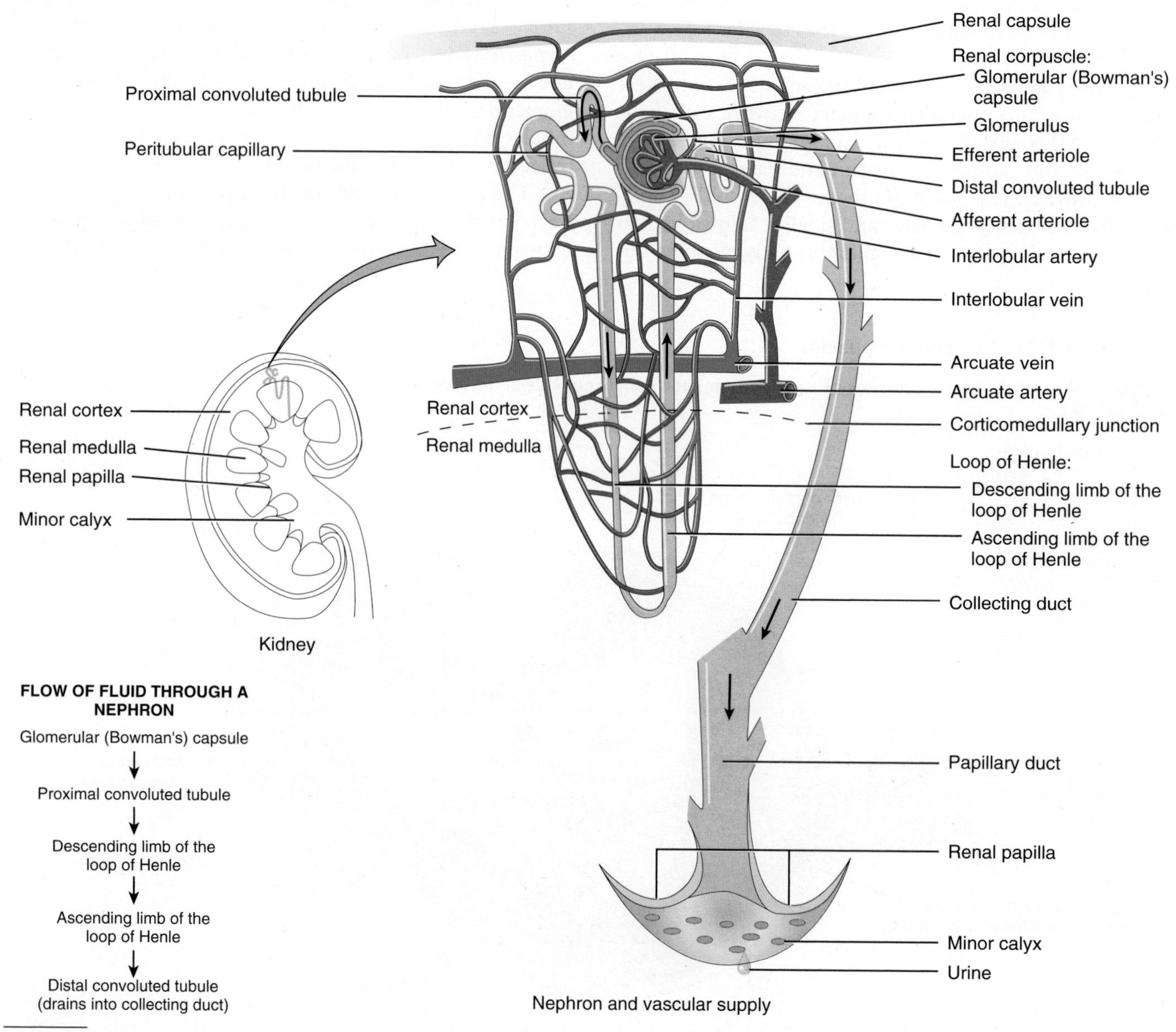

Renal capsule
Renal corpuscle:
  Glomerular (Bowman's) capsule
  Glomerulus
Efferent arteriole
Distal convoluted tubule
Afferent arteriole
Interlobular artery
Interlobular vein
Arcuate vein
Arcuate artery
Corticomedullary junction
Loop of Henle:
  Descending limb of the loop of Henle
  Ascending limb of the loop of Henle
Collecting duct
Papillary duct
Renal papilla
Minor calyx
Urine

Proximal convoluted tubule
Peritubular capillary

Renal cortex
Renal medulla

Renal cortex
Renal medulla
Renal papilla
Minor calyx

Kidney

Nephron and vascular supply

FLOW OF FLUID THROUGH A NEPHRON

Glomerular (Bowman's) capsule
↓
Proximal convoluted tubule
↓
Descending limb of the loop of Henle
↓
Ascending limb of the loop of Henle
↓
Distal convoluted tubule
(drains into collecting duct)

A water molecule has just entered the proximal convoluted tubule of the nephron. What parts of the nephron will it travel through (in order) to reach the renal pelvis as a component of urine?

are only several hundred large *papillary ducts,* which drain into the minor calyces at the renal papilla. The collecting ducts and papillary ducts extend from the renal cortex through the renal medulla to the renal pelvis. On average, there are about 30 papillary ducts per renal papilla.

Juxtaglomerular Apparatus

In each nephron, the final portion of the ascending limb of the loop of Henle makes contact with the afferent arteriole serving that renal corpuscle (see Figure 21.4). Because the ascending limb cells in this region are columnar and crowded together, they are known as the *macula densa* (*macula* = spot; *densa* = dense). The wall of the afferent arteriole adjacent to the macula densa contains modified smooth muscle fibers called *juxta-glomerular (JG) cells.* The JG cells and the macula densa make up the *juxtaglomerular apparatus (JGA),* which helps regulate renal blood pressure (described shortly).

FUNCTIONS OF THE KIDNEYS

Objective: • **Describe how the kidneys filter the blood and regulate its volume, chemical composition, and pH.**

While producing urine, nephrons and collecting ducts perform three basic processes—glomerular filtration, tubular reabsorption and tubular secretion (Figure 21.6):

1 **Glomerular filtration.** In the first step of urine production, water and most solutes in plasma pass from blood through the filtration membrane and into the glomerular capsule, which empties into the renal tubule.

2 **Tubular reabsorption.** As filtered fluid flows through the renal tubule and collecting duct, most filtered water and many useful solutes are absorbed by the cells lining these passageways and returned to the blood in the peritubular capillaries.

3 **Tubular secretion.** As fluid flows through the tubule and collecting duct, the tubule and duct cells secrete additional materials, such as wastes, drugs, and excess ions, into the fluid. Tubular secretion moves substances from the blood in peritubular capillaries into the fluid within the tubules.

Water and solutes in the fluid that drains into the renal pelvis remain in the urine and are excreted (eliminated from the body). The three activities of nephrons can be compared to a recycling center. Garbage trucks dump refuse into an input hopper, where the smaller refuse passes onto a conveyor belt (glomerular filtration of plasma). As the conveyor belt carries the garbage along, workers remove useful items, such as aluminum cans, plastics, and glass containers (reabsorption). Other workers place additional garbage left at the center and larger items onto the conveyor belt (secretion). At the end of the belt, all remaining garbage falls into a truck for transport to the landfill (excretion of wastes in urine). Each of the three steps that contribute to urine formation is described in more detail in the following sections.

Figure 21.6 ■ **Overview of functions of a nephron.** Excreted substances remain in the urine and subsequently leave the body.

Glomerular filtration occurs in the renal corpuscle; tubular reabsorption and tubular secretion occur all along the renal tubule and collecting duct.

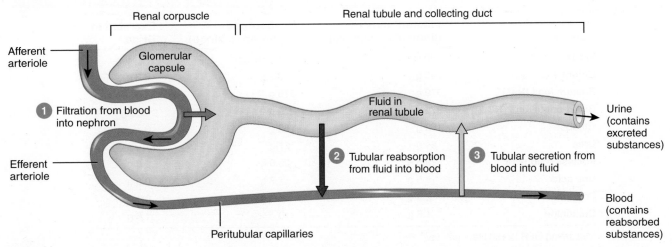

What is the name of the fluid that is formed as a result of glomerular filtration, tubular reabsorption, and tubular secretion?

Glomerular Filtration

Production of Glomerular Filtrate

The first step in the regulation of blood composition and volume by the kidneys is **glomerular filtration**. Filtration is the forcing of fluids and dissolved substances smaller than a certain size through a membrane by pressure. In the kidneys, filtration begins at the filtration membrane when blood enters the glomerulus. Blood pressure in the glomerulus forces water and most solutes from the plasma through the membrane. Blood cells and most proteins are too large to pass through. Table 21.1 compares the substances that are filtered, reabsorbed, and excreted in urine per day. Although the values shown are typical, they vary considerably according to diet.

Glomerular Filtration Rate

The amount of filtrate that forms in both kidneys every minute is called the **glomerular filtration rate (GFR)**. In adults, the GFR is about 105 mL/min in females and 125 mL/min in males. Thus, the daily volume of glomerular filtrate is about 150 liters in females and 180 liters in males. It is very important for the kidneys to maintain a constant GFR. If the GFR is too high, needed substances pass so quickly through the renal tubules that they are unable to be reabsorbed and pass out of the body as part of urine. On the other hand, if the GFR is too low, nearly all the filtrate is reabsorbed, and waste products are not adequately eliminated.

GFR is directly related to several factors. For example, changes in the size of afferent and efferent arterioles can change GFR. Constriction of the afferent arterioles decreases blood flow into the glomerulus, which decreases GFR. Constriction of the efferent arteriole, which takes blood out of the glomerulus, increases GFR. Obstructions, such as a kidney stone that blocks the ureter or an enlarged prostate gland that blocks the urethra in a male, can also decrease GFR.

Regulation of GFR

GFR is regulated by three main mechanisms: (1) renal autoregulation, (2) hormonal regulation, and (3) neural regulation.

RENAL AUTOREGULATION The ability of the kidneys to maintain a constant renal blood flow and GFR despite normal, everyday changes in systemic arterial pressure is called **renal autoregulation** (auto- = self). It consists of two mechanisms: the myogenic mechanism and tubuloglomerular feedback.

1. **Myogenic mechanism.** The stretching of smooth muscle is responsible for the **myogenic mechanism** (myo- = muscle; -genic = producing). GFR rises as systemic blood pressure increases because renal blood flow increases. However, the elevated blood pressure also stretches the walls of the afferent arterioles. In response, smooth muscle fibers in the walls of the afferent arterioles contract, the arterioles narrow, and renal blood flow decreases, which reduces GFR to its previous level. Conversely, when arterial blood pressure drops, the smooth muscle cells are stretched less and thus relax. The afferent arterioles dilate, renal blood flow increases, and GFR increases.

2. **Tubuloglomerular feedback.** In **tubuloglomerular feedback,** part of the renal tubule—the macula densa—provides feedback to the glomerulus. When GFR is above normal due to elevated systemic blood pressure (Figure 21.7), filtered fluid flows so rapidly through the renal tubules that they absorb less fluid. Macula densa cells of the juxtaglomerular apparatus (JGA) detect the increased load of Na^+, Cl^-, and water and stimulate the release of a chemical that causes afferent arterioles to constrict. Thus, less blood flows into the glomerular capillaries, and GFR decreases.

HORMONAL REGULATION Two hormones help regulate GFR. **Angiotensin II** is a very potent vasoconstrictor that reduces renal blood flow, thereby decreasing GFR. **Atrial natriuretic**

Table 21.1 / Substances Filtered, Reabsorbed, and Excreted in Urine Per Day

Substance	Filtered* (enters glomerular capsule)	Reabsorbed (returned to blood)	Excreted in Urine
Water	180 liters	178–179 liters	1–2 liters
Chloride ions (Cl^-)	640 g	633.7 g	6.3 g
Sodium ions (Na^+)	579 g	575 g	4 g
Bicarbonate ions (HCO_3^-)	275 g	275 g	0.03 g
Glucose	162 g	162 g	0
Urea	54 g	27 g	27 g[†]
Potassium ions (K^+)	29.6 g	29.6 g	2.0 g[‡]
Uric acid	8.5 g	7.7 g	0.8 g
Proteins	2.0 g	1.9 g	0.1 g
Creatinine	1.6 g	0	1.6 g

*Assuming GFR is 180 liters per day.

[†]In addition to being filtered and reabsorbed, urea is secreted.

[‡]After virtually all filtered K^+ is reabsorbed in the convoluted tubules and loop of Henle, a variable amount of K^+ is secreted in the collecting duct.

Figure 21.7 ■ Negative feedback regulation of glomerular filtration rate (GFR) by the juxtaglomerular apparatus (JGA).

The ability of the kidneys to maintain a constant GFR despite changes in systemic blood pressure is called renal autoregulation.

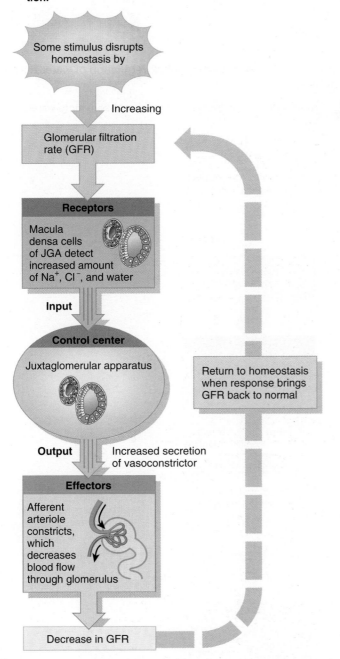

Some stimulus disrupts homeostasis by

Increasing

Glomerular filtration rate (GFR)

Receptors

Macula densa cells of JGA detect increased amount of Na⁺, Cl⁻, and water

Input

Control center

Juxtaglomerular apparatus

Return to homeostasis when response brings GFR back to normal

Output Increased secretion of vasoconstrictor

Effectors

Afferent arteriole constricts, which decreases blood flow through glomerulus

Decrease in GFR

What happens to GFR if blood pressure in the kidneys increases?

peptide (ANP) is secreted by cells in the atria of the heart. When blood volume increases, the atria are stretched, and ANP is secreted, which increases glomerular filtration rate.

NEURAL REGULATION Like most blood vessels of the body, those of the kidneys are supplied by sympathetic ANS fibers that cause vasoconstriction. At rest, sympathetic stimulation is moderately low, the afferent and efferent arterioles are dilated, and GFR is controlled by renal autoregulation. With moderate sympathetic stimulation, both afferent and efferent arterioles constrict to the same degree. Blood flow into and out of the glomerulus is restricted to the same extent, which slightly decreases GFR. With greater sympathetic stimulation, however, as occurs during exercise or hemorrhage, the afferent arterioles are constricted more than the efferent arterioles. As a result, blood flow into glomerular capillaries is greatly decreased, and GFR drops. This lowering of renal blood flow reduces urine output, which helps conserve blood volume and permits greater blood flow to other body tissues.

Tubular Reabsorption

About 99% of the glomerular filtrate, called *tubular fluid* once it enters the proximal convoluted tubule, is reabsorbed into the blood as it passes through the renal tubules and ducts. Thus, only about 1% of the filtrate actually leaves the body, about 1 to 2 liters per day. Tubular reabsorption occurs by osmosis, diffusion, and active transport.

Epithelial cells all along the renal tubule and collecting duct carry out reabsorption, but those of the proximal convoluted tubule reabsorb the bulk of the fluid. Reabsorption varies with the body's needs. Materials that are reabsorbed include water; glucose; amino acids; urea; and sodium, potassium, calcium, chloride, and bicarbonate ions (see Table 21.1). Tubular reabsorption allows the body to retain most of its nutrients.

Most of the energy used by the kidneys is for reabsorption of sodium ions (Na^+). The concentration of Na^+ in glomerular filtrate is high, the same as in blood plasma. As usual, the cytosol of the cells of the proximal convoluted tubule (PCT) has a much lower concentration of Na^+ that is maintained by active transport of Na^+ out of the PCT cells by sodium pumps (Figure 21.8). As a result of this difference in Na^+ concentration, Na^+ moves from the tubular fluid into PCT cells. Most of the Na^+ cross the membrane by means of *symporters* that simultaneously carry Na^+ and another substance into the cytosol. Sodium pumps located on the sides and at the base of the PCT cells actively transport Na^+ into interstitial fluid, and it then diffuses into a peritubular capillary (Figure 21.8). The substances that enter PCT cells along with Na^+ from the tubular fluid include glucose, amino acids, lactic acid, and bicarbonate ions (HCO_3^-). From the cytosol, these substances can move through the sides and base of the PCT cell by facilitated diffusion or another transport process.

Na^+ reabsorption also affects water reabsorption. The movement of Na^+ into peritubular capillaries decreases the solute concentration of the tubular fluid but increases the solute concentration in the peritubular capillaries, making the peritubular blood hypertonic with respect to the tubular fluid. As a result, water moves by osmosis into peritubular capillaries.

Water reabsorption is also regulated by ***antidiuretic hormone (ADH)*** in response to the body's state of hydration. Osmoreceptors in the hypothalamus detect a change in the concentration of water in the blood of as little as 1% (Figure 21.9). Nerve impulses from the osmoreceptors stimulate secretion of more ADH by hypothalamic cells and its release by the posterior pituitary into the blood. ADH causes cells in the last portion of the distal convoluted tubules and throughout the collecting ducts to become more permeable to water. As water reabsorption increases, the blood's water concentration rises, returning to normal. In a condition known as *diabetes insipidus*, ADH secretion is inadequate or the ADH receptors are faulty, and a person may excrete up to 20 liters of very dilute urine daily.

Diuretics are substances that slow reabsorption of water by the kidneys and thereby cause *diuresis*, an elevated urine flow rate. Naturally occurring diuretics include *caffeine* in coffee, tea, and cola sodas, which inhibits Na^+ reabsorption, and *alcohol* in beer, wine, and mixed drinks, which inhibits secretion of ADH.

Figure 21.8 ■ **Reabsorption of sodium ions (Na^+).**

Most of the energy used in the kidney is for Na^+ reabsorption.

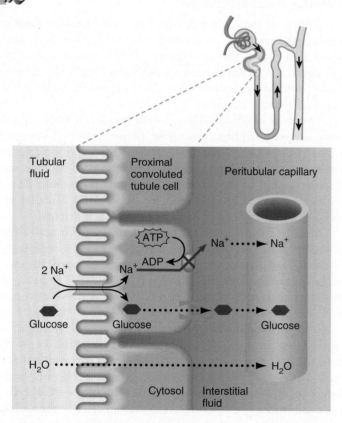

Key:

▱	Na^+– glucose symporter
⊃⊂	Glucose facilitated diffusion transporter
····▶	Diffusion of solutes and osmosis of water
⊗	Sodium pump (active transport)

How does filtered glucose move from tubular fluid into a PCT cell and how does it move from the cell into interstitial fluid?

Figure 21.9 ■ **Negative feedback regulation of water reabsorption by ADH.**

When ADH level is high, the kidneys reabsorb more water.

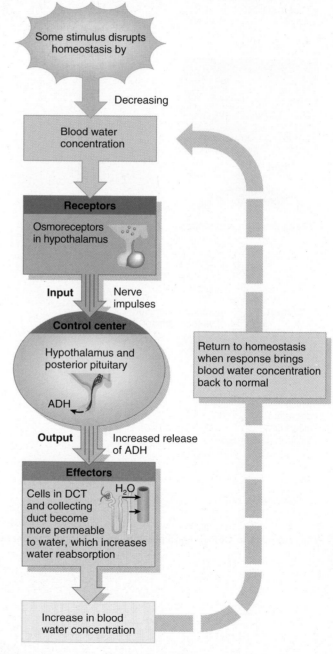

Would blood ADH level be higher or lower than normal in a person who has just completed a 5-km run without drinking any water?

Another consequence of Na^+ movement from cells of the PCT into peritubular capillaries is the reabsorption of chloride ions (Cl^-). Reabsorption of Cl^- occurs by diffusion due to a difference in the charges of Na^+ and Cl^-. The negatively charged Cl^- follows the positively charged Na^+ from the tubular fluid into PCT cells and then into peritubular capillaries. As water is reabsorbed by osmosis (following Na^+ and Cl^-), other ions become more concentrated in the remaining tubular fluid and diffuse down their concentration gradients into the peritubular capillaries. Overall, the proximal convoluted tubule Na^+ transporters cause reabsorption from the filtrate of 100% of the glucose, amino acids, and lactic acid; 80–90% of the HCO_3^-; 65% of the water, Na^+, and K^+; 50% of the Cl^-; and a variable amount of Ca^{2+}, Mg^{2+}, and HPO_4^{2-}.

Tubular Secretion

The third process involved in the regulation of blood composition and volume by the kidneys is ***tubular secretion,*** the addition of materials to the tubular fluid from the blood. Tubular secretion occurs in epithelial cells all along the renal tubules and collecting ducts. The secreted substances include K^+; hydrogen ions (H^+); ammonium ions (NH_4^+); creatinine; urea; some hormones; and certain drugs, such as penicillin. Secretion of excess K^+ for elimination in the urine is very important. If the level of K^+ in plasma rises above normal, serious disturbances in cardiac rhythm may develop; at higher levels, cardiac arrest may occur.

In addition to ridding the body of certain materials, tubular secretion helps control blood pH. A normal blood pH of 7.35 to 7.45 is maintained even though the typical high-protein diet in North America provides more acid-producing foods than alkali-producing foods. To eliminate acids, the cells of the renal tubules secrete H^+ into the tubular fluid, which helps maintain the pH of blood in the normal range. As a result of H^+ secretion, urine normally has a pH of 6.

Table 21.2 gives a summary of glomerular filtration, tubular reabsorption, and tubular secretion.

DIALYSIS

Objective: • **Explain the principle of dialysis and describe its importance.**

If a person's kidneys are so impaired by disease or injury that they are unable to function adequately, then blood must be cleansed artificially by ***dialysis,*** the removal of certain solutes from a solution through use of a selectively permeable membrane. One method of dialysis is the artificial kidney machine, which performs ***hemodialysis*** (*hemo-* = blood); it directly filters the patient's blood. As blood flows through tubing made of selectively permeable dialysis membrane, waste products diffuse from the blood in the tube into the dialysis solution surrounding the tubing (Figure 21.10). The machine continuously replaces the dialysis solution to maintain favorable concentration gradients for diffusion of wastes and excess solutes out of the blood.

Process	Region and Substances Involved
Table 21.2 / Summary of Filtration, Reabsorption, and Secretion	
Glomerular filtration	Renal corpuscle (filtration membrane): Blood in glomerular capillaries under pressure results in the formation of glomerular filtrate that contains water, Na^+, glucose, amino acids, lactic acid, water-soluble vitamins, K^+, Cl^-, HCO_3^-, urea, uric acid, creatinine, and other solutes in the same concentration as in blood plasma.
Tubular reabsorption	Renal tubule (mostly proximal convoluted tubule) and collecting duct: Materials absorbed from the tubular fluid include water, glucose, amino acids, urea, Na^+, K^+, Cl^-, and HCO_3^-.
Tubular secretion	Renal tubule and collecting duct: Materials secreted into the tubular fluid include K^+, H^+, NH_4^+, creatinine, urea, certain hormones, and some drugs (for example, penicillin).

Figure 21.10 ■ **Operation of an artificial kidney machine.** The blood route is indicated in red and blue. The route of the dialysis solution (dialysate) is indicated in gold.

Dialysis is the removal of solutes from solution through a selectively permeable membrane.

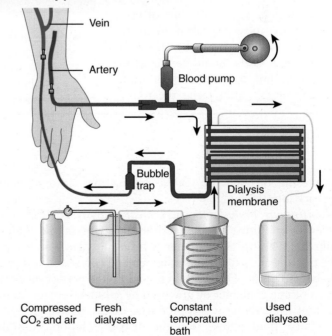

 What happens to plasma proteins during hemodialysis?

*U*rinary tract infections (UTIs) are the most common bacterial infections for women, and the second most common illness (after colds) for them. About 10–15% of women develop UTIs several times a month. Men get UTIs, too, but much less frequently. The female's shorter urethra allows bacteria to enter the urinary bladder more easily. In addition, the urethral and anal openings are closer in females. Most first-time UTIs are caused by *Escherichia coli (E. coli)* bacteria that have migrated to the urethra from the anal area. *E. coli* bacteria are necessary for proper digestion and are welcome in the intestinal tract, but they cause much pain and suffering if they infect the urinary system.

Infection Prevention

Personal hygiene is the first line of prevention. Care must be taken to avoid transporting bacteria from the anal area to the urethra. Girls should be taught to wipe from front to back and to wash hands thoroughly after using the toilet.

When bathing, women and girls should wash from front to back as well.

Menstrual blood provides an excellent growth medium for bacteria. Sanitary napkins and tampons should be changed often. Some women find that switching from tampons to napkins or from napkins to tampons reduces the frequency of UTIs. Deodorant tampons and napkins and superabsorbent tampons can increase irritation.

People who are prone to UTIs should drink at least 2 to 2.5 liters of fluid daily. Drinking cranberry and blueberry juice may help to decrease bacterial growth in the urinary bladder. Voiding frequently, every 2 to 3 hours, helps prevent recurrent UTIs because it expels bacteria and eliminates the urine needed for their growth.

Partners in Health

Sexual intercourse is frequently associated with the onset of UTIs in women. Women who find that sex brings on UTIs learn to develop stringent personal hygiene and teach their partners as well. Women should drink plenty of water be-

fore and after sex and urinate as soon afterward as possible. This flushes out bacteria that may have entered the urethra.

At times a woman's partner may be the source of bacterial transmission. When UTIs continue to recur, he should be tested for asymptomatic urethritis, which is the term for any bacterial infection of the urethra other than gonorrhea. Sometimes treating the partner with antibiotics cures both parties.

▶ *Think It Over*

▶ One of the basic tenets of the wellness philosophy is that the health-care system works best when patients work as partners with their providers to understand, treat, and prevent illness. Explain why treatment of recurrent UTIs is a good illustration of this belief.

After passing through the dialysis tubing, the cleansed blood flows back into the body.

Continuous ambulatory peritoneal dialysis (CAPD) uses the peritoneal lining of the peritoneal cavity as the dialysis membrane. The tip of a catheter is surgically placed in the patient's peritoneal cavity of the abdomen and connected to a sterile dialysis solution. The solution flows into the peritoneal cavity from its plastic container by gravity. The solution remains in the cavity until metabolic waste products, excess electrolytes, and extracellular fluid diffuse into the dialysis solution from the blood. The solution is then drained from the cavity by gravity into a sterile bag and discarded.

URINE TRANSPORTATION, STORAGE, AND ELIMINATION

Objective: • **Describe the structure and functions of the ureters, urinary bladder, and urethra.**

As you learned earlier in the chapter, urine produced by the nephrons drains through papillary ducts into the minor calyces, which join to become major calyces that unite to form the renal pelvis (see Figure 21.2). From the renal pelvis, urine first drains into the ureters and then into the urinary bladder; urine is then discharged from the body through the urethra (see Figure 21.1).

Ureters

Each of the two ***ureters*** (yoo-RĒ-ters) transports urine from the renal pelvis of one of the kidneys to the urinary bladder (see Figure 21.1). The ureters pass under the urinary bladder for several centimeters, causing the bladder to compress the ureters and thus prevent backflow of urine when pressure builds up in the bladder during urination. If this physiological valve is not operating, cystitis (urinary bladder inflammation) may develop into a kidney infection.

The wall of the ureter consists of three layers. The inner layer is a mucous membrane. Mucus secreted by the mucosa prevents the cells from coming in contact with urine. This barrier is important because the solute concentration and pH of urine differ drastically from the cytosol of the ureter wall cells. The middle layer consists of smooth muscle. Urine is transported primarily by peristaltic contractions of the smooth muscle, but the fluid pressure of the urine and gravity also contribute. The outer layer consists of areolar connective tissue containing blood vessels, lymphatic vessels, and nerves.

Urinary Bladder

The ***urinary bladder*** is a hollow muscular organ situated in the pelvic cavity behind the pubic symphysis. In males, it is directly in front of the rectum. In females, it is in front of the vagina and below the uterus. ...
toneum. The shape of ...
much urine it contains. When ...
loon. It becomes spherical when slig...
volume increases, becomes pear-shaped a...
dominal cavity. The average capacity of the u...
700 to 800 mL, about 3/4 quart.

Structure

At the base of the urinary bladder is a small area shaped like an upside-down triangle, the ***trigone*** (TRĪ-gōn = triangle; Figure 21.11). The opening into the urethra (*internal urethral orifice*) is in the apex of this triangle. At the two points of the base of the triangle, the ureters drain into the urinary bladder via the *ureteral openings*.

The mucous membrane of the urinary bladder contains transitional epithelium that can stretch, a marked advantage for an organ that stores a variable volume of fluid. Rugae (folds in the mucosa) are also present. The muscular layer of the bladder wall consists of three layers of smooth muscle called the ***detrusor muscle*** (de-TROO-ser = to push down). Around the opening to the urethra is an ***internal urethral sphincter*** composed of smooth muscle. The opening and closing of the internal urethral sphincter is involuntary. Below the internal sphincter is the ***external urethral sphincter,*** which is composed of skeletal

Figure 21.11 ■ **Ureters, urinary bladder, and urethra (female).**

🔑 **Urine is stored in the urinary bladder until it is expelled by an act called micturition.**

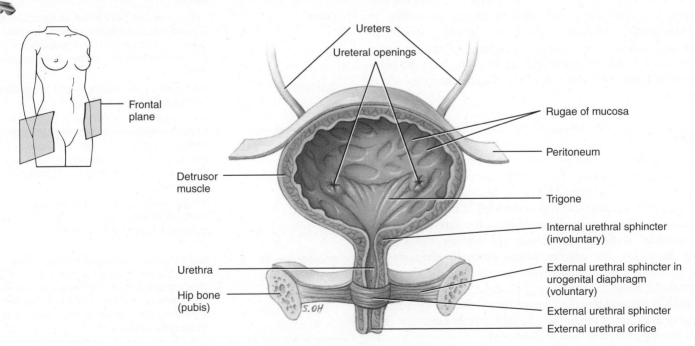

Anterior view of frontal section

 What is a lack of voluntary control over micturition called?

...under voluntary control. The peritoneum, which ...ers the superior surface of the bladder, forms a serous outer coat; the rest of the bladder has a fibrous outer covering.

Functions

The urinary bladder stores urine prior to its elimination and then expels urine into the urethra by an act called **micturition** (mik'-too-RI-shun = to urinate), commonly known as urination or voiding. Micturition requires a combination of involuntary and voluntary muscle contractions. When the volume of urine in the bladder exceeds 200 to 400 mL, pressure within the bladder increases considerably, and stretch receptors in its wall transmit nerve impulses into the spinal cord. These impulses propagate to the lower part of the spinal cord and trigger a reflex called the **micturition reflex.** In this reflex arc, parasympathetic impulses from the spinal cord cause *contraction* of the detrusor muscle and *relaxation* of the internal urethral sphincter muscle. Simultaneously, the spinal cord inhibits somatic motor neurons, causing relaxation of skeletal muscle in the external urethral sphincter. Upon contraction of the bladder wall and relaxation of the sphincters, urination takes place. Bladder filling causes a sensation of fullness that initiates a conscious desire to urinate before the micturition reflex actually occurs. Although emptying of the urinary bladder is a reflex, in early childhood we learn to initiate it and stop it voluntarily. Through learned control of the external urethral sphincter muscle and certain muscles of the pelvic floor, the cerebral cortex can initiate micturition or delay it for a limited period of time.

A lack of voluntary control over micturition is referred to as **incontinence.** Under about two years of age, incontinence is normal because neurons to the external urethral sphincter muscle are not completely developed. Infants void whenever the urinary bladder is sufficiently distended to trigger the reflex. In **stress incontinence,** physical stresses that increase abdominal pressure, such as coughing, sneezing, laughing, exercising, pregnancy, or simply walking, cause leakage of urine from the bladder. Smokers have twice the risk of developing incontinence as nonsmokers.

Urethra

The **urethra,** the terminal portion of the urinary system, is a small tube leading from the floor of the urinary bladder to the exterior of the body (Figure 21.11). In females, it lies directly behind the pubic symphysis and is embedded in the front wall of the vagina. The opening of the urethra to the exterior, the **external urethral orifice,** lies between the clitoris and vaginal opening. In males, the urethra passes vertically through the prostate gland, the urogenital diaphragm, and finally the penis (see Figures 23.1 and 23.6 on pages 547 and 552).

In both males and females, the urethra is the passageway for discharging urine from the body. The male urethra also serves as the duct through which semen is ejaculated.

URINALYSIS

Objective: • **List the normal and abnormal components of urine.**

An analysis of the volume and physical, chemical, and microscopic properties of urine, called a **urinalysis,** tells us much about the state of the body. The principal physical characteristics of urine are summarized in Table 21.3.

The volume of urine eliminated per day in a normal adult is 1 to 2 liters (about 1 to 2 qt). Urine volume is influenced by fluid intake, blood pressure, blood osmotic pressure, diet, temperature, diuretics, mental state, and general health. Water accounts for about 95% of the total volume of urine. The remaining 5% consists of solutes derived from cellular metabolism and outside sources such as drugs. Typical solutes normally present in urine are described in Table 21.4.

If the body's chemical processes are not operating efficiently, substances not normally present, such as glucose or proteins, may appear in the urine, or normal constituents may appear in abnormal amounts. Table 21.5 lists several abnormal constituents in urine that may be detected as part of a urinalysis.

Table 21.3 / Physical Characteristics of Normal Urine	
Characteristic	**Description**
Volume	One to two liters in 24 hours, but varies considerably.
Color	Yellow or amber, but varies with urine concentration and diet. Color is due to urobilin, a yellow pigment. Concentrated urine is darker in color. Diet, medications, and certain diseases affect color.
Turbidity	Transparent when freshly voided, but becomes turbid (cloudy) after a while.
Odor	Mildly aromatic but becomes ammonia-like after a time. Some people inherit the ability to form methylmercaptan from digested asparagus, which gives urine a characteristic odor. Urine of diabetics has a fruity odor due to presence of ketone bodies.
pH	Ranges between 4.6 and 8.0; average 6.0; varies considerably with diet. High-protein diets increase acidity; vegetarian diets increase alkalinity.
Specific gravity	Specific gravity (density) is the ratio of the weight of a volume of a substance to the weight of an equal volume of distilled water. Urine specific gravity ranges from 1.001 to 1.035. The higher the concentration of solutes, the higher the specific gravity.

Table 21.4 / Principal Solutes in Normal Urine

Solute	Comments
Organic Molecules	
Urea	Composes 60–90% of all nitrogen-containing material in urine; derived from ammonia produced by deamination of amino acids, which combines with carbon dioxide to form urea; amount excreted increases with increased dietary protein intake.
Creatinine	Normal constituent of blood. Derived primarily from breakdown of creatinine phosphate in muscle tissue.
Uric acid	Product of catabolism of nucleic acids (DNA and RNA) derived from food or cellular destruction. Because of its insolubility, uric acid tends to crystallize and is a common component of kidney stones.
Other substances	May be present in small quantities, depending on diet and general health. Include carbohydrates, pigments, fatty acids, mucin, enzymes, and hormones.
Ions	
Sodium (Na^+)	Amount excreted varies with dietary intake and level of aldosterone.
Potassium (K^+)	Amount excreted varies with dietary intake and level of aldosterone.
Chloride (Cl^-)	Amount excreted varies with dietary intake.
Magnesium (Mg^{2+})	Amount excreted varies with dietary intake; parathyroid hormone increases reabsorption and thus decreases urinary excretion.
Sulfates (SO_4^{2-})	Derived from amino acids. Amount excreted varies with dietary protein intake.
Phosphates ($H_2PO_4^-$, HPO_4^{2-}, PO_4^{3-})	Serve as buffers in blood and urine. Parathyroid hormone increases urinary excretion.
Ammonium (NH_4^+)	Derived from protein catabolism and from deamination of the amino acid glutamine in kidneys. Amount produced by kidneys may vary with need to produce HCO_3^- to offset acidity of blood and tissue fluids.
Calcium (Ca^{2+})	Amount excreted varies with dietary intake. Parathyroid hormone increases reabsorption, thus decreasing urinary excretion.

Table 21.5 / Summary of Abnormal Constituents in Urine

Abnormal Constituent	Comments
Albumin	Normal constituent of plasma; usually appears in only very small amounts in urine because it is too large to pass through the pores in capillary walls. The presence of excessive albumin in the urine, *albuminuria* (al′-byoo-mi-NOO-rē-a), indicates an increase in the permeability of filtering membranes due to injury or disease; increased blood pressure; or irritation of kidney cells by substances such as bacterial toxins, ether, or heavy metals.
Glucose	The presence of glucose in the urine, called *glucosuria* (gloo′-kō-SOO-rē-a), usually indicates diabetes mellitus. Occasionally, it may be caused by stress, which can cause excessive amounts of epinephrine to be secreted. Epinephrine stimulates the breakdown of glycogen and liberation of glucose from the liver.
Red blood cells (erythrocytes)	The presence of hemoglobin from ruptured red blood cells in the urine is called *hematuria* (hēm′-a-TOO-rē-a) and generally indicates a pathological condition. One cause is acute inflammation of the urinary organs as a result of disease or irritation from kidney stones. Other causes include tumors, trauma, and kidney disease. Hematuria is often misdiagnosed when a urine sample is contaminated with menstrual blood from the vagina.
White blood cells (leukocytes)	The presence of white blood cells and other components of pus in the urine, referred to as *pyuria* (pī-YOO-rē-a), indicates infection in the kidney or other urinary organs.
Ketone bodies	High levels of ketone bodies, called *ketosis* (kē-TŌ-sis), may indicate diabetes mellitus, anorexia, starvation, or simply too little carbohydrate in the diet.
Bilirubin	When red blood cells are destroyed by macrophages, the globin portion of hemoglobin is split off and the heme is converted to biliverdin. Most of the biliverdin is converted to bilirubin, which gives bile its major pigmentation. An above-normal level of bilirubin in urine is called *bilirubinuria* (bil′-ē-roo′-bi-NOO-rē-a).
Urobilinogen	The presence of urobilinogen (breakdown product of hemoglobin) in urine is called *urobilinogenuria* (yoo′-rō-bi-lin′-ō-jē-NOO-rē-a). Traces are normal, but increased urobilinogen may be due to hemolytic or pernicious anemia, infectious hepatitis, biliary obstruction, jaundice, cirrhosis, congestive heart failure, or infectious mononucleosis.
Casts	Tiny masses of material that have hardened and assumed the shape of the lumen of a tubule in which they formed. They are flushed out of the tubule when filtrate builds up behind them and are excreted in the urine. Casts are named after the cells or substances that compose them or on the basis of their appearance, for example, white blood cell casts, red blood cell casts, and epithelial cell casts.
Kidney stones	Insoluble stones occasionally formed from solidification of the crystals of urine salts. Can be caused by ingestion of excessive mineral salts, too low water intake, abnormally alkaline or acidic urine, or overactive parathyroid glands. Usually form in the renal pelvis, where they cause pain, hematuria, and pyuria.
Microbes	The number and type of bacteria vary with specific infections in the urinary tract. The most common fungus to appear in urine is *Candida albicans*, a cause of vaginitis. The most frequent protozoan seen is *Trichomonas vaginalis*, a cause of vaginitis in females and urethritis in males.

AGING AND THE URINARY SYSTEM

Objective: • **Describe the effects of aging on the urinary system.**

With aging, the kidneys shrink in size, have lowered blood flow, and filter less blood. The mass of the two kidneys decreases from an average of 260 g in 20 year olds to less than 200 g by age 80 due to a decrease in the total number of nephrons. Likewise, renal blood flow and filtration rate decline by 50% between ages 40 and 70. Kidney diseases that become more common with age include acute and chronic kidney inflammations and renal calculi (kidney stones). Because the sensation of thirst diminishes with age, older individuals are susceptible to dehy-

dration. Urinary tract infections are more common among the elderly, as are polyuria (excessive urine production), nocturia (excessive urination at night), increased frequency of urination, dysuria (painful urination), urinary retention or incontinence, and hematuria (blood in the urine).

• • •

To appreciate the many ways that the urinary system contributes to homeostasis of other body systems, examine Focus on Homeostasis: The Urinary System. Next, in Chapter 22, we will see how the kidneys and lungs contribute to maintenance of homeostasis of body fluid volume, electrolyte levels in body fluids, and acid–base balance.

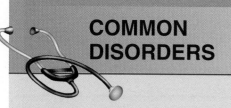

COMMON DISORDERS

Glomerulonephritis

Glomerulonephritis is an inflammation of the glomeruli of the kidney. One of the most common causes is an allergic reaction to the toxins produced by streptococcal bacteria that have recently infected another part of the body, especially the throat. The glomeruli become so inflamed, swollen, and engorged with blood that the filtration membranes allow blood cells and plasma proteins to enter the filtrate. As a result, the urine contains many erythrocytes (hematuria) and much protein.

Renal Failure

Renal failure is a decrease or cessation of glomerular filtration. In **acute renal failure (ARF)** the kidneys abruptly stop working entirely (or al-

most entirely). The main feature of ARF is the suppression of urine flow, usually characterized either by *oliguria* (olig- = scanty; -uria = urine production), which is daily urine output less than 250 mL, or by *anuria*, daily urine output less than 50 mL. Causes include low blood volume (for example, due to hemorrhage); decreased cardiac output; damaged renal tubules; kidney stones; or reactions to the dyes used to visualize blood vessels in angiograms, nonsteroidal anti-inflammatory drugs, and some antibiotic drugs.

Chronic renal failure (CRF) refers to a progressive and usually irreversible decline in glomerular filtration rate (GFR). CRF may result from chronic glomerulonephritis, pyelonephritis, polycystic kidney disease, or traumatic loss of kidney tissue. The final stage of CRF is called *end-stage renal failure* and occurs when about 90% of the nephrons have been lost. At this stage, GFR diminishes to 10–15% of normal, oliguria is present, and blood levels of nitrogen-containing wastes and creatinine increase further. People with end-stage renal failure require dialysis therapy (see page 521) and are possible candidates for a kidney transplant operation.

MEDICAL TERMINOLOGY AND CONDITIONS

Azotemia (az′-ō-TĒ-mē-a; *azot-* = nitrogen; *-emia* = condition of blood) Presence of urea or other nitrogenous solutes in the blood.

Cystocele (SIS-tō-sēl; *cysto-* = bladder; *-cele* = hernia or rupture) Hernia of the urinary bladder.

Diuresis (dī′-yoo-RĒ-sis; *dia-* = through; *-uresis* = urination) Increased excretion of urine.

Dysuria (dis-YOO-rē-a; *dys* = painful; *uria* = urine) Painful urination.

Enuresis (en′-yoo-RĒ-sis; = to void urine) Involuntary voiding of urine after the age at which voluntary control has typically been attained.

Intravenous pyelogram (in′-tra-VĒ-nus PĪ-e-lō-gram′; *intra-* = within; *veno-* = vein; *pyelo-* = pelvis of kidney; *-gram* = record), or **IVP** Radiograph (x-ray film) of the kidneys after venous injection of a dye.

Nocturnal enuresis (nok-TUR-nal en′-yoo-RĒ-sis) Discharge of urine during sleep, resulting in bed-wetting; occurs in about 15% of 5-year-old children and generally resolves spontaneously, afflict-

ing only about 1% of adults. Possible causes include smaller-than-normal bladder capacity, failure to awaken in response to a full bladder, and above-normal production of urine at night. Also termed **nocturia.**

Polyuria (pol′-ē-YOO-rē-a; *poly-* = too much) Excessive urine formation.

Stricture (STRIK-chur) Narrowing of the lumen of a canal or hollow organ, such as the ureter or urethra.

Uremia (yoo-RĒ-mē-a; *-emia* = condition of blood) Toxic levels of urea in the blood resulting from severe malfunction of the kidneys.

Urinary retention A failure to completely or normally void urine; may be due to an obstruction in the urethra or neck of the urinary bladder, to nervous contraction of the urethra, or to lack of the urge to urinate. In men, an enlarged prostate may constrict the urethra and cause urinary retention.

Body System	Contribution of Urinary System
For all body systems	Kidneys regulate volume, composition, and pH of body fluids by removing wastes and excess substances from blood and excreting them in urine; the ureter transports urine from the kidneys to the urinary bladder, which stores urine until it is eliminated through the urethra.
Integumentary system	Kidneys and skin both contribute to synthesis of calcitriol, the active form of vitamin D.
Skeletal system	Kidneys help adjust levels of blood calcium and phosphates, needed for building bone matrix.
Muscular system	Kidneys help adjust level of blood calcium, needed for contraction of muscle.
Nervous system	Kidneys perform gluconeogenesis, which provides glucose for ATP production in neurons.
Endocrine system	Kidneys participate in synthesis of calcitriol, the active form of vitamin D, and release erythropoietin, the hormone that stimulates production of red blood cells.
Cardiovascular system	By increasing or decreasing their reabsorption of water filtered from blood, the kidneys help adjust blood volume and blood pressure; renin released by kidneys raises blood pressure; some bilirubin from hemoglobin breakdown is converted to a yellow pigment (urobilin), which is excreted in urine.
Lymphatic and immune system 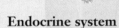	By increasing or decreasing their reabsorption of water filtered from blood, the kidneys help adjust volume of interstitial fluid and lymph; urine flushes microbes from urethra.
Respiratory system	Kidneys and lungs cooperate in adjusting pH of body fluids.
Digestive system	Kidneys help synthesize calcitriol, active form of vitamin D, which is needed for absorption of dietary calcium.
Reproductive systems	In males, the portion of the urethra that extends through the prostate gland and penis is a passageway for semen as well as urine.

STUDY OUTLINE

Introduction (p. 512)

1. The organs of the urinary system include the kidneys, ureters, urinary bladder, and urethra.

2. After the kidneys filter blood and return most water and many solutes to the bloodstream, the remaining water and solutes constitute urine.

Overview of Kidney Functions (p. 512)

1. The kidneys regulate blood ionic composition, blood volume, blood pressure, and blood pH.

2. The kidneys also release calcitriol and erythropoietin and excrete wastes and foreign substances.

Structure of the Kidneys (p. 513)

1. The kidneys are retroperitoneal organs attached to the posterior abdominal wall.

2. Each kidney is enclosed in a renal capsule, which is surrounded by adipose tissue.

3. Internally, the kidneys consist of a renal cortex, renal medulla, renal pyramids, renal papillae, renal columns, calyces, and a renal pelvis.

4. The nephron is the functional unit of the kidney. A nephron consists of a renal corpuscle (glomerulus and glomerular or Bowman's capsule) and a renal tubule (proximal convoluted tubule, descending limb of the loop of Henle, ascending limb of the loop of Henle, and distal convoluted tubule).

5. The filtering unit of a nephron is the filtration membrane.

6. Blood enters the kidney through the renal artery and leaves through the renal vein.

7. The juxtaglomerular apparatus (JGA) consists of the juxtaglomerular cells of an afferent arteriole and the macula densa of the final portion of the ascending limb of the loop of Henle.

Functions of the Kidneys (p. 517)

1. Nephrons perform three basic tasks: glomerular filtration, tubular secretion, and tubular reabsorption.

2. Blood pressure forces water and most dissolved blood components through the filtration membrane into the capsular space. Normally, blood cells and most proteins are not filtered.

3. The amount of filtrate that forms in both kidneys every minute is the glomerular filtration rate (GFR). It is regulated by renal autoregulation (myogenic mechanism and tubuloglomerular feedback), hormonal regulation (angiotensin II and atrial natriuretic peptide), and neural regulation (sympathetic division of the ANS).

4. Tubular reabsorption retains substances needed by the body, including water, glucose, amino acids, lactic acid, and ions (Na^+, K^+, Ca^{2+}, Mg^{2+}, Cl^-, HCO_3^-, and HPO_4^{2-}).

5. Most water is reabsorbed by osmosis together with Na^+ and other reabsorbed solutes in the proximal convoluted tubule; reabsorption of the remaining water is regulated by antidiuretic hormone (ADH) in the last part of the distal convoluted tubule and collecting duct.

6. Tubular secretion discharges chemicals not needed by the body into the urine. Included are excess ions, nitrogenous wastes, hormones, and certain drugs.

7. The kidneys help maintain blood pH by secreting H^+.

8. Tubular secretion also helps maintain proper levels of potassium in the blood.

Dialysis (p. 521)

1. Cleansing blood through an artificial kidney machine is called hemodialysis.

2. In continuous ambulatory peritoneal dialysis (CAPD), the peritoneal lining of the peritoneal cavity is used as the dialysis membrane.

Urine Transportation, Storage, and Elimination (p. 522)

1. The ureters are retroperitoneal and consist of a mucosa, muscularis, and adventitia. They transport urine from the renal pelvis to the urinary bladder, primarily by means of peristalsis.

2. The urinary bladder is posterior to the pubic symphysis. Its function is to store urine prior to micturition.

3. The mucous membrane of the urinary bladder contains transitional epithelium that can stretch. The muscular layer of the wall consists of three layers of smooth muscle together referred to as the detrusor muscle.

4. The micturition reflex discharges urine from the urinary bladder by means of parasympathetic impulses that cause contraction of the detrusor muscle and relaxation of the internal urethral sphincter muscle, and by inhibition of somatic motor neurons to the external urethral sphincter.

5. The urethra is a tube leading from the floor of the urinary bladder to the exterior. Its function is to discharge urine from the body.

Urinalysis (p. 524)

1. Urine volume is influenced by blood pressure, blood concentration, temperature, diuretics, and emotions.

2. The physical characteristics of urine evaluated in a urinalysis are color, odor, turbidity, pH, and specific gravity.

3. Chemically, normal urine contains about 95% water and 5% solutes. The main solutes are urea, creatinine, uric acid, and ions.

4. Abnormal constituents diagnosed through urinalysis include albumin, glucose, red blood cells, white blood cells, ketone bodies, bilirubin, urobilinogen, casts, kidney stones, and microbes.

Aging and the Urinary System (p. 526)

1. With aging, the kidneys shrink in size, have lowered blood flow, and filter less blood.

2. Common problems related to aging include urinary tract infections, increased frequency of urination, urinary retention or incontinence, and renal calculi (kidney stones).

SELF-QUIZ

1. Which of the following is NOT a function of the urinary system?
   a. regulation of blood volume and composition   **b.** stimulation of red blood cell production   **c.** regulation of body temperature   **d.** regulation of blood pressure   **e.** regulation of blood pH

2. One factor that contributes to high glomerular blood pressure is that
   a. the renal artery is smaller in diameter than the renal vein   **b.** the afferent arteriole is smaller in diameter than the efferent arteriole   **c.** the kidneys receive 25% of the body's blood   **d.** the glomeruli are located in the kidney's cortex   **e.** the efferent arteriole is smaller in diameter than the afferent arteriole

3. Which of the following increases water reabsorption in the distal convoluted tubules and collecting ducts?
   a. antidiuretic hormone (ADH)   **b.** angiotensin II   **c.** atrial natriuretic peptide (ANP)   **d.** diuretics   **e.** macula densa

4. The major openings located in the trigone of the bladder are the
   a. renal artery, renal vein, urethra   **b.** renal artery, renal vein, ureter   **c.** ureter, urethra, collecting tubes   **d.** urethra and two ureters   **e.** external urethral sphincter and papillary ducts

5. Which statement does NOT describe the kidneys?
   a. They are protected by the 11th and 12th pairs of ribs.   **b.** The average adult kidney is 10–12 cm (4–5 inches) long and 5–7 cm (2–3 inches) wide.   **c.** The left kidney is lower than the right to accommodate the large size of the liver.   **d.** Each kidney is surrounded by fat and connective tissue.   **e.** The kidneys are retroperitoneal.

6. Place the following structures in the correct order for the flow of urine:
   1. renal pyramids   **4.** major calyx
   2. minor calyx   **5.** collecting ducts
   3. renal pelvis   **6.** ureters
   a. 1, 2, 4, 3, 6, 5   **b.** 5, 1, 4, 2, 3, 6   **c.** 5, 1, 2, 4, 3, 6
   d. 3, 5, 1, 2, 4, 6   **e.** 1, 5, 2, 4, 3, 6

7. The functional unit of the kidney where urine is produced is the
   a. nephron   **b.** pyramid   **c.** juxtaglomerular apparatus
   d. glomerulus   **e.** calyx

8. What causes filtration of plasma across the filtration membrane?
   a. a full urinary bladder   **b.** control by the nervous system   **c.** water retention   **d.** the pressure of the blood   **e.** the pressure of urine in the glomerulus

9. Glomerular filtration rate (GFR) is the
   a. rate of bladder filling   **b.** amount of filtrate formed in both kidneys each minute   **c.** amount of filtrate reabsorbed at the collecting ducts   **d.** amount of blood delivered to the kidneys each minute   **e.** amount of urine formed per hour

10. Which of the following is secreted into the urine from the blood?
    a. hydrogen ions (H^+)   **b.** amino acids   **c.** glucose
    d. water   **e.** white blood cells

11. In the nephron, tubular fluid that is reabsorbed from the renal tubules enters the
    a. glomerulus   **b.** peritubular capillaries   **c.** efferent arteriole   **d.** afferent arteriole   **e.** renal vein

12. Place the following structures in the correct order as they are involved in the formation of urine in the nephrons.
    1. distal convoluted tubule
    2. renal corpuscle
    3. descending limb of loop of Henle
    4. proximal convoluted tubule
    5. collecting duct
    6. ascending limb of loop of Henle
    a. 4, 1, 6, 3, 2, 5   **b.** 2, 6, 3, 1, 5, 4   **c.** 2, 4, 3, 6, 5, 1
    d. 5, 1, 4, 3, 6, 2   **e.** 2, 4, 3, 6, 1, 5

13. Blood is carried out of the glomerulus by the
    a. renal arteries   **b.** afferent arterioles   **c.** peritubular venules   **d.** segmental arteries   **e.** efferent arterioles

14. Which of the following increases glomerular filtration rate (GFR)?
    a. atrial natriuretic peptide (ANP)   **b.** constriction of the afferent arterioles   **c.** increased sympathetic stimulation to the afferent arterioles   **d.** ADH   **e.** angiotensin II

15. Which of the following statements concerning tubular reabsorption is NOT true?
    a. Most reabsorption occurs in the proximal convoluted tubules.   **b.** Tubular reabsorption is a selective process.   **c.** Tubular reabsorption of excess potassium ions (K^+) maintains the correct blood level of K^+.   **d.** The reabsorption of water in the proximal convoluted tubules depends upon sodium ion (Na^+) reabsorption.   **e.** Tubular reabsorption allows the body to retain most filtered nutrients.

16. The micturition reflex
    a. is under the control of hormones   **b.** is activated by low pressure in the bladder   **c.** depends upon contraction of the internal urethral sphincter muscle   **d.** is an involuntary reflex over which normal adults have voluntary control   **e.** is also known as incontinence

17. Which of the following is NOT normally present in glomerular filtrate?
    a. blood cells   **b.** glucose   **c.** nitrogenous wastes such as urea   **d.** amino acids   **e.** water

18. Urine formation requires which of the following?
    a. glomerular filtration and tubular secretion only
    b. glomerular filtration and tubular reabsorption only
    c. glomerular filtration, tubular reabsorption, and tubular secretion   **d.** tubular reabsorption, tubular filtration, and tubular secretion   **e.** tubular secretion and tubular reabsorption only

19. The transport of urine from the renal pelvis into the urinary bladder is the function of the
    a. urethra   **b.** efferent arteriole   **c.** afferent arteriole
    d. renal pyramids   **e.** ureters

20. Incontinence is
    a. failure of the bladder to expel urine   **b.** a lack of voluntary control over the micturition reflex   **c.** an inability of the kidneys to produce urine   **d.** an ability to consciously control micturition   **e.** a form of kidney dialysis

CRITICAL THINKING APPLICATIONS

1. Yesterday, you attended a large, outdoor party where beer was the only beverage available. You remember having to urinate many, many times yesterday, and today you're very thirsty. What hormone is affected by alcohol, and how does this affect your kidney function?

2. Sarah is an "above average" one-year-old toddler whose parents would like her to be the first toilet-trained child in preschool. However, in this case at least, Sarah is average for her age and remains incontinent. Should her parents be concerned by this lack of success?

3. Ana visited the health center complaining of painful urination and blood in her urine. This was the second time she had experienced these symptoms this semester. Ana was given a prescription for antibiotics and some suggestions on how to avoid a recurrence of her problem. What are the proper terms for Ana's symptoms, and what is a likely cause? What are some suggestions that the health center may have given Ana?

4. Johann was being prepped to donate a kidney to his son, Johann Jr. Where are the kidneys located? What holds the kidneys in place?

ANSWERS TO FIGURE QUESTIONS

21.1 The kidneys do the major work of the urinary system.

21.2 Blood vessels, lymphatic vessels, and nerves enter the kidneys through the renal hilus.

21.3 About 1200 mL of blood enters the kidneys each minute.

21.4 Podocytes of the glomerular capsule and endothelium of the glomerulus make up the filtration membrane.

21.5 The water molecule will travel from the proximal convoluted tubule → descending limb of the loop of Henle → ascending limb of the loop of Henle → distal convoluted tubule → collecting duct → papillary duct → minor calyx → major calyx → renal pelvis.

21.6 Glomerular filtration, tubular reabsorption, and tubular secretion produce urine.

21.7 First, when blood pressure increases, GFR increases. Then, as the myogenic mechanism and tubuloglomerular feedback respond, the GFR returns to normal.

21.8 Glucose enters PCT cells by a Na^+–glucose symporter and then moves into interstitial fluid through a glucose facilitated diffusion transporter.

21.9 Due to loss of body water in sweat, blood level of ADH would be higher than normal after a 5-km run.

21.10 During hemodialysis, plasma proteins remain in the blood because they are too large to pass through the pores in the dialysis membrane.

21.11 A lack of voluntary control over micturition is termed incontinence.

Chapter **22**

Fluid, Electrolyte, and Acid–Base Balance

■ Student Learning Objectives

1. Compare the locations of intracellular fluid and extracellular fluid, and describe the various fluid compartments of the body. **532**
2. Describe the sources of water and solute gain and loss, and explain how each is regulated. **533**
3. Compare the electrolyte composition of the three major fluid compartments of the body. **536**

4. Discuss the functions of the various electrolytes, and explain the regulation of their concentrations. **537**
5. Compare the roles of buffers, exhalation of carbon dioxide, and kidney excretion of H^+ in maintaining pH of body fluids. **540**
6. Define acid–base imbalances, describe their effects on the body, and explain how they are treated. **541**

7. Describe the changes in fluid, electrolyte, and acid–base balance that may occur with aging. **542**

■ A Look Ahead

FLUID COMPARTMENTS AND FLUID BALANCE

Objectives: • **Compare the locations of intracellular fluid and extracellular fluid, and describe the various fluid compartments of the body.**

• **Describe the sources of water and solute gain and loss, and explain how each is regulated.**

*I*n Chapter 21 you learned how urine is formed by the kidneys to help maintain fluid balance in the body. The water and dissolved solutes in each of the body's fluid compartments constitute **body fluids.** Regulatory mechanisms normally maintain homeostasis of body fluids; malfunction in any or all of them may seriously endanger the functioning of organs throughout the body. In this chapter we will explore the mechanisms that regulate the volume and distribution of body fluids and examine how the concentrations of solutes and the pH of body fluids are determined.

In lean adults, body fluid constitutes about 55–60% of total body mass and is partitioned into two main compartments. **Intracellular fluid** (*intra-* = within) or **ICF**, fluid within the cells of the body, makes up about two-thirds of body fluid. The other third, called **extracellular fluid** (*extra-* = outside) or **ECF**, includes all body fluids outside the cells (Figure 22.1). About 80% of the ECF is **interstitial fluid** (*inter-* = between), which occupies the microscopic spaces between tissue cells, and 20% of the ECF is **plasma**, the liquid portion of the blood. The ECF also includes lymph in lymphatic vessels; cerebrospinal fluid in

Figure 22.1 ■ Body fluid compartments.

In lean adults, fluids constitute an average of 55–60% of body mass.

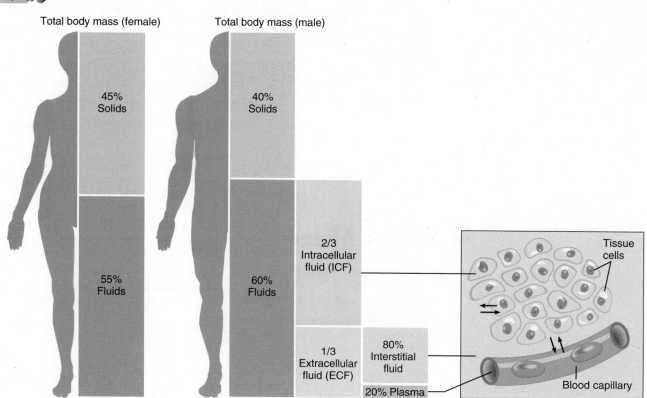

(a) Distribution of body water in an average lean, adult female and male

(b) Exchange of water among body fluid compartments

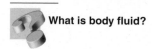 **What is body fluid?**

the nervous system; synovial fluid in joints; aqueous humor and vitreous body in the eyes; endolymph and perilymph in the ears; and pleural, pericardial, and peritoneal fluids between serous membranes.

Intracellular fluid is isolated from interstitial fluid by the plasma membranes of individual cells, and blood vessel walls separate interstitial fluid from plasma. Only the smallest blood vessels, the capillaries, have walls thin enough and leaky enough to permit the exchange of water and solutes between plasma and interstitial fluid.

The body is in *fluid balance* when the required amounts of water and solutes are present and correctly proportioned among the various compartments. Despite continual exchange of water and solutes between fluid compartments, the volume of fluid in each compartment remains fairly stable. Because osmosis is the primary means of water movement between intracellular fluid and interstitial fluid, the concentration of solutes in each fluid determines the *direction* of water movement. Most solutes in body fluids are *electrolytes,* inorganic compounds that dissociate into ions. Because intake of water and electrolytes rarely occurs in exactly the same proportions as their presence in body fluids, the ability of the kidneys to excrete excess water by producing dilute urine, or to excrete excess electrolytes by producing concentrated urine, is of utmost importance.

Avenues of Body Water Gain and Loss

Water is by far the largest single component of the body, constituting 45–75% of total body mass. Infants have the highest percentage of water, up to 75% of body mass; the percentage decreases until about two years of age. Until puberty, body mass of both boys and girls is about 60% water. Because adipose tissue contains almost no water, fatter people have a smaller proportion of water than leaner people. In lean adult males, water accounts for about 60% of body mass. Females, on average, have more subcutaneous fat than males, so their total body water is lower, accounting for about 55% of body mass.

The body can gain water by ingestion or by metabolic synthesis (Figure 22.2). The main sources of body water are ingested liquids and moist foods absorbed from the gastrointestinal tract, which total about 2300 mL/day. The other source of water is *metabolic water*—about 200 mL/day—which is produced during certain metabolic reactions, such as aerobic cellular respiration (see Figure 20.3 on page 497). Daily water gain totals about 2500 mL.

Normally, body fluid volume remains constant because water loss equals water gain. Water loss occurs in four ways (Figure 22.2). Each day the kidneys excrete about 1500 mL in urine, the skin evaporates about 600 mL, the lungs exhale about 300 mL as water vapor, and the gastrointestinal tract eliminates about 100 mL in feces, for a total daily loss of 2500 mL. In women of reproductive age, body water is also lost through menstrual flow.

Figure 22.2 ■ **Sources of daily water gain and loss under normal conditions.** Numbers are average volumes for adults.

 Normally, daily water loss equals daily water gain.

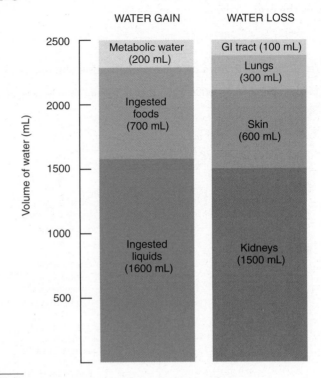

How does hyperventilation affect fluid balance?

Regulation of Body Water Gain

Adjusting the volume of water intake, mainly by drinking more or less fluid, is the principal method of regulating body water gain. An area in the hypothalamus known as the **thirst center** governs the urge to drink (Figure 22.3):

Figure 22.3 ■ **Pathways by which dehydration stimulates thirst.**

Dehydration occurs when water loss is greater than water gain.

1 Dehydration

2 Decreased flow of saliva | Increased blood osmotic pressure | Decreased blood volume

3 Dry mouth and pharynx | Stimulates osmoreceptors in hypothalamus | Decreased blood pressure

Increased renin release by juxta-glomerular cells of kidneys

Increased angiotensin II formation

4 Stimulate thirst center in hypothalamus

5 Increases thirst

6 Increases water intake

7 Increases body water to normal level and relieves dehydration

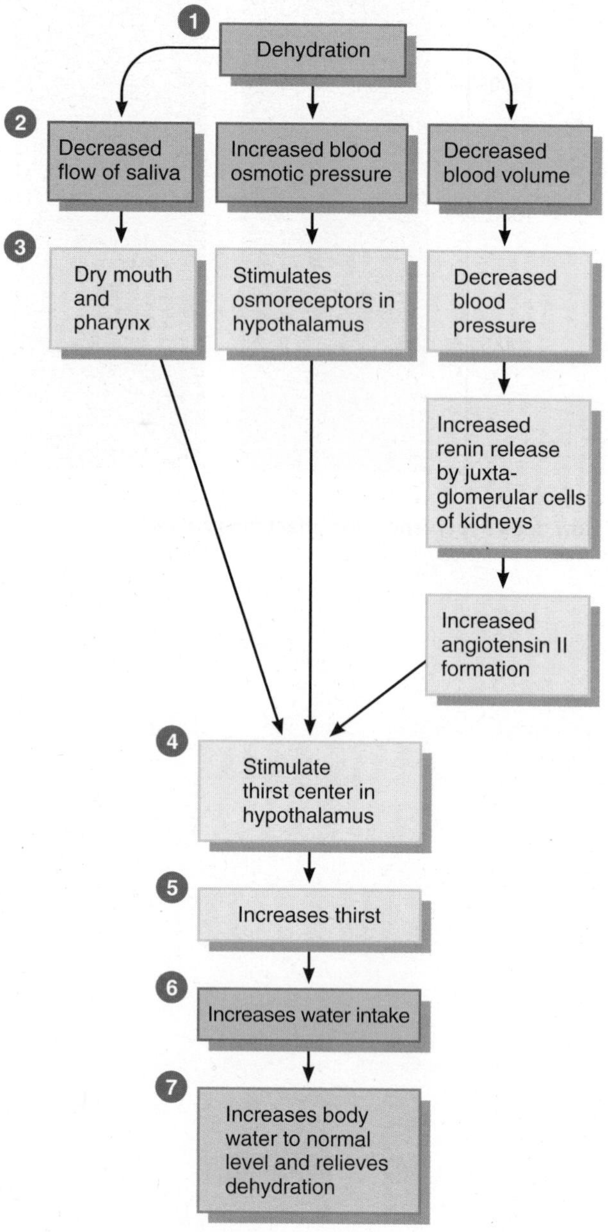

Does regulation of these pathways involve negative or positive feedback? Why?

1 When water loss is greater than water gain, the result is **dehydration,** a decrease in total body water content that may be moderate or severe.

2 Dehydration stimulates thirst by (1) decreasing production of saliva, (2) increasing blood osmotic pressure, and (3) decreasing blood volume.

3 Decreased production of saliva causes dryness of the mucosa of the mouth and pharynx. Increased blood osmotic pressure stimulates osmoreceptors in the hypothalamus. Decreased blood volume causes a drop in blood pressure, stimulating the kidneys to release renin, which promotes the formation of angiotensin II.

4 A dry mouth and pharynx, stimulation of osmoreceptors, and increased angiotensin II levels stimulate the thirst center in the hypothalamus.

5 The sensation of thirst increases.

6 If fluids are available, water intake increases.

7 Normal body water content is restored. The net effect is that water gain balances water loss thus relieving the dehydration.

If the sensation of thirst does not occur quickly enough or if access to fluids is restricted, significant dehydration can occur. In situations in which heavy sweating, diarrhea, or vomiting occurs, it is wise to start replacing body fluids by drinking even before thirst becomes apparent.

Regulation of Solute and Body Water Loss

As you learned in Chapter 21, elimination of *excess* body water or solutes occurs mainly via the urine and is controlled in part by hormones. **Angiotensin II** and **aldosterone** promote reabsorption (reduce urinary loss) of Na^+, Cl^-, and water, thereby increasing the volume of body fluids. **Antidiuretic hormone (ADH)** slows fluid loss in the urine by promoting increased water reabsorption. Yet another hormone—**atrial natriuretic peptide (ANP)**—promotes **natriuresis** (*natri-* = sodium; *-ure* = urine; *-sis* = the act of), elevated urinary excretion of Na^+ (and Cl^-) and water, which decreases blood volume.

Figure 22.4 shows the sequence of changes that occur after eating a salty meal:

1 The increased intake of NaCl produces an increase in plasma concentrations of Na^+ and Cl^-.

2 The increased plasma concentrations of Na^+ and Cl^- elevate the osmotic pressure of interstitial fluid, causing osmosis of water from intracellular fluid into interstitial fluid and then into plasma.

3 Blood volume is increased.

4 An increase in blood volume stretches the atria of the heart, which promotes release of atrial natriuretic peptide; the ensuing natriuresis decreases blood volume. An increase in blood volume also slows release of renin from the kidneys, so that less angiotensin II is formed. With less angiotensin II, glomerular filtration rate increases, and the secretion of aldosterone is reduced.

5 Na^+ and Cl^- reabsorption in the collecting ducts slows.

6 More sodium and chloride ions are excreted in the urine.

7 The excretion of more Na^+ and Cl^- is accompanied by increased loss of water in urine.

8 Blood volume is decreased.

An increase in the osmotic pressure of plasma and interstitial fluid also stimulates release of ADH, which increases the permeability of cells in the collecting ducts of the kidneys to water (see Figure 21.9 on page 520). Thus, water moves by osmosis from the tubular fluid back into the bloodstream, producing a small volume of very concentrated urine. By contrast, intake of a large amount of water decreases the osmotic pressure of blood and interstitial fluid. Within minutes, ADH secretion shuts down. When the cells of the collecting duct are not stimulated by ADH, their water permeability is very low, and more water is quickly lost in the urine.

Other factors besides blood osmotic pressure influence body water loss. An extreme decrease in blood volume also stimulates ADH release. In severe dehydration, glomerular filtration rate decreases because blood pressure falls, and less water is lost in the urine. The intake of too much water increases blood pressure, raises the rate of glomerular filtration, and causes greater water loss in the urine. Vomiting and diarrhea result in fluid loss from the gastrointestinal tract. Finally, fever, heavy sweating, and destruction of extensive areas of the skin from burns can all result in excessive water loss through the skin.

Movement of Water Between Fluid Compartments

Intracellular and interstitial fluids normally have the same osmotic pressure, so cells neither shrink nor swell. An increase in the osmotic pressure of interstitial fluid draws water out of cells, so they shrink slightly. A decrease in the osmotic pressure of interstitial fluid causes cells to swell. Changes in osmotic pressure most often result from changes in the concentration of Na^+.

A decrease in the osmotic pressure of interstitial fluid normally inhibits secretion of ADH. Normally functioning kidneys then excrete excess water in the urine, which raises the osmotic pressure of body fluids to normal. As a result, body cells swell only slightly, and only for a brief period of time. But when a person steadily consumes water faster than the kidneys can excrete it (the maximum urine flow rate is about 15 mL/min) or when kidney function is poor, the decreased Na^+ concentration of interstitial fluid causes water to move by osmosis from interstitial fluid into intracellular fluid. The result may be ***water intoxication,*** a state in which excessive body water causes cells to swell dangerously, producing convulsions, coma, and possibly death. To prevent this dire sequence of events, solutions given for intravenous or ***oral rehydration therapy*** (ORT) include a small amount of table salt (NaCl).

Figure 22.4 ■ **Hormonal changes following increased NaCl intake.** The resulting reduced blood volume is achieved by increasing urinary loss of Na^+ and Cl^-.

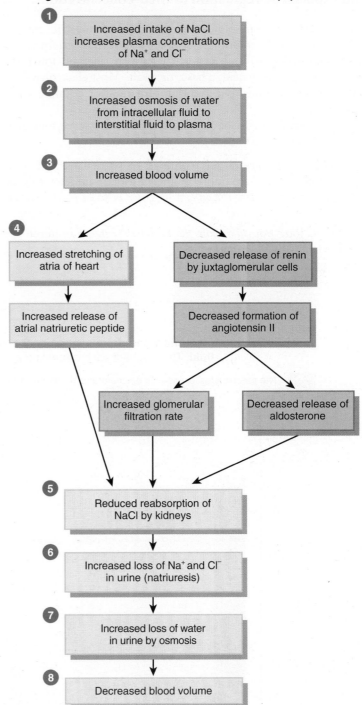

The three main hormones that regulate Na^+ and Cl^- reabsorption by the kidneys (and thus the amount lost in the urine) are angiotensin II, aldosterone, and atrial natriuretic peptide.

1 Increased intake of NaCl increases plasma concentrations of Na^+ and Cl^-

2 Increased osmosis of water from intracellular fluid to interstitial fluid to plasma

3 Increased blood volume

4 Increased stretching of atria of heart

Decreased release of renin by juxtaglomerular cells

Increased release of atrial natriuretic peptide

Decreased formation of angiotensin II

Increased glomerular filtration rate

Decreased release of aldosterone

5 Reduced reabsorption of NaCl by kidneys

6 Increased loss of Na^+ and Cl^- in urine (natriuresis)

7 Increased loss of water in urine by osmosis

8 Decreased blood volume

 How does excessive aldosterone secretion cause edema?

ELECTROLYTES IN BODY FLUIDS

Objectives: • **Compare the electrolyte composition of the three major fluid compartments of the body.** • **Discuss the functions of the various electrolytes, and explain the regulation of their concentrations.**

Electrolytes are compounds that conduct an electric current when dissolved. In solution, electrolytes split into ***cations*** (positively charged ions) and ***anions*** (negatively charged ions). An electrolyte can be an acid, a base, or a salt. Most electrolytes are inorganic compounds, but a few are organic (for example, most proteins).

The ions formed when electrolytes dissolve serve four general functions in the body:

1. Because they are largely confined to particular fluid compartments and are more numerous than nonelectrolytes, certain ions *control the osmosis of water between fluid compartments.*

2. Some ions *help maintain the acid–base balance* required for normal cellular activities.

3. The ability of ions to *carry electrical current* allows production of action potentials and graded potentials and controls secretion of some hormones and neurotransmitters.

4. Several ions are cofactors that are *needed for optimal activity of enzymes.*

Concentrations of Electrolytes in Body Fluids

The concentration of ions is typically expressed in ***milliequivalents per liter (mEq/liter),*** which indicates the total number of cations or anions (positive or negative electrical charges) in a given volume of solution. Figure 22.5 compares the concentrations of the main electrolytes and protein anions in extracellular fluid (plasma and interstitial fluid) and intracellular fluid. The chief difference between the two extracellular fluids is that plasma contains many protein anions, whereas interstitial fluid has very few. Because normal capillary membranes are virtually impermeable to proteins, only a few plasma proteins leak out of blood vessels into the interstitial fluid. This difference in protein concentration is largely responsible for the blood colloid osmotic pressure, the difference in osmotic pressure between

Figure 22.5 ■ **Electrolyte and protein anion concentrations in plasma, interstitial fluid, and intracellular fluid.** The height of each column represents the milliequivalents per liter (mEq/liter).

The electrolytes present in extracellular fluids are different from those present in intracellular fluid.

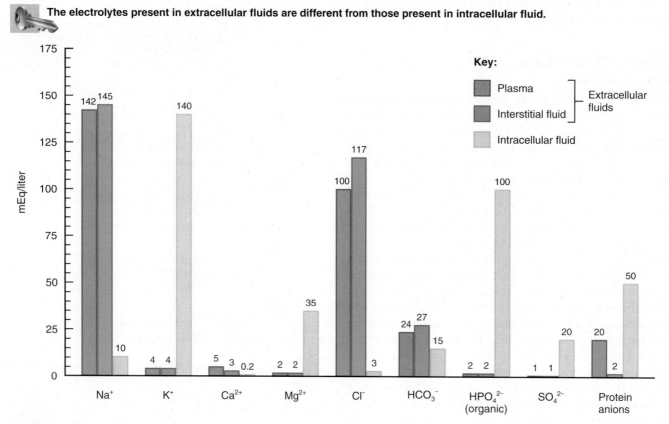

What is the major cation in ECF?

plasma and interstitial fluid. The other components of the two extracellular fluids are similar.

The electrolyte content of intracellular fluid differs considerably from that of extracellular fluid. The most abundant cation in extracellular fluid is Na^+, and the most abundant anion is Cl^-. The most abundant cation in intracellular fluid is K^+, and the three most abundant anions are proteins, phosphates (HPO_4^{2-}), and sulfates (SO_4^{2-}).

Sodium

Sodium ions (Na^+) are the most abundant extracellular ions, representing about 90% of extracellular cations. Normal plasma Na^+ concentration is 136 to 148 mEq/liter. Na^+ plays a pivotal role in fluid and electrolyte balance because it accounts for almost half of the osmotic pressure of extracellular fluid. As you learned in Chapter 9, the flow of Na^+ through voltage-gated channels in the plasma membrane is necessary for the generation and conduction of action potentials in neurons and muscle fibers. As you learned earlier in this chapter, the Na^+ level in the blood is controlled by aldosterone, antidiuretic hormone, and atrial natriuretic peptide.

If excess sodium ions remain in the body, water is also osmotically retained. The result is *edema,* an abnormal accumulation of interstitial fluid. Renal failure and excessive aldosterone secretion are two causes of Na^+ retention.

Chloride

Chloride ions (Cl^-) are the most prevalent extracellular anions. Normal plasma Cl^- concentration is 95 to 105 mEq/liter. Chloride ions are important to anion balance because they diffuse relatively easily between the extracellular and intracellular compartments. Chloride ions also are part of the hydrochloric acid secreted into gastric juice. Processes that increase or decrease renal reabsorption of sodium ions also affect reabsorption of chloride ions, which mainly follow sodium ions passively by electrical attraction.

Potassium

Potassium ions (K^+) are the most abundant cations in intracellular fluid (140 mEq/liter). K^+ plays a key role in producing action potentials in neurons and muscle fibers. Thus, abnormal plasma K^+ levels adversely affect neuromuscular and cardiac function. K^+ also helps maintain normal intracellular fluid volume. When K^+ moves in or out of cells, it often is exchanged for H^+. This shift of H^+ helps regulate pH.

Normal plasma K^+ concentration is 3.5 to 5.0 mEq/liter. The plasma level of K^+ is controlled mainly by aldosterone. When plasma K^+ concentration is high, more aldosterone is secreted into the blood. Aldosterone stimulates cells of the collecting ducts of the kidneys to secrete more K^+, which then is excreted in the urine. When plasma K^+ concentration is low, aldosterone secretion decreases and less K^+ is lost in urine.

Bicarbonate

Bicarbonate ions (HCO_3^-) are the second most prevalent extracellular anions. Normal plasma HCO_3^- concentration is 22 to 26 mEq/liter in systemic arterial blood and 23 to 27 mEq/liter in systemic venous blood. HCO_3^- concentration increases as blood flows through systemic capillaries because the carbon dioxide released by cells combines with water to form carbonic acid (H_2CO_3), which dissociates into H^+ and HCO_3^-. As blood flows through pulmonary capillaries, the concentration of HCO_3^- decreases as carbon dioxide is exhaled. Intracellular fluid also contains a significant amount of HCO_3^- (15 mEq/liter).

The kidneys are the main regulators of blood HCO_3^- concentration. Certain cells of the renal tubule can either form HCO_3^- and release it into the blood when the blood level of bicarbonate is low, or secrete excess HCO_3^- into the urine when the level in blood is too high.

Calcium

Because such a large amount of calcium is stored in bone, it is the most abundant mineral in the body. About 98% of the calcium in adults is in the skeleton and teeth, where it is combined with phosphates to form mineral salts. In body fluids, calcium is mainly an extracellular cation (Ca^{2+}). The normal concentration of free or unattached Ca^{2+} in plasma is 4.5 to 5.5 mEq/liter; about the same amount of Ca^{2+} is attached to various plasma proteins. Besides contributing to the hardness of bones and teeth, Ca^{2+} plays important roles in blood clotting, neurotransmitter release, maintenance of muscle tone, and excitability of nervous and muscle tissue.

The two main regulators of Ca^{2+} concentration in plasma are parathyroid hormone (PTH) and calcitriol, the form of vitamin D that acts as a hormone (see Figure 13.13 on page 317). A low plasma Ca^{2+} level promotes release of more PTH, which increases bone *resorption* by stimulating osteoclasts in bone tissue to release calcium (and phosphate) from mineral salts of bone matrix. PTH also enhances *reabsorption* of Ca^{2+} from glomerular filtrate back into blood and increases production of calcitriol (which in turn increases Ca^{2+} *absorption* from the gastrointestinal tract).

Phosphate

About 85% of the phosphate in adults is present as calcium phosphate salts, which are structural components of bone and teeth; the remaining 15% is ionized. Three phosphate ions ($H_2PO_4^-$, HPO_4^{2-}, and PO_4^{3-}) are important intracellular anions, but at the normal pH of body fluids, HPO_4^{2-} is the most prevalent form. Phosphates contribute about 100 mEq/liter of anions to intracellular fluid. HPO_4^{2-} is an important buffer of H^+, both in body fluids and in the urine. Although some are "free," most phosphate ions are covalently bound to organic molecules such as phospholipids, proteins, carbohydrates, nucleic acids (DNA and RNA), and adenosine triphosphate (ATP).

The normal plasma concentration of ionized phosphate is only 1.7 to 2.6 mEq/liter. The same two hormones that govern

*H*eavy or prolonged physical activity can lead to dehydration and disrupt fluid and electrolyte balance. During physical activity, muscle contraction generates a great deal of heat, up to 100 times more than when you are at rest. The body can get rid of this extra heat by increasing blood flow to the skin, where heat is given off by radiation, conduction, and convection, and by activating the sweat glands to increase heat loss by evaporation. Strenuous exercise in hot weather may cause the loss of over 2 liters (about 2 qt) of water per hour from the skin and lungs. Such losses can lead to dehydration and hyperthermia if fluids are not replaced.

Don't Sweat it

Dehydration is a loss of body fluid that amounts to 1% or more of total body weight. It is most common during physical activity at a high temperature but can also occur during strenuous exercise at lower temperatures. Fluid deficits of 5% are common in athletic events such as football, soccer, tennis, and long-distance running. Symptoms include irritability, fatigue, and loss of appetite.

With dehydration, water is lost from all body compartments. The decrease in blood volume impairs physical performance because it decreases the amount of blood the heart can pump per beat. Muscles need oxygen to work; as cardiac output is reduced, muscle performance declines. The body tries to maintain blood volume to the muscles by constricting blood vessels in the skin, so less heat is lost and body temperature rises.

► *Think It Over*

► Sports drinks contain electrolytes such as sodium and potassium. Why might such drinks help a dehydrated person regain normal hydration levels better than plain water?

Intracellular electrolyte changes may also occur.

Thirst is the body's signal that its water level is getting too low. Unfortunately, thirst is not a reliable indicator of fluid needs. People tend to drink just enough to relieve their parched throats. The thirst mechanism is especially unreliable in children and older adults. Aging decreases the kidneys' ability to retain water when the body needs fluids, which increases the susceptibility to dehydration.

---

calcium homeostasis—parathyroid hormone (PTH) and calcitriol—also regulate the level of HPO_4^{2-} in blood plasma. The resorption of bone matrix by osteoclasts caused by PTH releases both phosphate and calcium ions into the bloodstream. In the kidneys, however, PTH inhibits reabsorption of phosphate ions, increasing urinary excretion of phosphate and lowering blood phosphate level. As with calcium, calcitriol promotes absorption of phosphates from the digestive tract.

Magnesium

In adults, about 54% of the total body magnesium is deposited in bone matrix as magnesium salts; the remaining 46% occurs as magnesium ions (Mg^{2+}) in intracellular fluid (45%) and extracellular fluid (1%). Mg^{2+} is the second most common intracellular cation (35 mEq/liter). Mg^{2+} is needed for operation of the sodium pump and to assist some enzymes involved in the metabolism of carbohydrates and proteins. Mg^{2+} is also important in neuromuscular activity, nerve impulse transmission, and myocardial functioning, and it is needed for the secretion of parathyroid hormone.

Normal plasma Mg^{2+} concentration is low, only 1.3 to 2.1 mEq/liter, and is regulated by varying its excretion rate in urine. The kidneys increase urinary excretion of Mg^{2+} when extracellular fluid volume is increased; when levels of parathyroid hormone are decreased; and in disorders such as hypercalcemia, hypermagnesemia, and acidosis. Opposite conditions decrease urinary excretion of Mg^{2+}.

Table 22.1 describes the imbalances that result from the deficiency or excess of several electrolytes.

Table 22.1 / Blood Electrolyte Imbalances

Electrolyte*	Deficiency		Excess	
	Name and Causes	**Symptoms**	**Name and Causes**	**Symptoms**
Sodium (Na$^+$) 136–148 mEq/liter	**Hyponatremia** (hī´-pō-na-TRE-mē-a) may be due to decreased sodium intake; increased loss through vomiting, diarrhea, aldosterone deficiency, or use of diuretics; or excessive water intake.	Muscular weakness; dizziness, headache, and hypotension; tachycardia and shock; mental confusion, stupor, and coma.	**Hypernatremia** may occur with dehydration, water deprivation, or excessive sodium in diet or intravenous fluids.	Intense thirst, hypertension, edema, agitation, and convulsions.
Chloride (Cl$^-$) 95–105 mEq/liter	**Hypochloremia** (hī´-pō-klō-RE-mē-a) may be due to vomiting, overhydration, aldosterone deficiency, congestive heart failure, or therapy with some diuretics.	Muscle spasms, alkalosis, shallow respirations, hypotension, and tetany.	**Hyperchloremia** may result from dehydration due to water loss or water deprivation; excessive chloride intake; or renal failure, aldosterone excess, and some drugs.	Lethargy; weakness; acidosis; and rapid, deep breathing.
Potassium (K$^+$) 3.5–5.0 mEq/liter	**Hypokalemia** (hī´-pō-ka-LE-mē-a) may result from loss due to vomiting or diarrhea, decreased potassium intake, aldosterone excess, kidney disease, or therapy with some diuretics.	Muscle fatigue, flaccid paralysis, mental confusion, increased urine output, shallow respirations, and changes in the electrocardiogram.	**Hyperkalemia** may be due to excessive potassium intake, renal failure, aldosterone deficiency, or crushing injuries to body tissues.	Irritability, nausea, vomiting, diarrhea, muscular weakness; can cause death by inducing ventricular fibrillation.
Calcium (Ca^{2+}) 4.5–5.5 mEq/liter	**Hypocalcemia** (hī´-pō-kal-SE-mē-a) may be due to increased Ca^{2+} loss, reduced Ca^{2+} intake, elevated levels of phosphates, or parathyroid hormone deficiency.	Numbness and tingling of the fingers; hyperactive reflexes, muscle cramps, tetany, and convulsions; bone fractures; spasms of laryngeal muscles that can cause asphyxiation.	**Hypercalcemia** may result from hyperparathyroidism, some cancers, excessive intake of vitamin D, or Paget's disease of bone.	Lethargy, weakness, anorexia, nausea, vomiting, polyuria, itching, bone pain, depression, confusion, stupor, and coma.
Phosphate (HPO$_4^{2-}$) 1.7–2.6 mEq/liter	**Hypophosphatemia** (hī´-pō-fos´-fa-TE-mē-a) may occur through increased urinary losses or decreased intestinal absorption of phosphate.	Confusion, seizures, coma, chest and muscle pain, numbness and tingling of the fingers, memory loss, and lethargy.	**Hyperphosphatemia** occurs when the kidneys fail to excrete excess phosphate, as happens in renal failure; can also result from increased intake of phosphates or destruction of body cells, which releases phosphates into the blood.	Anorexia, nausea, vomiting, muscular weakness, hyperactive reflexes, tetany, and tachycardia.
Magnesium (Mg^{2+}) 1.3–2.1 mEq/liter	**Hypomagnesemia** (hī´-pō-mag´-ne-SE-mē-a) may be due to inadequate magnesium intake or excessive loss in urine or feces.	Weakness, tetany, delirium, convulsions, anorexia, nausea, vomiting, and cardiac arrhythmias.	**Hypermagnesemia** occurs in renal failure or with increased intake of Mg^{2+} such as magnesium-containing antacids.	Hypotension, muscular weakness or paralysis, nausea, vomiting, and altered mental functioning.

*Values are normal ranges of plasma levels in adults.

ACID–BASE BALANCE

Objectives: • **Compare the roles of buffers, exhalation of carbon dioxide, and kidney excretion of H$^+$ in maintaining pH of body fluids.**

• **Define acid–base imbalances, describe their effects on the body, and explain how they are treated.**

From our discussion thus far, it is clear that various ions play different roles in helping to maintain homeostasis. A major homeostatic challenge is keeping the H$^+$ level (pH) of body fluids in the appropriate range. This task—the maintenance of acid–base balance—is of critical importance because the three-dimensional shape of all body proteins, which enables them to perform specific functions, is very sensitive to the most minor changes in pH. When the diet contains a large amount of protein, as is typical in North America, cellular metabolism produces more acids than bases and thus tends to acidify the blood. (You may wish to review the discussion of acids, bases, and pH in Chapter 2 before proceeding with this section.)

In a healthy person, the pH of systemic arterial blood remains between 7.35 and 7.45. The removal of H$^+$ from body fluids and its subsequent elimination from the body depend on three major mechanisms: buffer systems, exhalation of carbon dioxide, and excretion of H$^+$ in the urine.

The Actions of Buffer Systems

Buffers are substances that act quickly to temporarily bind H$^+$, removing the highly reactive, excess H$^+$ from solution but not from the body. Buffers prevent rapid, drastic changes in the pH of a body fluid by converting strong acids and bases into weak acids and bases. Strong acids release H$^+$ more readily than weak acids and thus contribute more free hydrogen ions. Similarly, strong bases raise pH more than weak ones. The principal buffer systems of the body fluids are the protein buffer system, the carbonic acid–bicarbonate buffer system, and the phosphate buffer system.

Protein Buffer System

The *protein buffer system* is the most abundant buffer in intracellular fluid and plasma. Hemoglobin is an especially good buffer within red blood cells, and albumin is the main protein buffer in plasma. Recall that proteins are composed of amino acids, organic molecules that contain at least one carboxyl group (—COOH) and at least one amino group (—NH$_2$); it is the carboxyl groups and the amino groups that function in the protein buffer system. The carboxyl group releases H$^+$ when pH rises. The H$^+$ is then able to react with any excess OH$^-$ in the solution to form water. The amino group combines with H$^+$, forming an —NH$_3^+$ group, when pH falls. Thus, proteins can buffer both acids and bases.

Carbonic Acid–Bicarbonate Buffer System

The *carbonic acid–bicarbonate buffer system* is based on the *bicarbonate ion* (HCO$_3^-$), which can act as a weak base, and *carbonic*

acid (H$_2$CO$_3$), which can act as a weak acid. HCO$_3^-$ is a significant anion in both intracellular and extracellular fluids (see Figure 22.5). If there is an excess of H$^+$, HCO$_3^-$ can function as a weak base and remove the excess H$^+$, as follows:

$$\underset{\substack{\text{Hydrogen}\\\text{ion}}}{H^+} + \underset{\substack{\text{Bicarbonate ion}\\\text{(weak base)}}}{HCO_3^-} \longrightarrow \underset{\substack{\text{Carbonic acid}}}{H_2CO_3}$$

The H$_2$CO$_3$ then dissociates into water and carbon dioxide in the lungs, and the CO$_2$ is exhaled.

Conversely, if pH rises, H$_2$CO$_3$ can function as a weak acid and provide H$^+$ to lower the pH, as follows:

$$\underset{\substack{\text{Carbonic acid}\\\text{(weak acid)}}}{H_2CO_3} \longrightarrow \underset{\substack{\text{Hydrogen}\\\text{ion}}}{H^+} + \underset{\substack{\text{Bicarbonate ion}}}{HCO_3^-}$$

Phosphate Buffer System

The *phosphate buffer system* acts by essentially the same mechanism as the carbonic acid–bicarbonate buffer system. The components of the phosphate buffer system are the *dihydrogen phosphate ion* (H$_2$PO$_4^-$) and the *monohydrogen phosphate ion* (HPO$_4^{2-}$). Recall that phosphates are major anions in intracellular fluid and minor ones in extracellular fluids (see Figure 22.5). The dihydrogen phosphate ion acts as a weak acid and is capable of buffering strong bases such as OH$^-$, for example:

$$H_2PO_4^- + OH^- \longrightarrow HPO_4^{2-} + H_2O$$

The monohydrogen phosphate ion, in contrast, acts as a weak base and is capable of buffering the H$^+$ released by a strong acid:

$$HPO_4^{2-} + H^+ \longrightarrow H_2PO_4^-$$

Because the concentration of phosphates is highest in intracellular fluid, the phosphate buffer system is an important regulator of pH in the cytosol. It acts to a smaller degree in extracellular fluids, and buffers acids in urine. H$_2$PO$_4^-$ is formed when excess H$^+$ in the kidney tubule fluid combines with HPO$_4^{2-}$. The H$^+$ that becomes part of the H$_2$PO$_4^-$ passes into the urine. This reaction is one means by which the kidneys help maintain blood pH by excreting H$^+$ in the urine.

Exhalation of Carbon Dioxide

The pH of body fluids may be adjusted, usually in 1 to 3 minutes, by a change in the rate and depth of breathing affecting the following reversible reactions:

$$\underset{\substack{\text{Carbon}\\\text{dioxide}}}{CO_2} + \underset{\substack{\text{Water}}}{H_2O} \rightleftharpoons \underset{\substack{\text{Carbonic}\\\text{acid}}}{H_2CO_3} \rightleftharpoons \underset{\substack{\text{Hydrogen}\\\text{ion}}}{H^+} + \underset{\substack{\text{Bicarbonate}\\\text{ion}}}{HCO_3^-}$$

With increased ventilation, more carbon dioxide (CO$_2$) is exhaled. As the level of CO$_2$ in body fluids decreases, the reactions proceed from right to left, the concentration of H$^+$ in body fluids decreases, and blood pH rises. If ventilation slows, less CO$_2$ is exhaled, the concentration of CO$_2$ in body fluids increases, and the reactions proceed from left to right. As a result, the concentration of H$^+$ in body fluids increases and blood pH falls.

Figure 22.6 ■ **Negative feedback regulation of blood pH by the respiratory system.**

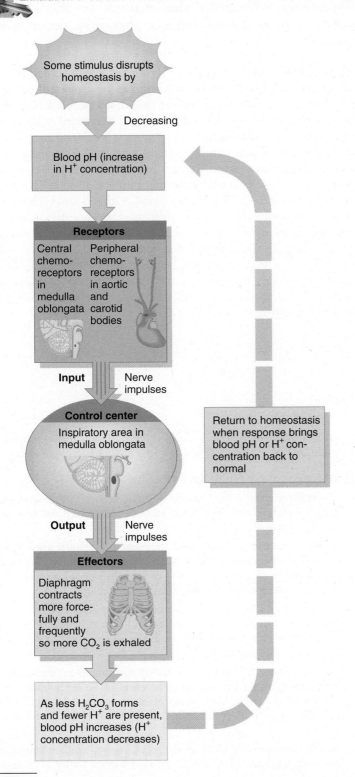

Exhalation of carbon dioxide lowers the H⁺ concentration of blood.

Some stimulus disrupts homeostasis by

Decreasing

Blood pH (increase in H⁺ concentration)

Receptors

Central chemoreceptors in medulla oblongata

Peripheral chemoreceptors in aortic and carotid bodies

Input Nerve impulses

Control center

Inspiratory area in medulla oblongata

Return to homeostasis when response brings blood pH or H⁺ concentration back to normal

Output Nerve impulses

Effectors

Diaphragm contracts more forcefully and frequently so more CO₂ is exhaled

As less H₂CO₃ forms and fewer H⁺ are present, blood pH increases (H⁺ concentration decreases)

If you hold your breath for 30 seconds, what is likely to happen to your blood pH?

Although this respiratory mechanism is powerful, it can only get rid of one acid, namely, carbonic acid.

The pH of body fluids and the rate and depth of breathing interact in a negative feedback system (Figure 22.6). When the blood becomes more acidic, the decrease in pH (increase in concentration of H⁺) is detected by central chemoreceptors in the medulla oblongata and peripheral chemoreceptors in the aortic and carotid bodies, both of which stimulate the inspiratory area in the medulla oblongata. As a result, the diaphragm and other breathing muscles contract more forcefully and frequently, so more CO₂ is exhaled. As less H₂CO₃ forms and H⁺ level drops, blood pH increases.

If the pH of the blood increases or the blood CO₂ level decreases, the respiratory center is inhibited and ventilation decreases. When ventilation decreases, the blood concentrations of both CO₂ and H⁺ increase.

Kidneys

The slowest mechanism for removal of acids is also the only way to eliminate most acids that form in the body: Cells of the renal tubules secrete H⁺, which then is excreted in urine. Also, because the kidneys synthesize new HCO₃⁻ and reabsorb filtered HCO₃⁻, this important buffer is not lost in the urine. Because of the contributions of the kidneys to acid–base balance, it's not surprising that renal failure can quickly cause death.

Table 22.2 summarizes the mechanisms that maintain pH of body fluids.

Acid-Base Imbalances

Acidosis is a condition in which arterial blood pH is below 7.35. The principal physiological effect of acidosis is depression of the central nervous system through depression of synaptic transmis-

Table 22.2 /	**Mechanisms That Maintain pH of Body Fluids**
Mechanism	**Comments**
Buffer Systems	Convert strong acids and bases into weak acids and bases, preventing drastic changes in body fluid pH.
Proteins	The most abundant buffers in body cells and blood. Hemoglobin is a buffer in the cytosol of red blood cells; albumin is a buffer in plasma.
Carbonic acid–bicarbonate	Important regulators of blood pH. The most abundant buffers in extracellular fluid.
Phosphates	Important buffers in intracellular fluid and in urine.
Exhalation of CO₂	With increased exhalation of CO₂, pH rises (fewer H⁺). With decreased exhalation of CO₂, pH falls (more H⁺).
Kidneys	Kidney tubules secrete H⁺ into the urine and reabsorb HCO₃⁻ so it is not lost in the urine.

sion. If the systemic arterial blood pH falls below 7, depression of the nervous system is so severe that the individual becomes disoriented, then becomes comatose, and may die.

In *alkalosis,* arterial blood pH is higher than 7.45. A major physiological effect of alkalosis is overexcitability in both the central nervous system and peripheral nerves. Neurons conduct impulses repetitively, even when not stimulated by normal stimuli; the results are nervousness, muscle spasms, and even convulsions and death.

A change in blood pH that leads to acidosis or alkalosis may be countered by *compensation,* the physiological response to an acid–base imbalance that acts to normalize arterial blood pH. Compensation may be either *complete,* if pH indeed is brought within the normal range, or *partial,* if systemic arterial blood pH is still lower than 7.35 or higher than 7.45. If a person has altered blood pH due to metabolic causes, hyperventilation or hypoventilation can help bring blood pH back toward the normal range; this form of compensation, termed *respiratory compensation,* occurs within minutes and reaches its maximum within hours. If, however, a person has altered blood pH due to respiratory causes, then *renal compensation*—changes in secretion of H^+ and reabsorption of HCO_3^- by the kidney tubules—can help reverse the change. Renal compensation may begin in minutes, but it takes days to reach maximum effectiveness.

AGING AND FLUID, ELECTROLYTE, AND ACID–BASE BALANCE

Objective: • **Describe the changes in fluid, electrolyte, and acid-base balance that may occur with aging.**

By comparison with children and younger adults, older adults often have an impaired ability to maintain fluid, electrolyte, and acid–base balance. With increasing age, many people have a decreased volume of intracellular fluid and decreased total body potassium due to declining skeletal muscle mass and increasing mass of adipose tissue (which contains very little water). Age-related decreases in respiratory and renal functioning may compromise acid–base balance by slowing the exhalation of CO_2 and the excretion of excess acids in urine. Other kidney changes, such as decreased blood flow, decreased glomerular filtration rate, and reduced sensitivity to antidiuretic hormone, have an adverse effect on the ability to maintain fluid and electrolyte balance. Due to a decrease in the number and efficiency of sweat glands, water loss from the skin declines with age. Because of these age-related changes, older adults are susceptible to several fluid and electrolyte disorders.

▮ STUDY OUTLINE

Fluid Compartments and Fluid Balance (p. 532)

1. Body fluid includes water and dissolved solutes.

2. About two-thirds of the body's fluid is located within cells and is called intracellular fluid (ICF). The other one-third, called extracellular fluid (ECF), includes all other body fluids. About 80% of the ECF is interstitial fluid, which occupies the microscopic spaces between tissue cells, and 20% of the ECF is plasma, the liquid portion of the blood.

3. Fluid balance means that the various body compartments contain the normal amount of water and solutes.

4. An electrolyte is an inorganic substance that dissociates into ions in solution. Fluid balance and electrolyte balance are interrelated.

5. Water is the largest single constituent in the body, about 55–60% of total body mass in lean adults.

6. Daily water gain and loss are each about 2500 mL. Sources of water gain are ingested liquids and foods and water produced by metabolic reactions (metabolic water). Water is lost from the body through urination, evaporation from the skin surface, exhalation of water vapor, and defecation. In women, menstrual flow is an additional route for loss of body water.

7. The main way to regulate body water gain is by adjusting the volume of water intake. The thirst center in the hypothalamus governs the urge to drink.

8. Angiotensin II and aldosterone reduce urinary loss of Na^+ and Cl^- and thereby increase the volume of body fluids. Antidiuretic hormone (ADH) decreases water loss in the urine, which also increases the volume of body fluids. ANP promotes natriuresis, elevated excretion of Na^+ (and Cl^-), which decreases blood volume.

Electrolytes in Body Fluids (p. 536)

1. Electrolytes control the osmosis of water between fluid compartments, help maintain acid–base balance, carry electrical current, and act as enzyme cofactors.

2. Sodium ions (Na^+) are the most abundant extracellular ions. They are involved in impulse transmission, muscle contraction, and fluid and electrolyte balance. Na^+ level is controlled by aldosterone, antidiuretic hormone, and atrial natriuretic peptide.

3. Chloride ions (Cl^-) are the major extracellular anions. They play a role in regulating osmotic pressure and forming HCl in gastric juice. Cl^- level is controlled by processes that increase or decrease kidney reabsorption of Na^+.

4. Potassium ions (K^+) are the most abundant cations in intracellular fluid. They play a key role in production of action potentials in neurons and muscle fibers, help maintain intracellular fluid volume, and contribute to regulation of pH. K^+ level is controlled by aldosterone.

5. Bicarbonate ions (HCO_3^-) are the second most abundant anions in extracellular fluid. They are the most important buffer in plasma.

6. Calcium is the most abundant mineral in the body. Calcium salts are structural components of bones and teeth. Ca^{2+}, which are principally extracellular cations, function in blood clotting, neurotransmitter release, and contraction of muscle. Ca^{2+} level is controlled mainly by parathyroid hormone and calcitriol.

7. Phosphate ions ($H_2PO_4^-$, HPO_4^{2-}, and PO_4^{3-}) are principally intracellular anions, and their salts are structural components of bones and teeth. They are also required for the synthesis of nucleic acids and ATP and participate in buffer reactions. Their level is controlled by parathyroid hormone and calcitriol.

8. Magnesium ions (Mg^{2+}) are primarily intracellular cations. They are needed for operation of the sodium pump and assist several enzymes.

9. Table 22.1 on page 000 describes the imbalances that result from deficiency or excess of important body electrolytes.

Acid–Base Balance (p. 540)

1. The normal pH of systemic arterial blood is 7.35 to 7.45.

2. Homeostasis of pH is maintained by buffer systems, by exhalation of carbon dioxide, and by kidney excretion of H^+ and reabsorption of HCO_3^-. Table 22.2 on page 000 summarizes the mechanisms that maintain pH of body fluids.

3. Acidosis is a systemic arterial blood pH below 7.35; its principal effect is depression of the central nervous system (CNS). Alkalosis is a systemic arterial blood pH above 7.45; its principal effect is overexcitability of the CNS.

Aging and Fluid, Electrolyte, and Acid–Base Balance (p. 542)

1. With increasing age, there is decreased intracellular fluid volume and decreased potassium due to declining skeletal muscle mass.

2. Decreased kidney function adversely affects fluid and electrolyte balance.

▪ SELF-QUIZ

1. Normally, most of the body's water is lost through

  a. the gastrointestinal tract **b.** cellular respiration **c.** exhalation by the lungs **d.** excretion of urine **e.** evaporation from the skin

2. Substances that dissociate into ions when dissolved in body fluids are

  a. neurotransmitters **b.** enzymes **c.** nonelectrolytes **d.** hormones **e.** electrolytes

3. Which of the following statements about sodium is NOT true?

  a. Sodium ions are the most abundant intracellular ions. **b.** Sodium is necessary for generating action potentials in neurons. **c.** Excess sodium ions can cause edema. **d.** Sodium levels are regulated by the kidneys. **e.** Aldosterone helps regulate the concentration of sodium in the blood.

4. Parathyroid hormone (PTH) controls blood level of

  a. magnesium **b.** sodium **c.** calcium **d.** potassium **e.** chloride

5. Fluid movement between intracellular fluid and extracellular fluid depends primarily on the concentration of which ion in extracellular fluid?

  a. sodium **b.** potassium **c.** calcium **d.** phosphate **e.** magnesium

6. Which of the following are mismatched?

  a. the most abundant extracellular anion, Cl^- **b.** the most abundant mineral in the body, Ca^{2+} **c.** the most abundant extracellular cation, Na^+ **d.** the most abundant intracellular cation, K^+ **e.** the most abundant intracellular anion, HCO_3^-

7. Which of the following statements concerning acid–base balance in the body is NOT true?

  a. An increase in respiration rate increases pH of body fluids. **b.** Normal pH of extracellular fluid is 7.35 to 7.45. **c.** Buffers are an important mechanism in the maintenance of pH balance. **d.** A blood pH of 7.2 is called alkalosis. **e.** Respiratory acidosis is characterized by a high level of CO_2 in body fluids.

8. Most human buffer systems consist of

  a. a weak acid and a weak base **b.** a strong acid and a strong base **c.** a strong acid such as HCl **d.** an electrolyte and nonelectrolyte **e.** a weak base and a gas

9. The most abundant buffer in body cells and plasma is the _____ buffer system.

  a. hemoglobin **b.** carbonic acid **c.** protein **d.** bicarbonate **e.** phosphate

10. Most (80%) of the extracellular fluid is part of the body's

  a. interstitial fluid **b.** lymph **c.** cerebrospinal fluid **d.** plasma **e.** synovial fluid

11. Which hormone stimulates the kidneys to secrete more K^+?

  a. atrial natriuretic peptide **b.** angiotensin **c.** aldosterone **d.** antidiuretic hormone **e.** parathyroid hormone

12. Most of the body's water comes from

  a. cellular respiration **b.** adipose tissue **c.** urine production **d.** water intoxication **e.** ingested liquids and foods

13. The thirst center can be activated by all of the following EXCEPT

  a. angiotensin II **b.** an increase in blood volume **c.** a decrease in flow from salivary glands **d.** a decrease in blood pressure **e.** an increase in blood osmotic pressure

14. The center for thirst is located in the

  a. kidneys **b.** adrenal cortex **c.** hypothalamus **d.** cerebral cortex **e.** liver

15. Which of the following is NOT one of the functions of electrolytes in the body?

  a. control of fluid movement between the extracellular and intracellular compartments **b.** regulation of pH **c.** enzyme cofactor **d.** energy source **e.** carrier of electric current

16. Aldosterone is secreted in response to

  a. increased blood pressure **b.** decreased blood volume **c.** increased calcium levels **d.** increased sodium levels **e.** increased water levels

17. What is the importance of buffer systems in the body?

  a. They help maintain the calcium and phosphate balances of bone. **b.** They control the body's water balance. **c.** They prevent drastic changes in the body's pH. **d.** They help regulate blood volume. **e.** They are responsible for the operation of the body's sodium pump.

18. Match the following:

_____ **a.** intracellular anion; structural component of bones and teeth

_____ **b.** most abundant ion in the body; controlled by parathyroid hormone (PTH)

_____ **c.** most abundant extracellular cation; needed for generation and conduction of action potentials

_____ **d.** intracellular ion that is an enzyme cofactor; needed for secretion of parathyroid hormone

_____ **e.** most abundant intracellular ion; involved in nerve and muscle homeostasis

_____ **f.** found in intracellular fluid; assists sodium in regulating osmotic pressure

A. calcium
B. chloride
C. potassium
D. sodium
E. phosphate
F. magnesium

CRITICAL THINKING APPLICATIONS

1. If anesthesia were improperly administered and a patient's breathing were depressed, what would be the effect on blood pH? Explain.

2. One-year-old Andrew had a busy morning at the "mom and tot" swim program. Today's lesson included lots of underwater exercises in blowing bubbles. After the lesson, Andrew seemed disoriented and then suffered a convulsion. The emergency room nurse thinks the swim class has something to do with Andrew's problem. What is wrong with Andrew?

3. Mike and Jennie are the same height and both weigh 150 lb, but when Mike and Jennie measured their blood alcohol after drinking

identical alcoholic beverages, Jennie's blood alcohol level was higher than Mike's. In the body, alcohol is transported in the body fluids. Use your knowledge about the differences in body water between males and females to explain the difference in alcohol level.

4. Alex was 15 minutes late for A & P class. While searching for his pen, he thought he heard the instructor say something about the heart affecting water balance, but he had thought it was the other way around. Alex just decided to ignore the whole thing. Bad move Alex! Explain the relationship of the heart to body fluid balance.

ANSWERS TO FIGURE QUESTIONS

22.1 The term body fluid refers to body water and its dissolved substances.

22.2 Hyperventilation increases loss of fluid from the body.

22.3 Negative feedback is in operation because the result (an increase in fluid intake) is opposite to the initiating stimulus (dehydration).

22.4 Elevated aldosterone promotes abnormally high renal reabsorption of NaCl and water, which expands blood volume and increases blood pressure. Increased blood pressure causes more fluid to filter out of capillaries and accumulate in the interstitial fluid, a condition called edema.

22.5 The major cation in ECF is Na^+.

22.6 Breath holding causes blood pH to decrease slightly as CO_2 and H^+ accumulate.

Chapter **23**

The Reproductive Systems

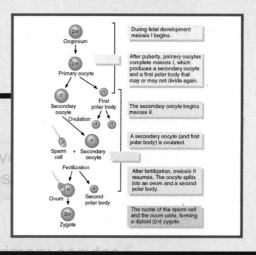

During fetal development meiosis I begins.

After puberty, primary oocytes complete meiosis I, which produces a secondary oocyte and a first polar body that may or may not divide again.

The secondary oocyte begins meiosis II.

A secondary oocyte (and first polar body) is ovulated.

After fertilization, meiosis II resumes. The oocyte splits into an ovum and a second polar body.

The nuclei of the sperm cell and the ovum unite, forming a diploid (2n) zygote.

■ Student Learning Objectives

1. Describe the location, structure, and functions of the organs of the male reproductive system. **546**
2. Describe how sperm cells are produced. **546**
3. Explain the roles of hormones in regulating male reproductive functions. **550**
4. Describe the location, structure, and functions of the organs of the female reproductive system. **555**

5. Describe how oocytes are produced. **556**
6. Explain the functions of the female reproductive hormones. **561**
7. Define the ovarian and menstrual cycles and explain how they are related. **561**

8. Compare the various types of birth control methods and outline the effectiveness of each. **565**
9. Describe the similarities and differences in the sexual responses of males and females. **567**
10. Describe the effects of aging on the reproductive systems. **568**

■ A Look Ahead

Sexual reproduction is a process in which organisms produce offspring by means of germ cells called *gametes* (GAM-ēts = spouses). After *fertilization,* when the male gamete (sperm cell) unites with the female gamete (secondary oocyte), the resulting cell contains one set of chromosomes from each parent. The organs that make up the male and female reproductive systems can be grouped by function. The *gonads*—testes in males and ovaries in females—produce gametes and secrete sex hormones. Various *ducts* then store and transport the gametes, and *accessory sex glands* produce substances that protect the gametes and facilitate their movement. Finally, *supporting structures,* such as the penis and the uterus, assist the delivery and joining of gametes. In addition, females have organs that sustain the growth of the embryo and fetus.

Gynecology (gī′-ne-KOL-ō-jē; *gynec-* = woman; *-ology* = study of) is the specialized branch of medicine concerned with the diagnosis and treatment of diseases of the female reproductive system. As noted in Chapter 21, *urology* (yoor-OL-ō-jē) is the study of the urinary system. Urologists also diagnose and treat diseases and disorders of the male reproductive system.

MALE REPRODUCTIVE SYSTEM

Objectives: • **Describe the location, structure, and functions of the organs of the male reproductive system.**

• **Describe how sperm cells are produced.**

• **Explain the roles of hormones in regulating male reproductive functions.**

The organs of the male reproductive system include the testes; a system of ducts (epididymis, ductus deferens, ejaculatory ducts, and urethra); accessory sex glands (seminal vesicles, prostate gland, and bulbourethral gland); and several supporting structures, including the scrotum and the penis (Figure 23.1). The testes produce sperm and secrete hormones. Sperm are transported and stored, helped to mature, and conveyed to the exterior by a system of ducts. Semen contains sperm plus the secretions provided by the testes and accessory sex glands.

Scrotum

The *scrotum* (SKRŌ-tum = bag) is a pouch that supports the testes; it consists of loose skin, superficial fascia, and smooth

muscle (Figure 23.1). Internally, a septum divides the scrotum into two sacs, each containing a single testis.

The production and survival of sperm require a temperature that is lower than normal body temperature. The location of the scrotum and contraction of its muscle fibers help regulate the temperature of the testes. Because the scrotum is outside the body cavities, its temperature normally is about 3°C below body temperature. On exposure to cold and during sexual arousal, skeletal muscles contract to elevate the testes, moving them closer to the pelvic cavity, where they can absorb body heat. Exposure to warmth causes relaxation of the skeletal muscles and descent of the testes, increasing the surface area exposed to the air, so that the testes can give off excess heat to the surrounding air.

Testes

The *testes* (TES-tēz; singular is *testis*), or *testicles,* are paired oval glands that develop on the embryo's posterior abdominal wall and usually begin their descent into the scrotum in the seventh month of fetal development. Failure of the testes to descend is termed *cryptorchidism* (krip-TOR-ki-dizm).

The testes are covered by a dense *white fibrous capsule* that extends inward and divides each testis into internal compartments called *lobules* (Figure 23.2a). Each of the 200 to 300 lobules contains one to three tightly coiled *seminiferous tubules* (*semin-* = seed; *fer-* = to carry) that produce sperm by a process called spermatogenesis. This process is considered shortly.

Seminiferous tubules are lined with spermatogenic (sperm-forming) cells in various stages of development (Figure 23.2b). Positioned against the basement membrane, toward the outside of the tubules, are the *spermatogonia* (sper-ma′-tō-GŌ-nē-a; *-gonia* = offspring), the stem cell precursors. Toward the lumen of the tubule are layers of cells in order of advancing maturity: primary spermatocytes, secondary spermatocytes, spermatids, and sperm. The mature *sperm cell,* or *spermatozoon* (sper-ma′-tō-ZŌ-on; *-zoon* = life), is released into the lumen of the tubule.

Large *Sertoli cells,* located between the developing sperm cells in the seminiferous tubules, support, protect, and nourish spermatogenic cells; phagocytize degenerating spermatogenic cells; secrete fluid for sperm transport; and release the hormone inhibin, which helps regulate sperm production. Between the seminiferous tubules are clusters of *Leydig cells.* These cells secrete the hormone testosterone, the most important *androgen* (AN-drō-jen). Androgens are substances that promote development of masculine characteristics.

Spermatogenesis

The process by which the seminiferous tubules of the testes produce sperm is called *spermatogenesis* (sper-ma′-tō-JEN-e-sis). It consists of three stages: meiosis I, meiosis II, and spermiogenesis. We begin with meiosis.

OVERVIEW OF MEIOSIS As you learned in Chapter 3, most body cells (somatic cells), such as brain cells, stomach cells, kidney cells, and so forth, contain 23 pairs of chromosomes, or a total of 46 chromosomes. One member of each pair is inherited from each parent. The chromosomes that make up each pair are called *homologous chromosomes* (hō-MOL-ō-gus; *homo-* = same) or *homologs;* they contain similar genes arranged in the same (or almost the same) order. Because somatic cells contain two sets of chromosomes, they are termed *diploid cells* (DIP-loyd; *dipl-* = double; *-oid* = form), symbolized as *2n*. Gametes differ from somatic cells in that they contain a single set of 23 chromosomes, symbolized as *n*; they are thus said to be *haploid* (HAP-loyd; *hapl-* = single).

In sexual reproduction, each new organism results from the fusion of two different gametes, one produced by each parent. If each gamete had the same number of chromosomes as somatic cells, then the number of chromosomes would double each time fertilization occurred. Instead, gametes receive a single set of chromosomes by means of a special type of cell division called *meiosis* (mī-Ō-sis; *mei-* = lessening; *-osis* = condition of). Meiosis occurs in two successive stages: *meiosis I* and *meiosis II.* First, we will examine how meiosis occurs during spermatogenesis. Later in the chapter, we will follow the steps of meiosis during oogenesis, the production of female gametes.

STAGES OF SPERMATOGENESIS Spermatogenesis begins during puberty and continues throughout life. The time from onset of

Figure 23.1 ■ Male organs of reproduction and surrounding structures.

Reproductive organs are adapted to produce new individuals and pass genetic material from one generation to the next.

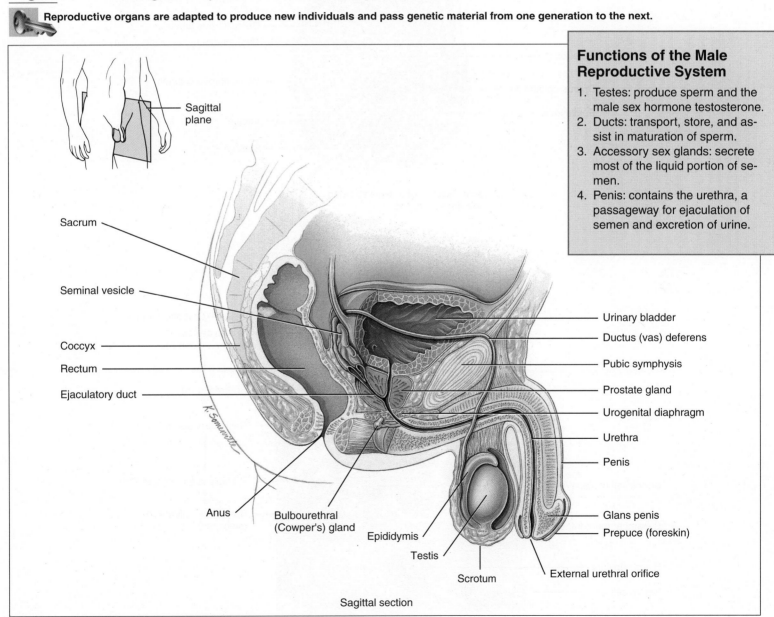

Functions of the Male Reproductive System

1. Testes: produce sperm and the male sex hormone testosterone.
2. Ducts: transport, store, and assist in maturation of sperm.
3. Accessory sex glands: secrete most of the liquid portion of semen.
4. Penis: contains the urethra, a passageway for ejaculation of semen and excretion of urine.

Sagittal plane

Sacrum

Seminal vesicle

Coccyx

Rectum

Ejaculatory duct

Anus

Bulbourethral (Cowper's) gland

Epididymis

Testis

Scrotum

Urinary bladder

Ductus (vas) deferens

Pubic symphysis

Prostate gland

Urogenital diaphragm

Urethra

Penis

Glans penis

Prepuce (foreskin)

External urethral orifice

Sagittal section

Among the male organs of reproduction, how is the penis classified functionally?

Figure 23.2 ■ Anatomy of the testes. The stages of spermatogenesis are shown in (b). Arrows in (b) indicate the progression from least to most mature spermatogenic cells. The (*n*) and (2*n*) refer to haploid and diploid chromosome number, to be described shortly.

The male gonads are the testes, which produce haploid sperm.

Sagittal plane

Spermatic cord

Blood vessels and nerves

Ductus (vas) deferens

Seminiferous tubule

Straight tubule

Rete testis

White fibrous capsule

Ductus epididymis

Epididymis

Lobule

Septum

(a) Sagittal section of a testis showing seminiferous tubules

SPERMATOGENIC CELLS:

Sperm cell or spermatozoon *(n)*

Spermatid *(n)*

Secondary spermatocyte *(n)*

Sertoli cell nucleus

Primary spermatocyte *(2n)*

Basement membrane

Spermatogonium *(2n)* (stem cell)

Blood capillary

Leydig cell

(b) Transverse section of a portion of a seminiferous tubule

Which spermatogenic cells in a seminiferous tubule are least mature?

cell division in a spermatogonium until sperm are released into the lumen of a seminiferous tubule is 65 to 75 days. The spermatogonia contain the diploid number of chromosomes (46). After a spermatogonium undergoes mitosis, one daughter cell stays near the basement membrane as a spermatogonium, so stem cells stays for future mitosis (Figure 23.3). The other daughter cell differentiates into a *primary spermatocyte* (sper-MA-tō-sīt'). Like spermatogonia, primary spermatocytes are diploid. Spermatogenesis proceeds as follows:

1. **Meiosis I.** During the interphase that precedes meiosis I, the chromosomes replicate, as also occurs in the interphase before mitosis in somatic cell division (see page 000). The 46 chromosomes, now each made up of two identical "sister" chromatids, line up as 23 pairs of homologs. (By contrast, pairing of homologs does not occur during mitosis.) The four chromatids of each homologous pair then twist around one another. At this time, portions of one chromatid may be exchanged with portions of another; such an exchange is termed *crossing-over.* Crossing-over results in *genetic recombination,* the formation of new combinations of genes. As a result, the sperm eventually produced are genetically unlike one another and unlike the parent cell that produced them.

   Next, the members of each homologous pair separate, with one member of each pair moving to opposite ends of the cell. The sister chromatids, held by a centromere, remain together. (During mitosis, the sister chromatids move to opposite ends of the cell.) The net effect of meiosis I is that each resulting daughter cell contains a haploid set of chromosomes, even though each one exists as two copies; each cell contains only one member of each pair of the homologous chromosomes present in the parent cell.

   The cells formed by meiosis I are haploid *secondary spermatocytes,* having 23 (duplicated) chromosomes. Each chromosome within a secondary spermatocyte, is made up of two chromatids (two identical copies of the DNA) still attached by a centromere. The genes on each chromatid may be rearranged as a result of crossing-over.

2. **Meiosis II.** In meiosis II there is no further replication of DNA. The chromosomes of the secondary spermatocytes line up in single file near the center of the nucleus, and this time the chromatids separate, as also occurs in mitosis. The cells formed from meiosis II, termed *spermatids,* contain 23 chromosomes, each of which is composed of a single chromatid.

3. **Spermiogenesis.** In the final stage of spermatogenesis, called *spermiogenesis* (sper'-mē-ō-JEN-e-sis), each haploid spermatid develops a head and a tail (see Figure 23.2b), and matures into a single haploid sperm cell.

Sperm

Sperm are produced at the rate of about 300 million per day. Once ejaculated, most do not survive more than 48 hours in the female reproductive tract. A sperm cell consists of structures

Figure 23.3 ■ **Spermatogenesis.** The designation *2n* means diploid (46 chromosomes); *n* means haploid (23 chromosomes).

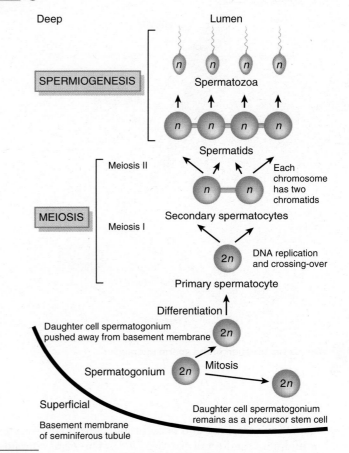

Spermiogenesis involves the maturation of spermatids into sperm.

 What is the significance of crossing-over?

Figure 23.4 ■ **Parts of a sperm cell.**

About 300 million sperm mature each day.

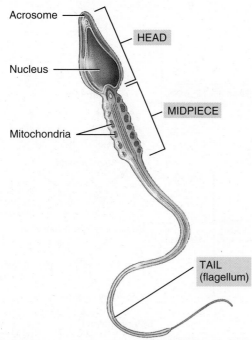

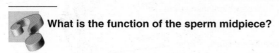

What is the function of the sperm midpiece?

highly adapted for reaching and penetrating a secondary oocyte: a head, a midpiece, and a tail (Figure 23.4). The *head* contains the nuclear material (DNA) and an *acrosome* (*acro-* = atop), a vesicle containing enzymes that aid penetration by the sperm cell into a secondary oocyte. Numerous mitochondria in the *midpiece* carry on the metabolism that provides ATP for locomotion. The *tail,* a typical flagellum, propels the sperm cell along its way.

Hormonal Control of Testicular Functions

At the onset of puberty, the anterior pituitary starts to secrete two hormones that have profound effects on male reproductive organs: LH (luteinizing hormone) and FSH (follicle-stimulating hormone). Their release is controlled by a releasing hormone from the hypothalamus called gonadotropin-releasing hormone (GnRH). LH stimulates Leydig cells to secrete the hormone testosterone. FSH and testosterone act on the seminiferous tubules to stimulate spermatogenesis.

Testosterone (tes-TOS-te-rōn) is synthesized from cholesterol in the testes. It has a number of effects on the male body:

- During prenatal development, it stimulates the development of male internal reproductive system structures.

- It stimulates the descent of the testes just before birth.

- At puberty, it brings about development and enlargement of the male sex organs and the development of male secondary

sex characteristics. These include pubic, axillary, facial, and chest hair; temporal hairline recession; thickening of the skin; increased sebaceous (oil) gland secretion; growth of skeletal muscles and bones; and enlargement of the larynx and deepening of the voice.

- It stimulates sex drive (libido).

- Testosterone is an anabolic hormone; that is, it stimulates protein synthesis. This effect is obvious in the heavier muscle and bone mass of most men as compared to women. It also stimulates closure of the epiphyseal plates, the structures responsible for bone growth in length.

A negative feedback system regulates testosterone production (Figure 23.5). Once testosterone concentration in the blood reaches a certain level, it inhibits the release of GnRH by cells in the hypothalamus. As a result, cells in the anterior pituitary release less LH, and the Leydig cells in the testes secrete less testosterone. If the testosterone concentration in the blood falls too low, however, GnRH is again released by the hypothalamus.

A protein hormone named *inhibin,* released by Sertoli cells, inhibits FSH secretion and thereby regulates the extent of spermatogenesis. When spermatogenesis is proceeding too quickly, Sertoli cells release more inhibin. If spermatogenesis is inadequate, less inhibin is released, which permits more FSH secretion and an increased pace of spermatogenesis.

Ducts

Ducts of the Testis

Following spermatogenesis, pressure generated by the continual release of sperm and fluid secreted by Sertoli cells propels sperm and fluid through the seminiferous tubules to the **straight tubules** (see Figure 23.2a). The straight tubules lead to a network of ducts in the testis called the **rete testis** (RĒ-tē = network). The sperm are next transported out of the testis into the epididymis (see Figure 23.2a)

Epididymis

The **epididymis** (ep′-i-DID-i-mis; *epi-* = over; *-didymis* = testis; plural is *epididymides*) is a comma-shaped organ that lies along the posterior border of the testis (see Figures 23.1 and 23.2a) and consists mostly of a tightly coiled tube, the **ductus epididymis.** Functionally, the ductus epididymis is the site where sperm motility increases over a 10- to 14-day period. The ductus epididymis also participates in the maturation of sperm, stores sperm, and helps propel them by peristaltic contraction of its smooth muscle into the ductus (vas) deferens. Sperm may remain in storage in the ductus epididymis for a month or more, after which they are expelled or reabsorbed.

Ductus (Vas) Deferens

At the end of the epididymis, the ductus epididymis becomes less convoluted, and its diameter increases. Beyond the epididymis, the duct is termed the **ductus deferens** or **vas deferens** (VAS DEF-er-enz; *vas* = vessel; *de-* = away). See Figure 23.2a.

Figure 23.5 ■ **Negative feedback control of blood level of testosterone.**

Testosterone production is controlled by a negative feedback system that involves the hypothalamus and anterior pituitary.

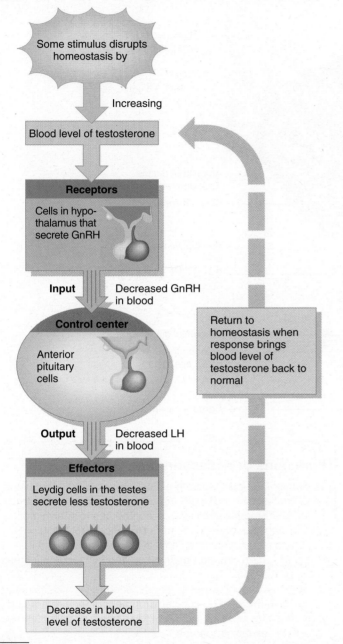

Which hormones inhibit secretion of LH and FSH by the anterior pituitary?

The vas deferens ascends along the posterior border of the epididymis and penetrates the inguinal canal, a passageway in the front abdominal wall. Then, it enters the pelvic cavity, where it loops over the side and down the posterior surface of the urinary bladder (see Figure 23.1). The vas deferens has a heavy coat of three layers of muscle. It stores sperm for up to several months and propels them toward the urethra during ejaculation by peristaltic contractions of its muscular coat. Sperm cells that are not ejaculated are reabsorbed.

Accompanying the vas deferens as it ascends in the scrotum are blood vessels, autonomic nerves, and lymphatic vessels that together make up the *spermatic cord*, a supporting structure of the male reproductive system (see Figure 23.2a).

Ejaculatory Ducts

The *ejaculatory ducts* (e-JAK-yoo-la-tō′-rē; *ejacul-* = to expel) (Figure 23.6) are formed by the union of the duct from the seminal vesicles (to be described shortly) and the vas deferens. The ejaculatory ducts eject sperm into the urethra.

Urethra

The *urethra* is the terminal duct of the male reproductive system, serving as a passageway for both sperm and urine. In the male, the urethra passes through the prostate gland, urogenital diaphragm (a fibrous membrane and muscle across the pelvic floor), and penis (see Figure 23.1). The opening of the urethra to the exterior is called the *external urethral orifice*.

Accessory Sex Glands

Whereas the ducts of the male reproductive system store and transport sperm cells, the *accessory sex glands* secrete most of the liquid portion of semen.

The paired *seminal vesicles* (VES-i-kuls) are pouchlike structures, lying at the base of the urinary bladder anterior to the rectum (Figures 23.1 and 23.6). They secrete an alkaline, viscous fluid that contains fructose, prostaglandins, and clotting proteins (unlike those found in blood). The alkaline nature of the fluid helps to neutralize the acidic environment of the male urethra and female reproductive tract that otherwise would inactivate and kill sperm. The fructose is used for ATP production by sperm. Prostaglandins contribute to sperm motility and viability and may also stimulate muscular contraction within the female reproductive tract. Clotting proteins help semen coagulate after ejaculation. Fluid secreted by the seminal vesicles normally constitutes about 60% of the volume of semen.

The *prostate gland* (PROS-tāt) is a single, doughnut-shaped gland about the size of a chestnut (Figure 23.6). It is inferior to the urinary bladder and surrounds the upper portion of the urethra. The prostate gland slowly increases in size from birth to puberty, and then it expands rapidly. The size attained by age 30 remains stable until about age 45, when further enlargement may occur. The prostate secretes a milky, slightly acidic fluid (pH about 6.5) that contains (1) *citric acid*, which can be used by sperm for ATP production via the Krebs cycle (see page 496); (2) acid phosphatase (the function of which is unknown); and (3) several

Figure 23.6 ■ **Male reproductive organs in relation to surrounding structures.**

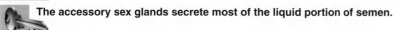

The accessory sex glands secrete most of the liquid portion of semen.

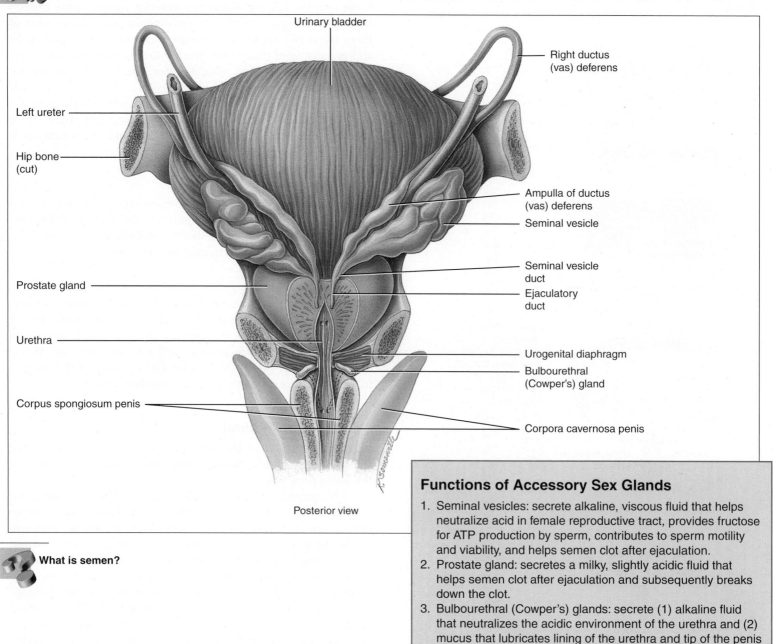

Posterior view

Functions of Accessory Sex Glands

1. Seminal vesicles: secrete alkaline, viscous fluid that helps neutralize acid in female reproductive tract, provides fructose for ATP production by sperm, contributes to sperm motility and viability, and helps semen clot after ejaculation.
2. Prostate gland: secretes a milky, slightly acidic fluid that helps semen clot after ejaculation and subsequently breaks down the clot.
3. Bulbourethral (Cowper's) glands: secrete (1) alkaline fluid that neutralizes the acidic environment of the urethra and (2) mucus that lubricates lining of the urethra and tip of the penis during sexual intercourse.

What is semen?

protein-digesting enzymes, such as *prostate-specific antigen (PSA)*, pepsinogen, and lysozyme. Prostatic secretions make up about 25% of the volume of semen.

The paired ***bulbourethral*** (bul′-bō-yoo-RĒ-thral) or ***Cowper's glands*** are about the size of peas. They are located inferior to the prostate gland on either side of the urethra within the urogenital diaphragm (Figure 23.6). During sexual arousal, the bulbourethral glands secrete an alkaline substance into the urethra that protects the passing sperm by neutralizing acids from urine in the urethra. At the same time, they secrete mucus that lubricates the end of the penis and the lining of the urethra,

thereby decreasing the number of sperm damaged during ejaculation.

Semen

Semen (= seed) is a mixture of sperm and the secretions of the Sertoli cells of the testes, seminal vesicles, prostate gland, and bulbourethral glands. The average volume of semen for each ejaculation is 2.5 to 5 mL, with 50 to 150 million sperm/mL. When the number falls below 20 million sperm/mL, the male is likely to be infertile. A very large number of sperm is required for fertilization because only a tiny fraction ever reach the secondary oocyte.

Despite the slight acidity of prostatic fluid, semen has a slightly alkaline pH of 7.2 to 7.7 due to the higher pH and larger volume of fluid from the seminal vesicles. The prostatic secretion gives semen a milky appearance, whereas fluids from the seminal vesicles and bulbourethral glands give it a sticky consistency. Semen also contains an antibiotic that can destroy certain bacteria. The antibiotic may help control the abundance of naturally occurring bacteria in the semen and in the lower female reproductive tract.

Once ejaculated, liquid semen clots within 5 minutes due to the presence of clotting proteins from the seminal vesicles. The functional role of semen clotting is not known, but the proteins involved are different from those that cause blood clotting. After about 10 to 20 minutes, semen reliquefies because enzymes pro-duced by the prostate gland break down the clot. Abnormal or delayed liquefaction of clotted semen may cause complete or partial immobilization of sperm, inhibiting their movement through the cervix of the uterus.

Penis

The **penis** contains the urethra and is a passageway for the ejaculation of semen and the excretion of urine (Figure 23.7). It is cylindrical in shape and consists of a root, a body, and the glans penis. The **root** is the proximal portion attached to the hip bones. The **body** is composed of three cylindrical masses of tissue. The two dorsolateral masses are called the **corpora cavernosa penis** (*corpora* = main bodies; *cavernosa* = hollow). The

Figure 23.7 ■ **Internal structure of the penis.** The inset in (b) shows details of the skin and fascia.

The penis contains a pathway for ejaculation of semen and excretion of urine.

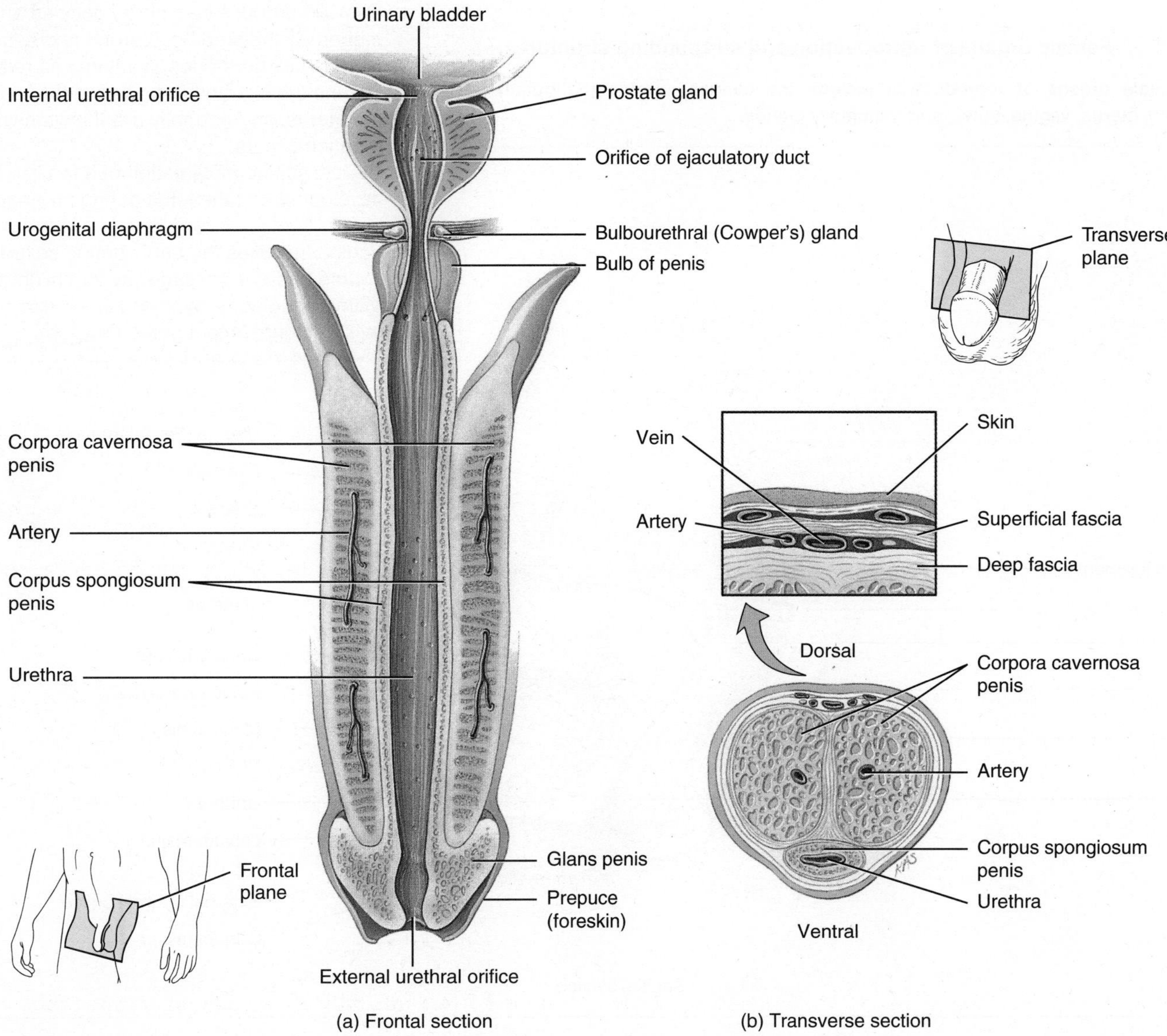

(a) Frontal section

(b) Transverse section

Which tissue masses form the erectile tissue in the penis, and why do they become rigid?

smaller midventral mass, the ***corpus spongiosum penis,*** contains the urethra. All three masses are enclosed by fascia (sheet of fibrous connective tissue) and skin and consist of erectile tissue permeated by blood sinuses.

Upon sexual stimulation, which may be visual, tactile, auditory, olfactory, or from the imagination, the arteries supplying the penis dilate, and large quantities of blood enter the blood sinuses. Expansion of these spaces compresses the veins draining the penis, so blood outflow is slowed. These vascular changes, due to a parasympathetic reflex, result in an ***erection,*** the enlargement and stiffening of the penis. The penis returns to its flaccid state when the arteries constrict and pressure on the veins is relieved.

Ejaculation is a sympathetic reflex. As part of the reflex, the smooth muscle sphincter at the base of the urinary bladder closes. Thus, urine is not expelled during ejaculation, and semen does not enter the urinary bladder. Even before ejaculation occurs, peristaltic contractions in the vas deferens, seminal vesicles, ejaculatory ducts, and prostate gland propel semen into the penile portion of the urethra. Typically, this leads to ***emission*** (ē-MISH-un), the discharge of a small volume of semen before ejaculation. Emission may also occur during sleep (nocturnal emission).

The distal end of the corpus spongiosum penis is a slightly enlarged region called the ***glans penis,*** which means shaped like an acorn. In the glans penis is the opening of the urethra (the ***external urethral orifice***) to the exterior. Covering the glans penis is the loosely fitting ***prepuce*** (PRĒ-pyoos), or ***foreskin.***

Circumcision (= to cut around) is a surgical procedure in which part or all of the prepuce is removed. Although not a

Figure 23.8 ■ **Female organs of reproduction and surrounding structures.**

The female organs of reproduction include the ovaries, uterine (Fallopian) tubes, uterus, vagina, vulva, and mammary glands.

Functions of the Female Reproductive System

1. Ovaries: produce secondary oocytes and hormones, including progesterone and estrogens (female sex hormones), inhibin, and relaxin.
2. Uterine tubes: transport a secondary oocyte to the uterus, and normally are the sites where fertilization occurs.
3. Uterus: site of implantation of a fertilized ovum, development of the fetus during pregnancy, and labor.
4. Vagina: receives the penis during sexual intercourse and is a passageway for childbirth.
5. Mammary glands: synthesize, secrete, and eject milk for nourishment of the newborn.

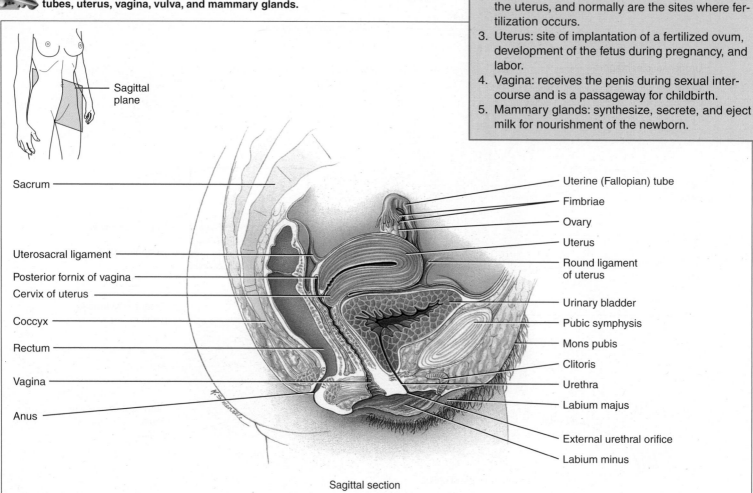

Sagittal section

What terms refer to the external genitalia of the female?

medically necessary procedure, it may be performed just after delivery, on the third or fourth day after birth, or on the eighth day as part of a Jewish religious rite.

FEMALE REPRODUCTIVE SYSTEM

Objectives: • **Describe the location, structure, and functions of the organs of the female reproductive system.**

• **Describe how oocytes are produced.**

The organs of the female reproductive system (Figure 23.8) include the ovaries; the uterine (Fallopian) tubes, or oviducts; the uterus; the vagina; and external organs collectively called the vulva, or pudendum. The mammary glands also are considered part of the female reproductive system.

Ovaries

The *ovaries* (= egg receptacles) are paired organs that produce secondary oocytes (cells that develop into mature ova, or eggs,

following fertilization) and hormones, such as progesterone and estrogens (the female sex hormones), inhibin, and relaxin. They arise from the same embryonic tissue as the testes, and they are the size and shape of unshelled almonds. One ovary lies on each side of the pelvic cavity, held in place by three ligaments: broad, ovarian, and suspensory (see Figure 23.11). Each contains a hilus where nerves, blood, and lymphatic vessels enter. They consist of the following parts (Figure 23.9):

1. The *germinal epithelium* is a layer of simple epithelium (low cuboidal or squamous) that covers the surface of the ovary.

2. The *ovarian cortex* is a region beneath the germinal epithelium that consists of dense connective tissue and contains ovarian follicles (described shortly).

3. The *ovarian medulla* is a region deep to the ovarian cortex that consists of loose connective tissue and contains blood vessels, lymphatic vessels, and nerves.

4. *Ovarian follicles* (*folliculus* = little bag) are in the cortex and consist of *oocytes* in various stages of development, plus the surrounding cells. The surrounding cells nourish the developing oocyte and begin to secrete estrogens as the follicle grows larger.

Figure 23.9 ■ **Histology of the ovary.** The arrows indicate the sequence of developmental stages that occur as part of the ovarian cycle.

The ovaries are the female gonads; they produce haploid oocytes.

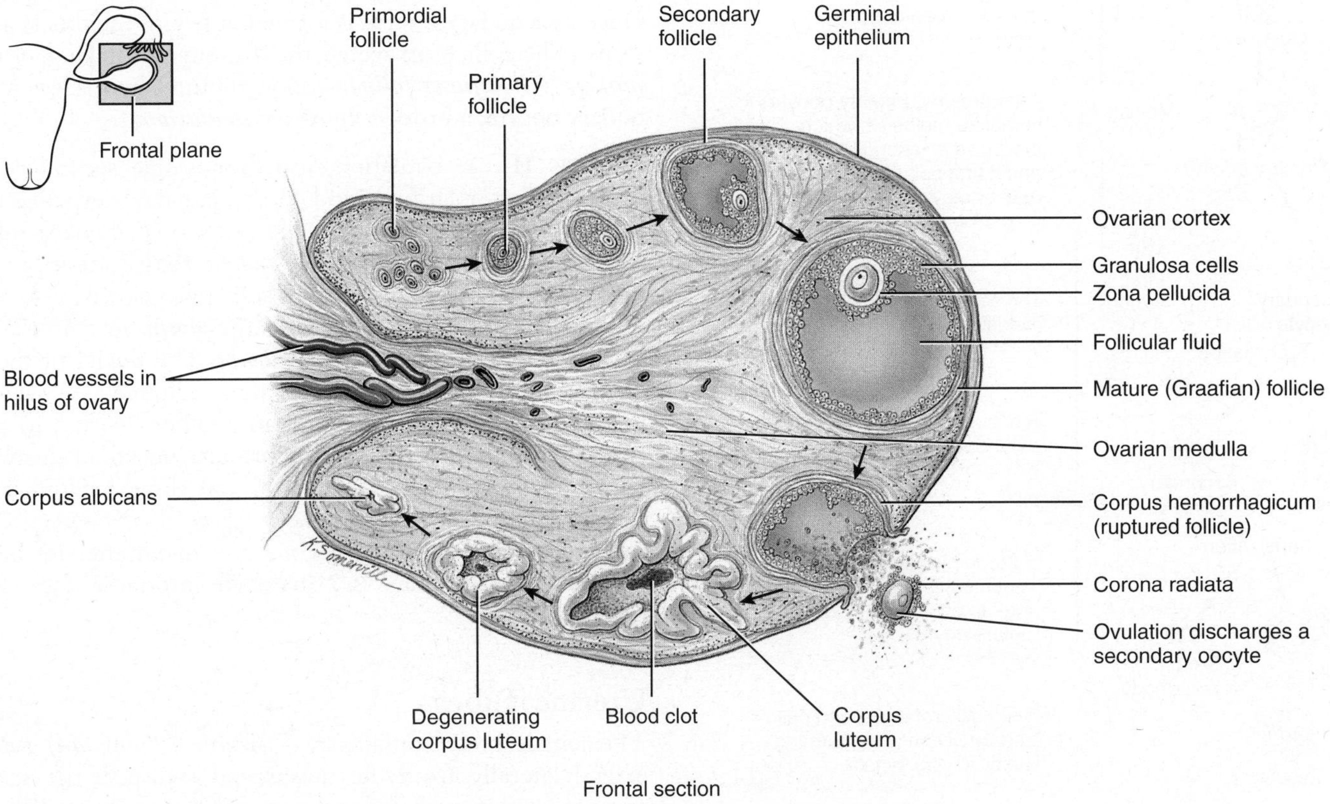

Frontal section

What structures in the ovary contain endocrine tissue, and what hormones do they secrete?

5. A *mature (Graafian) follicle* is a large, fluid-filled follicle that is preparing to rupture and expel a secondary oocyte, a process called *ovulation.*

6. A *corpus luteum* (= yellow body) contains the remnants of an ovulated mature follicle. The corpus luteum produces progesterone, estrogens, relaxin, and inhibin until it degenerates and turns into fibrous tissue called a *corpus albicans* (= white body).

Oogenesis

Formation of gametes in the ovaries is termed *oogenesis* (ō′-ō-JEN-e-sis; *oo-* = egg). Whereas spermatogenesis begins in males at puberty, oogenesis begins in females before they are even born. Oogenesis occurs in essentially the same manner as spermatogenesis. It involves meiosis and maturation.

MEIOSIS I During early fetal development, cells in the ovaries differentiate into *oogonia* (ō′-o-GŌ-nē-a), which can give rise to cells that develop into secondary oocytes (Figure 23.10). Even before birth, most of these cells degenerate. A few, however, develop into larger cells called *primary oocytes* (Ō-ō-sītz) that begin meiosis I during fetal development but do not complete it until after puberty. At birth, 200,000 to 2,000,000 oogonia and primary oocytes remain in each ovary. Of these, about 40,000 remain at puberty, but only 400 go on to mature and ovulate during a woman's reproductive lifetime. The remainder degenerate.

Each primary oocyte is surrounded by a single layer of follicular cells, and the entire structure is called a *primordial follicle* (see Figure 23.9). Primordial follicles develop into *primary follicles,* in which the oocyte is surrounded first by one layer of cuboidal-shaped follicular cells and later by six to seven layers of *granulosa cells.* As a follicle grows, it forms a clear glycoprotein layer, called the *zona pellucida* (pe-LOO-si-da), between the primary oocyte and the granulosa cells. The innermost layer of granulosa cells becomes firmly attached to the zona pellucida and is called the *corona radiata* (*corona* = crown; *radiata* = radiation). The follicle is called a *secondary follicle* after the granulosa cells begin to secrete follicular fluid, which builds up in the cavity of the follicle.

Each month after puberty, hormones secreted by the anterior pituitary stimulate the resumption of oogenesis (see Figure 23.15). Meiosis I resumes in several secondary oocytes, although only one follicle eventually reaches the maturity needed for ovulation in each cycle. The diploid primary oocyte completes meiosis I, resulting in two haploid cells of unequal size, both with 23 chromosomes (*n*) of two chromatids each. The smaller cell produced by meiosis I, called the *first polar body,* is essentially a packet of discarded nuclear material; the larger cell, known as the *secondary oocyte,* receives most of the cytoplasm. Once a secondary oocyte is formed, it begins meiosis II and then stops. The follicle in which these events are taking place—the *mature (Graafian) follicle*—soon ruptures and releases its secondary oocyte, a process known as *ovulation.*

MEIOSIS II At ovulation, usually a single secondary oocyte (with the first polar body and corona radiata) is expelled into the pelvic cavity and swept into the uterine (Fallopian) tube. If a sperm penetrates the secondary oocyte (fertilization), meiosis II resumes. The secondary oocyte splits into two haploid (*n*) cells of unequal size. The larger cell is the *ovum,* or mature egg; the smaller one is the *second polar body.* The nuclei of the sperm cell and the ovum then unite, forming a diploid (2*n*) *zygote.* The first polar body may also undergo another division to produce two polar bodies. If it does, the primary oocyte ultimately gives rise to a single haploid (*n*) ovum and three haploid (*n*) polar bodies. Thus each primary oocyte gives rise to a single gamete (secondary oocyte, which becomes an ovum after fertilization), whereas each primary spermatocyte produces four gametes (sperm).

Figure 23.10 ■ **Oogenesis.** Diploid cells (2*n*) have 46 chromosomes; haploid cells (*n*) have 23 chromosomes.

🔑 **In an oocyte, meiosis II is completed only if fertilization occurs.**

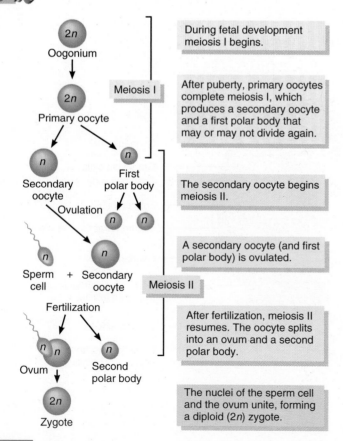

During fetal development meiosis I begins.

After puberty, primary oocytes complete meiosis I, which produces a secondary oocyte and a first polar body that may or may not divide again.

The secondary oocyte begins meiosis II.

A secondary oocyte (and first polar body) is ovulated.

After fertilization, meiosis II resumes. The oocyte splits into an ovum and a second polar body.

The nuclei of the sperm cell and the ovum unite, forming a diploid (2*n*) zygote.

 How does the age of a primary oocyte in a female compare with the age of a primary spermatocyte in a male?

Uterine Tubes

The female body contains two *uterine (Fallopian) tubes* that extend laterally from the uterus and transport the secondary oocytes from the ovaries to the uterus (Figure 23.11). The open, funnel-shaped end of each tube, the *infundibulum,* lies close to the ovary and is surrounded by fingerlike projections called *fim-*

briae (FIM-brē-ē = fringe). From the infundibulum, the uterine tubes extend medially, attaching to the upper and outer corners of the uterus.

Three layers of tissue compose the uterine tubes. The internal mucosa contains ciliated columnar epithelial cells and secretory cells, which have microvilli. The middle layer, the muscularis, is composed of smooth muscle. The outer layer of the uterine tubes is a serous membrane, the serosa.

After ovulation, local currents produced by movements of the fimbriae, which surround the surface of the mature follicle just before ovulation occurs, sweep the secondary oocyte into the uterine tube. The oocyte is then moved along the tube by the cilia of the tube's mucous lining and peristaltic contractions of the muscularis.

The usual site for fertilization of a secondary oocyte by a sperm cell is in the uterine tube. Fertilization may occur any time up to about 24 hours after ovulation. The fertilized ovum (zygote) descends into the uterus within seven days. Unfertilized secondary oocytes disintegrate.

Uterus

The **uterus** (womb) is the site of menstruation, implantation of a fertilized ovum, development of the fetus during pregnancy, and labor. It is situated between the urinary bladder and the rectum and is shaped like an inverted pear (see Figures 23.8 and 23.11).

Parts of the uterus include the dome-shaped portion superior to the uterine tubes called the **fundus,** the tapering central portion called the **body,** and the narrow portion opening into the vagina called the **cervix.** The interior of the body is called the **uterine cavity.**

Figure 23.11 ■ **Uterus and associated structures.** In the left side of the drawing, the uterine tube and uterus have been sectioned to show internal structures.

🔑 The uterus is the site of menstruation, implantation of a fertilized ovum, development of a fetus, and labor.

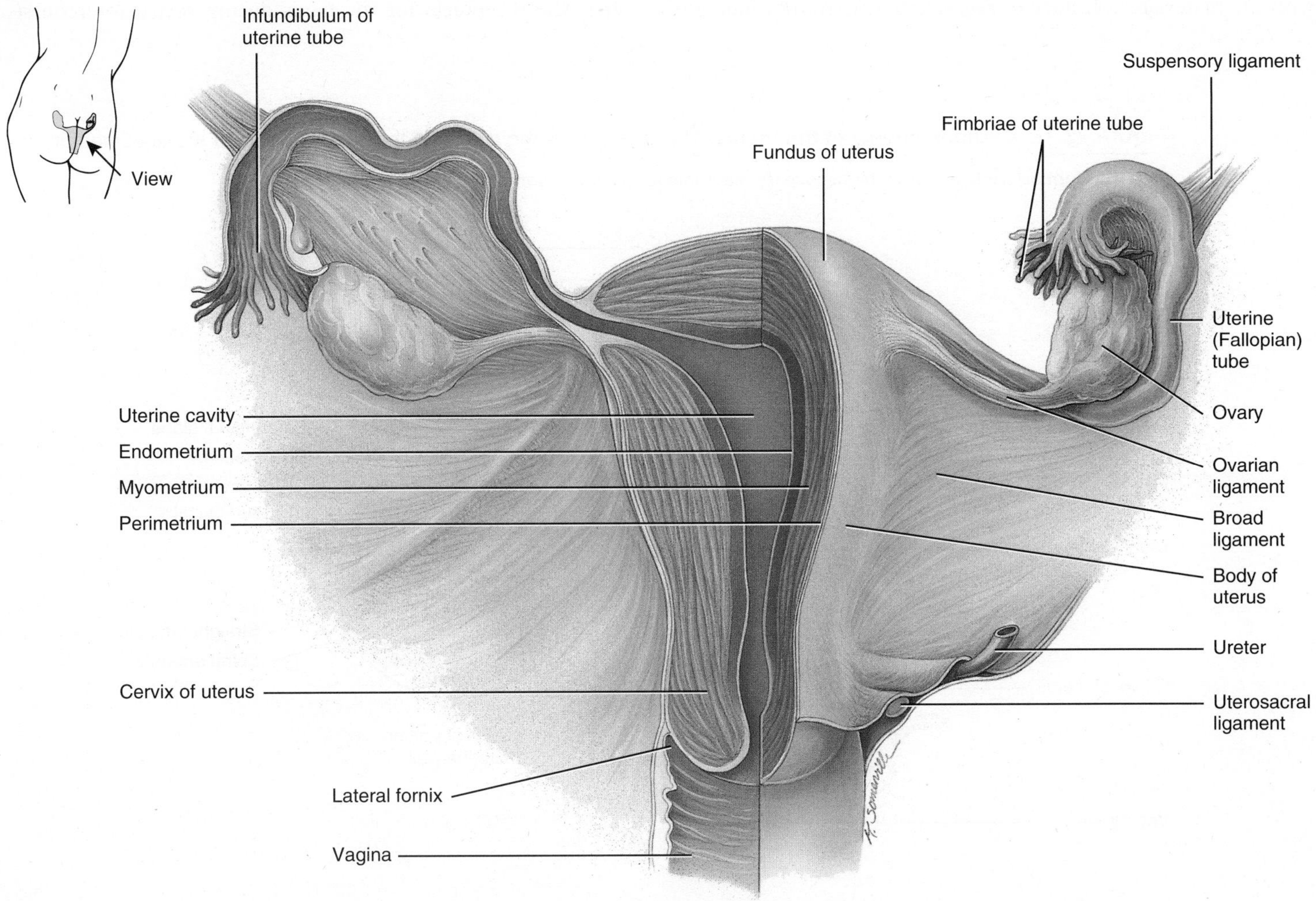

Posterior view

What ligaments hold the ovaries in position?

The uterus is supported and held in position by the broad, uterosacral, cardinal, and round ligaments (see Figures 23.8 and 23.11). The broad ligament also forms part of the outer layer of the uterus, the **perimetrium** (*peri-* = around).

The middle muscular layer of the uterus, the **myometrium** (*myo-* = muscle), consists of smooth muscle and forms the bulk of the uterine wall. During childbirth, coordinated contractions of uterine muscles help expel the fetus.

The innermost part of the uterine wall, the **endometrium** (*endo* = within), is a mucous membrane composed of two layers. The **stratum basalis** (STRAT-um ba-SĀ-lis; *stratum* = layer; *basalis* = base) is a permanent layer lying next to the myometrium. The **stratum functionalis** (fungk′-shun-A-lis) surrounds the uterine cavity. It nourishes a growing fetus or is shed each month during menstruation if fertilization does not occur. Following menstruation, it is replaced by the stratum basalis. The endometrium contains many *endometrial glands* whose secretions nourish sperm and the zygote.

Blood is supplied to the uterus by the **uterine arteries** (Figure 23.12). Branches of the uterine arteries penetrate deeply into the myometrium. Just before the branches enter the endometrium, they divide into two types of arterioles. The **straight arterioles** terminate in the stratum basalis and supply it with the materials necessary to regenerate the stratum functionalis. The **spiral arterioles** penetrate the functionalis and change markedly during the menstrual cycle. Blood leaves the uterus via the **uterine veins.**

Early stages of **cancer of the uterus** can be detected through the **Papanicolaou test** (pa′-pa-ni′-kō-LĀ-oo), or **Pap smear.** A few cells from the cervix and the part of the vagina surrounding the cervix are removed with a swab and examined microscopically. Malignant cells have a characteristic appearance that allows diagnosis even before symptoms occur. Estimates indicate that the Pap smear is more than 90% reliable in detecting cancer of the cervix.

Hysterectomy (hiss-ter-EK-tō-mē; *hyster-* = uterus), the surgical removal of the uterus, is the most common gynecological operation. It may be indicated in pelvic inflammatory disease; excessive uterine bleeding; or cancer of the cervix, uterus, or ovaries. In a partial or subtotal hysterectomy, the body of the uterus is removed but the cervix is left in place; a complete hysterectomy is the removal of the body and cervix of the uterus.

Vagina

The **vagina** (va-JĪ-na = sheath) is a tubular, fibromuscular canal that extends from the exterior of the body to the uterine cervix. It is the receptacle for the penis during sexual intercourse, the

Figure 23.12 ■. **Blood supply of the uterus.** The inset shows the details of the blood vessels of the endometrium.

🔑 **Straight arterioles supply the necessary materials for regeneration of the stratum functionalis.**

Perimetrium
Myometrium
Endometrium
Uterine cavity

Endometrial gland
Endometrium:
Stratum functionalis
Stratum basalis

Straight arteriole
Spiral arteriole

Uterine artery

Cervix of uterus

Uterine artery

Vagina

Details of portion of uterine wall

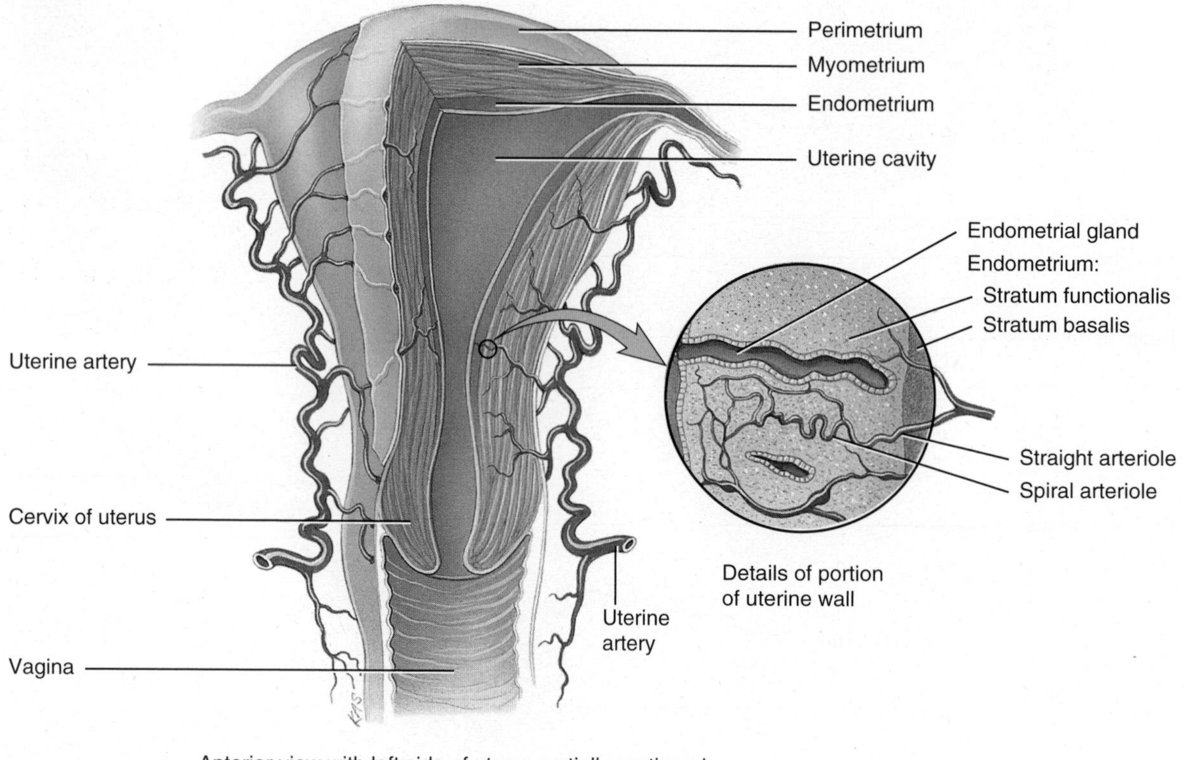

Anterior view with left side of uterus partially sectioned

What is the significance of the two layers of the endometrium?

outlet for menstrual flow, and the passageway for childbirth (see Figures 23.8 and 23.11). The vagina is situated between the urinary bladder and the rectum. A recess, called the *fornix* (= arch or vault), surrounds the cervix. When properly inserted, a contraceptive diaphragm rests on the fornix, covering the cervix.

The mucosa of the vagina is continuous with that of the uterus and cervix and lies in a series of transverse folds, the *rugae* (ROO-jē). The mucosa contains large stores of glycogen, the decomposition of which produces organic acids. The resulting acidic environment retards microbial growth, but it also is harmful to sperm. Alkaline components of semen, mainly from the seminal vesicles, neutralize the acidity of the vagina and increase viability of sperm. The muscular layer is composed of smooth muscle that can stretch to receive the penis during intercourse and allow for childbirth. There may be a thin fold of mucous membrane called the *hymen* (= membrane) partially covering the *vaginal orifice,* the vaginal opening (see Figure 23.13).

Vulva

The term *vulva* (VUL-va = to wrap around), or *pudendum* (pyoo-DEN-dum), refers to the external genitalia of the female (Figure 23.13). The *mons pubis* (MONZ PYOO-bis; *mons* = mountain) is an elevation of adipose tissue covered by coarse pubic hair, which cushions the pubic symphysis. From the mons pubis, two longitudinal folds of skin, the *labia majora* (LĀ-bē-a ma-JŌ-ra; *labia* = lips; *majora* = larger), extend down and back. In females the labia majora develops from the same embryonic tissue that the scrotum develops from in males. The labia majora contain adipose tissue and sebaceous (oil) and sudoriferous (sweat) glands. Like the mons pubis, they are covered by pubic hair. Inside the labia majora are two folds of skin called the *labia minora* (mī-NŌ-ra = smaller). The labia minora do not contain pubic hair or fat and have few sudoriferous (sweat) glands; they do, however, contain numerous sebaceous (oil) glands.

The *clitoris* (KLIT-o-ris) is a small, cylindrical mass of erectile tissue and nerves. It is located at the anterior junction of the labia minora. A layer of skin called the *prepuce* (foreskin) is formed at a point where the labia minora unite and cover the body of the clitoris. The exposed portion of the clitoris is the *glans.* Like the penis, the clitoris is capable of enlargement upon sexual stimulation.

The region between the labia minora is called the *vestibule.* In the vestibule are the hymen (if present); *vaginal orifice,* the opening of the vagina to the exterior; *external urethral orifice,* the opening of the urethra to the exterior; and on either side of the vaginal orifice, the openings of the ducts of the *paraurethral (Skene's) glands.* These glands in the wall of the urethra secrete

Figure 23.13 ■ **Components of the vulva.**

🔑 **Like the penis, the clitoris is capable of erection upon sexual stimulation.**

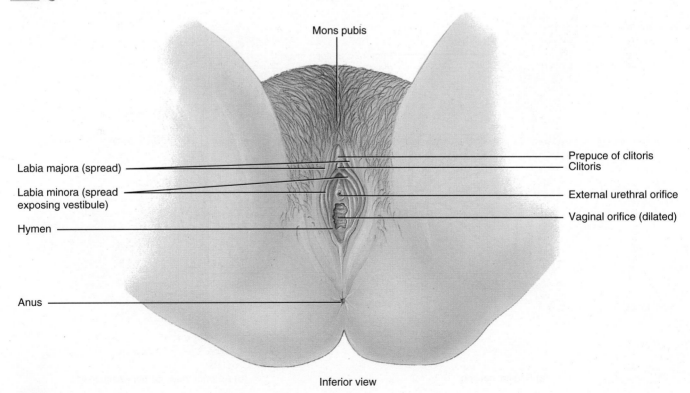

Mons pubis

Labia majora (spread)

Labia minora (spread exposing vestibule)

Hymen

Anus

Prepuce of clitoris

Clitoris

External urethral orifice

Vaginal orifice (dilated)

Inferior view

 What surface structures are anterior to the vaginal opening?

mucus. In males, the prostate gland develops from the same embryonic tissue as the paraurethral glands develop from in females. On either side of the vaginal orifice itself are the ***greater vestibular (Bartholin's) glands,*** which produce a small quantity of mucus during sexual arousal and intercourse that adds to cervical mucus and provides lubrication. In males, the bulbourethral glands are equivalent structures.

Perineum

The ***perineum*** (per′-i-NĒ-um) is the diamond-shaped area between the thighs and buttocks of both males and females (Figure 23.13). It contains the external genitals and anus. In the female, the region between the vagina and anus is known as the ***clinical perineum.*** During childbirth, if the vagina is too small to accommodate the head of an emerging fetus, the skin, vaginal epithelium, subcutaneous fat, and muscle of the clinical perineum may tear. Moreover, the tissues of the rectum may be damaged. To avoid such damage, a small incision called an ***episiotomy*** (e-piz-ē-OT-ō-mē; *episi-* = vulva or pubic region; *-otomy* = incision) is made in the perineal skin and underlying tissues just prior to delivery. After delivery, the episiotomy is sutured in layers.

Mammary Glands

The ***mammary glands*** (*mamma* = breast), located in the breasts, are modified sudoriferous (sweat) glands that produce milk. The breasts lie over the pectoralis major and serratus anterior muscles and are attached to them by a layer of connective tissue (Figure 23.14). Each breast has one pigmented projection, the ***nipple,*** with a series of closely spaced openings of ducts where milk emerges. The circular pigmented area of skin surrounding the nipple is called the ***areola*** (a-RĒ-ō-la = small space). This region appears rough because it contains modified sebaceous (oil) glands. Internally, each mammary gland consists of 15 to 20 ***lobes*** arranged radially and separated by adipose tissue and strands of connective tissue called ***suspensory ligaments of the breast (Cooper's ligaments),*** which support the breast. In each lobe are smaller ***lobules,*** in which milk-secreting glands called ***alveoli*** (= small cavities) are found. When milk is being produced, it passes from the alveoli into a series of tubules that drain toward the nipple.

At birth, both male and female mammary glands are undeveloped and appear as slight elevations on the chest. With the onset of puberty, under the influence of estrogens and progesterone, the female breasts begin to develop. The duct system matures and fat is deposited, which increases breast size. The

Figure 23.14 ■ **Mammary glands.**

The mammary glands function in the synthesis, secretion, and ejection of milk (lactation).

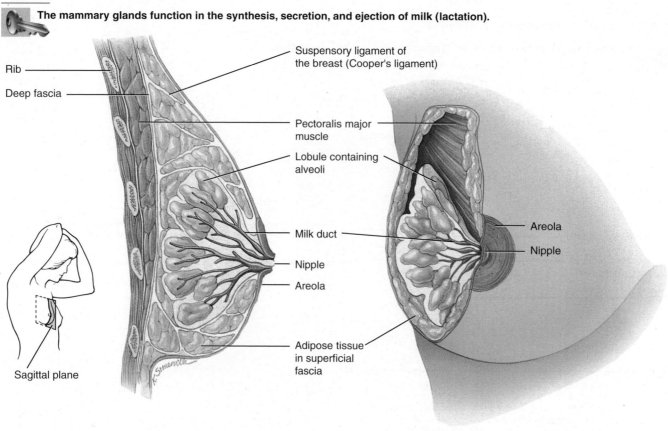

Rib

Deep fascia

Suspensory ligament of the breast (Cooper's ligament)

Pectoralis major muscle

Lobule containing alveoli

Milk duct

Nipple

Areola

Adipose tissue in superficial fascia

Areola

Nipple

Sagittal plane

(a) Sagittal section

(b) Anterior view, partially sectioned

 What hormones regulate the release of milk from the mammary glands?

areola and nipple also enlarge and become more darkly pigmented.

The essential functions of the mammary glands are the synthesis, secretion, and ejection of milk; these functions, called **lactation,** are associated with pregnancy and childbirth. Milk production is stimulated largely by the hormone prolactin, with contributions from progesterone and estrogens. The ejection of milk is stimulated by oxytocin, which is released from the posterior pituitary in response to the sucking of an infant on the mother's nipple (suckling).

FEMALE REPRODUCTIVE CYCLE

Objectives: • **Explain the functions of the female reproductive hormones.**
• **Define the ovarian and menstrual cycles and explain how they are related.**

The general term *female reproductive cycle* refers to the ovarian and menstrual cycles and the hormonal cycles that regulate them. The *ovarian cycle* is a monthly series of events associated with the maturation of an oocyte. Occurring at the same time is the *menstrual cycle* (*mens-* = monthly), a series of changes in the endometrium of the uterus to prepare it for the arrival and development of a fertilized ovum. If fertilization does not occur, the stratum functionalis of the endometrium is shed.

Hormonal Regulation

The ovarian and menstrual cycles are controlled by gonadotropin-releasing hormone (GnRH) from the hypothalamus (Figure 23.15). GnRH stimulates the release of follicle-stimulating hormone (FSH) and luteinizing hormone (LH) from the anterior pituitary. FSH stimulates the initial secretion of estrogens by growing follicles. LH stimulates further development of and estrogen secretion by ovarian follicles; brings about ovulation; promotes formation of the corpus luteum; and stimulates the production of estrogens, progesterone, relaxin, and inhibin by the corpus luteum.

Estrogens secreted by follicular cells have several important functions:

Figure 23.15 ■ **Secretion and physiological effects of estrogens, progesterone, relaxin, and inhibin.**

The uterine and ovarian cycles are controlled by GnRH and ovarian hormones.

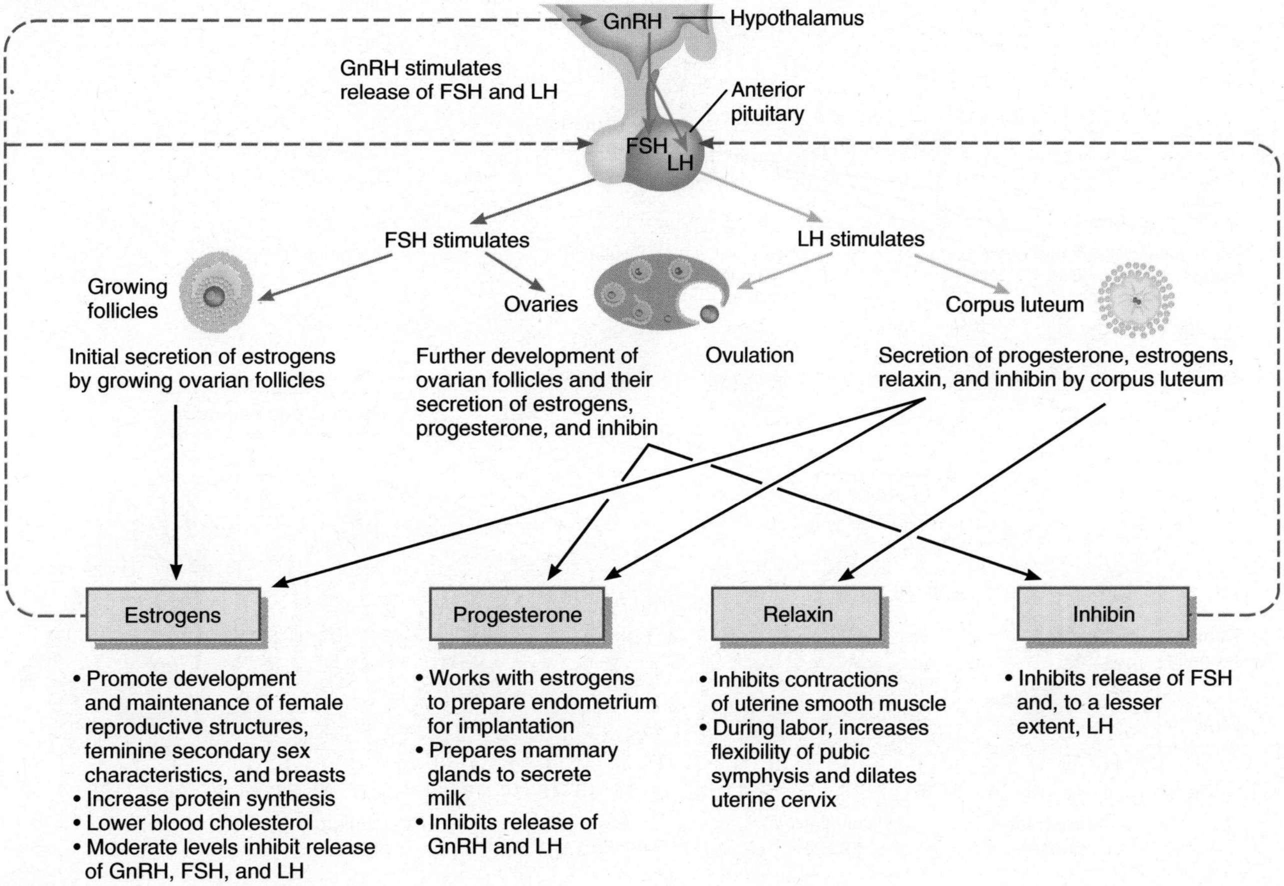

Which hormone helps ease delivery of a baby?

- Estrogens promote the development and maintenance of female reproductive structures, feminine secondary sex characteristics, and the mammary glands. The secondary sex characteristics include distribution of adipose tissue in the breasts, abdomen, mons pubis, and hips; a broad pelvis; and pattern of hair growth on the head and body.

- Estrogens stimulate protein synthesis, acting together with insulinlike growth factors, insulin, and thyroid hormones.

- Estrogens lower blood cholesterol level, which is probably the reason that women under age 50 have a much lower risk of coronary artery disease than do men of comparable age.

- Through negative feedback loops, moderate levels of estrogens in the blood inhibit both the release of GnRH by the hypothalamus and the secretion of LH and FSH by the anterior pituitary (Figure 23.15).

Progesterone, secreted mainly by cells of the corpus luteum, acts together with estrogens to prepare the endometrium for the implantation of a fertilized ovum and the mammary glands for milk secretion.

The small quantity of *relaxin* produced by the corpus luteum during each monthly cycle relaxes the uterus by inhibiting contractions; presumably, implantation of a fertilized ovum occurs more readily in a relaxed uterus. During pregnancy, the placenta produces additional relaxin, which continues to relax uterine smooth muscle. At the end of pregnancy, relaxin also increases the flexibility of the pubic symphysis and helps dilate the uterine cervix, both of which ease delivery of the baby.

Inhibin is secreted by granulosa cells of growing follicles and by the corpus luteum of the ovary. It contributes to regulating events of the female reproductive cycle by inhibiting secretion of FSH and, to a lesser extent, LH.

Phases of the Female Reproductive Cycle

The duration of the female reproductive cycle varies from 24 to 35 days. For this discussion we assume a duration of 28 days, divided into three phases: the menstrual phase, the preovulatory phase, and the postovulatory phase (Figure 23.16). Ovulation, the explusion of a secondary oocyte from an ovary, separates the

Figure 23.16 ■ Correlation of ovarian and uterine cycles with the hypothalamic and anterior pituitary hormones. In the cycle shown, fertilization and implantation have not occurred.

The preovulatory phase is more variable in length than the other phases.

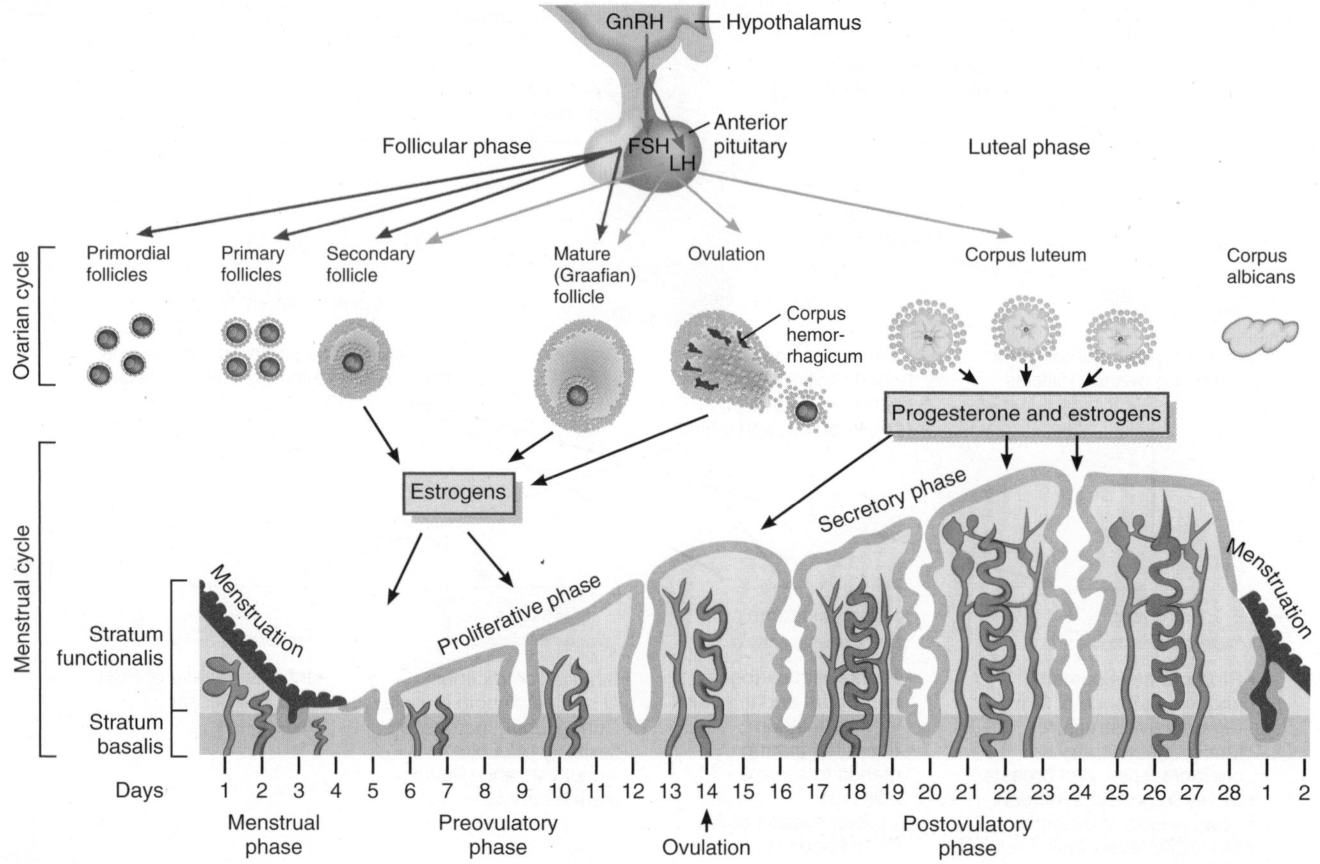

Which hormones directly stimulate proliferation of the endometrium?

pre- and postovulatory phases. Because they occur at the same time, the events of the ovarian cycle (events in the ovaries) and menstrual cycle (events in the uterus) will be discussed together.

Menstrual Phase

The *menstrual phase* (MEN-stroo-al), also called *menstruation* (men′-stroo-Ā-shun) or *menses* (= month), lasts for roughly the first five days of the cycle. (By convention, the first day of menstruation marks the first day of a new cycle.)

EVENTS IN THE OVARIES During the menstrual phase, 20 or so small secondary follicles, some in each ovary, begin to enlarge. Follicular fluid, secreted by the granulosa cells and oozing from blood capillaries, accumulates in the enlarging cavity in the follicle, while the oocyte remains near the edge of the follicle.

EVENTS IN THE UTERUS Menstrual flow from the uterus consists of 50 to 150 mL of blood and tissue cells from the endometrium. This discharge occurs because the declining level of estrogens and progesterone causes the uterine spiral arteries to constrict. As a result, the cells they supply become *ischemic* (deficient in blood) and start to die. Eventually, the entire stratum functionalis sloughs off so only the stratum basalis of the endometrium remains. The menstrual flow passes from the uterine cavity to the cervix and through the vagina to the exterior.

Preovulatory Phase

The *preovulatory phase,* the second phase of the female reproductive cycle, is the time between menstruation and ovulation. The preovulatory phase of the cycle accounts for most of the variation in cycle length. In a 28-day cycle, it lasts from days 6 to 13.

EVENTS IN THE OVARIES Under the influence of FSH, the group of about 20 secondary follicles continues to grow and begins to secrete estrogens and inhibin. By about day 6, a single follicle in one of the two ovaries has outgrown all the others to become the *dominant follicle.* Estrogens and inhibin secreted by the dominant follicle decrease the secretion of FSH (Figure 23.17, see days 8 to 11), which causes other, less well-developed follicles to stop growing and die.

The one dominant follicle becomes the *mature (Graafian) follicle.* The mature follicle continues to enlarge until it is ready for ovulation, forming a blisterlike bulge on the surface of the ovary. During maturation, the follicle continues to increase its production of estrogens under the influence of an increasing level of LH. Although estrogens are the primary ovarian hormones before ovulation, small amounts of progesterone are produced by the mature follicle a day or two before ovulation.

With reference to the ovaries, the menstrual phase and preovulatory phase together are termed the *follicular phase* (fō-LIK-yoo-lar) because ovarian follicles are growing and developing.

EVENTS IN THE UTERUS Estrogens liberated into the blood by growing ovarian follicles stimulate the repair of the endometrium; cells of the stratum basalis undergo mitosis and produce a new stratum functionalis. As the endometrium thickens,

Figure 23.17 ■ **Relative concentrations of anterior pituitary hormones (FSH and LH) and ovarian hormones (estrogens and progesterone) during a normal menstrual cycle.**

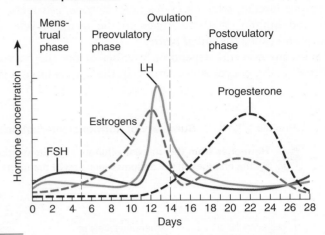

Estrogens are the primary ovarian hormones before ovulation; after ovulation, both progesterone and estrogens are secreted by the corpus luteum.

An over-the-counter home test that predicts ovulation detects the presence of which hormone?

the short, straight endometrial glands develop, and the spiral arterioles coil and lengthen as they penetrate the stratum functionalis. With reference to the menstrual cycle, the preovulatory phase is also termed the *proliferative phase* because the endometrium is proliferating, or growing.

Ovulation

Ovulation, the rupture of the mature follicle and the release of the secondary oocyte into the pelvic cavity, usually occurs on day 14 in a 28-day cycle. During ovulation, the secondary oocyte remains surrounded by its zona pellucida and corona radiata. Development of a secondary follicle into a fully mature follicle generally takes a total of about 20 days (spanning the last 6 days of the previous cycle and the first 14 days of the current cycle).

The high levels of estrogens during the last part of the preovulatory phase exert a positive feedback effect on both LH and GnRH. A high level of estrogens stimulates the hypothalamus to release more gonadotropin-releasing hormone (GnRH) and the anterior pituitary to produce more LH. GnRH then promotes the release of even more LH (Figure 23.17). The resulting surge of LH brings about rupture of the dominant follicle and expulsion of a secondary oocyte. An over-the-counter home test that detects the LH surge associated with ovulation can be used to predict ovulation a day in advance.

Postovulatory Phase

The *postovulatory phase* of the female reproductive cycle is the most constant in duration and lasts for 14 days, from days 15 to 28 in a 28-day cycle. It represents the time between ovulation and onset of the next menses.

EVENTS IN ONE OVARY After ovulation, the mature follicle collapses. The blood resulting from minor bleeding during rupture of the follicle clots, and the collapsed follicle is known as the **corpus hemorrhagicum** (hem-ō-RAJ-ik-um; *hemo-* = blood; *rrhagic-* = bursting forth) (see Figure 23.16). Stimulated by LH, the remaining follicular cells absorb the clot, enlarge, and form the corpus luteum, which secretes progesterone, estrogens, relaxin, and inhibin. With reference to the ovarian cycle, this phase is also called the **luteal phase**.

Subsequent events depend on whether or not the oocyte is fertilized. If the oocyte is not fertilized, the corpus luteum lasts for only two weeks, after which its secretory activity declines, and it degenerates into a corpus albicans (see Figure 23.16). As the levels of progesterone, estrogens, and inhibin decrease, release of GnRH, FSH, and LH rise due to loss of negative feedback suppression by the ovarian hormones. Then, follicular growth resumes and a new ovarian cycle begins.

If the secondary oocyte is fertilized and begins to divide, the corpus luteum persists past its normal two-week lifespan. It is "rescued" from degeneration by **human chorionic gonadotropin** (kōr′-ē-ON-ik) **(hCG),** a hormone produced by the chorion of the embryo beginning about eight days after fertilization. Like

Figure 23.18 ■ **Summary of hormonal interactions in the ovarian and menstrual cycles.**

Hormones from the anterior pituitary regulate ovarian function, and hormones from the ovaries regulate the changes in the endometrial lining of the uterus.

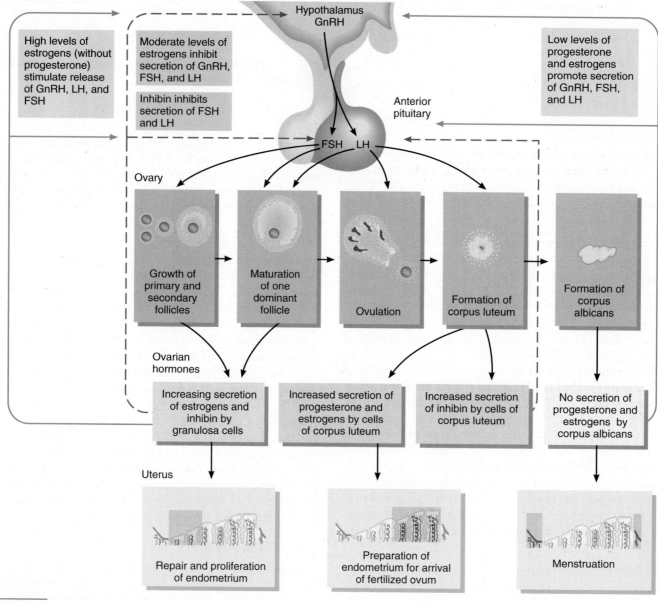

When declining levels of estrogens and progesterone stimulate secretion of GnRH, is this a positive or negative feedback effect? Why?

LH, hCG stimulates the secretory activity of the corpus luteum. The presence of hCG in maternal blood or urine is an indicator of pregnancy, and hCG is the hormone detected by home pregnancy tests.

EVENTS IN THE UTERUS Progesterone and estrogens produced by the corpus luteum promote growth of the endometrial glands, which begin to secrete glycogen, and vascularization and thickening of the endometrium. Because of the secretory activity, this phase of the uterine cycle is called the *secretory phase.* These preparatory changes peak about one week after ovulation, at the time a fertilized ovum might arrive at the uterus.

Figure 23.18 summarizes the hormonal interactions and cyclical changes in the ovaries and uterus during the ovarian and menstrual cycles.

BIRTH CONTROL METHODS AND ABORTION

Objective: • **Compare the various types of birth control methods and outline the effectiveness of each.**

No single, ideal method of **birth control** exists. The only method of preventing pregnancy that is 100% reliable is total **abstinence,** the avoidance of sexual intercourse. Several other methods are available, including surgical sterilization, hormonal methods, intrauterine devices, spermatocides, barrier methods, periodic abstinence, and coitus interruptus. Table 23.1 provides the failure rates for each method of birth control. We will also discuss induced abortion, the intentional termination of pregnancy.

Surgical Sterilization

Sterilization is a procedure that renders an individual incapable of reproduction. The most common means of sterilization of males is *vasectomy* (va-SEK-tō-mē; *-ectomy* = cut out), in which a portion of each vas deferens is removed. Even though sperm production continues in the testes after a vasectomy, sperm can no longer reach the exterior. Instead, they degenerate and are destroyed by phagocytosis. Blood testosterone level is normal, so a vasectomy has no effect on sexual desire or performance. Sterilization in females most often is achieved by performing a *tubal ligation* (lī-GĀ-shun), in which both uterine tubes are tied closed and then cut. As a result, the secondary oocyte cannot pass through the uterine tubes, and sperm cannot reach the oocyte.

Hormonal Methods

Aside from total abstinence or surgical sterilization, hormonal methods are the most effective means of birth control. By using *oral contraceptives* (*"the pill"*) to adjust hormone levels, it is possible to interfere with follicular development or implantation of a fertilized ovum in the uterus. "Combination pills" most often contain both a progestin (a substance similar to progesterone)

and an estrogen. These two hormones act by means of negative feedback on the anterior pituitary to decrease the secretion of FSH and LH, and on the hypothalamus to inhibit secretion of gonadotropin-releasing hormone. The low levels of FSH and LH usually prevent both follicular development and ovulation. Oral contraceptives also may be used for *emergency contraception (EC),* the so-called "morning-after pill." When two pills are taken within 72 hours after unprotected intercourse and another two pills are taken 12 hours later, the chance of pregnancy is reduced by 75%. Women who take the pill and smoke face far higher odds of having a heart attack or stroke than do nonsmoking pill users. Smokers seeking a reliable birth control method should use an alternative method of birth control, or even better—quit smoking.

Other hormonal methods of contraception include Norplant, Depo-Provera, and the vaginal ring. *Norplant* consists of six slender hormone-containing capsules that are surgically implanted under the skin of the arm using local anesthesia. They slowly and continually release a progestin, which inhibits ovulation and thickens the cervical mucus. The effects last for five years. Removing the Norplant capsules restores fertility. *Depo-provera,* which is given as an intramuscular injection once every three months, contains progestin that prevents maturation of the ovum and causes changes in the uterine lining that make

Table 23.1 / **Failure Rates of Birth Control Methods**		
Method	**Failure Rates***	
	Perfect Use†	**Typical Use‡**
None	85	85
Complete abstinence	0	0
Surgical sterilization		
Vasectomy	0.10	0.15
Tubal ligation	0.4	0.4
Hormonal methods		
Oral contraceptives	0.1	3
Norplant	0.1	0.1
Depo-Provera	0.05	0.05
Intrauterine device		
Copper T 380A	0.6	0.8
Spermatocides	6	26
Barrier methods		
Male condom	3	14
Vaginal pouch	5	21
Diaphragm	6	20
Periodic abstinence		
Sympto-thermal	2	20
Coitus interruptus	4	18

*Defined as percentage of women having an unintended pregnancy during the first year of use.

†Failure rate when the method is used correctly and consistently.

‡Includes couples who forgot to use the method.

The Female Athlete Triad—Disordered Eating, Amenorrhea, and Premature Osteoporosis

The female reproductive cycle can be disrupted by many factors, including weight loss, low body weight, disordered eating, and vigorous physical activity. Many athletes experience intense pressure from coaches, parents, peers, and themselves to lose weight to improve performance. Consequently, many develop disordered eating behaviors and engage in other harmful weight-loss practices in a struggle to maintain a very low body weight. The athletes with the highest rates of menstrual irregularity include runners, gymnasts, dancers, figure skaters, and divers.

Sticks and Stones and . . . the Female Athlete Triad?

Menstrual irregularity should never be ignored, because it may be caused by a serious underlying disorder for which the athlete should receive prompt medical treatment. Even when menstrual irregularity is apparently caused by disordered eating and physical training, and not as-sociated with another physical disorder, it is still a cause for concern. One reason is that women with amenorrhea, the absence of menstrual cycles, are at increased risk for premature osteoporosis. The observation that three conditions—disordered eating, amenorrhea, and osteoporosis—tend to occur together in female athletes led researchers to coin the term "female athlete triad."

Why osteoporosis? Remember that the ovarian follicles produce estrogen when stimulated by FSH and LH. If ovulation is not occurring, then the ovarian follicles, and later the corpus luteum, are not producing estrogens. Chronically low levels of estrogens are associated with loss of bone minerals, as estrogens help bones retain calcium. The loss of the protective effect of estrogens explains why many women experience a decline in bone density after menopause, when levels of estrogens drop. Amenorrheic runners have been shown to experience a similar effect. In one study, amenorrheic runners in their 20s had bone densities similar to those of postmenopausal women 50 to 70 years old. Whereas short periods of menstrual irregularity in young athletes may cause no lasting harm, long-term cessation of the menstrual cycle may be accompanied by a loss of bone mass or, in adolescent athletes, a failure to achieve an adequate bone mass, both of which can lead to premature osteoporosis and irreversible bone damage.

It is ironic that dedicated athletes should experience premature osteoporosis, because physical activity in general is associated with a *reduced* risk of osteoporosis. Exercise has been shown to increase bone density, especially if the exercise involves bone stress, such as running and aerobic dancing. However, in the presence of disordered eating and overtraining, exercise may simply add insult to injury.

▶ Think It Over

▶ Do you think that girls and women should be discouraged from participating in athletics and other forms of vigorous physical activities because of the female athlete triad?

---

pregnancy less likely. The **vaginal ring** is a doughnut-shaped ring that fits in the vagina and releases either a progestin alone or a progestin and an estrogen. It is worn for three weeks and then removed for one week to allow menstruation to occur.

The quest for an efficient oral contraceptive for males has been disappointing. The challenge is to find substances that block production of functional sperm without causing erectile dysfunction (impotence).

Intrauterine Devices

An **intrauterine device (IUD)** is a small object made of plastic, copper, or stainless steel that is inserted into the cavity of the uterus. IUDs cause changes in the uterine lining that prevent implantation of a fertilized ovum. The IUD most commonly used in the United States today is the Copper T 380A, which is approved for up to 10 years of use.

Spermatocides

Various foams, creams, jellies, suppositories, and douches that contain sperm-killing agents, or **spermatocides,** make the vagina and cervix unfavorable for sperm survival. The most widely used spermatocide is nonoxynol 9, which kills sperm by disrupting their plasma membrane. A spermatocide is more effective when used together with a barrier method such as a diaphragm or a condom.

Barrier Methods

Barrier methods are designed to prevent sperm from gaining access to the uterine cavity and uterine tubes. In addition to preventing pregnancy, some barrier methods provide protection against sexually transmitted diseases (STDs) such as AIDS. In contrast, oral contraceptives and IUDs confer no such protection. Among the barrier methods are use of a condom, a vaginal pouch, or a diaphragm.

A **condom** is a nonporous, latex covering placed over the penis that prevents deposition of sperm in the female reproductive tract. A **vaginal pouch,** sometimes called a female condom, is made of two flexible rings connected by a polyurethane sheath. One ring lies inside the sheath and is inserted to fit over the cervix; the other ring remains outside the vagina and covers the female's external genitals.

A **diaphragm** is a rubber, dome-shaped device that fits over the cervix and is used in conjunction with a spermatocide. It can be inserted up to six hours before intercourse. The diaphragm stops most sperm from passing into the cervix; the spermatocide kills most sperm that do get by.

Periodic Abstinence

A couple can use their knowledge of the functional changes that occur during the female reproductive cycle to decide either to abstain from intercourse on those days when pregnancy is a likely result, or to plan intercourse on those days if they wish to conceive a child.

One system is the **sympto-thermal method,** in which couples are instructed to know and understand certain signs of fertility and infertility. The signs of ovulation include increased basal body temperature; the production of clear, stretchy cervical mucus; abundant cervical mucus; and pain associated with ovulation. If a couple abstains from sexual intercourse when the signs of ovulation are present and for three days afterward, the chance of pregnancy is decreased. A problem with this method is that fertilization is likely to occur if intercourse took place up to two days *before* ovulation.

Coitus Interruptus

Coitus interruptus is withdrawal of the penis from the vagina just before ejaculation. Failures with this method are due either to failure to withdraw before ejaculation or to preejaculatory emission of sperm-containing fluid from the urethra. Like planned abstinence, this method offers no protection against transmission of STDs.

Abortion

Abortion is the premature expulsion from the uterus of the products of conception, usually before the 20th week of pregnancy. An abortion may be spontaneous (naturally occurring; also called a miscarriage) or induced (intentionally performed). Induced abortions may involve vacuum aspiration (suction), infusion of a saline solution, or surgical evacuation (scraping).

Certain drugs, most notably RU 486, can induce a so-called nonsurgical abortion. **RU 486 (Mifepristone)** blocks the action of progesterone. Progesterone maintains the uterine lining after implantation. If progesterone levels fall during pregnancy or if the action of the hormone is blocked, menstruation occurs, and the embryo is sloughed off along with the uterine lining. A form of prostaglandin E (misoprostol), which stimulates uterine contractions, is given after RU 486 to aid in expulsion of the endometrium. RU 486 can be taken up to five weeks after conception. It has been used for several years in France, Sweden, the United Kingdom, China, and other countries.

THE HUMAN SEXUAL RESPONSE

Objective: • **Describe the similarities and differences in the sexual responses of males and females.**

During heterosexual **sexual intercourse,** also called **coitus** (KŌ-i-tus = a coming together), sperm are ejaculated from the male urethra into the vagina. The similar sequence of physiological and emotional changes experienced by both males and females before, during, and after intercourse is termed the **human sexual response.**

Stages

The human sexual response has four stages: excitement, plateau, orgasm, and resolution. During sexual **excitement,** also known as **arousal,** various physical and psychological stimuli trigger parasympathetic reflexes. These reflexes cause nerve impulses to propagate from sacral segments of the spinal cord along the pelvic splanchnic nerves. One effect is relaxation of vascular smooth muscle, which allows **vasocongestion** (engorgement with blood) of genital tissues. Parasympathetic impulses also stimulate the secretion of lubricating fluids. Direct physical contact (as in kissing or touching), especially of the glans penis, clitoris, nipples of the breasts, and earlobes, is a potent initiator of excitement. Anticipation or fear; memories; visual, olfactory, and auditory sensations; and fantasies can enhance or diminish the chance that excitement will occur. Due to the tactile stimulation of the breast while nursing an infant, feelings of sexual arousal during breast-feeding are fairly common and should not be cause for concern.

The changes that begin during excitement are sustained at an intense level in the **plateau** stage, which may last for only a few seconds or for many minutes. At this stage, many females and some males begin to display a sex flush, a rashlike redness of the face and chest.

Generally, the briefest stage is *orgasm (climax)*. During orgasm, bursts of *sympathetic* nerve impulses leave the lumbar spinal cord and cause rhythmic contractions of smooth muscle in genital organs. At the same time, *somatic motor neurons* from lumbar and sacral segments of the spinal cord trigger powerful, rhythmic contractions of perineal skeletal muscles. The male ejaculates, and both sexes experience intense, pleasurable sensations.

The final stage, *resolution*, begins with a sense of profound relaxation. Genital tissues, heart rate, blood pressure, breathing, and muscle tone return to the unaroused state. The four phases of the human sexual response are not always clearly separated from one another. They may vary considerably among different people, and even in the same person at different times.

Changes in Males

Most of the time, the penis is flaccid (limp) because sympathetic impulses cause vasoconstriction of its arteries, which limits blood inflow. The first visible sign of sexual excitement is erection. Parasympathetic impulses cause release of neurotransmitters and local hormones, including the gas nitric oxide, which relaxes vascular smooth muscle in the penile arteries. The arteries of the penis dilate, and blood fills the blood sinuses of the three corpora. Expansion of these erectile tissues compresses the superficial veins that normally drain the penis, resulting in engorgement and rigidity. Mucus is secreted by the bulbourethral glands.

During the plateau stage, the head of the penis increases in diameter, and vasocongestion causes the testes to swell. As orgasm begins, rhythmic sympathetic impulses cause peristaltic contractions of smooth muscle in the walls of the ducts and glands. These contractions propel sperm and fluid into the urethra (emission). Then, peristaltic contractions in the ducts and urethra combine with rhythmic contractions of skeletal muscles in the perineum and at the base of the penis to ejaculate semen from the urethra to the exterior.

Erectile dysfunction (ED), previously termed *impotence*, is the consistent inability of an adult male to ejaculate or to attain or hold an erection long enough for sexual intercourse. Many cases of impotence are caused by insufficient release of nitric oxide. The drug sildenafil (Viagra) enhances the effect of nitric oxide.

Changes in Females

The first signs of sexual excitement in females are also due to vasocongestion. Although the vaginal mucosa lacks glands, engorgement of its connective tissue with blood during sexual excitement causes lubricating fluid to ooze from capillaries and seep through the epithelial layer. Glands within the cervical mucosa and the greater vestibular (Bartholin's) glands contribute a small quantity of lubricating mucus. During the excitement stage, parasympathetic impulses also trigger erection of the clitoris, vasocongestion of the labia, and relaxation of vaginal smooth muscle. Vasocongestion may also cause the breasts to swell and the nipples to become erect.

Late in the plateau stage, pronounced vasocongestion of the vagina swells the tissue and narrows the opening. Because of this response, the vagina grips the penis more firmly. If effective sexual stimulation continues, orgasm may occur, associated with 3 to 15 rhythmic contractions of the vagina, uterus, and perineal muscles.

AGING AND THE REPRODUCTIVE SYSTEMS

Objective: • **Describe the effects of aging on the reproductive systems.**

During the first decade of life, the reproductive system is in a juvenile state. At about age 10, hormone-directed changes start to occur in both sexes. *Puberty* (PYOO-ber-tē = a ripe age) is the period when secondary sexual characteristics begin to develop and the potential for sexual reproduction arises.

In females, the reproductive cycle normally occurs once each month from *menarche* (me-NAR-kē), the first menses, to *menopause,* the permanent cessation of menses. The female reproductive system has a time-limited span of fertility between menarche and menopause. Fertility declines with age, due to the small number of follicles left in the ovaries, less frequent ovulation, and possibly the declining ability of the uterine tubes and uterus to support the young embryo.

Between the ages of 40 and 50 the pool of remaining ovarian follicles becomes exhausted; as a result, the ovaries become less responsive to hormonal stimulation, and the production of estrogens declines, despite copious secretion of FSH and LH by the anterior pituitary. Changes in the pattern of gonadotropin-releasing hormone (GnRH) secretion may also contribute to menopausal changes. Some women experience hot flashes and night sweats, which appear to coincide with increased release of GnRH. Other symptoms of menopause include headache, hair loss, muscular pains, vaginal dryness, insomnia, depression, weight gain, and mood swings. Some atrophy of the ovaries, uterine tubes, uterus, vagina, external genitalia, and breasts occurs in postmenopausal women. Due to the diminished level of estrogens, osteoporosis can also develop. Sexual desire (libido) does not decline with the declining level of estrogens because it is maintained by adrenal androgens.

In males, declining reproductive function is much more subtle than in females. Healthy men often retain reproductive capacity into their 80s or 90s. At about age 55, a decline in testosterone synthesis leads to reduced muscle strength, fewer viable sperm, and decreased sexual desire. However, abundant sperm may be present even in old age.

Enlargement of the prostate gland to two to four times its normal size occurs in approximately one-third of all males over age 60. This condition, called *benign prostatic hyperplasia (BPH),* is characterized by frequent urination, nocturia (bedwetting), hesitancy in urination, decreased force of urinary stream, postvoiding dribbling, and a sensation of incomplete emptying.

COMMON DISORDERS

Reproductive System Disorders in Males

Testicular Cancer

Testicular cancer is the most common cancer, and also one of the most curable, in males between the ages of 20 and 35. An early sign of testicular cancer is a mass in the testis, often associated with a sensation of testicular heaviness or a dull ache in the lower abdomen; pain usually does not occur. All males should perform regular testicular self-examinations.

Prostate Disorders

Because the prostate surrounds a portion of the urethra, any prostatic infection, enlargement, or tumor can obstruct the flow of urine. Acute and chronic infections of the prostate gland are common in adult males, often in association with inflammation of the urethra. In **acute prostatitis,** the prostate gland becomes swollen and tender. **Chronic prostatitis** is one of the most common chronic infections in men of the middle and later years; on examination, the prostate gland feels enlarged, soft, and very tender, and its surface outline is irregular.

Prostate cancer is the leading cause of death from cancer in men in the United States. A blood test can measure the level of prostate-specific antigen (PSA) in the blood. The amount of PSA, which is produced only by prostate epithelial cells, increases with enlargement of the prostate gland and may indicate infection, benign enlargement, or prostate cancer. Males over the age of 40 should have an annual examination of the prostate gland. In a **digital rectal exam,** a physician palpates the gland through the rectum with the fingers (digits). Many physicians also recommend an annual PSA test for males over 50. Treatment for prostate cancer may involve surgery, radiation, hormonal therapy, and chemotherapy. Because many prostate cancers grow very slowly, some urologists recommend "watchful waiting" before treating small tumors in men over age 70.

Reproductive System Disorders in Females

Premenstrual Syndrome

Premenstrual syndrome (PMS) refers to severe physical and emotional distress that occurs during the postovulatory (luteal) phase of the female reproductive cycle. Signs and symptoms usually increase in severity until the onset of menstruation and then dramatically disappear. The signs and symptoms are highly variable from one woman to another and include edema, weight gain, breast swelling and tenderness, abdominal distention, backache, joint pain, constipation, skin eruptions, fatigue and lethargy, greater need for sleep, depression or anxiety, irritability, mood swings, headache, poor coordination and clumsiness, and cravings for sweet or salty foods. The cause of PMS is unknown. For some women, getting regular exercise; avoiding caffeine, salt, and alcohol; and eating a diet that is high in complex carbohydrates and lean proteins can bring considerable relief. Because PMS occurs only after ovulation, oral contraceptives are an effective treatment for women whose symptoms are incapacitating.

Endometriosis

Endometriosis (en′-dō-mē′-trē-Ō-sis; _endo-_ = within; _metri-_ = uterus; _-osis_ = condition of or disease) is characterized by the growth of endometrial tissue outside the uterus. The tissue enters the pelvic cavity via the open uterine tubes and may be found in any of several sites: on the ovaries, the outer surface of the uterus, the sigmoid colon, pelvic and abdominal lymph nodes, the cervix, the abdominal wall, the kidneys, and the urinary bladder. Endometrial tissue responds to hormonal fluctuations, whether it is inside or outside the uterus, by first proliferating and then breaking down and bleeding. When this occurs outside the uterus, it can cause inflammation, pain, scarring, and infertility. Symptoms include premenstrual pain or unusually severe menstrual pain.

Breast Cancer

One in eight women in the United States faces the prospect of **breast cancer,** the second-leading cause of female deaths from cancer. Early detection by breast self-examination and mammograms is the best way to increase the chance of survival.

The most effective technique for detecting tumors less than 1 cm (about 0.5 in.) in diameter is **mammography** (mam-OG-ra-fē; _-graphy_ = to record), a type of radiography using very sensitive x-ray film. The image of the breast, called a **mammogram,** is best obtained by compressing the breasts, one at a time, using flat plates. A supplementary procedure for evaluating breast abnormalities is **ultrasonography**. Although ultrasonography cannot detect tumors smaller than 1 cm in diameter, it can be used to determine whether a lump is a benign, fluid-filled cyst or a solid (and therefore possibly malignant) tumor.

Among the factors that increase the risk of developing breast cancer are (1) a family history of breast cancer, especially in a mother or sister; (2) never having borne a child or having a first child after age 35; (3) previous cancer in one breast; (4) exposure to ionizing radiation, such as x-rays; (5) excessive alcohol intake; and (6) cigarette smoking.

The American Cancer Society recommends the following steps to help diagnose breast cancer as early as possible:

- All women over 20 should develop the habit of monthly breast self-examination.

- A physician should examine the breasts every three years when a woman is between the ages of 20 and 40, and every year after age 40.

- A mammogram should be taken in women between the ages of 35 and 39, to be used later for comparison (baseline mammogram).

- Women with no symptoms should have a mammogram every year or two between ages 40 and 49, and every year after age 50.

- Women of any age with a history of breast cancer, a strong family history of the disease, or other risk factors should consult a physician to determine a schedule for mammography.

Treatment for breast cancer may involve hormone therapy, chemotherapy, radiation therapy, **lumpectomy** (removal of the tumor and the immediate surrounding tissue), a modified or radical mastectomy, or a combination of these approaches. A **radical mastectomy** (_mast-_ = breast) involves removal of the affected breast along with the underlying pectoral muscles and the axillary lymph nodes. (Lymph nodes are removed because metastasis of cancerous cells usually occurs through lymphatic or blood vessels.) Radiation treatment and

COMMON DISORDERS (CONTINUED)

chemotherapy may follow the surgery to ensure the destruction of any stray cancer cells. In some cases of metastatic (spreading) breast cancer, trastuzumab (Herceptin), a monoclonal antibody drug that targets an antigen on the surface of breast cancer cells, can cause regression of the tumors and retard progression of the disease. Finally, two promising drugs for breast cancer *prevention* are now on the market—tamoxifen (Nolvadex) and raloxifene (Evista).

Ovarian Cancer

Ovarian cancer is the sixth most common form of cancer in females, but the leading cause of death from all gynecological malignancies (excluding breast cancer) because it is difficult to detect before it metastasizes (spreads) beyond the ovaries. Risk factors associated with ovarian cancer include age (usually over age 50); race (whites are at highest risk); family history of ovarian cancer; more than 40 years of active ovulation; *nulliparity* (no pregnancies) or first pregnancy after age 30; a high-fat, low-fiber, vitamin A–deficient diet; and prolonged exposure to asbestos and talc. Early ovarian cancer may have no symptoms or mild ones such as abdominal discomfort, heartburn, nausea, loss of appetite, bloating, and flatulence. Later-stage signs and symptoms include an enlarged abdomen, abdominal and/or pelvic pain, persistent gastrointestinal disturbances, urinary complications, menstrual irregularities, and heavy menstrual bleeding.

Cervical Cancer

Cervical cancer, cancer of the uterine cervix, starts with **cervical dysplasia** (dis-PLĀ-zha), a change in the shape, growth, and number of cervical cells. The cells may either return to normal or progress to cancer. In most cases cervical cancer may be detected in its earliest stages by a Pap smear. Some evidence links cervical cancer to the virus that causes genital warts (human papilloma virus). Increased risk is associated with a large number of sexual partners, first intercourse at a young age, and smoking cigarettes.

Vulvovaginal Candidiasis

Candida albicans is a yeastlike fungus that commonly grows on mucous membranes of the gastrointestinal and genitourinary tracts. The organism is responsible for **vulvovaginal candidiasis** (vul′-vō-VAJ-i-nal can′-di-DĪ-a-sis), the most common form of **vaginitis** (vaj′-i-NĪ-tis), inflammation of the vagina. Candidiasis, commonly referred to as a yeast infection, is characterized by severe itching; a thick, yellow, cheesy discharge; a yeasty odor; and pain. The disorder, experienced at least once by about 75% of females, is usually a result of proliferation of the fungus following antibiotic therapy for another condition. Predisposing conditions include the use of oral contraceptives or cortisone-like medications, pregnancy, and diabetes.

Sexually Transmitted Diseases

A **sexually transmitted disease (STD)** is one that is spread by sexual contact. AIDS and hepatitis B, which are sexually transmitted diseases that also may be contracted in other ways, are discussed in Chapters 17 and 19, respectively.

Chlamydia

Chlamydia (kla-MID-ē-a) is a sexually transmitted disease caused by the bacterium *Chlamydia trachomatis* (*chlamy-* = cloak). This unusual bacterium cannot reproduce outside body cells; it "cloaks" itself inside cells, where it divides. At present, chlamydia is the most prevalent sexually transmitted disease. In most cases the initial infection is asymptomatic and thus difficult to recognize clinically. In males, urethritis is the principal result, causing a clear discharge and burning, frequent, and painful urination. Without treatment, the epididymides may also become inflamed, leading to male sterility. The uterine tubes may also become inflamed, which increases the risk of female infertility due to the formation of scar tissue. In 70% of females with chlamydia symptoms are absent, but chlamydia is the leading cause of pelvic inflammatory disease.

Gonorrhea

Gonorrhea (gon′-ō-Rē-a) is caused by the bacterium *Neisseria gonorrhoeae*. Discharges from infected mucus membranes are the source of transmission of the bacteria either during sexual contact or during the passage of a newborn through the birth canal. Males usually experience urethritis with profuse pus drainage and painful urination. In females, infection typically occurs in the vagina, often with a discharge of pus. In females, the infection and consequent inflammation can proceed from the vagina into the uterus, uterine tubes, and pelvic cavity. Thousands of women are made infertile by gonorrhea every year as a result of scar tissue formation that closes the uterine tubes. Transmission of bacteria in the birth canal to the eyes of a newborn can result in blindness.

Syphilis

Syphilis, caused by the bacterium *Treponema pallidum*, is transmitted through sexual contact or exchange of blood, or through the placenta to a fetus. The disease progresses through several stages. During the *primary stage*, the chief sign is a painless open sore, called a **chancre** (SHANG-ker), at the point of contact. The chancre heals within 1 to 5 weeks. From 6 to 24 weeks later, signs and symptoms such as a skin rash, fever, and aches in the joints and muscles usher in the *secondary stage*, which is systemic—the infection spreads to all major body systems. When signs of organ degeneration appear, the disease is said to be in the *tertiary stage*. If the nervous system is involved, the tertiary stage is called **neurosyphilis.** As motor areas become extensively damaged, victims may be unable to control urine and bowel movements; eventually they may become bedridden, unable even to feed themselves. Damage to the cerebral cortex produces memory loss and personality changes that range from irritability to hallucinations.

Genital Herpes

Genital herpes is caused by type 2 herpes simplex virus (HSV-2), producing painful blisters on the prepuce, glans penis, and penile shaft in males, and on the vulva or sometimes high up in the vagina in females. The blisters disappear and reappear in most patients, but the virus itself remains in the body; there is no cure. A related virus, type 1 herpes simplex virus (HSV-1), causes cold sores on the mouth and lips. Infected individuals typically experience recurrences of symptoms several times a year.

MEDICAL TERMINOLOGY AND CONDITIONS

Amenorrhea (ā-men′-ō-RĒ-a; *a-* = without; *men-* = month; *-rrhea* = a flow) The absence of menstruation; it may be caused by a hormone imbalance, obesity, extreme weight loss, or very low body fat as may occur during rigorous athletic training.

Colposcopy (kol-POS-kō-pē) A procedure used to directly examine the vaginal and cervical mucosa with a low-power binocular microscope called a colposcope, which magnifies the mucous membrane from about 6 to 40 times its actual size. This is often the first test done after an abnormal Pap smear.

Dysmenorrhea (dis-men′-ō-RĒ-a; *dys-* = difficult or painful) Painful menstruation; the term is usually reserved to describe menstrual symptoms that are severe enough to prevent a woman from functioning normally for one or more days each month. Some cases are caused by uterine tumors, ovarian cysts, pelvic inflammatory disease, or intrauterine devices.

Endocervical curettage (kū′-re-TAZH; *curette* = scraper) A procedure in which the cervix is dilated and the endometrium of the uterus is scraped with a spoon-shaped instrument called a curette; commonly called a D and C (dilation and curettage).

Hypospadias (hī′-pō-SPĀ-dē-as; *hypo-* = below) A displaced urethral opening. In males, the displaced opening may be on the underside of the penis, at the penoscrotal junction, between the scrotal folds, or in the perineum; in females, the urethra opens into the vagina.

Oophorectomy (ō′-of-ō-REK-tō-mē; *oophor-* = bearing eggs) Removal of the ovaries.

Ovarian cyst The most common form of ovarian tumor, in which a fluid-filled follicle or corpus luteum persists and continues growing.

Pelvic inflammatory disease (PID) A collective term for any extensive bacterial infection of the pelvic organs, especially the uterus, uterine tubes, or ovaries, which is characterized by pelvic soreness, lower back pain, abdominal pain, and urethritis. Often the early symptoms of PID occur just after menstruation. As infection spreads and cases advance, fever may develop, along with painful abscesses of the reproductive organs.

Salpingectomy (sal′-pin-JEK-tō-mē; *salpingo* = tube) Removal of a uterine tube (oviduct).

Smegma (SMEG-ma) The secretion, consisting principally of sloughed off epithelial cells, found chiefly around the external genitalia and especially under the foreskin of the male.

STUDY OUTLINE

Male Reproductive System (p. 546)

1. Sexual reproduction is the process of producing offspring by the union of gametes (oocytes and sperm).
2. The organs of reproduction are grouped as gonads (produce gametes), ducts (transport and store gametes), accessory sex glands (produce materials that support gametes), and supporting structures.
3. The male reproductive system includes the testes, ductus epididymis, ductus (vas) deferens, ejaculatory duct, urethra, seminal vesicles, prostate gland, bulbourethral (Cowper's) glands, and penis.
4. The scrotum is a sac that supports and regulates the temperature of the testes.
5. The male gonads are testes, oval-shaped organs in the scrotum that contain the seminiferous tubules, in which sperm cells develop; Sertoli cells, which nourish sperm cells and produce inhibin; and Leydig cells, which produce the male sex hormone testosterone.
6. Spermatogenesis occurs in the testes and consists of meiosis I, meiosis II, and spermiogenesis. It results in the formation of four haploid sperm cells from a primary spermatocyte.
7. Mature sperm consist of a head, midpiece, and tail. Their function is to fertilize a secondary oocyte.
8. At puberty, gonadotropin-releasing hormone (GnRH) stimulates anterior pituitary secretion of LH and FSH. LH stimulates Leydig cells to produce testosterone. FSH and testosterone initiate spermatogenesis.
9. Testosterone controls the growth, development, and maintenance of sex organs; stimulates bone growth, protein anabolism, and sperm maturation; and stimulates development of male secondary sex characteristics.
10. Inhibin is produced by Sertoli cells; its inhibition of FSH helps regulate the rate of spermatogenesis.
11. The duct system of the testes includes the seminiferous tubules, straight tubules, and rete testis.
12. Sperm are transported out of the testes into an adjacent organ, the epididymis, where their motility increases.
13. The ductus (vas) deferens stores sperm and propels them toward the urethra during ejaculation. Removing part of the vas deferens to prevent fertilization is called vasectomy.
14. The ejaculatory ducts are formed by the union of the ducts from the seminal vesicles and vas deferens, and they eject sperm into the urethra.
15. The male urethra passes through the prostate gland, urogenital diaphragm, and penis.
16. The seminal vesicles secrete an alkaline, viscous fluid that constitutes about 60% of the volume of semen and contributes to sperm viability.
17. The prostate gland secretes a slightly acidic fluid that constitutes about 25% of the volume of semen and contributes to sperm motility.
18. The bulbourethral (Cowper's) glands secrete mucus for lubrication and an alkaline substance that neutralizes acid.

19. Semen is a mixture of sperm and seminal fluid; it provides the fluid in which sperm are transported, supplies nutrients, and neutralizes the acidity of the male urethra and the vagina.

20. The penis consists of a root, a body, and a glans penis. It functions to introduce sperm into the vagina. Expansion of its blood sinuses under the influence of sexual excitation is called erection.

Female Reproductive System (p. 555)

1. The female organs of reproduction include the ovaries (gonads), uterine (Fallopian) tubes, uterus, vagina, and vulva.

2. The mammary glands are considered part of the reproductive system.

3. The female gonads are the ovaries, located in the upper pelvic cavity on either side of the uterus.

4. Ovaries produce secondary oocytes; discharge secondary oocytes (the process of ovulation); and secrete estrogens, progesterone, relaxin, and inhibin.

5. Oogenesis (production of haploid secondary oocytes) begins in the ovaries. The oogenesis sequence includes meiosis I and meiosis II. Meiosis II is completed only after an ovulated secondary oocyte is fertilized by a sperm cell.

6. The uterine (Fallopian) tube, which transports a secondary oocyte from an ovary to the uterus, is the normal site of fertilization.

7. The uterus is an organ the size and shape of an inverted pear that functions in menstruation, implantation of a fertilized ovum, development of a fetus during pregnancy, and labor. It also is part of the pathway for sperm to reach a uterine tube to fertilize a secondary oocyte.

8. The innermost layer of the uterine wall is the endometrium, which undergoes marked changes during the menstrual cycle.

9. The vagina is a passageway for the menstrual flow, the receptacle for the penis during sexual intercourse, and the lower portion of the birth canal. The smooth muscle of the vaginal wall makes it capable of considerable stretching.

10. The vulva, a collective term for the external genitals of the female, consists of the mons pubis, labia majora, labia minora, clitoris, vestibule, vaginal and urethral orifices, paraurethral (Skene's) glands, and greater vestibular (Bartholin's) glands.

11. An incision in the layers of the perineum, a diamond-shaped area at the inferior end of the trunk between the thighs and buttocks, prior to delivery is called an episiotomy.

12. The mammary glands of the female breasts are modified sweat glands located over the pectoralis major muscles. Their function is to secrete and eject milk (lactation).

13. Mammary gland development depends on estrogens and progesterone.

14. Milk production is stimulated by prolactin, estrogens, and progesterone; milk ejection is stimulated by oxytocin.

Female Reproductive Cycle (p. 561)

1. The female reproductive cycle includes the ovarian and menstrual cycles. The function of the ovarian cycle is development of a secondary oocyte, whereas that of the menstrual cycle is preparation of the endometrium each month to receive a fertilized egg.

2. The ovarian and menstrual cycles are controlled by GnRH from the hypothalamus, which stimulates the release of FSH and LH by the anterior pituitary.

3. FSH stimulates development of secondary follicles and initiates secretion of estrogens by the follicles. LH stimulates further development of the follicles, secretion of estrogens by follicular cells, ovulation, formation of the corpus luteum, and the secretion of progesterone and estrogens by the corpus luteum.

4. Estrogens stimulate the growth, development, and maintenance of female reproductive structures; the development of secondary sex characteristics; and protein synthesis.

5. Progesterone works with estrogens to prepare the endometrium for implantation and the mammary glands for milk synthesis.

6. Relaxin increases the flexibility of the pubic symphysis and helps dilate the uterine cervix to ease delivery of a baby.

7. During the menstrual phase, the stratum functionalis of the endometrium is shed, discharging blood and tissue cells.

8. During the preovulatory phase, a group of follicles in the ovaries begins to undergo final maturation. One follicle outgrows the others and becomes dominant while the others die. At the same time, endometrial repair occurs in the uterus. Estrogens are the dominant ovarian hormones during the preovulatory phase.

9. Ovulation is the rupture of the dominant mature (Graafian) follicle and the release of a secondary oocyte into the pelvic cavity. It is brought about by a surge of LH.

10. During the postovulatory phase, both progesterone and estrogens are secreted in large quantity by the corpus luteum of the ovary, and the uterine endometrium thickens in readiness for implantation.

11. If fertilization and implantation do not occur, the corpus luteum degenerates, and the resulting low level of progesterone allows discharge of the endometrium (menstruation) followed by the initiation of another reproductive cycle.

12. If fertilization and implantation occur, the corpus luteum is maintained by placental hCG and the corpus luteum.

Birth Control Methods and Abortion (p. 565)

1. Birth control methods include surgical sterilization (vasectomy, tubal ligation), hormonal methods, intrauterine devices, spermatocides, barrier methods (condom, vaginal pouch, diaphragm), periodic abstinence, and coitus interruptus. Table 23.1 on page 000 lists the failure rates for these methods. Abstinence is the only foolproof method of birth control.

2. Contraceptive pills of the combination type contain estrogens and progestins in concentrations that decrease the secretion of FSH and LH and thereby inhibit development of ovarian follicles and ovulation.

3. An abortion is the premature expulsion from the uterus of the products of conception; it may be spontaneous or induced. RU 486 can induce abortion by blocking the action of progesterone.

The Human Sexual Response (p. 567)

1. The similar sequence of changes experienced by both males and females before, during, and after intercourse is termed the human sexual response; it occurs in four stages: excitement (arousal), plateau, orgasm, and resolution.

2. During excitement and plateau, parasympathetic nerve impulses produce genital vasocongestion (engorgement of tissues with blood) and secretion of lubricating fluids. Heart rate, blood pressure, breathing rate, and muscle tone increase.

3. During orgasm, sympathetic and somatic motor nerve impulses cause rhythmic contractions of smooth and skeletal muscles.

4. During resolution, the body relaxes and returns to the unaroused state.

Aging and the Reproductive Systems (p. 568)

1. Puberty is the period of time when secondary sex characteristics begin to develop and the potential for sexual reproduction arises.

In older females, levels of progesterone and estrogens decrease, resulting in changes in menstruation and then menopause.

2. In older males, decreased levels of testosterone are associated with decreased muscle strength, waning sexual desire, and fewer viable sperm; prostate disorders are common.

■ SELF-QUIZ

1. The testes are located in the scrotum because

a. they must be separated from all other organs or sterility can occur   **b.** sperm and hormone production and survival require a temperature lower than the normal body temperature   **c.** the scrotum supplies the necessary hormones for sperm maturation   **d.** sperm in the testes cannot survive without the nutrients supplied by the scrotum   **e.** the scrotum produces alkaline fluids that neutralize the acids in the male urethra

2. Match the following:

____ **a.** cells that support, protect and nourish developing spermatogonia

____ **b.** contain developing oocytes

____ **c.** immature sperm cells

____ **d.** cells that secrete testosterone

____ **e.** produces progesterone and estrogens

A. corpus luteum
B. Leydig cells
C. Sertoli cells
D. follicles
E. spermatogonia

3. Removal of the prostate gland would

a. interfere with sperm production   **b.** inhibit testosterone release   **c.** decrease the volume of semen by about 75%   **d.** cause semen to become more acidic   **e.** affect semen clotting

4. Which of the following is true?

a. Meiosis is the process by which somatic (body) cells divide.   **b.** The haploid chromosome number is symbolized by $2n$.   **c.** Meiosis I results in diploid spermatocytes.   **d.** Gametes contain the haploid chromosome number.   **e.** Gametes contain 46 chromosomes in their nuclei.

5. The uterus is the site of all of the following except

a. menstruation   **b.** implantation of a fertilized ovum   **c.** ovulation   **d.** labor   **e.** development of the fetus

6. Menses is triggered by a

a. rapid rise in luteinizing hormone (LH)   **b.** rapid fall in luteinizing hormone (LH)   **c.** drop in estrogens and progesterone   **d.** rise in estrogens and progesterone   **e.** rise in inhibin

7. An inflammation of the seminiferous tubules would interfere with the ability to

a. secrete testosterone   **b.** produce sperm   **c.** void urine   **d.** make semen alkaline   **e.** regulate the temperature in the scrotum

8. The filling of blood sinuses in the penis under the influence of sexual stimulation causes

a. emission   **b.** orgasm   **c.** resolution   **d.** erection   **e.** ejaculation

9. Which of the following is NOT a function of semen?

a. transport sperm   **b.** lubricate the reproductive tract   **c.** provide an acidic environment needed for fertilization   **d.** provide nourishment for sperm   **e.** produce antibiotics to destroy some bacteria

10. Prior to ejaculation, sperm are stored in the

a. Leydig cells   **b.** scrotum   **c.** Sertoli cells   **d.** prostate gland   **e.** ductus (vas) deferens

11. Place the following in the correct order for the passage of sperm from the testes to the outside of the body.

1. ductus (vas) deferens   **4.** epididymis
2. rete testes   **5.** urethra
3. seminiferous tubules   **6.** straight tubules

a. 6, 3, 2, 4, 1, 5   **b.** 3, 2, 6, 4, 1, 5   **c.** 3, 6, 2, 4, 1, 5
d. 3, 6, 2, 4, 5, 1   **e.** 2, 4, 6, 1, 3, 5

12. In males, the gland that surrounds the urethra at the base of the bladder is the

a. glans penis   **b.** prostate gland   **c.** seminal vesicle   **d.** bulbourethral gland   **e.** greater vestibular gland

13. An oocyte is moved towards the uterus by

a. peristaltic contractions of the uterine (Fallopian) tubes   **b.** contraction of the uterus   **c.** gravity   **d.** swimming   **e.** flagella

14. Fertilization normally occurs in the

a. vagina   **b.** cervix   **c.** uterus   **d.** ovary   **e.** uterine (Fallopian) tube

15. In the female reproductive system, lubricating mucus is produced by the

a. vulva   **b.** clitoris   **c.** mons pubis   **d.** greater vestibular (Bartholin's) glands   **e.** sudoriferous glands

16. Ovarian follicles mature during

a. menses   **b.** ovulation   **c.** the postovulatory phase   **d.** the preovulatory phase   **e.** the secretory phase

17. Match the following:

____ **a.** initiated by sympathetic nerve impulses

____ **b.** return to unaroused state

____ **c.** initiated by parasympathetic impulses

____ **d.** stage in which a reddening of the face and chest may occur

A. excitement stage
B. plateau
C. orgasm
D. resolution

18. The portion of the uterus responsible for contraction is the

a. fundus   **b.** infundibulum   **c.** endometrium   **d.** myometrium   **e.** perimetrium

19. Match the following:

____ **a.** released by hypothalamus to regulate ovarian cycle

____ **b.** stimulates the initial secretion of estrogens by growing follicles

____ **c.** stimulates ovulation

____ **d.** stimulate growth, development, and maintenance of the female reproductive system

____ **e.** works with estrogens to prepare the uterus for implantation of a fertilized ovum

____ **f.** assists with labor by helping to dilate the cervix and increase flexibility of the pubic symphysis

____ **g.** inhibits release of FSH by the anterior pituitary

A. luteinizing hormone (LH)

B. gonadotropin-releasing hormone (GnRH)

C. relaxin

D. progesterone

E. inhibin

F. follicle-stimulating hormone (FSH)

G. estrogens

20. Birth control pills are a combination of ovarian hormones that prevent pregnancy by

a. neutralizing the pH of the vagina   **b.** inhibiting motility of the sperm   **c.** causing early ovulation, before the folicle is mature   **d.** preventing sperm from penetrating the zona pellucida **e.** inhibiting the secretion of LH and FSH from the pituitary gland

CRITICAL THINKING APPLICATIONS

1. Thirty-five-year-old Janelle has been advised to have a complete hysterectomy due to medical problems. She is worried that the procedure will cause menopause. Explain what is involved in the procedure and the likelihood that the procedure will result in menopause.

2. Phil has promised his wife that he will get a vasectomy after the birth of their next child. He is a little concerned, however, about the possible effects on his virility. What would you tell Phil about the procedure?

3. Maria was annoyed when her primary physician told her that she should schedule regular Pap smears. "I don't have any symptoms of an infection, so why should I?" Is Maria right to be annoyed?

4. An enlarged prostate gland (benign prostatic hyperplasia) is common in older men. What are the symptoms of this condition? If the prostate gland were removed, what would be the effect on the semen?

ANSWERS TO FIGURE QUESTIONS

23.1 Functionally, the penis is considered a supporting structure.

23.2 Spermatogonia (stem cells) are least mature.

23.3 Crossing-over permits recombination of genes.

23.4 The midpiece contains mitochondria, which produce ATP.

23.5 Testosterone inhibits secretion of LH and GnRH, and inhibin inhibits secretion of FSH.

23.6 Semen includes sperm and the combined secretions from the accessory sex glands and testes.

23.7 Two corpora cavernosa penis and one corpus spongiosum penis contain blood sinuses that fill with blood. Because the blood cannot flow out of the penis as quickly as it flows in, the trapped blood stiffens the tissue.

23.8 The female external genitalia are collectively referred to as the vulva or pudendum.

23.9 Broad, ovarian, and suspensory ligaments hold the ovaries in position.

23.10 Ovarian follicles secrete estrogens, and the corpus luteum secretes estrogens, progesterone, relaxin, and inhibin.

23.11 Primary oocytes are present in the ovary at birth, so they are as old as the woman is. In males, primary spermatocytes are continually being formed from spermatogonia and thus are only a few days old.

23.12 The stratum basalis provides cells to replace those of the stratum functionalis, which is shed during each menstruation.

23.13 The mons pubis, clitoris, prepuce, and external urethral orifice are anterior to the vagina.

23.14 Milk release is regulated by oxytocin.

23.15 Relaxin eases delivery.

23.16 Estrogens stimulate proliferation of the endometrium.

23.17 Ovulation predictor tests detect luteinizing hormone (LH).

23.18 This is negative feedback because the response is opposite to the stimulus. Decreasing estrogens and progesterone stimulate release of GnRH, which, in turn, increases production and release of estrogens.

Chapter 24

Development and Inheritance

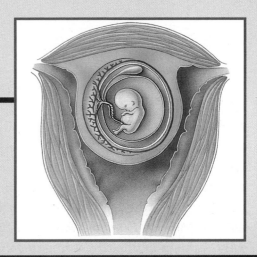

■ Student Learning Objectives

1. Explain the processes associated with fertilization, morula formation, blastocyst development, and implantation. **576**
2. Discuss the formation of the primary germ layers and extraembryonic membranes during the embryonic period. **579**
3. List representative body structures produced by the primary germ layers. **581**

4. Describe the formation of the placenta and umbilical cord. **582**
5. Describe the sources and functions of the hormones secreted during pregnancy. **586**
6. Describe the hormonal, anatomical, and physiological changes in the mother during pregnancy. **586**
7. Explain the effects of pregnancy on exercise, and of exercise on pregnancy. **588**

8. Outline the events associated with the three stages of labor. **588**
9. Define inheritance, and explain the inheritance of dominant, recessive, polygenic, and sex-linked traits. **589**

■ A Look Ahead

FROM FERTILIZATION TO IMPLANTATION

Objective: • **Explain the processes associated with fertilization, morula formation, blastocyst development, and implantation.**

Once sperm have been deposited in the vagina, pregnancy can occur. *Pregnancy* is a sequence of events that begins with fertilization; proceeds to implantation, embryonic development, and fetal development; and normally ends with birth about 38 weeks later.

Fertilization

During *fertilization* (fer′-ti-li-ZĀ-shun; *fertil-* = fruitful), the genetic material from a haploid sperm cell and a haploid secondary oocyte merges into a single diploid nucleus (Figure 24.1). Of the 300 to 500 million sperm introduced into the vagina, fewer than 1% reach the secondary oocyte. Fertilization normally occurs in the uterine (Fallopian) tube about 12 to 24 hours after ovulation. Sperm remain viable in the vagina for about 48 hours, and a secondary oocyte is viable for about 24 hours after ovulation. Thus, there is a three-day window during which pregnancy is most likely to occur—from two days before ovulation to one day after ovulation.

Sperm swim up the uterus and into the uterine tubes, propelled by the whiplike movements of their tails (flagella). In ad-

---

*D*evelopmental anatomy is the study of the sequence of events from the fertilization of a secondary oocyte to the formation of an adult organism. For the first two months after fertilization, the developing human is an *embryo* (*-bryo* = grow), and this is the period of *embryological development. Fetal development* begins at nine weeks and continues until birth; during this time the developing human is a *fetus* (*feo* = to bring forth). Together, embryological and fetal development constitute *prenatal* (before birth) *development,* a wonderfully complex and precisely coordinated series of events. *Embryology* (em′-brē-OL-ō-jē) is the study of development from the fertilized egg through the eighth week. *Obstetrics* (ob-STET-riks; *obstetrix* = midwife) is the branch of medicine that deals with the management of pregnancy, labor, and the *neonatal period,* the first 42 days after birth. In this chapter we focus on the developmental sequence from fertilization through implantation, embryonic and fetal development, labor, and birth.

Figure 24.1 ■ **Fertilization.** Sperm cell penetrating the corona radiata and zona pellucida around a secondary oocyte.

🔑 **During fertilization, genetic material from a sperm cell merges with that of a secondary oocyte to form a single diploid nucleus.**

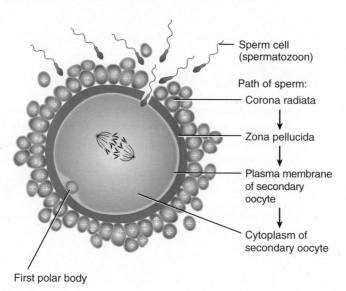

- Sperm cell (spermatozoon)
- Path of sperm:
- Corona radiata
- Zona pellucida
- Plasma membrane of secondary oocyte
- Cytoplasm of secondary oocyte
- First polar body

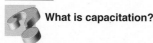 **What is capacitation?**

dition, prostaglandins in semen and oxytocin released from the posterior pituitary during female orgasm stimulate uterine contractions to further aid the movement of sperm toward the uterine tubes. Although 100 or so sperm reach the vicinity of the oocyte within minutes after ejaculation, they are not able to fertilize it until several hours later. During this time sperm undergo **capacitation,** a series of functional changes that cause the sperm's tail to beat even more vigorously and enable its plasma membrane to fuse with the oocyte's plasma membrane.

For fertilization to occur, a sperm cell first must penetrate the **corona radiata,** the cloud of granulosa cells that surround the oocyte, and then the **zona pellucida,** the clear glycoprotein layer between the corona radiata and the oocyte's plasma membrane (Figure 24.1). Glycoproteins in the zona pellucida act as sperm receptors by binding to specific membrane proteins in the sperm heads. This interaction triggers release of the contents of the acrosomes. The acrosomal enzymes digest a path through the zona pellucida as the lashing sperm tail pushes the sperm cell toward the oocyte's plasma membrane. Many sperm bind to the zona pellucida and release their enzymes, but only the first sperm cell to penetrate the entire zona pellucida and reach the oocyte's plasma membrane fuses with the oocyte. When fusion occurs, it triggers changes that block fertilization by other sperm. Within 1 to 3 seconds, the cell membrane of the oocyte depolarizes, which causes chemical changes that inactivate the sperm receptors and harden the entire zona pellucida.

Once a sperm cell enters a secondary oocyte, the oocyte completes meiosis II. It divides into a larger ovum (mature egg) and a smaller second polar body that fragments and disintegrates (see Figure 23.10 on page 556). The nucleus in the head of the sperm and the nucleus of the fertilized ovum fuse, producing a single diploid nucleus that contains 23 chromosomes from each parent, restoring the diploid number of 46 chromosomes. The fertilized ovum now is called a **zygote** (ZĪ-gōt; *zygosis* = a joining).

Dizygotic (fraternal) twins are produced from the independent release of two secondary oocytes and the subsequent fertilization of each by different sperm. Although fraternal twins look similar because they are the same age, they are genetically as dissimilar as siblings born at different times. Dizygotic twins may or may not be the same sex. Because **monozygotic (identical) twins** develop from a single fertilized ovum, they contain exactly the same genetic material and are always the same sex. Monozygotic twins arise from separation of cells of an embryo early in development into two clusters. The two clusters of embryonic cells then develop into two genetically identical fetuses.

Development of the Blastocyst

After fertilization, rapid mitotic cell divisions of the zygote, called **cleavage,** take place. The first division of the zygote begins about 24 hours after fertilization. Successive cleavages eventually produce a solid sphere of cells called the **morula** (MOR-yoo-la = mulberry), which is about the same size as the original zygote and is still surrounded by the zona pellucida. By the end of the fourth day, the number of cells in the morula in-

creases even more as it continues to move through the uterine tube toward the uterine cavity. At 4.5 to 5 days, the dense cluster of cells has developed into a hollow ball of cells, called a **blastocyst** (*-cyst* = bag), that enters the uterine cavity (Figure 24.2).

The blastocyst consists of an outer covering of cells called the **trophoblast** (TRŌF-ō-blast; *tropho-* = develop or nourish), an **inner cell mass,** and a fluid-filled **blastocyst cavity.** The trophoblast and part of the inner cell mass ultimately form the membranes composing the fetal portion of the placenta. The rest of the inner cell mass develops into the embryo.

Figure 24.2 ■ Blastocyst.

Cleavage refers to the early, rapid mitotic divisions in a zygote.

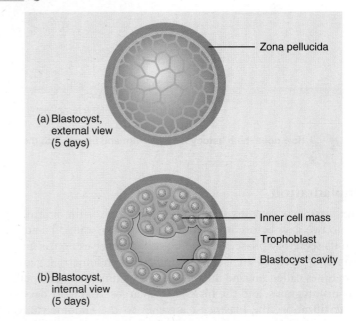

Zona pellucida

(a) Blastocyst, external view (5 days)

Inner cell mass

Trophoblast

Blastocyst cavity

(b) Blastocyst, internal view (5 days)

What is the difference between a morula and a blastocyst?

Figure 24.3 ■ **Implantation.** The blastocyst is shown in relation to the endometrium of the uterus at various times after fertilization.

 Implantation refers to the attachment of a blastocyst to the endometrium, which occurs about six days after fertilization.

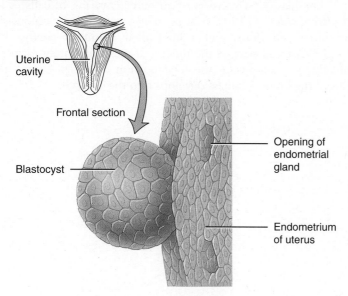

(a) External view, about 6 days after fertilization

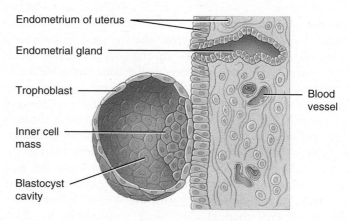

(b) Internal view, about 6 days after fertilization

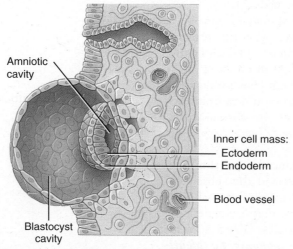

(c) Internal view, about 7 days after fertilization

How does the blastocyst merge with and burrow into the endometrium?

Implantation

The blastocyst remains free within the cavity of the uterus for one to two days before it attaches to the uterine wall. The uterus is in its secretory phase, and the blastocyst receives nourishment from the glycogen-rich secretions of the endometrial glands, sometimes called uterine milk. During this time, the zona pellucida disintegrates, and the blastocyst enlarges. About six days after fertilization, the blastocyst attaches to the endometrium, a process called *implantation* (Figure 24.3).

During implantation, the trophoblast secretes enzymes that digest the uterine lining and allow the blastocyst to burrow into the endometrium during the second week after fertilization.

The placenta develops between the inner cell mass and the endometrial wall to provide nutrients for the growth of the embryo (more detail on this shortly). By now, the trophoblast has begun to secrete *human chorionic gonadotropin (hCG),* a hormone that maintains the corpus luteum. Blood levels of hCG increase to a maximum during the ninth week of pregnancy, then the level decreases. Figure 24.4 summarizes the main events of fertilization and implantation.

In an *ectopic pregnancy* (ek-TOP-ik = out of place), the embryo implants in a location outside the uterine cavity, usually in the uterine tube. The usual cause of an ectopic pregnancy is

Figure 24.4 ■ Summary of events associated with fertilization and implantation.

Fertilization usually occurs in a uterine tube.

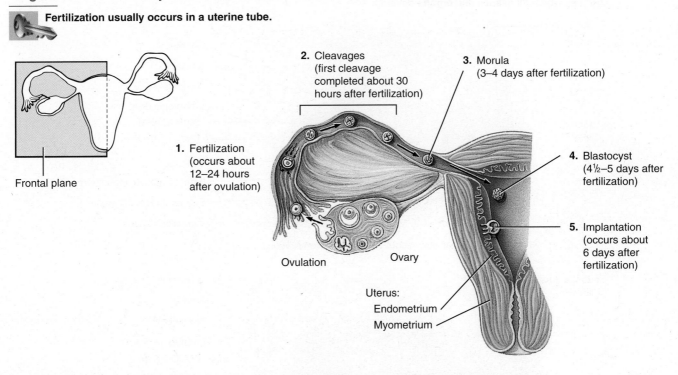

Frontal section through uterus, uterine tube, and ovary

 In what phase of the menstrual cycle is the uterus when implantation occurs?

impaired passage of the fertilized ovum through the uterine tube. Scar tissue from pelvic inflammatory disease or developmental abnormalities may block the tube. Also, nicotine in cigarette smoke slows passage of the ovum by paralyzing the cilia in the lining of the uterine tube. Ectopic pregnancy is characterized by one or two missed menses (periods), followed by vaginal bleeding and acute pelvic pain. Unless removed, the developing embryo can rupture the tube, often resulting in death of the mother.

EMBRYONIC AND FETAL DEVELOPMENT

Objectives: • **Discuss the formation of the primary germ layers and extraembryonic membranes during the embryonic period.**
• **List representative body structures produced by the primary germ layers.**
• **Describe the formation of the placenta and umbilical cord.**

The time span from fertilization to birth is the ***gestation period*** (jes-TĀ-shun; *gest-* = to bear). The human gestation period is about 38 weeks, counted from the estimated day of fertilization (or 2 weeks after the first day of the last menstruation). By the end of the ***embryonic period,*** the first two months of develop-

ment, the rudiments of all the principal adult organs are present, and the extraembryonic membranes are developed. During the ***fetal period,*** from the ninth week until birth, organs established by the primary germ layers grow rapidly, the placenta becomes functional, and the fetus takes on a human appearance.

The Beginnings of Organ Systems

The first major event of the embryonic period is differentiation of the inner cell mass of the blastocyst into three ***primary germ layers:*** ectoderm, endoderm, and mesoderm. These germ layers are the major embryonic tissues from which all tissues and organs of the body develop.

Within eight days after fertilization, cells of the inner cell mass proliferate and form a membrane, the ***amnion*** (AM-nē-on; *amnio-* = lamb), and a space, the ***amniotic cavity*** (am'-nē-OT-ik), adjacent to the inner cell mass (Figure 24.5). The layer of cells of the inner cell mass closest to the amniotic cavity develops into the ***ectoderm*** (*ecto-* = outside; *derm-* = skin). The inner cell mass layer that borders the blastocyst cavity develops into the ***endoderm*** (*endo-* = inside). As the amniotic cavity forms, the inner cell mass containing these two layers is called the ***embryonic disk.***

The cells of the endodermal layer divide rapidly. At the 12th day after fertilization, groups of endodermal cells extend around a hollow sphere, forming the ***yolk sac,*** another membrane (described shortly). Cells of the trophoblast give rise to a loose connective tissue, the ***extraembryonic mesoderm*** (*meso-* =

Figure 24.5 ■ Formation of the primary germ layers and associated structures.

The primary germ layers (ectoderm, mesoderm, and endoderm) are the embryonic tissues from which all tissues and organs develop.

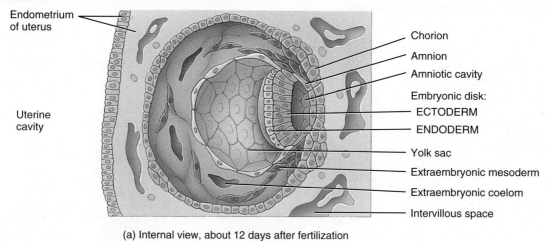

Endometrium of uterus

Uterine cavity

Chorion

Amnion

Amniotic cavity

Embryonic disk:
ECTODERM
ENDODERM

Yolk sac

Extraembryonic mesoderm

Extraembryonic coelom

Intervillous space

(a) Internal view, about 12 days after fertilization

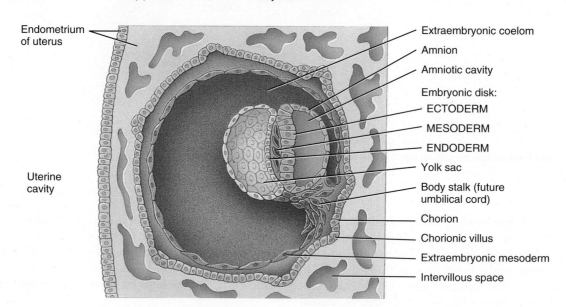

Endometrium of uterus

Uterine cavity

Extraembryonic coelom

Amnion

Amniotic cavity

Embryonic disk:
ECTODERM
MESODERM
ENDODERM

Yolk sac

Body stalk (future umbilical cord)

Chorion

Chorionic villus

Extraembryonic mesoderm

Intervillous space

(b) Internal view, about 14 days after fertilization

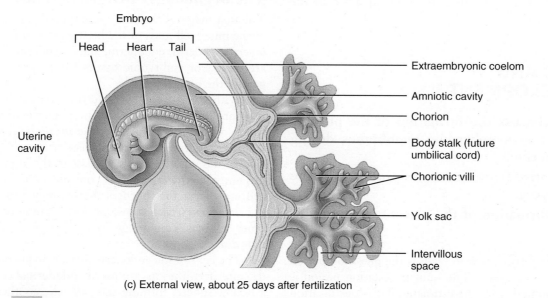

Embryo

Head   Heart   Tail

Uterine cavity

Extraembryonic coelom

Amniotic cavity

Chorion

Body stalk (future umbilical cord)

Chorionic villi

Yolk sac

Intervillous space

(c) External view, about 25 days after fertilization

Which structure will form the embryo?

middle), which completely fills the space between the trophoblast and yolk sac. Soon large spaces develop in the extraembryonic mesoderm and come together to form a single, larger cavity called the ***extraembryonic coelom*** (SĒ-lōm = cavity), the future ventral body cavity (Figure 24.5b).

By around day 14, the cells of the embryonic disk have differentiated into three distinct layers: ectoderm, mesoderm, and endoderm (Figure 24.5b). As the embryo develops, the endoderm becomes the epithelial lining of the gastrointestinal and respiratory tracts, and of several other organs. The mesoderm forms muscle, bone and other connective tissues, and the peritoneum. The ectoderm develops into the epidermis of the skin and the nervous system. Table 24.1 provides more details about the fates of these primary germ layers.

Formation of Extraembryonic Membranes

A second major event that occurs during the embryonic period is the formation of the ***extraembryonic membranes.*** These membranes, which lie outside the embryo, protect and nourish the embryo and, later, the fetus. (Recall that the developing embryo becomes a fetus after the second month.) The extraembryonic membranes are the yolk sac, the amnion, the chorion, and the allantois (Figure 24.6).

In species such as birds, the ***yolk sac*** is the primary source of blood vessels that transport nutrients to the embryo. Human embryos receive their nutrients from the endometrium. The human yolk sac remains small and functions as an early site of blood formation. It also contains cells that migrate into the gonads and differentiate into spermatogonia and oogonia.

Table 24.1 / Structures Produced by the Three Primary Germ Layers	
Germ Layer	**Structures Produced**
Endoderm	1. Epithelial lining of gastrointestinal tract (except the oral cavity and anal canal) and the epithelium of its glands.
	2. Epithelial lining of urinary bladder, gallbladder, and liver.
	3. Epithelial lining of pharynx, auditory (Eustachian) tubes, tonsils, larynx, trachea, bronchi, and lungs.
	4. Epithelium of thyroid gland, parathyroid glands, pancreas, and thymus.
	5. Epithelial lining of prostate and bulbourethral (Cowper's) glands, vagina, vestibule, urethra, and associated glands such as the greater vestibular (Bartholin's) and lesser vestibular glands.
Mesoderm	1. All skeletal, most smooth, and all cardiac muscle.
	2. Cartilage, bone, and other connective tissues.
	3. Blood, red bone marrow, and lymphatic tissue.
	4. Endothelium of blood vessels and lymphatic vessels.
	5. Dermis of skin.
	6. Fibrous tunic and vascular tunic of eye.
	7. Middle ear.
	8. Mesothelium of ventral body cavity.
	9. Epithelium of kidneys, ureters, adrenal cortex, gonads, and genital ducts.
Ectoderm	1. All nervous tissue.
	2. Epidermis of skin.
	3. Hair follicles, arrector pili muscles, nails, and epithelium of skin glands (sebaceous and sudoriferous).
	4. Lens, cornea, and internal eye muscles.
	5. Internal and external ear.
	6. Epithelium of sense organs.
	7. Epithelium of oral cavity, nasal cavity, paranasal sinuses, salivary glands, and anal canal.
	8. Epithelium of pineal gland, pituitary gland, and adrenal medullae.

Figure 24.6 ■ **Extraembryonic membranes.**

Extraembryonic membranes (outside the embryo) protect and nourish the embryo and, later, the fetus.

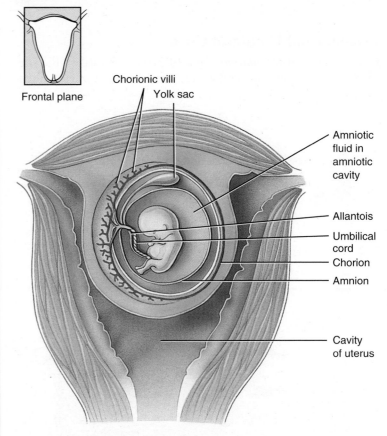

Frontal plane

Chorionic villi

Yolk sac

Amniotic fluid in amniotic cavity

Allantois

Umbilical cord

Chorion

Amnion

Cavity of uterus

Frontal section through uterus

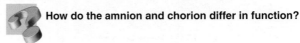

How do the amnion and chorion differ in function?

The *amnion* is a thin, protective membrane that forms by the eighth day after fertilization and initially overlies the embryonic disk (see Figure 24.5a,b). As the embryo grows, the amnion entirely surrounds the embryo, creating a cavity that becomes filled with amniotic fluid (Figure 24.6a), sometimes called the "bag of waters." Initially, amniotic fluid is filtered from maternal blood. Later, the fetus contributes to the fluid by excreting urine into the amniotic cavity. Amniotic fluid serves as a shock absorber for the fetus, helps regulate fetal body temperature, and prevents attachment between the skin of the fetus and surrounding tissues. Embryonic cells are sloughed off into amniotic fluid. The amnion usually ruptures just before birth, an event sometimes referred to as "water breaking."

The *chorion* (KŌR-ē-on) is derived from the trophoblast of the blastocyst and the mesoderm that lines the trophoblast. It surrounds the embryo and, later, the fetus. Eventually, the chorion becomes the principal embryonic part of the placenta. It also produces human chorionic gonadotropin (hCG). The inner layer of the chorion eventually fuses with the amnion.

The *allantois* (a-LAN-tō-is; *allant-* = sausage) is a small, vascularized structure that serves as another early site of blood formation.

Placenta and Umbilical Cord

Development of the *placenta* (pla-SEN-ta = flat cake), the site of exchange of nutrients and wastes between the mother and the fetus, occurs during the third month of pregnancy. The placenta forms from the chorion of the embryo and a portion of the endometrium of the mother.

Functionally, the placenta allows oxygen and nutrients to diffuse from maternal blood into fetal blood while carbon dioxide and wastes diffuse from fetal blood into maternal blood. The placenta also stores nutrients such as carbohydrates, proteins, calcium, and iron, which are released into the fetal circulation as required, and it produces several hormones that are needed during pregnancy. Because most microorganisms cannot pass through it, the placenta also is a protective barrier. However, certain viruses, such as those that cause AIDS, German measles, chickenpox, measles, encephalitis, and poliomyelitis, may cross the placenta. Many drugs, alcohol, and other substances that can cause birth defects can also pass freely through the placenta.

During embryonic life, fingerlike projections of the chorion, called *chorionic villi* (kō′-rē-ON-ik VIL-ī), grow into the endometrium of the uterus (Figure 24.7). These projections, which will contain fetal blood vessels, continue growing until they are bathed in maternal blood sinuses called *intervillous spaces* (in′-ter-VIL-us). Oxygen and nutrients from the mother's blood diffuse into capillaries of the villi. From the capillaries, the nutrients circulate into the *umbilical vein*. Wastes leave the fetus through the *umbilical arteries*, pass into the capillaries of the villi, and diffuse into the maternal blood. Thus maternal and fetal blood vessels are brought close together, but maternal and fetal blood do not normally mix. The *umbilical cord* (um-BIL-i-

Figure 24.7 ■ **Placenta and umbilical cord.**

🔑 The placenta is formed by the chorion of the embryo and part of the endometrium of the mother.

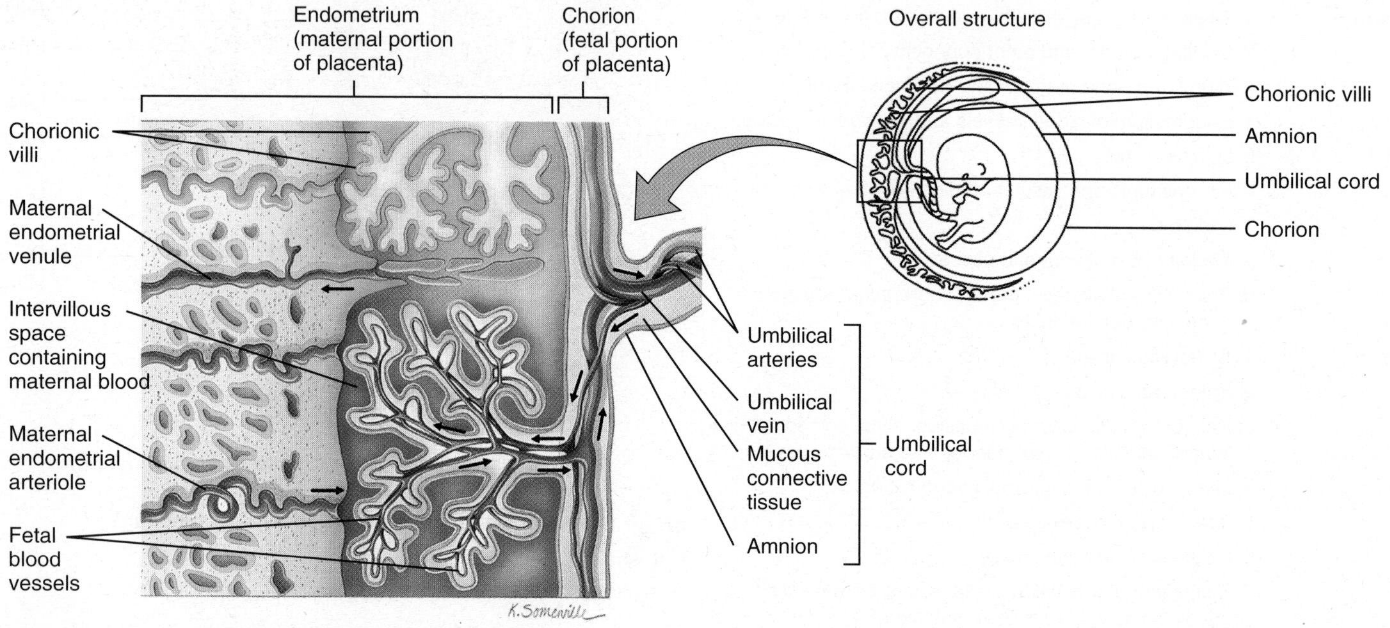

Details of placenta and umbilical cord

❓ **What is the function of the placenta?**

Table 24.2 / Changes Associated with Embryonic and Fetal Growth

End of Month	Approximate Size and Weight	Representative Changes
1	0.6 cm ($\frac{9}{16}$ in.)	Eyes, nose, and ears are not yet visible. Vertebral column and vertebral canal form. Limb buds form. Heart forms and starts beating. Body systems begin to form. The central nervous system appears at the start of the third week.
2	3 cm ($1\frac{1}{4}$ in.) 1 g ($\frac{1}{30}$ oz)	Eyes are far apart, eyelids fused. Nose is flat. Ossification begins. Limbs become distinct, and digits are well formed. Major blood vessels form. Many internal organs continue to develop.
3	$7\frac{1}{2}$ cm (3 in.) 30 g (1 oz)	Eyes are almost fully developed, but eyelids are still fused; nose develops a bridge; and external ears are present. Ossification continues. Limbs are fully formed and nails develop. Heartbeat can be detected. Urine starts to form. Fetus begins to move, but it cannot be felt by mother. Body systems continue to develop.
4	18 cm ($6\frac{1}{2}$–7 in.) 100 g (4 oz)	Head is large in proportion to rest of body. Face takes on human features, and hair appears on head. Many bones are ossified, and joints begin to form. Rapid development of body systems occurs.
5	25–30 cm (10–12 in.) 200–450 g ($\frac{1}{2}$–1 lb)	Head is less disproportionate to rest of body. Fine hair (lanugo) covers body. Fetal movements are commonly felt by mother (quickening). Rapid development of body systems occurs.
6	27–35 cm (11–14 in.) 550–800 g ($1\frac{1}{4}$–$1\frac{1}{2}$ lb)	Head becomes even less disproportionate to rest of body. Eyelids separate and eyelashes form. Substantial weight gain occurs. Skin is wrinkled. Type II alveolar cells begin to produce surfactant.
7	32–42 cm (13–17 in.) 110–1350 g ($2\frac{1}{2}$–3 lb)	Head and body are more proportionate. Skin is wrinkled. Seven-month fetus (premature baby) is capable of survival. Fetus assumes an upside-down position. Testes start to descend into scrotum.
8	41–45 cm ($16\frac{1}{2}$–18 in.) 2000–2300 g ($4\frac{1}{2}$–5 lb)	Subcutaneous fat is deposited. Skin is less wrinkled.
9	50 cm (20 in.) 3200–3400 g (7–$7\frac{1}{2}$ lb)	Additional subcutaneous fat accumulates. Lanugo is shed. Nails extend to tips of fingers and maybe even beyond.

kul) consists of the umbilical arteries, umbilical vein, and supporting connective tissue called mucous connective tissue (Figure 24.7).

After the birth of the baby, the placenta detaches from the uterus and is termed the *afterbirth*. At this time, the umbilical cord is severed, leaving the baby on its own. The small portion (about an inch) of the cord that remains still attached to the infant begins to wither and falls off, usually within 12 to 15 days after birth. The area where the cord was attached becomes covered by a thin layer of skin, and scar tissue forms. The scar is the *umbilicus (navel)*.

Pharmaceutical companies use human placentas as a source of hormones, drugs, and blood. Portions of placentas are also used to cover burns. The placental and umbilical cord veins can be used in blood vessel grafts, and cord blood can be frozen to preserve pluripotent stem cells for future use.

Table 24.2 lists the main changes that occur during embryonic and fetal growth.

Fetal Circulation

Some aspects of *fetal circulation* differ from blood circulation after birth because the lungs and gastrointestinal tract of a fetus are not functioning. The fetus derives its oxygen and nutrients from the maternal blood and eliminates its carbon dioxide and wastes into maternal blood.

The exchange of materials between fetal and maternal circulation occurs through the placenta. Blood passes from the fetus to the placenta via two *umbilical arteries* in the umbilical cord (Figure 24.8a). At the placenta, the blood picks up oxygen and nutrients and eliminates carbon dioxide and wastes. The oxygenated blood returns from the placenta via a single *umbilical vein,* which ascends to the liver of the fetus, where it divides into two branches. Some blood flows through the branch that joins the hepatic portal vein and enters the liver. The fetal liver manufactures red blood cells but does not function in digestion. Therefore, most of the blood bypasses the liver and flows into the second branch, the *ductus venosus* (DUK-tus ve-NŌ-sus). The ductus venosus eventually passes its blood to the inferior vena cava.

In general, circulation through other portions of the fetus is similar to circulation after birth. Deoxygenated blood returning from the lower limbs mixes with oxygenated blood from the ductus venosus in the inferior vena cava. This mixed blood then enters the right atrium. The circulation of blood through the upper portion of the fetus is also similar to circulation after birth. Deoxygenated blood returning from the upper regions of the fetus drains into the superior vena cava, and then into the right atrium.

Most of the blood does not flow from the right ventricle to the lungs, as it does after birth, because the fetal lungs do not operate. In the fetus, an opening called the *foramen ovale*

Figure 24.8 ■ **Fetal circulation and changes at birth.** The boxed areas indicate the fate of certain fetal structures once postnatal circulation is established.

 The lungs and gastrointestinal organs begin to function at birth.

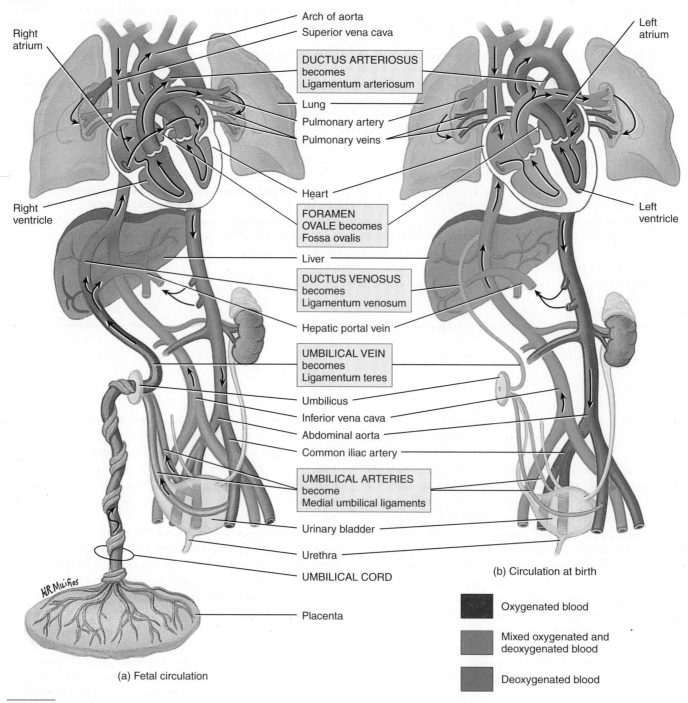

Arch of aorta
Superior vena cava
Right atrium
Left atrium

DUCTUS ARTERIOSUS
becomes
Ligamentum arteriosum

Lung
Pulmonary artery
Pulmonary veins

Right ventricle
Heart
Left ventricle

FORAMEN
OVALE becomes
Fossa ovalis

Liver

DUCTUS VENOSUS
becomes
Ligamentum venosum

Hepatic portal vein

UMBILICAL VEIN
becomes
Ligamentum teres

Umbilicus
Inferior vena cava
Abdominal aorta
Common iliac artery

UMBILICAL ARTERIES
become
Medial umbilical ligaments

Urinary bladder
Urethra
UMBILICAL CORD

Placenta

(a) Fetal circulation

(b) Circulation at birth

■ Oxygenated blood

■ Mixed oxygenated and deoxygenated blood

■ Deoxygenated blood

Which blood vessel carries blood with the highest level of oxygen in the fetal circulation?

(fō-RĀ-men ō-VA-lē) exists in the septum between the right and left atria. A valve in the inferior vena cava directs about a third of the blood through the foramen ovale so that it may be sent directly into systemic circulation. The blood that does descend into the right ventricle is pumped into the pulmonary trunk, but little of this blood actually reaches the lungs. Most is sent through the **ductus arteriosus** (ar-tē´-rē-Ō-sus), a small vessel connecting the pulmonary trunk with the aorta that allows most blood to bypass the fetal lungs. The blood in the aorta is carried to all parts of the fetus through its systemic branches. When the common iliac arteries branch into the external and internal iliacs, part of the blood flows into the internal iliacs. It then goes to the umbilical arteries and back to the placenta for another exchange of materials.

At birth, when the baby's respiratory, digestive, and liver functions are established, the special structures of fetal circulation are no longer needed. The foramen ovale normally closes shortly after birth to become the **fossa ovalis,** a depression in the interatrial septum. The ductus venosus, ductus arteriosus, and umbilical vessels constrict and become ligaments (Figure 24.8b).

Development and the Environment

Exposure of a developing embryo or fetus to certain environmental factors can damage the developing organism or even cause death. A **teratogen** (te-RAT-o-gen; *terato-* = monster; *-gen* = producing) is any agent or influence that causes developmental defects in the embryo. Alcohol is by far the number one fetal teratogen. Intrauterine exposure to alcohol may result in **fetal alcohol syndrome (FAS),** one of the most common causes of mental retardation and the most common preventable cause of birth defects. The symptoms of FAS may include slow growth before and after birth, characteristic facial features (short palpebral fissures, a thin upper lip, and sunken nasal bridge), defective heart and other organs, malformed limbs, genital abnormalities, and central nervous system damage. Behavioral problems, such as hyperactivity, extreme nervousness, reduced ability to concentrate, and an inability to appreciate cause-and-effect relationships, are common.

Other teratogens include pesticides; defoliants (chemicals that cause plants to shed their leaves prematurely); industrial chemicals; some hormones; antibiotics; oral anticoagulants, anticonvulsants, antitumor agents, thyroid drugs, thalidomide, diethylstilbestrol (DES), and numerous other prescription drugs; LSD; marijuana; and cocaine. A pregnant woman who uses cocaine, for example, subjects the fetus to higher risk of retarded growth, attention and orientation problems, hyperirritability, a tendency to stop breathing, malformed or missing organs, strokes, and seizures. The risks of spontaneous abortion, premature birth, and stillbirth also increase with fetal exposure to cocaine.

Cigarette smoking during pregnancy is linked to low infant birth weight and higher fetal and infant mortality. Cigarette smoke may be teratogenic and may cause cardiac abnormalities and anencephaly (a developmental defect characterized by the absence of a cerebrum). Infants nursing from smoking mothers have also been found to have an increased incidence of gastrointestinal disturbances. Even a mother's exposure to secondhand cigarette smoke (breathing air containing tobacco smoke) predisposes her baby to an increased incidence of respiratory problems, including bronchitis and pneumonia, during the first year of life.

Exposure of pregnant mothers to x-rays or radioactive isotopes during the embryo's susceptible period of development may cause microcephaly (small head size relative to the rest of the body), mental retardation, and skeletal malformations. Caution is advised, especially during the first three months of pregnancy.

Prenatal Diagnostic Tests

Several tests are available to detect genetic disorders and assess fetal well-being. Here we describe three of the most common tests: fetal ultrasonography, amniocentesis, and chorionic villi sampling (CVS).

Fetal Ultrasonography

If there is a question about the normal progress of a pregnancy, **fetal ultrasonography** (ul´-tra-son-OG-ra-fē) may be performed. By far the most common use of this procedure, also called *ultrasound*, is to determine a more accurate fetal age when the date of conception is unclear. It is also used to evaluate fetal viability and growth, determine fetal position, identify multiple pregnancies, identify fetal–maternal abnormalities, and guide special procedures such as amniocentesis. An instrument that emits high-frequency sound waves is passed back and forth over the mother's abdomen. The reflected sound waves from the developing fetus are detected by the same instrument and converted to an on-screen image called a **sonogram**.

Amniocentesis

Amniocentesis (am´-nē-ō-sen-TĒ-sis; *amnio-* = amnion; *-centesis* = puncture to remove fluid) involves withdrawing some of the amniotic fluid that bathes the developing fetus and analyzing the fetal cells and dissolved substances. It is performed for one of two purposes at one of two times during the pregnancy: (1) at 14 to 16 weeks to test for the presence of certain genetic disorders, such as Down syndrome (DS), spina bifida, hemophilia, Tay-Sachs disease, sickle cell disease, and certain muscular dystrophies, or (2) after week 35 to determine fetal maturity and well-being near the time of delivery. Because it involves examination of fetal chromosomes, amniocentesis can also reveal gender.

During amniocentesis, the position of the fetus and placenta is first identified using ultrasound and palpation. After the skin is prepared with an antiseptic and a local anesthetic is given, a hypodermic needle is inserted through the mother's abdominal wall and uterus into the amniotic cavity, and about 10 mL of fluid are withdrawn. The fluid and suspended cells are subjected to microscopic examination and biochemical testing. Amniocentesis is performed only when a risk for genetic defects is suspected because there is about a 0.5% chance of spontaneous abortion (miscarriage) after the procedure.

Chorionic Villi Sampling

In *chorionic villi sampling* (ko-rē-ON-ik VIL-ī) or *CVS,* a catheter is guided through the vagina and cervix of the uterus and then advanced to the chorionic villi under ultrasound guidance. About 30 mg of tissue are suctioned out and prepared for chromosomal analysis. Alternatively, the chorionic villi can be sampled by inserting a needle through the abdominal cavity, as in amniocentesis.

Although CVS can identify the same defects as amniocentesis because chorion cells and fetal cells contain the same genome, CVS offers some advantages over amniocentesis: It can be performed as early as eight weeks of gestation, and test results are available in only a few days, permitting an earlier decision on whether or not to continue the pregnancy. However, CVS is slightly riskier than amniocentesis; there is a 1–2% chance of spontaneous abortion after the procedure.

MATERNAL CHANGES DURING PREGNANCY

Objectives: • **Describe the sources and functions of the hormones secreted during pregnancy.**
• **Describe the hormonal, anatomical, and physiological changes in the mother during pregnancy.**

Hormones of Pregnancy

We have already discussed how trophoblasts and then the chorion secrete *human chorionic gonadotropin (hCG),* which stimulates the corpus luteum to continue production of *progesterone* and *estrogens* (Figure 24.9). Progesterone and estrogens maintain the lining of the uterus during pregnancy and prepare the mammary glands to secrete milk. Peak secretion of hCG occurs at about the ninth week of pregnancy. The hCG level decreases sharply during the fourth and fifth months and then levels off until childbirth. *Early pregnancy tests* can detect the tiny amounts of human chorionic gonadotropin (hCG) that appear in the blood and urine as early as eight days after fertilization. The high levels of human chorionic gonadotropin (hCG) and progesterone have been implicated as the cause of *morning sickness* or *emesis gravidarum* (EM-e-sis gra-VID-ar-um; *eme-* = to vomit; *gravida* = a pregnant woman), episodes of nausea and possibly vomiting that are most likely to occur in the morning during the early weeks of pregnancy.

During the first three to four months of pregnancy, the corpus luteum continues to secrete progesterone and estrogens. From the third month through the remainder of the pregnancy, the placenta itself produces progesterone and estrogens. A high level of progesterone ensures that the uterine myometrium is relaxed. After delivery, estrogens and progesterone in the blood decrease to nonpregnant levels.

Relaxin, a hormone produced first by the corpus luteum of the ovary and later by the placenta, increases the flexibility of the pubic symphysis and ligaments of the sacroiliac and sacro-coccygeal joints and helps dilate the uterine cervix during labor. All of these actions ease delivery of the baby.

Another hormone produced by the chorion of the placenta is *human chorionic somatomammotropin (hCS).* Secretion of hCS reaches a maximum after 32 weeks and then levels off. It helps prepare the mammary glands for lactation, enhances maternal growth by increasing protein synthesis, and decreases glucose use by the mother, making more glucose available for the fetus. Additionally, hCS promotes the release of fatty acids from adipose tissue, providing an alternative to glucose for the mother's ATP production.

A final important hormone produced by the placenta is *corticotropin-releasing hormone (CRH),* which in nonpregnant people is secreted only by neurosecretory cells in the hypothalamus. CRH is now thought to be the "clock" that establishes the timing of birth. Women who have higher levels of CRH earlier in pregnancy are more likely to deliver prematurely; those who have low levels are more likely to deliver after their due date. CRH also increases secretion of cortisol, which is needed for maturation of the fetal lungs and the production of surfactant, a substance that helps prevent collapse of the alveoli of the lungs.

Structural and Functional Changes During Pregnancy

By about the end of the third month of pregnancy, the uterus occupies most of the pelvic cavity. As the fetus continues to grow, the uterus extends higher and higher into the abdominal cavity. Toward the end of a full-term pregnancy, the uterus fills nearly all of the abdominal cavity, reaching nearly to the xiphoid

Figure 24.9 ■ Hormones of pregnancy.

Whereas the corpus luteum produces progesterone and estrogens during the first three to four months of pregnancy, the placenta assumes this function from the third month on.

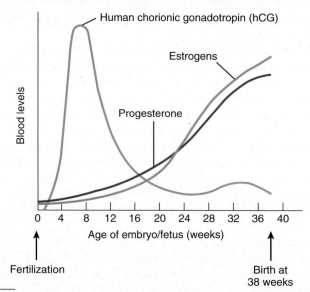

Which hormone serves as the basis for early pregnancy tests?

process of the sternum. It pushes the maternal intestines, liver, and stomach upward; elevates the diaphragm; and widens the thoracic cavity. Pressure on the stomach may force the stomach contents into the esophagus, resulting in heartburn. In the pelvic cavity, the uterus compresses the ureters and urinary bladder.

Changes in the skin during pregnancy include increased pigmentation around the eyes and cheekbones, in the areolae of the breasts, and in the linea alba of the lower abdomen. Striae (stretch marks) over the abdomen occur as the uterus enlarges, and hair loss increases.

Pregnancy also causes certain functional changes in the mother's body. Cardiac output rises 20–30% by the 27th week due to increased maternal blood flow to the placenta and increased metabolism. Pulmonary function is also altered: tidal volume increases, expiratory reserve decreases, and shortness of breath can occur during the last few months.

A general decrease in gastrointestinal tract motility can cause constipation and a delay in gastric emptying time. The pressure of the enlarging uterus on the urinary bladder can produce urinary symptoms, such as frequency, urgency, and stress incontinence.

EXERCISE AND PREGNANCY

Objective: • **Explain the effects of pregnancy on exercise, and of exercise on pregnancy.**

Only a few changes affect the mother's ability to exercise in early pregnancy. A pregnant woman may tire more easily than usual, or morning sickness may interfere with regular exercise. As the pregnancy progresses, weight is gained and posture changes, so more energy is needed to perform activities, and some maneuvers (sudden stopping, changes in direction, rapid movements) are more difficult to execute. In addition, increased levels of relaxin cause certain joints, especially the pubic symphysis, to become more flexible and thus less stable. To compensate, many mothers-to-be walk with widely spread legs and a shuffling gait.

Moderate physical activity does not endanger the fetuses of healthy women who have a normal pregnancy. Although blood shifts from viscera (including the uterus) to the muscles and skin during exercise, blood flow to the placenta does not appear to decrease. However, the heat generated during exercise may cause dehydration and further increase body temperature. During early pregnancy especially, excessive exercise and heat buildup should be avoided because elevated body temperature has been implicated in neural tube defects. Exercise has no known effect on lactation, provided a woman remains hydrated and wears a bra that provides good support. An overall benefit of exercise during pregnancy is a greater sense of well-being.

LABOR AND DELIVERY

Objective: • **Outline the events associated with the three stages of labor.**

Labor is the process by which the fetus is expelled from the uterus through the vagina to the outside environment. Progesterone inhibits uterine contractions. Toward the end of gestation, the levels of estrogens in the mother's blood rise sharply, producing changes that overcome the inhibiting effects of progesterone. Estrogens also stimulate the placenta to release prostaglandins. Prostaglandins induce production of enzymes that digest collagen fibers in the cervix, causing it to soften. High levels of estrogens cause uterine muscle fibers to display receptors for oxytocin, the hormone that stimulates uterine contractions. Relaxin assists by increasing the flexibility of the pubic symphysis and helping dilate the uterine cervix.

Uterine contractions occur in peristaltic waves that start at the top of the uterus and move downward, eventually expelling the fetus. Labor is divided into three stages (Figure 24.10)

❶ The time from the onset of labor to the complete dilation of the cervix is the *stage of dilation*. This stage typically lasts 6 to 12 hours. The uterus contracts at regular intervals, usually producing pain. As the interval between contractions shortens, the contractions intensify. The cervix dilates from 0 cm (completely closed) to 10 cm (fully dilated). If the amniotic sac does not rupture spontaneously, it is ruptured in-

Figure 24.10 ■ **Stages of labor.**

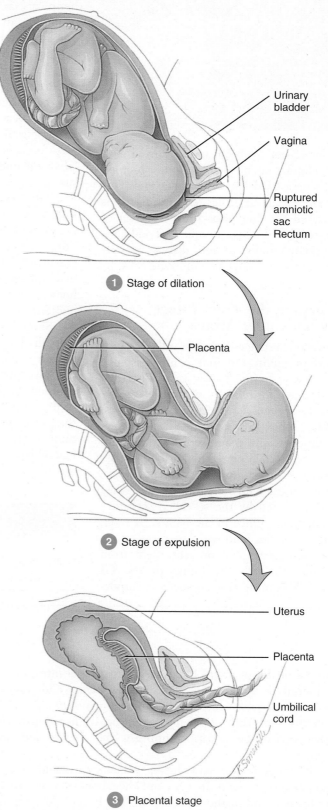

Labor is the process by which uterine contractions expel the fetus through the vagina to the outside.

Urinary bladder

Vagina

Ruptured amniotic sac

Rectum

❶ Stage of dilation

Placenta

❷ Stage of expulsion

Uterus

Placenta

Umbilical cord

❸ Placental stage

What event marks the beginning of the stage of expulsion?

tentionally. In addition, the cervix softens and becomes thinner (effaces).

2 The *stage of expulsion* is the time (10 minutes to several hours) from complete cervical dilation to delivery of the baby.

3 The *placental stage* is the time from delivery of the baby until the placenta or "afterbirth" is expelled by powerful uterine contractions (5 to 30 minutes or more). These contractions also constrict blood vessels that were torn during delivery, reducing the chance of hemorrhage.

During the six-week period after delivery of a baby, called the *puerperium* (pyoo'-er-PE-rē-um), the maternal reproductive organs and physiology return to the prepregnancy state.

INHERITANCE

Objective: • **Define inheritance, and explain the inheritance of dominant, recessive, polygenic, and sex-linked traits.**

The genetic material of a father and a mother unite when a sperm cell fuses with a secondary oocyte to form a zygote. Children resemble their parents because they inherit traits passed down from both parents, a process called *inheritance.* The branch of biology that deals with inheritance is called *genetics* (je-NET-iks). The area of health care that offers advice on genetic problems (or potential problems) is called *genetic counseling.*

Genotype and Phenotype

As you learned in Chapter 3, the nuclei of all of your cells except your gametes contain 23 pairs of chromosomes. One chromosome in each pair came from your mother, and the other came from your father. Each *homolog,* one of the two chromosomes that make up a pair, contains genes that control the same traits. Alternative forms of a gene that code for the same trait and are at the same location on homologous chromosomes are called *alleles* (ah-LĒLZ).

An allele that dominates or masks the presence of another allele is said to be a *dominant allele,* and the trait is called a dominant trait. The allele whose presence is completely masked is said to be a *recessive allele,* and the trait it controls is called a recessive trait. The symbols for genes are written in italics, with dominant alleles written in capital letters, for example, *P,* and recessive alleles in lowercase letters, for example, *p.* A person who has the same alleles on homologous chromosomes (for example, *PP* or *pp*) is **homozygous** for the trait. *PP* is homozygous dominant, and *pp* is homozygous recessive. A person who has different alleles on homologous chromosomes (for example, *Pp*) is **heterozygous** for the trait.

The relationship of genes to heredity is illustrated by examining the alleles involved in a disorder called *phenylketonuria* or *PKU.* People with PKU lack phenylalanine hydroxylase, an enzyme that converts the amino acid phenylalanine into tyrosine, another amino acid. If infants with PKU eat foods containing

phenylalanine, high levels of phenylalanine build up in the blood. The result is severe brain damage and mental retardation.

The allele that instructs the cell to make phenylalanine hydroxylase is symbolized as *P.* The mutated allele that fails to produce a functional enzyme is symbolized as *p.* The chart in Figure 24.11, which shows the possible combinations of gametes from two parents heterozygous for the PKU gene, is called a **Punnett square.** In constructing a Punnett square, the paternal alleles in sperm are written at the left side, and the maternal alleles in ova (or secondary oocytes) are written at the top. The four spaces on the chart show how the alleles can combine to produce the three different combinations of genes, or **genotypes** (JĒ-nō-tīps): *PP, Pp,* or *pp.* Notice from the Punnett square that an offspring has a 25% chance of inheriting the *PP* genotype, a 50% chance of inheriting the *Pp* genotype, and a 25% chance of inheriting the *PP* genotype. People who inherit *PP* or *Pp* genotypes do not have PKU; those with a *pp* genotype suffer from the disorder.

Phenotype (FĒ-nō-tīp; *pheno-* = showing) is the physical or outward expression of a gene. A person with *Pp* (a heterozygote) has a different genotype from a person with *PP* (a homozygote), but both have the same phenotype: normal production of phenylalanine hydroxylase. Heterozygous individuals, who carry a recessive gene but do not express it (*Pp*), can pass the gene on to their offspring. Such individuals are *carriers* of the recessive gene.

Figure 24.11 ▪ **Inheritance of phenylketonuria (PKU).**

🔑 Whereas genotype refers to genetic makeup, phenotype refers to the physical or outward expression of a gene.

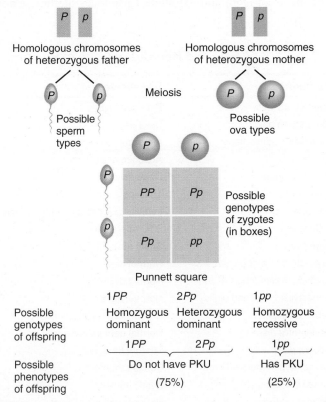

Punnett square

Possible genotypes of offspring	1*PP* Homozygous dominant	2*Pp* Heterozygous dominant	1*pp* Homozygous recessive
Possible phenotypes of offspring	1*PP* 2*Pp* Do not have PKU (75%)		1*pp* Has PKU (25%)

 If a couple has the genotypes shown here, what is the percent chance they will have a child with PKU?

Alleles that code for normal traits do not always dominate those that code for abnormal ones. Dominant alleles for severe disorders usually are lethal early in development—they cause death of the embryo or fetus. One exception is Huntington's disease (HD), in which symptoms typically do not appear for 30 or 40 years. Then, HD causes progressive degeneration of the nervous system and eventual death. By the time symptoms appear, many afflicted individuals have already passed the allele on to their children. Both homozygous dominant and heterozygous people exhibit HD. Homozygous recessive people do not suffer from the disorder.

Table 24.3 lists some dominant and recessive inherited traits in humans.

Variations on Dominant–Recessive Inheritance

Most patterns of inheritance are *not* as simple as that of PKU, in which a single pair of alleles determines genotype and phenotype. The expression of a particular gene may also be influenced by other genes and by the environment. Moreover, most inherited traits are influenced by more than one gene, and some genes can influence more than one trait.

Although a single individual inherits only two alleles for each gene, some genes may have more than two alternate forms, and this is the basis for *multiple-allele inheritance.* One example of multiple-allele inheritance in humans is the inheritance of the ABO blood group. The four blood types (phenotypes) of the ABO group—A, B, AB, and O—result from the inheritance of six combinations of three different alleles of a single gene called the *I* gene: (1) allele I^A produces the A antigen, (2) allele I^B produces the B antigen, and (3) allele i produces neither A nor B

antigen. Each person inherits two of the three *I*-gene alleles. The six possible genotypes produce four blood types, as follows:

Genotype	Blood type
$I^A I^A$ or $I^A i$	A
$I^B I^B$ or $I^B i$	B
$I^A I^B$	AB
$i\,i$	O

Notice that both I^A and I^B are inherited as dominant traits, whereas i is inherited as a recessive trait. Because type AB blood has characteristics of both type A blood and type B blood, alleles I^A and I^B are said to be *codominant.* In other words, both genes are expressed equally in the heterozygote.

Most inherited traits are controlled by the combined effects of several genes, a situation referred to as *polygenic inheritance* (*poly-* = many). A polygenic trait shows a continuous gradation of differences among individuals. Examples of polygenic traits include skin color, hair color, eye color, height, and body build. Suppose that skin color is controlled by three separate genes, each having two alleles: *A, a; B, b;* and *C, c.* A person with the genotype *AABBCC* is very dark skinned, an individual with the genotype *aabbcc* is very light skinned, and a person with genotype *AaBbCc* has an intermediate skin color.

A given phenotype often is the result of the interaction of genotype and environment. Factors in the environment seem to be particularly influential in the case of polygenic traits. For example, even though a person inherits several genes for tallness, full height potential may not be reached due to environmental factors, such as disease or malnutrition during the growth years.

Autosomes, Sex Chromosomes, and Sex Determination

When viewed under a microscope, the 46 human chromosomes in a normal somatic cell can be identified by their size, shape, and staining pattern to be a member of 23 different pairs of chromosomes (Figure 24.12a). In 22 of the pairs, the homologous chromosomes look alike and have the same appearance in both males and females; these 22 pairs are called *autosomes.* The two members of the 23rd pair are the *sex chromosomes.* In females the pair consists of two similar chromosomes called X chromosomes. In males the pair consists of one X chromosome and a much smaller Y chromosome.

When spermatocytes undergo meiosis, half of the resulting spermatids contain an X chromosome, and the other half contain a Y chromosome. Because females have two X chromosomes, oocytes have no Y chromosomes and produce only X-containing gametes. If the secondary oocyte is fertilized by an X-bearing sperm, the offspring is female (XX). Fertilization by a Y-bearing sperm produces a male (XY). Thus, gender is determined at the time of fertilization by the father's chromosomes (Figure 24.12b).

Both female and male embryos develop identically until about seven weeks after fertilization. At that point, one or more genes set into motion a cascade of events that leads to the development of a male. The prime male-determining gene is called *SRY (sex-determining region of the Y chromosome).* Only if the *SRY* gene is present and functional in a fertilized ovum can the

Table 24.3 / **Selected Hereditary Traits in Humans**	
Dominant	**Recessive**
Coarse body hair	Fine body hair
Male pattern baldness	Baldness
Normal skin pigmentation	Albinism
Freckles	Absence of freckles
Normal hearing	Deafness
Broad lips	Thin lips
Tongue roller	Inability to roll tongue into a U shape
PTC taster*	PTC nontaster
Large eyes	Small eyes
Polydactylism (extra digits)	Normal digits
Brachydactylism (short digits)	Normal digits
Syndactylism (webbed digits)	Normal digits
Diabetes insipidus	Normal ADH production
Huntington's disease	Normal nervous system
Widow's peak	Straight hairline
Curved (hyperextended) thumb	Straight thumb
Normal Cl⁻ transport	Cystic fibrosis
Hypercholesterolemia (familial)	Normal cholesterol level

*Ability to taste a chemical compound called phenylthiocarbamide (PTC).

fetus develop testes and differentiate into a male. In the absence of *SRY*, the fetus develops ovaries and differentiates into a female.

Sex-Linked Inheritance

The sex chromosomes also are responsible for the transmission of several nonsexual traits. Genes for many of these traits are present on X chromosomes but are absent from Y chromosomes. This feature produces a pattern of heredity that is different from the patterns described previously.

Consider the most common type of color blindness, called *red–green color blindness.* This condition is characterized by a deficiency in either red- or green-sensitive cone photoreceptors, so red and green are seen as the same color (either red or green,

depending on which cone is present). The allele that causes red–green color blindness is a recessive one designated *c*. Normal color vision, designated *C*, dominates. The *C/c* alleles are present only on the X chromosome. The possible combinations are as follows:

Genotype	Phenotype
$X^C X^C$	Normal female
$X^C X^c$	Normal female (but a carrier of the recessive allele)
$X^c X^c$	Red–green color-blind female
$X^C Y$	Normal male
$X^c Y$	Red–green color-blind male

Only females who have two X^c genes are red–green color blind. This rare situation can result only from the mating of a color-blind male and a color-blind or carrier female. Because males do not have a second X chromosome that could mask the trait, all males with an X^c gene are red–green color blind. Figure 24.13 shows an example of the inheritance of red–green color blindness in the offspring of a normal male and a carrier female.

Traits inherited in the manner just described are called *sex-linked traits.* The most common type of **hemophilia,** a condition in which the blood fails to clot or clots very slowly, is a sex-linked trait. Like the trait for red–green color blindness, hemophilia is caused by a recessive gene. A few other sex-linked traits in humans are nonfunctional sweat glands, certain forms of diabetes, some types of deafness, uncontrollable rolling of the eyeballs, absence of central incisors, night blindness, juvenile glaucoma, and juvenile muscular dystrophy.

Figure 24.12 ■ **Inheritance of gender (sex).** In (a) the sex chromosomes, pair 23, are indicated in the colored box.

Gender is determined at the time of fertilization by the sex chromosome of the sperm cell.

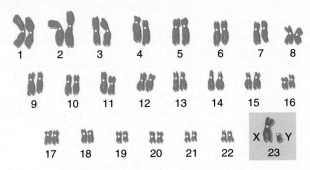

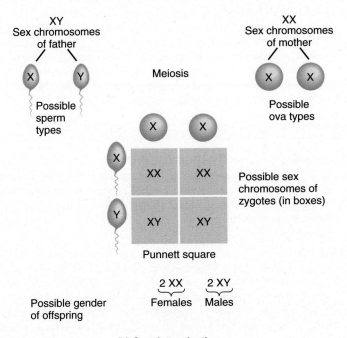

(a) Normal human male chromosomes

(b) Sex determination

What are chromosomes other than sex chromosomes called?

Figure 24.13 ■ **Inheritance of red–green color blindness.**

Red–green color blindness and hemophilia are examples of sex-linked traits.

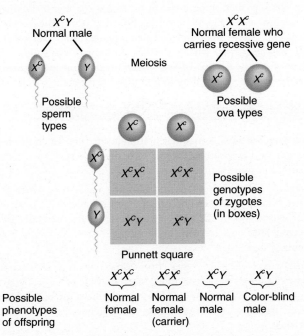

What would be the genotype of a red–green color-blind female?

COMMON DISORDERS

Infertility

Female infertility, or the inability to conceive, occurs in about 10% of all women of reproductive age in the United States. Female infertility may be caused by ovarian disease, obstruction of the uterine tubes, or conditions in which the uterus is not adequately prepared to receive a fertilized ovum. One cause of infertility in females is inadequate body fat. To begin and maintain a normal reproductive cycle, a female must have a minimum amount of body fat. Both dieting and intensive exercise may reduce body fat below the minimum amount and lead to infertility that is reversible, if weight gain or reduction of intensive exercise, or both, occur. Studies of very obese women indicate that they, like very lean ones, experience problems with amenorrhea (cessation of menstruation) and infertility.

Male infertility is an inability to fertilize a secondary oocyte; it does not imply impotence. Male fertility requires production of adequate quantities of viable, normal sperm by the testes; unobstructed transport of sperm though the ducts; and satisfactory deposition in the vagina. Males also experience reproductive problems in response to undernutrition and weight loss. For example, they may produce less prostatic fluid, reduced numbers of sperm, or sperm having decreased motility.

Many fertility-expanding techniques now exist for assisting infertile couples to have a baby. The birth of Louise Joy Brown on July 12, 1978, near Manchester, England, was the first recorded case of *in vitro fertilization (IVF)*—fertilization in a laboratory dish. In the IVF procedure, the mother-to-be is given follicle-stimulating hormone (FSH) soon after menstruation, so that several secondary oocytes, rather than the typical single oocyte, are produced (superovulation). When several follicles have reached the appropriate size, a small incision is made near the umbilicus, and the secondary oocytes are aspirated from the stimulated follicles. The oocytes then are transferred to a solution containing sperm, where they are fertilized. Alternatively, an oocyte may be fertilized in vitro by suctioning a sperm or even a spermatid obtained from the testis into a tiny pipette and then injecting it into the oocyte's cytoplasm. This procedure, termed *intracytoplasmic sperm injection (ICSI),* has been used when infertility is due to impairment of sperm motility or to the failure of spermatids to develop into spermatozoa. When the zygote achieved by IVF or ICSI reaches the 8-cell or 16-cell stage, it is introduced into the uterus for implantation and subsequent growth.

In *embryo transfer,* semen is used to artificially inseminate a fertile secondary oocyte donor. After fertilization in the donor's uterine tube, the morula or blastocyst is transferred from the donor to the infertile woman, who then carries it (and subsequently the fetus) to term. Embryo transfer is indicated for women who are infertile or who do not want to pass on their own genes because they are carriers of a serious genetic disorder.

In *gamete intrafallopian transfer (GIFT),* the goal is to mimic the normal process of conception by uniting sperm and secondary oocyte in the prospective mother's uterine tubes. In this procedure, a woman is given FSH and LH to stimulate the production of several secondary oocytes, which are then aspirated from the mature follicles, mixed outside the body with a solution containing sperm, and then immediately inserted into the uterine tubes.

Down Syndrome

Down syndrome is a disorder that most often results during meiosis when an extra chromosome 21 passes to one of the gametes. Most of the time the extra chromosome comes from the mother, a not too surprising finding given that all her oocytes began meiosis when she herself was a fetus. They may have been exposed to chromosome-damaging chemicals and radiation for years. (Sperm, by contrast, usually are less than 10 weeks old at the time they fertilize a secondary oocyte.) The chance of conceiving a baby with this syndrome, which is less than 1 in 3000 for women under age 30, increases to 1 in 300 in the 35 to 39 age group, and to 1 in 9 at age 48.

Down syndrome is characterized by mental retardation; retarded physical development (short stature and stubby fingers); distinctive facial structures (large tongue, flat profile, broad skull, slanting eyes, and round head); and malformations of the heart, ears, hands, and feet. Sexual maturity is rarely attained.

Fragile X Syndrome

Fragile X syndrome, a recently recognized disorder caused by a defective gene on the X chromosome, is so named because a small portion of the tip of the X chromosome seems susceptible to breakage. It is the leading cause of mental retardation among newborns. Fragile X syndrome affects males more than females and results in learning difficulties; mental retardation; and physical abnormalities such as oversized ears, elongated forehead, enlarged testes, and double-jointedness. The syndrome may also be involved in autism, in which the individual exhibits extreme withdrawal and refusal to communicate.

MEDICAL TERMINOLOGY AND CONDITIONS

Breech presentation   A malpresentation at birth in which the fetal buttocks or lower limbs extend into the maternal pelvis; the most common cause is prematurity.

Klinefelter's syndrome   A sex chromosome disorder in which there are two X chromosomes and one Y chromosome (XXY); occurs once in every 500 births and produces somewhat mentally disadvantaged, sterile males with undeveloped testes, scant body hair, and enlarged breasts.

Metafemale syndrome   A sex chromosome disorder characterized by the presence of at least three X chromosomes (XXX); occurs about once in every 700 births. These females have underdeveloped genital organs and limited fertility and generally are mentally retarded.

Preeclampsia   (prē'-e-KLAMP-sē-a) A syndrome of pregnancy characterized by sudden hypertension, large amounts of protein in urine, and generalized edema; possibly related to an autoimmune or allergic reaction to the presence of a fetus. When the condition is also associated with convulsions and coma, it is referred to as *eclampsia.*

Turner's syndrome   A sex chromosome abnormality caused by the presence of a single X chromosome (designated XO); occurs about once in every 5000 births and produces a sterile female with virtually no ovaries and limited development of secondary sex characteristics. Other features include short stature, webbed neck, underdeveloped breasts, and widely spaced nipples. Intelligence usually is normal.

STUDY OUTLINE

From Fertilization to Implantation (p. 576)

1. Pregnancy is a sequence of hormonally controlled events that starts with fertilization and normally ends 38 weeks later with birth.

2. Fertilization refers to the penetration of a secondary oocyte by a sperm cell and the subsequent formation of a zygote.

3. Penetration of the zona pellucida is facilitated by enzymes in the sperm's acrosome. Normally, only one sperm cell fertilizes a secondary oocyte.

4. Early rapid cell division of a zygote is called cleavage. The solid sphere of cells produced by cleavage is a morula. The morula develops into a blastocyst, a hollow ball of cells differentiated into a trophoblast and an inner cell mass.

5. The attachment of a blastocyst to the endometrium is called implantation; it occurs by means of enzymatic degradation of the endometrium.

Embryonic and Fetal Development (p. 579)

1. During embryonic growth, the primary germ layers and extraembryonic membranes are formed.

2. The primary germ layers—ectoderm, mesoderm, and endoderm—form all the tissues of the developing organism. Table 24.1 on page 581 summarizes the structures produced by the three germ layers.

3. The four extraembryonic membranes include the yolk sac, the amnion, the chorion, and the allantois.

4. Fetal and maternal materials are exchanged through the placenta.

5. During the fetal period, organs established by the primary germ layers grow rapidly.

6. The principal changes associated with embryonic and fetal growth are summarized in Table 24.2 on page 583.

7. The fetal circulation is designed to allow exchanges of materials between the fetus and the mother. At birth, when the baby's lung, digestive, and liver functions are established, the special structures of the fetal circulation are no longer needed.

8. Teratogens, which are agents that cause physical defects in developing embryos, include chemicals and drugs, alcohol, nicotine, and ionizing radiation.

9. Several prenatal diagnostic tests are used to detect genetic disorders and to assess fetal well-being, including fetal ultrasonography, in which an image of a fetus is displayed on a screen; amniocentesis, the withdrawal and analysis of amniotic fluid and the fetal cells within it; and chorionic villi sampling (CVS), which involves withdrawal of chorionic villi tissue for chromosomal analysis. CVS can be done earlier than amniocentesis, and the results are available more quickly, but it is also slightly more risky than amniocentesis.

Maternal Changes During Pregnancy (p. 586)

1. Pregnancy is maintained by human chorionic gonadotropin (hCG), estrogens, and progesterone.

2. Human chorionic somatomammotropin (hCS) helps prepare the mammary glands for lactation and plays roles in protein anabolism and in glucose and fatty acid catabolism.

3. Relaxin increases the flexibility of the pubic symphysis and helps dilate the uterine cervix near the end of pregnancy.

4. Corticotropin-releasing hormone, produced by the placenta, is thought to establish the timing of birth; it also stimulates the secretion of cortisol by the fetal adrenal gland.

5. During pregnancy, several structural and functional changes occur in the mother.

Exercise and Pregnancy (p. 588)

1. During pregnancy, some joints become less stable, and certain maneuvers are more difficult to execute.

2. Moderate physical activity does not endanger the fetus in a normal pregnancy.

Labor and Delivery (p. 588)

1. Labor is the process by which the fetus is expelled from the uterus through the vagina to the outside. It involves three stages: dilation of the cervix, expulsion of the fetus, and delivery of the placenta.

2. Oxytocin stimulates uterine contractions.

Inheritance (p. 589)

1. Inheritance is the passage of hereditary traits from one generation to the next.

2. The genetic makeup of an organism is called its genotype; the expressed traits are called its phenotype.

3. Dominant alleles control a particular trait; expression of recessive alleles is masked by dominant alleles.

4. In multiple-allele inheritance, genes have more than two alternate forms. An example is the inheritance of ABO blood groups.

5. In polygenic inheritance, an inherited trait is controlled by the combined effects of many genes. An example is skin color.

6. Each somatic cell has 46 chromosomes: 22 pairs of autosomes and 1 pair of sex chromosomes.

7. Females have two X chromosomes. Males have one X chromosome and a much smaller Y chromosome, which normally includes the prime male-determining gene, called *SRY*.

8. If the *SRY* gene is present and functional in a fertilized ovum, the fetus develops testes and differentiates into a male. In the absence of *SRY*, the fetus develops ovaries and differentiates into a female.

9. Red–green color blindness and hemophilia result from recessive genes located on the X chromosome; they occur primarily in males because of the absence of any alleles for these traits on the Y chromosome.

10. A given phenotype is the result of the interactions of genotype and the environment.

SELF-QUIZ

1. The change in a sperm that allows it to fertilize an egg is known as
   a. vasocongestion   b. capacitation   c. cleavage   d. gestation   e. meiosis

2. Match the following:
   ____ a. early division of the zygote that increases the cell number but not size
   ____ b. solid mass of cells 3 to 4 days following fertilization
   ____ c. a hollow ball of cells found in the uterine cavity about 5 days after fertilization
   ____ d. portion of the blastocyst that develops into the embryo
   ____ e. portion of the blastocyst that forms the chorion and placenta
   ____ f. results from fertilization of the sperm and egg

   A. blastocyst
   B. trophoblast
   C. morula
   D. inner cell mass
   E. zygote
   F. cleavage

3. The fetal structure that allows most blood to bypass the lungs is the
   a. ductus arteriosus   b. foramen ovale   c. fossa ovalis
   d. ductus venosus   e. umbilical vein

4. The placental hormone that appears to affect the timing of birth is
   a. cortisol   b. human chorionic somatomammotropin (hCS)
   c. relaxin   d. corticotropin-releasing hormone (CRH)
   e. human chorionic gonadotropin (hCG)

5. In fetal circulation,
   a. oxygenated blood is transported in the umbilical artery
   b. the gastrointestinal tract and lungs are not functioning
   c. the heart is not functioning
   d. the blood flow is reversed in the arteries and veins
   e. the amniotic fluid aids in the exchange of nutrients and oxygen

6. Which of the following is the embryonic membrane that most closely surrounds the developing fetus?
   a. amnion   b. umbilicus   c. chorion   d. placenta
   e. zona pellucida

7. The hormone that causes a home pregnancy test to show a positive result for pregnancy is
   a. follicle-stimulating hormone (FSH)   b. progesterone
   c. human chorionic somatomammotropin (hCS)   d. luteinizing hormone (LH)   e. human chorionic gonadotropin (hCG)

8. The foramen ovale is located
   a. between the atria
   b. between the ventricles
   c. between the right atrium and right ventricle
   d. between the right ventricle and pulmonary artery
   e. between the pulmonary artery and pulmonary vein

9. Match the following:
   ____ a. becomes part of the placenta
   ____ b. becomes muscle and bone
   ____ c. develops into the epidermis and nervous system
   ____ d. becomes the ventral body cavity
   ____ e. becomes the epithelial lining of the respiratory and gastrointestinal tracts

   A. extraembryonic coelom
   B. chorion
   C. endoderm
   D. ectoderm
   E. mesoderm

10. The period of time from conception of the zygote to delivery of the fetus is called
    a. fertilization   b. placentation   c. gestation   d. implantation   e. gastrulation

11. Homologous chromosomes
    a. contain genes that control the same trait
    b. contain genes that control different traits
    c. are inherited only from the mother
    d. are inherited only from the father
    e. contain all identical alleles

12. Sex-linked traits are carried on the
    a. autosomes   b. X and Y chromosomes   c. X chromosomes only   d. Y chromosomes only   e. Y chromosomes in males and X chromosomes in females

13. A person who is homozygous for a dominant trait on an autosome would have the genotype
    a. Aa b. AA c. aa d. $X^A X^A$ e. $X^a X^a$

14. The genotype of a normal female is
    a. XY and 44 autosomes
    b. 46 autosomes
    c. 46 X chromosomes
    d. XX and 44 autosomes
    e. XX and 46 autosomes

15. Which of the following is NOT a change that occurs in a female during pregnancy?
    a. increased cardiac output
    b. decreased pulmonary expiratory reserve volume
    c. decreased gastrointestinal tract motility
    d. increased frequency and urgency of urination
    e. decreased production of estrogen

16. The afterbirth is expelled from the uterus during the _____ stage of labor.
    a. parturition   b. placental   c. dilation   d. expulsion
    e. puerperium

CRITICAL THINKING APPLICATIONS

1. Your neighbors put up a sign to announce the birth of their twins, a girl and a boy. Another neighbor said, "Oh, how sweet! I wonder if they're identical twins." Without even seeing the twins, what can you tell her?

2. The science class was studying genetics at school, and Kendra came home in tears. She told her older sister, "We were doing our family tree, and when I filled in our traits, I discovered that Mom and Dad can't be my real parents 'cause the traits don't match!" It turns out that Mom and Dad can roll their tongues but Kendra can't. Could Mom and Dad still be Kendra's parents?

3. Baby Peterson was brought into the health clinic by a concerned grandparent. The alcoholic mother had abandoned the baby, and the grandparent was worried that the baby was growing slowly and not developing normally. One of the neighbors, a nurse, had told the grandparent that the baby's face was characteristic of a syndrome that she'd seen before. What is Baby Peterson's problem?

4. A friend told you that he was called a "blue baby" because he had a hole in his heart at birth. The hole was later closed surgically, and now your friend's color looks normal. Identify the hole in the heart of the fetus. Explain why the hole needs to close after birth. Why is a "blue baby" blue?

ANSWERS TO FIGURE QUESTIONS

24.1 Capacitation is the series of functional changes experienced by sperm while in the female reproductive tract that allows them to fertilize a secondary oocyte.

24.2 A morula is a solid ball of cells; a blastocyst is a ball of cells surrounding a cavity.

24.3 A blastocyst implants by secreting digestive enzymes that eat away the endometrial lining at the site of implantation.

24.4 The uterus is in the secretory phase at the time of implantation.

24.5 The embryo forms from the embryonic disk.

24.6 The amnion functions as a shock absorber, whereas the chorion forms the fetal portion of the placenta.

24.7 The placenta allows exchange of materials between fetus and mother.

24.8 The umbilical vein contains the most highly oxygenated blood.

24.9 Early pregnancy tests detect human chorionic gonadotropin (hCG).

24.10 Complete dilation of the cervix indicates the beginning of the stage of expulsion.

24.11 The chance of a PKU child is 25%.

24.12 Chromosomes other than sex chromosomes are called autosomes.

24.13 The genotype of a red–green color-blind female would be X^cX^c.

Self-Quiz Answers

CHAPTER 1

1. d 2. b 3a. G b. B c. E d. C e. F f. A g. D
4a. Nervous System b. brain, spinal cord, nerves, sense organs c. Lymphatic and Immune System d. Returns proteins & fluid to blood; sites of lymphocyte maturation and proliferation to protect against disease; carries lipids from digestive system to blood e. Respiratory System f. lungs, pharynx, larynx, trachea, bronchial tubes g. testes, ovaries, vagina, uterine tubes, uterus, penis h. reproduces the organism and releases hormones
5. d 6. a 7. c 8. e 9a. C b. D c. A d. B 10. a
11. a 12. d 13. c 14. e 15. b 16. c 17. a 18a. D
b. A c. H d. F e. G f. E g. C h. B 19. b 20. e

CHAPTER 2

1. d 2. c 3. a 4. e 5. d 6. b 7. b 8. c
9. a 10. d 11. e 12. d 13. c 14a. R, D b. D c. D
d. R e. R f. D g. D h. R i. R, D j. R, D 15. a
16. b 17. e 18. a 19. carbon, hydrogen, oxygen, nitrogen
20a. D b. C c. A d. E e. B

CHAPTER 3

1. b 2. e 3. a 4. a 5. c 6. d 7. b 8. d
9. a 10a. B b. F c. G d. H e. C f. E g. D h. A
11. e 12. c 13. a 14. d 15. b 16. b 17a. C b. E
c. F d. B e. D f. A 18. e 19. e 20. c

CHAPTER 4

1. c 2. d 3. b 4. d 5a. C b. G c. E d. D e. H
f. F g. B h. A 6. a 7. e 8. a 9. e 10. b 11. a
12. c 13. c 14. d 15. e 16. d 17. a 18. c 19. a
20. b

CHAPTER 5

1. d 2. d 3. a 4. e 5. e 6. a 7. c 8. c 9. a
10. a 11. b 12. a 13. c 14. b 15. d 16. e 17. b
18a. D b. F c. E d. G e. H f. A g. C h. B i. I
19. a 20. b

CHAPTER 6

1a. C b. E c. D d. A e. B 2. e 3. a 4a. E b. D
c. A d. C e. B 5. b 6. e 7. c 8. a 9. c 10. d
11. b 12a. C b. D c. B d. E e. A 13. e 14. b
15. e 16a. AX b. AP c. AP d. AX e. AP f. AP g. AP
h. AX i. AP j. AP k. AX l. AP m. AX n. AP o. AX

p. AP q. AX r. AX s. AP t. AX u. AX v. AP w. AX
x. AX y. AX z. AP aa. AP bb. AP cc. AX dd. AX

CHAPTER 7

1. d 2. a 3. c 4. b 5. d 6. e 7. b 8a. D b. A
c. E d. C e. B 9. a 10. c 11. e 12. a 13. b 14. e
15a. A b. G c. B d. C e. D f. E g. F h. J i. I
j. H

CHAPTER 8

1. e 2a. D b. E c. B d. A e. C 3. d 4a. C b. D
c. A d. E e. B 5. c 6. c 7. a 8a. SM, CA b. SK
c. SK, CA d. CA e. SK f. SK g. SM h. SM i. SK
j. CA 9. b 10. b 11. e 12. b 13. d 14. a 15. e
16. a 17. e 18a. C b. F c. D d. B e. G f. H g. A
h. E 19a. C b. E c. B d. D e. F f. A 20a. E
b. D c. G d. H e. B f. M g. O h. F i. J j. C
k. K l. A m. L n. I o. N

CHAPTER 9

1. c 2. e 3. a 4. d 5. a 6. e 7. c 8. a 9. c
10. e 11. b 12. d 13. b 14. d 15a. D b. E c. F
d. C e. A f. B 15a. I b. A c. D d. O e. E f. P
g. H h. J i. G j. C k. B l. L m. N n. K o. F
p. M

CHAPTER 10

1. e 2. a 3. c 4. c 5. d 6. e 7. c 8. a 9. c
10. a 11. b 12. e 13. e 14. a 15. b 16a. C b. D
c. A d. B 17. b 18. d 19. c, e 20a. G b. H c. K
d. J e. F f. I g. A h. D i. C j. B k. E

CHAPTER 11

1. c 2. a 3. c 4. e 5. d 6. a 7. d 8. b 9. d
10. e 11. b 12. d 13. b 14a. C b. F c. E d. B
e. A f. D 15a. S b. P c. P d. S e. P f. P g. S
h. S i. S

CHAPTER 12

1. b 2. b 3a. J b. E c. F d. I e. C f. G g. H
h. A i. B j. D 4. a 5. d 6. e 7. c
8. e 9. d 10. c 11. a 12. e 13a. D b. F c. G d. B
e. C f. A g. E 14. e 15. c 16. a 17. e 18. c
19. d 20. a

CHAPTER 13

1. d 2. e 3. c 4. a 5. d 6. a 7. e 8. a 9. b
10. c 11. c 12. a 13. b 14. d 15. b 16. a 17a. C
b. F c. G d. A e. D f. E g. B 18. b 19a. B b. C
c. E d. A e. F f. D 20a. R b. F c. F d. R e. F
f. E g. R h. E i. F

CHAPTER 14

1. c 2a. D b. G c. C d. A e. E f. F g. B 3. c
4. d 5. e 6. e 7. b 8. b 9. d 10. a 11. c 12. a
13a. E b. D c. C d. A e. B 14. e 15. b 16. a
17. e 18. a 19. b 20. c

CHAPTER 15

1a. D b. H c. F d. G e. C f. B g. A h. E 2. e
3. e 4. a 5. c 6. c 7. a 8. c 9. b 10. d 11. d
12. a 13. b 14. a 15. d 16. d 17. b 18. c 19a. A
b. A c. A d. B e. A f. B g. B h. A i. B j. A
k. B l. B m. A 20a. B b. G c. A d. F e. D f. C
g. E

CHAPTER 16

1. b 2. a 3. a 4. c 5. d 6. e 7a. E b. D c. A
d. B e. C 8. d 9. c 10. b 11. c 12a. H b. D c. E
d. F e. A f. B g. C h. I i. G 13a. A b. B c. B
d. A e. A f. A g. A h. A i. B 14. e 15. b 16. a
17. d

CHAPTER 17

1. c 2. d 3. e 4. c 5. a 6. b 7. b 8. a 9. d
10. e 11. b 12. c 13. d 14a. E b. B c. D d. A
e. C 15. c 16. b 17. c 18. b 19. e 20. a

CHAPTER 18

1. d 2. c 3. b 4. a 5. c 6. a 7. a 8. d 9a. B
b. D c. A d. C e. E 10. a 11a. F b. D c. H d. A
e. G f. C g. E h. B 12. b 13. c 14. e 15. a 16. b
17. c 18. e 19. c 20a. D b. A c. E d. C e. B

CHAPTER 19

1. e 2. b 3. b 4. c 5. d 6. a 7. c 8a. D b. G
c. A d. H e. B f. C g. E h. F 9. e 10. e 11. a
12. e 13. c 14. b 15. d 16. a 17. e 18. c 19. e
20. a

CHAPTER 20

1. b 2. c 3. a 4a. C b. A c. D d. B 5. c 6. a
7. b 8. e 9. d 10. a 11. a 12. b 13. e 14. b 15. c
16. c 17. d 18. a 19. d 20a. C b. D c. B d. E
e. A

CHAPTER 21

1. c 2. e 3. a 4. d 5. c 6. c 7. a 8. d 9. b
10. a 11. b 12. e 13. e 14. a 15. c 16. d 17. a
18. c 19. e 20. b

CHAPTER 22

1. d 2. e 3. a 4. c 5. a 6. e 7. d 8. a 9. c
10. a 11. c 12. e 13. b 14. c 15. d 16. b 17. c
18a. E b. A c. D d. F e. C f. B

CHAPTER 23

1. b 2a. C b. D c. E d. B e. A 3. e 4. d 5. c
6. c 7. b 8. d 9. c 10. e 11. c 12. b 13. a 14. e
15. d 16. d 17a. C b. D c. A d. B 18. d 19a. B
b. F c. A d. G e. D f. C g. E 20. e

CHAPTER 24

1. b 2a. F b. C c. A d. D e. B f. E 3. a 4. d
5. b 6. a 7. e 8. a 9a. B b. E c. D d. A e. C
10. c 11. a 12. c 13. b 14. d 15. e 16. b

Glossary

PRONUNCIATION KEY

1. The most strongly accented syllable appears in capital letters, for example, bilateral (bī-LAT-er-al) and diagnosis (dī′-ag-NŌ-sis).

2. If there is a secondary accent, it is noted by a prime (′), for example, physiology (fiz′-ē-OL-ō-jē). Any additional secondary accents are also noted by a prime, for example, decarboxylation (dē′-kar-bok′-si-LĀ-shun).

3. Vowels marked by a line above the letter are pronounced with the long sound, as in the following common words:
   ā as in māke   ē as in bē
   ī as in īvy   ō as in pōle

4. Vowels not so marked are pronounced with the short sound, as in the following words:
   a as in above   e as in bet
   i as in sip   o as in not
   u as in bud

5. Other phonetic symbols are used to indicate the following sounds:
   oo as in sue
   yoo as in cute
   oy as in oil

A

Abdomen (AB-dō-men) The area between the diaphragm and pelvis.

Abdominal (ab-DŌM-i-nal) **cavity** Superior portion of the abdominopelvic cavity that contains the stomach, spleen, liver, gallbladder, most of the small intestine, and part of the large intestine.

Abdominal thrust maneuver A first-aid procedure for choking. Employs a quick, upward thrust against the diaphragm that forces air out of the lungs with sufficient force to expel any lodged material. Also called the **Heimlich** (HĪM-lik) **maneuver.**

Abdominopelvic (ab-dom′-i-nō-PEL-vic) **cavity** Inferior component of the ventral body cavity that is subdivided into a superior abdominal cavity and an inferior pelvic cavity.

Abduction (ab-DUK-shun) Movement away from the axis or midline of the body.

Abortion (a-BOR-shun) The premature loss (spontaneous) or removal (induced) of the embryo or nonviable fetus; miscarriage due to a failure in the normal process of developing or maturing.

Abrasion (ah-BRĀ-zhun) A portion of skin that has been scraped away.

Abscess (AB-ses) A localized collection of pus and liquefied tissue in a cavity.

Absorption (ab-SORP-shun) Intake of fluids or other substances by cells of the skin or mucous membranes; the passage of digested foods from the gastrointestinal tract into blood or lymph.

Accommodation (ah-kom′-ō-DĀ-shun) A change in the curvature of the eye lens to adjust for vision at various distances.

Acetabulum (as′-e-TAB-yoo-lum) The rounded cavity on the external surface of the hip bone that receives the head of the femur.

Acetylcholine (as′-ē-til-KŌ-lēn) **(ACh)** A neurotransmitter liberated by many peripheral nervous system neurons and some central nervous system neurons. It is excitatory at neuromuscular junctions but inhibitory at some other synapses (for example, it slows heart rate).

Acid (AS-id) A proton donor, or a substance that dissociates into hydrogen ions (H^+) and anions; characterized by an excess of hydrogen ions and a pH less than 7.

Acidosis (as′-i-DŌ-sis) A condition in which blood pH is below 7.35.

Acini (AS-i-nē) Masses of cells in the pancreas that secrete digestive enzymes.

Acquired immunodeficiency syndrome (AIDS) A disorder caused by a virus called human immunodeficiency virus (HIV) and characterized by a positive HIV-antibody test, low T4 (helper) cell count, and certain indicator diseases (for example, Kaposi's sarcoma, *Pneumocystis carinii* pneumonia, tuberculosis, fungus diseases). Other symptoms include fever or night sweats, coughing, sore throat, fatigue, body aches, weight loss, and enlarged lymph nodes.

Acromegaly (ak′-rō-MEG-a-lē) Condition caused by hypersecretion of human growth hormone (hGH) during adulthood, characterized by thickened bones and enlargement of other tissues.

Acrosome (AK-rō-sōm) A dense lysosome-like body in the head of a sperm cell that contains enzymes that facilitate the penetration of a sperm cell into a secondary oocyte.

Actin (AK-tin) The contractile protein that is part of thin filaments in muscle fibers.

Action potential An electrical signal that propagates along the membrane of a neuron or muscle fiber (cell); a rapid change in membrane potential that involves a depolarization followed by a repolarization. Also called a **nerve action potential** or **nerve impulse** as it relates to a neuron, and a **muscle action potential** as it relates to a muscle fiber (cell).

Active transport The movement of substances across cell membranes against a concentration gradient, requiring the expenditure of cellular energy (ATP).

Acupuncture (AK-yoo-punk′-chur) The insertion of a needle into a tissue for the purpose of drawing fluid or relieving pain. It is also an ancient Chinese practice employed to cure illnesses by inserting needles into specific locations of the skin.

Acute (a-KYOOT) Having rapid onset, severe symptoms, and a short course; not chronic.

Adaptation (ad′-ap-TĀ-shun) The adjustment of the pupil of the eye to light variations. The property by which a neuron relays a decreased frequency of action potentials from a receptor, even though the strength of the stimulus remains constant; the decrease in perception of a sensation over time while the stimulus is still present.

Adduction (ad-DUK-shun) Movement toward the axis or midline of the body.

Adenohypophysis (ad′-e-nō-hī-POF-i-sis) The anterior portion of the pituitary gland.

Adenoids (AD-e-noyds) The pharyngeal tonsils.

Adenosine triphosphate (ah-DEN-ō-sēn trī-FOS-fāt) **(ATP)** The energy-carrying molecule manufactured in all living cells as a means of capturing and storing energy. ATP consists of the purine base adenine and the five-carbon sugar ribose, to which are added, in linear array, three phosphate groups.

Adipocyte (AD-i-pō-sīt) Fat cell, derived from a fibroblast.

Adipose (AD-i-pōz) **tissue** Tissue composed of adipocytes specialized for triglyceride storage and present in the form of soft pads between various organs for support, protection, and insulation.

Adrenal cortex (ah-DRĒ-nal KOR-teks) The outer portion of an adrenal gland, divided into three zones that secrete mineralocorticoids, glucocorticoids, and androgens.

Adrenal glands Two glands located superior to each kidney. Also called the **suprarenal** (soo′-pra-RĒ-nal) **glands.**

Adrenal medulla (me-DUL-a) The inner portion of an adrenal gland, consisting of cells that secrete epinephrine and norepinephrine (NE) in response to the stimulation of preganglionic sympathetic neurons.

Adrenergic (ad'-ren-ER-jik) **neuron** A neuron that when stimulated releases epinephrine (adrenaline) or norepinephrine (noradrenaline) at a synapse.

Adrenocorticotropic (ad-rē'-nō-kor'-ti-kō TRŌP-ik) **hormone (ACTH)** A hormone produced by the anterior pituitary that influences the production and secretion of certain hormones of the adrenal cortex.

Adventitia (ad'-ven-TISH-ya) The outermost covering of a structure or organ.

Aerobic (air-Ō-bik) Requiring molecular oxygen.

Afferent arteriole (AF-er-ent ar-TĒ-rē-ōl) A blood vessel of a kidney that divides into the capillary network called a glomerulus; there is one afferent arteriole for each glomerulus.

Agglutination (a-gloo'-ti-NĀ-shun) Clumping of microorganisms or blood cells, typically due to an antigen–antibody reaction.

Aggregated lymphatic follicles Clusters of lymph nodules that are most numerous in the ileum. Also called **Peyer's** (PĪ-erz) **patches.**

Albinism (AL-bin-izm) Abnormal, nonpathological, partial or total absence of melanin pigment in skin, hair, and eyes.

Albumin (al-BYOO-min) The most abundant (60%) and smallest of the plasma proteins, which is the main contributor to blood colloid osmotic pressure.

Aldosterone (al-DOS-ter-ōn) A mineralocorticoid produced by the adrenal cortex that brings about sodium and water reabsorption and potassium excretion.

Alkaline (AL-ka-līn) Containing more hydroxide ions (OH^-) than hydrogen ions (H^+); a pH higher than 7.

Alkalosis (al'-ka-LŌ-sis) A condition in which blood pH is higher than 7.45.

Allantois (a-LAN-tō-is) A small, vascularized membrane between the chorion and amnion of the fetus that serves as an early site for blood formation.

Alleles (a-LĒLZ) Alternate forms of a single gene that control the same inherited trait (such as height or eye color) and are located at the same position on homologous chromosomes.

Allergen (AL-er-jen) An antigen that evokes a hypersensitivity reaction.

Alpha (AL-fa) **cell** A cell in the pancreatic islets (islets of Langerhans) that secretes glucagon.

Alveolar (al-VĒ-ō-lar) **duct** Branch of a respiratory bronchiole around which alveoli and alveolar sacs are arranged.

Alveolar macrophage (MAK-rō-fāj) Highly phagocytic cell found in the alveolar walls of the lungs. Also called a **dust cell.**

Alveolar pressure Air pressure within the lungs.

Alveolar sac A collection or cluster of alveoli that share a common opening.

Alveolus (al-VĒ-ō-lus) A small hollow or cavity; an air sac in the lungs; milk-secreting portion of a mammary gland. Plural is **alveoli** (al-VĒ-ol-ī).

Alzheimer's (ALTZ-hī-merz) **disease (AD)** Disabling neurologic disorder characterized by dysfunction and death of specific cerebral neurons, resulting in widespread intellectual impairment, personality changes, and fluctuations in alertness.

Amenorrhea (ā-men'-ō-RĒ-a) Absence of menstruation.

Amino (a-MĒ-nō) **acid** An organic acid, containing an acidic carboxyl group (—COOH) and a basic amino group (—NH_2), that is the building unit from which proteins are formed.

Amnesia (am-NĒ-zē-a) A lack or loss of memory.

Amnion (AM-nē-on) The innermost extraembryonic membrane; a thin transparent sac that holds the fetus suspended in amniotic fluid. Also called the **"bag of waters."**

Amniotic (am'-nē-OT-ik) **fluid** Fluid in the amniotic cavity, the space between the developing embryo (or fetus) and amnion; the fluid is initially produced as a filtrate from maternal blood and later includes fetal urine.

Amphiarthrosis (am'-fē-ar-THRŌ-sis) A slightly movable articulation, in which the articulating bony surfaces are separated by fibrous connective tissue or fibrocartilage to which both are attached; types are syndesmosis and symphysis.

Anabolism (a-NAB-ō-lizm) Synthetic, energy-requiring reactions whereby small molecules are built up into larger ones.

Anaerobic (an'-a-RŌ-bik) Not requiring molecular oxygen.

Anal (Ā-nal) **canal** The terminal 2 or 3 cm (1 in.) of the rectum; opens to the exterior through the anus.

Anal triangle The subdivision of the female or male perineum that contains the anus.

Analgesia (an'-al-JĒ-zē-a) Pain relief.

Anaphase (AN-a-fāz) The third stage of mitosis in which the chromatids that have separated at the centromeres move to opposite poles of the cell.

Anaphylaxis (an'-a-fi-LAK-sis) Against protection; a hypersensitivity (allergic) reaction in which IgE antibodies attach to mast cells and basophils, causing them to produce mediators of anaphylaxis (histamine, leukotrienes, kinins, and prostaglandins) that bring about increased blood permeability, increased smooth muscle contraction, and increased mucus production. Examples are hay fever, hives, and anaphylactic shock.

Anatomical (an'-a-TOM-i-kal) **position** A position of the body universally used in anatomical descriptions in which the body is erect, facing the observer, the upper limbs are at the sides, the palms are facing forward, and the feet are flat on the floor.

Anatomic dead space Spaces of the nose, pharynx, larynx, trachea, bronchi, and bronchioles that contain 150 mL of tidal volume; the air does not reach the alveoli to participate in gas exchange.

Anatomy (a-NAT-ō-mē) The structure or study of structure of the body and the relation of its parts to each other.

Androgen (AN-drō-jen) Substance producing or stimulating masculine characteristics, such as the male hormone testosterone.

Anemia (a-NĒ-mē-a) Condition of the blood in which the number of functional red blood cells or their hemoglobin content is below normal.

Anesthesia (an'-es-THĒ-zē-a) A total or partial loss of feeling or sensation, usually defined with respect to loss of pain sensation; may be general or local.

Aneurysm (AN-yoo-rizm) A saclike enlargement of a blood vessel caused by a weakening of its wall.

Angina pectoris (an-JĪ-na or AN-ji-na PEK-tō-ris) A pain in the chest related to reduced coronary circulation that may or may not involve heart or artery disease.

Angiotensin (an'-jē-ō-TEN-sin) Either of two forms of a protein associated with regulation of blood pressure. Angiotensin I is produced by the action of renin on angiotensinogen and is converted by the action of ACE (angiotensin-converting enzyme) into angiotensin II, which stimulates aldosterone secretion by the adrenal cortex, stimulates the sensation of thirst, and causes vasoconstriction with resulting increase in systemic vascular resistance.

Anion (AN-ī-on) A negatively charged ion. An example is the chloride ion (Cl^-).

Anomaly (a-NOM-a-lē) An abnormality that may be a developmental (congenital) defect; a variant from the usual standard.

Anorexia nervosa (an'-ō-REK-sē-a ner-VŌ-sa) A chronic disorder characterized by self-induced weight loss, body-image and other perceptual disturbances, and physiologic changes that result from nutritional depletion.

Antagonist (an-TAG-ō-nist) A muscle that has an action opposite that of the prime mover (agonist) and yields to the movement of the prime mover.

Anterior (an-TĒR-ē-or) Nearer to or at the front of the body. Also called **ventral.**

Anterior pituitary Anterior lobe of the pituitary gland. Also called the **adenohypophysis** (ad'-e-nō-hī-POF-i-sis).

Anterolateral (an'-ter-ō-LAT-er-al) **pathway** Sensory pathway that conveys information related to pain, temperature, crude touch, pressure, tickle, and itch.

Antibody (AN-ti-bod'-ē) A protein produced by certain cells in response to a specific antigen; the antibody combines with that antigen to neutralize, inhibit, or destroy it. Also called an **immunglobulin** (im'-yoo-nō-GLOB-yoo-lin) or **Ig.**

Antibody-mediated immunity That component of immunity in which lymphocytes (B cells) develop into plasma cells that produce antibodies that destroy antigens. Also called **humoral** (HYOO-mor-al) **immunity.**

Anticoagulant (an'-tī-cō-AG-yoo-lant) A substance that can delay, suppress, or prevent the clotting of blood.

Antidiuretic (an'-ti-dī'-yoo-RET-ik) Substance that inhibits urine formation.

Antidiuretic hormone (ADH) Hormone produced by neurosecretory cells in the paraventricular and supraoptic nuclei of the hypothalamus that stimulates water reabsorption from kidney cells into the blood and vasoconstriction of arterioles. Also called **vasopressin** (vāz-ō-PRES-in).

Antigen (AN-ti-jen) A substance that has the ability to provoke an immune response and the ability to react with the antibodies or cells that result from the immune response.

Antigen-presenting cell (APC) Special class of

migratory cells that process and present antigens to T cells during an immune response; APCs include macrophages, B cells, and dendritic cells, which are present in the skin and in mucous membranes.

Anuria (a-NOO-rē-a) A daily urine output of less than 50 mL.

Anus (Ā-nus) The distal end and outlet of the rectum.

Aorta (ā-OR-ta) The main systemic trunk of the arterial system of the body that emerges from the left ventricle.

Aortic (ā-OR-tik) **body** Cluster of receptors on or near the arch of the aorta that respond to alterations in blood levels of oxygen, carbon dioxide, and hydrogen ions (H⁺).

Aortic reflex A reflex concerned with maintaining normal systemic blood pressure.

Aphasia (a-FĀ-zē-a) Loss of ability to express oneself properly through speech, or loss of verbal comprehension.

Apnea (AP-nē-a) Temporary cessation of breathing.

Apneustic (ap-NOOS-tik) **area** Portion of the respiratory center in the pons that sends stimulatory nerve impulses to the inspiratory area that activate and prolong inspiration and inhibit expiration.

Aponeurosis (ap′-ō-noo-RŌ-sis) A sheetlike tendon joining one muscle with another or with bone.

Apoptosis (ap′-ō-TŌ-sis) A normal type of cell death that removes unneeded cells during embryological development, regulates the number of cells in tissues, and eliminates many potentially dangerous cells such as cancer cells. During apoptosis, the DNA fragments and the cytoplasm shrinks, but the plasma membrane remains intact. Phagocytes engulf and digest the apoptotic cells, and an inflammatory response does not occur.

Aqueous humor (AK-wē-us HYOO-mur) The watery fluid, similar in composition to cerebrospinal fluid, that fills the anterior cavity of the eye.

Arachnoid (a-RAK-noyd) The middle of the three coverings (meninges) of the brain and spinal cord.

Arachnoid villus (VIL-us) Berrylike tuft of arachnoid that protrudes into the superior sagittal sinus and through which cerebrospinal fluid is reabsorbed into the bloodstream.

Areola (a-RĒ-ō-la) Any tiny space in a tissue. The pigmented ring around the nipple of the breast.

Arm The portion of the upper limb from the shoulder to the elbow.

Arrhythmia (a-RITH-mē-a) Irregular heart rhythm. Also called **dysrhythmia.**

Arteriole (ar-TĒR-ē-ōl) A small, almost microscopic artery that delivers blood to a capillary.

Arteriosclerosis (ar-tēr′-ē-ō-skle-RO-sis) Group of diseases characterized by thickening of the walls of arteries and loss of elasticity.

Artery (AR-ter-ē) A blood vessel that carries blood away from the heart.

Arthritis (ar-THRI-tis) Inflammation of a joint.

Arthrology (ar-THROL-ō-jē) The study or description of joints.

Arthroscopy (ar-THROS-cō-pē) A procedure for examining the interior of a joint, usually the knee, by inserting an arthroscope into a small incision; used to determine extent of damage, remove torn cartilage, repair cruciate ligaments, and obtain samples for analysis.

Arthrosis (ar-THRŌ-sis) A joint or articulation.

Articular (ar-TIK-yoo-lar) **capsule** Sleevelike structure around a synovial joint composed of a fibrous capsule and a synovial membrane.

Articular cartilage (KAR-ti-lij) Hyaline cartilage attached to articular bone surfaces.

Articulation (ar-tik′-yoo-LĀ-shun) A joint; a point of contact between bones, cartilage and bones, or teeth and bones.

Ascending colon (KŌ-lon) The portion of the large intestine that passes superiorly from the cecum to the inferior edge of the liver, where it bends at the right colic (hepatic) flexure to become the transverse colon.

Ascites (a-SĪ-tēz) Abnormal accumulation of serous fluid in the peroneal cavity.

Association area A portion of the cerebral cortex connected by many motor and sensory fibers to other parts of the cortex. The association areas are concerned with motor patterns, memory, concepts of word hearing and word seeing, reasoning, will, judgment, and personality traits.

Asthma (AZ-ma) Usually allergic reaction characterized by smooth muscle spasms in bronchi resulting in wheezing and difficult breathing. Also called **bronchial asthma.**

Astigmatism (a-STIG-ma-tizm) An irregularity of the lens or cornea of the eye causing the image to be out of focus and producing faulty vision.

Astrocyte (AS-trō-sīt) A neuroglial cell having a star shape that participates in brain development and the metabolism of neurotransmitters, helps form the blood–brain barrier and maintain the proper balance of K⁺ for generation of nerve impulses, and provides a link between neurons and blood vessels.

Ataxia (a-TAK-sē-a) A lack of muscular coordination, lack of precision.

Atherosclerotic (ath′-er-ō-skle-RO-tic) **plaque** (PLAK) A lesion that results from accumulated cholesterol and smooth muscle fibers (cells) of the tunica media of an artery; may become obstructive.

Atom Unit of matter that makes up a chemical element; consists of a nucleus and electrons.

Atomic mass (weight) Average mass of all stable atoms of an element, reflecting the relative proportion of atoms with different mass numbers.

Atomic number Number of protons in an atom.

Atrial fibrillation (Ā-trē-al fib′-ri-LĀ-shun) Asynchronous contraction of cardiac muscle fibers in the atria that results in the cessation of atrial pumping.

Atrial natriuretic (na′-trē-yoo-RET-ik) **peptide (ANP)** Peptide hormone, produced by the atria of the heart in response to stretching, that inhibits aldosterone production and thus lowers blood pressure.

Atrioventricular (AV) (ā′-trē-ō-ven-TRIK-yoo-lar) **bundle** The portion of the conduction system of the heart that begins at the atrioventricular (AV) node, passes through the cardiac skeleton separating the atria and the ventricles, then extends a short distance down the interventricular septum before splitting into right and left bundle branches. Also called the **bundle of His** (HISS).

Atrioventricular (AV) node The portion of the conduction system of the heart made up of a compact mass of conducting cells located in the septum between the two atria.

Atrioventricular (AV) valve A heart valve made up of membranous flaps or cusps that allows blood to flow in one direction only, from an atrium into a ventricle.

Atrium (Ā-trē-um) A superior chamber of the heart.

Atrophy (AT-rō-fē) Wasting away or decrease in size of a part, due to a failure, abnormality of nutrition, or lack of use.

Auditory ossicle (AW-di-tō-rē OS-si-kul) One of the three small bones of the middle ear called the **malleus, incus,** and **stapes.**

Auditory tube The tube that connects the middle ear with the nose and nasopharynx region of the throat. Also called the **Eustachian** (yoo-STĀ-kē-an) **tube.**

Auscultation (aws′-kul-TĀ-shun) Examination by listening to sounds in the body.

Autoimmunity An immunologic response against a person's own tissues.

Autolysis (aw-TOL-i-sis) Self-destruction of cells by their own lysosomal digestive enzymes after death or in a pathological process.

Autonomic ganglion (aw′-tō-NOM-ik GANG-lē-on) A cluster of sympathetic or parasympathetic cell bodies located outside the central nervous system.

Autonomic nervous system (ANS) Visceral sensory (afferent) and motor (efferent) neurons, both sympathetic and parasympathetic. Motor neurons conduct nerve impulses from the central nervous system to smooth muscle, cardiac muscle, and glands; so named because this portion of the nervous system was thought to be self-governing or spontaneous.

Autophagy (aw-TOF-a-jē) Process by which worn-out organelles are digested within lysosomes.

Autopsy (AW-top-sē) The examination of the body after death.

Autoregulation (aw′-tō-reg′-yoo-LĀ-shun) A local, automatic adjustment of blood flow in a given region of the body in response to tissue needs.

Autosome (AW-tō-sōm) Any chromosome other than the pair of sex chromosomes.

Axilla (ak-SIL-a) The small hollow beneath the arm where it joins the body at the shoulders. Also called the **armpit.**

Axon (AK-son) The usually single, long process of a nerve cell that propagates a nerve impulse toward the axon terminals.

B

B cell A lymphocyte that can develop into an antibody-producing plasma cell or a memory cell.

Babinski (ba-BIN-skē) **sign** Extension of the great toe, with or without fanning of the other toes, in response to stimulation of the outer margin of the sole of the foot; normal up to 18 months of age.

Back The posterior part of the body; the dorsum.

Ball-and-socket joint A synovial joint in which the rounded surface of one bone moves within a cup-shaped depression or fossa of another bone, as in the shoulder or hip joint. Also called a **spheroid** (SFĒ-roid) **joint.**

Baroreceptor (bar′-ō-re-SEP-tor) Nerve cell capable of responding to changes in blood, air, or fluid pressure. Also called a **pressoreceptor.**

Basal ganglia (GANG-glē-a) Paired clusters of cell bodies that make up the central gray matter in each cerebral hemisphere, including the caudate nucleus, lentiform nucleus, claustrum, and amygdala. Also called **cerebral nuclei** (SER-ebral NOO-klē-ī).

Basal metabolic (BĀ-sal met′-a-BOL-ik) **rate (BMR)** The rate of metabolism measured under standard or basal conditions (awake, at rest, fasting).

Base A nonacid or a proton acceptor, characterized by excess of hydroxide ions (OH^-) and a pH greater than 7. A ring-shaped, nitrogen-containing organic molecule that is one of the components of a nucleotide, namely, adenine, guanine, cytosine, thymine, and uracil; also known as nitrogenous base.

Basement membrane Thin, extracellular layer between epithelium and connective tissue consisting of a basal lamina and a reticular lamina.

Basilar (BĀS-i-lar) **membrane** A membrane in the cochlea of the internal ear that separates the cochlear duct from the scala tympani and on which the spiral organ (organ of Corti) rests.

Basophil (BĀ-sō-fil) A type of white blood cell characterized by a pale nucleus and large granules that stain blue-purple with basic dyes.

Belly The abdomen. The gaster or prominent, fleshy part of a skeletal muscle.

Benign (be-NĪN) Not malignant; favorable for recovery; a mild disease.

Beta (BĀ-ta) **cell** A cell in the pancreatic islets (islets of Langerhans) in the pancreas that secretes insulin.

Bicuspid (bī-KUS-pid) **valve** Atrioventricular (AV) valve on the left side of the heart. Also called the **mitral valve.**

Bilateral (bī-LAT-er-al) Pertaining to two sides of the body.

Bile (BĪL) A secretion of the liver consisting of water, bile salts, bile pigments, cholesterol, lecithin, and several ions; it emulsifies lipids prior to their digestion.

Bilirubin (bil′-ē-ROO-bin) An orange pigment that is one of the end products of hemoglobin breakdown in the hepatocytes and is excreted as a waste material in the bile.

Biopsy (BĪ-op-sē) Removal of tissue or other material from the living body for examination, usually microscopic.

Blastocyst (BLAS-tō-sist) In the development of an embryo, a hollow ball of cells that consists of a blastocele (the internal cavity), trophoblast (outer cells), and inner cell mass.

Blastomere (BLAS-tō-mēr) One of the cells resulting from the cleavage of a fertilized ovum.

Blind spot Area in the retina at the end of the optic (II) nerve in which there are no photoreceptors.

Blood The fluid that circulates through the heart, arteries, capillaries, and veins and that constitutes the chief means of transport within the body.

Blood–brain barrier (BBB) A barrier consisting of specialized brain capillaries and astrocytes that prevents the passage of materials from the blood to the cerebrospinal fluid and brain.

Blood pressure (BP) Force exerted by blood against the walls of blood vessels due to contraction of the heart and influenced by the elasticity of the vessel walls; clinically, a measure of the pressure in arteries during ventricular systole and ventricular diastole.

Blood–testis barrier (BTB) A barrier formed by Sertoli cells that prevents an immune response against antigens produced by spermatogenic cells by isolating the cells from the blood.

Body cavity A space within the body that contains various internal organs.

Body fluid Body water and its dissolved substances; constitutes about 60% of total body weight.

Bolus (BŌ-lus) A soft, rounded mass, usually food, that is swallowed.

Brachial plexus (BRĀ-kē-al PLEK-sus) A network of nerve axons of the anterior rami of spinal nerves C5, C6, C7, C8, and T1. The nerves that emerge from the brachial plexus supply the upper limb.

Bradycardia (brād′-ē-KAR-dē-a) A slow resting heart or pulse rate (under 60 beats/min).

Brain A mass of nervous tissue located in the cranial cavity.

Brain stem The portion of the brain immediately superior to the spinal cord, made up of the medulla oblongata, pons, and midbrain.

Brain waves Electrical activity produced as a result of action potentials of brain neurons.

Broad ligament A double fold of parietal peritoneum attaching the uterus to the side of the pelvic cavity.

Broca's (BRŌ-kahz) **area** Motor area of the brain in the frontal lobe that translates thoughts into speech. Also called the **motor speech area.**

Bronchi (BRONG-kē) Branches of the respiratory passageway including primary bronchi (the two divisions of the trachea), secondary or lobar bronchi (divisions of the primary bronchi that are distributed to the lobes of the lung), and tertiary or segmental bronchi (divisions of the secondary bronchi that are distributed to bronchopulmonary segments of the lung). Singular is **bronchus.**

Bronchial tree The trachea, bronchi, and their branching structures up to and including the terminal bronchioles.

Bronchiole (BRONG-kē-ōl) Branch of a tertiary bronchus further dividing into terminal bronchioles (distributed to lobules of the lung), which divide into respiratory bronchioles (distributed to alveolar sacs).

Bronchitis (brong-KI-tis) Inflammation of the bronchi characterized by hypertrophy and hyperplasia of seromucous glands and goblet cells that line the bronchi and which results in a productive cough.

Bronchoscopy (brong-KOS-kō-pē) Visual examination of the interior of the trachea and bronchi with a bronchoscope to biopsy a tumor, clear an obstruction, take cultures, stop bleeding, or deliver drugs.

Buccal (BUK-al) Pertaining to the cheek or mouth.

Buffer (BUF-er) **system** A pair of chemicals—one a weak acid and the other the salt of the weak acid, which functions as a weak base—that resists changes in pH.

Bulbourethral (bul′-bō-yoo-RĒ-thral) **gland** One of a pair of glands located inferior to the prostate gland on either side of the urethra that secretes an alkaline fluid into the cavernous urethra. Also called **Cowper's** (KOW-perz) **gland.**

Bulimia (boo-LIM-ē-a) A disorder characterized by overeating, at least twice a week, followed by purging by self-induced vomiting, strict dieting or fasting, vigorous exercise, or use of laxatives.

Bulk flow The movement of large numbers of ions, molecules, or particles in the same direction as a result of pressure differences (osmotic, hydrostatic, or air pressure).

Bundle branch One of the two branches of the atrioventricular (AV) bundle made up of specialized muscle fibers (cells) that transmit electrical impulses to the ventricles.

Bursa (BUR-sa) A sac or pouch of synovial fluid located at friction points, especially about joints.

Bursitis (bur-SĪ-tis) Inflammation of a bursa.

Buttocks (BUT-oks) The two fleshy masses on the posterior aspect of the inferior trunk, formed by the gluteal muscles.

C

Calcaneal (kal-KĀ-nē-al) **tendon** The tendon of the soleus, gastrocnemius, and plantaris muscles at the back of the heel. Also called the **Achilles** (ah-KIL-ēz) **tendon.**

Calcification (kal′-si-fi-KĀ-shun) Deposition of mineral salts, primarily hydroxyapatite, in a framework formed by collagen fibers in which the tissue hardens. Also called **mineralization** (min′-er-a-li-ZĀ-shun).

Calcitonin (kal′-si-TŌ-nin) **(CT)** A hormone produced by the thyroid gland that can lower the calcium and phosphate levels of the blood by inhibiting bone breakdown and accelerating calcium deposition into bone matrix.

Calculus (KAL-kyoo-lus) A stone, or insoluble mass of crystallized salts or other material, formed within the body, as in the gallbladder, kidney, or urinary bladder.

Callus (KAL-lus) A growth of new bone tissue in and around a fractured area, ultimately replaced by mature bone. An acquired, localized thickening.

Calorie (KAL-ō-rē) A unit of heat. A calorie (cal) is the standard unit and is the amount of heat necessary to raise the temperature of 1 g of water by 1°C. The Calorie, or kilocalorie (kcal), is used in metabolic and nutrition studies and is equal to 1000 cal.

Calyx (KĀL-iks) Any cuplike division of the kidney pelvis. Plural is **calyces** (KĀ-li-sēz).

Canal (ka-NAL) A narrow tube, channel, or passageway.

Canaliculus (kan′-a-LIK-yoo-lus) A small channel or canal, as in bones, where they connect lacunae. Plural is **canaliculi** (kan′-a-LIK-yoo-lī).

Capacitation (ka-pas′-i-TĀ-shun) The functional changes that sperm undergo in the female reproductive tract that allow them to fertilize a secondary oocyte.

Capillary (KAP-i-lar′-ē) A microscopic blood vessel located between an arteriole and venule through which materials are exchanged between blood and body cells.

Carbohydrate (kar′-bō-HĪ-drāt) An organic compound containing carbon, hydrogen, and oxygen in a particular amount and arrangement and composed of monosaccharide subunits; usually has the formula $(CH_2O)_n$.

Carcinogen (kar-SIN-ō-jen) A chemical agent or radiation that causes cancer.

Cardiac (KAR-dē-ak) **arrest** Cessation of an effective heartbeat in which the heart is completely stopped or in ventricular fibrillation.

Cardiac cycle A complete heartbeat consisting of systole (contraction) and diastole (relaxation) of both atria plus systole and diastole of both ventricles.

Cardiac muscle Striated muscle fibers (cells) that form the wall of the heart; stimulated by an intrinsic conduction system and autonomic motor neurons.

Cardiac notch An angular notch in the anterior border of the left lung into which a portion of the heart fits.

Cardiac output (CO) The volume of blood pumped from one ventricle of the heart (usually measured from the left ventricle) in 1 min; about 5.2 liters/min under normal resting conditions.

Cardiology (kar′-dē-OL-ō-jē) The study of the heart and diseases associated with it.

Cardiopulmonary resuscitation (rē-sus′-i-TA-shun) **(CPR)** A technique employed to restore life or consciousness to a person apparently dead or dying; includes external respiration (exhaled air respiration) and external cardiac massage.

Cardiovascular (kar′-dē-ō-VAS-kyoo-lar) **center** Groups of neurons scattered within the medulla oblongata that regulate heart rate, force of contraction, and blood vessel diameter.

Carotene (KAR-o-tēn) Antioxidant vitamin; yellow-orange pigment present in the stratum corneum of the epidermis. Accounts for the yellowish coloration of skin. Also termed **beta-carotene.**

Carotid (ka-ROT-id) **body** Cluster of chemoreceptors on or near the carotid sinus that respond to alterations in blood levels of oxygen, carbon dioxide, and hydrogen ions.

Carotid sinus A dilated region of the internal carotid artery immediately superior to the branching of the common carotid artery containing baroreceptors that monitor blood pressure.

Carotid sinus reflex A reflex concerned with maintaining normal blood pressure in the brain.

Carpal bones The eight bones of the wrist. Also called **carpals.**

Carpus (KAR-pus) Wrist.

Cartilage (KAR-ti-lij) A type of connective tissue consisting of chondrocytes in lacunae embedded in a dense network of collagen and elastic fibers and a matrix of chondroitin sulfate.

Cartilaginous (kar′-ti-LAJ-i-nus) **joint** A joint without a synovial (joint) cavity where the articulating bones are held tightly together by cartilage, allowing little or no movement.

Cast A small mass of hardened material formed within a cavity in the body and then discharged from the body; can originate in different areas and can be composed of various materials.

Catabolism (ka-TAB-ō-lizm) Chemical reactions that break down complex organic compounds into simple ones, with the net release of energy.

Catalyst (KAT-a-list) A substance that speeds up a chemical reaction without itself being altered; enzyme.

Cataract (KAT-a-rakt) Loss of transparency of the lens of the eye or its capsule, or both.

Cation (KAT-ī-on) A positively charged ion. An example is a sodium ion (Na^+).

Cauda equina (KAW-da ē-KWĪ-na) A tail-like array of roots of spinal nerves at the inferior end of the spinal cord.

Cecum (SĒ-kum) A blind pouch at the proximal end of the large intestine to which the ileum is attached.

Cell The basic structural and functional unit of all organisms; the smallest structure capable of performing all the activities vital to life.

Cell cycle Growth and division of a single cell into daughter cells; consists of interphase and cell division.

Cell division Process by which a cell reproduces itself that consists of a nuclear division (mitosis) and a cytoplasmic division (cytokinesis); types include somatic and reproductive cell division.

Cell-mediated immunity That component of immunity in which specially sensitized lymphocytes (T cells) attach to antigens to destroy them. Also called **cellular immunity.**

Cementum (se-MEN-tum) Calcified tissue covering the root of a tooth.

Center of ossification (os′-i-fi-KĀ-shun) An area in the cartilage model of a future bone where the cartilage cells hypertrophy and then secrete enzymes that result in the calcification of their matrix, resulting in the death of the cartilage cells, followed by the invasion of the area by osteoblasts that then lay down bone.

Central canal A microscopic tube running the length of the spinal cord in the gray commissure. A circular channel running longitudinally in the center of an osteon (Haversian system) of mature compact bone, containing blood and lymphatic vessels and nerves. Also called a **Haversian** (ha-VĒR-shun) **canal.**

Central nervous system (CNS) That portion of the nervous system that consists of the brain and spinal cord.

Centrioles (SEN-trē-ōlz) Paired, cylindrical structures within a centrosome, each consisting of a ring of microtubules and arranged at right angles to each other. They play a role in the formation or regeneration of flagella or cilia.

Centromere (SEN-trō-mēr) The constricted portion of a chromosome where the two chromatids are joined; serves as the point of attachment for microtubules that pull chromatids during anaphase of cell division.

Centrosome (SEN-trō-sōm) Organelle near the nucleus of a cell that consists of a pericentriolar area (dense region of small protein fibers) that contains a pair of centrioles. During prophase, the pericentriolar area forms the mitotic spindle.

Cephalic (se-FAL-ik) Pertaining to the head; superior in position.

Cerebellum (ser′-e-BEL-um) The portion of the brain lying posterior to the medulla oblongata and pons; governs balance and coordinates skilled movements.

Cerebral aqueduct (SER-e-bral AK-we-dukt) A channel through the midbrain connecting the third and fourth ventricles and containing cerebrospinal fluid. Also termed the **aqueduct of Sylvius.**

Cerebral arterial circle A ring of arteries forming an anastomosis at the base of the brain between the internal carotid and basilar arteries and arteries supplying the brain. Also called the **circle of Willis.**

Cerebral cortex The surface of the cerebral hemispheres, 2 to 4 mm thick, consisting of six layers of neuronal cell bodies (gray matter) in most areas.

Cerebral palsy (PAL-zē) A group of motor disorders resulting in muscular uncoordination and loss of muscle control and caused by damage to motor areas of the brain (cerebral cortex, basal ganglia, and cerebellum) during fetal life, birth, or infancy.

Cerebrospinal (se-rē′-brō-SPĪ-nal) **fluid (CSF)** A fluid produced by ependymal cells that cover choroid plexuses in the ventricles of the brain; the fluid circulates in the ventricles, the central canal, and the subarachnoid space around the brain and spinal cord.

Cerebrovascular (se-rē′-brō-VAS-kyoo-lar) **accident (CVA)** Destruction of brain tissue (infarction) resulting from disorders of blood vessels that supply the brain. Also called a **stroke.**

Cerebrum (SER-ē-brum or ser-Ē-brum) The two hemispheres of the forebrain, making up the largest part of the brain.

Cerumen (se-ROO-men) Waxlike secretion produced by ceruminous glands in the external auditory meatus (ear canal).

Ceruminous (se-ROO-mi-nus) **gland** A modified sudoriferous (sweat) gland in the external auditory meatus that secretes cerumen (ear wax).

Cervix (SER-viks) Neck; any constricted portion of an organ, such as the inferior cylindrical part of the uterus.

Cesarean (se-ZAR-ē-an) **section** Procedure in which a low, horizontal incision is made through the abdominal wall and uterus for removal of the baby and placenta. Also called a **C section.**

Chemical bond Force of attraction in a molecule or compound that holds its atoms together. Examples include ionic and covalent bonds.

Chemical element Unit of matter that cannot be decomposed into a simpler substance by ordinary chemical reactions. Examples include hydrogen (H), carbon (C), and oxygen (O).

Chemical reaction The combination or separation of atoms in which chemical bonds are formed or broken and new products with different properties are produced.

Chemoreceptor (kē′-mō-rē-SEP-tor) Receptor that detects the presence of specific chemicals.

Chemotaxis (kē′-mō-TAK-sis) Attraction of phagocytes to microbes by a chemical stimulus.

Chiasm (KĪ-azm) A crossing; especially the crossing of the optic (II) nerve fibers.

Chief cell The secreting cell of a gastric gland that produces pepsinogen, the precursor of the enzyme pepsin, and the enzyme gastric lipase.

Chiropractic (kī′-rō-PRAK-tik) A system of treating disease by using one's hands to manipulate body parts, mostly the vertebral column.

Chlamydia (kla-MID-ē-ah) Most prevalent sexually transmitted disease, characterized by frequent, painful, or burning urination and low back pain.

Cholecystectomy (kō′-lē-sis-TEK-tō-mē) Surgical removal of the gallbladder.

Cholesterol (kō-LES-ter-ol′) Classified as a lipid, the most abundant steroid in animal tissues; located in cell membranes and used for the synthesis of steroid hormones and bile salts.

Cholinergic (kō′-lin-ER-jik) **neuron** A neuron that liberates acetylcholine at its synapses.

Chondrocyte (KON-drō-sīt) Cell of mature cartilage.

Chondroitin (kon-DROY-tin) **sulfate** An amorphous matrix material found outside connective tissue cells.

Chordae tendineae (KOR-dē ten-DIN-ē-ē) Tendonlike, fibrous cords that connect heart AV valves with papillary muscles.

Chorion (KŌR-ē-on) The outermost extraembryonic membrane that becomes the principal embryonic portion of the placenta; serves a protective and nutritive function.

Chorionic villi (kōr′-ē-ON-ik VIL-ī) Fingerlike projections of the chorion that grow into the decidua basalis of the endometrium and contain fetal blood vessels.

Choroid (KŌR-oyd) Part of the vascular tunic of the eyeball; a thin membrane that lines most of the sclera.

Choroid plexus (PLEK-sus) A network of capillaries located in the roof of each of the four ventricles of the brain; ependymal cells around choroid plexuses produce cerebrospinal fluid.

Chromatid (KRŌ-ma-tid) One of a pair of identical connected nucleoprotein strands that are joined at the centromere and separate during cell division, each becoming a chromosome of one of the two daughter cells.

Chromatin (KRŌ-ma-tin) The threadlike mass of genetic material, consisting principally of DNA, which is present in the nucleus of a nondividing or interphase cell.

Chromatophilic substance Rough endoplasmic reticulum in the cell bodies of neurons that functions in protein synthesis. Also called **Nissl bodies.**

Chromosome (KRŌ-mō-sōm) One of the small, threadlike structures in the nucleus of a cell, normally 46 in a human diploid cell, that bears the genetic material; composed of DNA and proteins (histones) that form a delicate chromatin thread during interphase; becomes packaged into compact rodlike structures that are visible under the light microscope during cell division.

Chronic (KRON-ik) Long term or frequently recurring; applied to a disease that is not acute.

Chronic obstructive pulmonary disease (COPD) A disease, such as bronchitis or emphysema, in which there is some degree of obstruction of air passageways and consequently an increase in airway resistance.

Chyle (KĪL) The milky-appearing fluid found in the lacteals of the small intestine after absorption of lipids in food.

Chylomicron (kī′-lō-MĪ-kron) Protein-coated spherical structure that contains triglycerides, phospholipids, and cholesterol and is absorbed into the lacteal of a villus in the small intestine.

Chyme (KĪM) The semifluid mixture of partly digested food and digestive secretions found in the stomach and small intestine during digestion of a meal.

Ciliary (SIL-ē-ar′-ē) **body** One of the three portions of the vascular tunic of the eyeball, the others being the choroid and the iris; includes the ciliary muscle and the ciliary processes.

Cilium (SIL-ē-um) A hair or hairlike process projecting from a cell that may be used to move the entire cell or to move substances along the surface of the cell. Plural is **cilia.**

Circadian (ser-KĀ-dē-an) **rhythm** A cycle of active and nonactive periods in organisms determined by internal mechanisms and repeating about every 24 hours.

Circular folds Permanent, deep, transverse folds in the mucosa and submucosa of the small intestine that increase the surface area for absorption.

Circulation time Time required for blood to pass from the right atrium, through pulmonary circulation, back to the left ventricle, through systemic circulation to the foot, and back again to the right atrium; normally about 1 min.

Circumduction (ser′-kum-DUK-shun) A movement at a synovial joint in which the distal end of a bone moves in a circle while the proximal end remains relatively stable.

Circumvallate papilla (ser′-kum-VAL-at pah-PIL-a) One of the circular projections that is arranged in an inverted V–shaped row at the back of the tongue; the largest of the elevations on the upper surface of the tongue containing taste buds.

Cirrhosis (si-RO-sis) A liver disorder in which the parenchymal cells are destroyed and replaced by connective tissue.

Cisterna chyli (sis-TER-na KĪ-lē) The origin of the thoracic duct.

Climacteric (klī-MAK-ter-ik) Cessation of the reproductive function in the female or diminution of testicular activity in the male.

Climax The peak period or moments of greatest intensity during sexual excitement.

Clitoris (KLI-to-ris) An erectile organ of the female, located at the anterior junction of the labia minora, that is homologous to the male penis.

Clone (KLŌN) A population of identical cells.

Clot The end result of a series of biochemical reactions that changes liquid plasma into a gelatinous mass; specifically, the conversion of fibrinogen into a tangle of polymerized fibrin molecules.

Clot retraction (rē-TRAK-shun) The consolidation of a fibrin clot to pull damaged tissue together.

Clotting Process by which a blood clot is formed. Also known as **coagulation** (cō-ag′-yoo-LĀ-shun).

Coccyx (KOK-six) The fused bones at the inferior end of the vertebral column.

Cochlea (KŌK-lē-a) A winding, cone-shaped tube forming a portion of the inner ear and containing the spiral organ (organ of Corti).

Cochlear duct The membranous cochlea consisting of a spirally arranged tube enclosed in the bony cochlea and lying along its outer wall. Also called the **scala media** (SCĀ-la MĒ-dē-a).

Coenzyme A nonprotein organic molecule that is associated with and activates an enzyme; many are derived from vitamins. An example is nicotinamide adenine dinucleotide (NAD), derived from the B vitamin niacin.

Coitus (KŌ-i-tus) Sexual intercourse.

Collagen (KOL-a-jen) A protein that is the main organic constituent of connective tissue.

Colliculi (kō-LIK-yoo-lē) Four small elevations (superior and inferior) in the posterior portion of the midbrain concerned with visual and auditory reflexes.

Colon The division of the large intestine consisting of ascending, transverse, descending, and sigmoid portions.

Colostrum (kō-LOS-trum) A thin, cloudy fluid secreted by the mammary glands a few days prior to or after delivery before true milk is produced.

Column (KOL-um) Group of white matter tracts in the spinal cord.

Common bile duct A tube formed by the union of the common hepatic duct and the cystic duct that empties bile into the duodenum at the hepatopancreatic ampulla (ampulla of Vater).

Compact (dense) bone tissue Bone tissue that contains few spaces between osteons (Haversian systems); forms the external portion of all bones and the bulk of the diaphysis (shaft) of long bones; is found immediately deep to the periosteum and external to spongy bone.

Complement (KOM-ple-ment) A group of at least 20 normally inactive proteins found in plasma that forms a component of nonspecific resistance and immunity by bringing about cytolysis, inflammation, and opsonization.

Compound A substance that can be broken down into two or more other substances by chemical means.

Concha (KONG-ka) A scroll-like bone found in the skull. Plural is **conchae** (KONG-kē).

Concussion (kon-KUSH-un) Traumatic injury to the brain that produces no visible bruising but may result in abrupt, temporary loss of consciousness.

Conduction myofiber Muscle fiber (cell) in the ventricular tissue of the heart specialized for conducting an action potential to the myocardium; part of the conduction system of the heart. Also called a **Purkinje** (pur-KIN-jē) **fiber.**

Conduction system A series of autorhythmic cardiac muscle fibers that generates and distributes electrical impulses to stimulate coordinated contraction of the heart chambers; includes the sinoatrial (SA) node, the atrioventricular (AV) node, the atrioventricular (AV) bundle, the right and left bundle branches, and the conduction myofibers (Purkinje fibers).

Conductivity (kon′-duk-TIV-i-tē) The ability of a cell to conduct (propagate) action potentials along its plasma membrane; characteristic of neurons and muscle fibers (cells).

Condyloid (KON-di-loid) **joint** A synovial joint

structured so that an oval-shaped condyle of one bone fits into an elliptical cavity of another bone, permitting side-to-side and back-and-forth movements, such as the joint at the wrist between the radius and carpals. Also called an **ellipsoidal** (e′-lip-SOYD-al) **joint.**

Cone The type of photoreceptor in the retina that is specialized for highly acute, color vision in bright light.

Congenital (kon-JEN-i-tal) Present at the time of birth.

Conjunctiva (kon′-junk-TĪ-va) The delicate membrane covering the eyeball and lining the eyes.

Connective tissue The most abundant of the four basic tissue types in the body, performing the functions of binding and supporting; consists of relatively few cells in a generous matrix (the ground substance and fibers between the cells).

Consciousness (KON-shus-nes) A state of wakefulness in which an individual is fully alert, aware, and oriented, partly as a result of feedback between the cerebral cortex and reticular activating system.

Continuous conduction (kon-DUK-shun) Propagation of an action potential (nerve impulse) in a step-by-step depolarization of each adjacent area of an axon membrane.

Contraception (kon′-tra-SEP-shun) The prevention of fertilization or impregnation without destroying fertility.

Contractility (kon′-trak-TIL-i-tē) The ability of cells or parts of cells to actively generate force to undergo shortening for movements. Muscle fibers (cells) exhibit a high degree of contractility.

Control center The component of a feedback system, such as the brain, that determines the point at which a controlled condition, such as body temperature, is maintained.

Contusion (kon-TOO-zhun) Condition in which tissue below the skin is damaged, but the skin is not broken.

Conus medullaris (KŌ-nus med′-yoo-LAR-is) The tapered portion of the spinal cord inferior to the lumbar enlargement.

Convergence (con-VER-jens) The medial movement of the two eyeballs so that both are directed toward a near object being viewed in order to produce a single image.

Convulsion (con-VUL-shun) Violent, involuntary, tetanic contractions of an entire group of muscles.

Cor pulmonale (KOR pul′-mōn-AL-ē) (CP) Right ventricular hypertrophy from disorders that bring about hypertension in pulmonary circulation.

Cornea (KOR-nē-a) The nonvascular, transparent fibrous coat through which the iris can be seen.

Corona radiata The innermost layer of granulosa cells that is firmly attached to the zona pellucida around a secondary oocyte.

Coronary (KOR-ō-nar′-ē) **artery bypass grafting (CABG)** Surgical procedure in which a portion of a blood vessel is removed from another part of the body and grafted onto a coronary artery so as to bypass an obstruction in the coronary artery.

Coronary artery disease (CAD) A condition such as atherosclerosis that causes narrowing of coronary arteries so that blood flow to the heart is re-

duced. The result is **coronary heart disease (CHD),** in which the heart muscle receives inadequate blood flow due to an interruption of its blood supply.

Coronary circulation The pathway followed by the blood from the ascending aorta through the blood vessels supplying the heart and returning to the right atrium. Also called **cardiac circulation.**

Coronary sinus (SĪ-nus) A wide venous channel on the posterior surface of the heart that collects the blood from the coronary circulation and returns it to the right atrium.

Corpus albicans (KOR-pus AL-bi-kanz) A white fibrous patch in the ovary that forms after the corpus luteum regresses.

Corpus callosum (ka-LŌ-sum) The great commissure of the brain between the cerebral hemispheres.

Corpus luteum (KOR-pus LOO-tē-um) A yellow endocrine gland in the ovary formed when a follicle has discharged its secondary oocyte; secretes estrogens, progesterone, relaxin, and inhibin.

Corpuscle of touch The sensory receptor for the sensation of touch; found in the dermal papillae, especially in palms and soles. Also called a **Meissner** (MĪS-ner) **corpuscle.**

Cortex (KOR-teks) An outer layer of an organ. The convoluted layer of gray matter covering each cerebral hemisphere.

Costal (KOS-tal) Pertaining to a rib.

Costal cartilage (KAR-ti-lij) Hyaline cartilage that attaches a rib to the sternum.

Cramp A spasmodic, usually painful contraction of a muscle.

Cranial (KRĀ-nē-al) **cavity** A subdivision of the dorsal body cavity formed by the cranial bones and containing the brain.

Cranial nerve One of 12 pairs of nerves that leave the brain; pass through foramina in the skull; and supply sensory and motor neurons to the head, neck, part of the trunk, and viscera of the thorax and abdomen. Each is designated by a Roman numeral and a name.

Cranium (KRĀ-nē-um) The skeleton of the skull that protects the brain and the organs of sight, hearing, and balance; includes the frontal, parietal, temporal, occipital, sphenoid, and ethmoid bones.

Creatine phosphate (KRĒ-a-tin FOS-fāt) Molecule in skeletal muscle fibers that contains high-energy phosphate bonds; used to generate ATP rapidly from ADP by transfer of phosphate group. Also called **phosphocreatine** (fos′-fō-KRĒ-a-tin).

Crenation (kre-NĀ-shun) The shrinkage of red blood cells into knobbed, starry forms when they are placed in a hypertonic solution.

Crista (KRIS-ta) A crest or ridged structure. A small elevation in the ampulla of each semicircular duct that contains receptors for dynamic equilibrium.

Crossing-over The exchange of a portion of one chromatid with another during meiosis. It permits an exchange of genes among chromatids and is one factor that results in genetic variation of progeny.

Cryptorchidism (krip-TOR-ki-dizm) The condition of undescended testes.

Cupula (KUP-yoo-la) A mass of gelatinous material covering the hair cells of a crista; a receptor in the ampulla of a semicircular canal stimulated when the head moves.

Cushing's syndrome Condition caused by a hypersecretion of glucocorticoids characterized by spindly legs, "moon face," "buffalo hump," pendulous abdomen, flushed facial skin, poor wound healing, hyperglycemia, osteoporosis, weakness, hypertension, and increased susceptibility to disease.

Cutaneous (kyoo-TĀ-nē-us) Pertaining to the skin.

Cyanosis (sī′-a-NŌ-sis) A blue or dark purple discoloration, most easily seen in nail beds and mucous membranes, that results from an increased concentration of deoxygenated (reduced) hemoglobin (more than 5 g/dL).

Cyclic AMP (cyclic adenosine monophosphate) Molecule formed from ATP by the action of the enzyme adenylate cyclase; serves as an intracellular messenger (second messenger) for some hormones.

Cyst (SIST) A sac with a distinct connective tissue wall, containing a fluid or other material.

Cystic (SIS-tik) **duct** The duct that transports bile from the gallbladder to the common bile duct.

Cystic fibrosis (fī-BRŌ-sis) Inherited disease of secretory epithelia that affects the respiratory passageways, pancreas, salivary glands, and sweat glands; the most common lethal genetic disease among the white population.

Cystitis (sis-TĪ-tis) Inflammation of the urinary bladder.

Cystoscope (SIS-tō-skōp) An instrument used to examine the inside of the urinary bladder.

Cytolysis (sī-TOL-i-sis) The rupture of living cells in which the contents leak out.

Cytokines (SĪ-tō-kīnz) Small protein hormones produced by lymphocytes, fibroblasts, endothelial cells, and antigen-presenting cells that stimulate or inhibit cell growth and differentiation, regulate immune responses, or aid nonspecific defenses.

Cytokinesis (sī′-tō-ki-NĒ-sis) Distribution of the cytoplasm into two separate cells during cell division; coordinated with nuclear division (mitosis).

Cytology (sī-TOL-ō-jē) The study of cells.

Cytoplasm (SĪ-tō-plazm) Cytosol plus all organelles (except the nucleus).

Cytoskeleton Complex internal structure of cytoplasm consisting of microfilaments, microtubules, and intermediate filaments.

Cytosol (SĪ-tō-sol) Semifluid portion of cytoplasm in which organelles are suspended and solutes are dissolved. Also called **intracellular fluid.**

D

Deafness Lack of the sense of hearing or a significant hearing loss.

Decibel (DES-i-bel) **(dB)** A unit for expressing the relative intensity (loudness) of sound.

Deciduous (dē-SID-yoo-us) Falling off or being shed seasonally or at a particular stage of development. In the body, referring to the first set of teeth.

Deep Away from the surface of the body or an organ.

Deep fascia (FASH-ē-a) A sheet of connective tissue wrapped around a muscle to hold it in place.

Deep-venous thrombosis (DVT) The presence of a thrombus in a vein, usually a deep vein of the lower limbs.

Defecation (def′e-KĀ-shun) The discharge of feces from the rectum.

Deglutition (dē-gloo-TISH-un) The act of swallowing.

Dehydration (dē′-hī-DRĀ-shun) Excessive loss of water from the body or its parts.

Demineralization (de-min′-er-al-i-ZĀ-shun) Loss of calcium and phosphorus from bones.

Denaturation (dē-nā′-chur-Ā-shun) Disruption of the tertiary structure of a protein by agents such as heat, changes in pH, or other physical or chemical methods, in which the protein loses its physical properties and biological activity.

Dendrite (DEN-drīt) A neuronal process that carries electrical signals toward the cell body.

Dendritic (den-DRIT-ik) **cell** One type of antigen-presenting cell with long, branchlike projections that commonly is present in mucosal linings (such as the vagina) and in the skin (for example, Langerhans' cells in the epidermis).

Dental caries (KAR-ēz) Gradual demineralization of the enamel and dentin of a tooth that may invade the pulp and alveolar bone. Also called **tooth decay.**

Dentin (DEN-tin) The bony tissues of a tooth enclosing the pulp cavity.

Dentition (den-TI-shun) The eruption of teeth. The number, shape, and arrangement of teeth.

Deoxyribonucleic (dē-ok′-sē-ri′-bō-noo-KLĒ-ik) **acid (DNA)** A nucleic acid constructed of nucleotides consisting of one of four nitrogenous bases (adenine, cytosine, guanine, or thymine), deoxyribose, and a phosphate group; encoded in the nucleotides is genetic information.

Depression (dē-PRESH-un) Movement in which a part of the body moves inferiorly.

Dermal papilla (pah-PILL-a) Fingerlike projection of the papillary region of the dermis that may contain blood capillaries or corpuscles of touch (Meissner corpuscles).

Dermatology (der′-ma-TOL-ō-jē) The medical specialty dealing with diseases of the skin.

Dermatome (DER-ma-tōm) The cutaneous area developed from one embryonic spinal cord segment and receiving most of its sensory innervation from one spinal nerve. An instrument for incising the skin or cutting thin transplants of skin.

Dermis (DER-mis) A layer of dense irregular connective tissue lying deep to the epidermis.

Descending colon (KŌ-lon) The part of the large intestine descending from the left colic (splenic) flexure to the level of the left iliac crest.

Detrusor (dē-TROO-ser) **muscle** Muscle in the wall of the urinary bladder.

Developmental anatomy The study of development from the fertilized egg to the adult form. The branch of anatomy called embryology is generally restricted to the study of development from the fertilized egg through the eighth week in utero.

Diabetes insipidus (dī′-a-BĒ-tēz in-SIP-i-dus) Condition caused by hyposecretion of antidiuretic hormone (ADH) and characterized by thirst and excretion of large amounts of urine.

Diabetes mellitus (MEL-i-tus) Condition caused by hyposecretion of insulin and characterized by hyperglycemia, increased urine production, excessive thirst, and excessive eating.

Diagnosis (dī′-ag-NŌ-sis) Distinguishing one disease from another or determining the nature of a disease from signs and symptoms by inspection, palpation, laboratory tests, and other means.

Dialysis (dī-AL-i-sis) The removal of waste products from blood by diffusion through a selectively permeable membrane.

Diaphragm (DĪ-a-fram) Any partition that separates one area from another, especially the dome-shaped skeletal muscle between the thoracic and abdominal cavities. Also a dome-shaped device that is placed over the cervix, usually with a spermatocide, to prevent conception.

Diaphysis (dī-AF-i-sis) The shaft of a long bone.

Diarrhea (dī-a-RĒ-a) Frequent defecation of liquid feces caused by increased motility of the intestines.

Diarthrosis (dī-ar-THRŌ-sis) A freely movable joint; types are gliding, hinge, pivot, condyloid, saddle, and ball-and-socket.

Diastole (dī-AS-tō-lē) In the cardiac cycle, the phase of relaxation or dilation of the heart muscle, especially of the ventricles.

Diastolic (dī-as-TOL-ik) **blood pressure** The force exerted by blood on arterial walls during ventricular relaxation; the lowest blood pressure measured in the large arteries, about 80 mm Hg under normal conditions for a young adult.

Diencephalon (dī′-en-SEF-a-lon) A part of the brain consisting of the thalamus, hypothalamus, epithalamus, and subthalamus.

Digestion (dī-JES-chun) The mechanical and chemical breakdown of food to simple molecules that can be absorbed and used by body cells.

Diploid (DIP-loyd) Having the number of chromosomes characteristically found in the somatic cells of an organism. Symbolized as *2n*.

Direct motor pathways Collections of upper motor neurons with cell bodies in the motor cortex that project axons into the spinal cord, where they synapse with lower motor neurons or association neurons in the anterior horns. Also called the **pyramidal pathways.**

Disease Any change from a state of health.

Dislocation (dis′-lō-KĀ-shun) Displacement of a bone from a joint with tearing of ligaments, tendons, and articular capsules. Also called **luxation** (luk-SĀ-shun).

Dissect (di-SEKT) To separate tissues and parts of a cadaver (corpse) or an organ for anatomical study.

Distal (DIS-tal) Farther from the attachment of a limb to the trunk; farther from the point of origin or attachment.

Diuretic (dī-yoo-RET-ik) A chemical that inhibits sodium reabsorption, reduces antidiuretic hormone (ADH) concentration, and increases urine volume by inhibiting facultative reabsorption of water.

Diverticulum (dī-ver-TIK-yoo-lum) A sac or pouch in the wall of a canal or organ, especially in the colon.

Dominant allele An allele that overrides the influence of alternate alleles on the homologous chromosome; the allele that is expressed.

Dorsal body cavity Cavity near the dorsal (posterior) surface of the body that consists of a cranial cavity and vertebral canal.

Dorsiflexion (dor′-si-FLEK-shun) Bending the foot in the direction of the dorsum (upper surface).

Ductus arteriosus (DUK-tus ar-tēr′-ē-Ō-sus) A small vessel connecting the pulmonary trunk with the aorta; found only in the fetus.

Ductus (vas) deferens (DEF-er-ens) The duct that carries sperm from the epididymis to the ejaculatory duct. Also called the **seminal duct.**

Ductus epididymis (ep′-i-DID-i-mis) A tightly coiled tube inside the epididymis, distinguished into a head, body, and tail, in which sperm undergo maturation.

Ductus venosus (ve-NŌ-sus) A small vessel in the fetus that helps the circulation bypass the liver.

Duodenum (doo′-ō-DĒ-num or doo-OD-e-num) The first 25 cm (10 in.) of the small intestine, which connects the stomach and the ileum.

Dura mater (DOO-ra MĀ-ter) The outer membrane (meninx) covering the brain and spinal cord.

Dynamic equilibrium (ē′-kwi-LIB-rē-um) The maintenance of body position, mainly the head, in response to sudden movements such as rotation.

Dysfunction (dis-FUNK-shun) Absence of completely normal function.

Dyslexia (dis-LEK-sē-a) Impairment of the brain's ability to translate images received from the eyes into understandable language.

Dysmenorrhea (dis-men′-ō-RĒ-a) Painful menstruation.

Dysplasia (dis-PLĀ-zē-a) Change in the size, shape, and organization of cells due to chronic irritation or inflammation; may either revert to normal if stress is removed or progress to neoplasia.

Dyspnea (DISP-nē-a) Shortness of breath.

Dysuria (dis-YOO-rē-a) Painful urination.

E

Eardrum A thin, semitransparent partition of fibrous connective tissue between the external auditory meatus and the middle ear. Also called the **tympanic membrane.**

Ectoderm The primary germ layer that gives rise to the nervous system and the epidermis of skin and its derivatives.

Ectopic (ek-TOP-ik) Out of the normal location, as in ectopic pregnancy.

Edema (e-DĒ-ma) An abnormal accumulation of interstitial fluid.

Effector (e-FEK-tor) An organ of the body, either a muscle or a gland, that responds to a motor neuron impulse.

Efferent arteriole (EF-er-ent ar-TĒR-ē-ōl) A vessel of the renal vascular system that transports blood from a glomerulus to a peritubular capillary.

Efferent (EF-er-ent) **ducts** A series of coiled tubes that transport sperm cells from the rete testis to the epididymis.

Effusion (e-FYOO-zhun) The escape of fluid from the lymphatic vessels or blood vessels into a cavity or into tissues.

Eicosanoids (ī-KŌ-sa-noydz) Local hormones derived from a 20-carbon fatty acid (arachidonic acid); two important types are prostaglandins and leukotrienes.

Ejaculation (ē-jak′-yoo-LĀ-shun) The reflex ejection or expulsion of semen from the penis.

Ejaculatory (ē-JAK-yoo-la-tō′-rē) **duct** A tube that transports sperm cells from the ductus (vas) deferens to the prostatic urethra.

Elasticity (ē′-las-TIS-i-tē) The ability of tissue to return to its original shape after contraction or extension.

Electrocardiogram (ē-lek′-trō-KAR-dē-ō-gram′) **(ECG** or **EKG)** A recording of the electrical changes that accompany the cardiac cycle that can be detected at the surface of the body; may be resting, stress, or ambulatory.

Electroencephalogram (ē-lek′-trō-en-SEF-a-lō-gram′) **(EEG)** A recording of the electrical impulses of the brain from the scalp surface; used to diagnose certain diseases (such as epilepsy), furnish information regarding sleep and wakefulness, and confirm brain death.

Electrolyte (ē-LEK-trō-līt) Any compound that separates into ions when dissolved in water and can conduct electricity.

Electromyography (ē-lek′-trō-mī-OG-ra-fē) Evaluation of the electrical activity of resting and contracting muscle to ascertain causes of muscular weakness, paralysis, involuntary twitching, and abnormal levels of muscle enzymes; also used as part of biofeedback studies.

Electron transport chain A sequence of electron carrier molecules on the inner mitochondrial membrane that undergo oxidation and reduction as they synthesize ATP.

Elevation (el′-e-VĀ-shun) Movement in which a part of the body moves superiorly.

Embolism (EM-bō-lizm) Obstruction or closure of a vessel by an embolus.

Embolus (EM-bō-lus) A blood clot, bubble of air, fragment from broken bones, mass of bacteria, or other debris or foreign material transported by the blood.

Embryo (EM-brē-ō) The young of any organism in an early stage of development; in humans, the developing organism from fertilization to the end of the eighth week in utero.

Embryology (em′-brē-OL-ō-jē) The study of development from the fertilized egg to the end of the eighth week in utero.

Emesis (EM-e-sis) Vomiting.

Emigration (em′-e-GRĀ-shun) Process whereby white blood cells (WBCs) leave the bloodstream by slowing down, rolling along the endothelium, and squeezing between the endothelial cells. Adhesion molecules help WBCs stick to the endothelium. Also known as **migration** or **extravasation.**

Emission (ē-MISH-un) Propulsion of sperm into the urethra due to peristaltic contractions of the ducts of the testes, epididymides, and ductus (vas) deferens as a result of sympathetic stimulation.

Emmetropia (em′-e-TRŌ-pē-a) Normal vision in which light rays are focused exactly on the retina.

Emphysema (em′-fi-SĒ-ma) A lung disorder in which alveolar walls disintegrate, producing abnormally large air spaces and loss of elasticity in the lungs; typically caused by exposure to cigarette smoke.

Emulsification (ē-mul′-si-fi-KĀ-shun) The dispersion of large lipid globules to smaller, uniformly distributed particles in the presence of bile.

Enamel (ē-NAM-el) The hard, white substance covering the crown of a tooth.

Endocardium (en′-dō-KAR-dē-um) The layer of the heart wall, composed of endothelium and smooth muscle, that lines the inside of the heart and covers the valves and tendons that hold the valves open.

Endochondral ossification (en′-dō-KON-dral os′-i-fi-KĀ-shun) The replacement of cartilage by bone. Also called **intracartilaginous** (in′-tra-kar′-ti-LAJ-i-nus) **ossification.**

Endocrine (EN-dō-krin) **gland** A gland that secretes hormones into the blood; a ductless gland.

Endocrinology (en′-dō-kri-NOL-ō-jē) The science concerned with the structure and functions of endocrine glands and the diagnosis and treatment of disorders of the endocrine system.

Endocytosis (en′-dō-sī-TŌ-sis) The uptake into a cell of large molecules and particles in which a segment of plasma membrane surrounds the substance, encloses it, and brings it in; includes phagocytosis, pinocytosis, and receptor-mediated endocytosis.

Endoderm (EN-dō-derm) The primary germ layer of the developing embryo that gives rise to the gastrointestinal tract, urinary bladder and urethra, and respiratory tract.

Endodontics (en′-dō-DON-tiks) The branch of dentistry concerned with the prevention, diagnosis, and treatment of diseases that affect the pulp, root, periodontal ligament, and alveolar bone.

Endogenous (en-DOJ-e-nus) Growing from or beginning within the organism.

Endolymph (EN-dō-limf) The fluid within the membranous labyrinth of the inner ear.

Endometrium (en′-dō-MĒ-trē-um) The mucous membrane lining the uterus.

Endoplasmic reticulum (en′-dō-PLAZ-mik re-TIK-yoo-lum) **(ER)** A network of channels running through the cytoplasm of a cell that serves in intracellular transportation, storage, synthesis, and packaging of molecules. Portions of ER where ribosomes are attached to the outer surface are called **rough ER;** portions that have no ribosomes are called **smooth ER.**

Endorphin (en-DOR-fin) A neuropeptide in the central nervous system that acts as a painkiller.

Endosteum (en-DOS-tē-um) The membrane that lines the medullary (marrow) cavity of bones, consisting of osteoprogenitor cells and scattered osteoclasts.

Endothelium (en′-dō-THĒ-lē-um) The layer of simple squamous epithelium that lines the cavities of the heart, blood vessels, and lymphatic vessels.

Energy The capacity to do work.

Enkephalin (en-KEF-ah-lin) A peptide found in the central nervous system that acts as a painkiller.

Enteric (en-TER-ik) **nervous system** The part of the nervous system that is embedded in the submucosa and muscularis of the gastrointestinal (GI) tract; governs motility and secretions of the GI tract.

Enterogastric (en′-ter-ō-GAS-trik) **reflex** A reflex that inhibits gastric secretion; initiated by food in the small intestine.

Enzyme (EN-zīm) A substance that affects the speed of chemical changes; an organic catalyst, usually a protein.

Eosinophil (ē′-ō-SIN-ō-fil) A type of white blood cell characterized by granules that stain red or pink with acid dyes.

Ependymal (e-PEN-de-mal) **cells** Neuroglial cells that cover choroid plexuses and produce cerebrospinal fluid (CSF); they also line the ventricles of the brain and probably assist in the circulation of CSF.

Epicardium (ep′-i-KAHR-dē-um) The thin outer layer of the heart wall, composed of serous tissue and mesothelium. Also called the **visceral pericardium.**

Epidemiology (ep′-i-dē′-mē-OL-ō-jē) Medical science concerned with the occurrence and distribution of diseases and disorders in human populations.

Epidermis (ep′-i-DERM-is) The superficial, thinner layer of skin, composed of keratinized stratified squamous epithelium.

Epididymis (ep′-i-DID-i-mis) A comma-shaped organ that lies along the posterior border of the testis and contains the ductus epididymis, in which sperm undergo maturation. Plural is **epididymides** (ep′-i-DID-i-mi-dēz′).

Epidural (ep′-i-DOO-ral) **space** A space between the spinal dura mater and the vertebral canal, containing areolar connective tissue and a plexus of veins.

Epiglottis (ep′-i-GLOT-is) A large, leaf-shaped piece of cartilage lying on top of the larynx, with its "stem" attached to the thyroid cartilage and its "leaf" portion unattached and free to move up and down to cover the glottis (vocal folds and rima glottidis).

Epilepsy (EP-i-lep′-sē) Neurologic disorder characterized by short, periodic attacks of motor, sensory, or psychological malfunction.

Epinephrine (ep′-i-NEF-rin) Hormone secreted by the adrenal medulla that produces actions similar to those that result from sympathetic stimulation. Also called **adrenaline** (ah-DREN-ah-lin).

Epiphyseal (ep′-i-FIZ-ē-al) **line** The remnant of the epiphyseal plate in a long bone.

Epiphyseal plate The hyaline cartilage plate between the epiphysis and diaphysis that is responsible for the lengthwise growth of long bones.

Epiphysis (e-PIF-i-sis) The end of a long bone, usually larger in diameter than the shaft (diaphysis).

Episiotomy (eh-piz′-ē-OT-ō-mē) A cut made with surgical scissors to avoid tearing of the perineum at the end of the second stage of labor.

Epistaxis (ep′-i-STAK-sis) Loss of blood from the nose due to trauma, infection, allergy, neoplasm, and bleeding disorders. Also called **nosebleed.**

Epithelial (ep′-i-THĒ-lē-al) **tissue** The tissue that forms innermost and outermost surfaces of body structures and forms glands.

Erectile dysfunction Failure to maintain an erection long enough for sexual intercourse. Also known as **impotence** (IM-pō-tens).

Erection (ē-REK-shun) The enlarged and stiff state of the penis or clitoris resulting from the engorgement of the spongy erectile tissue with blood.

Erythema (er′-e-THĒ-mah) Skin redness usually caused by engorgement of the capillaries in the deeper layers of the skin.

Erythrocyte (e-RITH-rō-sīt) Red blood cell.

Erythropoiesis (e-rith′-rō-poy-Ē-sis) The process by which erythrocytes (red blood cells) are formed.

Erythropoietin (eh-rith′-rō-POY-eh-tin) A hormone released by the kidneys that stimulates erythrocyte (red blood cell) production.

Esophagus (e-SOF-a-gus) A hollow muscular tube connecting the pharynx and the stomach.

Essential amino acids Those 10 amino acids that cannot be synthesized by the human body at an adequate rate to meet its needs and therefore must be obtained from the diet.

Estrogens (ES-tro-jens) Female sex hormones produced by the ovaries concerned with the development and maintenance of female reproductive structures and secondary sex characteristics, fluid and electrolyte balance, and protein anabolism. Examples are estradiol, estrone, and estriol.

Etiology (ē′-tē-OL-ō-jē) The study of the causes of disease, including theories of the origin and organisms (if any) involved.

Euphoria (yoo-FOR-ē-a) A subjectively pleasant feeling of well-being marked by confidence and assurance.

Eupnea (yoop-NĒ-a) Normal quiet breathing.

Eversion (ē-VER-zhun) The movement of the sole laterally at the ankle joint or of an atrioventricular valve into an atrium during ventricular contraction.

Excitability (ek-sīt′-a-BIL-i-tē) The ability of muscle tissue to receive and respond to stimuli; the ability of nerve cells to respond to stimuli and convert them into nerve impulses.

Excrement (EKS-kreh-ment) Material eliminated from the body as waste, especially fecal matter.

Excretion (eks-KRĒ-shun) The process of eliminating waste products from the body; also the products excreted.

Exhalation Breathing out; expelling air from the lungs into the atmosphere. Also called **expiration** (ek′-spi-RĀ-shun).

Exocrine (EK-sō-krin) **gland** A gland that secretes substances via ducts into body cavities, the lining of an organ, or directly onto a free surface.

Exocytosis (ex′-ō-sī-TŌ-sis) A process of discharging large particles through the plasma membrane. The particles are enclosed in vesicles, that fuse with the plasma membrane, expelling the particles from the cell.

Exogenous (ek-SOJ-e-nus) Originating outside an organ or part.

Expiratory (ek-SPĪ-ra-tōr′-ē) **reserve volume** The volume of air in excess of tidal volume that can be exhaled forcibly; about 1200 mL.

Extensibility (ek-sten′-si-BIL-i-tē) The ability of muscle tissue to be stretched when pulled.

Extension (ek-STEN-shun) An increase in the angle between two bones; restoring a body part to its anatomical position after flexion.

External Located on or near the surface.

External auditory (AW-di-tōr′-ē) **canal** or **meatus** (mē-Ā-tus) A curved tube in the temporal bone that leads to the middle ear.

External ear The outer ear, consisting of the pinna, external auditory canal, and tympanic membrane (eardrum).

External nares (NA-rēz) The external nostrils, or the openings into the nasal cavity on the exterior of the body.

External respiration The exchange of respiratory gases between the lungs and blood. Also called **pulmonary respiration.**

Exteroceptor (eks′-ter-ō-SEP-tor) A receptor adapted for the reception of stimuli from outside the body.

Extracellular fluid (ECF) Fluid outside body cells, such as interstitial fluid and plasma.

Extrinsic (eks-TRIN-sik) Of external origin.

Exudate (EKS-yoo-dāt) Escaping fluid or semifluid material that oozes from a space and that may contain serum, pus, and cellular debris.

Eyebrow The hairy ridge superior to the eye.

F

Face The anterior aspect of the head.

Facilitated diffusion (fa-SIL-i-tā-ted di′-FYOO-zhun) Diffusion in which a substance not soluble by itself in lipids diffuses across a selectively permeable membrane with the help of a transporter (carrier) protein.

Fascia (FASH-ē-a) A fibrous membrane covering, supporting, and separating muscles.

Fascicle (FAS-i-kul) A small bundle or cluster, especially of nerve or muscle fibers (cells). Also called a **fasciculus** (fa-SIK-yoo-lus). Plural is **fasciculi** (fa-SIK-yoo-lī).

Fasciculation (fa-sik′-yoo-LĀ-shun) Abnormal, spontaneous twitch of all skeletal muscle fibers in one motor unit that is visible at the skin surface; not associated with movement of the affected muscle; present in progressive diseases of motor neurons, for example, poliomyelitis and amyotrophic lateral sclerosis (ALS).

Fauces (FAW-sēz) The opening from the mouth into the pharynx.

Feces (FĒ-sēz) Material discharged from the rectum and made up of bacteria, excretions, and food residue. Also called **stool.**

Feedback system A sequence of events in which information about the status of a situation is continually reported (fed back) to a control center.

Female reproductive cycle General term for the ovarian and menstrual cycles, the hormonal changes that accompany them, and cyclic changes in the breasts and cervix; includes changes in the endometrium of a nonpregnant female that prepares the lining of the uterus to receive a fertilized ovum.

Fertilization (fer′-ti-li-ZĀ-shun) Penetration of a secondary oocyte by a sperm cell, meiotic division of secondary oocyte to form an ovum, and subsequent union of the nuclei of the gametes.

Fetal (FĒ-tal) **alcohol syndrome (FAS)** Term applied to the effects of intrauterine exposure to alcohol, such as slow growth, defective organs, and mental retardation.

Fetal circulation The cardiovascular system of the fetus, including the placenta and special blood vessels involved in the exchange of materials between fetus and mother.

Fetus (FĒ-tus) In humans, the developing organism in utero from the beginning of the third month to birth.

Fever An elevation in body temperature above the normal temperature of 37°C (98.6°F) due to a resetting of the hypothalamic thermostat.

Fibrillation (fi-bri-LĀ-shun) Abnormal, spontaneous twitch of a single skeletal muscle fiber (cell) that can be detected with electromyography but is not visible at the skin surface; not associated with movement of the affected muscle; present in certain disorders of motor neurons, for example, amyotrophic lateral sclerosis (ALS). With reference to cardiac muscle, see **atrial fibrillation** and **ventricular fibrillation.**

Fibrin (FĪ-brin) An insoluble protein that is essential to blood clotting; formed from fibrinogen by the action of thrombin.

Fibrinogen (fī-BRIN-ō-jen) A clotting factor in blood plasma that by the action of thrombin is converted to fibrin.

Fibrinolysis (fī′-bri-NOL-i-sis) Dissolution of a blood clot by the action of a proteolytic enzyme, such as plasmin (fibrinolysin), that dissolves fibrin threads and inactivates fibrinogen and other blood clotting factors.

Fibroblast (FĪ-brō-blast) A large, flat cell that secretes most of the extracellular matrix material of areolar and dense connective tissues.

Fibrous (FĪ-brus) **joint** A joint that allows little or no movement, such as a suture or a syndesmosis.

Fibrous tunic (TOO-nik) The superficial coat of the eyeball, made up of the posterior sclera and the anterior cornea.

Fight-or-flight response The effects produced upon activation of the sympathetic division of the autonomic nervous system.

Filtrate (fil-TRĀT) The fluid produced when blood is filtered by the filtration membrane.

Filtration (fil-TRĀ-shun) The flow of a liquid through a filter (or membrane that acts like a filter) due to blood pressure, as occurs in capillaries.

Filtration membrane Site of blood filtration in nephrons of the kidneys, consisting of the endothelium and basement membrane of the glomerulus and the epithelium of the visceral layer of the glomerular (Bowman's) capsule.

Fissure (FISH-ur) A groove, fold, or slit that may be normal or abnormal.

Fixed macrophage (MAK-rō-fāj) Stationary phagocytic cell found in the liver, lungs, brain, spleen, lymph nodes, subcutaneous tissue, and red bone marrow. Also called a **histiocyte** (HIS-tē-ō-sīt′).

Flaccid (FLAS-id) Relaxed, flabby, or soft; lacking muscle tone.

Flagellum (fla-JEL-um) A hairlike, motile process on the extremity of a bacterium, protozoan, or sperm cell. Plural is **flagella** (fla-JEL-a).

Flatus (FLĀ-tus) Gas in the stomach or intestines; commonly used to denote expulsion of gas through the rectum.

Flexion (FLEK-shun) Movement in which there is a decrease in the angle between two bones.

Flexor reflex A protective reflex in which flexor muscles are stimulated while extensor muscles are inhibited.

Follicle (FOL-i-kul) A small secretory sac or cavity;

the group of cells that contains a developing oocyte in the ovaries.

Follicle-stimulating hormone (FSH) Hormone secreted by the anterior pituitary that initiates development of ova and stimulates the ovaries to secrete estrogens in females, and initiates sperm production in males.

Fontanel (fon'-ta-NEL) A membrane-covered spot where bone formation is not yet complete, especially between the cranial bones of an infant's skull.

Foot The terminal part of the lower limb, from the ankle to the toes.

Foramen (fō-RĀ-men) A passage or opening; a communication between two cavities of an organ, or a hole in a bone for passage of vessels or nerves. Plural is **foramina** (fō-RAM-i-na).

Foramen ovale (fō-RĀ-men ō-VAL-ē) An opening in the fetal heart in the septum between the right and left atria. A hole in the greater wing of the sphenoid bone that transmits the mandibular branch of the trigeminal (V) nerve.

Forearm (FOR-arm) The part of the upper limb between the elbow and the wrist.

Fossa (FOS-a) A furrow or shallow depression.

Fourth ventricle (VEN-tri-kul) A cavity filled with cerebrospinal fluid within the brain lying between the cerebellum and the medulla oblongata and pons.

Fovea (FŌ-vē-a) A cuplike depression in the center of the macula lutea of the retina, containing cones only; the area of clearest vision.

Frontal plane A plane at a right angle to a midsagittal plane that divides the body or organs into anterior and posterior portions. Also called a **coronal** (kō-RŌN-al) **plane.**

Functional residual (re-ZID-yoo-al) **capacity** The sum of residual volume plus expiratory reserve volume; about 2400 mL.

Fundus (FUN-dus) The part of a hollow organ farthest from the opening.

Fungiform papilla (FUN-ji-form pa-PIL-a) A mushroomlike elevation on the upper surface of the tongue appearing as a red dot; most contain taste buds.

G

Gallbladder A small pouch, located inferior to the liver, that stores bile and empties by means of the cystic duct.

Gallstone A solid mass, usually containing cholesterol, in the gallbladder or a bile-containing duct; formed anywhere between bile canaliculi in the liver and the hepatopancreatic ampulla (ampulla of Vater), where bile enters the duodenum. Also called a **biliary calculus.**

Gamete (GAM-ēt) A male or female reproductive cell; a sperm cell or secondary oocyte.

Gamete intrafallopian transfer (GIFT) A procedure in which aspirated secondary oocytes are combined with a solution containing sperm outside the body, and the mixture is then immediately inserted into the uterine (Fallopian) tubes.

Ganglion (GANG-glē-on) Usually, a group of nerve cell bodies lying outside the central nervous system (CNS); also used for one group of nerve cell bodies within the CNS—the basal ganglia. Plural is **ganglia** (GANG-glē-a).

Gastric (GAS-trik) **glands** Glands in the mucosa of the stomach composed of cells that empty their secretions into narrow channels called gastric pits. Types of cells are chief cells (secrete pepsinogen), parietal cells (secrete hydrochloric acid and intrinsic factor), mucous surface and mucous neck cells (secrete mucus), and G cells (secrete gastrin).

Gastroenterology (gas'-trō-en'-ter-OL-ō-jē) The medical specialty that deals with the structure, function, diagnosis, and treatment of diseases of the stomach and intestines.

Gastrointestinal (gas'-trō-in-TES-ti-nal) **(GI) tract** A continuous tube running through the ventral body cavity extending from the mouth to the anus. Also called the **alimentary** (al'-i-MEN-tar-ē) **canal.**

Gene (jēn) Biological unit of heredity; a segment of DNA located in a definite position on a particular chromosome; a sequence of DNA that codes for a particular mRNA, rRNA, or tRNA.

Generator potential The graded depolarization that results in a change in the resting membrane potential in a receptor (specialized neuronal ending); may trigger a nerve action potential (nerve impulse) if depolarization reaches threshold.

Genetic engineering The manufacture and manipulation of genetic material.

Genetics The study of genes and heredity.

Genital herpes (JEN-i-tal HER-pēz) A sexually transmitted disease caused by type 2 herpes simplex virus.

Genitalia (jen'-i-TĀL-ē-a) Reproductive organs.

Genome (JĒ-nōm) The complete set of genes of an organism.

Genotype (JĒ-nō-tīp) The total hereditary information carried by an individual; the genetic makeup of an organism.

Geriatrics (jer'-ē-AT-riks) The branch of medicine devoted to the medical problems and care of elderly persons.

Gestation (jes-TĀ-shun) The period of development from fertilization to birth.

Gingivae (JIN-ji-vē) Gums. They cover the alveolar processes of the mandible and maxilla and extend slightly into each socket.

Gland Specialized epithelial cell or cells that secrete substances.

Glans penis (glanz PĒ-nis) The slightly enlarged region at the distal end of the penis.

Glaucoma (glaw-KŌ-ma) An eye disorder in which there is increased intraocular pressure due to an excess of aqueous humor.

Gliding joint A synovial joint having articulating surfaces that are usually flat, permitting only side-to-side and back-and-forth movements, as between carpal bones, tarsal bones, and the scapula and clavicle. Also called an **arthrodial** (ar-THRŌ-dē-al) **joint.**

Glomerular (glō-MER-yoo-lar) **capsule** A double-walled globe at the proximal end of a nephron that encloses the glomerular capillaries. Also called **Bowman's** (BŌ-manz) **capsule.**

Glomerular filtrate (glō-MER-yoo-lar FIL-trāt) The fluid produced when blood is filtered by the filtration membrane in the glomeruli of the kidneys.

Glomerular filtration The first step in urine formation in which substances in blood are filtered at the filtration membrane, and the filtrate enters the proximal convoluted tubule of a nephron.

Glomerular filtration rate (GFR) The total volume of fluid that enters all the glomerular (Bowman's) capsules of the kidneys in 1 min; about 125 mL/min.

Glomerulus (glō-MER-yoo-lus) A rounded mass of nerves or blood vessels, especially the microscopic tuft of capillaries that is surrounded by the glomerular (Bowman's) capsule of each kidney tubule. Plural is **glomeruli.**

Glottis (GLOT-is) The vocal folds (true vocal cords) in the larynx plus the space between them (rima glottidis).

Glucagon (GLOO-ka-gon) A hormone, produced by the alpha cells of the pancreatic islets (islets of Langerhans), that increases blood glucose level.

Glucocorticoids (gloo'-kō-KOR-ti-koyds) Hormones secreted by the cortex of the adrenal gland, especially cortisol, that influence glucose metabolism.

Gluconeogenesis (gloo'-kō-nē'-ō-JEN-e-sis) The synthesis of glucose from glycerol, certain amino acids, or lactic acid.

Glucose (GLOO-kōs) A six-carbon sugar, $C_6H_{12}O_6$; the major energy source for the production of ATP by body cells.

Glucosuria (gloo'-kō-SOO-rē-a) The presence of glucose in the urine; may be temporary or pathological. Also called **glycosuria.**

Glycogen (GLĪ-kō-jen) A highly branched polymer of glucose containing thousands of subunits; functions as a compact store of glucose molecules in liver and muscle fibers (cells).

Glycogenesis (glī'-kō-JEN-e-sis) The process by which many molecules of glucose combine to form a molecule called glycogen.

Glycogenolysis (glī'-kō-je-NOL-i-sis) The breakdown of glycogen into glucose.

Glycolysis (glī-KOL-i-sis) Series of chemical reactions in the cytosol of a cell in which a molecule of glucose is split into two molecules of pyruvic acid with production of two molecules of ATP.

Goblet cell A goblet-shaped single cell gland that secretes mucus; present in epithelium of airways and intestines.

Goiter (GOY-ter) An enlargement of the thyroid gland.

Golgi (GOL-jē) **complex** An organelle in the cytoplasm of cells consisting of four to six flattened sacs (cisterns), stacked on one another, with expanded areas at their ends; functions in processing, sorting, packaging, and delivering proteins and lipids to the plasma membrane, lysosomes, and secretory vesicles.

Gomphosis (gom-FŌ-sis) A fibrous joint in which a cone-shaped peg fits into a socket.

Gonad (GŌ-nad) A gland that produces gametes and hormones; the ovary in the female and the testis in the male.

Gonadotropic hormone Anterior pituitary hormone affecting the gonads.

Gray matter Area in the central nervous system and ganglia consisting of nonmyelinated nerve tissue.

Greater omentum (ō-MEN-tum) A large fold in

the serosa of the stomach that hangs down like an apron anterior to the intestines.

Greater vestibular (ves-TIB-yoo-lar) **glands** A pair of glands on either side of the vaginal orifice that open by a duct into the space between the hymen and the labia minora. Also called **Bartholin's** (BAR-to-linz) **glands.**

Groin (GROYN) The depression between the thigh and the trunk; the inguinal region.

Gross anatomy The branch of anatomy that deals with structures that can be studied without using a microscope. Also called **macroscopic anatomy.**

Growth An increase in size due to an increase in (1) the number of cells, (2) the size of existing cells as internal components increase in size, or (3) the size of intercellular substances.

Gustatory (GUS-ta-tō′-rē) Pertaining to taste.

Gynecology (gī′-ne-KOL-ō-jē) The branch of medicine dealing with the study and treatment of disorders of the female reproductive system.

Gynecomastia (gī′-ne-kō-MAS-tē-a) Excessive growth (benign) of the male mammary glands due to secretion of sufficient estrogens by an adrenal gland tumor (feminizing adenoma).

Gyrus (JĪ-rus) One of the folds of the cerebral cortex of the brain. Plural is **gyri** (JĪ-rī). Also called a **convolution.**

H

Hair A threadlike structure, produced by hair follicles, that develops in the dermis. Also called **pilus** (PĪ-lus).

Hair follicle (FOL-li-kul) Structure, composed of epithelium and surrounding the root of a hair, from which hair develops.

Hair root plexus (PLEK-sus) A network of dendrites arranged around the root of a hair as free or naked nerve endings that are stimulated when a hair shaft is moved.

Hand The terminal portion of an upper limb, including the carpus, metacarpus, and phalanges.

Haploid (HAP-loyd) Having half the number of chromosomes characteristically found in the somatic cells of an organism; characteristic of mature gametes. Symbolized *n*.

Hard palate (PAL-at) The anterior portion of the roof of the mouth, formed by the maxillae and palatine bones and lined by mucous membrane.

Haustra (HAWS-tra) The sacculated elevations of the colon. Singular is **haustrum.**

Head The superior part of a human, cephalic to the neck. Also the superior or proximal part of a structure.

Heart A hollow muscular organ, lying slightly to the left of the midline of the chest, that pumps the blood through the cardiovascular system.

Heart block An arrhythmia (dysrhythmia) of the heart in which the atria and ventricles contract independently because of a blocking of electrical impulses through the heart at some point in the conduction system.

Heart murmur (MER-mer) An abnormal sound that consists of a flow noise that is heard before, between, or after the normal heart sounds, or that may mask normal heart sounds.

Heat exhaustion Condition characterized by cool, clammy skin; profuse perspiration; and fluid and electrolyte (especially salt) loss that results in

muscle cramps, dizziness, vomiting, and fainting. Also called *heat prostration*.

Heat stroke Condition produced when the body cannot easily lose heat and characterized by reduced perspiration and elevated body temperature. Also called *sunstroke*.

Hematocrit (hē-MAT-ō-krit) **(Hct)** The percentage of blood made up of red blood cells. Usually calculated by centrifuging a blood sample in a graduated tube, then reading the volume of red blood cells and dividing it by the total volume of blood in the sample.

Hematology (hē′-ma-TOL-ō-jē) The study of blood.

Hematoma (hē′-ma-TŌ-ma) A tumor or swelling filled with blood.

Hematuria (hē′-ma-TOOR-ē-a) Blood in the urine.

Hemiplegia (hem′-i-PLĒ-ja) Paralysis of the upper limb, trunk, and lower limb on one side of the body.

Hemoglobin (hē′-mō-GLŌ-bin) **(Hb)** A substance in red blood cells consisting of the protein globin and the iron-containing red pigment heme that transports oxygen and some carbon dioxide.

Hemolysis (hē-MOL-i-sis) The escape of hemoglobin from the interior of a red blood cell; results from disruption of the cell membrane by toxins or drugs, freezing or thawing, or hypotonic solutions.

Hemolytic disease of the newborn A hemolytic anemia of a newborn child that results from the destruction of the infant's erythrocytes (red blood cells) by antibodies produced by the mother; usually the antibodies are due to an Rh blood type incompatibility. Also called **erythroblastosis fetalis** (eh-rith′-rō-blas-TŌ-sis fe-TAL-is).

Hemophilia (hē′-mō-FĒL-ē-a) A hereditary blood disorder in which there is a deficient production of certain factors involved in blood clotting, resulting in excessive bleeding into joints, deep tissues, and elsewhere.

Hemopoiesis (hē′-mō-poy-Ē-sis) Blood cell production occurring in red bone marrow. Also called **hematopoiesis** (hē′-ma-tō-poy-Ē-sis).

Hemorrhage (HEM-o-rij) Bleeding; the escape of blood from blood vessels, especially when the loss is profuse.

Hemorrhoids (HEM-o-royds) Dilated or varicosed blood vessels (usually veins) in the anal region. Also called **piles.**

Hemostasis (hē′-mō-STĀ-sis) The stoppage of bleeding.

Heparin (HEP-a-rin) An anticoagulant given to slow the conversion of prothrombin to thrombin, thus reducing the risk of blood clot formation; also found naturally in basophils and most cells.

Hepatic (he-PAT-ik) Refers to the liver.

Hepatic duct A duct that receives bile from the bile capillaries. Small hepatic ducts merge to form the larger right and left hepatic ducts that unite to leave the liver as the common hepatic duct.

Hepatic portal circulation The flow of blood from the gastrointestinal organs to the liver before returning to the heart.

Hepatitis (hep′-a-TĪ-tis) Inflammation of the liver due to a virus, drugs, and chemicals.

Hepatocyte (he-PAT-ō-cyte) A liver cell.

Hepatopancreatic (hep′-a-tō-pan′-krē-AT-ik) **ampulla** A small, raised area in the duodenum where the combined common bile duct and main pancreatic duct empty into the duodenum. Also called the **ampulla of Vater** (VAH-ter).

Hernia (HER-nē-a) The protrusion or projection of an organ or part of an organ through a membrane or cavity wall, usually the abdominal cavity.

Herniated (her′-nē-ĀT-ed) **disc** A rupture of an intervertebral disk so that the nucleus pulposus protrudes into the vertebral cavity. Also called a **slipped disc.**

Heterozygous (het′-er-ō-ZĪ-gus) Possessing different alleles on homologous chromosomes for a particular hereditary trait.

Hiatus (hī-Ā-tus) An opening; a foramen.

Hilus (HĪ-lus) An area, depression, or pit where blood vessels and nerves enter or leave an organ. Also called a **hilum.**

Hinge joint A synovial joint in which a convex surface of one bone fits into a concave surface of another bone, such as the elbow, knee, ankle, and interphalangeal joints.

Hirsutism (HER-soot-izm) An excessive growth of hair in females and children, with a distribution similar to that in adult males, due to the conversion of vellus hairs into large terminal hairs in response to higher-than-normal levels of androgens.

Histamine (HISS-ta-mēn) Substance found in many cells, especially mast cells, basophils, and platelets, released when the cells are injured; results in vasodilation, increased permeability of blood vessels, and constriction of bronchioles.

Histology (hiss-TOL-ō-jē) Microscopic study of the structure of tissues.

Hives (HĪVZ) Condition of the skin marked by reddened elevated patches that are often itchy; may be caused by infections, trauma, emotional stress, or hypersensitivity to drugs or foods.

Homeostasis (hō′-mē-ō-STĀ-sis) The condition in which the body's internal environment remains relatively constant, within physiological limits.

Homologous chromosomes Two chromosomes that belong to a pair. Also called **homologues.**

Homozygous (hō′-mō-ZĪ-gus) Possessing the same alleles on homologous chromosomes for a particular hereditary characteristic.

Hormone (HOR-mōn) A secretion of endocrine cells that alters the physiological activity of target cells of the body.

Horn An area of gray matter (anterior, lateral, or posterior) in the spinal cord.

Human chorionic gonadotropin (hCG) (kōr′-ē-ON-ik gō-nad′-ō-TRŌ-pin) A hormone produced by the developing placenta that maintains the corpus luteum.

Human growth hormone (hGH) Hormone secreted by the anterior pituitary that stimulates release of insulinlike growth factors, which in turn stimulate growth of body tissues. Also known as **somatotropin** and **somatotropic hormone (STH).**

Hyaluronic (hī′-a-loo-RON-ik) **acid** A viscous, amorphous extracellular material that binds cells together, lubricates joints, and maintains the shape of the eyeballs.

Hymen (HĪ-men) A thin fold of vascularized mucous membrane at the vaginal orifice.

Hypercapnia (hī′-per-KAP-nē-a) An abnormal increase in the amount of carbon dioxide in the blood.

Hyperextension (hī′-per-ek-STEN-shun) Continuation of extension beyond the anatomical position, as in bending the head backward.

Hyperglycemia (hī′-per-glī-SĒ-mē-a) An elevated blood glucose level.

Hypermetropia (hī′-per-mē-TRŌ-pē-a) A condition in which visual images are focused behind the retina, with resulting defective vision of near objects; farsightedness.

Hyperplasia (hī′-per-PLĀ-zē-a) An abnormal increase in the number of normal cells in a tissue or organ, increasing its size.

Hyperpolarization (hī′-per-PŌ-lar-i-ZĀ-shun) Increase in the internal negativity across a cell membrane, thus increasing the voltage and moving it farther away from the threshold value.

Hypersecretion (hī′-per-sē-KRĒ-shun) Overactivity of glands resulting in excessive secretion.

Hypersensitivity (hī′-per-sen′-si-TIV-i-tē) Overreaction to an allergen that results in pathological changes in tissues. Also called **allergy.**

Hypertension (hī′-per-TEN-shun) High blood pressure.

Hyperthermia (hī′-per-THERM-ē-a) An elevated body temperature.

Hypertonia (hī-per-TŌ-nē-a) Increased muscle tone that is expressed as spasticity or rigidity.

Hypertonic (hī′-per-TON-ik) Solution that causes cells to shrink due to loss of water by osmosis.

Hypertrophy (hī-PER-trō-fē) An excessive enlargement or overgrowth of tissue without cell division.

Hyperventilation (hī′-per-ven′-ti-LĀ-shun) A rate of respiration higher than that required to maintain a normal level of plasma P_{CO_2}.

Hypoglycemia (hī′-pō-glī-SĒ-mē-a) An abnormally low concentration of glucose in the blood; can result from excess insulin (injected or secreted).

Hypophysis (hī-POF-i-sis) Pituitary gland.

Hyposecretion (hī′-pō-se-KRĒ-shun) Underactivity of glands resulting in diminished secretion.

Hypothalamus (hī′-pō-THAL-a-mus) A portion of the diencephalon, lying beneath the thalamus and forming the floor and part of the wall of the third ventricle.

Hypothermia (hī′-pō-THER-mē-a) Lowering of body temperature below 35°C (95°F); in surgical procedures, it refers to deliberate cooling of the body to slow down metabolism and reduce oxygen needs of tissues.

Hypotonia (hī′-pō-TŌ-nē-a) Decreased or lost muscle tone in which muscles appear flaccid.

Hypotonic (hī′-pō-TON-ik) Solution that causes cells to swell and perhaps rupture due to gain of water by osmosis.

Hypoventilation (hī′-pō-ven′-ti-LĀ-shun) A rate of respiration lower than that required to maintain a normal level of plasma P_{CO_2}.

Hypoxia (hī-PŎK-sē-a) Lack of adequate oxygen at the tissue level.

Hysterectomy (hiss′-ter-EK-tō-mē) The surgical removal of the uterus.

I

Ileocecal (il′-ē-ō-SĒ-kal) **sphincter** A fold of mucous membrane that guards the opening from the ileum into the large intestine. Also called the **ileocecal valve.**

Ileum (IL-ē-um) The terminal portion of the small intestine.

Immunity (i-MYOO-ni-tē) The state of being resistant to injury, particularly by poisons, foreign proteins, and invading pathogens.

Immunoglobulin (im′-yoo-nō-GLOB-yoo-lin) **(Ig)** An antibody synthesized by plasma cells, derived from B lymphocytes, in response to the introduction of antigen. Immunoglobulins are divided into five kinds (IgG, IgM, IgA, IgD, IgE) based primarily on the larger protein component present in the immunoglobulin.

Immunology (im′-yoo-NOL-ō-jē) The branch of science that deals with the responses of the body when challenged by antigens.

Implantation (im′-plan-TĀ-shun) The insertion of a tissue or a part into the body. The attachment of the blastocyst to the lining of the uterus 7 to 8 days after fertilization.

Incontinence (in-KON-ti-nens) Inability to retain urine, semen, or feces through loss of sphincter control.

Indirect motor pathways Motor tracts that convey information from the brain down the spinal cord for automatic movements, coordination of body movements with visual stimuli, skeletal muscle tone and posture, and balance. Also known as **extrapyramidal pathways.**

Infarction (in-FARK-shun) A localized area of necrotic tissue, produced by inadequate oxygenation of the tissue.

Infection (in-FEK-shun) Invasion and multiplication of microorganisms in body tissues, which may be inapparent or characterized by cellular injury.

Infectious mononucleosis (mon-ō-noo′-klē-Ō-sis) **(IM)** Contagious disease caused by the Epstein-Barr virus (EBV) and characterized by an elevated mononucleocyte and lymphocyte count, fever, sore throat, stiff neck, cough, and malaise.

Inferior (in-FĒR-ē-or) Away from the head or toward the lower part of a structure. Also called **caudad** (KAW-dad).

Inferior vena cava (VĒ-na KĀ-va) **(IVC)** Large vein that collects blood from parts of the body inferior to the heart and returns it to the right atrium.

Infertility Inability to conceive or to cause conception. Also called **sterility.**

Inflammation (in′-fla-MĀ-shun) Localized, protective response to tissue injury designed to destroy, dilute, or wall off the infecting agent or injured tissue; characterized by redness, pain, heat, swelling, and sometimes loss of function.

Inflation reflex Reflex that prevents overinflation of the lungs. Also called **Hering-Breuer reflex.**

Ingestion (in-JES-chun) The taking in of food, liquids, or drugs by mouth.

Inguinal (IN-gwi-nal) Pertaining to the groin.

Inhalation The act of drawing air into the lungs. Also termed **inspiration** (in′-spi-RĀ-shun).

Inheritance The acquisition of body characteristics and qualities by transmission of genetic information from parents to offspring.

Inhibin A hormone secreted by the gonads that inhibits release of follicle-stimulating hormone (FSH) by the anterior pituitary.

Inhibiting hormone Hormone secreted by the hypothalamus that can suppress secretion of hormones by the anterior pituitary.

Inner cell mass A region of cells of a blastocyst that differentiates into the three primary germ layers—ectoderm, mesoderm, and endoderm—from which all tissues and organs develop; also called an **embryoblast.**

Inorganic (in′-or-GAN-ik) **compound** Compound that usually lacks carbon, usually is small, and often contains ionic bonds. Examples include water and many acids, bases, and salts.

Insertion (in-SER-shun) The attachment of a muscle tendon to a movable bone or the end opposite the origin.

Inspiratory (in-SPĪ-ra-tor-ē) **capacity** Total inspiratory capacity of the lungs; the total of tidal volume plus inspiratory reserve volume; averages 3600 mL.

Inspiratory (in-SPĪ-ra-tor-ē) **reserve volume** Additional inspired air over and above tidal volume; averages 3100 mL.

Insula (IN-su-la) A triangular area of cerebral cortex that lies deep within the lateral cerebral fissue, under the parietal, frontal, and temporal lobes.

Insulin (IN-su-lin) A hormone produced by the beta cells of the pancreatic islets (islets of Langerhans) that decreases the blood glucose level by stimulating cellular uptake of glucose.

Insulinlike growth factor (IGF) Small protein, produced by the liver and other tissues in response to stimulation by human growth hormone (hGH), that mediates most of the effects of human growth hormone. Previously called **somatomedin** (sō′-ma-tō-MĒ-din).

Integumentary (in-teg′-yoo-MEN-tar-ē) Relating to the skin.

Intercalated (in-TER-ka-lāt′-ed) **disc** An irregular transverse thickening of sarcolemma that contains desmosomes that hold cardiac muscles fibers (cells) together and gap junctions that aid in conduction of muscle action potentials.

Intercostal (in′-ter-KOS-tal) **nerve** A nerve supplying a muscle located between the ribs.

Interferons (in′-ter-FĒR-ons) **(IFNs)** Three principal types of protein (alpha, beta, gamma) naturally produced by virus-infected host cells; induce uninfected host cells to synthesize antiviral proteins that inhibit intracellular viral replication.

Internal Away from the surface of the body.

Internal ear The inner ear or labyrinth, lying inside the temporal bone, containing the organs of hearing and balance.

Internal nares (NA-rēz) The two openings posterior to the nasal cavities leading into the nasopharynx. Also called the **choanae** (KŌ-a-nē).

Internal respiration The exchange of respiratory gases between blood and body cells. Also called **tissue respiration.**

Interneuron (in′-ter-NOO-ron) A nerve cell lying completely within the central nervous system. Also called an **association neuron.**

Interoceptor (in'-ter-ō-SEP-tor) Receptor located in blood vessels and viscera that provides information about the body's internal environment. Also termed **visceroceptor** (vis'-er-ō-SEP-tor).

Interphase (IN-ter-fāz) The period of the cell cycle between cell divisions, when the cell is engaged in growth, metabolism, and production of substances required for division; and during which chromosomes are replicated.

Interstitial (in'-ter-STISH-al) **fluid** The portion of extracellular fluid that fills the microscopic spaces between the cells of tissues; the internal environment of the body. Also called **intercellular** or **tissue fluid.**

Interstitial growth Growth from within, as in the growth of cartilage. Also called **endogenous** (en-DOJ-e-nus) **growth.**

Intervertebral (in'-ter-VER-te-bral) **disc** A pad of fibrocartilage located between the bodies of two vertebrae.

Intestinal gland A gland that opens onto the surface of the intestinal mucosa and secretes digestive enzymes. Also called a **crypt of Lieberkühn** (LĒ-ber-kyoon).

Intracellular (in'-tra-SEL-yoo-lar) **fluid (ICF)** Fluid located within cells. Also called **cytosol** (SĪ-tō-sol).

Intramembranous ossification (in'-tra-MEM-bra-nus os'-i-fi-KĀ-shun) The method of bone formation in which the bone is formed directly in membranous tissue.

Intraocular (in'-tra-OK-yoo-lar) **pressure (IOP)** Pressure in the eyeball, produced mainly by aqueous humor.

Intrapleural pressure Air pressure between the two pleural layers of the lungs, usually subatmospheric. Also called **intrathoracic pressure.**

Intrinsic (in-TRIN-sik) Of internal origin.

Intrinsic factor (IF) A glycoprotein, synthesized and secreted by the parietal cells of the gastric mucosa, that facilitates vitamin B₁₂ absorption in the small intestine.

In utero (YOO-ter-ō) Within the uterus.

In vitro (VĒ-trō) Literally, in glass; outside the living body and in an artificial environment such as a laboratory test tube.

In vivo (VĒ-vō) In the living body.

Ion (Ī-on) Any charged particle or group of particles; usually formed when a substance, such as a salt, dissolves and dissociates.

Ionization (ī'-on-i-ZĀ-shun) Separation of inorganic acids, bases, and salts into ions when dissolved in water. Also called **dissociation.**

Iris The colored portion of the vascular tunic of the eyeball seen through the cornea that contains circular and radial smooth muscle; the hole in the center of the iris is the pupil.

Ischemia (is-KĒ-mē-a) A lack of sufficient blood to a part due to obstruction or constriction of a blood vessel.

Isoantibody A specific antibody in blood plasma that reacts with specific isoantigens and causes the clumping of red blood cells. Also called an **agglutinin.**

Isoantigen A genetically determined antigen located on the surface of red blood cells; basis for the ABO grouping and Rh system of blood classification. Also called an **agglutinogen.**

Isometric contraction A muscle contraction in which tension on the muscle increases, but there is only minimal muscle shortening so that no movement is produced.

Isotonic (ī'-sō-TON-ik) Having equal tension or tone. A solution having the same concentration of impermeable solutes as cytosol.

Isotonic contraction Contraction in which the tension remains the same; occurs when a constant load is moved through the range of motions possible at a joint.

Isotopes (Ī-sō-tōps) Chemical elements that have the same number of protons but different numbers of neutrons.

Isthmus (IS-mus) A narrow strip of tissue or narrow passage connecting two larger parts.

J

Jaundice (JAWN-dis) A condition characterized by yellowness of skin, white of eyes, mucous membranes, and body fluids because of a buildup of bilirubin.

Jejunum (jē-JOO-num) The middle portion of the small intestine.

Joint kinesthetic (kin'-es-THET-ik) **receptor** A proprioceptive receptor located in a joint, stimulated by joint movement.

Juxtaglomerular (juks'-ta-glō-MER-yoo-lar) **apparatus (JGA)** Consists of the macula densa (cells of the distal convoluted tubule adjacent to the afferent and efferent arteriole) and juxtaglomerular cells (modified cells of the afferent and sometimes efferent arteriole); secretes renin when blood pressure starts to fall.

K

Keratin (KER-a-tin) An insoluble protein found in the hair, nails, and other keratinized tissues of the epidermis.

Keratinocyte (ke-RAT-in-ō-sīt) The most numerous of the epidermal cells; produces keratin.

Ketone (KĒ-tōn) **bodies** Substances produced primarily during excessive triglyceride catabolism, such as acetone, acetoacetic acid, and beta-hydroxybutyric acid.

Ketosis (kē-TŌ-sis) Abnormal condition marked by excessive production of ketone bodies.

Kidney (KID-nē) One of the paired reddish organs located in the lumbar region that regulates the composition, volume, and pressure of blood and produces urine.

Kidney stone A solid mass, usually consisting of calcium oxalate, uric acid, or calcium phosphate crystals, that may form in any portion of the urinary tract. Also called a **renal calculus.**

Kilocalorie (KIL-ō-kal'-ō-rē) **(kcal)** or **Calorie** The amount of heat required to raise the temperature of 1000 g of water 1°C from 14°C to 15°C; the unit used to express the energy content of foods and to measure metabolic rate.

Kinesiology (ki-nē'-sē-OL-ō-jē) The study of the movement of body parts.

Kinesthesia (kin'-es-THĒ-zē-a) Ability to perceive extent, direction, or weight of movement; muscle sense.

Korotkoff (korot-KOF) **sounds** The various sounds that are heard while taking blood pressure.

Krebs cycle A series of biochemical reactions that occurs in the matrix of mitochondria in which electrons are transferred to coenzymes and carbon dioxide is formed. The electrons carried by the coenzymes then enter the electron transport chain, which generates a large quantity of ATP. Also called the **citric acid cycle** or **tricarboxylic acid (TCA) cycle.**

Kyphosis (kī-FŌ-sis) An exaggeration of the thoracic curve of the vertebral column, resulting in a "round-shouldered" appearance. Also called **hunchback.**

L

Labia majora (LĀ-bē-a ma-JŌ-ra) Two longitudinal folds of skin extending downward and backward from the mons pubis of the female.

Labia minora (mi-NOR-a) Two small folds of mucous membrane lying medial to the labia majora of the female.

Labial frenulum (LĀ-bē-al FREN-yoo-lum) A medial fold of mucous membrane between the inner surface of the lip and the gums.

Labor The process by which a fetus is expelled from the uterus through the vagina.

Labyrinth (LAB-i-rinth) Intricate communicating passageway, especially in the internal ear.

Lacrimal canal (LAK-ri-mal) A duct, one on each eyelid, beginning at the punctum at the medial margin of an eyelid and conveying tears medially into the lacrimal sac.

Lacrimal gland Secretory cells, located at the superior anterolateral portion of each orbit, that secrete tears into excretory ducts that open onto the surface of the conjunctiva.

Lactation (lak-TĀ-shun) The secretion and ejection of milk by the mammary glands.

Lacteal (LAK-tē-al) One of many lymphatic vessels in villi of the intestines that absorb triglycerides and other lipids from digested food.

Lacuna (la-KOO-na) A small, hollow space, such as that found in bones in which the osteocytes lie. Plural is **lacunae** (la-KOO-nē).

Lamellae (la-MEL-ē) Concentric rings of hard, calcified matrix found in compact bone.

Lamellated corpuscle Oval-shaped pressure receptor located in subcutaneous tissue and consisting of concentric layers of connective tissue wrapped around a sensory nerve fiber. Also called a **Pacinian** (pa-SIN-ē-an) **corpuscle.**

Lamina propria (PRŌ-prē-a) The connective tissue layer of a mucous membrane.

Langerhans (LANG-er-hans) **cell** Epidermal dendritic cell that functions as an antigen-presenting cell (APC) during an immune response.

Lanugo (la-NOO-gō) Fine downy hairs that cover the fetus.

Large intestine The portion of the gastrointestinal tract extending from the ileum of the small intestine to the anus, divided structurally into the cecum, colon, rectum, and anal canal.

Laryngitis (lar'-in-JĪ-tis) Inflammation of the mucous membrane lining the larynx.

Laryngopharynx (lah-ring'-gō-FAR-inks) The inferior portion of the pharynx, extending downward from the level of the hyoid bone, that divides posteriorly into the esophagus and anteriorly into the larynx.

Larynx (LAR-inks) The voice box; a short passageway that connects the pharynx with the trachea.

Lateral (LAT-er-al) Farther from the midline of the body or a structure.

Lateral ventricle (VEN-tri-kul) A cavity within a cerebral hemisphere that communicates with the lateral ventricle in the other cerebral hemisphere and with the third ventricle by way of the interventricular foramen.

Leg The part of the lower limb between the knee and the ankle.

Lens A transparent organ constructed of proteins (crystallins), lying posterior to the pupil and iris of the eyeball and anterior to the vitreous body.

Lesion (LĒ-zhun) Any localized, abnormal change in a body tissue.

Lethargy (LETH-ar-jē) A condition of drowsiness or indifference.

Leukemia (loo-KĒ-mē-a) A malignant disease of the blood-forming tissues characterized by either uncontrolled production and accumulation of immature leukocytes in which many cells fail to reach maturity (acute), or an accumulation of mature leukocytes in the blood because they do not die at the end of their normal life span (chronic).

Leukocyte (LOO-kō-sīt) A white blood cell.

Leukocytosis (loo'-kō-sī-TŌ-sis) An increase in the number of white blood cells, characteristic of many infections and other disorders.

Leukopenia (loo'-kō-PĒ-nē-a) A decrease of the number of white blood cells below 5000 cells/μl.

Leydig (LĪ-dig) **cell** A type of cell that secretes testosterone; located in the connective tissue between seminiferous tubules in a mature testis. Also known as **interstitial cell of Leydig** or **interstitial endocrinocyte.**

Libido (li-BĒ-dō) Sexual desire.

Ligament (LIG-a-ment) Dense, regular, connective tissue that attaches bone to bone.

Ligand (LĪ-gand) A chemical substance that binds to a specific receptor.

Limbic system A portion of the forebrain, sometimes termed the visceral brain, concerned with various aspects of emotion and behavior.

Lipase An enzyme that splits fatty acids from triglycerides and phospholipids.

Lipid (LIP-id) An organic compound composed of carbon, hydrogen, and oxygen that is usually insoluble in water, but soluble in alcohol, ether, and chloroform; examples include triglycerides (fats and oils), phospholipids, steroids, and eicosanoids.

Lipid bilayer Arrangement of phospholipid, glycolipid, and cholesterol molecules in two parallel sheets in which the hydrophilic "heads" face outward and the hydrophobic "tails" face inward; found in cellular membranes.

Lipogenesis (lip'-ō-GEN-e-sis) The synthesis of triglycerides.

Lipolysis (li-POL-i-sis) The splitting of fatty acids from a triglyceride (fat) or phospholipid molecule.

Lipoprotein (lip'-ō-PRŌ-tēn) One of several types of particles containing lipids (cholesterol and triglycerides) and proteins that make it water soluble for transport in the blood; high levels of **low-density lipoproteins (LDLs)** are associated with increased risk of atherosclerosis, whereas high levels of **high-density lipoproteins (HDLs)** are associated with decreased risk of atherosclerosis.

Lithotripsy (LITH-ō-trip'-sē) A noninvasive procedure in which shock waves generated by a lithotriptor are used to pulverize kidney stones or gallstones.

Liver Large organ under the diaphragm that occupies most of the right hypochondriac region and part of the epigastric region. Functionally, it produces bile and synthesizes most plasma proteins; converts one nutrient into another; detoxifies substances; stores glycogen, minerals, and vitamins; carries on phagocytosis of worn-out blood cells and bacteria; and helps synthesize the active form of vitamin D.

Lobe (LŌB) A curved or rounded projection.

Lordosis (lor-DŌ-sis) An exaggeration of the lumbar curve of the vertebral column. Also called **swayback.**

Lower limb The appendage attached at the pelvic (hip) girdle, consisting of the thigh, knee, leg, ankle, foot, and toes. Also called **lower extremity.**

Lumbar (LUM-bar) Region of the back and side between the ribs and pelvis; loin.

Lumbar plexus (PLEK-sus) A network formed by the anterior (ventral) branches of spinal nerves L1 through L4.

Lumen (LOO-men) The space within an artery, vein, intestine, renal tubule, or other tubelike structure.

Lungs Main organs of respiration that lie on either side of the heart in the thoracic cavity.

Lunula (LOO-nyoo-la) The moon-shaped white area at the base of a nail.

Luteinizing (LOO-tē-i-nī'-zing) **hormone (LH)** A hormone secreted by the anterior pituitary that stimulates ovulation and progesterone secretion by the corpus luteum, and readies the mammary glands for milk secretion in females; stimulates testosterone secretion by the testes in males.

Lyme (LĪMZ) **disease** Disease caused by a bacterium (*Borrelia burgdorferi*) and transmitted to humans by ticks (mainly deer ticks); may be characterized by a bull's-eye rash. Symptoms include joint stiffness, fever and chills, headache, stiff neck, nausea, and low back pain. Later stages may involve cardiac and neurologic problems and arthritis.

Lymph (LIMF) Fluid confined in lymphatic vessels and flowing through the lymphatic system until it is returned to the blood.

Lymph node An oval or bean-shaped structure located along lymphatic vessels.

Lymphatic (lim-FAT-ik) **capillary** Closed-ended, microscopic lymphatic vessel that begins in spaces between cells and converges with other lymphatic capillaries to form lymphatic vessels.

Lymphatic tissue A specialized form of reticular tissue that contains large numbers of lymphocytes.

Lymphatic vessel A large vessel that collects lymph from lymphatic capillaries and converges with other lymphatic vessels to form the thoracic and right lymphatic ducts.

Lymphocyte (LIM-fō-sīt) A type of white blood cell, found in lymph nodes, associated with the immune system.

Lysosome (LĪ-sō-sōm) An organelle in the cytoplasm of a cell, enclosed by a single membrane and containing powerful digestive enzymes.

Lysozyme (LĪ-sō-zīm) A bactericidal enzyme found in tears, saliva, and perspiration.

M

Macrophage (MAK-rō-fāj) Phagocytic cell derived from a monocyte; may be fixed or wandering.

Macula (MAK-yoo-la) A discolored spot or a colored area. A small, thickened region on the wall of the utricle and saccule that contains receptors for static equilibrium.

Macula lutea (LOO-tē-a) The yellow spot in the center of the retina.

Major histocompatibility (MHC) antigens Surface proteins on white blood cells and other nucleated cells that are unique for each person (except for identical siblings) and are used to type tissues and help prevent rejection. Also known as **human lymphocyte antigens (HLA).**

Malignant (mah-LIG-nant) Referring to diseases that tend to become worse and cause death, especially the invasion and spreading of cancer.

Mammary (MAM-ar-ē) **gland** Modified sudoriferous (sweat) gland of the female that produces milk for the nourishment of the young.

Marrow (MAR-ō) Soft, spongelike material in the cavities of bone. Red bone marrow produces blood cells; yellow bone marrow, formed mainly of adipose tissue, has no blood-producing function.

Mass number The total number of protons and neutrons in an atom of a chemical element.

Mast cell A cell found in areolar connective tissue along blood vessels that produces histamine, a dilator of small blood vessels during inflammation.

Mastication (mas'-ti-KĀ-shun) Chewing.

Matrix (MĀ-triks) The ground substance and fibers between cells in a connective tissue.

Matter Anything that occupies space and has mass.

Mature follicle A relatively large, fluid-filled follicle containing a secondary oocyte and surrounding granulosa cells that secrete estrogens. Also called a **graafian** (GRAF-ē-an) **follicle.**

Meatus (mē-Ā-tus) A passage or opening, especially the external portion of a canal.

Mechanoreceptor (mek'-a-nō-rē-SEP-tor) Receptor that detects mechanical deformation of the receptor itself or adjacent cells; stimuli so detected include those related to touch, pressure, vibration, proprioception, hearing, equilibrium, and blood pressure.

Medial (MĒ-dē-al) Nearer the midline of the body or a structure.

Mediastinum (mē'-dē-as-TĪ-num) The broad, median partition between the pleurae of the lungs that extends from the sternum to the vertebral column in the thoracic cavity.

Medulla (me-DUL-a) An inner layer of an organ, such as the medulla of the kidneys.

Medulla oblongata (ob'-long-GAH-ta) The most inferior part of the brain stem.

Medullary (MED-yoo-lar'-ē) **cavity** The space within the diaphysis of a bone that contains yellow bone marrow. Also called the **marrow cavity.**

Medullary rhythmicity (rith-MIS-i-tē) **area** Portion of the respiratory center in the medulla oblongata that controls the basic rhythm of respiration.

Meiosis (mī-Ō-sis) A type of cell division that produces gametes; two successive nuclear divisions result in daughter cells with the haploid (*n*) number of chromosomes.

Melanin (MEL-a-nin) A dark black, brown, or yellow pigment found in some parts of the body such as the skin and hair.

Melanocyte (MEL-a-nō-sīt) A pigmented cell, located between or beneath cells of the deepest layer of the epidermis, that synthesizes melanin.

Melatonin (mel′-a-TŌ-nin) A hormone, secreted by the pineal gland, that helps set the timing of the body's biological clock.

Membrane A thin, flexible sheet of tissue composed of an epithelial layer and an underlying connective tissue layer, as in an epithelial membrane, or of areolar connective tissue only, as in a synovial membrane.

Membranous labyrinth (MEM-bra-nus LAB-i-rinth) The portion of the inner ear that is located inside the bony labyrinth and separated from it by the perilymph; made up of the membranous semicircular canals, the saccule and utricle, and the cochlear duct.

Menarche (me-NAR-kē) The first menses (menstrual flow) and beginning of ovarian and uterine cycles.

Ménière's (men-YAIRZ) **disease** A type of labyrinthine disease characterized by fluctuating loss of hearing, vertigo, and tinnitus due to an increased amount of endolymph that enlarges the labyrinth.

Meninges (me-NIN-jēz) Three membranes covering the brain and spinal cord, called the dura mater, arachnoid, and pia mater. Singular is **meninx** (MĒ-ninks).

Meningitis (men′-in-JĪ-tis) Inflammation of the meninges, most commonly the pia mater and arachnoid.

Meniscus (men-IS-cus) Fibrocartilage pad between articular surfaces of bones of some synovial joints.

Menopause (MEN-ō-pawz) The termination of the menstrual cycles.

Menstruation (men′-stroo-Ā-shun) Periodic discharge of blood, tissue fluid, mucus, and epithelial cells that usually lasts for five days; caused by a sudden reduction in estrogens and progesterone. Also called the **menstrual phase** or **menses.**

Merkel (MER-kel) **cell** Type of cell in the epidermis of hairless skin that makes contact with a tactile (Merkel) disk, which functions in touch.

Mesenchyme (MEZ-en-kīm) An embryonic connective tissue from which all other connective tissues arise.

Mesentery (MEZ-en-ter′-ē) A fold of peritoneum attaching the small intestine to the posterior abdominal wall.

Mesoderm The middle primary germ layer that gives rise to connective tissues, blood and blood vessels, and muscles.

Metabolism (me-TAB-ō-lizm) All the biochemical reactions that occur within an organism, including the synthetic (anabolic) reactions and decomposition (catabolic) reactions.

Metaphase (MET-a-fāz) The second stage of mitosis, in which chromatid pairs line up on the metaphase plate (equatorial plane) of the cell.

Metaphysis (me-TAF-i-sis) Growing portion of a bone.

Metastasis (me-TAS-ta-sis) The spread of cancer to surrounding tissues (local) or to other body sites (distant).

Micelle (mī-SEL) A spherical aggregate of bile salts that dissolves fatty acids and monoglycerides so that they can be absorbed into intestinal epithelial cells.

Microglia (mī′-krō-GLĒ-a) Neuroglial cells that carry on phagocytosis.

Microvilli (mī′-krō-VIL-ē) Microscopic, fingerlike projections of the plasma membranes of cells that increase surface area for absorption, especially in the small intestine and proximal convoluted tubules of the kidneys.

Micturition (mik′-too-RISH-un) The act of expelling urine from the urinary bladder. Also called **urination** (yoo′-ri-NĀ-shun).

Midbrain The part of the brain between the pons and the diencephalon. Also called the **mesencephalon** (mez′-en-SEF-a-lon).

Middle ear A small, epithelium-lined cavity hollowed out of the temporal bone, separated from the external ear by the eardrum and from the internal ear by a thin bony partition containing the oval and round windows; extending across the middle ear are the three auditory ossicles. Also called the **tympanic** (tim-PAN-ik) **cavity.**

Midline An imaginary vertical line that divides the body into equal left and right sides.

Midsagittal plane A vertical plane through the midline of the body that divides the body or organs into equal right and left sides. Also called a **median plane.**

Milk ejection reflex Contraction of alveolar cells to force milk into ducts of mammary glands, stimulated by oxytocin (OT), which is released from the posterior pituitary in response to suckling action. Also called the **milk letdown reflex.**

Mineral Inorganic, homogeneous solid substance that may perform a function vital to life; examples include calcium and phosphorus.

Mineralocorticoids (min′-er-al-ō-KOR-ti-koyds) A group of hormones of the adrenal cortex that help regulate sodium and potassium balance.

Minute ventilation (MV) Total volume of air inhaled and exhaled per minute; about 6000 mL.

Mitochondrion (mī′-tō-KON-drē-on) A double-membraned organelle that plays a central role in the production of ATP; known as the "powerhouse" of the cell.

Mitosis (mī-TŌ-sis) The orderly division of the nucleus of a cell that ensures that each new daughter nucleus has the same number and kind of chromosomes as the original parent nucleus. The process includes the replication of chromosomes and the distribution of the two sets of chromosomes into two separate and equal nuclei.

Mitotic spindle Collective term for a football-shaped assembly of microtubules that is responsible for the movement of chromosomes during cell division.

Mitral stenosis (MĪ-tral ste-NŌ-sis) Narrowing of the mitral valve by scar formation or a congenital defect.

Mitral valve prolapse (PRŌ-laps) or **MVP** An inherited disorder in which a portion of a mitral valve is pushed back too far (prolapsed) during ventricular contraction due to expansion of the cusps and elongation of the chordae tendineae.

Modality (mō-DAL-i-tē) Any of the specific sensory entities, such as vision, smell, taste, or touch.

Mole The weight, in grams, of a standard amount (6.0225×10^{23} molecules) of a substance.

Molecule (MOL-eh-kyool) The chemical combination of two or more atoms covalently bonded together.

Monocyte (MON-ō-sīt) A type of white blood cell characterized by agranular cytoplasm; the largest of the leukocytes.

Monounsaturated fat A fatty acid that contains one double covalent bond between two of its carbon atoms; it is not completely saturated with hydrogen atoms. Plentiful in triglycerides of olive and peanut oils.

Mons pubis (MONZ PYOO-bis) The rounded, fatty prominence over the pubic symphysis, covered by coarse pubic hair.

Morula (MOR-yoo-la) A solid sphere of cells produced by successive cleavages of a fertilized ovum about four days after fertilization.

Motor area The region of the cerebral cortex that governs muscular movement, particularly the precentral gyrus of the frontal lobe.

Motor end plate Portion of the sarcolemma of a muscle fiber (cell) that receives neurotransmitter liberated by an axon terminal.

Motor neuron (NOOR-on) A neuron that conducts nerve impulses from the brain and spinal cord to effectors that may be either muscles or glands. Also called an **efferent neuron.**

Motor unit A motor neuron together with the muscle fibers (cells) it stimulates.

Mucosa-associated lymphatic tissue (MALT) Lymphatic nodules scattered throughout the lamina propria (connective tissue) of mucous membranes lining the gastrointestinal tract, respiratory airways, urinary tract, and reproductive tract.

Mucous (MYOO-kus) **cell** A unicellular gland that secretes mucus. Two types are mucous neck cells and mucous surface cells in the stomach.

Mucous membrane A membrane that lines a body cavity that opens to the exterior. Also called the **mucosa** (myoo-KŌ-sa).

Mucus The thick fluid secretion of goblet cells, mucous cells, mucous glands, and mucous membranes.

Mumps Inflammation and enlargement of the parotid glands, accompanied by fever and extreme pain during swallowing.

Muscle An organ composed of one of three types of muscle tissue (skeletal, cardiac, or smooth), specialized for contraction to produce voluntary or involuntary movement of parts of the body.

Muscle action potential A stimulating impulse

that propagates along a sarcolemma and then into transverse tubules; in skeletal muscle, it is generated by acetylcholine, which alters permeability of the sarcolemma to cations, especially sodium ions (Na^+).

Muscle fatigue (fa-TĒG) Inability of a muscle to maintain its strength of contraction or tension; may be related to insufficient oxygen, depletion of glycogen, and/or lactic acid buildup.

Muscle tissue A tissue specialized to produce motion in response to muscle action potentials by its qualities of contractility, extensibility, elasticity, and excitability; types include skeletal, cardiac, and smooth.

Muscle tone A sustained, partial contraction of portions of a skeletal or smooth muscle in response to activation of stretch receptors or a baseline level of action potentials in the innervating motor neurons.

Muscular dystrophies (DIS-trō-fēz) Inherited muscle-destroying diseases, characterized by degeneration of the individual muscle fibers (cells), which leads to progressive atrophy of the skeletal muscle.

Muscularis (MUS-kyoo-la′-ris) A muscular layer (coat or tunic) of an organ.

Muscularis mucosae (myoo-KŌ-sē) A thin layer of smooth muscle fibers (cells) located in the most superficial layer of the mucosa of the gastrointestinal tract, underlying the lamina propria of the mucosa.

Mutation (myoo-TĀ-shun) Any change in the sequence of bases in a DNA molecule resulting in a permanent alteration in some inheritable characteristic.

Myasthenia (mī-as-THE-nē-a) **gravis** Weakness of skeletal muscles caused by antibodies directed against acetylcholine receptors that inhibit muscle contraction.

Myelin (MĪ-e-lin) **sheath** Multilayered lipid and protein covering, formed by Schwann cells and oligodendrocytes, around axons of many peripheral and central nervous system neurons.

Myocardial infarction (mī′-ō-KAR-dē-al in-FARK-shun) **(MI)** Gross necrosis of myocardial tissue due to interrupted blood supply. Also called **heart attack.**

Myocardium (mī′-ō-KAR-dē-um) The middle layer of the heart wall, made up of cardiac muscle tissue, lying between the epicardium and the endocardium and constituting the bulk of the heart.

Myofibril (mī′-ō-FĪ-bril) A threadlike structure, running longitudinally through a muscle fiber (cell), consisting mainly of thick filaments (myosin) and thin filaments (actin, troponin, and tropomyosin).

Myoglobin (mī′-ō-GLŌ-bin) The oxygen-binding, iron-containing protein present in the sarcoplasm of muscle fibers; contributes the red color to muscle.

Myogram (MĪ-ō-gram) The record or tracing produced by a myograph, an apparatus that measures and records the effects of muscular contractions.

Myology (mī-OL-ō-jē) The study of muscles.

Myometrium (mī′-ō-MĒ-trē-um) The smooth muscle layer of the uterus.

Myopathy (mī-OP-a-thē) Any abnormal condition or disease of muscle tissue.

Myopia (mī-O-pē-a) Defect in vision in which objects can be seen distinctly only when very close to the eyes; nearsightedness.

Myosin (MĪ-ō-sin) The contractile protein that makes up the thick filaments of muscle fibers (cells).

N

Nail A hard plate, composed largely of keratin, that develops from the epidermis of the skin to form a protective covering on the dorsal surface of the distal phalanges of the fingers and toes.

Nail matrix (MĀ-triks) The part of the nail beneath the body and root from which the nail is produced.

Nasal (NĀ-zal) **cavity** A mucosa-lined cavity on either side of the nasal septum that opens onto the face at the external nares and into the nasopharynx at the internal nares.

Nasal septum (SEP-tum) A vertical partition composed of bone (perpendicular plate of ethmoid and vomer) and cartilage, covered with a mucous membrane, separating the nasal cavity into left and right sides.

Nasolacrimal (nā′-zō-LAK-ri-mal) **duct** A canal that transports the lacrimal secretion (tears) from the nasolacrimal sac into the nose.

Nasopharynx (nā′-zō-FAR-inks) The superior portion of the pharynx, lying posterior to the nose and extending inferiorly to the soft palate.

Necrosis (ne-KRŌ-sis) A pathological type of cell death that results from disease, injury, or lack of blood supply in which many adjacent cells swell, burst, and spill their contents into the interstitial fluid, triggering an inflammatory response.

Negative feedback The principle governing most control systems; a mechanism of response in which a stimulus initiates actions that reverse or reduce the stimulus.

Neonatal (nē′-ō-NĀ-tal) Pertaining to the first four weeks after birth.

Neoplasm (NĒ-ō-plazm) A new growth that may be benign or malignant.

Nephron (NEF-ron) The functional unit of the kidney.

Nerve A cordlike bundle of nerve fibers (axons and/or dendrites) and its associated connective tissue coursing together outside the central nervous system.

Nerve impulse A wave of depolarization and repolarization that self-propagates along the plasma membrane of a neuron; also called a **nerve action potential.**

Nervous tissue Tissue containing neurons that initiate and conduct nerve impulses to coordinate homeostasis, and neuroglia that provide support and nourishment to neurons.

Net filtration pressure (NFP) Net pressure that promotes fluid outflow at the arterial end of a capillary, and fluid inflow at the venous end of a capillary; net pressure that promotes glomerular filtration in the kidneys.

Neuralgia (noo-RAL-jē-a) Attacks of pain along the entire course or branch of a peripheral sensory nerve.

Neuritis (noo-RĪ-tis) Inflammation of a single nerve, two or more nerves in separate areas, or many nerves simultaneously.

Neuroglia (noo-RŌG-lē-a) Cells of the nervous system that perform various supportive functions. The neuroglia of the central nervous system are the astrocytes, oligodendrocytes, microglia, and ependymal cells; neuroglia of the peripheral nervous system include the Schwann cells and the satellite cells. Also called **glial** (GLĒ-al) **cells.**

Neurolemma (noor′-ō-LEM-a) The peripheral, nucleated cytoplasmic layer of the Schwann cell. Also called **sheath of Schwann** (Schvon).

Neurology (noo-ROL-ō-jē) The branch of science that deals with the normal functioning and disorders of the nervous system.

Neuromuscular (noor′-ō-MUS-kyoo-lar) **junction** The area of contact between the axon terminal of a motor neuron and a portion of the sarcolemma of a muscle fiber (cell).

Neuron (NOOR-on) A nerve cell, consisting of a cell body, dendrites, and an axon.

Neurosecretory (noor′-ō-SE-kre-tō-rē) **cell** A neuron that secretes a hypothalamic releasing hormone, a hypothalamic inhibiting hormone, oxytocin, or antidiuretic hormone into blood capillaries of the hypothalamus or posterior pituitary.

Neurotransmitter One of a variety of molecules within axon terminals that are released into the synaptic cleft in response to a nerve impulse, and that affect the membrane potential of the postsynaptic neuron.

Neutrophil (NOO-trō-fil) A type of white blood cell characterized by granules that stain pale lilac with a combination of acidic and basic dyes.

Nipple A pigmented, wrinkled projection on the surface of the breast that is the location of the openings of the lactiferous ducts for milk release.

Nociceptor (nō′-se-SEP-tor) A free (naked) nerve ending that detects painful stimuli.

Node of Ranvier (ron-vē-Ā) A gap, along a myelinated nerve fiber, between the individual Schwann cells that form the myelin sheath. Also called **neurofibral node.**

Norepinephrine (nor′-ep-i-NEF-rin) **(NE)** A hormone secreted by the adrenal medulla that produces actions similar to those that result from sympathetic stimulation. Also called **noradrenaline** (nor′-a-DREN-a-lin).

Nuclear medicine The branch of medicine concerned with the use of radioisotopes in the diagnosis of disease and therapy.

Nuclease (NOO-klē-ās) An enzyme that breaks nucleic acids into nucleotides; examples are ribonuclease and deoxyribonuclease.

Nucleic (noo-KLĒ-ik) **acid** An organic compound that is a long polymer of nucleotides, with each nucleotide containing a pentose sugar, a phosphate group, and one of four possible nitrogenous bases (adenine, cytosine, guanine, and thymine or uracil).

Nucleolus (noo-KLĒ-ō-lus) Spherical body within the nucleus, composed of protein, DNA, and RNA, that functions in the synthesis and storage of ribosomal RNA.

Nucleus (NOO-klē-us) A spherical or oval organelle of a cell that contains the hereditary factors of the cell, called genes. A cluster of unmyelinated nerve cell bodies in the central nervous system. The central portion of an atom made up of protons and neutrons.

Nucleus pulposus (pul-PŌ-sus) A soft, pulpy, highly elastic substance in the center of an intervertebral disk; a remnant of the notochord.

Nutrient A chemical substance in food that provides energy, forms new body components, or assists in the functioning of various body processes.

O

Obesity (ō-BĒS-i-tē) Body weight more than 20% above a desirable standard due to excessive accumulation of fat.

Oblique (ō-BLĒK) **plane** A plane that passes through the body or an organ at an angle between the transverse plane and either the midsagittal, parasagittal, or frontal plane.

Obstetrics (ob-STET-riks) The specialized branch of medicine that deals with pmediately following delivery (about 42 days).

Occult (o-KULT) Obscure or hidden from view, as, for example, occult blood in stools or urine.

Olfactory (ōl-FAK-tō-rē) Pertaining to smell.

Olfactory bulb A mass of gray matter containing cell bodies of neurons that form synapses with neurons of the olfactory (I) nerve, lying inferior to the frontal lobe of the cerebrum on either side of the crista galli of the ethmoid bone.

Olfactory receptor A bipolar neuron with its cell body lying between supporting cells located in the mucous membrane lining the superior portion of each nasal cavity; transduces odors into neural signals.

Olfactory tract A bundle of axons that extends from the olfactory bulb posteriorly to the olfactory portion of the cerebral cortex.

Oligodendrocyte (ol′-i-gō-DEN-drō-sīt) A neuroglial cell that supports neurons and produces a myelin sheath around axons of neurons of the central nervous system.

Oligospermia (ol′-i-gō-SPER-mē-a) A deficiency of sperm cells in the semen.

Oliguria (ol′-i-GYOO-rē-a) Scanty daily urinary output, less than 250 mL.

Oncogene (ONG-kō-jēn) Gene that has the ability to transform a normal cell into a cancerous cell when it is activated.

Oncology (ong-KOL-ō-jē) The study of tumors.

Oogenesis (ō′-ō-JEN-e-sis) Formation and development of female gametes (oocytes).

Oophorectomy (ō′-of-ō-REK-tō-me) The surgical removal of the ovaries.

Ophthalmic (of-THAL-mik) Pertaining to the eye.

Ophthalmologist (of′-thal-MOL-ō-jist) A physician who specializes in the diagnosis and treatment of eye disorders using drugs, surgery, and corrective lenses.

Ophthalmology (of′-thal-MOL-ō-jē) The study of the structure, function, and diseases of the eye.

Opsonization (op′-so-ni-ZĀ-shun) The action of some antibodies that renders bacteria and other foreign cells more susceptible to phagocytosis.

Optic (OP-tik) Refers to the eye, vision, or properties of light.

Optic chiasm (KĪ-azm) A crossing point of the optic (II) nerves, anterior to the pituitary gland. Also called **optic chiasma.**

Optic disk A small area of the retina containing openings through which the axons of the ganglion cells emerge as the optic (II) nerve. Also called the **blind spot.**

Optician (op-TISH-an) A technician who fits, adjusts, and dispenses corrective lenses on prescription of an ophthalmologist or optometrist.

Optic tract A bundle of axons that transmits nerve impulses from the retina of the eye between the optic chiasm and the thalamus.

Optometrist (op-TOM-e-trist) Specialist with a doctorate in optometry who is licensed to examine and test the eyes and treat visual defects by prescribing corrective lenses.

Orbit (OR-bit) The bony, pyramid-shaped cavity of the skull that holds the eyeball.

Organ A structure composed of two or more different kinds of tissues with a specific function and usually a recognizable shape.

Organelle (or-gan-EL) A permanent structure within a cell with characteristic morphology that is specialized to serve a specific function in cellular activities.

Organic (or-GAN-ik) **compound** Compound that always contains carbon in which the atoms are held together by covalent bonds. Examples include carbohydrates, lipids, proteins, and nucleic acids (DNA and RNA).

Organism (OR-ga-nizm) A total living form; one individual.

Orgasm (OR-gazm) Sensory and motor events involved in ejaculation for the male and involuntary contraction of the perineal muscles in the female at the climax of sexual intercourse.

Orifice (OR-i-fis) Any aperture or opening.

Origin (OR-i-jin) The attachment of a muscle tendon to a stationary bone or the end opposite the insertion.

Oropharynx (or′-ō-FAR-inks) The intermediate portion of the pharynx, lying posterior to the mouth and extending from the soft palate to the hyoid bone.

Orthopedics (or′-thō-PĒ-diks) The branch of medicine that deals with the preservation and restoration of the skeletal system, articulations, and associated structures.

Osmoreceptor (oz′-mō-rē-SEP-tor) Receptor in the hypothalamus that is sensitive to changes in blood osmolarity and, in response to high osmolarity (low water concentration), causes synthesis and release of antidiuretic hormone (ADH).

Osmosis (oz-MŌ-sis) The net movement of water molecules through a selectively permeable membrane from an area of higher water concentration to an area of lower water concentration until equilibrium is reached.

Osseous (OS-ē-us) Bony.

Ossicle (OS-i-kul) Small bone, as in the middle ear (malleus, incus, stapes).

Ossification (os′-i-fi-KĀ-shun) Formation of bone. Also called **osteogenesis.**

Osteoblast (OS-tē-ō-blast′) Cell formed from an osteoprogenitor cell that participates in bone formation by secreting some organic components and inorganic salts.

Osteoclast (OS-tē-ō-clast′) A large, multinuclear cell that destroys or resorbs bone tissue.

Osteocyte (OS-tē-ō-sīt′) A mature bone cell that maintains the daily activities of bone tissue.

Osteology (os′-tē-OL-ō-jē) The study of bones.

Osteon (OS-tē-on) The basic unit of structure in adult compact bone, consisting of a central (haversian) canal with its concentrically arranged

lamellae, lacunae, osteocytes, and canaliculi. Also called a **Haversian** (ha-VER-zhun) **system.**

Osteoporosis (os′-tē-ō-pō-RŌ-sis) Age-related disorder characterized by decreased bone mass and increased susceptibility to fractures, often as a result of decreased levels of estrogens.

Otic (Ō-tik) Pertaining to the ear.

Otolith (Ō-tō-lith) A particle of calcium carbonate embedded in the otolithic membrane that functions in maintaining static equilibrium.

Otolithic (ō′-tō-LITH-ik) **membrane** Thick, gelatinous, glycoprotein layer located directly over hair cells of the macula in the saccule and utricle of the inner ear.

Otorhinolaryngology (ō′-tō-rī′-nō-lar′-in-GOL-ō-jē) The branch of medicine that deals with the diagnosis and treatment of diseases of the ears, nose, and throat.

Oval window A small, membrane-covered opening between the middle ear and inner ear into which the footplate of the stapes fits.

Ovarian (ō-VAR-ē-an) **cycle** A monthly series of events in the ovary associated with the maturation of an ovum.

Ovarian follicle (FOL-i-kul) A general name for oocytes (immature ova) in any stage of development, along with their surrounding epithelial cells.

Ovary (Ō-va-rē) Female gonad that produces ova and the hormones estrogen, progesterone, inhibin, and relaxin.

Ovulation (ov′-yoo-LĀ-shun) The rupture of a mature ovarian (Graafian) follicle with discharge of a secondary oocyte into the pelvic cavity.

Ovum (Ō-vum) The female reproductive or germ cell; an egg cell.

Oxidation (ok′-si-DĀ-shun) The removal of electrons from a molecule or, less commonly, the addition of oxygen to a molecule that results in a decrease in the energy content of the molecule. The oxidation of glucose in the body is called **cellular respiration.**

Oxyhemoglobin (ok′-sē-HĒ-mō-glō′-bin) **(Hb-O$_2$)** Hemoglobin combined with oxygen.

Oxytocin (ok′-sē-TŌ-sin) **(OT)** A hormone, secreted by neurosecretory cells in the paraventricular and supraoptic nuclei of the hypothalamus, that stimulates contraction of the smooth muscle fibers in the pregnant uterus and contractile cells around the ducts of mammary glands.

P

P wave The deflection wave of an electrocardiogram that reflects atrial depolarization.

Palate (PAL-at) The horizontal structure separating the oral and the nasal cavities; the roof of the mouth.

Palpate (PAL-pāt) To examine by touch; to feel.

Pancreas (PAN-krē-as) A soft, oblong organ lying along the greater curvature of the stomach and connected by a duct to the duodenum. It is both an exocrine gland (secreting pancreatic juice) and an endocrine gland (secreting insulin and glucagon).

Pancreatic (pan′-krē-AT-ik) **duct** A single large tube that unites with the common bile duct from the liver and gallbladder and drains pancreatic juice into the duodenum at the hepatopancreatic ampulla (ampulla of Vater). Also called the **duct of Wirsung.**

Pancreatic islet A cluster of endocrine gland cells in the pancreas that secretes insulin, glucagon, somatostatin, and pancreatic polypeptide. Also called an **islet of Langerhans** (LANG-er-hahnz).

Papanicolaou (pa′-pa-ni′-kō-LĀ-oo) **test** A cytological staining test for the detection and diagnosis of premalignant and malignant conditions of the female genital tract. Cells scraped from the epithelium of the cervix of the uterus are examined microscopically. Also called a **Pap smear.**

Papilla (pa-PIL-a) A small nipple-shaped projection or elevation.

Paralysis (pa-RAL-i-sis) Loss or impairment of motor function due to a lesion of nervous or muscular origin.

Paranasal sinus (par′-a-NĀ-zal SĪ-nus) A mucus-lined air cavity in a skull bone that communicates with the nasal cavity. Paranasal sinuses are located in the frontal, maxillary, ethmoid, and sphenoid bones.

Paraplegia (par′-a-PLĒ-jē-a) Paralysis of both lower limbs.

Parasagittal plane A vertical plane that does not pass through the midline and that divides the body or organs into unequal left and right portions.

Parasympathetic (par′-a-sim′-pa-THET-ik) **division** One of the two subdivisions of the autonomic nervous system, having cell bodies of preganglionic neurons in nuclei in the brain stem and in the lateral gray matter of the sacral portion of the spinal cord; primarily concerned with activities that conserve and restore body energy. Also called the **craniosacral** (krā′-nē-ō-SĀ-kral) **division.**

Parathyroid (par′-a-THI-royd) **gland** One of four (usually) small endocrine glands embedded in the posterior surfaces of the lateral lobes of the thyroid gland.

Parathyroid hormone (PTH) A hormone secreted by the parathyroid glands that increases blood calcium level and decreases blood phosphate level.

Parenchyma (pa-RENG-ki-ma) The functional parts of any organ, as opposed to tissue that forms its stroma or framework.

Parietal (pa-RĪ-e-tal) Pertaining to or forming the outer wall of a body cavity.

Parietal cell A type of secretory cell in gastric glands that produces hydrochloric acid and intrinsic factor.

Parietal pleura (PLOOR-a) The outer layer of the serous pleural membrane that encloses and protects the lungs; the layer that is attached to the wall of the pleural cavity.

Parkinson's disease Progressive degeneration of the basal ganglia and substantia nigra of the cerebrum, resulting in decreased production of dopamine (DA) that leads to tremor, slowing of voluntary movements, and muscle weakness.

Parotid (pa-ROT-id) **gland** One of the paired salivary glands located inferior and anterior to the ears; connected to the oral cavity by a duct (Stensen's) that opens into the inside of the cheek opposite the maxillary (upper) second molar.

Parturition (par′-tyoo-RI-shun) Act of giving birth; childbirth, delivery.

Patellar (pa-TEL-ar) **reflex** Extension of the leg, by contraction of the quadriceps femoris muscle in response to tapping the patellar ligament. Also called the **knee jerk reflex.**

Patent ductus arteriosus Congenital anatomical heart defect in which the fetal connection between the aorta and pulmonary trunk remains open instead of closing completely after birth.

Pathogen (PATH-ō-jen) A disease-producing microorganism.

Pathological (path′-ō-LOJ-i-kal) **anatomy** The study of structural changes caused by disease.

Pectoral (PEK-to-ral) Pertaining to the chest or breast.

Pediatrician (pē′-dē-a-TRISH-un) A physician who specializes in the care and treatment of children.

Pelvic (PEL-vik) **cavity** Inferior portion of the abdominopelvic cavity that contains the urinary bladder, sigmoid colon, rectum, and internal female and male reproductive structures.

Pelvic inflammatory disease (PID) Collective term for any extensive bacterial infection of the pelvic organs, especially the uterus, uterine (fallopian) tubes, and ovaries.

Pelvis The basinlike structure formed by the two hip bones, the sacrum, and the coccyx. The expanded, proximal portion of the ureter, lying within the kidney and into which the major calyces open.

Penis (PĒ-nis) The male copulatory organ, used to introduce semen into the female vagina.

Pepsin Protein-digesting enzyme secreted by chief (zymogenic) cells of the stomach in the inactive form pepsinogen, which is converted to active pepsin by hydrochloric acid.

Peptic ulcer An ulcer that develops in areas of the gastrointestinal tract exposed to hydrochloric acid; classified as a gastric ulcer if in the lesser curvature of the stomach, and as a duodenal ulcer if in the first part of the duodenum.

Percussion (per-KU-shun) The act of striking (percussing) an underlying part of the body with short, sharp blows as an aid in diagnosing the part by the quality of the sound produced.

Perforating canal A minute passageway by means of which blood vessels and nerves from the periosteum penetrate into compact bone. Also called **Volkmann's** (FŌK-manz) **canal.**

Pericardial (per′-i-KAR-dē-al) **cavity** Small potential space between the visceral and parietal layers of the serous pericardium that contains pericardial fluid.

Pericarditis (per′-i-KAR-dī-tis) Inflammation of the pericardium.

Pericardium (per′-i-KAR-dē-um) A loose-fitting membrane that encloses the heart, consisting of a superficial fibrous layer and a deep serous layer.

Perichondrium (per′-i-KON-drē-um) The membrane that covers cartilage.

Perilymph (PER-i-limf) The fluid contained between the bony and membranous labyrinths of the inner ear.

Perineum (per′-i-NĒ-um) The pelvic floor; the space between the anus and the scrotum in the male and between the anus and the vulva in the female.

Periodontal (per′-ē-ō-DON-tal) **disease** A collective term for conditions characterized by degeneration of gingivae, alveolar bone, periodontal ligament, and cementum.

Periodontal ligament The periosteum lining the alveoli (sockets) for the teeth in the alveolar processes of the mandible and maxillae.

Periosteum (per′-ē-OS-tē-um) The membrane that covers bone and consists of connective tissue, osteoprogenitor cells, and osteoblasts; essential for bone growth, repair, and nutrition.

Peripheral (pe-RIF-er-al) Located on the outer part or a surface of the body.

Peripheral nervous system (PNS) The part of the nervous system that lies outside the central nervous system, consisting of nerves and ganglia.

Peristalsis (per′-i-STAL-sis) Successive muscular contractions that propel materials along a hollow muscular structure.

Peritoneum (per′-i-tō-NĒ-um) The largest serous membrane of the body that lines the abdominal cavity and covers the viscera.

Peritonitis (per′-i-tō-NĪ-tis) Inflammation of the peritoneum.

Peroxisome (per-OK-si-sōm) Organelle similar in structure to a lysosome that contains enzymes that use molecular oxygen to oxidize various organic compounds. Such reactions produce hydrogen peroxide. Abundant in liver cells.

Perspiration Sweat; produced by sudoriferous (sweat) glands and containing water, salts, urea, uric acid, amino acids, ammonia, sugar, lactic acid, and ascorbic acid. Helps maintain body temperature and eliminate wastes.

pH A measure of the concentration of hydrogen ions (H^+) in a solution. The pH scale extends from 0 to 14, with a value of 7 expressing neutrality. Values lower than 7 indicate acidity, and values higher than 7 indicate alkalinity.

Phagocytosis (fag′-ō-sī-TŌ-sis) The process by which phagocytes ingest particulate matter; especially the ingestion and destruction of microbes, cell debris, and other foreign matter.

Phalanx (FĀ-lanks) The bone of a finger or toe. Plural is **phalanges** (fa-LAN-jēz).

Phantom pain A sensation of pain as originating in a limb that has been amputated.

Pharmacology (far′-ma-KOL-ō-jē) The science that deals with the effects and uses of drugs in the treatment of disease.

Pharynx (FAR-inks) The throat; a tube that starts at the internal nares and runs partway down the neck, where it opens into the esophagus posteriorly and the larynx anteriorly.

Phenotype (FĒ-nō-tīp) The observable expression of genotype; physical characteristics of an organism determined by genetic makeup and influenced by interaction between genes and internal and external environmental factors.

Phlebitis (fle-BĪ-tis) Inflammation of a vein, usually in a lower limb.

Photopigment A substance that can absorb light and undergo structural changes that can lead to the development of a receptor potential. An example is rhodopsin. Also called **visual pigment.**

Photoreceptor Receptor that detects light shining on the retina of the eye.

Physiology (fiz′-ē-OL-ō-jē) Science that deals with the functions of an organism or its parts.

Pia mater (PĪ-a MĀ-ter) The deep membrane (meninx) covering the brain and spinal cord.

Pineal (PIN-ē-al) **gland** The cone-shaped gland located in the roof of the third ventricle.

Pinna (PIN-a) The projecting part of the external ear, composed of elastic cartilage, covered by skin, and shaped like the flared end of a trumpet. Also called the **auricle** (OR-i-kul).

Pinocytosis (pi′-nō-sī-TŌ-sis) The process by which cells ingest liquid.

Pituitary (pi-TOO-i-tar′-ē) **dwarfism** Condition caused by hyposecretion of human growth hormone (hGH) during the growth years and characterized by childlike physical traits in an adult.

Pituitary gland A small endocrine gland lying in the sella turcica of the sphenoid bone and attached to the hypothalamus by the infundibulum. Also called the **hypophysis** (hī-POF-i-sis).

Pivot joint A synovial joint in which a rounded, pointed, or conical surface of one bone articulates with a ring formed partly by another bone and partly by a ligament. Examples include the joint between the atlas and axis and between the proximal ends of the radius and ulna.

Placenta (pla-SEN-ta) The special structure through which the exchange of materials between fetal and maternal circulations occurs. Also called the **afterbirth.**

Plantar flexion (PLAN-tar FLEK-shun) Bending the foot in the direction of the plantar surface (sole).

Plaque (PLAK) A mass of bacterial cells, dextran (polysaccharide), and other debris that adheres to teeth.

Plasma (PLAZ-ma) The extracellular fluid found in blood vessels; blood minus the formed elements.

Plasma cell Cell that produces antibodies and develops from a B cell (lymphocyte).

Plasma (cell) membrane Outer, limiting membrane that separates a cell's internal parts from extracellular fluid or the external environment.

Platelet (PLĀT-let) A fragment of cytoplasm enclosed in a cell membrane and lacking a nucleus, found in the circulating blood. Plays a role in blood clotting. Also called a **thrombocyte** (THROM-bō-sīt).

Platelet plug Aggregation of thrombocytes at a damaged blood vessel to prevent blood loss.

Pleura (PLOOR-a) The serous membrane that covers the lungs and lines the walls of the chest and the diaphragm.

Pleural cavity Small potential space between the visceral and parietal pleurae.

Plexus (PLEK-sus) A network of nerves, veins, or lymphatic vessels.

Pluripotent stem cell Immature stem cell in red bone marrow that gives rise to precursors of all the different mature blood cells. Previously called a **hemopoietic stem cell.**

Pneumotaxic (noo′-mō-TAK-sik) **area** Portion of the respiratory center in the pons that continually sends inhibitory nerve impulses to the inspiratory area, limiting inspiration and facilitating expiration.

Podiatry (pō-DĪ-a-trē) The diagnosis and treatment of foot disorders.

Polar body The smaller cell resulting from the unequal division of primary and secondary oocytes during meiosis. The polar body has no function and degenerates.

Polycythemia (pol′-ē-sī-THĒ-mē-a) Disorder characterized by a hematocrit above the normal level of 55% in which hypertension, thrombosis, and hemorrhage occur.

Polysaccharide (pol′-ē-SAK-a-rīd) A carbohydrate in which three or more monosaccharides are joined chemically.

Polyunsaturated fat A fatty acid that contains more than one double covalent bond between its carbon atoms; abundant in triglycerides of corn oil, safflower oil, and cottonseed oil.

Polyuria (pol′-ē-YOO-rē-a) An excessive production of urine.

Pons (PONZ) The portion of the brain stem that forms a "bridge" between the medulla oblongata and the midbrain, anterior to the cerebellum.

Positive feedback A feedback mechanism in which the response enhances the original stimulus.

Postcentral gyrus Gyrus of cerebral cortex located immediately posterior to the central sulcus; contains the primary somatosensory area.

Posterior (pos-TER-ē-or) Nearer to or at the back of the body.

Posterior column–medial lemniscus pathways Sensory pathways that carry information related to proprioception, fine touch, two-point discrimination, pressure, and vibration. First-order neurons project from the spinal cord to the ipsilateral (same side) medulla in the posterior columns. Second-order neurons project from the medulla to the contralateral (opposite side) thalamus in the medial lemniscus. Third-order neurons project from the thalamus to the somatosensory cortex (postcentral gyrus) on the same side.

Posterior pituitary Posterior lobe of the pituitary gland. Also called the **neurohypophysis** (noor′-ō-hī-POF-i-sis).

Postganglionic neuron (pōst′-gang-lē-ON-ik NOOR-on) The second visceral motor neuron in an autonomic pathway, having its cell body and dendrites located in an autonomic ganglion and its unmyelinated axon ending at cardiac muscle, smooth muscle, or a gland.

Postsynaptic (pōst-sin-AP-tik) **neuron** The nerve cell that is activated by the release of a neurotransmitter from another neuron and carries nerve impulses away from the synapse.

Precapillary sphincter (SFINGK-ter) A ring of smooth muscle fibers (cells) at the site of origin of true capillaries that regulate blood flow into true capillaries.

Precentral gyrus Gyrus of cerebral cortex located immediately anterior to the central sulcus; contains the primary motor area.

Preeclampsia (prē′-ē-KLAMP-sē-a) A syndrome characterized by sudden hypertension, large amounts of protein in urine, and generalized edema; it might be related to an autoimmune or allergic reaction due to the presence of a fetus.

Preganglionic (prē′-gang-lē-ON-ik) **neuron** The first visceral motor neuron in an autonomic pathway, with its cell body and dendrites in the brain or spinal cord and its myelinated axon ending at an autonomic ganglion, where it synapses with a postganglionic neuron.

Pregnancy Sequence of events that normally includes fertilization, implantation, embryonic growth, and fetal growth and terminates in birth.

Premenstrual syndrome (PMS) Severe physical and emotional stress occurring late in the postovulatory phase of the menstrual cycle, sometimes overlapping with menstruation.

Prepuce (PRĒ-pyoos) The loose-fitting skin covering the glans of the penis and clitoris. Also called the **foreskin.**

Presbyopia (prez′-bē-Ō-pē-a) A loss of elasticity of the lens of the eye due to advancing age, with a resulting inability to focus clearly on near objects.

Pressure sore Tissue destruction due to a constant deficiency of blood to tissues overlying a bony projection that has been subjected to prolonged pressure against an object such as a bed, cast, or splint. Also called **bedsore, decubitus** (dē-KYOO-bi-tus) **ulcer,** or **trophic ulcer.**

Presynaptic (prē-sin-AP-tik) **neuron** A neuron that propagates nerve impulses toward a synapse.

Prevertebral ganglion (prē-VER-te-bral GANG-glē-on) A cluster of cell bodies of postganglionic sympathetic neurons anterior to the spinal column and close to large abdominal arteries. Also called a **collateral ganglion.**

Primary germ layer One of three layers of embryonic tissue, called ectoderm, mesoderm, and endoderm, that give rise to all tissues and organs of the body.

Primary motor area A region of the cerebral cortex, in the precentral gyrus of the frontal lobe of the cerebrum, that controls specific muscles or groups of muscles.

Primary somatosensory area A region of the cerebral cortex, posterior to the central sulcus in the postcentral gyrus of the parietal lobe of the cerebrum, that localizes exactly the points of the body where somatic sensations originate.

Prime mover The muscle directly responsible for producing a desired motion. Also called an **agonist** (AG-ō-nist).

Principal cell Cell found in the parathyroid glands that secretes parathyroid hormone (PTH); cell type in the distal convoluted tubule and collecting duct of nephrons that is stimulated by aldosterone and antidiuretic hormone.

Proctology (prok-TOL-ō-jē) The branch of medicine that deals with the rectum and its disorders.

Progeny (PROJ-e-nē) Offspring or descendants.

Progesterone (prō-JES-te-rōn) A female sex hormone produced by the ovaries that helps prepare the endometrium of the uterus for implantation of a fertilized ovum and the mammary glands for milk secretion.

Prognosis (prog-NŌ-sis) A forecast of the probable results of a disorder; the outlook for recovery.

Prolactin (prō-LAK-tin) **(PRL)** A hormone secreted by the anterior pituitary that initiates and maintains milk secretion by the mammary glands.

Prolapse (PRŌ-laps) A dropping or falling down of an organ, especially the uterus or rectum.

Proliferation (prō-lif′-er-Ā-shun) Rapid and repeated reproduction of new parts, especially cells.

Pronation (prō-NĀ-shun) A movement of the hand or foot in which the palm or sole rotates medially to a more posterior or inferior orientation.

Prophase (PRŌ-fāz) The first stage of mitosis during which chromatid pairs are formed and aggregate around the metaphase plate of the cell.

Proprioception (prō'-prē-ō-SEP-shun) The receipt of information from muscles, tendons, and the labyrinth that enables the brain to determine movements and position of the body and its parts. Also called **kinesthesia** (kin'-es-THĒ-zē-a).

Proprioceptor (prō'-prē-ō-SEP-tor) A receptor located in muscles, tendons, or joints that provides information about body position and movements.

Prostaglandin (pros'-ta-GLAN-din) **(PG)** A membrane-associated lipid; released in small quantities and acts as a local hormone.

Prostate (PROS-tāt) **gland** A doughnut-shaped gland inferior to the urinary bladder that surrounds the superior portion of the male urethra and secretes a slightly acidic solution that contributes to sperm motility and viability.

Protein An organic compound consisting of carbon, hydrogen, oxygen, nitrogen, and sometimes sulfur; synthesized on ribosomes and made up of amino acids linked by peptide bonds.

Prothrombin (prō-THROM-bin) An inactive blood-clotting factor synthesized by the liver, released into the blood, and converted to active thrombin in the process of blood clotting by the activated enzyme prothrombinase.

Proto-oncogene (prō'-tō-ONG-kō-jēn) Gene responsible for some aspect of normal growth and development; it may transform into an oncogene, a gene capable of causing cancer.

Protraction (prō-TRAK-shun) The movement of the mandible or shoulder girdle forward on a plane parallel with the ground.

Proximal (PROK-si-mal) Used to describe the relative position of something that is nearer the attachment of a limb to the trunk or nearer to the point of origin or attachment.

Pseudopods (SOO-dō-pods) Temporary protrusions of the leading edge of a migrating cell, such as a phagocyte.

Ptosis (TŌ-sis) Drooping, as of the eyelid or the kidney.

Puberty (PYOO-ber-tē) The time of life during which the secondary sex characteristics begin to appear and sexual reproduction becomes possible; usually occurs between the ages of 10 and 17.

Pubic symphysis A slightly movable cartilaginous joint between the anterior surfaces of the hip bones.

Puerperium (pyoo'-er-PER-ē-um) The state immediately after childbirth, usually 4 to 6 weeks.

Pulmonary (PUL-mo-ner'-ē) Concerning or affected by the lungs.

Pulmonary circulation The flow of deoxygenated blood from the right ventricle to the lungs and the return of oxygenated blood from the lungs to the left atrium.

Pulmonary edema (e-DĒ-ma) An abnormal accumulation of interstitial fluid in the tissue spaces and alveoli of the lungs, due to increased pulmonary capillary permeability or increased pulmonary capillary pressure.

Pulmonary embolism (EM-bō-lizm) **(PE)** The presence of a blood clot or a foreign substance in a pulmonary arterial blood vessel that obstructs circulation to lung tissue.

Pulmonary ventilation The inflow (inspiration) and outflow (expiration) of air between the atmosphere and the lungs. Also called **breathing.**

Pulp cavity A cavity within the crown and neck of a tooth, which is filled with pulp, a connective tissue containing blood vessels, nerves, and lymphatic vessels.

Pulse (PULS) The rhythmic expansion and elastic recoil of a systemic artery after each contraction of the left ventricle.

Pulse pressure The difference between the maximum (systolic) and minimum (diastolic) pressures; normally about 40 mm Hg.

Pupil The hole in the center of the iris, the area through which light enters the posterior cavity of the eyeball.

Pus The liquid product of inflammation containing leukocytes or their remains and debris of dead cells.

Pyloric (pī-LOR-ik) **sphincter** A thickened ring of smooth muscle through which the pylorus of the stomach communicates with the duodenum. Also called the **pyloric valve.**

Pyogenesis (pī'-ō-JEN-e-sis) Formation of pus.

Pyramid (PIR-a-mid) A pointed or cone-shaped structure; one of two roughly triangular structures on the ventral side of the medulla oblongata composed of the largest motor tracts that run from the cerebral cortex to the spinal cord; a triangular structure in the renal medulla.

Q

QRS wave The deflection waves of an electrocardiogram that represent onset of ventricular depolarization.

Quadriplegia (kwod'-ri-PLĒ-jē-a) Paralysis of four limbs.

R

Radiographic (rā'-dē-ō-GRAF-ic) **anatomy** Diagnostic branch of anatomy that includes the use of x rays.

Rapid eye movement (REM) sleep Stage of sleep in which dreaming occurs, lasting for 5 to 10 minutes several times during a sleep cycle; characterized by rapid movements of the eyes beneath the eyelids.

Receptor A specialized cell or a distal portion of a neuron that responds to a specific sensory modality, such as touch, pressure, cold, light, or sound, and converts it to an electrical signal (generator or receptor potential). A specific molecule or cluster of molecules that recognizes and binds a particular ligand.

Receptor-mediated endocytosis A highly selective process whereby cells take up specific ligands, which usually are large molecules or particles, by enveloping them within a sac of plasma membrane. Ligands are eventually broken down by enzymes in lysosomes.

Recessive allele An allele whose presence is masked in the presence of a dominant allele on the homologous chromosome.

Recombinant DNA Synthetic DNA, formed by joining a fragment of DNA from one source to a portion of DNA from another.

Recovery oxygen consumption Elevated oxygen use following exercise, due to metabolic changes that start during exercise and continue after exercise. Previously called **oxygen debt.**

Recruitment (rē-KROOT-ment) The process of increasing the number of active motor units. Also called **motor unit summation.**

Rectum (REK-tum) The last 20 cm (8 in.) of the gastrointestinal tract, from the sigmoid colon to the anus.

Reduction The addition of electrons to a molecule or, less commonly, the removal of oxygen from a molecule, resulting in an increase in the energy content of the molecule.

Referred pain Pain that is felt at a site remote from the place of origin.

Reflex Fast response to a change (stimulus) in the internal or external environment that attempts to restore homeostasis; a reflex passes over a reflex arc.

Reflex arc The most basic conduction pathway through the nervous system, connecting a receptor and an effector and consisting of a receptor, a sensory neuron, an integrating center in the central nervous system, a motor neuron, and an effector.

Refraction (rē-FRAK-shun) The bending of light as it passes from one medium to another.

Refractory (re-FRAK-to-rē) **period** A period of time following depolarization during which an excitable cell (neuron or muscle fiber) cannot respond to a stimulus that is usually adequate to evoke an action potential.

Regional anatomy The division of anatomy dealing with a specific region of the body, such as the head, neck, chest, or abdomen.

Regurgitation (rē-gur'-ji-TĀ-shun) Return of solids or fluids to the mouth from the stomach; backward flow of blood through incompletely closed heart valves.

Relaxin A female hormone produced by the ovaries that increases flexibility of the pubic symphysis and helps dilate the uterine cervix to ease delivery of a baby.

Releasing hormone Hormone secreted by the hypothalamus that can stimulate secretion of hormones of the anterior pituitary.

Remodeling Replacement of old bone by new bone tissue.

Renal (RĒ-nal) Pertaining to the kidney.

Renal corpuscle (KOR-pus-el) A glomerular (Bowman's) capsule and its enclosed glomerulus.

Renal pelvis A cavity in the center of the kidney formed by the expanded, proximal portion of the ureter, into which the major calyces open.

Renal pyramid A triangular structure in the renal medulla containing the straight segments of renal tubules and the vasa recta.

Renin (REN-in) An enzyme released by the kidney into the plasma, where it converts angiotensinogen into angiotensin I.

Renin–angiotensin–aldosterone pathway A mechanism for the control of aldosterone secretion by angiotensin II, initiated by the secretion of renin by the kidney in response to low blood pressure.

Repolarization (rē-pō'-lar-i-ZĀ-shun) Restoration of a resting membrane potential following depolarization.

Reproduction (rē'-prō-DUK-shun) The formation of new cells for growth, repair, or replacement; the production of a new individual.

Reproductive cell division Type of cell division in which gametes (sperm and egg cells) are produced; consists of meiosis and cytokinesis.

Residual (re-ZID-yoo-al) **volume** The volume of air still contained in the lungs after a maximal expiration; about 1200 mL.

Resistance (re-ZIS-tans) Hindrance (impedance) to blood flow as a result of higher viscosity, longer total blood vessel length, and smaller blood vessel radius. Ability to ward off disease. The hindrance encountered by an electrical charge as it moves through a substance from one point to another. The hindrance encountered by air as it moves through the respiratory passageways.

Respiration (res′-pi-RĀ-shun) Overall exchange of gases between the atmosphere, blood, and body cells consisting of pulmonary ventilation, external respiration, and internal respiration.

Respiratory center Neurons in the pons and medulla oblongata of the brain stem that regulate the rate and depth of respiration.

Respiratory distress syndrome (RDS) of the newborn A disease of newborn infants, especially premature ones, in which insufficient amounts of surfactant are produced and breathing is labored. Also called **hyaline** (HI-ah-lin) **membrane disease (HMD).**

Respiratory membrane Structure in the lungs consisting of the alveolar wall and basement membrane and a capillary endothelium and basement membrane, through which the diffusion of respiratory gases occurs. Also called the **alveolar capillary membrane.**

Resting membrane potential The voltage difference between the inside and outside of a cell membrane when the cell is not responding to a stimulus; in many neurons and muscle fibers, it is −70 to −90 mV, with the inside of the cell negative relative to the outside.

Retention (rē-TEN-shun) A failure to void urine due to obstruction, nervous contraction of the urethra, or absence of sensation of desire to urinate.

Reticular (re-TIK-yoo-lar) **activating system (RAS)** A portion of the reticular formation that has many ascending connections with the cerebral cortex; when this area of the brain stem is active, nerve impulses pass to the thalamus and widespread areas of the cerebral cortex, resulting in generalized alertness or arousal from sleep.

Reticular formation A network of small groups of neural cell bodies scattered among bundles of axons (mixed gray and white matter) beginning in the medulla oblongata and extending superiorly through the central part of the brain stem.

Reticulocyte (re-TIK-yoo-lō-sīt′) An immature red blood cell.

Retina (RET-i-na) The deep coat of the posterior portion of the eyeball, consisting of nervous tissue (where the process of vision begins) and a pigmented layer of epithelial cells that contact the choroid.

Retinal (RE-ti-nal) A derivative of vitamin A that functions as the light-absorbing portion of all photopigments.

Retraction (rē-TRAK-shun) The movement of a protracted part of the body posteriorly on a plane parallel to the ground, as in pulling the lower jaw back in line with the upper jaw.

Retrograde degeneration (RE-trō-grād dē-jen′-er-Ā-shun) Changes that occur in the proximal portion of a damaged axon only as far as the first node of Ranvier; similar to changes that occur during wallerian degeneration.

Retroperitoneal (re′-trō-per′-i-tō-NĒ-al) Posterior to the peritoneal lining of the abdominal cavity.

Reye's (RĪZ) **syndrome** Disease, primarily of children and teenagers, that is characterized by vomiting and brain dysfunction and may progress to coma and death; seems to occur following a viral infection, particularly chickenpox or influenza, and aspirin is believed to be a risk factor.

Rh factor An antigen on the surface of erythrocytes (red blood cells). Among Rh⁺ individuals; missing among Rh⁻ individuals.

Rhinology (rī-NOL-ō-jē) The study of the nose and its disorders.

Rhodopsin (rō-DOP-sin) The photopigment in rods of the retina, consisting of a glycoprotein called opsin and a derivative of vitamin A called retinal.

Ribonucleic (rī′-bō-noo-KLĒ-ik) **acid (RNA)** A single-stranded nucleic acid constructed of nucleotides, each consisting of a nitrogenous base (adenine, cytosine, guanine, or uracil), ribose, and a phosphate group; three types are messenger RNA (mRNA), transfer RNA (tRNA), and ribosomal RNA (rRNA), each of which has a specific role during protein synthesis.

Ribosome (RĪ-bō-sōm) An organelle in the cytoplasm of cells, composed of ribosomal RNA and ribosomal proteins, that synthesizes proteins; nicknamed the "protein factory."

Rigidity (ri-JID-i-tē) Hypertonia characterized by increased muscle tone, but reflexes are not affected.

Rigor mortis State of partial contraction of muscles following death due to lack of ATP; myosin heads (cross bridges) remain attached to actin, thus preventing relaxation.

Rod One of two types of photoreceptor in the retina of the eye; specialized for vision in dim light.

Root canal A narrow extension of the pulp cavity lying within the root of a tooth.

Rotation (rō-TĀ-shun) Moving a bone around its own axis, with no other movement.

Round window A small opening between the middle and inner ear, directly inferior to the oval window, covered by the secondary tympanic membrane.

Rugae (ROO-jē) Large folds in the mucosa of an empty hollow organ, such as the stomach and vagina.

S

Saccule (SAK-yool) The inferior and smaller of the two chambers in the membranous labyrinth inside the vestibule of the internal ear, containing a receptor organ for static equilibrium.

Sacral plexus (PLEK-sus) A network formed by the ventral branches of spinal nerves L4 through S3.

Sacral promontory (PROM-on-tor′-ē) The superior surface of the body of the first sacral vertebra that projects anteriorly into the pelvic cavity; a line from the sacral promontory to the superior

border of the pubic symphysis divides the abdominal and pelvic cavities.

Saddle joint A synovial joint in which the articular surface of one bone is saddle shaped and the articular surface of the other bone is shaped like the legs of the rider sitting in the saddle, as in the joint between the trapezium and the metacarpal of the thumb.

Sagittal (SAJ-i-tal) **plane** A plane that divides the body or organs into left and right portions. Such a plane may be **midsagittal (median),** in which the divisions are equal, or **parasagittal,** in which the divisions are unequal.

Saliva (sah-LĪ-va) A clear, alkaline, somewhat viscous secretion produced mostly by the three pairs of salivary glands; contains various salts, mucin, lysozyme, salivary amylase, and lingual lipase (produced by glands in the tongue).

Salivary amylase (SAL-i-ver-ē AM-i-lās) An enzyme in saliva that initiates the chemical breakdown of starch.

Salivary gland One of three pairs of glands that lie external to the mouth and pour their secretory product (saliva) into ducts that empty into the oral cavity; the parotid, submandibular, and sublingual glands.

Salt A substance that, when dissolved in water, ionizes into cations and anions, neither of which are hydrogen ions (H⁺) or hydroxide ions (OH⁻).

Saltatory (sal′-ta-TŌ-rē) **conduction** The propagation of an action potential (nerve impulse) along the exposed portions of a myelinated nerve fiber. The action potential appears at successive nodes of Ranvier and therefore seems to jump or leap from node to node.

Sarcolemma (sar′-kō-LEM-a) The cell membrane of a muscle fiber (cell), especially of a skeletal muscle fiber.

Sarcomere (SAR-kō-mēr) A contractile unit in a striated muscle fiber extending from one Z disk to the next Z disk.

Sarcoplasm (SAR-kō-plazm) The cytoplasm of a muscle fiber (cell).

Sarcoplasmic reticulum (sar′-kō-PLAZ-mik re-TIK-yoo-lum) A network of saccules and tubes surrounding myofibrils of a muscle fiber (cell), comparable to endoplasmic reticulum; functions to reabsorb calcium ions during relaxation and to release them to cause contraction.

Satiety center A collection of neurons located in the ventromedial nuclei of the hypothalamus that, when stimulated, bring about the cessation of eating.

Saturated fat A fatty acid that contains only single bonds (no double bonds) between its carbon atoms; all carbon atoms are bonded to the maximum number of hydrogen atoms; prevalent in triglycerides of animal products such as meat, milk, milk products, and eggs.

Scala tympani (SKĀ-la TIM-pan-ē) The inferior spiral-shaped channel of the bony cochlea, filled with perilymph.

Scala vestibuli (ves-TI-byoo-lē) The superior spiral-shaped channel of the bony cochlea, filled with perilymph.

Schwann (SHVON) **cell** A neuroglial cell of the peripheral nervous system that forms the myelin sheath and neurolemma of a nerve fiber by wrapping around a nerve fiber in a jellyroll fashion. Also called a **neurolemmocyte.**

Sciatica (sī-AT-i-ka) Inflammation and pain along the sciatic nerve; felt along the posterior aspect of the thigh extending down the inside of the leg.

Sclera (SKLER-a) The white coat of fibrous tissue that forms the superficial protective covering over the eyeball except in the most anterior portion; the posterior portion of the fibrous tunic.

Scleral venous sinus A circular venous sinus located at the junction of the sclera and the cornea through which aqueous humor drains from the anterior chamber of the eyeball into the blood. Also called the **canal of Schlemm** (SHLEM).

Sclerosis (skle-RŌ-sis) A hardening with loss of elasticity of tissues.

Scoliosis (skō′-lē-Ō-sis) An abnormal lateral curvature, deviating from the normal vertical line, of the backbone.

Scrotum (SKRŌ-tum) A skin-covered pouch that contains the testes and their accessory structures.

Sebaceous (se-BĀ-shus) **gland** An exocrine gland in the dermis of the skin, almost always associated with a hair follicle, that secretes sebum. Also called an **oil gland.**

Sebum (SĒ-bum) Secretion of sebaceous (oil) glands.

Secondary response Accelerated, more intense cell-mediated or antibody-mediated immune response upon exposure to an antigen after the primary response.

Secondary sex characteristic A characteristic of the male or female body that develops at puberty under the influence of sex hormones but is not directly involved in sexual reproduction; examples are distribution of body hair, voice pitch, body shape, and muscle development.

Secretion (se-KRĒ-shun) Production and release of a fluid from a gland cell, especially a functionally useful product as opposed to a waste product. In the kidneys, movement of a substance from the blood, through a tubule cell, and into the urine.

Selective permeability (per′-mē-a-BIL-i-tē) The property of a membrane by which it permits the passage of certain substances but restricts the passage of others.

Sella turcica (SEL-a TUR-si-ka) A depression on the superior surface of the sphenoid bone that houses the pituitary gland.

Semen (SĒ-men) A fluid discharged at ejaculation by a male that consists of a mixture of sperm and the secretions of the seminiferous tubules, seminal vesicles, prostate gland, and bulbourethral (Cowper's) glands.

Semicircular canals Three bony channels (anterior, posterior, lateral), filled with perilymph, in which lie the membranous semicircular canals filled with endolymph. They contain receptors for equilibrium.

Semicircular ducts The membranous semicircular canals filled with endolymph and floating in the perilymph of the bony semicircular canals; they contain cristae that mediate dynamic equilibrium.

Semilunar (sem′-ē-LOO-nar) **valve** A value between the aorta or the pulmonary trunk and a ventricle of the heart.

Seminal vesicle (SEM-i-nal VES-i-kul) One of a pair of convoluted, pouchlike structures, lying posterior and inferior to the urinary bladder and anterior to the rectum, that secrete a component of semen into the ejaculatory ducts.

Seminiferous tubule (sem′-i-NIF-er-us TOO-by-ool) A tightly coiled duct, located in the testis, where sperm are produced.

Senescence (se-NES-ens) The process of growing old.

Sensation A state of awareness of external or internal conditions of the body.

Sensory area A region of the cerebral cortex concerned with the interpretation of sensory impulses.

Sensory neuron (NOOR-on) A neuron that conducts nerve impulses into the central nervous system. Also called an **afferent neuron.**

Septal defect An opening in the septum (interatrial or interventricular) between the left and right sides of the heart.

Septum (SEP-tum) A wall dividing two cavities.

Serosa (se-RŌ-sa) Any serous membrane. The external layer of an organ formed by a serous membrane. The membrane that lines the pleural, pericardial, and peritoneal cavities.

Serous (SĒR-us) **membrane** A membrane that lines a body cavity that does not open to the exterior. Also called the **serosa.**

Sertoli (ser-TŌ-lē) **cell** A supporting cell of seminiferous tubules that secretes fluid supplying nutrients to sperm and the hormone inhibin, phagocytizes excess cytoplasm from spermatogenic cells, and mediates the effects of FSH and testosterone on spermatogenesis. Also called a **sustentacular** (sus′-ten-TAK-yoo-lar) **cell.**

Serum Blood plasma minus its clotting proteins.

Sesamoid (SES-a-moyd) **bones** Small bones usually found in tendons.

Sex chromosomes The 23 pair of chromosomes, designated X and Y, which determine the genetic sex of an individual; in males, the pair is XY; in females, it is XX.

Sexual intercourse The insertion of the erect penis of a male into the vagina of a female. Also called **coitus** (KŌ-i-tus).

Sexually transmitted disease (STD) General term for any of a large number of diseases spread by sexual contact. Also called a **venereal disease (VD).**

Shin splints Soreness or pain along the tibia, probably caused by inflammation of the periosteum brought on by repeated tugging of the muscles and tendons attached to the periosteum. Also called **tibia stress syndrome.**

Shingles Acute infection of the peripheral nervous system caused by reactivation of the same virus that causes chickenpox.

Shivering Involuntary contraction of skeletal muscles that generates heat. Also called **involuntary thermogenesis.**

Shock Failure of the cardiovascular system to deliver adequate amounts of oxygen and nutrients to meet the metabolic needs of the body due to inadequate cardiac output. It is characterized by hypotension; clammy, cool, and pale skin; sweating; reduced urine formation; altered mental state; acidosis; tachycardia; weak, rapid pulse; and

thirst. Types include hypovolemic, cardiogenic, vascular, and obstructive.

Shoulder joint A synovial joint, where the humerus articulates with the scapula.

Sigmoid colon (SIG-moyd KŌ-lon) The S-shaped portion of the large intestine that begins at the level of the left iliac crest, projects medially, and terminates at the rectum at about the level of the third sacral vertebra.

Sign Any objective evidence of disease that can be observed or measured, such as a lesion, swelling, or fever.

Simple diffusion (di-FYOO-zhun) A passive process in which there is a net movement of molecules or ions from a region of high concentration to a region of low concentration until equilibrium is reached.

Sinoatrial (sī′-nō-Ā-trē-al) **(SA) node** A compact mass of cardiac muscle fibers (cells) specialized for conduction, located in the right atrium inferior to the opening of the superior vena cava. Normally initiates each heartbeat. Also called the **pacemaker.**

Sinus (SĪ-nus) A hollow in a bone (paranasal sinus) or other tissue; a channel for blood (vascular sinus); any cavity having a narrow opening.

Sinusoid (SĪ-nyoo-soyd) A microscopic space or passage for blood in certain organs, such as the liver or spleen.

Skeletal muscle An organ specialized for contraction, composed of striated muscle fibers (cells), supported by connective tissue, attached to a bone by a tendon or an aponeurosis, and stimulated by somatic motor neurons.

Skin The external covering of the body that consists of a superficial, thinner epidermis (epithelial tissue) and a deep, thicker dermis (connective tissue) that is anchored to the subcutaneous layer.

Skull The skeleton of the head, consisting of the cranial and facial bones.

Sleep A state of partial unconsciousness from which a person can be aroused; associated with a low level of activity in the reticular activating system.

Sliding-filament mechanism The explanation of how thick and thin filaments slide relative to one another during striated muscle contraction to decrease sarcomere length.

Small intestine A long tube of the gastrointestinal tract that begins at the pyloric sphincter of the stomach, coils through the central and inferior part of the abdominal cavity, and ends at the large intestine; divided into three segments: duodenum, jejunum, and ileum.

Smooth muscle A tissue specialized for contraction, composed of smooth muscle fibers (cells), located in the walls of hollow internal organs, and innervated by autonomic motor neurons.

Sodium pump An active transport pump located in the plasma membrane that transports sodium ions out of the cell and potassium ions into the cell at the expense of cellular ATP. It functions to keep the ionic concentrations of these elements at physiological levels. Also called **sodium–potassium ATPase.**

Soft palate (PAL-at) The posterior portion of the roof of the mouth, extending from the palatine bones to the uvula. It is a muscular partition lined with mucous membrane.

Solution A homogeneous molecular or ionic dispersion of one or more substances (solutes) in a dissolving medium (solvent) that is usually liquid.

Somatic (sō-MAT-ik) **cell division** Type of cell division in which a single parent cell duplicates itself to produce two identical daughter cells; consists of mitosis and cytokinesis.

Somatic nervous system (SNS) The portion of the peripheral nervous system consisting of somatic sensory (afferent) neurons and somatic motor (efferent) neurons.

Spasm (SPAZM) A sudden, involuntary contraction of large groups of muscles.

Spasticity (spas-TIS-i-tē) Hypertonia characterized by increased muscle tone, increased tendon reflexes, and pathological reflexes (Babinski sign).

Sperm cell A mature male gamete. Also termed **spermatozoon** (sper′-ma-tō-ZŌ-on).

Spermatic (sper-MAT-ik) **cord** A supporting structure of the male reproductive system, extending from a testis to the deep inguinal ring, that includes the ductus (vas) deferens, arteries, veins, lymphatic vessels, nerves, cremaster muscle, and connective tissue.

Spermatogenesis (sper′-ma-tō-JEN-e-sis) The formation and development of sperm in the seminiferous tubules of the testes.

Spermiogenesis (sper′-mē-ō-JEN-e-sis) The maturation of spermatids into sperm.

Sphincter (SFINGK-ter) A circular muscle that constricts an opening.

Sphincter of the hepatopancreatic ampulla A circular muscle at the opening of the common bile and main pancreatic ducts in the duodenum. Also called the **sphincter of Oddi** (OD-ē).

Sphygmomanometer (sfig′-mō-ma-NOM-e-ter) An instrument for measuring arterial blood pressure.

Spina bifida (SPĪ-na BIF-i-da) A congenital defect of the vertebral column in which the halves of the neural arch of a vertebra fail to fuse in the midline.

Spinal (SPĪ-nal) **cord** A mass of nerve tissue located in the vertebral canal from which 31 pairs of spinal nerves originate.

Spinal nerve One of the 31 pairs of nerves that originate on the spinal cord from posterior and anterior roots.

Spinal shock A period from several days to several weeks following transection of the spinal cord and characterized by the abolition of all reflex activity.

Spinothalamic (spī′-nō-tha-LAM-ik) **tracts** Sensory (ascending) tracts that convey information up the spinal cord to the thalamus for sensations of pain, temperature, crude touch, and deep pressure.

Spiral organ The organ of hearing, consisting of supporting cells and hair cells that rest on the basilar membrane and extend into the endolymph of the cochlear duct. Also called the **organ of Corti** (KOR-tē).

Spirometer (spī-ROM-e-ter) An apparatus used to measure lung volumes and capacities.

Spleen (SPLĒN) Large mass of lymphatic tissue between the fundus of the stomach and the diaphragm that functions in formation of blood cells during early fetal development, phagocytosis of worn-out blood cells, and proliferation of B cells during immune responses.

Spongy (cancellous) bone tissue Bone tissue that consists of an irregular latticework of thin plates of bone called trabeculae; spaces between trabeculae of some bones are filled with red bone marrow; found inside short, flat, and irregular bones and in the epiphyses (ends) of long bones.

Sprain Forcible wrenching or twisting of a joint with partial rupture or other injury to its attachments without dislocation.

Squamous (SKWĀ-mus) Flat or scalelike.

Starvation (star-VĀ-shun) The loss of energy stores in the form of glycogen, triglycerides, and proteins due to inadequate intake of nutrients or inability to digest, absorb, or metabolize ingested nutrients.

Static equilibrium (ē′-kwi-LIB-rē-um) The maintenance of posture in response to changes in the orientation of the body, mainly the head, relative to the ground.

Stellate reticuloendothelial (STEL-āt re-tik′-yoo-lō-en′-dō-THE-lē-al) **cell** Phagocytic cell bordering a sinusoid of the liver. Also called a **Kupffer's** (KOOP-ferz) **cell.**

Stenosis (ste-NŌ-sis) An abnormal narrowing or constriction of a duct or opening.

Sterile (STER-il) Free from any living microorganisms. Unable to conceive or produce offspring.

Sterilization (ster′-i-li-ZĀ-shun) Elimination of all living microorganisms. Any procedure that renders an individual incapable of reproduction (for example, castration, vasectomy, hysterectomy, oophorectomy).

Stimulus Any stress that changes a controlled condition; any change in the internal or external environment that excites a receptor, neuron, or muscle fiber.

Stomach The J-shaped enlargement of the gastrointestinal tract directly inferior to the diaphragm in the epigastric, umbilical, and left hypochondriac regions of the abdomen, between the esophagus and small intestine.

Stratum (STRĀ-tum) A layer.

Stratum basalis (ba-SAL-is) The superficial layer of the endometrium (next to the myometrium) that is maintained during menstruation and gestation, and that produces a new stratum functionalis following menstruation or parturition.

Stratum functionalis (funk′-shun-AL-is) The deep layer of the endometrium (next to the uterine cavity) that is shed during menstruation, and that forms the maternal portion of the placenta during gestation.

Stress response Wide-ranging set of bodily changes, triggered by a stressor, that gears the body to meet an emergency. Also termed **general adapation syndrome (GAS).**

Stressor A stress that is extreme, unusual, or long lasting and triggers the stress response.

Stretch receptor Receptor in the walls of blood vessels, airways, or organs that monitors the amount of stretching. Also termed **baroreceptor.**

Stretch reflex A monosynaptic reflex triggered by sudden stretching of muscle spindles that elicits contraction of that same muscle.

Stroke volume The volume of blood ejected by either ventricle in one systole; about 70 mL at rest.

Stromah (STRŌ-ma) The tissue that forms the ground substance, foundation, or framework of an organ, as opposed to its functional parts (parenchyma).

Subarachnoid (sub′-a-RAK-noyd) **space** A space between the arachnoid and the pia mater that surrounds the brain and spinal cord and through which cerebrospinal fluid circulates.

Subcutaneous (sub′-kyoo-TĀ-nē-us) Beneath the skin. Also called **hypodermic** (hi′-pō-DER-mik).

Subcutaneous layer A continuous sheet of areolar connective tissue and adipose tissue between the dermis of the skin and the deep fascia of the muscles. Also called the **superficial fascia** (FASH-ē-a).

Subdural (sub-DOO-ral) **space** A space between the dura mater and the arachnoid of the brain and spinal cord that contains a small amount of fluid.

Sublingual (sub-LING-gwal) **gland** One of a pair of salivary glands situated in the floor of the mouth deep to the mucous membrane and to the side of the lingual frenulum, with a duct that opens into the floor of the mouth.

Submandibular (sub′-man-DIB-yoo-lar) **gland** One of a pair of salivary glands found inferior to the base of the tongue under the mucous membrane in the posterior part of the floor of the mouth, posterior to the sublingual glands, with a duct situated to the side of the lingual frenulum. Also called the **submaxillary** (sub-MAK-si-ler-ē) **gland.**

Submucosa (sub′-myoo-KŌ-sa) A layer of connective tissue located deep to a mucous membrane, as in the gastrointestinal tract or the urinary bladder; the submucosa connects the mucosa to the muscularis layer.

Substrate A molecule upon which an enzyme acts.

Subthreshold stimulus A stimulus of such weak intensity that it cannot initiate an action potential (nerve impulse).

Sudden infant death syndrome (SIDS) Unexpected and unexplained death of an apparently well, or virtually well, infant; death usually occurs during sleep.

Sudoriferous (soo′-do-RIF-er-us) **gland** An exocrine gland in the dermis or subcutaneous layer that produces perspiration. Also called a **sweat gland.**

Sulcus (SUL-kus) A groove or depression between parts, especially between the convolutions of the brain. Plural is **sulci** (SUL-sī).

Summation (su-MĀ-shun) The addition of the excitatory and inhibitory effects of many stimuli applied to a neuron. The increased strength of muscle contraction that results when stimuli follow one another in rapid succession.

Superficial (soo′-per-FISH-al) Located on or near the surface of the body or an organ.

Superficial fascia (FASH-ē-a) A continuous sheet of fibrous connective tissue between the dermis of the skin and the deep fascia of the muscles. Also called **subcutaneous layer.**

Superior (soo-PĒR-ē-or) Indicating a location toward the head or upper part of a structure.

Superior vena cava (VĒ-na KĀ-va) **(SVC)** Large vein that collects blood from parts of the body superior to the heart and returns it to the right atrium.

Supination (soo′-pi-NĀ-shun) A movement of the hand or foot in which the palm or sole is turned anteriorly or superiorly by lateral rotation.

Surface anatomy The study of the structures that can be identified from the outside of the body.

Surfactant (sur-FAK-tant) Complex mixture of phospholipids and lipoproteins, produced by type II alveolar (septal) cells in the lungs, that decreases surface tension.

Susceptibility (su-sep′-ti-BIL-i-tē) Lack of resistance of a body to the deleterious or other effects of an agent such as a pathogen.

Suspensory ligament (sus-PEN-sor-ē LIG-a-ment) A fold of peritoneum extending laterally from the surface of the ovary to the pelvic wall.

Sutural (SOO-cher-al) **bone** A small bone located within a suture between certain cranial bones. Also called **Wormian** (WER-mē-an) **bone.**

Suture (SOO-cher) An immovable fibrous joint that joins skull bones.

Sympathetic (sim′-pa-THET-ik) **division** One of the two subdivisions of the autonomic nervous system, having cell bodies of preganglionic neurons in the lateral gray columns of the thoracic segment and the first two or three lumbar segments of the spinal cord; primarily concerned with processes involving the expenditure of energy. Also called the **thoracolumbar** (thor′-a-kō-LUM-bar) **division.**

Sympathetic trunk ganglion (GANG-glē-on) A cluster of cell bodies of postganglionic sympathetic neurons lateral to the vertebral column, close to the body of a vertebra. These ganglia extend inferiorly through the neck, thorax, and abdomen to the coccyx on both sides of the vertebral column and are connected to one another to form a chain on each side of the vertebral column. Also called **sympathetic chain** or **vertebral chain ganglia.**

Sympathomimetic (sim′-pa-thō-mī-MET-ik) Producing effects that mimic those brought about by the sympathetic division of the autonomic nervous system.

Symphysis (SIM-fi-sis) A line of union. A slightly movable cartilaginous joint such as the pubic symphysis.

Symport Process by which two substances, often Na⁺ and another substance, move in the same direction across a cell membrane. Also called **co-transport.**

Symptom (SIMP-tum) A subjective change in body function not apparent to an observer, such as pain or nausea, that indicates the presence of a disease or disorder of the body.

Synapse (SIN-aps) The functional junction between two neurons or between a neuron and an effector, such as a muscle or gland; may be electrical or chemical.

Synaptic (si-NAP-tik) **cleft** The narrow gap that separates the axon terminal of one neuron from another neuron or muscle fiber (cell), across which a neurotransmitter diffuses to affect the postsynaptic cell.

Synaptic end bulb Expanded distal end of an axon terminal that contains synaptic vesicles. Also called a **synaptic knob.**

Synaptic vesicle Membrane-enclosed sac in a synaptic end bulb that stores neurotransmitters.

Synarthrosis (sin′-ar-THRŌ-sis) An immovable joint; types are suture, gomphosis, and synchondrosis.

Synchondrosis (sin′-kon-DRŌ-sis) A cartilaginous joint in which the connecting material is hyaline cartilage.

Syndesmosis (sin′-dez-MŌ-sis) A slightly movable joint in which articulating bones are united by fibrous connective tissue.

Syndrome (SIN-drōm) A group of signs and symptoms that occur together in a pattern that is characteristic of a particular disease or abnormal condition.

Synergist (SIN-er-jist) A muscle that assists the prime mover by reducing undesired action or unnecessary movement.

Synovial (si-NŌ-vē-al) **cavity** The space between the articulating bones of a diarthrotic joint, filled with synovial fluid. Also called a **joint cavity.**

Synovial fluid Secretion of synovial membranes that lubricates joints and nourishes articular cartilage.

Synovial joint A fully movable or diarthrotic joint, in which a synovial (joint) cavity is present between the two articulating bones.

Synovial membrane The deeper of the two layers of the articular capsule of a synovial joint, composed of areolar connective tissue that secretes synovial fluid into the synovial (joint) cavity.

System A group of organs that have a common function.

Systemic (sis-TEM-ik) Affecting the whole body; generalized.

Systemic anatomy The anatomic study of particular systems of the body, such as the skeletal, muscular, nervous, cardiovascular, or urinary systems.

Systemic circulation The routes by which oxygenated blood flows from the left ventricle through the aorta to all the organs of the body and deoxygenated blood returns to the right atrium.

Systemic vascular resistance (SVR) All the vascular resistance offered by systemic blood vessels. Also called **total peripheral resistance.**

Systole (SIS-tō-lē) In the cardiac cycle, the phase of contraction of the heart muscle, especially of the ventricles.

Systolic (sis-TOL-ik) **blood pressure** The force exerted by blood on arterial walls during ventricular contraction; the highest pressure measured in the large arteries, about 120 mm Hg under normal conditions for a young adult.

T

T cell A lymphocyte that completes its development in the thymus gland and can differentiate into one of several types of effector cells that function in cell-mediated immunity.

T wave The deflection wave of an electrocardiogram that represents ventricular repolarization.

Tachycardia (tak′-i-KAR-dē-a) An abnormally rapid resting heartbeat or pulse rate (over 100 beats/min).

Tactile (TAK-tīl) Pertaining to the sense of touch.

Tactile disk Modified epidermal cell in the stratum basale of hairless skin that functions as a cutaneous receptor for discriminative touch. Also called a **Merkel** (MER-kel) **disk.**

Taenia coli (TĒ-nē-a KŌ-lī) One of three flat bands of thickened, longitudinal smooth muscle running the length of the large intestine.

Target cell A cell whose activity is affected by a particular hormone.

Tarsal gland Sebaceous (oil) gland that opens on the edge of each eyelid. Also called a **meibomian** (mī-BŌ-mē-an) **gland.**

Tarsals bones The seven bones of the ankle. Also called **tarsals.**

Tarsus (TAR-sus) The ankle.

Tay-Sachs (TĀ-SAKS) **disease** Inherited, progressive degeneration of the nervous system, due to a deficient lysosomal enzyme that causes excessive accumulations of a lipid called ganglioside.

Tectorial (tek-TŌ-rē-al) **membrane** A gelatinous membrane projecting over and in contact with the hair cells of the spiral organ (organ of Corti) in the cochlear duct.

Teeth Accessory structures of digestion, composed of calcified connective tissue and embedded in bony sockets of the mandible and maxillae; function to cut, shred, crush, and grind food. Also called **dentes** (DEN-tēz).

Telophase (TEL-ō-fāz) The final stage of mitosis in which the daughter nuclei become established.

Temporomandibular joint (TMJ) syndrome A disorder of the temporomandibular joint (TMJ) characterized by dull pain around the ear, tenderness of jaw muscles, a clicking or popping noise when opening or closing the mouth, limited or abnormal opening of the mouth, headache, tooth sensitivity, and abnormal wearing of the teeth.

Tendon (TEN-don) A white fibrous cord of dense, regularly arranged connective tissue that attaches muscle to bone.

Tendon organ A proprioceptive receptor, sensitive to changes in muscle tension and force of contraction, found chiefly near the junctions of tendons and muscles. Also called a **Golgi** (GOL-jē) **tendon organ.**

Teratogen (TER-a-tō-jen) Any agent or factor that causes physical defects in a developing embryo.

Testis (TES-tis) Male gonad that produces sperm and the hormones testosterone and inhibin. Also called a **testicle.**

Testosterone (tes-TOS-te-rōn) A male sex hormone (androgen) secreted by interstitial endocrinocytes (Leydig cells) of a mature testis; needed for development of sperm; together with a second androgen termed **dihydrotestosterone (DHT),** controls the growth and development of male reproductive organs, secondary sex characteristics, and body growth.

Tetanus (TET-a-nus) An infectious disease caused by the toxin of *Clostridium tetani*, characterized by tonic muscle spasms and exaggerated reflexes, lockjaw, and arching of the back; a smooth, sustained contraction produced by a series of very rapid stimuli to a muscle.

Tetany (TET-a-nē) Hyperexcitability of neurons and muscle fibers caused by hypocalcemia and characterized by intermittent or continuous tonic muscular contractions; may be due to hypoparathyroidism.

Thalamus (THAL-a-mus) A large, oval structure located superior to the midbrain, consisting of two masses of gray matter organized into nuclei; main relay center for sensory impulses ascending to cerebral cortex.

Thermoreceptor (ther′-mō-rē-SEP-tor) Receptor that detects changes in temperature.

Thigh The portion of the lower limb between the hip and the knee.

Third ventricle (VEN-tri-kul) A slitlike cavity between the right and left halves of the thalamus and between the lateral ventricles of the brain.

Thirst center A cluster of neurons in the hypothalamus that is sensitive to the osmotic pressure of extracellular fluid and brings about the sensation of thirst.

Thoracic (thō-RAS-ik) **cavity** Superior portion of the ventral body cavity that contains two pleural cavities, the mediastinum, and the pericardial cavity.

Thoracic duct A lymphatic vessel that begins as a dilation called the cisterna chyli. It receives lymph from the left side of the head, neck, and chest; the left arm; and the entire body below the ribs, and it empties into the left subclavian vein. Also called the **left lymphatic** (lim-FAT-ik) **duct.**

Thorax (THOR-aks) The chest.

Threshold potential The membrane voltage that must be reached to trigger an action potential.

Threshold stimulus Any stimulus strong enough to initiate an action potential or activate a sensory receptor.

Thrombin (THROM-bin) The active enzyme formed from prothrombin that acts to convert fibrinogen to fibrin during formation of a blood clot.

Thrombolytic (throm′-bō-LIT-ik) **agent** Chemical substance injected into the body to dissolve blood clots and restore circulation; mechanism of action is direct or indirect activation of plasminogen; examples include tissue plasminogen activator (t-PA), streptokinase, and urokinase.

Thrombosis (throm-BŌ-sis) The formation of a clot in an unbroken blood vessel, usually a vein.

Thrombus A stationary clot formed in an unbroken blood vessel, usually a vein.

Thymus (THĪ-mus) A bilobed organ, located in the superior mediastinum posterior to the sternum and between the lungs, that plays an essential role in immune responses.

Thyroglobulin (thī′-rō-GLOB-yoo-lin) **(TGB)** A large glycoprotein molecule produced by follicle cells of the thyroid gland in which iodine is combined with tyrosine to form thyroid hormones.

Thyroid cartilage (THĪ-royd KAHR-ti-lij) The largest single cartilage of the larynx, consisting of two fused plates that form the anterior wall of the larynx.

Thyroid gland An endocrine gland with right and left lateral lobes on either side of the trachea connected by an isthmus; located anterior to the trachea just inferior to the cricoid cartilage; secretes thyroxine (T$_4$), triiodothyronine (T$_3$), and calcitonin.

Thyroid-stimulating hormone (TSH) A hormone secreted by the anterior pituitary that stimulates the synthesis and secretion of thyroxine (T$_4$) and triiodothyronine (T$_3$).

Thyroxine (thī-ROK-sin) **(T$_4$)** A hormone secreted by the thyroid gland that regulates organic metabolism, growth and development, and the activity of the nervous system.

Tic Spasmodic twitching made involuntarily by muscles that are ordinarily under voluntary control.

Tidal volume The volume of air breathed in and out in any one breath; about 500 mL in quiet, resting conditions.

Tissue A group of similar cells and their intercellular substance joined together to perform a specific function.

Tissue factor (TF) A factor, or collection of factors, whose appearance initiates the blood clotting process. Also called **thromboplastin** (throm′-bō-PLAS-tin).

Tissue plasminogen activator (t-PA) An enzyme that dissolves small blood clots by initiating a process that converts plasminogen to plasmin, which degrades the fibrin of a clot.

Tongue A large skeletal muscle covered by a mucous membrane located on the floor of the oral cavity.

Tonicity (tō-NIS-i-tē) A measure of the concentration of impermeable solute particles in a solution relative to cytosol. When cells are bathed in an **isotonic solution,** they neither shrink nor swell.

Tonsil (TON-sil) An aggregation of large lymphatic nodules embedded in the mucous membrane of the throat.

Torn cartilage A tearing of an articular disk (meniscus) in the knee.

Total lung capacity The sum of tidal volume, inspiratory reserve volume, expiratory reserve volume, and residual volume; about 6000 mL in an average adult.

Trabecula (tra-BE-kyoo-la) Irregular latticework of thin plate of spongy bone. Fibrous cord of connective tissue serving as supporting fiber by forming a septum extending into an organ from its wall or capsule. Plural is **trabeculae** (tra-BEK-yoo-lē).

Trachea (TRĀ-kē-a) Tubular air passageway extending from the larynx to the fifth thoracic vertebra. Also called the **windpipe.**

Tracheostomy (trā′-kē-OS-tō-mē) Creation of an opening into the trachea through the neck (below the cricoid cartilage), with insertion of a tube, to facilitate passage of air or evacuation of secretions.

Tract A bundle of nerve fibers in the central nervous system.

Transcription (trans-KRIP-shun) The first step in the expression of genetic information in which a single strand of DNA serves as a template for the formation of an RNA molecule.

Transient ischemic (is-KĒ-mik) **attack (TIA)** Episode of temporary cerebral dysfunction caused by interference of the blood supply to the brain.

Translation (trans-LĀ-shun) The synthesis of a new protein on a ribosome as dictated by the sequence of codons in messenger RNA.

Transverse colon (trans-VERS KŌ-lon) The portion of the large intestine extending across the abdomen from right colic (hepatic) flexure to the left colic (splenic) flexure.

Transverse plane A plane that divides the body or organs into superior and inferior portions. Also called a **horizontal plane.**

Transverse tubules (TOO-byools) **(T tubules)** Small, cylindrical invaginations of the sarcolemma of striated muscle fibers (cells) that conduct muscle action potentials toward the center of the muscle fiber.

Trauma (TRAW-ma) An injury, either a physical wound or psychic disorder, caused by an external agent or force, such as a physical blow or emotional shock; the agent or force that causes the injury.

Tremor (TREM-or) Rhythmic, involuntary, purposeless contraction of opposing muscle groups.

Tricuspid (trī-KUS-pid) **valve** Atrioventricular (AV) valve on the right side of the heart.

Triglyceride (trī-GLIS-er-īd) A lipid formed from one molecule of glycerol and three molecules of fatty acids that may be either solid (fats) or liquid (oils) at room temperature; the body's most highly concentrated source of chemical potential energy. Found mainly within adipocytes. Also called a **neutral fat.**

Trigone (TRĪ-gōn) A triangular region at the base of the urinary bladder.

Triiodothyronine (trī-ī′-ō-dō-THĪ-rō-nēn) **(T$_3$)** A hormone produced by the thyroid gland that regulates organic metabolism, growth and development, and the activity of the nervous system.

Trophoblast (TRŌF-ō-blast) The superficial covering of cells of the blastocyst.

Tropic (TRŌ-pik) **hormone** A hormone whose target is another endocrine gland.

Trunk The part of the body to which the upper and lower limbs are attached.

Tubal ligation (lī-GĀ-shun) A sterilization procedure in which the uterine (Fallopian) tubes are tied and cut.

Tuberculosis (too-ber′-kyoo-LŌ-sis) An infection of the lungs and pleurae caused by *Mycobacterium tuberculosis*, resulting in destruction of lung tissue and its replacement by fibrous connective tissue.

Tubular reabsorption The movement of filtrate from renal tubules back into blood in response to the body's specific needs.

Tubular secretion The movement of substances in blood into renal tubular fluid in response to the body's specific needs.

Tumor suppressor gene A gene coding for a protein that normally inhibits cell division; loss or alteration of a tumor suppressor gene called *p53* is the most common genetic change in a wide variety of cancer cells.

Tunica externa (ek-STER-na) The superficial coat of an artery or vein, composed mostly of elastic and collagen fibers. Also called the **adventitia.**

Tunica interna (in-TER-na) The deep coat of an artery or vein, consisting of a lining of endothelium, basement membrane, and internal elastic lamina. Also called the **tunica intima** (IN-ti-ma).

Tunica media (MĒ-dē-a) The intermediate coat of an artery or vein, composed of smooth muscle and elastic fibers.

Twitch contraction Brief contraction of all muscle fibers in a motor unit triggered by a single action potential in its motor neuron.

Type II cutaneous mechanoreceptor A receptor embedded deeply in the dermis and deeper tissues that detects stretching of skin. Also called a **Ruffini corpuscle.**

U

Ulcer (UL-ser) An open lesion of the skin or a mucous membrane of the body with loss of substance and necrosis of the tissue.

Umbilical cord The long, ropelike structure containing the umbilical arteries and vein that connect the fetus to the placenta.

Umbilicus (um-BIL-i-kus) A small scar on the abdomen that marks the former attachment of the umbilical cord to the fetus. Also called the **navel.**

Upper limb The appendage attached at the shoulder gindle, consisting of the arm, forearm, wrist, hand, and fingers. Also called **upper extremity.**

Uremia (yoo-RĒ-mē-a) Accumulation of toxic levels of urea and other nitrogenous waste products in the blood, usually resulting from severe kidney malfunction.

Ureter (yoo-RĒ-ter) One of two tubes that connect the kidney with the urinary bladder.

Urethra (yoo-RĒ-thra) The duct from the urinary bladder to the exterior of the body that conveys urine in females and urine and semen in males.

Urinalysis The physical, chemical, and microscopic analysis or examination of urine.

Urinary (YOO-ri-ner′-ē) **bladder** A hollow, muscular organ situated in the pelvic cavity posterior to the pubic symphysis; recieves urine via two ureters and stores urine until it is excreted through the urethra.

Urine The fluid produced by the kidneys that contains wastes or excess materials; excreted from the body through the urethra.

Urology (yoo-ROL-ō-jē) The specialized branch of medicine that deals with the structure, function, and diseases of the male and female urinary systems and the male reproductive system.

Uterine (YOO-ter-in) **tube** Duct that transports ova from the ovary to the uterus. Also called the **Fallopian** (fa-LŌ-pē-an) **tube** or **oviduct.**

Uterus (YOO-ter-us) The hollow, muscular organ in females that is the site of menstruation, implantation, development of the fetus, and labor. Also called the **womb.**

Utricle (YOO-tri-kul) The larger of the two divisions of the membranous labyrinth located inside the vestibule of the inner ear, containing a receptor organ for static equilibrium.

Uvula (YOO-vyoo-la) A soft, fleshy mass, especially the V-shaped pendant part, descending from the soft palate.

V

Vagina (vah-JĪ-na) A muscular, tubular organ that leads from the uterus to the vestibule, situated between the urinary bladder and the rectum of the female.

Valence (VĀ-lens) The combining capacity of an atom; the number of deficit or extra electrons in the outermost electron shell of an atom.

Varicose (VAR-i-kōs) Pertaining to an unnatural swelling, as in the case of a varicose vein.

Vasa recta (VĀ-sa REK-ta) Extensions of the efferent arteriole of a juxtamedullary nephron that run alongside the loop of the nephron (Henle) in the medullary region of the kidney.

Vasa vasorum (va-SŌ-rum) Blood vessels that supply nutrients to the larger arteries and veins.

Vascular (VAS-kyoo-lar) Pertaining to or containing many blood vessels.

Vascular spasm Contraction of the smooth muscle in the wall of a damaged blood vessel to prevent blood loss.

Vascular tunic (TOO-nik) The middle layer of the eyeball, composed of the choroid, ciliary body, and iris. Also called the **uvea** (YOO-vē-a).

Vasectomy (va-SEK-tō-mē) A means of sterilization of males in which a portion of each ductus (vas) deferens is removed.

Vasoconstriction (vas′-ō-kon-STRIK-shun) A decrease in the size of the lumen of a blood vessel caused by contraction of the smooth muscle in the wall of the vessel.

Vasodilation (vas′-ō-dī-LĀ-shun) An increase in the size of the lumen of a blood vessel caused by relaxation of the smooth muscle in the wall of the vessel.

Vein A blood vessel that conveys blood from tissues back to the heart.

Vena cava (VĒ-na KĀ-va) One of two large veins that open into the right atrium, returning to the heart all of the deoxygenated blood from the systemic circulation, except that from the coronary circulation.

Ventral (VEN-tral) Pertaining to the anterior or front side of the body; opposite of dorsal.

Ventral body cavity Cavity near the ventral aspect of the body that contains viscera and consists of a superior thoracic cavity and an inferior abdominopelvic cavity.

Ventricle (VEN-tri-kul) A cavity in the brain or an inferior chamber of the heart.

Ventricular fibrillation (ven-TRIK-yoo-lar fib′-ri-LĀ-shun) Asynchronous ventricular contractions; unless reversed by defibrillation, results in heart failure.

Venule (VEN-yool) A small vein that collects blood from capillaries and delivers it to a vein.

Vermiform appendix (VER-mi-form a-PEN-diks) A twisted, coiled tube attached to the cecum.

Vertebral (VER-te-bral) **canal** A cavity within the vertebral column formed by the vertebral foramina of all the vertebrae and containing the spinal cord. Also called the **spinal canal.**

Vertebral column The 26 vertebrae of an adult or the 33 vertebrae of a child; encloses and protects the spinal cord and serves as a point of attachment for the ribs and back muscles. Also called the **backbone, spine,** or **spinal column.**

Vesicle (VES-i-kul) A small bladder or sac containing liquid.

Vestibular (ves-TIB-yoo-lar) **apparatus** Collective term for the organs of equilibrium, which includes the saccule, utricle, and semicircular ducts.

Vestibular membrane The membrane that separates the cochlear duct from the scala vestibuli.

Vestibule (VES-ti-byool) A small space or cavity at the beginning of a canal, especially the inner ear, larynx, mouth, nose, and vagina.

Villus (VIL-us) A projection of the intestinal mucosal cells containing connective tissue, blood vessels, and a lymphatic vessel; functions in the absorption of the end products of digestion. Plural is **villi** (VIL-ī).

Viscera (VIS-er-a) The organs inside the ventral body cavity. Singular is **viscus** (VIS-kus).

Visceral (VIS-er-al) Pertaining to the viscera or to the covering of an organ.

Visceral effectors (e-FEK-torz) Organs of the ventral body cavity that respond to neural stimulation, including cardiac muscle, smooth muscle, and glandular epithelium.

Visceral pleura (PLOOR-a) The deep layer of the serous membrane that covers the lungs.

Vital capacity The sum of inspiratory reserve volume, tidal volume, and expiratory reserve volume; about 4800 mL.

Vital signs Signs necessary to life that include temperature (T), pulse (P), respiratory rate (RR), and blood pressure (BP).

Vitamin An organic molecule necessary in trace amounts that is essential for normal metabolic processes in the body.

Vitiligo (vit′-i-LĪ-gō) Patchy, white spots on the skin due to partial or complete loss of melanocytes.

Vitreous (VIT-rē-us) **body** A soft, jellylike substance that fills the vitreous chamber of the eyeball, lying between the lens and the retina.

Vocal folds Pair of mucous membrane folds below the ventricular folds that function in voice production. Also called **true vocal cords.**

Voltage-gated channel An ion channel in a plasma membrane composed of integral proteins; functions like a gate to permit or restrict the movement of ions across the membrane in response to changes in membrane voltage.

Vulva (VUL-va) Collective designation for the external genitalia of the female. Also called the **pudendum** (pyoo-DEN-dum).

W

Wallerian (wahl-Ē-rē-an) **degeneration** Degeneration of the portion of the axon and mylin sheath of a neuron distal to the site of injury.

Wandering macrophage (MAK-rō-fāj) Phagocytic cell that develops from a monocyte, leaves the blood, and migrates to infected tissues.

Wave summation (su-MĀ-shun) The increased strength of muscle contraction that results when stimuli follow one another in rapid succession.

White matter Aggregations or bundles of myelinated axons located in the brain and spinal cord.

X

Xiphoid (ZIF-foyd) Sword-shaped. The inferior portion of the sternum is the **xiphoid process.**

Y

Yolk sac An extraembryonic membrane that connects with the midgut during early embryonic development but is nonfunctional in humans.

Z

Zona pellucida (pe-LOO-si-da) Clear glycoprotein layer between a secondary oocyte and the surrounding granulosa cells of the corona radiata.

Zygote (ZĪ-gōt) The single cell resulting from the union of male and female gametes; the fertilized ovum.

Credits

■

■ PHOTOS

Chapter 1

Page **3**: John Wilson White. Page **9**: Brian Bailey/Tony Stone Images/New York, Inc. Page **14** (top): Stephen A. Kieffer and E. Robert Heitzman, *An Atlas of Cross-Sectional Anatomy*. Harper & Row, Publishers, New York, 1979. Page **14** (center): Lester V. Bergman/Project Masters, Inc. Page **14** (bottom): Martin Rotker.

Chapter 2

Page **35**: William Whitehurst/The Stock Market.

Chapter 3

Page **44**: ©John Wiley & Sons. Page **63**: Courtesy Michael H. Ross. Page **64**: Ricardo Arias/Photo Researchers.

Chapter 4

Page **74** (top): Biophoto Associates/Photo Researchers. Pages **74** (bottom) and **75** (top): ©Ed Reschke. Page **75** (bottom): Douglas Merrill. Pages **76** and **77** (top): Biophoto Associates/Photo Researchers. Page **77** (bottom): ©Ed Reschke. Pages **78** and **79** (bottom): Bruce Iverson. Page **79** (top): Lester V. Bergman/Project Masters, Inc. Pages **83** and **84**: Courtesy Michael H. Ross. Page **85** (top): Biophoto Associates/Photo Researchers. Page **85** (bottom): Courtesy Andrew J. Kuntzman. Page **86** (top): ©Ed Reschke. Pages **86** (bottom) and **87** (top and bottom): Biophoto Associates/Photo Researchers. Page **88** (top): ©Biology Media/Photo Researchers. Pages **88** (center) and **89**: Courtesy Michael H. Ross. Pages **90** and **91**: Ed Rescke. Page **92**: P. Motta/Photo Researchers. Page **93**: ©Ed Reschke.

Chapter 5

Page **102**: Lester V. Bergman/Project Masters, Inc. Page **107**: Andrea Booher/Tony Stone Images/New York, Inc.

Chapter 6

Page **115**: Mark Nielson. Page **146**: Chris Sanders/Tony Stone Images/New York, Inc.

Chapter 7

Pages **159, 160** and **161**: John Wilson White. Page **162**: ©Corbis Images.

Chapter 8

Page **177**: Michael Kevin Daly/The Stock Market.

Chapter 9

Page **226**: Karen Leeds/The Stock Market.

Chapter 10

Page **239**: Courtesy Mark Nielsen. Page **243**: James D'Addio/The Stock Market.

Chapter 11

Page **266**: Jim Cummins/FPG International.

Chapter 12

Page **277**: Ken Reid/FPG International.

Chapter 13

Page **314**: Lester V. Bergman/Project Masters, Inc. Pages **316** and **318**: Courtesy Michael H. Ross. Page **321**: W. Marc Bernsau/The Image Works. Page **322**: Courtesy Jim Sheetz, University of Alabama. Page **328** (top left, top and center, and bottom left): Lester V. Bergman/Project Masters, Inc. Page **328** (top right): Martin Rotker/Phototake. Page **328** (bottom right): Biophoto Associates/Photo Researchers.

Chapter 14

Page **337**: Courtesy Michael H. Ross. Page **343**: ©Boehringer Ingelheim International. Photo by Lennart Nilsson/Albert Bonniers Publishers, THE INCREDIBLE MACHINE. Reproduced with permission. Page **346**: L. D. Gordon/The Image Bank.

Chapter 15

Page **361**: Bruce Ayres/Tony Stone Images/New York, Inc. Page **366** (left): ©Vu/Cabisco/Visuals Unlimited. Page **366** (right): W. Ober/Visuals Unlimited.

Chapter 16

Page **374**: Courtesy Michael H. Ross. Page **380**: Gary Nolton/Tony Stone Images/New York, Inc.

Chapter 17

Page **412**: National Cancer Institute/Photo Researchers. Page **417**: Ian Shaw/Tony Stone Images/New York, Inc.

Chapter 18

Page **433**: Courtesy Mark Nielson, University of Utah. Page **445**: Steve Taylor/Tony Stone Images/New York, Inc.

Chapter 19

Page **466**: Hessler/Visuals Unlimited. Page **474** (top): P. Motta and A. Familiari, Univerity of La Sapinza, Rome/SPL/Photo Researchers. Page **474** (bottom): Courtesy Mark Nielsen, University of Utah. Page **479**: P. Motta, University La

Sapienza, Rome/SPL/Photo Researchers. Page **481:** ©Telegraph Colour Library/FPG International.

Chapter 20

Page **504:** Lori Adamski Peek/Tony Stone Images/New York, Inc.

Chapter 21

Page **522:** Larry Gatz/The Image Bank.

Chapter 22

Page **538:** Lori Adamski Peek/Tony Stone Images/New York, Inc.

Chapter 23

Page **566:** DiMaggio/Kalish/The Stock Market.

Chapter 24

Page **587:** David Young Wolff/Tony Stone Images/New York, Inc.

■ LINE ART

Leonard Dank

Table 6.03, Table 6.04, 6.01, 6.02b-d, 6.04, 6.06, 6.07, 6.08, 6.09, 6.10, 6.11, 6.12, 6.13, 6.14a, 6.15, 6.16, 6.17, 6.18, 6.19, 6.20, 6.21, 6.22, 6.23, 6.24, 6.25, 6.26, 6.27, 6.28, 6.29, 7.01, 7.02, 7.03, 7.04, 7.11, 8.12, 8.13, 8.14, 8.15, 8.16, 8.17, 8.18, 8.19, 8.20, 8.21, 8.22, 8.23, 8.24, 8.25

Sharon Ellis

9.03, 9.04, 10.01, 10.02, 10.03, 10.04, 10.05, 10.06, 10.07a, 10.08, 10.09, 10.10, 10.11, 10.12, 12.04, 12.12-12.13 orientation figures, 17.01, 17.02, 17.04

Imagineering

2.01, 2.03, 2.04, 2.05, 2.09, 2.11, 2.15, 3.02 orientation diagram, 3.04, 3.05, 3.06, 3.09, 3.10, 3.11, 3.12, 3.17, 3.20, 3.21, 3.22, 3.23, 4.01, 5.02, 8.02b, 8.03, 8.06, 8.07, 8.08, 8.10, 9.02, 9.05, 9.06, 9.08, 14.08, 14.09, 15.07, 15.09, 17.08, 19.10d, 20.01, 20.03, 20.04, 20.05, 20.06, 20.07, 20.08, 21.05, 22.04, 22.05, Focus on Homeostasis icons.

Jean Jackson

Table 17.02

Lauren Keswick

3.24, 6.02a, Table 8.01

Hilda Muinos

8.04, 8.05, Table 10.02, 11.02, 14.01a, 15.05a, 15.08, 16.01, 16.08, 16.16, 17.07, 24.08

Tomo Narashima

1.01, Table 3.02, 3.01, 3.02, 3.13, 3.14, 3.15, 3.16, 3.18, 3.19, 12.02, 12.05a, 12.09, 12.10a, 12.10b, 12.11, 12.12, 12.13

Steve Oh

12.03b-c13.01, 19.01, 19.02, 19.03, 19.05, 19.07, 19.09, 21.02, 21.03, 21.11

Lynn O'Kelley

1.06, 12.03a, 12.05c, 13.05, 13.08, 13.10, 13.12, 13.14, 13.17, 18.01, 18.02, 18.03, 18.04, 18.05

Jared Schneidman Design

1.02, 1.03, 2.02, 2.06, 2.07, 2.08, 2.10, 2.12, 2.13, 2.14, 2.16, 3.07, 3.08, 5.05, 6.05, 8.09, 9.07, 11.01, 12.06, 12.07, 13.02, 13.03, 13.04, 13.06, 13.07, 13.09, 13.11, 13.13, 13.15, 13.16, 13.18, 14.04, 14.05, 15.05b, 16.03, 16.05, 16.06, 16.07, 17.09, 17.10, 17.11, 17.12, 18.08, 18.09, 18.12, 18.13, 18.14, 18.15, 18.16, 19.13, 20.02, 21.07, 21.08, 21.09, 22.01, 22.02, 22.03, 22.06, 23.03, 23.05, 23.10, 23.15, 23.16, 23.17, 23.18, 24.02, 24.09, 24.11, 24.12, 24.13

Nadine Sokol

Tables 4.01-4.05 orientation diagrams, Table 6.02, 6.14b, Table 10.01, 12.08, Table 14.02, 14.01b, 14.02a, 14.03, 14.06, 16.02a, 16.17b, 17.03, 17.05, 17.06, 17.13, 18.05 orientation diagrams, 18.11, 19.04, 19.06, 19.14, 21.06, 21.10, 24.01

Kevin Somerville

Table 1.01, 0.04, 1.05, 1.07, 1.08, 1.09, 1.10, 1.11, Table 4.01, Table 4.02, Table 4.03, 5.01, 5.03, 5.04, 6.03, Table 8.14, 8.01, 8.02a, Table 9.01, 9.01, 11.03, Table 12.02, Table 12.03, 12.01, 15.01, 15.02, 15.03, 15.04, 15.06, 16.09, 16.10, 16.11, 16.12, 16.13, 16.14, 16.15, 16.17a, 18.06, 18.07, 18.10, 19.08a, 19.08b, 19.10a-c, 19.11, 19.12a-b, 19.15, 21.01, 21.04, 23.01, 23.02, 23.04, 23.06, 23.07, 23.08, 23.09, 23.11, 23.12, 23.13, 23.14, 24.03, 24.04, 24.05, 24.06, 24.07, 24.10

Beth Willert

8.11

Index